STATISTICAL PRINCIPLES IN EXPERIMENTAL DESIGN

McGraw-Hill Series
in Psychology

Consulting Editors

NORMAN GARMEZY

RICHARD L. SOLOMON

LYLE V. JONES

HAROLD W. STEVENSON

Adams *Human Memory*
Beach, Hebb, Morgan, and Nissen *The Neuropsychology of Lashley*
Von Békésy *Experiments in Hearing*
Berkowitz *Aggression: A Social Psychological Analysis*
Berlyne *Conflict, Arousal, and Curiosity*
Blum *Psychoanalytic Theories of Personality*
Brown *The Motivation of Behavior*
Brown and Ghiselli *Scientific Method in Psychology*
Butcher *MMPI: Research Developments and Clinical Applications*
Campbell, Dunnette, Lawler, and Weick *Managerial Behavior, Performance, and Effectiveness*
Cofer *Verbal Learning and Verbal Behavior*
Cofer and Musgrave *Verbal Behavior and Learning: Problems and Processes*
Crafts, Schneirla, Robinson, and Gilbert *Recent Experiments in Psychology*
Crites *Vocational Psychology*
D'Amato *Experimental Psychology: Methodology, Psychophysics, and Learning*
Davitz *The Communication of Emotional Meaning*
Deese and Hulse *The Psychology of Learning*
Dollard and Miller *Personality and Psychotherapy*
Edgington *Statistical Inference: The Distribution-free Approach*
Ellis *Handbook of Mental Deficiency*
Epstein *Varieties of Perceptual Learning*
Ferguson *Statistical Analysis in Psychology and Education*
Forgus *Perception: The Basic Process in Cognitive Development*
Franks *Behavior Therapy: Appraisal and Status*
Ghiselli *Theory of Psychological Measurement*
Ghiselli and Brown *Personnel and Industrial Psychology*
Gilmer *Industrial Psychology*
Gray *Psychology Applied to Human Affairs*
Guilford *Fundamental Statistics in Psychology and Education*
Guilford *The Nature of Human Intelligence*
Guilford *Personality*
Guilford *Psychometric Methods*
Guilford and Hoepfner *The Analysis of Intelligence*
Guion *Personnel Testing*
Haire *Psychology in Management*
Hirsch *Behavior-genetic Analysis*

Hirsh *The Measurement of Hearing*
Hurlock *Adolescent Development*
Hurlock *Child Development*
Hurlock *Developmental Psychology*
Jackson and Messick *Problems in Human Assessment*
Karn and Gilmer *Readings in Industrial and Business Psychology*
Krech, Crutchfield, and Ballachey *Individual in Society*
Lawler *Pay and Organizational Effectiveness: A Psychological View*
Lazarus, A. *Behavior Therapy and Beyond*
Lazarus, R. *Adjustment and Personality*
Lazarus, R. *Psychological Stress and the Coping Process*
Lewin *A Dynamic Theory of Personality*
Lewin *Principles of Topological Psychology*
Maher *Principles of Psychopathology*
Marascuilo *Statistical Methods for Behavioral Science Research*
Marx and Hillix *Systems and Theories in Psychology*
Messick and Brayfield *Decision and Choice: Contributions of Sidney Siegel*
Miller *Language and Communication*
Morgan *Physiological Psychology*
Mulaik *The Foundations of Factor Analysis*
Nunnally *Psychometric Theory*
Overall and Klett *Applied Multivariate Analysis*
Rethlingshafer *Motivation as Related to Personality*
Rosenthal *Genetic Theory and Abnormal Behavior*
Robinson and Robinson *The Mentally Retarded Child*
Scherer and Wertheimer *A Psycholinguistic Experiment on Foreign Language Teaching*
Shaw *Group Dynamics: The Psychology of Small Group Behavior*
Shaw and Costanzo *Theories of Social Psychology*
Shaw and Wright *Scales for the Measurement of Attitudes*
Sidowski *Experimental Methods and Instrumentation in Psychology*
Siegel *Nonparametric Statistics for the Behavioral Sciences*
Spencer and Kass *Perspectives in Child Psychology*
Stagner *Psychology of Personality*
Townsend *Introduction to Experimental Methods for Psychology and the Social Sciences*
Vinacke *The Psychology of Thinking*
Wallen *Clinical Psychology: The Study of Persons*
Warren and Akert *The Frontal Granular Cortex and Behavior*
Waters, Rethlingshafer, and Caldwell *Principles of Comparative Psychology*
Winer *Statistical Principles in Experimental Design*
Zubek and Solberg *Human Development*

STATISTICAL PRINCIPLES IN EXPERIMENTAL DESIGN

Second Edition

B. J. WINER

Professor of Psychology
Purdue University

McGRAW-HILL BOOK COMPANY

NEW YORK ST. LOUIS SAN FRANCISCO
DÜSSELDORF JOHANNESBURG KUALA LUMPUR
LONDON MEXICO MONTREAL
NEW DELHI PANAMA RIO DE JANEIRO
SINGAPORE SYDNEY TORONTO

STATISTICAL PRINCIPLES IN EXPERIMENTAL DESIGN

Library of Congress Catalog Card Number 79-133392
07-070981-5

10 11 12 13 HDHD 7 8 0 9 8

This book was set in Times Roman, and was printed by The Halliday Lithograph Corporation, and bound by The Book Press, Inc. The designer was Edward Zytko; the drawings were done by John Cordes, J. & R. Technical Services, Inc. The editors were Walter Maytham, Albert Shapiro, and Joan Stern. Les Kaplan supervised production.

CONTENTS

Chapter 2
LINEAR MODELS 58

Chapter 3
DESIGN AND ANALYSIS OF SINGLE-FACTOR EXPERIMENTS 149

Chapter 4
SINGLE-FACTOR EXPERIMENTS HAVING REPEATED MEASURES ON THE SAME ELEMENTS 261

PREFACE

The second edition contains a great deal of new material as well as extensive revision of the material included in the first edition. However, the general objectives of this edition are fundamentally the same as those of the first edition.

Written primarily for students and research workers in the area of the behavioral and biological sciences, this book is meant to provide a text as well as a comprehensive reference source on statistical principles underlying experimental design. Particular emphasis is given to those designs that are likely to prove useful in research in the behavioral and biological sciences.

The logic basic to understanding principles underlying the statistical aspects of experimental design is emphasized, rather than the details of mathematical and statistical proofs. In the second edition there is somewhat more coverage of distribution theory associated with linear models. However, there is sufficient redundancy in the writing so that the more theoretical sections may be omitted without loss in the readability of the more applied sections.

The topics selected for inclusion are those which have been covered in courses taught by the author during the past several years. It has been the intention of the author to keep most of the applied chapters in the book at a readability level appropriate for students having a mathematical background equivalent to freshman college algebra.

The first course in design, as taught by the author, has as a prerequisite a basic course in statistical inference. Chapter 1 reviews the highlights of what is included in the prerequisite course. This chapter is not meant to provide a first exposure to the topics which are covered.

In general, the discussion of principles, interpretation of illustrative examples, and computational procedures are included in successive sections within the same chapter. However, to facilitate the use of the book as a reference source, this procedure is not followed in Chapters 5 and 6. Basic principles associated with a large class of designs for factorial experiments are discussed in Chapter 5. Detailed illustrative examples of these designs are presented in Chapter 6. For teaching purposes the author includes relevant material from Chapter 6 with corresponding material in Chapter 5.

Chapter 2, Linear Models, was not included in the first edition. It is a relatively long chapter. The use of matrix notation was found to be essential. However, the matrix algebra which is used is quite elementary. This chapter lays the foundation for much of the work that follows in later chapters. For an understanding of the statistical theory underlying the analysis-of-variance model, the contents of Chapter 2 are necessary. Some of the material in this chapter appears in modified form in various other chapters. It will be found that most of the contents of Chapter 2 may be omitted in a first course in experimental design.

Some topics from the multivariate analysis of variance have been included in Chapters 3 and 4. The latter chapter, in particular, is shown to be a special case of the multivariate analysis of variance. In Section 5.23 the topic of nonorthogonal factorial designs is covered in much greater detail than in the first edition. This extended coverage depends in part upon the material in Chapter 2.

To keep the size of the second edition within reasonable bounds, the chapter on incomplete-block designs included in the first edition has been deleted. The chapter on the analysis of covariance has been enlarged and rewritten to present this topic in a somewhat different light.

Some of the new material may be omitted without serious loss in readability of what remains. Those sections of the book which may be omitted in a first course in experimental design are as follows: 1.11 to 1.13, 2.3 to 2.9, 3.15 to 3.20, 4.9, 5.23 to 5.25, 6.14, 6.15, and 8.1 to 8.15.

The contents of Chapter 10 may be read independently of the contents of Chapter 9. From some points of view, Chapter 9, which deals with the Latin-square principle, is a special case of Chapter 8, which takes up the general principle of confounding in factorial experiments and partial replication of factorial experiments.

Appendix A, Random Variables, is new to the second edition. This appendix covers key sampling distributions underlying basic statistical theory. It is included to help provide background for reading the material on linear models in Chapters 2, 3, and 4.

Relatively complete tables for sampling distributions of statistics used in the analysis of experimental designs are included in Appendix C. Included in the second edition are tables of the noncentral t and F distributions. Ample references to source materials having mathematical proofs for the principles stated in the text are provided.

The author is indebted to E. S. Pearson and the trustees of *Biometrika* for permission to reproduce parts of Tables C.1, C.3, C.7, and C.9 from *Biometrika Tables for Statisticians*, vol. I, 2d ed. He wishes to thank H. L. Harter, D. S. Clem, and E. H. Guthrie for permission to reproduce Table C.4, which was taken from WADC Technical Report 58-484, vol. II, 1959. He appreciates the cooperation of the editor of the *Journal of the American Statistical Association* and of C. W. Dunnett for Table C.6, D. B. Owen for Table C.13, and M. L. Tiku for Table C.14.

The author is indebted to C. Eisenhart, M. W. Hastay, and W. A. Wallis for permission to reprint Table C.8, which appears in *Techniques of Statistical Analysis*, 1947. He wishes to thank L. S. Feldt and M. W. Mahmoud as well as the editor of *Psychometrika* for permission to reprint Table C.11.

Special thanks are due to Mrs. G. P. Lehman and Mrs. R. L. Smith for excellent secretarial assistance in preparing the manuscript for the first edition.

The author is particularly grateful to Dr. Dorothy C. Adkins for many reasons, and to Dr. A. Lubin, whose critical reading of the manuscript did much to help the author prepare the first edition.

Special thanks are due to Mrs. Carol S. Vester for her work in preparing the manuscript for the second edition.

B. J. Winer

STATISTICAL PRINCIPLES IN EXPERIMENTAL DESIGN

INTRODUCTION

The design of an experiment may be compared to an architect's plans for a structure, whether it be a giant skyscraper or a modest home. The basic requirements for the structure are given to the architect by the prospective owner. It is the architect's task to fill these basic requirements; yet the architect has ample room for exercising his ingenuity. Several different plans may be drawn up to meet all the basic requirements. Some plans may be more costly than others; given two plans having the same cost, one may offer potential advantages that the second does not.

In the design of an experiment, the designer has the role of the architect, the experimenter the role of the prospective owner. These two roles are not necessarily mutually exclusive—the experimenter may do a considerable portion of the design work. The basic requirements and primary objectives of the experiment are formulated by the experimenter; the experimenter may or may not be aware of the possible alternative approaches that can be followed in the conduct of his experiment. It is the designer's function to make the experimenter aware of these alternatives and to indicate the potential advantages and disadvantages of each of the alternative approaches. It is, however, the experimenter's task to reach the final decision about the conduct of the experiment.

The individual best qualified to design an experiment is the one who is (1) most familiar with the nature of the experimental material, (2) most familiar with the possible alternative methods for designing the experiment, (3) most capable of evaluating the potential advantages and disadvantages of the alternatives. Where an individual possesses all these qualifications, the roles of experimenter and designer are one. On some research problems in many experimental fields, the experimenter is capable of making all the necessary decisions without seeking extensive assistance. On more complex research problems, the experimenter may turn to colleagues who are equally or more familiar with the subject-matter area for assistance in formulating the basic requirements and primary objectives of his experiment. Problems on the design of the experiment may also be discussed with the subject-matter specialist, and considerable assistance on design problems may be obtained from this source. The experimenter may also turn to the individual whose specialized training is in the area of experimental design, just as the prospective builder turns to the architect for assistance on design problems. If the designer is familiar with the nature of the experimental material and the outcome of past experimentation in the general area of the experiment, he is in a better position to assist the experimenter in evaluating the possible choices as well as to suggest feasible alternative choices.

In the design of experiments there is ample opportunity for ingenuity in the method of attacking the basic problems. Two experiments having identical objectives may be designed in quite different ways: at the same cost in terms of experimental effort, one design may lead to unambiguous results no matter what the outcome, whereas the second design could potentially lead to ambiguous results no matter what the outcome. How good one design is relative to a second for handling the same general objective may be measured (1) in terms of the relative cost of the experimental effort and (2) in terms of the relative precision with which conclusions may be stated. More precise conclusions do not always demand the greater experimental effort, but they generally do demand more careful attention to experimental design.

Without an adequate experimental design, potentially fruitful hypotheses cannot be tested with any acceptable degree of precision. Before rejecting a hypothesis in a research field, one should examine the structure of the experiment to ascertain whether or not the experiment provided a real test of the hypothesis. On the other hand the most carefully planned experiment will not compensate for the lack of a fruitful hypothesis to be tested. In the latter case, the end product of this well-designed experiment can yield only relatively trivial results.

One of the primary objectives of this book is to provide the prospective experimenter with some of the basic principles used in the construction of experimental designs. These principles apply in all areas of experimental work. By the use of these principles, an extensive collection of relatively standard designs has been constructed to handle problems in design that have been encountered in a variety of experiments. These standard designs will be considered in detail, and their potential applications in research areas in the behavioral sciences will be indicated. Seldom does an experimenter have an experiment that is a perfect fit to a standard design. Some modification is frequently required; this is particularly true in experimental work in the area of the behavioral sciences. Careful planning by both the experimenter and the designer is often required in order to cast an experiment in a form that will permit the utilization of a standard design or to modify standard designs in a manner that will more closely meet the requirements of the experiment.

Principles of experimental design have their roots primarily in the logic of scientific method. Indeed logicians have made substantial contributions to the principles of experimental design. The steps from logic to mathematics are small ones. The now classic work on the basic statistical principles underlying experimental design is R. A. Fisher's *The Design of Experiments.* This work includes more than purely mathematical arguments—it probes into the basic logical structure of experiments and examines the manner in which experiments can provide information about problems put to experimental test. Depending upon how the experiment is conducted, it may or may not provide information about the issues at question. What has become standard working equipment for the individuals specializing in the area of experimental design stems in large measure from this and other works of R. A. Fisher.

What is perhaps the equivalent of a master collection of architect's plans is to be found in the work *Experimental Designs* by W. G. Cochran and G. M. Cox. This work is more than a mere collection of designs. It is a carefully prepared and well-organized text and reference book. Illustrative material is drawn from many different research areas, although most of the material is from the field of agriculture.

The statistical theory underlying major aspects of experimental design is by no means complete. The current literature in the area is extensive.

1

INFERENCE WITH RESPECT TO MEANS AND VARIANCES

1.1 Basic Terminology in Sampling

A statistical population is the collection of all elements about which one seeks information. Only a relatively small fraction, or *sample*, of the total number of elements in a statistical population can generally be observed. From data on the elements that are observed, conclusions or inferences are drawn about the characteristics of the entire population. In order to distinguish between quantities computed from observed data and quantities which characterize the population, the term *statistic* will be used to designate a quantity computed from sample data, and the term *parameter* will be used to designate a quantity characteristic of a population. Statistics are computed from sample data for two purposes: (1) to describe the data obtained in the sample, and (2) to estimate or test hypotheses about characteristics of the population.

If all the elements in a statistical population were measured on a characteristic of interest, and if the measurements were then tabulated in the form of a frequency distribution, the result would be the population distribution for the characteristic measured. A description of the population distribution is made in terms of parameters. The number of parameters necessary to describe the population depends on the form of the frequency distribution. If the form is that of the normal distribution, two parameters will completely specify the frequency distribution—the population mean, designated μ, and the population standard deviation, designated σ. If the form is not normal, the mean and the standard deviation may not be sufficient to specify the distribution. Indeed these two parameters may provide relatively little information about the distribution; other parameters may be required.

The sample mean, designated $\bar{X}$, generally provides an estimate of the population mean μ. In these same cases, the sample standard deviation, designated s, generally provides an estimate of the population standard deviation σ. The accuracy, or precision, of estimates of this kind depends upon the size of the sample from which such estimates are computed, the manner in which the sample was drawn from the population, the characteristics of the population from which the sample was drawn, and the principle used to estimate the parameter.

If a sample is drawn in such a way that (1) all elements in the population have an equal and constant chance of being drawn on all draws and (2) all possible samples have an equal (or a fixed and determinable) chance of being drawn, the resulting sample is a *random* sample from the specified population. By no means should a random sample be considered a haphazard, unplanned sample. Numerous other methods exist for drawing samples. Random samples have properties which are particularly important in statistical work. This importance stems from the fact that random sampling ensures constant and independent probabilities; the latter are relatively simple to handle mathematically.

Suppose that one were to draw a large number of samples (say, 100,000), each having n elements, from a specified population. Suppose further that the procedures by which the samples are drawn are comparable for all samples. For each of the samples drawn, suppose that the sample mean $\bar{X}$ and the sample variance s^2 are computed. The frequency distribution of the $\bar{X}$'s defines operationally what is meant by the *sampling distribution* of the sample mean. A distribution constructed in this way provides an *empirically determined* sampling distribution for the mean. The frequency distribution of the sample variances would provide an empirically determined sampling distribution for the variance. The sampling distribution of a statistic depends, in part, upon the way in which the samples are drawn.

Sampling distributions of statistics are generally tabulated in terms of cumulative frequencies, relative frequencies, or probabilities. The characteristics of sampling distributions are also described by parameters. Frequently the parameters of sampling distributions are related to the parameters of the population from which the samples are drawn. The mean of the sampling distribution is called the *expected value* of the statistic. The standard deviation of the sampling distribution is called the *standard error* of the statistic. The form of the sampling distribution as well as the magnitude of its parameters depends upon (1) the distribution of the measurements in the basic population from which the sample was drawn, (2) the sampling plan followed in drawing the samples, and (3) the number of elements in the sample.

Suppose that the basic population from which sample elements are drawn can be considered to be approximately normal in form, with mean equal to some value μ, and with standard deviation equal to some value σ. In other words, the frequency distribution of the measurements of interest is approximately normal in form, with specified values for the parameters. A normal distribution having a mean equal to μ and a standard deviation equal to σ is designated by $N(\mu, \sigma)$. If one were to draw a large number of random samples of size n from a population in which the measurements have the approximate form $N(\mu, \sigma)$, the sampling distribution of the statistic $\bar{X}$ would be approximately normal in form, with expected value approximately equal to μ, and with standard error approximately equal to $\sigma/\sqrt{n}$. Thus the sampling distribution of the mean of random samples of size n from the approximate population $N(\mu, \sigma)$ would be approximately $N(\mu, \sigma/\sqrt{n})$. This result may be verified by empirical sampling experiments.

The sampling distribution of the statistic, $\bar{X}$, assuming random sampling from the exact population $N(\mu, \sigma)$, can be derived mathematically from the properties of random samples; from purely mathematical considerations it can be shown that this sampling distribution is exactly $N(\mu, \sigma/\sqrt{n})$. Herein lies the importance of random samples—they have properties which permit the estimation of sampling distributions from purely mathematical considerations without the necessity for obtaining empirical sampling distributions. Estimates obtained from such samples have highly desirable properties—the latter will be discussed in a later section. Such purely mathematical considerations lead to scientifically useful results only when the experimental procedures adequately conform to the mathematical models used in predicting experimental results. Also, from purely mathematical considerations, it can be shown that the statistic $(n - 1)s^2/\sigma^2$ will have a sampling distribution that corresponds to the chi-square distribution which has $n - 1$ degrees of freedom. This last prediction may also be verified by sampling experiments.

If the population distribution is only approximately normal in form, the mathematical sampling distributions just discussed provide approximations to their operational counterparts; the larger the sample size, the better the approximation. One of the basic theorems in sampling theory, the *central-limit theorem*, states that the sampling distribution of the means of random samples will be approximately normal in form regardless of the form of the distribution in the population, provided that the sample size is sufficiently large and provided that the population variance is finite. The more the population distribution differs from a bell-shaped distribution, the larger the sample size must be for the theorem to hold.

Statistics obtained from samples drawn by means of sampling plans which are not random have sampling distributions which are either unknown or which can only be approximated with unknown precision.

Good approximations to sampling distributions of statistics are required if one is to evaluate the precision of the inferences made from sample data.

1.2 Basic Terminology in Statistical Estimation

Numerical values of parameters can be computed directly from observed data only when measurements on all elements in the population are available. Generally a parameter is estimated from statistics based upon one or more samples. Several criteria are used to evaluate how good a statistic is as an estimate of a parameter. One such criterion is lack of bias. A statistic is an *unbiased estimate* of a parameter if the expected value of the sampling distribution of the statistic is equal to the parameter of which it is an estimate. Thus the concept of unbiasedness is a property of the sampling distribution and not strictly a property of a single statistic. When one says that a given statistic is an unbiased estimate of a parameter, what one implies is that in the long run the mean of such statistics computed from a large number of samples of equal size will be equal to the parameter. If the statistic $\hat{\theta}$ is an estimator of the parameter θ, and if

$$E(\hat{\theta}) = \theta + c,$$

then the bias of the estimator is of magnitude c.

The mean $\bar{X}$ of a random sample from a normal population is an unbiased estimate of the population mean because the sampling distribution of $\bar{X}$ has an expected value equal to μ. Suppose that a random sample of size n is drawn from a specified normal population; suppose that the mean of this sample is 45. Then 45 is an unbiased estimate of the population mean. Suppose that a second random sample of size n is drawn from the same population; suppose that the mean of the second sample is 55. Then 55 is also an unbiased estimate of the population mean. Thus two random samples provide two unbiased estimates of the population mean; these estimates will not, in general, be equal to one another. There is no way of deciding which one, considered by itself, is the better estimate. The best single estimate of the population mean, given the two samples, is the average of the two sample means. This average is also an unbiased estimate of the population mean. It is a better estimate of μ in the sense that it has greater precision.

The *precision* of an estimator is generally measured by the standard error of its sampling distribution. The smaller the standard error, the greater the precision. Of two unbiased estimators whose sampling distributions have the same form, the better estimator is the one having the smaller standard error. The standard error of a sampling distribution is a good index of the precision only in those cases in which the form of the distribution approaches the normal distribution as the sample size increases. For statistics whose sampling distribution has this property, the *best* unbiased estimator is defined to be the one having the smallest standard error.

The *efficiency* of an unbiased estimator is measured relative to the square of the standard error of the best unbiased estimator. For example, if the squared standard error of one unbiased estimator is σ^2/n and the squared standard error of the best unbiased estimator is $\sigma^2/2n$, then the efficiency of the first estimator is defined to be

$$E_f = \frac{\sigma^2/2n}{\sigma^2/n} = \frac{1}{2}.$$

The concept of *consistency* in an estimator is in a sense related to that of unbiasedness. An estimator is a *consistent* estimate of a parameter if the probability that it differs from the parameter by any amount approaches zero as the sample size increases. In other words, a statistic is a consistent estimator if the bias tends toward zero as the sample size increases. An unbiased estimator is a consistent estimator. On the other hand, a consistent estimator may be biased for small samples.

Properties of estimators which hold as the sample size increases are called *asymptotic* properties. How large the sample size must be before asymptotic properties can be reasonably expected to hold varies as a function of the characteristics of the population and the method of sampling being used. Consistent estimators are asymptotically unbiased estimators. Where the bias of a consistent estimator is low but its precision is high, the consistent statistic may be used in preference to an unbiased estimator having less precision.

A parameter is, in most cases, a number. It may be estimated by a number, called a *point estimate* of the parameter. Another way of estimating a parameter is to specify a range of numbers, or an interval, within which the parameter lies. This latter type of estimate is known as an *interval estimate* of the parameter. The difference between the largest and smallest numbers of the interval estimate defines the range, or width, of the interval. The sampling distribution of a statistic obtained by means of purely mathematical considerations will provide information about the relative frequency (probability) of statistics in a given interval. Probabilities obtained directly from such sampling distributions provide predictions about the relative frequency with which statistics of given magnitudes will occur, assuming that conditions specified in the mathematical derivation are true in the empirical population. Thus knowledge of sampling distributions permits one to argue from a specified population to consequences in a series of samples drawn from this population.

In statistical estimation, the objective is to obtain estimates of the parameters in the population, given the observations in the sample. The parameters are unknown. Given the magnitude of certain statistics computed from the observed data, from which of several possible alternative populations was this sample drawn? Concepts of likelihood, confidence, inverse probability, and fiducial probability are used by some statisticians to evaluate the answer to this last question. This question can be rephrased in terms of two of these concepts.

1. Given a sample, what is the *likelihood* that it was drawn from a population having a specified set of parameters?
2. Given a sample, with what *confidence* can it be said that the population from which it was drawn has a specified parameter within a given range?

The likelihood of obtaining a given sample is the probability of obtaining the sample as a function of different values of the parameters underlying the population. Admittedly there is only a single set of parameters underlying a specified population. These values are, however, unknown. The relative frequency with which certain samples will occur depends upon the true values of these parameters. Under one set of assumptions about the parameter values, a given sample may have very high probability of occurring, whereas under a second set of assumptions the probability of the occurrence of a given sample may be very low.

R. A. Fisher introduced a widely used principle in statistical estimation: One selects as an estimator of a parameter that value which will maximize the likelihood of the sample that is actually observed to occur. Estimators having this property are known as *maximum-likelihood estimators.* In many areas of statistics, the principle of maximum likelihood provides estimators having maximum precision (i.e., minimum standard error).

The *least-squares principle* is another widely used principle of estimation. Gauss (among others) introduced this principle into statistics. If an observation X has the structure

$$X = f(\theta_1, \ldots, \theta_p) + \varepsilon,$$

where

$f(\theta_1, \ldots, \theta_p)$ is some known function of the parameters $\theta_1, \ldots, \theta_p$,
ε is a random variable, say "error,"

then the least-squares estimators, $\hat{\theta}_1, \ldots, \hat{\theta}_p$, make

$$\sum_i [X_i - f_i(\hat{\theta}_1, \ldots, \hat{\theta}_p)]^2 = \text{minimum}$$

for the sample data. If $f(\theta_1, \ldots, \theta_p)$ is a linear function of the θ_i, then, in the class of all estimators which are linear functions of the sample observations, the least-squares estimators are minimum-variance, unbiased estimators.

Yet another widely used principle of estimation yields what are called *Bayes estimators.* For large sample size, Bayes estimators tend to differ very little from maximum-likelihood estimators. Under the Bayes principle, the parameter being estimated (say θ) is considered to be a random variable with a known distribution function, called the prior distribution for θ. Associated with different estimators of θ (say $\hat{\theta}_i$) there is a loss function. In many applications this loss function is taken to be proportional to $(\hat{\theta}_i - \theta)^2$. The expected value of this loss function is called the *risk*. A Bayes estimator of θ is some function of the sample observations that minimizes the expected value of the risk.

An interval estimate is frequently referred to as a *confidence interval* for a parameter. The two extreme points in this interval, the upper and lower confidence bounds, define a range of values within which there is a specified likelihood (or level of confidence) that that parameter will fall. Given information from a single sample, the parameter either does or does not lie within this range. The procedure by which the upper and lower confidence bounds are determined will, in the long run (if the study is repeated many times), ensure that the proportion of correct statements is equal to the level of confidence for the interval. The numerical values of the upper and lower confidence bounds change from sample to sample, since these bounds depend in part upon statistics computed from the samples.

An interval estimate of a parameter provides information about the precision of the estimate; a point estimate does not include such information. The principles underlying interval estimation for a parameter are closely related to the principles underlying tests of statistical hypotheses.

1.3 Basic Terminology in Testing Statistical Hypotheses

A *statistical hypothesis* is a statement about a statistical population which, on the basis of information obtained from observed data, one seeks to support or refute. A *statistical test* is a set of rules whereby a decision about the hypothesis is reached. Associated with the decision rules is some indication of the accuracy of the decisions reached by following the rules. The measure of the accuracy is a probability statement about making the correct decision when various conditions are true in the population in which the hypothesis applies.

The design of an experiment has a great deal to do with the accuracy of the decisions based upon information supplied by an experiment. The decision rules depend in part upon what the experimenter considers critical bounds on arriving at the wrong decision. However, a statistical hypothesis does not become false when it exceeds such critical bounds, nor does the hypothesis become true when it does not exceed such bounds. Decision rules are guides in summarizing the results of a statistical test—following such guides enables the experimenter to attach probability statements to his decisions. In evaluating the outcome of a single experiment or in using the information in a single experiment as a basis for a course of action, whether an outcome exceeds an arbitrary critical value may or may not be relevant to the issue at hand. Probability statements that are associated with decision rules in a statistical test are predictions as to what may be expected to be the case if the conditions of the experiment were repeated a large number of times.

The logic of tests on statistical hypotheses is as follows: One assumes that the hypothesis that one desires to test is true. Then one examines the consequences of this assumption in terms of a sampling distribution which depends upon the truth of this hypothesis. If, as determined from the sampling distribution, observed data have relatively high probability of occurring, the decision is made that the data do not contradict the hypothesis. On the other hand, if the probability of an observed set of data is relatively low when the hypothesis is true, the decision is that the data tend to contradict the hypothesis. Frequently the hypothesis that is tested is stated in such a way that, when the data tend to contradict it, the experimenter is actually demonstrating what it is that he is trying to establish. In such cases the experimenter is interested in being able to reject or nullify the hypothesis being tested.

The *level of significance* of a statistical test defines the probability level that is to be considered too low to warrant support of the hypothesis being tested. If the probability of the occurrence of observed data (when the hypothesis being tested is true) is smaller than the level of significance, then the data are said to contradict the hypothesis being tested, and a decision is made to reject this hypothesis. Rejection of the hypothesis being tested is equivalent to supporting one of the possible alternative hypotheses which are not contradicted.

The hypothesis being tested will be designated by the symbol H_1. (In some notation systems this hypothesis has been designated by the symbol H_0.) The set of hypotheses that remain tenable when H_1 is rejected will be called the alternative hypothesis and will be designated by the symbol H_2. The decision rules in a statistical test are with respect to the rejection or nonrejection of H_1. The rejection of H_1 may be regarded as a decision to accept H_2; the nonrejection of H_1 may be regarded as a decision against the acceptance of H_2. If the decision rules reject H_1 when in fact H_1 is true, the rules lead to an erroneous decision. The probability of making this kind of error is at most equal to the level of significance of the test. Thus the level of significance sets an upper bound on the probability of making a decision to reject H_1 when in fact H_1 is true. This kind of erroneous decision is known as a *type* 1 *error;* the probability of making a type 1 error is controlled by the level of significance.

If the decision rules do not reject H_1, when in fact one of the alternative hypotheses is true, the rules also lead to an erroneous decision. This kind of error is known as a *type* 2 *error.* The potential magnitude of a type 2 error depends in part upon the level of significance and in part upon which one of the possible alternative hypotheses actually is true. Associated with each of the possible alternative hypotheses is a type 2 error of a different magnitude. The magnitude of a type 1 error is designated by the symbol α, and the magnitude of the type 2 error for a specified alternative hypothesis is designated by the symbol β. The definitions of type 1 and type 2 errors may be summarized as follows:

Decision	State of affairs in the population	
	H_1 true	H_1 false H_2 true
Reject H_1	Type 1 error (α)	No error
Do not reject H_1	No error	Type 2 error (β)

In this summary, rejection of H_1 is regarded as being equivalent to accepting H_2 and nonrejection of H_1 equivalent to not accepting H_2. The possibility of a type 1 error exists only when the decision is to reject H_1; the possibility of a type 2 error exists only when the decision is not to reject H_1.

The experimenter has the level of significance (type 1 error) directly under his control. Type 2 error is controlled indirectly, primarily through the design of the experiment. If possible, the hypothesis to be tested is stated in such a way that the more costly error is type 1 error. It is desirable to have both types of error small. However, the two types of error are not independent—the smaller numerically the type 1 error, the larger numerically the potential type 2 error.

To see the relationship between the two types of error, consider Fig. 1.3-1. In part *a* of this figure the left-hand curve represents the sampling

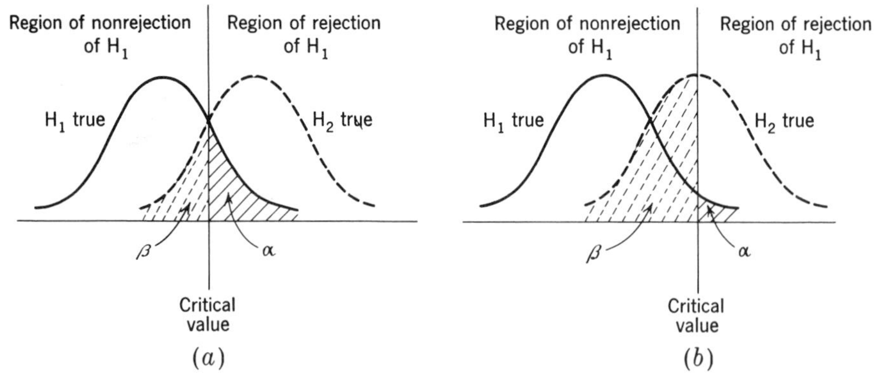

Figure 1.3-1

distribution of a relevant statistic when H_1 is true, and the right-hand curve represents the sampling distribution of the same statistic when a particular H_2 is true. The region of rejection of H_1 is defined with reference to the sampling distribution which assumes that H_1 is true. The decision rules specify that H_1 is to be rejected if an observed statistic has any value in the region of rejection. The probability of a statistic's falling in this region is equal to α when H_1 is true. The type 2 error associated with the particular H_2 represented in part *a* is numerically equal to the area under the right-hand curve which falls in the region of nonrejection of H_1.

In part *b* the numerical value of α is smaller than that in part *a*. This means that the decision rule has smaller type 1 error. The area under the right-hand curve in part *b* that falls in the region of nonrejection of H_1 is larger than the corresponding area in part *a*. Decreasing the numerical value of the type 1 error (level of significance) will increase the potential magnitude of the type 2 error.

The *power* of a test with respect to a specified alternative hypothesis is numerically equal to 1 minus the probability of a type 2 error. Represented geometrically, the power of a test is the area of the sampling distribution, when H_2 is true, that falls in the region of rejection of H_1. In part *a* of the figure this is the area under the right-hand curve that is to the right of the critical value. The power of a test decreases as the numerical value of α decreases.

The power of a test may be defined symbolically as

$$\text{Power} = P(\text{decision rejects } H_1 \mid H_2 \text{ true}).$$

In words, power is the probability that the decision rule rejects H_1 when a specified H_2 is true. Each of the possible hypotheses in H_2 has its own power. The closer an alternative hypothesis is to H_1, that is, the greater the overlap of the corresponding sampling distributions, the lower will be the power of the test with respect to that alternative. A well-designed experiment will have relatively high power with respect to all alternatives which are different in a practical sense from H_1. For example, if H_1 states that there is zero difference between two means, then one of the possible alternative hypotheses is that the difference is .001 unit. For all practical purposes this alternative may not be different from H_1; hence power with respect to this alternative need not be of concern to the experimenter. However, an alternative hypothesis which states that the difference is 5 units may have practically important consequences if true. Power with respect to this alternative would be a matter of concern to the experimenter.

In research in the area of the behavioral sciences, it is often difficult to evaluate the relative costs of type 1 and type 2 in terms of meaningful units. Both kinds of errors may be equally important, particularly in exploratory work. Too much emphasis has been placed upon the level of significance of a test and far too little emphasis upon the power of the test. In many cases where H_1 is not rejected, were the power of such tests studied carefully, the decisions might more appropriately have been that the experiment did not really provide an adequately sensitive (powerful) test of the hypothesis.

No absolute standards can be set up for determining the appropriate level of significance and power that a test should have. The level of significance used in making statistical tests should be gauged in part by the power of practically important alternative hypotheses at varying levels of significance. If experiments were conducted in the best of all possible worlds, the design of the experiment would provide adequate power for any predetermined level of significance that the experimenter were to set. However, experiments are conducted under the conditions that exist within the world in which one lives. What is needed to attain the demands of the well-designed experiment may not be realized. The experimenter must be satisfied with the best design feasible within the restrictions imposed by the working conditions. The frequent use of the .05 and .01 levels of significance is a matter of a convention having little scientific or logical basis. When the power of tests is likely to be low under these levels of significance, and when type 1 and type 2 errors are of approximately equal importance, the .30 and .20 levels of significance may be more appropriate than the .05 and .01 levels.

The evidence provided by a single experiment with respect to the truth or falsity of a statistical hypothesis is seldom complete enough to arrive at a decision which is free of all possible error. The potential risks in decisions based upon experimental evidence may in most cases be evaluated. What the magnitude of the risks should be before one takes a specified action in each case will depend upon existing conditions. The data from the statistical test will provide likelihoods associated with various actions.

1.4 Testing Hypotheses on Means—σ Assumed Known

To illustrate the basic procedures for making a statistical test, a highly simplified example will be used. Suppose that experience has shown that the form of the distribution of measurements on a characteristic of interest in a specified population is approximately normal. Further suppose, given data on a random sample of size 25 from this population, that information about the population mean μ is desired. In particular the experimenter is interested in finding out whether or not the data support the hypothesis that μ is greater than 50.

The first step in the test is to formulate H_1 and H_2. Suppose that an erroneous decision to reject the hypothesis that the population mean is 50 is more costly than an erroneous decision to reject the hypothesis that the mean is greater than 50. In this case H_1 is chosen to be $\mu = 50$; this choice for H_1 makes the more costly type of error the type 1 error, which is under the direct control of the experimenter. The alternative hypothesis in this case is $\mu > 50$. The decision rules for this test are to be formulated in such a way that rejection of H_1 is to provide evidence in favor of the tenability of H_2.

The choice for H_1 could also be $\mu \leq 50$. However, if the data tend to reject the hypothesis that $\mu = 50$ and support the hypothesis that $\mu > 50$, then the data will also tend to reject the hypothesis that $\mu < 50$. Thus, in formulating a decision rule which rejects H_1 only when the data support the hypothesis that $\mu > 50$, only the hypothesis that $\mu = 50$ need be considered. In essence the case $\mu < 50$ is irrelevant (inadmissible) in formulating the decision rule. However, nonrejection of H_1 would imply $\mu \leq 50$.

When it is true that $\mu = 50$, the sampling distribution of the mean of random samples from a normal population is normal in form, with expected value equal to 50 and standard error equal to the population standard deviation divided by the square root of the sample size. In practice the value of the population standard deviation will not be known, but to keep this example simple, suppose that the population standard deviation σ is equal to 10. Then the standard error of the sampling distribution of the mean for samples of size 25 is $\sigma/\sqrt{n} = 10/\sqrt{25} = 2$. Decision rules must now be formulated to indicate when observed data are consistent with H_1 and when observed data are not consistent with H_1. The decision rules must indicate a range of potentially observable values of $\bar{X}$ for which the decision will be to reject H_1. This range of values of $\bar{X}$ will be called the region of rejection for H_1. The probability of observing an $\bar{X}$ in this region is to be at most equal to the level of significance of the test, i.e., the magnitude of the type 1 error. This sets an upper bound on the probability of reaching the wrong decision when H_1 is true. In addition to satisfying this condition with respect to type 1 error, the region of rejection for H_1 must have relatively high probability for the observed $\bar{X}$ when H_2 is true. Hence the decision rules must specify a range of values of potentially observable $\bar{X}$ in which (1) the probability of an observed $\bar{X}$'s falling in this region is at most equal to the level of significance when H_1 is true and (2) the probability of an $\bar{X}$'s falling in this region is relatively high when H_2 is true. The latter condition is necessary to assure the power of the test.

Probabilities associated with the sampling distribution of $\bar{X}$ when H_1 is true are required in order to construct the decision rules. In addition, some knowledge about the relative location of the sampling distribution of $\bar{X}$ when each of the possible alternative hypotheses is true is required. Consider Fig. 1.4-1. When H_1 is true, the sampling distribution of $\bar{X}$ is given by (1). When H_2 is true (that is, μ is greater than 50), the sampling distribution of $\bar{X}$ will have an expected value somewhere to the right of 50. In particular, one possibility for this expected value is that $\mu =$

54. This possibility is represented by (2). Areas under these curves represent probabilities. The probability of observing an $\bar{X}$ in a range of

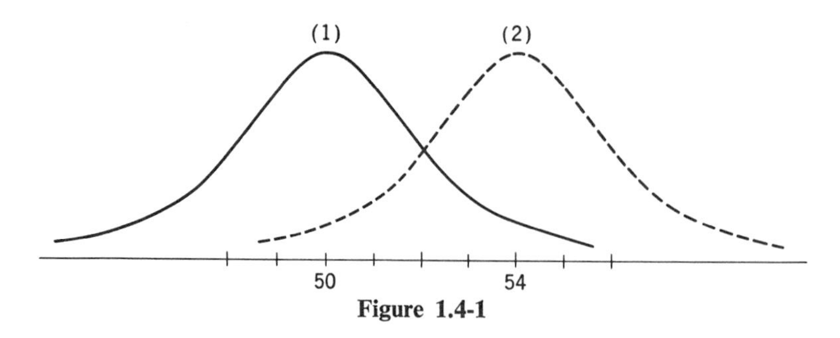

Figure 1.4-1

values covered by the extreme right-hand tail of (1) is relatively low when H_1 is true but relatively higher when the alternative hypothesis is true.

Suppose that the experimenter wants to formulate a set of decision rules which, in the long run, will make the probability of an erroneous decision when H_1 is true less than .01. This is another way of saying that the level of significance of the test is to be .01. Suppose that the mean of the potentially observable sample is designated by the symbol $\bar{X}_{\text{obs}}$. Then the decision rules will take the following form:

Reject H_1 when $\bar{X}_{\text{obs}}$ is greater than L.

Do not reject H_1 otherwise.

L is the critical value for $\bar{X}_{\text{obs}}$; L must have the property that

$$P(\bar{X}_{\text{obs}} > L \mid H_1 \text{ true}) = .01.$$

In words, the probability of drawing a sample whose mean is greater than L is to be .01 when H_1 is true.

Under the assumptions that have been made, when H_1 is true the form and parameter of the sampling distribution for sample means are known to be $N(50,2)$, that is, normal in form with expected value equal to 50 and standard error equal to 2. The tabulated values of the normal distribution are directly appropriate only for the standard normal, $N(0,1)$. From the table of the standard normal, the probability of observing a value 2.33 standard-error units or more above the mean of a population is .01. For the distribution $N(50,2)$, 2.33 standard-error units above the mean would be $50 + 2.33(2) = 54.66$. Therefore,

$$P(\bar{X} > 54.66 \mid \text{H}_1 \text{ true}) = .01.$$

Thus the region of rejection for H_1 is $\bar{X} > 54.66$. When H_1 is true, the probability that a random sample of size 25 from $N(50,10)$ will have a mean larger than 54.66 is less than .01. When one of the alternative hypotheses is true, i.e., when μ is greater than 50, the probability of a sample mean falling in this region will be higher than .01; the larger the difference between the true value of μ and 50, the higher the probability of an observed mean falling in the region of rejection.

The steps in the formulation of the decision rule have been as follows:

1. Basic population of measurements assumed to be normal in form, with $\sigma = 10$.
2. Random sample of size $n = 25$ elements to be drawn from this population.
3. $\bar{X}_{\text{obs}}$ to be computed from sample data.

The hypothesis being tested, the alternative hypothesis, the level of significance of the test, and the decision rule are as follows:

$$H_1: \quad \mu = 50.$$
$$H_2: \quad \mu > 50.$$
$$\alpha = .01.$$

Decision rules: Reject H_1 when $\bar{X}_{\text{obs}} > 54.66$.
Do not reject H_1 otherwise.

The region of rejection for H_1 may be represented geometrically as the right-hand tail of the sampling distribution for $\bar{X}$ which assumes H_1 to be true (see Fig. 1.4-2).

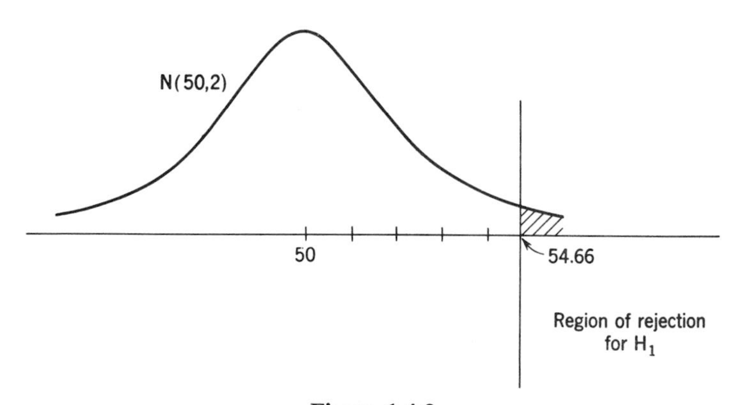

Figure 1.4-2

There are many regions in which the probability is .01 for observing a sample mean. The level of significance of a test does not determine where the region of rejection is to be located. The choice of the extreme right-hand tail of the sampling distribution which assumes H_1 to be true was necessary in order to minimize type 2 error (or, equivalently, to maximize the power). In general, the alternative hypothesis determines the *location* of the region of rejection, whereas the level of significance determines the *size* of the region of rejection. In this case the alternative hypothesis does not include the possibility that μ is less than 50. No matter how much smaller than 54.66 the observed sample mean is, H_1 is not rejected. Thus, if H_1 is not rejected, the evidence would indicate that μ is equal to or less than 50. On the other hand, if H_1 is rejected, the evidence would indicate that μ is greater than 50. Locating the region of rejection for H_1 in the right-hand tail provides maximum power with respect to the alternative hypothesis that μ is greater than 50.

The power of these decision rules with respect to various alternative hypotheses is readily computed. For example, the power with respect to the alternative hypothesis $\mu = 58$ is represented geometrically by the shaded area under curve (2) in Fig. 1.4-3. This area represents the probability of an observed mean's being greater than 54.66 when the true sampling distribution is $N(58,2)$. With reference to the latter sampling distribution, the point 54.66, which determines the region of rejection, is $(54.66 - 58.00)/2$ or 1.67 standard-error units below the mean. The area from the mean to 1.67 standard-error units below the mean is .45. Hence the total shaded area is $.45 + .50 = .95$. Thus the power of this test with respect to the alternative hypothesis $\mu = 58$ is .95. Conversely, the probability of a type 2 error when $\mu = 58$ is .05.

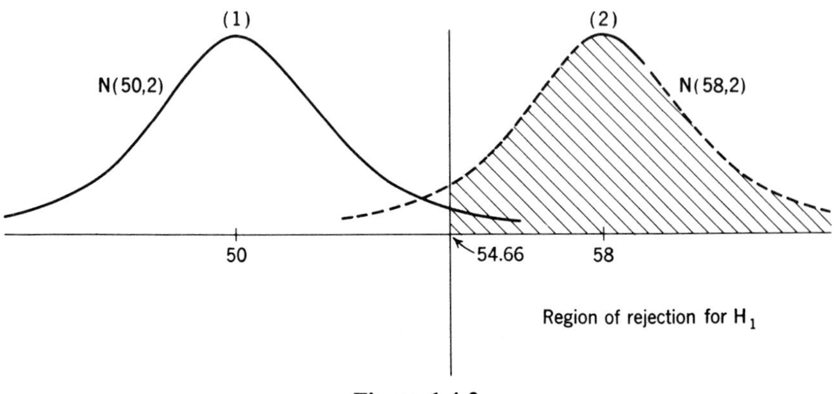

Figure 1.4-3

Instead of working directly with the sampling distribution of the statistic $\bar{X}$ and formulating the decision rules in terms of the statistic $\bar{X}$, it is more convenient to work with the statistic

$$z = \frac{\bar{X} - \mu_1}{\sigma/\sqrt{n}},$$

where μ_1 is the value specified by H_1. The sampling distribution of this z statistic is $N(0,1)$. Given the mean of an observed sample, $\bar{X}_{\text{obs}}$, the corresponding value of the z statistic, when H_1 is $\mu = 50$ and $\sigma/\sqrt{n} = 2$, is

$$z_{\text{obs}} = \frac{\bar{X}_{\text{obs}} - 50}{2}.$$

If the alternative hypothesis is H_2: $\mu > 50$, then the decision rules for a test having level of significance .01 are as follows:

Reject H_1 when $z_{\text{obs}} > 2.33$.
Do not reject H_1 otherwise.

The value 2.33 satisfies the condition that

$$P(z_{\text{obs}} > 2.33 \mid H_1 \text{ true}) = .01.$$

This numerical value actually is the 99 percentile point on $N(0,1)$ and will be designated by the symbol $z_{.99}$. Thus $z_{.99} = 2.33$. Since the level of significance for this test is $\alpha = .01$, the critical value for the decision rule can be designated $z_{1-\alpha}$, which in this case is $z_{.99}$.

For the general case in which the region of rejection for H_1 is the right-hand tail of $N(0,1)$ and the level of significance is equal to some value α, the decision rules take the following form:

Reject H_1 when $z_{\text{obs}} > z_{1-\alpha}$.
Do not reject H_1 otherwise.

Suppose that the mean for the sample observed actually is 60. Then the numerical value of the z statistic (when H_1 is that $\mu = 50$ and $\sigma/\sqrt{n} = 2$) is

$$z_{\text{obs}} = \frac{60 - 50}{2} = 5.00.$$

Since z_{obs} is larger than 2.33, H_1 is rejected. Hence the observed data do not support the hypothesis that the population mean is 50. The data indicate that the mean in the population is greater than 50.

If the alternative hypothesis has the form H_2: $\mu \neq 50$, then the region of rejection for H_1 usually has the form

$$z_{\text{obs}} < z_{\alpha/2} \qquad \text{and} \qquad z_{\text{obs}} > z_{1-(\alpha/2)}.$$

For this kind of alternative hypothesis, the region of rejection for H_1 includes both the left-hand and right-hand extreme tails of the sampling distribution associated with H_1. The two parts of the region of rejection for H_1 are sketched in Fig. 1.4-4. For example, if $\alpha = .01$, the two-tailed

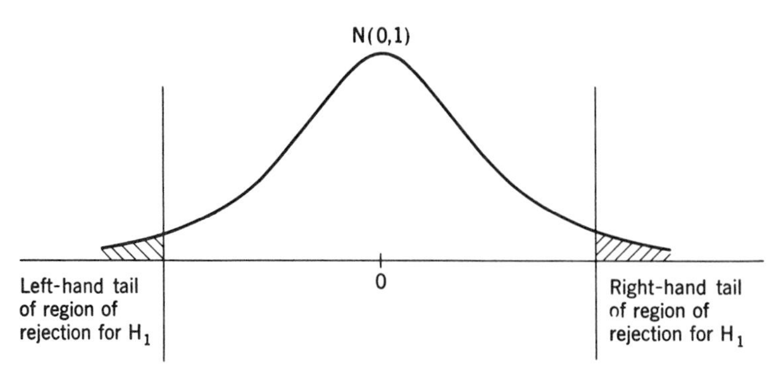

Figure 1.4-4

region of rejection for H_1 would be z_{obs} smaller than $z_{.005}$ and z_{obs} greater than $z_{.995}$. Locating the region of rejection for H_1 in this manner provides power with respect to the possibility that μ is less than 50, as well as to the possibility that μ is greater than 50. An alternative hypothesis of this form is called a two-tailed alternative hypothesis, and tests which admit to the possibility of a two-tailed alternative hypothesis are called two-tailed tests. The size of either tail of the region of rejection is equal to one-half the level of significance; the total size of the region of rejection is equal to the level of significance.

Dividing α into two equal parts in this way provides equal power with respect to alternative hypotheses equidistant (on the left and right) from H_1. If there is reason to want different power for such equidistant alternative hypotheses, α should not be divided into equal parts. For example, if it is desired to have greater power with respect to alternative hypotheses to the right of H_1, the size of the right-hand critical region may be $2\alpha/3$, whereas the size of the left-hand critical region may be $\alpha/3$.

In cases in which the experimenter is interested in rejecting H_1 only when the alternative hypothesis is one having a specified direction with respect to H_1, a one-tailed rather than a two-tailed alternative hypothesis is the more appropriate. Limiting the region of rejection to one tail of the sampling distribution for H_1 provides greater power with respect to an alternative hypothesis in the direction of that tail. This fact is illustrated geometrically in Fig. 1.4-5. The power under a two-tailed test with respect to a specified alternative hypothesis to the right of zero is shown by the shaded area in part *a*. The corresponding power with respect to a one-tailed test is shown in part *b*. Although the magnitude of the type 1 error is the same in both cases, the increased power in the one-tailed case is at the expense of zero power with respect to alternative hypotheses which are to the left of zero. In the latter case, all hypotheses corresponding to sampling distributions to the left of zero may be considered part of H_1.

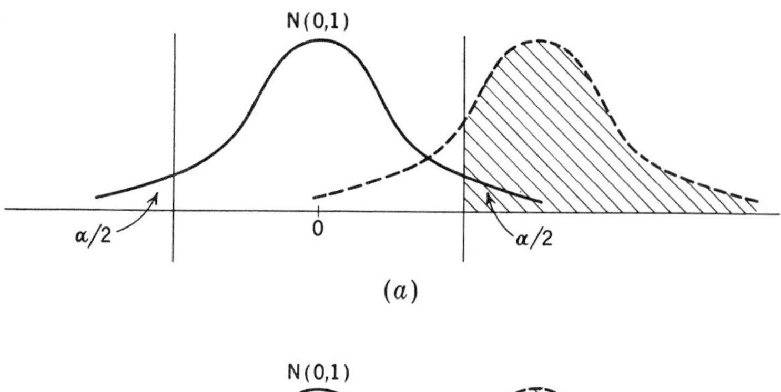

(a)

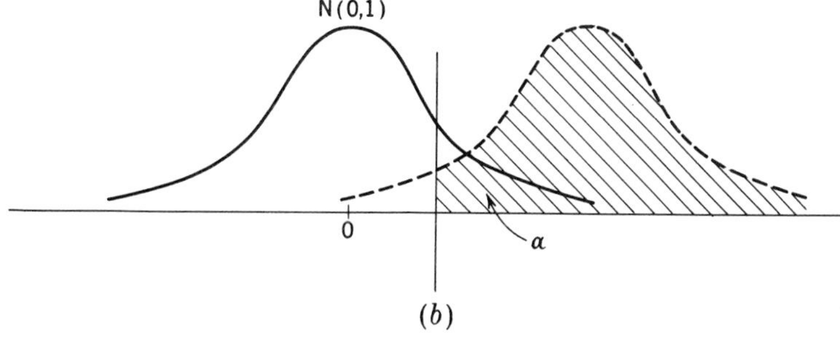

(b)

Figure 1.4-5

1.5 Testing Hypotheses on Means—σ Unknown

If $\bar{X}$ is distributed as $N(\mu, \sigma_X^2/n)$, the sampling distribution of

$$t = \frac{\bar{X} - \mu}{\sqrt{s^2/n}}, \qquad \text{where} \qquad s^2 = \frac{\Sigma(X_i - \bar{X})^2}{n - 1},$$

is Student's t with $n - 1$ degrees of freedom. The degrees of freedom for this distribution are those associated with s^2. If $\mu' \neq \mu$, and if $\bar{X}$ is distributed as $N(\mu', \sigma_X^2/n)$, then the statistic

$$t' = \frac{\bar{X} - \mu}{\sqrt{s^2/n}} = \frac{\bar{X} - \mu'}{\sqrt{s^2/n}} + \frac{\mu' - \mu}{\sqrt{\sigma^2/n}}\sqrt{\frac{\sigma^2}{s^2}}$$

$$= t + \delta\sqrt{\frac{\sigma^2}{s^2}},$$

where

$$\delta = (\mu' - \mu)\frac{\sqrt{n}}{\sigma},$$

is distributed as the *noncentral t* distribution with parameters $n - 1$ and δ, where δ is called the noncentrality parameter. Symbolically,

$$t' \text{ is distributed as } t(n - 1; \delta).$$

Note that when $\mu' = \mu$ then $\delta = 0$ and $t' = t$. Thus Student's t distribution is that special case of the noncentral t distribution for which the noncentrality parameter is zero.

It is noted that the normality assumption in the background of the t distribution is with respect to the sampling distribution of $\bar{X}$. Even though the distribution of X may not be normal, the distribution of $\bar{X}$ will tend to be normal in form asymptotically in n, that is, as n gets large. How large n has to be before this asymptotic property holds depends upon how far the distribution of X deviates from the normal distribution. If the latter distribution is normal in form, normality of the distribution of $\bar{X}$ holds for all n.

Since

$$t = \frac{\bar{X} - \mu}{\sqrt{s^2/n}}$$

contains the parameter μ, one may regard t as a parameter to be estimated. From data in a single sample, one obtains a set of numerical values for $\bar{X}$, s^2, and n. To estimate t one must specify a value for μ. The latter is given in terms of the hypothesis being tested. In testing hypotheses, Student's t distribution provides the statistical basis for setting up a region of rejection for H_1. The noncentral t distribution permits one to evaluate the power of the test with respect to alternative values for H_2.

For the one-tailed alternative hypothesis $\mu > \mu_1$, the region of rejection or H_1 is given by

$$t_{\text{obs}} > t_{1-\alpha}(n - 1),$$

where α is the level of significance and $t_{1-\alpha}(n - 1)$ is the $1 - \alpha$ percentile point on Student's t distribution having $n - 1$ degrees of freedom. An alternative notation system for this percentile point is

$$t_{1-\alpha;n-1}.$$

To illustrate these procedures, suppose that the data observed in a random sample from the population of interest are

$$n = 25, \qquad \bar{X} = 60, \qquad s = 15.$$

Suppose that the statement of the hypothesis to be tested and the alternative hypothesis are

$$H_1: \quad \mu = 50$$

$$H_2: \quad \mu > 50$$

and that the level of significance of the test is .01. From the table of the distribution of the t statistic having 24 degrees of freedom one finds that

$$P(t_{\text{obs}} > 2.49 \mid H_1 \text{ true}) = .01.$$

That is, the table of the t distribution indicates that $t_{.99}(24) = 2.49$. Hence the decision rules for this test are as follows:

Reject H_1 when t_{obs} is larger than 2.49.

Do not reject H_1 otherwise.

From the sample data, t_{obs} is found to be

$$t_{\text{obs}} = \frac{60 - 50}{15/\sqrt{25}} = 3.33.$$

Since t_{obs} is greater than the critical value 2.49, t_{obs} falls in the region of rejection for H_1. Hence the decision rules indicate that H_1 should be rejected.

The interpretation of this test is as follows: On the basis of the data in a random sample of size 25 from a population of interest, the hypothesis that $\mu = 50$ cannot be considered tenable when the test is made at the .01 level of significance. If this hypothesis were true, the probability of obtaining the data in the sample would be less than .01. The data obtained support the hypothesis that the mean of the population is greater than 50.

Having rejected the hypothesis that $\mu = 50$, suppose that the experimenter wanted to find the smallest value or μ_1 which would lead to nonrejection of H_1. The region of nonrejection for H_1 is defined by the inequality

$$\frac{\bar{X} - \mu_1}{s/\sqrt{n}} \leq t_{1-\alpha}(n - 1).$$

Solving this inequality for μ_1 gives

$$\mu_1 \geq \bar{X} - \frac{s}{\sqrt{n}} t_{1-\alpha}(n - 1).$$

Thus any value of μ_1 equal to or greater than $\bar{X} - (s/\sqrt{n})t_{1-\alpha}(n - 1)$ will yield a t statistic that will fall in the region of nonrejection for H_1. For the numerical example just considered, any H_1 that specifies μ_1 to be equal to or greater than

$$60 - (3.00)(2.49) = 52.53$$

would, on the basis of the single sample observed, lead to a decision not to reject H_1. Thus, on the evidence supplied by the single sample observed, any value for μ_1 equal to or greater than 52.53 would make t_{obs} smaller than the critical value of 2.49. Therefore the experimenter may conclude that the population mean is likely to be greater than 52.53. If the experimenter were to test hypotheses specifying that μ_1 is any value equal to or less than 52.53, the decision in every case (for the data in the given sample) would be to reject H_1. This conclusion may be expressed in the form of a one-tailed confidence interval on the population mean. This confidence interval takes the general form

$$C\left[\mu \geq \bar{X} - \frac{s}{\sqrt{n}} t_{1-\alpha}(n-1)\right] = 1 - \alpha.$$

The numerical values in terms of the observed sample data and $\alpha = .01$ are

$$C[\mu \geq 52.53] = .99.$$

The value 52.53 may be considered as the lower bound for μ. If one were to draw additional samples, the mathematical form of the lower bound would remain the same but its numerical value would change, since the numerical values of $\bar{X}$ and s would change. Once a sample has been drawn and numerical values for the confidence interval determined, the statement made in the confidence interval is either true or false. However, the procedure by which the confidence interval is constructed will, in the long run, lead to statements which are correct with probability equal to $1 - \alpha$.

The example that has just been considered involved a one-tailed alternative hypothesis. Suppose that the experimenter is willing to reject H_1 when μ is either smaller or larger than 50. In this case the alternative hypothesis takes the form $\mu \neq 50$. To provide power with respect to both tails of the alternative hypothesis, the decision rules for this case are as follows:

$$\text{Reject } H_1 \text{ when } t_{\text{obs}} \begin{cases} < t_{\alpha/2}(n-1). \\ > t_{1-(\alpha/2)}(n-1). \end{cases}$$

Otherwise do not reject H_1.

The region of rejection for H_1 for the case of a two-tailed alternative hypothesis is sketched in Fig. 1.5-1. The size of the region of rejection in

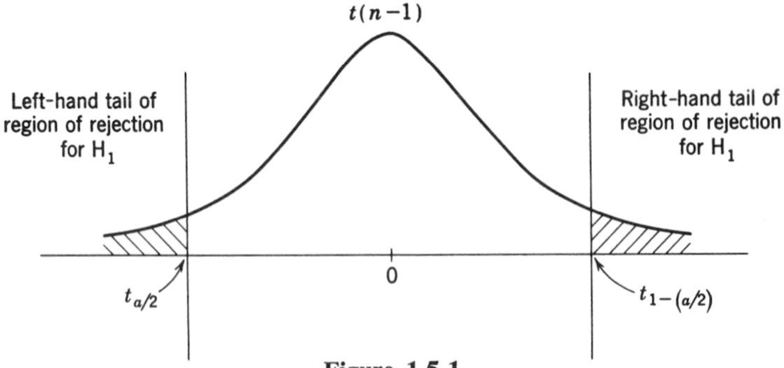

Figure 1.5-1

each tail is equal to $\alpha/2$. The left-hand tail of the region of rejection makes provision for power with respect to alternative hypotheses $\mu < 50$; the right-hand tail of the region of rejection makes provision for power with respect to alternative hypotheses $\mu > 50$.

For $n - 1 = 24$ and $\alpha = .01$,

$$t_{\alpha/2} = t_{.005} = -2.80,$$

and

$$t_{1-(\alpha/2)} = t_{.995} = 2.80.$$

(Since Student's t distribution is symmetrical, $t_{\alpha/2} = -t_{1-(\alpha/2)}$.) For this case the decision rules are

$$\text{Reject } H_1 \text{ when } t_{\text{obs}} \begin{cases} < -2.80. \\ > 2.80. \end{cases}$$

Otherwise do not reject H_1.

For $\bar{X}_{\text{obs}} = 60$ and $s_{\text{obs}} = 15$, $t_{\text{obs}} = 3.33$. Hence the decision rules lead to rejecting H_1. Having rejected the hypothesis that $\mu = 50$, the experimenter may be interested in determining the range of values of μ which, on the basis of the observed sample data, would not be rejected by these decision rules. This range of values is defined by a two-tailed confidence interval on μ, which is given by

$$C[\bar{X} - c \le \mu \le \bar{X} + c] = 1 - \alpha,$$

where

$$c = \frac{s}{\sqrt{n}} t_{1-(\alpha/2)}(n - 1).$$

The numerical value of this confidence interval for the observed sample and $\alpha = .01$ is

$$C[51.60 \le \mu \le 68.40] = .99.$$

To illustrate the fact that any value of μ in this range which is selected for H_1 leads to a decision not to reject H_1, consider the hypothesis that $\mu = 68$. For this H_1

$$t_{\text{obs}} = \frac{60 - 68}{3} = -2.33.$$

Since t_{obs} is greater than -2.80 the hypothesis that $\mu = 68$ is not rejected.

This relationship between confidence intervals and tests of hypotheses applies to many classes of tests. Given a set of statistics computed from a sample, and given a confidence interval of size $1 - \alpha$ on a parameter, then the range of values within this interval will provide a range for the parameter in H_1 which will lead to a decision not to reject H_1 at level of significance α. If H_1 specifies values of the parameter outside this range, the decision will be to reject H_1. Two-tailed confidence intervals are associated with two-tailed tests and one-tailed confidence intervals with one-tailed tests. Thus confidence intervals provide information about the potential outcomes of a series of individual tests .

The power of tests associated with Student's distribution is more difficult to compute than the power of tests associated with the normal distribution. Since the sampling distribution of the t statistic when H_1 is not true is a noncentral t distribution, computation of the power with respect to alternative hypotheses requires tables of the noncentral t distribution. Illustrations of the applications of this latter distribution are given in Johnson and Welch (1940).

Partial tables of the noncentral t distribution are given in Table C.13 in the Appendix. In the notation used in these tables the noncentrality parameter is denoted by the symbol δ, that is,

$$\delta = (\mu' - \mu)\frac{\sqrt{n}}{\sigma}.$$

In Table C.13, f represents the degrees of freedom associated with the central t statistic. To illustrate the use of these tables, consider the hypothesis

$$H_1: \quad \mu = 50,$$
$$H_2: \quad \mu > 50,$$
$$\alpha = .01.$$

If $n = 25$ and σ is assumed to be 5, then the noncentrality parameter corresponding to $\mu' = 53.36$ is

$$\delta = (53.36 - 50)\frac{\sqrt{25}}{5} = 3.36.$$

In this case $f = n - 1 = 24$. From Table C.13 corresponding to $\alpha = .01$, in row 24 one finds the entry 3.36 under the column corresponding to $\beta = .20$. Hence, the type 2 error associated with the alternative hypothesis $\mu' = 53.36$ is .20. The power is $1 - \beta = .80$. Had the level of significance of this test been $\alpha = .05$, the power with respect to $\mu' = 53.36$ would be approximately .94.

(In some widely used notation systems, μ_0 is the value of the parameter as specified by the hypothesis being tested, and μ_1 is the value of the parameter specified by an alternative hypothesis. In this notation system,

$$\mu_1 - \mu_0 \text{ corresponds to } \mu' - \mu.$$

Hence the notation for the noncentrality parameter as defined in Table C.13.)

1.6 Testing Hypotheses about the Difference between Two Means—Assuming Homogeneity of Variance

One problem common to many fields of research may be cast in the following form: Which one of two procedures will produce the better results when used in a specified population? To provide information relevant for an answer, the experimenter may draw two samples from the specified population. His experiment might consist of following procedure A in one of the samples and procedure B in the other sample. (These procedures will be referred to as treatments A and B.) The question, which one of the two is the better, requires some criterion on which to base the answer. Several criteria may be relevant—treatment A may be better with respect to some of these criteria, treatment B better with respect to others. Techniques exist for the evaluation of several criteria simultaneously, but in this section methods for evaluating a single criterion at a time will be considered.

Suppose that the experimenter measures each of the elements in the two samples on a single criterion. The results may be that some of the scores under treatment A are higher than those under B, and vice versa. If the distribution of the scores within each of the samples is approximately normal in form, then a comparison of the means of the criterion scores provides one kind of information about which of the two treatments gives the better results.

In some experimental situations, the variability on the criterion within the samples assigned to different treatments is primarily a function of (1) differences in the elements observed that existed before the start of the experiment—such differences are not directly related to the experimental treatment—and (2) uncontrolled sources of variability introduced during the course of the experiment which are in no way related to the treatment itself. In such cases one might reasonably expect that the criterion variance within each of the samples assigned to the experimental treatments is due to common sources of variance. In more technical language, one might reasonably expect homogeneity of variance, i.e., that the sources of variance within each of the samples are essentially the same and that the variances in the corresponding populations are equal.

It is convenient to formalize the arguments just given in more mathematical terms. The formal mathematical argument will serve to make explicit what it is that one assumes to arrive at the conclusions that have just been reached. Let the criterion measure on element i in the sample given experimental treatment j be designated by the symbol X_{ij}. In this case there are two experimental treatments; so j stands for either treatment A or treatment B. Suppose that this measurement may be expressed as the sum of a quantity τ_j, which represents the effect of experimental treatment j, and a quantity ε_{ij}, which is not directly related to experimental treatment j. That is, suppose that

$$X_{ij} = \tau_j + \varepsilon_{ij}. \tag{1}$$

The effect ε_{ij} includes all the unique characteristics associated with the element i as well as all uncontrolled effects associated with the experimental conditions under which the measurement is made. The effect τ_j

is assumed to be constant for all elements in the experimental group assigned to treatment j, whereas the effect ε_{ij} varies from element to element within the group and is in no direct way related to the experimental treatment. The effect ε_{ij} is frequently called the experimental error.

The term X_{ij} on the left-hand side of (1) represents a quantity that is observed. The terms on the right-hand side of (1) cannot be observed—they designate the variables that account for what is observed. The terms τ_j and ε_{ij} represent structural variables underlying the observed data; (1) is referred to as a structural model. The first basic assumption that has been made about the variables in the structural model is that they are uncorrelated. (Being uncorrelated is a less stringent assumption than statistical independence. The latter assumption is, however, required in making tests. For purposes of estimation, only the assumption of zero correlation is required.)

Let σ_a^2 and σ_b^2 designate the expected values of the criterion variance within the respective experimental groups. (That is, σ_a^2 represents the mean of the sampling distribution of the statistic s_a^2, the variance on the criterion for a sample of elements given treatment A.) Suppose that the experiment is designed in such a way that in the long run there will be no difference in the unique characteristics of the group of elements assigned to treatments A and B. Suppose also that the experiment is conducted in such a manner that the uncontrolled sources of variability are comparable in the two experimental groups. These latter assumptions imply homogeneity of experimental error; i.e., they imply that $\sigma_a^2 = \sigma_b^2$. These latter assumptions also imply that the expected value of the mean of the experimental error within treatment groups A and B will be equal. That is, if $\bar{\varepsilon}_a$ and $\bar{\varepsilon}_b$ represent the respective sample means of the experimental error within the treatment groups, the expected values of these quantities will be equal if the assumptions are true. The quantities $\bar{\varepsilon}_a$ and $\bar{\varepsilon}_b$ are the means of structural variables and cannot be computed directly from a single sample.

Experience has shown that the model (1) and the assumptions made about the variables in the model are appropriate for a large class of experimental situations. The tests to be considered in this section are suitable for this class. In terms of this structural model, the mean of the criterion scores for a sample of elements given treatment A may be represented as

$$\bar{X}_a = \tau_a + \bar{\varepsilon}_a. \tag{2}$$

Since the effect τ_a is assumed to be constant for all elements in the sample, its mean effect will be simply τ_a. The corresponding mean for a sample of elements given treatment B may be represented by

$$\bar{X}_b = \tau_b + \bar{\varepsilon}_b. \tag{3}$$

The difference between the two sample means has the form

$$(4) \qquad \bar{X}_a - \bar{X}_b = (\tau_a - \tau_b) + (\bar{\varepsilon}_a - \bar{\varepsilon}_b).$$

In words, (4) says that the observed difference between the criterion mean for the sample given treatment A and the criterion mean for the sample given treatment B is in part a function of the difference in effectiveness of the two treatments and in part a function of the difference between the average experimental error associated with each of the means. One purpose of a statistical test in this context is to find out whether the observed difference is of a magnitude that may be considered a function of experimental error alone or whether the observed difference indicates some effect larger than that due to experimental error.

The details in the analysis of an experiment designed to study the effects of treatments A and B on a specified criterion will now be considered. From the population of interest a random sample of n elements is drawn. The elements are subdivided at random into two subsamples—one of size n_a, the other of size n_b.

Note: In most cases n_a will be equal to n_b. The most sensitive design makes

$$n_a\sigma_b = n_b\sigma_a \qquad \text{or} \qquad \frac{n_a}{n_b} = \frac{\sigma_a}{\sigma_b} \qquad (n_a + n_b = n).$$

This choice for n_a and n_b makes

$$\frac{\sigma_a^2}{n_a} + \frac{\sigma_b^2}{n_b} = \text{minimum}.$$

The sample of size n_a is assigned to experimental treatment A; the sample of size n_b is assigned to experimental treatment B. After the administration of the treatments, each of the elements in the experiment is measured on a common criterion. Suppose that experience in related research indicates that the distribution of such criterion measures tends to be approximately normal in the population of interest. Suppose that inspection of the observed data does not contradict what past experimentation indicates about the form of the distribution of the criterion measures within each of the experimental conditions.

To summarize the information from the experiment, the mean and sample variance for each of the experimental groups are computed. The sample statistics $\bar{X}_a$ and s_a^2 provide, respectively, estimates of the parameters μ_a and σ_a^2. Similarly for treatment group B, $\bar{X}_b$ and s_b^2 provide, respectively, estimates of the parameters μ_b and σ_b^2. A schematic outline of this experiment is given in Table 1.6-1. If the form of the distribution of the measurements were not approximately bell-shaped, other statistics might be more appropriate to summarize the information in the samples.

Table 1.6-1 Outline of Steps in an Experiment

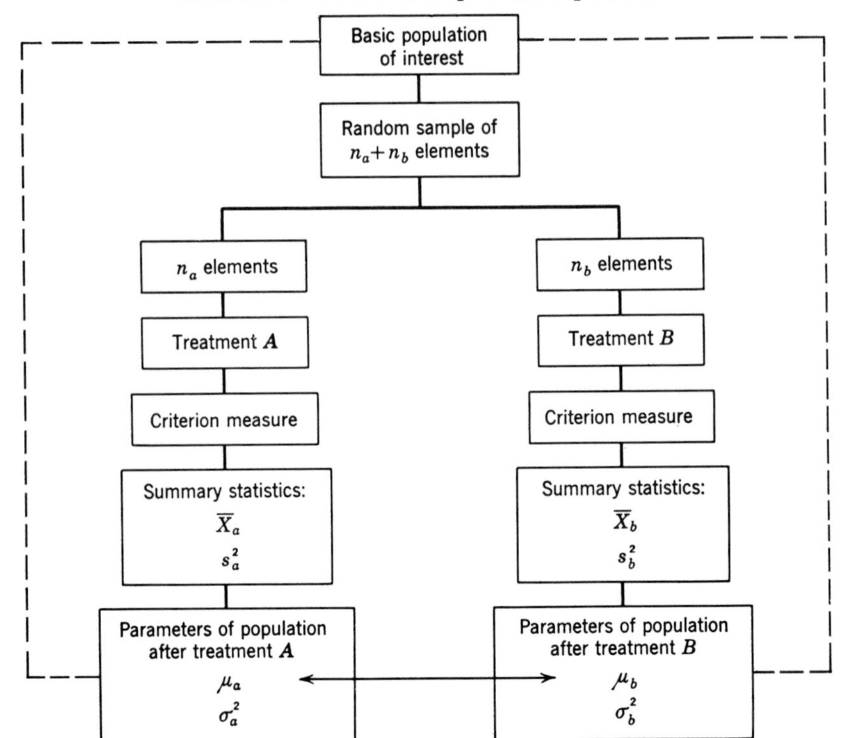

If one were to repeat this experiment a large number of times, each time starting with a different random sample, and if one were to compute the statistic $\bar{X}_a - \bar{X}_b$ for each experiment, the sampling distribution of this statistic would be found to be approximately normal in form. The expected value of the sampling distribution would be approximately $\mu_a - \mu_b$, and the standard error would be $\sqrt{(\sigma_a^2/n_a) + (\sigma_b^2/n_b)}$. Since σ_a^2 is assumed to be equal to σ_b^2, let the common value of the population variance be designated by σ_ε^2. In terms of σ_ε^2, the standard error of the sampling distribution would have the form

$$\text{(5)} \qquad \sigma_{\bar{X}_a - \bar{X}_b} = \sqrt{\sigma_\varepsilon^2\left(\frac{1}{n_a} + \frac{1}{n_b}\right)}.$$

The value of the common population variance σ_ε^2 is not known. Given s_a^2 and s_b^2 computed from sample data, the best estimate of σ_ε^2 is a weighted average of the sample variances, the weights being the respective degrees of freedom. Designating this weighted average by the symbol s_p^2,

$$\text{(6)} \qquad s_p^2 = \frac{(n_a - 1)s_a^2 + (n_b - 1)s_b^2}{(n_a - 1) + (n_b - 1)}.$$

This weighted average of the sample variances is known as the pooled within-class estimate of the common population variance. The degrees of freedom for s_p^2 are the sum of the respective degrees of freedom for the parts; i.e.,

$$(n_a - 1) + (n_b - 1) = n_a + n_b - 2.$$

The statistic used to test hypotheses about $\mu_a - \mu_b$ is

$$t = \frac{(\bar{X}_a - \bar{X}_b) - (\mu_a - \mu_b)}{\sqrt{s_p^2[(1/n_a) + (1/n_b)]}}. \tag{7}$$

The sampling distribution of this statistic, under the assumptions that have been made, is the t distribution having $n_a + n_b - 2$ degrees of freedom. Operationally, if this experiment were to be repeated a large number of times, each time with a different random sample from a specified population, and if one actually knew the value of $\mu_a - \mu_b$, then the distribution of the resulting t statistics could be approximated by the t distribution having $n_a + n_b - 2$ degrees of freedom. The latter degrees of freedom are those associated with s_p^2. In general one does not know the value of $\mu_a - \mu_b$. Its value is specified by H_1.

The denominator of the t statistic defined in (7) is often symbolized by $s_{\bar{X}_a - \bar{X}_b}$, that is,

$$s_{\bar{X}_a - \bar{X}_b} = \sqrt{s_p^2\left(\frac{1}{n_a} + \frac{1}{n_b}\right)}.$$

Under this notation system,

$$t = \frac{(\bar{X}_a - \bar{X}_b) - (\mu_a - \mu_b)}{s_{\bar{X}_a - \bar{X}_b}}.$$

The t distribution is used to test hypotheses about $\mu_a - \mu_b$. In terms of the right-hand side of the structural model in (4), the expected value of numerator of the t statistic in (7) is $E(\bar{\varepsilon}_a - \bar{\varepsilon}_b) = 0$, and the expected value of the denominator is $\sigma_{\bar{\varepsilon}_a - \bar{\varepsilon}_b}$. Hence, when the hypothecated value of $\mu_a - \mu_b$ is actually the true value of the difference between these parameters, the t statistic provides a standardized measure of the difference in the average experimental error for the two experimental conditions.

Suppose that the experimenter is interested in testing the following hypothesis:

$$H_1: \quad \mu_a - \mu_b = \delta.$$
$$H_2: \quad \mu_a - \mu_b \neq \delta.$$
$$\text{Level of significance} = \alpha.$$

The numerical value of δ is the smallest practically important difference of interest to the experimenter. (This value is often taken to be zero.)

Since the alternative hypothesis is two-tailed, a two-tailed region of rejection for H_1 is required. The region of rejection for H_1 is defined by the two tails of the sampling distribution which assumes H_1 to be true. The decision rules are as follows:

$$\text{Reject } H_1 \text{ when } t_{\text{obs}} \begin{cases} < t_{\alpha/2}(n_a + n_b - 2). \\ > t_{1-(\alpha/2)}(n_a + n_b - 2). \end{cases}$$

Do not reject H_1 otherwise.

The t statistic will be numerically large when (1) H_1 is not true or (2) H_1 is true but the difference between the mean experimental errors is unusually large relative to what is expected on the basis of the assumptions underlying the experimental design. The probability of rejecting H_1 when the latter contingency occurs is less than the level of significance of the test. The region of rejection for H_1 is sketched in Fig. 1.6-1.

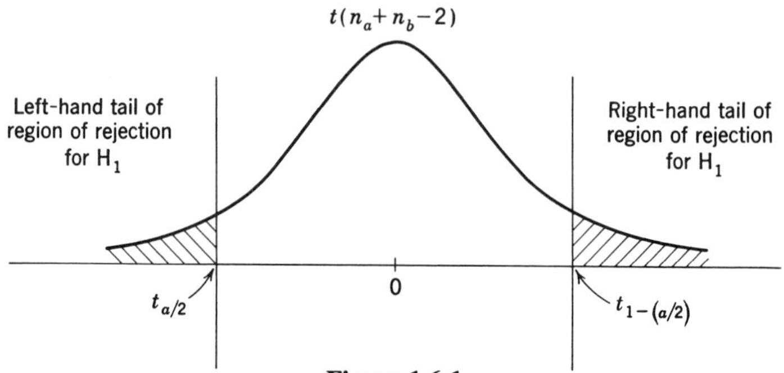

Figure 1.6-1

To illustrate the use of the t statistic and its sampling distribution in making the test about the difference $\mu_a - \mu_b$, suppose that the following data are obtained from an experiment designed and conducted under conditions such that the assumptions underlying the sampling distribution of the t statistic are satisfied:

Treatment	Sample size	Sample variance	Criterion mean
A	$n_a = 8$	$s_a^2 = 18$	$\bar{X}_a = 20$
B	$n_b = 10$	$s_b^2 = 12$	$\bar{X}_b = 25$

The experimenter is interested in making a two-tailed test on the hypothesis that $\mu_a - \mu_b = 0$. Since rejecting this hypothesis when it is true is considered by the experimenter to be a more costly error than not rejecting this hypothesis when it is false, H_1 has the form $\mu_a - \mu_b = 0$. The level of significance for this test is chosen to be .05. The pooled estimate of the common population variance will have 16 degrees of freedom. Hence the decision rules for this test will have the following form:

$$\text{Reject } H_1 \text{ when } t_{\text{obs}} \begin{cases} < -2.12. \\ > \quad 2.12. \end{cases}$$

$$\text{Do not reject } H_1 \text{ otherwise.}$$

The value 2.12 is the 97.5 percentile point on the t distribution having 16 degrees of freedom, that is, $t_{.975}(16) = 2.12$. Since the t distribution is symmetrical, $t_{.025}(16) = -2.12$.

For these data, using (6), the estimate of the population variance is

$$s_p^2 = \frac{7(18) + 9(12)}{16} = 14.62.$$

The value of t_{obs} is given by

$$t_{\text{obs}} = \frac{(20 - 25) - 0}{\sqrt{14.62(\frac{1}{8} + \frac{1}{10})}} = \frac{-5}{\sqrt{14.62(\frac{18}{80})}} = \frac{-5}{1.82} = -2.75.$$

Since t_{obs} is less than the critical value -2.12, H_1 is rejected. Thus the hypothesis that treatments A and B are equally effective with respect to the criterion measure is not supported by the experimental data. Inspection indicates that treatment B has the higher mean. Hence the data support the alternative hypothesis that $\mu_a - \mu_b < 0$, that is, $\mu_a < \mu_b$.

The hypothesis that the difference between the population means is zero having been rejected, the tenable values for this difference are given by a confidence interval for $\mu_a - \mu_b$. This interval has the general form

$$C[(\bar{X}_a - \bar{X}_b) - c \le \mu_a - \mu_b \le (\bar{X}_a - \bar{X}_b) + c] = 1 - \alpha,$$

where
$$c = t_{1-(\alpha/2)} s_{\bar{X}_a - \bar{X}_b}.$$

For the numerical data being considered,

$$c = 2.12(1.82) = 3.86.$$

Hence a .95 confidence interval on the difference between the two treatment means is

$$C[-8.86 \le \mu_a - \mu_b \le -1.14] = .95.$$

For these sample data, any H_1 which specifies that $\mu_a - \mu_b$ is within this interval will lead to nonrejection of H_1 when $\alpha = .05$.

Power of the t Test. Consider the one-tailed hypothesis

$$H_1: \quad \mu_a - \mu_b = 0,$$

$$H_2: \quad \mu_a - \mu_b > 0,$$

$$\alpha = .01.$$

Suppose one has the following sample data:

$$n_a = 10, \qquad s_a^2 = 15.33,$$

$$n_b = 15, \qquad s_b^2 = 21.00.$$

Hence,
$$s^2_{\text{pooled}} = \frac{(n_a - 1)s_a^2 + (n_b - 1)s_b^2}{n_a + n_b - 2} = 18.78,$$

$$\hat{\sigma} = s_{\text{pooled}} = \sqrt{s^2_{\text{pooled}}} = 4.33.$$

To obtain the power of this test with respect to the alternative hypothesis

$$(\mu_a - \mu_b)' = 6,$$

the noncentrality parameter is

$$\delta = \frac{6 - 0}{\sigma\sqrt{(1/n_a) + (1/n_b)}}.$$

One cannot evaluate this expression unless σ is known. If, however, one replaces σ by $\hat{\sigma}$, one obtains an approximate value for δ corresponding to $(\mu_a - \mu_b)' = 6$. Thus,

$$\delta \doteq \frac{6}{4.33\sqrt{\frac{1}{10} + \frac{1}{15}}} = 3.40.$$

The power of the test with respect to this value of δ may be obtained from Table C.13 for $\alpha = .01$. In this case $f = n_a + n_b - 2 = 23$.

Reading over on line 23, the value 3.40 lies between 3.84 and 3.37. Corresponding to 3.37, $\beta = .20$; corresponding to 3.84, $\beta = .10$. Using linear interpolation,

$$\beta = .10 + \frac{3.84 - 3.40}{3.84 - 3.37}(.20 - .10) = .194.$$

Hence, the power of this test when $\delta = 3.40$ is

$$\text{Power } (\delta = 3.40) = 1 - .194 = .806.$$

In Fig. 1.6-2, the power of the test when $\delta = 3.40$ is illustrated graphically. Note that $t_{.99}(23) = 2.50$.

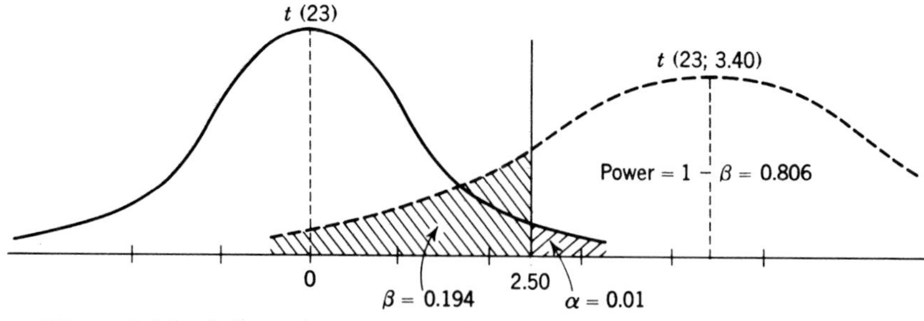

Figure 1.6-2 Schematic representation of power of t test when $\delta = 3.40$.

A relatively good approximation to percentile points on the noncentral t distribution may be obtained by the procedure illustrated in Fig. 1.6-3.

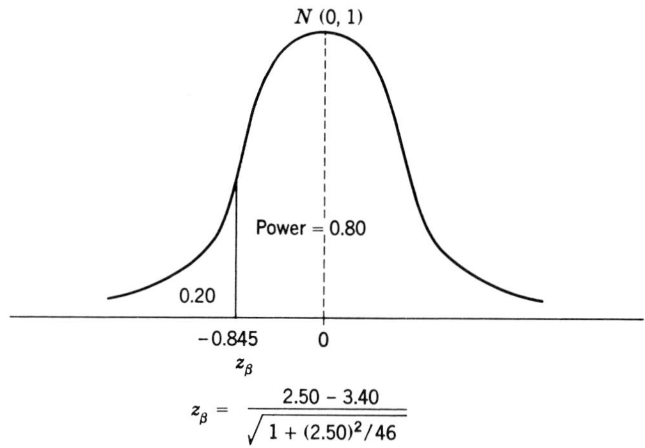

Figure 1.6-3 Normal approximation to noncentral t distribution.

If z is distributed as $N(0,1)$, then

$$\Pr\,[t(f,\delta) \geq t_{1-\alpha}(f)] \doteq \Pr\,(z \geq z_\beta),$$

where

$$z_\beta = \frac{t_{1-\alpha}(f) - \delta}{\sqrt{1 + [t^2_{1-\alpha}(f)/2f]}}\,.$$

In this case,

$$t_{1-\alpha}(f) = t_{.99}(23) = 2.50.$$

1.7 Computational Formulas for the t Statistic

For simplicity in computational work, the basic formula for the t statistic is not the most suitable for use. To avoid rounding error at intermediate stages of work, the best computational procedure puts off divisions and square roots until the final stages. For accurate computational work the general rule is to do all possible addition, subtraction, and multiplication before doing any division or taking any square root. To arrive at a computational formula for the t statistic, the following notation will be used:

ΣX_a = sum of the n_a observations in experimental group A.
ΣX_b = sum of the n_b observations in experimental group B.
$\Sigma X = \Sigma X_a + \Sigma X_b$ = sum of all observations in the experiment.
ΣX_a^2 = sum of the squares of the observations in experimental group A.
ΣX_b^2 = sum of the squares of the observations in experimental group B.
$\Sigma X^2 = \Sigma X_a^2 + \Sigma X_b^2$ = sum of the squares of all observations in the experiment.

The definition of the symbol L_a is

(1) $$L_a = n_a \Sigma X_a^2 - (\Sigma X_a)^2.$$

An analogous definition of the symbol L_b is

(2) $$L_b = n_b \Sigma X_b^2 - (\Sigma X_b)^2.$$

In terms of this notation, a computational formula for the square of the t statistic, when H_1 has the form $\mu_a - \mu_b = 0$, is

(3) $$t^2 = \frac{(n_a + n_b - 2)(n_b \Sigma X_a - n_a \Sigma X_b)^2}{(n_a + n_b)(n_b L_a + n_a L_b)}.$$

Taking the square root of t^2 yields the t statistic. The algebraic sign is given by the sign of $\bar{X}_a - \bar{X}_b$.

In terms of L_a and L_b computational formulas for variances are

(4) $$s_a^2 = \frac{L_a}{n_a(n_a - 1)},$$

(5) $$s_b^2 = \frac{L_b}{n_b(n_b - 1)}.$$

A computational formula for s_p^2, which minimizes rounding error at intermediate stages, is

(6) $$s_p^2 = \frac{n_a n_b \Sigma X^2 - n_b(\Sigma X_a)^2 - n_a(\Sigma X_b)^2}{n_a n_b(n_a + n_b - 2)}.$$

There is one disadvantage to these formulas: The numbers involved at intermediate steps tend to become quite large even for relatively small samples. Use of these formulas is illustrated in the numerical example given in Table 1.7-1.

The decision rules for H_1: $\mu_a - \mu_b = 0$ against a two-tailed alternative hypothesis may be stated in terms of either t_{obs} or t_{obs}^2. In terms of t_{obs} the decision rules are (assuming that $\alpha = .05$)

$$\text{Reject } H_1 \text{ if } t_{obs} \begin{cases} < -2.13. \\ > 2.13. \end{cases}$$

Otherwise do not reject H_1.

Since $t_{obs}^2 = -3.28$, H_1 is rejected. Inspection of the data indicates that treatment B gives the greater mean criterion score. In terms of t_{obs}^2, the decision rules are

$$\text{Reject } H_1 \text{ if } t_{obs}^2 > 4.54.$$

Otherwise do not reject H_1.

In this case $t_{obs}^2 = 10.76$; hence H_1 is rejected. The critical value for the t^2 statistic is $F_{1-\alpha}(1, n_a + n_b - 2)$ or equivalently $t_{1-(\alpha/2)}^2(n_a + n_b - 2)$.

Table 1.7-1 Numerical Example of Computation of *t* Statistic

	Treatment *A*	Treatment *B*	
	3	6	
	5	5	
$n_a = 7$	2	7	$n_b = 10$
	4	8	
	6	9	
	2	4	
	7	7	
		8	
		9	
		7	
	$\Sigma X_a = 29$	$\Sigma X_b = 70$	
	$\Sigma X_a^2 = 143$	$\Sigma X_b^2 = 514$	

$$L_a = 7(143) - (29)^2 = 160 \qquad L_b = 10(514) - (70)^2 = 240$$

$$s_a^2 = \frac{160}{7(6)} = 3.81 \qquad s_b^2 = \frac{240}{10(9)} = 2.67$$

$$\bar{X}_a = \tfrac{29}{7} = 4.14 \qquad \bar{X}_b = \tfrac{70}{10} = 7.00$$

$$t^2 = \frac{(n_a + n_b - 2)(n_b \Sigma X_a - n_a \Sigma X_b)^2}{(n_a + n_b)(n_b L_a + n_a L_b)}$$

$$= \frac{(15)[10(29) - 7(70)]^2}{(17)[10(160) + 7(240)]} = \frac{600{,}000}{55{,}760} = 10.760$$

$$F_{.95}(1,15) = 4.54$$

$$t = -\sqrt{10.760} = -3.28\dagger \qquad t_{.025}(15) = -2.13$$

† Negative sign is used because $\bar{X}_a - \bar{X}_b$ is negative.

1.8 Test for Homogeneity of Variance

The test on population means developed in Sec. 1.6 was based upon a structural model which assumed that $\sigma_a^2 = \sigma_b^2$. In the absence of extensive information from past experimentation in an area, the data obtained in the experiment are sometimes used to make preliminary tests on the model. Preliminary tests on structural models do not establish the appropriateness of the models; rather their appropriateness depends upon the design of the experiment and the nature of the sources of variation. The purpose of preliminary tests is to provide a partial check on whether or not the observed data tend to be consistent with the model. The observed data may actually be consistent with several models.

Moderate departures from the hypothesis that $\sigma_a^2 = \sigma_b^2$ do not seriously affect the accuracy of the decisions reached by means of the *t* test given in Sec. 1.6. In more technical language, the *t* test is *robust* with respect to moderate departures from the hypothesis of homogeneity of variance. An extensive investigation of the effect of unequal variances upon the *t* test and the corresponding F test is found in the work of Box (1954). The term *moderate* in this context is relative to the magnitude and difference in sample sizes. To illustrate the effect of unequal population variances upon the accuracy of the decision rule based upon the t test, which assumes equal population variances, consider the case in which $\sigma_a^2 = 2\sigma_b^2$. If $n_a = 5$ and $n_b = 5$, then the 95th percentile point on the F distribution which assumes the population variances equal is approximately equal to the 94th percentile point on the sampling distribution which actually takes into account the difference in the population variances. Thus for an .05-level test with these sample sizes, this violation of the assumption that $\sigma_a^2 = \sigma_b^2$ results in an error in the level of significance of approximately 1 percent (in the direction of rejecting H_1 more often than should be the case).

The work of Box (1954) also indicates that the t test is robust with respect to the assumption of normality of the distributions within the treatment populations. That is, the type 1 error of the decision rule is not seriously affected when the population distributions deviate from normality. Even when population distributions are markedly skewed, the sampling distribution of the t statistic, which assumes normality, provides a good approximation to the exact sampling distribution which takes into account the skewness. In summary, preliminary tests on the structural model for the t test, which assumes homogeneity of variance and normality of distributions, are not of primary importance with respect to type 1 error, particularly preliminary tests for normality of distribution.

The test of the hypothesis that $\sigma_a^2 = \sigma_b^2$ is useful in its own right, quite apart from its use in a preliminary test. Suppose that the following data are potentially available for random samples given treatment A and treatment B:

Treatment	Sample size	Sample variance
A	n_a	s_a^2
B	n_b	s_b^2

Assuming that the distribution of the measurements in the treatment populations is approximately normal, under the hypothesis that $\sigma_a^2 = \sigma_b^2$ the statistic

$$F = \frac{s_a^2}{s_b^2}$$

has a sampling distribution which is approximated by $F(n_a - 1, n_b - 1)$, that is, an F distribution with $n_a - 1$ degrees of freedom for the numerator and $n_b - 1$ degrees of freedom for the denominator. Operationally, if the experiment yielding the potential set of data given above were to be repeated a large number of times, and if for each experiment an F statistic were computed, then the resulting distribution of the F statistics would be approximately $F(n_a - 1, n_b - 1)$. The region of rejection for the hypothesis $\sigma_a^2 = \sigma_b^2$ against the alternative hypothesis $\sigma_a^2 \neq \sigma_b^2$ is sketched in Fig. 1.8-1. The F distribution is not symmetrical, but there is a relationship by means of which percentile points in the left tail may be obtained

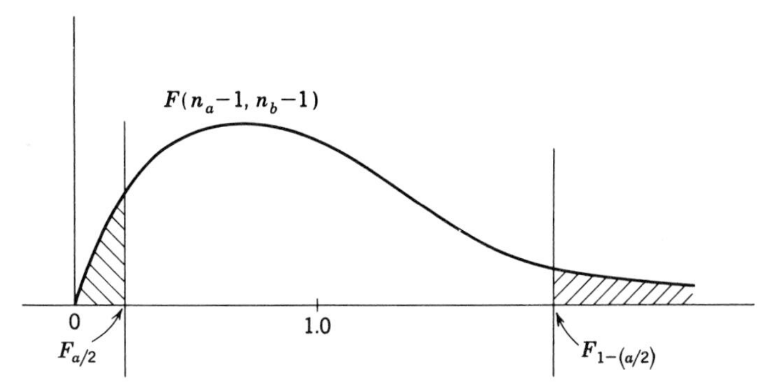

Figure 1.8-1

from points in the right tail. This relationship is

$$F_{\alpha/2}(n_a - 1, n_b - 1) = \frac{1}{F_{1-(\alpha/2)}(n_b - 1, n_a - 1)} \tag{1}$$

Most tables of the F distribution provide values only for the right-hand tail. Values for the left-hand tail are readily computed by means of the relation (1). There is, however, a procedure by which use of the left-hand tail may be avoided. It is actually immaterial which treatment is called A and which B. Suppose that F_{obs} is defined to be the ratio of the larger sample variance to the smaller sample variance. Hence F_{obs} will always tend to fall toward the right-hand tail of the F distribution for which the degrees of freedom for the numerator are equal to the degrees of freedom for the larger sample variance and the degrees of freedom for the denominator are equal to the degrees of freedom for the smaller sample variance.

To illustrate this test, suppose that the following data were obtained in an experiment:

Treatment	Sample size	Sample variance
A	$n_a = 10$	$s_a^2 = 15.58$
B	$n_b = 8$	$s_b^2 = 28.20$

To test the hypothesis $\sigma_a^2 = \sigma_b^2$ against the alternative hypothesis $\sigma_a^2 \neq \sigma_b^2$,

$$F_{\text{obs}} = \frac{28.20}{15.58} = 1.81.$$

If the level of significance for this test is chosen to be $\alpha = .10$, then the critical value for the region of rejection is $F_{.95}(7,9) = 3.29$. Since F_{obs}

does not exceed this value, the hypothesis that $\sigma_a^2 = \sigma_b^2$ is not contradicted by the observed data.

If one draws a large number of random samples of size n_a from an approximately normal population $N(\mu_a,\sigma_a^2)$, then the sampling distribution of the statistic

$$\chi_a^2 = \frac{(n_a - 1)s_a^2}{\sigma_a^2}$$

is approximated by the chi-square distribution having $n_a - 1$ degrees of freedom. If one were to draw similar samples from the population which is approximately $N(\mu_b,\sigma_b^2)$, then the sampling distribution of the statistic

$$\chi_b^2 = \frac{(n_b - 1)s_b^2}{\sigma_b^2}$$

is approximated by a chi-square distribution with $n_b - 1$ degrees of freedom. Assuming that the two populations have no elements in common, the two chi-square statistics will be independent. In an actual experiment there will be one sample from the A population and one sample from the B population.

The statistic

$$F = \frac{\chi_a^2/(n_a - 1)}{\chi_b^2/(n_b - 1)} = \frac{s_a^2/s_b^2}{\sigma_a^2/\sigma_b^2} \tag{2}$$

has a sampling distribution given by $F(n_a - 1, n_b - 1)$. This statistic may be used to test any hypothesis about the relationship between σ_a^2 and σ_b^2. Under the hypothesis that $\sigma_a^2 = \sigma_b^2$, the ratio $\sigma_a^2/\sigma_b^2 = 1.00$, and (2) assumes the form $F = s_a^2/s_b^2$. The ratio of two statistics whose sampling distributions are independent central chi squares will have a sampling distribution given by an F distribution.

The region of rejection for this test under the hypothesis that $\sigma_a^2/\sigma_b^2 = 1$ is defined by

$$\Pr\left[\frac{s_a^2}{s_b^2} > F_{1-\alpha}(n_a - 1, n_b - 1)\right].$$

The power of this test under the alternative hypothesis

$$\frac{\sigma_a^2}{\sigma_b^2} = \theta,$$

where $\theta > 1$, is

$$\Pr\left[\theta\frac{s_a^2}{s_b^2} > F_{1-\alpha}(n_a - 1, n_b - 1)\right].$$

Thus the power function for this test is given by a noncentrality parameter, θ, times a central F distribution. This is the power function that comes into play in the variance-component model of the analysis of variance.

Notation for the F Distribution. In one widely used notation system, the F distribution with parameters $n_a - 1$ and $n_b - 1$ is denoted by the symbol

$$F_{n_a-1,n_b-1}.$$

The $1 - \alpha$ percentile point on this distribution is denoted by the symbol

$$F_{\alpha;n_a-1,n_b-1}.$$

Thus, if an F statistic is distributed as F_{n_a-1,n_b-1}, then

$$\text{Prob}\,(F > F_{\alpha;n_a-1,n_b-1}) = \alpha.$$

1.9 Testing Hypotheses about the Difference between Two Means—Assuming Population Variances Not Equal

Suppose that independent random samples of size n_a and n_b are drawn from a common population and are assigned, respectively, to treatment conditions A and B. The elements are then measured on a common criterion. The following data may be obtained from this type of experiment:

Treatment	Sample size	Sample mean	Sample variance
A	n_a	$\bar{X}_a$	s_a^2
B	n_b	$\bar{X}_b$	s_b^2

The sample data for treatment A provide estimates of the parameters of a population A, a hypothetical population which consists of the collection of elements after the administration of treatment A. Thus $\bar{X}_a$ is an estimate of μ_a, and s_a^2 is an estimate of σ_a^2. If the distribution of the criterion measures is approximately normal in form, the population A may be designated as approximately $N(\mu_a,\sigma_a^2)$. The data from sample B are assumed to provide estimates of the parameters of a population which is approximately $N(\mu_b,\sigma_b^2)$. Thus populations A and B are potential populations that could arise from the basic common population from which the samples were drawn.

In Sec. 1.6, tests about $\mu_a - \mu_b$ were made under the assumption that $\sigma_a^2 = \sigma_b^2$. In this section, tests about $\mu_a - \mu_b$ make no assumption about the equality of the population variances. The sampling distribution of the statistic

$$z = \frac{(\bar{X}_a - \bar{X}_b) - (\mu_a - \mu_b)}{\sqrt{(\sigma_a^2/n_a) + (\sigma_b^2/n_b)}} \tag{1}$$

is $N(0,1)$. Two sets of parameters are involved in this statistic. Hypotheses about $\mu_a - \mu_b$ cannot be tested unless values for σ_a^2 and σ_b^2 are known. If n_a and n_b are both large (say larger than 30), then s_a^2 and s_b^2 may be substituted for the corresponding parameters and the resulting statistic has a sampling distribution approximated by $N(0,1)$. For small samples, however, the statistic

$$(1a) \qquad t^* = \frac{(\bar{X}_a - \bar{X}_b) - (\mu_a - \mu_b)}{\sqrt{(s_a^2/n_a) + (s_b^2/n_b)}}$$

has a sampling distribution which is neither the normal distribution nor Student's t distribution.

Several workers have attempted to obtain the exact sampling distribution of the t^* statistic. The end products have differed slightly because of differing assumptions that were made in the course of the derivations. Behrens made one attempt to obtain the sampling distribution of the t^* statistic; Fisher enlarged upon the work of Behrens, and the resulting distribution is called the Behrens-Fisher distribution. Tables of the Behrens-Fisher distribution are given in Fisher and Yates (1953, p. 52). A method of approximating critical values of the Behrens-Fisher distribution by Student's t distribution is given by Cochran and Cox (1957, p. 101).

Not all mathematical statisticians agree with the logic underlying the derivation of the Behrens-Fisher distribution. Using a somewhat different approach to a related problem, Satterthwaite (1946) derived an approximation for the t^* distribution which involves Student's t distribution having degrees of freedom approximated by the quantity

$$(2) \qquad f = \frac{U^2}{[V^2/(n_a - 1)] + [W^2/(n_b - 1)]},$$

where $\qquad V = \dfrac{s_a^2}{n_a}, \qquad W = \dfrac{s_b^2}{n_b}, \qquad$ and $\qquad U = V + W.$

The critical values for the t^* statistic are obtained from tables of Student's t distribution having degrees of freedom equal to the nearest integer to f. A slightly different (and perhaps closer) approximation for f (Welch, 1947) is given by

$$(3) \qquad f = \frac{U^2}{[V^2/(n_a + 1)] + [W^2/(n_b + 1)]} - 2.$$

Welch derived an exact sampling distribution for the t^* statistic. Tables for this distribution have been prepared by Aspin (1949).

A numerical example of the use of the t^* statistic appears in Table 1.9-1. The approximate degrees of freedom are obtained from (2). The value of f in (2) will be less than $f_a + f_b$ but greater than the smaller of f_a and f_b. If H_1 is rejected using the degrees of freedom equal to the smaller of f_a and f_b, the value of f need not be computed, since the decision reached by its use will be to reject H_1.

To test the hypothesis that $\mu_a - \mu_b = 0$ against a two-tailed alternative hypothesis, the decision rules for a test with level of significance equal to .05 are as follows:

$$\text{Reject } H_1 \text{ if } t^*_{\text{obs}} \begin{cases} < t_{.025}(f). \\ > t_{.975}(f). \end{cases}$$

Otherwise do not reject H_1.

The nearest integer to the numerical value of f computed in Table 1.9-1

Table 1.9-1 Numerical Example

	Treatments	
	A	B
Sample size	$n_a = 16$	$n_b = 8$
Sample mean	$\bar{X}_a = 30.00$	$\bar{X}_b = 21.00$
Sample variance	$s_a^2 = 32.00$	$s_b^2 = 80.00$

$H_1: \mu_a - \mu_b = 0$

$H_2: \mu_a - \mu_b \neq 0$

$\alpha = .05$

$$t^*_{\text{obs}} = \frac{(30.00 - 21.00) - 0}{\sqrt{(32.00/16) + (80.00/8)}} = \frac{9}{\sqrt{12}} = 2.60$$

$$f = \frac{(2 + 10)^2}{\frac{4}{15} + \frac{100}{7}} = 9.90$$

is 10. From tables of Student's t distribution, $t_{.975}(10) = 2.23$. The value of $t^*_{\text{obs}} = 2.60$ is greater than 2.23; hence the hypothesis that $\mu_a - \mu_b = 0$ is rejected. In this case it is not actually necessary to compute the value of f. The degrees of freedom for the variance associated with the smaller sample is 7; hence the numerical value of f will be larger than 7. Using $t_{.975}(7) = 2.36$ as a critical value leads to the decision to reject H_1. This critical value is larger than that based upon f. Hence H_1 will also be rejected when the proper number of degrees of freedom is used in determining the critical value.

The critical value obtained from tables of the Behrens-Fisher distribution is 2.20. The use of the Cochran and Cox approximation to the Behrens-Fisher test will also be illustrated for the data in Table 1.9-1. The critical value for t^*_{obs} is given by

$$t_{\text{critical}} = \frac{w_a t_a + w_b t_b}{w_a + w_b}, \tag{4}$$

where $w_a = s_a^2/n_a$, $w_b = s_b^2/n_b$, $t_a = t_{1-(\alpha/2)}(n_a - 1)$, and $t_b = t_{1-(\alpha/2)}(n_b - 1)$. For the data in Table 1.9-1 when $\alpha = .05$, $w_a = 2.00$, $w_b = 10.00$, $t_a = 2.13$, $t_b = 2.36$, and

$$t_{\text{critical}} = \frac{2.00(2.13) + 10.00(2.36)}{12.00} = 2.32.$$

In general t_{critical} computed in this manner will differ only slightly from the critical value obtained by means of the Satterthwaite approximation.

Formula (3) is actually a modified version of (2). Using (3) to compute f for the numerical data in Table 1.9-1 gives

$$f = \frac{(12.00)^2}{[(2.00)^2/17] + [(10.00)^2/9]} - 2 = \frac{144}{11.35} - 2 = 10.69.$$

Use of formula (2) gave 9.90.

1.10 Testing Hypotheses about the Difference between Two Means—Correlated Observations

In Secs. 1.6 and 1.9 the data from two independent samples were used as the basic data for tests. Elements in one of the samples were independent of (in no way related to) elements in the other sample. Hence the data in the two samples were uncorrelated. If the elements in single sample of size n are observed under both treatments A and B, then the resulting data are generally correlated. In this case two measurements are made on each element; hence the term repeated measurements to describe this type of design. If the order in which the treatments are administered has no effect upon the final outcome, then the difference between the two measures on the same element on a common criterion provides a measure of the relative effectiveness of the treatments. This measure is free of variation due to unique but systematic effects associated with the elements themselves. In this respect, each element serves as its own control.

The following structural model, which describes the sources of variation that underlie an observation, is applicable in a variety of experimental situations,

$$X_{ia} = \tau_a + \pi_i + \varepsilon_{ia}, \tag{1}$$

where X_{ia} = criterion measure for element i under treatment A;
τ_a = assumed magnitude of effect of treatment A;
π_i = unique systematic effect associated with element i;
ε_{ia} = uncontrolled sources of variance affecting observation X_{ia} (the experimental error).

Each of the structural variables on the right-hand side of (1) is assumed to be independent of the others. The corresponding structural model for an observation on element i under treatment B has the form

$$X_{ib} = \tau_b + \pi_i + \varepsilon_{ib}. \tag{2}$$

In Sec. 1.6, the structural model for an observation combined the term π_i with the experimental error, since π_i could not be distinguished from experimental error. In experimental designs having repeated measures, this source of variance may be eliminated from experimental error. The difference between two measures on the same criterion made on a randomly selected element i is

$$d_i = X_{ia} - X_{ib} = (\tau_a - \tau_b) + (\varepsilon_{ia} - \varepsilon_{ib}). \tag{3}$$

This difference does not involve the variable π_i. The term $\tau_a - \tau_b$ is assumed to be constant for all elements, whereas the term $\varepsilon_{ia} - \varepsilon_{ib}$ varies from element to element. Hence all variance of the d_i's is a function of $\varepsilon_{ia} - \varepsilon_{ib}$, which is assumed to be a measure of experimental error. The mean of the differences has the structural form

$$\bar{d} = \bar{X}_a - \bar{X}_b = (\tau_a - \tau_b) + (\bar{\varepsilon}_a - \bar{\varepsilon}_b). \tag{4}$$

Thus $\bar{d}$ consists of the sum of two components. One of the components is a measure of the difference between the treatment effects; the other is a measure of the difference between the average experimental error associated with each of the treatments.

If the experimental error for the observations made under treatment A differs in no systematic way from the experimental error for the observations made under treatment B, then it is reasonable to assume that

$$E(\bar{\varepsilon}_a - \bar{\varepsilon}_b) = 0. \tag{5}$$

Therefore,

$$E(\bar{d}) = \tau_a - \tau_b = \mu_a - \mu_b. \tag{6}$$

In words, $\bar{d}$ provides an unbiased estimate of $\mu_a - \mu_b$. If the distribution of the d's is approximately normal with variance equal to σ_ε^2, the expected value of the variance of the term $\varepsilon_{ia} - \varepsilon_{ib}$ in the structural model, then the sampling distribution of the statistic

$$t = \frac{\bar{d} - (\mu_a - \mu_b)}{\sqrt{s_d^2/n}} \tag{7}$$

may be approximated by Student's t distribution having $n - 1$ degrees of freedom, where n is the number of differences. In (7), s_d^2 is used as an estimate of σ_ε^2.

The application of this sampling distribution to testing hypotheses about $\mu_a - \mu_b$ will be illustrated by means of the data in Table 1.10-1. These

Table 1.10-1 Numerical Example

Person number	Before treatment	After treatment	Difference
1	3	6	3
2	8	14	6
3	4	8	4
4	6	4	−2
5	9	16	7
6	2	7	5
7	12	19	7
			$\Sigma d = 30$ $\quad \bar{d} = 4.29$
			$\Sigma d^2 = 188$

$$L_d = n\,\Sigma d^2 - (\Sigma d)^2$$
$$= 7(188) - (30)^2 = 416$$

$$s_d^2 = \frac{L_d}{n(n-1)} = \frac{416}{7(6)} = 9.90 \qquad \sqrt{s_d^2/n} = \sqrt{9.90/7} = 1.19$$

$$H_1\colon \mu_a - \mu_b = 0$$
$$H_2\colon \mu_a - \mu_b \neq 0 \qquad t_{\text{obs}} = \frac{4.29 - 0}{1.19} = 3.61$$
$$\alpha = .05$$

data represent the measures on a common criterion before and after the administration of a treatment. In this type of experiment the question of which treatment is to be administered first is not relevant. For a two-tailed test at the .05 level of significance of the hypothesis $\mu_a - \mu_b = 0$, the decision rules are as follows:

$$\text{Reject } H_1 \text{ if } t_{\text{obs}} \begin{cases} < t_{.025}(6) = -2.45. \\ > t_{.975}(6) = 2.45. \end{cases}$$

Do not reject H_1 otherwise.

The value of $t_{\text{obs}} = 3.61$ falls in the region of rejection for H_1. Inspection of the data indicates that the difference in the population means is probably greater than zero.

A confidence interval for the difference between the population means takes the following general form,

$$C(\bar{d} - c \leq \mu_a - \mu_b \leq \bar{d} + c) = 1 - \alpha,$$

where
$$c = t_{1-(\alpha/2)}(n-1)\sqrt{\frac{s_d^2}{n}}\,.$$

The symbol $t_{1-(\alpha/2)}(n-1)$ is to be interpreted as the $[1 - (\alpha/2)]$-percentile point on the t distribution with $n - 1$ degrees of freedom, not as $n - 1$

times this value. For the data in Table 1.10-1,

$$c = 2.45(1.19) = 2.92.$$

The confidence interval is

$$C(1.37 \le \mu_a - \mu_b \le 7.21) = .95.$$

Hypotheses which specify that the difference in the population means lies in this range will lead to a decision not to reject H_1 when the data in Table 1.10-1 are used to make the test.

If a single element is to be given two different treatments for the purpose of evaluating the difference in the effectiveness, there is the question of which treatment is to be given first. It may make some difference in the results if the treatments are administered in the order A followed by B or B followed by A. That is, the carry-over effect, if any, from the sequence A followed by B, or vice versa, may be different. This kind of effect is known as a treatment by order interaction. To check upon the possibility of this interaction in a repeated-measure design, half the elements in the study may be given one sequence, the other half the reverse sequence. This design will permit the evaluation of order effects as well as treatment by order interactions. Tests associated with the evaluation of interaction effects are discussed in Chap. 7. However, if no interaction of this type exists, the t statistic in (7) may be used to test the difference in treatment effects when the order of administration of the treatments is counterbalanced. In the absence of any information about this interaction, the method of attack in Chap. 7 is to be preferred to the use of the t statistic in (7).

The variance s_d^2 in the denominator of (7) may be expressed in terms of the variances of the individual measures, s_a^2 and s_b^2. This relationship is

$$(8) \qquad \begin{aligned} s_d^2 &= s_a^2 + s_b^2 - 2r_{ab}s_as_b \\ &= \text{var}_a + \text{var}_b - 2\,\text{cov}_{ab}, \end{aligned}$$

where r_{ab} is the product-moment correlation between the measures, and where cov_{ab} is the covariance between the measures. The following relationship holds:

$$\text{cov}_{ab} = r_{ab}s_as_b.$$

The notations var_a and var_b are sometimes used interchangeably with s_a^2 and s_b^2.

The computation of s_d^2 by means of this formula is illustrated by the numerical data in Table 1.10-1. A summary of the computational steps for the statistics needed is given in Table 1.10-2. Substituting the numerical values in Table 1.10-2 for the corresponding statistics in (8) gives

$$s_d^2 = 32.62 + 12.90 - 2(.868)(5.71)(3.59) = 9.91.$$

Within rounding error, this is the value obtained by working directly with the difference measures.

Table 1.10-2 Numerical Example (Continued)

Before treatment (B)	After treatment (A)	Product of treatments
$\Sigma X_b = 44$	$\Sigma X_a = 74$	
$\Sigma X_b^2 = 354$	$\Sigma X_a^2 = 978$	$\Sigma X_a X_b = 572$
$L_b = n\Sigma X_b^2 - (\Sigma X_b)^2$	$L_a = n\Sigma X_a^2 - (\Sigma X_a)^2$	$L_{ab} = n\Sigma X_a X_b - (\Sigma X_a)(\Sigma X_b)$
$= 7(354) - (44)^2$	$= 7(978) - (74)^2$	$= 7(572) - (74)(44)$
$= 542$	$= 1370$	$= 748$
$s_b^2 = \dfrac{L_b}{n(n-1)}$	$s_a^2 = \dfrac{L_a}{n(n-1)}$	$r_{ab} = \dfrac{L_{ab}}{\sqrt{L_a L_b}}$
$= 12.90$	$= 32.62$	$= \dfrac{748}{\sqrt{(1370)(542)}}$
$s_b = 3.59$	$s_a = 5.71$	$= .868$

One of the primary advantages of repeated measures is the potential reduction in variance due to experimental error. For this design

$$s_{\bar{d}}^2 = s_{\bar{X}_a - \bar{X}_b}^2 = \frac{s_d^2}{n} = \frac{s_a^2}{n} + \frac{s_b^2}{n} - \frac{2r_{ab}s_a s_b}{n}. \tag{9}$$

The corresponding estimate for the case of uncorrelated observations is

$$s_{\bar{X}_a - \bar{X}_b}^2 = \frac{s_a^2}{n} + \frac{s_b^2}{n}. \tag{10}$$

If the correlation is positive, and it will be positive if the model in (1) holds, then the estimate of the experimental error obtained from a design in which (9) is appropriate will be smaller than that obtained from a design involving uncorrelated observations by a factor of $2r_{ab}s_a s_b$. However, the degrees of freedom for the estimate in (9) are $n - 1$, whereas the degrees of freedom for the estimate in (10) are $2n - 2$. Before a repeated-measure design may be considered more efficient than a design which does not have repeated measures, the decrease in the experimental error must be sufficient to offset the reduction in degrees of freedom.

In areas of research in which the variance associated with the term π_i in the model in (2) is likely to be large, the control on this source of variation provided by a repeated-measure design will greatly increase the sensitivity of the experiment. The smaller the proportion of the total variance in an experiment due to effects which are not related to the treatments per se, the more sensitive the experiment.

1.11 Combining Several Independent Tests on the Same Hypothesis

A series of experiments designed to test the same hypothesis may be conducted at different places or at different times. Suppose that the samples used in each of these experiments are independent; i.e., different random samples are employed in each of the experiments. Suppose that the probabilities of the observed outcomes with respect to the common hypothesis tested are designated $P_1, P_2, \ldots, P_k$. If the experimenter desires to make an overall probability statement, the following statistic may be used:

$$\chi^2 = 2\Sigma u_i, \qquad \text{where } u_i = -\ln P_i. \tag{1}$$

Under the hypothesis that the observed probabilities are a random sample from a population of probabilities having a mean of .50, the χ^2 statistic in (1) has a sampling distribution which is approximated by the chi-square distribution having $2k$ degrees of freedom.

Use of this statistic is illustrated by the data in Table 1.11-1. These data give t statistics obtained in a series of five independent experiments

Table 1.11-1 Numerical Example

Experiment	t_{obs}†	Probability	$-\ln$ (probability)
1	.87	.20	1.61
2	.54	.30	1.20
3	1.10	.15	1.90
4	1.50	.07	2.66
5	1.30	.11	2.21
			9.58

$$\chi^2 = 2(9.58) = 19.16$$

$$\chi^2_{.95}(10) = 18.3; \qquad \chi^2_{.975}(10) = 20.5$$

† df = 14.

testing a specified hypothesis against a one-tailed alternative hypothesis. The probabilities of obtaining the observed t statistic or one more extreme in the direction of the alternative hypothesis is given in the column headed Probability. The natural logarithm of the probability with the sign changed is given in the next column. The numerical value for the chi-square statistic, 19.16, is twice the sum of this last column. k, the number of experiments, is 5; hence the degrees of freedom for this chi-square statistic are $2(5) = 10$. Since the probability of obtaining a chi-square statistic of 19.16 or larger is less than .05 when the mean probability in the population is .50, this latter hypothesis is rejected (if $\alpha = .05$). Thus the combined evidence from the series of experiments indicates that the common H_1 in each of the experiments should be rejected if inferences are to be made with respect to the combined populations represented by the series of experiments.

There is an alternative approach to combining the outcomes of several experiments involving the use of the t statistic based upon relatively large samples. Under the hypothesis that the mean value for the t statistic in the population is zero, the statistic

$$z = \frac{\Sigma t_j}{\sqrt{\Sigma[f_j/(f_j - 2)]}}, \tag{2}$$

where f_j represents the degrees of freedom associated with t_j, has a sampling distribution which is approximately $N(0,1)$. For the data in Table 1.11-1,

$$\sqrt{\Sigma \frac{f_j}{f_j - 2}} = \sqrt{\frac{5(14)}{12}} = 2.415.$$

Hence

$$z = \frac{.87 + .54 + 1.10 + 1.50 + 1.30}{2.415} = 2.20.$$

The probability of obtaining this value of z or one larger is

$$\Pr(z \geq 2.20) = .014.$$

Hence the combined evidence from the five experiments indicates that the H_1 common to each of the experiments should be rejected if the scope of the inference is with respect to the combined populations and if $\alpha = .05$.

One of the statistical principles underlying this last test procedure is that the statistic

$$\frac{t_j}{\sqrt{f_j/(f_j - 2)}},$$

is approximately distributed as $N(0,1)$. It is shown in Sec. A.5 that the variance of the t distribution having f_j degrees of freedom is $f_j/(f_j - 2)$. Hence the z statistic used in this test is the sum of independent, standardized t statistics. Except for extreme tails of the distribution, the approximation by the unit normal distribution is quite good when each f_j is 10 or more.

In some instances one could combine the data from all five experiments into a single larger experiment. This procedure might, however, tend to obscure rather than clarify the interpretation—the former would be the case if differences associated with the unique conditions of the individual experiments were allowed to augment the experimental error. This contingency can be avoided through use of designs which will be discussed in the next chapter.

1.12 Outliers and Winsorized t Statistic

One sometimes encounters samples having a few observations which are relatively extreme with respect to the bulk of the observations. It may be that these extreme observations are due to sources of error other than that attributable to sampling alone. For example, some subjects in an experiment may not be following instructions. As another example, if observations are being recorded automatically, it may be that the recorder did not function correctly for some of the observations. There can be many "after the fact" explanations for such deviant observations. Highly suspect observations in a sample are known in the statistical literature as "outliers" or "wild shots."

When one constructs confidence intervals on parameters of location, such as the mean, from a relatively small sample containing one or more observations suspect of being outliers, the resulting confidence bounds are likely to be unduly wide. Hence there is a real need for procedures which are efficient in guarding against undue influence of one or more highly extreme observations in relatively small samples.

One method of handling highly suspect extreme observations is that of *trimming*. Under this procedure equal numbers of the lowest and highest observations are removed from the sample; the resulting reduced (or trimmed) sample is treated as the sample data—with an appropriate adjustment made to correct the usual sampling distribution for the effect of the trimming. Some work has been done on determining the sampling distributions for statistics obtained from trimmed samples. This work is closely related to the problem of estimating parameters of location from order statistics obtained from doubly censored samples.

A second method for handling highly deviant outliers is that of *winsorization*. In the simplest case of winsorization ($g = 1$) the high and low extremes are replaced, respectively, by the next to the highest and next to the lowest extremes. The resulting data are processed as if it were the original data—approximate sampling distributions have been developed for statistics computed from winsorized samples. Under winsorization with $g = 2$, the highest two extremes are both replaced by the third highest; a similar replacement is made at the lower end of the ordered sample data. Winsorization with $g = 3$ operates in a similar way—the three highest and three lowest extremes are replaced by the next closest extreme.

Dixon and Tukey (1968) found that, when the basic population from which sample data have been drawn is normal in form, the usual t statistic computed from the winsorized data behaves, to a satisfactory approximation, like Student's t with modified degrees of freedom. Thus procedures for making tests or setting confidence intervals on a population mean are similar to those using the usual t statistic.

If g represents the number of elements in the basic sample that have been replaced at either extreme under winsorization, then the number of basic observations remaining is $n - 2g$. Let $h = n - 2g$. If the observations in the sample are ordered from low to high, one has

$$\underset{\text{lowest}}{X_1}, \ldots, X_g, \underline{X_{g+1}, \ldots, X_{g+h}}, X_{g+h+1}, \ldots, \underset{\text{highest}}{X_n}.$$

The winsorized data would have the form

$$X_{g+1}, \ldots, \underline{X_{g+1}, X_{g+1}, \ldots, X_{g+h}}, X_{g+h}, \ldots, X_{g+h}.$$

The underscored observations remain unchanged. The value of g should be tailored to the individual sample in the sense that intelligent "rejection" of raw observations should be guided by the configuration of the particular sample. Expertise with respect to the population from which the sample was drawn (or in experimental settings, with respect to the conditions under which the raw data have been obtained) can provide such guidance.

A confidence interval on μ based upon winsorized sample data has the form

$$C[\bar{X} - A \leq \mu \leq \bar{X} + A] = 1 - \alpha,$$

where $$A = \frac{n-1}{h-1} t_{1-(\alpha/2)}(h-1)s_{\bar{X}}, \qquad s_{\bar{X}} = \sqrt{\frac{\text{SS}}{n(n-1)}},$$

SS being computed from the winsorized sample data.

A numerical example is given in Table 1.12-1. In the first column of part i are the basic (ordered) sample data. ($g = 0$ corresponds to no winsorization.) The second column corresponds to winsorization with $g = 1$. One notes that the highest and lowest values of the sample data (68 and 10) have been replaced by the next most extreme values (46 and 31). In the third column, the sample data have been winsorized with $g = 2$. Here the two most extreme observations at both ends (68, 46 and 31, 10) are replaced by the next most extreme, namely, 45 at one end and 32 at the other.

In part ii of the table, quantities needed to obtain the confidence intervals are computed. In each case, computations are based upon the data in the corresponding column. For example, SS corresponding to $g = 2$ is computed from the third column:

$$\text{SS} = \Sigma X^2 - \frac{(\Sigma X)^2}{n} = 20{,}395 - \frac{(511)^2}{13} = 308.73.$$

Computations leading to the allowance factor, A, are outlined in part iii. Values for $t_{1-(\alpha/2)}$ with the appropriate degrees of freedom are obtained from tables of Student's t. These values are then multiplied by a factor of $(n-1)/(h-1)$ to approximate the corresponding value on the sampling

Table 1.12–1 Use of Winsorized t Test

Sample size $n = 13$

		$g = 0$, $h = 13$	$g = 1$, $h = 11$	$g = 2$, $h = 9$
(i)		68	$\underline{46}$	45
		46	46	$\underline{45}$
		45	45	45
		43	43	43
		43	43	43
		41	41	41
		40	40	40
		40	40	40
		38	38	38
		35	35	35
		32	32	$\underline{32}$
		31	$\underline{31}$	32
		10	31	32
(ii)	ΣX	512	511	511
	$\bar{X}$	39.38	39.31	39.31
	ΣX^2	22098	20451	20395
	SS	1933.08	364.77	308.73
	$\dfrac{SS}{n(n-1)}$	12.3915	2.3383	1.9790
	$\sqrt{\dfrac{SS}{n(n-1)}}$	3.520	1.529	1.407
(iii)	$t_{.975}(h-1)$	2.18	2.23	2.31
	$\dfrac{n-1}{h-1}\, t_{.975}(h-1)$	2.18	2.68	3.46
	A	7.67	4.10	4.87
(iv)	$g = 0$	$C(31.71 \le \mu \le 47.05) = .95$		
	$g = 1$	$C(35.21 \le \mu \le 43.41) = .95$		
	$g = 2$	$C(34.44 \le \mu \le 44.18) = .95$		

distribution of the winsorized t. The resulting confidence intervals are given in part iv. In this case the shortest confidence interval is that corresponding to $g = 1$.

If the degree of winsorization, g, in the sample is chosen a posteriori to correspond to the confidence interval having the smallest width, the resulting confidence statement will err on the side of having width smaller than should be the case; that is, a confidence interval which takes the a posteriori selection into account will have greater width. A conservative procedure which errs on the side of making the width of the confidence interval too wide, but does take the a posteriori decision into account, is obtained as follows:

Suppose one considered the following potential values for g:

$$g = 0, \quad g = 1, \quad \ldots, \quad g = k - 1.$$

The probability that any one of the k confidence intervals, each of size $1 - (\alpha/k)$, will fail to cover the mean of the population distribution is

$$\frac{\alpha}{k} + \frac{\alpha}{k} + \cdots + \frac{\alpha}{k} = \alpha.$$

Hence the probability that the shortest of the k confidence intervals will fail to cover μ is less than α. Thus, if one replaces α by α/k in determining the allowance factor, one obtains a confidence interval which permits the a posteriori selection of g and will not err on the side of producing confidence intervals having widths that will be too small with respect to replication under comparable a posteriori decisions.

For the numerical example in Table 1.12-1, suppose, a priori, one considered $g = 0, 1, 2$. That is, $k = 3$. The allowance factor for $\alpha = .05$ becomes

$$A = \frac{n-1}{h-1} t_{1-(\alpha/2k)}(h-1)s_{\bar{X}} = \frac{n-1}{h-1} t_{1-(.05/6)}(h-1)$$

$$s_{\bar{X}} = \frac{n-1}{h-1} t_{.9917}(h-1)s_{\bar{X}}.$$

For the case $g = 1$,

$$A = \frac{12}{10}(2.90)(1.529) = 5.32$$

The resulting, highly conservative, confidence interval is

$$C(33.99 \le \mu \le 44.63) = .95.$$

The width of this confidence interval (10.64) is still smaller than the interval corresponding to the unwinsorized data (15.34).

1.13 Multivariate Analog of Test on Difference between Two Means—Hotelling's T^2

In Sec. 1.7, to test the hypothesis

$$\mu_1 - \mu_2 = 0,$$

the statistic

$$(1) \qquad F = t^2 = \frac{(\bar{X}_1 - \bar{X}_2)^2}{s_p^2[(1/n_1) + (1/n_2)]} = \frac{n_1 n_2}{n_1 + n_2} \frac{(\bar{X}_1 - \bar{X}_2)^2}{s_p^2}$$

was used. For purposes of comparing this statistic with that which will be used in the multivariate case, the statistic in (1) may be written in the form

$$F = \frac{n_1 n_2}{n_1 + n_2}(\bar{X}_1 - \bar{X}_2)'(s_p^2)^{-1}(\bar{X}_1 - \bar{X}_2). \tag{2}$$

When the hypothesis being tested is true, the sampling distribution of the statistic in (2) is given by

$$F(1, n_1 + n_2 - 2).$$

Suppose one draws independent random samples of size n_1 and n_2 from the p-variate normal populations $N(\underline{\mu}_1,\Sigma_1)$ and $N(\underline{\mu}_2,\Sigma_2)$, respectively. It will be assumed that $\Sigma_1 = \Sigma_2 = \Sigma$, that is, homogeneity of covariances in the two populations. From data in a given sample one may compute the following statistics:

$$\underline{\bar{X}}_1 = \begin{bmatrix} \bar{X}_{11} \\ \vdots \\ \bar{X}_{p1} \end{bmatrix}, \qquad \underline{\bar{X}}_2 = \begin{bmatrix} \bar{X}_{12} \\ \vdots \\ \bar{X}_{p2} \end{bmatrix}, \qquad S_1, \quad S_2, \quad \text{and} \quad S,$$

where S_1 = covariance matrix obtained from sample 1,

S_2 = covariance matrix obtained from sample 2,

$$S = \text{pooled covariance matrix} = \frac{(n_1 - 1)S_1 + (n_2 - 1)S_2}{n_1 + n_2 - 2}.$$

$\underline{\bar{X}}_1$ provides an unbiased estimate of the mean vector $\underline{\mu}_1$; similarly $\underline{\bar{X}}_2$ provides an unbiased estimate of the mean vector $\underline{\mu}_2$. Under the hypothesis of homogeneity of covariance, S provides an unbiased estimate of Σ.

To test the p-variate hypothesis $\underline{\mu}_1 - \underline{\mu}_2 = \underline{0}$ against the alternative hypothesis $\underline{\mu}_1 - \underline{\mu}_2 \neq \underline{0}$, an appropriate test statistic is Hotelling's T^2. For this case,

$$T^2 = \frac{n_1 n_2}{n_1 + n_2}(\underline{\bar{X}}_1 - \underline{\bar{X}}_2)'S^{-1}(\underline{\bar{X}}_1 - \underline{\bar{X}}_2).$$

When the hypothesis being tested is true, the statistic

$$\frac{(n_1 + n_2 - 2) - p + 1}{(n_1 + n_2 - 2)p} T^2$$

is distributed as a central F distribution with degrees of freedom

$$p \qquad \text{and} \qquad (n_1 + n_2 - 2) - p + 1.$$

Equivalently, $100(1 - \alpha)$ percentile point on the distribution of the T^2

statistic is given by

$$T^2_{1-\alpha} = \frac{(n_1 + n_2 - 2)p}{(n_1 + n_2 - 2) - p + 1} F_{1-\alpha}(p, n_1 + n_2 - 2 - p + 1).$$

A decision rule with level of significance α is given by

Reject the hypothesis being tested if $T^2_{\text{obs}} > T^2_{1-\alpha}$.

For this special case, both the likelihood-ratio principle and the union-intersection principle for test construction lead to the use of Hotelling's T^2.

A numerical example of this test procedure is illustrated in Table 1.13-1. The observed data appear in part i. The sample mean vectors as well as the difference vector are given in part ii. The covariation matrices for the individual samples and the pooled covariation matrix appear in part iii. The inverse of the covariance matrix also is given in part iii. The T^2 statistic for the observed data is obtained in part iv. T^2_{obs} is 7.00. This value of T^2 is significant at the .10 level.

Often when the multivariate hypothesis $\underline{\mu}_1 - \underline{\mu}_2 = \underline{0}$ is rejected, one or more of the univariate hypotheses of the form $\mu_{j1} - \mu_{j2} = 0$ will also be rejected. However, it does *not* necessarily follow that one or more such univariate hypotheses will be rejected; in spite of a significant multivariate test it may be that no univariate test of the form indicated will be significant. Considered singly, the univariate tests disregard the covariances among the variables and hence use less of the information available about a set of observations.

In this case the univariate hypotheses (two-tailed) on the means may be tested through use of the statistics

$$t_1^2 = \frac{(-2.83)^2}{\frac{36.83}{10}\left(\frac{1}{6} + \frac{1}{6}\right)} = 6.52,$$

$$t_2^2 = \frac{(-3.00)^2}{\frac{183.00}{10}\left(\frac{1}{6} + \frac{1}{6}\right)} = 1.48.$$

For a test having level of significance .05 the critical value is

$$F_{.95}(1,10) = 5.12.$$

If the matrix of covariances is known to be Σ, then the statistic

$$\frac{n_1 n_2}{n_1 + n_2}(\underline{\bar{X}}_1 - \underline{\bar{X}}_2)'\Sigma^{-1}(\underline{\bar{X}}_1 - \underline{\bar{X}}_2)$$

is distributed as chi square with p degrees of freedom, under the hypothesis that $\underline{\mu}_1 = \underline{\mu}_2$. Hence this statistic can be used to test the hypothesis that $\underline{\mu}_1 = \underline{\mu}_2$ when the matrix Σ is known.

Table 1.13-1 Numerical Example of Hotelling's T^2

(i)

	Treatment 1		Treatment 2		
	X_1	X_2	X_1	X_2	
	3	10	7	14	
	6	18	9	22	
$n_1 = 6$	5	22	5	19	$n_2 = 6$
	8	20	10	24	
	4	16	10	26	
	7	19	9	18	
ΣX	33	105	50	123	
ΣX^2	199	1925	436	2617	
ΣX_1X_2	605		1053		

(ii)

$$\underline{\bar{X}}_1 = \begin{bmatrix} 5.50 \\ 17.50 \end{bmatrix} \qquad \underline{\bar{X}}_2 = \begin{bmatrix} 8.33 \\ 20.50 \end{bmatrix} \qquad \underline{d} = \underline{\bar{X}}_1 - \underline{\bar{X}}_2 = \begin{bmatrix} -2.83 \\ -3.00 \end{bmatrix}$$

(iii)

$$P_1 = (n_1 - 1)S_1 = \begin{bmatrix} 17.50 & 27.50 \\ 27.50 & 87.50 \end{bmatrix} \qquad P_2 = (n_2 - 1)S_2 = \begin{bmatrix} 19.33 & 28.00 \\ 28.00 & 95.50 \end{bmatrix}$$

$$P = (n_1 + n_2 - 2)S = P_1 + P_2 = \begin{bmatrix} 36.83 & 55.50 \\ 55.50 & 183.00 \end{bmatrix} \qquad S^{-1} = (n_1 + n_2 - 2)P^{-1} = \begin{bmatrix} .5001 & -.1517 \\ -.1517 & .1006 \end{bmatrix}$$

(iv)

$$T^2 = \frac{n_1 n_2}{n_1 + n_2} \underline{d}' S^{-1} \underline{d} = 9.00 \qquad T^2_{.90} = \frac{20}{9} F_{.90}(2,9) = 6.69$$

$$T^2_{.95} = \frac{20}{9} F_{.95}(2,9) = 9.47$$

2

LINEAR MODELS

Note: In this chapter statistical theory underlying principles given in later chapters is presented. Of necessity, the mathematical level of the writing is more demanding than the corresponding level of the more applied material in later chapters. Hence the reader may omit many or indeed all of the sections in this chapter without any serious loss in readability of what follows.

2.1 Linear Model—No Distribution Assumptions

Underlying the statistical aspects of the topic known as the analysis of variance is a series of linear models plus a set of distribution assumptions about the random variables in these models. In classic analysis of variance, the basic variables are what have sometimes been called *fixed* variables. In more general applications of the analysis of variance, the basic variables may be either fixed, random, or some mixture of fixed and random variables. If the basic variables are all random, the model is called the *variance-component model.* If the basic variables are a mixture of fixed and random variables, the model is called the *mixed model.*

There is no probability density associated with a fixed variable. Given the outcome of an experiment, if a specified treatment has been administered, the value of a fixed variable is an unknown constant associated with that treatment. On the other hand, associated with a random variable is a probability density having one or more parameters. The outcome of an experiment involving one or more random variables is a function of the form of the probability density and the values of the parameters.

The term *regression model* is used in two different senses. In the more classic sense, the basic variables are considered fixed. In this sense, the analysis-of-variance model for fixed variables is a special case of the regression model. In some of the more recent statistical literature (see Graybill, 1961), the term regression model refers to the case in which the basic variables are random (usually having a multivariate normal distribution).

In the models that are used in the analysis of variance, some of the variables are *counter* (or *indicator*) variables. This kind of variable takes the value one when some effect is present and zero when that effect is not present. In the classic regression model, the corresponding variables are continuous rather than dichotomous.

2.1-1 Univariate case

No assumptions will be made about the population distribution of the variables that are considered in this section. Attention will be directed only to the description of the sample data. Let the data in a sample from a specified population be represented as follows:

Sample element	X	Y
1	X_1	Y_1
2	X_2	Y_2
.	.	.
.	.	.
.	.	.
n	X_n	Y_n

Let $\bar{X} = \dfrac{\Sigma X}{n}$ = mean for X; $\bar{Y} = \dfrac{\Sigma Y}{n}$ = mean for Y;

$$SS_X = \Sigma(X - \bar{X})^2 = \Sigma X^2 - \frac{(\Sigma X)^2}{n} = \text{variation of } X;$$

$$SS_Y = \Sigma(Y - \bar{Y})^2 = \Sigma Y^2 - \frac{(\Sigma Y)^2}{n} = \text{variation of } Y;$$

$$SP_{XY} = \Sigma(X - \bar{X})(Y - \bar{Y})$$

$$= \Sigma XY - \frac{(\Sigma X)(\Sigma Y)}{n} = \text{covariation between } X \text{ and } Y.$$

Suppose it is desired to build a prediction system of the form

$$\hat{Y} = b_0 + b_1 X, \tag{1}$$

where b_0 and b_1 are to be determined in such a way as to "optimize" the prediction of Y in the sample. One criterion for optimization is to make

$$\Sigma(Y - \hat{Y})^2 = \text{minimum for sample data.} \tag{2}$$

Expression (2) represents the *least-squares* optimization criterion. b_0 and b_1 determined in such a way as to satisfy (2) are the least-squares definitions of b_0 and b_1.

Expression (2) may be written in the form

$$\Sigma(Y - b_0 - b_1X)^2 = \Sigma Y^2 + nb_0^2 + b_1^2(\Sigma X^2) - 2b_0(\Sigma Y) - 2b_1(\Sigma XY) + 2b_0b_1(\Sigma X) = \text{minimum.} \tag{3}$$

Expression (3) may be minimized by taking the partial derivations with respect to b_0 and b_1 and then setting what results equal to zero. The end product defines the set of *normal equations*. The latter are

$$\begin{aligned} nb_0 + b_1(\Sigma X) - \Sigma Y &= 0, \\ b_1(\Sigma X^2) - \Sigma(XY) + b_0(\Sigma X) &= 0. \end{aligned} \tag{4}$$

Solving the first equation for b_0 in terms of b_1 gives

$$b_0 = \bar{Y} - b_1\bar{X}. \tag{5}$$

Substituting the expression for b_0 given in (5) into the second equation in (4) yields, after some manipulation,

$$b_1(\text{SS}_X) = \text{SP}_{XY} \qquad \text{or} \qquad b_1 = \frac{\text{SP}_{XY}}{\text{SS}_X}. \tag{6}$$

The prediction equation, with b_0 and b_1 defined by the least-squares criterion, has the form

$$\hat{Y} = (\bar{Y} - b_1\bar{X}) + b_1X \qquad \text{or} \qquad \hat{Y} - \bar{Y} = b_1(X - \bar{X}). \tag{7}$$

If one sums both sides of the equation at the right over all sample elements, one obtains

$$\Sigma(\hat{Y} - \bar{Y}) = b_1\Sigma(X - \bar{X}) = 0,$$

since $\Sigma(X - \bar{X}) = 0$. Hence

$$\Sigma\hat{Y} = \Sigma Y \qquad \text{or} \qquad \bar{\hat{Y}} = \bar{Y}. \tag{8}$$

The following properties of variations are readily established. If

$$K = c_0 + c_1X, \quad \text{then} \quad \text{SS}_K = c_1^2\text{SS}_X.$$

If

$$U = V - W, \quad \text{then} \quad \text{SS}_U = \text{SS}_V + \text{SS}_W - 2\text{SP}_{UV}.$$

Since

$$\hat{Y} = b_0 + b_1X,$$

it follows that

$$\text{SS}_{\hat{Y}} = b_1^2\text{SS}_X = \frac{\text{SP}_{XY}^2}{\text{SS}_X^2}\text{SS}_X = \frac{\text{SP}_{XY}^2}{\text{SS}_X} = b_1\text{SP}_{XY}. \tag{9}$$

Also $$\mathrm{SS}_{Y-\hat{Y}} = \mathrm{SS}_Y + \mathrm{SS}_{\hat{Y}} - 2\mathrm{SP}_{Y\hat{Y}}.$$

But
$$\begin{aligned}\mathrm{SP}_{Y\hat{Y}} &= \Sigma Y\hat{Y} - \frac{(\Sigma Y)(\Sigma \hat{Y})}{n} = \Sigma Y\hat{Y} - \frac{(\Sigma Y)^2}{n} \\ &= \Sigma Y(\bar{Y} - b_1\bar{X} + b_1 X) - \frac{(\Sigma Y)^2}{n} \\ &= b_1\left[\Sigma XY - \frac{(\Sigma X)(\Sigma Y)}{n}\right] + \frac{(\Sigma Y)^2}{n} - \frac{(\Sigma Y)^2}{n} \\ &= b_1\mathrm{SP}_{XY} = \mathrm{SS}_{\hat{Y}}.\end{aligned}$$

Hence

$$\mathrm{SS}_{Y-\hat{Y}} = \mathrm{SS}_Y + \mathrm{SS}_{\hat{Y}} - 2\mathrm{SS}_{\hat{Y}} = \mathrm{SS}_Y - \mathrm{SS}_{\hat{Y}}. \tag{10}$$

Rearranging the terms in (10) one has

$$\mathrm{SS}_Y = \mathrm{SS}_{\hat{Y}} + \mathrm{SS}_{Y-\hat{Y}}. \tag{11}$$

Let the square of the product-moment correlation between X and Y, r^2_{XY}, be defined as

$$r^2_{XY} = \frac{\mathrm{SS}_{\hat{Y}}}{\mathrm{SS}_Y} = \frac{b_1\mathrm{SP}_{XY}}{\mathrm{SS}_Y} = \frac{\mathrm{SP}^2_{XY}}{\mathrm{SS}_X\mathrm{SS}_Y}. \tag{12}$$

That is, r^2_{XY} is the proportion of the total variation of Y in the sample that can be predicted by the prediction system in which b_0 and b_1 are defined by the least-squares criterion. In terms of r^2_{XY},

$$\mathrm{SS}_{\hat{Y}} = r^2_{XY}\mathrm{SS}_Y \quad \text{and} \quad \begin{aligned}\mathrm{SS}_{Y-\hat{Y}} &= \mathrm{SS}_Y - \mathrm{SS}_{\hat{Y}} \\ &= (1 - r^2_{XY})\mathrm{SS}_Y.\end{aligned} \tag{13}$$

A numerical example to illustrate the material discussed so far in this section is given in Table 2.1-1. The sample data are given at the left in part i. To the right of part i various summary statistics are computed. The sample values of b_0 and b_1 as defined by the least-squares criterion are obtained in part ii. The numerical value for r^2_{XY} is also computed in part ii. For this sample $r^2_{XY} = .980$, indicating that 98 percent of the total variation on Y can be predicted.

The prediction equation is given in part iii of the table. The predicted scores, $\hat{Y}$, as obtained from this prediction equation are given in part iv. For example, for element 1, $X = 3$,

$$\hat{Y}_1 = 5.853 + 1.693(3) = 10.902.$$

The error (or residual) associated with element 1 is

$$Y_1 - \hat{Y}_1 = 10 - 10.902 = -.902.$$

Table 2.1-1 Numerical Example of Construction of Prediction System for Sample Data

(i) Sample data:

Element	X	Y
1	3	10
2	5	14
3	8	20
4	12	28
$n = 5$	17	33
$\Sigma()$	45	105
$\Sigma()^2$	531	2569
ΣXY	1157	

$$\bar{X} = \frac{\Sigma X}{n} = 9 \qquad \bar{Y} = \frac{\Sigma Y}{n} = 21$$

$$\text{SS}_X = \Sigma X^2 - \frac{(\Sigma X)^2}{n} = 531 - 405 = 126$$

$$\text{SS}_Y = \Sigma Y^2 - \frac{(\Sigma Y)^2}{n} = 2569 - 2205 = 364$$

$$\text{SP}_{XY} = \Sigma XY - \frac{(\Sigma X)(\Sigma Y)}{n} = 1157 - 945 = 212$$

(ii)

$$b_1 = \frac{\text{SP}_{XY}}{\text{SS}_X} = \frac{212}{126} = 1.683 \qquad b_0 = \bar{Y} - b_1\bar{X} = 5.853$$

$$r^2_{XY} = \frac{\text{SP}^2_{XY}}{\text{SS}_X\text{SS}_Y} = \frac{(212)^2}{(126)(364)} = .980$$

(iii)

$$\hat{Y} = b_0 + b_1X = 5.853 + 1.683X$$

(iv)

Element	Y	$\hat{Y}$	$Y - \hat{Y}$
1	10	10.902	−0.902
2	14	14.268	−0.268
3	20	19.317	0.683
4	28	26.049	1.951
5	33	34.464	−1.464
$\Sigma()$	105	105.000	0.000
$\Sigma()^2$	2569	2561.89	7.30

$$\text{SS}_{\hat{Y}} = \Sigma\hat{Y}^2 - \frac{(\Sigma\hat{Y})^2}{n} = 2561.89 - 2205.00 = 356.89$$

$$= b_1\text{SP}_{XY} = 1.683(212) = 356.80$$

$$= \frac{\text{SP}^2_{XY}}{\text{SS}_X} = \frac{(212)^2}{126} = 356.70$$

$$= r^2_{XY}\text{SS}_Y = .980(364) = 356.72$$

(v) $\text{SS}_{Y-\hat{Y}} = \Sigma(Y - \hat{Y})^2 = 7.30 \quad \text{SS}_{Y-\hat{Y}} = (1 - r^2_{XY})\text{SS}_Y = .020(364) = 7.28$

The variation associated with the predicted Y scores, computed in four different (but equivalent) ways, is given in the bottom part of part iv. The computations agree to three significant digits. The residual or error variation is computed by two different methods in part v. Again the differences are due to rounding error.

A relationship that will be established in a later section is the following:

$$\Sigma \hat{Y}^2 = b_0 \Sigma Y + b_1 \Sigma XY$$
$$= 5.853(105) + 1.683(1157) = 2561.80.$$

As computed in part iv, $\Sigma \hat{Y}^2 = 2561.89$. The difference in the two numerical results is due to rounding error.

2.1-2 Univariate case—matrix notation

As in the preceding section, no assumption is made about the form of the distribution of the variables in the population. In order to be able to generalize the development in the preceding section to the multivariate case, it will be helpful to cast this material in matrix notation.

To introduce greater symmetry into the solution for b_0 and b_1 it is convenient to introduce a dummy variable $X_0 = 1$ for all sample elements. The sample data may now be represented as follows:

Sample element	X_0	X_1	Y
1	1	X_{11}	Y_1
2	1	X_{21}	Y_2
.	.	.	.
.	.	.	.
.	.	.	.
n	1	X_{n1}	Y_n

$(n > 2)$

Define

$$\underset{n,1}{\underline{Y}} = \begin{bmatrix} Y_1 \\ \cdot \\ \cdot \\ \cdot \\ Y_n \end{bmatrix}, \quad \underset{n,2}{X} = \begin{bmatrix} 1 & X_{11} \\ \cdot & \cdot \\ \cdot & \cdot \\ \cdot & \cdot \\ 1 & X_{n1} \end{bmatrix}, \quad \underset{2,1}{\underline{b}} = \begin{bmatrix} b_0 \\ b_1 \end{bmatrix}, \quad \underset{n,1}{\underline{\hat{Y}}} = \begin{bmatrix} \hat{Y}_1 \\ \cdot \\ \cdot \\ \cdot \\ \hat{Y}_n \end{bmatrix}.$$

In terms of these definitions, the prediction equation discussed in Sec. 2.1-1 takes the form

$$\underline{\hat{Y}} = X\underline{b}. \tag{1}$$

The error or residual is given by

$$\underline{e} = \underline{Y} - \underline{\hat{Y}} = \underline{Y} - X\underline{b}. \tag{2}$$

The least-squares criterion for goodness of fit of the prediction system makes

$$\Sigma e^2 = \underline{e}'\underline{e} = \underline{Y}'\underline{Y} + \underline{b}'X'X\underline{b} - 2'\underline{b}X'\underline{Y} = \text{minimum}. \tag{3}$$

Differentiation of (3) with respect to $\underline{b}$ and setting the result equal to zero yield the set of normal equations

$$(X'X)\underline{b} = X'\underline{Y}. \tag{4}$$

For the univariate case $X'X$ will be nonsingular, and the solution for $\underline{b}$ is

$$\underline{b} = (X'X)^{-1}X'\underline{Y} = C\underline{Y} \qquad \text{where} \qquad C = (X'X)^{-1}X'. \tag{5}$$

For the multivariate case, $X'X$ may not be nonsingular; in the latter case the solution for $\underline{b}$ will not be unique.

The prediction equation in (1) takes the form

$$\underline{\hat{Y}} = X\underline{b} = XC\underline{Y} = P\underline{Y}, \qquad \text{where} \qquad \underset{n,n}{P} = XC = X(X'X)^{-1}X'. \tag{6}$$

The vector of error (or residuals) is given by

$$\underline{e} = \underline{Y} - \underline{\hat{Y}} = \underline{Y} - P\underline{Y} = (I - P)\underline{Y} = M\underline{Y}, \qquad \text{where} \qquad M = I - P. \tag{7}$$

The matrices P and M have some very interesting properties. One notes that

$$P = P' \qquad \text{and} \qquad M = M'. \tag{8a}$$

That is, P and M are symmetric. Further,

$$PP = X(X'X)^{-1}X'X(X'X)^{-1}X = X(X'X)^{-1}X' = P. \qquad MM = (I - P)(I - P) = I + P - 2P = I - P = M. \tag{8b}$$

Property (8*b*) indicates that both P and M are *idempotent;* that is, $PP = P$ and $MM = M$. Another property is

$$PM = P(I - P) = P - P = 0 \qquad \text{and} \qquad MP = (I - P)P = 0. \tag{8c}$$

This last property indicates that P and M are *orthogonal.* One also has

$$PX = X(X'X)^{-1}X'X = X \qquad \text{and} \qquad MX = (I - P)X = X - X = 0. \tag{8d}$$

An additional property is

$$P + M = P + (I - P) = I. \tag{8e}$$

This property indicates that P and M are *complementary.* It may be verified that the following property also holds.

$$\underline{1}'P = \underline{1}' \qquad \text{and} \qquad \underline{1}'M = \underline{0}', \tag{8f}$$

where $\underline{1}$ is a vector of unities. Property (8*f*) indicates that each column in the matrix P sums to unity; also each column of the matrix M sums to zero.

The idempotent matrices P and M define what are called *projection* operators. The matrix operation

$$\underline{\hat{Y}} = P\underline{Y}$$

defines a projection of the vector $\underline{Y}$ onto what is called the prediction space. Similarly,

$$\underline{e} = \underline{Y} - \underline{\hat{Y}} = M\underline{Y}$$

defines a projection of the vector $\underline{Y}$ onto the error space. Since P and M are orthogonal, the prediction space and the error space are said to be orthogonal.

From (7) one has that

$$\Sigma e^2 = \underline{e}'\underline{e} = (M\underline{Y})'(M\underline{Y}) = \underline{Y}'MM\underline{Y} = \underline{Y}'M\underline{Y}. \tag{9}$$

From (6) one has

$$\Sigma \hat{Y}^2 = \underline{\hat{Y}}'\underline{\hat{Y}} = \underline{Y}'PP\underline{Y} = \underline{Y}'P\underline{Y}. \tag{10}$$

One notes that

$$\underline{b}'X'\underline{Y} = \underline{Y}'C'X'\underline{Y} = \underline{Y}'P'\underline{Y} = \underline{Y}'P\underline{Y} = \Sigma \hat{Y}^2. \tag{11}$$

The normal equations have the form

$$(X'X)\underline{b} = X'\underline{Y}.$$

From (11) and the normal equations, one notes the following extremely useful and general relationship: The sum of the squared predicted Y scores is obtained by premultiplying the vector on the right of the normal equations $(X'\underline{Y})$ by the transpose of vector $\underline{b}$.

An expression for the sum of the cross products $\Sigma Y\hat{Y}$ is given by

$$\Sigma Y\hat{Y} = \underline{Y}'\underline{\hat{Y}} = \underline{Y}'X\underline{b} = \underline{Y}'X(X'X)^{-1}X'\underline{Y} = \underline{Y}'P\underline{Y} = \Sigma \hat{Y}^2. \tag{12}$$

Thus the sum of the cross products $\Sigma Y\hat{Y} = \Sigma \hat{Y}^2$. Hence an alternative approach to the relationship in (9) is as follows:

$$\begin{aligned} \Sigma e^2 = \Sigma(Y - \hat{Y})^2 &= \Sigma Y^2 + \Sigma \hat{Y}^2 - 2\Sigma Y\hat{Y} \\ &= \Sigma Y^2 + \Sigma \hat{Y}^2 - 2\Sigma \hat{Y}^2 \\ &= \Sigma Y^2 - \Sigma \hat{Y}^2. \end{aligned} \tag{13}$$

But

$$\Sigma Y^2 - \Sigma \hat{Y}^2 = \underline{Y}'\underline{Y} - \underline{Y}'P\underline{Y} = \underline{Y}'(I - P)\underline{Y} = \underline{Y}'M\underline{Y}. \tag{13a}$$

Rearranging (13) one has

$$\Sigma Y^2 = \Sigma \hat{Y}^2 + \Sigma e^2. \tag{14}$$

If one subtracts $(\Sigma Y)^2/n$ from both sides of (14) one gets

$$\Sigma Y^2 - \frac{(\Sigma Y)^2}{n} = \Sigma \hat{Y}^2 - \frac{(\Sigma Y)^2}{n} + \Sigma e^2.$$

This expression may be written in the form

$$\text{(15)} \qquad \text{SS}_Y = \text{SS}_{\hat{Y}} + \text{SS}_{\text{error}}.$$

Expression (15) follows from the fact that $\Sigma Y = \Sigma \hat{Y}$ and $\Sigma e = 0$.

Since

$$\hat{Y} = b_0 X_0 + b_1 X_1, \qquad \text{where} \qquad b_0 X_0 = b_0,$$

using the fact that $b_1 = \text{SP}_{XY}/\text{SS}_X$, it follows that

$$\text{(16)} \qquad \begin{aligned} \text{SP}_{X\hat{Y}} &= b_1 \text{SS}_X = \text{SP}_{XY}, \\ \text{SS}_{\hat{Y}} &= b_1^2 \text{SS}_X = \frac{\text{SP}^2_{XY}}{\text{SS}_X}, \\ \text{SP}_{Y\hat{Y}} &= b_1 \text{SP}_{XY} = \frac{\text{SP}^2_{XY}}{\text{SS}_X} = \text{SS}_{\hat{Y}}. \end{aligned}$$

In terms of the relationships in (16), the square of the correlation is given by

$$\text{(17)} \qquad r^2_{XY} = \frac{\text{SS}_{\hat{Y}}}{\text{SS}_Y} = \frac{\text{SP}^2_{XY}/\text{SS}_X}{\text{SS}_Y} = \frac{\text{SP}^2_{XY}}{\text{SS}_X \text{SS}_Y}.$$

By analogy with the extreme right-hand expression in (17) one may define

$$\text{(18)} \qquad r^2_{Y\hat{Y}} = \frac{\text{SP}^2_{Y\hat{Y}}}{\text{SS}_Y \text{SS}_{\hat{Y}}} = \frac{\text{SS}^2_{\hat{Y}}}{\text{SS}_Y \text{SS}_{\hat{Y}}} = \frac{\text{SS}_{\hat{Y}}}{\text{SS}_Y} = r^2_{XY}.$$

Thus (18) indicates that the square of the correlation between X and Y is equal to the square of the correlation between Y and $\hat{Y}$. Since $\hat{Y}$ is a linear function of X, this last result could have been obtained directly from the fact that the correlation r_{XY} is invariant under linear transformations on either X or Y.

All the relationships developed in (1) through (15) can be generalized to the case in which the matrix X contains any number of variables, provided $X'X$ is nonsingular. None of the relationships has any constraint upon the size of the X matrix. However, if the number of rows of X is less than the number of columns, the matrix $X'X$ will be singular.

The numerical example in Table 2.1-1 is presented in matrix notation in Table 2.1-2. The data matrix appears at the left of part i. The first two columns of the data matrix define the matrix X. The third column defines the vector $\underline{Y}$. The inverse of $X'X$ is given in part ii. The components of the vector $\underline{b}$ are computed at the right of part ii. Within rounding error, the numerical values for b_0 and b_1 agree with those given in Table 2.1-1.

Table 2.1-2 Numerical Example

(i) Data matrix
$$X_{\text{aug}} = \begin{bmatrix} 1 & 3 & 10 \\ 1 & 5 & 14 \\ 1 & 8 & 20 \\ 1 & 12 & 28 \\ 1 & 17 & 33 \end{bmatrix}\ (\text{columns } X_0,\ X_1,\ Y); \quad X'X = \begin{bmatrix} 5 & 45 \\ 45 & 531 \end{bmatrix}; \quad X'\underline{Y} = \begin{bmatrix} 105 \\ 1157 \end{bmatrix}$$

(ii)
$$(X'X)^{-1} = \begin{bmatrix} .8429 & -.07143 \\ -.07143 & .007937 \end{bmatrix}; \quad \underline{b} = (X'X)^{-1}(X'\underline{Y}) = \begin{bmatrix} 5.860 \\ 1.683 \end{bmatrix}$$

(iii)
$$X'_{\text{aug}}X_{\text{aug}} = \begin{bmatrix} 5 & 45 & 105 \\ 45 & 531 & 1157 \\ 105 & 1157 & 2569 \end{bmatrix}; \quad (X'_{\text{aug}}X_{\text{aug}})^{-1} = \begin{bmatrix} 5.5392 & 1.2778 & -.8019 \\ 1.2778 & .3956 & -.2304 \\ -.8019 & -.2304 & .1369 \end{bmatrix}$$
(rows and columns of both matrices labelled X_0, X_1, Y)

(iv)
$$b_0 = -p^{X_0Y}/p^{YY} = -(-.8019)/.1369 = 5.856$$
$$b_1 = -p^{X_1Y}/p^{YY} = -(-.2304)/.1369 = 1.683$$
$$r^2_{XY} = 1 - [1/(p^{YY}\text{SS}_Y)] = 1 - [1/(.1369)(364.00)] = 1 - (1/49.83) = .97994$$

If one represents the data matrix by the symbol X_{aug}, the matrices $(X'_{\text{aug}}X_{\text{aug}})$ and $(X'_{\text{aug}}X_{\text{aug}})^{-1}$ are computed in part iii. Let an element of the latter inverse be denoted p^{ij}. Then the components of the vector $\underline{b}$ may be obtained as indicated in part iv. These relationships between the components of $\underline{b}$ and the elements in row (or column) Y of $(X'_{\text{aug}}X_{\text{aug}})^{-1}$ follow quite directly from the relationship between $(X'X)^{-1}$ and $(X'_{\text{aug}}X_{\text{aug}})^{-1}$. The elements of row Y of the latter inverse are proportional to $(X'X)^{-1}(X'\underline{Y})$.

The square of the correlation between X and Y may also be found from the element p^{YY} as indicated in part iv. The relationship follows from the fact that

$$\frac{1}{p^{YY}} = \text{SS}_{Y-\hat{Y}}.$$

Hence
$$1 - \frac{1}{p^{YY}\text{SS}_Y} = 1 - \frac{\text{SS}_{Y-\hat{Y}}}{\text{SS}_Y} = r^2_{XY}.$$

The matrix P for the data in Table 2.1-2 is given by

$$P = X(X'X)^{-1}X' = \begin{bmatrix} .4857 & .3905 & .2477 & .0571 & -.1809 \\ .3905 & .3271 & .2319 & .1048 & -.0539 \\ .2477 & .2319 & .2081 & .1762 & .1366 \\ .0571 & .1048 & .1762 & .2714 & .3906 \\ -.1809 & -.0539 & .1366 & .3906 & .7081 \end{bmatrix}.$$

It will be noted that the sum of the entries in each column is unity. Since the matrix is symmetric, the sum of the entries in each row will also be unity. It is of interest to note that

$$\text{trace } P = 2 = \text{rank } P.$$

The computations given in Table 2.1-2 may be simplified by applying what is called the SWP operator (which is described in Sec. 2.5) to the matrix $X'_{\text{aug}}X_{\text{aug}}$. For purposes of simplifying the notation let

$$M = X'_{\text{aug}}X_{\text{aug}}.$$

In part ii the matrix $M_{(1)}$ is generated by applying the operation SWP[1] to M. The last column of $M_{(1)}$ may be shown to contain $\bar{Y}$, SP_{X_1Y}, and SS_Y. In part iii the matrix $M_{(12)}$ is generated by applying the operation SWP[2] to $M_{(1)}$. The last column of $M_{(12)}$ contains the entries b_0, b_1, and $\text{SS}_{Y-\hat{Y}}$. The 2×2 submatrix in the upper left-hand corner of $M_{(12)}$ will be found to be $-(X'X)^{-1}$.

Table 2.1-3 Numerical Example Using SWP Operator

(i) $M = X'_{\text{aug}}X_{\text{aug}}$	5	45	105
	45	531	1157
	105	1157	2569
(ii) $M_{(1)} = \text{SWP}[1]\, M$	−.2000	9.0000	21.0000 $= \bar{Y}$
	9.0000	126.0000	212.0000 $= \text{SP}_{X_1Y}$
	21.0000	212.0000	364.0000 $= \text{SS}_Y$
(iii) $M_{(12)} = \text{SWP}[2]\, M_{(1)}$	−.84286	.071429	5.85714 $= b_0$
	.071429	−.0079365	1.68254 $= b_1$
	5.85714	1.68254	7.30159 $= \text{SS}_{Y-\hat{Y}}$
(iv) $M_{(123)} = \text{SWP}[3]\, M_{(12)}$	−5.54130	−1.27826	.80217
	−1.27826	−.39565	.23043
	.80217	.23043	−.13696

In part iv the matrix $M_{(123)}$ is generated by applying the operation SWP[3] to $M_{(12)}$. It will be found that

$$M^{-1} = -M_{(123)}.$$

Application of a Basic Theorem in Matrix Algebra. Let a nonsingular matrix M be partitioned as follows:

$$\underset{p,p}{M} = \begin{array}{c} \\ k \\ p-k \end{array}\!\!\begin{array}{c} \begin{array}{cc} k & p-k \end{array} \\ \begin{bmatrix} M_{11} & M_{12} \\ M_{21} & M_{22} \end{bmatrix} \end{array}.$$

The inverse of M in terms of submatrices will take the form

$$\underset{p,p}{M^{-1}} = \begin{matrix} & k & p-k \\ k & \multicolumn{2}{c}{\begin{bmatrix} C_{11} & C_{12} \\ C_{21} & C_{22} \end{bmatrix}} \\ p-k & & \end{matrix},$$

where

$$C_{11} = [M_{11} - M_{12}M_{22}^{-1}M_{21}]^{-1}, \qquad C_{22} = [M_{22} - M_{21}M_{11}^{-1}M_{12}]^{-1},$$
$$C_{12} = -M_{11}^{-1}M_{12}C_{22}, \qquad C_{21} = -M_{22}^{-1}M_{21}C_{11}.$$

This theorem in matrix algebra indicates how X_{aug} enters into the relationships discussed earlier in this section. One has

$$X'_{\text{aug}}X_{\text{aug}} = \begin{matrix} & p & 1 \\ p & X'X & X'\underline{Y} \\ 1 & \underline{Y}'X & \underline{Y}'\underline{Y} \end{matrix} = \begin{bmatrix} M_{11} & M_{12} \\ M_{21} & M_{22} \end{bmatrix}.$$

If

$$(X'_{\text{aug}}X_{\text{aug}})^{-1} = \begin{bmatrix} C_{11} & C_{12} \\ C_{21} & C_{22} \end{bmatrix},$$

then

$$C_{22} = [\underline{Y}'\underline{Y} - \underline{Y}'X(X'X)^{-1}X'\underline{Y}]^{-1}.$$

Hence

$$\frac{1}{C_{22}} = \underline{Y}'\underline{Y} - \underline{Y}'P\underline{Y}.$$

Further

$$C_{12} = -(X'X)^{-1}X'\underline{Y}C_{22} = -\underline{b}C_{22}.$$

Hence

$$\underline{b} = \frac{-C_{12}}{C_{22}},$$

since C_{22} is a scalar.

2.1-3 Multivariate case

Again, in this section, the discussion is limited to the development of a prediction system which optimizes (in a least-squares sense) the prediction within the sample. No assumptions are made about the joint distribution of the variables. Suppose the sample data are represented as follows:

Sample element	X_0	X_1	$\cdots$	X_p	Y
1	1	X_{11}	$\cdots$	X_{1p}	Y_1
2	1	X_{21}	$\cdots$	X_{2p}	Y_2
.	.	.		.	.
.	.	.		.	.
.	.	.		.	.
n	1	X_{n1}	$\cdots$	X_{np}	Y_n

There are $p + 1$ X variables, including a dummy variable X_0.

Let

$$\underset{n,p+1}{X} = \text{matrix of observations on } X_0, X_1, \ldots, X_p;$$

$$\underset{n,1}{\underline{Y}} = \text{vector of observations on } Y;$$

$$\underset{n,1}{\underline{\hat{Y}}} = \text{vector of predicted scores on } Y;$$

$$\underline{e} = \underline{Y} - \underline{\hat{Y}} = \text{vector of errors or residuals.}$$

The prediction system is to be linear in $X_0, X_1, \ldots, X_p$. That is,

$$\hat{Y} = b_0 X_0 + b_1 X_1 + \cdots + b_p X_p.$$

In some applications, a more complete notation system for the weights in the prediction system is needed. A widely used notation is one given below.

$$\hat{Y} = b_{Y0.12\cdots p} X_0 + \underline{b}_{Y1.02\cdots p} X_1 + b_{Y2.01\cdots p} X_2 + \cdots + b_{Yp.01\cdots(p-1)} X_p.$$

If, for example, $p = 3$, the prediction system in this notation has the form

$$\hat{Y} = b_{Y0.123} X_0 + b_{Y1.023} X_1 + b_{Y2.013} X_2 + b_{Y3.012} X_3.$$

Where the context makes clear just what predictors are in the system, the more complete notation will not be used. Where, however, one is using the complete set of predictors as well as subsets of the complete system, the more complete notation system will be needed.

One may represent the observed Y in the form

$$Y = \hat{Y} + e \qquad \text{or} \qquad e = Y - \hat{Y}.$$

Thus $\hat{Y}$ represents that part of Y which is a linear function of X, and e represents that part of Y which is *not* a linear function of X. In matrix notation,

$$\underline{\hat{Y}} = X\underline{b}, \qquad \underline{Y} = \underline{\hat{Y}} + \underline{e},$$

where $\underline{b}' = [b_0 \, b_1 \cdots b_p]$.

In Sec. 2.1-2, the relationships (1) through (15) were *not* restricted to a matrix X having only two columns. The only restriction on some of the relationships was that $X'X$ be nonsingular. Hence all the relationships obtained in (1) through (15) in Sec. 2.1-2 apply to the case in which the matrix X has $p + 1$ columns. The relationships in (16) through (18) in Sec. 2.1-2 also generalize to the multivariate case, but the generalization is somewhat less direct.

It will be found useful to have an expression for the components of the vector $\underline{b}$ in terms of the covariations of the variables. Toward this end, suppose the scale of measurement of the X variables is changed so that each variable has a mean of zero. That is, suppose $\bar{X}_j$ is subtracted from each X_{ij}. Under this transformation, the product $X'X$ takes the form

$$X'X = \begin{bmatrix} n & 0 & \cdots & 0 \\ 0 & \text{SS}_{X_1} & \cdots & \text{SP}_{X_1X_p} \\ \cdot & \cdot & & \cdot \\ \cdot & \cdot & & \cdot \\ \cdot & \cdot & & \cdot \\ 0 & \text{SP}_{X_pX_1} & \cdots & \text{SS}_{X_p} \end{bmatrix} = \begin{bmatrix} n & 0 & \cdots & 0 \\ 0 & & & \\ \cdot & & V_{XX} & \\ \cdot & & & \\ \cdot & & & \\ 0 & & & \end{bmatrix} = \begin{bmatrix} n & \underline{0}' \\ \underline{0} & V_{XX} \end{bmatrix},$$

where V_{XX} is the covariation matrix for the variables $X_1, \ldots, X_p$. If a similar transformation is made on the components of the vector $\underline{Y}$ one has

$$X'\underline{Y} = \begin{bmatrix} \Sigma Y \\ \text{SP}_{X_1Y} \\ \cdot \\ \cdot \\ \cdot \\ \text{SP}_{X_pY} \end{bmatrix} = \begin{bmatrix} \Sigma Y \\ \underline{v}_{XY} \end{bmatrix},$$

where $\underline{v}_{XY}$ = vector of covariations between Y and $X_1, \ldots, X_p$.

In terms of the transformed variables, the normal equations take the form

$$\begin{bmatrix} n & \underline{0}' \\ \underline{0} & V_{XX} \end{bmatrix} \underline{b} = \begin{bmatrix} \Sigma Y \\ \underline{v}_{XY} \end{bmatrix}. \tag{1}$$

Hence

$$\underline{b} = \begin{bmatrix} \frac{1}{n} & \underline{0}' \\ \underline{0} & V_{XX}^{-1} \end{bmatrix} \begin{bmatrix} \Sigma Y \\ \underline{v}_{XY} \end{bmatrix} = \begin{bmatrix} \bar{Y} \\ V_{XX}^{-1}\underline{v}_{XY} \end{bmatrix}. \tag{2}$$

From (2) one has

$$b_0 = \bar{Y}, \qquad \begin{bmatrix} b_1 \\ \cdot \\ \cdot \\ \cdot \\ b_p \end{bmatrix} = V_{XX}^{-1}\underline{v}_{XY}. \tag{3}$$

Thus, in terms of the transformed variables,

$$\underline{\hat{Y}} = X \begin{bmatrix} \bar{Y} \\ V_{XX}^{-1}\underline{v}_{XY} \end{bmatrix} \tag{4}$$

and

$$\Sigma \hat{Y}^2 = \underline{b}' \begin{bmatrix} n & \underline{0}' \\ \underline{0} & V_{XX} \end{bmatrix} \underline{b} = n\bar{Y}^2 + [b_1 \cdots b_p] V_{XX} \begin{bmatrix} b_1 \\ \cdot \\ \cdot \\ \cdot \\ b_p \end{bmatrix}.$$

Hence

$$\text{(5)} \qquad SS_{\hat{Y}} = \Sigma \hat{Y}^2 - n\bar{Y}^2 = [b_1 \cdots b_p]V_{XX}\begin{bmatrix} b_1 \\ \cdot \\ \cdot \\ \cdot \\ b_p \end{bmatrix} = \underline{v}'_{XY}V_{XX}^{-1}\underline{v}_{XY}$$

$$= [b_1 \cdots b_p]\underline{v}_{XY}.$$

The last expression in (5) represents a special case of (11) in Sec. 2.1-2. From relationship (15) in Sec. 2.1-2 one has, using (5),

$$\text{(6)} \qquad SS_{\text{error}} = SS_Y - SS_{\hat{Y}} = SS_Y - [b_1 \cdots b_p]\underline{v}_{XY}.$$

The square of the correlation between Y and $\hat{Y}$ is given by

$$\text{(7)} \qquad r^2_{Y\hat{Y}} = \frac{SS_{\hat{Y}}}{SS_Y} = \frac{[b_1 \cdots b_p]\underline{v}_{XY}}{SS_Y}.$$

Equivalently, using the relationship $SS_{\hat{Y}} = SS_Y - SS_{\text{error}}$,

$$\text{(8)} \qquad r^2_{Y\hat{Y}} = \frac{SS_Y - SS_{\text{error}}}{SS_Y} = 1 - \frac{SS_{\text{error}}}{SS_Y}.$$

The squared correlation for the case of multiple predictors is denoted by the symbol

$$\text{(9)} \qquad r^2_{Y\hat{Y}} = r^2_{Y(X_1X_2\cdots X_p)} = r^2_{Y(12\cdots p)}$$

and is called the squared *multiple* correlation. Thus, the squared multiple correlation represents the proportion of the total variation of Y in the sample that can be predicted from a linear function of $X_1, X_2, \ldots, X_p$, the coefficients in this linear function being determined by the least-squares criterion. It should be noted that b_0, the coefficient associated with the dummy variable X_0, has no effect on the correlation. The role of b_0 is that of determining the mean of the $\hat{Y}$; the least-squares solution for b_0 makes the mean of the $\hat{Y}$ equal to the mean of the Y. In terms of the squared multiple correlation,

$$\text{(10)} \qquad SS_{\hat{Y}} = r^2_{Y(12\cdots p)}SS_Y \qquad \text{and} \qquad SS_{\text{error}} = (1 - r^2_{Y(12\cdots p)})SS_Y.$$

A numerical example in which $p = 2$ and $n = 6$ appears in Table 2.1-4. At the right of part i, the variation of Y is given by

$$SS_Y = \Sigma Y^2 - \frac{(\Sigma Y)^2}{n} = \frac{2450 - (100)^2}{6} = 783.33.$$

The vector of covariations between Y and X_1 and Y and X_2 is also given

Table 2.1-4 Numerical Example

	Sample element	X_0	X_1	X_2	Y	
	1	1	3	6	10	$SS_Y = 783.33$
	2	1	5	1	5	
	3	1	7	2	15	
(i)	4	1	7	10	20	$\underline{v}_{XY} = \begin{bmatrix} 83.33 \\ 75.00 \end{bmatrix}$
	5	1	9	7	40	
	6	1	9	10	10	

(ii) $$X'X = \begin{bmatrix} 6 & 40 & 36 \\ 40 & 294 & 260 \\ 36 & 260 & 290 \end{bmatrix} \qquad X'\underline{Y} = \begin{bmatrix} 100 \\ 750 \\ 675 \end{bmatrix} \qquad \Sigma Y^2 = 2450$$

(iii) $$\underline{b} = (X'X)^{-1}X'\underline{Y} = \begin{bmatrix} 1.81375 & -.23006 & -.018900 \\ -.23006 & .045604 & -.012326 \\ -.018900 & -.012327 & .016845 \end{bmatrix} \begin{bmatrix} 100 \\ 750 \\ 675 \end{bmatrix} = \begin{bmatrix} -3.92326 \\ 2.87588 \\ .23622 \end{bmatrix}$$

(iv)
$$\Sigma \hat{Y}^2 = \underline{b}'X'\underline{Y} = [-3.92326 \quad 2.87588 \quad .23622] \begin{bmatrix} 100 \\ 750 \\ 675 \end{bmatrix} = 1924.03$$

$$\Sigma(Y - \hat{Y})^2 = \Sigma Y^2 - \Sigma \hat{Y}^2 = 525.97$$

$$r^2_{Y(12)} = 1 - \frac{\Sigma(Y - \hat{Y})^2}{SS_Y} = .32855$$

$$SS_{\hat{Y}} = [b_1\, b_2]\underline{v}_{XY} = [2.87588 \quad .23622] \begin{bmatrix} 83.33 \\ 75.00 \end{bmatrix} = 257.36$$

$$= r^2_{Y(12)} SS_Y = 257.36$$

$$SS_{\text{error}} = \Sigma(Y - \hat{Y})^2 = (1 - r^2_{Y(12)})SS_Y = 525.97$$

at the right of part i. For example,

$$SP_{X_1Y} = \Sigma X_1 Y - \frac{(\Sigma X_1)(\Sigma Y)}{n}$$

$$= 750 - \frac{(40)(100)}{6} = 83.33.$$

In part ii are $X'X$, $X'\underline{Y}$, and ΣY^2. The solution to the set of normal equations $(X'X)b = X'\underline{Y}$ appears in part iii. As a partial check on the arithmetic work

$$b_0 = \bar{Y} - b_1\bar{X}_1 - b_2\bar{X}_2.$$

The sum of the squares of the $\hat{Y}$, given by

$$\Sigma \hat{Y}^2 = \underline{b}'X'\underline{Y} = \underline{Y}'X(X'X)^{-1}X'\underline{Y},$$

is computed in part iv. One also has

$$\Sigma(Y - \hat{Y})^2 = \Sigma Y^2 + \Sigma \hat{Y}^2 - 2\Sigma Y\hat{Y}.$$

But $$\Sigma Y\hat{Y} = \underline{Y}'X\underline{b} = \underline{Y}'X(X'X)^{-1}X'\underline{Y} = \Sigma \hat{Y}^2.$$

Hence $$\Sigma(Y - \hat{Y})^2 = \Sigma Y^2 - \Sigma \hat{Y}^2.$$

This last quantity is computed in part iv. Since

$$\Sigma(Y - \hat{Y})^2 = \Sigma Y^2 - \Sigma \hat{Y}^2 = \left[\Sigma Y^2 - \frac{(\Sigma Y)^2}{n}\right] - \left[\Sigma \hat{Y}^2 - \frac{(\Sigma Y)^2}{n}\right],$$

one has

$$\Sigma(Y - \hat{Y})^2 = SS_Y - SS_{\hat{Y}}$$
$$= 783.33 - 257.36 = 525.97.$$

These relationships are illustrated numerically in part iv. Also in part iv, $SS_{\hat{Y}}$ is computed from the equivalent relationships

$$SS_{\hat{Y}} = [b_1 \; b_2]\underline{v}_{XY} = r^2_{Y(12)}SS_Y.$$

Table 2.1-5 Numerical Example (Continued)

(i)

$$X'_{\text{aug}}X_{\text{aug}} = \begin{bmatrix} & X_0 & X_1 & X_2 & Y \\ & 6 & 40 & 36 & 100 \\ & 40 & 294 & 260 & 750 \\ & 36 & 260 & 290 & 675 \\ & 100 & 750 & 675 & 2450 \end{bmatrix}$$

$$(X'_{\text{aug}}X_{\text{aug}})^{-1} = \begin{bmatrix} X_0 & X_1 & X_2 & Y \\ 1.84309 & -.25152 & -.020659 & .0074593 \\ -.25152 & .061328 & -.011034 & -.00546786 \\ -.020659 & -.011034 & .016951 & -.00044913 \\ .0074593 & -.00546786 & -.00044913 & .00190128 \end{bmatrix}$$

(ii)

$$b_0 = \frac{-p^{Y0}}{p^{YY}} = \frac{-.0074593}{.00190128} = -3.92326$$

$$b_1 = \frac{-p^{Y1}}{p^{YY}} = \frac{-(-.00546786)}{.00190128} = 2.87588$$

$$b_2 = \frac{-p^{Y2}}{p^{YY}} = \frac{-(-.00044913)}{.00190128} = .23622$$

$$r^2_{Y(12)} = 1 - \frac{1}{p^{YY}SS_Y} = 1 - \frac{1}{(.00190128)(783.33)} = .3286$$

(iii)

$$r^2_{1(Y2)} = 1 - \frac{1}{p^{11}SS_1} = 1 - \frac{1}{(.061328)(27.33)} = .4034$$

$$r^2_{2(Y1)} = 1 - \frac{1}{p^{22}SS_2} = 1 - \frac{1}{(.016951)(74.00)} = .2028$$

Table 2.1-6 Numerical Example—Applications of SWP Operator

	0	1	2	Y	
	6	40	36	100	
		294	260	750	
(i) $M = X'_{\text{aug}}X_{\text{aug}}$			290	675	
				2450	
	−.16667	6.66667	6.00000	16.66667	$= \bar{Y}$
		27.33333	20.00000	83.33333	$= SP_{X_1Y}$
(ii) $M_{(0)} = \text{SWP}[0]\,M$			74.00000	75.00000	$= SP_{X_2Y}$
				783.33333	$= SS_Y$
	−1.79269	.24390	1.12195	−3.65855	$= b_{Y0.1}$
		−.036585	.73171	3.04878	$= b_{Y1.0}$
(iii) $M_{(01)} = \text{SWP}[1]\,M_{(0)}$			59.36585	14.02439	
				529.26828	$= SS_{Y-\hat{Y}(1)}$
	−1.81389	.23007	.018899	−3.92360	$= b_{Y0.12}$
		−.045604	.012325	2.87592	$= b_{Y1.02}$
(iv) $M_{(012)} = \text{SWP}[2]\,M_{(01)}$			−.016845	.23624	$= b_{Y2.01}$
				525.95520	$= SS_{Y-\hat{Y}(12)}$
	−1.84316	.25152	.020661	−.0074600	
		−.061330	.011033	.0054680	$= -M^{-1}$
(v) $M_{(012Y)} = \text{SWP}[Y]\,M_{(012)}$			−.016951	.00044916	
				−.0019013	

In Table 2.1-5 computations are done through use of X_{aug}, where X_{aug} is the X matrix augmented by the Y column. In terms of the matrix in part i of Table 2.1-4, X'_{aug} includes columns X_0, X_1, X_2, and Y. Let p^{ij} denote the element in row i, column j of $(X'_{\text{aug}}X_{\text{aug}})^{-1}$. The components of the vector $\underline{b}$ are obtained as indicated in part ii. The square of the multiple correlation $r^2_{Y(12)}$ is also obtained in part ii. Other squared multiple correlations are computed in part iii. The rationale underlying these computations is discussed at the end of Sec. 2.1-1.

Use of the SWP operator (discussed in Sec. 2.5) is illustrated in Table 2.1-6. In part i is the upper half of the matrix

$$M = X'_{\text{aug}}X_{\text{aug}}.$$

Application of SWP[0] to the matrix M yields the matrix given in part ii. The 3×3 submatrix obtained by deleting the first row and first column of $M_{(0)}$ is a matrix of covariations. (Only the upper half of $M_{(0)}$ is given in part ii.)

In the Y column of $M_{(01)}$, given in part iii, are the coefficients in the prediction equation

$$Y_{(1)} = b_0X_0 + b_1X_1,$$

that is, the prediction equation based upon variable X_1. The entry 529.26828 in the Y column is that part of the total variation of Y, which is 783.33333, which cannot be predicted from X_1.

In the Y column of part iv are the coefficients in the prediction equation

$$Y_{(12)} = b_0X_0 + b_1X_1 + b_2X_2.$$

The entry in the Y column 525.95520 is that part of the total variation in Y which cannot be predicted from the variables X_1 and X_2. The difference

$$529.26828 - 525.95520 = 3.31308$$

represents the reduction in the residual variation brought about by adding X_2 to a prediction system already containing X_1.

In part iv of Table 2.1-6, a submatrix of $M_{(012)}$ is of special interest. It is readily shown that

$$-V_{XX}^{-1} = \begin{bmatrix} -.045604 & .012325 \\ .012325 & -.016845 \end{bmatrix}.$$

This is the submatrix in rows X_1, X_2 and columns X_1, X_2. Also the submatrix in rows X_1, X_2 and column Y, which is

$$\begin{bmatrix} 2.87592 \\ .23624 \end{bmatrix},$$

can be shown to be

$$V_{XX}^{-1}\underline{v}_{XY}.$$

Numerically,

$$V_{XX}^{-1}\underline{v}_{XY} = \begin{bmatrix} .045604 & -.012325 \\ -.012325 & .016845 \end{bmatrix}\begin{bmatrix} 83.33333 \\ 75.00000 \end{bmatrix} = \begin{bmatrix} 2.8759 \\ .2362 \end{bmatrix}.$$

The components of the $\underline{v}_{XY}$ vector are obtained from the Y column of $M_{(0)}$.

2.1-4 Multivariate case—singular matrix

In Sec. 2.1-3 the solution to the normal equations

$$(X'X)\underline{b} = X'\underline{Y} \tag{1}$$

was given as

$$\underline{b} = (X'X)^{-1}X'\underline{Y}. \tag{2}$$

This solution requires that $X'X$ be nonsingular. If $X'X$ is nonsingular, both $(X'X)^{-1}$ and $\underline{b}$ are unique.

In this section, the case in which $X'X$ is singular will be considered. It will be found that the components of the vector $\underline{b}$ will not be unique unless certain constraints are imposed above and beyond the set of normal equations.

The normal equations as given in (1) can be shown to be *consistent;* that is, at least one nontrivial solution for $\underline{b}$ will exist. Consider now the general solution to a set of consistent equations of the form

$$\underset{p,p}{A}\ \underset{p,1}{\underline{x}} = \underset{p,1}{\underline{c}}, \tag{3}$$

where the components of $\underline{x}$ are unknown, A is a known *symmetric* matrix, not necessarily nonsingular, and $\underline{c}$ is a known vector. All solutions to this system can be shown to have the form

$$\underline{x} = A^{-}\underline{c}, \tag{4}$$

where A^{-} is a matrix which has the property

$$AA^{-}A = A.$$

A matrix A^{-} having this property is called a *g*-inverse (*g* for generalized) of A. If A is singular, the *g*-inverse is not unique. To show that (4) does represent a solution, one has upon substituting in (3)

$$A(A^{-}\underline{c}) = \underline{c}.$$

Replacing $\underline{c}$ on the left by $A\underline{x}$ gives

$$AA^{-}A\underline{x} = \underline{c} \qquad \text{or} \qquad A\underline{x} = \underline{c}.$$

Thus $\underline{x} = A^{-}\underline{c}$ does represent a solution to the system (3). Conversely, it may be shown that, if there exists a matrix A^{-} such that $A^{-}\underline{c}$ is a solution for (3), then A^{-} has the property $AA^{-}A = A$.

From the property $AA^{-}A = A$, the following additional properties can be shown to hold:

(i) $A^{-}A = H$, where $H^2 = H$. (H is idempotent.)

(ii) $AH = A$.

When the matrix A is nonsingular, $H = I$. A general solution to the system $A\underline{x} = \underline{c}$ is given by

$$\underline{x} = A^{-}\underline{c} + (H - I)\underline{z},$$

where the components of $\underline{z}$ are completely arbitrary. To show that this is a solution, if one multiplies both sides of the last expression by A one has

$$\begin{aligned} A\underline{x} &= AA^{-}\underline{c} + (A - A)\underline{z} \\ &= AA^{-}\underline{c} = A\underline{x}, \end{aligned}$$

which is an identity.

A particularly useful form of a g-inverse of a *symmetric* matrix A may be obtained as follows: Partition the matrix A so that

$$\underset{p,p}{A} = \begin{array}{c} \\ r \\ p-r \end{array}\!\!\begin{array}{c} \begin{array}{cc} r & p-r \end{array} \\ \begin{bmatrix} A_{11} & A_{12} \\ A_{21} & A_{22} \end{bmatrix} \end{array}, \tag{5}$$

where A_{11} is the largest nonsingular submatrix contained in A. The rows and columns of A may be interchanged in order that r be a maximum. Under this type of partition, it can be shown that

$$A_{22} = A_{21}A_{11}^{-1}A_{12},$$

since A_{22} will be linearly dependent upon A_{21}, which in turn will be linearly dependent upon A_{11}. In applications, the task of finding the matrix A_{11} is usually relatively simple; A_{11} is not necessarily unique.

Consider now the matrix

$$A^- = \begin{bmatrix} A_{11}^{-1} & 0 \\ 0 & 0 \end{bmatrix}. \tag{6}$$

One notes that

$$AA^- = \begin{bmatrix} I & 0 \\ A_{21}A_{11}^{-1} & 0 \end{bmatrix} \quad \text{and} \quad A^-A = \begin{bmatrix} I & A_{11}^{-1}A_{12} \\ 0 & 0 \end{bmatrix}.$$

Hence

$$AA^-A = \begin{bmatrix} A_{11} & A_{12} \\ A_{21} & A_{21}A_{11}^{-1}A_{12} \end{bmatrix} = \begin{bmatrix} A_{11} & A_{12} \\ A_{21} & A_{22} \end{bmatrix} = A,$$

and

$$A^-AA^- = \begin{bmatrix} A_{11}^{-} & 0 \\ 0 & 0 \end{bmatrix} = A^-.$$

Also,

$$(AA^-)(AA^-) = AA^-,$$

and

$$(A^-A)(A^-A) = A^-A.$$

Thus the g-inverse defined by (6) has some useful special properties beyond the minimal property $AA^-A = A$.

For any matrix A, square or rectangular, singular or nonsingular, there exists a unique matrix K which has the following properties:

$$AKA = A, \qquad (KA)' = KA,$$
$$KAK = K, \qquad (AK)' = AK.$$

The matrix K is called the *Penrose inverse* of the matrix A. The Penrose inverse is a special case of a g-inverse. Another special case is the so-called *least-squares inverse.* The matrix L is a least-squares inverse if it satisfies the following properties:

$$ALA = A \qquad \text{and} \qquad AL = \text{symmetric.}$$

The Penrose inverse is a least-squares inverse. There are, however, least-squares inverses which are not Penrose inverses.

Returning to the normal equations as given in (1), if $X'X$ is singular then one of a family of solutions for $\underline{b}$ is given by

$$\underline{b} = (X'X)^{-}X'\underline{Y} = C\underline{Y}, \qquad \text{where} \qquad C = (X'X)^{-}X'. \tag{7}$$

Hence $\underline{\hat{Y}} = X\underline{b} = X(X'X)^{-}X'\underline{Y} = P\underline{Y}$, where $P = X(X'X)^{-}X'$.

If one selects a g-inverse having the property

$$(X'X)^{-}(X'X)(X'X)^{-} = (X'X)^{-},$$

then $$PP = X(X'X)^{-}X'X(X'X)^{-}X' = X(X'X)^{-}X' = P.$$

Thus, $$\Sigma \hat{Y}^2 = \underline{\hat{Y}}'\underline{\hat{Y}} = Y'PP\underline{Y} = \underline{Y}'P\underline{Y}.$$

Also $\underline{Y} - \underline{\hat{Y}} = \underline{Y} - P\underline{Y} = (I - P)\underline{Y} = M\underline{Y}$, where $M = I - P$.

$$\begin{aligned}\Sigma(Y - \hat{Y})^2 = (\underline{Y} - \underline{\hat{Y}})'(\underline{Y} - \underline{\hat{Y}}) &= \underline{Y}'MM\underline{Y} \\ &= \underline{Y}'(I - P)(I - P)\underline{Y} \\ &= \underline{Y}'(I + PP - 2P)\underline{Y} \\ &= \underline{Y}'(I - P)\,\underline{Y} = \underline{Y}'M\underline{Y}.\end{aligned}$$

The matrices P and M as defined above (in terms of a g-inverse having the special property) have all the properties of the matrices P and M defined in Sec. 2.1-2. Thus, all the major relationships that hold when $X'X$ is nonsingular also hold for the case in which $X'X$ is singular.

Although the solution for $\underline{b}$ is not unique when $X'X$ is singular, $\Sigma(Y - \hat{Y})^2 = \Sigma \hat{Y}^2 - \underline{b}'X'\underline{Y}$ will be unique since the normal equations represent both necessary and sufficient conditions for minimizing $\Sigma(Y - \hat{Y})^2$.

A numerical example of a set of normal equations in which $X'X$ is singular is given in Table 2.1-7. (In this example the first column of $X'X$ is the sum of the second and third columns; hence the matrix is singular.)

The 2×2 submatrix in the upper left-hand corner of $X'X$ is nonsingular. Hence a g-inverse of $X'X$ may be obtained from the regular inverse of the submatrix

$$\begin{bmatrix} 8 & 4 \\ 4 & 4 \end{bmatrix}, \qquad \text{which is} \qquad \begin{bmatrix} .25 & -.25 \\ -.25 & .50 \end{bmatrix}.$$

Table 2.1-7 Numerical Example of g-Inverse

Normal equations:

(i)

$$\underset{X'X}{\begin{bmatrix} 8 & 4 & 4 \\ 4 & 4 & 0 \\ 4 & 0 & 4 \end{bmatrix}} \underset{\underline{b}}{\begin{bmatrix} b_1 \\ b_2 \\ b_3 \end{bmatrix}} = \underset{X\underline{Y}}{\begin{bmatrix} 26 \\ 10 \\ 16 \end{bmatrix}}$$

$$(X'X)^- = \begin{bmatrix} .25 & -.25 & 0 \\ -.25 & .50 & 0 \\ 0 & 0 & 0 \end{bmatrix} \qquad \underline{b} = (X'X)^- X'\underline{Y} = \begin{bmatrix} 4.00 \\ -1.50 \\ 0 \end{bmatrix}$$

$$\Sigma \hat{Y}^2 = \underline{b}'X'\underline{Y} = [4.00 \quad -1.50 \quad 0] \begin{bmatrix} 26 \\ 10 \\ 16 \end{bmatrix} = 89$$

Normal equations (rearranged):

(ii)

$$\underset{(X'X)}{\begin{bmatrix} 4 & 0 & 4 \\ 0 & 4 & 4 \\ 4 & 4 & 8 \end{bmatrix}} \underset{\underline{b}}{\begin{bmatrix} b_2 \\ b_3 \\ b_1 \end{bmatrix}} = \underset{X'\underline{Y}}{\begin{bmatrix} 10 \\ 16 \\ 26 \end{bmatrix}}$$

$$(X'X)^- = \begin{bmatrix} .25 & 0 & 0 \\ 0 & .25 & 0 \\ 0 & 0 & 0 \end{bmatrix} \qquad \underline{b} = (X'X)^- X'\underline{Y} = \begin{bmatrix} 2.50 \\ 4.00 \\ 0 \end{bmatrix} = \begin{bmatrix} b_2 \\ b_3 \\ b_1 \end{bmatrix}$$

$$\Sigma \hat{Y}^2 = \underline{b}'X'\underline{Y} = [2.50 \quad 4.00 \quad 0] \begin{bmatrix} 10 \\ 16 \\ 26 \end{bmatrix} = 89$$

(iii)

$$\underline{b} = \begin{bmatrix} 4.00 \\ -1.50 \\ 0 \end{bmatrix}, \quad \begin{bmatrix} 0 \\ 2.50 \\ 4.00 \end{bmatrix}, \quad \begin{bmatrix} 2.50 \\ 0 \\ 1.50 \end{bmatrix}, \quad \begin{bmatrix} 5.50 \\ -3.00 \\ -1.50 \end{bmatrix}, \quad \begin{bmatrix} -12.00 \\ 14.50 \\ 16.00 \end{bmatrix}, \quad \begin{bmatrix} 3.25 \\ -0.75 \\ 0.75 \end{bmatrix}$$

Hence a g-inverse of $X'X$ is that matrix at the left of part i. The solution for $\underline{b}$ corresponding to this g-inverse appears at the right of part i. Also given in part i is $\Sigma \hat{Y}^2 = \underline{b}'X'\underline{Y}$ corresponding to this solution for $\underline{b}$.

For the g-inverse obtained in part i, one has

$$H = (X'X)^-(X'X) = \begin{bmatrix} 1 & 0 & 1 \\ 0 & 1 & -1 \\ 0 & 0 & 0 \end{bmatrix},$$

and
$$H - I = \begin{bmatrix} 0 & 0 & 1 \\ 0 & 0 & -1 \\ 0 & 0 & -1 \end{bmatrix}.$$

Hence a general solution for $\underline{b}$ has the form

$$\underline{b} = (X'X)^{-}X'\underline{Y} + (H - I)\underline{z}$$
$$= \begin{bmatrix} 4.00 \\ -1.50 \\ 0 \end{bmatrix} + \begin{bmatrix} 0 & 0 & 1 \\ 0 & 0 & -1 \\ 0 & 0 & -1 \end{bmatrix} \begin{bmatrix} z_1 \\ z_2 \\ z_3 \end{bmatrix} = \begin{bmatrix} 4.00 + z_3 \\ -1.50 - z_3 \\ 0 \quad - z_3 \end{bmatrix}.$$

The numerical value of z_3 is completely arbitrary. If, for example, one sets $z_3 = -4$, then

$$\underline{b} = \begin{bmatrix} 4.00 + z_3 \\ -1.50 - z_3 \\ 0 \quad - z_3 \end{bmatrix} = \begin{bmatrix} 0 \\ 2.50 \\ 4.00 \end{bmatrix} = \begin{bmatrix} b_1 \\ b_2 \\ b_3 \end{bmatrix}.$$

The last solution for $\underline{b}$ could have been obtained by rearranging the normal equations in part i of Table 2.1-7 so that they take the form shown in part ii. The 2×2 matrix in the upper left-hand corner of the rearranged normal equations is nonsingular. Hence a g-inverse has the form given in part ii. Since $\Sigma \hat{Y}^2$ is the same for *all* solutions to the normal equations, one notes that $\Sigma \hat{Y}^2$ as obtained from the solution in part ii is also 89.

A series of "different" solutions to the normal equations is shown in part iii. All these solutions are generated from the general solution given above. The solution given in part i may be obtained by adding the constraint $b_3 = 0$ to the original normal equations and then solving for b_1 and b_2. Analogously, the solution given in part ii may be obtained by adding the constraint $b_1 = 0$.

In some applications it will be found convenient to construct a g-inverse which is nonsingular. This can be done quite readily. If the matrix A is partitioned as in (5), the following matrix is a nonsingular g-inverse.

$$A^{-} = \begin{bmatrix} A_{11}^{-1} & -A_{11}^{-1}A_{12} \\ 0 & I \end{bmatrix}.$$

For the numerical example in Table 2.4-7, the g-inverse having the form given above, as obtained from part i, is

$$(X'X)^{-} = \begin{bmatrix} .25 & -.25 & -1 \\ -.25 & .50 & 1 \\ 0 & 0 & 1 \end{bmatrix}.$$

The solution for $\underline{b}$ corresponding to this g-inverse is

$$\underline{b} = \begin{bmatrix} -12.00 \\ 14.50 \\ 16.00 \end{bmatrix}.$$

No matter what g-inverse is used to obtain the $\underline{b}$ vector, one notes that the difference

$$b_2 - b_3 = -1.50.$$

Although the numerical values of b_2 and b_3 depend upon the choice of the g-inverse, the numerical value of the difference does not depend upon the choice. Since this difference is invariant under the choice of the g-inverse, the corresponding difference in the parameters estimated by $b_2 - b_3$ is said to be *estimable*.

Alternative Approach Used in the Analysis of Variance. An approach that has been found quite useful in the programming of the calculations in the analysis of variance has been suggested by Bock (1963). The basic problem here is equivalent to that of solving a set of normal equations involving a singular matrix. Let the model have the form

$$\underset{n,1}{\underline{Y}} = \underset{n,k}{A}\ \underset{k,1}{\underline{\beta}} + \underset{n,1}{\underline{\varepsilon}}, \qquad n > k.$$

Let the matrix A be factored as follows:

$$\underset{n,k}{A} = \underset{n,m}{K}\ \underset{m,k}{L},$$

where $\qquad \text{rank}\,(A) = m < k, \qquad K'K$ is nonsingular.

The model may be rewritten in the form

$$\begin{aligned} \underline{Y} &= KL\underline{\beta} + \underline{\varepsilon} \\ &= \underset{n,m}{K}\ \underset{m,1}{\underline{\theta}} + \underline{\varepsilon}, \qquad \text{where} \qquad \underline{\theta} = L\underline{\beta}. \end{aligned}$$

Thus the elements of the vector $\underline{\theta}$ are linear functions of the vector $\underline{\beta}$.

To obtain the least-squares estimate of $\underline{\theta}$, the normal equations have the form

$$(K'K)\underline{\theta} = K'\underline{Y}.$$

Hence $$\underline{\theta} = (K'K)^{-1}K'\underline{Y}.$$

Given L, the vector $\underline{\theta}$ is unique. However, there are many choices for L. The components of $\underline{\theta}$ define a set of estimable functions of the components of $\underline{\beta}$.

Premultiplying the equation $A = KL$ by $(K'K)^{-1}K'$, one obtains

$$\underset{m,k}{L} = (K'K)^{-1}K'A.$$

Thus the rows of the matrix L, which defines a set of estimable functions of $\underline{\beta}$, are linear functions of the columns of the matrix A. Hence the rows of the matrix L are said to belong to the column space of the matrix A.

As a numerical example, suppose

$$\underset{\underline{Y}}{\begin{bmatrix} 6 \\ 8 \\ 2 \\ 4 \end{bmatrix}} = \underset{A}{\begin{bmatrix} 1 & 1 & 0 \\ 1 & 1 & 0 \\ 1 & 0 & 1 \\ 1 & 0 & 1 \end{bmatrix}} \underset{\underline{\beta}}{\begin{bmatrix} \beta_0 \\ \beta_1 \\ \beta_2 \end{bmatrix}} + \underset{\underline{\varepsilon}}{\begin{bmatrix} \varepsilon_1 \\ \varepsilon_2 \\ \varepsilon_3 \\ \varepsilon_4 \end{bmatrix}}$$

Since the first column of A is the sum of the second and third columns, the rank of the matrix cannot exceed 2. Since the second and third columns of A are linearly independent, the rank of the matrix A is 2. One factorization of A is the following:

$$\underset{A}{\begin{bmatrix} 1 & 1 & 0 \\ 1 & 1 & 0 \\ 1 & 0 & 1 \\ 1 & 0 & 1 \end{bmatrix}} = \underset{K}{\begin{bmatrix} 1 & 0 \\ 1 & 0 \\ 0 & 1 \\ 0 & 1 \end{bmatrix}} \underset{L}{\begin{bmatrix} 1 & 1 & 0 \\ 1 & 0 & 1 \end{bmatrix}}.$$

In terms of this factorization,

$$K'K = \begin{bmatrix} 2 & 0 \\ 0 & 2 \end{bmatrix}, \qquad (K'K)^{-1} = \begin{bmatrix} \frac{1}{2} & 0 \\ 0 & \frac{1}{2} \end{bmatrix}, \qquad (K'K)^{-1}K' = \begin{bmatrix} \frac{1}{2} & \frac{1}{2} & 0 & 0 \\ 0 & 0 & \frac{1}{2} & \frac{1}{2} \end{bmatrix},$$

$$\underline{\theta} = L\underline{\beta} = \begin{bmatrix} 1 & 1 & 0 \\ 1 & 0 & 1 \end{bmatrix} \begin{bmatrix} \beta_0 \\ \beta_1 \\ \beta_2 \end{bmatrix} = \begin{bmatrix} \beta_0 + \beta_1 \\ \beta_0 + \beta_2 \end{bmatrix}.$$

The least-squares estimator of the vector $\underline{\theta}$ is given by

$$\hat{\underline{\theta}} = (K'K)^{-1}K'\underline{Y} = \begin{bmatrix} \frac{1}{2} & \frac{1}{2} & 0 & 0 \\ 0 & 0 & \frac{1}{2} & \frac{1}{2} \end{bmatrix} \begin{bmatrix} 6 \\ 8 \\ 2 \\ 4 \end{bmatrix} = \begin{bmatrix} 7 \\ 3 \end{bmatrix} = \begin{bmatrix} \widehat{\beta_0 + \beta_1} \\ \widehat{\beta_0 + \beta_2} \end{bmatrix}.$$

A second factorization of the matrix A in the example given above is the following:

$$\underset{A}{\begin{bmatrix} 1 & 1 & 0 \\ 1 & 1 & 0 \\ 1 & 0 & 1 \\ 1 & 0 & 1 \end{bmatrix}} = \underset{K}{\begin{bmatrix} 1 & 0 \\ 1 & 0 \\ 1 & -1 \\ 1 & -1 \end{bmatrix}} \underset{L}{\begin{bmatrix} 1 & 1 & 0 \\ 0 & 1 & -1 \end{bmatrix}}.$$

For this factorization

$$\underline{\theta} = L\underline{\beta} = \begin{bmatrix} \beta_0 + \beta_1 \\ \beta_1 - \beta_2 \end{bmatrix}.$$

Now $$K'K = \begin{bmatrix} 4 & -2 \\ -2 & 22 \end{bmatrix}, \qquad (K'K)^{-1} = \begin{bmatrix} \frac{1}{2} & \frac{1}{2} \\ \frac{1}{2} & 1 \end{bmatrix}.$$

Hence

$$\underline{\hat{\theta}} = (K'K)^{-1}K'\underline{Y} = \begin{bmatrix} \frac{1}{2} & \frac{1}{2} & 0 & 0 \\ 0 & 0 & -\frac{1}{2} & -\frac{1}{2} \end{bmatrix} \begin{bmatrix} 6 \\ 8 \\ 2 \\ 4 \end{bmatrix} = \begin{bmatrix} 7 \\ 4 \end{bmatrix} = \begin{bmatrix} \widehat{\beta_0 + \beta_1} \\ \widehat{\beta_1 - \beta_2} \end{bmatrix}.$$

The sum of squares of the $\hat{Y}$ is

$$\underline{\hat{Y}}'\underline{\hat{Y}} = \underline{\hat{\theta}}'K'\underline{Y} = \underline{Y}'K(K'K)^{-1}K'\underline{Y}$$
$$= \underline{Y}'P\underline{Y}, \qquad \text{where} \qquad P = K(K'K)^{-1}K'.$$

The sum of squares of the $Y - \hat{Y}$ is

$$(\underline{Y} - \underline{\hat{Y}})'(\underline{Y} - \underline{\hat{Y}}) = \underline{Y}'\underline{Y} - \underline{Y}'P\underline{Y} = \underline{Y}'(I - P)\underline{Y}.$$

Alternative Approach in Terms of Means. Suppose the model under consideration were in terms of means; that is,

$$\underset{r,1}{\underline{\bar{Y}}} = \begin{bmatrix} \bar{Y}_1 \\ \bar{Y}_2 \\ \cdot \\ \cdot \\ \cdot \\ \bar{Y}_r \end{bmatrix} = \underset{r,p}{A}\,\underset{p,1}{\underline{\beta}} + \underset{r,1}{\underline{\varepsilon}}, \qquad r > p,$$

where $\bar{Y}_j$ is a mean based upon n_j independent observations. Again let

$$\underset{r,p}{A} = \underset{r,m}{K}\,\underset{m,p}{L},$$

where $$\text{rank }(A) = m < p, \qquad K'K \text{ is nonsingular.}$$

Let $$\underset{m,1}{\underline{\theta}} = \underset{m,p}{L}\,\underset{p,1}{\underline{\beta}}.$$

The model may now be written in the form

$$\underline{\bar{Y}} = A\underline{\beta} + \underline{\varepsilon} = KL\underline{\beta} + \underline{\varepsilon} = K\underline{\theta} + \underline{\varepsilon}.$$

In this case the least-squares criterion for estimating $\underline{\theta}$ takes the form

$$\underline{\varepsilon}'D\underline{\varepsilon} = \text{minimum,}$$

where

$$\underset{r,r}{D} = \begin{bmatrix} n_1 & 0 & \cdots & 0 \\ 0 & n_2 & \cdots & 0 \\ \cdot & & & \cdot \\ \cdot & & & \cdot \\ \cdot & & & \cdot \\ 0 & 0 & \cdots & n_r \end{bmatrix}.$$

The D matrix enters into the picture because

$$\sigma^2_{\varepsilon_j} = \sigma^2_\varepsilon \frac{1}{n_j};$$

hence $$\operatorname{cov}(\underline{\varepsilon}) = \sigma^2_\varepsilon D^{-1}.$$

If the elements of the vector $\underline{\bar{Y}}$ were individual observations rather than means,

$$\operatorname{cov}(\underline{\varepsilon}) = \sigma^2_\varepsilon I.$$

The normal equations take the following form:

$$(K'DK)\underline{\theta} = K'D\underline{\bar{Y}}.$$

The least-squares estimator of $\underline{\theta}$ is

$$\underline{\hat{\theta}} = (K'DK)^{-1}K'D\underline{\bar{Y}} = C\underline{\bar{Y}} \qquad \text{where} \qquad C = (K'DK)^{-1}K'D.$$

In this case the predictable variation (including that due to the grand mean) is

$$\underline{\hat{\theta}}'K'D\underline{\bar{Y}} = \underline{\bar{Y}}'DK(K'DK)^{-1}K'D\underline{\bar{Y}} = \underline{\hat{\theta}}'K'DK\underline{\hat{\theta}}.$$

In this case the covariance matrix for the vector $\underline{\hat{\theta}}$ has the form

$$\operatorname{cov}(\underline{\hat{\theta}}) = C[\operatorname{cov}(\underline{\varepsilon})]C' = \sigma^2_\varepsilon CD^{-1}C' = \sigma^2_\varepsilon(K'DK)^{-1}.$$

2.2 Linear Model—Estimation in Univariate Case

2.2-1 *X* fixed, no distribution assumption on ε

The material that will be discussed in this section, although presented in terms of the univariate case, will be given in a form that may be generalized rather directly to the multivariate case.

Consider a population of elements in which the following model is assumed to hold:

$$Y = \beta_0 + \beta_1 X + \varepsilon. \tag{1}$$

Here X is a known constant associated with each element. ε is a random variable associated with each element. β_0 and β_1 are parameters of the model. Y is a potential observation on each element. No assumption will be made about the form of the distribution of the random variable.

It will, however, be assumed that $E(\varepsilon) = 0$ and that the variance of ε is some unknown value σ_ε^2, which is independent of X. σ_ε^2 is a parameter of the model.

Consider now the subpopulation of elements for which $X = X_i$. The model in (1) implies that the distribution of Y within this subpopulation has the parameters

$$E(Y \mid X_i) = \mu_{Y|X_i} = \mu_{Y_i} = \beta_0 + \beta_1 X_i,$$
$$\sigma_{Y|X_i}^2 = \sigma_{Y_i}^2 = \sigma_\varepsilon^2.$$

That is, the model in (1) implies that the expected value of the conditional distribution of Y for fixed X is a linear function of that X, and that the variance of this distribution is σ_ε^2 for all X. The form of the conditional distribution is not specified by the model.

Suppose one were to draw a random sample of size n from a population in which model (1) holds. Let the observations on the variable Y be denoted by the vector

$$\underline{Y} = \begin{bmatrix} Y_1 \\ \cdot \\ \cdot \\ \cdot \\ Y_n \end{bmatrix}.$$

In terms of the model in (1), the vector $\underline{Y}$ may be represented in the form

$$\underline{Y} = X\underline{\beta} + \underline{\varepsilon}, \tag{2}$$

where

$$X = \begin{bmatrix} 1 & X_1 \\ \cdot & \cdot \\ \cdot & \cdot \\ \cdot & \cdot \\ 1 & X_n \end{bmatrix}, \qquad \underline{\beta} = \begin{bmatrix} \beta_0 \\ \beta_1 \end{bmatrix}, \qquad \underline{\varepsilon} = \begin{bmatrix} \varepsilon_1 \\ \cdot \\ \cdot \\ \cdot \\ \varepsilon_n \end{bmatrix}.$$

The second column of the matrix X represents known constants associated with each element in the sample. Since the sample is assumed to be random, the components of the vector $\underline{\varepsilon}$ are independent random variables. Equivalently, the vector $\underline{\varepsilon}$ may be considered a vector variable with the properties

$$E(\underline{\varepsilon}) = \underline{0}, \qquad \operatorname{cov}(\underline{\varepsilon}) = \begin{bmatrix} \sigma_\varepsilon^2 & 0 & \cdots & 0 \\ 0 & \sigma_\varepsilon^2 & \cdots & 0 \\ \cdot & & & \cdot \\ \cdot & & & \cdot \\ \cdot & & & \cdot \\ 0 & 0 & \cdots & \sigma_\varepsilon^2 \end{bmatrix} = \sigma_\varepsilon^2 \underset{n,n}{I}.$$

Since $\sigma^2_{Y|X_i} = \sigma^2_\varepsilon$ for all X_i, one also has

$$\text{E}[\text{cov}\,(\underline{Y} \mid X_i)] = \text{E}[\text{cov}\,(\underline{Y})] = \sigma^2_\varepsilon I.$$

One may also write the model in (2) in the form

$$\underline{\varepsilon} = \underline{Y} - X\underline{\beta}. \tag{3}$$

The components of $\underline{\varepsilon}$ represent the errors or residuals in the sense that they give the extent to which the components of Y are not linear functions of X. From the assumptions underlying the model in (1), one has

$$E(\underline{Y}) = X\underline{\beta}.$$

The problem is to estimate $\underline{\beta}$ and σ^2_ε from the data in the sample. The least-squares principle will be used to obtain an estimate of $\underline{\beta}$. From the estimate of the latter, an estimate of σ^2_ε will be obtained.

Let $\underline{b}$ denote the least-squares estimate of $\underline{\beta}$, and let

$$\underline{e} = Y - X\underline{b}$$

for the sample data. To obtain the vector $\underline{b}$, one minimizes

$$\begin{aligned} \underline{e}'\underline{e} &= (\underline{Y} - X\underline{b})'(\underline{Y} - X\underline{b}) \\ &= \underline{Y}'\underline{Y} + \underline{b}'XX\underline{b} - 2\underline{b}'X'\underline{Y}. \end{aligned} \tag{4}$$

Upon differentiating (4) with respect to the components of $\underline{b}$ and setting the result equal to zero, one obtains the set of normal equations

$$(X'X)\underline{b} = X'\underline{Y}. \tag{5}$$

The solution to the system (5) was discussed in detail in Sec. 2.1-2. One has

$$\underline{b} = (X'X)^{-1}X'\underline{Y} = C\underline{Y} \qquad \text{where} \qquad \underset{2,n}{C} = (X'X)^{-1}X'. \tag{6}$$

From (6) one notes that $\underline{b}$ is a linear function of $\underline{Y}$. The component b_0 is that linear function defined by the first row of C; the component b_1 is that linear function defined by the second .ow of C.

To show that $\underline{b}$ is unbiased as an estimator of $\underline{\beta}$ one has

$$\begin{aligned} \text{E}(\underline{b}) = \text{E}(C\underline{Y}) = C\text{E}(\underline{Y}) &= CX\underline{\beta} \\ &= (X'X)^{-1}X'X\underline{\beta} \\ &= \underline{\beta}. \end{aligned} \tag{7}$$

To obtain the expected value of the covariance matrix associated with the sampling distribution of the vector $\underline{b}$, one uses the general principle that the covariance matrix of any vector variable $\underline{v}$ defined by

$$\underline{v} = A\underline{z},$$

where A is a matrix of known values and $\underline{z}$ is a vector variable having covariance matrix V, has the form

$$\text{cov}(\underline{v}) = AVA'.$$

Thus,

$$\text{E}[\text{cov}(\underline{b})] = \text{E}[\text{cov}(C\underline{Y})] = C\text{E}[\text{cov}(\underline{Y})]C' = C(\sigma_\varepsilon^2 I)C' = \sigma_\varepsilon^2 CC'. \tag{8}$$

But since $CC' = (X'X)^{-1}$, one has

$$\text{E}[\text{cov}(\underline{b})] = \sigma_\varepsilon^2 (X'X)^{-1}. \tag{9}$$

If one replaces $\underline{b}$ in (4) by its least-squares definition as given in (6), one obtains after some simplification

$$\text{Minimum} \sum_{i-1}^{n} e_i^2 = \underline{Y}'\underline{Y} - \underline{b}'X'\underline{Y} = \underline{Y}'\underline{Y} - X(X'X)^{-1}X'\underline{Y} = \underline{Y}'(I - P)\underline{Y} = \underline{Y}'M\underline{Y}, \tag{10}$$

where $\quad P = X(X'X)^{-1}X' \quad$ and $\quad M = I - P.$

The matrices P and M defined here are identical with the corresponding matrices in Sec. 2.1-2.

Since the vector $\underline{b}$ contains estimates of the parameters β_0 and β_1 as determined from the sample data, the degrees of freedom associated with the variance of the e_i are $n - 2$. Hence an estimate of σ_ε^2 is given by

$$\hat{\sigma}_\varepsilon^2 = \frac{\Sigma e_i^2}{n-2} = \frac{\underline{Y}'\underline{Y} - \underline{b}'X'\underline{Y}}{n-2}.$$

This estimate of σ_ε^2 may be shown to be unbiased. If one replaces σ_ε^2 in (9) by $\hat{\sigma}_\varepsilon^2$ one obtains an estimate of the covariance matrix for $\underline{b}$. Thus,

$$\text{est cov}(\underline{b}) = \hat{\sigma}_\varepsilon^2 (X'X)^{-1}. \tag{11}$$

If the elements of $(X'X)^{-1}$ are denoted by

$$(X'X)^{-1} = \begin{bmatrix} c^{00} & c^{01} \\ c^{10} & c^{11} \end{bmatrix} = \frac{1}{n\text{SS}_X} \begin{bmatrix} \Sigma X_i^2 & -\Sigma X_i \\ -\Sigma X_i & n \end{bmatrix},$$

then the estimate of the variance of b_0 is

$$\text{var}(b_0) = s_{b_0}^2 = \hat{\sigma}_\varepsilon^2 c^{00} = \hat{\sigma}_\varepsilon^2 \left(\frac{\Sigma X_i^2}{n\text{SS}_X}\right) = \hat{\sigma}_\varepsilon^2 \left(\frac{1}{n} + \frac{\bar{X}^2}{\text{SS}_X}\right).$$

The estimate of the variance of b_1 is

$$\text{var}(b_1) = s_{b_1}^2 = \hat{\sigma}_\varepsilon^2 c^{11} = \hat{\sigma}_\varepsilon^2 \left(\frac{n}{n\text{SS}_X}\right) = \hat{\sigma}_\varepsilon^2 \left(\frac{1}{\text{SS}_X}\right),$$

and the estimate of the covariance between b_0 and b_1 is

$$\text{cov}\,(b_0,b_1) = \hat{\sigma}_\varepsilon^2 c^{01} = \hat{\sigma}_\varepsilon^2\left(\frac{-\Sigma X_i}{n\text{SS}_X}\right) = \hat{\sigma}_\varepsilon^2\left(\frac{-\bar{X}}{\text{SS}_X}\right).$$

The Gauss-Markov theorem says the following about least-squares estimators obtained under the model as stated in (1): In the class of all unbiased estimators of $\underline{\beta}$ which are linear functions of $\underline{Y}$, the least-squares are best in the sense of having minimum variance. Hence the least-squares estimator $\underline{b}$ has some highly desirable properties.

From the last two results one has (since $\hat{Y} = b_0 + b_1X$)

$$s_{\hat{Y}}^2 = \text{var}\,(\hat{Y}) = \text{var}\,(b_0) + \text{var}\,(b_1X) + 2\,\text{cov}\,(b_0, b_1X)$$

$$= \hat{\sigma}_\varepsilon^2\left(\frac{1}{n} + \frac{\bar{X}^2}{\text{SS}_X} + \frac{X^2}{\text{SS}_X} + 2\,\frac{-\bar{X}X}{\text{SS}_X}\right) = \hat{\sigma}_\varepsilon^2\left(\frac{1}{n} + \frac{(X - \bar{X})^2}{\text{SS}_X}\right),$$

where var $(\hat{Y})$ is the estimate of the variance of the sampling distribution of $\hat{Y}$.

2.2-2 X fixed, ε normally distributed

In Sec. 2.2-1 no assumption was made about the form of the distribution of the random variable ε. If one assumes that ε is $N(0,\sigma_\varepsilon^2)$, then the form of the sampling distributions of the estimators, under random sampling, is also specified.

It should be noted that the estimators of $\underline{\beta}$ and σ_ε^2 are not affected by the form of distribution specified for ε. Under the distribution assumption made about ε, the conditional distribution of Y for $X = X_i$ is normal in form with

$$\text{(1)} \qquad \text{E}(Y\,|\,X_i) = \beta_0 + \beta_1X_i \qquad \text{and} \qquad \sigma_{Y|X_i}^2 = \sigma_\varepsilon^2.$$

The sampling distribution of the statistic b_1, since it is a linear function of Y, will also be normally distributed with

$$\text{(2)} \qquad \text{E}(b_1) = \beta_1 \qquad \text{and} \qquad \sigma_{b_1}^2 = \sigma_\varepsilon^2 c^{11} = \sigma_\varepsilon^2\frac{1}{\text{SS}_X}.$$

The sampling distribution of b_0 will also be normally distributed with

$$\text{(3)} \qquad \text{E}(b_0) = \beta_0 \qquad \text{and} \qquad \sigma_{b_0}^2 = \sigma_\varepsilon^2 c^{00} = \frac{\sigma_\varepsilon^2(\Sigma X_i^2)}{n\text{SS}_X}.$$

If one defines (for fixed X_i)

$$\hat{Y}_i = b_0 + b_1X_i,$$

then $\hat{Y}_i$ is a linear function of b_0 and b_1. Hence the sampling distribution of $\hat{Y}_i$ is normal in form with

$$\text{(4)} \qquad E(\hat{Y}_i) = \beta_0 + \beta_1 X_i$$

and with variance

$$\text{(5)} \qquad \sigma^2_{\hat{Y}_i} = \sigma^2_{b_0} + X_i^2\sigma^2_{b_1} + 2X_i \operatorname{cov}(b_0,b_1)$$
$$= \sigma^2_\varepsilon(c^{00} + X_i^2c^{11} + 2X_ic^{01}).$$

Since

$$\begin{bmatrix} c^{00} & c^{01} \\ c^{10} & c^{11} \end{bmatrix} = \frac{1}{n\text{SS}_X}\begin{bmatrix} \Sigma X_i^2 & -\Sigma X_i \\ -\Sigma X_i & n \end{bmatrix},$$

(5) becomes

$$\text{(6)} \qquad \sigma^2_{\hat{Y}_i} = \frac{\sigma^2_\varepsilon}{n\text{SS}_X}(\Sigma X_i^2 + nX_i^2 - 2X_i\Sigma X_i)$$
$$= \sigma^2_\varepsilon\left(\frac{\Sigma X_i^2 - n\bar{X}^2}{n\text{SS}_X} + \frac{nX_i^2 - 2nX_i\bar{X} + n\bar{X}^2}{n\text{SS}_X}\right)$$
$$= \sigma^2_\varepsilon\left(\frac{1}{n} + \frac{(X_i - \bar{X})^2}{\text{SS}_X}\right).$$

If σ^2_ε is replaced by $\hat{\sigma}^2_\varepsilon$ one obtains the corresponding estimators of the variances of the sampling distributions.

It will also be of interest to consider the sampling distribution of the statistic $Y_i - \hat{Y}_i$ for fixed X_i. This sampling distribution will be normal in form with

$$E(Y_i - \hat{Y}_i) = 0.$$

Since, for random samples, Y_i and $\hat{Y}_i$ will be uncorrelated

$$\text{(7)} \qquad \sigma^2_{Y_i - \hat{Y}_i} = \sigma^2_{Y_i} + \sigma^2_{\hat{Y}_i}$$
$$= \sigma^2_\varepsilon + \sigma^2_\varepsilon\left(\frac{1}{n} + \frac{(X_i - \bar{X}_i)^2}{\text{SS}_X}\right)$$
$$= \sigma^2_\varepsilon\left(1 + \frac{1}{n} + \frac{(X_i - \bar{X}_i)^2}{\text{SS}_X}\right).$$

2.2-3 X and Y bivariate normal

The probability density function for the bivariate normal distribution is defined by

$$\Pr(\underline{u}') = \frac{1}{(2\pi)\,|\Sigma|^{\frac{1}{2}}} \exp\left[-\tfrac{1}{2}(\underline{u} - \underline{\mu})'\Sigma^{-1}(\underline{u} - \underline{\mu})\right],$$

where $\underline{u}' = [X \quad Y]$, $\quad \underline{\mu}' = [\mu_X \quad \mu_Y]$, $\quad \Sigma = \begin{bmatrix} \sigma^2_X & \sigma_{XY} \\ \sigma_{YX} & \sigma^2_Y \end{bmatrix}.$

The parameters of this distribution are $\underline{\mu}$, the vector of means, and Σ, the matrix of covariances. The marginal distributions of the bivariate normal are $N(\mu_X,\sigma_X^2)$ and $N(\mu_Y,\sigma_Y^2)$.

The conditional distribution of Y for $X = X_i$ is normal in form, with mean

$$\mu_{Y|X_i} = \mu_Y - \beta_1\mu_X + \beta_1 X_i, \qquad \text{where} \qquad \beta_1 = \frac{\sigma_{XY}}{\sigma_X^2}.$$

If one lets

$$\beta_0 = \mu_Y - \beta_1\mu_X,$$

then the mean of the conditional distribution of Y for $X = X_i$ is

$$\mu_{Y|X_i} = \beta_0 + \beta_1 X_i.$$

The variance of this conditional distribution is

$$\sigma_{Y|X_i}^2 = (1 - \rho_{XY}^2)\sigma_Y^2, \qquad \text{where} \qquad \rho_{XY}^2 = \frac{\sigma_{XY}^2}{\sigma_X^2\sigma_Y^2}.$$

This conditional distribution may be represented in the form

$$Y_i = \beta_0 + \beta_1 X_i + \varepsilon_i,$$

where the distribution of the random variable ε_i is $N(0,\sigma_{Y|X_i}^2)$. Since $\sigma_{Y|X_i}^2$ does not depend upon X, one has

$$\sigma_{\varepsilon_i}^2 = \sigma_{Y|X_i}^2 = \sigma_\varepsilon^2 = (1 - \rho_{XY}^2)\sigma_Y^2.$$

Thus the random variable ε associated with the conditional distribution of $Y|\,X_i$ is $N(0,\sigma_\varepsilon^2)$. Hence the conditional distribution of Y for $X = X_i$ has

$$\text{E}(Y_i) = \beta_0 + \beta_1 X_i \qquad \text{and} \qquad \sigma_{Y|X_i}^2 = \sigma_\varepsilon^2.$$

One notes that the conditional distribution of Y for $X = X_i$ has all the properties of the linear model defined by (1) in Sec. 2.2-1, with the additional property that ε is normally distributed. Since β_0 and β_1 as well as σ_ε^2 are defined in terms of the parameters of the bivariate distribution, it would appear reasonable that the problem of estimating β_0, β_1, and σ_ε^2 would reduce to the problem of estimating the parameters $\underline{\mu}$ and Σ of the bivariate distribution. Such is indeed the case.

Estimates of $\underline{\mu}$ and Σ are readily obtained through use of the maximum-likelihood principle. Given a random sample of size n from the population $N(\underline{\mu},\Sigma)$, the natural logarithm of the likelihood of the sample data is

$$\phi = L(\text{sample}) = -\tfrac{1}{2}n \ln (2\pi) - \frac{n}{2} \ln |\Sigma| - \tfrac{1}{2}\sum_{i=1}^{n} (\underline{u}_i - \underline{\mu})'\Sigma^{-1}(\underline{u}_i - \underline{\mu}),$$

where

$$U = \begin{bmatrix} X_1 & Y_1 \\ \cdot & \cdot \\ \cdot & \cdot \\ \cdot & \cdot \\ X_i & Y_i \\ \cdot & \cdot \\ \cdot & \cdot \\ \cdot & \cdot \\ X_n & Y_n \end{bmatrix}, \qquad \text{and} \qquad \underline{u}_i' = [X_i \quad Y_i].$$

To find the maximum-likelihood estimators $\underline{\tilde{\mu}}$ and $\tilde{\Sigma}$, one maximizes ϕ with respect to $\underline{\mu}$ and Σ. To do this, one takes the partial derivative of ϕ with respect to $\underline{\mu}$ and Σ and sets these partial derivatives equal to zero. Setting the partial derivative with respect to $\underline{\mu}$ equal to zero gives

$$\frac{\partial \phi}{\partial \underline{\mu}} = \Sigma^{-1} \sum_{i=1}^{n} (\underline{u}_i - \underline{\tilde{\mu}}) = 0 \qquad \text{or} \qquad \sum_{i=1}^{n} \underline{u}_i = n\underline{\tilde{\mu}}.$$

Equivalently,

$$\sum_{i=1}^{n} \underline{u}_i = \begin{bmatrix} \sum_{i=1}^{n} X_i \\ \sum_{i=1}^{n} Y_i \end{bmatrix} = n \begin{bmatrix} \tilde{\mu}_X \\ \tilde{\mu}_Y \end{bmatrix}.$$

The maximum-likelihood estimator $\underline{\tilde{\mu}}$ is taken to be

$$\underline{\bar{u}} = \underline{\tilde{\mu}} = \begin{bmatrix} \tilde{\mu}_X \\ \tilde{\mu}_Y \end{bmatrix} = \frac{1}{n} \begin{bmatrix} \sum_{i=1}^{n} X_i \\ \sum_{i=1}^{n} Y_i \end{bmatrix} = \begin{bmatrix} \bar{X} \\ \bar{Y} \end{bmatrix}.$$

Taking the partial derivative of ϕ with respect to the matrix Σ and then setting this partial derivative equal to zero yields the set of normal equations

$$\frac{\partial \phi}{\partial \Sigma} = -n\tilde{\Sigma}^{-1} + \tilde{\Sigma}^{-1} B \tilde{\Sigma}^{-1} = 0,$$

where

$$\underset{2,2}{B} = \sum_{i=1}^{n} (\underline{u}_i - \underline{\mu})(\underline{u}_i - \underline{\mu})'$$

$$= \begin{bmatrix} \Sigma(X_i - \mu_X)^2 & \Sigma(X_i - \mu_X)(Y_i - \mu_Y) \\ \Sigma(Y_i - \mu_Y)(X_i - \mu_X) & \Sigma(Y_i - \mu_Y)^2 \end{bmatrix}.$$

Let the matrix A be defined as

$$\underset{2,2}{A} = \sum_{i=1}^{n} (\underline{u}_i - \underline{\tilde{\mu}})(\underline{u}_i - \underline{\tilde{\mu}})' = \begin{bmatrix} \Sigma(X_i - \bar{X})^2 & \Sigma(X_i - \bar{X})(Y_i - \bar{Y}) \\ \Sigma(Y_i - \bar{Y})(X_i - \bar{X}) & \Sigma(Y_i - \bar{Y})^2 \end{bmatrix};$$

that is, $\underline{\mu}$ in $\underline{B}$ is replaced by its maximum-likelihood estimator. If one replaces B by A in the normal equations, one has

$$-n\tilde{\Sigma}^{-1} + \tilde{\Sigma}^{-1}A\tilde{\Sigma}^{-1} = 0.$$

Pre- and post-multiplying this equation by $\tilde{\Sigma}$ gives

$$-n\tilde{\Sigma} + A = 0.$$

Solving for $\tilde{\Sigma}$, one obtains

$$\tilde{\Sigma} = \frac{1}{n} A$$

$$= \begin{bmatrix} \Sigma(X_i - \bar{X})^2/n & \Sigma(X_i - \bar{X})(Y_i - \bar{Y})/n \\ \Sigma(Y_i - \bar{Y})(X_i - \bar{X})/n & \Sigma(Y_i - \bar{Y})^2/n \end{bmatrix}$$

$$= \frac{n}{n-1} \begin{bmatrix} s_X^2 & \text{cov}_{XY} \\ \text{cov}_{YX} & s_Y^2 \end{bmatrix} = \frac{n}{n-1} S.$$

The maximum-likelihood estimator $\tilde{\Sigma}$ can be shown to be biased as an estimator of Σ. The unbiased estimator is given by

$$\hat{\Sigma} = \frac{n-1}{n} \tilde{\Sigma} = S = \begin{bmatrix} s_X^2 & \text{cov}_{XY} \\ \text{cov}_{YX} & s_Y^2 \end{bmatrix}.$$

Under random sampling with samples of size n from the bivariate normal population $N(\underline{\mu},\Sigma)$, the sampling distributions of $\underline{\tilde{\mu}}$ and $\hat{\Sigma}$ can be shown to be independent. The sampling distribution of $\underline{\tilde{\mu}}$ will be bivariate normal with parameters

$$\text{E}(\underline{\tilde{\mu}}) = \underline{\mu} \qquad \text{and} \quad \sigma^2_{\underline{\tilde{\mu}}} = \frac{1}{n} \Sigma.$$

The joint sampling distribution of the elements of the matrix $(n-1)\hat{\Sigma}$ will be the Wishart distribution with parameters $p = 2$, $n - 1$, and Σ. (The Wishart distribution reduces to the chi-square distribution if one is dealing with the case $p = 1$. Conversely, the Wishart distribution may be regarded as a multivariate analog of the chi-square distribution.)

From the maximum-likelihood estimators of $\underline{\mu}$ and Σ one obtains

$$b_1 = \frac{s_{XY}}{s_X^2}, \qquad b_0 = \bar{Y} - b_1\bar{X}, \qquad r_{YX}^2 = \frac{s_{XY}^2}{s_X^2 s_Y^2} = \frac{b_1 s_{XY}}{s_Y^2},$$

$$\hat{\sigma}_\varepsilon^2 = \frac{n-1}{n-2} s_Y^2(1 - r_{XY}^2) = \frac{n-1}{n-2}(s_Y^2 - b_1 s_{XY}).$$

Obtaining the maximum-likelihood estimators of the conditional distribution by using the corresponding estimators of the unconditional distribution is a relatively simple operation. It is this simple only because the maximum-likelihood estimators of $\underline{\mu}$ and Σ can be shown to be consistent, efficient, complete, and sufficient. The estimators b_1, b_0, and $\hat{\sigma}_\varepsilon^2$ can be shown to be unbiased as estimators of β_1, β_0, and σ_ε^2. However, r_{YX}^2 is not unbiased as an estimator of ρ_{YX}^2.

The maximum-likelihood estimators of the parameters of the conditional distribution may be obtained more directly by starting with the probability density function of the conditional distribution for Y. The latter can be shown to be

$$\Pr(Y \mid X = X_i) = \frac{1}{(2\pi)^{\frac{1}{2}}\sigma_{Y|X_i}} \exp\left[-\frac{\frac{1}{2}(Y - \beta_0 - \beta_1 X_i)^2}{\sigma_{Y|X_i}^2}\right],$$

where $\sigma_{Y|X_i}^2 = \sigma_Y^2 - \beta_1\sigma_{XY}.$

By taking the partial derivatives of the likelihood function with respect to β_0 and β_1, and then setting the results equal to zero, one obtains

$$b_1 = \frac{\sum_i (X_i - \bar{X})(Y_i - \bar{Y})}{\Sigma(X_i - \bar{X})^2}, \qquad b_0 = \bar{Y} - b_1\bar{X}.$$

The estimate of error variance (corrected for bias) is given by

$$\begin{aligned}\hat{\sigma}_\varepsilon^2 = s_\varepsilon^2 &= \frac{1}{n-2}\sum_i (Y_i - b_0 - b_1 X_i)^2 \\ &= \frac{1}{n-2}(\mathrm{SS}_Y - b_1\mathrm{SP}_{XY}).\end{aligned}$$

These estimators are identical to those obtained earlier.

2.3 Linear Model—Multivariate Case with Distribution Assumptions

2.3-1 $X_1, \ldots, X_p$ fixed; ε independently and identically distributed

The discussion in this section follows very closely a corresponding discussion in Sec. 2.2-1.

Consider a population of elements in which the following model holds.

$$\begin{aligned}Y &= \beta_0 + \beta_1 X_1 + \cdots + \beta_p X_p + \varepsilon \\ &= \underline{X}'\underline{\beta} + \varepsilon,\end{aligned} \tag{1}$$

where $\underline{X}' = [1 \quad X_1 \quad \cdots \quad X_p]$ and $\underline{\beta} = \begin{bmatrix} \beta_0 \\ \beta_1 \\ \cdot \\ \cdot \\ \cdot \\ \beta_p \end{bmatrix}$.

For a specific element i,

$$\underline{X}' = \underline{X}_i' = [1 \quad X_{i1} \quad \cdots \quad X_{ip}],$$

and the model takes the form

$$Y_i = \underline{X}_i'\underline{\beta} + \varepsilon_i.$$

The components of the vector $\underline{X}_i'$ are a set of fixed constants associated with the element i. The error ε_i is a random variable.

Let the data given by a random sample of n elements from a population in which model (1) holds be denoted

$$\begin{bmatrix} Y_1 & 1 & X_{11} & \cdots & X_{1p} \\ \cdot & \cdot & \cdot & & \cdot \\ \cdot & \cdot & \cdot & & \cdot \\ \cdot & \cdot & \cdot & & \cdot \\ Y_i & 1 & X_{i1} & \cdots & X_{ip} \\ \cdot & \cdot & \cdot & & \cdot \\ \cdot & \cdot & \cdot & & \cdot \\ \cdot & \cdot & \cdot & & \cdot \\ Y_n & 1 & X_{n1} & \cdots & X_{np} \end{bmatrix} = \underset{n,1}{[Y} \quad \underset{n,\, p+1}{X\,]}.$$

The model for the sample has the form

$$\underset{n,1}{\underline{Y}} = \underset{n,\, p+1}{X} \quad \underset{p+1,\, 1}{\underline{\beta}} + \underset{n,1}{\underline{\varepsilon}}. \tag{2}$$

In form, (2) above is identical with (2) in Sec. 2.2-1. As part of the model it is assumed that

$$E(\underline{\varepsilon}) = \underline{0}, \qquad \text{hence} \qquad E(\underline{Y}) = X\underline{\beta}.$$

It is also assumed that the components of $\underline{\varepsilon}$ are identically and independently distributed random variables. Hence the covariance matrix for the vector variable $\underline{\varepsilon}$ is

$$E(\underline{\varepsilon}\,\underline{\varepsilon}') = \sigma_\varepsilon^2 \underset{n,n}{I}. \tag{3}$$

The right-hand side of (2) contains only one random variable, namely, ε. Hence the least-squares principle may be used to obtain estimators of the parameters $\underline{\beta}$ and σ_ε^2. The model represented by (2) is precisely the set of conditions under which the Gauss-Markov theorem holds. This theorem states that the least-squares estimators have minimum sampling variance in the class of all unbiased estimators which are linear functions of $\underline{Y}$. To obtain the least-squares estimators, one first notes that

$$\underline{\varepsilon} = \underline{Y} - X\underline{\beta}. \tag{4}$$

Hence one minimizes

$$\sum_{i=1}^{n} \varepsilon_i^2 = \underline{\varepsilon}'\underline{\varepsilon} \qquad \text{with respect to } \underline{\beta}. \tag{5}$$

The resulting normal equations are identical in form with those given in (6) in Sec. 2.2-1. That is,

$$X'X\underline{b} = X'\underline{Y} \qquad \text{or} \tag{6}$$

$$\underset{p+1,\,1}{\underline{b}} = (X'X)^{-1}X'\underline{Y} = C\underline{Y}, \qquad \text{where} \qquad C = (X'X)^{-1}X,$$

provided $(X'X)$ is nonsingular.

All the developments made in both Sec. 2.1-2 and Sec. 2.2-1 generalize quite directly to the present case. Thus if

$$\underline{\hat{Y}} = X\underline{b}, \tag{7}$$

then

$$\underline{\hat{Y}} = XCY = P\underline{Y}, \qquad \text{where} \qquad P = XC = X(X'X)^{-1}X', \tag{8}$$

and

$$\underline{Y} - \underline{\hat{Y}} = \underline{Y} - \underline{P}Y = (I - P)Y = MY \qquad \text{where} \qquad M = I - P. \tag{9}$$

One also has

$$\underline{\hat{Y}}'\underline{\hat{Y}} = \underline{b}'X'X\underline{b} = \underline{Y}'P\underline{Y} = \underline{b}'X'\underline{Y} \qquad (\text{since } X'X\underline{b} = X'\underline{Y}) \tag{10}$$

$$(\underline{Y} - \underline{\hat{Y}})'(\underline{Y} - \underline{\hat{Y}}) = \underline{Y}'\underline{Y} - \underline{Y}'P\underline{Y} = \underline{Y}'M\underline{Y}. \tag{11}$$

Further,

$$\mathrm{E}(\underline{b}) = \mathrm{E}(C\underline{Y}) = C\mathrm{E}(\underline{Y}) = CX\underline{\beta} = (X'X)^{-1}X'X\underline{\beta} = \underline{\beta}, \tag{12}$$

$$\hat{\sigma}_\varepsilon^2 = \frac{\sum_{i=1}^{n} \varepsilon_i^2}{n - p - 1} = \frac{\underline{Y}'\underline{Y} - \underline{b}'X'\underline{Y}}{n - p - 1}, \tag{13}$$

$$\mathrm{E}[\mathrm{cov}\,(\underline{b})] = \sigma_\varepsilon^2(X'X)^{-1}. \tag{14}$$

If one lets

$$\hat{Y}_i = \underline{X}_i'\underline{b} \qquad \text{where} \qquad \underline{X}_i' = [1 \quad X_{i1} \quad \cdots \quad X_{ip}],$$

then $\hat{Y}_i$ is a linear function of the components of $\underline{b}$. Hence

(15) $$\sigma^2_{\hat{Y}_i} = \mathrm{E}[\mathrm{var}\,(\hat{Y}_i)] = \underline{X}_i'\mathrm{E}[\mathrm{cov}\,(\underline{b})]\underline{X}_i = \sigma^2_\varepsilon \underline{X}_i'(X'X)^{-1}\underline{X}_i.$$

Consider the statistic $Y_i - \hat{Y}_i$ for fixed i. The expected value of this statistic is

(16) $$\mathrm{E}(Y_i - \hat{Y}_i) = 0.$$

The expected value of the variance of this statistic is

(17) $$\begin{aligned}\mathrm{E}[\mathrm{var}\,(Y_i - \hat{Y}_i)] &= \sigma^2_{Y_i} + \sigma^2_{\hat{Y}_i} \\ &= \sigma^2_\varepsilon + \sigma^2_\varepsilon \underline{X}_i'(X'X)^{-1}\underline{X}_i \\ &= \sigma^2_\varepsilon[1 + \underline{X}_i'(X'X)^{-1}\underline{X}_i].\end{aligned}$$

Expression (12) indicates that $\underline{b}$ is unbiased as an estimator of $\underline{\beta}$. Expression (13) provides an estimator of the parameter σ^2_ε. Expression (14) gives the estimators of the variances and covariances of the components of $\underline{b}$. Thus, if c^{jj} denotes a diagonal element of the matrix $(X'X)^{-1}$, then

$$\hat{\sigma}^2_{b_j} = \hat{\sigma}^2_\varepsilon c^{jj}.$$

From (14) one also has

$$\hat{\sigma}^2_{b_j - b_k} = \hat{\sigma}^2_\varepsilon(c^{jj} + c^{kk} - 2c^{jk}).$$

Expression (15) provides an estimate of the variance of the predicted Y corresponding to a specified set of $\underline{X}$. On the other hand, (17) provides an estimate of the variance of the difference between an observed Y and a Y predicted from the set of $\underline{X}$ associated with that Y. One notes that the variance of the sampling distribution of the statistic $Y_i - \hat{Y}_i$ is the sum of the conditional variance of Y for fixed X (which is σ^2_ε) and the variance of $\hat{Y}_i$ [which is $\sigma^2_\varepsilon \underline{X}_i'(X'X)^{-1}\underline{X}_i$]. For fixed $\underline{X}_i$, Y_i and $\hat{Y}_i$ are uncorrelated.

2.3-2 $X_1, \ldots, X_p$ fixed; distribution of ε normal

If one adds to the model in Sec. 2.3-1 the assumption that $\underset{n,1}{\underline{\varepsilon}}$ is multivariate normal, then the maximum-likelihood principle may be used to obtain estimators of the parameters in the linear model. The maximum-likelihood estimators (adjusted for bias) will be identical to those obtained in Sec. 2.3-1.

Under the normality assumption on $\underline{\varepsilon}$, the conditional distribution of Y for $\underline{X}' = \underline{X}_i' = [1 \quad \underline{X}_{i1} \quad \cdots \quad X_{ip}]$ will be univariate normal with

(1) $$\mathrm{E}(Y_i) = \mathrm{E}(\underline{X}_i'\underline{b}) = \underline{X}_i'\underline{\beta}$$

and with

(2) $$\sigma^2_{Y_i} = \sigma^2_\varepsilon.$$

Further the sampling distribution of $\hat{Y}_i = \underline{X}_i'\underline{b}$, which will be a function of the components of $\underline{b}$, will also be normal in form with

(3) $$E(\hat{Y}_i) = \underline{X}_i\underline{\beta}$$

and with

(4) $$\sigma^2_{\hat{Y}_i} = \underline{X}_i'E(\text{cov}\ \underline{b})\underline{X}_i = \sigma^2_\varepsilon \underline{X}_i'(X'X)^{-1}\underline{X}_i.$$

The sampling distribution of the statistic $Y_i - \hat{Y}_i$ will also be normal in form with

(5) $$E(Y_i - \hat{Y}_i) = 0$$

and with

(6) $$\begin{aligned}\sigma^2_{Y_i-\hat{Y}_i} &= \sigma^2_{Y_i} + \sigma^2_{\hat{Y}_i} \\ &= \sigma^2_\varepsilon + \sigma^2_\varepsilon \underline{X}_i'(X'X)^{-1}\underline{X}_i \\ &= \sigma^2_\varepsilon[1 + \underline{X}_i'(X'X)^{-1}\underline{X}_i].\end{aligned}$$

The expression in (6) follows from the fact that, for fixed $\underline{X}_i$, Y_i and $\hat{Y}_i$ will be uncorrelated.

An unbiased estimate of σ^2_ε is given by

(7) $$\hat{\sigma}^2_\varepsilon = \frac{\underline{Y}'\underline{Y} - \underline{b}'X'\underline{Y}}{n - p - 1}.$$

The statistic $(n - p - 1)\hat{\sigma}^2_\varepsilon/\sigma^2_\varepsilon$ will be distributed as chi square with $n - p - 1$ degrees of freedom. Hence the statistic

(8) $$t = \frac{Y_i - \underline{X}_i\underline{\beta}}{\hat{\sigma}_{\hat{Y}_i}},$$

where $\hat{\sigma}_{\hat{Y}_i}$ is obtained from (4) by replacing σ^2_ε by $\hat{\sigma}^2_\varepsilon$, is distributed as Student's t with parameter $n - p - 1$.

If $\underline{c}'\underline{\beta}$ represents a linear function of the components of $\underline{\beta}$, then the sampling distribution of the statistic $\underline{c}'\underline{b}$ will be univariate normal with

(9) $$E(\underline{c}'\underline{b}) = \underline{c}'\underline{\beta}$$

and with

(10) $$\sigma^2_{\underline{c}'\underline{b}} = \underline{c}'E[\text{cov}\ (\underline{b})]\underline{c} = \sigma^2_\varepsilon \underline{c}'(X'X)^{-1}\underline{c}.$$

Hence the statistic

(11) $$t = \frac{\underline{c}'(\underline{b} - \underline{\beta})}{\hat{\sigma}_{\underline{c}'\underline{b}}}$$

is distributed as Student's t with parameter $n - p - 1$. $\hat{\sigma}_{\underline{c}'\underline{b}}$ is obtained by replacing σ^2_ε by $\hat{\sigma}^2_\varepsilon$ in (10).

2.3-3 $X_1, \ldots, X_p$, Y multivariate normal

In this section the variates $X_1, \ldots, X_p$, Y are assumed to have a $(p+1)$-variate normal distribution. Let the parameters of this distribution be denoted

$$\underline{\mu} = \begin{bmatrix} \mu_1 \\ \cdot \\ \cdot \\ \cdot \\ \mu_p \\ \mu_Y \end{bmatrix} = \begin{bmatrix} \underline{\mu}_X \\ \mu_Y \end{bmatrix}, \qquad \Sigma = \begin{matrix} p \\ 1 \end{matrix}\overset{\begin{matrix} p & \quad 1 \end{matrix}}{\begin{bmatrix} \Sigma_{XX} & \underline{\sigma}_{XY} \\ \underline{\sigma}'_{XY} & \sigma_Y^2 \end{bmatrix}}.$$

Let $\underline{u}' = [X_1 \cdots X_p \quad Y]$. The probability density for the $(p+1)$-variate normal distribution has the form

$$\text{(1)} \qquad \Pr(\underline{u}) = \frac{1}{(2\pi)^{(p+1)/2}\,|\Sigma|^{\frac{1}{2}}} \exp\left[-\tfrac{1}{2}(\underline{u} - \underline{\mu})'\Sigma^{-1}(\underline{u} - \underline{\mu})\right].$$

Suppose a random sample of size n is drawn from this population. Let the sample data be denoted

$$U = \begin{bmatrix} X_{11} & \cdots & X_{1p} & Y_1 \\ \cdot & & \cdot & \cdot \\ \cdot & & \cdot & \cdot \\ \cdot & & \cdot & \cdot \\ X_{i1} & \cdots & X_{ip} & Y_i \\ \cdot & & \cdot & \cdot \\ \cdot & & \cdot & \cdot \\ \cdot & & \cdot & \cdot \\ X_{n1} & \cdots & X_{np} & Y_n \end{bmatrix} = {}_n\overset{\begin{matrix} p & 1 \end{matrix}}{[X \quad \underline{Y}]}.$$

Let the data on element i in the sample be denoted

$$\underline{u}_i = [\underline{X}'_i \quad Y_i], \qquad \text{where} \qquad \underline{X}'_i = [X_{i1} \quad \cdots \quad X_{ip}].$$

The likelihood of the sample data is given by

$$L(\text{sample}) = \frac{1}{(2\pi)^{n(p+1)/2}\,|\Sigma|^{n/2}} \exp\left[-\tfrac{1}{2}\sum_{i=1}^{n}(\underline{u}_i - \underline{\mu})'\Sigma^{-1}(\underline{u}_i - \underline{\mu})\right].$$

The ln likelihood of the sample is given by

$$\text{(2)} \quad \phi = \ln L(\text{sample}) = -\tfrac{1}{2}n(p+1)\ln(2\pi) - \frac{n}{2}\ln|\Sigma| - \tfrac{1}{2}\sum_{i=1}^{n}(\underline{u}_i - \underline{\mu})'\Sigma^{-1}(\underline{u}_i - \underline{\mu}).$$

The marginal distribution of each X_j $(j = 1, \ldots, p)$ is $N(\mu_j,\sigma_j^2)$. The marginal distribution of Y is $N(\mu_Y,\sigma_Y^2)$. To characterize the conditional distribution of Y when $\underline{X}' = \underline{X}_i' = [X_{i1} \cdots X_{ip}]$, let

$$\underline{\beta} = \begin{bmatrix} \beta_1 \\ \cdot \\ \cdot \\ \cdot \\ \beta_p \end{bmatrix} = \Sigma_{XX}^{-1}\underline{\sigma}_{XY}, \qquad \beta_0 = \mu_Y - \underline{\beta}'\underline{\mu}_X, \qquad \rho^2_{Y(1\ldots p)} = \frac{\underline{\beta}'\underline{\sigma}_{XY}}{\sigma_Y^2}.$$

The conditional distribution of Y for $\underline{X} = \underline{X}_i$ is univariate normal in form with

$$E(Y \mid \underline{X}_i) = \beta_0 + \underline{\beta}'\underline{X}_i \qquad \text{and} \qquad \sigma^2_{Y|\underline{X}_i} = \sigma_Y^2(1 - \rho^2_{Y(1\ldots p)}).$$

One notes that when $\rho^2_{Y(1\ldots p)} = 0$, or equivalently when $\underline{\beta} = \underline{0}$, the conditional distribution of Y is identical with the marginal distribution of Y.

The conditional distribution may be represented in the form

$$\begin{aligned} Y \mid \underline{X}_i = Y_i &= \beta_0 + \beta_1 X_{i1} + \cdots + \beta_p X_{ip} + \varepsilon_i \\ &= \beta_0 + \underline{\beta}'\underline{X}_i + \varepsilon_i, \end{aligned}$$

where ε is normally distributed with

$$E(\varepsilon) = 0 \qquad \text{and} \qquad \sigma_\varepsilon^2 = \sigma^2_{Y|\underline{X}_i} = \sigma_Y^2(1 - \rho^2_{Y(1\ldots p)}).$$

The problem of estimating the parameters of this conditional distribution may be approached directly or indirectly by first estimating the parameters of the complete distribution.

Taking the partial derivative of the ln L(sample) with respect to $\underline{\mu}$ and setting the result equal to zero yield the maximum-likelihood estimator

$$\hat{\underline{\mu}} = \begin{bmatrix} \bar{\underline{X}} \\ \bar{Y} \end{bmatrix}, \qquad \text{where} \qquad \bar{\underline{X}} = \begin{bmatrix} \bar{X}_1 \\ \cdot \\ \cdot \\ \cdot \\ \bar{X}_p \end{bmatrix}.$$

A similar operation with respect to Σ on ln L(sample) yields, after adjustment for bias, the maximum-likelihood estimator

$$\hat{\Sigma} = \frac{1}{n-1}\begin{bmatrix} V_{XX} & \underline{v}_{XY} \\ \underline{v}'_{XY} & v_{YY} \end{bmatrix} = \begin{bmatrix} \hat{\Sigma}_{XX} & \hat{\underline{\sigma}}_{XY} \\ \hat{\underline{\sigma}}'_{XY} & \hat{\sigma}_Y \end{bmatrix},$$

where
$$\frac{1}{n-1}\underset{p,p}{V_{XX}} = \text{sample covariance matrix} = \hat{\Sigma}_{XX} = S_{XX},$$

$$\frac{1}{n-1}\underline{v}_{XY} = \text{vector of sample covariances } \mathrm{cov}_{X_jY},$$

$$\frac{1}{n-1}v_{YY} = \text{sample variance of } Y = \hat{\sigma}_Y^2 = s_Y^2.$$

Upon replacing parameters by their maximum-likelihood estimators, from the definitions of β_0, $\underline{\beta}$, and σ_ε^2 one obtains

$$\underline{b} = V_{XX}^{-1}\underline{v}_{XY}, \qquad b_0 = \bar{Y} - \underline{b}'\underline{\bar{X}}, \qquad r_{Y(1\ldots p)}^2 = \frac{\underline{b}'\underline{v}_{XY}}{v_{YY}},$$

$$s_\varepsilon^2 = \frac{v_{YY}}{n-p-1}(1 - r_{Y(1\ldots p)}^2).$$

For the case of fixed X, the joint distribution of the components of the random vector $\underline{b}$ is multivariate normal. For the case in which Y, $X_1, \ldots, X_p$ are multivariate normal, the distribution of the random vector $\underline{b}$ is not multivariate normal. Further the distribution of the components of the vector $\underline{b}$ in the latter case is not univariate normal. However, asymptotically in n, the distributions of the components of $\underline{b}$ will be univariate normal. The sampling distribution of the vector $\underline{b}$ for the random model is given by Seal (1951, p. 142).

The probability density function for the conditional distribution of $Y \mid \underline{X}_i$ is given by

$$\Pr(Y \mid \underline{X}_i) = \frac{\Pr(\underline{X}_i, Y)}{\Pr(\underline{X}_i)},$$

that is, the ratio of the joint density $(\underline{X}_i, Y)$ to the density $\underline{X}_i$. The numerator is the density for the $(p+1)$-variate normal; the denominator is the density for the p-variate normal. Explicitly,

$$\Pr(Y \mid \underline{X}_i) = \frac{1}{(2\pi)\sigma_{Y|X_i}} \exp\left[-\frac{1}{2}\frac{(Y - \mu_{Y|X_i})^2}{\sigma_{Y|X_i}^2}\right]$$

where $\sigma_{Y|X_i}^2 = \sigma_Y^2 - \underline{\beta}'\underline{\sigma}_{XY}, \qquad \underline{\beta} = \Sigma_{XX}^{-1}\underline{\sigma}_{XY},$

$$\mu_{Y|X_i} = \mu_y + \underline{\beta}'(\underline{X}_i - \underline{\mu}_X) = \mu_Y - \underline{\beta}'\underline{\mu}_X + \underline{\beta}'\underline{X}_i$$
$$= \beta_0 + \underline{\beta}'\underline{X}_i.$$

From the density function for an individual element, the likelihood function for a random sample of elements is readily obtained. From the likelihood function (or, more conveniently, from its logarithm) one may derive the maximum-likelihood estimators of the conditional distribution.

Returning to the estimators of the parameters of the parent distribution, the sampling distribution of the maximum-likelihood estimator $\underline{\bar{X}}' = [\bar{X}_1 \cdots \bar{X}_p]$ is p-variate normal with mean vector

$$E(\underline{\bar{X}}) = \underline{\mu}_X$$

and covariance matrix

$$\sigma^2(\underline{\bar{X}}) = \frac{1}{n}\Sigma_{XX}.$$

The joint sampling distribution of the elements of the matrix

$$(n - 1)\hat{\Sigma}_{XX} = V_{XX}$$

is the Wishart distribution with parameters p, $n - 1$, and Σ_{XX}.

The sampling distribution of the statistic

$$T^2 = n(\underline{\bar{X}} - \underline{\mu})' S_{XX}^{-1} (\underline{\bar{X}} - \underline{\mu})$$

is Hotelling's T^2 distribution with parameters n and p. For this special case

$$\frac{n-p}{p(n-1)} T^2 \text{ is distributed as } F(p, n - p).$$

When $p = 1$,

$$T^2 = \frac{n(\bar{X} - \mu)^2}{s_x^2},$$

$$\frac{n-p}{p(n-1)} T^2 = T^2 \qquad \text{is distributed as} \qquad F(1, n - 1).$$

Thus, when $p = 1$, this special case of Hotelling's T^2 reduces to the square of Student's t, which has an F distribution with one degree of freedom for the numerator and $n - 1$ degrees of freedom for the denominator.

2.3-4 $X'X$ singular and estimable parametric functions

In the first part of this section, the model will be

$$(1) \qquad Y = \beta_0 + \beta_1 X_1 + \cdots + \beta_p X_p + \varepsilon = \underline{X}'\underline{\beta} + \varepsilon,$$

where $\underset{1,\, p+1}{\underline{X}'} = [1 \quad X_1 \quad \cdots \quad X_p]$ is a vector of fixed variables, and ε is $N(0,\sigma_\varepsilon^2)$. The variance of ε is assumed to be independent of $\underline{X}$. Let

$$\underset{n,\, p+1}{X} = \text{matrix of sample values for } \underline{X}' = [1 \quad X_1 \quad \cdots \quad X_p].$$

It will be assumed that $X'X$ is singular.

The normal equations for obtaining the least-squares estimators of $\underline{\beta}$ have the form

$$(2) \qquad (X'X)\underline{\beta} = X'\underline{Y}.$$

Corresponding to a generalized inverse, say $(X'X)_u^-$, one may obtain the solution

$$\underline{b}_u = (X'X)_u^- X'\underline{Y}.$$

Corresponding to a different generalized inverse, say $(X'X)_v^-$, the solution is

$$\underline{b}_v = (X'X)_v^- X'\underline{Y}.$$

In general,

$$\underline{b}_u \neq \underline{b}_v.$$

Hence, when $X'X$ is singular, $\underline{\beta}$ is said to be *nonestimable.*

Since the normal equations in (2) represent both necessary and sufficient conditions that $\Sigma\varepsilon_i^2$ be minimized for the sample data,

$$\begin{aligned}\text{Minimum } \Sigma\varepsilon_i^2 &= \underline{Y}'\underline{Y} - \underline{b}_u'X'\underline{Y} \\ &= \underline{Y}'\underline{Y} - \underline{b}_v'X'\underline{Y} = \Sigma(Y_i - \hat{Y}_i)^2, \\ \underline{b}_u'X'\underline{Y} &= \underline{b}_v'X'\underline{Y} = \Sigma\hat{Y}_i^2.\end{aligned}$$

Thus, although the solution for $\underline{b}$ is not unique, $\Sigma(Y_i - \hat{Y}_i)^2$ and $\Sigma\hat{Y}_i^2$ are unique. Hence

$$\hat{\sigma}_\varepsilon^2 = \frac{\underline{Y}'\underline{Y} - \underline{b}'X'\underline{Y}}{n - p - 1}$$

is a unique estimate of σ_ε^2, where $\underline{b}$ is any solution to the normal equations.

The solution to the system represented by (2) may be made unique by imposing an appropriate number of side conditions (or constraints) upon the components of $\underline{\beta}$. This procedure is equivalent to adding an appropriate number of rows and columns to the matrix $X'X$ so that the augmented matrix is nonsingular, the vector $X'\underline{Y}$ being augmented appropriately.

It is, however, of considerable interest to find the answer to the following question: What linear functions of the components of $\underline{\beta}$ can be estimated? If the components of $\underline{c}'$ are known constants, the parametric function $\underline{c}'\underline{\beta}$ is said to be *estimable* if $\underline{c}'\underline{b}$ is unique for all solutions $\underline{b}$ to the normal equations in (2). It has already been pointed out that the individual components of $\underline{\beta}$ are not estimable in the sense defined here.

Both a necessary and sufficient condition that $\underline{c}'\underline{\beta}$ be estimable is that $\underline{c}'$ have the form

$$\underline{c}' = \underline{\lambda}'(X'X), \tag{3}$$

where $\underline{\lambda}$ is any vector. That is, the components of $\underline{c}'$ must be a linear function of the columns of $X'X$. In geometric language, the vector $\underline{c}'$ must lie in the space defined by the column vectors of $X'X$. If $\underline{c}'$ has the form given in (3), then

$$\underline{c}'\underline{b} = \underline{\lambda}'(X'X)\underline{b} = \underline{\lambda}'X'\underline{Y} \qquad \text{since} \qquad (X'X)\underline{b} = X'\underline{Y}.$$

The expected value of $\underline{c}'\underline{b}$, if $\underline{c}' = \underline{\lambda}'(X'X)$, is

$$\text{E}(\underline{c}'\underline{b}) = \text{E}(\underline{\lambda}'X'\underline{Y}) = \underline{\lambda}\text{E}(X'\underline{Y}) = \underline{\lambda}'(X'X)\underline{\beta} = \underline{c}'\underline{\beta}.$$

A generalization of the Gauss-Markov theorem asserts that if $\underline{c}'\underline{\beta}$ is an estimable parametric function, then $\underline{c}'\underline{b}$, where $\underline{b}$ is any solution to the normal equations, is a unique, unbiased, minimum-variance, linear (in

$\underline{Y}$) estimator of $\underline{c}'\underline{\beta}$. The variance of the estimator $\underline{c}'\underline{b}$ is

$$\begin{aligned}\sigma^2_{\underline{c}'\underline{b}} = \underline{c}'\mathrm{E}[\mathrm{cov}\,(\underline{b})]\underline{c} &= \sigma^2_{\varepsilon}\underline{c}'(X'X)^{-}\underline{c} \\ &= \sigma^2_{\varepsilon}\underline{\lambda}'(X'X)(X'X)^{-}(X'X)\underline{\lambda} \\ &= \sigma^2_{\varepsilon}\underline{\lambda}'(X'X)\underline{\lambda}.\end{aligned}$$

An estimate of the variance of $\underline{c}'\underline{b}$ is obtained by replacing σ^2_ε by $\hat{\sigma}^2_\varepsilon$.

Numerical Example. A numerical example of some of the points discussed in this section appears in Table 2.3-1. In part i is a set of normal

Table 2.3-1 Numerical Example

(i)	Normal equations: $\underset{X'X}{\begin{bmatrix} 6 & 3 & 3 \\ 3 & 3 & 0 \\ 3 & 0 & 3 \end{bmatrix}} \underset{\underline{b}}{\begin{bmatrix} b_0 \\ b_1 \\ b_2 \end{bmatrix}} = \underset{X'\underline{Y}}{\begin{bmatrix} 42 \\ 15 \\ 27 \end{bmatrix}}$
(ii)	$(X'X)^{-}_{u} = \begin{bmatrix} \frac{1}{3} & -\frac{1}{3} & 0 \\ -\frac{1}{3} & \frac{2}{3} & 0 \\ 0 & 0 & 0 \end{bmatrix}$ $\quad \underline{b}_u = (X'X)^{-}_{u}X'\underline{Y} = \begin{bmatrix} 9 \\ -4 \\ 0 \end{bmatrix}$
(iii)	$(X'X)^{-}_{v} = \begin{bmatrix} 0 & 0 & 0 \\ 0 & \frac{1}{3} & 0 \\ 0 & 0 & \frac{1}{3} \end{bmatrix}$ $\quad \underline{b}_v = (X'X)^{-}_{v}X'\underline{Y} = \begin{bmatrix} 0 \\ 5 \\ 9 \end{bmatrix}$
(iv)	$\Sigma\hat{Y}^2 = \underline{b}'_u X'\underline{Y} = [9 \quad -4 \quad 0]\begin{bmatrix} 42 \\ 15 \\ 27 \end{bmatrix} = 318$ $= \underline{b}'_v X'\underline{Y} = [0 \quad 5 \quad 9]\begin{bmatrix} 42 \\ 15 \\ 27 \end{bmatrix} = 318$
(v)	$(a)\ \underline{\lambda}' = [\frac{1}{3} \quad 0 \quad -\frac{1}{3}] \qquad \underline{c}' = \underline{\lambda}'(X'X) = [1 \quad 1 \quad 0]$ $(b)\ \underline{\lambda}' = [\frac{1}{3} \quad -\frac{1}{3} \quad 0] \qquad \underline{c}' = \underline{\lambda}'(X'X) = [1 \quad 0 \quad 1]$ $(c)\ \underline{\lambda}' = [0 \quad \frac{1}{3} \quad -\frac{1}{3}] \qquad \underline{c}' = \underline{\lambda}'(X'X) = [0 \quad 1 \quad -1]$ $(d)\ \underline{\lambda}' = [1 \quad 1 \quad 1] \qquad \underline{c}' = \underline{\lambda}'(X'X) = [12 \quad 6 \quad 6]$

equations in which $X'X$ is singular. One of the generalized inverses of $X'X$ is illustrated at the left of part ii; the corresponding solution $\underline{b}_u$ appears at the right. A different generalized inverse of $X'X$ is shown at the left of part iii; the corresponding solution $\underline{b}_v$ is given at the right. The numerical value of $\Sigma\hat{Y}^2$ corresponding to $\underline{b}_u$ and $\underline{b}_v$ is given in part iv. In both cases $\Sigma\hat{Y}^2 = 318$. This illustrates the principle that $\Sigma\hat{Y}^2$ is invariant for any solution to the normal equations.

In part v, $\underline{c}'$ corresponding to various values for the vector $\underline{\lambda}'$ are indicated. Since a necessary and sufficient condition that the parametric function $\underline{c}'\underline{\beta}$ be estimable is that $\underline{c}' = \underline{\lambda}'(X'X)$, some of the estimable parametric functions associated with the normal equations in part i are illustrated in part v. For example, the parametric function

(*a*) $\beta_0 + \beta_1$ is estimable with estimator $b_0 + b_1$,

(*b*) $\beta_0 + \beta_2$ is estimable with estimator $b_0 + b_2$,

(*c*) $\beta_1 - \beta_2$ is estimable with estimator $b_1 - b_2$,

(*d*) $12\beta_0 + 6\beta_1 + 6\beta_2$ is estimable with estimator $12b_0 + 6b_1 + 6b_2$.

In each case, the numerical value of the estimator is the same no matter which one of the solutions to the normal equations is chosen. For example, if $\underline{b}_u$ is chosen to estimate the parametric function represented by (*d*),

$$12(9) + 6(-4) + 6(0) = 84.$$

If $\underline{b}_v$ is chosen, the parametric function defined by (*d*) is estimated by

$$12(0) + 6(5) + 6(9) = 84.$$

2.4 Correlations

2.4-1 Multiple correlation

Consider the $p + 1$ random variables $\underline{X}' = [X_1 \cdots X_p]$ and Y. Let the covariance matrix be denoted

$$\underset{p+1,\,p+1}{\Sigma} = \begin{matrix} & p & 1 \\ p & \Sigma_{XX} & \underline{\sigma}_{XY} \\ 1 & \underline{\sigma}_{YX} & \sigma_Y \end{matrix}.$$

No assumption about the form of the joint distribution of these variables will be made at this stage of the discussion.

Suppose one defines the following arbitrary linear function of $X_1, \ldots, X_p$:

$$L = L(\underline{X}) = a_1X_1 + \cdots + a_pX_p = \underline{a}'\underline{X}. \tag{1}$$

The variance of the random variable L is

$$\sigma_L^2 = \underline{a}'\Sigma_{XX}\underline{a}, \tag{2}$$

and the covariance of the random variable L with the random variable Y is

$$\sigma_{YL} = \underline{a}'\underline{\sigma}_{XY}. \tag{3}$$

The square of the product-moment correlation between Y and L is

$$\rho_{YL}^2 = \frac{\sigma_{YL}^2}{\sigma_Y^2\sigma_L^2} = \frac{(\underline{a}'\underline{\sigma}_{XY})^2}{\sigma_Y^2(\underline{a}'\Sigma_{XX}\underline{a})}. \tag{4}$$

Suppose one were to choose

(5) $$\underline{a} = \underline{\beta} = \Sigma_{XX}^{-1}\underline{\sigma}_{XY}.$$

Let

(6) $$L^* = \underline{\beta}'\underline{X}.$$

Then

(7) $$\sigma_{L^*}^2 = \underline{\beta}'\Sigma_{XX}\underline{\beta} = \underline{\beta}'\underline{\sigma}_{XY} \qquad [\text{since from (5) } \Sigma_{XX}\underline{\beta} = \underline{\sigma}_{XY}],$$

(8) $$\sigma_{YL^*} = \underline{\beta}'\underline{\sigma}_{XY} = \sigma_{L^*}^2.$$

(9) $$\rho_{YL^*}^2 = \frac{(\sigma_{YL^*})^2}{\sigma_Y^2\sigma_{L^*}^2} = \frac{(\sigma_{L^*}^2)^2}{\sigma_Y^2\sigma_{L^*}^2} = \frac{\sigma_{L^*}^2}{\sigma_Y^2}.$$

Thus when $L = L^*$ the squared product-moment correlation between Y and L reduces to the ratio of the variance of L^* to the variance of Y.

Suppose one were to choose

(10) $$\underline{a} = \underline{\alpha} \neq \underline{\beta}.$$

Let $$L^{**} = \underline{\alpha}'\underline{X}.$$

That is, L^{**} is any linear function other than L^*. One has

(11) $$\sigma_{L^{**}}^2 = \underline{\alpha}'\Sigma_{XX}\underline{\alpha},$$

(12) $$\sigma_{YL^{**}} = \underline{\alpha}'\underline{\sigma}_{XY} = \underline{\alpha}'\Sigma_{XX}\underline{\beta} \qquad [\text{since from (5) } \Sigma_{XX}\underline{\beta} = \underline{\sigma}_{XY}].$$

Thus

(13) $$\rho_{YL^{**}}^2 = \frac{(\sigma_{YL^{**}})^2}{\sigma_Y^2\sigma_{L^{**}}^2} = \frac{(\underline{\alpha}'\Sigma_{XX}\underline{\beta})^2}{\sigma_Y^2(\underline{\alpha}'\Sigma_{XX}\underline{\alpha})},$$

or

(14) $$\sigma_Y^2\rho_{YL^{**}}^2 = \frac{(\underline{\alpha}'\Sigma_{XX}\underline{\beta})^2}{\underline{\alpha}'\Sigma_{XX}\underline{\alpha}}.$$

The Cauchy-Schwartz inequality states that

$$(\Sigma UV)^2 \leq (\Sigma U)^2(\Sigma V)^2.$$

The analog of this inequality in the present case is

(15) $$(\underline{\alpha}'\Sigma_{XX}\underline{\beta})^2 \leq (\underline{\alpha}'\Sigma_{XX}\underline{\alpha})(\underline{\beta}'\Sigma_{XX}\underline{\beta}),$$

or

(16) $$\frac{(\underline{\alpha}'\Sigma_{XX}\underline{\beta})^2}{\underline{\alpha}'\Sigma_{XX}\underline{\alpha}} \leq \underline{\beta}'\Sigma_{XX}\underline{\beta} = \sigma_{L^*}^2.$$

Hence from (14)

$$\sigma_Y^2 \rho_{YL^{**}}^2 = \frac{(\underline{\alpha}'\Sigma_{XX}\underline{\beta})^2}{\underline{\alpha}'\Sigma_{XX}\underline{\alpha}} \leq \sigma_{L^*}^2 = \sigma_Y^2 \rho_{YL^*}^2. \tag{17}$$

The relationship in (17) implies that

$$\rho_{YL^{**}}^2 \leq \rho_{YL^*}^2. \tag{18}$$

That is, $\rho_{YL^*}^2$ is the maximum squared product-moment correlation between Y and any linear function L.

The maximum squared product-moment correlation is called the squared multiple correlation and is denoted

$$\rho_{YL^*}^2 = \rho_{Y(1\cdots p)}^2 = \frac{\sigma_{YL^*}}{\sigma_Y^2} = \frac{\underline{\beta}'\underline{\sigma}_{XY}}{\sigma_Y^2}. \tag{19}$$

To obtain another expression for the squared multiple correlation, define

$$\varepsilon = Y - L^*.$$

Hence

$$\begin{aligned} \sigma_\varepsilon^2 &= \sigma_Y^2 + \sigma_{L^*}^2 - 2\sigma_{YL^*} \\ &= \sigma_Y^2 - \sigma_{L^*}^2 \qquad (\text{since } \sigma_{YL^*} = \sigma_{L^*}^2) \\ &= \sigma_Y^2 - \sigma_Y^2\rho_{YL^*}^2 = \sigma_Y^2(1 - \rho_{YL^*}^2). \end{aligned} \tag{20}$$

Solving (20) for $\rho_{YL^*}^2$ gives

$$\rho_{YL^*}^2 = 1 - \frac{\sigma_\varepsilon^2}{\sigma_Y^2}. \tag{21}$$

Expression (21) is sometimes taken as the definition of the squared multiple correlation.

Some Computational Details. Let

$$\Sigma_{p+1,\,p+1}^{-1} = \begin{bmatrix} \sigma^{11} & \cdots & \sigma^{1p} & \sigma^{1Y} \\ \cdot & & \cdot & \cdot \\ \cdot & & \cdot & \cdot \\ \cdot & & \cdot & \cdot \\ \sigma^{Y1} & \cdots & \sigma^{Yp} & \sigma^{YY} \end{bmatrix} = \begin{bmatrix} \Sigma_{XX} & \underline{\sigma}_{XY} \\ \underline{\sigma}_{YX} & \sigma_Y^2 \end{bmatrix}^{-1}.$$

From the relationship between Σ^{-1} and Σ_{XX}^{-1}, it may be shown that

$$\beta_j = -\frac{\sigma^{jY}}{\sigma^{YY}} \qquad \text{and} \qquad \rho_{Y(1\ldots p)}^2 = 1 - \frac{1}{\sigma^{YY}\sigma_Y^2}.$$

Let the matrix of intercorrelations be denoted

$$\underset{p+1,\,p+1}{R} = D^{-1}\Sigma D^{-1} \qquad \text{where} \qquad D = \begin{bmatrix} \sigma_1 & \cdots & 0 \\ \cdot & & \cdot \\ \cdot & & \cdot \\ \cdot & & \cdot \\ 0 & \cdots & \sigma_Y \end{bmatrix}$$

$$= [\rho_{jk}].$$

Then

$$R^{-1} = D\Sigma^{-1}D = [\rho^{jk}] = [\sigma_j\sigma_k\sigma^{jk}].$$

Hence

$$\beta_j = -\frac{\sigma^{jY}}{\sigma^{YY}} = -\frac{\rho^{jY}/\sigma_j\sigma_Y}{\rho^{YY}/\sigma_Y^2} = -\frac{\rho^{jY}}{\rho^{YY}}\frac{\sigma_Y}{\sigma_j},$$

$$\rho^2_{Y(1\ldots p)} = 1 - \frac{1}{\sigma^{YY}\sigma_Y^2} = 1 - \frac{1}{\rho^{YY}}.$$

This last relationship indicates that the multiple correlation is a function of the elements of the matrix R. Equivalently, the multiple correlation is a function of the intercorrelations among X_j and X_k (for all possible j and k) as well as the correlations between each X_j and Y.

Heuristic Estimation. Since (as yet) no assumption has been made about the joint distribution of $\underline{X}$ and Y, the maximum-likelihood principle of estimation (or for that matter any other estimation principle that depends upon the joint distribution) cannot be used to obtain estimators. If one adopts the heuristic principle of replacing individual parameters in definitions by their corresponding sample estimators, the resulting estimators will be consistent but will *not* necessarily be unbiased or have minimum variance.

Suppose the sample covariance matrix is denoted

$$\underset{p+1,\,p+1}{S} = \begin{bmatrix} S_{XX} & \underline{s}_{XY} \\ \underline{s}'_{XY} & s_Y^2 \end{bmatrix}.$$

Let

$$S^{-1} = \begin{bmatrix} s^{11} & \cdots & s^{1p} & s^{1Y} \\ \cdot & & \cdot & \cdot \\ \cdot & & \cdot & \cdot \\ \cdot & & \cdot & \cdot \\ s^{p1} & \cdots & s^{pp} & s^{pY} \\ s^{Y1} & \cdots & s^{Yp} & s^{YY} \end{bmatrix}.$$

For the sample estimate of $\underline{\beta}$ suppose one takes

$$b_j = \frac{-s^{jY}}{s^{YY}}, \quad j = 1, \ldots, p.$$

For the sample estimate of $\rho^2_{Y(1\ldots p)}$ suppose one takes

$$r^2_{Y(1\ldots p)} = 1 - \frac{1}{s^{YY}s^2_Y}.$$

Equivalently, one may take

$$\underline{b} = S^{-1}_{XX}\underline{s}_{XY} \qquad \text{and} \qquad r^2_{Y(1\ldots p)} = \frac{\underline{b}'\underline{s}_{XY}}{s^2_Y}.$$

For the case in which $\underline{X}$ and Y have a joint multivariate normal distribution, $\underline{b}$ will turn out to be the maximum-likelihood estimator of $\underline{\beta}$, but $r^2_{Y(1\cdots p)}$ will not be unbiased as an estimator of $\rho^2_{Y(1\cdots p)}$. One maximum-likelihood estimator of the latter is obtained from (21) by replacing the variances by their unbiased maximum-likelihood estimators. One has

$$\hat{\sigma}^2_\varepsilon = \frac{\mathrm{SS}_Y(1 - r^2_{Y(1\ldots p)})}{n - p - 1},$$

$$\hat{\sigma}^2_Y = \frac{\mathrm{SS}_Y}{n - 1}.$$

Thus, $$\hat{\rho}^2_{Y(1\ldots p)} = 1 - \frac{\hat{\sigma}^2_\varepsilon}{\hat{\sigma}^2_Y} = 1 - \frac{n-1}{n-p-1}(1 - r^2_{Y(1\ldots p)}).$$

Although $\hat{\sigma}^2_\varepsilon$ and $\hat{\sigma}^2_Y$ are unbiased as estimators of their respective parameters, the ratio $\hat{\sigma}^2_\varepsilon/\hat{\sigma}^2_Y$ will not be an unbiased estimator of the ratio $\sigma^2_\varepsilon/\sigma^2_Y$. Hence $\hat{\rho}^2_{Y(1\ldots p)}$ as defined above will be biased.

For the case in which $\underline{X}$ and Y have a $(p + 1)$-variate normal distribution, Olkin and Pratt (1958) obtained an unbiased estimator of ρ^2. If one neglects terms of order $1/n^2$, the unbiased estimator reduces to

$$\text{est } \rho^2_{Y(1\ldots p)} = 1 - \frac{n-3}{n-p-1}(1 - r^2_{Y(1\ldots p)}) - \frac{n-3}{n-p-1}\frac{2}{n-p+1}(1 - r^2_{Y(1\ldots p)})^2.$$

2.4-2 Correlations related to the multiple correlation

Let the set of variables $X_1, \ldots, X_p$ be divided into two subsets $X_1, \ldots, X_k$ and $X_{k+1}, \ldots, X_p$. Let $\underline{X}'_a = [X_1 \cdots X_k]$ and $\underline{X}'_b = [X_{k+1} \cdots X_p]$. The covariance matrix Σ defined earlier now has the form

$$\Sigma = \begin{bmatrix} \Sigma_{XX} & \underline{\sigma}_{XY} \\ \underline{\sigma}_{YX} & \sigma^2_Y \end{bmatrix} = \begin{array}{c} \\ k \\ p-k \\ 1 \end{array}\!\!\begin{array}{c} \begin{array}{ccc} k & p-k & 1 \end{array} \\ \left[\begin{array}{cc|c} \Sigma_{aa} & \Sigma_{ab} & \underline{\sigma}_{aY} \\ \Sigma_{ba} & \Sigma_{bb} & \underline{\sigma}_{bY} \\ \hline \underline{\sigma}_{Ya} & \underline{\sigma}_{Yb} & \sigma^2_Y \end{array}\right] \end{array}.$$

Let

$$L_a = \underline{\beta}_a'\underline{X}_a \qquad \text{where} \qquad \underline{\beta}_a = \Sigma_{aa}^{-1}\underline{\sigma}_{aY},$$

$$L_b = \underline{\beta}_b'\underline{X}_b \qquad \text{where} \qquad \underline{\beta}_b = \Sigma_{bb}^{-1}\underline{\sigma}_{bY},$$

$$L_{a,b} = L = \underline{\beta}'\underline{X} \qquad \text{where} \qquad \underline{\beta} = \Sigma_{XX}^{-1}\underline{\sigma}_{XY}.$$

Hence

$$\rho_{YL_a}^2 = \rho_{Y(1\ldots k)}^2 = \frac{\underline{\beta}_a'\underline{\sigma}_{aY}}{\sigma_Y^2} = \frac{\sigma_{L_a}^2}{\sigma_Y^2},$$

$$\rho_{YL_b}^2 = \rho_{Y(k+1\ldots p)}^2 = \frac{\underline{\beta}_b'\underline{\sigma}_{bY}}{\sigma_Y^2} = \frac{\sigma_{L_b}^2}{\sigma_Y^2},$$

$$\rho_{YL}^2 = \rho_{Y(1\ldots p)}^2 = \frac{\underline{\beta}'\underline{\sigma}_{XY}}{\sigma_Y^2} = \frac{\sigma_L^2}{\sigma_Y^2}.$$

One notes that

$$\underline{\beta}_a'\underline{\sigma}_{aY} = \underline{\beta}_a'\Sigma_{aa}\underline{\beta}_a = \sigma_{L_a}^2 = \sigma_{YL_a} = \sigma_Y^2\rho_{YL_a}^2.$$

Thus $\rho_{Y(1\ldots k)}^2$ is the maximum squared product-moment correlation between a linear function of $X_1, \ldots, X_k$ and Y. Similarly $\rho_{Y(k+1\ldots p)}^2$ is the maximum squared product-moment correlation between a linear function of $X_{k+1}, \ldots, X_p$ and Y. Alternatively, the latter is the multiple correlation between Y and $X_{k+1}, \ldots, X_p$.

The squared *semipartial* (or *part*) correlation between Y and L_b is defined by

$$\rho_{Y(L_b \cdot L_a)}^2 = \rho_{YL}^2 - \rho_{YL_a}^2 = \rho_{Y(1\ldots p)}^2 - \rho_{Y(1\ldots k)}^2. \tag{1}$$

The square of this semipartial measures the increase in the squared multiple correlation achieved by adding the set of variables $\underline{X}_b$ to a prediction system already containing the set $\underline{X}_a$. The linear prediction system defined by the combined sets is L. One notes that

$$\rho_{YL}^2 = \rho_{YL_a}^2 + \rho_{Y(L_b \cdot L_a)}^2.$$

An alternative notation system for this squared semipartial correlation is

$$\rho_{Y(L_b \cdot L_a)}^2 = \rho_{Y(L_b|L_a)}^2 = \rho_{Y(k+1\ldots p|1\ldots k)}^2.$$

This notation system is more indicative of the conditional distributions that are involved.

The square of the semipartial correlation between Y and L_a is defined by

$$\begin{aligned}\rho_{Y(L_a \cdot L_b)}^2 = \rho_{Y(L_a|L_b)}^2 &= \rho_{YL}^2 - \rho_{YL_b}^2 \\ &= \rho_{Y(1\ldots p)}^2 - \rho_{Y(k+1\ldots p)}^2.\end{aligned} \tag{2}$$

This squared semipartial correlation measures the increase in the squared multiple correlation that is achieved by adding the set $\underline{X}_a$ to the set of variables $\underline{X}_b$.

The square of the *partial* correlation between Y and L_b is defined by

$$\rho^2_{Y L_b \cdot L_a} = \rho^2_{Y L_b | L_a} = \frac{\rho^2_{Y(L_b|L_a)}}{1 - \rho^2_{Y L_a}}. \tag{3}$$

To interpret this squared partial correlation, $1 - \rho^2_{Y L_a}$ is a measure of the residual nonpredictability in Y when the optimum linear prediction system (based upon the set $\underline{X}_a$ alone) is used. The squared partial correlation in (3) represents the proportion of this residual that can be predicted by adding the set $\underline{X}_b$ to the linear prediction system, in an optimal way.

The square of the partial correlation between Y and L_a is defined by

$$\rho^2_{Y L_a \cdot L_b} = \rho^2_{Y L_a | L_b} = \frac{\rho^2_{Y(L_a|L_b)}}{1 - \rho^2_{Y L_b}}. \tag{4}$$

If $1 - \rho^2_{Y L_b}$ is considered a measure of the residual unpredictability in Y from a linear prediction system based upon $\underline{X}_b$, then the squared partial correlation defined in (4) represents the proportion of this residual that can be predicted by adding the set $\underline{X}_a$ to the prediction system. An alternative notation for the squared partial correlation in (4) is

$$\rho^2_{Y L_a | L_b} = \rho^2_{Y(1\ldots k)|(k+1\ldots p)}.$$

There is another approach to semipartial and partial correlations that provides somewhat different interpretations. Define the random variable $L_{b|a}$ as follows:

$$L_{b|a} = L - L_a = L_{a,b} - L_a, \qquad \text{where} \qquad L = L_{a,b}. \tag{5}$$

One notes that the covariance between L and L_a is given by

$$\sigma_{L L_a} = \underline{\beta}' \Sigma_{XX} \begin{bmatrix} \underline{\beta}_a \\ \underline{0} \end{bmatrix} = \underline{\sigma}'_{XY} \begin{bmatrix} \underline{\beta}_a \\ \underline{0} \end{bmatrix} = \sigma^2_Y \rho^2_{Y L_a}.$$

Hence the variance of $L_{b|a}$ is

$$\begin{aligned} \sigma^2_{L_{b|a}} &= \sigma^2_L + \sigma^2_{L_a} - 2\sigma_{L L_a} \\ &= \sigma^2_Y \rho^2_{Y L} + \sigma^2_Y \rho^2_{Y L_a} - 2\sigma^2_Y \sigma^2_{Y L_a} \\ &= \sigma^2_Y(\rho^2_{Y L} - \rho^2_{Y L_a}). \end{aligned}$$

Consider now the random variable

$$U_{Y|a} = Y - L_a. \tag{6}$$

The variance of this variable is

$$\begin{aligned} \sigma^2_{U_{Y|a}} &= \sigma^2_Y + \sigma^2_{L_a} - 2\sigma_{Y L_a} \\ &= \sigma^2_Y + \sigma^2_Y \rho^2_{Y L_a} - 2\sigma^2_Y \rho^2_{Y L_a} \\ &= \sigma^2_Y(1 - \rho^2_{Y L_a}). \end{aligned}$$

The covariance between Y and $L_{b|a}$ is

$$\sigma_{YL_{b|a}} = \sigma_{YL} - \sigma_{YL_a} = \sigma_Y^2(\rho_{YL}^2 - \rho_{YL_a}^2).$$

Hence the squared product-moment correlation between Y and $L_{b|a}$ is

$$(7) \qquad \rho_{YL_{b|a}}^2 = \frac{(\sigma_{YL_{b|a}})^2}{\sigma_Y^2 \sigma_{L_{b|a}}^2} = \frac{[\sigma_Y^2(\rho_{YL}^2 - \rho_{YL_a}^2)]^2}{\sigma_Y^2[\sigma_Y^2(\rho_{YL}^2 - \rho_{YL_a}^2)]}$$
$$= \rho_{YL}^2 - \rho_{YL_a}^2 = \rho_{Y(L_b|L_a)}^2.$$

Thus, the squared semipartial correlation is the squared product-moment correlation between Y and $L - L_a$. The latter random variable is the difference between two linear functions of $\underline{X}$—that is,

$$L_{b|a} = L - L_a = \underline{\beta}'\underline{X} - \underline{\beta}_a'\underline{X}_a$$
$$= [\underline{\beta}' - (\underline{\beta}_a'\ \underline{0}')]\underline{X}.$$

One notes that

$$\underline{\beta}_a'\underline{X}_a = (\underline{\beta}_a'\ \underline{0}')\underline{X}.$$

The semipartial correlation is

$$\pm\sqrt{\rho_{YL_{b|a}}^2};$$

the algebraic sign is that of $\sigma_{YL_{b|a}}$.

The covariance between $U_{Y|a} = Y - L_a$ and $L_{b|a} = L - L_a$ is given by

$$\sigma_{U_{Y|a}L_{b|a}} = \sigma_{YL} - \sigma_{YL_a} - \sigma_{LL_a} + \sigma_{L_aL_a}$$
$$= \sigma_Y^2(\rho_{YL}^2 - \rho_{YL_a}^2 - \rho_{YL_a}^2 + \rho_{YL_a}^2)$$
$$= \sigma_Y^2(\rho_{YL}^2 - \rho_{YL_a}^2).$$

Hence the square of the product-moment correlation between $U_{Y|a}$ and $L_{b|a}$ is

$$(8) \qquad \rho_{U_{Y|a}L_{b|a}}^2 = \frac{(\sigma_{U_{Y|a}L_{b|a}})^2}{\sigma_{U_{Y|a}}^2\sigma_{L_{b|a}}^2} = \frac{[\sigma_Y^2(\rho_{YL}^2 - \rho_{YL_a}^2)]^2}{\sigma_Y^2(1 - \rho_{YL_a}^2)\sigma_Y^2(\rho_{YL}^2 - \rho_{YL_a}^2)}$$
$$= \frac{\rho_{YL}^2 - \rho_{YL_a}^2}{1 - \rho_{YL_a}^2} = \rho_{YL_b|L_a}^2.$$

Thus the partial correlation is the product-moment correlation between $Y - L_a$ and $L - L_a$. This correlation may be viewed as the correlation between two sets of residuals. $Y - L_a$ represents the residual in Y from a prediction system based upon the set $\underline{X}_a$. $L - L_a$ represents the difference between a prediction of Y based upon $\underline{X}$ (which includes $\underline{X}_a$ and $\underline{X}_b$) and a prediction of Y based upon $\underline{X}_a$ alone.

Suppose now one defines the random variables

$$U_{Y|b} = Y - L_b,$$
$$L_{a|b} = L - L_b = L_{a.b} - L_b.$$

By analogy with what has just been demonstrated,

$$\rho^2_{YL_a|_b} = \rho^2_{YL} - \rho^2_{YL_b} = \rho^2_{Y(L_a|L_b)}, \tag{9}$$

$$\rho^2_{U_{Y|b}L_a|_b} = \frac{\rho^2_{YL} - \rho^2_{YL_b}}{1 - \rho^2_{YL_b}} = \rho^2_{YL_a|L_b}. \tag{10}$$

2.4-3 Correlation ratio and related correlations

Consider the problem of predicting the random variable Y from the set of random variables $X_1, \ldots, X_p$. Let the predicted value of Y be defined by

$$\tilde{Y} = f(\underline{X}),$$

where $f(\underline{X})$ is some function of $X_1, \ldots, X_p$. The function need not be linear. Suppose it is desired to have a prediction system in which

$$E(Y - \tilde{Y})^2 = \text{minimum}.$$

If $f^*(\underline{X}) = \mu_{Y|X}$, then for $\underline{X} = \underline{X}_j$,

$$\underset{i}{E}(Y_{ij} - \mu_{Y|\underline{X}_j})^2 = \text{minimum} \qquad (j \text{ fixed}).$$

This last relationship follows from the fact that the sum of the squared deviations about the mean of a distribution is a minimum. Hence one has

$$\underset{j}{E}\left[\underset{i}{E}(Y_{ij} - \mu_{Y|X_j})^2\right] = \text{minimum},$$

since the sum over a series of local minima is an overall minimum. If one lets

$$f^*(\underline{X}_j) = \mu_{Y|\underline{X}_j} = M_j,$$

the last expectation may be written in the form

$$E(Y_j - M_j)^2 = \text{minimum}.$$

The function M_j is called the *regression* function of Y on $X_1, \ldots, X_p$. For the special case in which Y and $X_1, \ldots, X_p$ are jointly multivariate normal, the regression function is a linear function of $X_1, \ldots, X_p$. In general, the regression function need not be a linear function of the components of $\underline{X}$.

Let

$$f(\underline{X}_j) = f_j$$

be an arbitrary function of $\underline{X}_j$. Consider a finite population in which $\underline{X}$ can assume only the two values $\underline{X} = \underline{X}_1$ and $\underline{X} = \underline{X}_2$. The conditional distribution of Y for these values of $\underline{X}$ may be represented as follows:

$Y \mid \underline{X}_1$	$Y \mid \underline{X}_2$	
Y_{11}	Y_{12}	$N = N_1 + N_2$
$\cdot$	$\cdot$	
$\cdot$	$\cdot$	
$\cdot$	$\cdot$	
$\underline{Y_{N_1 1}}$	$\underline{Y_{N_2 2}}$	
$E(Y \mid \underline{X}_1) = M_1$	$E(Y \mid \underline{X}_2) = M_2$	$M = E(Y) = \dfrac{N_1 M_1 + N_2 M_2}{N}$
$f(\underline{X}_1) = f_1$	$f(\underline{X}_2) = f_2$	$f = E(f_j) = \dfrac{N_1 f_1 + N_2 f_2}{N}$

The variance of M_j is given by

$$\sigma_M^2 = \tfrac{1}{2} \sum_{i,j} (M_j - M)^2 = \frac{1}{2}\left[\sum_{i,j} M_j^2 - \frac{(\sum_{i,j} M_j)^2}{N}\right]$$

$$= \frac{1}{2}\left[N_1 M_1^2 + N_2 M_2^2 - \frac{(N_1 M_1 + N_2 M_2)^2}{N}\right]$$

$$= \tfrac{1}{2}[N_1 M_1^2 + N_2 M_2^2 - N M^2].$$

The covariance between Y_{ij} and f_j is given by

$$\text{(1)}\quad \operatorname{cov}(Y_{ij}, f_j) = \frac{1}{N}\left[f_1 N_1 M_1 + f_2 N_2 M_2 - \frac{(N_1 f_1 + N_2 f_2)(N_1 M_1 + N_2 M_2)}{N}\right]$$

$$= \frac{1}{N}[f_1 N_1 M_1 + f_2 N_2 M_2 - f N_1 M_1 - f N_2 M_2]$$

$$= \frac{1}{N}[(f_1 - f) N_1 M_1 + (f_2 - f) N_2 M_2]$$

$$= \frac{1}{N} \sum_{ij} (f_j - f) M_j.$$

One notes that the covariance between f_j and M_j is

$$\operatorname{cov}(f_j, M_j) = \frac{1}{N}\left[f_1 N_1 M_1 + f_2 N_2 M_2 - \frac{(N_1 f_1 + N_2 f_2)(N_1 M_1 + N_2 M_2)}{N}\right]$$

$$= \operatorname{cov}(Y_{ij}, f_j).$$

For the case of an infinite population, expression (1) generalizes to

$$\text{(2)}\qquad \operatorname{cov}(Y_{ij}, f_j) = E[(f_j - f) M_j] = \operatorname{cov}(M_j, f_j).$$

Equivalently,

$$\text{(3)}\qquad \sigma_{Yf} = \sigma_{Mf}.$$

When, in particular, $f_j = M_j$ [that is, when $f^*(X_j) = \mu_{Y|\underline{X}_j}$], then (3) becomes

$$\sigma_{YM} = \sigma_{MM} = \sigma_M^2, \tag{3a}$$

where $$\sigma_M^2 = \text{E}[M - \text{E}(M)]^2.$$

The square of the product-moment correlation between Y_{ij} and M_j is given by

$$\rho_{YM}^2 = \rho^2(Y_{ij}, M_j) = \frac{(\sigma_{YM})^2}{\sigma_Y^2 \sigma_M^2} = \frac{(\sigma_M^2)^2}{\sigma_Y^2 \sigma_M^2} = \frac{\sigma_M^2}{\sigma_Y^2}. \tag{4}$$

This squared product-moment correlation is called the *correlation ratio* and is denoted by the symbol η_{YX}^2.

Consider now the problem of finding the maximum squared product-moment correlation between Y_{ij} and f_j. One notes that

$$\rho^2(Y_{ij}, f_j) = \frac{(\sigma_{Yf})^2}{\sigma_Y^2 \sigma_f^2}. \tag{5}$$

By using (3), one may rewrite (5) in the form

$$\begin{aligned} \rho_{Yf}^2 = \rho^2(Y_{ij}, f_j) &= \frac{(\sigma_{Mf})^2}{\sigma_f^2 \sigma_M^2} \frac{\sigma_M^2}{\sigma_Y^2} \\ &= \rho_{Mf}^2 \rho_{YM}^2. \end{aligned} \tag{6}$$

Clearly, from (6),

$$\rho_{Yf}^2 = \rho_{Mf}^2 \rho_{YM}^2 \le \rho_{YM}^2 = \eta_{YX}^2. \tag{7}$$

When $f_j = M_j$, $\rho_{Mf}^2 = 1$ and ρ_{Yf}^2 will assume its maximum value, which is $\eta_{Y\underline{X}}^2$. Equivalently, if f_j is some linear function of M_j, then ρ_{Yf}^2 will also assume its maximum value. Thus the correlation ratio is the maximum possible squared product-moment correlation between Y and $f(\underline{X})$. This maximum is achieved if $f(\underline{X})$ is the regression function of Y on $X_1, \ldots, X_p$, that is, $f(\underline{X}) = f^*(\underline{X}) = \mu_{Y|X}$.

Let $$\underline{X} = [\underline{X}_a \quad \underline{X}_b]$$

where $\underline{X}_a' = [X_1 \cdots X_k]$ and $\underline{X}_b' = [X_{k+1} \cdots X_p]$.

Suppose one defines the following regression functions:

$$\begin{aligned} f^*(\underline{X}) &= M = \mu_{Y|\underline{X}}, \\ f^*(\underline{X}_a) &= M_a = \mu_{Y|\underline{X}_a}, \\ f^*(\underline{X}_b) &= M_b = \mu_{Y|\underline{X}_b}. \end{aligned}$$

The square of the maximum product-moment correlation between Y and any $f(\underline{X}_a)$, where $f(\underline{X}_a)$ is arbitrary, is given by

$$\eta^2_{Y\underline{X}_a} = \frac{\sigma^2_{M_a}}{\sigma^2_Y}. \tag{8}$$

Similarly, the square of the maximum product-moment correlation between Y and any $f(\underline{X}_b)$ is

$$\eta^2_{Y\underline{X}_b} = \frac{\sigma^2_{M_b}}{\sigma^2_Y}. \tag{9}$$

(8) and (9) represent the correlation ratios associated with prediction systems based upon subsets of the entire set.

The squared semipartial correlation ratio is defined by

$$\eta^2_{Y(\underline{X}_b|\underline{X}_a)} = \rho^2_{Y(M_b|M_a)} = \eta^2_{Y\underline{X}} - \eta^2_{Y\underline{X}_a} = \rho^2_{YM} - \rho^2_{YM_a}. \tag{10}$$

This is the square of the semipartial correlation ratio between Y and that part of $\underline{X}_b$ having $\underline{X}_a$ partialled out. The squared partial correlation ratio is

$$\eta^2_{Y\underline{X}_b|\underline{X}_a} = \rho^2_{Y(M_b|M_a)} = \frac{\eta^2_{Y\underline{X}} - \eta^2_{Y\underline{X}_a}}{1 - \eta^2_{Y\underline{X}_a}} = \frac{\rho^2_{YM} - \rho^2_{YM_a}}{1 - \rho^2_{YM_a}}. \tag{11}$$

By analogy,

$$\eta^2_{YX_a|X_b} = \rho^2_{Y(M_a|M_b)} = \frac{\eta^2_{Y\underline{X}} - \eta^2_{Y\underline{X}_b}}{1 - \eta^2_{Y\underline{X}_b}} = \frac{\rho^2_{YM} - \rho^2_{YM_b}}{1 - \rho^2_{YM_b}}. \tag{12}$$

The semipartial and partial correlation ratios may be related to squared product-moment correlations among residuals in the same way that corresponding semipartial and partial correlations were related to residuals. If one lets

$$M_{b|a} = M - M_a = f^*(\underline{X}) - f^*(\underline{X}_a),$$

$$W_{Y|a} = Y - M_a = Y - f^*(\underline{X}_a),$$

then it may be shown that

$$\eta^2_{Y(X_b|X_a)} = \rho^2(Y, M_{b|a}) = \rho^2_{YM} - \rho^2_{YM_a}. \tag{13}$$

That is, the squared semipartial correlation ratio is equal to the square of the product-moment correlation between Y and $M_{b|a}$. The squared partial correlation ratio is

$$\eta^2_{Y\underline{X}_b|\underline{X}_a} = \rho^2(W_{Y|a}, M_{b|a}) = \frac{\operatorname{cov}(W_{Y|a}, M_{b|a})}{\sigma^2_{W_{Y|a}}\sigma^2_{M_{b|a}}}. \tag{14}$$

The extreme right-hand side of (14) reduces to

$$\frac{\rho^2_{YM} - \rho^2_{YM_a}}{1 - \rho^2_{YM_a}} .$$

If one defines

$$M_{a|b} = M - M_b = f^*(\underline{X}) - f^*(\underline{X}_b),$$
$$W_{Y|b} = Y - M_b = Y - f^*(\underline{X}_b),$$

then one has the following squared semipartial and partial correlation ratios:

$$\eta^2_{Y(\underline{X}_a|\underline{X}_b)} = \rho^2(Y, M_{a|b}), \tag{15}$$

$$\eta^2_{Y\underline{X}_a|\underline{X}_b} = \rho^2(W_{Y|b}, M_{a|b}). \tag{16}$$

In order to obtain an expression for the correlation ratio which will be useful in tackling the estimation problem, define

$$\varepsilon^* = Y - M = Y - \mu_{Y|\underline{X}}.$$

The variance of ε^* is

$$\begin{aligned}
\sigma^2_{\varepsilon^*} &= \sigma^2_Y + \sigma^2_M - 2\sigma_{YM} \\
&= \sigma^2_Y + \sigma^2_M - 2\sigma^2_M \qquad (\text{since } \sigma^2_{YM} = \sigma^2_M) \\
&= \sigma^2_Y - \sigma^2_M \\
&= \sigma^2_Y - \sigma^2_Y \eta^2_{Y\underline{X}} = \sigma^2_Y(1 - \eta^2_{Y\underline{X}}).
\end{aligned}$$

Solving the last expression for $\eta^2_{Y\underline{X}}$ gives

$$\eta^2_{Y\underline{X}} = 1 - \frac{\sigma^2_{\varepsilon^*}}{\sigma^2_Y} . \tag{17}$$

Estimation. Since no assumptions have been made about the form of the distribution of the random variables, only a heuristic approach to the problem of estimation may be taken. This approach will lead to consistent estimators.

For the sample data, define

$$\text{SS}_{\text{within class}} = \Sigma(Y_{ij} - \bar{Y}_j)^2,$$

where $\bar{Y}_j$ is the mean of the subsample in which $\underline{X} = \underline{X}_j$. Also define

$$\text{SS}_Y = \Sigma(Y_{ij} - \bar{Y})^2,$$

where $\bar{Y}$ is the mean of the entire sample. Let $n = \Sigma n_j$, where n_j is the size of the subsample for which $\underline{X} = \underline{X}_j$. Let

$$\text{MS}_{\text{within class}} = \frac{\text{SS}_{\text{within class}}}{\Sigma n_j - k} ,$$

where k is the number of subsamples,

$$\mathrm{MS}_Y = \frac{\mathrm{SS}_Y}{\Sigma n_j - 1}.$$

One estimator of the correlation ratio is given by

$$e_{YX}^2 = 1 - \frac{\mathrm{SS}_{\text{within class}}}{\mathrm{SS}_Y}.$$

A second estimator of the correlation ratio is given by

$$\hat{\eta}_{YX}^2 = 1 - \frac{\mathrm{MS}_{\text{within class}}}{\mathrm{MS}_Y}.$$

Since

$$\frac{\mathrm{MS}_{\text{within class}}}{\mathrm{MS}_Y} = \frac{\Sigma n_j - 1}{\Sigma n_j - k}\,\frac{\mathrm{SS}_{\text{within class}}}{\mathrm{SS}_Y} = \frac{\Sigma n_j - 1}{\Sigma n_j - k}(1 - e_{YX}^2),$$

the first and second estimators are related as follows:

$$\hat{\eta}_{YX}^2 = 1 - \frac{\Sigma n_j - 1}{\Sigma n_j - k}(1 - e_{YX}^2).$$

The second estimator tends to be less biased than the first estimator. From the definitions, it follows that (for finite samples)

$$\hat{\eta}_{YX}^2 \le e_{YX}^2.$$

For example, if

$$n = 51 \qquad \mathrm{SS}_Y = 100$$

$$k = 11 \qquad \mathrm{SS}_{\text{within class}} = 30$$

then

$$e_{YX}^2 = 1 - \frac{30}{100} = .70,$$

$$\hat{\eta}_{YX}^2 = 1 - \frac{30/40}{100/50} = .62.$$

The difference between the square of the sample multiple correlation and the square of the sample correlation ratio is an index of the extent to which the relationship between Y and the best-fitting $f(\underline{X})$ deviates from a linear function. This index of "lack of linear fit" for the sample data is

$$e_{Y\underline{X}}^2 - r_{Y\underline{X}}^2 = e_{Y(1\ldots p)}^2 - r_{Y(1\ldots p)}^2.$$

Numerical Example. A numerical example of the computation of the correlation ratio is given in Table 2.4-1. It is assumed that random samples of size n_1 and n_2 have been drawn, respectively, from the conditional

Table 2.4-1 Numerical Example—Correlation Ratio

(i) Basic data:

	$Y \mid X_1$	$Y \mid X_2$		
	3	8		
	5	14		
$n_1 = 4$	7	10	$n_1 = 5$	
	9	12		
		16		
$\Sigma()$:	24	60		84
	T_1	T_2		G
Mean:	6	12		
	M_1	M_2		
$\Sigma()^2$:	164	760		924
	U_1	U_2		U

(ii)

$$SS_{\text{within class}} = U - \frac{T_1^2}{n_1} - \frac{T_2^2}{n_2}$$

$$= 924 - \frac{(24)^2}{4} - \frac{(60)^2}{5} = 60.00$$

$$SS_Y = U - \frac{G^2}{n_{..}} = 924 - \frac{(84)^2}{9} = 140.00$$

$$e_{YX}^2 = 1 - \frac{SS_{\text{within class}}}{SS_Y}$$

$$= 1 - \frac{60.00}{140.00} = .5714$$

(iii)

$$\Sigma M = n_1M_1 + n_2M_2 = 4(6) + 5(12) = 84$$

$$\Sigma M^2 = n_1M_1^2 + n_2M_2^2 = 4(36) + 5(144) = 864$$

$$\Sigma YM = T_1M_1 + T_2M_2 = 24(6) + 60(12) = 864$$

$$SS_M = \Sigma M^2 - \frac{(\Sigma M)^2}{n_{..}} = 864 - \frac{(84)^2}{9} = 80.00$$

$$SP_{YM} = \Sigma YM - \frac{(\Sigma Y)(\Sigma M)}{n_{..}} = 864 - \frac{(84)(84)}{9} = 80.00$$

$$r_{YM}^2 = \frac{SP_{YM}^2}{SS_Y SS_M} = .5714$$

distributions of Y corresponding to $\underline{X} = \underline{X}_1$ and $\underline{X} = \underline{X}_2$. In part i are the basic observations. For these data,

$$\Sigma(Y \mid \underline{X}_1) = T_1 = 24, \qquad M_1 = \bar{Y}_1 = 6;$$

$$\Sigma(Y \mid \underline{X}_2) = T_2 = \underline{60}, \qquad M_2 = \bar{Y}_2 = 12;$$

$$\Sigma Y = G = 84.$$

Also,

$$\Sigma(Y^2 \mid \underline{X}_1) = U_1 = 164,$$

$$\Sigma(Y^2 \mid \underline{X}_2) = U_2 = \underline{760},$$

$$\Sigma Y^2 = U = 924.$$

In part ii the correlation ratio, e^2_{YX}, as obtained from the sample data is computed.

The squared product-moment correlation between Y and M is computed in part iii. One notes that

$$\Sigma M_{ij} = \Sigma Y_{ij} = 84,$$

$$\Sigma Y_{ij} M_{ij} = \Sigma M^2_{ij} = 864.$$

The equalities on the left hold in general. From these relationships it follows that

$$\text{SS}_M = \text{SP}_{YM},$$

or in this case

$$\text{SS}_M = \text{SP}_{YM} = 864.$$

Hence

$$r^2_{YM} = \frac{\text{SP}^2_{YM}}{\text{SS}_Y \text{SS}_M} = \frac{\text{SS}^2_M}{\text{SS}_Y \text{SS}_M} = \frac{\text{SS}_M}{\text{SS}_Y}$$

$$= \frac{80.00}{140.00} = .5714.$$

2.5 Dwyer and SWP Algorithms for the Inverse of a Symmetric Matrix

The solution to a set of normal equations generally requires finding the inverse of a symmetric matrix. For matrices of high order (say 10×10 or higher) the computations become relatively difficult without the use of an electronic computer. For relatively small matrices, an algorithm suggested by Dwyer (among others) is readily adapted to the desk calculator. This algorithm is based upon the following principles.

Let M be a symmetric nonsingular matrix. This matrix may be factored into two triangular matrices:

$$M = TT',$$

where T is a lower triangular matrix. (A lower triangular matrix is a square matrix having zeros everywhere above the main diagonal.) The inverse of the matrix M may be expressed in the form

$$M^{-1} = (TT')^{-1} = (T')^{-1}T^{-1}$$
$$= U'U, \qquad \text{where} \qquad U = T^{-1}.$$

The inverse of the matrix T is relatively easy to obtain. The Dwyer algorithm obtains the T and U matrices simultaneously. The inverse of a lower triangular matrix will be a lower triangular matrix.

Before going into the details of the algorithm, consider the problem of finding the matrix T for the case in which M is a 2×2 matrix. For this case,

$$\underset{T}{\begin{bmatrix} t_{11} & 0 \\ t_{21} & t_{22} \end{bmatrix}} \underset{T'}{\begin{bmatrix} t_{11} & t_{21} \\ 0 & t_{22} \end{bmatrix}} = \underset{M}{\begin{bmatrix} m_{11} & m_{12} \\ m_{21} & m_{22} \end{bmatrix}} .$$

Hence

$$t_{11}^2 + 0 = m_{11} \qquad \text{or} \qquad t_{11} = \sqrt{m_{11}},$$

$$t_{11}t_{21} + 0 = m_{12} \qquad \text{or} \qquad t_{21} = \frac{m_{12}}{t_{11}},$$

$$t_{21}^2 + t_{22}^2 = m_{22} \qquad \text{or} \qquad t_{22} = \sqrt{m_{22} - t_{21}^2}.$$

To find U, the inverse of T,

$$\underset{T}{\begin{bmatrix} t_{11} & 0 \\ t_{21} & t_{22} \end{bmatrix}} \underset{U}{\begin{bmatrix} u_{11} & u_{12} \\ u_{21} & u_{22} \end{bmatrix}} = \underset{I}{\begin{bmatrix} 1 & 0 \\ 0 & 1 \end{bmatrix}}.$$

Hence

$$t_{11}u_{11} = 1 \qquad \text{or} \qquad u_{11} = \frac{1}{t_{11}},$$

$$t_{11}u_{12} = 0 \qquad \text{or} \qquad u_{12} = 0,$$

$$t_{21}u_{11} + t_{22}u_{21} = 0 \qquad \text{or} \qquad u_{21} = \frac{-t_{21}u_{11}}{t_{22}},$$

$$t_{21}u_{12} + t_{22}u_{22} = 1 \qquad \text{or} \qquad u_{22} = \frac{1}{t_{22}}.$$

Details in the computations are given in Table 2.5-1. A numerical example for the case of a 3×3 matrix appears in Table 2.5-2. At the left of part i of Table 2.5-2 is the upper half of the entries in a 3×3 symmetric matrix. At the right of part i is an identity matrix of the same order as M; only the lower half of this identity matrix is recorded. The

Table 2.5-1 Dwyer Algorithm for the Inverse of a Symmetric Matrix

(i)				M	I			
				T'	U			
					$M^{-1} = U'U$			
(ii)		m_{11}	m_{12}	m_{13}	d_{11}			
			m_{22}	m_{23}	d_{21}	d_{22}		
				m_{33}	d_{31}	d_{32}	d_{33}	Check
	t_1	t_{11}	t_{12}	t_{13}	u_{11}			$m_{11}u_{11} = t_{11}$
	$t_{2.1}$		t_{22}	t_{23}	u_{21}	u_{22}		$m_{11}\mu_{21} + m_{12}u_{22} = 0$
	$t_{3.12}$			t_{33}	u_{31}	u_{32}	u_{33}	$m_{11}u_{31} + m_{12}u_{32} + m_{13}u_{33} = 0$

(iii)

$$t_{11} = \sqrt{m_{11}}; \qquad t_{1j} = m_{1j}/t_{11}; \qquad u_{11} = d_{11}/t_{11} \qquad j = 2, 3$$

$$t_{22} = \sqrt{m_{22} - t_{12}^2}; \qquad t_{2j} = (m_{2j} - t_{12}t_{1j})/t_{22}; \qquad j = 3$$

$$u_{2k} = (d_{2k} - t_{12}u_{2k})/t_{22} \qquad k = 1, 2$$

$$t_{33} = \sqrt{m_{33} - t_{13}^2 - t_{23}^2}; \qquad u_{3k} = (d_{3k} - t_{13}u_{1k} - t_{23}u_{2k})/t_{33} \quad k = 1, 2, 3$$

(iv)

$$t_{pp} = \sqrt{m_{pp} - t_{1p}^2 - t_{2p}^2 - \cdots - t^2_{(p-1)p}}$$

$$t_{pj} = (m_{pj} - t_{1p}t_{1j} - t_{2p}t_{2j} - \cdots - t_{(p-1)p}t_{(p-1)j})/t_{pp} \qquad j > p$$

$$u_{pk} = (d_{pk} - t_{1p}u_{1k} - t_{2p}u_{2k} - \cdots - t_{(p-1)p}u_{(p-1)k})/t_{pp} \qquad k \leq p$$

Table 2.5-2 Numerical Example of Dwyer Algorithm

		1	2	3	1	2	3	
(i)	M	16	8	4	1			
			29	12	0	1		
				86	0	0	1	Check
(ii)	T'	4.0000	2.0000	1.0000	.2500			4.0000
			5.0000	2.0000	−.1000	.2000		.0000
				9.0000	−.005555	−.04444	.1111	0000
(iii)					.07253	−.01975	−.0006173	1.0000
					−.01975	.041975	−.004938	.0000
					−.0006173	−.004938	.012346	.0000

computational formulas for obtaining the entries in the T matrix in part ii of Table 2.5-2 are given in part iii of Table 2.5-1. Thus,

$$t_{11} = \sqrt{m_{11}} = \sqrt{16} = 4.0000,$$

$$t_{12} = \frac{m_{12}}{t_{11}} = \frac{8}{4.0000} = 2.0000,$$

$$t_{13} = \frac{m_{13}}{t_{11}} = \frac{4}{4.0000} = 1.0000.$$

The first entry in the U matrix (at the right of part ii) is

$$u_{11} = \frac{d_{11}}{t_{11}} = \frac{1}{4} = .2500.$$

The entries in the second row of the matrices in part ii of Table 2.5-2 are given by

$$t_{22} = \sqrt{m_{22} - t_{12}^2} = \sqrt{29 - (2.0000)^2} = \sqrt{25.0000} = 5.0000,$$

$$t_{23} = \frac{m_{23} - t_{12}t_{13}}{t_{22}} = \frac{12 - (2.0000)(1.0000)}{5.0000} = 2.0000,$$

$$u_{21} = \frac{d_{21} - t_{12}u_{13}}{t_{22}} = \frac{0 - (2.0000)(.2500)}{5.0000} = -.1000,$$

$$u_{22} = \frac{1}{t_{22}} = \frac{1}{5.0000} = .2000.$$

As a partial check on the computations,

$$m_{11}u_{21} + m_{12}u_{22} = 16(-.1000) + 8(.2000) = 0.0000 \text{ (within rounding error).}$$

The entries in the third row of part ii of Table 2.5-2 are obtained as follows:

$$t_{33} = \sqrt{m_{33} - t_{13}^2 - t_{23}^2} = \sqrt{86 - (1)^2 - (2)^2} = 9.0000,$$

$$u_{31} = \frac{d_{31} - t_{13}u_{11} - t_{23}u_{21}}{t_{33}}$$

$$= \frac{0 - (1)(.2500) - (2)(-1.000)}{9} = -.005555,$$

$$u_{32} = \frac{d_{32} - t_{13}u_{12} - t_{23}u_{22}}{t_{33}}$$

$$= \frac{0 - (1)(0) - (2)(.2000)}{9} = -.04444,$$

$$u_{33} = \frac{1}{t_{33}} = .1111.$$

It is noted that an entry of the form u_{kk} will always be $1/t_{kk}$. Again as a partial check on the computations,

$$m_{11}u_{31} + m_{12}u_{32} + m_{13}m_{33} = 0 \qquad \text{(within rounding error).}$$

As another computational check, the sum of the squares of the entries in each column of the T matrix should be the entry in the corresponding main diagonal of the matrix M. For example,

$$(2.0000) + (5.0000)^2 = 29,$$
$$(1.0000)^2 + (2.0000)^2 + (9.0000)^2 = 86.$$

The inverse of the matrix M is shown in part iii of Table 2.5-2. It is obtained by premultiplying the matrix at the right of part ii by its transpose. This can be done conveniently as a column-by-column multiplication. For example, column 1 times itself is

$$(.2500)^2 + (-.1000)^2 + (-.005555)^2 = .07253.$$

Column 1 times column 2 is

$$(-.1000)(.2000) + (-.005555)(-.04444) = -.01975.$$

A final check is given by

$$M^{-1}M = I \qquad \text{(within rounding error).}$$

The general computational formula for any diagonal element of the T matrix appears in the first row of part iv in Table 2.5-1. A similar formula for any off-diagonal element of the T matrix appears in the second row. In the third row is the general formula for any element in the U matrix. With regard to the latter formula it may be noted that

$$d_{pk} = 0 \qquad \text{if } p \neq k,$$
$$d_{pk} = 1 \qquad \text{if } p = k,$$
$$u_{jk} = 0 \qquad \text{if } j < k.$$

Further u_{kk} will always have the form $1/t_{kk}$.

It should be noted that rounding error cumulates very rapidly in this type of computation. If four-significant-digit accuracy is wanted in the inverse at least six significant digits should be carried at all intermediate states of the calculations. It should also be noted that, if the matrix M is singular, a diagonal entry of the T matrix will become zero. Hence the algorithm will collapse since it requires division by such diagonal elements.

If the matrix M is a covariation matrix, the elements of the matrix T will turn out to be of interest in their own right—quite apart from being a step along the way of getting the inverse matrix. The elements on the main diagonal of the T matrix will be found to be partial variations, and the elements of the main diagonal will be related to partial covariations.

SWP Operator. What may be considered a modified version of the Dwyer algorithm is illustrated in Table 2.5-3. At the right of part i, m_{ij} is an element of the matrix M, and n_{ij} is an element in the matrix $M_{(1)}$.

Table 2.5-3 SWP Algorithm

(i)	M	16 8 4	8 29 12	4 12 86	$m_{11} = 16 \quad n_{11} = -1/16$ $n_{i1} = m_{i1}/16 \quad n_{1j} = m_{1j}/16$ $n_{ij} = m_{ij} - (m_{i1}m_{1j}/16)$
(ii)	$M_{(1)}$	−.06250 .5000 .2500	.5000 25.0000 10.0000	.2500 10.0000 85.0000	$m_{22} = 25 \quad n_{22} = -1/25$ $n_{i2} = m_{i2}/25 \quad n_{2j} = m_{2j}/25$ $n_{ij} = m_{ij} - (m_{i2}m_{2j}/25)$
(iii)	$M_{(12)}$	−.07250 .02000 .05000	.02000 −.04000 .4000	.05000 .4000 81.0000	$m_{33} = 81 \quad n_{33} = -1/81$ $n_{i3} = m_{i3}/81 \quad n_{3j} = m_{3j}/81$ $n_{ij} = m_{ij} - (m_{i3}m_{3j}/81)$
(iv)	$M_{(123)}$	−.07253 .01975 .0006173	.01975 −.041975 .004938	.0006173 .004938 −.012345	$M^{-1} = -M_{(123)}$
(v)		$M_{(1)} = \text{SWP}[1]\, M$ $M_{(12)} = \text{SWP}[2]\, M_{(1)} = \text{SWP}[1,2]\, M$ $M_{(123)} = \text{SWP}[3]\, M_{(2)} = \text{SWP}[1,2,3]\, M$			

The elements of $M_{(1)}$ are obtained from the elements of the matrix M by means of the following set of operations:

$$n_{11} = \frac{-1}{m_{11}}, \qquad n_{i1} = \frac{m_{i1}}{m_{11}}, \qquad n_{1j} = \frac{m_{1j}}{m_{11}},$$

$$n_{ij} = m_{ij} - \frac{m_{i1}m_{1j}}{m_{11}}, \qquad \text{for } i, j \neq 1.$$

These operations are indicated at the right of part i.

If now the elements of $M_{(1)}$ are denoted m_{ij}, and the elements of $M_{(12)}$ are denoted n_{ij}, then the elements of $M_{(12)}$ are obtained from the elements of $M_{(1)}$ by means of the following set of operations:

$$n_{22} = \frac{-1}{m_{22}}, \qquad n_{i2} = \frac{m_{i2}}{m_{22}}, \qquad n_{2j} = \frac{m_{2i}}{m_{22}},$$

$$n_{ij} = m_{ij} - \frac{m_{i2}m_{2j}}{m_{22}}, \qquad \text{for } i, j \neq 2.$$

These operations are indicated at the right of part ii. A similar set of operations is used to obtain $M_{(123)}$ from $M_{(12)}$, as illustrated in part iii. It will be found that

$$-M_{(123)} = M^{-1}.$$

The set of operations by which $M_{(1)}$ is obtained from M defines the matrix operation SWP[1], called the *sweep-out* operation, applied to row and column 1. In symbols,

$$\text{SWP}[1]\, M = M_{(1)}.$$

The set of operations by which $M_{(12)}$ is obtained from $M_{(1)}$ may be used to define the matrix operation

$$\text{SWP}[2]\, M_{(1)} = M_{(12)}.$$

Similarly,

$$\text{SWP}[3]\, M_{(12)} = M_{(123)}.$$

The following matrix operations are readily verified:

$$\begin{aligned}\text{SWP}[1,2]\, M &= \text{SWP}[2]\, M_{(1)} \\ &= \text{SWP}[1]\, M_{(2)},\end{aligned}$$

where

$$M_{(2)} = \text{SWP}[2]\, M.$$

Also

$$\begin{aligned}\text{SWP}[2,3]\, M &= \text{SWP}[3]\, M_{(2)} \\ &= \text{SWP}[2]\, M_{(3)},\end{aligned}$$

where

$$M_{(3)} = \text{SWP}[3]\, M.$$

A very useful property of the SWP operator is the following: Suppose the symmetric matrix M is partitioned

$$\underset{p,p}{M} = \begin{matrix} & n & p-n \\ n & \multicolumn{2}{c}{\multirow{2}{*}{}} \end{matrix} \begin{bmatrix} M_{11} & M_{12} \\ M_{21} & M_{22} \end{bmatrix}.$$

It may be verified that

$$\text{SWP}[1 \cdots n]\, M = \begin{bmatrix} -M_{11}^{-1} & M_{11}^{-1}M_{12} \\ M_{21}M_{11}^{-1} & M_{22} - M_{21}M_{11}^{-1}M_{12} \end{bmatrix}.$$

This last relationship may be illustrated by the numerical example in Table 2.5-3. In part iii the 2×2 matrix in the upper left-hand corner is the inverse of the corresponding matrix in M, except for algebraic sign. That is,

$$\begin{bmatrix} 16 & 8 \\ 8 & 29 \end{bmatrix}^{-1} = -\begin{bmatrix} -.0725 & .0200 \\ .0200 & -.0400 \end{bmatrix}.$$

2.6 Transformations Yielding Uncorrelated Variables

The discussion in this section is cast in terms of a problem in which $p = 3$; however, the generalization to arbitrary p is quite direct.

Consider the variables X_1, X_2, X_3. No assumptions will be made about the form of the joint distribution. Consider the following prediction equations in which the weights are defined by the least-squares criterion.

$$(1) \qquad \hat{X}_2 = b_{20} + b_{21}X_1,$$
$$\hat{X}_3 = b_{30} + b_{31}X_1 + b_{32}X_2.$$

Define a new set of variables as follows:

$$(2) \qquad \begin{aligned} X_1 &= X_1, \\ X_{2.1} &= X_2 - \hat{X}_2, \\ X_{3.12} &= X_3 - \hat{X}_3. \end{aligned}$$

Thus, variable $X_{2.1}$ represents the residual from the prediction equation in which X_2 is predicted from X_1; equivalently, $X_{2.1}$ represents that part of X_2 from which X_1 has been partialled out. Similarly, the variable $X_{3.12}$ represents that part of X_3 from which X_1 and X_2 have been partialled out.

The variables in the set (2) will be uncorrelated. To show this,

$$r_{X_1X_{2.1}} = 0,$$

since $X_{2.1}$ has all information that is a linear function of X_1 partialled out. Similarly,

$$r_{X_1X_{3.12}} = 0,$$

since the linear information on X_1 (as well as X_2) has been partialled out of $X_{3.12}$.

Consider now a prediction system in which Y is to be predicted from variables in the set (2). For the *standardized* variables the linear prediction system has the form

$$(3) \qquad \hat{Y}^* = b_1^*X_1^* + b_{2.1}^*X_{2.1}^* + b_{3.12}^*X_{3.12}^*.$$

The vector of weights $\underline{b}^*$ is obtained from the solution to the normal equations

$$R\underline{b}^* = \underline{r}$$

or

$$\begin{bmatrix} 1 & 0 & 0 \\ 0 & 1 & 0 \\ 0 & 0 & 1 \end{bmatrix} \begin{bmatrix} b_1^* \\ b_{2.1}^* \\ b_{3.12}^* \end{bmatrix} = \begin{bmatrix} r_{Y1} \\ r_{Y(2.1)} \\ r_{Y(3.12)} \end{bmatrix}.$$

Hence

$$\begin{bmatrix} b_1^* \\ b_{2.1}^* \\ b_{3.12}^* \end{bmatrix} = \begin{bmatrix} r_{Y1} \\ r_{Y(2.1)} \\ r_{Y(3.12)} \end{bmatrix}.$$

The square of the multiple correlation associated with the prediction system given in (3) is

$$r_{Y(123)}^2 = \underline{b}^{*\prime}\underline{r} = r_{Y1}^2 + r_{Y(2.1)}^2 + r_{Y(3.12)}^2.$$

It should be noted that the set of variables in (2) is equivalent to the set X_1, X_2, and X_3 insofar as problems of linear prediction are concerned. This is so because the set in (2) contains all the linear information in the original set in the sense that

X_2 is completely predictable from X_1 and $X_{2.1}$,
X_3 is completely predictable from X_1, $X_{2.1}$, and $X_{3.12}$.

Hence the multiple correlation associated with the prediction equation in (3) is the same as the multiple correlation associated with the prediction system

$$\hat{Y}^* = b_1^* X_1^* + b_2^* X_2^* + b_3^* X_3^*.$$

The correlation between Y and the variable $X_{2.1}$, denoted $r_{Y(2.1)}$, is called a semipartial (or part) correlation and measures the correlation between Y and that part of X_2 from which X_1 has been partialled out. Similarly, $r_{Y(3.12)}$ denotes the product-moment correlation between Y and $X_{3.12}$. In terms of these semipartial correlations one has

$$r_{Y(12)}^2 = r_{Y1}^2 + r_{Y(2.1)}^2,$$

$$r_{Y(123)}^2 = r_{Y(12)}^2 + r_{Y(3.12)}^2.$$

If one had a set of four variables,

$$r_{Y(1234)}^2 = r_{Y(123)}^2 + r_{Y(4.123)}^2.$$

From a computational point of view, semipartial correlations are readily obtained from the Dwyer algorithm for computing the inverse of a symmetric matrix. Suppose the matrix M in this algorithm were the following:

$$\begin{bmatrix} 1 & r_{12} & r_{13} & r_{Y1} \\ r_{21} & 1 & r_{23} & r_{Y2} \\ r_{31} & r_{32} & 1 & r_{Y3} \\ r_{Y1} & r_{Y2} & r_{Y3} & 1 \end{bmatrix}$$

The T' matrix in the algorithm will turn out to be

$$T' = \begin{bmatrix} 1 & r_{12} & r_{13} & r_{Y1} \\ 0 & r_{2(2.1)} & r_{3(2.1)} & r_{Y(2.1)} \\ 0 & 0 & r_{3(3.12)} & r_{Y(3.12)} \\ 0 & 0 & 0 & r_{Y(Y.123)} \end{bmatrix}.$$

The entries below the main diagonal are actually semipartial correlations of the form $r_{1(2.1)}$, $r_{1(3.12)}$, $r_{2(3.12)}$, $r_{2(3.12)}$, etc., which are all equal to zero.

The squared correlation $r_{Y(Y.123)}^2$ represents the proportion of the variation in Y that cannot be predicted from a linear function X_1, X_2, and

X_3. Hence the proportion of the variation of Y that can be predicted from this set of variables is

$$r^2_{Y(123)} = 1 - r^2_{Y(Y.123)}.$$

If the M matrix in the Dwyer algorithm is the covariation matrix, including a Y variable, then the entries in the Y column of the T matrix would be the following:

$$u_1 = \frac{\text{SP}_{Y1}}{\sqrt{\text{SS}_1}},$$

$$u_{2.1} = \frac{\text{SP}_{Y(2.1)}}{\sqrt{\text{SS}_{2.1}}},$$

$$u_{3.12} = \frac{\text{SP}_{Y(3.12)}}{\sqrt{\text{SS}_{3.12}}},$$

$$u_{Y.123} = \sqrt{\text{SS}_{Y.123}}.$$

These entries may be converted into correlations as follows:

$$r_{Y1} = \frac{u_1}{\sqrt{\text{SS}_Y}} = \frac{\text{SP}_{Y1}}{\sqrt{\text{SS}_Y}\sqrt{\text{SS}_1}},$$

$$r_{Y(2.1)} = \frac{u_{2.1}}{\sqrt{\text{SS}_Y}} = \frac{\text{SP}_{Y(2.1)}}{\sqrt{\text{SS}_Y}\sqrt{\text{SS}_{2.1}}},$$

$$r_{Y(3.12)} = \frac{u_{3.12}}{\sqrt{\text{SS}_Y}} = \frac{\text{SP}_{Y(3.12)}}{\sqrt{\text{SS}_Y}\sqrt{\text{SS}_{3.12}}},$$

$$r_{Y(Y.123)} = \frac{u_{Y.123}}{\sqrt{\text{SS}_Y}} = \frac{\text{SP}_{Y(Y.123)}}{\sqrt{\text{SS}_Y}\sqrt{\text{SS}_{Y.123}}}.$$

One will find that

$$\text{SS}_Y = \underbrace{u_1^2 + u_{2.1}^2 + u_{3.12}^2}_{} + u_{Y.123}^2,$$

$$= \quad \text{SS}_{\hat{Y}} \quad + \text{SS}_{Y-\hat{Y}}.$$

That is,

$$R(X_1,X_2,X_3) = u_1^2 \quad + u_{2.1}^2 \quad + u_{3.12}^2$$
$$= R(X_1) + R(X_2 \mid X_1) + R(X_3 \mid X_1,X_2),$$

where $R(X_1)$ = predictable variation due to X_1,

$R(X_2 \mid X_1)$ = increase in predictable variation due to adding X_2 to a prediction system already containing X_1,

$R(X_3 \mid X_1,X_2)$ = increase in predictable variation due to adding X_3 to a prediction system already containing X_1 and X_2.

Numerical Example. A numerical example of what has been discussed in this section is presented in Table 2.6-1. Basic data on the variables X_1, X_2, and X_3 are shown at the left of part i for a sample of size $n = 5$. The transformed variables are shown at the right. The prediction equations used to obtain the transformed data are given at the right of part ii.

To obtain the first prediction equation, one has

$$b_{21} = \frac{\text{SP}_{21}}{\text{SS}_1} = \frac{12.00}{40.00} = .30,$$

$$b_{20} = \bar{X}_2 - b_{21}\bar{X}_1 = 4.40 - .30(7.00) = 2.30.$$

To obtain the first entry in the column $X_{2.1}$ in part i,

$$X_{2.1} = X_2 - \hat{X}_2 = 4 - [2.30 + .30(3.00)] = .80.$$

The second prediction equation in part i is obtained from

$$\begin{bmatrix} b_{31} \\ b_{32} \end{bmatrix} = \begin{bmatrix} 40.00 & 12.00 \\ 12.00 & 11.20 \end{bmatrix}^{-1} \begin{bmatrix} 38.00 \\ 1.20 \end{bmatrix} = \begin{bmatrix} 1.35 \\ -1.34 \end{bmatrix},$$

$$b_{30} = \bar{X}_3 - b_{31}\bar{X}_1 - b_{32}\bar{X}_3 = 1.85.$$

It is readily verified that the variables X_1, $X_{2.1}$, and $X_{3.12}$ have zero covariation. For example, since $\Sigma X_{2.1} = 0$,

$$\text{SP}_{1(2.1)} = \Sigma X_1 X_{2.1} = 3(.80) + \cdots + 11(.40) = 0.00.$$

If one starts the Dwyer algorithm with the covariation matrix at the left of part ii, the resulting T' matrix is shown at the left of part iii. The entries on the main diagonal of this T' matrix are the square roots of the variations of the transformed variables. For example, by direct computation from part i one has

$$\text{SS}_{2.1} = (.80)^2 + (-1.80)^2 + \cdots + (.40)^2 = 7.60,$$

$$\sqrt{\text{SS}_{2.1}} = \sqrt{7.60} = 2.7568.$$

This last entry is the second entry on the main diagonal of the T' matrix in part iii. Similarly,

$$\text{SS}_{3.12} = (.46)^2 + (-.92)^2 + \cdots + (-.66)^2 = 3.4108,$$

$$\sqrt{\text{SS}_{3.12}} = \sqrt{3.4108} = 1.8468.$$

The correlations at the right of part iii are obtained by dividing the entries in the Y column of the T' matrix by $\sqrt{\text{SS}_Y}$. Thus,

$$r_{Y(3.12)} = \frac{6.6694}{\sqrt{2800}} = .1260.$$

That part of the total variation of Y which is predictable from a linear function of the variables X_1, X_2, and X_3 (or, equivalently, from the

Table 2.6-1 Numerical Example

		Original				Transformed		
	Element	X_1	X_2	X_3	Y	X_1	$X_{2.1}$	$X_{3.12}$
(i)	1	3	4	1	30	3	.80	.46
	2	5	2	5	10	5	−1.80	−.92
	3	7	6	3	30	7	1.60	−.26
	4	9	4	10	50	9	−1.00	1.36
	5	11	6	8	80	11	.40	−.66
	Σ()	35	22	27	200	35	.00	−.02
	Mean	7.00	4.40	5.40		7.00	.00	.00

(ii) Covariation matrix:

$$
\begin{array}{cccc}
1 & 2 & 3 & Y \\
\end{array}
$$

$$
\begin{bmatrix}
40.00 & 12.00 & 38.00 & 280.00 \\
 & 11.20 & 1.20 & 120.00 \\
 & & 53.20 & 230.00 \\
\text{Symmetric} & & & 2800.00
\end{bmatrix}
$$

Prediction equations:

$$\hat{X}_2 = 2.30 + .30X_1$$

$$\hat{X}_3 = 1.85 + 1.35X_1 - 1.34X_2$$

(iii) T' matrix in Dwyer algorithm:

$$
\begin{array}{cccc}
1 & 2 & 3 & Y \\
\end{array}
$$

$$
\begin{bmatrix}
6.3246 & 1.8974 & 6.0083 & 44.2719 \\
0 & 2.7568 & -3.7000 & 13.0582 \\
0 & 0 & 1.8467 & 6.6694 \\
0 & 0 & 0 & 25.0000
\end{bmatrix}
\qquad
\begin{array}{l}
.8367 = r_{Y1} \\
.2468 = r_{Y(2.1)} \\
.1260 = r_{Y(3.12)} \\
.4725 = r_{Y(Y.123)}
\end{array}
$$

(iv) Correlation matrix:

$$
\begin{array}{cccc}
1 & 2 & 3 & Y \\
\end{array}
$$

$$
\begin{bmatrix}
1.00000 & .56696 & .82377 & .83666 \\
 & 1.00000 & .04916 & .67764 \\
\text{Symmetric} & & 1.00000 & .59593 \\
 & & & 1.00000
\end{bmatrix}
$$

$$
\begin{bmatrix}
1.00000 & .56696 & .82377 & .83666 \\
0 & .82374 & -.50730 & .24679 \\
0 & 0 & .25308 & .12609 \\
0 & 0 & 0 & .47244
\end{bmatrix}
\begin{array}{l}
= r_{Y1} \\
= r_{Y(2.1)} \\
= r_{Y(3.12)} \\
= r_{Y(Y.123)}
\end{array}
$$

(v)

$$
\begin{aligned}
r^2_{Y(123)} &= r^2_{Y1} + r^2_{Y(2.1)} + r^2_{Y(3.12)} = .7768 \\
&= 1 - r^2_{Y(Y.123)} \qquad\qquad = .7768
\end{aligned}
$$

variables X_1, $X_{2.1}$, and $X_{3.12}$) may be obtained from the Y column of the T' matrix in part iii.

$$R(X_1,X_2,X_3) = \text{SS}_{\hat{Y}} = (44.2719)^2 + (13.0582)^2 + (6.6694)^2 = 2175$$
$$= R(X_1) + R(X_2 \mid X_1) + R(X_3 \mid X_1,X_2).$$

That part of the total variation of Y which is not predictable is

$$\text{SS}_{\text{error}} = \text{SS}_{Y-\hat{Y}} = (25.0000)^2 = 625.00.$$

Hence $$r^2_{Y(123)} = 1 - \frac{\text{SS}_{\text{error}}}{\text{SS}_Y} = 1 - \frac{625.00}{2800.00} = .7768.$$

In the upper half of part iv is the correlation matrix for X_1, X_2, X_3, and Y. The corresponding T' matrix is in the lower half of part iv. On the main diagonal of this T' matrix are semipartial correlations of the form

$$r_{2(2.1)}, \qquad r_{3(3.12)}, \qquad r_{Y(3.12)}.$$

This may be verified by computing these correlations directly from what appears in part i. For example,

$$\Sigma X_2 X_{2.1} = 7.60, \qquad \text{SS}_{X_2} = 11.20, \qquad \text{SS}_{X_{2.1}} = \Sigma X^2_{2.1} = 7.60.$$

Hence $$r_{X_2 X_{2\cdot 1}} = r_{2(2.1)} = \frac{\text{SP}_{X_2 X_{2\cdot 1}}}{\sqrt{\text{SS}_{X_2}}\sqrt{\text{SS}_{X_{2.1}}}} = \frac{7.60}{\sqrt{11.20}\sqrt{7.60}}$$
$$= .82.$$

2.6-1 Gram-Schmidt orthogonalization process

The transformation defined by (2) is known as the Gram-Schmidt orthogonalization process. The algebraic manipulation involved may be simplified in terms of matrix operations. Let

$\underset{p,p}{M}$ = covariance matrix for original variables $X_1, \ldots, X_p$;

$\underset{n,p}{X}$ = matrix of observations on original variables scaled so that $\frac{1}{n-1} X'X = M$;

$\underset{n,p}{Z}$ = matrix of observations in terms of the transformed variables $Z_1, \ldots, Z_p$.

In terms of the Dwyer algorithm, the matrices T and U are defined as

$\underset{p,p}{T}$ = lower triangular matrix such that $M = TT'$;

$\underset{p,p}{U}$ = lower triangular matrix such that $U = T^{-1}$.

Hence $$U(M)U' = U(TT')U' = I.$$

Under the Gram-Schmidt orthogonalization process,

$$\underset{n,p}{Z} = \underset{n,p}{X} \; \underset{p,p}{U'} \tag{4}$$

defines the transformed set of variables $Z_1, \ldots, Z_p$. The matrix U' is called the matrix of the transformation. The covariance matrix for the transformed set is

$$\frac{1}{n-1} Z'Z = \frac{1}{n} UX'XU' = UMU' = I.$$

Except for a scaling factor the transformation defined by (4) is equivalent to the transformation defined by (2). A set of variables whose covariance matrix is diagonal is said to form an orthogonal set.

2.6-2 Principal-components transformation

The principal-components transformation represents a particularly useful way of summarizing the information in a set of variables $X_1, \ldots, X_p$. (Assume $X_1, \ldots, X_p$ are scaled so that $\bar{X}_j = 0$, $j = 1, \ldots, p$.) The transformation from the original set $X_1, \ldots, X_p$ to the new set $Z_1, \ldots, Z_p$ is defined by

$$\underset{n,p}{Z} = \underset{n,p}{X} \; \underset{p,p}{A}, \tag{5}$$

where $\underset{p,p}{M} = \frac{1}{n-1} X'X =$ nonsingular covariance matrix for the original set of variables;

$$A'A = I;$$

$$A'MA = D_\lambda = \begin{bmatrix} \lambda_1 & 0 & \cdots & 0 \\ 0 & \lambda_2 & \cdots & 0 \\ \cdot & \cdot & & \cdot \\ \cdot & \cdot & & \cdot \\ \cdot & \cdot & & \cdot \\ 0 & 0 & \cdots & \lambda_p \end{bmatrix}.$$

From the last two relationships it follows that $M = AD_\lambda A'$. Under the transformation defined by the matrix A, the covariance matrix for the variables $Z_1, \ldots, Z_p$ is

$$\frac{1}{n-1} Z'Z = \frac{1}{n-1} A'X'XA = A'MA = D_\lambda.$$

That is, under the principal-components transformation the set $Z_1, \ldots, Z_p$ are orthogonal variables.

The column vectors of the matrix A are called the eigenvectors (or latent vectors) of the matrix M. The diagonal elements of the matrix D_λ are called the eigenvalues (eigenroots or latent roots) corresponding to the eigenvectors.

Let

$$A = [\underline{a}_1 \quad \underline{a}_2 \quad \cdots \quad \underline{a}_p],$$

where $\underset{p,1}{\underline{a}_j} = j\text{th column vector of the matrix } A.$

Similarly, let

$$Z = [\underline{z}_1 \quad \underline{z}_2 \quad \cdots \quad \underline{z}_p],$$

where $\underset{n,1}{\underline{z}_j} = j\text{th column vector of the matrix } Z.$

The components of the vector $\underline{z}_1$ are the "observed" scores on the variable Z_1. From (5) one has

$$\underset{n,1}{\underline{z}_1} = \underset{n,p}{X}\ \underset{p,1}{\underline{a}_1}.$$

The variance of the variable Z_1 is given by

$$s^2_{Z_1} = \underline{a}_1' M \underline{a}_1 = \lambda_1, \qquad \text{where} \qquad \underline{a}_1'\underline{a}_1 = 1.$$

λ_1 is called the root corresponding to the first eigenvector. It can be shown that the vector $\underline{a}_1$ makes $s^2_{Z_1}$ = maximum. Thus, the variable Z_1 is that normalized linear function of $X_1, \ldots, X_p$ that has maximum variance, where a *normalized* linear function is one for which $\underline{a}_1'\underline{a}_1 = 1$. Symbolically, the variable Z_1 is given by

$$Z_1 = a_{11}X_1 + a_{21}X_2 + \cdots + a_{p1}X_p,$$

where a_{j1} is an element of the vector $\underline{a}_1$.

The jth column vector of the matrix Z is given by

$$\underset{n,1}{\underline{z}_j} = \underset{n,p}{X}\ \underset{p,1}{\underline{a}_j},$$

where

$$\underline{a}_j'\underline{a}_{j-1} = 0, \qquad \underline{a}_j'\underline{a}_{j-2} = 0, \ldots, \qquad \underline{a}_j'\underline{a}_1 = 0, \qquad \underline{a}_j'\underline{a}_j = 1.$$

The variance of the variable Z_j is

$$s^2_{Z_j} = \underline{a}_j' M \underline{a}_j = \lambda_j.$$

The variable Z_j is that normalized linear function of $X_1, \ldots, X_p$ that has maximum variance subject to the constraints indicated.

The *generalized variance* of the set of variables $X_1, \ldots, X_p$ is, by definition,

$$\text{gen var}\,(X_1, \ldots, X_p) = |M|.$$

Since $$M = AD_\lambda A',$$

it follows that

$$\begin{aligned}|M| &= |A|\,|D_\lambda|\,|A'| \\ &= |D_\lambda|\,|AA'| \\ &= |D_\lambda|\,|I| \\ &= \lambda_1\lambda_2 \cdots \lambda_p.\end{aligned}$$

That is, the generalized variance is the product of the eigenroots of the matrix M. If M is nonsingular, then all the eigenroots will be greater than zero. (In this context, each λ_j is a variance.)

A numerical example of the principal-components transformation appears in Table 2.6-2. The matrix of observations, X, appears at the left of part i. At the right of part i is the covariance matrix associated

Table 2.6-2 Principal-components Transformation

	Element	X_1	X_2		
(i)	1	3	4	$M = \begin{bmatrix} 6.67 & 4.67 \\ 4.67 & 8.67 \end{bmatrix}$	$\lvert M \rvert = 36.00$
	2	5	1		trace $M = 15.34$
	3	7	3		
	$n = 4$	9	8		
(ii)	$A = \begin{bmatrix} .6283 & .7780 \\ .7780 & -.6283 \end{bmatrix}$			$A'MA = D_\lambda = \begin{bmatrix} 12.41 & 0 \\ 0 & 2.90 \end{bmatrix}$	
	Element	Z_1	Z_2		
(iii)	1	5.00	−0.18		
	2	3.92	3.26	$s^2_{Z_1} = 12.43$	
	3	6.73	3.55	$s^2_{Z_2} = 2.89$	
	$n = 4$	11.88	1.95	$\text{cov}_{Z_1Z_2} = -0.06$	
(iv)				$\lambda_1 + \lambda_2 = 15.31 = \text{trace } M$	
				$\lambda_1\lambda_2 = 35.99 = \lvert M \rvert$	

with the variables X_1, X_2. In part ii is the matrix A which defines the principal-components transformation. The computational details for finding the matrix A are not given. Essentially, the computational procedure is an iterative one based upon the relationship

$$M\underline{a}_j = \lambda_j\underline{a}_j.$$

This algorithm is carried out rather readily on an electronic computer but is not easily done on a desk calculator because of the accumulation of rounding error.

Given the matrix A, the matrix of the eigenroots is obtained at the right of part ii. The matrix of "observations" on the transformed variables Z_1 and Z_2 is obtained from the relationship

$$Z = XA.$$

From the columns of the Z matrix the variances and covariance between the Z variables are obtained and are given at the right of part iii. Within rounding error, the variances are equal to the corresponding eigenroots, and the covariance is zero. Two additional relationships between the eigenroots and the matrix M are noted in part iv.

It is of interest to note that

$$\underset{p,p}{A'}\ \underset{p,p}{A} = \underset{p,p}{I} \qquad \text{implies} \qquad AA' = I.$$

Hence from

$$M = AD_\lambda A'$$

one has

$$M^2 = (AD_\lambda A')(AD_\lambda A') = AD_\lambda^2 A'.$$

In general

$$M^k = AD_\lambda^k A'.$$

This last relationship implies that the eigenvectors are invariant under the operation of powering the matrix M. However, the eigenroots of a powered matrix are equal to the corresponding power of the original matrix. This relationship is used in many computing algorithms.

A covariance matrix is said to have *compound symmetry* if it has the following form:

$$\underset{p,p}{\Sigma} = \sigma^2 \begin{bmatrix} 1 & \rho & \cdots & \rho \\ \rho & 1 & \cdots & \rho \\ \cdot & \cdot & & \cdot \\ \cdot & \ddots & & \cdot \\ \cdot & \cdot & & \cdot \\ \rho & \rho & \cdots & 1 \end{bmatrix}.$$

A patterned matrix of this type occurs in an important class of models in the analysis of variance. The largest eigenroot of this type of matrix is given by

$$\lambda_1 = \sigma^2[1 + (p - 1)\rho].$$

All other roots will be equal to

$$\lambda_j = \sigma^2(1 - \rho), \qquad j = 2, \ldots, p.$$

For the special case $p = 2$,

$$\Sigma = \sigma^2 \begin{bmatrix} 1 & \rho \\ \rho & 1 \end{bmatrix}.$$

Hence $\quad |\Sigma| = \sigma^4(1 - \rho^2), \qquad \lambda_1 = \sigma^2(1 + \rho), \qquad \lambda_2 = \sigma^2(1 - \rho).$

Thus
$$\lambda_1 + \lambda_2 = \sigma^2(1 + \rho) + \sigma^2(1 - \rho) = 2\sigma^2 = \text{trace } \Sigma;$$
$$\lambda_1\lambda_2 = \sigma^2(1 + \rho)\sigma^2(1 - \rho) = \sigma^4(1 - \rho^2) = |\Sigma|.$$

2.7 Two Sets of Predictors

Suppose the set of variables $X_1, \ldots, X_p$ is divided into two subsets as follows:

Subset 1: $\quad \underline{X}_1 = [X_1 \cdots X_k]', \qquad k < p$

Subset 2: $\quad \underline{X}_2 = [X_{k+1} \cdots X_p]'.$

The subdivision is completely arbitrary. Let the covariation matrix and vector for the complete set be partitioned in accordance with the subsets.

$$\underset{p,p}{V} = \begin{array}{c} \\ k \\ p-k \end{array}\!\!\begin{array}{c} \begin{array}{cc} k & p-k \end{array} \\ \begin{bmatrix} V_{11} & V_{12} \\ V_{21} & V_{22} \end{bmatrix} \end{array}, \qquad \underset{p,1}{\underline{v}_{XY}} = \begin{array}{c} k \\ p-k \end{array}\!\!\begin{bmatrix} \underline{v}_{1Y} \\ \underline{v}_{2Y} \end{bmatrix}.$$

Based upon the variables in subset 1, one may construct the following linear prediction equation:

$$\hat{Y}_1 = a_1 + \underline{b}_1'\underline{X}_1, \tag{1}$$

where $\quad \underline{b}_1 = V_{11}^{-1}\underline{v}_{1Y}, \qquad a_1 = \bar{Y} - \underline{b}_1'\underline{\bar{X}}_1,$

where $\underline{\bar{X}}_1$ is the vector of means for the variables in subset 1. Based upon the variables in subset 2, one has

$$\hat{Y}_2 = a_2 + \underline{b}_2'\underline{X}_2, \tag{2}$$

where $\quad \underline{b}_2 = V_{22}^{-1}\underline{v}_{2Y}, \qquad a_2 = \bar{Y} - \underline{b}_2'\underline{\bar{X}}_2.$

The prediction equation based upon the complete set has the form

$$\hat{Y} = a + \underline{b}'\underline{X}, \tag{3}$$

where $\quad \underline{b} = V^{-1}\underline{v}_{XY}, \qquad a = \bar{Y} - \underline{b}'\underline{\bar{X}}.$

The predictable variation associated with each of the prediction equations may be summarized as follows:

Subset 1: $\quad R(X_1, \ldots, X_k) = \underline{b}_1'\underline{v}_{1Y} = r^2_{Y(1\ldots k)}\text{SS}_Y,$

Subset 2: $\quad R(X_{k+1}, \ldots, X_p) = \underline{b}_2'\underline{v}_{2Y} = r^2_{Y(k+1\ldots p)}\text{SS}_Y,$

Complete set: $\quad R(X_1, \ldots, X_p) = \underline{b}'\underline{v}_{XY} = r^2_{Y(1\ldots p)}\text{SS}_Y.$

The increase in predictability achieved by adding the second subset of predictors to a prediction system already containing the first subset is

$$\begin{aligned} R(X_{k+1}, \ldots, X_p \mid X_1, \ldots, X_k) &= R(X_1, \ldots, X_p) - R(X_1, \ldots, X_k) \\ &= \underline{b}'\underline{v}_{XY} - \underline{b}_1'\underline{v}_{1Y} \\ &= (r^2_{Y(1\ldots p)} - r^2_{Y(1\ldots k)})\text{SS}_Y. \end{aligned}$$

If one defines the squared semipartial correlation as

$$r^2_{Y(k+1\ldots p|1\ldots k)} = r^2_{Y(1\ldots p)} - r^2_{Y(1\ldots k)},$$

then $$R(X_{k+1}, \ldots, X_p \mid X_1, \ldots, X_k) = r^2_{Y(k+1\ldots p|1\ldots k)}\text{SS}_Y.$$

By analogy, the increase in predictability achieved by adding the first subset to a system already containing the second subset is

$$\begin{aligned} R(X_1, \ldots, X_k \mid X_{k+1}, \ldots, X_p) &= (r^2_{Y(1\ldots p)} - r^2_{Y(k+1\ldots p)})\text{SS}_Y \\ &= r^2_{Y(1\ldots k|k+1\ldots p)}\text{SS}_Y. \end{aligned}$$

If $X_1, \ldots, X_p$ represent a set of fixed variables, the following notation system is sometimes used.

$$R(\underline{\beta}) = R(\underline{\beta}_1,\underline{\beta}_2) = \underline{b}'\underline{v}_{XY},$$

$$R(\underline{\beta}_1) = \underline{b}_1'\underline{v}_{1Y}, \qquad R(\underline{\beta}_1 \mid \underline{\beta}_2) = R(\underline{\beta}_1,\underline{\beta}_2) - R(\underline{\beta}_2),$$

$$R(\underline{\beta}_2) = \underline{b}_2'\underline{v}_{2Y}, \qquad R(\underline{\beta}_2 \mid \underline{\beta}_1) = R(\underline{\beta}_1,\underline{\beta}_2) - R(\underline{\beta}_1).$$

$R(\underline{\beta})$ is read "The reduction in sum of squares due to $\underline{\beta}$."

Suppose the matrix V for the complete set has the following structure:

$$V = \begin{bmatrix} V_{11} & 0 \\ 0 & V_{22} \end{bmatrix}.$$

Then $$V^{-1} = \begin{bmatrix} V_{11}^{-1} & 0 \\ 0 & V_{22}^{-1} \end{bmatrix}.$$

For this special case,

$$\underline{b} = \begin{bmatrix} V_{11}^{-1}\underline{v}_{1Y} \\ V_{22}^{-1}\underline{v}_{2Y} \end{bmatrix} = \begin{bmatrix} \underline{b}_1 \\ \underline{b}_2 \end{bmatrix}.$$

That is, the weights associated with prediction systems based upon the subsets are actually subsets of the weights associated with the prediction system associated with the complete set. Thus,

$$\begin{aligned} \hat{Y} &= a + \underline{b}'\underline{X} \\ &= a + \underline{b}_1'\underline{X}_1 + \underline{b}_2'\underline{X}_2 \\ &= \hat{Y}_1 + \hat{Y}_2 - \bar{Y} \end{aligned}$$

Also for this special case,

$$R(X_1, \ldots, X_p) = R(X_1, \ldots, X_k) + R(X_{k+1}, \ldots, X_p),$$

$$r^2_{Y(1\ldots p)} = r^2_{Y(1\ldots k)} + r^2_{Y(k+1\ldots p)}.$$

Given the subset 1, it is always possible to find a transformation for the subset 2 such that the covariation matrix for the complete set $X_1, \ldots, X_k$

$U_{k+1}, \ldots, U_p$ (where $U_{k+1}, \ldots, U_p$ represent the transformed subset) has the form

$$V = \begin{bmatrix} V_{11} & 0 \\ 0 & V_{22} \end{bmatrix}.$$

In proving certain relationships it is sometimes convenient to carry out this transformation. The transformation will not necessarily be unique.

A question that is often asked is this: Within the complete system of predictor variables $X_1, \ldots, X_p$, what is the contribution of the subset $X_{k+1}, \ldots, X_p$? There are actually two answers to this question. One answer is given by the magnitude of $R(X_{k+1}, \ldots, X_p)$ or the magnitude of $r^2_{Y(k+1,\ldots,p)}$. A second answer is given by the magnitude of

$$R(X_{k+1}, \ldots, X_p \mid X_1, \ldots, X_p)$$

or the corresponding semipartial correlation. Each of these answers presents a different point of view. If the covariation matrix $V_{12} = 0$, then both answers reduce to a single answer.

2.8 Testing Statistical Hypotheses—Fixed Model

2.8-1 Testing the hypothesis $\underline{\beta} = \underline{\beta}^*$

Let

$\underset{n,p}{X}$ = matrix of sample observations on variables $X_1, \ldots, X_p$,

$\underset{n,1}{\underline{Y}}$ = vector of sample observations on Y.

Consider the fixed model

$$\underline{Y} = X\underline{\beta} + \underline{\varepsilon},$$

where

$$\underline{\varepsilon}: N(\underline{0}, \sigma^2_\varepsilon I).$$

Least-squares estimation procedures give

$$\underline{\hat{Y}} = X\underline{b},$$

where

$$\underline{b} = (X'X)^{-1}X'\underline{Y},$$

$$\Sigma \hat{Y}^2 = \underline{Y}'X(X'X)^{-1}X'\underline{Y} = \underline{Y}'P\underline{Y}, \quad \text{where } P = X(X'X)^{-1}X',$$

$$Q_0 = \Sigma(\underline{Y} - \underline{\hat{Y}})^2 = \underline{Y}'\underline{Y} - \underline{Y}'P\underline{Y}.$$

Suppose one wants to test the hypothesis that $\underline{\beta} = \underline{\beta}^*$, where $\underline{\beta}^*$ is a specified vector. When this hypothesis is true, the estimate of variation due to error is

$$Q_1 = \Sigma(Y - X\underline{\beta}^*)^2 = (\underline{Y} - X\underline{\beta}^*)'(\underline{Y} - X\underline{\beta}^*)$$
$$= \underline{Y}'\underline{Y} - 2\underline{\beta}^{*\prime}X'\underline{Y} + \underline{\beta}^{*\prime}(X'X)\underline{\beta}^*.$$

A measure of deviation from hypothesis is given by

$$Q_2 = Q_1 - Q_0 = \underline{\beta}^{*\prime}(X'X)\underline{\beta}^* - 2\underline{\beta}^{*\prime}X'\underline{Y} + \underline{Y}'P\underline{Y}$$
$$= (\underline{Y} - X\underline{\beta}^*)'P(\underline{Y} - X\underline{\beta}^*).$$

The last line in the above expression follows from the line above by rearrangement of the terms.

It may be shown that Q_2 and Q_0 are independently distributed as chi square with respective degrees of freedom

$$n - (n - p) = p \quad \text{and} \quad n - p.$$

Hence the ratio

$$F = \frac{(Q_1 - Q_0)/p}{Q_0/(n - p)}$$

is distributed as $F(p, n - p)$ when the hypothesis being tested is true. The test procedure is summarized in Table 2.8-1.

Table 2.8-1 Testing the Hypothesis $\underline{\beta} = \underline{\beta}^*$

Source of variation	SS	df	MS
Residual assuming $\underline{\beta} = \underline{\beta}^*$	Q_1	n	
Residual if $\underline{\beta}$ is estimated by $\underline{b}$	Q_0	$n - p$	$Q_0/(n - p)$
Deviation from hypothesis	$Q_2 = Q_1 - Q_0$	p	$(Q_1 - Q_0)/p$

$$F = \frac{(Q_1 - Q_0)/p}{Q_0/(n - p)}$$

For the special case $\underline{\beta}^* = \underline{0}$, one has the following:

$$Q_1 = \Sigma Y^2 = \underline{Y}'\underline{Y},$$
$$Q_0 = \Sigma(Y - \hat{Y})^2 = \underline{Y}'\underline{Y} - \underline{Y}'P\underline{Y},$$
$$Q_2 = Q_1 - Q_0 = \underline{Y}'P\underline{Y}.$$

Hence $$F = \frac{(Q_1 - Q_0)/p}{Q_0/(n - p)} = \frac{\underline{Y}'P\underline{Y}/p}{(\underline{Y}'\underline{Y} - \underline{Y}'P\underline{Y})/(n - p)}.$$

2.8-2 Testing the hypothesis $\underline{\beta}_j = \underline{0}$

Suppose that one has the following sample data:

$$\underset{n,1}{\underline{Y}} \quad \underset{n,p}{X} \qquad (n = \text{sample size}).$$

Let the matrix X be partitioned as follows:

$$\underset{n,p}{X} = [\underset{n,k}{X_1} \quad \underset{n,p-k}{X_2}], \qquad \text{where } k < p.$$

Consider the model

$$(1) \qquad \underline{Y} = X\underline{\beta} + \underline{\varepsilon}_{\beta},$$

where $X_1, \ldots, X_p$ are fixed variables and $\underline{\varepsilon}_{\beta}$: $N(\underline{0},\sigma^2_{\varepsilon_\beta}I)$. If one lets

$$\underline{\beta} = \begin{matrix} k \\ p-k \end{matrix}\begin{bmatrix} \underline{\beta}_1 \\ \underline{\beta}_2 \end{bmatrix},$$

then (1) takes the form

$$\underline{Y} = X_1\underline{\beta}_1 + X_2\underline{\beta}_2 + \underline{\varepsilon}_{\beta}.$$

Consider the following models:

$$(2) \qquad \underline{Y} = X_1\underline{\alpha}_1 + \underline{\varepsilon}_{\alpha_1}, \qquad \text{where } \varepsilon_{\alpha_1}: N(\underline{0},\sigma^2_{\varepsilon_{\alpha_1}}I);$$

$$(3) \qquad \underline{Y} = X_2\underline{\alpha}_2 + \underline{\varepsilon}_{\alpha_2}, \qquad \text{where } \varepsilon_{\alpha_2}: N(\underline{0},\sigma^2_{\varepsilon_{\alpha_2}}I).$$

Using least-squares estimation procedures, the variation due to error under models (1) through (3) is summarized in part i of Table 2.8-2. Thus

Table 2.8-2 Testing the Hypothesis $\underline{\beta}_j = \underline{0}$

	Source of variation	df
(i)	(1) $Q_\beta = \underline{Y}'\underline{Y} - \underline{Y}'P_\beta\underline{Y} = \Sigma Y^2 - \hat{\underline{\beta}}'X'\underline{Y}$	$n - p$
	(2) $Q_{\alpha_1} = \underline{Y}'\underline{Y} - \underline{Y}'P_{\alpha_1}\underline{Y} = \Sigma Y^2 - \hat{\underline{\alpha}}_1'X_1'\underline{Y}$	$n - k$
	(3) $Q_{\alpha_2} = \underline{Y}'\underline{Y} - \underline{Y}'P_{\alpha_2}\underline{Y} = \Sigma Y^2 - \hat{\underline{\alpha}}_2'X_2'\underline{Y}$	$n - (p - k)$
(ii)	$Q_{\alpha_2} - Q_\beta = \underline{Y}'P_\beta\underline{Y} - \underline{Y}'P_{\alpha_2}\underline{Y} = R(\underline{\beta}) - R(\underline{\alpha}_2) = R(\underline{\beta}_1 \mid \underline{\beta}_2)$	$n - (p - k) - (n - p) = k$
	$Q_{\alpha_1} - Q_\beta = \underline{Y}'P_\beta\underline{Y} - \underline{Y}'P_{\alpha_1}\underline{Y} = R(\underline{\beta}) - R(\underline{\alpha}_1) = R(\underline{\beta}_2 \mid \underline{\beta}_1)$	$n - k - (n - p) = p - k$
(iii)	$F_{\beta_1\mid\beta_2} = \dfrac{(Q_{\alpha_2} - Q_\beta)/k}{Q_\beta/(n-p)}$	$F_{\beta_2\mid\beta_1} = \dfrac{(Q_{\alpha_1} - Q_\beta)/(p-k)}{Q_\beta/(n-p)}$

Q_β is the variation due to error as obtained under model (1). In the expression Q_β,

$$P_\beta = X(X'X)^{-1}X'.$$

Similarly,

$$P_{\alpha_1} = X_1(X_1'X_1)^{-1}X_1' \qquad \text{and} \qquad P_{\alpha_2} = X_2(X_2'X_2)^{-1}X_2'$$

in Q_{α_1} and Q_{α_2}.

The difference between Q_{α_2} and Q_β may be expressed in the following form:

$$Q_{\alpha_2} - Q_\beta = (\underline{Y}'\underline{Y} - \underline{Y}'P_{\alpha_2}\underline{Y}) - (\underline{Y}'\underline{Y} - \underline{Y}'P_\beta\underline{Y})$$
$$= \underline{Y}'P_\beta\underline{Y} - \underline{Y}'P_{\alpha_2}\underline{Y}$$
$$= R(\underline{\beta}) - R(\underline{\alpha}_2) = R(\underline{\beta}_1 \mid \underline{\beta}_2),$$

where $R(\underline{\beta})$ is that part of the total sum of squares for Y that can be predicted under model (1), and $R(\underline{\alpha}_2)$ is that part of the total sum of squares for Y that can be predicted under model (2). $R(\underline{\beta}_1 \mid \underline{\beta}_2)$ represents that part of the variation due to $\underline{\beta}_1$ that is orthogonal to $\underline{\beta}_2$.

The difference

$$Q_{\alpha_1} - Q_\beta = \underline{Y}'P_\beta\underline{Y} - \underline{Y}'P_{\alpha_1}\underline{Y}$$
$$= R(\underline{\beta}) - R(\underline{\alpha}_1) = R(\underline{\beta}_2 \mid \underline{\beta}_1)$$

represents that part of the variation due to $\underline{\beta}_2$ which is orthogonal to $\underline{\beta}_1$. Hence one has the following relationships:

$$R(\underline{\beta}) = R(\underline{\beta}_1 \mid \underline{\beta}_2) + R(\underline{\alpha}_2)$$
$$= R(\underline{\beta}_2 \mid \underline{\beta}_1) + R(\underline{\alpha}_1).$$

Under the hypothesis that $\underline{\beta}_2 = \underline{0}$, a measure of deviation from hypothesis is given by

$$Q_{\alpha_1} - Q_\beta = R(\underline{\beta}_2 \mid \underline{\beta}_1).$$

When the hypothesis that $\underline{\beta}_2 = \underline{0}$ is true,

$$\mathrm{E}[R(\underline{\beta}_2 \mid \underline{\beta}_1)] = (p - k)\sigma_\varepsilon^2,$$
$$\mathrm{E}(Q_\beta) = (n - p)\sigma_\varepsilon^2.$$

Further, $Q_{\alpha_1} - Q_\beta$ and Q_β are independently distributed as chi square with respective degrees of freedom $p - k$ and $n - p$. Hence the appropriate F ratio for testing the hypothesis that $\underline{\beta}_2 = \underline{0}$ is

$$\frac{(Q_{\alpha_1} - Q_\beta)/(p - k)}{Q_\beta/(n - p)} = \frac{R(\underline{\beta}_2 \mid \underline{\beta}_1)/(p - k)}{Q_\beta/(n - p)}.$$

By analogy, to test the hypothesis that $\underline{\beta}_1 = \underline{0}$

$$\frac{(Q_{\alpha_2} - Q_\beta)/k}{Q_\beta/(n - p)} = \frac{R(\underline{\beta}_1 \mid \underline{\beta}_2)/k}{Q_\beta/(n - p)}.$$

When $\underline{\beta}_1 = \underline{0}$, the sampling distribution of the latter ratio is $F(k, n - p)$.

Consider the special case in which

$$X'X = \begin{bmatrix} X_1'X_1 & 0 \\ 0 & X_2'X_2 \end{bmatrix};$$

that is, the case in which $X_1'X_2 = 0$. For this special case

$$\begin{aligned} P_\beta &= X(X'X)^{-1}X' = [X_1 \quad X_2]\begin{bmatrix}(X_1'X_1)^{-1} & 0 \\ 0 & (X_2'X_2)^{-1}\end{bmatrix}\begin{bmatrix}X_1' \\ X_2'\end{bmatrix} \\ &= X_1(X_1'X_1)^{-1}X_1' + X_2(X_2'X_2)^{-1}X_2' \\ &= P_{\alpha_1} + P_{\alpha_2}. \end{aligned}$$

Hence $$\underline{Y}'P_\beta\underline{Y} = \underline{Y}'(P_{\alpha_1} + P_{\alpha_2})\underline{Y} = \underline{Y}'P_{\alpha_1}\underline{Y}' + \underline{Y}'P_{\alpha_2}\underline{Y}.$$

Thus $$R(\underline{\beta}) = R(\underline{\alpha}_1) + R(\underline{\alpha}_2).$$

From this last relationship it follows that

$$\begin{aligned} R(\underline{\beta}_1 \mid \underline{\beta}_2) &= R(\underline{\alpha}_1) = R(\underline{\beta}_1), \\ R(\underline{\beta}_2 \mid \underline{\beta}_1) &= R(\underline{\alpha}_2) = R(\underline{\beta}_2). \end{aligned}$$

For this special case it also follows that

$$\underline{\beta}_1 = \underline{\alpha}_1 \qquad \text{and} \qquad \underline{\beta}_2 = \underline{\alpha}_2.$$

For the general case (in which $X_1'X_2 \neq 0$), one may partition the inverse of $X'X$ as follows:

$$(X'X)^{-1} = \begin{array}{c} \\ k \\ p-k \end{array}\begin{array}{c} \begin{array}{cc} k & p-k \end{array} \\ \begin{bmatrix} C_{11} & C_{12} \\ C_{21} & C_{22} \end{bmatrix} \end{array},$$

where
$$\begin{aligned} C_{11} &= [X_1'X_1 - X_1'P_{\alpha_2}X_1]^{-1}, \\ C_{22} &= [X_2'X_2 - X_2'P_{\alpha_1}X_2]^{-1}, \\ C_{12} &= C_{21}' = -(X_1'X_1)^{-1}X_1'X_2[X_2'X_2 - X_2'P_{\alpha_1}X_2]^{-1}. \end{aligned}$$

Hence,

$$\begin{aligned} \underline{\hat{\beta}} &= (X'X)^{-1}X'\underline{Y} = (X'X)^{-1}\begin{bmatrix}X_1' \\ X_2'\end{bmatrix}\underline{Y} = (X'X)^{-1}\begin{bmatrix}X_1'\underline{Y} \\ X_2'\underline{Y}\end{bmatrix} \\ &= \begin{bmatrix}C_{11} & C_{12} \\ C_{21} & C_{22}\end{bmatrix}\begin{bmatrix}X_1'\underline{Y} \\ X_2'\underline{Y}\end{bmatrix} \\ &= \begin{bmatrix}C_{11}X_1'\underline{Y} + C_{12}X_2'\underline{Y} \\ C_{21}X_1'\underline{Y} + C_{22}X_2'\underline{Y}\end{bmatrix} = \begin{bmatrix}\underline{\hat{\beta}}_1 \\ \underline{\hat{\beta}}_2\end{bmatrix}. \end{aligned}$$

This last relationship indicates that $\underline{\hat{\beta}}_1$ is in part a function of the variables in X_2 as well as the variables in X_1. However, when $X_1'X_2 = 0$, $C_{12} = 0$, and $C_{11} = (X_1'X_1)^{-1}$; hence $\underline{\hat{\beta}}_1$ in this special case will be a function of only those variables in X_1.

Another special case of considerable interest is that in which $k = 1$ and $X_1 = 1$ for all elements. For this case,

$$\begin{aligned} X_1'X_1 &= n, \\ P_{\alpha_1} &= X_1(X_1'X_1)^{-1}X_1' = \frac{1}{n}\underset{n,n}{U}, \end{aligned}$$

where U is a matrix of unities. Further,

$$\underline{Y}'P_{\alpha_1}\underline{Y} = \frac{1}{n}(\Sigma Y)^2.$$

Hence, $$Q_{\alpha_1} = \underline{Y}'\underline{Y} - \underline{Y}'P_{\alpha_1}\underline{Y} = \Sigma Y^2 - \frac{1}{n}(\Sigma Y)^2 = SS_Y.$$

In this case, to test the hypothesis that $\underline{\beta}_2 = \underline{0}$, one has

$$\begin{aligned} R(\underline{\beta}_2 \mid \underline{\beta}_1) &= Q_{\alpha_1} - Q_\beta \\ &= SS_Y - SS_{Y-\hat{Y}} \\ &= SS_{\hat{Y}}. \end{aligned}$$

Hence the F ratio takes the form

$$F = \frac{SS_{\hat{Y}}/(p-1)}{SS_{Y-\hat{Y}}/(n-p)}.$$

When $\underline{\beta}_2 = \underline{0}$, this F ratio is distributed as $F(p-1, n-p)$.

As an application of the general case of the test procedure, consider the numerical example in Table 2.1-4. In this example, the X variables are X_0, X_1, X_2. For this example,

$$n = 6 \qquad \text{and} \qquad p = 3.$$

The numerical value of Q_β as obtained from part iv of Table 2.1-6 is

$$Q_\beta = \Sigma(Y - \hat{Y})^2 = 525.96.$$

Let $$\underset{6,3}{X} = [\underline{X}_0 \quad \underline{X}_1 \mid \underline{X}_2] = [\underset{6,2}{X_1} \quad \underset{6,1}{X_2}].$$

That is,

$$X_1 = [\underline{X}_0 \quad \underline{X}_1] \qquad \text{and} \qquad X_2 = [\underline{X}_2].$$

From part iii of Table 2.1-6, one finds that

$$Q_{\alpha_1} = 529.27.$$

Hence $$R(\beta_2 \mid \beta_0, \beta_1) = Q_{\alpha_1} - Q_\beta = 3.29.$$

That is, that part of reduction in the sum of squares due to β_2 which is orthogonal to both β_0 and β_1 is 3.30. Under the hypothesis that $\beta_2 = 0$, $R(\beta_2 \mid \beta_0, \beta_1)$ measures the deviation from hypothesis. To test the hypothesis that $\beta_2 = 0$, the appropriate F ratio is

$$F = \frac{(Q_{\alpha_1} - Q_\beta)/1}{Q_\beta/3} = \frac{3.29}{525.96/3} = .0189.$$

The critical value for a .05-level test is $F_{.95}(1,3) = 10.1$. Hence these data do not contradict the hypothesis that $\beta_2 = 0$. The power of this test is extremely low since the degrees of freedom for the denominator of the F ratio are so small.

As another application of the general test procedure, again consider the data in Table 2.1-4. This time suppose one wants to test the hypothesis that

$$\begin{bmatrix}\beta_1\\ \beta_2\end{bmatrix} = \begin{bmatrix}0\\ 0\end{bmatrix}.$$

From part ii and part iv of Table 2.1-6, one has

$$Q_{\alpha_1} = 783.33,$$
$$Q_{\beta} = 525.96.$$

Hence the measure of deviation from hypothesis is

$$R(\beta_1,\beta_2 \mid \beta_0) = Q_{\alpha_1} - Q_{\beta} = 257.37.$$

In this case the appropriate F ratio is

$$F = \frac{257.37/2}{525.96/3} = .73.$$

For a .05-level test, $F_{.95}(2,3) = 9.55$. Again because of the very small number of degrees of freedom for the denominator of the F ratio, this test has extremely low power.

2.9 Regression of Regression Coefficients on Supplementary Variables

In this section the *direct* or *Kronecker* product of two matrices is used. This type of product is defined by

$$\underset{r,s}{A} \times \underset{t,u}{B} = \underset{rt,\ su}{\begin{bmatrix} Ab_{11} & Ab_{12} & \cdots & Ab_{1u} \\ Ab_{21} & Ab_{22} & \cdots & Ab_{2u} \\ \cdot & & & \cdot \\ \cdot & & & \cdot \\ \cdot & & & \cdot \\ Ab_{t1} & Ab_{t2} & \cdots & Ab_{tu} \end{bmatrix}}$$

For example,

$$\underset{2,2}{\begin{bmatrix}2 & 1\\ 0 & 3\end{bmatrix}} \times \underset{1,3}{[4 \quad 5 \quad 6]} = \underset{2,6}{\left[\begin{array}{cc:cc:cc} 8 & 4 & 10 & 5 & 12 & 6\\ 0 & 12 & 0 & 15 & 0 & 18\end{array}\right]}.$$

The following results with the direct product operator are readily verified.

(i) $$(A \times B)' = A' \times B'.$$

(ii) $$(\underset{p,q}{A} \times \underset{r,s}{B})(\underset{q,u}{C} \times \underset{s,v}{E}) = \underset{p,u}{(AC)} \times \underset{r,v}{(BE)}.$$

As a special case of (ii),

$$(A \times B)'(A \times B) = (A'A) \times (B'B).$$

If A and B are square, nonsingular matrices,

(iii) $$[A \times B]^{-1} = A^{-1} \times B^{-1}.$$

As a special case of (iii), if $X'X$ and $Z'Z$ are nonsingular, then

$$[(X'X) \times (Z'Z)]^{-1} = (X'X)^{-1} \times (Z'Z)^{-1}.$$

Suppose that one has the following data:

Element	$Y_1 \quad Y_2 \quad \cdots \quad Y_m$	$X_1 \quad X_2 \quad \cdots \quad X_p$
1		
2		
.	$\underset{n,m}{Y}$	$\underset{n,p}{X}$
.		
.		
n		

From these data one may compute the regression of each Y_j ($j = 1, 2, \ldots, m$) on $X_1, X_2, \ldots, X_p$. The m sets of regression weights would be the column vectors of the matrix B, where

(1) $$\underset{p,m}{B} = (X'X)^{-1}X'Y.$$

Suppose that one also has, in addition to the B matrix, supplementary information $Z_1, Z_2, \ldots, Z_q$ on each Y_j. The combined data may be represented as follows:

	$b_1 \quad b_2 \quad \cdots \quad b_p$	$Z_1 \quad Z_2 \quad \cdots \quad Z_q$
Y_1		
Y_2		
.		
.	$\underset{m,p}{B'}$	$\underset{m,q}{Z}$
.		
Y_m		

From these data one may obtain the regression of each b_k ($k = 1, 2, \ldots, p$) on $Z_1, Z_2, \ldots, Z_q$. The normal equations take the form

(2) $$(Z'Z)C = Z'B',$$

where $$\underset{q,p}{C} = [\underline{c}_1 \quad \underline{c}_2 \quad \cdots \quad \underline{c}_k \quad \cdots \quad \underline{c}_p],$$

$$\underset{q,1}{\underline{c}_k} = k\text{th column vector of } C.$$

The regression of b_k on $Z_1, Z_2, \ldots, Z_q$ takes the form

$$b_k = \underline{c}_k'\underline{Z} = c_{1k}Z_1 + c_{2k}Z_2 + \cdots + c_{qk}Z_q.$$

The solution for the elements of C is

$$C = (Z'Z)^{-1}Z'B' = (Z'Z)^{-1}Z'Y'X(X'X)^{-1}. \tag{3}$$

An equivalent expression for the normal equations represented by (2) is

$$(Z'Z)C(X'X) = Z'Y'X. \tag{4}$$

Expression (4) is obtained from (3) by pre- and post-multiplying (3) by $(Z'Z)$ and $(X'X)$, respectively.

An alternative approach to formulating the normal equations represented by (4) permits one to interpret the problem of regression of regression coefficients in another way. Let the elongated vector $C^{(1)}$ be defined by

$$\underset{pq,1}{C^{(1)}} = \begin{matrix} q \\ q \\ \\ \\ \\ q \end{matrix}\overset{1}{\begin{bmatrix} \underline{c}_1 \\ \underline{c}_2 \\ \cdot \\ \cdot \\ \cdot \\ \underline{c}_p \end{bmatrix}}.$$

That is, $C^{(1)}$ is the vector formed by laying the column vectors of C end to end. Similarly, if

$$\underset{n,m}{Y} = [\underline{Y}_1 \quad \underline{Y}_2 \quad \cdots \quad \underset{n,1}{\underline{Y}_j} \quad \cdots \quad \underline{Y}_m],$$

let $Y^{(1)}$ be defined by

$$\underset{mn,1}{Y^{(1)}} = \begin{matrix} n \\ n \\ \\ \\ \\ n \end{matrix}\overset{1}{\begin{bmatrix} \underline{Y}_1 \\ \underline{Y}_2 \\ \cdot \\ \cdot \\ \cdot \\ \underline{Y}_m \end{bmatrix}}.$$

In terms of these elongated vectors and the direct product operator, (4) may be written

$$[\underset{pq,pq}{(X'X) \times (Z'Z)}]\underset{pq,1}{C^{(1)}} = \underset{pq,mn}{(X' \times Z')}\underset{mn,1}{Y^{(1)}}. \tag{5}$$

Now suppose that a simple regression problem is started with the following basic data:

Dependent variable	Independent variables
$\underset{mn,1}{Y^{(1)}}$	$\underset{mn,pq}{X \times Z}$

The normal equations associated with this problem have the form

$$(X \times Z)'(X \times Z)C^{(1)} = (X \times Z)'Y^{(1)}$$

or

$$[(X'X) \times (Z'Z)]C^{(1)} = (X' \times Z')Y^{(1)}. \tag{6}$$

Expression (6) is the same as (5). Hence the problem of the regression of b_k on $Z_1, Z_2, \ldots, Z_q$ is equivalent to a simple regression problem starting with the data organized as given above. The solution to (6) is

$$\begin{aligned} C^{(1)} &= [(X'X) \times (Z'Z)]^{-1}(X' \times Z')Y^{(1)} \\ &= [(X'X)^{-1} \times (Z'Z)^{-1}](X' \times Z')Y^{(1)} \\ &= [(X'X)^{-1}X' \times (Z'Z)^{-1}Z']Y^{(1)}. \end{aligned} \tag{7}$$

An interesting application of the principles discussed in this section will be found in Cornish (1957).

3

DESIGN AND ANALYSIS OF SINGLE-FACTOR EXPERIMENTS

3.1 Introduction

In testing statistical hypotheses or in setting confidence bounds on estimates, one uses sampling distributions determined by purely mathematical considerations. One postulates a model, imposes various conditions upon this model, and then derives the consequences in terms of sampling distributions which are valid for the mathematical system. To the extent that the model and the conditions imposed upon it approximate an actual experiment, the model can be used as a guide in drawing inferences from the data.

If use is to be made of existing, well-developed mathematical models, experiments must be designed in a way that makes these models appropriate. If an experiment cannot meet the specifications in existing models, the experimenter may be able to develop a model tailored to the specific needs of his experiment. But the problem of the analysis of the resulting data must still be solved. If the mathematics needed to derive sampling distributions having known characteristics is manageable, the specially tailored model leads to inferences of known precision. Otherwise inferences drawn from an experiment of this kind will have unknown precision.

The analysis of data obtained from an experiment is dependent upon its design and the sampling distributions appropriate for the underlying population distributions. The design, in part, determines what the sampling distributions will be. Associated with standard designs are analyses justified by the mathematical models which led to the construction of these designs. Alternative designs are often available for an experiment having stated objectives. Depending upon the specific situations, one design may be more efficient than another for the same amount of experimental effort—more efficient in terms of the power of the resulting tests and the narrowness of the resulting confidence intervals. A major problem in planning an experiment is to find or develop that design which is most efficient per unit of experimental effort in reaching the primary objectives of the experiment.

The most efficient design from a purely mathematical point of view may be so costly in terms of time, money, and effort as to render it unworkable. In general, the smaller the variation due to experimental error, other factors being constant, the more efficient the design. This source of variation may be reduced by introducing various kinds of controls or by increasing the sample size. Both methods of reducing experimental error may be used, but which one provides the greater reduction per unit of cost depends upon features unique to the experimental conditions.

The study of the statistical aspects of experimental design will assist the experimenter in finding the most adequate and feasible model for his experiment. This model must permit the experimenter to reach decisions with respect to all the objectives of his experiment. Whether or not a model adequately represents the experimental situation calls for expert knowledge of the subject-matter field. In appraising the adequacy of alternative models, the experimenter is often made aware of sources of variation and possible implications that he had not thoroughly considered. Thus design problems often force the experimenter to formulate his experiment in terms of variables that will lead to clear-cut interpretations for the end product.

Some of the criteria for good experimental designs are as follows:

1. The analyses resulting from the design should provide unambiguous information on the primary objectives of the experiment. In particular the design should lead to unbiased estimates.
2. The model and its underlying assumptions should be appropriate for the experimental material.
3. The design should provide maximum information with respect to the major objectives of the experiment per minimum amount of experimental effort.
4. The design should provide some information with respect to all the objectives of the experiment.
5. The design must be feasible within the working conditions that exist for the experimenter.

The designs that will be considered in this chapter are appropriate for what have been called single-factor experiments. The primary objective of this kind of experiment is to compare the relative effectiveness of two or more treatments on a common criterion. The term single-factor in this context is used in contrast to the term multifactor. In this latter type of experiment the primary objective is to compare the effect of combinations of treatments acting simultaneously on each of the elements. There are some instances in which the distinction between single-factor and multi-factor experiments is difficult to make.

In this chapter only designs involving independent observations will be discussed; corresponding designs having correlated observations are considered in the next chapter. The designs in this chapter form a special case of what are called *completely randomized* designs. These form the building blocks for many other designs; they also provide a standard against which the efficiency of other types of designs is measured.

In some of the work that follows, hypothetical "treatment" populations are formulated to correspond to the design of an experiment. The data-analysis phase consists of estimating parameters of these populations and testing statistical hypotheses about such parameters, as well as all other appropriate operations useful in interpreting experimental data.

Statistical models are introduced to define and summarize the assumptions that have been made about such populations. Models serve as guides in formalizing statistical bases of the data analysis. Models are also useful tools in guiding test procedures. The terms which do or do not appear in a model, however, must reflect corresponding effects that are present in an experiment as it actually is conducted. Further, the distribution assumptions on such terms must also reflect, realistically, corresponding distributions in the actual experiment.

The scope of inferences from an experiment stems not from any formal model but rather from its design and the procedures followed in its conduct. In this sense, the role of the model is secondary rather than primary.

There is a duality between sampling operations and what are called fixed and random variables in a model. Consider, for example, the operation of drawing random samples from the population $N(\mu,\sigma^2)$. A model for an observation in the sample is given by

$$X_{ij} = \mu + \varepsilon_{ij},$$

where μ = fixed but unknown constant,

ε_{ij} = random variable distributed as $N(0,\sigma^2)$.

Thus a potential observation from the population $N(\mu,\sigma^2)$ may be represented by a fixed constant plus a random variable.

In much of the material that follows, the covariance structure that is assumed about terms in a model is meant to correspond to the sampling plan followed in the conduct of an experiment. The sampling plan is part of the experimental design. An excellent summary of the relationship between models and the analysis of experimental data appears in Wilk and Kempthorne (1955, pp. 1160–1161).

3.2 Definitions and Numerical Example

A numerical example will illustrate the definitions of terms used in the analysis of a single-factor experiment. The actual analysis of this example will be given in detail. The rationale justifying this analysis will be discussed in the next section.

An experimenter is interested in evaluating the effectiveness of three methods of teaching a given course. A group of 24 subjects is available to the experimenter. This group is considered by the experimenter to be the equivalent of a random sample from the population of interest. Three subgroups of 8 subjects each are formed at random; the subgroups are then taught by one of the three methods. Upon completion of the course, each of the subgroups is given a common test covering the material in the course. The resulting test scores are given in part i of Table 3.2-1. The symbol n designates the number of subjects in a subgroup and k the number of methods.

In part ii of this table, the symbol T_j designates the sum of the test scores for the subjects who were taught by method j. For example, T_1 designates the sum of the test scores for the subjects taught by method 1. The symbol G designates the grand total of all observations in the experiment. G is most readily obtained by summing the T_j's; that is, $G = 38 + 37 + 62 = 137$. The symbol ΣX_j^2 designates the sum of the squares of the observations on the subjects taught by method j. For example, $\Sigma X_1^2 = 3^2 + 5^2 + \cdots + 9^2 = 224$.

In part iv, $\bar{T}_j$ designates the mean of the test scores for subjects taught by method j. For example, $\bar{T}_1 = 4.75$ is the mean test score for the subjects under method 1. The symbol $\bar{G}$ designates the grand mean of all test scores. When there is an equal number of subjects in each of the subgroups, $\bar{G} = (\Sigma\bar{T}_j)/k$, where k is the number of subgroups; this relationship does not hold if the number of subjects within each subgroup varies.

The symbol SS_j in part iii designates the sum of squares, or *variation*, of the test scores within the group under method j. By definition, the variation of the observations within method j is

$$(1) \qquad SS_j = \sum_i (X_{ij} - \bar{T}_j)^2,$$

i.e., the sum of the squared deviations of the test scores under method j about the mean of subgroup j. This definition is algebraically equivalent to

$$(2) \qquad SS_j = \Sigma X_j^2 - \frac{T_j^2}{n},$$

Table 3.2-1

	Method 1	Method 2	Method 3	
(i)	3	4	6	
	5	4	7	$n = 8$
	2	3	8	
	4	8	6	$k = 3$
	8	7	7	
	4	4	9	
	3	2	10	
	9	5	9	
(ii)†	$T_1 = 38$	$T_2 = 37$	$T_3 = 62$	$G = \Sigma T_j = 137$
	$\Sigma X_1^2 = 224$	$\Sigma X_2^2 = 199$	$\Sigma X_3^2 = 496$	$\Sigma(\Sigma X_j^2) = 919$
(iii)	$SS_1 = \Sigma X_1^2 - \frac{T_1^2}{n}$	$SS_2 = \Sigma X_2^2 - \frac{T_2^2}{n}$	$SS_3 = \Sigma X_3^2 - \frac{T_3^2}{n}$	$SS_w = \Sigma SS_j$
	$= 224 - \frac{38^2}{8}$	$= 199 - \frac{37^2}{8}$	$= 496 - \frac{62^2}{8}$	
	$= 43.50$	$= 27.88$	$= 15.50$	$SS_w = 86.88$
(iv)	$\bar{T}_1 = T_1/n$	$\bar{T}_2 = T_2/n$	$\bar{T}_3 = T_3/n$	$\bar{G} = G/nk$
	$= 38/8 = 4.75$	$= 37/8 = 4.62$	$= 62/8 = 7.75$	$= 137/24 = 5.71$

† The symbol T_1 is an abbreviation for the more complete notation given by

$$T_1 = \sum_i X_{i1}.$$

Similarly,

$$\Sigma X_1^2 = \sum_i X_{i1}^2.$$

which is a more convenient computational formula. Use of this formula is illustrated in part iii. The symbol SS_w designates the pooled within-method variation (or sum of squares). By definition, SS_w is the sum of the variation within each of the methods,

$$SS_w = \Sigma SS_j. \tag{3}$$

A computationally more convenient formula for SS_w is

$$SS_w = \Sigma(\Sigma X_j^2) - \frac{\Sigma T_j^2}{n}. \tag{4}$$

For the data in the table, (4) is

$$SS_w = 919 - \frac{(38)^2 + (37)^2 + (62)^2}{8}$$

$$= 919 - \frac{6657}{8} = 919 - 832.12 = 86.88.$$

This value for SS_w is the same as that obtained in the table using (3).

The variation (or sum of squares) due to the methods of training is by definition

$$\text{(5)} \qquad SS_{\text{methods}} = n\Sigma(\bar{T}_j - \bar{G})^2.$$

This statistic measures the extent to which the means for the subgroups differ from the grand mean. It is also a measure of the extent to which the subgroup means differ from one another. In terms of this latter interpretation, (5) may be shown to be algebraically equivalent to

$$\text{(6)} \qquad SS_{\text{methods}} = \frac{n\Sigma(\bar{T}_j - \bar{T}_{j'})^2}{k},$$

where the symbol $\bar{T}_j - \bar{T}_{j'}$ designates the difference between a pair of means. For the data in the table, (6) is

$$SS_{\text{methods}} = \frac{8[(4.75 - 4.62)^2 + (4.75 - 7.75)^2 + (4.62 - 7.75)^2]}{3}$$

$$= 50.17.$$

Using (5),

$$SS_{\text{methods}} = 8[(4.75 - 5.71)^2 + (4.62 - 5.71)^2 + (7.75 - 5.71)^2]$$

$$= 50.17.$$

Neither (5) nor (6) is a convenient computational formula for SS_{methods}: the latter is given by

$$\text{(7)} \qquad SS_{\text{methods}} = \frac{\Sigma T_j^2}{n} - \frac{G^2}{nk}.$$

For the data in the table, (7) is numerically

$$SS_{\text{methods}} = \frac{(38)^2 + (37)^2 + (62)^2}{8} - \frac{(137)^2}{24}$$

$$= 832.12 - 782.04 = 50.08.$$

The numerical value for SS_{methods} computed from (5) and (6) involves more rounding errors than does the computational formula (7).

The variation due to experimental error is, by definition, the pooled within-method variation,

$$\text{(8)} \qquad SS_{\text{error}} = SS_w = \Sigma SS_j.$$

This statistic measures the sum of the variation within each of the subgroups. Its computational formula is given by (4). The total variation, or total sum of squares, is

$$\text{(9)} \qquad SS_{\text{total}} = \Sigma\Sigma(X_{ij} - \bar{G})^2,$$

the sum of the squared deviation of each observation in the experiment about the grand mean. Its computational formula is

$$\text{(10)} \qquad SS_{\text{total}} = \Sigma(\Sigma X_j^2) - \frac{G^2}{nk}.$$

For the data in the table,

$$SS_{\text{total}} = 919 - \frac{(137)^2}{24}$$

$$= 136.96.$$

From these definitions of SS_{total}, SS_{methods}, and SS_{error}, it may be shown algebraically that

$$\text{(11)} \qquad \begin{aligned} SS_{\text{total}} &= SS_{\text{methods}} + SS_{\text{error}}, \\ \Sigma\Sigma(X_{ij} - \bar{G})^2 &= n\Sigma(\bar{T}_j - \bar{G})^2 + \Sigma\Sigma(X_{ij} - \bar{T}_j)^2. \end{aligned}$$

The relation (11) describes a *partition*, or division, of the total variation into two additive parts. One part is a function of differences between the mean scores made by the subgroups having different methods of training; the other part is the sum of the variation of scores within subgroups. Numerically,

$$136.96 = 50.08 + 86.88.$$

The partition represented by (11) is basic to the analysis of variance. Its derivation is not difficult. Let

$$a_{ij} = X_{ij} - \bar{T}_j \qquad \text{and} \qquad b_j = \bar{T}_j - \bar{G}.$$

Then
$$a_{ij} + b_j = X_{ij} - \bar{G},$$

and
$$\sum_i\sum_j(X_{ij} - \bar{G})^2 = \sum_i\sum_j(a_{ij} + b_j)^2 = \sum_i\sum_j a_{ij}^2 + \sum_i\sum_j b_j^2 + 2\sum_i\sum_j a_{ij}b_j.$$

The term at the extreme right is

$$\sum_i\sum_j a_{ij}b_j = \sum_j b_j(\sum_i a_{ij}) = 0,$$

since $\Sigma_i a_{ij} = 0$ for each j. (That is, $\Sigma_i a_{ij}$ is the sum of deviations about the mean of observations in class j.) Since b_j^2 is a constant for all i's in the same class,

$$\sum_i\sum_j b_j^2 = n\sum_j b_j^2.$$

Hence
$$\begin{aligned} \sum_i\sum_j(X_{ij} - \bar{G})^2 &= \sum_i\sum_j a_{ij}^2 + n\sum_j b_j^2 \\ &= \sum_i\sum_j(X_{ij} - \bar{T}_j)^2 + n\sum_j(\bar{T}_j - \bar{G})^2. \end{aligned}$$

A variance, in the terminology of analysis of variance, is more frequently called a mean square (abbreviated MS). By definition

$$\text{(12)} \qquad \text{Mean square} = \frac{\text{variation}}{\text{degrees of freedom}} = \frac{\text{SS}}{\text{df}}.$$

In words, a mean square is the average variation per degree of freedom; this is also the basic definition for a variance. The term mean square is a more general term for the average of squared measures. Hence a variance is actually a special case of a mean square.

The term *degrees of freedom* originates from the geometric representation of problems associated with the determination of sampling distributions for statistics. In this context the term refers to the dimension of the geometric space appropriate in the solution of the problem. The following definition permits the computation of the degrees of freedom for any source of variation:

(13)

$$\text{Degrees of freedom} = \begin{pmatrix}\text{no. of independent}\\ \text{observations on}\\ \text{source of variation}\end{pmatrix} - \begin{pmatrix}\text{no. of independent}\\ \text{parameters estimated}\\ \text{in computing variation}\end{pmatrix}.$$

In this context, a statistic may be used either as an estimate of a parameter or as a basic observation in estimating a source of variation. The source of variation being estimated will indicate what role a particular statistic will have in a specified context. More accurately,

(13*a*) df = (no. independent observations) − (no. linear restraints).

Substituting an $\bar{X}_j$ for a μ_j in the computation of a mean square is equivalent to imposing a linear restraint upon the estimation procedure. The substitution restricts the sum of a set of observations to be a specified number. For example, if the mean of four scores is required to be 10, and if the first three observations are

$$3, \; -5, \text{ and } 20,$$

then the fourth score must be 22; that is, the total must be 40. Under this restraint on the mean, only three of the four scores are free to vary. Hence the term *freedom*.

In the computation of SS_{methods}, the $\bar{T}_j$'s are considered to be the basic observations; SS_{methods} is a measure of the variation of the $\bar{T}_j$'s. Thus, there are k independent observations in SS_{methods}. In the computation of this source of variation, $\bar{G}$ is used as a parameter estimating the mean of the $\bar{T}_j$'s. Hence one estimate of a parameter is used in the computation of SS_{methods}. Therefore, by (13), the degrees of freedom for this source of variation are $k - 1$.

An alternative, computational definition of the degrees of freedom for a source of variation is

$$\text{(13}b\text{)}\quad \text{Degrees of freedom} = \begin{pmatrix}\text{no. squared}\\ \text{deviations}\end{pmatrix} - \begin{pmatrix}\text{no. independent points}\\ \text{about which deviations}\\ \text{are taken}\end{pmatrix}.$$

For example, $SS_{\text{methods}} = n\Sigma(\bar{T}_j - \bar{G})^2$ involves k squared deviations all taken about the single point $\bar{G}$. Hence the degrees of freedom are $k - 1$. As another example, $SS_{\text{error}} = \Sigma SS_j$ involves n squared deviations for each of the k subgroups, or a total of kn squared deviations. Within each subgroup the deviations are taken about the mean $\bar{T}_j$ of that subgroup. Since there are k subgroups, there are k different points about which the deviations are taken. Hence the degrees of freedom are $nk - k$ for SS_{error}.

SS_{error} is the pooled variation within each of the subgroups. The variation within the subgroup j is a measure of the extent to which each of the n observations deviates from the mean of the subgroup, $\bar{T}_j$. For this source of variation $\bar{T}_j$ is used as an estimate of the mean for population j. Hence the number of degrees of freedom for the variation within subgroup j is $n - 1$. The degrees of freedom for the pooled within-subgroup variation are the sum of the degrees of freedom for each of the subgroups. If the variation within each of k subgroups has $n - 1$ degrees of freedom, then the total degrees of freedom for k subgroups is $k(n - 1)$, that is, the sum $(n - 1) + (n - 1) + \cdots + (n - 1)$ for k terms.

SS_{total} is the variation of the nk independent observations about the grand mean $\bar{G}$. Here $\bar{G}$ is an estimate of the overall population mean. Hence SS_{total} is based upon $nk - 1$ degrees of freedom. Corresponding to the partition of the total variation in (11), there is a partition of the total degrees of freedom

$$\text{(14)}\quad \begin{aligned} df_{\text{total}} &= df_{\text{methods}} + df_{\text{error}},\\ kn - 1 &= (k - 1) + (kn - k). \end{aligned}$$

A summary of the statistics used in the analysis of a single-factor experiment is given in Table 3.2-2. The numerical entries are those computed from the data in Table 3.2-1.

Table 3.2-2 Summary of Analysis of Variance

Source of variation	Sum of squares	Degrees of freedom	Mean square
Between methods	$SS_{\text{methods}} = 50.08$	$k - 1 = 2$	$MS_{\text{methods}} = 25.04$
Experimental error	$SS_{\text{error}} = 86.88$	$kn - k = 21$	$MS_{\text{error}} = 4.14$
Total	$SS_{\text{total}} = 136.96$	$kn - 1 = 23$	

Assuming that there is no difference in the effectiveness of the methods of training, as measured by the mean scores on the test, and making additional assumptions which will become explicit in the next section, the statistic

$$F = \frac{\text{MS}_{\text{methods}}}{\text{MS}_{\text{error}}} \tag{15}$$

has a sampling distribution which is approximated by an F distribution having $k - 1$ degrees of freedom for the numerator and $kn - k$ degrees of freedom for the denominator. Thus the F statistic may be used to test hypotheses about the equality of the population means for the methods. To test the hypothesis that the population means for the test scores are equal, that is, $\mu_1 = \mu_2 = \mu_3$, against a two-tailed alternative hypothesis, the decision rules are as follows:

Reject H_1 when $F_{\text{obs}} > F_{1-\alpha}(k - 1, kn - k)$.
Otherwise do not reject H_1.

For the data in Table 3.2-2, $F_{\text{obs}} = 25.04/4.14 = 6.05$. Critical values for $\alpha = .05$ and $\alpha = .01$ are, respectively, $F_{.95}(2,21) = 3.47$ and $F_{.99}(2,21) = 5.78$. In this case F_{obs} exceeds the critical value for $\alpha = .01$. Hence the data do not support the hypothesis that the population means are equal. Inspection of the means in Table 3.2-1 indicate that method 3 has the largest mean.

The experiment represented by the data in Table 3.2-1 is a special case of a single-factor experiment. For this case, the experimental variable is the method of training. In the general case the term treatment will be used interchangeably with the terms experimental variable, experimental condition, or whatever it is that distinguishes the manner in which the subgroups are handled (treated) in the experiment. The elements assigned to a treatment constitute what will be called a treatment class. A general notation for the observed data in a single-factor experiment having n observations in each treatment class is given in part i of Table 3.2-3. For example, an observation on the element i in treatment class j is designated X_{ij}. Notation for the totals required in the computation of the sums of squares appears in part ii. For example, T_j designates the sum of all observations in treatment class j, and ΣX_j^2 designates the sum of the squares of the observations in treatment class j.

In part iii the computational formulas for the sums of squares used in the analysis of variance and the associated degrees of freedom are given. A convenient method for summarizing the computational formulas is given in part iv in terms of what may be called computational symbols. For example, the symbol (2) designates the numerical value of $\Sigma(\Sigma X_j^2)$.

Table 3.2-3 General Notation

	Treatment 1	...	Treatment j	...	Treatment k	
(i)	X_{11} X_{21} . . . X_{i1} . . . X_{n1}		X_{1j} X_{2j} . . . X_{ij} . . . X_{nj}		X_{1k} X_{2k} . . . X_{ik} . . . X_{nk}	
(ii)	T_1 ΣX_1^2 $\bar{T}_1$	...	T_j ΣX_j^2 $\bar{T}_j$	...	T_k ΣX_k^2 $\bar{T}_k$	G $\Sigma(\Sigma X_j^2)$ $\bar{G}$

(iii)

$$SS_{treat} = \frac{\Sigma T_j^2}{n} - \frac{G^2}{kn} \qquad df_{treat} = k - 1$$

$$SS_{error} = \Sigma(\Sigma X_j^2) - \frac{\Sigma T_j^2}{n} \qquad df_{error} = kn - k$$

$$SS_{total} = \Sigma(\Sigma X_j^2) - \frac{G^2}{kn} \qquad df_{total} = kn - 1$$

(iv)

Computational symbols	Sums of squares in terms of computational symbols
$(1) = G^2/kn$	$SS_{treat} = (3) - (1)$
$(2) = \Sigma(\Sigma X_j^2)$	$SS_{error} = (2) - (3)$
$(3) = (\Sigma T_j^2)/n$	$SS_{total} = (2) - (1)$

The degrees of freedom for a sum of squares may be computed directly from the computational formula by means of the following rule: Count the number of quantities which are squared in a term. Then replace this term by this number in the computational formula. For example, in the term $(\Sigma T_j^2)/n$ there are k quantities that are squared ($T_1, T_2, \ldots, T_k$). In the term G^2/kn there is just one term which is squared, namely, G. Hence the degrees of freedom of the sum of squares defined by $[(\Sigma T_j^2)/n - G^2/kn]$ are $k - 1$. As another example, in the term $\Sigma(\Sigma X_j^2)$ there are kn terms that are squared, namely, each of the kn individual observations. Hence the sum of squares defined by $[\Sigma(\Sigma X_j^2) - G^2/kn]$ has $kn - 1$ degrees of freedom.

The general form used in summarizing the analysis of variance for a single-factor experiment is given in Table 3.2-4. The F statistic is used in testing the hypothesis that $\mu_1 = \mu_2 = \cdots = \mu_k$ against the equivalent of a two-tailed alternative hypothesis. If this hypothesis is rejected, additional tests are required for more detailed information about which means are different from the others. Specialized tests for comparing individual means with each other are discussed in later sections.

Table 3.2-4 General Form of Summary Data

Source	SS	df	MS	F
Treatments	SS_{treat}	$k - 1$	MS_{treat}	$F = \frac{MS_{treat}}{MS_{error}}$
Experimental error	SS_{error}	$kn - k$	MS_{error}	
Total	SS_{total}	$kn - 1$		

The formal method for testing statistical hypotheses requires that the level of significance of a test be set in advance of obtaining the data. Convention in the analysis of variance is somewhat opposed to this procedure. The value of F_{obs} is generally compared with tabled critical values, and the outcome is described in terms of the statement: F_{obs} exceeds a specified percentile point (usually the 95 or the 99 percentile points). The choice of the level of significance is thus in part determined by the observed data. This procedure is not objectionable for purposes of estimating the probability of the observed outcome in terms of an assumed underlying sampling distribution.

3.3 Structural Model for Single-factor Experiment—Model I

Suppose all elements in a specified basic population are given treatment 1. After the treatment, the elements are measured on a criterion related to the effectiveness of the treatment. Assume that the distribution of the resulting measurements is approximately normal in form, with parameters μ_1 and σ_1^2. Thus, μ_1 and σ_1^2 are the parameters of a population of measurements that would exist if treatment 1 were administered. This potential population will be designated as that corresponding to treatment 1. Suppose, instead, that treatment 2 is administered to all the elements and then measurements on the same criterion of effectiveness are made. Assume that the resulting distribution is approximately normal in form, with parameters μ_2 and σ_2^2.

Corresponding to each treatment about which the experimenter seeks to make inferences, there is assumed to be a population of approximately normally distributed criterion measures. The number of such populations is equal to the number of treatments. Assume that the number of treatments is K. In the experiment, data are obtained on k of the possible K treatments. If $k = K$, then observed data are available on all treatments in the domain to which inferences are to be made. If k is less than K, then observed data are available on only some of the treatments about which inferences are to be drawn. In this section the case in which $k = K$ will be considered. Other cases are considered in the next section.

The parameters defining the treatment populations are summarized in Table 3.3-1. When $k = K$, the grand mean μ is

$$\mu = \frac{\Sigma \mu_j}{k}. \tag{1}$$

Table 3.3-1 Parameters of Populations Corresponding to Treatments in the Experiment

Treatment	Population mean	Population variance	Treatment effect
1	μ_1	σ_1^2	$\tau_1 = \mu_1 - \mu$
2	μ_2	σ_2^2	$\tau_2 = \mu_2 - \mu$
.	.	.	.
.	.	.	.
.	.	.	.
j	μ_j	σ_j^2	$\tau_j = \mu_j - \mu$
.	.	.	.
.	.	.	.
.	.	.	.
j'	$\mu_{j'}$	$\sigma_{j'}^2$	$\tau_{j'} = \mu_{j'} - \mu$
.	.	.	.
.	.	.	.
.	.	.	.
$k = K$	μ_k	σ_k^2	$\tau_k = \mu_k - \mu$
	Grand mean $= \mu$		$\bar{\tau} = 0$

The effect of treatment j, designated τ_j, is the difference between the mean for treatment j and the grand mean of the population means,

$$\tau_j = \mu_j - \mu. \tag{2}$$

Thus τ_j is a parameter which measures the degree to which the mean for treatment j differs from the mean of all other relevant population means. Since the sum of the deviations about the mean of a set is zero, $\Sigma\tau_j = 0$; hence the mean of the τ_j's, designated $\bar{\tau}$, is equal to zero.

Let X_{ij} be the criterion measure on a randomly selected element i in treatment population j. The following structural model is assumed for this measurement,

$$X_{ij} = \mu + \tau_j + \varepsilon_{ij}, \tag{3}$$

where μ = grand mean of treatment populations,

τ_j = effect of treatment j,

ε_{ij} = experimental error.

The term μ is constant for all measurements in all treatment populations. The effect τ_j is constant for all measurements within population j; however, a different value, say $\tau_{j'}$, is associated with population j', where j' represents some treatment other than j. The experimental error ε_{ij} represents all the uncontrolled sources of variance affecting individual measurements; this effect is unique for each of the elements i in the basic population. This effect is further assumed to be independent of τ_j.

Since both μ and τ_j are constant for all measurements within population j, the only source of variance for these measurements is that due to experimental error. Thus

(4) $$\sigma_j^2 = \sigma_{\varepsilon_j}^2,$$

where $\sigma_{\varepsilon_j}^2$ designates the variance due to experimental error for the measurements within treatment population j. If $X_{ij'}$ represents a measurement in population j', then (3) takes the form

(5) $$X_{ij'} = \mu + \tau_{j'} + \varepsilon_{ij'}.$$

The variance within population j' is due solely to the experimental error; hence

(6) $$\sigma_{j'}^2 = \sigma_{\varepsilon_{j'}}^2.$$

If the elements (subjects, animals) that are observed under each of the treatment conditions are assigned at random to these conditions, one has some degree of assurance that the error effects will be independent of the treatment effects. Hence the importance of randomization in design problems. The elements, in this context, are called the experimental units. Random assignment of the experimental units to the experimental conditions tends to make the unique effects of the units per se independent of the treatment effects.

The experimental error measures all uncontrolled effects which are not related to the treatments. As such, the experimental error is the combined effect of many random variables that are independent of the treatment effect. Under these conditions, it is reasonable to assume that the distribution of the ε_{ij}'s within population j will be approximately normal in form, with expected value $\mu_{\varepsilon_j} = 0$ and variance $\sigma_{\varepsilon_j}^2$. If the sources of experimental error are comparable in each of the treatment populations, it is also reasonable to assume that

(7) $$\sigma_{\varepsilon_j}^2 = \sigma_{\varepsilon_{j'}}^2 .$$

This last relationship may be written in more general form as

(8) $$\sigma_{\varepsilon_1}^2 = \sigma_{\varepsilon_2}^2 = \cdots = \sigma_{\varepsilon_k}^2 = \sigma_{\varepsilon}^2,$$

where σ_ε^2 is the variance due to experimental error within any of the treatment populations.

To summarize the assumptions underlying the structural model (3), a measurement X_{ij} is expressed as the sum of three components: (1) a component μ which is constant for all treatments and all elements; (2) a component τ_j which is constant for all elements within a treatment population but may differ for different treatment populations; (3) a component ε_{ij}, independent of τ_j, and distributed as $N(0,\sigma_\varepsilon^2)$ within each treatment population. This structural model is called the *fixed-constants model*, or model I. The component μ in model I is actually an unknown constant; the component τ_j is a systematic or fixed component which depends upon the difference between the means of the treatment populations; the component ε_{ij} is a random component (or a random variable) depending upon uncontrolled sources of variances assumed to be drawn randomly from a population in which the distribution is $N(0,\sigma_\varepsilon^2)$.

A parameter indicating the extent to which the treatment effects differ is

$$\sigma_\tau^2 = \frac{\Sigma\tau_j^2}{k-1} = \frac{\Sigma(\mu_j - \mu)^2}{k-1}. \tag{9}$$

An equivalent definition in terms of differences between treatment effects is

$$\sigma_\tau^2 = \frac{\Sigma(\tau_j - \tau_{j'})^2}{k(k-1)} = \frac{\Sigma(\mu_j - \mu_{j'})^2}{k(k-1)}, \tag{10}$$

where the summation is over all the different possible pairs of means; there are $k(k-1)/2$ such distinct pairs. When the treatment effects are equal, i.e., when $\tau_1 = \tau_2 = \cdots = \tau_k$, σ_τ^2 will be zero. The larger the differences between the τ's the larger will be σ_τ^2. Thus the hypothesis specifying that $\sigma_\tau^2 = 0$ is equivalent to the hypothesis that specifies $\tau_1 = \tau_2 = \cdots = \tau_k$ or $\mu_1 = \mu_2 = \cdots = \mu_k$.

Statistics useful in estimating the parameters in Table 3.3-1 are summarized in Table 3.3-2. Since model I assumes that the population variances are all equal to σ_ε^2, the best estimate of this parameter is the pooled within-class sample variance

$$s_{\text{pooled}}^2 = \frac{\Sigma s_j^2}{k} = \text{MS}_{\text{error}}. \tag{11}$$

For this design the pooled within-class variance is designated MS_{error}. The latter is an unbiased estimate of σ_ε^2; that is,

$$E(\text{MS}_{\text{error}}) = \sigma_\varepsilon^2.$$

In terms of the structural model in (3), the treatment means for the samples of n observations may be expressed as

$$\begin{aligned} \bar{T}_1 &= \mu + \tau_1 + \bar{\varepsilon}_1, \\ \bar{T}_2 &= \mu + \tau_2 + \bar{\varepsilon}_2, \\ &\cdots \\ \bar{T}_k &= \mu + \tau_k + \bar{\varepsilon}_k, \end{aligned} \tag{12}$$

Table 3.3-2 Estimates of Parameters of Treatment Populations

Sample size	Treatment	Sample mean	Sample variance	Treatment effect
n	1	$\bar{T}_1$	s_1^2	$t_1 = \bar{T}_1 - \bar{G}$
n	2	$\bar{T}_2$	s_2^2	$t_2 = \bar{T}_2 - \bar{G}$
.	.	.	.	.
.	.	.	.	.
.	.	.	.	.
n	j	$\bar{T}_j$	s_j^2	$t_j = \bar{T}_j - \bar{G}$
.	.	.	.	.
.	.	.	.	.
.	.	.	.	.
n	j'	$\bar{T}_{j'}$	$s_{j'}^2$	$t_{j'} = \bar{T}_{j'} - \bar{G}$
.	.	.	.	.
.	.	.	.	.
.	.	.	.	.
n	k	$\bar{T}_k$	s_k^2	$t_k = \bar{T}_k - \bar{G}$
		Grand mean $= \bar{G}$		$\bar{t} = 0$

where $\bar{\varepsilon}_j$ is the mean experimental error for a sample of n observations within treatment class j. For random samples, it is reasonable to assume that $E(\bar{\varepsilon}_j) =$ constant. Without any loss in generality, this constant may be assumed to be zero. Therefore,

$$
\begin{aligned}
E(\bar{T}_1) &= \mu + \tau_1, \\
E(\bar{T}_2) &= \mu + \tau_2, \\
&\cdots \\
E(\bar{T}_k) &= \mu + \tau_k.
\end{aligned} \tag{13}
$$

The first relation in (13) may be interpreted as follows: If an infinite number of random samples of size n are given treatment 1 and the statistic $\bar{T}_1$ is computed for each of the samples, the distribution of the resulting statistic would have an expected value (or mean) equal to $\mu + \tau_1$.

The statistic t_j in Table 3.3-2 may be represented as

$$
\begin{aligned}
t_j = \bar{T}_j - \bar{G} &= (\mu + \tau_j + \bar{\varepsilon}_j) - (\mu + \bar{\varepsilon}) \\
&= \tau_j + \bar{\varepsilon}_j - \bar{\varepsilon}.
\end{aligned} \tag{14}
$$

where $\bar{\varepsilon}$ is the mean of the $\bar{\varepsilon}_j$'s. The expected value of t_j is equal to τ_j, since the expected value of the quantity $\bar{\varepsilon}_j - \bar{\varepsilon}$ equals 0.

A measure of the degree to which the sample means for the various treatments differ is provided by the statistic

$$
\text{MS}_{\text{treat}} = \frac{n\Sigma(\bar{T}_j - \bar{G})^2}{k - 1} = \frac{n\Sigma t_j^2}{k - 1}. \tag{15}
$$

Inspection of the right-hand side of the relation (14) indicates that differences among the $\bar{T}$'s depend upon differences among the τ_j's and differences among the $\bar{\varepsilon}_j$'s. Since the τ_j's and the $\bar{\varepsilon}_j$'s are assumed to be independent, the variance of the $\bar{T}$'s is the sum of the variance of the τ's and the variance of the $\bar{\varepsilon}$'s. Therefore n times the variance of the $\bar{T}$'s has the expected value

$$(16) \qquad E(MS_{treat}) = \frac{n\Sigma\tau_j^2}{k-1} + \frac{n\Sigma\bar{\varepsilon}_j^2}{k-1} = n\sigma_\tau^2 + n\sigma_{\bar{\varepsilon}}^2.$$

Since the variance of the mean of n observations is $1/n$ times the variance of the individual observations, using the relationship that $\sigma_\varepsilon^2 = n\sigma_{\bar{\varepsilon}}^2$ permits (16) to be written as

$$(17) \qquad E(MS_{treat}) = n\sigma_\tau^2 + \sigma_\varepsilon^2.$$

Thus, when $\tau_1 = \tau_2 = \cdots = \tau_k$, the term σ_τ^2 is equal to zero and the expected value of MS_{treat} is σ_ε^2. This implies that MS_{treat} is an unbiased estimate of the variance due to experimental error when there are no differences among the treatment effects.

Note: For the case in which the sample size under treatment j is n_j, where n_j is not the same for all treatments, it can be shown that

$$E(MS_{treat}) = \frac{N^2 - \Sigma n_j^2}{N(k-1)}\sigma_\tau^2 + \sigma_\varepsilon^2,$$

where $N = \Sigma n_j$. (See Graybill, 1961, pp. 353–354.) When $n_j = n$ for all treatments,

$$N = kn \qquad \text{and} \qquad \Sigma n_j^2 = kn^2.$$

Hence for the case of equal sample sizes,

$$\frac{N^2 - \Sigma n_j^2}{N(k-1)} = \frac{k^2n^2 - kn^2}{kn(k-1)} = \frac{kn^2(k-1)}{kn(k-1)} = n.$$

A more general approach to the result obtained in (17) is to start from the relation

$$(18) \qquad \bar{T}_j - \bar{T}_{j'} = (\tau_j - \tau_{j'}) + (\bar{\varepsilon}_j - \bar{\varepsilon}_{j'}).$$

This relation is obtained from (12). In this form, a difference between two treatment means estimates a difference between two treatment effects and a difference between two average error effects. Further, $\bar{T}_j - \bar{T}_{j'}$ provides an unbiased estimate of $\tau_j - \tau_{j'}$. MS_{treat} is readily expressed in terms of all possible pairs of differences among the $\bar{T}$'s. Similarly σ_τ^2 and σ_ε^2 may be expressed in terms of the differences on the right-hand side of (18). In the derivation of the expected values for the mean square due to treatments, the more direct approach is by means of relations analogous to (18).

The expected values of the mean squares which are used in the analysis of variance are summarized in Table 3.3-3. The ratio of the expected values has the form

$$\frac{\text{E}(\text{MS}_{\text{treat}})}{\text{E}(\text{MS}_{\text{error}})} = \frac{\sigma_\varepsilon^2 + n\sigma_\tau^2}{\sigma_\varepsilon^2}.$$

When $\sigma_\tau^2 = 0$, the expected value of the numerator is σ_ε^2. Thus, on condition that $\sigma_\tau^2 = 0$, numerator and denominator are unbiased, independent estimates of σ_ε^2. The independence of the estimators follows from the fact that the sampling distribution of the within-class variance is independent of the class means. MS_{error} is obtained from the within-class data, whereas MS_{treat} is obtained from the class means.

Table 3.3-3 Expected Values of Mean Squares (n Observations per Treatment)

Source	df	MS	E(MS)
Treatments	$k - 1$	MS_{treat}	$\sigma_\varepsilon^2 + n\sigma_\tau^2$
Experimental error	$kn - k$	MS_{error}	σ_ε^2

On condition that $\sigma_\tau^2 = 0$, under relatively weak assumptions, the sampling distribution of the statistic

$$F = \frac{\text{MS}_{\text{treat}}}{\text{MS}_{\text{error}}} \tag{19}$$

can be shown to be that of the F distribution having $k - 1$ degrees of freedom for the numerator and $kn - k$ degrees of freedom for the denominator. It should be noted that

$$\text{E}(F) \neq \frac{\text{E}(\text{MS}_{\text{treat}})}{\text{E}(\text{MS}_{\text{error}})}, \tag{20}$$

since, in general, the expected value of any ratio is not equal to the ratio of expected values, even though the latter may be equal. Actually,

$$\text{E}(F) = \frac{\text{df}_{\text{denominator}}}{\text{df}_{\text{denominator}} - 2} \qquad \text{on condition that } \sigma_\tau^2 = 0. \tag{21}$$

Thus, on condition that $\sigma_\tau^2 = 0$,

$$\frac{\text{E}(\text{MS}_{\text{treat}})}{\text{E}(\text{MS}_{\text{error}})} = 1 \qquad \text{but} \qquad \text{E}(F) > 1.$$

This F statistic may be used to test the hypothesis that $\sigma_\tau^2 = 0$, which is equivalent to the hypothesis that $\tau_1 = \tau_2 = \cdots = \tau_k$. The assumptions under which this test is valid are those underlying model I; these assumptions were discussed in connection with the structural model. The magnitude of type 1 error is not seriously affected if the distributions depart moderately from normality or if the population variances depart moderately from equality; i.e., the test is robust with respect to the assumptions of normality of distribution and homogeneity of error variance. A study of the effect of lack of homogeneity of error variance is given in the work of Box (1954).

When $\sigma_\tau^2 \neq 0$, the expected value of the F statistic will be greater than 1.00 by an amount which depends in part upon the magnitude of σ_τ^2. Thus, if the F ratio is larger than 1.00 by an amount having low probability when $\sigma^2 = 0$, the inference is that $\sigma_\tau^2 \neq 0$.

The notation ave(MS), read average value of a mean square, is used by some authors in place of the notation E(MS). Also the notation σ_τ^2 is sometimes reserved for use in connection with model II, and no special symbol is used to designate the quantity $n(\Sigma\tau_j^2)/(k-1)$. Thus, for model I

$$E(MS_{treat}) = \sigma_\varepsilon^2 + \frac{n\Sigma\tau_j^2}{k-1}.$$

As long as the assumptions underlying the model are made explicit, no ambiguity will result in the use of the symbol σ_τ^2 in model I.

3.4 Structural Model for Single-factor Experiment—Model II (Variance-component Model)

One of the basic assumptions underlying model I is that all treatments about which inferences are to be made are included in the experiment. Thus, if the experiment were to be replicated, the same set of treatments would be included in each of the replications. Model I is usually the most appropriate for a single-factor experiment. If, however, the k treatments that are included in a given experiment constitute a random sample from a collection of K treatments, where k is small relative to K, then upon replication a different random sample of k treatments will be included in the experiment.

Model II covers this latter case. The structural equation for model II has the same form as that of model I, namely,

$$X_{ij} = \mu + \tau_j + \varepsilon_{ij}. \tag{1}$$

However, the assumptions underlying this model are different. The term μ is still assumed to be constant for all observations; the term ε_{ij} is still assumed to have the distribution $N(0,\sigma_\varepsilon^2)$ for all treatments; but the term τ_j is now considered to be a random variable. The distribution of τ_j is assumed to be $N(0,\sigma_\tau^2)$. In the variance-component model, the primary objective is to estimate σ_τ^2 rather than the individual τ_j.

From the point of view of the computation of estimates of the various effects, the two models do not lead to different procedures. For samples of size n, the expected values of the mean squares of the estimators are as follows:

Source	df	MS	E(MS)
Treatments	$k - 1$	MS_{treat}	$\sigma_\varepsilon^2 + n\sigma_\tau^2$
Experimental error	$kn - k$	MS_{error}	σ_ε^2

Here the expected value of the mean square for treatments is the sum of the variances of two independent variables, that is, $\sigma_\varepsilon^2 + n\sigma_\tau^2$. This latter value was the one that was obtained under model I. The test on the hypothesis $\sigma_\tau^2 = 0$ is identical with that made under model I. In more complex designs, the F ratios for analogous tests do not have identical forms.

A more comprehensive treatment of the variance-component model is given in Sec. 3.17. The distribution theory underlying model II is actually quite different from the corresponding theory underlying model I.

3.5 Methods for Deriving Estimates and Their Expected Values

Under model I, the structural equation for an observation has the form

$$X_{ij} = \mu + \tau_j + \varepsilon_{ij}, \qquad (1)$$

where μ is a constant for all observations, τ_j is a constant (or fixed variable) for observations under treatment j, and ε_{ij} is a random variable independent of τ_j having the distribution $N(0,\sigma_\varepsilon^2)$ for all treatments. Thus, there are k fixed variables and one random variable in the model. This model is similar to that underlying a k-variate (fixed) multiple regression equation; hence the methods for the solution of estimation problems in multiple regression are applicable. The method of least squares, which is used in the solution of multiple regression problems, will yield unbiased minimum-variance estimators when applied to estimation problems posed by model I. (The usual multiple regression problem involves correlated predictors. In this case, the predictors are uncorrelated. The latter restriction simplifies the computational procedures, but the underlying principles are identical.)

The generalization of model I forms a major part of what is known as classic analysis of variance; estimation problems in this area may be handled by means of least-squares analysis or by means of maximum-likelihood methods. The two approaches lead to identical results for this model.

In this section only a restricted version of model I will be considered—namely, that special case in which the constraint $\Sigma\tau_j = 0$ is imposed upon the model. Without this constraint it is not possible to obtain unbiased estimators of the individual τ_j. The more general approach to model I, which does not impose this constraint on the τ_j, will be considered in a later section. Under the more general approach unbiased estimators can be obtained only for certain linear functions of the parameters in the linear model. The latter are known as *estimable* parametric functions.

To illustrate the least-squares approach to (1), let $m, t_1, t_2, \ldots, t_k$, and e_{ij} designate the respective least-squares estimators of $\mu, \tau_1, \tau_2, \ldots, \tau_k$, and ε_{ij}. The method of least squares minimizes the quantity

$$\Sigma\Sigma e_{ij}^2 = \sum_i \sum_j (X_{ij} - m - t_j)^2 = \text{minimum}. \tag{2}$$

By taking partial derivatives with respect to the estimators $m, t_1, t_2, \ldots, t_k$ (this is the procedure used to obtain the normal equations in the multiple regression problem for fixed variables) and setting these derivatives equal to zero, one obtains the equations

$$\begin{aligned} knm + n\Sigma t_j - G &= 0, \\ nm + nt_1 - T_1 &= 0, \\ nm + nt_2 - T_2 &= 0, \\ &\cdots \\ nm + nt_k - T_k &= 0. \end{aligned} \tag{3}$$

There are $k + 1$ equations and $k + 1$ unknowns; however, the $k + 1$ equations are not independent, since the first equation may be obtained by summing the last k equations. A unique solution to these equations requires a restriction on the variables. Since $\Sigma\tau_j = 0$, a reasonable restriction is that $\Sigma t_j = 0$. The solutions to (3) now take the form

$$\begin{aligned} m &= \frac{G}{kn} = \bar{G}, \\ t_1 &= \frac{T_1 - n\bar{G}}{n} = \bar{T}_1 - \bar{G}, \\ t_2 &= \frac{T_2 - n\bar{G}}{n} = \bar{T}_2 - \bar{G}, \\ &\cdots \\ t_k &= \frac{T_k - n\bar{G}}{n} = \bar{T}_k - \bar{G}. \end{aligned} \tag{4}$$

To show that these estimators are unbiased, one has

$$\begin{aligned} \mathrm{E}(m) = \mathrm{E}(\bar{G}) &= \mathrm{E}\left(m + \frac{\Sigma t_j}{k} + \frac{\Sigma\varepsilon_{ij}}{kn}\right) \\ &= \mu + 0 + 0 \qquad = \mu, \end{aligned}$$

$$\begin{aligned} \mathrm{E}(t_j) = \mathrm{E}(\bar{T}_j - \bar{G}) &= \mathrm{E}(\bar{T}_j) - \mathrm{E}(\bar{G}) \\ &= \mathrm{E}(\mu + \tau_j + \bar{\varepsilon}_j) - \mu \qquad = \tau_j. \end{aligned}$$

Although the expected value of t_j is the parameter τ_j, in terms of the parameters of the model, t_j has the form

$$(5) \qquad t_j = \bar{T}_j - \bar{G} = \left(\mu + \tau_j + \frac{\Sigma_i \varepsilon_{ij}}{n}\right) - \left(\mu + \frac{n\Sigma\tau_j}{kn} + \frac{\Sigma\Sigma\varepsilon_{ij}}{kn}\right)$$

$$= \tau_j + \bar{\varepsilon}_j - \bar{\varepsilon},$$

where $\bar{\varepsilon}_j = (\Sigma_i \varepsilon_{ij})/n$ and $\bar{\varepsilon} = (\Sigma\Sigma\varepsilon_{ij})/kn$. Thus,

$$(6) \qquad \text{MS}_{\text{treat}} = \frac{n\Sigma t_j^2}{k-1} = \frac{n\Sigma(\tau_j + \bar{\varepsilon}_j - \bar{\varepsilon})^2}{k-1}.$$

Since the τ's and the ε's are independent, the expected value of MS_{treat} will be n times the sum of the mean squares of the parameters on the right-hand side of (6), that is, $n(\sigma_\tau^2 + \sigma_{\bar{\varepsilon}}^2) = n\sigma_\tau^2 + \sigma_\varepsilon^2$. This is a general property of least-squares estimators: the expected value of the mean square of an estimator is the sum of the respective mean squares of the parameters estimated. The term

$$(7) \qquad \begin{aligned} e_{ij} &= X_{ij} - m - t_j = X_{ij} - \bar{G} - (\bar{T}_j - \bar{G}) \\ &= X_{ij} - \bar{T}_j = (\mu + \tau_j + \varepsilon_{ij}) - (\mu + \tau_j + \bar{\varepsilon}_j) \\ &= \varepsilon_{ij} - \bar{\varepsilon}_j. \end{aligned}$$

Thus

$$(8) \qquad \text{E}(e_{ij}^2) = \text{E}\left\{\frac{\Sigma\Sigma(X_{ij} - \bar{T}_j)^2}{kn - k}\right\} = \sigma_\varepsilon^2.$$

Model II, which has more than one random variable, does not permit the direct application of least-squares methods. Instead, maximum-likelihood methods are used to obtain estimators and the expected values of the mean squares of these estimators. General methods for obtaining the expected values for mean squares under model II are discussed by Crump (1951).

In experimental designs discussed in later chapters, a mixture of models I and II often occurs. The expected values of the mean squares in designs having a mixed model depend upon the randomization restrictions which are imposed upon the variables. When different sets of restrictions are imposed, different expected values for the mean squares are obtained.

3.6 Comparisons among Treatment Means

A *comparison* or *contrast* between two treatment means is by definition the difference between the two means, with appropriate algebraic sign. Thus, $\bar{T}_1 - \bar{T}_2$ defines a comparison between these two means. Comparisons among three treatment means can be made in several different ways. Each mean may be compared or contrasted with each of the other means, giving rise to the following differences:

$$\bar{T}_1 - \bar{T}_2, \qquad \bar{T}_1 - \bar{T}_3, \qquad \bar{T}_2 - \bar{T}_3.$$

Another set of comparisons may be obtained by averaging two of the three means and then comparing this average with the third mean. This procedure gives rise to the following differences:

$$\frac{\bar{T}_1 + \bar{T}_2}{2} - \bar{T}_3, \qquad \frac{\bar{T}_1 + \bar{T}_2}{2} - \bar{T}_3, \qquad \frac{\bar{T}_2 + \bar{T}_3}{2} - \bar{T}_1.$$

In general, a *comparison* or *contrast* among three means is an expression of the form

$$(1) \qquad c_1\bar{T}_1 + c_2\bar{T}_2 + c_3\bar{T}_3, \qquad \text{where } \Sigma c_j = 0.$$

[Any expression having the form

$$w_1\bar{T}_1 + w_2\bar{T}_2 + w_3\bar{T}_3,$$

where there is no restriction on the w's, is called a *linear* function of the $\bar{T}$'s. A comparison, as defined in (1), is a specialized linear function of the $\bar{T}$'s.]

For example, if $c_1 = 1$, $c_2 = -1$, and $c_3 = 0$, (1) becomes

$$1\bar{T}_1 + (-1)\bar{T}_2 + (0)\bar{T}_3 = \bar{T}_1 - \bar{T}_2.$$

If $c_1 = 0$, $c_2 = 1$, and $c_3 = -1$, then (1) becomes

$$\bar{T}_2 - \bar{T}_3.$$

If $c_1 = \frac{1}{2}$, $c_2 = -1$, and $c_3 = \frac{1}{2}$, then (1) becomes

$$(\tfrac{1}{2})\bar{T}_1 + (-1)\bar{T}_2 + (\tfrac{1}{2})\bar{T}_3 = \frac{\bar{T}_1 + \bar{T}_3}{2} - \bar{T}_2.$$

The general definition of a comparison or contrast among k means has the form

$$(2) \qquad c_1\bar{T}_1 + c_2\bar{T}_2 + \cdots + c_k\bar{T}_k, \qquad \text{where } \Sigma c_j = 0.$$

The expression (2) is called a linear combination or weighted sum of the means. If the sum of the weights in this expression is equal to zero, then the linear combination is a comparison or contrast among the means. When the number of observations in each treatment class is equal to n, it is more convenient to work with totals rather than means. A comparison among the treatment totals is defined by

$$(3) \qquad c_1T_1 + c_2T_2 + \cdots + c_kT_k, \qquad \text{where } \Sigma c_j = 0.$$

A comparison between the first and second treatments is obtained by setting $c_1 = 1$, $c_2 = -1$, and setting all other c's equal to zero.

Orthogonal comparisons will be defined by means of a numerical example. Consider the following comparisons:

$$\begin{aligned} C_1 &= (-3)T_1 + (-1)T_2 + (1)T_3 + (3)T_4, \\ C_2 &= (1)T_1 + (-1)T_2 + (-1)T_3 + (1)T_4, \\ C_3 &= (1)T_1 + (1)T_2 + (-1)T_3 + (-1)T_4. \end{aligned}$$

For comparisons C_1 and C_2, the sum of the products of corresponding coefficients is

$$(-3)(1) + (-1)(-1) + (1)(-1) + (3)(1) = 0.$$

For comparisons C_1 and C_3, the sum of the products of corresponding coefficients is

$$(-3)(1) + (-1)(1) + (1)(-1) + (3)(-1) = -8.$$

Two comparisons are orthogonal if the sum of the products of corresponding coefficients is equal to zero. Thus, the comparisons C_1 and C_2 are orthogonal, but the comparisons C_1 and C_3 are not orthogonal. To find out whether or not C_2 is orthogonal to C_3, one forms the sum of products of corresponding coefficients,

$$(1)(1) + (-1)(1) + (-1)(-1) + (1)(-1) = 0.$$

Since this sum is equal to zero, the two comparisons are orthogonal. The concept of orthogonality in this context is analogous to the concept of nonoverlapping or uncorrelated sources of variation.

A *component* of the sum of squares for treatments is defined by the expression

$$\frac{(c_1T_1 + c_2T_2 + \cdots + c_kT_k)^2}{n(c_1^2 + c_2^2 + \cdots + c_k^2)}, \qquad \text{where } \Sigma c_j = 0.$$

(It is assumed that there are n observations in each treatment total.) In words, a component of a sum of squares is the square of a comparison divided by n times the sum of the squares of the coefficients in the comparison. A component of a sum of squares has one degree of freedom because it basically represents the squared difference between two observations, each observation being the weighted average of treatment means. A sum of squares based upon two basic observations has one degree of freedom.

Two components of the sum of squares for treatments are orthogonal if the comparisons in these components are orthogonal. Any treatment sum of squares having $k - 1$ degrees of freedom can be divided into $k - 1$ orthogonal components; there are usually many ways in which this can be done. A treatment sum of squares can also be subdivided into a relatively large number of nonorthogonal components. Orthogonal components are additive; i.e., the sum of the parts equals the whole. Each component in an orthogonal set covers a different portion of the total variation. Nonorthogonal components are not additive in this sense.

The computation and interpretation of the components of the sums of squares for treatments will be illustrated by means of the numerical example used in Table 3.2-1. For these data,

$$n = 8, \quad k = 3, \quad T_1 = 38, \quad T_2 = 37, \quad T_3 = 62,$$
$$\text{SS}_{\text{methods}} = 50.08, \quad \text{MS}_{\text{error}} = 4.14.$$

Consider the following comparisons:

$$C_1 = (1)T_1 + (-1)T_2 + (0)T_3 = T_1 - T_2 = 38 - 37 = 1,$$
$$C_2 = (1)T_1 + (1)T_2 + (-2)T_3 = T_1 + T_2 - 2T_3 = 38 + 37 - 124 = -49.$$

C_1 represents a comparison between methods 1 and 2, method 3 being disregarded. This comparison is used in testing the hypothesis that $\mu_1 - \mu_2 = 0$. C_2 represents a comparison between method 3 and the average of methods 1 and 2. This comparison is used in testing the hypothesis that $[(\mu_1 + \mu_2)/2] - \mu_3 = 0$. C_1 and C_2 are orthogonal, since the sum of the products of corresponding coefficients is zero; that is,

$$(1)(1) + (-1)(1) + (0)(-2) = 0.$$

The component of the sum of squares corresponding to C_1 is

$$\text{SS}_{C_1} = \frac{(T_1 - T_2)^2}{n[(1)^2 + (-1)^2 + (0)^2]} = \frac{(1)^2}{8(2)} = .0625.$$

This component is interpreted as follows: Of the total variation among the methods of training, $\text{SS}_{\text{methods}} = 50.08$, that part which is due to the difference between methods 1 and 2 is $\text{SS}_{C_1} = .0625$. The component corresponding to C_2 is

$$\text{SS}_{C_2} = \frac{(T_1 + T_2 - 2T_3)^2}{8[(1)^2 + (1)^2 + (-2)^2]} = \frac{(-49)^2}{8(6)} = 50.02.$$

Thus, of the total variation between the methods, that part which is due to the difference between method 3 and the other two methods combined is $\text{SS}_{C_2} = 50.02$. Method 3 appears to be clearly different from the other two methods; there is very little difference between methods 1 and 2.

Since a component of a sum of squares has one degree of freedom, there is no difference between a sum of squares or a mean square for a component. The expected value of the sum of squares for C_1 is

$$\text{E}(\text{SS}_{C_1}) = \sigma_\varepsilon^2 + \frac{n}{2}(\tau_1 - \tau_2)^2.$$

Under the hypothesis that $\tau_1 - \tau_2 = 0$, $E(SS_{C_1}) = \sigma_\varepsilon^2$. Hence the ratio

$$\frac{E(SS_{C_1})}{E(MS_{error})} = \frac{\sigma_\varepsilon^2}{\sigma_\varepsilon^2} = 1.00, \qquad \text{when } \tau_1 - \tau_2 = 0.$$

Thus the statistic

$$F = \frac{SS_{C_1}}{MS_{error}}$$

may be used in a test of the hypothesis that $\tau_1 - \tau_2 = 0$. This F ratio has one degree of freedom for the numerator and $kn - k$ degrees of freedom for the denominator. For these data $F_{.95}(1,21) = 4.32$. The numerical value of F_{obs} is

$$F_{obs} = \frac{.0625}{4.14}.$$

This F ratio is less than 1.00 and hence does not exceed the critical value 4.32. Thus, the observed data do not contradict the hypothesis that $\tau_1 - \tau_2 = 0$.

The hypothesis that $[(\tau_1 + \tau_2)/2] - \tau_3 = 0$ is tested by use of the statistic

$$F = \frac{SS_{C_2}}{MS_{error}} = \frac{50.02}{4.14} = 12.08.$$

The critical value is again $F_{.95}(1,21) = 4.32$. Thus, the observed data contradict the hypothesis that method 3 is no different from the average effect of methods 1 and 2.

The comparison used in a test of the hypothesis that $\tau_1 - \tau_3 = 0$ is

$$T_1 - T_3 = 38 - 62 = -24.$$

The component of variation corresponding to this comparison is

$$\frac{(T_1 - T_3)^2}{n[(1)^2 + (-1)^2]} = \frac{(-24)^2}{8(2)} = 36.00.$$

The F ratio for this test is

$$F = \frac{36.00}{4.14} = 8.70.$$

The critical value for a .05-level test is $F_{.95}(1,21) = 4.32$. Since the observed value of F exceeds 4.32, the hypothesis that $\tau_1 - \tau_3 = 0$ is rejected.

The sum of squares for methods, $SS_{methods} = 50.08$, has two degrees of freedom. Since SS_{C_1} and SS_{C_2} are orthogonal, they represent nonoverlapping (additive) parts of $SS_{methods}$. Hence

$$SS_{methods} = SS_{C_1} + SS_{C_2},$$
$$50.08 = .06 + 50.02.$$

C_1 and C_2 are not the only pair of orthogonal comparisons that may be formed. For example,

$$C_3 = (1)T_1 + (0)T_2 + (-1)T_3,$$
$$C_4 = (-1)T_1 + (2)T_2 + (-1)T_3$$

are also orthogonal. The components corresponding to these comparisons will also sum to $SS_{methods}$. In practice the comparisons that are constructed are those having some meaning in terms of the experimental variables; whether these comparisons are orthogonal or not makes little or no difference.

The general form of the hypothesis tested by a comparison is

$$H_1: \quad \Sigma c_j\mu_j = 0, \qquad \Sigma c_j = 0$$
$$H_2: \quad \Sigma c_j\mu_j \neq 0,$$

where the c_j's are the coefficients in the comparison. The statistic used in making the test has the general form

$$F = \frac{(\Sigma c_j T_j)^2}{(n\Sigma c_j^2)(MS_{error})}.$$

The numerator of this F ratio has one degree of freedom. The degrees of freedom for the denominator are those of MS_{error}.

When a large number of comparisons are made following a significant overall F, some of the decisions which reject H_1 may be due to type 1 error. For example, if 5 independent tests are each made at the .05 level, the probability of a type 1 error in one or more of the five decisions is $1 - (.95)^5 = .23$. If 10 independent tests are made, the probability of a type 1 error's occurring in one or more of the decisions is $1 - (.95)^{10} = .40$. In general, if $\alpha = .05$ and a series of m independent tests are made, then the probability of one or more type 1 errors in the series of m tests is obtained by subtracting the last term of the expansion of $(.05 + .95)^m$ from unity. Equivalently, one may add all terms in the expansion except the last one.

Thus, when the number of comparisons is large, the number of decisions that can potentially be wrong owing to type 1 error can be relatively large. Tukey and Scheffé developed methods for constructing simultaneous confidence intervals which avoid the pitfall of permitting the type 1 error to become excessively large. These methods as well as others for dealing with multiple comparisons are discussed in Secs. 3.8 and 3.9.

Interpretation of Overall *F* Test in Terms of Comparisons. Consider the following comparison:

$$C = c_1\bar{T}_1 + c_2\bar{T}_2 + \cdots + c_k\bar{T}_k, \qquad \Sigma c_j = 0.$$

Assuming equal sample sizes under each of the treatments, the variation due to this comparison is

$$SS_C = \frac{C^2}{(\Sigma c_j^2)/n}.$$

The answer to the following question is of considerable interest: What choice of $c_1, c_2, \ldots, c_k$ will make SS_C a maximum? The maximum value of SS_C will be attained (see Scheffé, 1959, p. 118) if one sets

$$c_j = \frac{n(\bar{T}_j - \bar{G})}{\sqrt{SS_{\text{treat}}}}.$$

For this choice of c_j, one notes that

$$\frac{\Sigma c_j^2}{n} = \frac{\Sigma n(\bar{T}_j - \bar{G})^2}{SS_{\text{treat}}} = 1.$$

A comparison for which the denominator of SS_C is equal to unity is called a *normalized* comparison.

If one lets

$$C_{\max} = \text{normalized comparison for which } SS_C \text{ is maximum},$$

then

$$C_{\max} = \frac{\Sigma n(\bar{T}_j - \bar{G})}{\sqrt{SS_{\text{treat}}}}\bar{T}_j = \frac{1}{\sqrt{SS_{\text{treat}}}}\Sigma n(\bar{T}_j - \bar{G})\bar{T}_j$$

$$= \sqrt{SS_{\text{treat}}},$$

since $n\Sigma(\bar{T}_j - \bar{G})\bar{T}_j = SS_{\text{treat}}$. Hence

$$SS_{C_{\max}} = C_{\max}^2 = SS_{\text{treat}}.$$

For example, suppose

$$n = 5, \qquad \bar{T}_1 = 5, \qquad \bar{T}_2 = 10, \qquad \bar{T}_3 = 30, \qquad \bar{G} = 15.$$

For these data,

$$SS_{\text{treat}} = n\Sigma(\bar{T}_j - \bar{G})^2 = 5(100 + 25 + 225) = 1750.$$

To find $C_{\max}$, let

$$c_1 = \frac{5(-10)}{\sqrt{1750}}, \qquad c_2 = \frac{5(-5)}{\sqrt{1750}}, \qquad c_3 = \frac{5(15)}{\sqrt{1750}}.$$

Hence $$C_{\max} = c_1(5) + c_2(10) + c_3(30) = \frac{1750}{\sqrt{1750}} = \sqrt{1750}$$

Thus $$SS_{C_{\max}} = C_{\max}^2 = 1750 = SS_{\text{treat}}.$$

If one were to test the hypothesis that the comparison of which $C_{\max}$ is an estimate is zero, for the case of equal sample size, the region of rejection for this test (as obtained from the Scheffé simultaneous-confidence-interval approach, which is discussed in detail in Sec. 3.9) is

$$F = \frac{SS_{C_{max}}}{MS_{error}} > (k-1)F_{1-\alpha}(k-1, kn-k),$$

or, equivalently,

$$F = \frac{SS_{C_{max}}/(k-1)}{MS_{error}} = \frac{MS_{treat}}{MS_{error}} > F_{1-\alpha}(k-1, kn-k).$$

In the latter form, one notes that the test of the hypothesis that the comparison estimated by C_{max} is zero is identical to the test of no variation due to treatments. If the hypothesis being tested in terms of the comparison is not rejected, the implication is the corresponding population comparison is not significantly different from zero. If the comparison having maximum sum of squares is not significantly different from zero then no other comparison will be significantly different from zero when tested by means of the Scheffé approach. If the hypothesis being tested is rejected, the implication is that the corresponding comparison (and possibly others) is significantly different from zero.

For the collection of all tests on comparisons, considered as a single test, the Scheffé approach has level of significance α. That is, the probability of a type 1 error on the entire collection is α.

For the case of unequal sample sizes,

$$SS_C = \frac{C^2}{(c_1^2/n_1) + (c_2^2/n_2) + \cdots + (c_k^2/n_k)}.$$

To maximize SS_C in this case one sets

$$c_j = \frac{n_j(\bar{T}_j - \bar{G})}{M}, \qquad M = \sqrt{\Sigma n_j(\bar{T}_j - \bar{G})^2}.$$

The maximum value of SS_C will be found to be

$$SS_{C_{max}} = \Sigma n_j(\bar{T}_j - \bar{G})^2 = SS_{treat}.$$

3.7 Use of Orthogonal Components in Tests for Trend

The total variation of the treatments may be subdivided in many different ways. Depending upon the nature of the experimental variables and the purpose of the experiment, some subdivisions may be meaningful and others may not. The meaningful comparisons need not be the orthogonal ones. If the treatments form a series of equal steps along an ordered scale, i.e., increasing dosage, intensity, complexity, time, etc., then treatment variation may be subdivided into linear, quadratic, cubic, etc., trend components through the use of orthogonal polynomials. This method of subdivision provides information about the form of the relationship between the criterion and the steps along the treatment scale.

The use of orthogonal polynomials as described in this section requires that the sample size n be constant for all treatments and that the treatments represent equally spaced quantitative steps along an underlying continuum. For example, if the treatments represent dosage of a drug, the treatments may be dosages of 1 cc, 3 cc, 5 cc, 7 cc. Alternatively, the dosages may be equally spaced in terms of a log scale; that is, log dosage may be c, $c + k$, $c + 2k$, etc. If these conditions are not met, one may use the usual regression techniques for polynomial regression in order to achieve the same result. When one does not have equally spaced quantitative steps a slight modification of the procedure to be described is required; this modification is given by Robson (1959).

Many different curves may be found to fit a given set of empirical data. The use of polynomials of varying degree may or may not be the most appropriate choice for the form of the curve. From one point of view this choice simplifies certain of the statistical problems. From another point of view, the use of higher-degree polynomials may be scientifically meaningless. Polynomials are, however, quite useful in showing the general form of relationships. Within a limited range of values, polynomials can be used to approximate curves which are actually exponential or logarithmic in form.

Caution is required in the use of the methods to be described in this section. The nature of the data and the purpose of the experiment must be considered in evaluation of the results.

In regression analysis, a first-degree (linear) equation is used as the first approximation to the relationship between two variables. Second-degree (quadratic), third-degree (cubic), and higher-degree equations are used if the fit of the linear relationship does not prove to be satisfactory. When there are equal numbers of observations in each of the treatment classes, and when the treatment classes form equal steps along an ordered scale, the work of finding the degree of the best-fitting curve is simplified by use of comparisons corresponding to these curves. Once the degree of the best-fitting polynomial is found, the regression coefficients for this curve are also readily obtained. Extensive tables exist giving the coefficients for the comparisons corresponding to regression equations of varying degree. The most complete set of coefficients is in the work of Anderson and Houseman (1942). The Fisher and Yates tables (1953) also have an adequate set. A numerical example will be used to illustrate the application of individual components of variation to the problem of finding the degree of the best-fitting curve. (Tests for the best-fitting curve are also called tests for trend.)

Suppose that an experimenter is interested in approximating the form of the relationship between the degree of complexity of a visual display (i.e., an instrument panel containing a series of dials) and the reaction time of subjects in responding to the displacement of one or more of the dials. After deciding upon a definition of the complexity scale, the experimenter constructs six displays representing six equal steps along this scale. The treatments in this experiment correspond to the degree of complexity of the displays. Random samples of 10 subjects are assigned to each of the displays; each sample is observed under only one of the displays. The criterion used in the analysis is the mean reaction time (or some transformation thereof) for each subject to a series of trials which are as comparable as the displays permit.

Summary data for this experiment are presented in Table 3.7-1. Each T_j in this table represents the sum over the 10 criterion scores for scale j. The sum of the squares of the 10 criterion scores under each scale is also given in part i of this table. Steps in the computation of the sums of squares are given in part ii.

Table 3.7-1 Numerical Example

	Complexity of display	1	2	3	4	5	6	$k = 6;\ n = 10$ Total
(i)	T_j	100	110	120	180	190	210	$910 = G$
	ΣX_j^2	1180	1210	1600	3500	3810	4610	$15{,}910 = \Sigma\Sigma X^2$
	$\bar{T}_j$	10.0	11.0	12.0	18.0	19.0	21.0	

(ii)

$$(1) = G^2/kn = (910)^2/60 = 13{,}801.67$$
$$(2) = \Sigma\Sigma X^2 = 15{,}910$$
$$(3) = (\Sigma T_j^2)/n = (100^2 + 110^2 + \cdots + 210^2)/10 = 14{,}910.00$$

$$SS_{\text{displays}} = (3) - (1) = 1108.33$$
$$SS_{\text{error}} = (2) - (3) = 1000.00$$
$$SS_{\text{total}} = (2) - (1) = 2108.33$$

The analysis of variance appears in Table 3.7-2. Since there are six displays, the degrees of freedom for the displays are $6 - 1 = 5$. Ten

Table 3.7-2 Summary of Analysis of Variance

Source of variation	SS	df	MS	F
Displays	1108.33	5	221.67	11.97**
Experimental error	1000.00	54	18.52	
Total	2108.33	59		

** $F_{.99}(5,54) = 3.38$

subjects were used under each display. Hence the sum of squares within a single display has degrees of freedom equal to 9; the pooled within-display variation has $6 \times 9 = 54$ degrees of freedom. The pooled within-display variation defines the experimental error. To test the hypothesis that the mean reaction times (or the means of the transformed measures) for the displays are equal,

$$F = \frac{\text{MS}_{\text{displays}}}{\text{MS}_{\text{error}}} = \frac{221.67}{18.52} = 11.97.$$

The critical value for this test, using the .01 level of significance, is

$$F_{.99}(5,54) = 3.38.$$

Clearly the data indicate that reaction times for the displays differ. Inspection of the T_j's in Table 3.7-1 shows that the greater the degree of complexity the slower the reaction time. The means corresponding to the T_j's are plotted in Fig. 3.7-1. This figure shows a strong linear relationship between display complexity and reaction time. There is also the

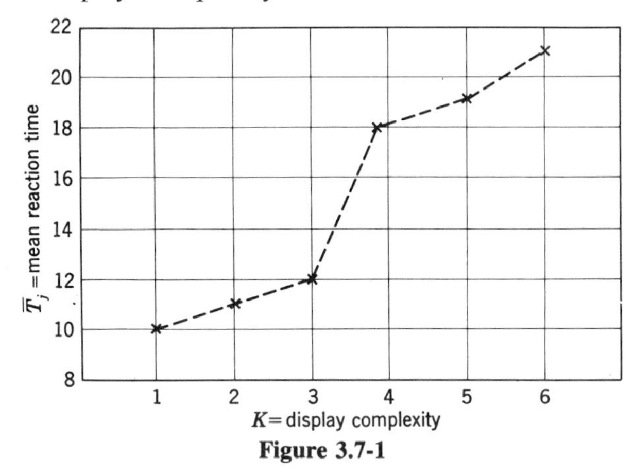

Figure 3.7-1

suggestion that an S-shaped curve might be the best-fitting curve; an S-shaped curve corresponds to an equation having degree 3 (cubic).

In the upper left part of Table 3.7-3 are the coefficients for linear, quadratic, and cubic comparisons for $k = 6$. These coefficients were obtained from Table C.10. Notice that the sum of the linear coefficients is zero, that each coefficient differs by 2 units from its neighbor, and that only once in the sequence from -5 to $+5$ does the sign change. For the coefficients defining the quadratic comparison, the signs change twice in the sequence from $+5$ to $+5$; in the cubic comparison there are three changes of sign in the sequence from -5 to $+5$. The number of times the signs change determines the degree of the polynomial.

The numerical value of the linear comparison for these data is

$$C_{\text{lin}} = (-5)(100) + (-3)(110) + (-1)(120) + (1)(180) + (3)(190) + (5)(210)$$

$$= 850.$$

Table 3.7-3 Tests for Trend

	Complexity T_j	1 100	2 110	3 120	4 180	5 190	6 210	Σc^2	C	$D = n\Sigma c^2$	C^2/D
(i)	Linear	−5	−3	−1	1	3	5	70	850	700	1032.14
	Quadratic	5	−1	−4	−4	−1	5	84	50	840	2.98
	Cubic	−5	7	4	−4	−7	5	180	−250	1800	34.72
											1069.84

(ii)

Test for linear trend: $F = \dfrac{1032.14}{18.52} = 55.73^{**}$

Test for quadratic trend: $F = \dfrac{2.98}{18.52} = .16$

Test for cubic trend: $F = \dfrac{34.72}{18.52} = 1.87$

** $F_{.99}(1,54) = 7.14$

Since $D = n\Sigma c^2$, the component of variation corresponding to this comparison has the form

$$\frac{C_{\text{lin}}^2}{D_{\text{lin}}} = \frac{850^2}{10[(-5)^2 + (-3)^2 + (-1)^2 + 1^2 + 3^2 + 5^2]}$$

$$= \frac{850^2}{10(70)} = 1032.14.$$

Thus the variation due to the linear trend in the data is 1032.14 units. From Table 3.7-2 it is seen that the total variation in the displays is 1108.33 units. Hence 1032.14/1108.33 = 93 percent of the variation in the reaction time for the displays may be predicted from a linear regression equation. The test for linear trend is given by

$$F = \frac{\text{linear component}}{\text{MS}_{\text{error}}} = \frac{1032.14}{18.52} = 55.73.$$

The critical value for a .01-level test is $F_{.99}(1,f)$, where f is the degrees of freedom for MS_{error}. In this case $f = 54$, and the critical value is 7.14. Clearly the linear trend is statistically significant. However,

$$1108.33 - 1032.14 = 76.19$$

units of reaction-time variation are not predicted by the linear regression equation. There is still the possibility that the quadratic or cubic trend might be statistically significant.

An overall measure of deviations from linearity is given by

$$\text{SS}_{\text{nonlin}} = \text{SS}_{\text{displays}} - \text{SS}_{\text{lin}} = 1108.33 - 1032.14 = 76.19.$$

The degrees of freedom for this source of variation are $k - 2$. An overall test for nonlinearity is given by

$$F = \frac{\text{SS}_{\text{nonlin}}/(k-2)}{\text{MS}_{\text{error}}} = \frac{76.19/4}{18.52} = \frac{19.05}{18.52} = 1.03.$$

For a .01-level test, the critical value is $F_{.99}(4.54) = 3.68$. Since the F ratio does not exceed this critical value, the data do not indicate that there are significant deviations from linearity. However, when the degrees of freedom for $\text{MS}_{\text{nonlin}}$ are large, the overall test may mask a significant higher-order component. In spite of the fact that the overall test indicates no significant deviations from linearity, if there is a priori evidence that a higher-order component might be meaningful, tests may continue beyond the linear component. Caution is required in the interpretation of significant higher-order trend components when the overall test for nonlinearity indicates no significant nonlinear trend.

The quadratic comparison is orthogonal to the linear comparison, since the sum of the products of corresponding coefficients is zero, i.e.,

$$(-5)(5) + (-3)(-1) + (-1)(-4) + (1)(-4) + (3)(-1) + (5)(5) = 0.$$

Therefore the quadratic component of variation is part of the 76.19 units of variation which is not predicted by the linear component. The computation of this component is similar to the computation of the linear component, the quadratic coefficients replacing the linear coefficients in the calculation of C and D. The numerical value of the quadratic component is 2.98. This is the increase in predictability that would accrue for the sample data by using a second-degree instead of a first-degree equation. A test on whether or not this increase in predictability is significantly greater than zero uses the statistic

$$F = \frac{\text{quadratic component}}{\text{MS}_{\text{error}}} = \frac{2.98}{18.52} = .16.$$

The critical value for a .01-level test is $F_{.99}(1,f)$; in this case the critical value is 7.14. The F statistic for quadratic trend is .16. Hence the increase in predictability due to the quadratic component is not significantly different from zero.

It is readily verified that the cubic comparison is orthogonal to both the linear and quadratic components. Thus, the cubic component is part of the $1108.33 - 1032.14 - 2.98 = 73.21$ units of variation not predicted by the linear or quadratic trends. The component of variation corresponding to the cubic trend is 34.72. The F statistic for this component is 1.87; the critical value for a .01-level test is 7.14. Thus, the sample data indicate that the cubic component does not increase the predictability by an amount which is significantly different from zero.

Of the total of 1108.33 units of variation due to differences in complexity of the displays, 1069.84 units are predictable from the linear, quadratic, and cubic components. The remaining variation is due to higher-order components, none of which would be significantly different from zero. In summary, the linear trend appears to be the only trend that is significantly greater than zero. Hence a first-degree equation (linear equation) is the form of the best-fitting curve.

In this case the linear equation will have the form

$$X = bK + a,$$

where X = predicted reaction time,
K = degree of complexity of display,
b, a = regression coefficients.

Since the degree of complexity of the displays is assumed to be equally spaced along a complexity dimension, the degree of complexity may conveniently be indicated by the integers 1, 2, . . . , 6. The sum of squares for the complexity variable is given by

$$SS_K = \frac{n(k^3 - k)}{12},$$

where k is the number of displays and n is the number of observations under each display. The entry 12 is constant for all values of n and k. The regression coefficient b is given by the relation

$$b = \sqrt{\frac{\text{linear component}}{SS_K}} = \sqrt{\frac{SS_{\text{lin}}}{SS_K}}$$

Equivalently,

$$b = \frac{\lambda_1 C_{\text{lin}}}{D_{\text{lin}}},$$

where λ_1 is a constant which depends upon k; tables of coefficients for the orthogonal polynomials will give values of λ_1.

The symbol SS_{lin} is generally used in regression analysis to indicate the variation predictable from the linear equation. The numerical value of the regression coefficient a is given by the relation

$$a = \bar{X} - b\bar{K}.$$

Computation of the coefficients for the regression equation is summarized in Table 3.7-4.

Table 3.7-4 Computation of Regression Equation

X = reaction time $\qquad K$ = display complexity

$\bar{X} = G/kn = 15.17 \qquad \bar{K} = (k + 1)/2 = 3.50$

$$SS_K = \frac{n(k^3 - k)}{12} = \frac{(10)(216 - 6)}{12} = 175$$

$$b = \sqrt{\frac{SS_{\text{lin}}}{SS_K}} = \sqrt{\frac{1032.14}{175}} = 2.43$$

$$a = \bar{X} - b\bar{K} = 15.17 - (2.43)(3.50) = 6.67$$

Regression equation: $X = bK + a = 2.43K + 6.67$

The linear correlation between degree of complexity and reaction time is given by

$$r = \sqrt{\frac{SS_{\text{lin}}}{SS_{\text{total}}}} = \sqrt{\frac{1032.11}{2108.33}} = .70.$$

The numerical value for SS_{total} is obtained from Table 3.7-2. The actual fit of the regression equation to the points in Fig. 3.7-1 is shown in Fig.

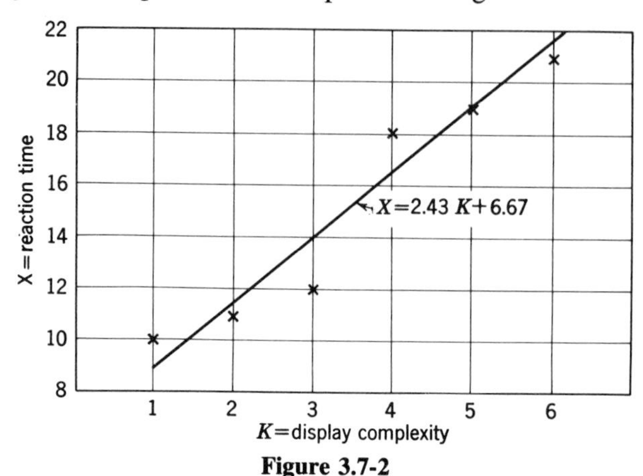

Figure 3.7-2

3.7-2. The correlation associated with the cubic relationship is

$$r = \sqrt{\frac{SS_{\text{lin}} + SS_{\text{quad}} + SS_{\text{cubic}}}{SS_{\text{total}}}}$$

$$= \sqrt{\frac{1032.14 + 2.98 + 34.72}{2108.33}} = .71.$$

The tests that have been made indicate that the cubic correlation does not differ significantly from the linear correlation.

Another measure of correlation associated with an analysis of variance (usually the variance-component model) is the *intraclass* correlation. Borrowing this measure for descriptive purposes, one defines

$$\rho_{\text{intraclass}} = \frac{\sigma^2_{\text{displays}}}{\sigma_\varepsilon^2 + \sigma^2_{\text{displays}}}.$$

For this example,

$$E(MS_{\text{displays}}) = \sigma_\varepsilon^2 + 10\sigma^2_{\text{displays}},$$

$$E(MS_{\text{error}}) = \sigma_\varepsilon^2.$$

Thus,

$$\hat{\sigma}_\varepsilon^2 = 18.52,$$

$$\hat{\sigma}_\varepsilon^2 + 10\hat{\sigma}^2_{\text{displays}} = 221.67 \quad \text{or} \quad \hat{\sigma}^2_{\text{displays}} = \frac{221.67 - 18.52}{10}$$

$$= 20.31.$$

An estimate of $\rho_{\text{intraclass}}$ is given by

$$\hat{\rho}_{\text{intraclass}} = \frac{\hat{\sigma}^2_{\text{displays}}}{\hat{\sigma}^2_\varepsilon + \hat{\sigma}^2_{\text{displays}}} = \frac{20.31}{18.52 + 20.31} = .5299.$$

(This estimate of $\rho_{\text{intraclass}}$ is biased.) One has

$$\sqrt{\hat{\rho}_{\text{intraclass}}} = .728.$$

The square root of the intraclass correlation is an upper bound for the correlation associated with any polynomial regression.

In working with higher-order regression equations, in predicting V from U it is convenient to work with the equation in the second of the following forms:

$$\begin{aligned} V &= a_0 + a_1U + a_2U^2 + a_3U^3 \\ &= A_0'\xi_0' + A_1'\xi_1' + A_2'\xi_2' + A_3'\xi_3'. \end{aligned}$$

In this latter form,

$$\begin{aligned} A_0' &= \bar{V}, & \xi_0' &= 1, \\ A_1' &= \frac{C_{\text{lin}}}{D_{\text{lin}}}, & \xi_1' &= \lambda_1(U - \bar{U}), \\ A_2' &= \frac{C_{\text{quad}}}{D_{\text{quad}}}, & \xi_2' &= \lambda_2\left[(U - \bar{U})^2 - \frac{k^2 - 1}{12}\right], \\ A_3' &= \frac{C_{\text{cubic}}}{D_{\text{cubic}}}, & \xi_3' &= \lambda_3\left[(U - \bar{U})^3 - (U - \bar{U})\frac{3k^2 - 7}{20}\right]. \end{aligned}$$

The numerical values of the λ_j's depend upon the value of k, the number of treatments. Numerical values for λ_j will be found in tables of the coefficients.

3.8 Use of the Studentized Range Statistic

Although the hypothesis $\tau_1 = \tau_2 = \cdots = \tau_k$ is generally tested by means of an F ratio, there are other methods for testing this same hypothesis. One method is through use of the studentized range statistic, defined by

$$q = \frac{\bar{T}_{\text{largest}} - \bar{T}_{\text{smallest}}}{\sqrt{\text{MS}_{\text{error}}/n}}, \tag{1}$$

where n is the number of observations in each $\bar{T}$. In words, the studentized range statistic is the difference between the largest and smallest treatment means (the range of the treatment means) divided by the square root of the quantity mean-square experimental error over n. A numerically equivalent form of the q statistic is given by

$$(1a) \qquad q = \frac{T_{\text{largest}} - T_{\text{smallest}}}{\sqrt{n\text{MS}_{\text{error}}}},$$

where a T represents a sum of n observations. Computationally it is more convenient to compute first the following statistic:

$$(2) \qquad F_{\text{range}} = \frac{(T_{\text{largest}} - T_{\text{smallest}})^2}{2n\text{MS}_{\text{error}}}.$$

Then q may readily be shown to be equal to

$$(3) \qquad q = \sqrt{2F_{\text{range}}}.$$

Under the hypothesis that $\tau_1 = \tau_2 = \cdots = \tau_k$ and under all the assumptions underlying either model I or model II, the sampling distribution of the q statistic is approximated by the studentized range distribution having parameters k = number of treatments and f = degrees of freedom for MS_{error}. The symbol $q_{.99}(k,f)$ designates the 99th percentile point on the q distribution. Tables of the latter are given in Table C.4. In contrast to the overall F ratio, which uses all the k treatment means to obtain the numerator, the q statistic uses only the two most extreme means in the set of k. Thus, the q statistic uses less of the information from the experiment than does the F statistic. Use of the F statistic generally leads to a more powerful test with respect to a broader class of alternative hypotheses than does the use of the q statistic. However, there are some alternative hypotheses for which the q statistic leads to a more powerful test. Whether or not the two tests lead to the same decision with respect to the hypothesis being tested depends upon the distributions in the populations from which the means were obtained. If the distributions are each $N(\mu,\sigma^2)$, the use of either the F or the q statistic will lead to the same decision. In other cases the decisions reached may differ.

Table 3.8-1 Numerical Example

Treatments . . .	a	b	c	d	e	
T_j	10	14	18	14	14	$k = 5;\ n = 4$
$\bar{T}_j$	2.50	3.50	4.50	3.50	3.50	

Source	SS	df	MS	F
Treatments	8.00	4	2.00	4.00
Experiment error	7.50	15	0.50	
Total	15.50	19		

$F_{.99}(4,15) = 4.89$

The numerical example summarized in Table 3.8-1 provides an illustration of a case in which the two tests lead to different decisions. Each T_j in this table represents the sum over 4 observations ($n = 4$) under each of 5 treatments ($k = 5$). Since the observed overall F ratio (4.00) does not exceed the critical value (4.89) for a .01-level test, the data do not contradict the hypothesis that the treatment effects are all equal. The q statistic as computed from (1) is

$$q = \frac{4.50 - 2.50}{\sqrt{.50/4}} = 5.67.$$

The critical value for a .01-level test is $q_{.99}(5,15) = 5.56$. Since the observed q statistic exceeds the critical value, the data contradict the hypothesis that the treatment effects are all equal. In this case the F statistic and the q statistic lead to conflicting decisions. This is so because the means for treatments b, d, and e fall at a single point.

In contrast to the example in Table 3.8-1, consider the example in Table 3.8-2. In this table the means tend to be more evenly distributed between

Table 3.8-2 Numerical Example

Treatments . . .	a	b	c	d	e	
T_j	10	12	18	16	14	$k = 5;\ n = 4$
$\bar{T}_j$	2.50	3.00	4.50	4.00	3.50	

Source	SS	df	MS	F
Treatments	10.00	4	2.50	5.00**
Experimental error	7.50	15	0.50	
Total	17.50	19		

** $F_{.99}(4,15) = 4.89$

the highest and lowest values. SS_{treat} for this case is larger than it is for the case in which the means concentrate at a single point within the range. The experimental error in this example is numerically equal to that in Table 3.8-1. The overall F value exceeds the critical value for a .01-level test. Since the ranges are the same in the two examples and since the error is also the same, there is no numerical change in the value of the q statistic. If, in practice, the overall F leads to nonrejection of the hypothesis being tested but the q test leads to rejection of the hypothesis being tested, the experimenter should examine his data quite carefully before attempting any interpretation. In most cases, the F test is more powerful than the corresponding q test.

A numerical example will be used to illustrate the marked difference in sampling distributions for statistics constructed in a way which takes into account the order of magnitude of the observed outcomes. Consider a population having the following five elements:

Element	X
1	10
2	15
3	20
4	25
5	30

The collection of all possible samples of size $n = 3$, the sequence within the sample being disregarded, that can be constructed by sampling without replacement after each draw is enumerated in part i of Table 3.8-3.

Table 3.8-3 Sampling Distributions for Ordered Differences

	Sample	d_2	d_3	Sample	d_2	d_3
(i)	10, 15, 20	5	10	10, 25, 30	15	20
	10, 15, 25	5	15	15, 20, 25	5	10
	10, 15, 30	5	20	15, 20, 30	5	15
	10, 20, 25	10	15	15, 25, 30	10	15
	10, 20, 30	10	20	20, 25, 30	5	10

	d_2	Cumulative rel. frequency	d_3	Cumulative rel. frequency
(ii)	20	1.00	20 III	1.00
	15 I	1.00	15 IIII	.70
	10 III	.90	10 III	.30
	5 ~~IIII~~ I	.60	5	.00

The statistic d_3 is defined to be the difference between the largest and the smallest of the sample values. The statistic d_2 is the difference between the next largest and the smallest. Under a sampling plan which gives each of the samples in part i an equal chance of being drawn, the sampling distributions for d_2 and d_3 are the cumulative relative-frequency distributions in part ii. From the latter, one obtains the following probability statements:

$$P(d_2 > 10) = .10,$$
$$P(d_3 > 10) = .70.$$

In words, the probability of large values for d_3 is larger than the corresponding probability for d_2.

The principle that emerges is this: When one attempts to draw inferences from statistics determined by taking into account the order of magnitude of the observations within an experiment, the appropriate sampling distribution depends upon the number of ordered steps between the observations from which the statistic was computed.

As a further illustration of the nature of the sampling distribution of what are called order statistics, consider random samples of size n from a normal population in which $\mu = 0$ and $\sigma = 1$. Suppose that one arranges the observations within these samples in rank order from low to high as illustrated below.

(1)	(2)	...	(k)	...	($n - 1$)	(n)
smallest						largest

The expected values of the (k)th-order statistic for various values of n are given in Table 3.8-4. Standard errors are also given.

Table 3.8-4 Expected Value and Standard Error of Order Statistics from Unit Normal Distribution

n	(k)	E(k)	$\sigma_{(k)}$	n	(k)	E(k)	$\sigma_{(k)}$
3	(3)	0.85	.75	8	(8)	1.42	.61
	(2)	0.00	.67		(7)	0.85	.49
					(6)	0.47	.45
4	(4)	1.03	.70		(5)	0.15	.43
	(3)	0.30	.60				
				9	(9)	1.49	.60
5	(5)	1.16	.67		(8)	0.93	.48
	(4)	0.50	.56		(7)	0.57	.43
	(3)	0.00	.54		(6)	0.27	.41
					(5)	0.00	.41
6	(6)	1.27	.64				
	(5)	0.64	.53	10	(10)	1.54	.59
	(4)	0.20	.50		(9)	1.00	.46
					(8)	0.66	.42
7	(7)	1.35	.63		(7)	0.38	.40
	(6)	0.76	.51		(6)	0.12	.39
	(5)	0.35	.47				
	(4)	0.00	.46				

Note: For $n = 10$, E(5) $= -.012$, E(4) $= -0.38$, etc.
For $n = 9$, E(4) $= -0.27$, E(3) $= -0.57$, etc.

From this table one is led to the following conclusion: The expected value of an order statistic depends upon both k and n. Further, the difference between the expected value of the (k)th- and the ($k + 1$)th-order statistic depends upon k—the larger k, the larger the magnitude of the difference between expected values. Still further, one notes that the standard error of the sampling distribution depends upon k as well as n—in general, the larger k, the larger $\sigma_{(k)}$. As one might expect, the larger n, the smaller $\sigma_{(k)}$.

For the case $n = 9$, when $\sigma_X = 1$,

$$\sigma_{\bar{X}} = \frac{\sigma_X}{\sqrt{n}} = \frac{1}{\sqrt{9}} = .33.$$

However, for the case $n = 9$, from the table one finds

$$\sigma_{(5)} = .41.$$

But (5) is actually the median value of a sample of size 9. For random samples from a normal distribution, in general one has

$$\sigma_{\text{median}} > \sigma_{\text{mean}},$$

$$\sigma_{\text{median}} = \sqrt{\frac{\pi}{2}\,\sigma^2_{\text{mean}}} = 1.253\sigma_{\text{mean}}.$$

A comparison of the standard errors of various order statistics with the standard error of the mean is shown in Fig. 3.8-1. One notes that the

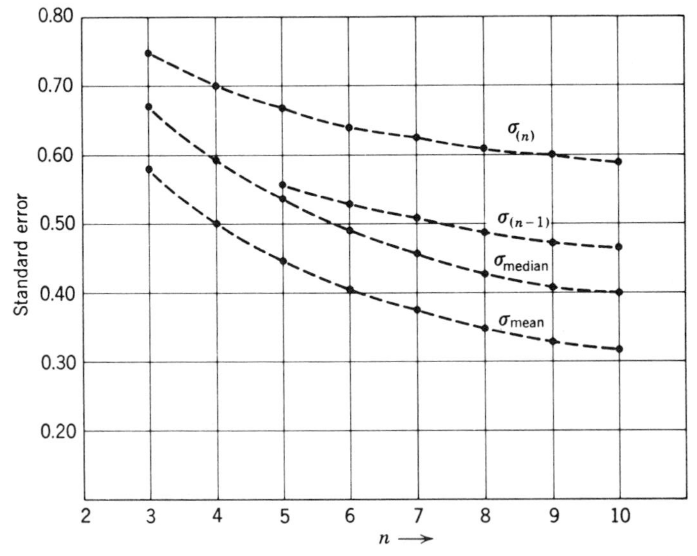

Figure 3.8-1 Comparison of standard error of order statistics with standard error of mean.

curves for σ_{mean} and σ_{median} are parallel. The larger n, the larger is the vertical distance between σ_{mean} and $\sigma_{(k)}$. From Table 3.8-4 one notes that the largest relative difference in standard error is between $\sigma_{(n-1)}$ and $\sigma_{(n)}$.

The direct use of the sampling distribution of the (k)th-order statistic in tests of statistical hypotheses about sets of means in the analysis-of-variance setting is of somewhat limited utility, since the standard error depends upon k and the sampling distribution of the (k)th-order statistic is not independent of the $(k + i)$th-order statistic, where $i \geq 1$. (However the covariance between such order statistics can be obtained.)

A more useful approach to the problem of tests on a set of means obtained in the analysis of variance is through use of the studentized range statistic. The test procedure which is about to be described is known as the Newman-Keuls test. The basic strategy underlying the Newman-Keuls approach is this: The set of ranked treatment means (or totals) is divided into subsets which are consistent with the hypothesis of no differences. Within any specified subset no tests are made unless the range of the set containing the specified subset is statistically different from zero. The test procedure focuses upon a series of ranges rather than a collection of differences between the expected values of order statistics.

For this purpose, the q_r statistic will be used, where r is the number of steps two means (or totals) are apart on an ordered scale. Consider the following treatment totals arranged in increasing order of magnitude:

Order . . .	1	2	3	4	5	6	7
T_j	$T_{(1)}$	$T_{(2)}$	$T_{(3)}$	$T_{(4)}$	$T_{(5)}$	$T_{(6)}$	$T_{(7)}$

In this notation $T_{(1)}$ designates the smallest treatment total and $T_{(7)}$ the largest treatment total. $T_{(7)}$ is defined as being seven steps from $T_{(1)}$, $T_{(6)}$ is six steps from $T_{(1)}$, etc.; $T_{(7)}$ is six steps from $T_{(2)}$, $T_{(6)}$ is five steps from $T_{(2)}$, etc. In general, the number of steps between the ordered totals $T_{(j)}$ and $T_{(i)}$ is $j - i + 1$. Thus,

$$q_7 = \frac{T_{(7)} - T_{(1)}}{\sqrt{n\text{MS}_{\text{error}}}};$$

there is only one q_7 statistic. There are, however, two q_6 statistics,

$$q_6 = \frac{T_{(7)} - T_{(2)}}{\sqrt{n\text{MS}_{\text{error}}}} \quad \text{and} \quad q_6 = \frac{T_{(6)} - T_{(1)}}{\sqrt{n\text{MS}_{\text{error}}}}.$$

q_7 corresponds to the ordinary studentized range statistic when $k = 7$. q_r, where r is less than k, is a modified or truncated studentized range statistic. Critical values for q_r are obtained from tables of the studentized range statistic by setting r equal to the range. For example, the critical value for q_6 in a .01-level test is $q_{.99}(6,f)$; the corresponding critical value for q_4 is $q_{.99}(4,f)$, where f is the degrees of freedom for MS_{error}. (Tables of the q statistic are given in Table C.4.)

In terms of treatment means,

$$q_7 = \frac{\bar{T}_{(7)} - \bar{T}_{(1)}}{\sqrt{\text{MS}_{\text{error}}/n}}.$$

Also,

$$q_6 = \frac{\bar{T}_{(7)} - \bar{T}_{(2)}}{\sqrt{\text{MS}_{\text{error}}/n}}, \qquad q_6 = \frac{\bar{T}_{(6)} - \bar{T}_{(1)}}{\sqrt{\text{MS}_{\text{error}}/n}}.$$

When the sample sizes are unequal, an approximate q_r statistic may be obtained by setting n equal to the harmonic mean of the sample sizes. An alternative approximation is to set n equal to the harmonic mean of the largest and smallest of the sample sizes. The latter approximation will generally be the more conservative in the sense of keeping the actual level of significance somewhat smaller than the nominal level, where the nominal level is the value of α for which the tables of the q_r statistic are entered.

Rather than working directly with the critical value for the q_r statistic, it is more convenient, in making a large number of tests, to obtain a critical value for the difference between two totals. For example,

$$T_{(7)} - T_{(2)} = q_6\sqrt{n\text{MS}_{\text{error}}}.$$

Since the critical value for q_6 is $q_{1-\alpha}(6,f)$, the critical value for $T_{(7)} - T_{(2)}$ will be $q_{1-\alpha}(6,f)\sqrt{n\text{MS}_{\text{error}}}$. In general, the critical value for the difference between two treatment totals which are r steps apart on an ordered scale will be $q_{1-\alpha}(r,f)\sqrt{n\text{MS}_{\text{error}}}$.

The use of the q_r statistic in testing the difference between all pairs of means following a significant overall F will be illustrated by the numerical example in Table 3.8-5. The summary of the analysis of variance appears in part i. Since the observed F ratio (9.40) exceeds the critical value for a .01-level test, the hypothesis that the effects of the seven treatments are all equal is rejected.

The seven treatments are designated by the symbols a through g. The totals of the five observations under each of the treatments are arranged in increasing order of magnitude in part ii of this table. For example, treatment c has the smallest total; treatment f has the largest total. A table of differences between the treatment totals also appears in part ii. For example, the entry in column a, row c, is $T_a - T_c = 12 - 10 = 2$; the entry in column g, row a, is $T_g - T_a = 22 - 12 = 10$. In general, the entry in column j, row i, is $T_j - T_i$. Only the entries shown in this table need be computed.

The critical values for the q_r statistic (when $\alpha = .01$) are given in part iii of Table 3.8-5. Since mean square for experimental error has 28 degrees of freedom, $f = 28$. The entry 3.91 is obtained from tables of the studentized range statistic in which $k = 2$ and $f = 28$. The entry 4.48 is obtained from the same row of the tables, but from the column in which $k = 3$. Similarly the entry 4.84 is obtained from the column in which $k = 4$. These entries are critical values for q_r; to obtain the critical value for the difference between treatment totals which are r steps apart, the critical values are multiplied by

$$\sqrt{n\text{MS}_{\text{error}}} = \sqrt{5(.80)} = \sqrt{4.00} = 2.00.$$

Table 3.8-5 Tests on All Ordered Pairs of Means

(i)

Source of variation	SS	df	MS	F
Treatments	45.09	6	7.52	9.40
Experimental error	22.40	28	0.80	
Total	67.49	34		

Order	1	2	3	4	5	6	7
Treat. in order of T_j	c	a	d	b	g	e	f

(ii)

T_j	10	12	13	18	22	24	25		
	c	a	d	b	g	e	f	r	$q_{.99}(r,28)\sqrt{n\text{MS}_{\text{error}}}$
c	—	2	3	8	12	14	15	7	10.90
a		—	1	6	10	12	13	6	10.56
d			—	5	9	11	12	5	10.18
b				—	4	6	7	4	9.68
g					—	2	3	3	8.96
e						—	1	2	7.82

(iii)

Truncated range r	2	3	4	5	6	7
$q_{.99}(r,28)$	3.91	4.48	4.84	5.09	5.28	5.45
$q_{.99}(r,28)\sqrt{n\text{MS}_{\text{error}}}$	7.82	8.96	9.69	10.18	10.56	10.90

(iv)

	c	a	d	b	g	e	f
c					**	**	**
a					**	**	**
d					**	**	**

Thus, the entry $7.82 = q_{.99}(2,28)\sqrt{n\text{MS}_{\text{error}}} = (3.91)(2.00)$. Similarly the entry $10.90 = q_{.99}(7,28)\sqrt{n\text{MS}_{\text{error}}} = (5.45)(2.00)$. Hence the critical value for the difference between two treatment totals that are $r = 2$ steps apart is 7.82, whereas the critical value for the difference which is $r = 7$ steps apart is 10.90. Differences between totals an intermediate number of steps apart have critical values between these two limits.

The farther apart two means (or totals) are on an ordered scale, the larger the difference between them must be before their range exceeds its critical value. Thus, if one examines the data obtained from an experiment and decides to test the difference between the largest and the smallest mean in a set of k means, the critical value for a test of this kind is larger than the critical value for two means which are adjacent to each other on an ordered scale. The larger critical value is required because the sampling distribution of the difference between the largest and smallest means in a set of k will have a greater relative frequency of more extreme values than will the sampling distribution of two adjacent means in a set of k.

In part ii of the table of ordered differences, entries on a diagonal running from upper left to lower right have the same r value. For example, for the diagonal having the entries 2, 1, 5, 4, 2, 1, one has $r = 2$. Also for the diagonal having the entries 3, 6, 9, 6, 3, one has $r = 3$. The critical value for these entries appears at the extreme right. In order to avoid a type of contradiction that is explained later in this section, there is a prescribed sequence in which tests on differences between the ordered totals should be made.

1. The first test is made on the difference in the upper right-hand corner. This is the difference for which r has its maximum value, which in this case is $r = 7$. The critical value for a .01-level test is 10.90. Since the observed difference exceeds this critical value, the hypothesis that $\tau_{(7)} - \tau_{(1)} = \tau_f - \tau_c = 0$ is rejected. Two asterisks are placed in the cell (c,f) in part iv to indicate that this hypothesis is rejected at the .01 level of significance. If this hypothesis had not been rejected, no further tests would be made. If there is no significant difference between the two most extreme treatment totals, the implication is that there is no significant difference between two totals which are less extreme.

2. Tests are now made on all differences for which $r = 6$. The critical value for these tests is 10.56. The entries on the diagonal for which $r = 6$ are 14 and 13. Both entries exceed the appropriate critical value. Hence two asterisks are placed in the cells (c,e) and (a,f) in part iv.

3. Tests are next made on all differences for which $r = 5$. In this case the critical value is 10.18. The diagonal corresponding to this value of r has the entries 12, 12, 12. Since all entries on this diagonal exceed this critical value, double asterisks are entered in the corresponding cells in part iv.

4*a*. Tests are now made on all differences for which $r = 4$. The appropriate critical value is 9.68. The entries for which $r = 4$ are 8, 10, 11, 7.

4*b*. The entry 8 does not exceed the critical value. No further tests are made in the triangular region defined by the rows and columns of which the entry 8 forms the upper right-hand corner.

4*c*. The entry 7 does not exceed the critical value. No further tests are made in the triangular region defined by the rows and columns of which 7 forms the upper right-hand corner.

4*d*. The entries 10 and 11 do exceed the critical value. Hence double asterisks are placed in the corresponding cells in part iv.

5. Tests are now made on entries on the diagonal corresponding to $r = 3$ which are not in the triangular regions defined by 4*b* and 4*c*. There is just one entry to be checked, namely, 9. The critical value is 8.96. Hence the corresponding ordered difference is statistically significant. No additional tests are made.

The information with respect to the significant differences summarized in part iv may be presented schematically as follows:

$$\underline{c\ a\ d\ b}\ \ g\ e\ f. \qquad \text{(second line: } \underline{b\ g\ e\ f}\text{)}$$

Treatments underlined by a common line do not differ from each other; treatments not underlined by a common line do differ. Thus, treatment *f* differs from treatments *c*, *a*, and *d*, but treatment *f* does not differ from treatments *e*, *g*, and *b*. Similarly treatment *e* differs from *c*, *a*, and *d* but does not differ from *b*, *g*, *f*.

This sequence for making the tests prevents one from arriving at contradictory decisions of the following type: Suppose there are four means in an ordered set and that the difference $T_{(4)} - T_{(1)}$ is close to being significant but does not quite "make it." Further suppose that the difference $T_{(3)} - T_{(1)}$ is just larger than the appropriate critical value. In this case one might be tempted to conclude that there is no significant difference between the largest and the smallest means in the set but that there is a significant difference between the next to the largest and the smallest. Geometrically this conclusion would be equivalent to inferring that the distance between the largest and smallest of four means is zero but that the distance from the smallest to the next to the largest is greater than zero. Yet the latter distance has to be smaller than the former. Clearly this kind of inference leads to a contradiction.

In general, if the sequence indicated above is followed in making tests on all possible pairs of ordered means, the patterns of significant differences indicated below will *not* occur:

(1)	(2)	(3)	(4)	(5)
	*	—	*	*
		—	*	*
			*	—
			—	—

(1)	(2)	(3)	(4)	(5)
	*	*	—	*
		—	*	—
		*	—	*
			*	—

That is, between any two asterisks in the same row or column there can be no gaps (nonsignificant differences). Further, if the extreme position at the right of a row is a gap, then there can be no asterisks in that row or any row below that row.

It is of interest to compare the critical values for tests on ordered means with tests on individual components made through use of the *F* statistic. Since $q = \sqrt{2F}$, the critical value for a test on individual components for the example in Table 3.8-5 would be (in comparable units)

$$\sqrt{2F_{.99}(1,28)} = \sqrt{2(7.64)} = 3.91.$$

This is the critical value for totals which are two steps apart. Hence use of the F statistic is equivalent to using the 3.91 as the critical value for all q's, or using the value 7.82 as the critical value for differences between treatment totals. The critical values for the Newman-Keuls procedure range from 7.82 to 10.90. Hence the use of the F statistic leads to more "significant" results than does the use of the q_r statistic.

If the meaningful comparisons are relatively few in number and are planned before the data are obtained, the F test associated with individual components of variation should be used. This type of comparison is called an a priori comparison in contrast to comparisons made after inspection of the experimental data; the latter are called a posteriori or post-mortem comparisons. The a priori type is always justified whether or not the overall F is significant. If the k treatments fall into one or more natural groupings in terms of the treatments, tests on ordered differences may be made separately within each of the groupings. Other procedures for making a posteriori comparisons are discussed in the next section.

3.9 Alternative Procedures for Making A Posteriori Tests

The Newman-Keuls procedure, described in the last section, keeps the level of significance at most α for sets of ranges and subsets of ranges within an overall inclusive range. However, the level of significance with respect to the collection of all tests made, considered as a single test, is considerably lower than α. Thus, the power of the collection of all tests made is less than that associated with an ordinary α-level test. Duncan (1955) has developed a procedure which uses a *protection level* of α for the collection of tests, rather than an α level for the individual tests.

[Scheffé (1959, p. 78) takes issue with the principles underlying the development of the sampling distributions which Duncan uses in obtaining the critical values for tests.]

In terms of individual tests, a protection level with $\alpha = .01$ provides a level of significance equal to .01 for means which differ by two ordered steps; for means differing by three steps, the level of significance is equal to $1 - (.99)^2 = .02$; for means differing by four steps, the level of significance is equal to $1 - (.99)^3 = .03$. Thus, with protection level equal to α, the level of significance for individual tests is numerically higher than α when the pairs are more than two steps apart.

The statistic used in the Duncan procedure is the same as that used in the Newman-Keuls test, namely, q_r, where r is the number of steps apart two means or totals are in an ordered sequence. The steps followed in using the Duncan procedure are identical to those followed in the Newman-Keuls procedure. However, the critical values for the q_r statistic are obtained from tables prepared by Duncan for use with protection levels.

For the case $k = 7$ and degrees of freedom for experimental error equal to 28, critical values for the .01-level Newman-Keuls tests and the .01-level protection level are as follows:

k	2	3	4	5	6	7
Newman-Keuls	3.91	4.48	4.84	5.09	5.28	5.45
Duncan	3.91	4.08	4.18	4.28	4.34	4.39

When $k = 2$, the two procedures have identical critical values. For values of k larger than 2, the Duncan procedure has the smaller critical value. The larger the value of r, the larger the difference between the critical values for the two procedures. Thus, on the average, a larger difference between two means (or two totals) is required for statistical significance under the Newman-Keuls procedure.

To illustrate the difference in size of the type 1 error associated with individual tests, which disregards the order aspect, the q_r statistic may be transformed into an F statistic by means of the relation

$$q_r = \sqrt{2F} \qquad \text{or} \qquad F = \frac{q_r^2}{2}.$$

For the case discussed in the preceding paragraph, 5.45 is equivalent to an F of 14.85, and 4.39 is equivalent to an F of 8.63. Relative to the critical value for an individual comparison made by means of an F statistic, the type 1 error associated with the difference between two totals which are seven steps apart is as follows:

	Critical F value	"Actual" α
Individual comparison	7.64	.01
Newman-Keuls	14.85	.0005
Duncan	8.63	.007

In other words, if a .01-level test were made on the difference between two means that are seven steps apart on an ordered scale, assuming that MS_{error} has 28 degrees of freedom, and if the order of the means in the sample were disregarded, the critical value would be $F = 7.64$. If, however, the order were taken into account, the equivalent critical value would be 14.85. With a .01 protection level, the equivalent critical value would be 8.63.

A considerably more conservative procedure in terms of keeping the type 1 error small is the use of $q_{1-\alpha}(k,f)$ as the critical value for all tests, no matter how many steps apart the means may be. Thus, instead of changing the critical value as a function of the number of steps two means are apart on an ordered scale, the critical value for the maximum number of steps is used for all tests. This approach, suggested by Fisher, has been studied and extended by Tukey and will be called the Tukey (*a*) procedure. [This procedure has also been called the honestly significant difference (hsd) procedure.] Compared with the Newman-Keuls and Duncan approaches, fewer significant differences will be obtained. For the example considered in Table 3.8-5, the critical value for the q_r statistic would be 5.45 for all tests; equivalently, the critical value for the difference between two treatment totals would be 10.90 for all differences. If the Tukey (*a*) test were used with the data in Table 3.8-5, part iv of this table would have the following form:

	c	*a*	*d*	*b*	*g*	*e*	*f*
c					**	**	**
a						**	**
d						**	**

The power of Tukey (*a*) tests is lower than those of the Newman-Keuls and the Duncan procedures. The Tukey (*a*) procedure has this general property: All tests on differences between pairs have a level of significance which is at most equal to α.

Tukey also proposed a second procedure, which is a compromise between the Tukey (*a*) tests and the Newman-Keuls procedure. The statistic employed in making what will be called Tukey (*b*) tests is again the q_r statistic. However, the critical value is the average for the corresponding value in the Newman-Keuls tests and the critical value for the Tukey (*a*) tests. In symbols, this critical value is

$$\frac{q_{1-\alpha}(k,f) + q_{1-\alpha}(r,f)}{2},$$

where k is the number of means in the set and r is the number of steps between the two means being compared. For example, if $k = 7$ and $f = 28$, the critical value for a .01-level test for two means which are four steps apart is

$$\frac{5.45 + 4.84}{2} = 5.14.$$

A procedure for making all possible comparisons, not specifically comparisons involving two means, has been developed by Scheffé. An F statistic corresponding to a component of variation is computed, but the critical value for this component is $(k - 1)F_{1-\alpha}(k - 1, f)$. All tests use this critical value. For example, if $k = 7$, $f = 28$, and $\alpha = .01$, the critical value for Scheffé tests is $6[F_{.99}(6,28)] = 6(3.53) = 21.18$. In terms of the q statistic this is equivalent to a critical value of $\sqrt{2(21.18)} = 6.51$. If Scheffé tests were made on the data in Table 3.8-5, the critical value would be $(6.51)\sqrt{n\text{MS}_{\text{error}}} = 13.02$ for all differences between treatment totals. The statistically significant differences using this critical value are as follows:

	c	*a*	*d*	*b*	*g*	*e*	*f*
c						**	**

Thus, the Scheffé method applied to testing differences between all possible pairs is even more conservative with respect to type 1 errors than is the Tukey (*a*) method.

The Scheffé approach has this optimum property: The type 1 error is at most α for any of the possible comparisons. In the original development, Scheffé was concerned with constructing a set of simultaneous confidence intervals on all the comparisons within a subdivision of the entire experiment. Before a set of simultaneous confidence intervals is considered to be true, each of the separate statements must be true. If any one of the confidence intervals in the set is false, then the confidence statement is considered to be false. The simultaneous-confidence-interval approach can be translated into a procedure for making all possible tests on comparisons. This translation has been outlined above.

Another approach to the problem of regulating and apportioning the type 1 error rate is a method suggested by R. A. Fisher and called the modified *least-significant difference* (lsd) approach. The usual t statistic is computed, i.e.,

$$t = \frac{\bar{T}_j - \bar{T}_{j'}}{\sqrt{2\text{MS}_{\text{error}}/n}}.$$

If there are k treatments, there are $k(k-1)/2$ unique differences of the form $\bar{T}_j - \bar{T}_{j'}$. If the level of significance for a test on a specified difference is denoted by α, let

$$\alpha_{\text{E}} = \frac{\alpha}{k(k-1)/2}.$$

If α_{E} is taken as the level of significance for the t test, then the probability of making a type 1 error in testing all $k(k-1)/2$ possible differences will be at most

$$\frac{k(k-1)}{2}\alpha_{\text{E}} = \frac{k(k-1)}{2}\frac{\alpha}{k(k-1)/2} = \alpha.$$

Thus, if α_E is taken to be the level of significance for the t test, then

$$\alpha = \frac{\text{number of type 1 errors with respect to tests on simple differences}}{\text{number of experiments}}.$$

A level of significance defined in this way is called a *per experiment* level of significance. If one uses the critical values

$$\pm t_{1-(\alpha_E/2)}(f), \qquad \text{where } f = \text{degrees of freedom for } \text{MS}_{\text{error}},$$

for the t statistic defined above, then the per experiment type 1 error rate will be equal to α.

In terms of a q statistic, the critical value for a .01-level (per experiment) test for the numerical data in Table 3.8-5 (where $k = 7$, $f = 28$, $\sqrt{n\text{MS}_{\text{error}}} = 2$) is

$$\begin{aligned} \sqrt{2}\, t_{1-(\alpha_E/2)}(f) &= 1.4141t_{.99976}(28) \\ &= 1.4141(3.96) = 5.60. \end{aligned}$$

Note that

$$1 - \frac{\alpha_E}{2} = 1 - \frac{\alpha/2}{k(k-1)/2} = 1 - \frac{.01}{42} = .99976.$$

The critical value (per experiment) for the difference between two treatment totals is

$$t_{.99976}(28)\sqrt{2n\text{MS}_{\text{error}}} = 3.96\sqrt{2(5)(.80)} = 11.20.$$

In contrast to the per experiment error rate for tests on differences as represented by the lsd approach, the Tukey hsd approach represents what may be called an *experimentwise* error rate. The latter may be represented as

$$\alpha' = \frac{\text{number of experiments having one or more type 1 errors with respect to tests on simple differences}}{\text{number of experiments}}.$$

For a fixed set of outcomes, the numerator of the experimentwise error rate will, in general, be smaller than the numerator for the per experiment error rate. Hence, in terms of comparable units, for fixed α the experimentwise critical value will be somewhat smaller in absolute value than the per experiment critical value.

The data in Table 3.8-5 will be used to summarize the various methods for making a posteriori tests. For $\alpha = .01$, the critical values for differences between pairs of ordered totals are as follows:

Method	k: 2	3	4	5	6	7
Scheffé	13.02	13.02	13.02	13.02	13.02	13.02
Modified lsd	11.20	11.20	11.20	11.20	11.20	11.20
Tukey (*a*) (hsd)	10.90	10.90	10.90	10.90	10.90	10.90
Tukey (*b*)	9.36	9.93	10.29	10.54	10.74	10.90
Newman-Keuls	7.82	8.96	9.68	10.18	10.56	10.90
Duncan	7.82	8.16	8.36	8.56	8.68	8.78
Individual comparisons	7.82	7.82	7.82	7.82	7.82	7.82

The last method is primarily for meaningful comparisons planned prior to inspection of the data; it is included in this summary only for purposes of illustration. The Scheffé method is clearly the most conservative with respect to type 1 error; this method will lead to the smallest number of significant differences. In making tests on differences between all possible pairs of means it will yield too few significant results. The Tukey (*a*) test will also tend to yield too few significant results. Because the Tukey (*a*) test is applicable in a relatively broad class of situations, and because it is simple to apply, there is much to recommend the Tukey (*a*) test for general use in making a posteriori tests.

A confidence interval on the difference between two means using the Tukey (*a*) procedure has the form (for all pairs, j, j')

$$C(\bar{T}_j - \bar{T}_{j'} - A \leq \mu_j - \mu_{j'} \leq \bar{T}_j - \bar{T}_{j'} + A) = 1 - \alpha,$$

where
$$A = q_{1-\alpha}[k, k(n-1)]s_{\bar{T}}.$$

In this case the confidence coefficient, $1 - \alpha$, is with respect to the collection of all possible pairs in the set of k considered simultaneously. Thus, with simultaneous confidence $1 - \alpha$, the differences $\mu_j - \mu_{j'}$ lie in the intervals specified. A set of confidence intervals is true simultaneously if and only if all the individual statements in the set are true. A simultaneous confidence interval of size $1 - \alpha$ implies that no fewer than $100(1 - \alpha)$ percent of all sets of statements made will be true.

3.10 Comparing All Means with a Control

If one of the k treatments in an experiment represents a control condition, the experimenter is generally interested in comparing each treatment with the control condition, regardless of the outcome of the overall F. There are $k - 1$ comparisons of this kind. Rather than setting a level of significance equal to α for each of the tests, the experimenter may want to have a level equal to α for the collection of the $k - 1$ decisions, considered as a single decision summarizing the outcomes.

Since each of the tests uses the same information on the control condition and a common estimate of experimental error, the tests are not independent. Dunnett (1955) has derived the sampling distribution for a t statistic appropriate for use when level of significance α is desired for the set of all comparisons between several treatments and a control. The parameters of Dunnett's distribution for the t statistic are:

$$k = \text{number of treatments (including the control)},$$

$$\text{df} = \text{degrees of freedom for } \text{MS}_{\text{error}}.$$

The approach used by Dunnett is a special case of the problem of handling multiple comparisons by constructing a joint confidence interval on the set of all relevant comparisons.

The numerical example in Table 3.10-1 will be used to illustrate the application of Dunnett's t statistic. In this experiment there are four treatment conditions ($k = 4$); three observations ($n = 3$) are made under each of the treatment conditions. One of the treatments represents a standard manufacturing process; the other three represent different methods for manufacturing the same product. The criterion measure in each case is an index of quality of the manufactured product. In this case the overall $F = 7.05$ exceeds the critical value for a .05-level test. The t statistic for the difference between method j and the standard method is

$$t = \frac{\bar{T}_j - \bar{T}_0}{\sqrt{2\text{MS}_{\text{error}}/n}}.$$

The critical value for the collection of $k - 1$ statistics of this form that may be computed is obtained from the Dunnett tables given in Table C.6. For the data in Table 3.10-1 the critical value for a two-tailed .05-level test is ± 2.88 [that is, $t_{.975}(4,8) = 2.88$]. This value is found under the column headed 4 and the row corresponding to degrees of freedom equal to 8. For example, in comparing method I with the standard,

$$t = \frac{61 - 50}{\sqrt{2(19)/3}} = \frac{11}{3.56} = 3.09.$$

Since the observed t statistic exceeds the critical value, the hypothesis that the two methods of manufacturing yield products having equal average quality index is rejected. The level of significance for this single test is not .05; it is approximately .02. The critical value of 2.88 is associated with the collection of the three tests that would be made. The other two tests are

$$t = \frac{52 - 50}{3.56} = .56,$$

$$t = \frac{45 - 50}{3.56} = -1.41.$$

These three tests may be summarized in the statement that method I differs significantly from the standard but methods II and III do not. This summary decision has significance level equal to .05. In the long run the summary decision reached by the procedures followed will have the equivalent of a type 1 error equal to, at most, .05.

Rather than working with a series of tests of significance, the experimenter may construct a series of confidence intervals. For this purpose it is convenient to introduce the concept of what Tukey has called an allowance, designated by the symbol A. In this context an allowance is defined by

$$A = t_{1-(\alpha/2)}\sqrt{\frac{2\text{MS}_{\text{error}}}{n}},$$

where $t_{1-(\alpha/2)}$ is a value obtained from the Dunnett tables. For the data in Table 3.10-1, with $\alpha = .05$, $A = 2.88(3.56) = 10.25$. The general form for the lower and upper confidence limits is

$$(\bar{T}_j - \bar{T}_0) \pm A.$$

Table 3.10-1 Numerical Example

	Methods				
	Standard	I	II	III	
	55	55	55	50	
	47	64	49	44	$n = 3;\ k = 4$
	48	64	52	41	
T_j	150	183	156	135	$G = 624$
$\Sigma(X_j^2)$	7538	11,217	8130	6117	$\Sigma(\Sigma X^2) = 33{,}002$
$\bar{T}_j$	50	61	52	45	

$$(1) = G^2/kn = (624)^2/12 = 32{,}448.00$$
$$(2) = \Sigma(\Sigma X^2) = 33{,}002$$
$$(3) = (\Sigma T_j^2)/n = (150^2 + 183^2 + 156^2 + 135^2)/3 = 32{,}850.00$$

$$\text{SS}_{\text{methods}} = (3) - (1) = 402.00$$
$$\text{SS}_{\text{error}} = (2) - (3) = 152.00$$
$$\text{SS}_{\text{total}} = (2) - (1) = 554.00$$

Source of variation	SS	df	MS	F
Methods	402.00	3	134.00	7.05
Experimental error	152.00	8	19.00	
Total	554.00	11		

$F_{.95}(3,8) = 4.07$

For the data in Table 3.10-1, 95 percent confidence limits for the collection of confidence statements on the difference $\mu_j - \mu_0$ are as follows:

For method I: $(61 - 50) \pm 10.25 = .75$ and 21.25,

For method II: $(52 - 50) \pm 10.25 = -8.25$ and 12.25,

For method III: $(45 - 50) \pm 10.25 = -15.25$ and 5.25.

The joint (simultaneous) confidence coefficient for the intervals defined by these limits is .95. The confidence interval for method I takes the form

$$.75 \leq \mu_{\text{I}} - \mu_0 \leq 21.25.$$

In words, the difference between the mean quality indices for method I and the standard method is between .25- and 21.25-quality index units. This statement and the additional confidence statements for methods II and III have a joint confidence coefficient of .95.

It is of interest to compare the distribution of Dunnett's t statistic with that of Student's t statistic. For the case in which $k = 2$, the two distributions are identical. For k greater than 2, corresponding critical values in the Dunnett tables are larger. For example, with the degrees of freedom for error equal to 10 and $k = 7$, two-tailed tests with joint significance level .05 are equivalent to individual two-tailed tests at the .01 level. It is also of interest to compare the critical values for this type of comparison with those made by means of the Tukey (a) test. The two critical values may be cast in terms of comparable units of measurement by means of the relationship

$$q = \sqrt{2}\, t.$$

Corresponding critical values for .05 joint two-tailed significance levels are as follows:

df for error	k	Tukey (a)	Dunnett
10	8	5.30	4.58
20	6	4.45	3.86
∞	10	4.47	3.80
10	2	3.15	3.15

For values of k greater than 2, the Tukey (a) values are higher than the corresponding Dunnett values. Since the Tukey (a) test was designed to be appropriate for all possible differences, of which comparisons of each treatment with the control are a subset, it is reasonable to expect that the Tukey (a) tests would require the larger critical values.

3.11 Tests for Homogeneity of Variance

One of the basic assumptions underlying both models I and II is that the variance due to experimental error within each of the treatment populations be homogeneous, that is $\sigma^2_{\varepsilon_1} = \sigma^2_{\varepsilon_2} = \cdots = \sigma^2_{\varepsilon_k}$. Moderate departures from this assumption do not, however, seriously affect the sampling distribution of the resulting F statistic. That is, when the variances in the population are not equal, the F statistic using a pooled variance has approximately the same distribution as the F statistic which takes the differences in the population variances into account. The following examples, taken from Box (1954, p. 299), illustrate the effect of lack of homogeneity of variance. In these examples $n = 5$ and $k = 3$.

	Populations			Probability of F exceeding $F_{.95}$
	1	2	3	
(*a*) Variances	1	1	1	.050
(*b*) Variances	1	2	3	.058
(*c*) Variances	1	1	3	.059

In (*a*) all variances are equal; hence an F statistic which pools variances has probability of .05 of exceeding $F_{.95}$ when $\tau_1 = \tau_2 = \tau_3$. In (*b*) the variances have the ratio 1:2:3; that is, the variance for the second population is twice the variance of the first population, and the variance of the third population is three times the variance of the first population. For this case, the exact sampling distribution (assuming $\tau_1 = \tau_2 = \tau_3$) for the F statistic shows probability equal to .058 of exceeding $F_{.95}$ obtained from the F statistic which assumes $\sigma^2_{\varepsilon_1} = \sigma^2_{\varepsilon_2} = \sigma^2_{\varepsilon_3}$. Using the ordinary F test when the population variances are in the ratio 1:2:3 gives a test having a small positive bias, since relatively more significant results will be obtained than the exact sampling distribution warrants. In (*c*) the ratio of the variances is 1:1:3; the F test in this case would also have a small positive bias, since the probability of exceeding $F_{.95}$ is .059 rather than .050. When the number of observations in each treatment class varies considerably, Box indicates that the bias becomes somewhat larger and the direction of the bias is not always positive. Also, the greater the skewness in the distribution of the population variances, the more bias in the resulting tests.

Even if the variances are not homogeneous, the sampling distribution of the statistic

$$F = \frac{\text{MS}_{\text{treat}}}{\text{MS}_{\text{error}}}$$

may be approximated by an F distribution having parameters

$$F[(k-1)\varepsilon', k(n-1)\varepsilon],$$

where $$\varepsilon' = \left[1 + \frac{k-2}{k-1}c^2\right]^{-1}, \qquad \varepsilon = (1 + c^2)^{-1},$$

$$c^2 = \frac{1}{k}\frac{\Sigma(\sigma_j^2 - \bar{\sigma}^2)^2}{(\bar{\sigma}^2)^2}, \qquad \bar{\sigma}^2 = \frac{1}{k}\Sigma\sigma_j^2.$$

The value of c^2 may be shown to lie in the interval

$$0 \le c^2 \le k - 1.$$

If c^2 assumes its maximum value, then

$$\varepsilon' = \frac{1}{k-1} \qquad \text{and} \qquad \varepsilon = \frac{1}{k}.$$

Hence, when $c^2 = k - 1$, then

$$F = \frac{MS_{treat}}{MS_{error}} \text{ is approximately distributed as } F[1, n-1].$$

The decision rule

$$\text{Reject } H_1 \text{ when } F_{obs} > F_{1-\alpha}[1, n-1]$$

will be extremely conservative in the sense that the critical value will probably be larger than it should be to represent level of significance α. This conservatism follows from the fact that the degrees of freedom for the distribution used to approximate the exact distribution are set equal to their lower bounds. This approximation is one suggested by Box (1954, p. 300).

In cases where the experimenter has no knowledge about the effect of the treatments upon the variance, tests for homogeneity of variances may be appropriate as preliminary tests on the model underlying the analysis. There is no need, however, for a high degree of sensitivity in such tests, because F tests are robust with respect to departures from homogeneity of variance. The experimenter need be concerned about only relatively large departures from the hypothesis of equal population variances.

A relatively simple, but adequate, test of the hypothesis that

$$\sigma_1^2 = \sigma_2^2 = \cdots = \sigma_k^2$$

is one proposed by Hartley. When n is constant for all the k treatments in an experiment, this hypothesis may be tested by means of the statistic

$$F_{max} = \frac{\text{largest of } k \text{ treatment variances}}{\text{smallest of } k \text{ treatment variances}}$$

$$= \frac{s^2_{largest}}{s^2_{smallest}}.$$

Under the hypothesis that $\sigma_1^2 = \sigma_2^2 = \cdots = \sigma_k^2$, the sampling distribution of the F_{max} statistic (assuming independent random samples from normal populations) has been tabulated by Hartley. This distribution is given in Table C.7. The parameters for this distribution are k, the number of treatments, and $n - 1$, the degrees of freedom for each of the treatment class variances. If the observed F_{max} is greater than the tabled value associated with an α-level test, then the hypothesis of homogeneity of variance is rejected. This test will be illustrated by use of the data in Table 3.11-1.

Table 3.11-1 Numerical Example
($n = 10$)

	Treatment 1	Treatment 2	Treatment 3	Treatment 4
T_j	140	95	83	220
ΣX_j^2	2320	1802	869	6640
T_j^2/n	1960.00	902.50	688.90	4840.00
SS_j	360.00	899.50	180.10	1800.00
s_j^2	40.00	99.94	20.01	200.00
				$\Sigma SS_j = 3239.60$
				$\Sigma s_j^2 = 359.95$

In this table $k = 4$, and $n = 10$. SS_j is the variation within treatment class j and is given by

$$SS_j = \Sigma X_j^2 - \frac{T_j^2}{n}.$$

The variance within treatment class j is given by

$$s_j^2 = \frac{SS_j}{n - 1}.$$

The largest of the within-class variances is 200.00; the smallest is 20.01. Thus, the numerical value of the F_{max} statistic for these data is

$$F_{max} = \frac{200.00}{20.01} = 10.0.$$

From the tables of the F_{max} distribution, $F_{max,.99}(4,9) = 9.9$. Since the observed value of the F_{max} statistic is greater than the critical value for a .01-level test, the hypothesis of homogeneity of variance is rejected. This test, referred to as Hartley's test, is in practice sufficiently sensitive for use as a preliminary test in situations where such a test is in order. When the number of observations in each of the treatment classes is not constant, but the n_j's are relatively close to being equal, the largest of the sample sizes may be used instead of n in obtaining the degrees of freedom required for use in the Hartley tables. This procedure leads to a slight positive bias in the test, i.e., rejecting H_1 more frequently than should be the case.

Another relatively simple test for homogeneity of variance, developed by Cochran, uses the statistic

$$C = \frac{s^2_{\text{largest}}}{\Sigma s_j^2}.$$

The parameters of the sampling distribution of this statistic are k, the number of treatments, and $n - 1$, the degrees of freedom for each of the variances. Tables of the 95th and 99th percentile points of the distribution of the C statistic are given in Table C.8. For the data in Table 3.11-1,

$$C = \frac{200.00}{359.95} = .56.$$

For a .01-level test the critical value is $C_{.99}(4,9) = .57$. The observed value of C is quite close to the critical value but does not exceed it. However, the experimenter should on the basis of this result seriously question the tenability of the hypothesis of homogeneity of variance.

In most situations encountered in practice, the Cochran and Hartley tests will lead to the same decisions. Since the Cochran test uses more of the information in the sample data, it is generally somewhat more sensitive than is the Hartley test. In cases where n_j, the number of observations in each treatment class, is not constant but is relatively close, the largest of the n_j's may be used in place of n in determining the degrees of freedom needed to enter the tables.

Bartlett's test for homogeneity of variance is perhaps the most widely used test. The routine use of Bartlett's test as a preliminary test on the model underlying the analysis of variance is not, however, recommended. Only in relatively few cases is Bartlett's test useful. From the computational point of view it is more complex than is either the Hartley test or the Cochran test. In Bartlett's test the n_j's in each of the treatment classes need not be equal; however, no n_j should be smaller than 3, and most n_j's should be larger than 5. The statistic used in Bartlett's test is

$$\chi^2 = \frac{2.303}{c}(f \log \text{MS}_{\text{error}} - \Sigma f_j \log s_j^2),$$

where

$$f_j = n_j - 1 = \text{degrees of freedom for } s_j^2, \qquad j = 1, \ldots, k,$$

$$f = \Sigma f_j = \text{degrees of freedom for MS}_{\text{error}},$$

$$c = 1 + \frac{1}{3(k-1)}\left(\Sigma \frac{1}{f_j} - \frac{1}{f}\right),$$

$$\text{MS}_{\text{error}} = \frac{\Sigma \text{SS}_j}{\Sigma f_j}.$$

When $\sigma_1^2 = \sigma_2^2 = \cdots = \sigma_k^2$, the sampling distribution of the χ^2 statistic is approximated by the χ^2 distribution having $k - 1$ degrees of freedom.

The data in Table 3.11-1 will be used to illustrate the computation of the χ^2 statistic. Computations will be indicated for the case in which the n_j's are not assumed to be equal. For this case MS_{error} is most readily obtained from

$$MS_{error} = \frac{\Sigma SS_j}{\Sigma f_j} = \frac{3239.60}{36} = 89.99,$$

since $\Sigma f_j = 9 + 9 + 9 + 9 = 36$. Other items required for the computation of the χ^2 statistic are

$$\begin{aligned}
f \log MS_{error} &= 36 \log \ 89.99 = 36(1.954) = \underline{70.344} \\
f_1 \log s_1^2 &= 9 \log \ 40.00 = 9(1.602) = 14.418 \\
f_2 \log s_2^2 &= 9 \log \ 99.94 = 9(1.999) = 17.991 \\
f_3 \log s_3^2 &= 9 \log \ 20.01 = 9(1.301) = 11.709 \\
f_4 \log s_4^2 &= 9 \log 200.00 = 9(2.301) = \underline{20.709} \\
\Sigma f_j \log s_j^2 & \qquad\qquad\qquad\qquad\qquad\quad 64.827
\end{aligned}$$

$$\begin{aligned}
c &= 1 + \frac{1}{3(3)}\left(\frac{1}{9} + \frac{1}{9} + \frac{1}{9} + \frac{1}{9} - \frac{1}{36}\right) \\
&= 1 + \tfrac{1}{9}\left(\tfrac{15}{36}\right) \\
&= 1.046.
\end{aligned}$$

From these terms, the χ^2 statistic in Bartlett's test is

$$\chi^2 = \frac{2.303}{1.046}(70.344 - 64.827) = 12.14.$$

The larger the variation between the s_j^2's, the larger will be the value of the χ^2 statistic. For a .01-level test of the hypothesis that $\sigma_1^2 = \sigma_2^2 = \cdots = \sigma_k^2$ the critical value is $\chi^2_{.99}(3) = 11.3$. Since the observed value of the χ^2 statistic is larger than the critical value, the experimental data do not support the hypothesis being tested.

For the data in Table 3.11-1, the Hartley, Cochran, and Bartlett tests give comparable results. The Hartley test uses what is equivalent to the range of the sample variances as a measure of heterogeneity, whereas the Bartlett test uses what is equivalent to the ratio of the arithmetic mean to the geometric mean of the variances. The sampling distribution of the latter measure has a smaller standard error and hence provides a more powerful test of the hypothesis being tested. For purposes of detecting large departures from the hypothesis of homogeneity of variance, either the Hartley or the Cochran test is adequate in most cases occurring in practice.

There is some evidence that all the tests for homogeneity of variance that have been discussed above are oversensitive to departures from normality of the distributions of the basic observations. Bartlett and Kendall have proposed a test for homogeneity of variance which is less sensitive to nonnormality than any of the above tests. A detailed description and applications of the Bartlett and Kendall test will be found in the work of Odeh and Olds (1959).

Another test for homogeneity of variance which is relatively insensitive to departure from normality is one proposed by Scheffé. In essence this procedure represents an analysis of variance on the logarithm of a set of variances. The approach to the Scheffé test is described in detail in Sec. 3.12.

3.12 Unequal Sample Sizes

The plan of an experiment may call for an equal number of observations under each treatment, but the completed experiment may not meet this objective. For comparable precision in the evaluation of each treatment effect this objective is a highly desirable one, assuming that the variances for the treatment classes are equal. Circumstances not related to the experimental treatments often prevent the experimenter from having an equal number of observations under each treatment. For example, in animal research, deaths may occur from causes in no way related to the experimental treatments. In areas of research in which people are the subjects, it may be that only intact groups can be handled; such intact groups may vary in size.

In the earlier sections of this chapter it is generally assumed that random samples of size n were assigned at random to each of the treatments. In this section it will be assumed that a random sample of size n_1 is assigned to treatment 1, a random sample of size n_2 is assigned to treatment 2, etc. The size of the random sample is not assumed to be constant for all treatments. The form of the definitions of the sums of squares is different from those appropriate for the case in which the sample size is constant. The notation that will be used is outlined in Table 3.12-1. The number of treatments in the experiment is k. The treatments are designated by the symbols $1, 2, \ldots, j, \ldots, k$, where the symbol j represents any treatment within the set. The size of the sample observed under treatment j is designated by the symbol n_j. The total number of elements in the experiment is

$$n_1 + n_2 + \cdots + n_k = N.$$

To obtain estimates of the variation due to error and treatment effects, one starts with the model

$$X_{ij} = \mu + \tau_j + \varepsilon_{ij}.$$

The usual assumptions underlying the general linear model apply here.

Table 3.12-1 Notation

	Treatment 1	Treatment 2	$\cdots$	Treatment j	$\cdots$	Treatment k	
Number of observations	n_1	n_2	$\cdots$	n_j	$\cdots$	n_k	$N = \Sigma n_j$
Sum of observations	T_1	T_2	$\cdots$	T_j	$\cdots$	T_k	$G = \Sigma T_j$
Mean of observations	$\bar{T}_1$	$\bar{T}_2$	$\cdots$	$\bar{T}_j$	$\cdots$	$\bar{T}_k$	$\bar{G} = G/N$
Sum of squares of observations	ΣX_1^2	ΣX_2^2	$\cdots$	ΣX_j^2	$\cdots$	ΣX_k^2	$\Sigma(\Sigma X_j^2)$
T_j^2/n_j	T_1^2/n_1	T_2^2/n_2	$\cdots$	T_j^2/n_j	$\cdots$	T_k^2/n_k	
Within-class variation	SS_1	SS_2	$\cdots$	SS_j	$\cdots$	SS_k	
Within-class variance	$s_1^2 = \dfrac{SS_1}{n_1 - 1}$	$s_2^2 = \dfrac{SS_2}{n_2 - 1}$	$\cdots$	$s_j^2 = \dfrac{SS_j}{n_j - 1}$	$\cdots$	$s_k^2 = \dfrac{SS_k}{n_k - 1}$	

Least-squares estimates of the parameters μ and τ_j are obtained by making

$$\Sigma\hat{\varepsilon}_{ij}^2 = \Sigma(X_{ij} - \hat{\mu} - \hat{\tau}_j)^2 = \text{minimum}.$$

The normal equations for this case have the form

$$\begin{aligned} N\hat{\mu} + \Sigma n_j\hat{\tau}_j &= G, \\ n_j\hat{\mu} + n_j\hat{\tau}_j &= T_j, \end{aligned} \qquad j = 1, \ldots, k.$$

Under the constraint $\Sigma n_j\hat{\tau}_j = 0$, the least-squares estimators of the parameters are

$$\begin{aligned} \hat{\mu} &= \frac{G}{N} = \bar{G}, \\ \hat{\tau}_j &= \bar{T}_j - \bar{G}, \end{aligned} \qquad j = 1, \ldots, k.$$

The total sum of squares due to μ and the τ_j is given by

$$\begin{aligned} R(\mu,\tau) &= \bar{G}G + \Sigma\hat{\tau}_j T_j \\ &= \frac{G^2}{N} + \Sigma\frac{T_j^2}{n_j} - \frac{G^2}{N} = \Sigma\frac{T_j^2}{n_j}. \end{aligned}$$

The variation due to treatment is

$$\begin{aligned} R(\tau \mid \mu) = SS_{\text{treat}} &= R(\mu,\tau) - R(\mu) \\ &= \Sigma\frac{T_j^2}{n_j} - \frac{G^2}{N} \\ &= \Sigma n_j(\bar{T}_j - \bar{G})^2. \end{aligned}$$

The variation due to error is

$$\begin{aligned} \Sigma X_{ij}^2 - R(\tau,\mu) &= \Sigma X_{ij}^2 - \Sigma\frac{T_j^2}{n_j} \\ &= \Sigma(X_{ij} - \bar{T}_j)^2 = SS_{\text{error}}. \end{aligned}$$

The same estimates of variation would result under maximum-likelihood procedures assuming the τ_j were random variables having a joint multivariate normal distribution.

Computational formulas are summarized in Table 3.12-2. Symbols (1) and (2) have the same general form as they do for the case of equal n's. However, symbol (3) is different; that is,

Table 3.12-2 Computational Formulas

$(1) = \dfrac{G^2}{N}$	$(2) = \Sigma(\Sigma X_j^2)$	$(3) = \Sigma\left(\dfrac{T_j^2}{n_j}\right)$
$SS_{treat} = (3) - (1)$		$df_{treat} = k - 1$
$SS_{error} = (2) - (3)$		$df_{error} = N - k$
$SS_{total} = (2) - (1)$		$df_{total} = N - 1$

$$(3) = \frac{T_1^2}{n_1} + \frac{T_2^2}{n_2} + \cdots + \frac{T_k^2}{n_k}.$$

Thus each T_j^2 must be divided by its n_j before the summation is made. The degrees of freedom for each of the sums of squares are also shown in Table 3.12-2. For SS_{error} the number of degrees of freedom is the pooled degrees of freedom for the variation within each of the treatments, i.e.,

$$\begin{aligned} df_{error} &= (n_1 - 1) + (n_2 - 1) + \cdots + (n_k - 1) \\ &= N - k. \end{aligned}$$

The computational formulas given in Table 3.12-2 become those for the case of equal sample sizes when n_j is constant for all treatments. Hence the computational formulas for equal n's are simplified special cases of those given in Table 3.12-2.

The basic partition of the overall variation now has the form

$$(1) \qquad \Sigma\Sigma(X_{ij} - \bar{G})^2 = \Sigma\Sigma(X_{ij} - \bar{T}_j)^2 + \Sigma n_j(\bar{T}_j - \bar{G})^2,$$

$$SS_{total} \quad = \quad SS_{error} \quad + \quad SS_{treat}.$$

Thus in the computation of SS_{treat} each $(\bar{T}_j - \bar{G})^2$ is multiplied by n_j. Hence the variation due to the treatments is a *weighted* sum of the squared deviations of the treatment means about the grand mean; squared deviations based upon a large number of observations are given greater weight than those based upon a small number of observations. (A definition of SS_{treat} which assigns equal weight to the squared deviations is given at the end of this section. The latter definition is to be preferred in cases where differences in the n_j's have no direct meaning in terms of the population about which inferences are being drawn.)

A numerical example is given in Table 3.12-3. Data in part i represent the basic observations. Detailed summary data are given in part ii. The

Table 3.12-3 Numerical Example

	Treatment 1	Treatment 2	Treatment 3	Treatment 4	
(i)	3	7	3	10	
	2	8	2	12	
	4	4	1	8	
	3	10	2	5	
	1	6	4	12	
	5		2	10	
			3	9	
			1		
(ii)	$n_1 = 6$	$n_2 = 5$	$n_3 = 8$	$n_4 = 7$	$N = 26$
	$T_1 = 18$	$T_2 = 35$	$T_3 = 18$	$T_4 = 66$	$G = 137$
	$\Sigma X_1^2 = 64$	$\Sigma X_2^2 = 265$	$\Sigma X_3^2 = 48$	$\Sigma X_4^2 = 658$	$\Sigma(\Sigma X_j^2) = 1035$
	$\frac{T_1^2}{n_1} = 54.00$	$\frac{T_2^2}{n_2} = 245.00$	$\frac{T_3^2}{n_3} = 40.50$	$\frac{T_4^2}{n_4} = 622.29$	$\Sigma\left(\frac{T_j^2}{n_j}\right) = 961.79$
	$SS_1 = 10.00$	$SS_2 = 20.00$	$SS_3 = 7.50$	$SS_4 = 35.71$	$\Sigma SS_j = 73.21$
	$s_1^2 = 2.00$	$s_2^2 = 5.00$	$s_3^2 = 1.07$	$s_4^2 = 5.95$	
	$\bar{T}_1 = 3.00$	$\bar{T}_2 = 7.00$	$\bar{T}_3 = 2.25$	$\bar{T}_4 = 9.43$	$\bar{G} = \frac{137}{26} = 5.27$

(iii)

$(1) = G^2/N = (137)^2/26 = 721.88$ $\quad (2) = \Sigma\Sigma X^2 = 1035$ $\quad (3) = \Sigma(T_j^2/n_j) = 961.79$

$$SS_{treat} = (3) - (1) = 239.91$$

$$SS_{error} = (2) - (3) = 73.21$$

$$SS_{total} = (2) - (1) = 313.12$$

symbols used in part ii are defined in Table 3.12-1. For example,

$$SS_1 = \Sigma X_1^2 - \frac{T_1^2}{n_1} = 64 - 54.00 = 10.00.$$

As a rough check for homogeneity of variance,

$$F_{max} = \frac{s^2_{largest}}{s^2_{smallest}} = \frac{5.95}{1.07} = 5.56.$$

The largest of the n_j's is 8. For a .05-level test, an approximate critical value is $F_{max.95}(k = 4, df = 8 - 1 = 7) = 8.44$. The data do not contradict the hypothesis of homogeneity of variance.

Use of the computational formulas is illustrated in part iii. There are alternative methods for computing these sums of squares. SS_{error} may be obtained from

$$SS_{error} = \Sigma SS_j = 73.21.$$

From the definition,

$$\begin{aligned} SS_{treat} &= \Sigma n_j(\bar{T}_j - \bar{G})^2 \\ &= 6(3.00 - 5.27)^2 + 5(7.00 - 5.27)^2 + 8(2.25 - 5.27)^2 \\ &\quad + 7(9.43 - 5.27)^2 \\ &= 239.95. \end{aligned}$$

For purposes of showing just what it is that forms the sum of squares for error and treatments, these alternative computational methods are more revealing, but they also involve more computational effort.

Table 3.12-4 Analysis of Variance

Source of variation	SS	df	MS	F
Treatments	239.91	3	79.97	24.02**
Experimental error	73.21	22	3.33	
Total	313.12	25		

**$F_{.99}(3,22) = 4.82$

The analysis of variance is summarized in Table 3.12-4. A test on the hypothesis that all the treatment effects are equal is given by the F ratio,

$$F = \frac{MS_{treat}}{MS_{error}} = \frac{79.97}{3.33} = 24.02.$$

The critical value for a .01-level test is $F_{.99}(3,22) = 4.82$. The data contradict the hypothesis of no differences in treatment effects. Inspection of the treatment means indicates that treatments 2 and 4 are quite different from treatments 1 and 3.

A comparison has the same form as that for the case of equal n's, that is,

$$C = c_1\bar{T}_1 + c_2\bar{T}_2 + \cdots + c_k\bar{T}_k, \qquad \text{where } \Sigma c_j = 0.$$

A component of variation corresponding to a comparison is in this case

$$\text{SS}_C = \frac{C^2}{(c_1^2/n_1) + (c_2^2/n_2) + \cdots + (c_k^2/n_k)}.$$

Two comparisons,

$$C_1 = c_{11}\bar{T}_1 + c_{12}\bar{T}_2 + \cdots + c_{1k}\bar{T}_k,$$
$$C_2 = c_{21}\bar{T}_1 + c_{22}\bar{T}_2 + \cdots + c_{2k}\bar{T}_k,$$

are orthogonal if

$$\frac{c_{11}c_{21}}{n_1} + \frac{c_{12}c_{22}}{n_2} + \cdots + \frac{c_{1k}c_{2k}}{n_k} = 0.$$

These definitions reduce to those given for equal n's when all the n_j's are equal.

Computational procedures for comparisons will be illustrated through use of the data in Table 3.12-3. The component of variation corresponding to the difference between treatments 2 and 4 is

$$\text{SS}_C = \frac{(\bar{T}_2 - \bar{T}_4)^2}{[(1)^2/n_2] + [(-1)^2/n_4]} = \frac{(7.00 - 9.43)^2}{\frac{1}{5} + \frac{1}{7}} = 17.20.$$

To test the hypothesis that $\tau_2 = \tau_4$, the statistic used is

$$F = \frac{\text{SS}_C}{\text{MS}_{\text{error}}} = \frac{17.20}{3.33} = 5.17.$$

The numerator of this statistic has 1 degree of freedom, the denominator 22 degrees of freedom (the degrees of freedom for MS_{error}). For a .01-level test the critical value is $F_{.99}(1,22) = 7.94$. Since the observed F statistic does not exceed the critical value, the data do not contradict the hypothesis that $\tau_2 = \tau_4$.

As another example, suppose that a meaningful comparison planned in advance of the experiment is that between treatment 4 and all others combined; i.e., does treatment 4 differ from the average of all other treatment effects? This comparison is given by

$$\text{SS}_C = \frac{(3\bar{T}_4 - \bar{T}_1 - \bar{T}_2 - \bar{T}_3)^2}{(c_4^2/n_4) + (c_1^2/n_1) + (c_2^2/n_2) + (c_3^2/n_3)}$$
$$= \frac{[3(9.43) - (3.00) - (7.00) - (2.25)]^2}{\frac{9}{7} + \frac{1}{6} + \frac{1}{5} + \frac{1}{8}} = 144.54.$$

To test the hypothesis that $\tau_4 = (\tau_1 + \tau_2 + \tau_3)/3$, the statistic used is

$$F = \frac{SS_C}{MS_{error}} = \frac{144.54}{3.33} = 43.40.$$

The critical value for a .01-level test is $F_{.99}(1,22) = 7.94$. Hence the data clearly contradict the hypothesis that the effect of treatment 4 is equal to the average effect of the other three treatments.

When the n_j's do not differ markedly, the Newman-Keuls, the Duncan, or either of the Tukey methods may be adapted for use in making tests on differences between all pairs of means. The Newman-Keuls method will be used to illustrate the principles involved. With unequal sample sizes it is convenient to work with the treatment means. (For the case of equal sample sizes it is more convenient to work with the treatment totals.) The example in Table 3.12-3 will be used to illustrate the numerical operations. Part i of Table 3.12-5 gives the treatment means arranged in order of increasing magnitude. The differences between all possible pairs of means are shown. For example, the entry 7.18 in the first row is the difference 9.43 − 2.25. The entry 4.75 is the difference 7.00 − 2.25. In general an entry in this table is the difference between the mean at the top of the column and the mean at the left of the row.

The statistic to be used in making tests on these differences is q_r,

$$q_r = \frac{\bar{T}_j - \bar{T}_{j'}}{\sqrt{MS_{error}/n}},$$

where r is the number of steps the two means are apart on an ordered scale. The n in the expression $\sqrt{MS_{error}/n}$ refers to the number of observations in each of the means and is assumed to be constant. If the n_j's do not differ markedly from each other, the harmonic mean of the n_j's may be used instead of n in this expression. The harmonic mean $\tilde{n}$ is defined as

$$\tilde{n} = \frac{k}{(1/n_1) + (1/n_2) + \cdots + (1/n_k)}.$$

For the numerical example,

$$\tilde{n} = \frac{4}{\frac{1}{6} + \frac{1}{5} + \frac{1}{8} + \frac{1}{7}} = 6.30.$$

Note: An alternative procedure is to replace n by the harmonic mean of the sample sizes corresponding to the two extreme means. That is, n is replaced by $\dot{n}$, where

$$\dot{n} = \frac{2}{(1/n_{(1)}) + (1/n_{(k)})},$$

where $n_{(1)}$ = sample size corresponding to smallest treatment mean,
$n_{(k)}$ = sample size corresponding to largest treatment mean.

This alternative procedure tends to be conservative in the sense of erring on the side of reducing the nominal value of the level of significance. That is, a test having nominal value $\alpha = .05$ relative to equal n tends to have actual $\alpha = .04$ (approximately) for unequal n, provided the sample sizes are relatively close to each other.

Since the degrees of freedom for MS_{error} are 22, the critical values for the q_r statistic are found in the tables of the studentized range statistic in the row corresponding to 22 degrees of freedom. The critical values for a .01-level test are given in part ii of Table 3.12-5. Thus 3.99 is the critical

Table 3.12-5 Tests on Differences between All Pairs of Means

	Treatments		3	1	2	4
		Means	2.25	3.00	7.00	9.43
(i)	3	2.25	—	0.75	4.75	7.18
	1	3.00		—	4.00	6.43
	2	7.00			—	2.43
	4	9.43				—
				$r = 2$	$r = 3$	$r = 4$
(ii)			$q_{.99}(r,22)$	3.99	4.59	4.96
(iii)			$\sqrt{MS_{error}/\tilde{n}}\, q_{.99}(r,22)$	2.90	3.34	3.61

(iv)		3	1	2	4
	3			**	**
	1			**	**
	2				
	4				

value for the q_r statistic when $r = 2$, that is, when the means are two steps apart; 4.59 is the critical value for q_r when $r = 3$. In making several tests it is convenient to work with the critical value of the difference between a pair of means rather than the critical value of q_r. Since

$$\sqrt{\frac{MS_{error}}{\tilde{n}}}\, q_r = \bar{T}_j - \bar{T}_{j'},$$

the critical value for the difference between two means is

$$q_{1-\alpha}(r,\text{df}) \sqrt{\frac{MS_{error}}{\tilde{n}}}\,.$$

The numerical value of $\sqrt{MS_{error}/\tilde{n}}$ is in this case $\sqrt{3.33/6.30} = .727$. Hence the critical values for .01-level tests on the differences between pairs of means are given by multiplying the entries in part ii of Table 3.12-5 by .727. These values are given in part iii. For example, the entry 2.90 = (.727)(3.99).

The sequence in which the tests must be made is given in Sec. 3.8. This sequence must be followed here. The sequence indicated in the following steps is equivalent to that followed in Sec. 3.8.

1. The first test made is on the difference 7.18 in the upper right of part i. Since this entry is the difference between two means that are four steps apart, the critical value is 3.61. Hence the hypothesis that $\tau_3 = \tau_4$ is contradicted by the experimental data.
2. The next test is on the entry 4.75, the difference between two means which are three steps apart. The critical value for this test is 3.34. Hence the data contradict the hypothesis that $\tau_2 = \tau_3$.
3. The entry .75, which is the difference between two means that are two steps apart, is tested next. The critical value is 2.90. Hence the data do not contradict the hypothesis that $\tau_1 = \tau_3$.
4. The entry 6.43 is tested against the critical value 3.34, since this entry is the difference between two means which are three steps apart. Hence the data contradict the hypothesis that $\tau_1 = \tau_4$.
5. The entry 4.00 is tested against the critical value 2.90. The data contradict the hypothesis that $\tau_1 = \tau_2$.
6. The entry 2.43 is tested against the critical value 2.90. The data do not contradict the hypothesis that $\tau_2 = \tau_4$.

A summary of the tests is given in part iv. The cells with asterisks indicate that the corresponding differences are statistically significant at the .01 level. Schematically this summary may be represented as follows:

$$\underline{3 \quad 1} \qquad\qquad \underline{2 \quad 4}$$

Treatments underlined by a common line do not differ; treatments not underlined by a common line do differ. Hence treatments 2 and 4 differ from treatments 3 and 1, but there is no difference between treatments 2 and 4 and no difference between 3 and 1.

In adapting the Duncan method or either of the Tukey methods to the case of unequal sample sizes, the harmonic mean $\tilde{n}$ is used in place of n. For example, in the Tukey (*a*) test the critical value for .01-level tests on all differences would be $\sqrt{\text{MS}_{\text{error}}/\tilde{n}}\, q_{.99}(k,\text{df})$, where k is the number of treatments and df is the degrees of freedom for MS_{error}. For the numerical example, this critical value is 3.61. In this case the Tukey (*a*) test would give outcomes identical with those obtained by means of the Newman-Keuls test.

Unweighted-means Analysis. The definition of SS_{treat} given earlier in this section requires that each $(\bar{T}_j - \bar{G})^2$ be weighted by n_j. If the error variances for each of the treatments are equal, this weighting procedure gives each squared deviation a weight which is proportional to the reciprocal of its squared standard error, which is n_j. If, however, the n_j's are in no way related to the hypothesis being tested and it is desired to give each treatment mean a numerically equal weight in determining SS_{treat}, then the latter source of variation may be defined as

$$SS_{treat} = \tilde{n}\Sigma(\bar{T}_j - \bar{G})^2,$$

where
$$\bar{G} = \frac{\Sigma\bar{T}_j}{k}.$$

This definition of $\bar{G}$ differs from that used earlier in this section. If the latter definition of SS_{treat} is used, $SS_{treat} + SS_{error}$ will not be numerically equal to SS_{total}.

Scheffé Test for Homogeneity of Variance. A test for homogeneity of variance which is relatively insensitive to departure from normality of the basic variable is one proposed by Scheffé (1959, pp. 83–87). The hypothesis being tested under the procedure to be described is $\sigma_1^2 = \sigma_2^2 = \cdots = \sigma_k^2 = \sigma^2$. The computations to be described are essentially those of Table 3.12-2, with the exception of a system of weights.

Let the sample sizes under the treatments be denoted

$$n_1, n_2, \ldots, n_j, \ldots, n_k.$$

Suppose the n_j elements under treatment j are subdivided into subsamples of size

$$n_{1j}, n_{2j}, \ldots, n_{ij}, \ldots, n_{p_j j}.$$

(See part i of Table 3.12-6.) The number of subsamples p_j is arbitrary. As a rule of thumb each of the subsamples should be of approximately equal size; each subsample should be of size larger than three. Also, as a

Table 3.12-6 Scheffé Test for Homogeneity of Variance

	Treatment 1	Treatment 2	$\cdots$	Treatment k	
(i)	n_{11} y_{11}	n_{12} y_{12}	$\cdots$	n_{1k} y_{1k}	
	n_{21} y_{21}	n_{22} y_{22}	$\cdots$	n_{2k} y_{2k}	
	$\cdot$ $\cdot$	$\cdot$ $\cdot$		$\cdot$ $\cdot$	
	$\cdot$ $\cdot$	$\cdot$ $\cdot$		$\cdot$ $\cdot$	
	$\cdot$ $\cdot$	$\cdot$ $\cdot$		$\cdot$ $\cdot$	
	$n_{p_1 1}$ $y_{p_1 1}$	$n_{p_2 2}$ $y_{p_2 2}$	$\cdots$	$n_{p_k k}$ $y_{p_k k}$	
Total	n_1	n_2	$\cdots$	n_k	
(ii)	$f_{i1} = n_{i1} - 1$	$f_{i2} = n_{i2} - 1$	$\cdots$	$f_{ik} = n_{ik} - 1$	
	$f_1 = \Sigma f_{i1}$	$f_2 = \Sigma f_{i2}$	$\cdots$	$f_k = \Sigma f_{ik}$	$f = \Sigma f_j$
	$T_1 = \Sigma f_{i1} y_{i1}$	$T_2 = \Sigma f_{i2} y_{i2}$	$\cdots$	$T_k = \Sigma f_{ik} y_{ik}$	$G = \Sigma T_j$
	$\bar{T}_1 = T_1/f_1$	$\bar{T}_2 = T_2/f_2$	$\cdots$	$\bar{T}_k = \Sigma T_k/f_k$	$\bar{G} = G/f$
(iii)	$(1) = G^2/f$	$(2) = \Sigma\Sigma f_{ij} y_{ij}^2$		$(3) = \Sigma(T_j^2/f_j)$	

rule of thumb, $\Sigma(p_j - 1)$ should be 10 or more to assure reasonable power for the test.

For each of the subsamples compute

$$s_{ij}^2 = \frac{\Sigma(X_{ijk} - \bar{X}_{ij})^2}{n_{ij} - 1} = \frac{\sum_k X_{ijk}^2 - \left[\sum_k (X_{ijk})\right]^2 / n_{ij}}{n_{ij} - 1}$$

that is, the variance for the subsample of size n_{ij}. Let

$$y_{ij} = \ln s_{ij}^2.$$

The data for the analysis of variance are shown in part i of Table 3.12-6. In essence one now carries out an analysis of variance with the y_{ij} as the basic observations. Each y_{ij} is considered to have $f_{ij} = n_{ij} - 1$ degrees of freedom. Hence one carries out a weighted analysis of variance, the weights being the f_{ij}. The y_{ij} are assumed to be approximately normally distributed.

Additional notation is defined in part ii. One has

$$f_j = \sum_i f_{ij}, \qquad T_j = \sum_i f_{ij} y_{ij}.$$

Convenient computational symbols are defined in part iii. The analysis of variance is

$$\begin{aligned} SS_{\text{treat}} &= \Sigma f_j(\bar{T}_j - \bar{G})^2 = (3) - (1), \\ SS_{\text{error}} &= \Sigma\Sigma f_{ij}(y_{ij} - \bar{T}_j)^2 = (2) - (3). \end{aligned}$$

To test the hypothesis $\sigma_1^2 = \sigma_2^2 = \cdots = \sigma_k^2$, one uses the test statistic

$$F = \frac{SS_{\text{treat}}/(k - 1)}{SS_{\text{error}}/\Sigma(p_j - 1)}.$$

When the hypothesis being tested is true this F statistic is approximately an F distribution with degrees of freedom $k - 1$ and $\Sigma(p_j - 1)$.

As an example of the application of this test, suppose that $k = 5$ and that there are $n = 12$ observations under each treatment. One may subdivide (at random) each set of 12 observations into $p_j = 3$ subsamples, each of size $n_{ij} = 4$. Under each treatment condition there will be three variances. The degrees of freedom for the resulting F ratio will be $k - 1 = 4$ and $\Sigma(p_j - 1) = 10$.

3.13 Power and Determination of Sample Size—Fixed Model

In terms of a single-factor experiment (fixed model) having k treatments and n experimental units per treatment, the expected values of the mean squares for the overall F ratio have the form

$$\frac{E(MS_{\text{treat}})}{E(MS_{\text{error}})} = \frac{\sigma_\varepsilon^2 + n(\Sigma\tau_j^2)/(k - 1)}{\sigma_\varepsilon^2} = 1 + \frac{n(\Sigma\tau_j^2)/(k - 1)}{\sigma_\varepsilon^2} = 1 + \frac{\lambda}{k - 1},$$

where $$\lambda = \frac{n\Sigma\tau_j^2}{\sigma_\varepsilon^2}.$$

(The parameters τ_j are assumed to be scaled so that $\Sigma\tau_j = 0$.)

λ is called the noncentrality parameter. Under the hypothesis that $\tau_1 = \tau_2 = \cdots = \tau_k = 0$, it follows that $\lambda = 0$, and the sampling distribution of the overall F ratio has the *central* F distribution $F[(k-1), k(n-1)]$. When $\lambda \neq 0$, the sampling distribution of the F ratio is a *noncentral* F distribution $F[(k-1), k(n-1); \lambda]$. That is, the distribution depends upon the parameter λ.

To compute the power of the overall F test with respect to alternative hypotheses specified by the noncentrality parameter λ, tables of the noncentral F distribution corresponding to specified values of λ are needed. Such tables are available for a limited number of values for λ. (See Tiku, 1967.) The central F distribution is that special case of the noncentral F distribution for which $\lambda = 0$. Most tables of the noncentral F distribution are tabulated in terms of the noncentrality parameter

$$\phi = \sqrt{\frac{n\Sigma\tau_j^2}{k\sigma_\varepsilon^2}} = \sqrt{\frac{\lambda}{k}}.$$

The corresponding noncentral F distribution is symbolized $F[(k-1), k(n-1); \phi]$. The noncentral F distribution is given in Table C.14 in the Appendix.

The power of an overall F test (having $\alpha = .05$) with respect to an alternative hypothesis specified by $\phi = 1.00$ is indicated graphically in Fig. 3.13-1. The magnitude of the type 2 error is represented by the area β. The power of the test is numerically equal to $1 - \beta$.

Since the noncentrality parameter ϕ depends in part upon n, the power of the overall F test also depends upon n. By making n suitably large, the power of the overall F test can be made suitably large for any nonzero

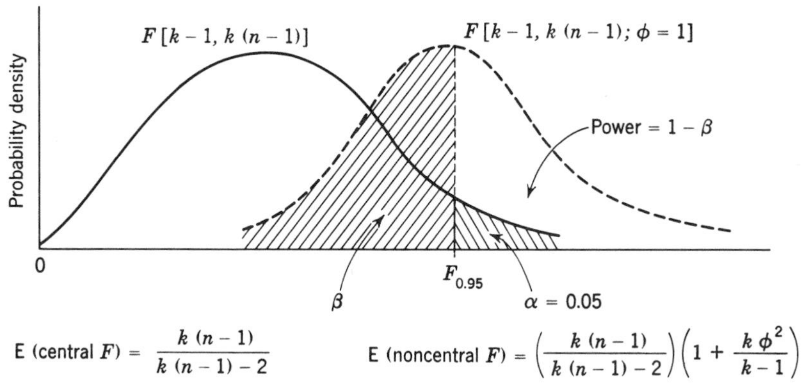

Figure 3.13-1 Power of F test under fixed model.

specified value of $\Sigma\tau_j^2/\sigma_\varepsilon^2$. That is, for a fixed value of $\Sigma\tau_j^2/\sigma_\varepsilon^2$, the larger n, the larger ϕ becomes. The larger ϕ, for a fixed level of significance, the smaller β; hence the larger is the power.

Once $\Sigma\tau_j^2/\sigma_\varepsilon^2$ and k have been specified, the power of the overall F test becomes a function solely of n. An estimate of σ_ε^2 can be obtained either from a pilot study or from previous experimentation. The problem of specifying $\Sigma\tau_j^2$ is not a simple one. For purposes of what follows, $\Sigma\tau_j^2$ should be taken as the smallest value that is considered of practical importance. (Expert judgment backed by a reasonable rationale is needed here.) The sample size n will then be determined so as to make the power of the overall F test adequately high (say power equal to .80 or .90) with respect to this specified value for $\Sigma\tau_j^2$. Equivalently, the sample size will be determined so as to make the experiment sensitive enough to detect differences in the parameters τ_j that are considered large enough to be of practical importance—if indeed such differences exist.

Alternative Approach to Noncentral *F*. The approach taken earlier in this section to justify the use of the noncentral distribution in terms of the expected value of the mean squares can be somewhat misleading. (It will be found that the power function for the variance-component model involves the central F distribution rather than the noncentral F. Yet approached from the expected values of the mean squares, the two models appear quite similar.)

Under the model,

$$X_{ij} = \mu + \varepsilon_{ij},$$

the estimate of the variation due to error is

$$SS' = SS_{\text{total}} = \Sigma(X_{ij} - \bar{G})^2.$$

Under the model, assuming that the τ_j are fixed constants,

$$X_{ij} = \mu + \tau_j + \varepsilon_{ij},$$

the estimate of the variation due to error is

$$SS'' = SS_{\text{error}} = \Sigma(X_{ij} - \bar{T}_j)^2.$$

A measure of deviation from the hypothesis that all $\tau_j = 0$ is

$$SS' - SS'' = SS_{\text{total}} - SS_{\text{error}} = SS_{\text{treat}} = n\Sigma(\bar{T}_j - \bar{G})^2.$$

It may be shown (Scheffé, 1959, p. 38) that

$$SS_{\text{total}} - SS_{\text{error}} = \sum_{j=1}^{k-1} z_j^2,$$

where the z_j are independently distributed as $N(\xi_j, \sigma^2/n)$. Hence

$$\frac{n\Sigma z_j^2}{\sigma^2} \quad \text{is distributed as} \quad \chi^2\left(k - 1; \frac{n\Sigma\xi_j^2}{\sigma^2}\right).$$

That is,
$$\frac{(k-1)\text{MS}_{\text{treat}}}{\sigma^2}$$
is distributed as a noncentral chi square with noncentrality parameter
$$\frac{n\Sigma\xi_j^2}{\sigma^2} = \frac{n\Sigma\tau_j^2}{\sigma_\varepsilon^2}.$$

Numerical Example. Consider the case in which $k = 2$ and the overall F test is to be made at level of significance $\alpha = .05$. From tables of the noncentral F distribution the power of the overall test may be evaluated for various combinations of ϕ and n. A summary of the latter is presented in Table 3.13-1.

Table 3.13-1 Power of the Overall F Test in the Analysis of Variance When $k = 2$ and $\alpha = .05$†

ϕ	$n = 5$ $F(1,8;\phi)$	$n = 10$ $F(1,18;\phi)$	$n = 15$ $F(1,28;\phi)$	$n = 20$ $F(1,38;\phi)$	$n = 25$ $F(1,48;\phi)$
0.00	.05	.05	.05	.05	.05
0.50	.10	.10	.10	.11	.11
1.00	.24	.27	.28	.28	.29
1.20	.32	.36	.37	.38	.38
1.40	.41	.47	.48	.49	.49
1.60	.51	.57	.59	.60	.60
1.80	.61	.67	.69	.70	.70
2.00	.70	.76	.78	.79	.79
2.20	.78	.84	.85	.86	.86
2.60	.89	.94	.94	.95	.95
3.00	.96	.98	.98	.99	.999

† The value of n, in part, determines the degrees of freedom for MS_{error}, which is $k(n-1)$. These values are obtained from the Tiku (1967) tables. Part of these tables appears as Table C.14 in the Appendix.

If the minimum value of τ_j to be considered of practical importance is 3, then $\Sigma\tau_j^2 = 9 + 9 = 18$. If σ_ε^2 is estimated to be equal to 20, then
$$\frac{\Sigma\tau_j^2}{\sigma_\varepsilon^2} = \frac{18}{20} = .90.$$

Hence
$$\phi = \sqrt{\frac{n\Sigma\tau_j^2}{k\sigma_\varepsilon^2}} = \sqrt{.45n}\,.$$

For this special case, the value of ϕ associated with various values of n and the corresponding power are as follows:

n:	5	10	15	20	25
ϕ:	1.50	2.12	2.60	3.25	3.35
Power:	.46	.81	.94	.99	.999

If power .80 is considered to be appropriate for $\Sigma\tau_j^2 = 18$, then letting $n = 10$ will provide a test having this power. If power .90 is desired for this alternative hypothesis, then n should be between 10 and 15. It will be found that, when $n = 13$,

$$F(1,24;\ 2.42) = .90.$$

To avoid the necessity of having to prepare summary tables such as Table 3.13-1 from tables of the noncentral F distribution, charts like those in Table C.11 in the Appendix have been prepared. Let

$$\phi' = \sqrt{\frac{\Sigma\tau_j^2}{k\sigma_\varepsilon^2}}.$$

For the special case $k = 2$, $\Sigma\tau_j^2 = 18$, and $\sigma_\varepsilon^2 = 20$,

$$\phi' = .67.$$

From Table C.11 ($k = 2$), for $\alpha = .05$ and $\phi' = .67$, to find n required for power .90, one reads up from the point $\phi' = .67$ until one intersects the curve corresponding to $P = .9$. The point of intersection corresponds approximately to $n = 13$ on the scale for n at the left. If the overall test were to be made with level of significance $\alpha = .01$, one reads up on the same chart at $\phi' = .67$ on the scale for $\alpha = .01$. For power .90 in this case, n must be approximately 18.

As a second numerical example, suppose $k = 3$. Assume $\sigma_\varepsilon^2 = 50$ and that the minimal practically important value for $\Sigma\tau_j^2 = 75$. Hence the value of ϕ corresponding to $\Sigma\tau_j^2 = 75$ is

$$\phi = \sqrt{\frac{n\Sigma\tau_j^2}{k\sigma_\varepsilon^2}} = \sqrt{\frac{n(75)}{3(50)}} = \sqrt{.50n}.$$

In Table 3.13-2 are tabulated the powers corresponding to various combinations of n and ϕ obtained from tables of the noncentral F distributions.

Given below are the values of ϕ for different values of n corresponding to the alternative hypothesis $\Sigma\tau_j^2 = 75$ for $k = 3$ and $\sigma_\varepsilon^2 = 50$. Associated with these values of ϕ and n is the power for specified level of significance.

n:	5	10	15	20
ϕ:	1.58	2.24	2.74	3.16
Power ($\alpha = .01$):	.29	.73	.92	.99
Power ($\alpha = .05$):	.58	.91	.98	.99

Table 3.13-2 Power of the Overall F Test in the Analysis of Variance When $k = 3$

ϕ	$n = 5$ $F(2,12;\phi)$ $\alpha = .01$	$\alpha = .05$	$n = 10$ $F(2,27;\phi)$ $\alpha = .01$	$\alpha = .05$	$n = 15$ $F(2,42;\phi)$ $\alpha = .01$	$\alpha = .05$	$n = 20$ $F(2,57;\phi)$ $\alpha = .01$	$\alpha = .05$
.00	.01	.05	.01	.05	.01	.05	.01	.05
.50	.02	.10	.03	.11	.03	.11	.03	.11
1.00	.09	.26	.11	.29	.12	.30	.13	.31
1.20	.14	.36	.18	.40	.19	.42	.20	.43
1.40	.20	.47	.27	.53	.29	.54	.31	.55
1.60	.29	.58	.38	.65	.41	.66	.43	.68
1.80	.39	.69	.50	.75	.53	.77	.55	.78
2.00	.49	.78	.62	.84	.65	.86	.68	.87
2.20	.60	.86	.73	.91	.76	.92	.78	.92
2.60	.78	.95	.89	.98	.91	.98	.93	.98
3.00	.91	.99	.97	.99	.98	.99	.98	.99

The power for specified α is obtained from the appropriate column of Table 3.13-2.

Thus, if $n = 10$ is chosen, the power corresponding to the alternative hypothesis $\Sigma\tau_j^2 = 75$ is .73 when $\alpha = .01$ and .91 when $\alpha = .05$. If $n = 15$, the corresponding power is .92 when $\alpha = .01$ and .98 when $\alpha = .05$.

If one were to use the charts in Table C.11,

$$\phi' = \sqrt{\frac{\Sigma\tau_j^2}{k\sigma_\varepsilon^2}} = \sqrt{\frac{75}{3(50)}} = \sqrt{.50} = .7.$$

On the chart for $k = 3$, on the scale for which $\alpha = .01$, corresponding to power .90, n is 15. For power .70, the approximate n is 10. The values of n determined from these charts will, except for interpolation errors, be the same as those obtained directly from the noncentral F distributions—indeed the charts were prepared from the latter.

Approximating Percentile Points on the Noncentral F. Very close approximations to percentile points on the noncentral F distribution may be obtained from tables of the central F distribution—provided relatively complete tables of the latter are available. The approximation to be outlined here is one suggested by Patnaik (1949).

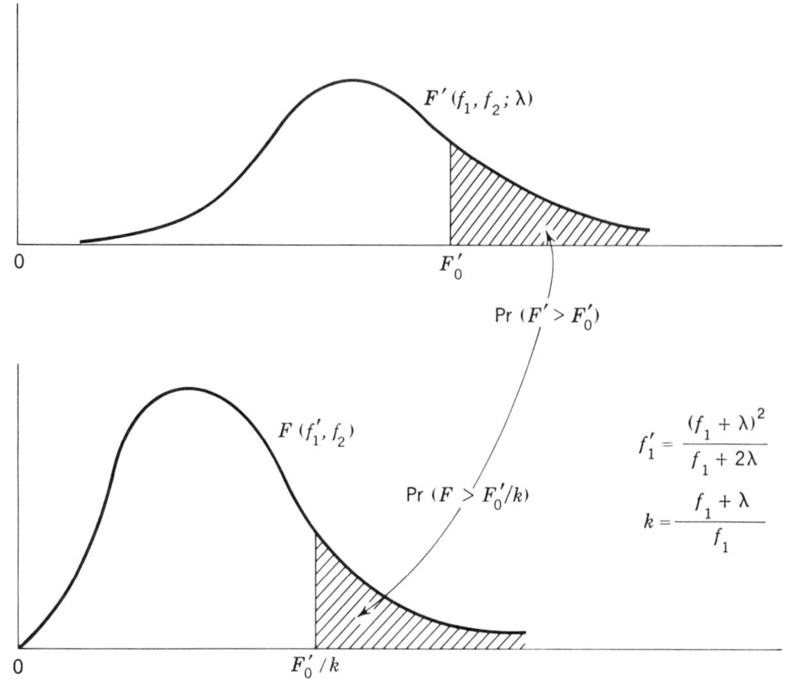

Figure 3.13-2 Approximating noncentral F.

At the top of Fig. 3.13-2 is a sketch of a noncentral F distribution with parameters f_1, f_2, and λ, denoted $F'(f_1,f_2;\lambda)$. The area to the right of the point F_0' is the probability

$$\Pr(F' \geq F_0'),$$

that is, the probability of obtaining an F' statistic larger than F_0'. This probability is approximated by

$$\Pr\left(F \geq \frac{F_0'}{k}\right)$$

from the F distribution specified in the lower half of Fig. 3.13-2. In general, f_1' will not be an integer. Hence, interpolation in tables of the central F distribution will be necessary.

To illustrate this approximation procedure, suppose one desires

$$\Pr(F' \geq 6.55)$$

for the distribution $F'(3,10;4)$. For this case,

$$f_1' = \frac{(3+4)^2}{3+8} = 4.45; \qquad k = \frac{3+4}{3} = 2.33; \qquad \frac{F'}{k} = 2.81.$$

Hence, $$\Pr(F' > 6.55) \doteq \Pr(F > 2.81),$$

where F has the central F distribution $F(4.45,10)$. (See Fig. 3.13-3.) By interpolation in tables of the central F,

$$\Pr(F \geq 2.81) \doteq .081.$$

From tables of the noncentral F, the exact probability is .082.

To illustrate the degree of accuracy of the approximation procedure, from Patnaik (1949) one has the following:

				$\Pr(F' \geq F'_0)$	
f_1	f_2	λ	F'_0	Approx.	Exact
3	10	4	3.708	.248	.254
3	10	4	6.552	.081	.082
8	10	9	5.057	.091	.092
8	30	9	3.173	.185	.187

A more accurate approximation of the noncentral F from the central F distribution may be obtained from a method given by Tiku (1967). The approximation method used by Patnaik has the first two moments of the central F in common with the noncentral F; the Tiku approximation

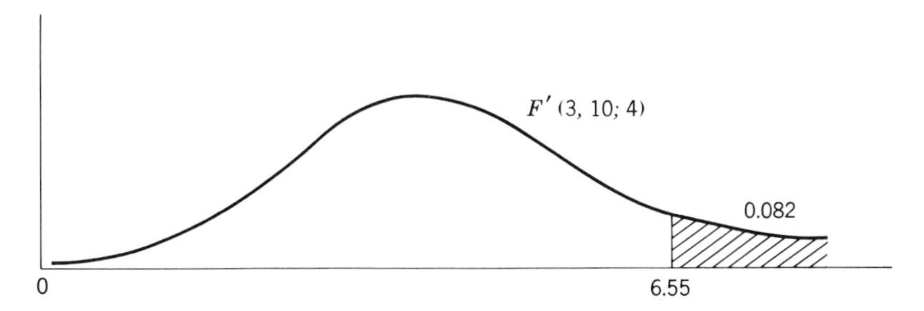

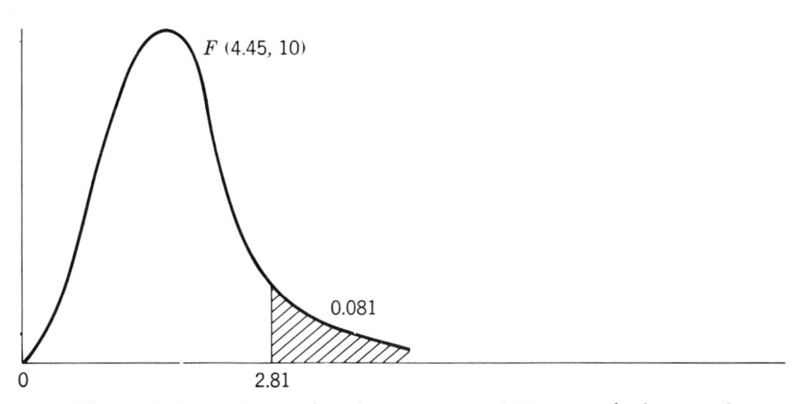

Figure 3.13-3 Approximating noncentral F—numerical example.

method has the first three moments of the central F in common with the noncentral F.

3.14 Linear Model with Fixed Variables

The model in Sec. 3.5 may be cast in a form which more closely resembles the models discussed in Chap. 2 through the use of indicator (or dummy) variables. This model has the general form

$$Y_{ij} = \beta_0 X_0 + \beta_1 X_{i1} + \cdots + \beta_k X_{ik} + \varepsilon_{ij}. \tag{1}$$

In this notation system,

Y_{ij} = observation on element i under treatment j,
X_0 = 1 for all elements,

$$X_{i1} = \begin{cases} 1 \text{ if } j = 1, \\ 0 \text{ if } j \neq 1, \end{cases}$$

$$\vdots$$

$$X_{ik} = \begin{cases} 1 \text{ if } j = k, \\ 0 \text{ if } j \neq k, \end{cases}$$

ε_{ij} is a random variable distributed as $N(0,\sigma_\varepsilon^2)$,
$\beta_0, \beta_1, \ldots, \beta_k$ are parameters to be estimated.

If, for example, $j = 3$, then $X_{i1} = 0$, $X_{i2} = 0$, $X_{i3} = 1, \ldots, X_{ik} = 0$.

For the case $k = 2$, the model in (1) has the matrix form

$$\begin{bmatrix} Y_{11} \\ \cdot \\ \cdot \\ \cdot \\ Y_{n_1 1} \\ \hline Y_{12} \\ \cdot \\ \cdot \\ \cdot \\ Y_{n_2 2} \end{bmatrix} = \begin{bmatrix} 1 & 1 & 0 \\ \cdot & \cdot & \cdot \\ \cdot & \cdot & \cdot \\ \cdot & \cdot & \cdot \\ 1 & 1 & 0 \\ \hline 1 & 0 & 1 \\ \cdot & \cdot & \cdot \\ \cdot & \cdot & \cdot \\ \cdot & \cdot & \cdot \\ 1 & 0 & 1 \end{bmatrix} \begin{bmatrix} \beta_0 \\ \beta_1 \\ \beta_2 \end{bmatrix} + \begin{bmatrix} \varepsilon_{11} \\ \cdot \\ \cdot \\ \cdot \\ \varepsilon_{n_1 1} \\ \hline \varepsilon_{12} \\ \cdot \\ \cdot \\ \cdot \\ \varepsilon_{n_2 2} \end{bmatrix}, \tag{2a}$$

or in general

$$\underset{n.,k+1}{\underline{Y}} = \underset{n.,k+1}{X} \underset{k+1,1}{\underline{\beta}} + \underset{n.,1}{\underline{\varepsilon}}, \qquad \text{where } n. = \Sigma n_j. \tag{2b}$$

The matrix X is called the *design* matrix. Entries in this matrix are either 0 or 1. The experimental design determines the number of zeros or ones in each column beyond the first.

The vector $\underline{\varepsilon}$ in (2*b*) represents the experimental error or residual. The sum of the squared residuals is

$$\Sigma\Sigma\varepsilon_{ij}^2 = \underline{\varepsilon}'\underline{\varepsilon} = (\underline{Y} - X\underline{\beta})'(\underline{Y} - X\underline{\beta}) \tag{3}$$
$$= \underline{Y}'\underline{Y} + \underline{\beta}'X'X\underline{\beta} - 2\underline{\beta}'X'\underline{Y}.$$

The least-squares estimators of the components of the vector $\underline{\beta}$ are obtained by taking the derivative of (3) with respect to the components of $\underline{\beta}$, setting these derivatives equal to zero, and solving the resulting system of equations. The latter system is given by

$$(X'X)\underline{\hat{\beta}} = X'\underline{Y}. \tag{4}$$

For the special case $k = 2$, the system of equations has the form

$$\begin{bmatrix} n. & n_1 & n_2 \\ n_1 & n_1 & 0 \\ n_2 & 0 & n_2 \end{bmatrix} \begin{bmatrix} \hat{\beta}_0 \\ \hat{\beta}_1 \\ \hat{\beta}_2 \end{bmatrix} = \begin{bmatrix} Y_{..} \\ Y_{.1} \\ Y_{.2} \end{bmatrix}, \tag{5}$$

where $\quad Y_{..} = \Sigma\Sigma Y_{ij}, \qquad Y_{.1} = \Sigma Y_{i1}, \qquad Y_{.2} = \Sigma Y_{i2}.$

In this context, the matrix $X'X$ will be singular. Hence the components of the vector $\underline{\hat{\beta}}$ will not be unique. One solution to the system given in (5) is

$$\underline{\hat{\beta}} = \begin{bmatrix} 0 & 0 & 0 \\ 0 & 1/n_1 & 0 \\ 0 & 0 & 1/n_2 \end{bmatrix} \begin{bmatrix} Y_{..} \\ Y_{.1} \\ Y_{.2} \end{bmatrix} + \begin{bmatrix} z \\ -z \\ -z \end{bmatrix} \tag{6}$$
$$= \begin{bmatrix} 0 + z \\ \bar{Y}_1 - z \\ \bar{Y}_2 - z \end{bmatrix},$$

where z is any real number. A general solution for $\underline{\hat{\beta}}$ in (2*b*) has the form

$$\underline{\hat{\beta}} = \begin{bmatrix} 0 + z \\ \bar{Y}_1 - z \\ \vdots \\ \bar{Y}_k - z \end{bmatrix}. \tag{7}$$

The solution to the estimation problem given in Sec. 3.5 is that special case of (7) in which $\Sigma n_j\hat{\beta}_j = 0$ or, equivalently, $z = \bar{Y}$.

From (7) one notes that $\hat{\beta}_0 + \hat{\beta}_1 = \bar{Y}_1$ for all possible choices of z. In general, $\hat{\beta}_0 + \hat{\beta}_j = \bar{Y}_j$ for all possible choices of z. One also notes that $\hat{\beta}_j - \hat{\beta}_k$ does not depend upon the choice of z.

Any function of $\beta_0, \beta_1, \ldots, \beta_k$ of the form $\Sigma c_j\beta_j$ is said to be *estimable* if the corresponding function $\Sigma c_j\hat{\beta}_j$ does not depend upon the choice of z.

Estimability in this sense requires that

$$c_0 = \sum_{j=1}^{k} c_j \qquad \text{or} \qquad c_0 - \sum_{j=1}^{k} c_j = 0.$$

For the case $k = 2$, the following functions are estimable:

(i) $\beta_0 + \beta_1,$

(ii) $\beta_0 + \beta_2,$

(iii) $\beta_1 - \beta_2;$

the following functions are not estimable:

(i) $\beta_0,$

(ii) $\beta_2,$

(iii) $\beta_1 + \beta_2.$

In the general case, $\beta_0 + \beta_j$ will be estimable. Further, any comparison among the β_j will be estimable; that is,

$$c_1\beta_1 + c_2\beta_2 + \cdots + c_k\beta_k \qquad \text{where } \Sigma c_j = 0.$$

If the components of the vector $\underline{c}$ can be expressed in the form

$$\underline{c} = \underline{\lambda}'(X'X)$$

then $\underline{c}'\underline{\beta}$ will be estimable. The corresponding estimator will be $\underline{\lambda}'X'\underline{Y}$. Conversely, for each choice of $\underline{\lambda}'$ there is a $\underline{c}$ which defines an estimable parametric function.

For example, for the case $k = 2$, suppose

$$\underline{\lambda}' = [1 \quad 2 \quad 0],$$

then

$$\underline{c} = \underline{\lambda}'(X'X) = [1 \quad 2 \quad 0]\begin{bmatrix} n. & n_1 & n_2 \\ n_1 & n_1 & 0 \\ n_2 & 0 & n_2 \end{bmatrix} = \begin{bmatrix} n. + 2n_1 \\ 3n_1 \\ n_2 \end{bmatrix} = \begin{bmatrix} c_0 \\ c_1 \\ c_2 \end{bmatrix}.$$

Note that

$$c_0 = c_1 + c_2,$$

$$n_. + 2n_1 = 3n_1 + n_2,$$

since

$$n_1 + n_2 = n_. .$$

The estimator of $\underline{c}'\underline{\beta}$ is

$$\underline{\lambda}'X'\underline{Y} = [1 \quad 2 \quad 0]\begin{bmatrix} Y_. \\ Y_{1.} \\ Y_{2.} \end{bmatrix} = Y_. + 2Y_{1.}.$$

In general, one notes that if

$$(X'X)\underline{\beta} = X'\underline{Y} \qquad \text{then} \qquad \underline{\lambda}'(X'X)\underline{\beta} = \underline{\lambda}'X'\underline{Y}.$$

Hence, if $\underline{c} = \underline{\lambda}'(X'X)$, then the expression at the right becomes an identity.

An equivalent condition for estimability may be obtained as follows: Let the design matrix X be partitioned

$$X = [A_0 \quad A_1],$$

where A_1 is the largest matrix in X that has full column rank. That is, A_1 contains all the linearly independent columns in X. Let $\underline{\beta}' = [\underline{\beta}_0' \quad \underline{\beta}_1']$, where $\underline{\beta}_0$ corresponds to A_0 and $\underline{\beta}_1$ corresponds to A_1. The parametric function

$$\underline{c}'\underline{\beta} = \underline{c}_0'\underline{\beta}_0 + \underline{c}_1'\underline{\beta}_1$$

will be estimable if

$$\underline{c}_0' = \underline{c}_1'(A_1'A_1)^{-1}A_1'A_0.$$

For the case $k = 2$, A_0 may be taken to be the first column of the X matrix given in (2*b*), and A_1 may be taken to be the second and third columns. Hence

$$A_1'A_1 = \begin{bmatrix} n_{1.} & 0 \\ 0 & n_{2.} \end{bmatrix} \qquad \text{and} \qquad (A_1'A_1)^{-1} = \begin{bmatrix} 1/n_{1.} & 0 \\ 0 & 1/n_{2.} \end{bmatrix}.$$

Thus

$$\underline{c}_0' = \underline{c}_1'\begin{bmatrix} 1/n_{1.} & 0 \\ 0 & 1/n_{2.} \end{bmatrix}\begin{bmatrix} n_{1.} \\ n_{2.} \end{bmatrix} = \underline{c}_1'\begin{bmatrix} 1 \\ 1 \end{bmatrix},$$

or

$$c_0 = c_1 + c_2.$$

The following correspondence exists between terminology in regression analysis with dichotomous variables $X_1, \ldots, X_k$ and the analysis of variance. Corresponding sums of squares are equal.

Analysis of variance			Multiple regression		
Source	SS	df	df	SS	Source
Treatment	SS_{treat}	$k - 1$	$k - 1$	SS_{pred}	Predictable
Error	SS_{error}	$N - k$	$N - k$	SS_{res}	Residual
Total	SS_{total}	$N - 1$	$N - 1$	SS_Y	Total

It should be noted that the analysis-of-variance model for fixed variables is generally defined to be that special case of the linear model defined by (1) which includes the side condition

$$\Sigma n_j\beta_j = 0.$$

With this side condition as part of the model, $\beta_0, \beta_1, \ldots, \beta_k$ are given by

$$\hat{\beta}_0 = \bar{Y}, \qquad \hat{\beta}_1 = \bar{Y}_1 - \bar{Y}, \ldots, \hat{\beta}_k = \bar{Y}_k - \bar{Y}.$$

In terms of the notation system used in earlier sections,

$$\bar{Y}_j = \bar{T}_j \qquad \text{and} \qquad \bar{Y} = \bar{G}.$$

3.15 Multivariate Analysis of Variance

In the *univariate* analysis of variance, the effect of a series of treatments on a single criterion variable X is observed. In the *multivariate* analysis of variance each experimental unit is observed on several criterion variables, namely, $X_1, X_2, \ldots, X_p$. A treatment may have an effect upon several characteristics of an experimental unit. For example, the effect of a drug may be on blood pressure and heart rate, as well as temperature, simultaneously. In the multivariate analysis of variance, the effect of the treatment on all criteria is observed simultaneously. Each observation is a vector rather than a scalar.

By analyzing the effects of the treatments on $X_1, X_2, \ldots, X_p$ separately, that is, as a series of univariate analyses, one fails to take into account the correlations among the variables. The simultaneous response of the experimental units to all variables, considered as a single response, generally contains more information about the total effect of the treatment than does the series of responses considered singly. The latter is what happens when one treats the multivariate data as a series of independent univariate data.

The model for the single-factor, multivariate analysis of variance has the form

$$\begin{bmatrix} X_{i1j} \\ X_{i2j} \\ \cdot \\ \cdot \\ \cdot \\ X_{ipj} \end{bmatrix} = \begin{bmatrix} \mu_1 \\ \mu_2 \\ \cdot \\ \cdot \\ \cdot \\ \mu_p \end{bmatrix} + \begin{bmatrix} \tau_{1j} \\ \tau_{2j} \\ \cdot \\ \cdot \\ \cdot \\ \tau_{pj} \end{bmatrix} + \begin{bmatrix} \varepsilon_{i1j} \\ \varepsilon_{i2j} \\ \cdot \\ \cdot \\ \cdot \\ \varepsilon_{ipj} \end{bmatrix}$$

$$\underset{p,1}{\underline{X}'_{ij}} = \underbrace{\underset{p,1}{\underline{\mu}} + \underset{p,1}{\underline{\tau}_j}} + \underset{p,1}{\underline{\varepsilon}_{ij}}$$

$$= \qquad \underline{\mu}_j \qquad + \underline{\varepsilon}_{ij},$$

where $\underline{X}_{ij}$ = an observation vector on experimental unit i under treatment j, where $\underline{X}_{ij}$ is a row vector,

$\underline{\tau}_j$ = vector of effects of treatment j on variables $X_1, X_2, \ldots, X_p$,

$\underline{\varepsilon}_{ij}$ = vector of experimental error.

As in the case of univariate analysis of variance, $\underline{\tau}$ and $\underline{\varepsilon}$ are assumed to be distributed independently of each other. Under model I, the components of $\underline{\tau}_j$ are fixed constants associated with treatment j and variables $X_1, \ldots, X_p$. The components of the vector $\underline{\varepsilon}$ are assumed to have a joint multivariate normal distribution with mean vector $\underline{0}$ and arbitrary covariance matrix Σ. Each row of the multivariate model considered singly represents the model for a univariate analysis of variance associated with a specified variable X_m.

In the univariate analysis of variance, one of the basic hypotheses tested is

$$\mu_1 = \mu_2 = \cdots = \mu_k. \tag{1}$$

The corresponding multivariate hypothesis is

$$\begin{bmatrix} \mu_{11} \\ \mu_{21} \\ \cdot \\ \cdot \\ \cdot \\ \mu_{p1} \end{bmatrix} = \begin{bmatrix} \mu_{12} \\ \mu_{22} \\ \cdot \\ \cdot \\ \cdot \\ \mu_{p2} \end{bmatrix} = \cdots = \begin{bmatrix} \mu_{1k} \\ \mu_{2k} \\ \cdot \\ \cdot \\ \cdot \\ \mu_{pk} \end{bmatrix}, \tag{2}$$

$$\underline{\mu}_1 = \underline{\mu}_2 = \cdots = \underline{\mu}_k .$$

For the case in which there are two criterion variables, say X_1 and X_2, the basic data for the multivariate analysis of variance would take the form given in Table 3.15-1. From the data of the type given in Table 3.15-1

Table 3.15-1 Multivariate Analysis of Variance, $p = 2$

	Treatment 1		Treatment 2		$\cdots$	Treatment k	
	X_1	X_2	X_1	X_2	$\cdots$	X_1	X_2
	X_{111}	X_{121}	X_{112}	X_{122}	$\cdots$	X_{11k}	X_{12k}
	X_{211}	X_{221}	X_{212}	X_{222}	$\cdots$	X_{21k}	X_{22k}
	$\cdot$	$\cdot$	$\cdot$	$\cdot$		$\cdot$	$\cdot$
	$\cdot$	$\cdot$	$\cdot$	$\cdot$		$\cdot$	$\cdot$
	$\cdot$	$\cdot$	$\cdot$	$\cdot$		$\cdot$	$\cdot$
	X_{n11}	X_{n21}	X_{n12}	X_{n22}	$\cdots$	X_{n1k}	X_{n2k}
Total	T_{11}	T_{21}	T_{12}	T_{22}	$\cdots$	T_{1k}	T_{2k}
	$G_1 = \Sigma T_{1j}$			$G_2 = \Sigma T_{2j}$			

one may compute the following matrices:

$$W = \begin{bmatrix} W_{11} & W_{12} \\ W_{21} & W_{22} \end{bmatrix}, \qquad B = \begin{bmatrix} B_{11} & B_{12} \\ B_{21} & B_{22} \end{bmatrix},$$

where $W_{11} = \Sigma X_{i1j}^2 - \dfrac{\Sigma T_{1j}^2}{n}, \qquad W_{22} = \Sigma X_{i2j}^2 - \dfrac{\Sigma T_{2j}^2}{n},$

$$B_{11} = \frac{\Sigma T_{1j}^2}{n} - \frac{G_1^2}{kn}, \qquad B_{22} = \frac{\Sigma T_{2j}^2}{n} - \frac{G_2^2}{kn},$$

$$W_{12} = \Sigma X_{i1j} X_{i2j} - \frac{\Sigma T_{1j} T_{2j}}{n},$$

$$B_{12} = \frac{\Sigma T_{1j} T_{2j}}{n} - \frac{G_1 G_2}{kn}.$$

Note that W_{11} and B_{11} represent, respectively, the within-treatment (or error) sum of squares and the between-treatment (or treatment) sum of squares in the univariate analysis of variance for X_1. Similarly, W_{22} and B_{22} are the corresponding sum of squares in the univariate analysis of variance for X_2. W_{12} and E_{12} represent the error and between-treatment sum of squares associated with the cross products; these terms take the covariance between the criteria into account. The W and B matrices are symmetric. In this context, the W matrix is said to be that "due to experimental error" and the B matrix that "due to treatments."

For the case of p variates,

$$W = \begin{bmatrix} W_{11} & \cdots & W_{1p} \\ \cdot & & \cdot \\ \cdot & & \cdot \\ \cdot & & \cdot \\ W_{p1} & \cdots & W_{pp} \end{bmatrix} \quad \text{and} \quad B = \begin{bmatrix} B_{11} & \cdots & B_{1p} \\ \cdot & & \cdot \\ \cdot & & \cdot \\ \cdot & & \cdot \\ B_{p1} & \cdots & B_{pp} \end{bmatrix}.$$

The elements on the main diagonal of both matrices are those corresponding to what would be obtained in the univariate analysis associated with the variable under consideration. (In order that W be nonsingular, n must be greater than p.)

The F statistic used in testing the *univariate* hypothesis in (1) can be expressed in the form

$$F = \frac{1 - \Lambda}{\Lambda} \frac{kn - k}{k - 1},$$

where $\Lambda = \dfrac{W_{11}}{W_{11} + B_{11}}, \qquad W_{11} = \text{SS}_{\text{error}}, \qquad B_{11} = \text{SS}_{\text{treat}}.$

To show this,

$$F = \frac{1 - W_{11}/(W_{11} + B_{11})}{W_{11}/(W_{11} + B_{11})} \frac{kn - k}{k - 1} = \frac{B_{11}/(k - 1)}{W_{11}/(kn - k)}$$

$$= \frac{\text{SS}_{\text{treat}}/(k - 1)}{\text{SS}_{\text{error}}/(kn - k)}.$$

There are several methods of testing the hypothesis represented by (2); these methods do not necessarily lead to the same conclusions since they focus upon somewhat different ways of formulating the alternative hypothesis and the level of significance. The likelihood-ratio approach to testing the hypothesis (2) uses the statistic

$$\Lambda = \frac{|W|}{|W + B|}.$$

For the special case $p = 2$, the sampling distribution of the statistic

$$\frac{1 - \sqrt{\Lambda}}{\sqrt{\Lambda}} \frac{kn - k - 1}{k - 1}, \tag{3}$$

assuming (2) is true, is

$$F[2(k - 1), 2(kn - k - 1)]. \tag{4}$$

In the general case, the statistic

$$\frac{1 - \Lambda^{1/s}}{\Lambda^{1/s}} \frac{ms + 2\lambda}{2r}, \tag{5}$$

assuming (2) is true, is approximately

$$F[2r, ms + 2\lambda], \tag{6}$$

where

$$m = kn - 1 - \tfrac{1}{2}(p + k),$$

$$r = \tfrac{1}{2}p(k - 1),$$

$$s = \left[\frac{p^2(k - 1)^2 - 4}{p^2 + (k - 1)^2 - 5}\right]^{\frac{1}{2}},$$

$$\lambda = \frac{-1}{4}[p(k - 1) - 2].$$

This approximation neglects additive terms of order $1/m^4$. Another approximation to the null distribution of the test statistic is given by (7).

$$-m \ln \Lambda \text{ is approximately } \chi^2[p(k - 1)]. \tag{7}$$

The approximation in (7) is appropriate when m is large, say over 50.

When $p = 2$, (6) reduces to (4). Also, when $p = 2$, the sampling distribution of the statistic in (5) is exact rather than approximate. A summary of the special cases of (5) that reduce to exact F distributions, under the assumption that (2) is true, is given in Table 3.15-2. A more complete discussion of these sampling distributions will be found in Rao (1952, pp. 260–263, and 1965, pp. 471–472).

An alternative test statistic for use in testing the hypothesis in (2) is

$$\frac{\rho_{\max}}{1 + \rho_{\max}},$$

where $\rho_{\max}$ = largest characteristic root to BW^{-1}.

Table 3.15-2 Special Cases of Distribution of Likelihood-ratio Statistic That Reduce to F Distributions

Parameters	F statistic	df
$k = 2$, any p	$\frac{1-\Lambda}{\Lambda} \cdot \frac{2k-p-1}{p}$	$p, 2k-p-1$
$k = 3$, any p	$\frac{1-\sqrt{\Lambda}}{\sqrt{\Lambda}} \cdot \frac{3k-p-2}{p}$	$2p, 2(kn-p-2)$
$p = 1$, any k	$\frac{1-\Lambda}{\Lambda} \cdot \frac{kn-k}{k-1}$	$k-1, kn-k$
$p = 2$, any k	$\frac{1-\sqrt{\Lambda}}{\sqrt{\Lambda}} \cdot \frac{kn-k-1}{k-1}$	$2(k-1), 2(kn-k-1)$

This statistic is obtained from applying what is called Roy's union-intersection principle. The latter is a generalization of the univariate simultaneous-confidence-interval approach. Still a third statistic useful in testing the hypothesis in (2) is

$$\text{trace } BW^{-1}.$$

In terms of the characteristic roots of the matrix BW^{-1}, the likelihood-ratio test statistic involves the product of the roots, whereas trace BW^{-1} is actually the sum of the roots. There is some evidence to indicate that some symmetric function of all the roots is a more appropriate test statistic than the single largest root. When BW^{-1} has only one root (i.e., when $p = 1$), all test statistics indicated above reduce to the usual univariate F statistic.

A numerical example of the multivariate analysis of variance for the case $k = 3$ and $p = 2$ is given in Table 3.15-3. The entry 3, 10 under treatment 1 represents the observation vector on the first experimental unit assigned to treatment 1. The entry 5, 16 represents the observation vector on the second experimental unit assigned to treatment 1. Similarly, 1, 9 represents the observation vector on the last experimental unit assigned to treatment 1.

The entries 18 and 65 under treatment 1 are the column totals T_{11} and T_{21}, respectively. Under treatment 1, one also has the totals

$$\Sigma X_1^2 = 76, \qquad \Sigma X_2^2 = 889, \qquad \Sigma X_1 X_2 = 255.$$

Table 3.15-3 Numerical Example
$k = 3, p = 2, n = 5$

	Treatment 1		Treatment 2		Treatment 3			
	X_1	X_2	X_1	X_2	X_1	X_2		
	3	10	8	12	10	16		
	5	16	4	8	4	10		
	5	16	4	6	10	18		
(i)	4	14	2	6	4	14		
	1	9	9	14	10	16		
T_{mj}	18	65	27	46	38	74	$G_1 = 83$	$G_2 = 185$
$\Sigma(\ )^2$	76	889	181	476	332	1132	$\Sigma X_1^2 = 589$	$\Sigma X_2^2 = 2497$
$\Sigma X_{1j}X_{2j}$	255		290		596		1141	

Variable X_1:

(ii)

$$SS_{11} = SS_{\text{total}} = \Sigma X_1^2 - G_1^2/kn = 589.00 - 459.27 = 129.73$$
$$B_{11} = SS_{\text{treat}} = \Sigma T_{1j}^2/n - G_1^2/kn = 499.40 - 459.27 = 40.13$$
$$W_{11} = SS_{\text{w.cell}} = \Sigma X_1^2 - \Sigma T_{1j}^2/n = 589.00 - 499.40 = 89.60$$

Variable X_2:

(iii)

$$SS_{22} = SS_{\text{total}} = \Sigma X_2^2 - G_2^2/kn = 2497.00 - 2281.67 = 215.33$$
$$B_{22} = SS_{\text{treat}} = \Sigma T_{2j}^2/n - G_2^2/kn = 2363.40 - 2281.67 = 81.73$$
$$W_{22} = SS_{\text{w.cell}} = \Sigma X_2^2 - \Sigma T_{2j}^2/n = 2497.00 - 2363.40 = 133.60$$

Cross product X_1X_2:

(iv)

$$SS_{12} = SS_{\text{total}} = \Sigma X_1X_2 - G_1G_2/n = 1141.00 - 1023.67 = 117.33$$
$$B_{12} = SS_{\text{treat}} = \Sigma T_{1j}T_{2j}/n - G_1G_2/n = 1044.80 - 1023.67 = 21.13$$
$$W_{12} = SS_{\text{w.cell}} = \Sigma X_1X_2 - \Sigma T_{1j}T_{2j}/n = 1141.00 - 1044.80 = 96.20$$

In part ii one has the sums of squares that would be obtained in the univariate analysis of variance for X_1. A similar univariate analysis of variance for X_2 appears in part iii. The sums of squares that would be obtained in a univariate analysis of variance of the cross products appears in part iv.

The matrices W, B, and $B + W$ are given in Table 3.15-4. For the case of a single-factor design with n observations per treatment

$$W + B = T,$$

where $$T = \text{matrix of total variations.}$$

The statistic Λ for the basic data is equal to .1917. Since $p = 2$, under the hypothesis that

$$\underline{\mu}_1 = \underline{\mu}_2 = \underline{\mu}_3,$$

the F statistic shown in Table 3.15-4 has an F distribution with degrees of freedom indicated. Since

$$F_{\text{obs}} > F_{.99}(4,22),$$

the experimental data tend to contradict the hypothesis of no treatment effects, the level of significance being .01.

Table 3.15-4 Numerical Example (Continued)

$$[W] = \begin{bmatrix} 89.60 & 96.20 \\ 96.20 & 133.60 \end{bmatrix} \qquad [B] = \begin{bmatrix} 40.13 & 21.13 \\ 21.13 & 81.73 \end{bmatrix}$$

$$[W + B] = \begin{bmatrix} 129.73 & 117.33 \\ 117.33 & 215.33 \end{bmatrix}$$

$$\Lambda = \frac{|W|}{|W + B|} = \frac{2716.12}{14168.43} = .1917$$

$$F = \frac{1 - \sqrt{\Lambda}}{\sqrt{\Lambda}} \frac{nk - k - 1}{k - 1} = \frac{1 - .4379}{.4379} \frac{11}{2} = 7.06$$

$$F_{.95}[2(k - 1), 2(nk - k - 1)] = F_{.95}(4,22) = 2.82 \qquad F_{.99}(4,22) = 4.31$$

The means for each of the variates under each of the treatments are summarized in the following vectors:

$$\underline{\bar{T}}_1 = \begin{bmatrix} 3.60 \\ 13.00 \end{bmatrix}, \qquad \underline{\bar{T}}_2 = \begin{bmatrix} 5.40 \\ 9.20 \end{bmatrix}, \qquad \underline{\bar{T}}_3 = \begin{bmatrix} 7.60 \\ 14.80 \end{bmatrix}.$$

The F test in Table 3.15-4 contradicts the hypothesis that

$$\underline{\mu}_1 = \underline{\mu}_2 = \underline{\mu}_3,$$

where $\underline{\mu}_j$ is estimated by $\underline{\bar{T}}_j$. To test the hypothesis (a priori)

$$H_1: \qquad \underline{\mu}_2 - \underline{\mu}_3 = \underline{0},$$

$$H_2: \qquad \underline{\mu}_2 - \underline{\mu}_3 \neq \underline{0},$$

the multivariate analog of Student's t statistic may be used, namely, Hotelling's T^2 statistic. Let

$$\underline{d}_{23} = \underline{\bar{T}}_2 - \underline{\bar{T}}_3 = \begin{bmatrix} -2.80 \\ -5.60 \end{bmatrix}.$$

Then Hotelling's T^2 takes the form

$$T^2 = \left(\frac{fn}{2}\right) \underline{d}'_{23} W^{-1} \underline{d}_{23},$$

where f = degrees of freedom associated with an element of W, in this case $kn - k$.

For the numerical example,

$$T^2 = \frac{(12)(5)}{2}[-2.80 \quad -5.60] \begin{bmatrix} .04919 & -.03542 \\ -.03542 & .03299 \end{bmatrix} \begin{bmatrix} -2.80 \\ -5.60 \end{bmatrix}$$

$$= 9.285.$$

When the hypothesis being tested is true, the statistic

$$F = \frac{f - p + 1}{fp} T^2$$

is distributed as

$$F(p, f - p + 1).$$

For the numerical example,

$$F = \frac{12 - 2 + 1}{(12)(2)}(9.285) = 4.26.$$

The critical values are

$$F_{.95}(2,11) = 3.98 \qquad \text{and} \qquad F_{.99}(2,11) = 7.21.$$

With level of significance .05, the experimental data contradict the hypothesis that $\underline{\mu}_2 - \underline{\mu}_3 = \underline{0}$.

The univariate analysis for the $p = 2$ variates is shown in Table 3.15-5.

Table 3.15-5 Univariate Analyses for Example in Table 3.15-3

Variable X_1:		SS	df	MS	F
	Treatment	40.13	2	20.06	2.69
	Error	89.60	12	7.47	
	Total	129.73	14		

$F_{.90}(2,12) = 2.81$

Variable X_2:		SS	df	MS	F
	Treatment	81.73	2	40.87	3.67
	Error	133.60	12	11.13	
	Total	215.33	14		

$F_{.90}(2,12) = 2.81, \qquad F_{.95}(2,12) = 4.75$

Although the overall F test in the multivariate analysis of variance is highly significant (actually beyond the .001 level of significance), neither of the overall univariate tests exceeds the .05 level. This result, on the surface, may appear to be contradictory. However, this contradiction disappears when one notes the high within-class correlation between X_1 and X_2 $[r = 96.20/\sqrt{(89.60)(133.60)} = .879]$. In the multivariate analysis this correlation plays an important role in determining the F ratio; in the univariate analyses this correlation is disregarded.

In one sense the multivariate test can be considered as a univariate test on an optimally weighted composite of the individual variates—optimal in the sense of maximizing the ratio of the between-treatment variance to the within-treatment variance. More realistically, the multivariate test incorporates a series of univariate tests involving a series of optimally weighted composites of the variates.

3.16 Randomized Complete-block Designs

Consider a design in which there are $k = 3$ treatments arranged as follows:

Block 1	Block 2
Treatment 3	Treatment 2
Treatment 1	Treatment 3
Treatment 2	Treatment 1

In the agricultural setting, from which the terminology originates, a "block" often corresponds to an area of land. Each block is divided into k subblocks of equal size. Within each block the k treatments are assigned at random to the subblocks. Each block is complete in the sense that each contains all treatments.

In general, a block may correspond to a repetition of an experiment under essentially comparable conditions. Each block is divided into k subblocks; the number of blocks n will be equal to the number of repetitions called for by the design. The primary purpose for arranging the treatments in blocks is to eliminate variation due to differences between blocks from the experimental error. Thus the principle underlying "blocking" of treatments is to provide a certain degree of control over the heterogeneity of the experimental units.

Although the blocks are often considered to represent a random factor (in the sense that the blocks actually included in the experiment are a random sample from a population of blocks), in what follows in this section the blocks will be considered to represent a fixed factor. As long as a primary interest lies in the elimination of block effects from experimental error and as long as statistical tests are restricted to treatment effects, handling blocks as a fixed factor will not alter the principal features of the analysis.

Suppose the observed data from a design in which treatments are arranged in n complete blocks are symbolized as follows:

	Treatment 1	Treatment 2	$\cdots$	Treatment k	Total
Block 1	X_{11}	X_{12}	$\cdots$	X_{1k}	B_1
Block 2	X_{21}	X_{22}	$\cdots$	X_{2k}	B_2
$\vdots$	$\vdots$	$\vdots$		$\vdots$	$\vdots$
Block n	X_{n1}	X_{n2}	$\cdots$	X_{nk}	B_n
Total	T_1	T_2	$\cdots$	T_k	G

$$\bar{B}_i = B_i/k, \qquad \bar{T}_j = T_j/n, \qquad \bar{G} = G/kn$$

Suppose the following model is appropriate for an observation on treatment j in block i.

$$X_{ij} = \mu + \beta_i + \tau_j + \varepsilon_{ij}, \qquad i = 1, \ldots, n; j = 1, \ldots, k. \tag{1}$$

The least-squares estimators of the parameters in (1) are obtained by solving the following set of normal equations:

$$\begin{aligned} \mu: &\quad knm + k\Sigma b_i + n\Sigma t_j = G, \\ \beta_i: &\quad km + kb_i + \Sigma t_j = B_i, \\ \tau_j: &\quad nm + \Sigma b_i + nt_j = T_j. \end{aligned} \tag{2}$$

For the special case $n = 2$, $k = 3$, the normal equations may be written schematically as follows:

(3)

	m	b_1	b_2	t_1	t_2	t_3	
m:	6	3	3	2	2	2	G
b_1:	3	3	0	1	1	1	B_1
b_2:	3	0	3	1	1	1	B_2
t_1:	2	1	1	2	0	0	T_1
t_2:	2	1	1	0	2	0	T_2
t_3:	2	1	1	0	0	2	T_3

The first line in the above schematic representation corresponds to the equation

$$6m + 3b_1 + 3b_2 + 2t_1 + 2t_2 + 2t_3 = G.$$

The set of equations (3) represents six equations in six unknowns. One will find, however, that

(i) Sum of equations b_1 and b_2 = sum of equations t_1, t_2, and t_3,

(ii) Equation m = sum of equations b_1, b_2, t_1, t_2, and t_3.

Thus there are two linear relationships among the $k + n + 1$ normal equations in which there are $k + n + 1$ unknowns. Hence, two linearly independent constraints on the unknowns will be needed before the system (2) will have a unique solution.

A relatively simple solution to the system (2) is obtained if one imposes the constraints

$$\Sigma b_i = 0 \qquad \text{and} \qquad \Sigma t_j = 0.$$

The solution corresponding to these constraints is

$$\hat{\mu} = m = \frac{G}{kn} = \bar{G},$$

$$(4) \qquad \hat{\beta}_i = b_i = \frac{B_i - km}{k} = \bar{B}_i - \bar{G},$$

$$\hat{\tau}_j = t_j = \frac{T_j - nm}{n} = \bar{T}_j - \bar{G}.$$

By a direct extension of the principles in Sec. 3.14, the reduction in sums of squares due to $\mu, \beta_1, \ldots, \beta_n, \tau_1, \ldots, \tau_k$ is given by

$$(5) \qquad R(\mu,\beta,\tau) = mG + \Sigma b_i B_i + \Sigma t_j T_j.$$

Although the estimators in (4) do depend upon the choice of the constraints, the estimate of the variation due to the corresponding parameters does not depend upon the constraints chosen.

If one uses the models indicated below, least-squares procedures will provide the variations indicated.

Disregarding treatments:

$$X_{ij} = \mu + \beta_i + \varepsilon_{ij}, \qquad R(\mu,\beta) = mG + \Sigma b_i B_i.$$

Disregarding blocks:

$$X_{ij} = \mu + \tau_j + \varepsilon_{ij}, \qquad R(\mu,\tau) = mG + \Sigma t_j T_j.$$

Disregarding treatments and blocks:

$$X_{ij} = \mu + \varepsilon_{ij}, \qquad R(\mu) = mG.$$

If the constraints $\Sigma b_i = 0$ and $\Sigma t_j = 0$ are used in solving the normal equations associated with these models, then

$$m = \bar{G}, \qquad b_i = \bar{B}_i - \bar{G}, \qquad t_j = \bar{T}_j - \bar{G}.$$

Hence, one has the following relationships for complete-block designs:

$$\begin{aligned} R(\beta \mid \mu) &= SS_{\text{blocks}} = R(\mu,\beta,\tau) - R(\mu,\tau) \\ &= \Sigma b_i B_i \\ &= k\Sigma(\bar{B}_i - \bar{G})^2 = \frac{\Sigma B_i^2}{k} - \frac{G^2}{kn}. \end{aligned}$$

$$
\begin{aligned}
R(\tau \mid \mu) = \text{SS}_{\text{treat}} &= R(\mu,\beta,\tau) - R(\mu,\beta) \\
&= \Sigma t_j T_j \\
&= n\Sigma(\bar{T}_j - \bar{G})^2 = \frac{\Sigma T_j^2}{n} - \frac{G^2}{kn}.
\end{aligned}
$$

$$
\begin{aligned}
\text{SS}_{\text{error}} = \Sigma X_{ij}^2 - R(\mu,\beta,\tau) &= \Sigma X_{ij}^2 - mG - \Sigma b_i B_i - t_j T_j \\
&= \Sigma(X_{ij} - \bar{G})^2 - \text{SS}_{\text{blocks}} - \text{SS}_{\text{treat}} \\
&= \text{SS}_{\text{total}} - \text{SS}_{\text{blocks}} - \text{SS}_{\text{treat}}.
\end{aligned}
$$

It should be noted that

$$
\begin{aligned}
R(\tau,\beta \mid \mu) = R(\mu,\beta,\tau) - R(\mu) &= \Sigma b_i B_i + \Sigma t_j T_j \\
&= R(\beta \mid \mu) + R(\tau \mid \mu).
\end{aligned}
$$

In general, when

$$R(u,v) = R(u) + R(v),$$

the design is said to be *orthogonal* with respect to u and v. In a complete-block design, the block effects and the treatment effects are orthogonal. (In what are called incomplete-block designs, the block effects are not orthogonal to the treatment effects.)

The analysis of variance associated with a randomized, complete-block design is summarized in Table 3.16-1. The degrees of freedom for

Table 3.16-1 Analysis of Variance for Complete-block Design

Source of variation	SS	df	MS	E(MS)
Blocks	$\text{SS}_{\text{blocks}}$	$n-1$	$\text{MS}_{\text{blocks}}$	$\sigma_\varepsilon^2 + k\sigma_\beta^2$
Treatments	SS_{treat}	$k-1$	MS_{treat}	$\sigma_\varepsilon^2 + n\sigma_\tau^2$
Error	SS_{error}	$(k-1)(n-1)$	MS_{error}	σ_ε^2
Total	SS_{total}	$kn-1$		

$(1) = G^2/kn$ $(3) = (\Sigma T_j^2)/n$

$(2) = \Sigma X^2$ $(4) = (\Sigma B_i^2)/k$

$\text{SS}_{\text{blocks}} = (4) - (1)$ $\text{SS}_{\text{error}} = (2) - (3) - (4) + (1)$

$\text{SS}_{\text{treat}} = (3) - (1)$ $\text{SS}_{\text{total}} = (2) - (1)$

SS_{error} are obtained as follows: There are kn observations in all. There are $(k + n + 1) - 2$ linearly independent normal equations. Hence, the number of linearly independent estimators obtained from the normal equations is $(k + n + 1) - 2$. The degrees of freedom for SS_{error} are given by

$$
\begin{aligned}
kn - [(k + n + 1) - 2] &= kn - k - n + 1 \\
&= (k - 1)(n - 1).
\end{aligned}
$$

An alternative rationale for obtaining these degrees of freedom is as follows:

$$\begin{aligned}\mathrm{df}_{\text{error}} &= \mathrm{df}_{\text{total}} - \mathrm{df}_{\text{blocks}} - \mathrm{df}_{\text{treat}} \\ &= (kn - 1) - (n - 1) - (k - 1) = (k - 1)(n - 1).\end{aligned}$$

A development which has many points in common with that used in this section will be found in Sec. 4.4. In the latter section, a block corresponds to a person. The equivalent of a block by treatment interaction is included in one of the models discussed in Sec. 4.4.

3.17 Some Special Features of the Variance-component Model

In the variance-component model,

(1) $$X_{ij} = \mu + \tau_j + \varepsilon_{ij}, \qquad j = 1, \ldots, k;\ i = 1, \ldots, n,$$

μ is considered to be a fixed constant. τ_j and ε_{ij} are independently distributed random variables, with respective distributions $N(0,\sigma_\tau^2)$ and $N(0,\sigma_\varepsilon^2)$. Thus, as part of the model,

$$\underset{i,j}{\mathrm{E}}(X_{ij}) = \mathrm{E}(\mu) + \mathrm{E}(\tau_j) + \mathrm{E}(\varepsilon_{ij})$$

$$= \mu,$$

and

$$\begin{aligned}\operatorname{var}(X_{ij}) &= \operatorname{var}(\mu) + \operatorname{var}(\tau_j) + \operatorname{var}(\varepsilon_{ij}) \\ &= \quad 0 \quad + \quad \sigma_\tau^2 \quad + \quad \sigma_\varepsilon^2 .\end{aligned}$$

The covariance structure of this model may be represented as follows:

If $i = i'$,

$$\operatorname{cov}(X_{ij}, X_{i'j}) = \operatorname{var}(X_{ij}) = \sigma_\varepsilon^2 + \sigma_\tau^2.$$

If $i \neq i'$,

$$\operatorname{cov}(X_{ij}, X_{i'j}) = \sigma_\tau^2.$$

The *intraclass* correlation is, by definition,

$$\rho = \frac{\operatorname{cov}(X_{ij}, X_{i'j})}{\operatorname{var}(X_{ij})} = \frac{\sigma_\tau^2}{\sigma_\varepsilon^2 + \sigma_\tau^2}.$$

In the numerator of the expression for the intraclass correlation, $i \neq i'$.

If the n observations under treatment j are considered to be the components of a vector variable

$$\underline{X}_j = \begin{bmatrix} X_{1j} \\ \cdot \\ \cdot \\ \cdot \\ X_{nj} \end{bmatrix},$$

then the variance-component model assumes that the random variable $\underline{X}_j$

is multivariate normal with parameters

$$E(\underline{X}_j) = \begin{bmatrix} \mu \\ \mu \\ \cdot \\ \cdot \\ \cdot \\ \mu \end{bmatrix}, \qquad \underset{n,n}{\Sigma} = \begin{bmatrix} \sigma_\varepsilon^2 + \sigma_\tau^2 & \sigma_\tau^2 & \cdots & \sigma_\tau^2 \\ \sigma_\tau^2 & \sigma_\varepsilon^2 + \sigma_\tau^2 & \cdots & \sigma_\tau^2 \\ \cdot & \cdot & & \cdot \\ \cdot & \cdot & & \cdot \\ \cdot & \cdot & & \cdot \\ \sigma_\tau^2 & \sigma_\tau^2 & \cdots & \sigma_\varepsilon^2 + \sigma_\tau^2 \end{bmatrix}.$$

The corresponding matrix of intercorrelations is

$$\underset{n,n}{P} = \begin{bmatrix} 1 & \rho & \cdots & \rho \\ \rho & 1 & \cdots & \rho \\ \cdot & \cdot & & \cdot \\ \cdot & \cdot & & \cdot \\ \cdot & \cdot & & \cdot \\ \rho & \rho & \cdots & 1 \end{bmatrix},$$

where ρ is the intraclass correlation.

The matrix Σ may be written in the form

$$\Sigma = \sigma_\varepsilon^2 I + \sigma_\tau^2 J,$$

where I = diagonal matrix with unity in the main diagonal,
J = square matrix whose elements are all unity.

The inverse of the matrix Σ may be shown to be

$$\Sigma^{-1} = \alpha_1 I + \alpha_2 J,$$

where

$$\alpha_1 = \frac{1}{\sigma_\varepsilon^2}, \qquad \alpha_2 = \frac{-\sigma_\tau^2}{\sigma_\varepsilon^2(\sigma_\varepsilon^2 + n\sigma_\tau^2)}.$$

This inverse has a role in obtaining the standard error of estimators. Under this type of covariance structure, the method of maximum likelihood may be used to obtain estimates of the parameters in the model.

An alternative path toward obtaining estimates of the parameters σ_ε^2 and σ_τ^2 is to set mean squares equal to their corresponding expected values and then solve the resulting equations. Thus, let

$$\text{MS}_{\text{treat}} = \hat{\sigma}_\varepsilon^2 + n\hat{\sigma}_\tau^2,$$
$$\text{MS}_{\text{error}} = \hat{\sigma}_\varepsilon^2.$$

Substituting the last result in the first expression gives

$$\hat{\sigma}_\tau^2 = \frac{1}{n}(\text{MS}_{\text{treat}} - \text{MS}_{\text{error}}).$$

In matrix notation,

$$\begin{bmatrix} 1 & n \\ 1 & 0 \end{bmatrix} \begin{bmatrix} \hat{\sigma}_\varepsilon^2 \\ \hat{\sigma}_\tau^2 \end{bmatrix} = \begin{bmatrix} \text{MS}_{\text{treat}} \\ \text{MS}_{\text{error}} \end{bmatrix}.$$

Hence
$$\begin{bmatrix} \hat{\sigma}_\varepsilon^2 \\ \hat{\sigma}_\tau^2 \end{bmatrix} = \begin{bmatrix} 0 & 1 \\ \frac{1}{n} & -\frac{1}{n} \end{bmatrix} \begin{bmatrix} \text{MS}_{\text{treat}} \\ \text{MS}_{\text{error}} \end{bmatrix},$$

or
$$\begin{bmatrix} \hat{\sigma}_\varepsilon^2 \\ \hat{\sigma}_\tau^2 \end{bmatrix} = \begin{bmatrix} \text{MS}_{\text{error}} \\ \frac{1}{n}(\text{MS}_{\text{treat}} - \text{MS}_{\text{error}}) \end{bmatrix}.$$

It can be shown (Graybill, 1961, pp. 338–347) that these estimators are the unbiased, maximum-likelihood estimators of the corresponding parameters. (The constant μ in the model is estimated by $\bar{G}$, the mean of all observations.) From the assumptions underlying the model

(2)
$$\frac{(k-1)\text{MS}_{\text{treat}}}{\sigma_\varepsilon^2 + n\sigma_\tau^2} \text{ is distributed as } \chi^2(k-1),$$
$$\frac{[k(n-1)]\text{MS}_{\text{error}}}{\sigma_\varepsilon^2} \text{ is distributed as } \chi^2[k(n-1)].$$

Further, these distributions are independent. Hence from (2) one has

(3)
$$F = \frac{\text{MS}_{\text{treat}}}{\text{MS}_{\text{error}}} \text{ is distributed as } \frac{\sigma_\varepsilon^2 + n\sigma_\tau^2}{\sigma_\varepsilon^2} \frac{\chi^2(k-1)/(k-1)}{\chi^2[k(n-1)]/k(n-1)}$$
$$= \frac{\sigma_\varepsilon^2 + n\sigma_\tau^2}{\sigma_\varepsilon^2} F[(k-1), k(n-1)].$$

One may rewrite (3) in the form

(4)
$$F = (1 + n\theta)F[k-1, k(n-1)], \qquad \text{where } \theta = \frac{\sigma_\tau^2}{\sigma_\varepsilon^2};$$

that is, the F statistic $\text{MS}_{\text{treat}}/\text{MS}_{\text{error}}$ is distributed as a multiple of the central F distribution. Under the hypothesis that σ_τ^2 is zero, this multiple is unity. Under the hypothesis that $\sigma_\tau^2 > 0$, this multiple is not unity.

Thus, whether or not θ is zero, the F statistic has a central F distribution. Hence the power of the test $\sigma_\tau^2 = 0$ with respect to nonzero θ is given by

$$\Pr\left\{F > \frac{F_{1-\alpha}[k-1, k(n-1)]}{1 + n\theta}\right\}.$$

(See Fig. 3.17-1.)

Since
$$F = (1 + n\theta)F[k-1, k(n-1)],$$

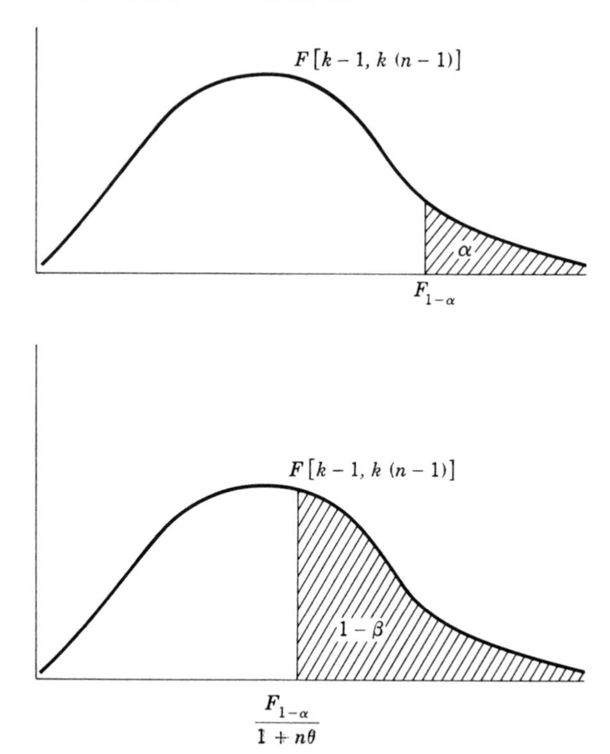

Figure 3.17-1 Power of F test under variance-component model.

one may test any hypothesis of the form $\theta = \theta_1$ by means of the following decision rule:

$$\text{Reject the hypothesis that } \theta = \theta_1 \text{ if}$$
$$F_{\text{obs}} > (1 + n\theta_1)F_{1-\alpha}[k - 1, k(n - 1)].$$

From (4) one has

$$\text{(5)} \quad \Pr\left\{F_\alpha[k - 1, k(n - 1)] \leq \frac{F}{1 + n\theta} \leq F_{1-\alpha}[k - 1, k(n - 1)]\right\} = 1 - \alpha.$$

One may rewrite (5) in the form

$$\text{(6)} \quad \Pr\left\{\frac{F}{F_{1-\alpha}[k - 1, k(n - 1)]} \leq 1 + n\theta \leq \frac{F}{F_\alpha[k - 1, k(n - 1)]}\right\} = 1 - \alpha.$$

From (6) one obtains the following confidence interval on θ:

$$\text{(7)} \qquad C(A \leq \theta \leq B) = 1 - \alpha,$$

where
$$A = \frac{1}{n}\left(\frac{\text{MS}_{\text{treat}}/\text{MS}_{\text{error}}}{F_{1-\alpha}[k-1, k(n-1)]} - 1\right),$$
$$B = \frac{1}{n}\left(\frac{\text{MS}_{\text{treat}}/\text{MS}_{\text{error}}}{F_{\alpha}[k-1, k(n-1)]} - 1\right).$$

In terms of the parameter θ, the intraclass correlation is given by

$$\rho = \frac{\theta}{1+\theta} = \frac{\sigma_\tau^2/\sigma_\varepsilon^2}{1 + (\sigma_\tau^2/\sigma_\varepsilon^2)} = \frac{\sigma_\tau^2}{\sigma_\varepsilon^2 + \sigma_\tau^2}.$$

From (7) one obtains the following confidence interval on ρ:

$$C\left(\frac{A}{1+A} \le \rho \le \frac{B}{1+B}\right) = 1 - \alpha, \tag{8}$$

where A and B are defined in (7). A point estimate of the intraclass correlation is given by

$$\hat{\rho} = \frac{\hat{\sigma}_\tau^2}{\hat{\sigma}_\varepsilon^2 + \hat{\sigma}_\tau^2} = \frac{(\text{MS}_{\text{treat}} - \text{MS}_{\text{error}})/n}{\text{MS}_{\text{error}} + (\text{MS}_{\text{treat}} - \text{MS}_{\text{error}})/n}$$
$$= \frac{\text{MS}_{\text{treat}} - \text{MS}_{\text{error}}}{\text{MS}_{\text{treat}} + (n-1)\text{MS}_{\text{error}}}.$$

This estimate will be biased.

In problems involving variance components, it is of interest to obtain unbiased estimates of the ratio of variance components. Toward this end consider the problem of finding an unbiased estimate of the ratio $\theta = \sigma_\tau^2/\sigma_\varepsilon^2$. One will find that

$$\text{E}\left(\frac{\hat{\sigma}_\tau^2}{\hat{\sigma}_\varepsilon^2}\right) \ne \frac{\sigma_\tau^2}{\sigma_\varepsilon^2},$$

where $\hat{\sigma}_\tau^2$ and $\hat{\sigma}_\varepsilon^2$ are the estimators obtained earlier in this section.

From the relationship

$$\frac{\text{MS}_{\text{treat}}}{\text{MS}_{\text{error}}} = (1 + n\theta)F[k-1, k(n-1)],$$

one has

$$\text{E}\left(\frac{\text{MS}_{\text{treat}}}{\text{MS}_{\text{error}}}\right) = (1 + n\theta)\frac{k(n-1)}{k(n-1) - 2}, \tag{9}$$

since
$$\text{E}\{F[k-1, k(n-1)]\} = \frac{k(n-1)}{k(n-1)-2}.$$

Solving (9) for θ gives

$$\text{E}\left\{\frac{1}{n}\left[\frac{\text{MS}_{\text{treat}}}{\text{MS}_{\text{error}}}\frac{k(n-1)-2}{k(n-1)} - 1\right]\right\} = \theta. \tag{10}$$

Hence an unbiased estimate of θ is

$$\hat{\theta} = \frac{1}{n}\left[\frac{\text{MS}_{\text{treat}}}{\text{MS}_{\text{error}}}\frac{k(n-1)-2}{k(n-1)} - 1\right]$$

$$= \frac{[k(n-1)-2]\text{MS}_{\text{treat}} - k(n-1)\text{MS}_{\text{error}}}{kn(n-1)\text{MS}_{\text{error}}}$$

$$= \frac{\text{MS}_{\text{treat}} - m\text{MS}_{\text{error}}}{nm\text{MS}_{\text{error}}} \qquad \text{where} \quad m = \frac{k(n-1)}{k(n-1)-2}.$$

By way of contrast,

$$\frac{\hat{\sigma}_\tau^2}{\hat{\sigma}_\varepsilon^2} = \frac{\text{MS}_{\text{treat}} - \text{MS}_{\text{error}}}{n\text{MS}_{\text{error}}}.$$

To indicate the order of the magnitude of the bias in the estimator $\hat{\sigma}_\tau^2/\hat{\sigma}_\varepsilon^2$, consider the following numerical example:

$$k = 4 \qquad \text{MS}_{\text{treat}} = 250$$

$$n = 10 \qquad \text{MS}_{\text{error}} = 50$$

For this example,

$$\hat{\sigma}_\tau^2 = \tfrac{1}{10}(250 - 50) = 20, \qquad \hat{\sigma}_\varepsilon^2 = 50.$$

Hence $$\frac{\hat{\sigma}_\tau^2}{\hat{\sigma}_\varepsilon^2} = .400.$$

Whereas $$\hat{\theta} = \frac{250 - \frac{36}{34}(50)}{\frac{360}{34}(50)} = .372.$$

An estimate of ρ in terms of the unbiased estimate of θ is given by

$$\hat{\rho} = \frac{\hat{\theta}}{1 + \hat{\theta}}.$$

For the numerical example in the preceding paragraph,

$$\hat{\rho} = \frac{.372}{1 + .372} = .271.$$

An estimate of ρ in terms of the estimators of σ_ε^2 and σ_τ^2 is

$$\hat{\rho}' = \frac{\hat{\sigma}_\tau^2}{\hat{\sigma}_\varepsilon^2 + \hat{\sigma}_\tau^2} = \frac{20}{50 + 20} = .286.$$

To obtain the variance of an estimate of a variance component, one notes that

$$\text{E}[\chi^2(k-1)] = k - 1, \qquad \text{var}\,[\chi^2(k-1)] = 2(k-1);$$

$$\text{E}\{\chi^2[k(n-1)]\} = k(n-1), \qquad \text{var}\,\{\chi^2[k(n-1)]\} = 2k(n-1).$$

Since $$k(n-1)\frac{\hat{\sigma}_\varepsilon^2}{\sigma_\varepsilon^2} = \chi^2[k(n-1)],$$

it follows that $$\hat{\sigma}_\varepsilon^2 = \frac{\sigma_\varepsilon^2}{k(n-1)}\chi^2[k(n-1)].$$

Hence

$$\text{var}\,(\hat{\sigma}_\varepsilon^2) = \text{var}\,(\text{MS}_{\text{error}}) = \frac{\sigma_\varepsilon^4}{k^2(n-1)^2}2k(n-1) = \frac{2\sigma_\varepsilon^4}{k(n-1)}.$$

Since $$\frac{(k-1)\text{MS}_{\text{treat}}}{\sigma_\varepsilon^2 + n\sigma_\tau^2} = \chi^2(k-1),$$

it follows that $$\text{MS}_{\text{treat}} = \frac{\sigma_\varepsilon^2 + n\sigma_\tau^2}{k-1}\chi^2(k-1).$$

Hence $$\text{var}\,(\text{MS}_{\text{treat}}) = \frac{(\sigma_\varepsilon^2 + n\sigma_\tau^2)^2}{(k-1)^2}2(k-1)$$

$$= \frac{2(\sigma_\varepsilon^2 + n\sigma_\tau^2)^2}{k-1}.$$

Since MS_{error} and MS_{treat} are distributed independently,

$$\text{var}\,(\hat{\sigma}_\tau^2) = \text{var}\,\frac{1}{n}(\text{MS}_{\text{treat}} - \text{MS}_{\text{error}})$$

$$= \frac{1}{n^2}[\text{var}\,(\text{MS}_{\text{treat}}) + \text{var}\,(\text{MS}_{\text{error}})].$$

Estimates of these variances are obtained by replacing parameters by their estimators. Thus,

$$\text{est var}\,(\text{MS}_{\text{error}}) = \frac{2\hat{\sigma}_\varepsilon^4}{k(n-1)} = \frac{2\text{MS}_{\text{error}}^2}{k(n-1)}.$$

It may happen that estimates of variance components will turn out to be negative. One may avoid using a negative number as an estimate of a variance component by defining the estimate as being zero whenever the estimator is less than or equal to zero. However, if one adopts the latter procedure, the estimators given above will no longer be unbiased. Further, the distribution theory used to obtain the variance of the estimators will no longer be valid.

Alternative Approach to Variance-component Model. The E(MS) for the variance-component model may be obtained by constructing the analysis-of-variance table corresponding to the fixed model. That is, the sources of variation and mean squares are computed as though one were working with model I. The E(MS) are then obtained by assuming that

1. The τ_j are uncorrelated random variables with

$$E(\tau_j) = 0 \qquad \text{and} \qquad \text{var}\,(\tau_j) = \sigma_\tau^2,$$

2. The ε_{ij} are uncorrelated with the τ_j and uncorrelated with each other with

$$\underset{i}{E}(\varepsilon_{ij}) = 0 \qquad \text{and} \qquad \underset{i}{\text{var}}\,(\varepsilon_{ij}) = \sigma_\varepsilon^2 \qquad \text{for all } j.$$

The estimates of σ_τ^2 and σ_ε^2 that are obtained under this procedure are identical with those obtained under the variance-component model. Without any distribution assumptions, however, no F tests may be constructed. (Randomization tests on the experimental data can, of course, be made without prior distribution assumptions.) Even without distribution assumptions on the random variables, it may be shown that the estimators are unbiased and have minimum variance in the class of estimators that are quadratic functions of the basic observations.

From some points of view, the fixed model may be considered a limiting case of the variance-component model. It is that limiting case in which one is dealing with a finite population of size $k = K$. Methods of obtaining the E(MS) that are used in later chapters consider the fixed model as a limiting case of this type.

3.18 Maximum-likelihood Estimation and Likelihood-ratio Test

3.18-1 Maximum-likelihood estimation—univariate normal case

If the probability density function of a random variable X is $f(X)$, then the joint density of a random sample $X_1, X_2, \ldots, X_n$ is given by

$$\Pr\,(X_1,X_2,\ldots,X_n) = f(X_1)f(X_2)\cdots f(X_n) = \prod_{i=1}^{n} f(X_i).$$

If the density function $f(X)$ contains the parameters $\theta_1, \theta_2, \ldots, \theta_k$, then this joint probability depends upon these parameters. Considered as a function of the parameters of the probability distribution, the joint probability is called the *likelihood* of the parameters, given the sample, and is denoted by the symbol $L(\theta_1,\theta_2,\ldots,\theta_k)$. Thus,

$$\begin{aligned} L(\theta_1,\theta_2,\ldots,\theta_k) &= \Pr\,(X_1,X_2,\ldots,X_n) \\ &= f(X_1)f(X_2)\cdots f(X_n). \end{aligned}$$

Assume that the probability density of the random variable X is $N(\mu,\sigma^2)$. That is,

$$f(X;\mu,\sigma^2) = \left(\frac{1}{2\pi\sigma^2}\right)^{\frac{1}{2}} \exp\left[-\frac{1}{2\sigma^2}(X-\mu)^2\right].$$

For a random sample of size n, the likelihood of the parameters has the form

$$L(\mu,\sigma^2) = \left(\frac{1}{2\pi\sigma^2}\right)^{\frac{n}{2}} \exp\left[-\frac{1}{2\sigma^2}\Sigma(X_i - \mu)^2\right]. \tag{1}$$

The *maximum-likelihood estimators* of the parameters in (1), say $\hat{\mu}$ and $\hat{\sigma}^2$, are those values which make

$$L(\mu,\sigma^2) = \text{maximum}.$$

Equivalently, $\hat{\mu}$ and $\hat{\sigma}^2$ make

$$L^* = \ln L(\mu,\sigma^2) = \text{maximum}. \tag{2}$$

In this case,

$$L^* = -\frac{n}{2}\ln(2\pi) - \frac{n}{2}\ln\sigma^2 - \frac{1}{2\sigma^2}\Sigma(X_i - \mu)^2.$$

This maximization problem is solved by setting the partial derivatives of L^* (with respect to μ and σ^2) equal to zero, and then solving the resulting set of simultaneous equations. The partial derivatives of (2) are given in (3).

$$\frac{\partial L^*}{\partial\mu} = \frac{1}{\sigma^2}\Sigma(X_i - \mu).$$

(3)

$$\frac{\partial L^*}{\partial\sigma^2} = -\frac{n}{2\sigma^2} + \frac{1}{2\sigma^4}\Sigma(X_i - \mu)^2.$$

The equations in (3) represent the so-called normal equations. Setting the first equation in (3) equal to zero and replacing μ by $\hat{\mu}$ give

$$\Sigma(X_i - \hat{\mu}) = 0 \qquad \text{or} \qquad \hat{\mu} = \frac{\Sigma X_i}{n} = \bar{X}.$$

Replacing μ in the second equation by $\bar{X}$ and then setting the result equal to zero give

$$-\frac{n}{2\hat{\sigma}^2} + \frac{1}{2\hat{\sigma}^4}\Sigma(X_i - \bar{X})^2 = 0.$$

Hence
$$\frac{n}{2\hat{\sigma}^2} = \frac{1}{2\hat{\sigma}^4}\Sigma(X_i - \bar{X})^2,$$

or
$$n\hat{\sigma}^2 = \Sigma(X_i - \bar{X})^2.$$

Thus
$$\hat{\sigma}^2 = \frac{\Sigma(X_i - \bar{X})^2}{n}.$$

In general, maximum-likelihood estimators, if they exist, will be found to have the following highly desirable properties. They are:

1. Asymptotically efficient
2. The best asymptotically normal estimators
3. Consistent
4. A function of minimal sufficient statistics, if such exist

These properties hold provided the distribution function is regular with respect to the first two derivatives, the derivatives being taken with respect to the parameters being estimated, and provided the maximum-likelihood estimators are unique. Maximum-likelihood estimators are not necessarily unbiased.

3.18-2 Likelihood-ratio test

The end product of a single-factor (univariate) experiment, with k treatments and n independent experimental units per treatment, may be regarded as k random samples of size n each. Assume that the populations from which the samples were drawn are specified by

$$X_{ij}: N(\mu_j, \sigma^2) \qquad i = 1, \ldots, n \quad j = 1, \ldots, k.$$

That is, the n experimental units under treatment j constitute a random sample from a population in which X_{ij} is normally distributed with mean μ_j and variance σ^2, where σ^2 is the same for all populations.

The parameters in this model are

$$\mu_j \, (j = 1, \ldots, k) \qquad \text{and} \qquad \sigma^2.$$

Two sets of estimators of these parameters will be obtained. One set will be obtained under a set of assumptions labeled Ω, the second set under a set of assumptions labeled ω.

The set Ω is defined as follows:

$$\Omega: \{-\infty < \mu_j < \infty, j = 1, \ldots, k; 0 < \sigma^2 < \infty\}.$$

Under Ω, no restrictions are imposed upon the values that the parameters may assume and still satisfy the distribution assumptions. The maximum-likelihood estimators of the parameters under Ω will be

$$\hat{\mu}_j = \bar{T}_j, \qquad \hat{\sigma}^2 = \frac{\Sigma\Sigma(X_{ij} - \bar{T}_j)^2}{kn} = \frac{SS_{\text{error}}}{kn}.$$

The set ω is defined as follows:

$$\omega: \{-\infty < \mu_j < \infty; \mu_1 = \cdots = \mu_k = \mu; 0 < \sigma^2 < \infty\}.$$

These restrictions define the hypothesis $H_1: \mu_1 = \cdots = \mu_k$. Under ω

the maximum-likelihood estimators of the parameters are

$$\hat{\mu}_j = \bar{G}, \qquad \hat{\sigma}^2 = \frac{\Sigma\Sigma(X_{ij} - \bar{G})^2}{kn} = \frac{SS_{total}}{kn}.$$

The expression for the likelihood of the sample data is

$$(1) \qquad L(\text{sample} \mid \mu_j, \sigma^2) = \left(\frac{1}{2\pi\sigma^2}\right)^{kn/2} \exp\left[-\frac{1}{2\sigma^2}\Sigma\Sigma(X_{ij} - \mu_j)^2\right].$$

If one replaces the parameters in (1) by their maximum-likelihood estimators under Ω, one obtains

$$(2) \qquad L(\hat{\Omega}) = \left(\frac{kn}{2\pi SS_{error}}\right)^{kn/2} \exp\left[-\frac{kn}{2}\frac{SS_{error}}{SS_{error}}\right] = \left(\frac{kn}{2\pi SS_{error}}\right)^{kn/2} \exp\left[-\frac{kn}{2}\right].$$

On the other hand, if one replaces the parameters in (1) by their maximum-likelihood estimators under ω, one obtains

$$(3) \qquad L(\hat{\omega}) = \left(\frac{kn}{2\pi SS_{total}}\right)^{kn/2} \exp\left[-\frac{kn}{2}\frac{SS_{total}}{SS_{total}}\right] = \left(\frac{kn}{2\pi SS_{total}}\right)^{kn/2} \exp\left[-\frac{kn}{2}\right].$$

The likelihood ratio is, by definition,

$$(4) \qquad \lambda = \frac{L(\hat{\omega})}{L(\hat{\Omega})} = \left(\frac{SS_{error}}{SS_{total}}\right)^{kn/2} = \left(\frac{SS_{error}}{SS_{error} + SS_{treat}}\right)^{kn/2} = \left[\frac{1}{1 + (SS_{treat}/SS_{error})}\right]^{kn/2}$$

Hence

$$(5) \qquad \lambda^{2/kn} = \frac{1}{1 + (SS_{treat}/SS_{error})} = \frac{1}{1 + [(k-1)/k(n-1)](MS_{treat}/MS_{error})} = \frac{1}{1 + [(k-1)/k(n-1)]F}.$$

Under the distribution assumptions that have been made, when the hypothesis that $\mu_1 = \mu_2 = \cdots = \mu_k$ is true (that is, under the conditions defined by ω), the statistic defined by (5) has a central beta distribution, since SS_{treat} and SS_{error} are independently distributed as chi square. When the conditions in ω do not hold, (5) has a noncentral beta distribution. From (4) one notes that the smaller the numerical value of λ the more the data tend to contradict the hypothesis of no differences among the population means.

Probabilities associated with the beta distribution are readily obtained from tables of the F distribution since the F statistic is a relatively simple transformation on a statistic having a beta distribution. One may, however, work directly with the F statistic as defined in (5).

The principles leading to the likelihood ratio in (4) represent an approach to hypothesis testing that is readily generalized to the multivariate case. Also, most of the approaches to hypothesis testing that have been discussed in earlier chapters may be shown to be equivalent to likelihood-ratio tests.

A general form of F as used in (5) is

$$F = \frac{k(n-1)(Q_\omega - Q_\Omega)}{(k-1)Q_\Omega}, \tag{6}$$

where

$$Q_\omega = \Sigma\Sigma(X_{ij} - \bar{G})^2 = \text{SS}_{\text{total}},$$
$$Q_\Omega = \Sigma\Sigma(X_{ij} - \bar{T}_j)^2 = \text{SS}_{\text{error}}.$$

Q is a symbol representing a quadratic form. In the chapters that follow, the structure of F ratios that have an F distribution (both central and noncentral) can almost always be represented in the form (6).

3.19 General Principle in Hypothesis Testing

What is discussed in this section is a special case of the principle in Sec. 2.8. This principle is the basis for a broad class of tests, particularly in the analysis of covariance.

Let Ω represent the following set of conditions:

$$\Omega\colon X_{ij} = \mu_j + \varepsilon_{ij}, \qquad i = 1, \ldots, n; j = 1, \ldots, k,$$
$$\varepsilon_{ij} \text{ independently distributed as } N(0,\sigma_\varepsilon^2),$$
$$\mu_j = \text{set of constants associated with the treatments.}$$

Thus Ω summarizes the model and its underlying assumptions. Under Ω, the least-squares estimators of the μ_j make

$$Q = \Sigma(X_{ij} - \mu_j)^2 = \text{minimum.} \tag{1}$$

In earlier sections it was found that the least-squares estimators of the μ_j were

$$\hat{\mu}_j = \bar{T}_j \qquad j = 1, \ldots, k.$$

Under Ω, the minimum value of Q is

$$Q_\Omega = \Sigma(X_{ij} - \bar{T}_j)^2 = \text{SS}_{\text{error}}. \tag{2}$$

Let ω represent the following set of conditions:

$$\omega\colon X_{ij} = \mu_j + \varepsilon_{ij}, \qquad i = 1, \ldots, n; j = 1, \ldots, k,$$
$$\varepsilon_{ij} \text{ independently distributed as } N(0,\sigma_\varepsilon^2),$$
$$\mu_1 = \mu_j = \cdots = \mu_k = \mu.$$

Symbolically,

$$\omega = \Omega + H,$$

where H represents the hypothesis that all the μ_j are equal. H actually represents a constraint on Ω. Under ω, the least-squares estimator of μ makes

$$Q = \Sigma(X_{ij} - \mu)^2 = \text{minimum}. \tag{3}$$

The least-squares estimator of μ is

$$\hat{\mu} = \bar{G}.$$

Hence the minimum value of Q under ω is

$$Q_\omega = \Sigma(X_{ij} - \bar{G})^2 = SS_{\text{total}}. \tag{4}$$

Thus Q_ω is the variation due to error under a model which imposes the restriction that the hypothesis $\mu_1 = \mu_2 = \cdots = \mu_k$ is true. On the other hand, Q_Ω is the variation due to error under a model which imposes no restrictions on the μ_j. The difference

$$\begin{aligned} Q_\omega - Q_\Omega &= \text{measure of deviation from hypothesis,} \\ &= \text{variation due to the } \mu_j, \qquad j = 1, \ldots, k. \end{aligned}$$

One has

$$Q_\omega - Q_\Omega = SS_{\text{total}} - SS_{\text{error}} = SS_{\text{treat}}.$$

The degrees of freedom associated with Q_ω and Q_Ω in this case are

$$f_\omega = kn - 1, \qquad f_\Omega = k(n - 1).$$

The degrees of freedom associated with the difference $Q_\omega - Q_\Omega$ are

$$f_\omega - f_\Omega = (kn - 1) - k(n - 1) = k - 1.$$

Under random assignment of the sampling units to the treatments, it may be shown that $Q_\omega - Q_\Omega$ and Q_Ω are independently distributed as chi-square variables. Hence

$$F = \frac{(Q_\omega - Q_\Omega)/(f_\omega - f_\Omega)}{Q_\Omega/f_\Omega} \tag{5}$$

has an F distribution—central when the hypothesis in ω is true, noncentral when this hypothesis is not true.

Scheffé (1959, p. 39) gives the following rule for determining the noncentrality parameter associated with (5):

> Replace all terms in $Q_\omega - Q_\Omega$ by their expectations under Ω. The results will be $\sigma_\varepsilon^2\lambda$, where λ is the noncentrality parameter.

For the case of a single-factor experiment having n observations per treatment

$$Q_\omega - Q_\Omega = n\Sigma(\bar{T}_j - \bar{G})^2.$$

Using this rule, since $E(\bar{T}_j) = \mu_j$ and $E(\bar{G}) = \mu$,

$$n\Sigma(\bar{T}_j - \bar{G})^2 \quad \text{becomes} \quad n\Sigma(\mu_j - \mu)^2 = \sigma_\varepsilon^2\lambda.$$

Hence
$$\lambda = \frac{n\Sigma(\mu_j - \mu)^2}{\sigma_\varepsilon^2}.$$

The extension of this principle to the case in which the sample sizes under each of the treatments are not equal is direct. The test procedure outlined in Sec. 2.8 utilizes the basic principle in this section, and the F ratio has the same structure as that represented by (5).

3.20 Testing the Hypothesis of Equality of a Subset of τ_j (Fixed Model)

In a single-factor experiment, the overall F test is on the hypothesis that

$$H_1: \qquad \tau_1 = \tau_2 = \cdots = \tau_k.$$

If this hypothesis is rejected, it is sometimes of interest to test a hypothesis of the following form:

$$H_1: \qquad \tau_1 = \tau_2 = \cdots = \tau_m, \qquad \text{where } m < k.$$

A general procedure for testing the latter type of hypothesis is outlined in this section.

A numerical example using the computational procedures outlined earlier in this chapter is summarized in Table 3.20-1. The F test here leads

Table 3.20-1 Numerical Example

	Treatment 1	Treatment 2	Treatment 3	
(i)	4	4	12	
	8	10	14	
	6	10	14	
	10	12	16	
			12	
			16	
	$28 = T_1$	$36 = T_2$	$84 = T_3$	$148 = G$
	$4 = n_1$	$4 = n_2$	$6 = n_3$	$14 = n$

(ii)
$(1) = G^2/n = (148)^2/14 = 1565.57$
$(2) = \Sigma X^2 = 1768$
$(3) = \Sigma(T_j^2/n_j) = (28^2/4) + (36^2/4) + (84^2/6) = 1696$

	Source of variation	SS	df	MS	F
(iii)	Treatments	$(3) - (1) = 130.43$	2	65.21	9.96**
	Error	$(2) - (3) = 72.00$	11	6.55	
(iv)	Mean	$(1) = 1565.57$	1		
	Mean + treatments	$(3) = 1696.00$	3		

$^{**}F_{.99}(2,11) = 7.21$

to rejecting the hypothesis that

$$H_1: \qquad \tau_1 = \tau_2 = \tau_3.$$

Suppose now it is desired to test the hypothesis

$$H_1: \qquad \tau_1 = \tau_2.$$

In this special case, this hypothesis may be tested by means of the comparison corresponding to $\tau_1 - \tau_2 = 0$. That is,

$$C = \bar{T}_1 - \bar{T}_2 = 7 - 9 = -2.$$

$$\text{MS}_C = \frac{C^2}{(1/n_1) + (1/n_2)} = \frac{4}{\frac{2}{4}} = 8.$$

Hence, to test the hypothesis that $\tau_1 = \tau_2$,

$$F = \frac{\text{MS}_C}{\text{MS}_{\text{error}}} = \frac{8}{6.55} = 1.22, \qquad F_{.95}(1,11) = 4.84.$$

The experimental data do not reject this hypothesis.

To make the approach to testing this type of hypothesis general enough to cover the case

$$\tau_1 = \tau_2 = \cdots = \tau_m, \qquad \text{where } 2 < m < k,$$

consider the model

$$X_{ij} = \mu + \tau_j + \varepsilon_{ij} \qquad j = 1, 2, \ldots, k,$$

where the τ_j are unknown constants. The normal equations for the data in part i of Table 3.20-1 are as follows:

$$(1) \qquad \begin{bmatrix} 14 & 4 & 4 & 6 \\ 4 & 4 & 0 & 0 \\ 4 & 0 & 4 & 0 \\ 6 & 0 & 0 & 6 \end{bmatrix} \begin{bmatrix} \hat{\mu} \\ \hat{\tau}_1 \\ \hat{\tau}_2 \\ \hat{\tau}_3 \end{bmatrix} = \begin{bmatrix} 148 = G \\ 28 = T_1 \\ 36 = T_2 \\ 84 = T_3 \end{bmatrix}.$$

Under the side condition

$$4\hat{\tau}_1 + 4\hat{\tau}_2 + 6\hat{\tau}_3 = 0,$$

the system of equations (1) has the following solution:

$$\begin{aligned} 14 &= 148 & &\text{or} & \hat{\mu} &= 10.5714, \\ 4\hat{\tau}_1 &= 28 - 4\mu & &\text{or} & \hat{\tau}_1 &= -3.5714, \\ 4\hat{\tau}_2 &= 36 - 4\mu & &\text{or} & \hat{\tau}_2 &= -1.5714, \\ 6\hat{\tau}_3 &= 84 - 8\mu & &\text{or} & \hat{\tau}_3 &= 3.4286. \end{aligned}$$

Hence, the sum of squares due to prediction (including the mean) is

$$R(\mu,\tau_1,\tau_2,\tau_3) = \hat{\mu}G + \Sigma\hat{\tau}_j T_j = 1696.$$

The sum of squares due to error is

$$\begin{aligned} SS_{\text{error}} &= \Sigma X^2 - R(\mu,\tau_1,\tau_2,\tau_3) \\ &= 1768 - 1696 \\ &= 72. \end{aligned}$$

Now consider the model

$$X_{ij} = \mu + \tau_j + \varepsilon_{ij} \qquad j = 1, 2, \ldots, k,$$
$$\tau_1 = \tau_2 = \beta.$$

Under this restricted model, the normal equations for the data in part i of Table 3.20-1 are as follows:

$$\text{(2)} \qquad \begin{bmatrix} 14 & 8 & 6 \\ 8 & 8 & 0 \\ 6 & 0 & 6 \end{bmatrix} \begin{bmatrix} \hat{\mu} \\ \hat{\beta} \\ \hat{\tau}_3 \end{bmatrix} = \begin{bmatrix} 148 = G \\ 64 = T_1 + T_2 \\ 84 = T_3 \end{bmatrix}.$$

Under the side condition

$$8\hat{\beta} + 6\hat{\tau}_3 = 0,$$

the solution to the system of equations (2) is as follows:

$$\begin{aligned} 14\hat{\mu} &= 148 & &\text{or} & \hat{\mu} &= 10.5714, \\ 8\hat{\beta} &= 64 - 8\mu & &\text{or} & \hat{\beta} &= -2.5714, \\ 6\hat{\tau}_3 &= 84 - 6\mu & &\text{or} & \hat{\tau}_3 &= 3.4286. \end{aligned}$$

Hence the predictable sum of squares under this restricted model is

$$\begin{aligned} R(\mu, \tau_1 = \tau_2, \tau_3) &= \hat{\mu}G + \hat{\beta}(T_1 + T_2) + \hat{\tau}_3 T_3 \\ &= 1688. \end{aligned}$$

The sum of squares due to error under the restricted model is

$$\begin{aligned} SS'_{\text{error}} &= \Sigma X^2 - R(\mu, \tau_1 = \tau_2, \tau_3) \\ &= 1768 - 1688 \\ &= 80. \end{aligned}$$

A measure of deviation from hypothesis is

$$\begin{aligned} SS'_{\text{error}} - SS_{\text{error}} &= R(\mu,\tau_1,\tau_2,\tau_3) - R(\mu, \tau_1 = \tau_2, \tau_3) \\ &= 1696 - 1688 \\ &= 8. \end{aligned}$$

To test the hypothesis that $\tau_1 = \tau_2$, the F ratio is given by

$$F = \frac{MS_{\text{dev hypothesis}}}{MS_{\text{error}}} = \frac{8}{6.55} = 1.22.$$

A summary of the test procedure is given in Table 3.20-2.

Table 3.20-2 Summary

Source	SS	df	MS
Total (including mean)	$\Sigma X^2 = 1768$	14	
$R(\mu,\tau_1,\tau_2,\tau_3)$	1696	3	
$R(\mu,\tau_1 = \tau_2,\tau_3)$	1688	2	
Deviation from hypothesis	8	1	8
Error $= \Sigma X^2 - R(\mu,\tau_1,\tau_2,\tau_3)$	72	11	6.55

The test procedure that has just been outlined is equivalent to testing (simultaneously) the following hypothesis in terms of orthogonal comparisons:

$$\begin{bmatrix} \tau_1 - \tau_2 \\ \tau_1 + \tau_2 - 2\tau_3 \\ \tau_1 + \tau_2 + \tau_3 - 3\tau_4 \\ \cdots \\ \sum_{j=1}^{m-1} \tau_j - (m-1)\tau_m \end{bmatrix} = 0.$$

For the general case, the computational procedures outlined above lead to

$$SS_{\text{dev hypothesis}} = \left[\sum_{j=1}^{m}\left(\frac{T_j^2}{n_j}\right)\right] - \frac{\left(\sum_{j=1}^{m} T_j\right)^2}{\sum_{j=1}^{m} n_j}.$$

For the numerical example,

$$\sum_{j=1}^{2}\frac{T_j^2}{n_j} = \frac{28^2}{4} + \frac{36^2}{4} = 520,$$

$$\frac{\left(\sum_{j=1}^{2} T_j\right)^2}{\sum_{j=1}^{2} n_j} = \frac{(28+36)^2}{4+4} = 512.$$

Hence $\quad SS_{\text{dev hypothesis}} = 520 - 512 = 8.$

4

SINGLE-FACTOR EXPERIMENTS HAVING REPEATED MEASURES ON THE SAME ELEMENTS

4.1 Purpose

In experimental work in the behavioral sciences the elements forming the statistical population are frequently people. Because of large differences in experience and background, the responses of people to the same experimental treatment may show relatively large variability. In many cases, much of this variability is due to differences between people existing prior to the experiment. If this latter source of variability can be separated from treatment effects and experimental error, then the sensitivity of the experiment may be increased. If this source of variability cannot be estimated, it remains part of the uncontrolled sources of variability and is thus automatically part of the experimental error.

One of the primary purposes of experiments in which the same subject is observed under each of the treatments is to provide a control on differences between subjects. In this type of experiment, treatment effects for subject i are measured relative to the average response made by subject i on all treatments. In this sense each subject serves as his own control—responses of individual subjects to the treatments are measured in terms of deviations about a point which measures the average responsiveness of that individual subject. Hence variability due to differences in the average responsiveness of the subjects is eliminated from the experimental error (if an additive model is appropriate).

Experiments in which the same elements are used under all the k treatments require k observations on each element. Hence the term *repeated measurements* to describe this kind of design. To the extent that unique characteristics of the individual elements remain constant under the different treatments, pairs of observations on the same elements will tend to be positively correlated. More generally, the observations will be *dependent* rather than independent. If the population distributions involved are multivariate normal, the terms *dependent* and *correlated* are synonymous; analogously, the terms independent and uncorrelated are synonymous in this context. Since the models that will be used are assumed to have underlying multivariate normal distributions, correlated measurements imply statistically dependent measurements. The designs in this chapter may be said to involve correlated, or dependent, observations.

The notation to be used and general computational procedures to be followed are given in the next section. The rationale underlying the analysis and special uses of these designs are presented in later sections.

4.2 Notation and Computational Procedures

Notation for this type of design will be illustrated in terms of people as the elements of the statistical population. However, the notation is not

Table 4.2-1 Notation

	Treatment							
Person	1	2	...	j	...	k	Total	Mean
1	X_{11}	X_{12}		X_{1j}		X_{1k}	P_1	$\bar{P}_1$
2	X_{21}	X_{22}		X_{2j}		X_{2k}	P_2	$\bar{P}_2$
.	.	.		.		.	.	.
.	.	.		.		.	.	.
.	.	.		.		.	.	.
i	X_{i1}	X_{i2}		X_{ij}		X_{ik}	P_i	$\bar{P}_i$
.	.	.		.		.	.	.
.	.	.		.		.	.	.
.	.	.		.		.	.	.
n	X_{n1}	X_{n2}		X_{nj}		X_{nk}	P_n	$\bar{P}_n$
Total	T_1	T_2	...	T_j	...	T_k	G	
Mean	$\bar{T}_1$	$\bar{T}_2$	...	$\bar{T}_j$	...	$\bar{T}_k$		$\bar{G}$

restricted to this case. In Table 4.2-1 the symbol X_{11} represents the measurement on person 1 under treatment 1, X_{12} the measurement on person 1 under treatment 2, X_{1j} the measurement of person 1 under treatment j. In general the first subscript to an X indicates the person observed and the second subscript the treatment under which the observation is made.

The symbol P_1 represents the sum of the k observations on person 1, P_2 the sum of the k observations on person 2, P_i the sum of the k observations on person i. In summation notation,

$$P_i = \sum_j X_{ij};$$

that is, P_i is the sum of the k entries in row i. Summation over the subscript j is equivalent to summing over all columns within a single row. The mean of the observations on person i is

$$\bar{P}_i = \frac{P_i}{k}.$$

The symbol T_1 represents the sum of the n observations under treatment 1, T_2 the sum of the n observations under treatment 2, T_j the sum of the n observations under treatment j. In summation notation,

$$T_j = \sum_i X_{ij}.$$

Summation over the subscript i is equivalent to summing all entries in a single column. The mean of the n observations under treatment j, designated $\bar{T}_j$, is

$$\bar{T}_j = \frac{T_j}{n}.$$

The sum of the kn observations in the experiment, designated G, is

$$G = \Sigma P_i = \Sigma T_j = \Sigma\Sigma X_{ij}.$$

The symbol $\Sigma\Sigma X_{ij}$ represents the sum over all observations in the experiment. The grand mean of all observations, designated $\bar{G}$, is

$$\bar{G} = \frac{G}{kn} = \frac{\Sigma \bar{P}_i}{n} = \frac{\Sigma \bar{T}_j}{k}.$$

In the analysis of this type of experiment, the total variation is divided into two parts: One part is a function of differences between the means of the people; the other part is a function of the pooled variation within individuals. The total variation is

$$\text{(1)} \qquad SS_{\text{total}} = \Sigma\Sigma(X_{ij} - \bar{G})^2,$$

the sum of the squared deviations of each observation about the grand mean. This source of variation has $kn - 1$ degrees of freedom. That part of the total variation due to differences between the means of the people is

$$\text{(2)} \qquad SS_{\text{b. people}} = k\Sigma(\bar{P}_i - \bar{G})^2.$$

In words, the between-people variation is a function of the squared deviations of the means for the people about the grand mean. Alternatively, this source of variation may be viewed as due to the differences between all possible pairs of $\bar{P}_i$; the larger such differences, the larger this source of variation. Since there are n means, this source of variation has $n - 1$ degrees of freedom.

The variation within person i is

$$SS_{\text{w. person } i} = \sum_j (X_{ij} - \bar{P}_i)^2,$$

the sum of the squared deviations of the observations on person i about the mean for person i. This source of variation has $k - 1$ degrees of freedom. The pooled within-person variation, designated $SS_{\text{w. people}}$, is

(3) $$SS_{\text{w. people}} = \sum_i SS_{\text{w. person } i} = \Sigma\Sigma(X_{ij} - \bar{P}_i)^2.$$

Since the variation within each person has $k - 1$ degrees of freedom, the pooled within-person variation will have $n(k - 1)$ degrees of freedom. It is readily shown that the between- and within-people sources of variation are statistically independent and that

$$SS_{\text{total}} = SS_{\text{b. people}} + SS_{\text{w. people}}.$$

The degrees of freedom corresponding to these sources of variation are also additive,

$$kn - 1 = (n - 1) + n(k - 1).$$

To show this partition of SS_{total} algebraically, let

$$b_{ij} = X_{ij} - \bar{P}_i,$$
$$a_i = \bar{P}_i - \bar{G}.$$

Then, $\sum_j b_{ij} = 0$ for all i, $\quad \sum_i a_i = 0, \quad \sum_j a_i = ka_i.$

Hence, $$\sum_i \sum_j a_i b_{ij} = \sum_i a_i \left(\sum_j b_{ij} \right) = \sum_j (0) = 0.$$

From the definitions of b_{ij} and a_i, it follows that

$$X_{ij} - \bar{G} = b_{ij} + a_i.$$

Hence, $$\begin{aligned} SS_{\text{total}} = \sum_i \sum_j (X_{ij} - \bar{G})^2 &= \sum_i \sum_j (b_{ij} + a_i)^2 \\ &= \sum_i \sum_j b_{ij}^2 + \sum_i \sum_j a_i^2 + 2\sum_i \sum_j a_i b_{ij} \\ &= \sum_i \sum_j b_{ij}^2 + k\sum_i a_i^2 + 2(0) \\ &= SS_{\text{w. people}} + SS_{\text{b. people}}. \end{aligned}$$

The difference between two observations on the same person depends in part upon the difference in treatment effects and in part upon uncontrolled or residual sources of variation. Hence the pooled within-person variation may be divided into two parts: one part which depends upon differences between the treatment means, and a second part which consists of residual variation. That part which depends upon differences between treatment effects is defined as

$$\text{(4)} \qquad SS_{\text{treat}} = n\Sigma(\bar{T}_j - \bar{G})^2.$$

Alternatively, this source of variation may be expressed as

$$SS_{\text{treat}} = \frac{n\Sigma(\bar{T}_j - \bar{T}_{j'})^2}{k}.$$

The expression $\bar{T}_j - \bar{T}_{j'}$ represents the difference between a pair of treatment means; the summation is with respect to all possible pairs of treatment means, order within the pair being disregarded. For example, if $k = 3$,

$$SS_{\text{treat}} = \frac{n[(\bar{T}_1 - \bar{T}_2)^2 + (\bar{T}_1 - \bar{T}_3)^2 + (\bar{T}_2 - \bar{T}_3)^2]}{3}.$$

This source of variation has $k - 1$ degrees of freedom.

The residual variation is

$$\text{(5)} \qquad SS_{\text{res}} = \Sigma\Sigma[(X_{ij} - \bar{G}) - (\bar{P}_i - \bar{G}) - (\bar{T}_j - \bar{G})]^2.$$

The terms that are subtracted from $X_{ij} - \bar{G}$ are, respectively, the person and treatment effects so that the residual variation represents those sources of variation in the total that cannot be accounted for by differences between the people and differences between the treatments. The degrees of freedom for the residual variation are

$$\begin{aligned} df_{\text{res}} &= df_{\text{total}} - df_{\text{b. people}} - df_{\text{treat}} \\ &= (kn - 1) - (n - 1) - (k - 1) \\ &= kn - n - k + 1 = n(k - 1) - (k - 1) \\ &= (k - 1)(n - 1). \end{aligned}$$

It is readily shown that SS_{treat} and SS_{res} are statistically independent and that

$$SS_{\text{w. people}} = SS_{\text{treat}} + SS_{\text{res}}.$$

The degrees of freedom for the corresponding variations are also additive, i.e.,

$$n(k - 1) = (k - 1) + (n - 1)(k - 1).$$

The analysis of the sources of variation and the corresponding degrees of freedom are shown schematically in Fig. 4.2-1.

The definitions of the sources of variation do not provide the most convenient formulas for their computation. Formulas for this purpose are summarized in Table 4.2-2. The symbols (1), (2), and (3) are identical to those used in the case of single-factor experiments which do not have repeated measures. Symbol (4) occurs only in experiments having repeated measures. In each case the divisor in a term is the number of observations that are summed to obtain an element in the numerator. For example, G is the sum of kn observations; T_j is the sum of n observations; P_i is the sum of k observations. A summary of the analysis of variance

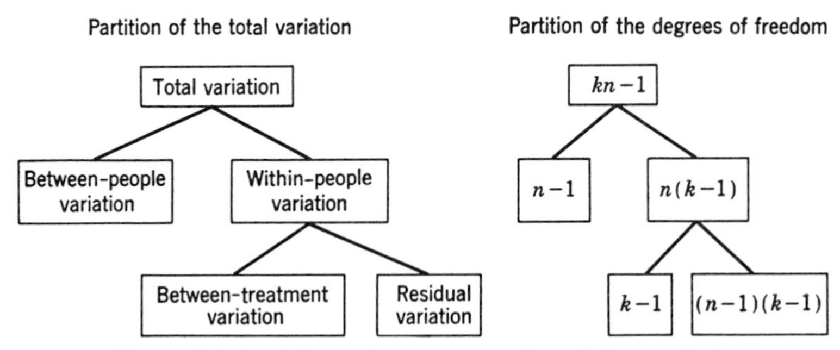

Figure 4.2-1 Schematic representation of the analysis.

appropriate for this design is given in part ii of this table. Mean squares are obtained from corresponding sums of squares by dividing the latter by their respective degrees of freedom.

The F ratio

$$F = \frac{\text{MS}_{\text{treat}}}{\text{MS}_{\text{res}}}$$

provides a test of the hypothesis that $\tau_1 = \tau_2 = \cdots = \tau_k$, where the τ's represent treatment effects and are defined in the same manner as they were for the case of designs not having repeated measures. The rationale underlying the use of this statistic for this test is discussed in Sec. 4.4.

Under one set of assumptions (made explicit in Sec. 4.4) about the underlying sources of variation, the F ratio has a sampling distribution which is approximated by the F distribution having $k - 1$ and $(n - 1) \times (k - 1)$ degrees of freedom. This is the usual test. Under less restrictive assumptions about the relations between the underlying sources of vari-

Table 4.2-2 Summary of Computational Procedures

(i) $(1) = G^2/kn$ $\quad$ $(2) = \Sigma\Sigma X^2$ $\quad$ $(3) = (\Sigma T_j^2)/n$ $\quad$ $(4) = (\Sigma P_i^2)/k$

	Source of variation	SS		df
(ii)	Between people	$\text{SS}_{\text{b.people}}$	$= (4) - (1)$	$n - 1$
	Within people	$\text{SS}_{\text{w. people}}$	$= (2) - (4)$	$n(k - 1)$
	Treatments	SS_{treat}	$= (3) - (1)$	$k - 1$
	Residual	SS_{res}	$= (2) - (3) - (4) + (1)$	$(n - 1)(k - 1)$
	Total	SS_{total}	$= (2) - (1)$	$kn - 1$

ation, Box (1954) has shown that the F ratio in the last paragraph has a sampling distribution (assuming that all $\tau_j = 0$) which is approximated by the F distribution having $(k - 1)\theta$ and $(n - 1)(k - 1)\theta$ degrees of freedom, where θ is a quantity which depends upon a set of homogeneity assumptions. The maximum value of θ is 1.00, and the minimum value is $1/(k - 1)$. The maximum value of θ is attained when the homogeneity assumptions underlying the usual test are met. Use of the minimum value of θ provides a conservative test. Thus, if the ratio

$$F = \frac{\text{MS}_{\text{treat}}}{\text{MS}_{\text{res}}}$$

is assumed to be distributed as an F distribution with 1 and $n - 1$ degrees of freedom (assuming that all $\tau_j = 0$), one has a conservative test relative to the usual test. However, the assumptions underlying this test are much weaker than those underlying the usual test. (*Conservative* in this context implies that a larger value of the F ratio is required for statistical significance at a specified level of α.)

4.3 Numerical Example

The computational procedures described in the last section will be illustrated by means of the numerical example in Table 4.3-1. The statistical basis for the analysis is discussed in the next section. The purpose of this experiment was to study the effects of four drugs upon reaction time to a series of standardized tasks. All subjects had been given extensive training on these tasks prior to the experiment. The five subjects used in the experiment are a random sample from a population of interest to the experimenter.

Each subject was observed under each of the drugs; the order in which a subject was administered a given drug was randomized. (In designs considered in later chapters, the order in which treatments are given to the same subject is either controlled or counterbalanced.) A sufficient time was allowed between the administration of the drugs to avoid the effect of one drug upon the effects of subsequent drugs, i.e., an interaction effect. The numerical entries in Table 4.3-1 represent the score (mean reaction time) on the series of standardized tasks. Thus person 1 had scores of 30, 28, 16, and 34 under the respective drug conditions. The total of these scores is 108; thus the numerical value of P_1 is 108. The other values for the P's are obtained by summing the entries in the respective rows in part i. The numerical values for the T's are obtained by summing the columns. For example, T_1 is the sum of the five entries under drug 1. The grand total, G, is obtained either by summing the P's or by summing the T's. A check on the arithmetic work is provided by computing G by both methods.

Quantities required in the computation of the sums of squares are given in part ii. The first three of these quantities are identical to those com-

Table 4.3-1 Numerical Example

	Person	Drug 1	Drug 2	Drug 3	Drug 4	Total
	1	30	28	16	34	$108 = P_1$
	2	14	18	10	22	$64 = P_2$
(i)	3	24	20	18	30	$92 = P_3$
	4	38	34	20	44	$136 = P_4$
	5	26	28	14	30	$98 = P_5$
		132	128	78	160	$498 = G$
		T_1	T_2	T_3	T_4	

(ii)

$$(1) = \frac{G^2}{kn} = \frac{(498)^2}{4(5)} = \frac{248{,}004}{20} = 12{,}400.20$$

$$(2) = \Sigma\Sigma X^2 = 13{,}892$$

$$(3) = \frac{\Sigma T_j^2}{n} = \frac{132^2 + 128^2 + 78^2 + 160^2}{5} = \frac{65{,}492}{5} = 13{,}098.40$$

$$(4) = \frac{\Sigma P_i^2}{k} = \frac{108^2 + 64^2 + 92^2 + 136^2 + 98^2}{4} = \frac{52{,}324}{4} = 13{,}081.00$$

(iii)

$$SS_{\text{b. people}} = (4) - (1) = 13{,}081.00 - 12{,}400.20 = 680.80$$

$$SS_{\text{w. people}} = (2) - (4) = 13{,}892 - 13{,}081.00 = 811.00$$

$$SS_{\text{drugs}} = (3) - (1) = 13{,}098.40 - 12{,}400.20 = 698.20$$

$$SS_{\text{res}} = (2) - (3) - (4) + (1)$$

$$= 13{,}892 - 13{,}098.40 - 13{,}081.00 + 12{,}400.20 = 112.80$$

$$SS_{\text{total}} = (2) - (1) = 13{,}892 - 12{,}400.20 = 1491.80$$

puted for designs which do not involve repeated measures. Symbol (4) is obtained from the P's. Each P is the sum over $k = 4$ drugs; hence the divisor associated with the symbol (4) is 4. The computation of the sums of squares required in the analysis of variance is illustrated in part iii. An alternative method for computing SS_{res} is

$$SS_{\text{res}} = SS_{\text{w. people}} - SS_{\text{drugs}}$$
$$= 811.00 - 698.20 = 112.80.$$

The latter method is actually simpler than the method used in part iii; however, the method in part iii provides a partial check on the numerical work, since the sum of SS_{drugs} and SS_{res} should total $SS_{\text{w. people}}$.

The analysis of variance is summarized in Table 4.3-2. The F ratio

$$F = \frac{MS_{\text{treat}}}{MS_{\text{res}}} = \frac{232.73}{9.40} = 24.76$$

is used in testing hypotheses about reaction time as a function of the effects of the drugs. For a .01-level test on the hypothesis that $\tau_1 = \tau_2 = \tau_3 = \tau_4$, the critical value for the F ratio is $F_{.99}(3,12) = 5.95$. The experimental data contradict this hypothesis. Inspection of the totals for the drugs in Table 4.3-1 indicates that drug 3 is associated with the fastest reaction.

Suppose that it had been anticipated before the experiment had been conducted that drug 3 would have a different effect from all others. The comparison that would be used in testing this hypothesis is

$$C = 3T_3 - T_1 - T_2 - T_4 = 3(78) - 132 - 128 - 160 = -186.$$

The component of variation corresponding to this comparison is

$$SS_C = \frac{C^2}{n\Sigma c^2} = \frac{(-186)^2}{5[3^2 + (-1)^2 + (-1)^2 + (-1)^2]} = 576.60.$$

The F statistic

$$F = \frac{SS_C}{MS_{res}} = \frac{576.60}{9.40} = 61.34$$

is used to test the hypothesis that $\tau_3 = (\tau_1 + \tau_2 + \tau_4)/3$. The critical value for a .01-level test of this hypothesis $F_{.99}(1,12) = 9.33$. The observed data contradict this hypothesis. If this comparison were suggested by inspection of the data, the procedure given by Scheffé (described in Sec. 3.9) would be used to obtain the critical value. The latter critical value for a .01-level test is $(k - 1)F_{.99}(k - 1, \text{df}_{res}) = 3F_{.99}(3,12) = 3(5.95) = 17.85$. Even with this critical value, the data indicate that drug 3 is different in its effect on reaction time from the effects of the other three drugs.

Table 4.3-2 Analysis of Variance

Source of variation		SS		df	MS	F
Between people		680.80		4		
Within people		811.00		15		
Drugs	698.20		3		232.73	24.76**
Residual	112.80		12		9.40	
Total		1491.80		19		

** $F_{.99}(3,12) = 5.95$

To test the hypothesis that $\tau_1 = \tau_2 = \tau_4$, the sum of squares for these three drugs is given by

$$SS_{\text{drugs } 1,2,4} = \frac{T_1^2 + T_2^2 + T_4^2}{n} - \frac{(T_1 + T_2 + T_4)^2}{3n}$$

$$= \frac{132^2 + 128^2 + 160^2}{5} - \frac{(132 + 128 + 160)^2}{15}$$

$$= 121.60.$$

The mean square corresponding to this sum of squares is

$$MS_{\text{drugs } 1,\,2,\,4} = \frac{121.60}{2} = 60.80.$$

The statistic used in the test is

$$F = \frac{MS_{\text{drugs } 1,\,2,\,4}}{MS_{\text{res}}} = \frac{60.80}{9.40} = 6.47.$$

For a .01-level test, the critical value of this statistic is $F_{.99}(2,12) = 6.93$. Although the observed F statistic does not exceed the critical value for a .01-level test, the observed F is large enough to question the hypothesis that the drugs 1, 2, and 4 are equally effective with respect to reaction time. Inspection of the drug totals in Table 4.3-1 indicates that drug 4 has a somewhat longer reaction time, but the evidence is not quite strong enough to establish this conclusion at the .01 level of significance.

The data, in this case, can be adequately summarized in terms of a few selected comparisons. Analogous conclusions can be reached by other, somewhat more systematic probing procedures. Any of the methods discussed in Secs. 3.8 and 3.9 may be used to test the difference between all possible pairs of means. In such tests MS_{res} has the role of MS_{error}. Application of the Newman-Keuls method is illustrated in Table 4.3-3.

Table 4.3-3 Tests on Differences between Pairs of Means

	Drugs		3	2	1	4
		Totals	78	128	132	160
(i)	3	78	—	50	54	82
	2	128		—	4	32
	1	132			—	28
	4	160				—

(ii)	$q_{.99}(r,12)$	4.32	5.04	5.50
(iii)	$\sqrt{nMS_{\text{res}}}\, q_{.99}(r,12)$	29.64	34.57	37.73

		3	2	1	4
(iv)	3		**	**	**
	2		—	—	—
	1			—	—
	4				—

With repeated measures, barring missing data, the numbers of observations under each treatment will be equal. In this case treatment totals may be used rather than treatment means. The drug totals, in increasing order of magnitude, are given in part i. The entry in a cell of part i is the difference between a total at the head of a column and a total to the left of a row. Critical values for the statistic

$$q_r = \frac{T_j - T_{j'}}{\sqrt{n\text{MS}_{\text{res}}}},$$

where r is the number of steps two totals are apart on an ordered scale, are given in part ii. These values are obtained from the first three columns of tables for the 99th percentile point for the q statistic; the degrees of freedom for this q statistic are the degrees of freedom of MS_{res}. Critical values for

$$T_j - T_{j'} = q_r\sqrt{n\text{MS}_{\text{res}}}$$

are given in part iii. In this case

$$\sqrt{n\text{MS}_{\text{res}}} = \sqrt{5(9.40)} = \sqrt{47.00} = 6.86.$$

Thus the entries in part iii are 6.86 times the corresponding entries in part ii.

The order in which tests are made is given in Sec. 3.8. The critical value for the difference $T_4 - T_3 = 82$ is 37.73. Hence the data contradict the hypothesis that $\tau_4 = \tau_3$. The difference $T_1 - T_3 = 54$ has the critical value 34.57, and the difference $T_2 - T_3 = 50$ has the critical value 29.64. The difference $T_4 - T_2 = 32$ has the critical value 34.57; this difference does not quite exceed the critical value. No further tests are made. The tests which yield statistically significant results are summarized in part iv. Drug 3 appears to be different from the other drugs in its effect on reaction time. Although the differences between drug 4 and drugs 2 and 1 are relatively large, the differences do not exceed critical values of a .01-level test. This latter result is consistent with the outcome of the test of the hypothesis that $\tau_1 = \tau_2 = \tau_4$. This hypothesis was not rejected at the .01 level of significance, but the observed F statistic was close to the critical value.

The computational formula for MS_{res} is algebraically equivalent to the expression

$$\text{MS}_{\text{res}} = \overline{\text{var}} - \overline{\text{cov}},$$

where $\overline{\text{var}}$ is the mean of the variances within each of the drug conditions and $\overline{\text{cov}}$ is the mean of the covariances between the pairs of observations under any two drug conditions. To show this equivalence for the numerical data in Table 4.3-1, the variance-covariance matrix for these data is given in Table 4.3-4.

Table 4.3-4 Variance-Covariance Matrix

	Drug 1	Drug 2	Drug 3	Drug 4
Drug 1	76.80	53.20	29.20	69.00
Drug 2		42.80	15.80	47.00
Drug 3			14.80	27.00
Drug 4				64.00

The variance of the observations made under each of the drugs appears along the main diagonal of this table. For example, the variance of the observations made under drug 1 is

$$\text{var}_{X_1} = \frac{\Sigma X_1^2 - (T_1^2/n)}{n-1} = \frac{3792 - (132^2/5)}{4} = 76.80.$$

The covariances appear above the main diagonal. For example, the covariance between the observations under drug 1 and those made under drug 2 is

$$\text{cov}_{X_1X_2} = \frac{\Sigma(X_{i1}X_{i2}) - (T_1T_2/n)}{n-1}$$

$$= \frac{(30)(28) + \cdots + (26)(28) - [(132)(128)/5]}{4} = 53.20.$$

The mean of the variances is 49.60; the mean of the covariances is 40.20. Thus,

$$\overline{\text{var}} - \overline{\text{cov}} = 49.60 - 40.20 = 9.40.$$

The numerical value of MS_{res} obtained by the computational formula is also 9.40. It is considerably more work to obtain MS_{res} from the variance-covariance matrix than it is to obtain MS_{res} by means of the computational formula. However, in order to check certain of the assumptions underlying the F test, computation of the variance-covariance matrix is sometimes required and is often enlightening in its own right.

$\text{MS}_{\text{b. people}}$ is also related to $\overline{\text{var}}$ and $\overline{\text{cov}}$. This relationship is

$$\text{MS}_{\text{b. people}} = \overline{\text{var}} + (k-1)\,\overline{\text{cov}}.$$

From Table 4.3-2,

$$\text{MS}_{\text{b. people}} = \frac{\text{SS}_{\text{people}}}{n-1} = \frac{680.80}{4} = 170.20.$$

In terms of the average variance and the average covariance,

$$\text{MS}_{\text{b. people}} = 49.60 + 3(40.20) = 170.20.$$

The matrix of intercorrelations for the data in Table 4.3-1 is given in Table 4.3-5. The average intercorrelation of the off-diagonal elements is $\bar{r} = .86$. Had a common estimate of the population variance been used in computing these intercorrelations, this average would have been $\bar{r}' = \overline{\text{cov}}/\overline{\text{var}} = .8104$. In terms of this latter average,

$$\text{MS}_{\text{res}} = \overline{\text{var}}\,(1 - \bar{r}') = 49.60(1 - .8104) = 9.40.$$

Had the data been uncorrelated, MS_{res} would have been equal to 49.60. The larger the average intercorrelation, the smaller will be MS_{res}.

Table 4.3-5 Intercorrelation Matrix

	Drug 1	Drug 2	Drug 3	Drug 4
Drug 1	1.000	.928	.866	.984
Drug 2		1.000	.628	.898
Drug 3			1.000	.877
Drug 4				1.000

4.4 Statistical Basis for the Analysis

The validity of tests in the last section rests upon a set of assumptions about the nature of the underlying sources of variation. The case in which the number of treatments is equal to 2 will be considered first. The assumptions that will be made for this case are actually stronger than those required; however, all the assumptions will be required for the case in which the number of treatments is greater than 2. Suppose that each of the observations may be expressed in terms of the strictly additive model given below:

Person	Treatment 1	Treatment 2
1	$X_{11} = \mu + \pi_1 + \tau_1 + \varepsilon_{11}$	$X_{12} = \mu + \pi_1 + \tau_2 + \varepsilon_{12}$
.	.	.
.	.	.
.	.	.
n	$X_{n1} = \mu + \pi_n + \tau_1 + \varepsilon_{n1}$	$X_{n2} = \mu + \pi_2 + \tau_2 + \varepsilon_{n2}$
	$T_1 = n\mu + \Sigma\pi_i + n\tau_1 + \Sigma\varepsilon_{i1}$	$T_2 = n\mu + \Sigma\pi_i + n\tau_2 + \Sigma\varepsilon_{i2}$
	$\bar{T}_1 = \mu + \bar{\pi} + \tau_1 + \bar{\varepsilon}_1$	$\bar{T}_2 = \mu + \bar{\pi} + \tau_2 + \bar{\varepsilon}_2$

The notation is defined below:

X_{ij} = an observation on person i under treatment j.
μ_1 = mean of all potential observations under treatment 1, that is, mean for treatment 1 if the entire population of people were observed under treatment 1.
μ_2 = mean of all potential observations under treatment 2.
π_i = a constant associated with person i. In the population of people the mean of the π_i is assumed to be zero.
$\mu = (\mu_1 + \mu_2)/2$ = grand mean of all potential observations.
$\tau_1 = \mu_1 - \mu$ = main effect of treatment 1.
$\tau_2 = \mu_2 - \mu$ = main effect of treatment 2.
ε_{ij} = experimental error associated with X_{ij}.
= all sources of variation in X_{ij} except those accounted for by the τ's and the π's.

Treatments 1 and 2 will be assumed to constitute the population of treatments. From the definition of τ_1 and τ_2, it follows that $\tau_1 + \tau_2 = 0$. Since the n people in the experiment are assumed to be a random sample from a potentially infinite population of people, π_i is a random variable. In the population of people, π_i will be assumed to be normally distributed with mean zero and variance σ_π^2.

Within the population of potential observations under treatment 1, the experimental error ε_{i1} is assumed to be normally distributed, with mean equal to zero and variance equal to $\sigma_{\varepsilon_1}^2$. Within the corresponding population under treatment 2, the experimental error is also assumed to be normally distributed, with mean equal to zero and variance equal to $\sigma_{\varepsilon_2}^2$. The two distributions of experimental error will be assumed to be independent; further it will be assumed that

$$\sigma_{\varepsilon_1}^2 = \sigma_{\varepsilon_2}^2 = \sigma_\varepsilon^2;$$

i.e., the error variances are homogeneous.

Since μ and τ_1 are constant for all observations under treatment 1, the variance of the X's within treatment population 1 is a function of the variance due to π_i and ε_{i1}. Assuming π_i and ε_{i1} uncorrelated,

$$\sigma_{X_1}^2 = \sigma_\varepsilon^2 + \sigma_\pi^2.$$

In words, under the assumptions that have been made, the variance of the potential observations under treatment 1 is the sum of the variance due to experimental error and the variance due to differences between the π's. Similarly the variance due to the potential observations under treatment 2 is

$$\sigma_{X_2}^2 = \sigma_\varepsilon^2 + \sigma_\pi^2.$$

Since the term π_i is common to two measurements on the same person, the covariance between X_1 and X_2 will not in general be equal to zero.

All covariance between X_1 and X_2 is assumed to be due to the term π_i. If this covariance is denoted by the symbol $\sigma_{X_1X_2}$, then

$$\sigma_{X_1X_2} = \sigma_\pi^2.$$

Under the assumptions that have been made,

$$\frac{\sigma_{X_1}^2 + \sigma_{X_2}^2}{2} = \sigma_\varepsilon^2 + \sigma_\pi^2.$$

Hence
$$\frac{\sigma_{X_1}^2 + \sigma_{X_2}^2}{2} - \sigma_{X_1X_2} = \sigma_\varepsilon^2.$$

The computational formula for MS_{res}, given in Sec. 4.2, is equivalent to the left-hand side of the above expression if statistics are substituted for corresponding parameters. MS_{res} provides an unbiased estimate of σ_ε^2; in symbols, this last statement is expressed by

$$\text{E}(\text{MS}_{\text{res}}) = \sigma_\varepsilon^2.$$

The expectation in this case is with respect to random samples of size n. Under the assumptions made, the sampling distribution of $(n - 1)\text{MS}_{\text{res}}/\sigma_\varepsilon^2$ is a chi-square distribution having $n - 1$ degrees of freedom.

From the structural model it is seen that

$$\bar{T}_1 - \bar{T}_2 = (\tau_1 - \tau_2) + (\bar{\varepsilon}_1 - \bar{\varepsilon}_2).$$

Since the same people are observed under both the treatments, this difference is free of any effects associated with the $\bar{\pi}$'s. The variance of the quantity $\bar{T}_1 - \bar{T}_2$, when the experiment is replicated with random samples of size n people, has the form

$$\sigma_{\bar{T}_1-\bar{T}_2}^2 = \sigma_{\tau_1-\tau_2}^2 + \sigma_{\bar{\varepsilon}_1-\bar{\varepsilon}_2}^2.$$

This expression assumes that $\tau_1 - \tau_2$ and $\bar{\varepsilon}_1 - \bar{\varepsilon}_2$ are uncorrelated. Implied by the expression for $\sigma_{\bar{T}_1-\bar{T}_2}^2$ is

$$\sigma_{\bar{T}}^2 = \sigma_\tau^2 + \sigma_{\bar{\varepsilon}}^2.$$

This last expression is implied by the previous expression since the variance of a variable is a function of the differences between all possible pairs of the variables. Multiplying both sides of this last expression by n gives

$$n\sigma_{\bar{T}}^2 = n\sigma_\tau^2 + n\sigma_{\bar{\varepsilon}}^2.$$

For random samples of size n, it has been shown in earlier chapters that $n\sigma_{\bar{\varepsilon}}^2 = \sigma_\varepsilon^2$. It may be further shown that MS_{treat} is an unbiased estimate of $n\sigma_{\bar{T}}^2$. Hence

$$\text{E}(\text{MS}_{\text{treat}}) = n\sigma_{\bar{T}}^2 = n\sigma_\tau^2 + \sigma_\varepsilon^2.$$

Under the assumptions that have been made, when $\sigma_\tau^2 = 0$ the sampling distribution of $\text{MS}_{\text{treat}}/\sigma_\varepsilon^2$ will be a chi-square distribution having one degree of freedom. Further the sampling distribution of MS_{treat} is independent of the sampling distribution of MS_{res}. Thus the statistic

$$F = \frac{MS_{treat}}{MS_{res}},$$

under the hypothesis that $\sigma_\tau^2 = 0$, represents the ratio of mean squares having independent chi-square distributions. Further, when $\sigma_\tau^2 = 0$, the numerator and denominator have expected values equal to σ_ε^2. Hence the F statistic has a sampling distribution which is an F distribution having one degree of freedom for the numerator and $n - 1$ degrees of freedom for the denominator. More explicitly,

$$\frac{E(MS_{treat})}{E(MS_{res})} = \frac{\sigma_\varepsilon^2 + n\sigma_\tau^2}{\sigma_\varepsilon^2} = \frac{\sigma_\varepsilon^2}{\sigma_\varepsilon^2} = 1.00,$$

when $\sigma_\tau^2 = 0$.

Since the logic underlying the design discussed in this section is basic to a broad class of designs to be discussed in later chapters, two alternative general approaches will be outlined. The second approach will be what is known as the mixed model with a person by treatment interaction. The first approach is that special case of the mixed model in which there is no interaction. However, certain developments in the first approach will be unique.

First Approach. Consider the model

$$(1) \qquad X_{ij} = \mu + \pi_i + \tau_j + \varepsilon_{ij}, \qquad i = 1, \ldots, n; j = 1, \ldots, k.$$

The following constraint is imposed upon the model:

$$\sum_j \tau_j = 0.$$

One assumes that

π_i are independent and distributed as $N(0,\sigma_\pi^2)$,
ε_{ij} are independent and distributed independently of the π_i as $N(0,\sigma_\varepsilon^2)$.

The τ_j are constants associated with the observations under treatment j.

Assume that the potential observations under treatment j have variance

$$\sigma_{X_j}^2 = \sigma_X^2 \qquad \text{for all } j.$$

Also assume that the covariances $(j \neq j')$

$$\sigma_{X_jX_{j'}} = \text{constant} \qquad \text{for all pairs } j, j'.$$

Define

$$\rho = \frac{\sigma_{X_jX_{j'}}}{\sigma_X^2}.$$

Thus $$\sigma_{X_jX_{j'}} = \rho\sigma_X^2.$$

The covariance matrix for the potential observations under treatments $1, \ldots, k$ is thus assumed to have the form

$$(2) \qquad \underset{k,k}{\Sigma} = \begin{bmatrix} \sigma_X^2 & \rho\sigma_X^2 & \cdots & \rho\sigma_X^2 \\ \rho\sigma_X^2 & \sigma_X^2 & \cdots & \rho\sigma_X^2 \\ \cdot & \cdot & & \cdot \\ \cdot & \cdot & & \cdot \\ \cdot & \cdot & & \cdot \\ \rho\sigma_X^2 & \rho\sigma_X^2 & \cdots & \sigma_X^2 \end{bmatrix} = \sigma_X^2 \begin{bmatrix} 1 & \rho & \cdots & \rho \\ \rho & 1 & \cdots & \rho \\ \cdot & \cdot & & \cdot \\ \cdot & \cdot & & \cdot \\ \cdot & \cdot & & \cdot \\ \rho & \rho & \cdots & 1 \end{bmatrix}.$$

This matrix is said to have *compound symmetry*.

In terms of the assumptions underlying the model given in (1),

$$\sigma_{X_j}^2 = \sigma_X^2 = \sigma_\varepsilon^2 + \sigma_\pi^2, \qquad \text{for all } j,$$

since τ_j is a constant for all observations under treatment j. Also,

$$\sigma_{X_jX_{j'}} = \sigma_\pi^2, \qquad \text{for all pairs } j, j'.$$

Hence $$\rho = \frac{\sigma_{X_jX_{j'}}}{\sigma_X^2} = \frac{\sigma_\pi^2}{\sigma_\varepsilon^2 + \sigma_\pi^2}.$$

Also, $$\rho\sigma_X^2 = \sigma_{X_jX_{j'}} = \sigma_\pi^2.$$

From these last relationships, it follows that

$$(3) \qquad \sigma_X^2(1 - \rho) = \sigma_X^2 - \rho\sigma_X^2 = (\sigma_\varepsilon^2 + \sigma_\pi^2) - \sigma_\pi^2 = \sigma_\varepsilon^2.$$

Also,

$$(4) \qquad \begin{aligned} \sigma_X^2[1 + (k-1)\rho] &= \sigma_X^2 + (k-1)\rho\sigma_X^2 \\ &= (\sigma_\varepsilon^2 + \sigma_\pi^2) + (k-1)\sigma_\pi^2 = \sigma_\varepsilon^2 + k\sigma_\pi^2. \end{aligned}$$

In terms of the model in (1),

$$\bar{P}_i = \frac{1}{k}\sum_j X_{ij} = \mu + \pi_i + \bar{\varepsilon}_i.$$

Hence

$$\begin{aligned} kE[\text{var}(\bar{P}_i)] &= E(MS_{\text{b. people}}) \\ &= k(\sigma_{\bar{\varepsilon}}^2 + \sigma_\pi^2) = \sigma_\varepsilon^2 + k\sigma_\pi^2 = \sigma_X^2[1 + (k-1)\rho]. \end{aligned}$$

The relationship on the extreme right follows from (4).

From the model in (1),

$$\bar{T}_j - \bar{T}_{j'} = \tau_j - \tau_{j'} + \bar{\varepsilon}_{.j} - \bar{\varepsilon}_{.j'}.$$

Since MS_{treat} is a function of $n(\bar{T}_j - \bar{T}_{j'})^2/2$, it follows that

$$E(MS_{\text{treat}}) = \sigma_\varepsilon^2 + n\sigma_\tau^2 = \sigma_X^2(1 - \rho) + n\sigma_\tau^2.$$

The expression on the extreme right follows from the relationship given in (3).

Also from (1), one has

$$X_{ij} - \bar{P}_i - \bar{T}_j + \bar{G} = \varepsilon_{ij} - \bar{\varepsilon}_{i.} - \bar{\varepsilon}_{.j} + \bar{\varepsilon}_{..}.$$

Since $$\text{MS}_{\text{res}} = \frac{1}{(n-1)(k-1)} \Sigma(X_{ij} - \bar{P}_i - \bar{T}_j + \bar{G})^2,$$

it follows that

$$\text{E}(\text{MS}_{\text{res}}) = \sigma_\varepsilon^2 = \sigma_X^2(1 - \rho).$$

To summarize, under the model in (1) one has the following E(MS):

$$\begin{aligned}
\text{MS}_{\text{b. people}}&: \quad \sigma_\varepsilon^2 + k\sigma_\pi^2 = \sigma_X^2[1 + (k-1)\rho], \\
\text{MS}_{\text{treat}}&: \quad \sigma_\varepsilon^2 + n\sigma_\tau^2 = \sigma_X^2(1 - \rho) + n\sigma_\tau^2, \\
\text{MS}_{\text{res}}&: \quad \sigma_\varepsilon^2 \qquad\quad = \sigma_X^2(1 - \rho).
\end{aligned}$$

The E(MS) at the extreme right are in terms of the parameters of the variance-covariance matrix for the X_j. The correspondence between the parameters of this covariance matrix and the parameters in the model (1) was established by a set of assumptions on the parameters in the model. It will be found that the E(MS) at the extreme right hold for a class of models of which that in (1) is a special case.

One notes that

$$\frac{\text{E}(\text{MS}_{\text{treat}})}{\text{E}(\text{MS}_{\text{res}})} = \frac{\sigma_\varepsilon^2 + n\sigma_\tau^2}{\sigma_\varepsilon^2} = 1 \qquad \text{when } \sigma_\tau^2 = 0.$$

Further, the ratio

$$F = \frac{\text{MS}_{\text{treat}}}{\text{MS}_{\text{res}}}$$

is distributed as a central F distribution when $\sigma_\tau^2 = 0$. Hence, to test the hypothesis that $\sigma_\tau^2 = 0$, the F ratio indicated is appropriate. Since τ_j corresponds to a fixed factor, it is understood that

$$\sigma_\tau^2 = \frac{\Sigma\tau_j^2}{k-1}.$$

When $\sigma_\tau^2 > 0$, the F ratio indicated has a noncentral F distribution with noncentrality parameter

$$\lambda = \frac{n\Sigma\tau_j^2}{\sigma_\varepsilon^2}.$$

Second Approach. The model that will be used in the approach that follows is called the mixed model with a person by treatment interaction. Consider the model

$$(5) \qquad X_{ij} = \mu + \pi_i + \tau_j + \pi\tau_{ij} + \varepsilon_{ij}, \qquad \begin{matrix} i = 1, \ldots, n. \\ j = 1, \ldots, k. \end{matrix}$$

Assume that τ_j corresponds to a fixed factor with

$$\Sigma\tau_j = 0 \qquad \text{and} \qquad \sigma_\tau^2 = \frac{\Sigma\tau_j^2}{k-1}.$$

The terms π_i, $\pi\tau_{ij}$, and ε_{ij} correspond to random variables with

$$E(\pi_i) = 0, \qquad E(\pi\tau_{ij}) = 0, \qquad E(\varepsilon_{ij}) = 0.$$

The variables ε_{ij} are assumed to be independent and distributed as

$$\varepsilon_{ij}: N(0,\sigma_\varepsilon^2).$$

Further, the ε_{ij} are assumed to be distributed independently of π_i and $\pi\tau_{ij}$.

The random variables π_i are assumed to be independent and distributed as $N(0,\sigma_\pi^2)$. Further, the π_i are assumed to be independent of the $\pi\tau_{ij}$.

The variables $\pi\tau_{ij}$ have the constraint $\sum_j \pi\tau_{ij} = 0$. The distribution of the $\pi\tau_{ij}$ is

$$\pi\tau_{ij}: N\left(0, \frac{k-1}{k}\,\sigma_{\pi\tau}^2\right).$$

The $\pi\tau_{ij}$ are not, however, independent. The covariance between $\pi\tau_{ij}$ and $\pi\tau_{ij'}$ is

$$\text{cov}(\pi\tau_{ij},\pi\tau_{ij'}) = -\frac{1}{k}\,\sigma_{\pi\tau}^2.$$

However, $$\text{cov}(\pi\tau_{ij},\pi\tau_{i'j}) = 0, \qquad i \neq i'.$$

Note that under this covariance structure on the $\pi\tau_{ij}$ it follows that

$$\begin{aligned}\text{var}\,(\pi\tau_{i1} + \cdots + \pi\tau_{ik}) &= k\left(\frac{k-1}{k}\right)\sigma_{\pi\tau}^2 + k(k-1)\left(-\frac{1}{k}\,\sigma_{\pi\tau}^2\right)\\ &= 0.\end{aligned}$$

Under the constraint

$$\sum_j \pi\tau_{ij} = 0$$

this sum will always be zero. Hence the constraint implies the covariance structure, and the covariance structure implies that this sum will be constant.

For the model in (5),

$$\bar{P}_i = \frac{1}{k}\sum_j X_{ij} = \mu + \pi_i + \bar{\varepsilon}_{i.}.$$

Hence $$k\text{E}[\text{var}\,(\bar{P}_i)] = k(\sigma_{\bar{\varepsilon}}^2 + \sigma_\pi^2) = \sigma_\varepsilon^2 + k\sigma_\pi^2.$$

Also from (5),

$$\bar{T}_j - \bar{T}_{j'} = (\tau_j - \tau_{j'}) + (\overline{\pi\tau}_{.j} - \overline{\pi\tau}_{.j'}) + (\bar{\varepsilon}_{.j} - \bar{\varepsilon}_{.j'}).$$

Note that

$$\overline{\pi\tau}_{.j} - \overline{\pi\tau}_{.j'} = \frac{1}{n}[(\pi\tau_{1j} - \pi\tau_{1j'}) + \cdots + (\pi\tau_{nj} - \pi\tau_{nj'})].$$

$$\text{Hence} \quad \text{E}[\text{var}\,(\overline{\pi\tau}_{.j} - \overline{\pi\tau}_{.j'})] = \frac{n}{n^2}\left[2\left(\frac{k-1}{k}\right)\sigma^2_{\pi\tau} - 2\left(-\frac{1}{k}\,\sigma^2_{\tau\pi}\right)\right]$$

$$= \frac{2}{n}\,\sigma^2_{\pi\tau}.$$

Since MS_{treat} is a function of $n(\bar{T}_j - \bar{T}_{j'})^2/2$, one has

$$\text{E}(\text{MS}_{\text{treat}}) = \sigma^2_\varepsilon + \sigma^2_{\pi\tau} + n\sigma^2_\tau.$$

The person by treatment interaction may be defined as that part of an observation which is not an additive function of the person and treatment main effects. In this model

$$\widehat{\tau\pi}_{ij} = X_{ij} - \bar{P}_i - \bar{T}_j + \bar{G},$$

and

$$\text{MS}_{\text{people}\times\text{treat}} = \frac{\sum_i\sum_j(\widehat{\pi\tau}_{ij})^2}{(n-1)(k-1)}.$$

But this was the definition of MS_{res} in the analysis of variance. Hence, for this model,

$$\text{MS}_{\text{res}} = \text{MS}_{\text{person}\times\text{treat}}.$$

Under the model in (5)

$$\widehat{\pi\tau}_{ij} = \pi\tau_{ij} - \overline{\pi\tau}_j + \varepsilon_{ij} - \bar{\varepsilon}_{i.} - \bar{\varepsilon}_{.j} + \bar{\varepsilon}_{..}$$

Thus

$$\text{E}(\widehat{\pi\tau}_{ij}) = \pi\tau_{ij},$$

$$\text{E}(\text{MS}_{\text{person}\times\text{treat}}) = \sigma^2_\varepsilon + \sigma^2_{\tau\pi}.$$

The expected values of the mean squares are summarized in Table 4.4-1.

Table 4.4-1 E(MS) for Mixed Model as Defined in (5)

Source of variation	df	E(MS)
Between people	$n-1$	$\sigma^2_\varepsilon + k\sigma^2_\pi$
Treatments	$k-1$	$\sigma^2_\varepsilon + \sigma^2_{\pi\tau} + n\sigma^2_\tau$
Person × treatment (res.)	$(n-1)(k-1)$	$\sigma^2_\varepsilon + \sigma^2_{\pi\tau}$

The $k \times k$ variance-covariance matrix Σ which was made explicit in the first approach is actually in the background of the present approach.

If the matrix Σ has the symmetry properties indicated under the first approach, then the F ratio for testing the hypothesis $\sigma_\tau^2 = 0$ suggested by Table 4.4-1, namely,

$$F = \frac{\text{MS}_{\text{treat}}}{\text{MS}_{\text{person} \times \text{treat}}},$$

will have the distribution $F[k - 1, (n - 1)(k - 1)]$ when $\sigma_\tau^2 = 0$. If the matrix Σ does not have the symmetry properties indicated, then the test based upon the F distribution is an approximate one. Monte Carlo studies indicate that the approximation is a relatively good one.

Summary Remarks on F Test. The assumption of compound symmetry on the matrix Σ, which underlies the usual F test on the treatment effects in this design, is highly restrictive. That is, the experimental data must conform to a prescribed pattern of variances and covariances before the statistical test can be considered exact. Box (1954) has indicated that the usual F test in case of uncorrelated data is relatively robust (insensitive) with respect to violation of the assumption of homogeneity of variance. That is, for uncorrelated data, violation of the assumption of homogeneity of variance does not seriously bias the final F test.

Box has shown, however, that heterogeneity of both the variances and covariances in a design having correlated observations will generally result in a positive bias in the usual F test. That is, the critical value as obtained from an F table tends to be too low relative to a critical value appropriate for an arbitrary variance-covariance matrix. (The usual F test is appropriate only for a very special variance-covariance matrix.) For the case in which the variance-covariance matrix is arbitrary in form, an approximate test may be made through use of the usual F statistic, but the degrees of freedom are taken to be $(k - 1)\theta$ and $(k - 1)(n - 1)\theta$, where θ is a number that measures the extent to which the covariance matrix in the underlying model deviates from a pattern that is associated with an F distribution for the F ratio. For $\theta = 1$, the covariance matrix either has compound symmetry or some other pattern for which the F ratio has an F distribution. The upper bound for $\theta = 1$; the lower bound is $\theta = 1/(k - 1)$.

If one assumes that θ is actually equal to this lower bound, the resulting test will tend to err on the conservative side. That is, the F value determined from an F table will tend to be somewhat larger than the exact value. (This kind of test is said to be negatively biased.) The degrees of freedom for the numerator of the F ratio, assuming that $\theta = 1/(k - 1)$, are

$$(k - 1)\theta = \frac{k - 1}{k - 1} = 1.$$

The degrees of freedom for the denominator, assuming that $\theta = 1/(k - 1)$, are

$$(n - 1)(k - 1)\theta = \frac{(n - 1)(k - 1)}{k - 1} = n - 1.$$

Hence the conservative test assumes that the F ratio has one and $n - 1$ degrees of freedom. When the assumptions of homogeneity of variance and homogeneity of covariances are questionable, the conservative test indicated in this paragraph provides an approximate test.

For the case in which $k \geq 2$ and $n > k$, Hotelling's T^2 statistic may be used to test the hypothesis that $\sigma_\tau^2 = 0$. This test is exact if the underlying distribution is multivariate normal. Use of Hotelling's T^2 statistic requires no assumptions of homogeneity on the covariance matrix. Computation of Hotelling's T^2 is illustrated in Sec. 4.9.

Procedures for testing homogeneity hypotheses about population covariance matrices are described in Sec. 7.7. In particular χ_2^2 as defined in Sec. 7.7 may be used to test the hypothesis that the population covariance matrix has the form

$$\begin{bmatrix} \sigma^2 & \rho\sigma^2 & \rho\sigma^2 \\ \rho\sigma^2 & \sigma^2 & \rho\sigma^2 \\ \rho\sigma^2 & \rho\sigma^2 & \sigma^2 \end{bmatrix}.$$

This is the form specified by the model used in this section to justify the subsequent F tests. In the definition of χ_2^2 in Sec. 7.7, q corresponds to the number of treatments; in the notation of this section, $k = q$, and $N = nk$.

Comment on Compound Symmetry. Compound symmetry of the matrix Σ is a sufficient condition for the ratio $F = \text{MS}_{\text{treat}}/\text{MS}_{\text{res}}$ to have an F distribution under the hypothesis of no treatment effects. It is not, however, a necessary condition. That is, Σ may have other patterns and the ratio in question will still have an F distribution. If Σ has a pattern such that

$$\sigma^2_{\bar{T}_i - \bar{T}_j} = \text{constant} \qquad \text{for all } i \text{ and } j,$$

the ratio will have an F distribution.

For example, consider the matrix

$$\Sigma = \begin{bmatrix} \sigma_{11} & \sigma_{12} & \sigma_{13} \\ \sigma_{21} & \sigma_{22} & \sigma_{23} \\ \sigma_{31} & \sigma_{32} & \sigma_{33} \end{bmatrix} = \begin{bmatrix} 5 & 2.5 & 5 \\ 2.5 & 10 & 7.5 \\ 5 & 7.5 & 15 \end{bmatrix}.$$

Assume $n = 5$. Then one has

$$\sigma^2_{\bar{T}_1 - \bar{T}_2} = \frac{1}{n}(\sigma_{11} + \sigma_{22} - 2\sigma_{12}) = \tfrac{1}{5}(5 + 10 - 5) = 2,$$

$$\sigma^2_{\bar{T}_1 - \bar{T}_3} = \frac{1}{n}(\sigma_{11} + \sigma_{33} - 2\sigma_{13}) = \tfrac{1}{5}(5 + 15 - 10) = 2,$$

$$\sigma^2_{\bar{T}_2-\bar{T}_3} = \frac{1}{n}(\sigma_{22} + \sigma_{33} - 2\sigma_{23}) = \tfrac{1}{5}(10 + 15 - 15) = 2.$$

This matrix, although it does not have compound symmetry, has the property that $\sigma^2_{\bar{T}_i-\bar{T}_j}$ = constant. A matrix which does have compound symmetry will also have this property. Thus matrices having compound symmetry belong to a larger class of matrices for which the variance of a difference between treatment means will be a constant for all treatments.

The parameter θ, which measures the departure of the matrix Σ from a pattern for which the F ratio has an F distribution, is defined by

$$\theta = \frac{k^2(\bar{\sigma}_{ii} - \bar{\sigma}_{..})^2}{(k-1)(\Sigma\Sigma\sigma^2_{ij} - 2k\Sigma\bar{\sigma}^2_{i.} + k^2\bar{\sigma}^2_{..})},$$

where $\bar{\sigma}_{ii}$ = mean of entries on main diagonal of Σ,

$\bar{\sigma}_{..}$ = mean of all entries in Σ,

$\bar{\sigma}_{i.}$ = mean of entries in row i of Σ.

θ will be equal to unity when Σ has compound symmetry. θ will also be equal to unity $\sigma^2_{\bar{T}_i-\bar{T}_j}$ = constant. For the numerical example given here,

$$\theta = \frac{9(10 - 6.67)^2}{2[525 - 6(4.17^2 + 6.67^2 + 9.17^2) + 9(6.67)^2]} = 1.$$

Although the statistic $F = \text{MS}_{\text{treat}}/\text{MS}_{\text{res}}$ will have an F distribution for a broader class of matrices than those having compound symmetry, the logic underlying the linear model for a univariate analysis of variance will often rule out the possibility of any patterned form of Σ other than that of compound symmetry. For most situations in which Σ does not have compound symmetry, the multivariate analysis of variance will often be a more informative approach than the univariate approach. It is, however, difficult to formulate a set of general rules whereby one type of analysis will be preferred to the other. The univariate approach pools information from repeated measures—this kind of pooling usually carries with it a set of homogeneity assumptions.

4.5 Use of Analysis of Variance to Estimate Reliability of Measurements

Given a person possessing a magnitude π of a specified characteristic. In appraising this characteristic with some measuring device, the observed score may have the magnitude $\pi + \eta$. The quantity η is the error of measurement; all measurement has some of this kind of error. The latter is due in part to the measuring device itself and in part to the conditions surrounding the measurement. In the development that follows, it will be assumed that the magnitude of the error of measurement is uncorrelated with π. A measurement on person i with measuring instrument

j may be represented as

(1) $$X_{ij} = \pi_i + \eta_{ij},$$

where X_{ij} = observed measurement,
π_i = true magnitude of characteristic being measured,
η_{ij} = error of measurement.

Upon repeated measurement with the same or comparable instruments, π_i is assumed to remain constant, whereas η_{ij} is assumed to vary. The mean of k such repeated measures may be represented as

(2) $$\frac{\sum_j X_{ij}}{k} = \bar{P}_i = \pi_i + \bar{\eta}_{i\cdot}$$

A schematic representation of a random sample of k measurements on the same or comparable measuring instruments is shown in Table 4.5-1.

Table 4.5-1 Estimation of Reliability

Person	Comparable measurements						Total	Mean
	1	2	...	j	...	k		
1	X_{11}	X_{12}		X_{1j}		X_{1k}	P_1	$\bar{P}_1$
2	X_{21}	X_{22}		X_{2j}		X_{2k}	P_2	$\bar{P}_2$
.	.	.	.	.	.	.	.	.
.	.	.	.	.	.	.	.	.
.	.	.	.	.	.	.	.	.
i	X_{i1}	X_{i2}		X_{ij}		X_{ik}	P_i	$\bar{P}_i$
.	.	.	.	.	.	.	.	.
.	.	.	.	.	.	.	.	.
.	.	.	.	.	.	.	.	.
n	X_{n1}	X_{n2}		X_{nj}		X_{nk}	P_n	$\bar{P}_n$
Total	T_1	T_2	...	T_j	...	T_k	G	

If π_i remains constant for such measurement, the variance within person i is due to error of measurement, and the pooled within-person variance also estimates variance due to error of measurement. On the other hand, the variance in the $\bar{P}$'s is in part due to differences between the true magnitudes of the characteristic possessed by the n people and in part due to differences in the average error of measurement for each person. The analysis of variance and the expected values for the mean squares for data of the type shown in Table 4.5-1 are given in Table 4.5-2. $MS_{b.\ people}$ is defined as

$$MS_{b.\ people} = \frac{k\Sigma(\bar{P}_i - \bar{G})^2}{n-1},$$

Table 4.5-2 Analysis of Variance for Model in (1)

Source of variation	MS	E(MS)
Between people	$MS_{b.\ people}$	$\sigma_\eta^2 + k\sigma_\pi^2$
Within people	$MS_{w.\ people}$	σ_η^2

whereas the variance of the $\bar{P}$'s is given by

$$s_{\bar{P}}^2 = \frac{\Sigma(\bar{P}_i - \bar{G})^2}{n - 1}.$$

Thus
$$MS_{b.\ people} = ks_{\bar{P}}^2.$$

In terms of (2), the expected value of the variance of the $\bar{P}$'s is

$$E(s_{\bar{P}}^2) = \sigma_{\bar{\eta}}^2 + \sigma_\pi^2.$$

The quantity σ_π^2 is the variance of the true measures in the population of which the n people in the study represent a random sample. From the relationship between $MS_{b.\ people}$ and $s_{\bar{P}}^2$,

$$E(MS_{b.\ people}) = k\sigma_{\bar{\eta}}^2 + k\sigma_\pi^2 = \sigma_\eta^2 + k\sigma_\pi^2,$$

since
$$k\sigma_{\bar{\eta}}^2 = \sigma_\eta^2.$$

The reliability of $\bar{P}_i$, the mean of k measurements, is defined as

$$\rho_k = \frac{\sigma_\pi^2}{\sigma_\pi^2 + \sigma_{\bar{\eta}}^2} = \frac{\sigma_\pi^2}{\sigma_\pi^2 + (\sigma_\eta^2/k)}. \tag{3}$$

In words, the reliability of the mean of k measurements is the variance due to true scores divided by the sum of the variance due to true scores and the variance due to the mean of the errors of measurement. If one defines

$$\theta = \frac{\sigma_\pi^2}{\sigma_\eta^2},$$

then the expression for ρ_k may be cast in the form

$$\rho_k = \frac{\sigma_\pi^2}{\sigma_\pi^2 + (\sigma_\eta^2/k)} = \frac{\sigma_\pi^2/\sigma_\eta^2}{(\sigma_\pi^2/\sigma_\eta^2) + (\sigma_\eta^2/k\sigma_\eta^2)} \tag{4}$$

$$= \frac{k(\sigma_\pi^2/\sigma_\eta^2)}{1 + k(\sigma_\pi^2/\sigma_\eta^2)}$$

$$= \frac{k\theta}{1 + k\theta}.$$

When $k = 1$, (4) becomes

(5) $$\rho_1 = \frac{\theta}{1 + \theta} = \frac{\sigma_\pi^2}{\sigma_\pi^2 + \sigma_\eta^2},$$

which, by definition, is the reliability of a single measurement. Within the context of the variance-component model of the analysis of variance, (5) represents the *intraclass* correlation.

The reliability of the mean of k measurements, ρ_k, may be expressed in terms of the reliability of a single measurement, ρ_1, as follows:

(6) $$\rho_k = \frac{k\rho_1}{1 + (k - 1)\rho_1}.$$

To establish (6), replace ρ_1 in (6) by the expression for ρ_1 in (5). Thus,

$$\rho_k = \frac{k[\theta/(1 + \theta)]}{1 + (k - 1)[\theta/(1 + \theta)]} = \frac{k\theta}{1 + k\theta}.$$

Hence the expression on the right-hand side of (6) is identical to the expression on the right-hand side of the expression given in (4).

In the psychometric literature, (6) is known as the Spearman-Brown prediction formula. The assumptions underlying this formula are those underlying the model in (1), namely, that the error of measurement is uncorrelated with the true score, that the sample of n people on whom the observations are made is a random sample from a population of people to which inferences are to be made, that the sample of k measuring instruments used is a random sample from a population of comparable measuring instruments, and that the within-person variance may be pooled to provide an estimate of σ_η^2.

The expected values of the mean squares associated with the model in (1) are given in Table 4.5-2. From these expected values one has the following estimator for σ_η^2:

$$\hat{\sigma}_\eta^2 = \text{MS}_{\text{w. people}}.$$

If one equates $\text{MS}_{\text{b. people}}$ to its expectation and then replaces parameters by their estimators, one has

$$\begin{aligned} \text{MS}_{\text{b. people}} &= \hat{\sigma}_\eta^2 + k\hat{\sigma}_\pi^2 \\ &= \text{MS}_{\text{w. people}} + k\hat{\sigma}_\pi^2. \end{aligned}$$

Hence an estimate of the variance component σ_π^2 is given by

$$\hat{\sigma}_\pi^2 = \frac{\text{MS}_{\text{b. people}} - \text{MS}_{\text{w. people}}}{k}.$$

Under the assumption that the random variables π and η in (1) are independently and normally distributed, the estimators for $\hat{\sigma}_\pi^2$ and $\hat{\sigma}_\eta^2$ can be shown to be the unbiased maximum-likelihood estimators.

From these estimators, one logical estimate for θ would seem to be

$$\hat{\theta} = \frac{\hat{\sigma}_\pi^2}{\hat{\sigma}_\eta^2} = \frac{\text{MS}_{\text{b. people}} - \text{MS}_{\text{w. people}}}{k\text{MS}_{\text{w. people}}}.$$

This estimator will, however, be biased. In general, the ratio of two unbiased estimators will provide a biased estimator of the ratio. In terms of this estimator of θ, an estimate of the reliability, ρ_k, is given by

$$(7) \quad r_k = \hat{\rho}_k = \frac{k\hat{\theta}}{1 + k\hat{\theta}} = \frac{(\text{MS}_{\text{b. people}} - \text{MS}_{\text{w. people}})/\text{MS}_{\text{w. people}}}{1 + [(\text{MS}_{\text{b. people}} - \text{MS}_{\text{w. people}})/\text{MS}_{\text{w. people}}]}$$

$$= \frac{\text{MS}_{\text{b. people}} - \text{MS}_{\text{w. people}}}{\text{MS}_{\text{b. people}}}$$

$$= 1 - \frac{\text{MS}_{\text{w. people}}}{\text{MS}_{\text{b. people}}} = \frac{F - 1}{F},$$

where $F = \text{MS}_{\text{b. people}}/\text{MS}_{\text{w. people}}$. The reliability of a single measurement is estimated by

$$(8) \quad r_1 = \hat{\rho}_1 = \frac{\hat{\theta}}{1 + \hat{\theta}} = \frac{\text{MS}_{\text{b. people}} - \text{MS}_{\text{w. people}}}{\text{MS}_{\text{b. people}} + (k - 1)\text{MS}_{\text{w. people}}}.$$

It is possible to obtain an unbiased estimator of θ; indeed the unbiased estimator of θ was obtained in Sec. 3.17. In terms of the notation used in this section, the unbiased estimator obtained in Sec. 3.17 is given by

$$(9) \quad \hat{\theta}' = \frac{\text{MS}_{\text{b. people}} - \{n(k-1)/[n(k-1) - 2]\}\text{MS}_{\text{w. people}}}{\{kn(k-1)/[n(k-1) - 2]\}\text{MS}_{\text{w. people}}}$$

$$= \frac{\text{MS}_{\text{b. people}} - m\text{MS}_{\text{w. people}}}{km\text{MS}_{\text{w. people}}} \qquad \text{where } m = \frac{n(k-1)}{n(k-1) - 2}.$$

In terms of the estimator in (1), the estimator of the reliability of the mean of k measurements is

$$(10) \quad r_k' = \frac{k\hat{\theta}'}{1 + k\hat{\theta}'}.$$

In general $\qquad \hat{\theta}' < \hat{\theta}.$

Hence $\qquad r_k' < r_k.$

Numerical Example. To illustrate the material that has been discussed up to this point, a numerical example is given in Table 4.5-3. In this example $n = 6$ people are rated by $k = 4$ judges on a specified characteristic. The data in the columns represent the rating given by the individual judges. The data within a row represent the rating received by a person. For example, person 1 received the ratings 2, 4, 3, and 3. The row totals are designated by the symbols P_i; the column totals are designated by the

Table 4.5-3 Numerical Example

	Person	Judge 1	Judge 2	Judge 3	Judge 4	Total
(i)	1	2	4	3	3	$12 = P_1$
	2	5	7	5	6	$23 = P_2$
	3	1	3	1	2	$7 = P_3$
	4	7	9	9	8	$33 = P_4$
	5	2	4	6	1	$13 = P_5$
	6	6	8	8	4	$26 = P_6$
	Total	23	35	32	24	$114 = G$
		T_1	T_2	T_3	T_4	

(ii) $(1) = \dfrac{G^2}{kn} = 541.50$ $\quad (2) = \Sigma(\Sigma X^2) = 700$ $\quad (3) = \dfrac{\Sigma T_j^2}{n} = 559.00$ $\quad (4) = \dfrac{\Sigma P_i^2}{k} = 664.00$

(iii)

$$
\begin{aligned}
SS_{\text{b. people}} &= (4) - (1) &&= 122.50\\
SS_{\text{w. people}} &= (2) - (4) &&= 36.00\\
SS_{\text{b. judges}} &= (3) - (1) &&= 17.50\\
SS_{\text{res}} &= (2) - (3) - (4) + (1) &&= 18.50\\
SS_{\text{total}} &= (2) - (1) &&= 158.50
\end{aligned}
$$

symbols T_j. In the computational work in parts ii and iii, the judges play the role that treatments had in earlier sections.

A summary of the analysis of variance appears in Table 4.5-4. Only

Table 4.5-4 Analysis of Variance

Source of variation	SS		df		MS
Between people		122.50		5	24.50
Within people		36.00		18	2.00
Between judges	17.50		3		5.83
Residual	18.50		15		1.23
Total		158.50		23	

the between- and within-subject mean squares are needed to estimate the reliability as given by (7). The additional mean squares given in Table 4.5-4 will be used in connection with an alternative definition of reliability, which will be discussed later in this section.

From the data in Table 4.5-4, the estimate of θ is

$$\hat{\theta} = \frac{MS_{\text{b. people}} - MS_{\text{w. people}}}{kMS_{\text{w. people}}} = \frac{24.50 - 2.00}{4(2.00)} = 2.8125.$$

Hence the estimate of the reliability of the mean of the $k = 4$ judges, as obtained from (7), is

$$r_4 = \frac{4\hat{\theta}}{1 + 4\hat{\theta}} = \frac{4(2.8125)}{1 + 4(2.8125)} = .9184.$$

The estimate of the reliability of a single judge, as obtained from (8), is

$$r_1 = \frac{\hat{\theta}}{1 + \hat{\theta}} = \frac{2.8125}{1 + 2.8125} = .7377.$$

To illustrate the application of the Spearman-Brown prophecy formula, one may obtain r_4 from the relationship

$$r_4 = \frac{4r_1}{1 + 3r_1} = \frac{4(.7377)}{1 + 3(.7377)} = .9184.$$

The unbiased estimator of θ as obtained from the data in Table 4.5-4, as obtained from (9), is

$$\hat{\theta}' = \frac{24.50 - \{6(3)/[6(3) - 2]\}(2.00)}{\{4(6)(3)/[6(3) - 2]\}(2.00)} = 2.4722.$$

In terms of this estimator, the reliability of the mean of $k = 4$ judges, as obtained from (10), is

$$r_4' = \frac{4\hat{\theta}'}{1 + 4\hat{\theta}'} = \frac{4(2.4722)}{1 + 4(2.4722)} = .9082.$$

Alternative Model—Adjustment for Anchor Points. A more general (in one sense) model than that used in (1) has the form

$$X_{ij} = \pi_i + \alpha_j + \eta_{ij} \qquad \begin{cases} i = 1, \ldots, n, \\ j = 1, \ldots, k. \end{cases} \tag{11}$$

where X_{ij} = observed measurement,
π_i = true magnitude of characteristic measured,
α_j = anchor point (main effect) of measuring instrument,
η_{ij} = error of measurement.

In terms of the representation in Table 4.5-1, differences due to anchor points are due to differences in the T_j.

Under the model in (11), the analysis of variance has the form given in Table 4.5-5. One should note that the term η_{ij} in the model represented by (1) includes the term α_j that appears in model (11). That is, the term η_{ij} in model (1) is more inclusive than the corresponding term in model (11).

Table 4.5-5 Analysis of Variance for Model in (11)

Source of variation	SS	df	MS	E(MS)
Between people	$SS_{b.\,people}$	$n - 1$	$MS_{b.\,people}$	$\sigma_\eta^2 + k\sigma_\pi^2$
Within people	$SS_{w.\,people}$	$n(k - 1)$		
Judges (anchor points)	SS_{judges}	$k - 1$	MS_{judges}	$\sigma_\eta^2 + n\sigma_\alpha^2$
Residual	SS_{res}	$(n - 1)(k - 1)$	MS_{res}	σ_η^2

In terms of the model in (11), the reliability of the mean of k observations is defined by

$$(12) \qquad \rho_k = \frac{k\theta}{1 + k\theta}, \qquad \text{where } \theta = \frac{\sigma_\pi^2}{\sigma_\eta^2}.$$

In form, (12) is identical to that given earlier in this section; however, the model on which (12) is based is different from the earlier model. From the E(MS) in Table 4.5-5,

$$\hat{\sigma}_\eta^2 = \text{MS}_{\text{res}}.$$

(From the model used earlier in this section, $\hat{\sigma}_\eta^2 = \text{MS}_{\text{w. people}}$.) Also

$$\hat{\sigma}_\pi^2 = \frac{\text{MS}_{\text{b. people}} - \text{MS}_{\text{res}}}{k}.$$

Hence

$$\hat{\theta} = \frac{\hat{\sigma}_\pi^2}{\hat{\sigma}_\eta^2} = \frac{\text{MS}_{\text{b. people}} - \text{MS}_{\text{res}}}{k\text{MS}_{\text{res}}}.$$

The unbiased estimate of θ is given by

$$\hat{\theta}' = \frac{\text{MS}_{\text{b. people}} - \{(n-1)(k-1)/[(n-1)(k-1)-2]\}\text{MS}_{\text{res}}}{\{k(n-1)(k-1)/[(n-1)(k-1)-2]\}\text{MS}_{\text{res}}}$$

$$= \frac{\text{MS}_{\text{b. people}} - m'\text{MS}_{\text{res}}}{km'\text{MS}_{\text{res}}} \qquad \text{where } m' = \frac{(n-1)(k-1)}{(n-1)(k-1)-2}.$$

In terms of the numerical example summarized in Table 4.5-4,

$$\hat{\theta} = \frac{24.500 - 1.233}{4(1.233)} = 4.7176,$$

$$\hat{\theta}' = \frac{24.500 - \{5(3)/[5(3)-2]\}(1.233)}{\{4(5)(3)/[5(3)-2]\}(1.233)} = 4.0552.$$

Thus, when differences due to anchor points are not considered part of the error of measurement, the estimate of reliability of the mean of $k = 4$ observations, using $\hat{\theta}$, is

$$r_4 = \frac{4\hat{\theta}}{1 + 4\hat{\theta}} = \frac{18.8704}{19.8704} = .9497,$$

$$r_1 = \frac{\hat{\theta}}{1 + \hat{\theta}} = \frac{4.7176}{5.7176} = .8251.$$

The corresponding estimate using $\hat{\theta}'$ is

$$r_4' = \frac{4\hat{\theta}'}{1 + 4\hat{\theta}'} = \frac{16.2208}{17.2208} = .9419.$$

In terms of the $k \times k$ variance-covariance matrix of the measuring instruments, it will be found that

$$r_1 = \frac{\overline{\text{cov}}}{\overline{\text{var}}}; \tag{13}$$

that is, the estimate of reliability for a single measurement, when differences due to anchor points are not considered part of error of measurement, is the average covariance among the measurements divided by the average variance. The relationship in (13) follows from the fact that

$$MS_{\text{b. people}} = \overline{\text{var}} + (k-1)\overline{\text{cov}},$$
$$MS_{\text{res}} = \overline{\text{var}} - \overline{\text{cov}}.$$

Hence
$$\hat{\theta} = \frac{[\overline{\text{var}} + (k-1)\overline{\text{cov}}] - [\overline{\text{var}} - \overline{\text{cov}}]}{k(\overline{\text{var}} - \overline{\text{cov}})}$$
$$= \frac{\overline{\text{cov}}}{\overline{\text{var}} - \overline{\text{cov}}}.$$

Thus
$$r_1 = \frac{\hat{\theta}}{1+\hat{\theta}} = \frac{\overline{\text{cov}}/(\overline{\text{var}} - \overline{\text{cov}})}{1 + \overline{\text{cov}}/(\overline{\text{var}} - \overline{\text{cov}})},$$
$$= \frac{\overline{\text{cov}}}{\overline{\text{var}}}.$$

To illustrate the relationship in (13) the variance-covariance matrix for the numerical data given in Table 4.5-3 is given in Table 4.5-6. For this

Table 4.5-6 Variance-Covariance Matrix for Numerical Example in Table 4.5-3

	Judge 1	Judge 2	Judge 3	Judge 4
Judge 1	6.1660	6.1660	6.4660	5.6000
Judge 2		6.1660	6.4660	5.6000
Judge 3			9.0660	4.6000
Judge 4				6.8000

matrix the mean of the entries on the main diagonal is

$$\overline{\text{var}} = \frac{6.1660 + 6.1660 + 9.0660 + 6.800}{4} = 7.0495.$$

The mean of the entries off the main diagonal is

$$\overline{\text{cov}} = 5.8163.$$

Hence
$$r_1 = \frac{\overline{\text{cov}}}{\overline{\text{var}}} = \frac{5.8163}{7.0495} = .8251.$$

Since neither the variance within a judge nor the covariance between any two judges depends upon the anchor points, any correlational measure for reliability will not depend upon differences due to anchor points.

Direct Adjustment for Anchor Points—Numerical Example. Consider the numerical data in Table 4.5-3. The totals and means for each of the columns (judges) are as follows:

	Judge 1	Judge 2	Judge 3	Judge 4	
Total	23	35	32	24	114
Mean	3.83	5.83	5.33	4.00	4.75 $= \bar{G}$

The anchor points for the judges are given by

$$\hat{\alpha}_1 = \bar{T}_1 - \bar{G} = -0.99, \qquad \hat{\alpha}_3 = \bar{T}_3 - \bar{G} = 0.59,$$

$$\hat{\alpha}_2 = \bar{T}_2 - \bar{G} = 1.08, \qquad \hat{\alpha}_4 = \bar{T}_4 - \bar{G} = -0.75.$$

To adjust the data obtained from judge j for his anchor point, one subtracts $\hat{\alpha}_j$ from his ratings. For example, the adjusted data for judge 1 is obtained by subtracting $\hat{\alpha}_1 = -0.99$ from all entries in the first column of the data in part i of Table 4.5-3. If similar adjustments are made for the other columns, the data in part i of Table 4.5-7 will be obtained. The

Table 4.5-7 Numerical Example (Adjusted Data)

	Person	Judge 1	Judge 2	Judge 3	Judge 4	Total
(i)	1	2.92	2.92	2.41	3.75	12
	2	5.92	5.92	4.41	6.75	23
	3	1.92	1.92	.41	2.75	7
	4	7.92	7.92	8.41	8.75	33
	5	2.92	2.92	5.41	1.75	13
	6	6.92	6.92	7.41	4.75	26
	Total	28.52	28.52	28.46	28.50	114

(ii)

$$(1) = \frac{G^2}{kn} = 541.50 \qquad (2) = \Sigma\Sigma X^2 = 682.50 \qquad (3) = \frac{\Sigma T_j^2}{n} = 541.50$$

$$(4) = \frac{\Sigma P_i^2}{k} = 664.00$$

(iii)

$$
\begin{aligned}
SS_{\text{b. people}} &= (4) - (1) &&= 122.50\\
SS_{\text{w. people}} &= (2) - (4) &&= 18.50\\
SS_{\text{b. judges}} &= (3) - (1) &&= .00\\
SS_{\text{res}} &= (2) - (3) - (4) + (1) &&= 18.50\\
SS_{\text{total}} &= (2) - (1) &&= 141.00
\end{aligned}
$$

model for the adjusted data may be represented as follows:

$$X'_{ij} = X_{ij} - \hat{\alpha}_j = \pi_i + \eta_{ij}. \tag{14}$$

Note that the adjustment process has no effect upon the row totals. The adjustment process does, however, make all the column totals equal, within rounding error.

In parts ii and iii of Table 4.5-7 the sums of squares for the various sources of variation are computed by means of the usual computational formulas. A summary of the analysis of variance is given in Table 4.5-8.

Table 4.5-8 Analysis of Variance (Adjusted Data)

Source of variation	SS	df	MS
Between people	122.50	5	24.500
Within people (adj.)	18.50	15	1.233
Total	141.00	20	

Since $k - 1$ independent parameters are estimated in order to obtain the adjustments, the degrees of freedom associated with $SS_{w.\ people}$(adjusted) are

$$n(k-1) - (k-1) = (n-1)(k-1) \qquad \text{rather than} \qquad n(k-1).$$

One notes that $MS_{w.\ people}$(adjusted) in this analysis is equal numerically to MS_{res} in the analysis-of-variance Table 4.5-4. Thus replacing $MS_{w.\ people}$ by $MS_{w.\ people}$(adjusted) is equivalent to replacing $MS_{w.\ people}$ by MS_{res}. In essence, this is what was done in the preceding subsection.

The right-hand side of the model in (14) has the same form as the model in (1), but in (1) η_{ij} includes differences due to anchor points. The assumptions underlying the model in (1) also apply to the model in (14). However, in estimating the parameters in the model, under the model in (14) $MS_{w.\ people}$(adjusted) replaces $MS_{w.\ people}$. For example, under the model in (14), the (biased) estimator of θ is given by

$$\hat{\theta} = \frac{MS_{b.\ people} - MS_{w.\ people}(\text{adjusted})}{kMS_{w.\ people}(\text{adjusted})}$$

$$= \frac{MS_{b.\ people} - MS_{res}}{kMS_{res}}.$$

Reliability of a Test of k Items. Table 4.5-9 represents data that would be obtained from the administration of a test of k items. The totals $P_1, P_2, \ldots, P_n$ represent the test scores for the people taking the tests, provided that the score on the test is obtained by summing the scores on the individual items. An estimate of the reliability of the test may be obtained from (12). The reliability of a total is the same as that of a mean score. The computational procedures used in Sec. 4.2, with the test items having the role of the treatments, will provide the quantities required in (12). If the test items are scored 1 for correct and 0 for incorrect,

Table 4.5-9 Representation of Test Scores

Person	Item 1	Item 2	. . .	Item k	Test score
1	X_{11}	X_{12}		X_{1k}	P_1
2	X_{21}	X_{22}		X_{2k}	P_2
.	.	.		.	.
.	.	.		.	.
.	.	.		.	.
n	X_{n1}	X_{n2}		X_{nk}	P_n
	T_1	T_2	. . .	T_k	G

Table 4.5-10 Reliability of a Test of $k = 5$ Dichotomous Items

Person	Items						
	1	2	3	4	5		
1	1	1	1	1	1	$5 = P_1$	
2	1	1	1	1	0	$4 = P_2$	
3	1	1	1	0	1	4	
4	1	1	0	1	0	3	$k = 5$
5	1	1	1	0	0	3	$n = 10$
6	1	1	0	0	1	3	
7	1	1	0	0	0	2	
8	0	1	1	0	0	2	
9	1	0	1	0	0	2	
10	1	0	0	0	0	1	
Total	9	8	6	3	3	$29 = G$	

$(1) = G^2/kn = 16.82$ $\qquad$ $(3) = \Sigma T^2/n = 19.90$

$(2) = \Sigma X^2 = 29$ $\qquad$ $(4) = \Sigma P^2/k = 19.40$

$SS_{b.\,people} = (4) - (1) = 2.58 \qquad MS_{b.\,people} = .2867$

$SS_{res} = (2) - (3) - (4) + (1) = 6.52 \qquad MS_{res} = .1811$

$$\hat{\theta} = \frac{MS_{b.\,people} - MS_{res}}{kMS_{res}} = .1166$$

$$r_5 = \frac{5\hat{\theta}}{1 + 5\hat{\theta}} = \frac{.5831}{1.5831} = .3683$$

the only information required to obtain an estimate of the reliability of the test is the scores on the tests and the number of correct responses to each item. (In cases where the observed data consist of 0's and 1's, $\Sigma X = \Sigma X^2$.) The error of measurement in this case may be interpreted (in part) as a measure of the extent to which people having the same test scores do not have identical profiles of correct responses. For person 1, the profile of responses is $X_{11}, X_{12}, \ldots, X_{1k}$.

Two numerical examples using dichotomous items are given in this subsection. In both examples the distribution of the test scores is the same. However in the second example individuals having the same scores have identical profiles. In Table 4.5-10, the steps in the computation of the reliability are given in detail. The reliability for the test is $r_5 = .3683$.

The covariance matrix for the items is given in Table 4.5-11. From $\overline{\text{var}}$ and $\overline{\text{cov}}$ in this table one obtains r_1. From the latter, by means of the Spearman-Brown relationship, one may compute r_5. Within rounding error, r_5 obtained here is the same as r_5 obtained in Table 4.5-10.

Table 4.5-11 Variance-Covariance Matrix for Items in Table 4.5-10

		1	2	3	4	5
(i)	1	.1000	−.0222	−.0444	.0333	.0333
	2		.1778	.0222	.0667	.0667
	3			.2667	.0222	.0667
	4				.2333	.0111
	5					.2333

(ii)

$$\overline{\text{var}} = .2021 \qquad \overline{\text{cov}} = .0211$$

$$r_1 = \frac{\overline{\text{cov}}}{\overline{\text{var}}} = .1044$$

$$r_5 = \frac{5r_1}{1 + 4r_1} = .3682$$

In Table 4.5-12, the row totals are the same as those in Table 4.5-10. However in Table 4.5-12 individuals having the same row totals have the same row profiles. MS_{res} in this table is reduced from .1811 (in Table 4.5-10) to .1033. The resulting reliability is raised from .3683 to .6397. One notes that, in spite of the fact that all individuals having the same score have identical row profiles, the reliability is not unity.

Under this method of defining reliability, MS_{res} is a measure of what is called the subject by item interaction. As long as $MS_{b.\ people} > 0$, with dichotomous items $MS_{res} > 0$. Hence the reliability cannot reach unity.

Table 4.5-12 Reliability of a Test of $k = 5$ Dichotomous Items—Comparable Profiles

Person	Items					Total
	1	2	3	4	5	
1	1	1	1	1	1	5
2	1	1	1	1	0	4
3	1	1	1	1	0	4
4	1	1	1	0	0	3
5	1	1	1	0	0	3
6	1	1	1	0	0	3
7	1	1	0	0	0	2
8	1	1	0	0	0	2
9	1	1	0	0	0	2
10	1	0	0	0	0	1
Total	10	9	6	3	1	29

$MS_{b.\ people} = .2867$ $\qquad$ $MS_{res} = .1033$

$$\hat{\theta} = .1033 \qquad r_5 = \frac{5\hat{\theta}}{1 + 5\hat{\theta}} = .6397$$

Summary. One additional point about the model underlying reliability should be made explicit and emphasized. Just as in the basic model for correlated observations discussed in Sec. 4.4, one assumes that the π_i's are constant under all measurements made. This assumption implies that the correlation between the judges (tests, items) be constant. In particular the analysis-of-variance model cannot be used to estimate reliability when the true score changes irregularly from one measurement to the next, as, for example, when practice effects are present in some non-systematic manner. If, however, changes in the underlying true score are systematic and constant for all subjects, then adjustments for this change may be made by eliminating variation due to change from the within-subject variation.

4.6 Tests for Trend

The methods of curve fitting discussed in Sec. 3.7 may be adapted for use in connection with experiments involving repeated measures. These methods will be illustrated by means of a learning experiment.

A sample of 10 subjects is used in the experiment to be described. Each subject is given 28 trials in which to learn a discrimination problem. The trials are grouped into blocks of 4 trials each. The block is considered as the observational unit. Hence the data are analyzed as if there were 7 observations on each subject, an observation being the outcome of a series of 4 trials. The degrees of freedom for the between-subject variation are 9; the degrees of freedom for the within-subject variation are 60, 6 degrees of freedom for each of the 10 subjects.

Table 4.6-1 Analysis of Learning Data

Source of variation	SS	df	MS	F
Between subjects	90.00	9	10.00	
Within subjects	155.24	60		
Blocks	103.94	6	17.32	18.23
Residual	51.30	54	.95	

A summary of the overall analysis appears in Table 4.6-1. For $\alpha = .01$ the critical value for a test of the hypothesis that there is no difference in the block means is $F_{.99}(6,54) = 3.28$. The data indicate that there are significant differences between the block means. A graph of the block totals is given in Fig. 4.6-1. Inspection of this figure indicates that a

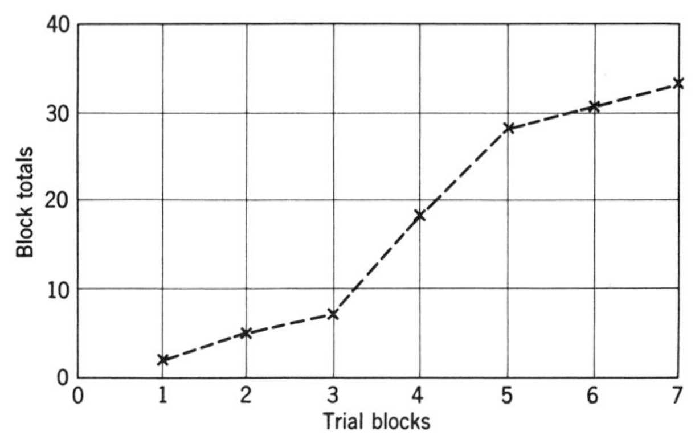

Figure 4.6-1

straight line would provide a good fit to the points, but there is also some evidence to indicate that an S-shaped (cubic) curve would provide a better fit. The nature of the subject matter also suggests that a cubic curve would be more appropriate for these data. Hence a point to be investigated is whether or not a cubic curve provides a better fit than a straight line (within the range of blocks included in this study).

Toward this end, the mean squares corresponding to the linear and cubic trends are computed in Table 4.6-2. The block totals for these

Table 4.6-2 Tests for Trends

	Blocks:	1	2	3	4	5	6	7	Σc^2	C	MS
	Block totals:	2	5	7	18	28	31	33			
Linear		−3	−2	−1	0	1	2	3	28	166	98.41**
Quadratic		5	0	−3	−4	−3	0	5	84	−2	.00
Cubic		−1	1	1	0	−1	−1	1	6	−16	4.27*
											102.68

experimental data are given near the top of Table 4.6-2. Each of these totals is the sum over 10 observations. Since there are seven blocks, the coefficients corresponding to the linear, quadratic, and cubic trends are obtained from the set of coefficients for which $k = 7$ in Table C.10. The entries under the column headed Σc^2 represent the sums of the squares of the coefficients in the corresponding rows. The entries under the column headed C represent the numerical value of the comparisons. For example, the linear comparison is

$$(-3)(2) + (-2)(5) + (-1)(7) + (0)(18) + (1)(28) + (2)(31) + (3)(33) = 166.$$

The mean square corresponding to the linear comparison is

$$\text{MS}_{\text{lin}} = \frac{C^2_{\text{lin}}}{n\Sigma c^2} = \frac{(166)^2}{10(28)} = 98.41.$$

A test on the significance of the linear trend is given by the F ratio

$$F = \frac{\text{MS}_{\text{lin}}}{\text{MS}_{\text{res}}} = \frac{98.41}{.95} = 103.59.$$

The sampling distribution of this statistic (assuming no linear trend) may be approximated by an F distribution having degrees of freedom (1,54). The linear trend is significant beyond the .01 level.

For these data the numerical value of the quadratic comparison is zero to two decimal places. The mean square corresponding to the cubic comparison is

$$\text{MS}_{\text{cubic}} = \frac{C^2_{\text{cubic}}}{n\Sigma c^2} = \frac{(-16)^2}{10(6)} = 4.27.$$

A test on whether or not the cubic trend adds significant predictability beyond that already given by the linear and quadratic trends employs the statistic

$$F = \frac{\text{MS}_{\text{cubic}}}{\text{MS}_{\text{res}}} = \frac{4.27}{.95} = 4.49.$$

The sampling distribution of this statistic (assuming no cubic trend) may be approximated by an F distribution having (1,54) degrees of freedom. The critical value when $\alpha = .05$ is 4.03. Hence the data indicate that, within the range of blocks included in the experiment, the cubic comparison does add significant predictability to that given by the linear trend ($\alpha = .05$).

The total variation between blocks (as given in Table 4.6-1) is 103.94. Of this total, 102.68 is accounted for by the linear and cubic trends. The remaining between-block variation appears negligible relative to experimental error. The between-block variation due to higher-order trend components is

$$\begin{aligned} SS_{\text{higher order}} &= SS_{\text{blocks}} - SS_{\text{lin}} - SS_{\text{quad}} - SS_{\text{cubic}} \\ &= 1.23. \end{aligned}$$

The corresponding mean square is

$$MS_{\text{higher order}} = \frac{1.23}{3} = .41.$$

The F ratio in the test for trend components higher than the third is

$$F = \frac{MS_{\text{higher order}}}{MS_{\text{res}}} = \frac{.41}{.95} < 1.$$

Since this ratio is less than unity, the data indicate that no component higher than the third is relevant.

An alternative, more widely used testing procedure for trends is possible. In this procedure the error term in the test for linear trend is obtained from

$$\begin{aligned} SS_{\text{dev lin}} &= SS_{\text{res}} + (SS_{\text{blocks}} - SS_{\text{lin}}) \\ &= 51.30 + (103.94 - 98.41) = 56.83. \end{aligned}$$

The degrees of freedom for $SS_{\text{dev lin}}$ in this type of design are $n(k - 1) - 1$. For the case under consideration, the degrees of freedom are $60 - 1 = 59$. The mean square for deviations from linearity is

$$MS_{\text{dev lin}} = \frac{SS_{\text{dev lin}}}{59} = .96.$$

Under this testing procedure the statistic used in the test for linear trend is

$$F = \frac{MS_{\text{lin}}}{MS_{\text{dev lin}}} = \frac{98.41}{.96} = 102.61.$$

The sampling distribution for this statistic may be approximated by an F distribution having degrees of freedom (1,59). This test procedure has a slight negative bias; i.e., if $F_{.95}(1,59)$ is used as a critical value for a test having $\alpha = .05$, this critical value will tend to be slightly larger than the exact critical value if the trend is actually different from linear.

Table 4.6-3 Summary of Alternative Tests for Trend

Source of variation	SS	df	MS	F
Linear trend	98.41	1	98.41	102.51**
Dev from lin	56.83	59	.96	
Quadratic trend	.00	1		
Dev from quad	56.83	58		
Cubic trend	4.27	1	4.27	4.64*
Dev from cubic	52.56	57	.92	

Under the alternative test procedure, the test for cubic trend uses a denominator obtained from

$$\begin{aligned} SS_{\text{dev cubic}} &= SS_{\text{res}} + (SS_{\text{blocks}} - SS_{\text{lin}} - SS_{\text{quad}} - SS_{\text{cubic}}) \\ &= 51.30 + (103.94 - 98.41 - .00 - 4.27) = 52.56. \end{aligned}$$

The degrees of freedom for this source of variation are $n(k - 1) - 3$, which in this case is $60 - 3 = 57$. The mean square corresponding to this source of variation is

$$MS_{\text{dev cubic}} = \frac{SS_{\text{dev cubic}}}{57} = \frac{52.56}{57} = .92.$$

The test for cubic trend uses the statistic

$$F = \frac{MS_{\text{cubic}}}{MS_{\text{dev cubic}}} = \frac{4.27}{.92} = 4.64.$$

The sampling distribution of this statistic may be approximated by an F distribution having (1,57) degrees of freedom. The critical value for this test ($\alpha = .05$) is 4.01. Hence the data contradict the hypothesis that the cubic trend adds no predictability to the linear and quadratic trends.

From some points of view this alternative test procedure is to be preferred to that which has a constant denominator. Under this procedure, the denominator of F ratios tends to err on the side of being too large. When the degrees of freedom for MS_{res} are large (say over 30), the first approach presented differs only slightly from the alternative approach.

In those cases where the experimental data provide a direct estimate of σ_ε^2, tests on trend may be viewed as a special case of testing a priori hypotheses. Tests on trend of the kind discussed here focus on individual comparisons that go to make up the overall main effect.

4.7 Analysis of Variance for Ranked Data

Suppose that an experimenter is interested in determining whether or not there is any difference between various methods of packaging a product. In an experiment, subjects are asked to rank the methods in order of preference. A numerical example of this kind of experiment is given in Table 4.7-1. Person 1, for example, assigned rank 1 to method 3, rank 2 to method 2, rank 3 to method 1, and rank 4 to method 4.

Table 4.7-1 Numerical Example

Person	Method 1	Method 2	Method 3	Method 4	Total
1	3	2	1	4	10
2	4	3	1	2	10
3	2	4	1	3	10
4	1	3	2	4	10
5	2	3	1	4	10
6	1	4	2	3	10
7	2	3	1	4	10
$n = 8$	1	4	2	3	10
Total	16	26	11	27	80

$$(1) = \frac{G^2}{kn} = 200.00 \qquad (2) = \Sigma\Sigma X^2 = 240 \qquad (3) = \frac{\Sigma T_j^2}{n} = \frac{1782}{8} = 222.75$$

$$(4) = \frac{\Sigma P_i^2}{k} = \frac{800}{4} = 200.00$$

$$SS_{\text{methods}} = (3) - (1) = 22.75$$
$$SS_{\text{res}} = (2) - (3) - (4) + (1) = 17.25$$
$$SS_{\text{w. people}} = (2) - (4) = 40.00$$

The computational formulas in Sec. 4.2 may be used to obtain sums of squares required for the analysis of variance. $SS_{\text{b. people}}$ will always be zero. Upon replicating the experiment with random samples of subjects, SS_{total} will remain constant provided that no tied ranks are permitted. Rather than an F statistic, the chi-square statistic

$$(1) \qquad \chi^2_{\text{ranks}} = \frac{SS_{\text{methods}}}{(SS_{\text{methods}} + SS_{\text{res}})/n(k-1)} = \frac{SS_{\text{methods}}}{MS_{\text{w. people}}} = \frac{n(k-1)SS_{\text{methods}}}{SS_{\text{w. people}}}$$

is used to test the hypothesis of no difference in mean rank for the methods. The higher the agreement between people in ranking the methods, the larger $SS_{methods}$ will be. For the data in Table 4.7-1,

$$\chi^2_{ranks} = \frac{8(3)(22.75)}{40.00} = 13.65.$$

The critical value of this statistic for a .01-level test is

$$\chi^2_{.99}(k - 1) = \chi^2_{.99}(3) = 11.3.$$

Since the observed χ^2 exceeds the critical value, the data contradict the hypothesis of no difference between the mean ranks for the different methods of packaging. Inspection of Table 4.7-1 indicates that methods 1 and 3 have the smaller means and methods 2 and 4 the larger means.

When no tied ranks are permitted,

$$SS_{w.\ people} = \frac{nk(k^2 - 1)}{12},$$

and the χ^2 statistic becomes

$$\chi^2_{ranks} = \frac{12}{nk(k + 1)}(\Sigma T_j^2) - 3n(k + 1). \tag{2}$$

The expression for χ^2 in (2) is algebraically equivalent to (1) when no ties are permitted; (1) may be used whether or not ties are present. For the data in Table 4.7-1, expression (2) gives

$$\chi^2_{ranks} = \frac{12}{8(4)(5)}(1782) - 3(8)(5)$$
$$= 13.65.$$

Use of this statistic in testing the hypothesis on the mean ranks defines what is called the Friedman test.

The rationale underlying the use of the chi-square distribution rather than an F distribution for making this type of test is (nonrigorously) as follows: If no ties are permitted, upon replication of the experiment, $MS_{w.\ subj}$ is not subject to sampling variation. Hence $MS_{w.\ subj}$ may be regarded as a parameter rather than as a statistic. Thus

$$\chi^2_{ranks} = \frac{SS_{methods}}{MS_{w.\ subj}} = \frac{SS_{methods}}{\sigma^2_{w.\ subj}}.$$

Under the hypothesis of no differences among the methods

$$\chi^2_{ranks} \quad \text{is distributed as} \quad \chi^2(k - 1).$$

An index of the extent to which people agree in their preferences is given by the coefficient of concordance, which is defined as

$$W = \frac{SS_{methods}}{SS_{total}}. \tag{3}$$

This coefficient is related to the average intercorrelation between the rankings assigned by the people; this relationship is

$$(4) \qquad \bar{r} = \frac{nW - 1}{n - 1}.$$

The test in (1) may be regarded as a test of the hypothesis that the coefficient of concordance in the population of people is zero. For the data in Table 4.7-1,

$$W = \frac{22.75}{40.00} = .569.$$

The average intercorrelation between the people is

$$\bar{r} = \frac{8(.569) - 1}{8 - 1} = .507.$$

The index W corresponds to a correlation ratio, whereas $\bar{r}$ corresponds to a product-moment correlation (or equivalently a rank-difference correlation).

The average intercorrelation between the people may also be computed through use of what is equivalent to the model in (11) in Sec. 4.5. In this case

$$\hat{\theta} = \frac{\hat{\sigma}^2_{\text{methods}}}{\hat{\sigma}^2_{\text{res}}} = \frac{(\text{MS}_{\text{methods}} - \text{MS}_{\text{res}})/n}{\text{MS}_{\text{res}}}$$

$$= \frac{[(22.75/3) - (17.25/21)]/8}{17.25/21} = 1.0290.$$

Thus

$$\bar{r} = \frac{\hat{\theta}}{1 + \hat{\theta}} = \frac{1.0290}{2.0290} = .507.$$

Alternatively,

$$\bar{r} = \frac{\hat{\sigma}^2_{\text{methods}}}{\hat{\sigma}^2_{\text{res}} + \hat{\sigma}^2_{\text{methods}}} = \frac{\hat{\theta}}{1 + \hat{\theta}},$$

which is a biased estimator of the intraclass correlation.

4.8 Dichotomous Data

Observed data may in some cases be classified into one of two classes; for example, a characteristic may be present or absent, a response is either yes or no, a drug is either fatal or it is not. Such data are said to be dichotomous. One of the two dichotomies may conveniently be designated by a 0, the other by a 1. Which category is assigned the zero is arbitrary.

Consider an experiment designed to study the effects of an advertising campaign upon the attitude of a potential population of buyers toward a

Table 4.8-1 Numerical Example

Subject	Time 1	Time 2	Time 3	Time 4	Time 5	Total
1	0	0	0	0	0	$0 = P_1$
2	0	0	1	1	0	$2 = P_2$
3	0	0	1	1	1	3
4	0	1	1	1	1	4
5	0	0	0	0	1	1
6	0	1	0	1	1	3
7	0	0	1	1	1	3
8	1	0	0	1	1	3
9	1	1	1	1	1	5
10	1	1	1	1	1	5
Total	3	4	6	8	8	29

$$(1) = \frac{G^2}{kn} = \frac{29^2}{50} = 16.82 \qquad (2) = \Sigma\Sigma X^2 = 29 \qquad (3) = \frac{\Sigma T_j^2}{n} = \frac{189}{10} = 18.90$$

$$(4) = \frac{\Sigma P_i^2}{k} = \frac{107}{5} = 21.40$$

			df	
$SS_{\text{b. people}}$	$= (4) - (1)$	$= 4.58$		
$SS_{\text{w. people}}$	$= (2) - (4)$	$= 7.60$		
SS_{time}	$= (3) - (1)$	$= 2.08$	4	$MS_{\text{time}} = .520$
SS_{res}	$= (2) - (3) - (4) + (1)$	$= 5.52$	36	$MS_{\text{res}} = .153$
SS_{total}	$= (2) - (1)$	$= 12.18$		

product. The data may take the form given in Table 4.8-1. In this table a 0 is used to indicate an unfavorable attitude, and a 1 is used to indicate a favorable attitude. Suppose that a random sample of 10 subjects is selected for the study. Each subject is interviewed at the end of five different time periods. For example, subject 1 was not favorable at any time; subject 5 was not favorable on the first four interviews but was favorable on the fifth interview.

To test the hypothesis of no change in the percentage of favorable replies over the time periods, the statistic

$$Q = \frac{n(k-1)SS_{\text{time}}}{SS_{\text{w. people}}} \tag{1}$$

may be used. The Q statistic has the same form as χ^2_{ranks}, which was used in Sec. 4.7. Under the hypothesis of no change in the percentage of favorable replies, Cochran (1950) has shown that the sampling distribution of the Q statistic is approximated by a chi-square distribution with $k - 1$ degrees of freedom, when n is reasonably large. For an α-level test the critical value for the Q statistic is $\chi^2_{1-\alpha}(k-1)$.

For the data in Table 4.8-1,

$$Q = \frac{10(5-1)(2.08)}{7.60} = 10.95.$$

The critical value for a .05-level test is $\chi^2_{.95}(5 - 1) = 9.5$. Hence the experimental data contradict the hypothesis of no change in the percentage of favorable replies. Examination of the data indicates a systematic increase in the percentage of favorable replies. Cochran (1950) has indicated that the F statistic computed by treating the data as if the measurements were normally distributed variables will yield probability statements which are relatively close to those obtained by use of the Q statistic. For the data in Table 4.8-1,

$$F = \frac{MS_{\text{time}}}{MS_{\text{res}}} = \frac{.520}{.153} = 3.40.$$

For a .05-level test the critical value for the F statistic is $F_{.95}(4,36) = 2.63$. Thus use of the F statistic also leads to the rejection of the hypothesis of no change in the percentage of favorable replies.

4.9 Hotelling's T^2

In Sec. 4.4 it was indicated that tests in the analysis of variance were exact only if the population covariance matrix Σ had a specified pattern. Exact procedures for testing the hypothesis that $\sigma_\tau^2 = 0$, when Σ is arbitrary, will be described in this section. Here the basic assumption is that the variables have a joint multivariate normal distribution with arbitrary covariance matrix Σ. If the pattern assumption on Σ is appropriate, the test to be described in this section will be less powerful than the corresponding test in the analysis of variance.

When the *sample* covariance matrix is replaced by a matrix which has compound symmetry, the T^2 statistic to be computed in this section will actually be equal to $(k - 1)F_a$, where F_a is the F ratio obtained in the analysis of variance. For small deviations from compound symmetry, T^2 will be quite close to $(k - 1)F_a$.

In what follows it is assumed that k, the number of treatments, is four. Generalization to k equal to any number is direct. Throughout it will be assumed that $n > k$. Let

$$\underset{4,4}{S} = \text{covariance matrix for observations,}$$

$$\underline{\bar{T}}' = [\bar{T}_1 \quad \bar{T}_2 \quad \bar{T}_3 \quad \bar{T}_4] = \text{vector of treatment means,}$$

$$\underset{3,4}{C} = \begin{bmatrix} 1 & 0 & 0 & -1 \\ 0 & 1 & 0 & -1 \\ 0 & 0 & 1 & -1 \end{bmatrix},$$

$$\underset{3,3}{S_y} = CSC',$$

$$\underset{3,1}{\underline{\bar{y}}} = C\underline{\bar{T}}.$$

There are many possible choices for the matrix C. All that is required is that the matix C have rank $k - 1$, and, for each row, the sum must be zero. For example, an alternative choice for C might be

$$C^{(1)} = \begin{bmatrix} 1 & -1 & 0 & 0 \\ 0 & 1 & -1 & 0 \\ 0 & 0 & 1 & -1 \end{bmatrix}.$$

It will be found that the T^2 statistic will be invariant for all choices of C.

One notes that

$$\underline{\bar{y}} = C\underline{\bar{T}} = \begin{bmatrix} \bar{T}_1 - \bar{T}_4 \\ \bar{T}_2 - \bar{T}_4 \\ \bar{T}_3 - \bar{T}_4 \end{bmatrix}.$$

Under the hypothesis that $\tau_1 = \tau_2 = \tau_3 = \tau_4$,

$$\text{E}(\underline{\bar{y}}) = \begin{bmatrix} 0 \\ 0 \\ 0 \end{bmatrix}.$$

One also notes that

$$\underline{\bar{y}}^{(1)} = C^{(1)}\underline{\bar{T}} = \begin{bmatrix} \bar{T}_1 - \bar{T}_2 \\ \bar{T}_2 - \bar{T}_3 \\ \bar{T}_3 - \bar{T}_4 \end{bmatrix},$$

and under the hypothesis specified above,

$$\text{E}(\underline{\bar{y}}^{(1)}) = \begin{bmatrix} 0 \\ 0 \\ 0 \end{bmatrix}.$$

The covariance matrix associated with $\underline{\bar{y}}$ is $(1/n)CSC' = (1/n)S_y$.

Under the hypothesis indicated, the statistic

$$T^2 = n\underline{\bar{y}}'S_y^{-1}\underline{\bar{y}}$$

is distributed as Hotelling's (central) T^2. Equivalently, under the hypothesis that $\sigma_\tau^2 = 0$,

$$\frac{n - k + 1}{(n - 1)(k - 1)} T^2 \qquad \text{is distributed as} \qquad F(k - 1, n - k + 1).$$

One should note that in the analysis of variance the F distribution involved is $F[k - 1, (n - 1)(k - 1)]$.

What has been done here is to transform the hypothesis

$$\tau_1 = \tau_2 = \cdots = \tau_k$$

into the equivalent hypothesis

$$\underline{\mu} = \begin{bmatrix} \tau_1 - \tau_k \\ \tau_2 - \tau_k \\ \cdot \\ \cdot \\ \cdot \\ \tau_{k-1} - \tau_k \end{bmatrix} = \begin{bmatrix} 0 \\ 0 \\ \cdot \\ \cdot \\ \cdot \\ 0 \end{bmatrix}.$$

Given a $k - 1$ multivariate normal population, the T^2 distribution represents the sampling distribution for the vector of sample means $\underline{\bar{y}}$ when $\underline{\mu} = \underline{0}$ in the population. The covariance matrix Σ in this population is estimated by S_y. Percentile points on the T^2 distribution can be obtained from the F distribution.

The computation of Hotelling's T^2 for the numerical data in Table 4.3-1 is illustrated in Table 4.9-1. In part i, the components of the vector $\underline{\bar{T}}'$ are those obtained from Table 4.9-1. S_y is obtained from the covariance matrix in Table 4.3-4. The T^2 statistic is computed in part ii. Converted to an F statistic, $F = 28.39$. The sampling distribution of this F statistic has $k - 1$ and $n - k + 1$ degrees of freedom, or three and two degrees of freedom. With only two degrees of freedom for the denominator, the resulting test will have very low power. Whenever k is large relative to n,

Table 4.9-1 Computation of Hotelling's T^2
$(n = 5, k = 4)$

(i)

$$\bar{T}' = [26.40 \quad 25.60 \quad 15.60 \quad 32.00]$$

$$\underline{\bar{y}}' = \underline{\bar{T}}'C' = [-5.60 \quad -6.40 \quad -16.40]$$

$$S_y = CSC'$$

$$= \begin{bmatrix} 1 & 0 & 0 & -1 \\ 0 & 1 & 0 & -1 \\ 0 & 0 & 1 & -1 \end{bmatrix} \begin{bmatrix} 76.80 & 53.20 & 29.20 & 69.00 \\ 53.20 & 42.80 & 15.80 & 47.00 \\ 29.20 & 15.80 & 14.80 & 27.00 \\ 69.00 & 47.00 & 27.00 & 64.00 \end{bmatrix} \begin{bmatrix} 1 & 0 & 0 \\ 0 & 1 & 0 \\ 0 & 0 & 1 \\ -1 & -1 & -1 \end{bmatrix}$$

$$= \begin{bmatrix} 2.80 & 1.20 & -2.80 \\ 1.20 & 13.00 & 5.80 \\ -2.80 & 5.80 & 24.80 \end{bmatrix}$$

$$S_y^{-1} = \begin{bmatrix} .4569 & -.07280 & .06861 \\ -.07280 & .09749 & -.03102 \\ .06861 & -.03102 & .05532 \end{bmatrix}$$

(ii)

$$T^2 = n\underline{\bar{y}}'S_y^{-1}\underline{\bar{y}} = 170.36$$

$$F = \frac{n - k + 1}{(n - 1)(k - 1)} T^2 = 28.39 \qquad F_{.95}(3,2) = 19.2 \qquad F_{.99}(3,2) = 99.2$$

the power of a test involving the T^2 statistic will tend to be low. The resulting T^2 statistic exceeds the critical value for a test having level of significance .05. (The analysis-of-variance test exceeded the .01 level, but the latter test had a larger number of degrees of freedom in the denominator—resulting from a pooling under the assumption of compound symmetry.) The model under which the test using Hotelling's T^2 statistic is used is a more general one in the sense that it involves fewer assumptions about symmetry in the underlying parameters.

It will be instructive to compute Hotelling's T^2 statistic for the case in which the sample covariance matrix is replaced by a matrix which has compound symmetry. Toward this end, suppose the S matrix in Table 4.9-1 is replaced by the S^* matrix in Table 4.9-2. The latter has $\overline{\text{var}}$ on

Table 4.9-2 Computation of Hotelling's T^2—Assuming Σ Has Compound Symmetry

(i)

$$\bar{y} = [-5.60 \quad -6.40 \quad -16.40]$$

$$S^* = \begin{bmatrix} 49.60 & 40.20 & 40.20 & 40.20 \\ 40.20 & 49.60 & 40.20 & 40.20 \\ 40.20 & 40.20 & 49.60 & 40.20 \\ 40.20 & 40.20 & 40.20 & 49.60 \end{bmatrix}$$

$$S_y^* = (CSC')^{-1} = \begin{bmatrix} .07980 & -.02662 & -.02662 \\ -.02662 & .07980 & -.02662 \\ -.02662 & -.02662 & .07980 \end{bmatrix}$$

(ii)

$$T^2 = n\underline{\bar{y}}' S_y^{*-1} \underline{\bar{y}} = 74.29$$

$$F_a = \frac{1}{k-1} T^2 = \tfrac{1}{3}(74.29) = 24.76$$

the main diagonal and $\overline{\text{cov}}$ off the main diagonal. This matrix would be an estimate of Σ under the hypothesis of compound symmetry. The resulting T^2 statistic is shown in part ii. One notes that $T^2 = 74.29$. From Table 4.3-2 one has $F_a = 24.76$. Hence

$$T^2 = (k-1)F_a = 3(24.76) = 74.28.$$

Under the hypothesis of compound symmetry,

$$\frac{1}{k-1} T^2 \qquad \text{is distributed as} \qquad F[k-1, (n-1)(k-1)].$$

A more detailed discussion of the use of the T^2 statistic will be found in T. W. Anderson (1958, pp. 107–112) and Morrison (1967, pp. 133–141).

5

DESIGN AND ANALYSIS OF FACTORIAL EXPERIMENTS

5.1 General Purpose

Factorial experiments permit the experimenter to evaluate the combined effect of two or more experimental variables when used simultaneously. Information obtained from factorial experiments is more complete than that obtained from a series of single-factor experiments, in the sense that factorial experiments permit the evaluation of *interaction* effects. An interaction effect is an effect attributable to the combination of variables above and beyond that which can be predicted from the variables considered singly.

For example, many of the properties of the chemical substance H_2O (water) cannot be predicted from the properties of oxygen and the properties of hydrogen studied in isolation. Most of the properties of water are attributable to the effect of the interaction between oxygen and hydrogen. The compound formed by this interaction has properties which are not given by simply adding the properties of oxygen to the properties of hydrogen.

At the end of a factorial experiment, the experimenter has information which permits him to make decisions which have a broad range of applicability. In addition to information about how the experimental variables operate in relative isolation, the experimenter can predict what will happen when two or more variables are used in combination. Apart from the information about interactions, the estimates of the effects of the individual variables is, in a sense, a more practically useful one; these estimates are obtained by averaging over a relatively broad range of other relevant experimental variables. By contrast, in a single-factor experiment some relevant experimental variables may be held constant, while others may be randomized. In the case of a factorial experiment, the population to which inferences can be made is more inclusive than the corresponding population for a single-factor experiment.

In working with factorial experiments in the behavioral science area, a sharp distinction must be drawn between experiments involving repeated measures on the same elements and those which do not involve repeated measures. The material in this chapter will be concerned primarily with experiments which do not involve repeated measures. However, many of the basic principles to be developed in this chapter will be applicable, with only slight modification, to the case in which there are repeated measures.

The term factor will be used in a broad sense. For some purposes a distinction will be made between treatment and classification factors. The latter group the experimental units into classes which are homogeneous with respect to what is being classified. In contrast, treatment factors define the experimental conditions applied to an experimental unit. The administration of the treatment factors is under the direct control of the experimenter, whereas classification factors are not, in a sense. The effects of the treatment factors are of primary interest to the experimenter, whereas classification factors are included in an experiment to reduce experimental error and clarify interpretation of the effects of the treatment factors.

The design of factorial experiments is concerned with answering the following questions:

1. What factors should be included?
2. How many levels of each factor should be included?
3. How should the levels of the factors be spaced?
4. How should the experimental units be selected?
5. How many experimental units should be selected for each treatment combination?
6. What steps should be taken to control experimental error?
7. What criterion measures should be used to evaluate the effects of the treatment factors?
8. Can the effects of primary interest be estimated adequately from the experimental data that will be obtained?

The answers to these questions will be considered in some detail in this chapter and the chapters that follow. An excellent and readable overview of the planning of factorial experiments will be found in Cox (1958, chap. 7).

The estimation and hypothesis-testing problems in experimental design work may be approached in two ways. Under one approach, a hypothetical population of elements is postulated. If the treatments were administered to all elements of the population, the resulting population could be described in terms of distributions having specified forms with specified sets of parameters. The experiment consists of drawing a sample from this hypothetical population. The problem is to make inferences about the population from the experiment. In this context, a given experiment may be considered as a sample from a potential population of replications of the experiment.

On the other hand, one may approach the same problems in terms of a (linear) model for a basic observation. Associated with this model are one or more random variables and a set of distribution assumptions on the random variables. In the latter case one is concerned with estimating and testing hypotheses about the parameters in the model. In most cases the two approaches lead to similar results. However, in the approach postulating a hypothetical population of elements, certain constraints appear as a natural part of the system. Both approaches are used in the discussion that follow in this chapter.

5.2 Terminology and Notation

The term *factor* will be used interchangeably with the terms treatment and experimental variable. More specifically, a factor is a series of related treatments or related classifications. The related treatments making up a factor constitute the *levels* of that factor. For example, a factor *color* may consist of three levels: red, green, and yellow. A factor *size* may consist of two levels: small and large. A factor *dosage* may consist of four levels: 1 cc, 3 cc, 5 cc, and 7 cc. The number of levels within a factor is determined largely by the thoroughness with which an experimenter desires to investigate the factor. Alternatively, the levels of a factor are determined by the kind of inferences the experimenter desires to make upon conclusion of the experiment. The levels of a factor may be quantitative variations in an essentially quantitative variable, or they may be qualitatively different categories within an essentially qualitative variable. Basically a factor is a qualitative variable; in special cases it becomes a quantitative variable.

The *dimensions* of a factorial experiment are indicated by the number of factors and the number of levels of each factor. For example, a factorial experiment in which there are two factors, one having three levels and the other having four levels, is called a 3×4 (read "three by four") factorial experiment. In a $2 \times 3 \times 5$ factorial experiment, there are three factors, having respective levels of two, three, and five. (Many different designs may be constructed for a given factorial experiment.) The treatment combinations in a 2×3 factorial experiment may be represented schematically as follows:

		Levels of factor B		
		b_1	b_2	b_3
Levels of factor A	a_1	ab_{11}	ab_{12}	ab_{13}
	a_2	ab_{21}	ab_{22}	ab_{23}

In this schema, a_1 and a_2 designate the levels of factor A; b_1, b_2, and b_3 designate the levels of factor B. In a 2×3 factorial experiment six possible combinations of treatments may be formed. Level a_1 may be used in combination with each of the three levels of factor B; level a_2 may also be used in combination with each of the three levels of factor B. The resulting treatment combinations are labeled in the cells of the schema. For example, the symbol ab_{12} represents the experimental condition resulting when factor A is at level a_1 and factor B is at level b_2.

For the case of a $p \times q$ factorial experiment, pq different treatment combinations are possible. In a $p \times q \times r$ factorial experiment there are pqr treatment combinations. As the number of factors increases, or as the number of levels within a factor increases, the number of treatment combinations in a factorial experiment increases quite rapidly. For example, a $5 \times 5 \times 5$ factorial experiment has 125 treatment combinations.

In an experiment, the elements observed under each of the treatment combinations will generally be a random sample from some specified population. This specified population will, in most cases of interest, contain a potentially infinite number of elements. If n elements are to be observed under each treatment combination in a $p \times q$ factorial experiment, a random sample of npq elements from the population is required (assuming no repeated measurements on the same elements). The npq elements are then subdivided at random into pq subsamples of size n each. These subsamples are then assigned at random to the treatment combinations.

The potential (or population) levels of factor A will be designated by the symbols $a_1, a_2, \ldots, a_I, \ldots, a_P$. The number of such potential levels, P, may be quite large. The experimenter may group the P potential levels into p levels ($p < P$) by either combining adjoining levels or deliberately selecting what are considered to be representative levels. For example, if factor A represents the dimension of age, the experimenter may choose to group the levels into 1-year intervals; alternatively, the experimenter may choose to group the levels into 2-year intervals. On the other hand, if factor A represents a dosage dimension, the experimenter may deliberately select a series of representative dosages; i.e., in terms of previous research the levels selected may be representative of low, middle, and high dosages.

When p, the number of levels of factor A included in the experiment, is equal to P, then factor A is called a *fixed* factor. When the selection of the p levels from the potential P levels is determined by some systematic, nonrandom procedure, then factor A is also considered a fixed factor. In this latter case, the selection procedure reduces the potential P levels to p *effective* levels. Under this type of selection procedure, the effective, potential number of levels of factor A in the population may be designated $P_{\text{effective}}$, and $P_{\text{effective}} = p$.

In contrast to this systematic selection procedure, if the p levels of factor A included in the experiment represent a random sample from the potential P levels, then factor A is considered to be a *random* factor. For example, in an experiment designed to test the effectiveness of various drugs upon categories of patients, the factor A may be the hospitals in which the patients are located. Potentially the number of different hospitals may be quite large. If a random sample of p of the P potential hospitals are included in the experiment, then factor A is a random factor. (If, further, the sampling within each of the hospitals selected is proportional to the number of patients within the hospital, then conclusions drawn from this kind of experiment will be relevant to the domain of all patients and not just to the domain of all hospitals.) In most practical situations in which random factors are encountered, p is quite small relative to P, and the ratio p/P is quite close to zero.

Similar definitions apply to factor B. Let the potential level of factor B be designated $b_1, b_2, \ldots, b_J, \ldots, b_Q$. Of these Q potential levels, let the number of levels actually in an experiment be q. If $q = Q$, or if the effective number of levels of factor B is reduced from Q to q by some systematic, nonrandom procedure, then factor B is considered fixed. (The reduction from Q potential levels to q effective levels is a function of the experimental design.) The actual levels of factor B included in an experiment will be designated by the notation $b_1, b_2, \ldots, b_j, \ldots, b_q$. If the q levels in an experiment are a random sample from the Q potential levels, then factor B is considered a random factor. In most practical cases in which factor B is a random factor, Q will be quite large relative to q, and the ratio q/Q will be close to zero.

The ratio of the number of levels of a factor in an experiment to the potential number of levels in the population is called the *sampling fraction* for a factor. In terms of this sampling fraction, the definitions of fixed and random factors may be summarized as follows:

Sampling fraction	Factor
$p/P = 1$ or $p/P_{\text{effective}} = 1$	A is a fixed factor
$p/P \doteq 0$	A is a random factor
$q/Q = 1$ or $q/Q_{\text{effective}} = 1$	B is a fixed factor
$q/Q \doteq 0$	B is a random factor

Cases in which the sampling fraction assumes a value between 0 and 1 do occur in practice. However, cases in which the sampling fraction is either 1 or very close to 0 are encountered more frequently.

The symbol a_I will be used to refer to an arbitrary level of factor A when the frame of reference is the potential P levels of factor A. The symbol a_i will be used to refer to an arbitrary level of factor A when the frame of reference is the p levels actually included in the experiment. A similar distinction will be made between the symbols b_J and b_j. The first symbol has as its frame of reference the potential Q levels of factor B, whereas the second symbol has as its frame of reference the q levels of factor B in the actual experiment. In those cases in which factor A is a random factor, the symbol a_1, when the frame of reference is the potential population levels, refers to a specified population level. However, when the frame of reference is the experiment, the symbol a_1 refers to a specified level in the experiment. Thus the symbol a_1 may refer to different levels of the same factor, depending upon the frame of reference. The context in which the notation is used will clarify the ambiguity of the symbol considered in isolation.

To illustrate additional notation, assume that all the potential elements are included in all the potential cells of a factorial experiment. (Generally only a random sample of the potential elements are included in a cell of a factorial experiment.) Assume further that, after the treatment combinations have been administered, each of the elements is measured (observed) on the characteristic being studied (the *criterion* or dependent variable). The mean of the observations made under each treatment combination will be denoted by the following notation:

	b_1	b_2	$\cdots$	b_J	$\cdots$	b_Q
a_1	μ_{11}	μ_{12}	$\cdots$	μ_{1J}	$\cdots$	μ_{1Q}
a_2	μ_{21}	μ_{22}	$\cdots$	μ_{2J}	$\cdots$	μ_{2Q}
$\vdots$	$\vdots$	$\vdots$		$\vdots$		$\vdots$
a_I	μ_{I1}	μ_{I2}	$\cdots$	μ_{IJ}	$\cdots$	μ_{IQ}
$\vdots$	$\vdots$	$\vdots$		$\vdots$		$\vdots$
a_P	μ_{P1}	μ_{P2}	$\cdots$	μ_{PJ}	$\cdots$	μ_{PQ}

In this notation, μ_{IJ} denotes the mean on the criterion for the potential population of elements under treatment combination ab_{IJ}. Equivalently, μ_{IJ} represents the population mean for the dependent variable in cell ab_{IJ}. It will be assumed that the potential number of elements in each of the cells is the same for all cells.

The average of the cell means appearing in row I is

$$\mu_{I.} = \frac{\sum_J \mu_{IJ}}{Q}.$$

In words, $\mu_{I.}$ is the mean of the dependent variable averaged over all potential treatment combinations in which factor A is at level a_I. Similarly, the mean of the dependent variable averaged over all potential treatment combinations in which factor B is at level b_J is

$$\mu_{.J} = \frac{\sum_I \mu_{IJ}}{P}.$$

The grand mean on the dependent variable is

$$\mu_{..} = \frac{\sum_I \sum_J \mu_{IJ}}{PQ} = \frac{\sum_I \mu_{I.}}{P} = \frac{\sum_J \mu_{.J}}{Q}.$$

The grand mean may also be defined as the mean of the dependent variable for all potential observations under all potential treatment combinations.

Notation for the statistics obtained from actual experimental data is as follows. (It is assumed that each cell mean is based upon an independent random sample of size n.)

	b_1	b_2	$\cdots$	b_j	$\cdots$	b_q	
a_1	$\overline{AB}_{11}$	$\overline{AB}_{12}$	$\cdots$	$\overline{AB}_{1j}$	$\cdots$	$\overline{AB}_{1q}$	$\bar{A}_1$
a_2	$\overline{AB}_{21}$	$\overline{AB}_{22}$	$\cdots$	$\overline{AB}_{2j}$	$\cdots$	$\overline{AB}_{2q}$	$\bar{A}_2$
$\vdots$	$\vdots$	$\vdots$		$\vdots$		$\vdots$	$\vdots$
a_i	$\overline{AB}_{i1}$	$\overline{AB}_{i2}$	$\cdots$	$\overline{AB}_{ij}$	$\cdots$	$\overline{AB}_{iq}$	$\bar{A}_i$
$\vdots$	$\vdots$	$\vdots$		$\vdots$		$\vdots$	$\vdots$
a_p	$\overline{AB}_{p1}$	$\overline{AB}_{p2}$	$\cdots$	$\overline{AB}_{pj}$	$\cdots$	$\overline{AB}_{pq}$	$\bar{A}_p$
	$\bar{B}_1$	$\bar{B}_2$	$\cdots$	$\bar{B}_j$	$\cdots$	$\bar{B}_q$	$\bar{G}$

The symbol $\overline{AB}_{ij}$ represents the mean of the measurements on the dependent variable for the n elements under treatment combination ab_{ij}. The average of all observations at level a_i is

$$\bar{A}_i = \frac{\sum_j \overline{AB}_{ij}}{q}.$$

Similarly, the average of all observations at level b_j is

$$\bar{B}_j = \frac{\sum_i \overline{AB}_{ij}}{p}.$$

The grand mean $\bar{G}$ is the mean of all means. Thus

$$\bar{G} = \frac{\sum_i \sum_j \overline{AB}_{ij}}{pq} = \frac{\sum_i \bar{A}_i}{p} = \frac{\sum_j \bar{B}_j}{q}.$$

In most designs for factorial experiments, these statistics are unbiased estimates of corresponding parameters. That is,

$$\mathrm{E}(\overline{AB}_{ij}) = \mu_{ij}, \qquad \mathrm{E}(\bar{A}_i) = \mu_{i.}, \qquad \text{and} \qquad \mathrm{E}(\bar{B}_j) = \mu_{.j}.$$

5.3 Main Effects

Main effects are defined in terms of parameters. Direct estimates of these parameters will, in most cases, be obtainable for corresponding statistics. The main effect of level a_I of factor A is by definition

$$\alpha_I = \mu_{I.} - \mu_{..}.$$

In words, the main effect for level a_I is the difference between the mean of all potential observations on the dependent variable at level a_I and the grand mean of all potential observations. The main effect for level a_I may be either positive or negative. The main effect for level $a_{I'}$, where I' designates a level of factor A different from I, is

$$\alpha_{I'} = \mu_{I'.} - \mu_{..}.$$

For most practical purposes, only the difference between two main effects will be needed. The *differential* main effect is defined to be

$$\alpha_I - \alpha_{I'} = \mu_{I.} - \mu_{I'.}.$$

In terms of what are called *estimable parametric functions* in the general linear model, differential main effects are estimable whereas the individual main effects are not estimable.

Analogous definitions hold for the main effects of the levels of factor B. Thus, the main effect for level b_J is

$$\beta_J = \mu_{.J} - \mu_{..}.$$

The differential main effect for levels b_J and $b_{J'}$ is

$$\beta_J - \beta_{J'} = \mu_{.J} - \mu_{.J'}.$$

A differential main effect measures the extent to which criterion means for two levels within the same factor differ. Thus, if all the means for the various levels of a factor are equal, all the differential main effects for the levels of that factor will be zero. It should be pointed out that equality of the population means does not imply equality of the sample main effects. Hence estimates of differential main effects may not be zero even though the population values are zero. However, when the population differential main effects are zero, their sample estimates will differ from zero by amounts which are functions of experimental error.

The variance of the main effects due to factor A is, by definition,

$$\sigma_\alpha^2 = \frac{\sum_I (\mu_{I.} - \mu_{..})^2}{P - 1} = \frac{\sum_I \alpha_I^2}{P - 1}.$$

An equivalent definition in terms of the differential main effects is

$$\sigma_\alpha^2 = \frac{\Sigma(\alpha_I - \alpha_{I'})^2}{P(P - 1)}, \qquad I < I'.$$

The symbol $I < I'$ indicates that the summation is over different pairs of α's; that is, $\alpha_1 - \alpha_2$ is not considered to be different from $\alpha_2 - \alpha_1$. The variance of the main effects of factor A measures the extent to which the criterion means for the various levels of factor A differ. Equivalently, the variance of the main effects may be regarded as an overall measure of the differential main effects for that factor. Thus, σ_α^2 will be small when all the differential main effects are small; σ_α^2 will be large when one or more of the differential main effects are large. When all the $\mu_{I.}$ are equal, σ_α^2 will be zero.

Analogous definitions hold for the variance of the main effects of factor B.

$$\sigma_\beta^2 = \frac{\sum_J (\mu_{.J} - \mu_{..})^2}{Q - 1} = \frac{\sum_J \beta_J^2}{Q - 1}.$$

In terms of the differential main effects,

$$\sigma_\beta^2 = \frac{\sum_J (\beta_J - \beta_{J'})^2}{Q(Q - 1)}, \qquad J < J'.$$

To illustrate the definitions of main effects and their variance, suppose the population means on the dependent (criterion) variable are those given in the following table. Here $P = 2$ and $Q = 3$.

	b_1	b_2	b_3	Mean
a_1	10	5	15	10
a_2	20	5	5	10
Mean	15	5	10	10

For the data in this table, $\mu_{1.} = 10$, $\mu_{2.} = 10$, and $\mu_{..} = 10$. Hence $\alpha_1 = 0$, and $\alpha_2 = 0$. It is also noted that $\mu_{.1} = 15$, $\mu_{.2} = 5$, $\mu_{.3} = 10$. Hence

$$\beta_1 = 15 - 10 = 5,$$

$$\beta_2 = 5 - 10 = -5,$$

$$\beta_3 = 10 - 10 = 0.$$

The differential main effects for factor B are

$$\beta_1 - \beta_2 = 10, \qquad \beta_1 - \beta_3 = 5, \qquad \beta_2 - \beta_3 = -5.$$

[In the general case there will be $Q(Q - 1)/2$ distinct differential main effects for factor B.]

The variance due to the main effects of factor B is

$$\sigma_\beta^2 = \frac{\sum_J \beta_J^2}{Q - 1} = \frac{5^2 + (-5)^2 + 0^2}{3 - 1} = 25.$$

In terms of the differential main effect of factor B, the variance is

$$\sigma_\beta^2 = \frac{\Sigma(\beta_J - \beta_{J'})^2}{Q(Q - 1)}$$

$$= \frac{10^2 + 5^2 + (-5)^2}{3(3 - 1)} = 25.$$

In summary, main effects as well as differential main effects are defined in terms of a specified population and not in terms of individual treatments within the population. Thus a main effect is *not* a parameter associated with a specified level of a specified factor; rather, a main effect depends upon all the other factors that may be present, as well as the number of levels assumed for the specified factor. It should also be noted that main effects are computed from means which are obtained by averaging over the totality of all other factors present in the population to which inferences are to be made.

5.4 Interaction Effects

The interaction between level a_I and level b_J, designated by the symbol $\alpha\beta_{IJ}$, is a measure of the extent to which the criterion mean for treatment combination ab_{IJ} can*not* be predicted from the sum of the corresponding main effects. From many points of view, interaction is a measure of the nonadditivity of the main effects. To some extent the existence or nonexistence of interaction depends upon the scale of measurement. For example, in terms of a logarithmic scale of measurement, interaction may not be present, whereas in terms of some other scale of measurement an interaction effect may be present. The choice of a scale of measurement for the dependent variable is generally at the discretion of the experimenter. If alternative choices are available, then that scale which leads to the simplest additive model will generally provide the most complete and adequate summary of the experimental data.

In terms of the population means and the population main effects,

$$\alpha\beta_{IJ} = \mu_{IJ} - (\alpha_I + \beta_J + \mu_{..}).$$

From the definition of the main effects, the above expression may be shown to be algebraically equivalent to

$$\alpha\beta_{IJ} = \mu_{IJ} - \mu_{I.} - \mu_{.J} + \mu_{..}.$$

From the way in which $\mu_{I.}$ and $\mu_{.J}$ were defined, it follows that

$$\begin{aligned}\sum_I \alpha\beta_{IJ} &= \sum_I \mu_{IJ} - \sum_I \mu_{I.} - \sum_I \mu_{.J} + \sum_I \mu_{..} \\ &= P\mu_{.J} - P\mu_{..} - P\mu_{.J} + P\mu_{..} \\ &= 0.\end{aligned}$$

It also follows that

$$\sum_J \alpha\beta_{IJ} = 0.$$

In words, the sum of the interaction effects within any row or any column of the population cells is zero.

Differential interaction effects are defined by the following examples:

(i) $\alpha\beta_{12} - \alpha\beta_{34} = \mu_{12} - \mu_{34} - (\alpha_1 - \alpha_3) - (\beta_2 - \beta_4),$

(ii) $\alpha\beta_{12} - \alpha\beta_{13} = \mu_{12} - \mu_{13} - (\beta_2 - \beta_3),$

(iii) $\alpha\beta_{12} - \alpha\beta_{32} = \mu_{12} - \mu_{32} - (\alpha_1 - \alpha_3).$

In (i) the differential interaction effect depends upon two differential main effects as well as the difference between cell means. In (ii) and (iii) only a single differential main effect is involved; (ii) depends upon the differential main effect associated with two levels of factor B, whereas (iii) depends upon a differential main effect associated with two levels of factor A. Relative to (i), the differential interaction effects represented by (ii) and (iii) are classified as *simple* interaction effects.

The variance due to interaction effects in the population is, by definition,

$$\sigma^2_{\alpha\beta} = \frac{\Sigma\Sigma(\alpha\beta)^2_{IJ}}{(P-1)(Q-1)}.$$

An equivalent definition can also be given in terms of the differential interaction effects. Under the hypothesis that the individual cell means may be predicted from a knowledge of corresponding main effects, $\sigma^2_{\alpha\beta} = 0$; that is, under the hypothesis of additivity of main effects, the variance due to the interaction effect is equal to zero.

For the numerical data given in the last section,

$$\begin{aligned}\alpha\beta_{11} &= 10 - 10 - 15 + 10 = -5, \\ \alpha\beta_{12} &= \ \ 5 - 10 - \ \ 5 + 10 = 0, \\ \alpha\beta_{13} &= 15 - 10 - 10 + 10 = 5.\end{aligned}$$

It is also noted that $\sum_J \alpha\beta_{1J} = -5 + 0 + 5 = 0.$

5.5 Experimental Error and Its Estimation

All uncontrolled sources of variance influencing an observation under a specified treatment combination contribute to what is known as the variance due to experimental error. The greater the number of relevant sources of variance which are controlled and measured, the smaller the variance due to experimental error. In a population in which there are N potential observations (i.e., measurements on a criterion) under a specified treatment combination, the variance of these observations defines the variance due to experimental error. Since all the observations within this cell have been made under the same treatment combination, effects associated with the treatment per se have been held constant. Hence the treatment effects per se are not contributing to the within-cell variance. Differences between units of the experimental material existing prior to the experimental treatment, variance introduced by inaccuracies or uncontrolled changes in experimental techniques, possible unique interactions between the material and the treatments—all these sources contribute to the within-cell variance.

If an observation on element K under treatment combination ab_{IJ} is designated by the symbol X_{IJK}, then the variance of the N potential elements under this treatment combination is given by

$$\sigma_{IJ}^2 = \frac{\sum_K (X_{IJK} - \mu_{IJ})^2}{N - 1}.$$

[In the usual definition of a population variance, N rather than $N - 1$ appears as the divisor. However, using $N - 1$ as the divisor simplifies the notation that follows. This definition of σ_{IJ}^2 is the one adopted by Cornfield and Tukey (1956, p. 916). As N approaches infinity, $N - 1$ approaches N.] Thus, σ_{IJ}^2, the within-cell variance for the population cell ab_{IJ}, is the variance due to experimental error in this cell. For purposes of making tests of significance and obtaining confidence bounds on parameters, it must be assumed that the variance due to experimental error is constant for all cells in the experiment. This is the assumption of homogeneity of error variance; this assumption may be represented symbolically by

$$\sigma_{IJ}^2 = \sigma_\varepsilon^2 \qquad \text{for all } IJ\text{'s},$$

where σ_ε^2 is the variance due to experimental error within any cell in the population. For the same purposes, it will generally be assumed that X_{IJK} is normally distributed within each of the cells.

In an experiment, assume that there are n independent observations within each of the treatment combinations included in the experiment.

The variance of the observations in the cell ab_{ij} is given by

$$s_{ij}^2 = \frac{\sum_k (X_{ijk} - \overline{AB}_{ij})^2}{n - 1}.$$

Assuming that the experimental error in the population is σ_ε^2 for all cells, s_{ij}^2 provides an estimate of σ_ε^2. A better estimate of σ_ε^2 is obtained by averaging the within-cell variances for each of the pq cells in the experiment. This average within-cell variance, denoted by the symbol s_{pooled}^2, may be represented as

$$s_{\text{pooled}}^2 = \frac{\Sigma\Sigma s_{ij}^2}{pq}.$$

Under the assumptions made, s_{pooled}^2 will be an unbiased estimate of σ_ε^2. Further, the sampling distribution of $pq(n - 1)s_{\text{pooled}}^2/\sigma_\varepsilon^2$ will be a chi-square distribution having $pq(n - 1)$ degrees of freedom. [Since the variance within each of the pq cells in the experiment is based upon n independent observations, each of the within-cell variances has $n - 1$ degrees of freedom. The pooled variance, s_{pooled}^2, has degrees of freedom equal to the combined degrees of freedom for each of the within-cell variances, that is, $pq(n - 1)$.] Because the observations within each of the cells are independent, the degrees of freedom for each of the variances that are averaged are additive.

The statistic s_{pooled}^2 is called the within-cell mean square, abbreviated $MS_{\text{w. cell}}$. Alternatively, this source of variance is designated as MS_{error} in designs which do not have repeated measures.

5.6 Estimation of Mean Squares Due to Main Effects and Interaction Effects

Some of the assumptions that will be made for purposes of estimation and analysis in a two-factor experiment are summarized by the following structural model:

$$X_{ijk} = \mu_{..} + \alpha_i + \beta_j + \alpha\beta_{ij} + \varepsilon_{ijk}. \quad (1)$$

In this model, X_{ijk} is an observation made in the experiment on element k under treatment combination ab_{ij}. On the right-hand side of (1) are the factorial effects and the experimental error. This model assumes that the factorial effects as well as the experimental error are additive, i.e., that an observation is a linear function of the factorial effects and the experimental error. Expression (1) is a special case of the general linear hypothesis. Each of the terms on the right-hand side is assumed to be statistically independent of the others.

The left-hand side of (1) represents an observation on the dependent variable made in the experiment. The terms on the right-hand side of (1) are parameters underlying the independent variables. The terms on the right-hand side cannot be observed directly; however, data from an experiment will give unbiased estimators of these parameters, provided a set of constraints are added to the model. If these constraints are not included, only certain linear functions of the parameters can be estimated. Tests of hypotheses are possible only with respect to estimable functions of the parameters. In the analysis of variance, the constraints necessary to obtain unbiased estimators of the individual parameters are usually considered part of the model.

For all observations made under treatment combination ab_{ij}, the terms α_i, β_j, and $\alpha\beta_{ij}$ are constant. ($\mu_{..}$ is a constant for all observations.) Hence the only source of variation for observations made under treatment combination ab_{ij} is that due to experimental error. If s_{ij}^2 is the variance of the observations within cell ab_{ij}, then s_{ij}^2 is an estimate of σ_ε^2.

In terms of (1), the mean of the observations within cell ab_{ij} may be expressed as

$$(2) \qquad \overline{AB}_{ij} = \frac{\sum_k X_{ijk}}{n} = \mu_{..} + \alpha_i + \beta_j + \alpha\beta_{ij} + \bar{\varepsilon}_{ij}.$$

The notation $\bar{\varepsilon}_{ij}$ denotes the mean experimental error for the n observations in cell ab_{ij}. If the observations in cell ab_{ij} were to be replicated with an independent sample of size n, the term $\bar{\varepsilon}_{ij}$ would not remain constant.

Table 5.6-1 Structural Parameters Estimated by Various Statistics

Statistic	Structural parameters estimated
$\bar{A}_i - \bar{A}_{i'}$	$(\alpha_i - \alpha_{i'}) + (\overline{\alpha\beta}_i - \overline{\alpha\beta}_{i'}) + (\bar{\varepsilon}_i - \bar{\varepsilon}_{i'})$
$\bar{B}_j - \bar{B}_{j'}$	$(\beta_j - \beta_{j'}) + (\overline{\alpha\beta}_j - \overline{\alpha\beta}_{j'}) + (\bar{\varepsilon}_j - \bar{\varepsilon}_{j'})$
$\overline{AB}_{ij} - \overline{AB}_{i'j'}$	$(\alpha_i - \alpha_{i'}) + (\beta_j - \beta_{j'}) + (\alpha\beta_{ij} - \alpha\beta_{i'j'}) + (\bar{\varepsilon}_{ij} - \bar{\varepsilon}_{i'j'})$
$(\overline{AB}_{ij} - \bar{A}_i - \bar{B}_j) - (\overline{AB}_{i'j'} - \bar{A}_{i'} - \bar{B}_{j'})$	$(\alpha\beta_{ij} - \alpha\beta_{i'j'}) + (\bar{\varepsilon}_{ij} - \bar{\varepsilon}_{i'j'})$

However, all the other terms on the right-hand side of (2) would remain constant for this kind of replication. Thus, for a large number of replications of the observations within cell ab_{ij}, each replication with an independent sample of size n,

$$(3) \qquad \begin{aligned} E(\overline{AB}_{ij}) &= E(\mu_{..}) + E(\alpha_i) + E(\beta_j) + E(\alpha\beta_{ij}) + E(\bar{\varepsilon}_{ij}) \\ &= \mu_{..} + \alpha_i + \beta_j + \alpha\beta_{ij} + 0. \end{aligned}$$

In this context, the expected value of terms on the right-hand side is the average value of a large number of replications for cell ab_{ij} with independent samples. Although the mean experimental error $\bar{\varepsilon}_{ij}$ has expected value equal to zero, it varies from one replication to the next. Hence the variance of $\bar{\varepsilon}_{ij}$ is not zero.

From (1) and the definitions of the various parameters, the statistics summarized in Table 5.6-1 have the expressions indicated. The right-hand side of this table gives the structural variables estimated by the statistics on the left-hand side.

The notation $\overline{\alpha\beta}_i$ denotes the average for level a_i of the $\alpha\beta_{ij}$ effects over the levels of factor B included in the experiment. If $q = Q$ (that is, if factor B is fixed), then $\overline{\alpha\beta}_i$ is zero. If $q \neq Q$, $\overline{\alpha\beta}_i$ need not be zero for any single experiment. To illustrate this point, suppose that there are six levels of factor B in the population of such levels, and suppose that the interaction effects associated with level a_i are those given below:

	b_1	b_2	b_3	b_4	b_5	b_6
a_i	$\alpha\beta_{i1} = 3$	$\alpha\beta_{i2} = 2$	$\alpha\beta_{i3} = -3$	$\alpha\beta_{i4} = -2$	$\alpha\beta_{i5} = 5$	$\alpha\beta_{i6} = -5$

In this case $Q = 6$. Note that $\sum_J \alpha\beta_{iJ} = 0$. Suppose that only a random sample of $q = 3$ levels of factor B is included in any single experiment. Suppose that the levels b_2, b_4, and b_5 are included in an experiment which is actually conducted. For this experiment

$$\overline{\alpha\beta}_i = \frac{\sum_j \alpha\beta_{ij}}{q} = \frac{\alpha\beta_{i2} + \alpha\beta_{i4} + \alpha\beta_{i5}}{3} = \frac{5}{3}.$$

For a large number of random replications of this experiment, where an independent random sample of the levels of factor B is drawn for each replication, the expected value of $\overline{\alpha\beta}_i$ will be zero. The variance of the distribution of $\overline{\alpha\beta}_i$ generated by this kind of sampling procedure depends upon n, q, Q, and the variance $\sigma^2_{\alpha\beta}$.

The notation $\overline{\alpha\beta}_j$ denotes the average of the $\alpha\beta_{ij}$ effects over the levels of factor A present in an experiment at level b_j of factor B. If factor A is a fixed factor, $\overline{\alpha\beta}_j = 0$. However, if factor A is not a fixed factor, $\overline{\alpha\beta}_j$ is not necessarily equal to zero for any single experiment. Over a large number of random replications of the experiment, the expected value of $\overline{\alpha\beta}_j$ will be equal to zero. If factor A is a fixed factor, $\sigma^2_{\overline{\alpha\beta}_j}$ will be zero, since $\overline{\alpha\beta}_j$ will be zero for all replications. If factor A is a random factor, $\sigma^2_{\overline{\alpha\beta}_j}$ will be a function of n, p, P, and $\sigma^2_{\alpha\beta}$.

The mean square due to the main effects of factor A in the experiment is defined to be

$$\text{MS}_a = \frac{nq\Sigma(\bar{A}_i - \bar{G})^2}{p - 1}.$$

An equivalent definition in terms of differences between pairs of means is

$$\text{MS}_a = \frac{nq\Sigma(\bar{A}_i - \bar{A}_{i'})^2}{p(p - 1)}.$$

The summation in this last expression is with respect to all distinct pairs of means, no pair being included twice. The multiplier nq is the number of observations in each $\bar{A}_i$. The expected value of $\bar{A}_i - \bar{A}_{i'}$ (for independent, random replications of the experiment) is

$$\begin{aligned} E(\bar{A}_i - \bar{A}_{i'}) &= E(\alpha_i - \alpha_{i'}) + E(\overline{\alpha\beta}_i - \overline{\alpha\beta}_{i'}) + E(\bar{\varepsilon}_i - \bar{\varepsilon}_{i'}) \\ &= \alpha_i - \alpha_{i'} \quad + 0 \quad + 0. \end{aligned}$$

The expected value of MS_a may be shown to be

$$E(MS_a) = \left(\frac{N - n}{N}\right)\sigma_\varepsilon^2 + \left(\frac{Q - q}{Q}\right)n\sigma_{\alpha\beta}^2 + nq\sigma_\alpha^2.$$

Detailed steps in the derivation of this latter expected value are given by Cornfield and Tukey (1956).

The mean square due to factor B is defined to be

$$\begin{aligned} MS_b &= \frac{np\Sigma(\bar{B}_j - \bar{G})^2}{q - 1} \\ &= \frac{np\Sigma(\bar{B}_j - \bar{B}_{j'})^2}{q(q - 1)}. \end{aligned}$$

The multiplier np represents the number of observations in each $\bar{B}_j$. The expected value of this mean square in terms of the parameters in (1) is given in Table 5.6-2.

Table 5.6-2 Expected Values of Mean Squares

Mean square as obtained from experimental data	Expected value of mean square in terms of parameters of (1)
MS_a	$(1 - n/N)\sigma_\varepsilon^2 + n(1 - q/Q)\sigma_{\alpha\beta}^2 + nq\sigma_\alpha^2$
MS_b	$(1 - n/N)\sigma_\varepsilon^2 + n(1 - p/P)\sigma_{\alpha\beta}^2 + np\sigma_\beta^2$
MS_{ab}	$(1 - n/N)\sigma_\varepsilon^2 + n\sigma_{\alpha\beta}^2$
MS_{error}	$(1 - n/N)\sigma_\varepsilon^2$

The mean square due to interaction effects in the experiment is defined as

$$MS_{ab} = \frac{n\Sigma\Sigma(\overline{AB}_{ij} - \bar{A}_i - \bar{B}_j + \bar{G})^2}{(p - 1)(q - 1)}.$$

An equivalent definition can be given in terms of differential interaction effects. The multiplier n is the number of observations in each $\overline{AB}_{ij}$. The expected value of this mean square represents the average of the MS_a. computed from a large number of independent, random replications of the experiment; this average value is expressed in terms of the parameters in the general linear model (1), which is assumed to represent the sources of variance underlying an observation.

Certain of the coefficients in Table 5.6-2 are either zero or unity, depending upon whether a factor is fixed or random. It will generally be assumed that the number of elements observed in an experiment is small relative to the number of potential elements in the population of elements, i.e., that the ratio n/N for all practical purposes is equal to zero. Hence the coefficient $1 - n/N$ is assumed to be equal to unity.

If factor A is fixed, the ratio p/P will be equal to unity and the coefficient $1 - p/P$ will be equal to zero. If, on the other hand, factor A is random, the ratio p/P will be equal to zero and the coefficient $1 - p/P$ will be equal to unity. In an analogous manner, the coefficient $1 - q/Q$ is equal to zero when factor B is a fixed factor and equal to unity when factor B is a random factor.

Special cases of the expected values of the mean squares are summarized in Table 5.6-3. Each of these cases is obtained from the general values given in Table 5.6-2 by evaluating the coefficients which depend upon the ratios n/N, p/P, and q/Q. Several different approaches may be used to obtain the special cases given in Table 5.6-3. Specialization of the generalized approach represented by Table 5.6-2 provides the simplest method of attack on the evaluation of more complex experimental designs.

Table 5.6-3 Special Cases of Expected Values of Mean Squares

Mean squares	Case 1 Factor A fixed Factor B fixed	Case 2 Factor A fixed Factor B random	Case 3 Factor A random Factor B random
MS_a	$\sigma_\varepsilon^2 + nq\sigma_\alpha^2$	$\sigma_\varepsilon^2 + n\sigma_{\alpha\beta}^2 + nq\sigma_\alpha^2$	$\sigma_\varepsilon^2 + n\sigma_{\alpha\beta}^2 + nq\sigma_\alpha^2$
MS_b	$\sigma_\varepsilon^2 + np\sigma_\beta^2$	$\sigma_\varepsilon^2 \qquad + np\sigma_\beta^2$	$\sigma_\varepsilon^2 + n\sigma_{\alpha\beta}^2 + np\sigma_\beta^2$
MS_{ab}	$\sigma_\varepsilon^2 + n\sigma_{\alpha\beta}^2$	$\sigma_\varepsilon^2 + n\sigma_{\alpha\beta}^2$	$\sigma_\varepsilon^2 + n\sigma_{\alpha\beta}^2$
MS_{error}	σ_ε^2	σ_ε^2	σ_ε^2

Case 1, in which both factors are fixed, has been designated by Eisenhart (1947) as model I. Case 2, in which one factor is fixed and the second is random, is called the mixed model. Case 3, in which both factors are random, is called model II, or the *variance-component* model. Model I has been more extensively studied than the other two models. In its most general form, the statistical principles underlying model I are identical to those underlying the general regression model having any number of fixed variates and one random variate. As such, the best estimates of various parameters can readily be obtained by the method of least squares. For the case of the generalized model I, application of the method of least squares is straightforward and leads to no difficulties. For the generalized mixed model, application of the principles of maximum likelihood are more direct.

Since the statistical tests made on the experimental data depend upon what these expected values are assumed to be, it is particularly important to specify the conditions under which these expected values are derived. To obtain the general expected values given in Table 5.6-2, the following assumptions are made:

1. There is a population of size P and variance σ_α^2 of main effects of factor A, of which the effects $(\alpha_1, \alpha_2, \ldots, \alpha_p)$ occurring in the experiment constitute a random sample (sampling without replacement) of size p. The sample may include all the levels of factor A in the population; that is, p may be equal to P.

2. There is a population of size Q and variance σ_β^2 of main effects of factor B, of which the effects $(\beta_1, \beta_2, \ldots, \beta_q)$ occurring in the experiment constitute a random sample of size q. The sample may include all the levels of factor B in the population; that is, q may be equal to Q.

3. There is a population of interaction effects of size PQ and variance $\sigma_{\alpha\beta}^2$; the $\alpha\beta_{ij}$'s which occur in the experiment correspond to the combinations of the levels of factor A and factor B that occur in the experiment. That is, one does not have a random sample of pq interaction effects; rather, the interaction effects in the experiment are tied to the levels of factor A and factor B that occur in the experiment. It is assumed that the average (in the population) of the interaction effects over all levels of one factor is independent of the main effects of the other factor; that is, $\overline{\alpha\beta}_i$ is independent of α_i, and $\overline{\alpha\beta}_j$ is independent of β_j.

4. The sampling of the levels of factor A is independent of the sampling of the levels of factor B.

5. The experimental error is independent of all main effects and all interactions. Further, within each cell in the population, ε, the experimental error, is assumed to be normally distributed, with mean equal to zero and variance equal to σ_ε^2 for all cells in the population.

6. The n observations within each cell of the experiment constitute a random sample of size n from a population of size N (assumed infinite in most cases). The n observations within each cell constitute independent random samples from a random sample of npq independent elements drawn from the basic population.

For purposes of deriving the expected values of mean squares, some of these assumptions may be relaxed. The assumption of normality of the distribution of the experimental error is not required for the derivation of the expected values. However, all these assumptions are needed for the validity of the tests involving the use of F ratios, which are based upon the expected values of the mean squares. In particular, the assumption that the distribution of the experimental error is normal is required in order that the sampling distributions of mean squares be chi-square distributions.

Under the conditions that have been stated, the mean squares computed in the analysis of variance have the following sampling distributions:

Statistic	Sampling distribution
$(p-1)\text{MS}_a/\text{E}(\text{MS}_a)$	Chi square with $p-1$ df
$(q-1)\text{MS}_b/\text{E}(\text{MS}_b)$	Chi square with $q-1$ df
$(p-1)(q-1)\text{MS}_{ab}/\text{E}(\text{MS}_{ab})$	Chi square with $(p-1)(q-1)$ df

Principles Underlying Derivation of Expected Values for Mean Squares. To provide some insight into the principles underlying the derivation of the expected values for mean squares, a nonrigorous derivation will be outlined. From finite sampling theory, one has the following theorem:

$$(1) \qquad \sigma_{\bar{X}}^2 = \frac{N-n}{N-1}\frac{\sigma_X^2}{n}.$$

Under random sampling without replacement after each draw, (1) relates the square of the standard error of a mean, $\sigma_{\bar{X}}^2$, to the population size N, the sample size n, and the variance σ_X^2 of the variable X in the population.

In (1), the population variance is defined to be

$$\sigma_X^2 = \frac{\Sigma(X-\mu)^2}{N}.$$

If one uses as the definition of the population variance

$$\sigma_X^2 = \frac{\Sigma(X-\mu)^2}{N-1},$$

then (1) has the form

$$(2) \qquad \sigma_{\bar{X}}^2 = \frac{N-n}{N-1}\frac{\Sigma(X-\mu)^2}{nN}\frac{N-1}{N-1}$$

$$= \frac{N-n}{N}\frac{\Sigma(X-\mu)^2}{n(N-1)}$$

$$= \left(1-\frac{n}{N}\right)\frac{\sigma_X^2}{n} \qquad \text{where} \qquad \sigma_X^2 = \frac{\Sigma(X-\mu)^2}{N-1}.$$

To simplify the notation, the present development will define a population variance using $N-1$ as the divisor. This definition is consistent with that used by Cornfield and Tukey (1956).

In terms of the right-hand side of the structural model given in (1) at the beginning of this section, the mean of all observations made at level a_i in an experiment is

$$(3) \qquad \bar{A}_i = \mu_{..} + \alpha_i + \bar{\beta} + \overline{\alpha\beta}_i + \bar{\varepsilon}_{i.}$$

In this notation $\bar{\beta}$ represents the average effect of all levels of factor B included in the experiment; $\bar{\beta}$ is constant for all levels of factor A. The notation $\overline{\alpha\beta}_i$ represents the average interaction effect associated with level a_i; $\overline{\alpha\beta}_i$ may differ for the various levels of factor A. The notation $\bar{\varepsilon}_i$ denotes the average experimental error associated with each $\bar{A}_i$.

The variance of $\bar{A}_i$ is defined to be

$$s_{\bar{A}}^2 = \frac{\Sigma(\bar{A}_i - \bar{G})^2}{p - 1}.$$

Under the assumption that the terms on the right-hand side of (3) are statistically independent, the expected value of the variance of the left-hand side of (3) will be equal to the sum of the variances of the terms on the right-hand side. Terms which are constants have zero variance. Thus,

$$\text{(4)} \qquad \mathrm{E}(s_{\bar{A}}^2) = \sigma_{\bar{\alpha}}^2 + \sigma_{\overline{\alpha\beta}}^2 + \sigma_{\bar{\varepsilon}}^2.$$

The mean square due to the main effect of factor A may be written as

$$\mathrm{MS}_a = nqs_{\bar{A}}^2.$$

Hence (4) becomes

$$\text{(5)} \qquad \mathrm{E}(\mathrm{MS}_a) = \mathrm{E}(nqs_{\bar{A}}^2) = nq\mathrm{E}(s_{\bar{A}}^2) = nq\sigma_{\bar{\alpha}}^2 + nq\sigma_{\overline{\alpha\beta}}^2 + nq\sigma_{\bar{\varepsilon}}^2.$$

The analog of the theorem stated in (1) may now be applied to each of the variances in (5) which have a mean as a subscript. Thus, $\bar{\varepsilon}$ is the mean of the experimental error associated with the nq observations from which $\bar{A}_i$ is computed. Therefore,

$$nq\sigma_{\bar{\varepsilon}}^2 = \frac{nq(1 - n/N)\sigma_\varepsilon^2}{nq} = \left(1 - \frac{n}{N}\right)\sigma_\varepsilon^2.$$

(The experimental error in the basic linear model is associated with each element in the basic population from which the elements are drawn. Hence the sampling fraction associated with the experimental error is n/N.)

The mean interaction effect in (4) represents the average over the q values of $\alpha\beta_{ij}$ present at level a_i. Hence $\overline{\alpha\beta}_i$ may be considered to be the mean of a sample of size q levels from a population of size Q levels of factor B. The analog of the theorem in (1) now takes the form

$$nq\sigma_{\overline{\alpha\beta}}^2 = \frac{nq(1 - q/Q)\sigma_{\alpha\beta}^2}{q} = n\left(1 - \frac{q}{Q}\right)\sigma_{\alpha\beta}^2.$$

Since the main effects and interaction effects are assumed to be independent, restricting the sample of q values of $\alpha\beta_{ij}$ to level a_i does not invalidate the theorem summarized in (1).

The expression for the expected value of the mean square of the main effect of factor A may now be written as

$$\text{(6)} \qquad \mathrm{E}(\mathrm{MS}_a) = \left(1 - \frac{n}{N}\right)\sigma_\varepsilon^2 + n\left(1 - \frac{q}{Q}\right)\sigma_{\alpha\beta}^2 + nq\sigma_\alpha^2.$$

Each of the variances in (6) is associated with the parameters in the general linear model, whereas in (5) some of the variances were in terms of parameters which are not explicitly in the general linear model. The purpose of defining MS_a as $nqs_{\bar{A}}^2$ now becomes clear. By adding the multiplier nq, one may conveniently express each of the variances in (5) in terms of parameters which are explicit in the linear model.

Derivation of the expected values of the mean squares for the main effects of B and for the AB interaction follows the same general line of reasoning. The algebraic procedures whereby the various means are obtained from the experimental data are carried through in terms of the right-hand side of the basic structural model for an observation. Then, using the assumptions underlying the model, one obtains the expected values of the mean squares by the principles that have just been outlined.

The principles underlying the derivation of expected values of the mean squares will be illustrated for the case of a 2 × 2 factorial experiment having n observations in each cell. In terms of a general linear model, the cell means and the marginal means for the levels of factor B may be represented as follows:

b_1	b_2
$\overline{AB}_{11} = \mu_{..} + \alpha_1 + \beta_1 + \alpha\beta_{11} + \bar{\varepsilon}_{11}$	$\overline{AB}_{12} = \mu_{..} + \alpha_1 + \beta_2 + \alpha\beta_{12} + \bar{\varepsilon}_{12}$
$\overline{AB}_{21} = \mu_{..} + \alpha_2 + \beta_1 + \alpha\beta_{21} + \bar{\varepsilon}_{21}$	$\overline{AB}_{22} = \mu_{..} + \alpha_2 + \beta_2 + \alpha\beta_{22} + \bar{\varepsilon}_{22}$
$\bar{B}_1 = \mu_{..} + \bar{\alpha} + \beta_1 + \overline{\alpha\beta}_{.1} + \bar{\varepsilon}_{.1}$	$\bar{B}_2 = \mu_{..} + \bar{\alpha} + \beta_2 + \overline{\alpha\beta}_{.2} + \bar{\varepsilon}_{.2}$

The difference between the two marginal means for factor B estimates the following parameters:

$$\bar{B}_1 - \bar{B}_2 = (\beta_1 - \beta_2) + (\overline{\alpha\beta}_{.1} - \overline{\alpha\beta}_{.2}) + (\bar{\varepsilon}_{.1} - \bar{\varepsilon}_{.2}).$$

Multiplying each side of the above expression by np gives

$$np(\bar{B}_1 - \bar{B}_2) = np(\beta_1 - \beta_2) + np(\overline{\alpha\beta}_{.1} - \overline{\alpha\beta}_{.2}) + np(\bar{\varepsilon}_{.1} - \bar{\varepsilon}_{.2}).$$

Since the terms on the right-hand side of the above expression are assumed to be statistically independent, the variance of the term on the left-hand side will estimate the sum of the variances of the terms on the right-hand side. Hence,

$$\text{E(MS}_b) = np\sigma_\beta^2 + np\sigma_{\overline{\alpha\beta}}^2 + np\sigma_{\bar{\varepsilon}}^2.$$

Using the analog of the relation given in (1),

$$np\sigma_{\overline{\alpha\beta}}^2 = n\left(1 - \frac{p}{P}\right)\sigma_{\alpha\beta}^2, \qquad \text{and} \qquad np\sigma_{\bar{\varepsilon}}^2 = \left(1 - \frac{n}{N}\right)\sigma_\varepsilon^2.$$

Thus, $$\text{E(MS}_b) = np\sigma_\beta^2 + n\left(1 - \frac{p}{P}\right)\sigma_{\alpha\beta}^2 + \left(1 - \frac{n}{N}\right)\sigma_\varepsilon^2.$$

In most experimental designs the potential number of experimental units, N, is considered to be infinite. Hence $1 - n/N = 1$.

Least-squares Estimates of Parameters in Model. In a $p \times q$ factorial experiment having n observations per cell, assume factors A and B to be fixed. The least-squares principle leads quite directly to estimators of the parameters in a linear model, provided a set of constraints are imposed upon the model. Let the linear model have the form

$$X_{ijk} = \mu + \alpha_i + \beta_j + \alpha\beta_{ij} + \varepsilon_{ijk}, \tag{7}$$

with the constraints

$$\sum_i \alpha_i = 0, \qquad \sum_j \beta_j = 0, \qquad \sum_i \alpha\beta_{ij} = 0, \qquad \sum_j \alpha\beta_{ij} = 0.$$

It is also assumed that

$$\underset{k}{\mathrm{E}}(\varepsilon_{ijk}) = 0 \qquad \text{for all } i, j,$$

$$\underset{k}{\mathrm{E}}(\varepsilon_{ijk}^2) = \sigma_\varepsilon^2 \qquad \text{for all } i, j.$$

That is, σ_ε^2 does not depend upon i or j.

If the least-squares estimators are denoted $\hat{\mu}$, $\hat{\alpha}_i$, $\hat{\beta}_j$, and $\widehat{\alpha\beta}_{ij}$, the least-squares criterion makes

$$\sum \hat{\varepsilon}_{ijk}^2 = \sum(X_{ijk} - \hat{\mu} - \hat{\alpha}_i - \hat{\beta}_j - \widehat{\alpha\beta}_{ij})^2 = \text{minimum} \tag{8}$$

for the experimental data. Taking the partial derivative of (8) with respect to each of the estimators and then setting each of the partial derivatives equal to zero yield the following set of normal equations:

$$\begin{aligned}
\mu&: \quad npq\hat{\mu} + nq\sum_i\hat{\alpha}_i + np\sum_j\hat{\beta}_j + n\sum_i\sum_j\widehat{\alpha\beta}_{ij} = G,\\
\alpha_i&: \quad nq\hat{\mu} + nq\hat{\alpha}_i + n\sum_j\hat{\beta}_j + n\sum_j\widehat{\alpha\beta}_{ij} = A_i, \qquad i = 1, \ldots, p,\\
\beta_j&: \quad np\hat{\mu} + n\sum_i\hat{\alpha}_i + np\hat{\beta}_j + n\sum_i\widehat{\alpha\beta}_{ij} = B_j, \qquad j = 1, \ldots, q,\\
\alpha\beta_{ij}&: \quad n\hat{\mu} + n\hat{\alpha}_i + n\hat{\beta}_j + n\widehat{\alpha\beta}_{ij} = AB_{ij}.
\end{aligned} \tag{9}$$

The set of linear equations represented by (9) contains

$$1 + p + q + pq$$

individual equations in the same number of unknowns. However, the equations are not linearly independent. To illustrate one linear dependency in the set, one notes that the sum of the p equations represented by α_i is the equation represented by μ. Another linear dependency in the set is noted by the fact that the sum over j of the pq $\alpha\beta_{ij}$ equations yields the equation corresponding to α_i. Because of these and other linear dependencies in the set (9), there is no unique solution to the system without imposing a set of constraints on the unknowns.

If one imposes upon the estimators the same set of constraints that are imposed upon the parameters in (7), namely,

$$\sum_i \hat{\alpha}_i = 0, \qquad \sum_j \hat{\beta}_j = 0, \qquad \sum_i \widehat{\alpha\beta}_{ij} = 0, \qquad \sum_j \widehat{\alpha\beta}_{ij} = 0,$$

then the solution to the set of equations in (9) has the following relatively simple solution:

$$\begin{aligned} \hat{\mu} = \bar{G}, \qquad \hat{\alpha}_i &= \bar{A}_i - \bar{G} && i = 1, \ldots, p, \\ \hat{\beta}_j &= \bar{B}_j - \bar{G} && j = 1, \ldots, q, \\ \widehat{\alpha\beta}_{ij} &= \overline{AB}_{ij} - \bar{A}_i - \bar{B}_j + \bar{G}. \end{aligned}$$

From the Gauss-Markov theorem, it follows that these estimators are unbiased and are the *best* unbiased estimators, which are linear in the X_{ijk}, in the sense of having minimum standard errors.

The estimators obtained here are identical to those that are obtained if one sets up a hypothetical population of elements and levels of factors A and B. In this hypothetical population, the key constructs were the cell variances and the cell and marginal means. Estimates of the parameters in the linear model were expressed in terms of corresponding estimates of the cell and marginal means obtained from the experimental data.

From the properties of the general linear model, the variation due to the mean is

$$R(\mu) = \hat{\mu}G = \frac{G^2}{npq}.$$

The sum of squares due to the main effects of factor A, adjusted for the mean, is

$$\begin{aligned} SS_a = R(\alpha \mid \mu) &= \Sigma \hat{\alpha}_i A_i = \Sigma(\bar{A}_i - \bar{G})A_i \\ &= \frac{\Sigma A_i^2}{nq} - \frac{G^2}{npq}. \end{aligned}$$

Similarly, the sum of squares due to the main effects of factor B, adjusted for the mean, is

$$\begin{aligned} SS_b = R(\beta \mid \mu) &= \Sigma \hat{\beta}_j B_j \\ &= \frac{\Sigma B_j^2}{np} - \frac{G^2}{npq}. \end{aligned}$$

The sum of squares due to the interaction effects, adjusted for the mean, is

$$\begin{aligned} SS_{ab} = R(\alpha\beta \mid \mu) &= \Sigma \widehat{\alpha\beta}_{ij} AB_{ij} \\ &= \frac{\Sigma AB_{ij}^2}{n} - \frac{G^2}{npq} - SS_a - SS_b. \end{aligned}$$

If the cell frequencies are not equal, the solution to the normal equations and the estimation of the variation due to various effects are somewhat more complex. The case of unequal cell frequencies is considered in detail in Sec. 5.23.

5.7 Principles for Constructing F Ratios

For either model I (all factors assumed fixed) or model II (all factors assumed random) the sampling distributions of the mean squares for main effects and interactions are independent chi-square distributions. For model III (some factors fixed, some factors random) all required sampling distributions are independent chi-square distributions only if highly restrictive assumptions on covariances are made. The principles for testing hypotheses to be presented in this section hold rigorously for models I and II. In practice, tests under model III follow the principles presented here; however, interpretations under model III require special care. Scheffé (1959, pp. 264, 288), in particular, has questioned principles for constructing F tests of the type to be presented here for special cases of model III. Under the mixed model, it is the test on the main effects of the fixed factor that is only approximate unless a set of homogeneity assumptions on covariances are satisfied. The latter are made explicit in Sec. 4.4. Indications are, however, that the operating characteristics of the approximate tests are quite close to the exact tests under relatively weak assumptions on the relevant covariances.

The hypothesis that $\alpha_1 = \alpha_2 = \cdots = \alpha_P$ (that is, the hypothesis of no differences between the main effects of factor A) is equivalent to the hypothesis that $\sigma_\alpha^2 = 0$. This hypothesis is in turn equivalent to the following hypotheses:

1. All possible comparisons (or contrasts) among the main effects of factor A are equal to zero.

2. $$\mu_{1.} = \mu_{2.} = \cdots = \mu_{P.} = \mu_{..}.$$

To test this hypothesis against the alternative hypothesis that $\sigma_\alpha^2 > 0$ requires the construction of an F ratio. In terms of the expected value of mean squares, the F ratio for this test has the general form

$$\frac{\text{E(numerator)}}{\text{E(denominator)}} = \frac{u + c\sigma_\alpha^2}{u},$$

where u is some linear function of the variances of other parameters in the model and c is some coefficient. In words, $E(MS_{\text{numerator}})$ must be equal to $E(MS_{\text{denominator}})$ when $\sigma_\alpha^2 = 0$.

For the test under consideration, the mean square in the numerator of the F ratio must be MS_a. The mean square that is in the denominator depends upon the expected value of MS_a under the proper model. For model I, the appropriate denominator for this F ratio is MS_{error}; for model II, the appropriate denominator (in a two-factor factorial experiment) is MS_{ab}. Thus, in order to form an F ratio in the analysis of variance, knowledge of expected value of mean squares under the appropriate model is needed. This, in essence, implies that the F ratio depends upon the design of the experiment.

If the numerator and denominator of an F ratio satisfy the structural requirements in terms of the expected values of the mean squares, and if the sampling distributions of these mean squares are independent chi squares when the hypothesis being tested is true, then the resulting F ratio will have a sampling distribution given by an F distribution. The degrees of freedom for the resulting F distribution are, respectively, the degrees of freedom for the numerator mean square and the degrees of freedom for the denominator mean square. General principles for setting up F ratios are illustrated in Table 5.7-1.

Table 5.7-1 Tests of Hypotheses under Model II

Source of variation	E(MS)	Hypothesis being tested	F ratio
Main effect of factor A	$\sigma_\varepsilon^2 + n\sigma_{\alpha\beta}^2 + nq\sigma_\alpha^2$	$H_1\colon \sigma_\alpha^2 = 0$	$F = \text{MS}_a/\text{MS}_{ab}$
Main effect of factor B	$\sigma_\varepsilon^2 + n\sigma_{\alpha\beta}^2 + np\sigma_\beta^2$	$H_1\colon \sigma_\beta^2 = 0$	$F = \text{MS}_b/\text{MS}_{ab}$
$A \times B$ interaction	$\sigma_\varepsilon^2 + n\sigma_{\alpha\beta}^2$	$H_1\colon \sigma_{\alpha\beta}^2 = 0$	$F = \text{MS}_{ab}/\text{MS}_{\text{error}}$
Error	σ_ε^2		

The expected values in this table are those appropriate for model II. In terms of these expected values the F ratio used to test the hypothesis that $\sigma_\alpha^2 = 0$ has the form

$$\frac{\text{E(numerator)}}{\text{E(denominator)}} = \frac{\sigma_\varepsilon^2 + n\sigma_{\alpha\beta}^2 + nq\sigma_\alpha^2}{\sigma_\varepsilon^2 + n\sigma_{\alpha\beta}^2}.$$

When the hypothesis being tested is true (that is, when $\sigma_\alpha^2 = 0$), the expected value of the numerator is equal to the expected value of the denominator. Thus $\text{E}(\text{MS}_a) = \text{E}(\text{MS}_{ab})$ when $\sigma_\alpha^2 = 0$. When σ_α^2 is greater than zero, the expected value of the numerator is greater than the expected value of the denominator by an amount which depends upon the term $nq\sigma_\alpha^2$. From the structure of the F ratio, the numerator can be less than the denominator only because of sampling error associated with the estimation of MS_a and MS_{ab}; for any single experiment each of these statistics may be independently either less than or greater than its expected value. Alternatively, the F ratio may be less than unity when some of the assumptions about the model do not hold.

The F ratio appropriate for testing the hypothesis $\sigma_{\alpha\beta}^2 = 0$ has the structure

$$\frac{\text{E(numerator)}}{\text{E(denominator)}} = \frac{\sigma_\varepsilon^2 + n\sigma_{\alpha\beta}^2}{\sigma_\varepsilon^2}.$$

When the hypothesis being tested is true, numerator and denominator have the same expected value. When $\sigma_{\alpha\beta}^2 > 0$, the expected value of this ratio will be greater than unity by an amount which depends in part upon the term $n\sigma_{\alpha\beta}^2$. The ratio obtained from the experimental data has the form

$$F = \frac{\text{MS}_{ab}}{\text{MS}_{\text{error}}}.$$

When the hypothesis being tested is true, the sampling distribution of this F ratio is the F distribution having $(p - 1)(q - 1)$ degrees of freedom for the numerator and $pq(n - 1)$ degrees of freedom for the denominator.

Power of Tests. Under model I, the sampling distribution of the F statistic when the hypothesis being tested is not true is the *noncentral F* with the appropriate noncentrality parameter. For example, in testing the hypothesis that $\sigma_\alpha^2 = 0$, the noncentrality parameter is

$$\lambda = \frac{nq\sigma_\alpha^2}{\sigma_\varepsilon^2}.$$

Tables of the noncentral F are usually oriented in terms of $\phi = \sqrt{\lambda/p}$.

Under model II, the sampling distribution of the F statistic when the hypothesis being tested is not true is a multiple of a central F distribution. For example, the statistic

$$\frac{\text{MS}_a}{\text{MS}_{ab}} \quad \text{is distributed as} \quad cF[p - 1, (p - 1)(q - 1)],$$

where

$$c = \frac{\sigma_\varepsilon^2 + n\sigma_{\alpha\beta}^2 + nq\sigma_\alpha^2}{\sigma_\varepsilon^2 + n\sigma_{\alpha\beta}^2} = 1 + \frac{nq\sigma_\alpha^2}{\sigma_\varepsilon^2 + n\sigma_{\alpha\beta}^2}.$$

Under the hypothesis that $\sigma_\alpha^2 = 0$, then $c = 1$; under the hypothesis that $\sigma_\alpha^2 > 0$, then $c > 1$.

Under model III, if factor A is a fixed factor and B is a random factor, the E(MS) lead to the ratio

$$F = \frac{\text{MS}_a}{\text{MS}_{ab}}.$$

Treating this F statistic as if it had the central F distribution when the hypothesis being tested is true requires a set of highly restrictive assumptions about the covariance structure of the parameters in the model. Some of these assumptions were discussed in Chap. 4. Evaluation of power in this case involves the use of the noncentral F distribution.

Monte Carlo studies as well as randomization tests indicate that treating the F ratio suggested by the E(MS) as if it had an F distribution results in a good approximation to these tests, even though the homogeneity assumptions on the covariance structure may be violated.

5.8 Higher-order Factorial Experiments

When a factorial experiment includes three or more factors, different orders of interaction are possible. For example, in a $2 \times 3 \times 5$ factorial experiment, having 10 independent observations in each cell, the analysis of variance generally has the form given in Table 5.8-1.

Table 5.8-1 Analysis of Variance for $2 \times 3 \times 5$ Factorial Experiment Having 10 Observations per Cell

Source of variation	Sum of squares	df	df (general)
A main effects	SS_a	1	$p - 1$
B main effects	SS_b	2	$q - 1$
C main effects	SS_c	4	$r - 1$
AB interaction	SS_{ab}	2	$(p - 1)(q - 1)$
AC interaction	SS_{ac}	4	$(p - 1)(r - 1)$
BC interaction	SS_{bc}	8	$(q - 1)(r - 1)$
ABC interaction	SS_{abc}	8	$(p - 1)(q - 1)(r - 1)$
Experimental error (within cell)	SS_{error}	270	$pqr(n - 1)$
Total	SS_{total}	299	$npqr - 1$

In a three-factor experiment there are three interactions which involve two factors: $A \times B$, $A \times C$, $B \times C$. There is one three-factor interaction. The $A \times B \times C$ interaction represents the unique effects attributable to the combination of the three factors, i.e., the effects that cannot be predicted from a knowledge of the main effects and two-factor interactions. The notation that was introduced for the case of a two-factor experiment can be extended as follows:

	Levels in population	Levels in experiment
Factor A	$a_1, a_2, \ldots, a_I, \ldots, a_P$	$a_1, a_2, \ldots, a_i, \ldots, a_p$
Factor B	$b_1, b_2, \ldots, b_J, \ldots, b_Q$	$b_1, b_2, \ldots, b_j, \ldots, b_q$
Factor C	$c_1, c_2, \ldots, c_K, \ldots, c_R$	$c_1, c_2, \ldots, c_k, \ldots, c_r$

The definitions of fixed and random factor given in Sec. 5.2 also apply to factor C. If $r = R$, then factor C is a fixed factor. If the r levels of factor C in the experiment are a random sample from the R levels in the population, and if R is quite large relative to r, then factor C is a random factor. If the R levels are reduced to $R_{\text{effective}}$ levels by some systematic, nonrandom procedure, then factor C is considered fixed when $r = R_{\text{effective}}$.

The notation for cell means used for the case of a two-factor experiment may also be extended. An observation on element m under treatment combination abc_{ijk} is designated by X_{ijkm}. Notation for cell means is summarized in Table 5.8-2. In this notation system, μ_{ijk} designates the

Table 5.8-2 Notation for Means in a Three-factor Experiment

	Population mean	Experiment mean
Elements in cell abc_{ijk}	μ_{ijk}	$\overline{ABC}_{ijk}$
Elements under ab_{ij}	$\mu_{ij.}$	$\overline{AB}_{ij}$
Elements under ac_{ik}	$\mu_{i.k}$	$\overline{AC}_{ik}$
Elements under bc_{jk}	$\mu_{.jk}$	$\overline{BC}_{jk}$
Elements under a_i	$\mu_{i..}$	$\bar{A}_i$
Elements under b_j	$\mu_{.j.}$	$\bar{B}_j$
Elements under c_k	$\mu_{..k}$	$\bar{C}_k$

mean of the N potential observations that could be made under treatment combination abc_{ijk}. The notation $\mu_{ij.}$ designates the mean of the NR potential observations that could be made under the treatment combinations $abc_{ij1}, abc_{ij2}, \ldots, abc_{ijR}$ (N potential observations under each of the R treatment combinations). In terms of symbols,

$$\mu_{ij.} = \frac{\sum_K \mu_{ijK}}{R}.$$

(The subscript K is used here to indicate that the average is over all potential levels of factor C and not just those in any single experiment.)

The notation $\mu_{i..}$ deisgnates the mean of the NQR potential observations that could be made under the treatment combinations in which factor A is at level a_i. Thus,

$$\mu_{i..} = \frac{\sum_J \sum_K \mu_{iJK}}{QR}.$$

(The subscripts J and K indicate that the average is over all levels of factors B and C, not just those included in the experiment.) Similarly, $\mu_{..k}$ denotes the mean of the potential NPQ observations that could be made under level c_k, that is,

$$\mu_{..k} = \frac{\sum_I \sum_J \mu_{IJk}}{PQ}.$$

The notation k refers to a level of factor C actually included in an experiment. If all the factors in an experiment are fixed factors, there is no need to make the distinction between I, J, and K and i, j, and k.

The numerical data given in Table 5.8-3 will be used to illustrate the definitions of main effects and interactions in a three-factor experiment. The data in this table include all levels of each of the factors, and the entries in the cells are the means of all the potential observations that could be in each of the cells. Thus the numerical entries represent the parameters for a specified population.

Table 5.8-3 Population Means for 2 × 3 × 2 Factorial Experiment

(i)

	b_1		b_2		b_3	
	c_1	c_2	c_1	c_2	c_1	c_2
a_1	20	0	30	10	70	50
a_2	60	40	40	20	50	30
Mean	40	20	35	15	60	40

(ii)

	c_1	c_2	Mean
a_1	40	20	30
a_2	50	30	40
Mean	45	25	35

(iii)

	b_1	b_2	b_3	Mean
a_1	10	20	60	30
a_2	50	30	40	40
Mean	30	25	50	35

(ia)

	b_1		b_2		b_3		
	c_1	c_2	c_1	c_2	c_1	c_2	Mean
a_1	μ_{111}	μ_{112}	μ_{121}	μ_{122}	μ_{131}	μ_{132}	$\mu_{1..}$
a_2	μ_{211}	μ_{212}	μ_{221}	μ_{222}	μ_{231}	μ_{232}	$\mu_{2..}$
Mean	$\mu_{.11}$	$\mu_{.12}$	$\mu_{.21}$	$\mu_{.22}$	$\mu_{.31}$	$\mu_{.32}$	

(iia)

	c_1	c_2	Mean
a_1	$\mu_{1.1}$	$\mu_{1.2}$	$\mu_{1..}$
a_2	$\mu_{2.1}$	$\mu_{2.2}$	$\mu_{2..}$
Mean	$\mu_{..1}$	$\mu_{..2}$	$\mu_{...}$

(iiia)

	b_1	b_2	b_3	Mean
a_1	$\mu_{11.}$	$\mu_{12.}$	$\mu_{13.}$	$\mu_{1..}$
a_2	$\mu_{21.}$	$\mu_{22.}$	$\mu_{23.}$	$\mu_{2..}$
Mean	$\mu_{.1.}$	$\mu_{.2.}$	$\mu_{.3.}$	$\mu_{...}$

In part i the cell entries may be designated by the symbol μ_{IJK}. For example, $\mu_{211} = 60$. In symbols,

$$\mu_{211} = \frac{\sum_M X_{211M}}{N} = 60.$$

The entries along the lower margin of part i represent the means of the respective columns. Thus the entry 40 at the extreme left represents the mean of all potential observations in which factor B is at level b_1 and factor C is at level c_1; in symbols, this mean is $\mu_{.11}$. Thus,

$$\mu_{.11} = \frac{\sum_I \mu_{I11}}{P} = \frac{20 + 60}{2} = 40.$$

Part i*a* summarizes the symbols for the entries in part i.

In part ii each of the numerical entries in the cells represents a mean which has the general symbol $\mu_{I.K}$. For example, the entry in cell ac_{21} is

$$\mu_{2.1} = \frac{\sum_J \mu_{2J1}}{Q} = \frac{60 + 40 + 50}{3} = 50.$$

Each of the marginal entries to the right of the cells in part ii has the general symbol $\mu_{I..}$. For example, the entry 30 is the mean of the entries in row a_1; in symbols, $\mu_{1..} = 30$. Thus,

$$\mu_{1..} = \frac{\sum_K \mu_{1.K}}{R} = \frac{40 + 20}{2} = 30.$$

The entries along the bottom margin of part ii represent the means of all potential observations at specified levels of factor C. Thus,

$$\mu_{..2} = \frac{\sum_I \mu_{I.2}}{P} = \frac{20 + 30}{2} = 25.$$

Part ii*a* summarizes the symbols for corresponding entries in part ii.

In part iii each of the cell entries has the general designation $\mu_{IJ.}$. The marginal entries at the right may be designated $\mu_{I..}$; the marginal entries at the bottom may be designated $\mu_{.J.}$. Thus the entry 50 at the bottom of part iii is

$$\mu_{.3.} = \frac{\sum_I \mu_{I3.}}{P} = \frac{60 + 40}{2} = 50.$$

The main effects and interaction effects for a three-factor experiment will be defined in terms of the data in Table 5.8-3. The main effect due to level a_1 is

$$\alpha_1 = \mu_{1..} - \mu_{...} = 30 - 35 = -5.$$

In words, the main effect of level a_1 is a measure of the extent to which the mean of all potential observations at level a_1, averaged over all potential levels of factors B and C, differs from the grand mean of all potential observations. In general the main effect of level a_i is

$$\alpha_i = \mu_{i..} - \mu_{...}.$$

The main effect of level b_3 of factor B is by definition

$$\beta_3 = \mu_{.3.} - \mu_{...} = 50 - 35 = 15.$$

In general the main effect of level b_j is

$$\beta_j = \mu_{.j.} - \mu_{...}.$$

The main effect for level c_k of factor C has the general definition

$$\gamma_k = \mu_{..k} - \mu_{...}.$$

The interaction effect of level a_i with level b_j is the definition

$$\begin{aligned} \alpha\beta_{ij} &= \mu_{ij.} - \mu_{...} - \alpha_i - \beta_j = \mu_{ij.} - (\mu_{...} + \alpha_i + \beta_j) \\ &= \mu_{ij.} - \mu_{i..} - \mu_{.j.} + \mu_{...}. \end{aligned}$$

In words, $\alpha\beta_{ij}$ measures the extent to which the mean of all potential observations differs from the sum of the main effects of a_i and b_j and the grand mean. For example,

$$\alpha\beta_{13} = 60 - 30 - 50 + 35 = 15.$$

The interaction effect of level a_i with c_k is

$$\begin{aligned} \alpha\gamma_{ik} &= \mu_{i.k} - \mu_{...} - \alpha_i - \gamma_k \\ &= \mu_{i.k} - \mu_{i..} - \mu_{..k} + \mu_{...}. \end{aligned}$$

The interaction effect of level b_j with c_k is

$$\begin{aligned} \beta\gamma_{jk} &= \mu_{.jk} - \mu_{...} - \beta_j - \gamma_k \\ &= \mu_{.jk} - \mu_{.j.} - \mu_{..k} + \mu_{...}. \end{aligned}$$

For example, the interaction of level b_3 with level c_1 is

$$\beta\gamma_{31} = 60 - 50 - 45 + 35 = 0.$$

The interaction of a_i, b_j, and c_k is defined to be

$$\begin{aligned} \alpha\beta\gamma_{ijk} &= \mu_{ijk} - \mu_{...} - (\alpha\beta_{ij} + \alpha\gamma_{ik} + \beta\gamma_{jk} + \alpha_i + \beta_j + \gamma_k) \\ &= \mu_{ijk} - \mu_{ij.} - \mu_{i.k} - \mu_{.jk} + \mu_{i..} + \mu_{.j.} + \mu_{..k} - \mu_{...}. \end{aligned}$$

In words, the interaction of three factors measures the difference between the mean of all potential observations under a specified combination, μ_{ijk}, and the sum of two-factor interactions, main effects, and the grand mean. The three-factor interaction is in essence a measure of the nonadditivity of two-factor interactions and main effects. For example, for the data in Table 5.8-3, the interaction of a_1, b_3, and c_2 is

$$\begin{aligned} \alpha\beta\gamma_{132} &= \mu_{132} - \mu_{13.} - \mu_{1.2} - \mu_{.32} + \mu_{1..} + \mu_{.3.} + \mu_{..2} - \mu_{...} \\ &= 50 - 60 - 20 - 40 + 30 + 50 + 25 - 35 \\ &= 0. \end{aligned}$$

Equivalently,
$$\begin{aligned}\alpha\beta\gamma_{132} &= \mu_{132} - \mu_{...} - (\alpha\beta_{13} + \alpha\gamma_{12} + \beta\gamma_{32} + \alpha_1 + \beta_3 + \gamma_2)\\ &= 50 - 35 - [15 + 0 + 0 + (-5) + 15 + (-10)]\\ &= 0.\end{aligned}$$

From the definitions of these effects it follows that

$$\sum_I \alpha_I = 0, \qquad \sum_J \beta_J = 0, \qquad \sum_K \gamma_K = 0;$$

$$\sum_I \alpha\beta_{IJ} = \sum_J \alpha\beta_{IJ} = \sum_I \alpha\gamma_{IK} = \sum_K \alpha\gamma_{IJ} = \sum_J \beta\gamma_{JK} = \sum_K \beta\gamma_{JK} = 0;$$

$$\sum_I \alpha\beta\gamma_{IJK} = \sum_J \alpha\beta\gamma_{IJK} = \sum_K \alpha\beta\gamma_{IJK} = 0.$$

In each case the summation is over all the potential levels of the factors. If, for example, $k \neq K$, then

$$\sum_k \gamma_k \neq 0.$$

The variance due to the main effects of factor A is

$$\sigma_\alpha^2 = \frac{\sum_I \alpha_I^2}{P - 1}.$$

Similarly, the variances due to the main effects of factors B and C are, respectively,

$$\sigma_\beta^2 = \frac{\sum_J \beta_J^2}{Q - 1},$$

$$\sigma_\gamma^2 = \frac{\sum_K \gamma_K^2}{R - 1}.$$

When the main effects within any factor are all equal, the variance corresponding to these main effects will be zero. Hence the equality of main effects implies that the variance corresponding to these main effects is zero.

The variance due to two-factor interactions is defined as follows:

$$\sigma_{\alpha\beta}^2 = \frac{\sum_I\sum_J (\alpha\beta_{IJ})^2}{(P - 1)(Q - 1)},$$

$$\sigma_{\alpha\gamma}^2 = \frac{\sum_I\sum_K (\alpha\gamma_{IK})^2}{(P - 1)(R - 1)},$$

$$\sigma_{\beta\gamma}^2 = \frac{\sum_J\sum_K (\beta\gamma_{JK})^2}{(Q - 1)(R - 1)}.$$

The variance due to the three-factor interaction is

$$\sigma^2_{\alpha\beta\gamma} = \frac{\sum_I\sum_J\sum_K(\alpha\beta\gamma_{IJK})^2}{(P-1)(Q-1)(R-1)}.$$

The variance due to the experimental error has the same general definition as that given for two-factor experiments; it is the variance of the measurements on the N potential elements within each potential cell of the experiment. Thus for cell abc_{ijk},

$$\sigma^2_{\varepsilon_{ijk}} = \frac{\sum_M(X_{ijkM} - \mu_{ijk})^2}{N-1},$$

where the subscript M represents a potential element in the cell specified. Assuming that the variance due to experimental error is equal for all potential cells in the experiment, the overall variance due to experimental error is

$$\sigma^2_\varepsilon = \frac{\sum_I\sum_J\sum_K\sigma^2_{\varepsilon_{IJK}}}{PQR}.$$

Mean squares as computed from the actual experimental data are estimates of linear functions of these population variances. The mean squares of main effects, as obtained from observed data, will estimate a linear function of the variance due to the specified main effects, variance due to interactions effects, and variance due to experimental error.

The extension of the notation and definitions to four-factor experiments is direct. For example, the main effect of level d_m of factor D is

$$\delta_m = \mu_{...m} - \mu_{....}.$$

This effect is estimated by

$$\text{est}\,(\delta_m) = \bar{D}_m - \bar{G}.$$

The notation est (δ_m) denotes an estimate of the parameter δ_m obtained from the data in the experiment. The interaction effect associated with levels a_i and d_m is

$$\alpha\delta_{im} = \mu_{i..m} - \mu_{....} - (\alpha_i + \delta_m).$$

This interaction effect is estimated by

$$\text{est}\,(\alpha\delta_{im}) = \overline{AD}_{im} - \bar{A}_i - \bar{D}_m + \bar{G}.$$

The interaction of levels a_i, b_j, and d_m is defined to be

$$\alpha\beta\delta_{ijm} = \mu_{ij.m} - \mu_{....} - (\alpha\beta_{ij} + \alpha\delta_{im} + \beta\delta_{jm}) - (\alpha_i + \beta_j + \delta_m).$$

This interaction effect is estimated by

$$\text{est}\ (\alpha\beta\delta_{ijm}) = \overline{ABD}_{ijm} - \overline{AB}_{ij} - \overline{AD}_{im} - \overline{BD}_{jm} + \bar{A}_i + \bar{B}_j + \bar{D}_m - \bar{G}.$$

This last estimate has the general form

$$(\text{3-factor mean}) - \Sigma(\text{2-factor means}) + \Sigma(\text{1-factor means}) - \bar{G}.$$

The estimate of a four-factor interaction has the general form

$$\begin{aligned}(\text{4-factor mean}) &- \Sigma(\text{3-factor means}) + \Sigma(\text{2-factor means})\\ &- \Sigma(\text{1-factor means}) + \bar{G}.\end{aligned}$$

For example, the interaction of levels a_1, b_2, c_3, d_4 is estimated by

$$\begin{aligned}\text{est}\ (\alpha\beta\gamma\delta_{1234}) = \overline{ABCD}_{1234} &- (\overline{ABC}_{123} + \overline{ABD}_{124} + \overline{ACD}_{134} + \overline{BCD}_{234})\\ &+ (\overline{AB}_{12} + \overline{AC}_{13} + \overline{AD}_{14} + \overline{BC}_{23} + \overline{BD}_{24} + \overline{CD}_{34})\\ &- (\bar{A}_1 + \bar{B}_2 + \bar{C}_3 + \bar{D}_4) + \bar{G}.\end{aligned}$$

The term Σ(3-factor means) in the general expression for the estimate of a four-factor interaction effect includes all possible means of the form UVW_{rst}, where r, s, and t are the subscripts for corresponding terms in the interaction effect being estimated. In a four-factor experiment the number of terms in Σ(3-factor means) is equal to the number of combinations of four things taken three at a time, which is symbolized ${}_4C_3$. This number is

$${}_4C_3 = \frac{4 \cdot 3 \cdot 2}{1 \cdot 2 \cdot 3} = 4.$$

For a four-factor experiment the number of terms in the summation Σ(2-factor means) is

$${}_4C_2 = \frac{4 \cdot 3}{1 \cdot 2} = 6.$$

In a k-factor experiment, the estimate of the k-factor interaction effect has the general form

$$\begin{aligned}(k\text{-factor mean}) &- \Sigma[(k-1)\text{-factor means}] + \Sigma[(k-2)\text{-factor means}]\\ &- \Sigma[(k-3)\text{-factor means}] + \Sigma[(k-4)\text{-factor means}]\\ &- \cdots.\end{aligned}$$

The last term is $\pm[(k-k)\text{-factor means}] = \pm\bar{G}$. If k is an even number, the last term is $+\bar{G}$; if k is an odd number, the last term is $-\bar{G}$. The number of terms in the summation $\Sigma[(k-1)$-factor means] is ${}_kC_{k-1}$; the number of terms in the summation $\Sigma[(k-2)$-factor means] is ${}_kC_{k-2}$. For example, if $k = 5$,

$${}_kC_{k-2} = {}_5C_3 = \frac{5 \cdot 4 \cdot 3}{1 \cdot 2 \cdot 3} = 10; \qquad {}_kC_{k-1} = {}_5C_4 = \frac{5 \cdot 4 \cdot 3 \cdot 2}{1 \cdot 2 \cdot 3 \cdot 4} = 5.$$

5.9 Estimation and Tests of Significance for Three-factor Experiments

For purposes of demonstrating the principles underlying the analysis that will be made, it is convenient to formulate a structural model for an observation. For a three-factor experiment, the structural model has the form

$$X_{ijkm} = f(abc_{ijk}) + \varepsilon_{ijkm}. \tag{1}$$

The observation on element m under treatment combination abc_{ijk} is designated by the symbol X_{ijkm}. The symbol $f(abc_{ijk})$ denotes the hypothetically true effect of the treatment combination abc_{ijk}. The symbol ε_{ijkm} is the experimental error associated with the measurement on element m. In this context the experimental error is considered to be the difference between the observed measurement and the predicted measurement given by $f(abc_{ijk})$.

For purposes of the analysis that follows it will be assumed that the predicted measurement is a linear function of the main effects and interaction effects. Specifically,

$$f(abc_{ijk}) = \mu_{...} + \alpha_i + \beta_j + \gamma_k + \alpha\beta_{ij} + \alpha\gamma_{ik} + \beta\gamma_{jk} + \alpha\beta\gamma_{ijk}. \tag{2}$$

If (1) and (2) are combined, the resulting structural model is a generalization of the model given in Sec. 5.6 for a two-factor experiment. The assumptions that will be made about this model in the course of the analysis are direct generalizations of those summarized in Sec. 5.6.

It will be assumed that p levels of factor A are selected at random from a population of P levels. It will further be assumed that an independent random sample of q levels of factor B is selected from a population of Q levels and that a third independent random sample of size r is selected from a population of R levels of factor C. The treatments in the experiment are the pqr combinations that result when each of the selected levels of one factor is combined with each of the selected levels from the other factors. For example, if level a_1 is used in combination with b_2 and c_3, the resulting treatment is designated abc_{123}. In a $p \times q \times r$ factorial experiment there are pqr treatment combinations.

It will also be assumed that a random sample of $npqr$ element is drawn from a specified population. Random subsamples of size n each are assigned to each of the pqr treatment combinations to be studied in the experiment. After administration of the treatments, each of the elements is measured on a criterion of effectiveness (the dependent variable). The scale of measurement for the criterion is assumed to be given in terms of an experimentally meaningful unit.

From the data obtained in the experiment, mean squares are computed to estimate variances due to the structural variables on the right-hand side of the model. For a three-factor experiment, the definitions of the mean squares for main effects, interactions, and experimental error are summarized in Table 5.9-1. With the exception of multipliers and the ranges

Table 5.9-1 Definition of Mean Squares

A main effect	$MS_a = nqr\Sigma(\bar{A}_i - \bar{G})^2/(p-1)$
B main effect	$MS_b = npr\Sigma(\bar{B}_j - \bar{G})^2/(q-1)$
C main effect	$MS_c = npq\Sigma(\bar{C}_k - \bar{G})^2/(r-1)$
AB interaction	$MS_{ab} = nr\sum_i\sum_j(\overline{AB}_{ij} - \bar{A}_i - \bar{B}_j + \bar{G})/(p-1)(q-1)$
AC interaction	$MS_{ac} = nq\sum_i\sum_k(\overline{AC}_{ik} - \bar{A}_i - \bar{C}_k + \bar{G})^2/(p-1)(r-1)$
BC interaction	$MS_{bc} = np\sum_j\sum_k(\overline{BC}_{jk} - \bar{B}_j - \bar{C}_k + \bar{G})^2/(q-1)(r-1)$
ABC interaction	$MS_{abc} = n\Sigma\Sigma\Sigma(\overline{ABC}_{ijk} - \overline{AB}_{ij} - \overline{AC}_{ik} - \overline{BC}_{jk}$ $+ \bar{A}_i + \bar{B}_j + \bar{C}_k - \bar{G})^2/(p-1)(q-1)(r-1)$
Experimental error	$MS_{\text{error}} = \sum_i\sum_j\sum_k\sum_m(X_{ijkm} - \overline{ABC}_{ijk})^2/pqr(n-1)$

of summation, these definitions carry out on the experimental data the same operations that would be carried out on the population to obtain the variance due to main effects, interactions, and experimental error.

For example, in the population the variance due to the main effects of factor A is

$$\sigma_\alpha^2 = \frac{\sum_I(\mu_{I..} - \mu_{...})^2}{P-1}.$$

The mean square due to the main effects of factor A as computed from the data in the experiment is

$$MS_a = \frac{nqr\sum_i(\bar{A}_i - \bar{G})^2}{p-1}.$$

As another example, in the population the variance due to the BC interaction is

$$\sigma_{\beta\gamma}^2 = \frac{\sum_J\sum_K(\mu_{.JK} - \mu_{.J.} - \mu_{..K} + \mu_{...})^2}{(Q-1)(R-1)}.$$

The mean square due to the BC interaction as obtained from the data in the experiment is

$$MS_{bc} = \frac{np\sum_j\sum_k(\overline{BC}_{jk} - \bar{B}_j - \bar{C}_k + \bar{G})^2}{(q-1)(r-1)}.$$

In each case, with the exceptions noted, the mean square duplicates for the data obtained from the experiment the definitions for the variance due to main effects and interactions in the population.

The expected values of the mean squares computed from the experimental data are not, however, equal to a simple multiple of the corresponding variance in the population. For example,

$$E(MS_a) \neq nqr\sigma_\alpha^2.$$

In general, the expected value of the mean square due to the main effects of factor A will depend upon variance due to interactions with the levels of factor A, variance due to experimental error, and the variance due to the main effects of factor A. Similarly,

$$E(MS_{bc}) \neq np\sigma_{\beta\gamma}^2.$$

In general the expected value of MS_{bc} will depend upon the variance due to higher-order interactions which involve factors B and C, variance due to experimental error, and the variance due to the BC interaction. The expected values for the mean squares defined in Table 5.9-1 are summarized in Table 5.9-2. Part i of this table gives general values; part ii specializes these general values for specific cases. The last column is one of several possible special cases of the mixed model; for example, two factors may be random and one factor fixed; or one factor may be random, and two factors may be fixed. In all these cases it is assumed that the sample of elements observed under any treatment combination is a random sample of size n from a potentially infinite population of elements.

Tests of significance in a three-factor experiment follow the same general rules as those for a two-factor experiment. Appropriate F ratios are determined from the structure of the expected values of the mean squares which correspond to the design of the experiment. For the case of model I, all F ratios have MS_{error} for a denominator. For the case of model III given in Table 5.9-2, F ratios for main effects have the following structure:

$$H_1: \quad \sigma_\alpha^2 = 0, \qquad F = \frac{MS_a}{MS_{error}};$$

$$H_1: \quad \sigma_\beta^2 = 0, \qquad F = \frac{MS_b}{MS_{ab}};$$

$$H_1: \quad \sigma_\gamma^2 = 0, \qquad F = \frac{MS_c}{MS_{ac}}.$$

Tests on main effects for model II are considered in Sec. 5.15.

Under model III, the appropriate F ratio to test the hypothesis $\sigma_{\beta\gamma}^2 = 0$ is

$$F = \frac{MS_{bc}}{MS_{abc}}.$$

However, the appropriate F ratio for a test on the hypothesis that $\sigma_{\alpha\beta}^2 = 0$ has the form

$$F = \frac{MS_{ab}}{MS_{error}}.$$

Table 5.9-2 Expected Value for Mean Squares in a Three-factor Experiment Having n Observations per Cell

(i)

MS_a	$(1 - n/N)\sigma_\varepsilon^2 + n(1 - q/Q)(1 - r/R)\sigma_{\alpha\beta\gamma}^2 + nq(1 - r/R)\sigma_{\alpha\gamma}^2 + nr(1 - q/Q)\sigma_{\alpha\beta}^2 + nqr\sigma_\alpha^2$
MS_b	$(1 - n/N)\sigma_\varepsilon^2 + n(1 - p/P)(1 - r/R)\sigma_{\alpha\beta\gamma}^2 + np(1 - r/R)\sigma_{\beta\gamma}^2 + nr(1 - p/P)\sigma_{\alpha\beta}^2 + npr\sigma_\beta^2$
MS_c	$(1 - n/N)\sigma_\varepsilon^2 + n(1 - p/P)(1 - q/Q)\sigma_{\alpha\beta\gamma}^2 + np(1 - q/Q)\sigma_{\beta\gamma}^2 + nq(1 - p/P)\sigma_{\alpha\gamma}^2 + npq\sigma_\gamma^2$
MS_{ab}	$(1 - n/N)\sigma_\varepsilon^2 + n(1 - r/R)\sigma_{\alpha\beta\gamma}^2 + nr\sigma_{\alpha\beta}^2$
MS_{ac}	$(1 - n/N)\sigma_\varepsilon^2 + n(1 - q/Q)\sigma_{\alpha\beta\gamma}^2 + nq\sigma_{\alpha\gamma}^2$
MS_{bc}	$(1 - n/N)\sigma_\varepsilon^2 + n(1 - p/P)\sigma_{\alpha\beta\gamma}^2 + np\sigma_{\beta\gamma}^2$
MS_{abc}	$(1 - n/N)\sigma_\varepsilon^2 + n\sigma_{\alpha\beta\gamma}^2$
MS_{error}	$(1 - n/N)\sigma_\varepsilon^2$

Special Cases ($n/N = 0$ in all cases)

(ii)

Mean square	Model I All factors fixed	Model II All factors random	Model III Factor A random, all others fixed
MS_a	$\sigma_\varepsilon^2 + nqr\sigma_\alpha^2$	$\sigma_\varepsilon^2 + n\sigma_{\alpha\beta\gamma}^2 + nq\sigma_{\alpha\gamma}^2 + nr\sigma_{\alpha\beta}^2 + nqr\sigma_\alpha^2$	$\sigma_\varepsilon^2 + nqr\sigma_\alpha^2$
MS_b	$\sigma_\varepsilon^2 + npr\sigma_\beta^2$	$\sigma_\varepsilon^2 + n\sigma_{\alpha\beta\gamma}^2 + np\sigma_{\beta\gamma}^2 + nr\sigma_{\alpha\beta}^2 + npr\sigma_\beta^2$	$\sigma_\varepsilon^2 + nr\sigma_{\alpha\beta}^2 + npr\sigma_\beta^2$
MS_c	$\sigma_\varepsilon^2 + npq\sigma_\gamma^2$	$\sigma_\varepsilon^2 + n\sigma_{\alpha\beta\gamma}^2 + nq\sigma_{\alpha\gamma}^2 + np\sigma_{\beta\gamma}^2 + npq\sigma_\gamma^2$	$\sigma_\varepsilon^2 + nq\sigma_{\alpha\gamma}^2 + npq\sigma_\gamma^2$
MS_{ab}	$\sigma_\varepsilon^2 + nr\sigma_{\alpha\beta}^2$	$\sigma_\varepsilon^2 + n\sigma_{\alpha\beta\gamma}^2 + nr\sigma_{\alpha\beta}^2$	$\sigma_\varepsilon^2 + nr\sigma_{\alpha\beta}^2$
MS_{ac}	$\sigma_\varepsilon^2 + nq\sigma_{\alpha\gamma}^2$	$\sigma_\varepsilon^2 + n\sigma_{\alpha\beta\gamma}^2 + nq\sigma_{\alpha\gamma}^2$	$\sigma_\varepsilon^2 + nq\sigma_{\alpha\gamma}^2$
MS_{bc}	$\sigma_\varepsilon^2 + np\sigma_{\beta\gamma}^2$	$\sigma_\varepsilon^2 + n\sigma_{\alpha\beta\gamma}^2 + np\sigma_{\beta\gamma}^2$	$\sigma_\varepsilon^2 + n\sigma_{\alpha\beta\gamma}^2 + np\sigma_{\beta\gamma}^2$
MS_{abc}	$\sigma_\varepsilon^2 + n\sigma_{\alpha\beta\gamma}^2$	$\sigma_\varepsilon^2 + n\sigma_{\alpha\beta\gamma}^2$	$\sigma_\varepsilon^2 + n\sigma_{\alpha\beta\gamma}^2$
MS_{error}	σ_ε^2	σ_ε^2	σ_ε^2

The F ratios constructed for tests on fixed effects under model III are only approximately distributed as the corresponding F distributions. The distribution theory in this case is exact only under a set of homogeneity conditions on a set of covariances—or equivalently on a set of intraclass correlations.

5.10 Simple Effects and Their Tests

The definitions of simple effects will be cast in terms of a three-factor factorial experiment. However, the generalization to higher- or lower-order factorial experiments is direct. Simple effects are associated with both main effects and interaction effects. The former are called simple main effects, the latter simple interaction effects.

A $p \times q \times r$ factorial experiment may be considered from different points of view. For some purposes, it can be considered as a set of qr single-factor experiments on factor A, each experiment being for a different combination of b_j and c_k. For other purposes, it can be considered as a set of r experiments, each experiment being a $p \times q$ factorial experiment for a different level of factor C. It will be convenient, for the present, to adopt this latter point of view. The analysis of the experiment for level c_k takes the following form:

Source of variation	df
A (for c_k)	$p - 1$
B (for c_k)	$q - 1$
AB (for c_k)	$(p - 1)(q - 1)$

The main effects for this single experiment, relative to the entire set of experiments, are called the simple main effects for level c_k. The interaction in this experiment is called the simple AB interaction for level c_k. The variation due to the simple main effects of factor A, $SS_{a \text{ for } c_k}$, is related to the variation of the overall main effect of factor A and the overall AC interaction. Specifically,

$$\sum_k SS_{a \text{ for } c_k} = SS_a + SS_{ac}.$$

In words, the sum of the variation due to the simple main effects of A at the various levels of factor C is equal to the variation due to the overall main effect of factor A and the overall AC interaction. When the variation due to the AC interaction is zero, the sum of the variation for these simple effects is equal to the overall main effects. Analogously, it can be shown that

$$\sum_k SS_{ab \text{ for } c_k} = SS_{ab} + SS_{abc}.$$

Table 5.10-1 Definition of Effects

Effect	Definition
Overall main effect of a_i	$\alpha_i = \mu_{i..} - \mu_{...}$
Simple main effect of a_i for c_k	$\alpha_{i(c_k)} = \mu_{i.k} - \mu_{..k}$
Overall main effect of b_j	$\beta_j = \mu_{.j.} - \mu_{...}$
Simple main effect of b_j for c_k	$\beta_{j(c_k)} = \mu_{.jk} - \mu_{..k}$
Overall interaction effect of ab_{ij}	$\alpha\beta_{ij} = \mu_{ij.} - \mu_{...} - \alpha_i - \beta_j$
Simple interaction effect of ab_{ij} for c_k	$\alpha\beta_{ij(c_k)} = \mu_{ijk} - \mu_{..k} - \alpha_{i(c_k)} - \beta_{j(c_k)}$

In words, the sum of the variation of the simple two-factor interactions is equal to the variation of the overall two-factor interaction plus the three-factor interaction.

The definitions of the simple effects, in terms of population means, are given in Table 5.10-1. It will be noted that simple effects have the same general form as overall factorial effects; simple effects, however, are restricted to a single level of one or more of the factors. The degree to which overall effects approximate simple effects depends upon magnitudes of interactions. In the absence of interaction, overall effects will be equal to corresponding simple effects.

The definition of simple effects will be illustrated by the numerical data given in Table 5.10-2. The data in this table represent population means. An entry in a cell of part i has the general symbol μ_{ijk}; an entry in a cell of part ii has the general symbol $\mu_{ij.}$; an entry in a cell of part iii has the

Table 5.10-2 Population Means for 2 × 2 × 2 Factorial Design

(i)

	c_1			c_2		
	b_1	b_2	Mean	b_1	b_2	Mean
a_1	60	20	40	80	40	60
a_2	0	0	0	20	20	20
Mean	30	10	20	50	30	40

(ii)

	b_1	b_2	Mean
a_1	70	30	50
a_2	10	10	10
Mean	40	20	30

(iii)

	c_1	c_2	Mean
a_1	40	60	50
a_2	0	20	10
Mean	20	40	30

(iv)

	c_1	c_2	Mean
b_1	30	50	40
b_2	10	30	20
Mean	20	40	30

general symbol $\mu_{i.k}$. It is assumed that $P = Q = R = 2$. The simple main effects for factor A at levels c_1 and c_2 are most readily obtained from part iii. It will be found that

$$\alpha_1 = 50 - 30 = 20, \qquad \alpha_2 = 10 - 30 = -20;$$
$$\alpha_{1(c_1)} = 40 - 20 = 20, \qquad \alpha_{2(c_1)} = 0 - 20 = -20;$$
$$\alpha_{1(c_2)} = 60 - 40 = 20, \qquad \alpha_{2(c_2)} = 20 - 40 = -20.$$

In each case the simple main effect is equal to the corresponding overall main effect. This finding indicates that the two-factor interaction effects $\alpha\gamma_{ik}$ are all zero and hence that $\sigma^2_{\alpha\gamma} = 0$. The $\alpha\gamma_{ik}$'s are most readily obtained from part iii.

$$\alpha\gamma_{11} = 40 - 50 - 20 + 30 = 0,$$
$$\alpha\gamma_{12} = 60 - 50 - 40 + 30 = 0,$$
$$\alpha\gamma_{21} = 0 - 10 - 20 + 30 = 0,$$
$$\alpha\gamma_{22} = 20 - 10 - 40 + 30 = 0.$$

Conversely, when two-factor interaction effects are found to be zero, corresponding simple main effects will be equal to overall main effects.

The simple main effects for factor A at levels b_1 and b_2 are most readily obtained from part ii. These overall and simple main effects are

$$\alpha_1 = 50 - 30 = 20, \qquad \alpha_2 = 10 - 30 = -20;$$
$$\alpha_{1(b_1)} = 70 - 40 = 30, \qquad \alpha_{2(b_1)} = 10 - 40 = -30;$$
$$\alpha_{1(b_2)} = 30 - 20 = 10, \qquad \alpha_{2(b_2)} = 10 - 20 = -10.$$

In this case, simple main effects are not equal to corresponding overall main effects. This finding indicates that the two-factor interactions $\alpha\beta_{ij}$ are not all zero and hence that $\sigma^2_{\alpha\beta} \neq 0$. The $\alpha\beta_{ij}$'s are obtained from part ii.

$$\alpha\beta_{11} = 70 - 50 - 40 + 30 = 10,$$
$$\alpha\beta_{12} = 30 - 50 - 20 + 30 = -10,$$
$$\alpha\beta_{21} = 10 - 10 - 40 + 30 = -10,$$
$$\alpha\beta_{22} = 10 - 10 - 20 + 30 = 10.$$

The fact that the $\alpha\beta_{ij}$'s are not all equal to zero implies that the simple main effects are not equal to the corresponding overall main effects.

The overall two-factor interaction effects are related to corresponding simple effects. Specifically,

$$\alpha\beta_{ij} = \alpha_{i(b_j)} - \alpha_i. \tag{1}$$

For example,

$$\alpha\beta_{12} = \alpha_{1(b_2)} - \alpha_1 = 10 - 20 = -10,$$
$$\alpha\beta_{21} = \alpha_{2(b_1)} - \alpha_2 = -30 - (-20) = -10,$$

Similarly,

$$\alpha\beta_{ij} = \beta_{j(a_i)} - \beta_j. \tag{2}$$

These last relationships indicate that two-factor interaction effects will be zero when simple main effects are equal to overall main effects and, conversely, that when two-factor interaction effects are zero corresponding simple main effects will be equal to overall main effects.

The simple two-factor interaction effects $\alpha\beta_{ij(c_k)}$ are most readily obtained from part i. For example,

$$\begin{aligned} \alpha\beta_{11(c_1)} &= 60 - 40 - 30 + 20 = 10 \\ &= \mu_{111} - \mu_{1.1} - \mu_{.11} + \mu_{..1}, \\ \alpha\beta_{11(c_2)} &= 80 - 60 - 50 + 40 = 10 \\ &= \mu_{112} - \mu_{1.2} - \mu_{.12} + \mu_{..2}. \end{aligned}$$

Thus it is noted that

$$\alpha\beta_{11(c_k)} = \alpha\beta_{11} = 10, \qquad k = 1, 2.$$

In general, the overall three-factor interaction effect is related to simple two-factor interaction effects by means of the relation

$$\alpha\beta\gamma_{ijk} = \alpha\beta_{ij(c_k)} - \alpha\beta_{ij}. \tag{3}$$

Thus, when the simple interaction effect is equal to the overall interaction effect, the corresponding three-factor interaction effect will be zero.

For the data in part i in Table 5.10-2,

$$\alpha\beta_{11(c_1)} = \alpha\beta_{11}.$$

Hence from relation (3) it follows that $\alpha\beta\gamma_{111} = 0$. In general, for the data in part i it will be found that

$$\alpha\beta_{ij(c_k)} = \alpha\beta_{ij}, \qquad \text{for all } k\text{'s}.$$

From relation (3) it follows that all $\alpha\beta\gamma_{ijk}$'s are equal to zero. Hence $\sigma^2_{\alpha\beta\gamma} = 0$. Analogous to (3) are the relations

$$\alpha\beta\gamma_{ijk} = \alpha\gamma_{ik(b_j)} - \alpha\gamma_{ik}, \tag{4}$$

$$\alpha\beta\gamma_{ijk} = \beta\gamma_{jk(a_i)} - \beta\gamma_{jk}. \tag{5}$$

The relation (3) implies (4) and (5) is illustrated geometrically in the next section.

Estimates of simple effects are in general obtained in the same manner as corresponding overall effects. For example,

$$\text{est } \alpha_i = \bar{A}_i - \bar{G},$$

whereas

$$\text{est } \alpha_{i(c_k)} = \overline{AC}_{ik} - \bar{C}_k.$$

Similarly,

$$\text{est } \beta_{j(c_k)} = \overline{BC}_{jk} - \bar{C}_k,$$

$$\text{est } \alpha_{i(b_j)} = \overline{AB}_{ij} - \bar{B}_j.$$

The overall interaction effect is estimated by

$$\text{est } \alpha\beta_{ij} = \overline{AB}_{ij} - \bar{A}_i - \bar{B}_j + \bar{G},$$

whereas a simple interaction effect is estimated by

$$\text{est } \alpha\beta_{ij(c_k)} = \overline{ABC}_{ijk} - \overline{AC}_{ik} - \overline{BC}_{jk} + \bar{C}_k.$$

Variation due to simple effects is given by the general formula

$$\Sigma[\text{est (simple effect)}]^2.$$

Computational formulas for such variation are given in Chap. 6.

Tests on simple effects depend upon expected values of mean squares for such effects. In general, appropriate expected values of mean squares for simple effects may be obtained from the expected values appropriate for experiments of lower dimension than the original experiment, i.e., by considering the simple effects as overall effects of experiments of lower dimension. F ratios are constructed from expected values thus obtained. However, data from the entire experiment may be used to estimate mean squares, where such data are relevant.

5.11 Geometric Interpretation of Higher-order Interactions

Various orders of interactions may be represented geometrically. In terms of such representation, interesting aspects of the meaning of interactions will become clearer. In particular, it will be noted how simple interactions are related to overall interactions. It will also be noted how simple interactions are related to higher-order interactions. The numerical data in Table 5.11-1 will be used for this purpose. It will be assumed that these data represent population means.

Table 5.11-1 Population Means

(i)	c_1					c_2			
	b_1	b_2	b_3	Mean		b_1	b_2	b_3	Mean
a_1	20	30	70	40	a_1	0	10	50	20
a_2	60	40	50	50	a_2	40	20	30	30
Mean	40	35	60	45	Mean	20	15	40	25

(ii)	b_1	b_2	b_3	Mean	(iii)	b_1	b_2	b_3	Mean
a_1	10	20	60	30	c_1	40	35	60	45
a_2	50	30	40	40	c_2	20	15	40	25
Mean	30	25	50	35	Mean	30	25	50	35

Data necessary to plot the profiles for the BC interactions for the two levels of factor A are given in part i. The left-hand side of Fig. 5.11-1 represents the profiles of means which are in the a_1 row of part i. The dotted line represents the means of the form μ_{1j1}; the solid line represents means of the form μ_{1j2}. Thus the two profiles on the left represent the *BC* means for level a_1. These two profiles have the same shape (i.e., are parallel). This finding implies that all simple interaction effects $\beta\gamma_{jk(a_1)}$ are zero.

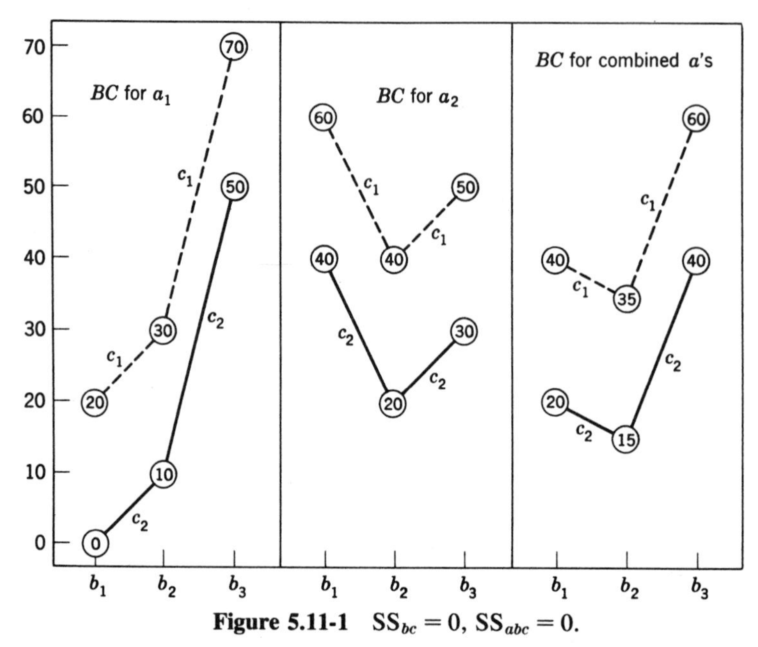

Figure 5.11-1 $SS_{bc} = 0$, $SS_{abc} = 0$.

In general, the simple two-factor interaction at level a_i has the definition

$$\beta\gamma_{jk(a_i)} = \mu_{ijk} - \mu_{ij.} - \mu_{i.k} + \mu_{i..}. \tag{1}$$

For the data in Table 5.11-1,

$$\beta\gamma_{11(a_1)} = 20 - 10 - 40 + 30 = 0,$$
$$\beta\gamma_{21(a_1)} = 30 - 20 - 40 + 30 = 0,$$
$$\beta\gamma_{31(a_1)} = 70 - 60 - 40 + 30 = 0.$$

The observation that the profiles on the left are parallel has the algebraic counterpart

$$\mu_{1j1} - \mu_{1j2} = \text{constant} \qquad \text{for all } j\text{'s}.$$

Hence the variation arising from such differences will be zero. This source of variation is that due to the *BC* interaction for a_1.

It is noted that the profiles of the *BC* means for level a_2 are parallel. This finding implies that all $\beta\gamma_{jk(a_2)}$'s are zero. Hence

$$SS_{bc \text{ for } a_1} + SS_{bc \text{ for } a_2} = 0 + 0 = 0.$$

In words, the sum of variations due to the simple interaction effects for the levels of factor A is zero. Since

$$\sum_i SS_{bc \text{ for } a_i} = SS_{bc} + SS_{abc},$$

when $\sum_i SS_{bc \text{ for } a_i} = 0$, both SS_{bc} and SS_{abc} must be equal to zero.

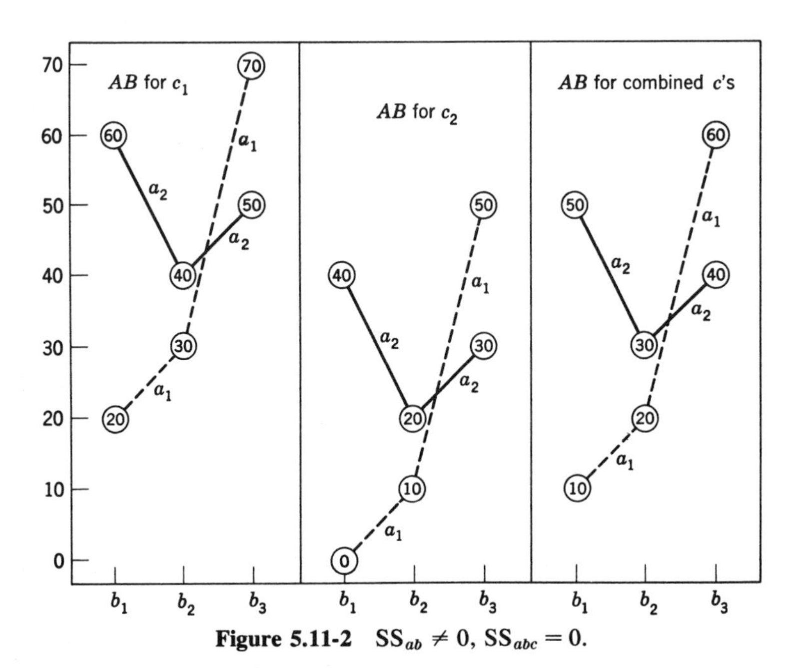

Figure 5.11-2 $SS_{ab} \neq 0$, $SS_{abc} = 0$.

The profiles of the overall BC means, obtained from part iii of Table 5.11-1, are given at the right in Fig. 5.11-1. The two profiles here are also parallel. This finding implies that

$$\beta\gamma_{jk} = 0, \qquad \text{for all } j\text{'s and } k\text{'s}.$$

Hence SS_{bc} must be zero. This latter result was implied by the fact that all the simple interaction effects of the $\beta\gamma_{jk(a_i)}$'s are zero. The fact that the BC profiles for a_1 and a_2 are parallel actually implies that the BC profiles for the combined levels of factor A are also parallel. To summarize, when the simple interactions of two factors at various levels of a third factor are all zero, the corresponding two-factor and three-factor interactions will also be zero.

It is, however, possible for two-factor interactions to be nonzero and yet have zero three-factor interaction. The profiles in Fig. 5.11-2 illustrate this case. These profiles represent the AB means for the two levels of factor C (Table 5.11-1). The profiles of AB for c_1 are not parallel; the profiles AB for c_2 are not parallel. Hence

$$\sum_k \text{SS}_{ab \text{ for } c_k} \neq 0.$$

Although the profiles within each level of factor C are not parallel, the a_1 profile for c_1 is parallel to the a_1 profile for the combined levels of factor C. Similarly, the a_1 profile for c_2 is parallel to the a_1 profile for the combined levels of factor C. The a_2 profiles for each level of factor C are also parallel to the a_2 profile for the combined data. This finding implies that SS_{abc} will be zero. The fact that the AB profiles for the combined data are not parallel indicates that SS_{ab} will not be zero.

To summarize the implications of the profiles with respect to the three-factor interaction, the latter will be zero when (1) the profiles of the two-factor means are parallel within each level of the third factor or when (2) the pattern of profiles for the two-factor means is geometrically similar to the pattern for the combined levels. In order that patterns be geometrically similar, corresponding profiles must be parallel.

A set of profiles in which the three-factor interaction is nonzero but the two-factor interaction is zero is given in Fig. 5.11-3. (These profiles are *not* based upon the data in Table 5.11-1.) The profiles within level c_1 are not parallel to each other, nor is the profile at a_1 parallel to the a_1 profile for the combined levels of factor C. Hence the three-factor interaction is nonzero. However, the AB profiles for the combined levels of factor C are parallel. Thus $\text{SS}_{ab} = 0$.

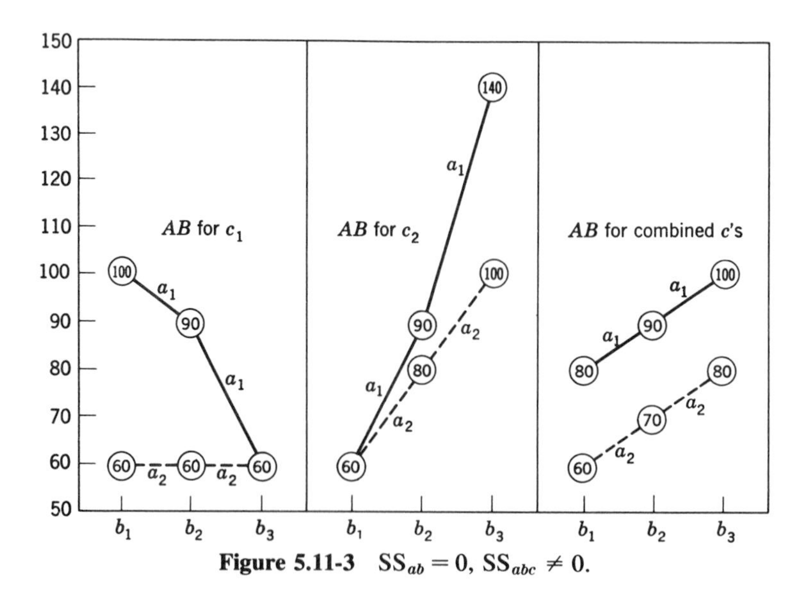

Figure 5.11-3 $\text{SS}_{ab} = 0$, $\text{SS}_{abc} \neq 0$.

The geometric relationships between the simple two-factor interactions, the three-factor interaction, and the overall two-factor interaction are seen more clearly by drawing three-dimensional profiles. The following data will be used for illustrative purposes:

	c_1		c_2	
	b_1	b_2	b_1	b_2
a_1	60	20	80	40
a_2	0	0	20	20

(These data are the same as those in part i of Table 5.10-2.) These data represent the population means for a $2 \times 2 \times 2$ factorial experiment. A

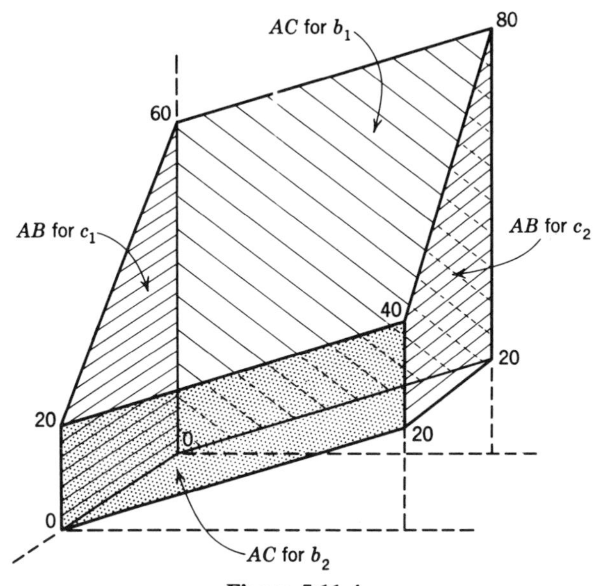

Figure 5.11-4

geometric representation of the patterns formed by these means is given in Fig. 5.11-4. The left-hand panel represents the four means in which factor C is at level c_1. This panel is denoted AB for c_1. The right-hand panel represents the four means in which factor C is at level c_2. The line (60,20) in the left panel is parallel to the line (80,40) in the right panel. The line (0,0) in the left panel is parallel to the line (20,20) in the right panel. The vertical lines in these panels are automatically parallel by the method of construction. Since corresponding sides of the left and right panels are parallel, the two panels are geometrically similar.

Geometric similarity of the left and right panels forces similarity of the front and back panels as well as similarity of the top and bottom panels. In analysis-of-variance terms, similarity of panels implies that the profiles of the simple two-factor interactions have the same patterns for all levels of a third factor. Thus the fact that the panel AB for c_1 is similar to the panel AB for c_2 implies that the simple AB interactions have the same pattern for c_1 as they do for c_2. This in turn implies that $SS_{abc} = 0$.

The fact that $SS_{abc} = 0$ is implied by the similarity of any two opposite panels—in turn, similarity of one pair of opposite panels implies the similarity of all other pairs. This geometric fact illustrates why one cannot distinguish between the interactions $(AB) \times C$, $(AC) \times B$, and $(BC) \times A$. Similarity of the left and right panels actually implies that $(AB) \times C$ is zero. But similarity of the left and right panels forces similarity of the front and back panels. The latter similarity implies that $(AC) \times B$ is zero. Thus, when $(AB) \times C$ is zero, $(AC) \times B$ must also be zero; one implies the other. More emphatically, one is not distinguishable from the other.

When the three-factor interaction is zero, inspection of individual panels will provide information about the two-factor interactions. In Fig.

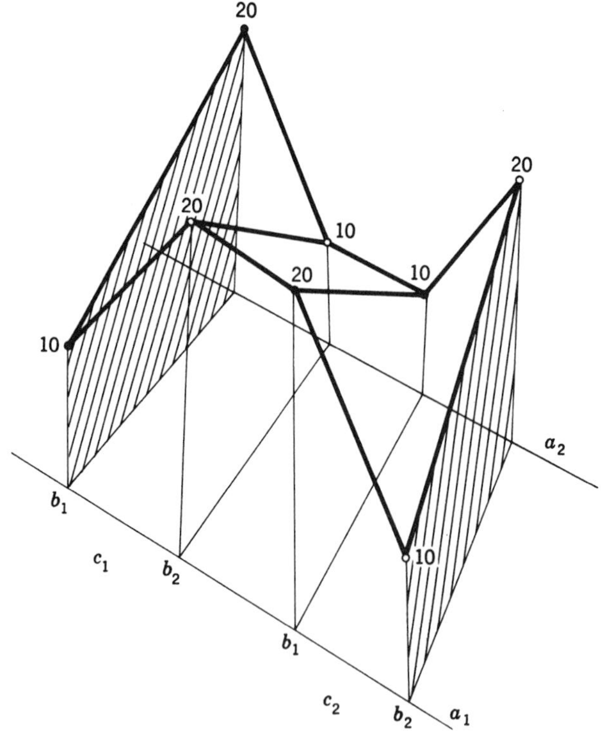

Figure 5.11-5

5.11-4 the line (60,20) is not parallel to the line (0,0). This implies that the variance due to the AB interaction is not zero. In the front panel, the line (20,40) is parallel to the line (0,20). This implies that the variance due to the AC interaction is zero. In the top panel, the line (60,80) is parallel to the line (20,40). This implies that $SS_{bc} = 0$. When the three-factor interaction is not zero, inspection of the individual panels does not provide information with respect to the two-factor interactions. The individual panels in pairs must be averaged in order to obtain information relevant to the overall two-factor interaction.

An example of a set of cell means which defines a population in which all main effects and all two-factor interactions are zero but the three-factor interaction is not zero is given below. A three-dimensional plot of

	c_1		c_2	
	b_1	b_2	b_1	b_2
a_1	10	20	20	10
a_2	20	10	10	20

these means appears in Fig. 5.11-5. Another view of the configuration in Fig. 5.11-5 is shown in Fig. 5.11-6. In the latter figure, the information

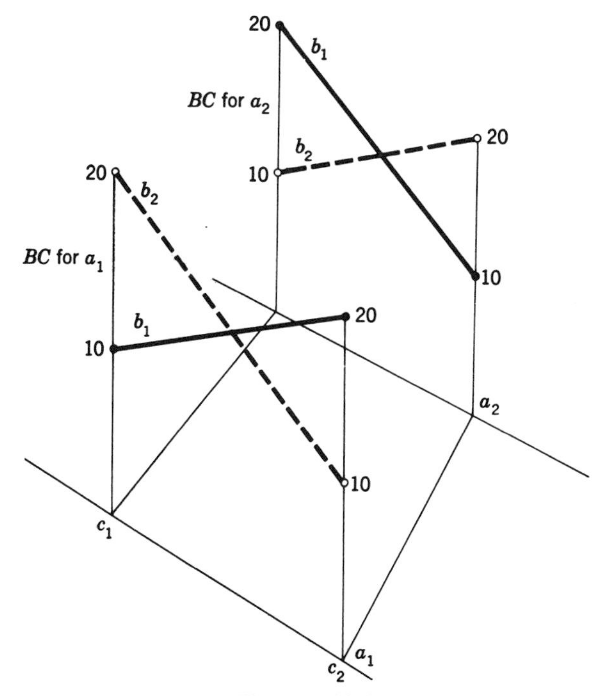

Figure 5.11-6

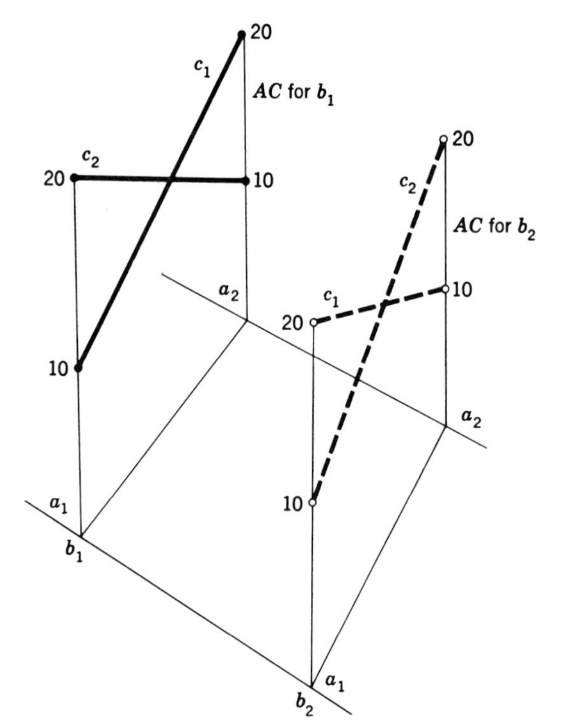

Figure 5.11-7

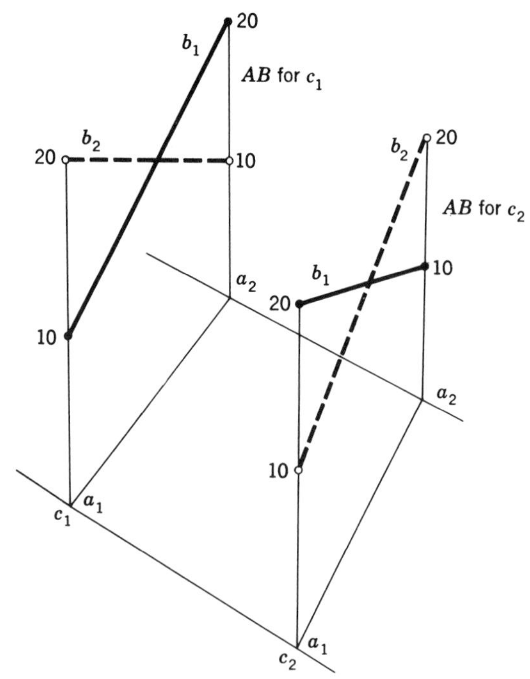

Figure 5.11-8

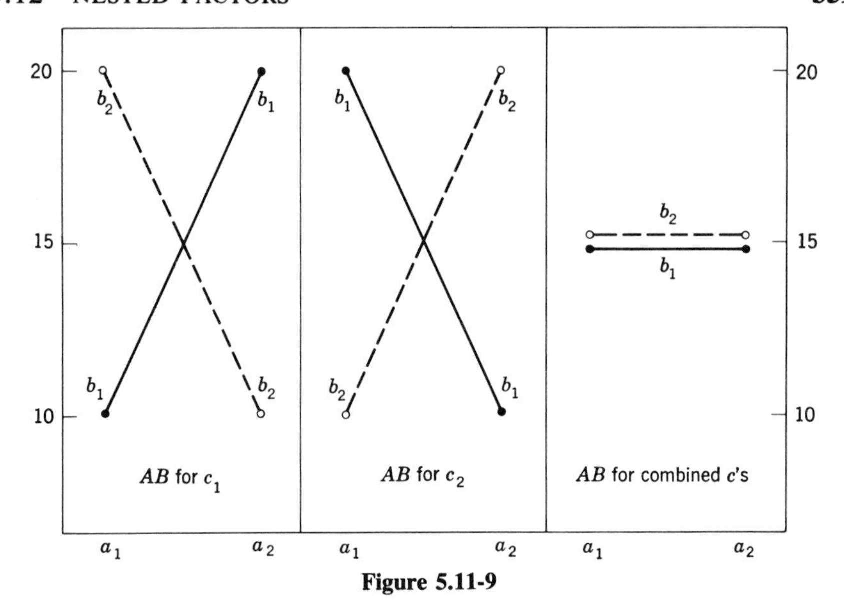

Figure 5.11-9

on the levels of factor B appears as profiles. Figures 5.11-7 and 5.11-8 present somewhat different views of Fig. 5.11-5. In Fig. 5.11-8, the left-hand panel represents the relationship between the responses to combinations of factors A and B for level c_1; the right-hand panel represents the corresponding relationships for level c_2.

Another way of representing Fig. 5.11-8 appears in Fig. 5.11-9. Here the left- and right-hand panels of Fig. 5.11-8 appear side by side in two dimensions. The combined profiles appear at the extreme right. The fact that the combined profiles are parallel (actually identical in this case) indicates that the variation due to the AB interaction is zero. The fact that the combined profiles are identical and horizontal indicates that variation due to the main effects of factors A and B are both zero.

5.12 Nested Factors (Hierarchal Designs)

Consider an experiment conducted to evaluate the relative effectiveness of two drugs with respect to some specified criterion. Suppose that the design calls for the administration of drug 1 to n patients from each of hospitals 1, 2, and 3; drug 2 is to be administered to n patients from each of hospitals 4, 5, and 6. This design can be represented schematically as follows:

Drug 1			Drug 2		
Hosp. 1	Hosp. 2	Hosp. 3	Hosp. 4	Hosp. 5	Hosp. 6
n	n	n	n	n	n

The difference between the mean effect of drug 1 and the mean effect of drug 2 will be due in part to differences between the unique effects associated with hospitals 1, 2, and 3 and the unique effects associated with hospitals 4, 5, and 6.

The unique effects associated with hospitals 1, 2, and 3 are confined to drug 1 whereas the unique effects associated with hospitals 4, 5, and 6 are confined to drug 2. Effects which are restricted to a single level of a factor are said to be *nested* within that factor. In the experimental design being considered, the hospital effects are nested under the drug factor. Since a given hospital appears only under one of the two drugs, there is no way of evaluating the interaction effect between the hospital and the drug. Before such an interaction effect can be evaluated, each hospital must appear under both levels of the drug factor.

Thus, in a two-factor experiment having one factor nested under the other, the interaction effect cannot be evaluated. For the general case of a two-factor experiment in which factor B is nested under factor A, the structural model is

$$\overline{AB}_{ij} = \mu_{..} + \alpha_i + \beta_{j(i)} + \bar{\varepsilon}_{ij}.$$

The notation $\beta_{j(i)}$ indicates that the effect of level b_j is nested under level a_i. Note that no interaction term of the form $\alpha\beta_{ij(i)}$ appears in the model. Inferences made from this type of design assume implicitly that the variation associated with this latter interaction is either zero or negligible relative to the variation associated with the main effects.

The analysis of variance for the design outlined at the beginning of this section takes the following form:

	Factor	df	df for general case
A	Drug	1	$p - 1$
B (w. a_1)	Hospitals within drug 1	2	$q - 1$
B (w. a_2)	Hospitals within drug 2	2	$q - 1$
	Within hospital	$6(n - 1)$	$pq(n - 1)$

The expected values of the mean squares in this analysis are as follows:

Source of variation	df	E(MS)
A	$p - 1$	$\sigma_\varepsilon^2 + nD_q\sigma_\beta^2 + nq\sigma_\alpha^2$
B (pooled)	$p(q - 1)$	$\sigma_\varepsilon^2 + n\sigma_\beta^2$
Experimental error (within cell)	$pq(n - 1)$	σ_ε^2

The symbol D_q is used to designate the expression $1 - q/Q$. Numerically $D_q = 0$ when $q = Q$ (that is, when factor B is fixed), and $D_q = 1$ when $q/Q = 0$ (that is, when factor B is random). To test the hypothesis that $\sigma_\alpha^2 = 0$,

$$F = \frac{\text{MS}_a}{\text{MS}_b}, \qquad \text{when factor } B \text{ is random;}$$

$$F = \frac{\text{MS}_a}{\text{MS}_{\text{error}}}, \qquad \text{when factor } B \text{ is fixed.}$$

By way of contrast, in a two-factor factorial experiment each level of one of the factors is associated with each level of the second factor. If the design outlined at the beginning of this section were changed to a two-factor factorial experiment, the new design could be represented schematically as follows:

	Hosp. 1	Hosp. 2	Hosp. 3	Hosp. 4	Hosp. 5	Hosp. 6
Drug 1	$n/2$	$n/2$	$n/2$	$n/2$	$n/2$	$n/2$
Drug 2	$n/2$	$n/2$	$n/2$	$n/2$	$n/2$	$n/2$

This factorial experiment requires n subjects from each of the hospitals, but $n/2$ of the subjects from each hospital are given drug 1, and $n/2$ of the subjects are given drug 2. In many cases the two-factor factorial experiment is to be preferred to the two-factor design in which the hospital factor is nested under the drug factor—particularly in those cases in which an interaction effect might be suspected. However, there are some instances in which the experimenter may be forced to use a design in which one factor is nested within another.

As another illustration of nested effects, consider the following experimental design:

	Drug 1			Drug 2		
	Hosp. 1	Hosp. 2	Hosp. 3	Hosp. 4	Hosp. 5	Hosp. 6
Category 1	n	n	n	n	n	n
Category 2	n	n	n	n	n	n

This design calls for a sample of n patients in category 1 and n patients in category 2 (random samples) from each of the hospitals. Patients from hospitals 1, 2, and 3 receive drug 1; patients from hospitals 4, 5, and 6 receive drug 2. In this design the hospital factor is nested under the drug factor. Since some patients from each category receive drug 1 and some patients from each category receive drug 2, the category factor is not nested under the drug factor. Further, since patients from each of the categories are obtained from each hospital, the category factor is not nested under the hospital factor.

In this case the hospital factor is nested under the drug factor but the hospital factor is crossed with the category factor. The drug factor is also crossed with the category factor. One may indicate the nesting and crossing schematically as follows:

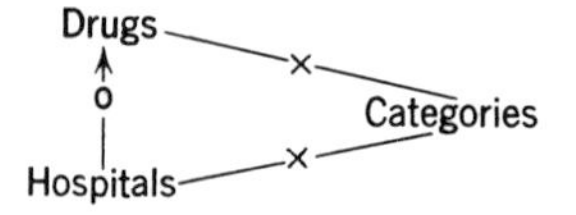

The design that has just been sketched may be considered as a three-factor experiment, the factors being drugs (A), hospitals (B), and categories (C). In this case, factor B is nested under factor A, but all other relationships are those of a bona fide factorial experiment, namely, crossing. The model for this type of experiment has the following form:

$$\overline{ABC}_{ijk} = \mu + \alpha_i + \beta_{j(i)} + \gamma_k + \alpha\gamma_{ik} + \beta\gamma_{j(i)k} + \bar{\varepsilon}_{ijk}.$$

No interaction in which the subscript i appears twice occurs in this model. That is, the interactions $\alpha\beta_{ij(i)}$ and $\alpha\beta\gamma_{ij(i)k}$ do not appear. Hence the utility of this type of design is limited to situations in which such interactions are either zero or negligible relative to other sources of variation of interest.

The analysis of variance for this design takes the form given in Table 5.12-1. In making tests by means of the F ratio, the appropriate ratios are determined from the expected values of the mean squares. The latter

Table 5.12-1 Analysis of Three-factor Experiment in Which Factor B is Nested under Factor A

Source of variation		df	df for general case
A	Drugs	1	$p - 1$
B (w. a_1)	Hospital w. drug 1	2 } 4	$q - 1$ } $p(q-1)$
.		.	.
.		.	.
.		.	.
B(w. a_p)	Hospital w. drug 2	2	$q - 1$
C	Categories	1	$r - 1$
AC	Drug × category	1	$(p-1)(r-1)$
(B w. A) × C	Hospital × category	4	$p(q-1)(r-1)$
Error	Within cell	$12(n-1)$	$pqr(n-1)$
	Total	$11 + 12(n-1)$	$npqr - 1$

are in part a function of whether the factors in the experiment are fixed or random. General expressions for the expected values of mean square

Table 5.12-2 Expected Values of Mean Squares, Factor *B* Nested under Factor *A*

Source of variation	df	Expected value of mean square
A	$p - 1$	$\sigma_\varepsilon^2 + nD_qD_r\sigma_{\beta\gamma}^2 + nqD_r\sigma_{\alpha\gamma}^2 + nrD_q\sigma_\beta^2 + nqr\sigma_\alpha^2$
B w. A	$p(q - 1)$	$\sigma_\varepsilon^2 + nD_r\sigma_{\beta\gamma}^2 + nr\sigma_\beta^2$
C	$r - 1$	$\sigma_\varepsilon^2 + nD_q\sigma_{\beta\gamma}^2 + nqD_p\sigma_{\alpha\gamma}^2 + npq\sigma_\gamma^2$
AC	$(p - 1)(r - 1)$	$\sigma_\varepsilon^2 + nD_q\sigma_{\beta\gamma}^2 + nq\sigma_{\alpha\gamma}^2$
$(B$ w. $A)C$	$p(q - 1)(r - 1)$	$\sigma_\varepsilon^2 + n\sigma_{\beta\gamma}^2$
Within cell	$pqr(n - 1)$	σ_ε^2

appropriate for this type of design are given in Table 5.12-2. In this table the symbol D_p is used to designate $1 - p/P$, D_q is used to designate $1 - q/Q$, and D_r is used to designate $1 - r/R$. Each of these D's is either 0 or 1 depending, respectively, on whether the corresponding factor is fixed or random.

In this table, the source of variation B w. A (the main effects due to factor B, which is nested within factor A) is actually the sum of the following main effects:

		df
	B w. a_1	$q - 1$
	. . .	. . .
	B w. a_p	$q - 1$
Sum =	B w. A	$p(q - 1)$

Similarly, the interaction $(B$ w. $A) \times C$ represents a pooling of the following interactions:

		df
	$(B$ w. $a_1)C$	$(q - 1)(r - 1)$
	. . .	. . .
	$(B$ w. $a_p)C$	$(q - 1)(r - 1)$
Sum =	$(B$ w. $A)C$	$p(q - 1)(r - 1)$

If this pooled interaction is used in the denominator of an F ratio, the variations which are pooled must be homogeneous if the resulting F ratio is to have an F distribution when the hypothesis being tested is true.

As another illustration of nested effects, consider an experiment in which the subjects have been classified as follows:

Company 1				Company 2			
Dept. 1		Dept. 2		Dept. 3		Dept. 4	
Job 1	Job 2	Job 3	Job 4	Job 5	Job 6	Job 7	Job 8
n	n	n	n	n	n	n	n

Suppose that n people from each job are included in an experiment in which attitude toward a retirement plan is being studied. This design may be considered as a three-factor experiment in which the department factor (B) is nested under the company factor (A). The job factor (C) is nested under both factors B and A. This type of design is referred to as a *hierarchal* design. In a three-factor hierarchal experiment, factor B is nested under factor A, and factor C is nested under both factors B and A. (The design in Table 5.12-1 is only partially hierarchal, since factor C was not nested under either factor A or factor B.)

The model for a three-factor hierarchal experiment has the form

$$\overline{ABC}_{ijk} = \mu_{...} + \alpha_i + \beta_{j(i)} + \gamma_{k(ij)} + \bar{\varepsilon}_{ijk}.$$

The notation $\gamma_{k(ij)}$ indicates that factor C is nested under both factors A and B. It should be noted that no interaction terms appear explicitly in this model. The expected values of the mean squares for this design are summarized in Table 5.12-3. The numerical values of the D's in these expected values depend upon the respective sampling fractions for the levels of the factors.

Table 5.12-3 Expected Values of Mean Squares for Three-factor Hierarchal Experiment

Source of variation	df	Expected value of mean square
A	$p - 1$	$\sigma_\varepsilon^2 + nD_r\sigma_\gamma^2 + nrD_q\sigma_\beta^2 + nqr\sigma_\alpha^2$
B w. A	$p(q - 1)$	$\sigma_\varepsilon^2 + nD_r\sigma_\gamma^2 + nr\sigma_\beta^2$
C w. (A and B)	$pq(r - 1)$	$\sigma_\varepsilon^2 + n\sigma_\gamma^2$
Experimental error	$pqr(n - 1)$	σ_ε^2
Total	$npqr - 1$	

The expected values of the mean squares for a completely hierarchal design are readily obtained, once the expected value of the mean square for the factor within which all other factors are nested is determined. The expected value for each succeeding factor is obtained from the one above by dropping the last term and making D in the next to the last term unity.

The following design represents a partially hierarchal design:

	City 1		City 2	
	School 1	School 2	School 3	School 4
Method 1	n	n	n	n
Method 2	n	n	n	n

Suppose that the purpose of this experiment is to evaluate the relative effectiveness of two different methods of teaching a specified course. Since both methods of training are given within each of the schools and within each of the cities, there is no nesting with respect to the methods factor. The school factor is, however, nested within the city factor. The expected values of the mean squares for this design have the form given in Table 5.12-4. This design enables the experimenter to eliminate systematic

Table 5.12-4 Expected Values of Mean Squares for Three-factor Partially Hierarchal Experiment

Source of variation	df	Expected value of mean square
A Methods	$p - 1$	$\sigma_\varepsilon^2 + nD_r\sigma_{\alpha\gamma}^2 + nrD_q\sigma_{\alpha\beta}^2 + nqr\sigma_\alpha^2$
B Cities	$q - 1$	$\sigma_\varepsilon^2 + nD_rD_p\sigma_{\alpha\gamma}^2 + nrD_p\sigma_{\alpha\beta}^2 + npD_r\sigma_\gamma^2 + npr\sigma_\beta^2$
C w. B Schools within cities	$q(r - 1)$	$\sigma_\varepsilon^2 + nD_p\sigma_{\alpha\gamma}^2 + np\sigma_\gamma^2$
AB	$(p - 1)(q - 1)$	$\sigma_\varepsilon^2 + nD_r\sigma_{\alpha\gamma}^2 + nr\sigma_{\alpha\beta}^2$
$A \times (C$ w. $B)$	$q(p - 1)(r - 1)$	$\sigma_\varepsilon^2 + n\sigma_{\alpha\gamma}^2$
Within cell	$pqr(n - 1)$	σ_ε^2

sources of variation associated with differences between cities and differences between schools within cities from the experimental error—at the cost, however, of reduced degrees of freedom for experimental error.

As still another example of a useful design involving a nested factor, consider an experiment with the following schematic representation:

	Method 1			Method 2		
	Person 1	Person 2	Person 3	Person 4	Person 5	Person 6
Period 1						
Period 2						
Period 3						
Period 4						

In this design persons 1, 2, and 3 are observed under method of training 1; criterion measures are obtained at four different periods during the training process. Persons 4, 5, and 6 are observed at comparable periods under training method 2. This design may be considered as a three-factor experiment in which the person factor is nested under the methods factor. If methods is considered to be factor A, persons factor B, and periods factor C, the model for this design is

$$ABC_{ijk} = \mu_{..} + \alpha_i + \beta_{j(i)} + \gamma_k + \alpha\gamma_{ik} + \beta\gamma_{j(i)k} + \varepsilon_{ijk}.$$

For this design there is only one observation in each cell of the experiment. Hence there is no within-cell variation. The analysis of variance is identical in form with that given in Table 5.12-2 if the within-cell variation is deleted. The expected values of the mean squares are also identical to those given in Table 5.12-2, with n set equal to unity. Specializing this design to the case in which factors A (methods) and C (periods) are fixed and factor B (persons) is random, one obtains the expected values given in Table 5.12-5. In this table, the source of variation due to B w. A

Table 5.12-5 Three-factor Partially Hierarchal Design
(Factors A and C fixed, B random)

Source of variation		df	Expected value of mean square
A	Methods	$p - 1$	$\sigma_\varepsilon^2 + r\sigma_\beta^2 + qr\sigma_\alpha^2$
B w. A	People within methods	$p(q - 1)$	$\sigma_\varepsilon^2 + r\sigma_\beta^2$
C	Periods	$r - 1$	$\sigma_\varepsilon^2 + \sigma_{\beta\gamma}^2 + pq\sigma_\gamma^2$
AC	Method × period	$(p - 1)(r - 1)$	$\sigma_\varepsilon^2 + \sigma_{\beta\gamma}^2 + q\sigma_{\alpha\gamma}^2$
$(B$ w. $A)C$		$p(q - 1)(r - 1)$	$\sigma_\varepsilon^2 + \sigma_{\beta\gamma}^2$

represents the pooled variation of people within methods. A homogeneity assumption is required for this pooling. The interaction term $(B$ w. $A)C$ also represents a pooling of different sources of variation. The homogeneity assumption required for pooling in this case is equivalent to the assumption that the correlation between periods be constant within each of the methods. The F tests for the analysis in Table 5.21-5 have the following form:

$$H_1: \quad \sigma_\alpha^2 = 0, \qquad F = \frac{\text{MS}_{\text{methods}}}{\text{MS}_{\text{people w. methods}}};$$

$$H_1: \quad \sigma_\gamma^2 = 0, \qquad F = \frac{\text{MS}_{\text{periods}}}{\text{MS}_{(B\text{ w. }A)C}};$$

$$H_1: \quad \sigma_{\alpha\gamma}^2 = 0, \qquad F = \frac{\text{MS}_{\text{method}\times\text{period}}}{\text{MS}_{(B\text{ w. }A)C}}.$$

Rather than considering this last experiment as three-factor partially hierarchal, there are other ways of classifying this type of experiment. In particular, it can be considered as a two-factor experiment in which there are repeated measurements on one of the factors. This latter type of classification will receive more extensive treatment in Chap. 7.

5.13 Split-plot Designs

The *split-plot* design has much in common with the partially hierarchal design. The term split-plot comes from agricultural experimentation in which a single level of one treatment is applied to a relatively large plot of ground (the whole plot) but all levels of a second treatment are applied to subplots within the whole plot. For example, consider the following design, in which the levels of factor A are applied to the whole plots and the levels of factor B are applied to the subplots:

a_1 Plot 1	a_2 Plot 2	a_1 Plot 3	a_2 Plot 4
b_2	b_1	b_2	b_3
b_3	b_3	b_1	b_2
b_1	b_2	b_3	b_1

In this design, differences between the levels of factor A cannot be estimated independently of differences between groups of plots. That is,

$$\begin{aligned} 6 \text{ est } (\alpha_1 - \alpha_2) &= (\text{plot 1} + \text{plot 3}) - (\text{plot 2} + \text{plot 4}) \\ &= \qquad A_1 \qquad\qquad - \qquad\qquad A_2. \end{aligned}$$

For this reason the variation due to the levels of factor A is part of the between-plot effects. However, comparisons among the levels of factor B are part of the within-plot variation. From the information within each of the plots, estimates of the main effects due to factor B may be obtained. These estimates are free of variation due to whole plots. In the analysis of this type of design, sources which are part of the whole-plot variation are usually grouped separately from those which are part of the within-plot variation.

The model for this experiment is

$$X_{ijk} = \mu_{...} + \alpha_i + \pi_{k(i)} + \beta_j + \alpha\beta_{ij} + \pi'_{k(ij)} + \varepsilon_{ijk}.$$

The notation $\pi_{k(i)}$ designates the effect of plot k within level a_i. (This notation indicates that the plot effects are nested within the levels of factor A.) The notation $\pi'_{k(ij)}$ designates residual subplot effects. For the special, but frequently occurring, case in which A and B are fixed factors and the plots are a random sample from a specified population of plots, the analysis of variance assumes the form given in Table 5.13-1. In this design, each of the p levels of factor A is assigned at random to n plots. Within each plot, the levels of factor B are assigned at random to the subplots. The expected values of the mean squares in this table are actually identical in form with those given in Table 5.12-5 if plots are assigned the role of factor B in this latter design.

There is a distinction between the designs usually placed in the hierarchal category and designs in the split-plot category. In the hierarchal designs, generally (but not always) all except one of the factors are modes of classifying the experimental units rather than treatments which are administered to the units by the experimenter. Such modes of classification are set up primarily to eliminate, in part, differences among the experimental units from the experimental error. Interaction between the

Table 5.13-1 Expected Values of Mean Squares for Split-plot Design
(A and B fixed, plots random)

Source of variation	df	Expected value of mean square
Between plots	$np - 1$	
A	$p - 1$	$\sigma_\varepsilon^2 + q\sigma_\pi^2 + nq\sigma_\alpha^2$
Plots w. a_1 ... Plots w. a_p	$p(n - 1)$	$\sigma_\varepsilon^2 + q\sigma_\pi^2$
Within plots	$np(q - 1)$	
B	$q - 1$	$\sigma_\varepsilon^2 + \sigma_{\pi'}^2 + np\sigma_\beta^2$
AB	$(p - 1)(q - 1)$	$\sigma_\varepsilon^2 + \sigma_{\pi'}^2 + n\sigma_{\alpha\beta}^2$
$B \times$ plots w. a_1 ... $B \times$ plots w. a_p	$p(q - 1)(n - 1)$	$\sigma_\varepsilon^2 + \sigma_{\pi'}^2$

treatment and the classifications is generally considered to be negligible. The usual hierarchal experiment may be regarded as a single-factor experiment with controls on the grouping of the experimental units.

In contrast, the split-plot design has two treatment factors. Whole plots are the experimental units for one of the factors, whereas the subplots are the experimental units for the second factor. If this distinction is made, the final design considered in the last section should be placed in the split-plot rather than the partially hierarchal category.

By inspection of the design outlined at the beginning of this section it is not obvious that the AB interaction is free of variation due to whole-plot effects. To demonstrate that such is the case, consider only the whole-plot effects in each of the following means:

Mean	Whole-plot effects
$\bar{A}_1$	$\pi_1 + \pi_3$
$\bar{B}_1$	$\pi_1 + \pi_2 + \pi_3 + \pi_4$
$\overline{AB}_{11}$	$\pi_1 + \pi_3$
$\bar{G}$	$\pi_1 + \pi_2 + \pi_3 + \pi_4$

An estimate of the interaction effect associated with treatment combination ab_{11} is

$$\text{est } \alpha\beta_{11} = \overline{AB}_{11} - \bar{A}_1 - \bar{B}_1 + \bar{G}.$$

The whole-plot effects associated with est $\alpha\beta_{11}$ are

$$(\pi_1 + \pi_3) - (\pi_1 + \pi_3) - (\pi_1 + \pi_2 + \pi_3 + \pi_4) + (\pi_1 + \pi_2 + \pi_3 + \pi_4) = 0.$$

Thus the whole-plot effects for this interaction sum to zero. In general, the whole-plot effects for est $\alpha\beta_{ij}$ will sum to zero for all i's and j's. Hence the variation due to whole-plot effects does not influence variation due to the AB interaction.

It should be noted, however, that the variation due to the AB interaction is not free of variation, if any, associated with the $B \times$ plot interaction. This latter source of variation may be analyzed as follows:

$B \times$ plot	$(q - 1)(np - 1)$
AB	$(p - 1)(q - 1)$
$B \times$ plots w. a_1 . . . $B \times$ plots w. a_p	$p(q - 1)(n - 1)$

This analysis of the $B \times$ plot interaction shows the AB interaction as part of this source of variation. This actually follows directly from the fact that the effects of factor A are confounded with groups of plots.

Several variations of the split-plot design are possible. One such variation is a double-split or split-split-plot design. In this type of design each level of factor A is assigned at random to n whole plots. (A total of np whole plots is required for this design.) Each of the whole plots is divided into q subplots. The q levels of factor B are then assigned at random to the subplots within each whole plot, and each subplot is divided into r sub-subplots. The r levels of factor C are then assigned at random to each of the sub-subplots. Part of a split-split-plot design is illustrated schematically below:

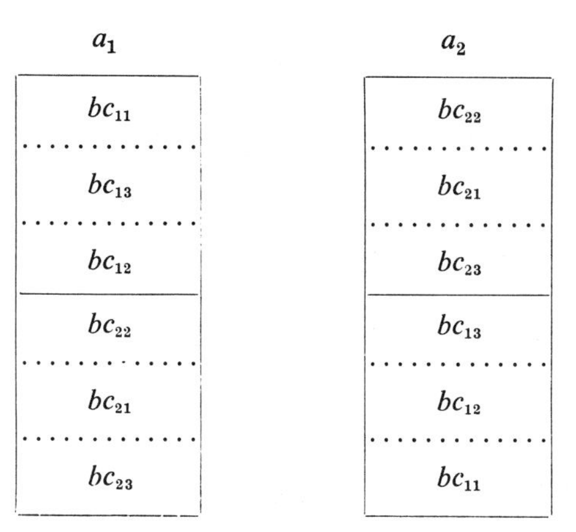

Thus the experimental unit for factor A is the whole plot; the experimental unit for factor B is the subplot; and the experimental unit for factor C

is the sub-subplot. Since the sub-subplots are nested within the subplots and the latter are nested within the whole plots, factor C is nested under the subplots and factor B is nested under the whole plots. Factor A is partially confounded with groups of whole plots.

The model for this type of design may be written

$$X_{ijkm} = \mu_{...} + \alpha_i + \pi_{m(i)} + \beta_j + \alpha\beta_{ij} + \pi'_{m(ij)} + \gamma_k + \alpha\gamma_{ik} + \beta\gamma_{jk} + \alpha\beta\gamma_{ijk} + \pi''_{m(ijk)} + \varepsilon_{ijkm}.$$

The notation $\pi''_{m(ij)}$ designates the residual sub-subplot effect. (This latter may also be regarded as the pooled $\gamma\pi_{km(i)}$ and $\beta\gamma\pi_{jkm(i)}$ interaction effects.) The analysis of this type of experiment takes the form given in Table 5.13-2. The expected values of the mean squares given in this table

Table 5.13-2 Expected Values of Mean Squares for Split-split-plot Design
(A, B, C fixed, plots random)

Source of variation	df	Expected value of mean square
Between whole plots	$np - 1$	
A	$p - 1$	$\sigma_\varepsilon^2 + qr\sigma_\pi^2 + nqr\sigma_\alpha^2$
Whole-plot residual	$p(n - 1)$	$\sigma_\varepsilon^2 + qr\sigma_\pi^2$
Within subplots	$np(q - 1)$	
B	$q - 1$	$\sigma_\varepsilon^2 + r\sigma_{\pi'}^2 + npr\sigma_\beta^2$
AB	$(p - 1)(q - 1)$	$\sigma_\varepsilon^2 + r\sigma_{\pi'}^2 + nr\sigma_{\alpha\beta}^2$
Subplot residual	$p(n - 1)(q - 1)$	$\sigma_\varepsilon^2 + r\sigma_{\pi'}^2$
Within sub-subplots	$npq(r - 1)$	
C	$r - 1$	$\sigma_\varepsilon^2 + \sigma_{\pi''}^2 + npq\sigma_\gamma^2$
AC	$(p - 1)(r - 1)$	$\sigma_\varepsilon^2 + \sigma_{\pi''}^2 + nq\sigma_{\alpha\gamma}^2$
BC	$(q - 1)(r - 1)$	$\sigma_\varepsilon^2 + \sigma_{\pi''}^2 + np\sigma_{\beta\gamma}^2$
ABC	$(p - 1)(q - 1)(r - 1)$	$\sigma_\varepsilon^2 + \sigma_{\pi''}^2 + n\sigma_{\alpha\beta\gamma}^2$
Sub-subplot residual	$pq(n - 1)(r - 1)$	$\sigma_\varepsilon^2 + \sigma_{\pi''}^2$

are for the special case in which factors A, B, and C are fixed and plots are random. The numbers of levels of factor A, B, and C are, respectively, p, q, and r; each level of factor A is assigned to n whole plots.

The expected values of the mean squares indicate the structure of appropriate F ratios. It will be noted that the sub-subplot residual mean square is the appropriate denominator for the main effect and all interactions involving factor C. More extensive consideration of split-plot designs is given in connection with designs having repeated measures on the same people. These designs form the subject matter of Chap. 7.

The design summarized in Table 5.13-2 involves pqr cells. Within each cell there are n independent observations. Hence the within-cell variation has $pqr(n - 1)$ degrees of freedom. It is of interest to note that the sum of the degrees of freedom for the residuals is the degrees of freedom for the within-cell variation.

Source	df
Whole-plot residual	$p(n - 1)$
Subplot residual	$p(n - 1)(q - 1)$
Sub-subplot residual	$pq(n - 1)(r - 1)$
Total	$pqr(n - 1)$

5.14 Rules for Deriving the Expected Values of Mean Squares

Given an experiment in which the underlying variables can be assumed to satisfy the conditions of the general linear model, the expected values of the mean squares computed from the experimental data can be obtained by means of a relatively simple set of rules. Although these rules lead to an end product which has been proved to be statistically correct when the assumptions underlying the general linear model are met, the rules themselves provide little insight into the mathematical rationale underlying the end product. The assumptions that underlie the general linear model under consideration have been stated in Sec. 5.6. The rules which will be outlined in this section are those developed by Cornfield and Tukey (1956). A similar set of rules will also be found in Bennett and Franklin (1954).

As the rules will be given, no distinction will be made between random and fixed factors. However, certain of the terms become either 0 or 1 depending upon whether an experimental variable corresponds to a fixed or random factor. For purposes of simplifying the expressions that will result, the following notation will be used:

$$D_p = 1 - \frac{p}{P}, \qquad i = 1, \ldots, p.$$

$$D_q = 1 - \frac{q}{Q}, \qquad j = 1, \ldots, q.$$

$$D_r = 1 - \frac{r}{R}, \qquad k = 1, \ldots, r.$$

If $p = P$, that is, if factor A is fixed, then D_p is zero. On the other hand, if factor A is random, D_p is unity. Similarly, the other D's are either 0 or 1 depending upon whether the corresponding factor is fixed or random. In the application of these rules to designs of special interest, the appropriate evaluation of the D's should be used rather than the D's themselves. The general statement of the rules is followed by a series of examples.

Rule 1. Write the appropriate model for the design, making explicit in the notation those effects which are nested.

Rule 2. Construct a two-way table in which the terms in the model (except the grand mean) are the row headings and the subscripts appearing in the model are the column headings. The number of columns in this table will be equal to the number of different subscripts in the model. The number of rows will be equal to the number of terms in the model which have subscripts. The row headings should include all subscripts associated with terms in the model.

Rule 3. To obtain the entries in column i,
enter D_p in those rows having headings containing an i which is not nested,
enter unity in those rows having headings containing an i which is nested,
enter p in those rows having headings which do not contain an i.

Rule 4. To obtain entries in column j,
enter D_q in those rows having headings containing a j which is not nested,
enter unity in those rows having headings containing a j which is nested,
enter q in those rows having headings which do not contain a j.

Rule 5. Entries in all other columns follow the general pattern outlined in rules 3 and 4. For example, the possible entries in column k would be D_r, unity, and r.

Rule 6. The expected value of the mean square for the main effect of factor A is a weighted sum of the variances due to all effects which contain the subscript i. If a row heading contains a subscript i, then the weight for the variance due to this row effect is the product of all entries in this row, the entry in column i being omitted. (For nested effects, see rule 10.)

Rule 7. The expected value of the mean square for the main effect of factor B is a weighted sum of the variances due to all effects which contain the subscript j. If a row heading contains a j, the weight for the variance due to this effect is the product of all entries in this row, the entry in column j being omitted. (See rule 10.)

Rule 8. The expected value of the mean square for the AB interaction is a weighted sum of the variances due to all effects which contain both the subscripts i and j. If a row heading contains both the subscripts i and j, then the weight for the variance corresponding to this effect is the product of all entries in this row, the entries in both columns i and j being omitted.

Rule 9. In general, the expected value of a mean square for an effect which has the general representation XYZ_{uvw} is a weighted sum of the variances due to all effects in the model which contain all the subscripts u, v, and w (and possibly other subscripts). If a row heading does contain all three of the three subscripts u, v, and w, then the weight for the variance due to the corresponding row effects is the product of all entries in this row, the entries in columns u, v, and w being omitted.

Rule 10. If an effect is nested, the expected value of its mean square is a weighted sum of variances corresponding to all effects containing the same subscripts as the nested effect. For example, if the main effect of factor B appears as $\beta_{j(i)}$ in the model, then the relevant effects are those which contain both the subscripts i and j. Similarly, if the term $\beta\gamma_{j(i)k}$ appears in the model, in considering the set of relevant variances, row headings must contain all three of the subscripts i, j, and k.

The application of these rules will be illustrated by means of a $p \times q \times r$ partially hierarchal factorial design having n observations per cell. In this design, it will be assumed that factor B is nested under factor A. This type of design may be represented schematically as follows:

	a_1			a_2		
	$b_{1(1)}$	$b_{2(1)}$	$b_{3(1)}$	$b_{1(2)}$	$b_{2(2)}$	$b_{3(2)}$
c_1						
c_2	n observations in each cell					
c_3						
c_4						

The notation $b_{j(i)}$ indicates that factor B is nested under factor A. The structural model for this design may be written as

$$X_{ijkm} = \mu_{...} + \alpha_i + \beta_{j(i)} + \gamma_k + \alpha\gamma_{ik} + \beta\gamma_{j(i)k} + \varepsilon_{m(ijk)}.$$

In accordance with rule 1, the notation in the structural model makes explicit those effects which are nested. Thus, the notation $\beta_{j(i)}$ indicates that the levels of factor B are nested under the levels of factor A. The notation $\varepsilon_{m(ijk)}$ indicates that the unique effects associated with an observation on element m in cell abc_{ijk} are nested under all effects; i.e., the experimental error is nested under all factors.

The two-way table called for by rule 2 is given in Table 5.14-1. The row headings are the terms in the model, the term $\mu_{...}$ being omitted. The column headings are the different subscripts that appear in the model.

The entries that appear in column i of this table were obtained in accordance with rule 3. Since rows α_i and $\alpha\gamma_{ik}$ contain a subscript i which is not in parentheses, the entry in column i for each of these rows is D_p. Since rows $\beta_{j(i)}$, $\beta\gamma_{j(i)k}$, and $\varepsilon_{m(ijk)}$ each contains the subscript i in parentheses, the entry in column i for each of these rows is unity. All other entries in column i are p, since none of the remaining row headings contain the subscript i.

The entries that appear in column j were obtained in accordance with rule 4. The entries that appear in column m follow from rule 5—with one exception. From rule 5 the entry in row $\varepsilon_{m(ijk)}$ would be D_n. Since the experimental error will always be considered a random variable,

Table 5.14-1 Expected Value of Mean Squares for Partially Hierarchal Factorial Design

Effect	i	j	k	m	E(MS)
α_i	D_p	q	r	n	$\sigma_\varepsilon^2 + nD_qD_r\sigma_{\beta\gamma}^2 + nqD_r\sigma_{\alpha\gamma}^2 + nrD_q\sigma_\beta^2 + nqr\sigma_\alpha^2$
$\beta_{j(i)}$	1	D_q	r	n	$\sigma_\varepsilon^2 + nD_r\sigma_{\beta\gamma}^2 + nr\sigma_\beta^2$
γ_k	p	q	D_r	n	$\sigma_\varepsilon^2 + nD_q\sigma_{\beta\gamma}^2 + nqD_p\sigma_{\alpha\gamma}^2 + npq\sigma_\gamma^2$
$\alpha\gamma_{ik}$	D_p	q	D_r	n	$\sigma_\varepsilon^2 + nD_q\sigma_{\beta\gamma}^2 + nq\sigma_{\alpha\gamma}^2$
$\beta\gamma_{j(i)k}$	1	D_q	D_r	n	$\sigma_\varepsilon^2 + n\sigma_{\beta\gamma}^2$
$\varepsilon_{m(ijk)}$	1	1	1	1	σ_ε^2

$D_n = 1 - (n/N)$ will always be considered equal to unity. Thus the n observations that appear within a cell in this design are considered to be a random sample from a potentially infinite number of observations that could be made within a cell.

In accordance with rule 6, the expected value of the mean square for the main effect of factor A is a weighted sum of the variances due to all row effects which contain the subscript i. Thus E(MS$_a$) is a weighted sum of the following variances:

$$\sigma_\varepsilon^2, \quad \sigma_{\beta\gamma}^2, \quad \sigma_{\alpha\gamma}^2, \quad \sigma_\beta^2, \quad \sigma_\alpha^2.$$

The weight for σ_α^2, which is given by the product of all terms in row α_i, the term in column i being omitted, is nqr. The weight for σ_β^2, which is the product of all terms in row $\beta_{j(i)}$, the term in column i being omitted, turns out to be nrD_q. The weight for $\sigma_{\alpha\gamma}^2$ is the product of all terms in row $\alpha\gamma_{ik}$, the term in column i being omitted; this weight is nqD_r. The properly weighted sum for E(MS$_a$) appears to the right of row α_i. Thus,

$$\text{E(MS)}_a = \sigma_\varepsilon^2 + nD_qD_r\sigma_{\beta\gamma}^2 + nqD_r\sigma_{\alpha\gamma}^2 + nrD_q\sigma_\beta^2 + nqr\sigma_\alpha^2.$$

In words, the expected value of the mean square for the main effect of factor A is a weighted sum of a set of variances; the variances included in this set and the weights for each are determined by application of rule 6.

Since, in this design, factor B is nested under factor A, the main effect of factor B is denoted by the symbol $\beta_{j(i)}$. In this case the expected value of the mean square for the main effect of factor B is a weighted sum of the variances of terms containing both the subscripts i and j. These variances are

$$\sigma_\varepsilon^2, \quad \sigma_{\beta\gamma}^2, \quad \text{and} \quad \sigma_\beta^2.$$

The weight for σ_β^2 is the product of all terms in row $\beta_{j(i)}$, the terms in columns i and j being omitted; this weight is nr. The weight for $\sigma_{\beta\gamma}^2$ is the product of all terms in row $\beta\gamma_{j(i)k}$, the terms in columns i and j being omitted. This weight is nD_r. The expected value for the mean square of the main effect of factor B (which is nested under factor A) is given at the right of the row $\beta_{j(i)}$.

The other terms under the heading E(MS) in Table 5.14-1 were obtained by application of rules 9 and 10. For example, the variances that enter into the expected value of MS_{ac} are

$$\sigma_\varepsilon^2, \quad \sigma_{\beta\gamma}^2, \quad \text{and} \quad \sigma_{\alpha\gamma}^2.$$

These are the variances corresponding to those effects which contain the subscripts i and k.

The expected values for the mean squares in Table 5.14-1 may be specialized to cover specific designs by evaluating the D's. For example, if factors A and C are fixed and factor B is random, $D_p = 0$, $D_r = 0$, and $D_q = 1$. For this case,

$$\text{E}(\text{MS}_a) = \sigma_\varepsilon^2 + nr\sigma_\beta^2 + nqr\sigma_\alpha^2.$$

This expected value was obtained from the corresponding expected value, in Table 5.14-1 by evaluating the D_p, D_q, and D_r. By a similar procedure

$$\text{E}(\text{MS}_b) = \sigma_\varepsilon^2 + nr\sigma_\beta^2.$$

The expected values of mean squares for specialized designs may be obtained by evaluating the D's prior to the application of the rules. For the special case under consideration, the expected values of the mean squares are derived in Table 5.14-2. The expected values of the mean squares

Table 5.14-2 Expected Value of Mean Squares for Partially Hierarchal Factorial Design
(Factors A and C fixed, factor B random)

Effect	i	j	k	m	E(MS)
α_i	0	q	r	n	$\sigma_\varepsilon^2 + nr\sigma_\beta^2 + nqr\sigma_\alpha^2$
$\beta_{j(i)}$	1	1	r	n	$\sigma_\varepsilon^2 + nr\sigma_\beta^2$
γ_k	p	q	0	n	$\sigma_\varepsilon^2 + n\sigma_{\beta\gamma}^2 + npq\sigma_\gamma^2$
$\alpha\gamma_{ik}$	0	q	0	n	$\sigma_\varepsilon^2 + n\sigma_{\beta\gamma}^2 + nq\sigma_{\alpha\gamma}^2$
$\beta\gamma_{j(i)k}$	1	1	0	n	$\sigma_\varepsilon^2 + n\sigma_{\beta\gamma}^2$
$\varepsilon_{m(ijk)}$	1	1	1	1	σ_ε^2

shown on the right are identical to those obtained by specializing the E(MS) given at the right of Table 5.14-1. For special cases of interest to the experimenter, the simplest method of deriving the expected values of the mean squares is that illustrated in Table 5.14-2.

5.15 Quasi F Ratios

In some cases the appropriate F ratio cannot be constructed by direct application of the rules based upon expected values of mean squares. For example, consider the case of a $p \times q \times r$ factorial experiment having n observations per cell. If all factors are random, the expected values of the mean squares are those given in Table 5.15-1. (These expected values correspond to model II in Table 5.9-2.)

Table 5.15-1 Expected Values of Mean Squares
(Model II)

Source of variation	E(MS)
A	$\sigma_\varepsilon^2 + n\sigma_{\alpha\beta\gamma}^2 + nq\sigma_{\alpha\gamma}^2 + nr\sigma_{\alpha\beta}^2 + nqr\sigma_\alpha^2$
B	$\sigma_\varepsilon^2 + n\sigma_{\alpha\beta\gamma}^2 + np\sigma_{\beta\gamma}^2 + nr\sigma_{\alpha\beta}^2 + npr\sigma_\beta^2$
C	$\sigma_\varepsilon^2 + n\sigma_{\alpha\beta\gamma}^2 + np\sigma_{\beta\gamma}^2 + nq\sigma_{\alpha\gamma}^2 + npq\sigma_\gamma^2$
AB	$\sigma_\varepsilon^2 + n\sigma_{\alpha\beta\gamma}^2 + nr\sigma_{\alpha\beta}^2$
AC	$\sigma_\varepsilon^2 + n\sigma_{\alpha\beta\gamma}^2 + nq\sigma_{\alpha\gamma}^2$
BC	$\sigma_\varepsilon^2 + n\sigma_{\alpha\beta\gamma}^2 + np\sigma_{\beta\gamma}^2$
ABC	$\sigma_\varepsilon^2 + n\sigma_{\alpha\beta\gamma}^2$
Experimental error	σ_ε^2

In practice preliminary tests on the model (which are discussed in Sec. 5.16) are made on the higher-order interactions before proceeding with the tests on main effects. (In many cases which arise in practice, tests on main effects may be relatively meaningless when interactions are significantly different from zero.) If none of the interaction terms may be dropped from the model, then no single mean square can serve as a denominator in a test on main effects due to factor A. The proper denominator for a test in this case should have the expected value

$$\sigma_\varepsilon^2 + n\sigma_{\alpha\beta\gamma}^2 + nq\sigma_{\alpha\gamma}^2 + nr\sigma_{\alpha\beta}^2.$$

None of the individual mean squares in Table 5.15-1 has this expected value. However, by adding and subtracting certain of the mean squares one may obtain a composite mean square which has the required expected value. One such composite, assuming each of the mean squares is independent of the others, may be constructed as follows:

$$\begin{aligned} E(MS_{ac}) &= \sigma_\varepsilon^2 + n\sigma_{\alpha\beta\gamma}^2 + nq\sigma_{\alpha\gamma}^2 \\ E(MS_{ab}) &= \sigma_\varepsilon^2 + n\sigma_{\alpha\beta\gamma}^2 \qquad\qquad + nr\sigma_{\alpha\beta}^2 \\ -E(MS_{abc}) &= -\sigma_\varepsilon^2 - n\sigma_{\alpha\beta\gamma}^2 \\ \hline E(MS_{ac} + MS_{ab} - MS_{abc}) &= \sigma_\varepsilon^2 + n\sigma_{\alpha\beta\gamma}^2 + nq\sigma_{\alpha\gamma}^2 + nr\sigma_{\alpha\beta}^2 \end{aligned}$$

A quasi F ratio, which has the proper structural requirements in terms of expected values of mean squares, for a test on the main effect of factor A is

$$F' = \frac{MS_a}{MS_{ac} + MS_{ab} - MS_{abc}}.$$

The symbol F' is used for this ratio rather than the symbol F. Since the denominator is a composite of different sources of variation, the sampling distribution of the F' ratio is not the usual F distribution, although the latter distribution may be used as an approximation. The denominator of this F' ratio calls for subtracting a mean square. This could lead to the possibility of obtaining a negative denominator. According to the population model, it is not possible to have a negative denominator in terms of parameters. However, in terms of the estimates of these parameters it is possible to have a negative denominator.

The following quasi F ratio avoids the possibility of a negative denominator and still satisfies the structural requirements for a test on the main effect of factor A:

$$F'' = \frac{\text{MS}_a + \text{MS}_{abc}}{\text{MS}_{ac} + \text{MS}_{ab}}.$$

In terms of expected values of the mean squares, the ratio of the expectations has the form

$$\frac{2\sigma_\varepsilon^2 + 2n\sigma_{\alpha\beta\gamma}^2 + nq\sigma_{\alpha\gamma}^2 + nr\sigma_{\alpha\beta}^2 + nqr\sigma_\alpha^2}{2\sigma_\varepsilon^2 + 2n\sigma_{\alpha\beta\gamma}^2 + nq\sigma_{\alpha\gamma}^2 + nr\sigma_{\alpha\beta}^2}.$$

Under the hypothesis that $\sigma_\alpha^2 = 0$, E(numerator) = E(denominator). Similarly, the following F'' ratio has the structural requirements in terms of the expected valus of mean squares for a test of the hypothesis that $\sigma_\beta^2 = 0$:

$$F'' = \frac{\text{MS}_b + \text{MS}_{abc}}{\text{MS}_{ab} + \text{MS}_{bc}}.$$

Although these F'' ratios satisfy the structural requirements in terms of expected values, the sampling distributions of these F'' ratios can only be roughly approximated by the usual F distributions, provided that special degrees of freedom are used for numerator and denominator. Suppose that the F'' ratio has the following general form,

$$F'' = \frac{u + v}{w + x},$$

where u, v, w, and x are appropriate mean squares. Let respective degrees of freedom for these mean squares be f_u, f_v, f_w, and f_x. Then the degrees of freedom for the numerator are approximated by the nearest integral value to

$$\frac{(u + v)^2}{(u^2/f_u) + (v^2/f_v)}.$$

The degrees of freedom for the denominator are approximated by

$$\frac{(w + x)^2}{(w^2/f_w) + (x^2/f_x)}.$$

This approximation to the F distribution and the associated degrees of freedom are those suggested by Satterthwaite (1946).

The F' ratio has the following general form:

$$F' = \frac{u}{w + x - v}.$$

If the F distribution is used to approximate the sampling distribution of the F' statistic, the degrees of freedom for the denominator are

$$\frac{(w + x - v)^2}{(w^2/f_w) + (x^2/f_x) + (v^2/f_v)}.$$

The numerator of this expression may be represented as

$$(\Sigma a_i \text{MS}_i)^2 \qquad \text{where } a_1 = 1, \qquad a_2 = 1, \qquad \text{and} \qquad a_3 = -1.$$

The denominator has the form

$$\sum_i \frac{(a_i \text{MS}_i)^2}{f_i}.$$

When one or more of the a_i are negative, the Satterthwaite procedure may lead to relatively poor estimates of the appropriate degrees of freedom. A study of the consequences of negative a_i appears in Gaylor and Hopper (1969).

5.16 Preliminary Tests on the Model and Pooling Procedures

In deriving the expected values for the mean squares, extensive use is made of a model which is assumed to be appropriate for an experiment. The model indicates the relevant sources of variability. A question might be raised about why certain terms appear in the model and why other possible terms are omitted. If in fact there is no interaction effect of a given kind in the population of interest, why should such an interaction term be included in the model? Including such an interaction term in the model can potentially affect the expected values of several mean squares. The latter in turn determine the structure of F ratios.

Decisions about what terms should appear in the model and what terms should be omitted are generally based upon experience in an experimental area and knowledge about what are reasonable expectations with respect to underlying sources of variation—in short, subject-matter information. All sources of variation not specifically included in the model are in reality classified as part of the experimental error. In most cases the latter variation is the residual variation after all controlled sources have been estimated. Previous experimentation in an area may indicate that no interaction between two factors is to be expected; hence in designing a new experiment in this area such an interaction term may be omitted from the model. However, any variation due to this interaction, if it exists, is automatically included as part of the experimental error or automatically confounds other estimates, depending upon the experimental design.

Lacking knowledge about interaction effects from past experimentation, one might ask whether or not data obtained in a given factorial experiment could be used as a basis for revising an initial model. The specification of the parameters in the initial model could be considered incomplete or left open to more complete specification. Tests designed to revise or complete the specification of parameters to be included in the model are called *preliminary* tests on the model. Such tests are particularly appropriate when one is dealing with experiments in which interactions between fixed and random factors or interactions between random factors are potentially in the model. Such terms may turn out to be denominators for F ratios. If such terms have a relatively small number of degrees of freedom, corresponding F ratios will have very low power. Since the mean squares for interactions between two or more fixed factors can never form the denominator in an F ratio, preliminary tests on such interactions are not generally required.

The procedure of making preliminary tests on higher-order interactions before proceeding with tests of lower-order interactions and main effects may be regarded as a multistage decision rule. Depending upon the outcome of a sequence of tests, the parameters in the model become more completely specified; in turn the expected values for the mean squares are revised sequentially.

If a preliminary test does not reject the hypothesis that the variance due to an interaction effect is zero, one proceeds as if this variance were actually zero and drops the corresponding term from the model. The expected values of the mean squares are then revised in accordance with the new model, and additional preliminary tests, if necessary, are made. Care must be taken in such tests to avoid type 2 error, i.e., accepting the hypothesis of zero interaction when it should be rejected. Type 2 error can be kept numerically small by making preliminary tests at a numerically high type 1 error, that is, $\alpha = .20$ or $.30$.

The sequence in which preliminary tests on the model are made will be illustrated for the case of a $p \times q \times r$ factorial experiment in which factor A is considered fixed and factors B and C are considered random. The initial model for this experiment is the following:

$$(1) \quad X_{ijkm} = \mu + \alpha_i + \beta_j + \gamma_k + \alpha\beta_{ij} + \alpha\gamma_{ik} + \beta\gamma_{jk} + \alpha\beta\gamma_{ijk} + \varepsilon_{m(ijk)}.$$

The question of whether or not all the interactions between random factors and between fixed and random factors should be included in the model is left unanswered for the time being. Assuming the complete model, the expected values of the mean squares are those given in Table 5.16-1. Suppose that the experimental work is completed and the analysis-of-variance table prepared. Tests of hypotheses depend upon the model and associated expected values. The model as given in (1) is tentative and subject to change. Associated with this model are the expected values given in Table 5.16-1. Inspection of this model indicates that the appropriate denominator for tests on the hypotheses $\sigma^2_{\alpha\beta\gamma} = 0$ and $\sigma^2_{\beta\gamma} = 0$ is $\text{MS}_{\text{w. cell}}$. Suppose both of these tests are made at the 25 percent level of significance. Suppose that these tests do not reject the respective hypotheses and that a priori information indicates no good basis for expecting that such interactions exist. On these grounds $\alpha\beta\gamma_{ijk}$ and $\beta\gamma_{jk}$ are now dropped from the model.

The revised model now has the form

$$(2) \qquad X_{ijkm} = \mu + \alpha_i + \beta_j + \gamma_k + \alpha\beta_{ij} + \alpha\gamma_{ik} + \varepsilon_{ijkm}.$$

The term ε now includes the interaction terms which were dropped from the model in (1). The revised expected values of the mean squares are obtained by dropping $\sigma^2_{\alpha\beta\gamma}$ and $\sigma^2_{\beta\gamma}$ from the terms in Table 5.16-1. When

Table 5.16-1 Expected Values of Mean Squares for Model in (1)
(A fixed, B and C random)

Source of variation	df	E(MS)
A	$p-1$	$\sigma^2_\varepsilon + n\sigma^2_{\alpha\beta\gamma} + nq\sigma^2_{\alpha\gamma} + nr\sigma^2_{\alpha\beta} + nqr\sigma^2_\alpha$
B	$q-1$	$\sigma^2_\varepsilon + np\sigma^2_{\beta\gamma} + npr\sigma^2_\beta$
C	$r-1$	$\sigma^2_\varepsilon + np\sigma^2_{\beta\gamma} + npq\sigma^2_\gamma$
AB	$(p-1)(q-1)$	$\sigma^2_\varepsilon + n\sigma^2_{\alpha\beta\gamma} + nr\sigma^2_{\alpha\beta}$
AC	$(p-1)(r-1)$	$\sigma^2_\varepsilon + n\sigma^2_{\alpha\beta\gamma} + nq\sigma^2_{\alpha\gamma}$
BC	$(q-1)(r-1)$	$\sigma^2_\varepsilon + np\sigma^2_{\beta\gamma}$
ABC	$(p-1)(q-1)(r-1)$	$\sigma^2_\varepsilon + n\sigma^2_{\alpha\beta\gamma}$
Within cell	$pqr(n-1)$	σ^2_ε

this is done, MS_{bc}, MS_{abc}, and $\text{MS}_{\text{w. cell}}$ are all estimates of σ^2_ε. These three mean squares may be pooled to provide a single estimate of σ^2_ε as shown in Table 5.16-2. The pooled estimate is

$$\text{MS}_{\text{res}} = \frac{\text{SS}_{bc} + \text{SS}_{abc} + \text{SS}_{\text{w. cell}}}{(q-1)(r-1) + (p-1)(q-1)(r-1) + pqr(n-1)}.$$

The degrees of freedom for the denominator are the sum of the degrees of freedom for the sources of variation which are pooled. This sum is equal to $npqr - pq - pr + p$.

Inspection of Table 5.16-2 indicates that the hypotheses $\sigma^2_{\alpha\gamma} = 0$ and $\sigma^2_{\alpha\beta} = 0$ may both be tested with MS_{res} as a denominator. Suppose that these tests are made at the 25 percent level of significance. Suppose that the outcome of these tests does not reject the hypothesis that $\sigma^2_{\alpha\gamma} = 0$ but

Table 5.16-2 Expected Values of Mean Squares for Model in (2)

Source of variation	df	E(MS)
A	$p - 1$	$\sigma^2_\varepsilon + nq\sigma^2_{\alpha\gamma} + nr\sigma^2_{\alpha\beta} + nqr\sigma^2_\alpha$
B	$q - 1$	$\sigma^2_\varepsilon + npr\sigma^2_\beta$
C	$r - 1$	$\sigma^2_\varepsilon + npq\sigma^2_\gamma$
AB	$(p - 1)(q - 1)$	$\sigma^2_\varepsilon + nr\sigma^2_{\alpha\beta}$
AC	$(p - 1)(r - 1)$	$\sigma^2_\varepsilon + nq\sigma^2_{\alpha\gamma}$
BC, ABC, Within cell } residual	$npqr - pq - pr + p$	σ^2_ε

that the hypothesis that $\sigma^2_{\alpha\beta} = 0$ is rejected. The revised model now has the form

$$X_{ijkm} = \mu + \alpha_i + \beta_j + \gamma_k + \alpha\beta_{ij} + \varepsilon_{ijkm}. \tag{3}$$

The experimental error in (3) includes variation due to $\alpha\gamma$ as well as the interaction terms included in model (2). The expected values associated with (3) are given in Table 5.16-3. These expected values may be taken as those appropriate for final tests.

Table 5.16-3 Expected Values of Mean Squares for Model in (3)

Source of variation	df	E(MS)
A	$p - 1$	$\sigma^2_\varepsilon + nr\sigma^2_{\alpha\beta} + nqr\sigma^2_\alpha$
B	$q - 1$	$\sigma^2_\varepsilon + npr\sigma^2_\beta$
C	$r - 1$	$\sigma^2_\varepsilon + npq\sigma^2_\gamma$
AB	$(p - 1)(q - 1)$	$\sigma^2_\varepsilon + nr\sigma^2_{\alpha\beta}$
Residual (pooled AC, BC, ABC, and within cell)	$npqr - pq - r + 1$	σ^2_ε

If there is a priori evidence to indicate interaction between factors, preliminary tests on such interactions should in general be avoided and tests should be made in accordance with the original formulation. When preliminary tests are made, only interactions in which random factors appear are considered in such tests. (Only interactions with random factors can potentially form the denominator of an F ratio.) If an adequate number of degrees of freedom (say 20 or more) is available for the denominator of F ratios constructed in terms of the original model, preliminary tests should also be avoided. However, if the denominator of an F ratio constructed in accordance with the original model has relatively few degrees of freedom (say less than 10), in the absence of a priori knowledge preliminary tests are in order.

A numerical example will be used to illustrate the pooling procedures associated with Tables 5.16-1, 5.16-2, and 5.16-3. Part i of Table 5.16-4 represents the analysis of variance for a $4 \times 2 \times 3$ factorial experiment having two observations per cell. Assume factor A fixed and factors B

Table 5.16-4 Numerical Example of Pooling Procedures
($n = 2, p = 4, q = 2, r = 3$)

First-stage preliminary tests: $H_1\colon \sigma^2_{\alpha\beta\gamma} = 0$; $H_1\colon \sigma^2_{\beta\gamma} = 0$

	Source	SS	df	MS	F	
	A	120.00	3	40.00		
	B	60.00	1	60.00		
	C	40.00	2	20.00		
	AB	96.00	3	32.00		
(i)	AC	72.00	6	12.00		
	BC	18.00	2	9.00	0.90	$F_{.75}(2,24) = 1.47$
	ABC	72.00	6	12.00	1.20	$F_{.75}(6,24) = 1.41$
	Within cell	240.00	24	10.00		

Second-stage preliminary tests: $H_1\colon \sigma^2_{\alpha\beta} = 0$; $H_1\colon \sigma^2_{\alpha\gamma} = 0$

	Source	SS	df	MS	F	
	A	120.00	3	40.00		
	B	60.00	1	60.00		
	C	40.00	2	20.00		
(ii)	AB	96.00	3	32.00	3.10	$F_{.75}(3,32) = 1.44$
	AC	72.00	6	12.00	1.16	$F_{.75}(6,32) = 1.39$
	Residual (ii)	330.00	32	10.31		

Analysis for final tests:

	Source	SS	df	MS	F	
	A	120.00	3	40.00	1.25	$F_{.95}(3,3) = 9.28$
	B	60.00	1	60.00	5.67	$F_{.95}(1,38) = 4.10$
(iii)	C	40.00	2	20.00	1.89	$F_{.95}(2,38) = 3.25$
	AB	96.00	3	32.00	3.02	$F_{.95}(3,38) = 2.85$
	Residual (iii)	402.00	38	10.58		

and C random. If the model in (1) is assumed, the expected values of the mean squares given in Table 5.16-1 are appropriate. Preliminary tests on $\sigma^2_{\alpha\beta\gamma}$ and $\sigma^2_{\beta\gamma}$ are made in part i. These tests indicate that the hypotheses that $\sigma^2_{\alpha\beta\gamma} = 0$ and $\sigma^2_{\beta\gamma} = 0$ cannot be rejected at the 25 percent level of significance. Hence the corresponding terms are dropped from the model in (1).

The residual term in part ii is obtained as follows:

Source	SS	df	MS
BC	18.00	2	
ABC	72.00	6	
Within cell	240.00	24	
Residual (ii)	330.00	32	10.31

The variation due to residual ii corresponds to variation due to experimental error in the model given in (2). The expected values for the mean squares in part ii are those given in Table 5.16-2. Second-stage tests are made in accordance with the latter expected values. As a result of the tests in part ii, the variation associated with the AC interaction is pooled with the experimental error, but the variation due to the AB interaction is not pooled.

The residual term in part iii is obtained as follows:

Source	SS	df	MS
AC	72.00	6	
Residual (ii)	330.00	32	
Residual (iii)	402.00	38	10.58

Statistical tests in part iii are based upon the expected values in Table 5.16-3. The denominator for the F ratio in the test on the main effect of factor A is MS_{ab}; all other F ratios have MS_{res} as a denominator. As distinguished from parts i and ii, the tests in part iii are final tests rather than preliminary tests; hence the difference in the level of significance.

The sampling distributions of the statistics used in tests made following preliminary tests are actually different from the sampling distributions associated with tests which are not preceded by preliminary tests. What is really required in the second stage of a sequential decision procedure is the sampling distribution of the statistic in question under the condition that specified decisions have been made in the first stage. Using sampling distributions which do not have such conditions attached generally introduces a slight bias into the testing procedure. Specified percentile points on the unconditional sampling distribution are probably slightly lower than corresponding points on the conditional sampling distribution. That is, a statistic which falls at the 95th percentile point when referred to the unconditional distribution may fall at only the 92d percentile point when referred to the conditional distribution. Hence use of the unconditional distributions for sequential tests probably gives tests which have a slight positive bias; i.e., the type 1 error is slightly larger than the specified level of significance.

By way of summary, it should be noted that there is no widespread agreement among statisticians on the wisdom of the pooling procedures which have been discussed in this section. Those statisticians who adhere to the "never pool" rule demand a completely specified model prior to the analysis of the experimental data. This position has much to recommend it. The inferences obtained from adopting this point of view will be based upon exact sampling distributions, provided that the model that has been specified is appropriate for the experiment.

Using data from the experiment to revise the model introduces contingencies which are difficult to evaluate statistically. However, working from a revised model which more adequately fits the data may potentially provide more powerful tests than those obtained from the "never pool" rule. Admittedly the change in power cannot be evaluated with precision. The conservative attitude toward pooling adopted in this section attempts to take middle ground: One departs from the initial model only if the experimental data strongly suggest that the initial model is not appropriate. The position taken by the author in this section is quite close to that adopted by Green and Tukey (1960). It is also in line with the point of view developed by Bozivich et al. (1956).

5.17 Individual Comparisons

Procedures discussed in Chap. 3 for making individual and multiple comparisons among means can be extended rather directly to factorial experiments. A significant overall F test on a main effect, for example, indicates that one or more of a multitude of possible comparisons is significant. (In particular, a significant overall main effect implies that the maximum normalized comparison among the relevant marginal means is statistically significant.)

The specific comparisons which are built into the design or suggested by the theoretical basis for the experiment can and should be made individually, regardless of the outcome of the corresponding overall F test. Seldom, if ever, should a posteriori comparisons be made when the overall F is nonsignificant. (In this context an a posteriori comparison is one suggested by deliberate inspection of the data.) Should such comparisons be made, statistically significant outcomes should be interpreted with extreme caution. The experimenter should not hesitate to describe fully all aspects of his experimental results.

The procedure for making individual comparisons will be illustrated for the case of a $p \times q$ factorial experiment having n observations per cell. To test the hypothesis that $\alpha_i = \alpha_{i'}$ against the two-tailed alternative hypothesis $\alpha_i \neq \alpha_{i'}$, one may use the test statistic

$$F = \frac{(\bar{A}_i - \bar{A}_{i'})^2}{\text{MS}_{\bar{A}_i - \bar{A}_{i'}}}.$$

The best estimate of $\text{MS}_{\bar{A}_i - \bar{A}_{i'}}$ depends upon whether factor B is fixed or random. For the case in which factor B is fixed,

$$\text{MS}_{\bar{A}_i - \bar{A}_{i'}} = \frac{2\text{MS}_{\text{w.cell}}}{nq}.$$

For the case in which factor B is random and $\sigma^2_{\alpha\beta} \neq 0$,

$$\text{MS}_{\bar{A}_i - \bar{A}_{i'}} = \frac{2\text{MS}_{ab}}{nq}.$$

(When the AB interaction is significant, one is generally interested only in the simple main effects. Hence tests on the overall main effects in the presence of significant interaction seldom are made in practice.) For the first case, the degrees of freedom for $\text{MS}_{\bar{A}_i - \bar{A}_{i'}}$ are $pq(n - 1)$; for the second case, the degrees of freedom are $(p - 1)(q - 1)$. In either case

$$t = \sqrt{F} = \frac{\bar{A}_i - \bar{A}_{i'}}{\sqrt{\text{MS}_{\bar{A}_i - \bar{A}_{i'}}}}.$$

Either one-tailed or two-tailed tests may be made by using the t statistic. An equivalent but computationally simpler form of the F statistic for the case in which factor B is fixed, in terms of totals rather than means, is

$$F = \frac{(A_i - A_{i'})^2}{2nq\text{MS}_{\text{w. cell}}}.$$

For the case in which factor B is random,

$$F = \frac{(A_i - A_{i'})^2}{2nq\text{MS}_{ab}}.$$

Regardless of whether factors A or B are fixed or random, a test on the difference between cell means has the form

$$F = \frac{(AB_{ij} - AB_{km})^2}{2n\text{MS}_{\text{w. cell}}}.$$

In making the tests indicated in this section, the denominators use data from all cells in the experiment, not just those from which the means have been computed. If the experimental error is homogeneous, use of information from all cells to estimate experimental error is justified. If, however, there is a sufficient number of degrees of freedom (say over 30) for the estimation of experimental error from only those cells which are used in the estimation of the means being compared, this latter estimate is to be preferred to the pooled estimate from all cells.

A test on the difference between two means is a special case of a comparison among several means. For $p \times q$ factorial experiments having n observations in each cell, the mean square for a comparison among the main effects of factor A has the general form

$$\frac{(c_1A_1 + c_2A_2 + \cdots + c_pA_p)^2}{nq\Sigma c_i^2},$$

where $\Sigma c_i = 0$. The case in which $c_1 = 1$, $c_2 = -1$, and all other c's are zero defines the mean square

$$\frac{(A_1 - A_2)^2}{2nq}.$$

The case in which $c_1 = 1$, $c_2 = 1$, and $c_3 = -2$ defines the mean square

$$\frac{(A_1 + A_2 - 2A_3)^2}{6nq}.$$

The general mean square corresponding to a comparison among the main effects of factor B has the form

$$\frac{(c_1B_1 + c_2B_2 + \cdots + c_qB_q)^2}{np\Sigma c_j^2},$$

where $\Sigma c_j = 0$. Similarly, the mean square for a comparison among the simple effects of factor A for level b_j has the form

$$\frac{(c_1AB_{1j} + c_2AB_{2j} + \cdots + c_pAB_{pj})^2}{n\Sigma c_i^2}.$$

A comparison involving marginal totals (or means) may always be expressed in terms of cell totals (or means). For example, the comparison

$$\frac{(AB_{11} + AB_{12} + \cdots + AB_{1q} - AB_{21} - AB_{22} - \cdots - AB_{2q})^2}{n(2q)}$$

reduces to

$$\frac{(A_1 - A_2)^2}{2nq}.$$

For the case in which A and B are fixed factors, comparisons planned before the data were obtained (and assuming that the number of such comparisons is small relative to the total number possible) may be made by use of the following statistic:

$$F = \frac{\text{MS}_{\text{comparison}}}{\text{MS}_{\text{w. cell}}}.$$

Under the assumption that the hypothesis being tested is true, this statistic has a sampling distribution which is given by an F distribution. The critical value for this test is $F_{1-\alpha}[1, pq(n-1)]$. In cases where the comparisons are large in number or of the a posteriori type, the appropriate critical value, as suggested by Scheffé (1953), for tests on main effects due to factor A is

$$(p-1)F_{1-\alpha}[(p-1), pq(n-1)].$$

An analogous critical value for comparisons among the main effects of factor B is

$$(q-1)F_{1-\alpha}[(q-1), pq(n-1)].$$

The critical value for comparisons of the a posteriori type among cell means is

$$(pq-1)F_{1-\alpha}[(pq-1), pq(n-1)].$$

To illustrate the magnitude of the difference between the two types of critical values, consider the case in which $p = 10$ and $pq(n-1) = 40$. For $\alpha = .01$,

$$F_{.99}(1,40) = 7.31$$

is the critical value for an a priori type of comparison. For an a posteriori type

$$9F_{.99}(9,40) = 9(2.8876) = 25.99.$$

The difference between the two critical values is quite marked—but then the difference in the underlying logic is also quite marked. Whenever a relatively large number of tests of significance are to be made, or whenever comparisons suggested by the data are made, the usual sampling distributions (that is, t or F) associated with tests of significance no longer apply. The critical value for the a posteriori comparison is much larger because the sampling distribution appropriate for this type of comparison must take into account sources of variation which are not relevant to the a priori comparison.

In making comparisons between all possible pairs of ordered means within a logical grouping, the Tukey or the Newman-Keuls procedures may be adapted for use. For example, all pairs of means within a single row or column of a $p \times q$ factorial experiment may be compared by means of the Tukey procedure. If the number of degrees of freedom for estimating the standard error is sufficiently large (say over 30), a single row or a single column of the $p \times q$ factorial experiment may be considered as a single-factor experiment and the methods in Chap. 3 may be applied directly to this part of the factorial experiment. Alternatively, if the assumption of homogeneity of within-cell variance is met, an estimate of the standard error of a cell mean is given by

$$s_{\overline{AB}} = \sqrt{\frac{\text{MS}_{\text{w. cell}}}{n}}.$$

The degrees of freedom for this estimate are $pq(n - 1)$.

If all possible pairs of means corresponding to the main effects of factor A are to be compared, and if factor B is fixed, then

$$s_{\bar{A}} = \sqrt{\frac{\text{MS}_{\text{w. cell}}}{nq}}.$$

Numerical applications of comparisons of this kind are given in Sec. 6.2.

5.18 Partition of Main Effects and Interaction into Trend Components

In some cases arising in practice it is desirable to divide main effects as well as interactions into components associated with functional forms assumed to account for trends in the criterion responses. As an example, consider a 3 × 4 factorial experiment in which the levels of both factors may be regarded as steps along an essentially underlying continuum. The magnitudes of the criterion scores within each of the cells may be considered to define a response surface. It is frequently of interest to explore regions on this response surface, particularly when one is seeking an optimum combination of treatments.

Main effects and interactions in a 3 × 4 factorial experiment may be subdivided as indicated below:

Source	df	
A		2
Linear	1	
Quadratic	1	
B		3
Linear	1	
Quadratic	1	
Cubic	1	

AB				6
Linear × linear	1	Quadratic × linear	1	
Linear × quadratic	1	Quadratic × quadratic	1	
Linear × cubic	1	Quadratic × cubic	1	

The components of the variation due to the main effects of factor A can be expressed in the form

$$\text{SS}_{A\text{ component}} = \frac{(\Sigma c_i A_i)^2}{nq\Sigma c_i^2},$$

where the c_i are the coefficients of trend comparisons corresponding to the main effects of factor A. The computational work is simplified if the levels of factor A are equally spaced along the underlying continuum; in this case the coefficients are defined by appropriate entries in Table C.10, the so-called coefficients of orthogonal polynomials. In the case of unequal spacing, these coefficients may be modified by using methods described by Robson (1959).

In a similar manner, the components of the variation due to the main effects of factor B may be expressed in the form

$$SS_{B\ \text{component}} = \frac{(\Sigma c_j B_j)^2}{np\Sigma c_j^2}.$$

The trend components of the variation due to the interaction have the general form

$$SS_{AB\ \text{component}} = \frac{(\Sigma c_{ij} AB_{ij})^2}{n\Sigma c_{ij}^2},$$

where $c_{ij} = c_i c_j$ and c_i and c_j are the respective trend components of the main effects. (See Table 5.18-1.)

Table 5.18-1 Some Trend Comparisons in a 3 × 4 Factorial Experiment

A_{lin}

	b_1	b_2	b_3	b_4	Total
a_1	−1	−1	−1	−1	−4
a_2	0	0	0	0	0
a_3	1	1	1	1	4
Total	0	0	0	0	0

B_{quad}

	b_1	b_2	b_3	b_4	Total
a_1	1	−1	−1	1	0
a_2	1	−1	−1	1	0
a_3	1	−1	−1	1	0
Total	3	−3	−3	3	0

$A_{\text{lin}} \times B_{\text{quad}}$

	b_1	b_2	b_3	b_4	Total
a_1	−1	1	1	−1	0
a_2	0	0	0	0	0
a_3	1	−1	−1	1	0
Total	0	0	0	0	0

The response surface defined by such trend components is that described by a polynomial in which the independent variates are mutually uncorrelated. For the case of a 3 × 4 factorial experiment, an exact fit to the surface on which the cell means lie can always be obtained by a polynomial having the following form:

$$\overline{AB} = b_0 + b_{1,0}P_{1,0} + b_{2,0}P_{2,0} + b_{0,1}P_{0,1} + b_{0,2}P_{0,2} + b_{0,3}P_{0,3}$$
$$+ b_{1,1}P_{1,1} + b_{1,2}P_{1,2} + b_{1,3}P_{1,3}$$
$$+ b_{2,1}P_{2,1} + b_{2,2}P_{2,2} + b_{2,3}P_{2,3},$$

where each $P_{i,j}$ is uncorrelated with each of the others. The $P_{i,j}$ actually correspond to polynomials of degree i in a variate Z_1 and of degree j in a variate Z_2. However, for prediction purposes the $P_{i,j}$ may be regarded as a set of independent variates. Because the $P_{i,j}$ are uncorrelated, the sources of variation associated with each of the $b_{i,j}$ are orthogonal.

The variation due to the regression coefficient

$b_{1,0}$ corresponds to that due to A_{lin},
$b_{2,0}$ corresponds to that due to A_{quad},
$b_{0,1}$ corresponds to that due to B_{lin},
$b_{0,2}$ corresponds to that due to B_{quad},
$b_{1,1}$ corresponds to that due to $A_{\text{lin}} \times B_{\text{lin}}$,
$b_{2,3}$ corresponds to that due to $A_{\text{quad}} \times B_{\text{cubic}}$.

In terms of the $P_{i,j}$ variates, the components of the regression analysis and the components of the analysis of variance are identical. However, this identity is not readily established if the regression analysis is carried out in terms of a correlated set of regression variates.

The components of the interaction obtained in this way describe features of a response surface that might be described in a more parsimonious form by means of functions other than polynomials. More extensive methods for exploring response surfaces are available. A survey of some of these methods is given in Cochran and Cox (1957, chap. 8A).

If only one of the factors, say, factor B, is continuous, then the following analysis is possible:

Source	df	
A		2
B		3
Linear	1	
Quadratic	1	
Cubic	1	
AB		6
Difference in lin trend	2	
Difference in quad trend	2	
Difference in cubic trend	2	

This method of partitioning the interaction is particularly appropriate in attempting to interpret differences in shapes of profiles. Computational details for this kind of partition are given in Secs. 6.9 and 7.6. In Sec. 7.6, computational steps are given in terms of a design calling for repeated measures. If the design does not have repeated measures, computational procedures for treatment effects remain the same but no partition of the error variation is required.

For the case of a 2×2 factorial experiment in which both factors represent what can be considered an underlying continuum, an exact fit to the surface defined by the four cell means is given by the polynomial

$$\overline{AB} = b_0 + b_{1,0}P_{1,0} + b_{0,1}P_{0,1} + b_{1,1}P_{1,1},$$

where the $P_{i,j}$ represent a mutually uncorrelated set of variates. For the case of a 2×3 factorial experiment, an exact fit to the surface defined by the six cell means is given by the polynomial

$$\overline{AB} = b_0 + b_{1,0}P_{1,0} + b_{0,1}P_{0,1} + b_{0,2}P_{0,2} + b_{1,1}P_{1,1} + b_{1,2}P_{1,2}.$$

5.19 Replicated Experiments

A replication of an experiment is an independent repetition under as nearly identical conditions as the nature of the experimental material will permit. *Independent* in this context implies that the experimental units in the repetitions are independent samples from the population being studied. That is, if the elements are people, in a replication of an experiment an independent sample of people is used in the replication. Inferences made from replicated experiments are with respect to the outcomes of a series of replications of the experiment. In a sense that will be indicated, inferences from replicated experiments have a broader scope than do inferences from nonreplicated experiments.

An experiment in which there are n observations per cell is to be distinguished from an experiment having n replications with one observation per cell. The total number of observations per treatment is the same, but the manner in which the two types of experiments are conducted differs. Consequently the relevant sources of variation will differ; hence a change in the model for the experiment is needed. Associated with the latter change is a different set of expected values for the mean squares.

In conducting an experiment having n observations per cell, all observations in a single cell are generally made within the same approximate time interval or under experimental conditions that can be considered to differ only as a function of experimental error. In a $p \times q$ factorial experiment having n observations per cell (no repeated measures) the total variation may be subdivided as follows.

Source	df
Between cells	$pq - 1$
Within cells	$pq(n - 1)$

In this partition it is assumed that differences between observations within the same cell are attributable solely to experimental error. On the other hand, differences among cells are attributed to treatment effects and experimental error. Should there be systematic sources affecting between-cell variation, other than treatment effects and experimental error, such variation is completely or partially confounded with the treatment effects.

The purpose of a replicated experiment, in contrast to an experiment having n observations per cell, is to permit the experimenter to maintain more uniform conditions within each cell of the experiment, as well as to eliminate possible irrelevant sources of variation between cells. For example, the replications may be repetitions of the experiment at different times or at different places. In this case, sources of variation which are functions of time or place are eliminated from both between- and within-cell variation.

Instead of having n replications with one observation per cell, one may have two replications in which there are $n/2$ observations per cell, one may have three replications with $n/3$ observations per cell, etc. The number of observations per cell for any single replication should be the maximum that will permit uniform conditions within all cells of the experiment and, at the same time, reduce between-cell sources of variation which are not directly related to the treatment effects.

As an example of a replicated experiment, suppose that it is desired to have 15 observations within each cell of a 2×3 factorial experiment. Suppose that the experimental conditions are such that only 30 observations can be made within a given time period; suppose further that there are, potentially, differences in variation associated with the time periods. To eliminate sources of variation directly related to the time dimension, the experiment may be set up in three replications as follows (cell entries indicate the number of observations):

	Replication 1				Replication 2				Replication 3		
	b_1	b_2	b_3		b_1	b_2	b_3		b_1	b_2	b_3
a_1	5	5	5	a_1	5	5	5	a_1	5	5	5
a_2	5	5	5	a_2	5	5	5	a_2	5	5	5

If the replications are disregarded, there are 15 observations under each treatment combination. Within each replication, conditions may be relatively uniform with respect to uncontrolled sources of variation; between replications, conditions may not be uniform. In essence, a replicated experiment adds another dimension or factor to the experiment—a replication factor. The latter is a random factor.

In Table 5.19-1 the analysis of a $p \times q$ factorial experiment having n observations per cell is contrasted with a $p \times q$ factorial experiment in which there are r replications of a $p \times q$ factorial experiment having n/r observations per cell. In both designs there are n observations under treatment combination and a total of $npq - 1$ degrees of freedom. The

Table 5.19-1 Comparison of Analysis for Replicated and Nonreplicated Factorial Experiments

	$p \times q$ factorial experiment with n observations per cell		r replications of $p \times q$ factorial experiment with n/r observations per cell	
		df		df
(i)	A	$p - 1$	A	$p - 1$
	B	$q - 1$	B	$q - 1$
	AB	$(p - 1)(q - 1)$	AB	$(p - 1)(q - 1)$
			Reps	$r - 1$
			$A \times$ rep	$(p - 1)(r - 1)$
			$B \times$ rep	$(q - 1)(r - 1)$
			$AB \times$ rep	$(p - 1)(q - 1)(r - 1)$
	Within cell	$pq(n - 1)$	Within cell	$pqr[(n/r) - 1]$
	Total	$npq - 1$	Total	$npq - 1$
(ii)	Between cells	$pq - 1$	Between cells	$pqr - 1$
			Treatments	$pq - 1$
			Reps	$r - 1$
			Treat $\times$ rep	$(pq - 1)(r - 1)$
	Within cell	$pq(n - 1)$	Within cell	$pqr[(n/r) - 1]$

degrees of freedom in braces in part i are often pooled in a replicated experiment. Part ii gives an alternative partition of the variation. In the nonreplicated experiment the between-cell variation defines the estimates of variation due to treatment effects. On the other hand, in the replicated experiment part of the between-cell variation is due to replication effects as well as interactions with replications. The within-cell variations in both designs is considered to be due to experimental error. Since observations in the replicated experiment are made under somewhat more controlled conditions, the within-cell variation in a replicated experiment is potentially smaller than the corresponding variation in a nonreplicated experiment.

The partition of the total variation given in part ii may be illustrated numerically. Consider a 3×4 factorial experiment in which there are to be 10 observations under each treatment combination. The design calling for 10 observations per cell is contrasted with the design calling for five replications with 2 observations per cell in the following partition:

10 observations per cell		5 reps, 2 observations per cell	
Between cells	11	Between cells	59
		Treatments	11
		Reps	4
		Treat $\times$ rep	44
Within cell	108	Within cell	60

In a replicated design of this kind interactions with replications are often considered to be part of the experimental error. However, preliminary tests on the model may be used to check upon whether or not such pooling is justified.

There are both advantages and disadvantages to replicated experiments when contrasted with those in which there are n observations per cell. The advantages of being able to eliminate sources of variation associated with replications depend upon the magnitude of the latter relative to treatment effects and effects due to experimental error. One possible disadvantage of a replicated design might arise in situations requiring precise settings of experimental equipment. Considerable time and effort may be lost in resetting equipment for each of the replications, rather than making all observations under a single set of experimental conditions before moving to the next experimental condition. However, the inferences which can be drawn from the replicated experiment are stronger when possible variations in the resettings are considered as a relevant source of variation in the conduct of the experiment.

5.20 The Case $n = 1$ and a Test for Nonadditivity

If there is only one observation in each cell of a $p \times q$ factorial experiment, there can be no within-cell variation and hence no direct estimate of experimental error. Among other models, the following two may be postulated to underlie the observed data:

(i) $$X_{ij} = \mu + \alpha_i + \beta_j + \varepsilon_{ij}.$$

(ii) $$X_{ij} = \mu + \alpha_i + \beta_j + \alpha\beta_{ij} + \varepsilon_{ij}.$$

In (i) no interaction effect is postulated; hence all sources of variation other than main effects are considered to be part of the experimental error. In (ii) an interaction term is postulated. From some points of view the interaction term may be considered as a measure of nonadditivity of the main effects. In this context, (i) will be considered the additive model, whereas (ii) will be considered the nonadditive model.

The choice of the scale of measurement for the basic criterion measurement will to some extent determine whether (i) or (ii) is the more appropriate model. The degree of appropriateness is gauged in terms of the degree of heterogeneity of experimental error under model (i). Subject-matter knowledge and experience gained from past experimentation about the functional form of the underlying sources of variation are the best guides for deciding between models (i) and (ii). To supplement these sources of information, the experimenter may want to use information provided by preliminary tests before specifying the model. So far in this chapter such tests have been considered only for the case in which direct estimates of experimental error were available from the experimental data. Tukey (1949) developed a test applicable to the case in which there is a single observation per cell. This test is discussed in detail by Scheffé (1959, pp. 130–134) and by Rao (1965, pp. 207–209).

Tukey's test is called a test for *nonadditivity*. Its purpose is to help in the decision between models (i) and (ii). This test has also been used to choose between alternative scales of measurement, the decision being made in favor of the scale for which model (i) is the more appropriate.

In Tukey's approach to the problem, he starts with the model

(iii) $$X_{ij} = \mu + \alpha_i + \beta_j + \lambda\alpha_i\beta_j + \varepsilon_{ij},$$

where $\alpha_i\beta_j$ is the product of the main effects and λ is a regression coefficient. Here the product term is that part of the interaction which can be expressed as the product of main effects. A test on the hypothesis that $\lambda = 0$ is equivalent to a test on the hypothesis that product terms of this form do not contribute to the prediction of the X_{ij}.

Assuming that the α_i and the β_j are known, a least-squares estimate of λ is obtained by making

$$\sum\hat{\varepsilon}_{ij}^2 = \sum(X_{ij} - \mu - \alpha_i - \beta_j - \hat{\lambda}\alpha_i\beta_j)^2 = \text{minimum}.$$

Solving the normal equation which results yields the following estimate for λ:

$$\hat{\lambda} = \frac{\Sigma\alpha_i\beta_j(X_{ij} - \mu - \alpha_i - \beta_j)}{\Sigma\alpha_i^2\beta_j^2},$$

all summations being over i and j. If one now replaces μ, α_i, and β_j by their respective least-squares estimates,

$$\hat{\mu} = \bar{G}, \qquad \hat{\alpha}_i = \bar{A}_i - \bar{G}, \qquad \hat{\beta}_j = \bar{B}_j - \bar{G},$$

one obtains, after simplification,

$$\hat{\lambda} = \frac{\Sigma\hat{\alpha}_i\hat{\beta}_j X_{ij}}{\Sigma\hat{\alpha}_i^2\hat{\beta}_j^2}.$$

If one lets

$$k_{ij} = \hat{\alpha}_i\hat{\beta}_j,$$

from the definitions of $\hat{\alpha}_i$ and $\hat{\beta}_j$, one finds that

$$\Sigma k_{ij} = 0 \qquad \text{and} \qquad \Sigma k_{ij}^2 = \Sigma\hat{\alpha}_i^2\hat{\beta}_j^2,$$

where all summations are over both i and j. In terms of the k_{ij} the expression for $\hat{\lambda}$ takes the form

$$\hat{\lambda} = \frac{\Sigma k_{ij}X_{ij}}{\Sigma k_{ij}^2}.$$

Let $$D = \hat{\lambda}\Sigma k_{ij}^2 = \Sigma k_{ij}X_{ij}.$$

Since $\Sigma k_{ij} = 0$, D represents a comparison or contrast among the X_{ij}. The component of variation corresponding to this comparison has the form

$$SS_D = \frac{D^2}{\Sigma k_{ij}^2} = \hat{\lambda}^2(\Sigma k_{ij}^2) = \frac{(\Sigma k_{ij} X_{ij})^2}{\Sigma k_{ij}^2}.$$

This source of variation Tukey calls the nonadditivity component. Upon replacing the k_{ij} by their basic definitions one has

$$SS_{\text{nonadd}} = \hat{\lambda}^2(\Sigma k_{ij}^2) = \frac{(\Sigma \hat{\alpha}_i \hat{\beta}_j X_{ij})^2}{(\Sigma \hat{\alpha}_i^2)(\Sigma \hat{\beta}_j^2)} = \frac{pq(\Sigma \hat{\alpha}_i \hat{\beta}_j X_{ij})^2}{SS_a SS_b},$$

since $SS_a \doteq q\Sigma\hat{\alpha}_i^2$ and $SS_b = p\Sigma\hat{\beta}_j^2$.

In the application of Tukey's test for nonadditivity to the case of a $p \times q$ factorial experiment having one observation per cell, the analysis of variance takes the following form:

Source of variation	SS	df	MS
A	SS_a	$p - 1$	MS_a
B	SS_b	$q - 1$	MS_b
Residual	SS_{res}	$(p - 1)(q - 1)$	
Nonadditivity	SS_{nonadd}	1	MS_{nonadd}
Balance	SS_{bal}	$(p - 1)(q - 1) - 1$	MS_{bal}

In this table

$$SS_{\text{bal}} = SS_{\text{res}} - SS_{\text{nonadd}},$$

and

$$SS_{\text{res}} = SS_{\text{total}} - SS_a - SS_b.$$

The test for nonadditivity is given by

$$F = \frac{MS_{\text{nonadd}}}{MS_{\text{bal}}}.$$

When this F ratio exceeds the critical value defined by the level of significance of the test, the hypothesis that model (i) is appropriate is rejected. Tukey's test for nonadditivity is sensitive to only one source of nonadditivity—that associated with a component of the interaction represented by $\alpha_i\beta_j$. (This component is somewhat related to the linear $\times$ linear component of interaction in the analysis of trend.) In working with this component there is an implicit assumption that, the larger the main effects of the individual levels in a treatment combination, the larger the potential interaction effect for the treatment combination, if this does exist. This assumption appears reasonable in some cases. In other cases it might be that the equivalent of the linear $\times$ quadratic or the quadratic $\times$ quadratic component could more appropriately be used as a measure of nonadditivity. A numerical example is given in Sec. 6.8.

The principles underlying Tukey's test for a two-factor experiment can be extended to higher-order factorial experiments. For the three-factor case, the comparison for nonadditivity is given by

$$D = \Sigma c_{ijk} X_{ijk}$$

where $$c_{ijk} = (\bar{A}_i - \bar{G})(\bar{B}_j - \bar{G})(\bar{C}_k - \bar{G}).$$

The corresponding component of variation is

$$\text{SS}_D = \text{SS}_{\text{nonadd}} = \frac{D^2}{\Sigma c_{ijk}^2} = \frac{pqrD^2}{\text{SS}_a \text{SS}_b \text{SS}_c}.$$

In this case the residual variation is partitioned as follows:

Source of variation	df
$\text{SS}_{\text{residual}}$	$(p - 1)(q - 1)(r - 1)$
$\text{SS}_{\text{nonadd}}$	1
$\text{SS}_{\text{balance}}$	$(p - 1)(q - 1)(r - 1) - 1$

5.21 The Choice of a Scale of Measurement and Transformations

In the analysis of variance of a factorial experiment, the total variation of the criterion variable is subdivided into nonoverlapping parts which are attributable to main effects, interactions, and experimental error. The relative magnitude of each of the corresponding variances depends upon the scale of measurement as well as the spacing of the levels of the factors used in the experiment. In cases where alternative choices of a scale of measurement appear equally justifiable on the basis of past experience and theory, analysis in terms of each of the alternative scales is warranted provided that each of the scales satisfies the assumptions underlying the respective analyses.

It may happen that within-cell variances will be homogeneous in terms of one scale of measurement but heterogeneous in terms of a second scale. The within-cell distributions may be highly skewed in terms of one scale but approximately normal in terms of a second scale. In terms of one scale of measurement, an additive (i.e., no interaction terms) model may be appropriate, whereas in terms of a second scale the additive model will not be appropriate.

In determining the choice of a scale of measurement for the observed data, two cases will be contrasted. In one case, a priori theory and experience determine the appropriate model as well as the appropriate scale. In the second case, where there is neither adequate theory nor experience to serve as guides, the appropriate model and the proper scale of measurement are determined only after the experimental data have been partially analyzed. In the latter case the design of the experiment should provide the experimenter with sufficient data to permit the evaluation of alternative formulations of the model.

A readable summary of methods for determining appropriate transformations will be found in Olds et al. (1956). A series of alternative methods is also given by Tukey (1949). A transformation in this context is a change in the scale of measurement for the criterion. For example rather than time in seconds the scale of measurement may be logarithm time in seconds; rather than number of errors, the square root of the number of errors may be used as the criterion score. There are different reasons for making such transformations. Some transformations have as their primary purpose the attainment of homogeneity of error variance. The work of Box (1953) has shown that the distribution of the F ratio in the analysis of variance is affected relatively little by inequalities in the variances which are pooled into the experimental error. Transformations which have homogeneity of error variance as their primary purpose are relatively less important than they were formerly considered to be. With regard to the usual tests for homogeneity of error variance, Box (1953, p. 333) says, "To make the preliminary tests on variances is rather like putting to sea in a rowing boat to find out whether conditions are sufficiently calm for an ocean liner to leave port."

Another reason for using transformations is to obtain normality of within-cell distributions. Often non-normality and heterogeneity of variance occur simultaneously. The same transformation will sometimes normalize the distributions as well as make the variances more homogeneous. The work of Box (1953) has shown that the sampling distribution of the F ratio is relatively insensitive to moderate departures from normality. Hence transformations whose primary purpose is to attain normal within-cell distributions are now considered somewhat less important than was the case previously.

A third reason for transformations is to obtain additivity of effects. In this context *additivity of effects* implies a model which does not contain interaction terms. In some of the designs which are discussed in later chapters, a strictly additive model is required. In designs of this type certain interaction effects are completely confounded with experimental error. For those designs which do permit independent estimation of interaction effects and error effects, the strictly additive model is not essential. There are, however, advantages to the strictly additive model, if it is appropriate, over the nonadditive model—particularly in cases in which fixed and random factors are in the same experiment. The interaction of the random factors with the fixed factors, if these exist, will tend to increase the variance of the sampling distribution of the main effects. Tukey (1949) has pointed out rather vividly the influence of the scale of measurement upon existence or nonexistence of interaction effects.

The use of transformations in order to obtain additivity has received more attention in recent works than it has in the past. Tukey's test for nonadditivity has been used, in part, as a guide for deciding between alternative possible transformations. It is not possible to find transformations which will eliminate nonadditivity in all cases. In some cases there is an intrinsic interaction between the factors which cannot be considered a function of the choice of the scale of measurement. These cases are not always easily distinguished from cases in which the interaction is essentially an artifact of the scale of measurement.

A *monotonic* transformation is one which leaves ordinal relationships (i.e., greater than, equal to, or less than) unchanged. If the means for the levels of factor A have the same rank for all levels of factor B, then a monotonic transformation can potentially remove the $A \times B$ interaction. When such rank order is not present, a monotonic transformation cannot remove the $A \times B$ interaction. Only monotonic transformations will be discussed in this section. Nonmonotonic transformations in this connection would be of extremely limited utility; the author is aware of no studies in which the latter class of transformation has been used.

There are some overall guides in selecting a scale of measurement which will satisfy the assumptions of homogeneity of error variance. These guides will be considered in terms of the relationship between the cell means and the cell variances. The latter relationship will be presented in terms of a $p \times q$ factorial experiment; the principles to be discussed hold for all designs.

Case (i): $\sigma_{ij}^2 = c^2\mu_{ij}$. In this case the cell variances tend to be functions of the cell means: the larger the mean, the larger the variance. This kind of relationship exists when the within-cell distribution is Poisson in form. For this case, a square-root transformation will tend to make the variances more homogeneous. This transformation has the form

$$X'_{ijk} = \sqrt{X_{ijk}},$$

where X is the original scale and X' is the transformed scale. If X is a frequency, i.e., number of errors, number of positive responses, and if X is numerically small in some cases (say less than 10), then a more appropriate transformation is

$$X'_{ijk} = \sqrt{X_{ijk}} + \sqrt{X_{ijk} + 1}.$$

The following transformation is also used for frequency data in which some of the entries are numerically small:

$$X'_{ijk} = \sqrt{X_{ijk} + \tfrac{1}{2}}.$$

Either of the last two transformations is suitable for the stated purpose.

Case (ii): $\sigma_{ij}^2 = \mu_{ij}(1 - \mu_{ij})$. This case occurs in practice when the basic observations have a binomial distribution. For example, if the basic observations are proportions, variances and means will be related in the manner indicated. The following transformation is effective in stabilizing the variances,

$$X'_{ijk} = 2 \arcsin \sqrt{X_{ijk}},$$

where X_{ijk} is a proportion. In many cases only a single proportion appears in a cell. Tables are available for this transformation. Numerically, X'_{ijk} is an angle measured in radians. For proportions between .001 and .999, X'_{ijk} assumes values between .0633 and 3.0783. The notation $\sin^{-1}$ (read inverse sine) is equivalent to the notation arcsin. For values of X close to zero or close to unity, the following transformation is recommended:

$$X'_{ijk} = 2 \arcsin \sqrt{X_{ijk} \pm [1/(2n)]},$$

where n is the number of observations on which X is based. The plus sign is used for X_{ijk} close to zero; the minus sign is used for X_{ijk} close to unity.

Case (iii): $\sigma^2_{ij} = k^2\mu^2_{ij}$. In this case the logarithmic transformation will stabilize the variances.

$$X'_{ijk} = \log X_{ijk}.$$

To avoid values of X close to zero, an alternative transformation

$$X'_{ijk} = \log (X_{ijk} + 1)$$

is often used when some of the measurements are equal to or close to zero. The logarithmic transformation is particularly effective in normalizing distributions which have positive skewness. Such distributions occur in psychological research when the criterion is in terms of a time scale, i.e., number of seconds required to complete a task.

The rationale underlying the choice of a transformation is the following: Let m be a cell mean in terms of an original scale of measurement. Let the cell variance be

$$\sigma^2_m = f(m).$$

That is, the cell variances are some function of the cell means. Let $\phi(m)$ define a transformation. In terms of the transformed scale, the cell variances will be approximately

$$\sigma^2_\phi = \left(\frac{d\phi}{dm}\right)^2 f(m) \qquad \text{or} \qquad d\phi = \frac{\sigma_\phi}{\sqrt{f(m)}}\, dm.$$

In order that the cell variances be some constant, say k^2, in terms of the transformed scale, one must have

$$\phi(m) = \int \frac{k\, dm}{\sqrt{f(m)}}.$$

For example, suppose

$$\sigma^2_m = f(m) = c^2 m.$$

Then $$\frac{df}{dm} = c^2 \qquad \text{and} \qquad dm = \frac{df}{c^2}.$$

Hence $$\phi(m) = \int \frac{k}{\sqrt{c^2 m}} \frac{df}{c^2} = \frac{2k}{c^3} \int \frac{1}{2\sqrt{m}}\, df$$
$$= \frac{2k}{c^3} \sqrt{m}.$$

If all that is wanted is a transformation that will make σ_m^2 some constant, then one may use

$$\phi(m) = \sqrt{m}.$$

In terms of the basic observations, this transformation takes the form

$$\phi(X_{ijk}) = X'_{ijk} = \sqrt{X_{ijk}}.$$

Use of the Range in Deciding between Alternative Transformations. In deciding which one of several possible transformations to use in a specific problem, one may investigate several before deciding which one puts the data in a form that most nearly satisfies the basic assumptions underlying the analysis of variance. The use of the range statistic (or the truncated range statistic) provides a relatively simple method for inspecting the potential usefulness of several transformations with a minimum of computational effort. An example given by Rider et al. (1956, pp. 47–55) will be used to illustrate the method.

In this example, eight operators individually measured the resistance of each of four propeller blades with each of two instruments. Order was randomized. The range of the 16 measurements on each blade in terms of the original scale as well as in terms of transformed scales is given below. (Only the end points of the ranges in terms of the original scale need to be transformed in order to obtain the following data.)

Blade	Scale of measurement			
	Original	Square root	Logarithm	Reciprocal
1	3.10	0.61	0.21	0.077
2	0.10	0.08	0.12	0.833
3	0.15	0.12	0.17	1.111
4	11.00	1.01	0.16	0.015

The logarithmic transformation is seen to make the ranges more uniform. In many practical cases the range tends to be proportional to the variance. A transformation which tends to make the ranges uniform will also tend to make the variances more uniform. As a further step in checking on the adequacy of the logarithmic transformation, the authors applied

Tukey's test for nonadditivity to each of the four interaction terms. On the original scale of measurement, two of the four F ratios for nonadditivity were significant at the 5 percent level. None of the F ratios for nonadditivity was significant at the 5 percent level in terms of the logarithmic scale.

5.22 Unequal Cell Frequencies

Although an experimental design in its initial planning phases may call for an equal number of observations per cell, the completed experiment may not have an equal number of observations in all cells. There may be many reasons for such a state of affairs. The experimenter may be forced to work with intact groups having unequal size; the required number of individuals in a given category may not be available to the experimenter at a specified time; subjects may not show up to complete their part in an experiment; laboratory animals may die in the course of an experiment. If the original plan for an experiment calls for an equal number of observations in each cell, and if the loss of observations in cells is essentially random (in no way directly related to the experimental variables), then the experimental data may appropriately be analyzed by the method of unweighted means. In essence the latter method considers each cell in the experiment as if it contained the same number of observations as all other cells (at least with regard to the computation of main effects and interaction effects).

Under the conditions that have been specified, the number of observations within each cell will be of the same order of magnitude. The procedures for an unweighted-means analysis will be described in terms of a 2×3 factorial experiment. These procedures may be generalized to higher-order factorial experiments. The number of observations in each cell may be indicated as follows:

	b_1	b_2	b_3
a_1	n_{11}	n_{12}	n_{13}
a_2	n_{21}	n_{22}	n_{23}

The harmonic mean of the number of observations per cell is

$$\bar{n}_h = \frac{pq}{(1/n_{11}) + (1/n_{12}) + \cdots + (1/n_{pq})}.$$

In the computation of main effects and interactions each cell is considered to have $\bar{n}_h$ observations. (The harmonic mean rather than the arithmetic mean is used here because the standard error of a mean is proportional to $1/n_{ij}$ rather than n_{ij}.)

The mean for each of the cells may be represented as follows:

	b_1	b_2	b_3
a_1	$\overline{AB}_{11}$	$\overline{AB}_{12}$	$\overline{AB}_{13}$
a_2	$\overline{AB}_{21}$	$\overline{AB}_{22}$	$\overline{AB}_{23}$

The estimate of the mean $\mu_{1.}$ is

$$\bar{A}_1 = \frac{\sum_j \overline{AB}_{1j}}{q}.$$

That is, $\bar{A}_1$ is the mean of the means in row a_1, not the mean of all observations at level a_1. These two means will differ when each cell does not have the same number of observations. The estimate of $\mu_{.1}$ is

$$\bar{B}_1 = \frac{\sum_i \overline{AB}_{i1}}{p}.$$

Again there will be a difference between the mean of the means within a column and the mean of all observations in a column. The grand mean in the population is estimated by

$$\bar{G} = \frac{\Sigma \bar{A}_i}{p} = \frac{\Sigma \bar{B}_j}{q} = \frac{\Sigma\Sigma \overline{AB}_{ij}}{pq}.$$

Variation due to main effects and interactions are estimated by the following sums of squares:

$$\begin{aligned}
SS_a &= \bar{n}_h q \Sigma(\bar{A}_i - \bar{G})^2, \\
SS_b &= \bar{n}_h p \Sigma(\bar{B}_j - \bar{G})^2, \\
SS_{ab} &= \bar{n}_h \Sigma(\overline{AB}_{ij} - \bar{A}_i - \bar{B}_j + \bar{G})^2.
\end{aligned}$$

These sums of squares have the same form as corresponding sums of squares for the case of equal cell frequencies. However, $\bar{A}_i$, $\bar{B}_j$, and $\bar{G}$ are computed in a different manner. If all cell frequencies were equal, both computational procedures would lead to identical results.

The variation within cell ij is

$$SS_{ij} = \sum_m X_{ijm}^2 - \frac{\left(\sum_m X_{ijm}\right)^2}{n_{ij}}.$$

The pooled within-cell variation is

$$SS_{\text{w. cell}} = \Sigma\Sigma SS_{ij}.$$

The degrees of freedom for this latter source of variation are

$$df_{\text{w. cell}} = (\Sigma\Sigma n_{ij}) - pq.$$

Other methods are available for handling the analysis for unequal cell frequencies. If, however, the differences in cell frequencies are primarily functions of sources of variation irrelevant to the experimental variables, there are no grounds for permitting such frequencies to influence the estimation of population means. On the other hand, should the cell frequencies be directly related to the size of corresponding population strata, then such frequencies should be used in estimating the mean of the population composed of such strata.

5.23 Unequal Cell Frequencies—Least-squares Estimation

As long as the cell frequencies in a factorial experiment are equal (or proportional in a sense that is defined later in this section) the variations due to overall main effects and interactions are additive. That is, the joint variation due to the main effects is the sum of the variations due to the separate main effects:

$$\mathrm{SS}_{a+b} = \mathrm{SS}_a + \mathrm{SS}_b.$$

Additivity in this sense corresponds to orthogonality of the separate variations. Further, for the case of equal or proportional cell frequencies,

$$\mathrm{SS}_{a+b+ab} = \mathrm{SS}_a + \mathrm{SS}_b + \mathrm{SS}_{ab}.$$

That is, the interaction is orthogonal to the main effects.

Orthogonality of the various sources of variation simplifies the interpretation of the outcome of an experiment; because of orthogonality one may interpret one source of variation independently of the others. This simplicity of interpretation disappears (in part) in a nonorthogonal design. Disproportionate cell frequencies will give rise to nonorthogonality.

Throughout this section it is assumed that the factors are fixed. Computational details for what is presented in this section will be found in Sec. 6.14.

An interesting discussion of some possible interpretations associated with the various methods discussed in this section will be found in Overall and Spiegel (1969).

5.23-1 Notation

In this section the following notation system will be used for the cell and marginal frequencies:

	b_1	$\cdots$	b_q	
a_1	n_{11}	$\cdots$	n_{1q}	$n_{1.}$
$\vdots$	$\vdots$		$\vdots$	$\vdots$
a_p	n_{p1}	$\cdots$	n_{pq}	$n_{p.}$
	$n_{.1}$	$\cdots$	$n_{.q}$	$n_{..}$

The cell and marginal totals will be denoted as follows:

	b_1	$\cdots$	b_q	
a_1	AB_{11}	$\cdots$	AB_{1q}	A_1
$\cdot$	$\cdot$		$\cdot$	$\cdot$
$\cdot$	$\cdot$		$\cdot$	$\cdot$
$\cdot$	$\cdot$		$\cdot$	$\cdot$
a_p	AB_{p1}	$\cdots$	AB_{pq}	A_p
	B_1	$\cdots$	B_q	G

Cell and marginal means will be defined as follows unless indicated otherwise.

$$\overline{AB}_{ij} = \frac{AB_{ij}}{n_{ij}}, \qquad \bar{A}_i = \frac{A_i}{n_{i.}}, \qquad \bar{B}_j = \frac{B_j}{n_{.j}}, \qquad \bar{G} = \frac{G}{n_{..}}.$$

The observations within cell ab_{ij} will be denoted

$$Y_{ij1} \quad Y_{ij2} \quad \cdots \quad Y_{ijn_{ij}}.$$

5.23-2 Principles from general linear model

Consider the following linear function of the variables $X_0, X_1, \ldots, X_p$:

$$\hat{Y} = b_0 X_0 + b_1 X_1 + \cdots + b_p X_p = \underline{b}'\underline{X}. \tag{1}$$

In earlier sections it was indicated that the least-squares definition of the components of $\underline{b}' = [b_0 \quad b_1 \quad \cdots \quad b_p]$ were obtained by solving the system of nomal equations

$$(X'X)\underline{b} = X'\underline{Y}. \tag{2}$$

The solution to (2) has the form

$$\underline{b} = (X'X)^{-1}X'\underline{Y} \tag{3}$$

if $X'X$ is nonsingular. If $X'X$ is singular, then a generalized inverse replaces $(X'X)^{-1}$; a generalized inverse is not unique. That part of ΣY^2 which can be predicted from the linear system (1) is given by

$$\Sigma \hat{Y}^2 = \underline{b}'X'\underline{Y}, \tag{4}$$

where $\underline{b}$ is any solution to the system (2). It is noted that (2) represents both a necessary and sufficient condition for minimizing $\Sigma(Y - \hat{Y})^2$. Hence any solution for (2) will minimize $\Sigma(Y - \hat{Y})^2$. It was found that

$$\Sigma Y^2 = \Sigma \hat{Y}^2 + \Sigma(Y - \hat{Y})^2.$$

Hence any solution which minimizes $\Sigma(Y - \hat{Y})^2$ will maximize $\Sigma \hat{Y}^2$. For a given set of sample data both $\Sigma(Y - \hat{Y})^2$ and $\Sigma \hat{Y}^2$ will be unique.

In applying the linear function represented by (1) to experimental designs, $X_0, X_1, \ldots, X_p$ are *indicator* (or counter) variables which are defined as follows:

$$X_0 = 1 \qquad \text{for all observations,}$$

$$X_j = \begin{cases} 1 & \text{if effect } j \text{ is present,} \\ 0 & \text{if effect } j \text{ is not present,} \end{cases} \qquad j = 1, \ldots, p.$$

5.23-3 Estimation of parameters under different (fixed) models

Consider the model

$$(1) \qquad Y_{ijk} = \mu + \varepsilon_{ijk}.$$

In terms of the general linear model, (1) is equivalent to

$$(1a) \qquad Y = \mu X_0 + \varepsilon_{ijk},$$

where μ has the role of β_0. In terms of matrix notation,

$$\underset{n_{..},1}{\underline{Y}} = \begin{bmatrix} Y_{111} \\ \cdot \\ \cdot \\ \cdot \\ Y_{pqn_{pq}} \end{bmatrix}, \qquad \underset{n_{..},1}{\underline{X_0}} = \begin{bmatrix} 1 \\ \cdot \\ \cdot \\ \cdot \\ 1 \end{bmatrix}.$$

The normal equations for this model have the form

$$(1b) \qquad (\underline{X}_0'\underline{X}_0)\hat{\mu} = \underline{X}_0'\underline{Y} \qquad \text{or} \qquad n_{..}\hat{\mu} = G.$$

Hence the least-squares estimate of μ is

$$(1c) \qquad \hat{\mu} = \frac{G}{n_{..}} = \bar{G}.$$

That part of ΣY^2 which can be predicted from the model in (1) is

$$(1d) \qquad R(\mu) = \hat{\mu}(\underline{X}_0'\underline{Y}) = \bar{G}G = \frac{G^2}{n_{..}}.$$

Consider the model

$$(2) \qquad Y_{ijk} = \mu + \alpha_i + \varepsilon_{ijk}.$$

This model is equivalent to

$$(2a) \qquad Y = \mu X_0 + \alpha_1 X_1 + \cdots + \alpha_p X_p + \varepsilon_{ijk},$$

where $X_0, X_1, \ldots, X_p$ are indicator variables. For this case the normal equations have the form

$$
\text{(2b)} \qquad \underset{(X'X)}{\begin{bmatrix} n_{..} & n_{1.} & n_{2.} & \cdots & n_{p.} \\ n_{1.} & n_{1.} & 0 & \cdots & 0 \\ n_{2.} & 0 & n_{2.} & \cdots & 0 \\ \cdot & \cdot & \cdot & & \cdot \\ \cdot & \cdot & \cdot & & \cdot \\ \cdot & \cdot & \cdot & & \cdot \\ n_{p.} & 0 & 0 & \cdots & n_{p.} \end{bmatrix}} \underset{\underline{\beta}}{\begin{bmatrix} \mu \\ \alpha_1 \\ \alpha_2 \\ \cdot \\ \cdot \\ \cdot \\ \alpha_p \end{bmatrix}} = \underset{X'\underline{Y},}{\begin{bmatrix} G \\ A_1 \\ A_2 \\ \cdot \\ \cdot \\ \cdot \\ A_p \end{bmatrix}}.
$$

In this case $X'X$ is singular. If one imposes the constraint

$$n_{1.}\hat{\alpha}_1 + \cdots + n_{p.}\hat{\alpha}_p = 0$$

upon the system in (2*b*), one obtains the following solution:

$$\hat{\mu} = \bar{G}, \qquad \hat{\alpha}_i = \bar{A}_i - \bar{G}.$$

That part of ΣY^2 which is predictable from the model in (2) is thus

$$
\begin{aligned}
\text{(2c)} \qquad \Sigma \hat{Y}^2 &= R(\mu,\underline{\alpha}) = \hat{\mu}G + \hat{\alpha}_1 A_1 + \cdots + \hat{\alpha}_p A_p \\
&= \bar{G}G + (\bar{A}_1 - \bar{G})A_1 + \cdots + (\bar{A}_i - \bar{G})A_p \\
&= \frac{G^2}{n_{..}} + \frac{A_1^2}{n_{1.}} + \cdots + \frac{A_p^2}{n_{p.}} - \bar{G}(A_1 + \cdots + A_p) \\
&= \frac{A_1^2}{n_{1.}} + \cdots + \frac{A_p^2}{n_{p.}}.
\end{aligned}
$$

Had the constraint $\hat{\mu} = 0$ been imposed upon the system (2*b*) the following solution would have been obtained:

$$\hat{\mu} = 0, \qquad \hat{\alpha}_i = \bar{A}_i.$$

For this solution,

$$
\begin{aligned}
R(\mu,\underline{\alpha}) &= 0G + \bar{A}_i A_i + \cdots + \bar{A}_p A_p \\
&= \frac{A_1^2}{n_{1.}} + \cdots + \frac{A_p^2}{n_{p.}}.
\end{aligned}
$$

This result is identical with that in (2*c*) since $R(\mu,\underline{\alpha})$ is a constant for all solutions to the system (2*b*).

Consider now the linear model

$$\text{(3)} \qquad Y_{ijk} = \mu + \beta_j + \varepsilon_{ijk}$$

or the equivalent form

$$\text{(3a)} \qquad Y = \mu X_0 + \beta_1 X_1 + \cdots + \beta_q X_q + \varepsilon.$$

Here the indicator variable X_j indicates the presence or absence of the treatment effect corresponding to the parameter β_j (the main effect of treatment b_j). In this case the normal equations are

$$\text{(3b)} \qquad \overset{(X'X)}{\begin{bmatrix} n_{..} & n_{.1} & n_{.2} & \cdots & n_{.q} \\ n_{.1} & n_{.1} & 0 & \cdots & 0 \\ n_{.2} & 0 & n_{.2} & \cdots & 0 \\ \cdot & \cdot & \cdot & & \cdot \\ \cdot & \cdot & \cdot & & \cdot \\ \cdot & \cdot & \cdot & & \cdot \\ n_{.q} & 0 & 0 & \cdots & n_{.q} \end{bmatrix}} \overset{\underline{\beta}}{\begin{bmatrix} \mu \\ \beta_1 \\ \beta_2 \\ \cdot \\ \cdot \\ \cdot \\ \beta_q \end{bmatrix}} = \overset{X'\underline{Y},}{\begin{bmatrix} G \\ B_1 \\ B_2 \\ \cdot \\ \cdot \\ \cdot \\ B_q \end{bmatrix}}.$$

Under the constraint

$$n_{.1}\hat{\beta}_1 + \cdots + n_{.q}\hat{\beta}_q = 0,$$

a solution to the system (3*b*) is given by

$$\hat{\mu} = \bar{G}, \qquad \hat{\beta}_j = \bar{B}_j - \bar{G}.$$

That part of ΣY^2 which is predictable from the model (3) is

$$\text{(3c)} \qquad \Sigma \hat{Y}^2 = R(\mu, \underline{\beta}) = \hat{\mu}G + \hat{\beta}_1 B_1 + \cdots + \hat{\beta}_q B_q$$
$$= \frac{B_1^2}{n_{.1}} + \cdots + \frac{B_q^2}{n_{.q}}.$$

Consider now the model

$$\text{(4)} \qquad Y_{ijk} = \mu + \tau_{ij} + \varepsilon_{ijk},$$

where τ_{ij} is an effect associated with cell ab_{ij}. In terms of main effects and interactions,

$$\tau_{ij} = \alpha_i + \beta_j + \alpha\beta_{ij}.$$

In terms of indicator variables, (4) has the form

$$\text{(4a)} \qquad Y = \mu X_0 + \tau_{11}X_{11} + \tau_{12}X_{12} + \cdots + \tau_{pq}X_{pq} + \varepsilon.$$

For the linear function in (4*a*) the normal equations have the form

$$\text{(4b)} \qquad \overset{(X'X)}{\begin{bmatrix} n_{..} & n_{11} & n_{12} & \cdots & n_{pq} \\ n_{11} & n_{11} & 0 & \cdots & 0 \\ n_{12} & 0 & n_{12} & \cdots & 0 \\ \cdot & \cdot & \cdot & & \cdot \\ \cdot & \cdot & \cdot & & \cdot \\ \cdot & \cdot & \cdot & & \cdot \\ n_{pq} & 0 & 0 & \cdots & n_{pq} \end{bmatrix}} \overset{\underline{\beta}}{\begin{bmatrix} \mu \\ \tau_{11} \\ \tau_{12} \\ \cdot \\ \cdot \\ \cdot \\ \tau_{pq} \end{bmatrix}} = \overset{X'\underline{Y},}{\begin{bmatrix} G \\ AB_{11} \\ AB_{12} \\ \cdot \\ \cdot \\ \cdot \\ AB_{pq} \end{bmatrix}}.$$

Under the constraint

$$n_{11}\hat{\tau}_{11} + \cdots + n_{pq}\hat{\tau}_{pq} = 0,$$

a solution to the system (4*b*) is given by

$$\hat{\mu} = \bar{G}, \qquad \hat{\tau}_{ij} = \overline{AB}_{ij} - \bar{G}.$$

The part of ΣY^2 which is predictable from the model in (4) is

$$(4c) \qquad R(\mu,\underline{\tau}) = R(\mu,\underline{\alpha},\underline{\beta},\underline{\alpha\beta}) = \hat{\mu}G + \hat{\tau}_{11}AB_{11} + \cdots + \hat{\tau}_{pq}AB_{pq} = \frac{AB_{11}^2}{n_{11}} + \cdots + \frac{AB_{pq}^2}{n_{pq}}.$$

In terms of predictable *variation* rather than the uncorrected sum of squares one has

$$(2d) \qquad R(\underline{\alpha} \mid \mu) = R(\mu,\underline{\alpha}) - R(\mu) = \Sigma\left(\frac{A_i^2}{n_{i.}}\right) - \frac{G^2}{n_{..}} = SS_a(\text{unadjusted}),$$

$$(3d) \qquad R(\underline{\beta} \mid \mu) = R(\mu,\underline{\beta}) - R(\mu) = \Sigma\left(\frac{B_j^2}{n_{.j}}\right) - \frac{G^2}{n_{..}} = SS_b(\text{unadjusted}),$$

$$(4d) \qquad R(\underline{\tau} \mid \mu) = R(\mu,\underline{\tau}) - R(\mu) = \Sigma\left(\frac{AB_{ij}^2}{n_{ij}}\right) - \frac{G^2}{n_{..}} = SS_{\text{b. cell}}.$$

Consider now a fifth model given by

$$(5) \qquad Y_{ijk} = \mu + \alpha_i + \beta_j + \varepsilon_{ijk},$$

or, in terms of indicator variables,

$$(5a) \qquad Y = \mu X_0 + \alpha_1 X_1 + \cdots + \alpha_p X_p + \beta_1 X_{p+1} + \cdots + \beta_q X_{p+q} + \varepsilon.$$

This model is somewhat more complex than those considered up to this point. Hence a more elaborate notation system will be needed in order to obtain the type of solution desired. The notation system that will be used is summarized in Table 5.23-1. In this table one notes that

$\underset{p,q}{C}$ = matrix of cell frequencies.

$\underset{p,1}{\underline{u}_p}$ = vector of unities.

$\underset{q,1}{\underline{u}_q}$ = vector of unities.

$C\underline{u}_q$ = vector of row totals of $C = [n_{i.}]$.

$C'\underline{u}_p$ = vector of column totals of $C = [n_{.j}]$.

D_a = diagonal matrix with elements $n_{i.}$.

D_b = diagonal matrix with elements $n_{.j}$.

Table 5.23-1 Notation and Relationships

(i)	$\underset{p,q}{C} = \begin{bmatrix} n_{11} & \cdots & n_{1q} \\ \cdot & & \cdot \\ \cdot & & \cdot \\ \cdot & & \cdot \\ n_{p1} & \cdots & n_{pq} \end{bmatrix}$	$\underset{p,1}{\underline{u}_p} = \begin{bmatrix} 1 \\ \cdot \\ \cdot \\ \cdot \\ 1 \end{bmatrix} \quad \underset{q,1}{\underline{u}_q} = \begin{bmatrix} 1 \\ \cdot \\ \cdot \\ \cdot \\ 1 \end{bmatrix}$
	$\underset{p,p}{D_a} = \begin{bmatrix} n_{1.} & & & 0 \\ & \cdot & & \\ & & \cdot & \\ & & & \cdot \\ 0 & & & n_{p.} \end{bmatrix}$	$\underset{q,q}{D_b} = \begin{bmatrix} n_{.1} & & & 0 \\ & \cdot & & \\ & & \cdot & \\ & & & \cdot \\ 0 & & & n_{.q} \end{bmatrix}$
(ii)	$C\underline{u}_q = \begin{bmatrix} n_{1.} \\ \cdot \\ \cdot \\ \cdot \\ n_{p.} \end{bmatrix}$	$C'\underline{u}_p = \begin{bmatrix} n_{.1} \\ \cdot \\ \cdot \\ \cdot \\ n_{.q} \end{bmatrix}$
(iii)	$CD_b^{-1} = \begin{bmatrix} \frac{n_{11}}{n_{.1}} & \cdots & \frac{n_{1q}}{n_{.q}} \\ \cdot & & \cdot \\ \cdot & & \cdot \\ \cdot & & \cdot \\ \frac{n_{p1}}{n_{.1}} & \cdots & \frac{n_{pq}}{n_{.q}} \end{bmatrix}$	$(CD_b^{-1})(C'\underline{u}_p) = \begin{bmatrix} n_{1.} \\ \cdot \\ \cdot \\ \cdot \\ n_{p.} \end{bmatrix} = C\underline{u}_q$
(iv)	$C'D_a^{-1} = \begin{bmatrix} \frac{n_{11}}{n_{1.}} & \cdots & \frac{n_{p1}}{n_{p.}} \\ \cdot & & \cdot \\ \cdot & & \cdot \\ \cdot & & \cdot \\ \frac{n_{1q}}{n_{1.}} & \cdots & \frac{n_{pq}}{n_{p.}} \end{bmatrix}$	$(C'D_a^{-1})(C\underline{u}_q) = \begin{bmatrix} n_{.1} \\ \cdot \\ \cdot \\ \cdot \\ n_{.q} \end{bmatrix} = C'\underline{u}_p$

One notes that

$$(CD_b^{-1})(C'\underline{u}_p) = \left[\frac{n_{ij}}{n_{.j}}\right][n_{.j}] = [n_{i.}] = C\underline{u}_q.$$

Similarly,

$$(CD_a^{-1})(C\underline{u}_q) = \left[\frac{n_{ji}}{n_{i.}}\right][n_{i.}] = [n_{.j}] = C'\underline{u}_p.$$

In terms of the notation system that has just been defined, the normal equations corresponding to the system in (5a) take the form

$$(5b)\qquad (X'X)\quad \underline{\gamma} = X'\underline{Y},$$

$$\begin{matrix}\text{(i)}\\ \text{(ii)}\\ \text{(iii)}\end{matrix}\quad \begin{bmatrix} n_{..} & \underline{u}_q'C' & \underline{u}_p'C \\ C\underline{u}_q & D_a & C \\ C'\underline{u}_p & C' & D_b \end{bmatrix} \begin{bmatrix} \mu \\ \underline{\alpha} \\ \underline{\beta} \end{bmatrix} = \begin{bmatrix} G \\ \underline{a} \\ \underline{b} \end{bmatrix},$$

where $\underline{a}' = [A_1 \cdots A_p]$ and $\underline{b}' = [B_1 \cdots B_q]$. If one multiplies row (iii) in (5b) by CD_b^{-1} and then subtracts the product from row (ii) one will obtain the following system:

$$(D_a - CD_b^{-1}C')\underline{\alpha} = \underline{a} - CD_b^{-1}\underline{b},$$

or, equivalently,

$$(5c)\qquad M_a\underline{\alpha} = \underline{v}_a,$$

where

$$M_a = D_a - CD_b^{-1}C',$$

$$\underline{v}_a = \underline{a} - CD_b^{-1}\underline{b}.$$

The system in (5c) is called the *reduced* set of normal equations for $\underline{\alpha}$. The matrix M_a will be symmetric; it will also be singular since each row will sum to zero.

The reduced set of normal equations corresponding to $\underline{\beta}$ is obtained by multiplying row (ii) in (5b) by $C'D_a^{-1}$ and then subtracting the product from row (iii). The result is

$$(5d)\qquad M_b\underline{\beta} = \underline{v}_b,$$

where

$$M_b = D_b - C'D_a^{-1}C,$$

$$\underline{v}_b = \underline{b} - C'D_a^{-1}\underline{a}.$$

The matrix M_b will be symmetric; it will also be singular since the row sums will all be zero.

A convenient generalized inverse of M_a is obtained by first forming the matrix M_a^*, given by

$$M_a^* = \begin{bmatrix} M_a & \underline{u}_p \\ \underline{u}_p' & 0 \end{bmatrix}.$$

Forming the matrix M_a^* is equivalent to imposing the constraint

$$\hat{\alpha}_1 + \cdots + \hat{\alpha}_p = 0$$

on the system. The matrix M_a^* will be nonsingular. Suppose its regular inverse is represented schematically as follows:

$$(M_a^*)^{-1} = \begin{bmatrix} E_a & \underline{e}_a \\ \underline{e}_a' & 0 \end{bmatrix}.$$

The matrix E_a will be a generalized inverse of M_a. To show this, from the properties of the regular inverse of M_a^* one has

$$\begin{bmatrix} M_aE_a + \underline{u}_p\underline{e}_a' = I & M_a\underline{e}_a + \underline{u}_p0 = \underline{0} \\ \underline{u}_p'E_a + 0\underline{e}_a' = \underline{0}' & \underline{u}_p'\underline{e}_a = I \end{bmatrix}.$$

Hence
$$M_aE_a = I - \underline{u}_p\underline{e}_a'.$$

Postmultiplying both sides of the last expression by M_a gives

$$M_aE_aM_a = M_a - \underline{u}_p\underline{e}_a'M_a = M_a$$

since $\underline{e}_a'M_a = \underline{0}'$.

A solution to the system in (5*c*) is given by

$$\hat{\underline{\alpha}} = E_a\underline{v}_a. \tag{5e}$$

The corresponding predictable sum of squares is

$$R(\underline{\alpha} \mid \mu,\underline{\beta}) = \hat{\underline{\alpha}}'\underline{v}_a = \underline{v}_a'E_a\underline{v}_a. \tag{5f}$$

In the course of obtaining the reduced set of normal equations the estimators were "adjusted" for the effects of μ and $\underline{\beta}$; hence (5*f*) represents $R(\underline{\alpha} \mid \mu,\underline{\beta})$ rather than $R(\underline{\alpha})$.

A solution to the system (5*d*) is given by

$$\hat{\underline{\beta}} = E_b\underline{v}_b, \tag{5g}$$

where
$$M_b^* = \begin{bmatrix} M_b & \underline{u}_q \\ \underline{u}_q' & 0 \end{bmatrix}, \qquad (M_b^*)^{-1} = \begin{bmatrix} E_b & \underline{e}_b \\ \underline{e}_b' & 0 \end{bmatrix}.$$

The corresponding predictable variation associated with the reduced system in (5*d*) is

$$R(\underline{\beta} \mid \mu, \underline{\alpha}) = \hat{\underline{\beta}}'\underline{v}_b = \underline{v}_b'E_b\underline{v}_b. \tag{5h}$$

One may now combine the results from models (1) through (5) to obtain the following relationships:

$$\begin{aligned} R(\mu,\underline{\alpha},\underline{\beta}) &= R(\mu,\underline{\alpha}) + R(\underline{\beta} \mid \mu,\underline{\alpha}) \\ &= R(\mu,\underline{\beta}) + R(\underline{\alpha} \mid \mu,\underline{\beta}). \end{aligned} \tag{6}$$

$$\begin{aligned} R(\underline{\alpha},\underline{\beta} \mid \mu) &= R(\underline{\alpha} \mid \mu) + R(\underline{\beta} \mid \mu,\underline{\alpha}) \\ &= R(\underline{\beta} \mid \mu) + R(\underline{\alpha} \mid \mu,\underline{\beta}). \end{aligned} \tag{7}$$

Relationship (7) expresses the joint predictability due to $\underline{\alpha}$ and $\underline{\beta}$ in terms of the predictability due to the parts. In terms of variation, (7) has the form

$$\begin{aligned} SS_{a+b} &= SS_a \text{ (unadjusted)} + SS_b \text{ (adjusted for } A) \\ &= SS_b \text{ (unadjusted)} + SS_a \text{ (adjusted for } B). \end{aligned} \tag{7a}$$

If one combines the results in (7) and (7*a*) with those associated with the model in (4) one obtains

(8) $$R(\underline{\alpha\beta} \mid \mu,\underline{\alpha},\underline{\beta}) = R(\mu,\underline{\alpha},\underline{\beta},\underline{\alpha\beta}) - R(\mu,\underline{\alpha},\underline{\beta}) = R(\underline{\alpha},\underline{\beta},\underline{\alpha\beta} \mid \mu) - R(\underline{\alpha},\underline{\beta} \mid \mu).$$

In terms of variation, (8) has the form

(8a) $$SS_{ab} \text{ (adjusted for } A \text{ and } B) = SS_{b.\ cell} - SS_{a+b}.$$

Table 5.23-2 Summary of Analyses for Various Models

Model	Predictable sum of squares	Estimator
(1) $Y_{ijk} = \mu + \varepsilon_{ijk}$	$R(\mu) = \hat{\mu}G$	$\hat{\mu} = \bar{G}$
(2) $Y_{ijk} = \mu + \alpha_i + \varepsilon_{ijk}$	$R(\mu,\underline{\alpha}) = \Sigma\hat{\alpha}_i A_i$	$\hat{\alpha}_i = \bar{A}_i + \hat{\mu}G$
(3) $Y_{ijk} = \mu + \beta_j + \varepsilon_{ijk}$	$R(\mu,\underline{\beta}) = \Sigma\hat{\beta}_j B_j$	$\hat{\beta}_j = \bar{B}_j + \hat{\mu}G$
(4) $Y_{ijk} = \mu + \tau_{ij} + \varepsilon_{ijk}$	$R(\mu,\underline{\tau}) = R(\mu,\underline{\alpha},\underline{\beta},\underline{\alpha\beta}) = \Sigma\hat{\tau}_{ij}AB_{ij}$	$\hat{\tau}_{ij} = \overline{AB}_{ij} + \hat{\mu}G$
(5) $Y_{ijk} = \mu + \alpha_i + \beta_j + \varepsilon_{ijk}$	$R(\mu,\underline{\alpha},\underline{\beta}) = \underline{\gamma}'(X'\underline{Y})$	
	$R(\underline{\alpha} \mid \mu,\underline{\beta}) = \hat{\underline{\alpha}}'\underline{v}_a$	$\hat{\underline{\alpha}} = E_a\underline{v}_a$
	$R(\underline{\beta} \mid \mu,\underline{\alpha}) = \hat{\underline{\beta}}'\underline{v}_b$	$\hat{\underline{\beta}} = E_b\underline{v}_b$

The procedures that have been discussed in this subsection have been called the *method of fitting constants*. The approach used here shows that this method is a direct application of the least-squares criterion to the general linear model, the variables being indicator variables for the levels of the factors (or the presence or absence of an effect associated with a cell). Which one of the several models is appropriate depends upon the purpose of the analysis. Under the hypothesis of no interaction, the model given in (5) is appropriate for estimation of variation due to main effects. A test for interaction requires an estimate of the variation due to interaction adjusted for main effects. The latter is obtained from (8). If interaction is present, procedures for testing main effects are discussed in a later subsection.

A readable summary of some of the problems encountered in the analysis of variance of disproportionate data when interaction is present is contained in Gosslee and Lucas (1965). An approach to the problem of handling unequal cell frequencies in a mixed model is considered in Mielke and McHugh (1965).

Some numerical details of the principles discussed in this section appear in Sec. 6.14.

5.23-4 Special case when one factor has only two levels

If one factor (say factor A) has only two levels, the computation of $R(\underline{\alpha} \mid \mu,\underline{\beta})$ and $R(\underline{\alpha\beta} \mid \mu,\underline{\alpha},\underline{\beta})$ may be obtained in a relatively simple manner.

Let

$$d_j = \overline{AB}_{1j} - \overline{AB}_{2j},$$
$$d_{j'} = \overline{AB}_{1j'} - \overline{AB}_{2j'}.$$

If one assumes the model

$$\overline{AB}_{ij} = \mu + \alpha_i + \beta_j + \alpha\beta_{ij} + \bar{\varepsilon}_{ij},$$

then
$$d_j = (\alpha_1 - \alpha_2) + (\alpha\beta_{1j} - \alpha\beta_{2j}) + (\bar{\varepsilon}_{1j} - \bar{\varepsilon}_{2j}),$$
$$d_{j'} = (\alpha_1 - \alpha_2) + (\alpha\beta_{1j'} - \alpha\beta_{2j'}) + (\bar{\varepsilon}_{1j'} - \bar{\varepsilon}_{2j'}).$$

Since $(\alpha_1 - \alpha_2)$ is a constant for all d_j, the variation due to the d_j will be a function only of interaction and error.

The square of the standard error of d_j will be proportional to

$$\frac{1}{w_j} = \frac{1}{n_{1j}} + \frac{1}{n_{2j}} = \frac{n_{1j} + n_{2j}}{n_{1j}n_{2j}}.$$

If one weights each d_j by w_j (a factor inversely proportional to the squared standard error of d_j) in computing the variation of the d_j one will obtain

$$R(\underline{\alpha\beta} \mid \mu,\underline{\alpha},\underline{\beta}) = \Sigma w_j d_j^2 - \frac{(\Sigma w_j d_j)^2}{\Sigma w_j}.$$

It will also be found that

$$R(\underline{\alpha} \mid \mu,\underline{\beta}) = \frac{(\Sigma w_j d_j)^2}{\Sigma w_j}.$$

Hence
$$\Sigma w_j d_j^2 = R(\underline{\alpha\beta} \mid \mu,\underline{\alpha},\underline{\beta}) + R(\underline{\alpha} \mid \mu,\underline{\beta})$$
$$= R(\underline{\alpha},\underline{\alpha\beta} \mid \mu,\underline{\beta}).$$

This last result is one that might have been anticipated from the expression of d_j in terms of $(\alpha_1 - \alpha_2)$ and $(\alpha\beta_{1j} - \alpha\beta_{2j})$.

5.23-5 Tests on interaction and main effects

The procedure for testing interaction is outlined in Table 5.23-3. To obtain an estimate of error one uses the model

$$Y_{ijk} = \mu + \tau_{ij} + \varepsilon_{ijk},$$

where
$$\tau_{ij} = \alpha_i + \beta_j + \alpha\beta_{ij}.$$

Variation due to error is given by

$$SS_{\text{error}} = \Sigma Y^2 - R(\mu,\underline{\tau})$$
$$= \Sigma Y^2 - \Sigma \frac{AB_{ij}^2}{n_{ij}} = SS_{\text{w. cell}}.$$

Variation due to interaction adjusted for main effects, $R(\underline{\alpha\beta} \mid \mu,\underline{\alpha},\underline{\beta})$, is obtained from relationship (8) in Sec. 5.23-3.

A test on the hypothesis of no interaction is given by

$$F = \frac{\text{MS}_{ab}}{\text{MS}_{\text{error}}}.$$

If, either on a priori grounds or on the basis of the test indicated above, it is reasonable to assume that no interaction exists, variation due to main effects may be estimated from a model which does not include an interaction term. Since the variation due to main effects are not orthogonal in this case, the adjusted variation due to main effects are obtained as indicated in Sec. 5.23-3. The latter are summarized in Table 5.23-3.

One form of test on main effects is

$$F = \frac{\text{MS}_a}{\text{MS}_{\text{error}}}, \qquad F = \frac{\text{MS}_b}{\text{MS}_{\text{error}}},$$

where MS_{error} is that in Table 5.23-3. An alternative form of the test on main effects uses the error estimated from the model which does not include an interaction term. This estimate of error is obtained from

$$\Sigma Y^2 - R(\mu,\underline{\alpha},\underline{\beta}).$$

This error may be estimated by pooling the variation due to error and the variation due to interaction in Table 5.23-3.

If interaction is present, it generally makes more sense to test simple main effects rather than main effects, since interaction implies that the effect of one factor is dependent upon the level of the other factors. However, if tests on main effects are wanted even though interaction is present, the variation due to main effects as given in Table 5.23-3 cannot be used. Since these sources of variation were estimated from a model which did *not* include an interaction term, they will be biased as estimates of variation due to main effects in a model which does include an interaction term.

Table 5.23-3 Test on Interaction

Source of variation	SS	df	MS	F
Mean	$R(\mu)$	1		
A (adj. for B)	$R(\underline{\alpha} \mid \mu,\underline{\beta}) = R(\mu,\underline{\alpha},\underline{\beta}) - R(\mu,\underline{\beta})$	$p - 1$		
B (adj. for A)	$R(\underline{\beta} \mid \mu,\underline{\alpha}) = R(\mu,\underline{\alpha},\underline{\beta}) - R(\mu,\underline{\alpha})$	$q - 1$		
AB (adj. for A and B)	$R(\underline{\alpha\beta} \mid \mu,\underline{\alpha},\underline{\beta}) = R(\mu,\underline{\tau}) - R(\mu,\underline{\alpha},\underline{\beta})$	$(p-1)(q-1)$	MS_{ab}	$\text{MS}_{ab}/\text{MS}_{\text{error}}$
Error	$\Sigma Y^2 - R(\mu,\underline{\tau})$	$n_{..} - pq$	MS_{error}	

One possible approach to testing variation due to main effects when interaction is present is to obtain

$$R(\underline{\alpha} \mid \mu,\underline{\beta},\underline{\alpha\beta}) = R(\mu,\underline{\alpha},\underline{\beta},\underline{\alpha\beta}) - R(\mu,\underline{\beta},\underline{\alpha\beta}),$$

where $R(\mu,\underline{\beta}, \underline{\alpha\beta})$ is obtained by using the model

$$X_{ijk} = \mu + \beta_j + \alpha\beta_{ij} + \varepsilon_{ijk}.$$

In this context, $R(\underline{\alpha} \mid \mu,\underline{\beta},\underline{\alpha\beta})$ represents that part of the variation due to the main effects of factor A which are orthogonal to the main effects of factor B as well as the AB interaction. Similarly, one may obtain

$$R(\underline{\beta} \mid \mu,\underline{\alpha},\underline{\alpha\beta}) = R(\mu,\underline{\alpha},\underline{\beta},\underline{\alpha\beta}) - R(\mu,\underline{\alpha},\underline{\alpha\beta}).$$

An alternative method for making a test on main effects is described in Sec. 5.23-6. Under this approach the variation due to the main effects is not orthogonal in the sense that

$$SS_{a+b} \neq SS_a + SS_b.$$

The test procedure on interaction outlined earlier in this section may be cast in the form of the general test principle given in Sec. 3.20. Let Ω be the set of conditions associated with the linear model

$$X_{ijk} = \mu + \alpha_i + \beta_j + \alpha\beta_{ij} + \varepsilon_{ijk}. \tag{1}$$

Under least-squares estimation procedures, the variation due to error (under Ω) is

$$\begin{aligned} Q_\Omega &= \Sigma(X_{ijk} - \overline{AB}_{ij})^2 \\ &= SS_{\text{total}} - R(\underline{\alpha},\underline{\beta},\underline{\alpha\beta} \mid \mu) \\ &= SS_{\text{w. cell}}. \end{aligned}$$

Let ω be the set of conditions associated with the model

$$X_{ijk} = \mu + \alpha_i + \beta_j + \varepsilon_{ijk}. \tag{2}$$

Under the hypothesis that all $\alpha\beta_{ij} = 0$, (1) reduces to (2). Hence

$$\omega = \Omega + (H\text{: all } \alpha\beta_{ij} = 0).$$

Under least-squares estimation procedures, variation due to error in this case is

$$Q_\omega = SS_{\text{total}} - R(\underline{\alpha},\underline{\beta} \mid \mu).$$

Thus,

$$\begin{aligned} F &= \frac{(Q_\omega - \Omega_\Omega)/(f_{\omega-\Omega})}{Q_\Omega/f_\Omega} \\ &= \frac{[R(\underline{\alpha},\underline{\beta},\underline{\alpha\beta} \mid \mu) - R(\underline{\alpha},\underline{\beta} \mid \mu)]/(p-1)(q-1)}{SS_{\text{w. cell}}/(n_{..} - pq)}. \end{aligned}$$

5.23-6 Method of weighted squares of means

The estimation procedures to be described in this subsection lead to unbiased estimates of the variation due to main effects even though interaction may be present. The method of weighted squares of means has many features in common with the method used in Sec. 5.23-3.

Under the method of weighted squares of means, the estimators of the main effects are

$$\hat{\alpha}_i = \bar{A}_i - \bar{G} \qquad \text{and} \qquad \hat{\beta}_j = \bar{B}_j - \bar{G}$$

where $$\bar{A}_i = \frac{\sum_j \overline{AB}_{ij}}{q}, \qquad \bar{B}_j = \frac{\sum_i \overline{AB}_{ij}}{p}, \qquad \bar{G} = \frac{\sum_i \sum_j \overline{AB}_{ij}}{pq}.$$

That is, $\bar{A}_i$ is an unweighted mean of the cell means at level a_i. Let

$$h_i = \frac{q}{\sum_j (1/n_{ij})} = \text{harmonic mean of cell frequencies at level } a_i,$$

$$h_j = \frac{p}{\sum_i (1/n_{ij})} = \text{harmonic mean of cell frequencies at level } b_j.$$

Also let $$w_i = \frac{h_i}{q} \qquad \text{and} \qquad w_j = \frac{h_j}{p}.$$

The variation due to the main effects of factor A is defined as

$$\text{(1)} \qquad \text{SS}_a = q\left[\Sigma h_i \hat{\alpha}_i^2 - \frac{(\Sigma h_i \hat{\alpha}_i)^2}{\Sigma h_i}\right]$$

$$= q^2\left[\Sigma w_i \bar{A}_i^2 - \frac{(\Sigma w_i \bar{A}_i)^2}{\Sigma w_i}\right].$$

It will be noted that, when $n_{ij} = n$ for all cell frequencies, $h_i = n$. Hence in (1) $\Sigma h_i \hat{\alpha}_i = n\Sigma\hat{\alpha}_i = 0$, and (1) becomes

$$nq\Sigma\hat{\alpha}_i^2,$$

which is the least-squares estimator of SS_a when the cell frequencies are equal.

Under the method of weighted squares of means,

$$\text{(2)} \qquad \text{SS}_b = p\left[\Sigma h_j \hat{\beta}_j^2 - \frac{(\Sigma h_j \hat{\beta}_j)^2}{\Sigma h_j}\right]$$

$$= p^2\left[\Sigma w_j \bar{B}_j^2 - \frac{(\Sigma w_j \bar{B}_j)^2}{\Sigma w_j}\right].$$

For the case of equal cell frequencies, (2) reduces to

$$\text{SS}_b = np\Sigma\hat{\beta}_j^2.$$

By way of contrast, in the method of *unweighted* means (or more accurately *equally* weighted means)

$$h_i \text{ is replaced by } \bar{n}_h \qquad \text{and} \qquad h_j \text{ is replaced by } \bar{n}_h,$$

where $\bar{n}_h$ is the harmonic mean of all cell frequencies and not just those frequencies in a single row or a single column. Under both methods,

$$SS_{error} = SS_{w.\ cell}.$$

For the special case in which factor A has only two levels, the method of weighted squares of means reduces to constructing that comparison among the cell means corresponding to the main effect of factor A. For the case of a 2×3 factorial, the variation due to this comparison is given by

$$SS_a = \frac{(\overline{AB}_{11} + \overline{AB}_{12} + \overline{AB}_{13} - \overline{AB}_{21} - \overline{AB}_{22} - \overline{AB}_{23})^2}{\Sigma(1/n_{ij})}$$

In general, if

$$\bar{A}_i = \frac{1}{q}\sum_j \frac{AB_{ij}}{n_{ij}},$$

then a comparison among the $\bar{A}_i$ has the form

$$C = \sum_i c_i \bar{A}_i = \sum_i \sum_j u_{ij} \overline{AB}_{ij},$$

where

$$\sum_i c_i = 0, \qquad u_{ij} = \frac{c_i}{q}.$$

The variation due to this comparison is

$$SS_C = \frac{C^2}{\Sigma(u_{ij}^2/n_{ij})}.$$

If one defines

$$u_{ij}^* = \frac{u_{ij}}{k}, \qquad \text{where } k = \sqrt{\sum_{ij} \frac{u_{ij}^2}{n_{ij}}},$$

and if one defines

$$C^* = \Sigma\Sigma u_{ij}^* \overline{AB}_{ij},$$

then

$$SS_{C^*} = \frac{(C^*)^2}{\Sigma(u_{ij}^{*2}/n_{ij})} = (C^*)^2$$

since

$$\frac{\Sigma u_{ij}^{*2}}{n_{ij}} = 1.$$

C^* is called a normalized comparison.

It may be shown that the weighted sum of squares of means corresponds to the variation due to the maximum normalized comparison (Scheffé, 1959, p. 118).

5.23-7 Standard errors of estimators

In Sec. 5.23-3 when considering the model

$$Y_{ijk} = \mu + \alpha_i + \beta_j + \varepsilon_{ijk}$$

it was shown that

$$\hat{\underline{\alpha}} = E_a \underline{v}_a, \qquad \Sigma\hat{\alpha}_i = 0.$$

Here E_a is the generalized inverse of M_a obtained from the regular inverse of M_a^*. From the general property of least-squares estimation in a model assuming var $(\underline{\varepsilon}) = \sigma_\varepsilon^2 I$, the covariance matrix associated with the vector of estimators is given by

(1) $$\text{cov}\,(\hat{\underline{\alpha}}) = \sigma_\varepsilon^2 E_a.$$

Thus,

(2) $$\sigma_{\hat{\alpha}_i}^2 = \sigma_\varepsilon^2 e_{ii}, \qquad \text{where } e_{ii} \text{ is a diagonal element of } E_a.$$

Also

(3) $$\sigma_{\hat{\alpha}_i - \hat{\alpha}_k}^2 = \sigma_\varepsilon^2(e_{ii} + e_{kk} - 2e_{ik}),$$

where e_{ik} is the element in row i, column k of E_a. If σ_ε^2 is replaced by $\hat{\sigma}_\varepsilon^2$, then one has an estimate of $\sigma_{\hat{\alpha}_i - \hat{\alpha}_k}^2$. One estimate of σ_ε^2 is given by

$$\hat{\sigma}_\varepsilon^2 = \frac{\Sigma Y^2 - R(\mu, \underline{\alpha}, \underline{\beta})}{n_{..} - p - q + 1}.$$

Alternatively, one may use

$$\hat{\sigma}_\varepsilon^2 = \frac{SS_{\text{w. cell}}}{n_{..} - pq}.$$

The latter estimator will have fewer degrees of freedom than the former since it does not include the variation due to interaction, which in a model that does not include an interaction term is considered part of error.

By analogy,

(4) $$\text{cov}\,(\hat{\underline{\beta}}) = \sigma_\varepsilon^2 E_b,$$

(5) $$\sigma_{\hat{\beta}_j - \hat{\beta}_k}^2 = \sigma_\varepsilon^2(e_{jj} + e_{kk} - 2e_{jk}),$$

where e_{jk} is the element in row j, column k of E_b.

5.23-8 Special case of proportional cell frequencies

That special case in which the cell frequencies are proportional (in a sense that will be defined later) has a particularly elegant solution. It will be found that

$$SS_{a+b} = SS_a + SS_b;$$

that is, the variations due to the main effects of factors A and B are additive. This kind of additivity holds for equal cell frequencies. In this context, equal cell frequencies are a special case of proportional cell frequencies.

Consider the following numerical example:

	b_1	b_2	b_3	Total
a_1	5	15	10	$30 = n_{1.}$
a_2	10	30	20	$60 = n_{2.}$
Total	15	45	30	$90 = n_{..}$
	$n_{.1}$	$n_{.2}$	$n_{.3}$	

Let $$r_1 = \frac{n_{1.}}{n_{1.}} = 1, \qquad r_2 = \frac{n_{2.}}{n_{1.}} = 2.$$

Also let
$$s_1 = \frac{n_{.1}}{n_{.1}} = 1, \qquad s_2 = \frac{n_{.2}}{n_{.1}} = 3, \qquad s_3 = \frac{n_{.3}}{n_{.1}} = 2.$$

If one lets
$$k = \frac{n_{1.}\,n_{.1}}{n_{..}} = \frac{(30)(15)}{90} = 5,$$

one notes that the cell frequencies given above may be expressed as follows:

$$n_{11} = kr_1s_1 \qquad n_{12} = kr_1s_2 \qquad n_{13} = kr_1s_3$$

$$n_{21} = kr_2s_1 \qquad n_{22} = kr_2s_2 \qquad n_{23} = kr_2s_3$$

In general, if cell frequencies are such that

$$n_{ij} = kr_is_j = \frac{n_{i.}n_{.j}}{n_{..}},$$

the cell frequencies are said to be *proportional*. If the cell frequencies are proportional, then

$$n_{i.} = \sum_j kr_is_j = kr_i\sum_j s_j = kSr_i \qquad \text{where } S = \sum_j s_j,$$

(1) $$n_{.j} = \sum_i kr_is_j = ks_j\sum_i r_i = kRs_j \qquad \text{where } R = \sum_i r_i,$$

$$n_{..} = \sum_i n_{i.} = kS\sum_i r_i = kRS.$$

In terms of the numerical example given above,

$$R = \Sigma r_i = 3, \qquad S = \Sigma s_j = 6, \qquad kRS = 5(3)(6) = 90 = n_{..}.$$

Under the model

$$X_{ijk} = \mu + \alpha_i + \beta_j + \varepsilon_{ijk},$$

the normal equations take the following form:

$$
\begin{aligned}
&\mu: && kRS\hat{\mu} + kS\sum_i r_i\hat{\alpha}_i + kR\sum_j s_j\hat{\beta}_j = G, \\
(2)\quad &\alpha_i: && kSr_i\hat{\mu} + kSr_i\hat{\alpha}_i + kr_i\sum_j s_j\hat{\beta}_j = A_i, \qquad i = 1, \dots, p, \\
&\beta_j: && kRs_j\hat{\mu} + ks_j\sum_i s_i\hat{\alpha}_i + kRs_j\hat{\beta}_j = B_j, \qquad j = 1, \dots, q.
\end{aligned}
$$

Here $A_i = \sum_j AB_{ij}$, $B_j = \sum_i AB_{ij}$, and $G = \sum_i\sum_j AB_{ij}$.

If one imposes the constraints

$$\sum_i r_i\hat{\alpha}_i = 0, \qquad \sum_j s_j\hat{\beta}_j = 0,$$

one obtains the following solutions:

$$
\begin{aligned}
&\hat{\mu} = \frac{G}{kRS} = \bar{G}, \\
(3)\quad &\hat{\alpha}_i = \frac{A_i - kSr_i\hat{\mu}}{kSr_i} = \bar{A}_i - \bar{G}, \qquad \text{where } \bar{A}_i = \frac{A_i}{kSr_i}, \\
&\hat{\beta}_j = \frac{B_j - kRs_j\hat{\mu}}{kRs_j} = \bar{B}_j - \bar{G}, \qquad \text{where } \bar{B}_j = \frac{B_j}{kRs_j}.
\end{aligned}
$$

Hence

$$
\begin{aligned}
(4)\quad R(\mu, \underline{\alpha}, \underline{\beta}) &= \hat{\mu}G + \Sigma\hat{\alpha}_i A_i + \Sigma\hat{\beta}_j B_j \\
&= \frac{G^2}{n_{..}} + \Sigma A_i(\bar{A}_i - \bar{G})^2 + \Sigma B_j(\bar{B}_j - \bar{G})^2 \\
&= \frac{G^2}{n_{..}} + \Sigma\frac{A_i^2}{n_{i.}} + \Sigma\frac{B_j^2}{n_{.j}} - \frac{2G^2}{n_{..}} \\
&= \Sigma\frac{A_i^2}{n_{i.}} + \Sigma\frac{B_j^2}{n_{.j}} - \frac{G^2}{n_{..}}.
\end{aligned}
$$

From Sec. 5.23-3 one has the following:

$$
\begin{aligned}
R(\mu) &= \frac{G^2}{n_{..}}, \\
R(\underline{\alpha} \mid \mu) &= \Sigma\frac{A_i^2}{n_{i.}} - \frac{G^2}{n_{..}} = \Sigma n_i(\bar{A}_i - \bar{G})^2, \\
R(\underline{\beta} \mid \mu) &= \Sigma\frac{B_j^2}{n_{.j}} - \frac{G^2}{n_{..}} = \Sigma n_j(\bar{B}_j - \bar{G})^2.
\end{aligned}
$$

Combining (3) and (4) one has for the case of proportional cell frequencies

$$(5) \qquad \begin{aligned} R(\underline{\alpha},\underline{\beta} \mid \mu) &= R(\mu,\underline{\alpha},\underline{\beta}) - R(\mu) \\ &= \Sigma \frac{A_i^2}{n_{i.}} + \Sigma \frac{B_j^2}{n_{.j}} - \frac{G^2}{n_{..}} - \frac{G^2}{n_{..}} \\ &= R(\underline{\alpha} \mid \mu) + R(\underline{\beta} \mid \mu) \\ &= \text{SS}_a + \text{SS}_b. \end{aligned}$$

The analysis of variance for the case of proportional cell frequencies is summarized in Table 5.23-4.

It is readily verified that the variation due to interaction which is defined by

$$\text{SS}_{ab} = \text{SS}_{\text{b.cell}} - \text{SS}_a - \text{SS}_b$$

may be expressed in the algebraically equivalent form

$$\text{SS}_{ab} = \Sigma n_{ij}(\overline{AB}_{ij} - \bar{A}_i - \bar{B}_j + \bar{G})^2.$$

Table 5.23-4 Analysis of Variance (Least Squares) for Proportional Cell Frequencies

$(1) = G^2/n_{..} \quad (2) = \Sigma X_{ijk}^2 \quad (3) = \Sigma(A_i^2/n_{i.})$
$(4) = \Sigma(B_j^2/n_{.j}) \quad (5) = \Sigma[(AB_{ij})^2/n_{ij}]$

Source of variation	SS	df
A	$(3) - (1)$	$p - 1$
B	$(4) - (1)$	$q - 1$
AB	$(5) - (3) - (4) + (1)$	$(p - 1)(q - 1)$
Within cell	$(2) - (5)$	$n_{..} - pq$

5.24 Estimability in a General Sense

The material in this section will be presented in the context of a 2×2 factorial experiment having $n = 2$ observations per cell. The factors will be assumed fixed. Generalization to the case of a $p \times q$ factorial experiment having n observations per cell is direct.

Let the observations in the experiment be the components of the vector Y.

$$\underline{Y}' = [Y_{111} Y_{112} \quad Y_{121} Y_{122} \quad Y_{211} Y_{212} \quad Y_{221} Y_{222}].$$

Thus, $\quad Y_{ijk}$ = observation on element k under treatment ab_{ij}.

The linear model for an observation may be written in the alternative forms

$$(1) \quad Y_{ijk} = \mu + \alpha_i + \beta_j + \alpha\beta_{ij} + \varepsilon_{ijk}, \qquad i = 1, 2; j = 1, 2; k = 1, 2;$$

$$(2) \quad \underset{8,1}{\underline{Y}} = \underset{8,9}{X} \underset{9,1}{\underline{\beta}} + \underset{8,1}{\underline{\varepsilon}},$$

where

$$X = \begin{bmatrix} 1 & 1 & 0 & 1 & 0 & 1 & 0 & 0 & 0 \\ 1 & 1 & 0 & 1 & 0 & 1 & 0 & 0 & 0 \\ 1 & 1 & 0 & 0 & 1 & 0 & 1 & 0 & 0 \\ 1 & 1 & 0 & 0 & 1 & 0 & 1 & 0 & 0 \\ 1 & 0 & 1 & 1 & 0 & 0 & 0 & 1 & 0 \\ 1 & 0 & 1 & 1 & 0 & 0 & 0 & 1 & 0 \\ 1 & 0 & 1 & 0 & 1 & 0 & 0 & 0 & 1 \\ 1 & 0 & 1 & 0 & 1 & 0 & 0 & 0 & 1 \end{bmatrix}, \qquad \underline{\beta} = \begin{bmatrix} \mu \\ \alpha_1 \\ \alpha_2 \\ \beta_1 \\ \beta_2 \\ \alpha\beta_{11} \\ \alpha\beta_{12} \\ \alpha\beta_{21} \\ \alpha\beta_{22} \end{bmatrix}.$$

The matrix X is called the design matrix. The least-squares estimators of the components of $\underline{\beta}$ are obtained from the solution to the following system of equations:

$$(X'X)\hat{\underline{\beta}} = X'\underline{Y}. \tag{3}$$

For the example under consideration, (3) may be written schematically as follows:

	$\hat{\mu}$	$\hat{\alpha}_1$	$\hat{\alpha}_2$	$\hat{\beta}_1$	$\hat{\beta}_2$	$\widehat{\alpha\beta}_{11}$	$\widehat{\alpha\beta}_{12}$	$\widehat{\alpha\beta}_{21}$	$\widehat{\alpha\beta}_{22}$	
μ:	8	4	4	4	4	2	2	2	2	G
α_1:	4	4	0	2	2	2	2	0	0	A_1
α_2:	4	0	4	2	2	0	0	2	2	A_2
β_1:	4	2	2	4	0	2	0	2	0	B_1
β_2:	4	2	2	0	4	0	2	0	2	B_2
$\alpha\beta_{11}$:	2	2	0	2	0	2	0	0	0	AB_{11}
$\alpha\beta_{12}$:	2	2	0	0	2	0	2	0	0	AB_{12}
$\alpha\beta_{21}$:	2	0	2	2	0	0	0	2	0	AB_{21}
$\alpha\beta_{22}$:	2	0	2	0	2	0	0	0	2	AB_{22}
					$X'X$					$X'\underline{Y}$

In the above schematic form, the normal equation corresponding to μ is

$$8\hat{\mu} + 4\hat{\alpha}_1 + 4\hat{\alpha}_2 + 4\hat{\beta}_1 + 4\hat{\beta}_2 + 2\widehat{\alpha\beta}_{11} + 2\widehat{\alpha\beta}_{12} + 2\widehat{\alpha\beta}_{21} + 2\widehat{\alpha\beta}_{22} = G.$$

Another representation of (3) is as follows:

$$\begin{aligned} \mu: &\quad 8\hat{\mu} + 4\hat{\alpha}_. + 4\hat{\beta}_. + 2\widehat{\alpha\beta}_{..} = G, & & \\ \alpha_i: &\quad 4\hat{\mu} + 4\hat{\alpha}_i + 2\hat{\beta}_. + 2\widehat{\alpha\beta}_{i.} = A_i, & & i = 1, 2; \\ \beta_j: &\quad 4\hat{\mu} + 2\hat{\alpha}_. + 4\hat{\beta}_j + 2\widehat{\alpha\beta}_{.j} = B_j, & & j = 1, 2; \\ \alpha\beta_{ij}: &\quad 2\hat{\mu} + 2\hat{\alpha}_i + 2\hat{\beta}_j + 2\widehat{\alpha\beta}_{ij} = AB_{ij}. & & \end{aligned}$$

Here

$$\hat{\alpha}_. = \sum_i \hat{\alpha}_i, \quad \hat{\beta}_. = \sum_j \hat{\beta}_j, \quad \widehat{\alpha\beta}_{i.} = \sum_j \widehat{\alpha\beta}_{ij}, \quad \widehat{\alpha\beta}_{.j} = \sum_i \widehat{\alpha\beta}_{ij}, \quad \widehat{\alpha\beta}_{..} = \sum_i \sum_j \widehat{\alpha\beta}_{ij}.$$

Since the columns of the matrix X are linearly dependent, the matrix $X'X$ will be singular, and (3) will not have a unique solution. Suppose one augments the matrix $X'X$ by a set of rows and columns which are equivalent to the following side conditions:

$$\sum_i \hat{\alpha}_i = 0, \qquad \sum_j \hat{\beta}_j = 0, \qquad \sum_i \widehat{\alpha\beta}_{ij} = 0, \qquad \sum_j \widehat{\alpha\beta}_{ij} = 0.$$

The resulting augmented matrix will be nonsingular, and there will be a unique solution for $\underline{\beta}$.

If the model which *defines* the general mean μ, the main effects α_i and β_j, and the interactions $\alpha\beta_{ij}$ includes the conditions

$$\sum_i \alpha_i = 0, \qquad \sum_j \beta_j = 0, \qquad \sum_i \alpha\beta_{ij} = 0, \qquad \sum_j \alpha\beta_{ij} = 0,$$

then the solution to the least-squares estimation problem includes corresponding constraints as part of the normal equations. The linear model which includes these constraints may be called a restricted model. This restricted model is the one which is generally assumed to underlie the analysis of variance when the factors are fixed and cell frequencies are equal.

However, these side conditions are not part of a more general linear model which has no restrictions on the parameters. In this more general case, $\underline{c}'\underline{\beta}$ has a unique estimator if and only if (see Sec. 2.3-4)

$$\underline{c} = \lambda'(X'X). \tag{4}$$

Without going into details, what is implied by (4) with regard to estimability of "main" effects may be seen by expressing the marginal means in terms of the model in (1).

$$\bar{A}_1 = \mu + \alpha_1 + \frac{\beta_1 + \beta_2}{2} + \overline{\alpha\beta}_{1.} + \bar{\varepsilon}_{1..},$$

$$\bar{A}_2 = \mu + \alpha_2 + \frac{\beta_1 + \beta_2}{2} + \overline{\alpha\beta}_{2.} + \bar{\varepsilon}_{2..}.$$

Thus $$\bar{A}_1 - \bar{A}_2 = \alpha_1 - \alpha_2 + \overline{\alpha\beta}_{1.} - \overline{\alpha\beta}_{2.} + \bar{\varepsilon}_{1..} - \bar{\varepsilon}_{2..}.$$

Hence

$$\mathrm{E}(\bar{A}_1 - \bar{A}_2) = \alpha_1 - \alpha_2 + \overline{\alpha\beta}_{1.} - \overline{\alpha\beta}_{2.}, \qquad \text{assuming } \mathrm{E}(\bar{\varepsilon}_{1..} - \bar{\varepsilon}_{2..}) = 0.$$

If interaction is present, the parametric function $\alpha_1 - \alpha_2$ will not be estimable. Further, any hypothesis about $\alpha_1 - \alpha_2$ will not be testable. The parametric function

$$\alpha_1 - \alpha_2 + \overline{\alpha\beta}_{1.} + \overline{\alpha\beta}_{2.}$$

will be estimable, the estimator being $\bar{A}_1 - \bar{A}_2$.

By analogy,

$$E(\bar{B}_1 - \bar{B}_2) = \beta_1 - \beta_2 + \overline{\alpha\beta}_{.1} - \overline{\alpha\beta}_{.2}.$$

If interaction is present, the parametric function $\beta_1 - \beta_2$ will not be estimable. The parametric function

$$\beta_1 - \beta_2 + \overline{\alpha\beta}_{.1} - \overline{\alpha\beta}_{.2}$$

is estimated by $\bar{B}_1 - \bar{B}_2$.

Under the definition of an estimable parametric function as given by (4), the class of estimable parametric functions will include all linear functions of the marginal means (that is, linear functions of the $\mu_{i.}$ or $\mu_{.j}$). The expected values of estimators of these means, in terms of the parameters in the linear model, depend upon the restrictions, if any, that are imposed upon the linear model. Following Scheffé (1959, p. 92), the restrictions indicated have been considered part of the model in earlier sections of this chapter. Hence main effects and linear functions thereof were considered estimable (in the restricted sense) and therefore testable, whether or not interaction was present. Without the set of restrictions, the parameters in (1) should not really be called the general mean, main effects, and interactions since these latter terms imply that a set of side conditions is satisfied.

5.25 Estimation of Variance Components

For the special case of a single-factor experiment, the problem of estimation of variance components was discussed in Sec. 3.17. Estimation of variance components for a simple case of the mixed model was discussed in Sec. 4.5. The discussion in this section is limited largely to the case of a balanced, two-factor, variance-component model.

The estimation problems that are encountered in unbalanced designs having one or more random factors have been summarized by Searle (1968). In the statistical literature there are three methods for estimating components of variance for the case of disproportional cell frequencies. All three methods were studied in some detail by Henderson (1953). Most of the difficulties that are encountered occur in the mixed model. When the cell frequencies are equal, all three methods of estimation are equivalent.

Consider a two-factor experiment in which factors A and B are both random. The expected values of the mean squares for main effects and interactions (when cell frequencies are equal) have the following form:

$$\begin{aligned} E(\text{MS}_a) &= \sigma_\varepsilon^2 + n\sigma_{\alpha\beta}^2 + nq\sigma_\alpha^2, \\ E(\text{MS}_b) &= \sigma_\varepsilon^2 + n\sigma_{\alpha\beta}^2 + np\sigma_\beta^2, \\ E(\text{MS}_{ab}) &= \sigma_\varepsilon^2 + n\sigma_{\alpha\beta}^2, \\ E(\text{MS}_{\text{w. cell}}) &= \sigma_\varepsilon^2. \end{aligned} \tag{1}$$

(1) may be written in matrix form as

$$(2) \qquad \mathrm{E}\begin{bmatrix} \mathrm{MS}_a \\ \mathrm{MS}_b \\ \mathrm{MS}_{ab} \\ \mathrm{MS}_{\text{w. cell}} \end{bmatrix} = \begin{bmatrix} 1 & n & 0 & nq \\ 1 & n & np & 0 \\ 1 & n & 0 & 0 \\ 1 & 0 & 0 & 0 \end{bmatrix} \begin{bmatrix} \sigma_\varepsilon^2 \\ \sigma_{\alpha\beta}^2 \\ \sigma_\beta^2 \\ \sigma_\alpha^2 \end{bmatrix}$$

$$\mathrm{E}(\underline{s}^2) = M \quad \underline{\sigma}^2 .$$

Estimators of the components of $\underline{\sigma}^2$ may be obtained by equating mean squares to their expectations. If this is done, (2) becomes

$$(3) \qquad \begin{bmatrix} \mathrm{MS}_a \\ \mathrm{MS}_b \\ \mathrm{MS}_{ab} \\ \mathrm{MS}_{\text{w. cell}} \end{bmatrix} = \begin{bmatrix} 1 & n & 0 & nq \\ 1 & n & np & 0 \\ 1 & n & 0 & 0 \\ 1 & 0 & 0 & 0 \end{bmatrix} \begin{bmatrix} \hat\sigma_\varepsilon^2 \\ \hat\sigma_{\alpha\beta}^2 \\ \hat\sigma_\beta^2 \\ \hat\sigma_\alpha^2 \end{bmatrix}$$

$$\underline{s}^2 = M \quad \underline{\hat\sigma}^2 .$$

For the case of balanced designs (i.e., equal cell frequencies), solution of the system (3) will lead to unbiased estimators.

The general solution to the system (3) is given by

$$(4) \qquad \underline{\hat\sigma}^2 = M^{-1}\underline{s}^2$$

provided M is nonsingular. For the special case represented by (3), the solution is readily obtained by inspection. From the last equation in (3) one has

$$\hat\sigma_\varepsilon^2 = \mathrm{MS}_{\text{w. cell}}.$$

Substituting this result in the equation corresponding to MS_{ab} gives

$$\hat\sigma_{\alpha\beta}^2 = \frac{\mathrm{MS}_{ab} - \mathrm{MS}_{\text{w. cell}}}{n} .$$

If one subtracts the equation corresponding to MS_{ab} from the equation corresponding to MS_b, one obtains

$$\frac{\mathrm{MS}_b - \mathrm{MS}_{ab}}{np} = \frac{\hat\sigma_\varepsilon^2 + n\hat\sigma_{\alpha\beta}^2 + np\hat\sigma_\beta^2 - \hat\sigma_\varepsilon^2 - n\hat\sigma_{\alpha\beta}^2}{np} = \hat\sigma_\beta^2.$$

Similarly,

$$\frac{\mathrm{MS}_a - \mathrm{MS}_{ab}}{nq} = \hat\sigma_\alpha^2.$$

This procedure for estimating variance components may lead to negative estimators for variance components. Estimates of the variances and covariances of the estimators are readily obtained by methods developed by Tukey (1956). If the assumption is made that the random effects are independently and normally distributed, the variance and covariance of the estimators are obtained by a direct extension of the methods used in Sec. 3.17. Similar procedures are given in Scheffé (1959, pp. 228–229).

To illustrate the nature of the problem of estimating variance components for the case of disproportionate cell frequencies, consider an experiment for which the following additive model is appropriate:

$$X_{ijk} = \mu + \alpha_i + \beta_j + \varepsilon_{ijk},$$

where factors A and B are random. Suppose that there are n_{ij} observations in the cell corresponding to treatment ab_{ij}.

Let the total variation be partitioned as follows:

$$\Sigma(X_{ijk} - \bar{G})^2 = \left[\Sigma\left(\frac{A_i^2}{n_{i.}}\right) - \frac{G^2}{n_{..}}\right] + \left[\Sigma\left(\frac{B_j^2}{n_{.j}}\right) - \frac{G^2}{n_{..}}\right] + \text{SS}_{\text{rem}},$$

$$\text{SS}_{\text{total}} = \text{SS}_a + \text{SS}_b + \text{SS}_{\text{rem}},$$

where

$$\text{SS}_{\text{rem}} = \text{SS}_{\text{total}} - \text{SS}_a - \text{SS}_b.$$

(It is possible that SS_{rem} will be negative.) The expected values of the mean squares corresponding to this partition may be shown to be those given in Table 5.25-1. If one equates the expected values of the mean squares to their expectations, replacing parameters by their estimators, one obtains the following system of equations:

$$\begin{bmatrix} \text{MS}_a \\ \text{MS}_b \\ \text{MS}_{\text{rem}} \end{bmatrix} = \begin{bmatrix} 1 & c_{ab} & c_{aa} \\ 1 & c_{ba} & c_{bb} \\ 1 & c_{ea} & c_{eb} \end{bmatrix} \begin{bmatrix} \hat{\sigma}_\varepsilon^2 \\ \hat{\sigma}_\beta^2 \\ \hat{\sigma}_\alpha^2 \end{bmatrix}. \tag{5}$$

Table 5.25-1 Expected Value of Mean Squares

	Source of variation	df	E(MS)
(i)	A	$p - 1$	$\sigma_\varepsilon^2 + c_{ab}\sigma_\beta^2 + c_{aa}\sigma_\alpha^2$
	B	$q - 1$	$\sigma_\varepsilon^2 + c_{bb}\sigma_\beta^2 + c_{ba}\sigma_\alpha^2$
	Remainder	$n_{..} - p - q + 1$	$\sigma_\varepsilon^2 + c_{eb}\sigma_\beta^2 + c_{ea}\sigma_\alpha^2$

(ii)

$$c_{aa} = \frac{1}{p-1}\left(n_{..} - \sum_i \frac{n_{i.}^2}{n_{..}}\right) \qquad c_{ab} = \frac{1}{p-1}\left(\sum_{i,j} \frac{n_{ij}^2}{n_{i.}} - \sum_j \frac{n_{.j}^2}{n_{..}}\right)$$

$$c_{bb} = \frac{1}{q-1}\left(n_{..} - \sum_j \frac{n_{.j}^2}{n_{..}}\right) \qquad c_{ba} = \frac{1}{q-1}\left(\sum_{i,j} \frac{n_{ij}^2}{n_{.j}} - \sum_i \frac{n_{i.}^2}{n_{..}}\right)$$

$$c_{eb} = \frac{1}{n_{..} - p - q + 1}\left(\sum_j \frac{n_{.j}^2}{n_{..}} - \sum_{i,j} \frac{n_{ij}^2}{n_{i.}}\right)$$

$$c_{ea} = \frac{1}{n_{..} - p - q + 1}\left(\sum_i \frac{n_{i.}^2}{n_{..}} - \sum_{i,j} \frac{n_{ij}^2}{n_{.j}}\right)$$

Solution of the system (5) provides a set of estimators for the variance components σ_ε^2, σ_β^2, and σ_α^2. For the unbalanced case, just what optimum properties the estimators have has not been studied in detail. The estimators will be consistent.

5.26 Estimation of the Magnitude of Experimental Effects

The topic that is discussed in this section is closely related to the problem of estimating components of variance in the variance-component model. The problem that will be considered in this section, however, deals only with the fixed model.

The "magnitude" of an experimental effect is usually measured relative to some base. Depending upon what base is chosen, the relative magnitude of an effect will vary. The discussion which follows will be cast in terms of a 3×6 factorial experiment having $n = 5$ independent observations per cell. Both factors are assumed fixed. Suppose that the analysis of variance is that given in Table 5.26-1.

Table 5.26-1 Numerical Example

Source of variation	SS	df	MS	E(MS)	F
A	96.00	2	48.00	$\sigma_\varepsilon^2 + nq\sigma_\alpha^2$	$F_a = 12.00$
B	200.00	5	40.00	$\sigma_\varepsilon^2 + np\sigma_\beta^2$	$F_b = 10.00$
AB	60.00	10	6.00	$\sigma_\varepsilon^2 + n\sigma_{\alpha\beta}^2$	$F_{ab} = 1.50$
Error	288.00	72	4.00	σ_ε^2	

Let the parameter θ_α^2 be defined as follows:

$$\theta_\alpha^2 = \frac{\Sigma\alpha_i^2}{p}.$$

It is convenient to use θ_α^2 as an index of the overall effects of the levels of factor A. Since

$$\sigma_\alpha^2 = \frac{\Sigma\alpha_i^2}{p-1},$$

one has

$$\theta_\alpha^2 = \frac{p-1}{p}\sigma_\alpha^2.$$

In terms of the parameter θ_α^2,

$$E(MS_a) = \sigma_\varepsilon^2 + nq\sigma_\alpha^2 = \sigma_\varepsilon^2 + \frac{npq}{p-1}\theta_\alpha^2.$$

Similarly, if one defines

$$\theta_\beta^2 = \frac{\Sigma\beta_j^2}{q}, \qquad \text{then} \qquad \theta_\beta^2 = \frac{q-1}{q}\sigma_\beta^2.$$

Thus $$E(MS_b) = \sigma_\varepsilon^2 + np\sigma_\beta^2 = \sigma_\varepsilon^2 + \frac{npq}{q-1}\theta_\beta^2.$$

Similarly,

$$E(MS_{ab}) = \sigma_\varepsilon^2 + n\sigma_{\alpha\beta}^2 = \sigma_\varepsilon^2 + \frac{npq}{(p-1)(q-1)}\theta_{\alpha\beta}^2,$$

where $$\theta_{\alpha\beta}^2 = \frac{(p-1)(q-1)}{pq}\sigma_{\alpha\beta}^2 = \frac{\Sigma(\alpha\beta_{ij})^2}{pq}.$$

One has

$$\hat{\sigma}_\varepsilon^2 = MS_{error}.$$

Using this estimate of σ_ε^2 in $E(MS_a)$, one has

$$\hat{\theta}_\alpha^2 = \frac{(p-1)(MS_a - MS_{error})}{npq} = \frac{(p-1)(F_a - 1)MS_{error}}{npq},$$

where $F_a = MS_a/MS_{error}$. Similarly, one has

$$\hat{\theta}_\beta^2 = \frac{(q-1)(MS_b - MS_{error})}{npq} = \frac{(q-1)(F_b - 1)MS_{error}}{npq}$$

and

$$\hat{\theta}_{\alpha\beta}^2 = \frac{(p-1)(q-1)(MS_{ab} - MS_{error})}{npq} = \frac{(p-1)(q-1)(F_{ab} - 1)MS_{error}}{npq}.$$

One index (of several possible) of the relative magnitude of the effect of treatment A is

(1*a*) $$\omega_\alpha^2 = \frac{\theta_\alpha^2}{\theta_\alpha^2 + \theta_\beta^2 + \theta_{\alpha\beta}^2 + \sigma_\varepsilon^2}.$$

The numerator represents a measure of the main effects due to factor A; the denominator represents the sum of all effects underlying an observation. A heuristic principle for estimating this ratio is to replace each parameter in (1*a*) by its estimator. The resulting estimator of ω_α^2, although biased, will be consistent. In this case

(2*a*) $$\widehat{\omega_\alpha^2} = \frac{(p-1)(F_a - 1)}{(p-1)(F_a - 1) + (q-1)(F_b - 1) + (p-1)(q-1)(F_{ab} - 1) + npq}.$$

For the numerical data in Table 5.26-1,

$$\widehat{\omega_\alpha^2} = \frac{2(12-1)}{2(12-1) + 5(10-1) + 2(5)(1.50-1) + 5(3)(6)} = \frac{22}{162} = .1358.$$

In terms of this index, 13.58 percent of the total variance is due to the main effects of factor A.

If one defines

$$(1b) \qquad \omega_\beta^2 = \frac{\theta_\beta^2}{\theta_\alpha^2 + \theta_\beta^2 + \theta_{\alpha\beta}^2 + \sigma_\varepsilon^2},$$

then

$$(2b) \qquad \widehat{\omega_\beta^2} = \frac{(q-1)(F_b - 1)}{(p-1)(F_a - 1) + (p-1)(F_b - 1) + (p-1)(q-1)(F_{ab} - 1) + npq}.$$

For the numerical data in Table 5.26-1,

$$\widehat{\omega_\beta^2} = \frac{5(9)}{162} = .2778.$$

The index $\omega_{\alpha\beta}^2$ is defined in an analogous manner. One may also define

$$\omega_{\text{all effects}}^2 = \frac{\theta_\alpha^2 + \theta_\beta^2 + \theta_{\alpha\beta}^2}{\theta_\alpha^2 + \theta_\beta^2 + \theta_{\alpha\beta}^2 + \sigma_\varepsilon^2}.$$

For the numerical example,

$$\widehat{\omega_{\text{all effects}}^2} = \frac{162 - 90}{162} = .4444.$$

Thus 44.44 percent of the variance is due to treatment effects, and 55.56 percent of the variance is due to error variance.

A discussion of the advantages and disadvantages of some of the other indices that have been proposed as measures of the magnitude of a treatment effect will be found in Fleiss (1969).

6

FACTORIAL EXPERIMENTS—COMPUTATIONAL PROCEDURES AND NUMERICAL EXAMPLES

6.1 General Purpose

In this chapter the principles discussed in Chap. 5 will be illustrated by numerical examples; detailed computational procedures will be given for a variety of factorial experiments. It should be noted that the formulas convenient for computational work are not necessarily those which are most directly interpretable. For purposes of interpretation the basic definitions given in Chap. 5 are the important sources for reference; however, in all cases the computational formulas are algebraically equivalent to the basic definitions. The algebraic proofs underlying this equivalence are not difficult. Factorial experiments in which there are repeated measures are discussed in Chap. 7.

6.2 $p \times q$ Factorial Experiment Having n Observations per Cell

The treatment combinations in this type of experiment are represented in the cells of the following table:

	b_1	b_2	$\cdots$	b_j	$\cdots$	b_q
a_1	ab_{11}	ab_{12}	$\cdots$	ab_{1j}	$\cdots$	ab_{1q}
a_2	ab_{21}	ab_{22}	$\cdots$	ab_{2j}	$\cdots$	ab_{2q}
$\vdots$	$\vdots$	$\vdots$		$\vdots$		$\vdots$
a_i	ab_{i1}	ab_{i2}	$\cdots$	ab_{ij}	$\cdots$	ab_{iq}
$\vdots$	$\vdots$	$\vdots$		$\vdots$		$\vdots$
a_p	ab_{p1}	ab_{p2}	$\cdots$	ab_{pj}	$\cdots$	ab_{pq}

The cell in row a_i and column b_j corresponds to that part of the experiment in which treatment a_i is used in combination with b_j to yield the treatment combination ab_{ij}. The first subscript in this notation system refers to the level of factor A, the second subscript to the level of factor B.

Within each cell of the experiment there are n observations. The observations made under treatment combination ab_{ij} may be symbolized as follows:

	b_j					
a_i	X_{ij1}	X_{ij2}	$\cdots$	X_{ijk}	$\cdots$	X_{ijn}

The symbol X_{ij1} denotes a measurement on the first element in this cell. The symbol X_{ij2} denotes a measurement on the second element in this cell. The measurement on element k is denoted by the symbol X_{ijk}. The subscript k assumes the values $1, 2, \ldots, n$ within each of the cells. The sum of the n observations within cell ab_{ij} will be denoted by the symbol AB_{ij}; thus,

$$AB_{ij} = \sum_k X_{ijk}.$$

The following table summarizes the notation that will be used for sums of basic measurements:

	b_1	b_2	$\cdots$	b_j	$\cdots$	b_q	Row sum $= \sum_j$
a_1	AB_{11}	AB_{12}	$\cdots$	AB_{1j}	$\cdots$	AB_{1q}	$A_1 = \sum AB_{1j}$
a_2	AB_{21}	AB_{22}	$\cdots$	AB_{2j}	$\cdots$	AB_{2q}	A_2
$\vdots$	$\vdots$	$\vdots$		$\vdots$		$\vdots$	$\vdots$
a_i	AB_{i1}	AB_{i2}	$\cdots$	AB_{ij}	$\cdots$	AB_{iq}	A_i
$\vdots$	$\vdots$	$\vdots$		$\vdots$		$\vdots$	$\vdots$
a_p	AB_{p1}	AB_{p2}	$\cdots$	AB_{pj}	$\cdots$	AB_{pq}	A_p
Column sum	B_1	B_2	$\cdots$	B_j	$\cdots$	B_q	G

The sum of the nq measurements in row i, that is, the sum of all measurements made at level a_i, is denoted by the symbol A_i. Thus,

$$A_i = \sum_j AB_{ij} = \sum_j \sum_k X_{ijk}.$$

The double summation symbol $\sum_j \sum_k$ indicates that one sums within each cell as well as across all cells in row i. (Summing over the subscripts

j and k is equivalent to summing over all observations within a given row.) The sum of the np measurements in column j, that is, the sum of all measurements made under level b_j, is denoted by the symbol B_j. Thus,

$$B_j = \sum_i AB_{ij} = \sum_i \sum_k X_{ijk}.$$

The grand total of all measurements is denoted by the symbol G. Thus,

$$G = \sum_i A_i = \sum_j B_j = \sum_i \sum_j AB_{ij} = \sum_i \sum_j \sum_k X_{ijk}.$$

The mean of all measurements under treatment combination ab_{ij} is

$$\overline{AB}_{ij} = \frac{AB_{ij}}{n}.$$

The mean of all measurements at level a_i is

$$\bar{A}_i = \frac{A_i}{nq}.$$

The mean of all measurements at level b_j is

$$\bar{B}_j = \frac{B_j}{np}.$$

The grand mean of all observations in the experiment is

$$\bar{G} = \frac{G}{npq}.$$

To summarize the notation, a capital letter with a subscript ijk represents an individual observation; a pair of capital letters with the subscript ij represents the sum over the n observations represented by the subscript k. A capital letter with the subscript i represents the sum of nq observations within a row of the experimental plan; a capital letter with the subscript j represents the sum of the np observations within a column of the experimental plan. A widely used equivalent notation system is summarized below:

Notation	Equivalent notation
X_{ijk}	X_{ijk}
AB_{ij}	$X_{ij.}$
A_i	$X_{i..}$
B_j	$X_{.j.}$
G	$X_{...}$

In the equivalent notation system, the periods indicate the subscript over which the summation has been made.

An alternative notation system for the levels of the factors and the combinations of the levels is the following:

A:	a_1	a_2	$\cdots$	a_i	$\cdots$	a_I
B:	b_1	b_2	$\cdots$	b_j	$\cdots$	b_J
AB:	ab_{11}	ab_{12}	$\cdots$	ab_{ij}	$\cdots$	ab_{IJ}

Thus $\quad i = 1, \ldots, I \quad$ and $\quad j = 1, \ldots, J.$

In this notation system, the number of levels of factor A in the experiment is I, and no distinction is made between the number of levels of factor A in the experiment and the number of levels of A in the population of levels.

In some notation systems, an individual observation is denoted by Y_{ijk}, and the observations within cell ab_{ij} take the form

	b_j				
a_i	Y_{ij1} Y_{ij2}	$\cdots$	Y_{ijk}	$\cdots$	$Y_{ijn_{ij}}$,

where n_{ij} is the number of observations in cell ab_{ij}. For the case of equal cell frequencies

$$n_{ij} = n \qquad \text{for all } i, j.$$

Definition of Computational Symbols. In order to simplify the writing of the computational formulas for the sums of squares needed in the analysis of variance, it is convenient to introduce a set of computational symbols. The symbols appropriate for a $p \times q$ factorial experiment having n observations per cell are defined in Table 6.2-1. The computational procedures for this case are more elaborate than they need be, but the procedures to be developed here can be readily extended to more complex experimental plans in which they are not more elaborate than they need be.

In using the summation notation, where the index of summation is not indicated it will be understood that the summation is over all possible subscripts. For example, the notation $\sum_i \sum_j \sum_k X_{ijk}^2$ will be abbreviated ΣX_{ijk}^2; similarly the notation $\sum_i \sum_j (AB_{ij})^2$ will be abbreviated $\Sigma(AB_{ij})^2$. Where the summation is restricted to a single level of one of the subscripts, the index of summation will be indicated. Thus

$$\sum_i (AB_{ij})^2 = (AB_{1j})^2 + (AB_{2j})^2 + \cdots + (AB_{ij})^2 + \cdots + (AB_{pj})^2,$$

$$\sum_j (AB_{ij})^2 = (AB_{i1})^2 + (AB_{i2})^2 + \cdots + (AB_{ij})^2 + \cdots + (AB_{iq})^2.$$

In the definitions of the computational formulas given in part i of Table 6.2-1, the divisor in each case is the number of observations summed to obtain one of the terms that is squared. For example, n observations are summed to obtain an AB_{ij}. Hence the denominator n in the term

Table 6.2-1 Definition of Computational Symbols

(i)	$(1) = G^2/npq$	$(3) = (\Sigma A_i^2)/nq$
	$(2) = \Sigma X_{ijk}^2$	$(4) = (\Sigma B_j^2)/np$
	$(5) = [\Sigma(AB_{ij})^2]/n$	

	Source of variation	Computational formula for SS
(ii)	A	$SS_a = (3) - (1)$
	B	$SS_b = (4) - (1)$
	AB	$SS_{ab} = (5) - (3) - (4) + (1)$
	Experimental error (within cell)	$SS_{error} = (2) - (5)$
	Total	$SS_{total} = (2) - (1)$

$[\Sigma(AB_{ij})^2]/n$. There are nq observations summed to obtain an A_i. Hence the denominator nq in the term $(\Sigma A_i^2)/nq$.

An estimate of the variation due to the main effects of factor A is given by

$$SS_a = nq\Sigma(\bar{A}_i - \bar{G})^2.$$

This is not a convenient computational formula for this source of variation. An algebraically equivalent form is

$$SS_a = \frac{\Sigma A_i^2}{nq} - \frac{G^2}{npq} = (3) - (1).$$

The other sources of variation in the analysis of variance are as follows:

$$\begin{aligned}
SS_b &= np\Sigma(\bar{B}_j - \bar{G})^2 &&= (4) - (1),\\
SS_{ab} &= n(\Sigma\overline{AB}_{ij} - \bar{A}_i - \bar{B}_j + \bar{G})^2 &&= (5) - (3) - (4) + (1),\\
SS_{w.\ cell} &= \Sigma(X_{ijk} - \overline{AB}_{ij})^2 &&= (2) - (5),\\
SS_{total} &= \Sigma(X_{ijk} - \bar{G})^2 &&= (2) - (1).
\end{aligned}$$

These computational formulas are summarized in part ii of Table 6.2-1.

Should the interaction term in the analysis of variance prove to be statistically significant, it is generally desirable to analyze the simple main effects rather than the overall main effects. Computational symbols for variation due to simple main effects are summarized in Table 6.2-2. By definition the variation due to the simple main effect of factor A for level b_1 is

$$\begin{aligned}
SS_{a\ \text{for}\ b_1} &= n\Sigma(\overline{AB}_{i1} - \bar{B}_1)^2\\
&= \frac{\Sigma(AB_{i1})^2}{n} - \frac{B_1^2}{np} = (5b_1) - (4b_1).
\end{aligned}$$

By definition, the variation due to the simple main effect of factor B for level a_1 is

$$SS_{b \text{ for } a_1} = n\Sigma(\overline{AB}_{1j} - \bar{A}_1)^2$$
$$= \frac{\Sigma(AB_{1j}^2)^2}{n} - \frac{A_1^2}{nq} = (5a_1) - (3a_1).$$

Table 6.2-2 Definition of Computational Symbols for Simple Effects

(i)

$(3a_1) = A_1^2/nq$	$(4b_1) = B_1^2/np$
$(3a_2) = A_2^2/nq$	$(4b_2) = B_2^2/np$
...	...
$(3a_p) = A_p^2/nq$	$(4b_q) = B_q^2/np$
$(3) = (\Sigma A_i^2)/nq$	$(4) = (\Sigma B_j^2)/np$
$(5a_1) = [\sum_j (AB_{1j})^2]/n$	$(5b_1) = [\sum_i (AB_{i1})^2]/n$
$(5a_2) = [\sum_j (AB_{2j})^2]/n$	$(5b_2) = [\sum_i (AB_{i2})^2]/n$
...	...
$(5a_p) = [\sum_j (AB_{pj})^2]/n$	$(5b_q) = [\sum_i (AB_{iq})^2]/n$
$(5) = [\Sigma(AB_{ij})^2]/n$	$(5) = [\Sigma(AB_{ij})^2]/n$

(ii)

Source of variation	Computational formula for SS
Simple effects for A:	
For level b_1	$(5b_1) - (4b_1)$
For level b_2	$(5b_2) - (4b_2)$
...	...
For level b_q	$(5b_q) - (4b_q)$
Simple effects for B:	
For level a_1	$(5a_1) - (3a_1)$
For level a_2	$(5a_2) - (3a_2)$
...	...
For level a_p	$(5a_p) - (3a_p)$

Numerical Example. A numerical example of a 2×3 factorial experiment having three observations per cell will be used to illustrate the computational procedures. Suppose that an experimenter is interested in evaluating the relative effectiveness of three drugs (factor B) in bringing about behavioral changes in two categories, schizophrenics and depressives, of patients (factor A). What is considered to be a random sample of nine patients belonging to category a_1 (schizophrenics) is divided at random into three subgroups, with three patients in each subgroup.

Each subgroup is then assigned to one of the drug conditions. An analogous procedure is followed for a random sample of nine patients belonging to category a_2 (depressives). Criterion ratings are made on each patient before and after the administration of the drugs. The numerical entries in part i of Table 6.2-3 represent the difference between

Table 6.2-3 Numerical Example

Observed data:

		Drug b_1			Drug b_2			Drug b_3	
(i) Category a_1	8	4	0	10	8	6	8	6	4
Category a_2	14	10	6	4	2	0	15	12	9

AB summary table:

(ii)		b_1	b_2	b_3	Total
	a_1	12	24	18	$54 = A_1$
	a_2	30	6	36	$72 = A_2$
	Total	42	30	54	$126 = G$
		B_1	B_2	B_3	

(iii)

$$\begin{aligned}
(1) &= (126)^2/18 &&= 882.00\\
(2) &= (8^2 + 4^2 + 0^2 + \cdots + 15^2 + 12^2 + 9^2) &&= 1198\\
(3) &= (54^2 + 72^2)/9 &&= 900.00\\
(4) &= (42^2 + 30^2 + 54^2)/6 &&= 930.00\\
(5) &= (12^2 + 24^2 + 18^2 + 30^2 + 6^2 + 36^2)/3 &&= 1092.00
\end{aligned}$$

(iv)

$$\begin{aligned}
SS_a &= (3) - (1) = 900.00 - 882.00 &&= 18.00\\
SS_b &= (4) - (1) = 930.00 - 882.00 &&= 48.00\\
SS_{ab} &= (5) - (3) - (4) + (1) &&\\
&= 1092.00 - 900.00 - 930.00 + 882.00 &&= 144.00\\
SS_{w.\,cell} &= (2) - (5) = 1198 - 1092.00 &&= 106.00\\
SS_{total} &= (2) - (1) = 1198 - 882.00 &&= 316.00
\end{aligned}$$

the two ratings on each of the patients. (An analysis of covariance might be more appropriate for this plan; covariance analysis is discussed in Chap. 10.)

As a first step in the analysis, the AB summary table in part ii is obtained. The entry in row a_1, column b_1 is

$$AB_{11} = 8 + 4 + 0 = 12.$$

The entry in row a_2, column b_3 is

$$AB_{23} = 15 + 12 + 9 = 36.$$

Data for all computational symbols except (2) are in part ii; computational symbol (2) is obtained from data in part i. The analysis of variance is summarized in Table 6.2-4.

Table 6.2-4 Summary of Analysis of Variance

Source of variation	SS	df	MS	F
A (category of patient)	18.00	1	18.00	2.04
B (drug)	48.00	2	24.00	2.72
AB	144.00	2	72.00	8.15
Within cell	106.00	12	8.83	
Total	316.00	17		

The structure of the F ratios used in making tests depends upon the expected values of mean squares appropriate for the experimental data. If the categories and drugs are fixed factors, i.e., if inferences are to be made only with respect to the two categories of patients represented in the experiment and only with respect to the three drugs included in the experiment, then the appropriate expected values of the mean squares are given in Table 6.2-5. The structure of F ratios is determined in accordance with the principles given in Sec. 5.7.

Table 6.2-5 Expected Values for Mean Squares

Source	MS	E(MS)	F
Main effect of A	MS_a	$\sigma_\varepsilon^2 + 9\sigma_\alpha^2$	$MS_a/MS_{w.\ cell}$
Main effect of B	MS_b	$\sigma_\varepsilon^2 + 6\sigma_\beta^2$	$MS_b/MS_{w.\ cell}$
AB interaction	MS_{ab}	$\sigma_\varepsilon^2 + 3\sigma_{\alpha\beta}^2$	$MS_{ab}/MS_{w.\ cell}$
Within cell	$MS_{w.\ cell}$	σ_ε^2	

If tests are made at the .05 level of significance, the critical value for the test of the hypothesis that the action of the drugs is independent of the category of patient (i.e., zero interaction) is $F_{.95}(2,12) = 3.89$. In this case $F_{\text{obs}} = 8.15$ exceeds the critical value. Hence the experimental data do not support the hypothesis of zero interaction. The data indicate that the effect of a drug differs for the two types of patients—the effect of a drug depends upon the category of patient to which it is administered. A significant interaction indicates that a given drug has different effects for one category of patient from what it has for a second category. The nature of the interaction effects is indicated by inspecting of the cell means. These means are given below:

	Drug 1	Drug 2	Drug 3
Category 1	4	8	6
Category 2	10	2	12

A geometric representation of these means is given in Fig. 6.2-1. This figure represents the profiles corresponding to the simple effects of the drugs (factor B) for each of the categories (factor A). A test for

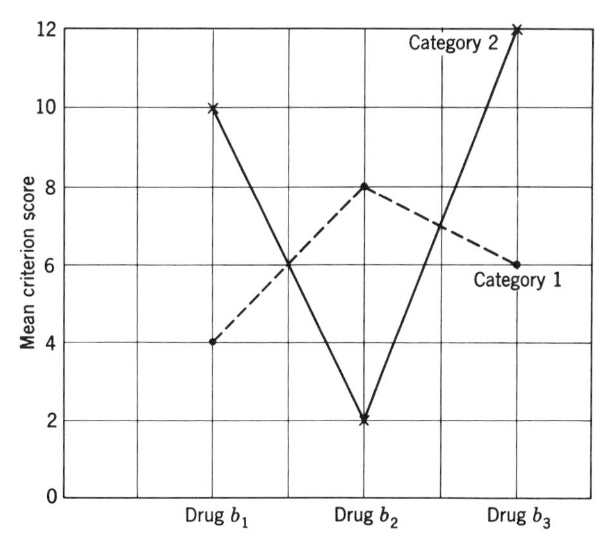

Figure 6.2-1 Profiles of simple effects for drugs.

the presence of interaction is equivalent to a test on the difference in the shapes of the profiles of these simple effects. An equivalent geometric representation of the table of means is given in Fig. 6.2-2. This figure represents the profiles corresponding to the simple effects of the cate-

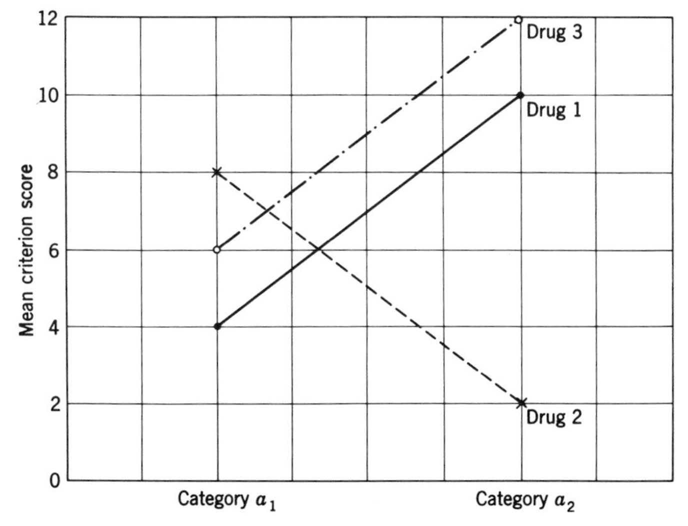

Figure 6.2-2 Profiles of simple effects for categories.

gories for each of the drugs. The profile for drug 2 appears to have a slope which is different from the slopes of the profiles for the other drugs. As an aid in the interpretation of interactions, geometric representation of the profiles corresponding to the means is generally of considerable value.

Tests on differences between means within the same profile are given by tests on simple effects. Computational procedures for obtaining the variation due to simple effects are summarized in Table 6.2-6. Data from which the symbols in part ii are obtained are given in part i; the latter is the AB summary of Table 6.2-3.

The analysis of variance for the simple effects of the drugs for each of the categories is summarized in Table 6.2-7. The structure of the F ratios for simple effects is dependent upon appropriate expected values for the mean squares. Assuming that factors A and B are fixed factors, expected values for the mean squares of simple effects are given in Table 6.2-8. It should be noted, in terms of the general linear model, that a simple main effect is actually a sum of an overall main effect and an interaction. For example,

$$\overline{AB}_{ij} - \bar{B}_j \qquad \text{estimates} \qquad \alpha_i + \alpha\beta_{ij} = \alpha_i \text{ for } b_j.$$

A test on the simple effect of factor A for level b_j is equivalent to a test that $\sigma^2_{\alpha \text{ for } b_j} = 0$. The appropriate F ratio for this test would have an estimate of σ_ε^2 as denominator; the latter estimate is given by $MS_{w.cell}$.

By using the .05 level of significance for the data in Table 6.2-7, the critical value for an F ratio is $F_{.95}(2,12) = 3.89$. The experimental data indicate no difference between the drugs for category a_1 ($F_{obs} = 1.36$). The experimental data indicate that there are differences between the drugs for category a_2 ($F_{obs} = 9.51$). Inspection of the profile of the drugs for category a_2 in Fig. 6.2-2 indicates that the effect of drug b_2 is different from the effects of the other drugs.

Returning to the analysis in Table 6.2-4, neither of the main effects is statistically significant. However, in the presence of interaction, inferences made with respect to main effects must be interpreted with caution. The data indicate that the main effects due to drugs do not differ. However, this does not mean that the drugs are equally effective for each of the categories considered separately. Because of the significant interaction effect, conclusions with respect to main effects due to drugs cannot be applied separately to each of the categories.

The drugs are differentially effective only when one considers comparisons within each of the categories separately. It must be noted and emphasized that statistical inference is a tool to help the experimenter in drawing scientifically meaningful conclusions from his experiment. Too frequently the tool is allowed to become the master rather than the servant. The primary objective of the experiment should determine which tests

Table 6.2-6 Computation of Simple Effects

AB summary table:

(i)

	b_1	b_2	b_3	Total
a_1	12	24	18	54
a_2	30	6	36	72
Total	42	30	54	126

(ii)

$(3a_1) = (54^2)/9 = 324.00$
$(3a_2) = (72^2)/9 = 576.00$

$(3) = 900.00$

$(4b_1) = (42^2)/6 = 294.00$
$(4b_2) = (30^2)/6 = 150.00$
$(4b_3) = (54^2)/6 = 486.00$

$(4) = 930.00$

$(5a_1) = (12^2 + 24^2 + 18^2)/3 = 348.00$
$(5a_2) = (30^2 + 6^2 + 36^2)/3 = 744.00$

$(5) = 1092.00$

$(5b_1) = (12^2 + 30^2)/3 = 348.00$
$(5b_2) = (24^2 + 6^2)/3 = 204.00$
$(5b_3) = (18^2 + 36^2)/3 = 540.00$

$(5) = 1092.00$

(iii)

Simple effects of A:

For level b_1 $\quad SS_{a \text{ for } b_1} = (5b_1) - (4b_1) = 54.00$
For level b_2 $\quad SS_{a \text{ for } b_2} = (5b_2) - (4b_2) = 54.00$
For level b_3 $\quad SS_{a \text{ for } b_3} = (5b_3) - (4b_3) = 54.00$

162.00

Simple effects of B:

For level a_1 $\quad SS_{b \text{ for } a_1} = (5a_1) - (3a_1) = 24.00$
For level a_2 $\quad SS_{b \text{ for } a_2} = (5a_2) - (3a_2) = 168.00$

192.00

Check: $SS_a + SS_{ab} = \Sigma SS_{a \text{ for } b_j}$
$18.00 + 144.00 = 162.00$

Check: $SS_b + SS_{ab} = \Sigma SS_{b \text{ for } a_i}$
$48.00 + 144.00 = 192.00$

Table 6.2-7 Analysis of Variance for Simple Effects

Source of variation	SS	df	MS	F
B for a_1 (drugs for category a_1)	24.00	2	12.00	1.36
B for a_2 (drugs for category a_2)	168.00	2	84.00	9.51
Within cell	106.00	12	8.83	

Table 6.2-8 Expected Values for Mean Squares of Simple Effects

Source of variation	E(MS)
Simple effects of factor A:	
For level b_1	$\sigma_\varepsilon^2 + 3\sigma_{\alpha \text{ for } b_1}^2$
For level b_2	$\sigma_\varepsilon^2 + 3\sigma_{\alpha \text{ for } b_2}^2$
For level b_3	$\sigma_\varepsilon^2 + 3\sigma_{\alpha \text{ for } b_3}^2$
Simple effects for factor B:	
For level a_1	$\sigma_\varepsilon^2 + 3\sigma_{\beta \text{ for } a_1}^2$
For level a_2	$\sigma_\varepsilon^2 + 3\sigma_{\beta \text{ for } a_2}^2$

are to be made and which tests are to be avoided for lack of meaning. Statistical elegance does not necessarily imply scientifically meaningful inferences.

Individual Comparisons. To make comparisons between two or more means, the procedures given in Sec. 5.17 may be used. For example, to test the hypothesis that drugs 1 and 2 are equally effective for category 2, one may use the following statistic (data from the last numerical example will be used for illustrative purposes):

$$F = \frac{n(\overline{AB}_{21} - \overline{AB}_{22})^2}{2\text{MS}_{\text{w.cell}}} = \frac{(AB_{21} - AB_{22})^2}{2n\text{MS}_{\text{w.cell}}} = \frac{(30 - 6)^2}{6(8.83)} = 10.87.$$

For a .05-level test, the critical value is $F_{.95}(1,12) = 4.75$. Hence the hypothesis that $\mu_{21} = \mu_{22}$ is not supported by the experimental data.

To test differences between all possible pairs of means in a *logical grouping* of means, the procedures given in Sec. 3.9 may be adapted for use. The Newman-Keuls procedure will be illustrated here. Suppose that it is desired to test the differences between all possible pairs of means for category a_2. The procedures for such multiple comparisons are outlined in Table 6.2-9.

In part i the means to be compared are first arranged in rank order, from low to high. Then differences between all possible pairs which give a positive value are obtained. An estimate of the standard error of a single mean is computed in part ii. This estimate is based upon data in all cells in the experiment, not just those from which the means are obtained. (If there is any real question about homogeneity of variance, only cells from which the means are obtained may be used in estimating the standard error of a mean.) If the within-cell variance from all cells in the experiment is used, degrees of freedom for $s_{\overline{AB}}$ are $pq(n - 1)$, which in this case is equal to 12.

To obtain critical values for a .05-level test, one obtains values of the $q_{.95}(r,12)$ statistic, where 12 is the degrees of freedom for $s_{\overline{AB}}$ and r is the number of steps two means are apart in an ordered sequence.

Table 6.2-9 Comparisons between Means for Category a_2

		$\overline{AB}_{22}$	$\overline{AB}_{21}$	$\overline{AB}_{23}$		Critical
	Ordered means:	2	10	12	r	value
(i)		2 —	8*	10*	3	6.45
		10	—	2	2	5.27

(ii) $s_{\overline{AB}} = \sqrt{\text{MS}_{\text{w. cell}}/n} = \sqrt{8.83/3} = 1.71$

	r:	2	3
	$q_{.95}(r,12)$:	3.08	3.77
(iii)	Critical value $= s_{\overline{AB}}q_{.95}(r,12)$:	5.27	6.45

These values are obtained from tables of the q statistic in Table C.4. The actual critical values are $s_{\overline{AB}}q_{.95}(r,12)$. For example, the difference

$$\overline{AB}_{23} - \overline{AB}_{22} = 10.$$

These means are three steps apart in the ordered sequence ($r = 3$); hence the critical value is 6.45. The difference

$$\overline{AB}_{23} - \overline{AB}_{21} = 2,$$

for which $r = 2$, has the critical value 5.27. Tests of this kind must be made in a sequence which is specified in Sec. 3.9.

From the outcome of the tests in part iii one concludes that, for category a_2, drugs 1 and 3 differ from drug 2 but that there is no statistically significant difference (on the criterion used) between drugs 1 and 3.

Test for Homogeneity of Error Variance. Although F tests in the analysis of variance are robust with respect to the assumption of homogeneity of error variance, a rough but simple check may be made on this assumption through use of the $F_{\max}$ statistic. Apart from the use of this statistic, the variances of the individual cells should be inspected for any kind of systematic pattern between treatments and variances. In cases having a relatively large number of observations within each cell, within-cell distributions should also be inspected.

Use of the $F_{\max}$ test for homogeneity of variance will be illustrated for the data in part i of Table 6.2-3. For this purpose the within-cell variation for each of the cells is required. In the computational procedures given in Table 6.2-3, the pooled within-cell variation from all cells is computed. The variation within cell ij has the form

$$\text{SS}_{ij} = \sum_k (X_{ijk} - \overline{AB}_{ij})^2$$

$$= \Sigma X_{ijk}^2 - \frac{(AB_{ij})^2}{n}.$$

For example,

$$\text{SS}_{11} = (8^2 + 4^2 + 0^2) - \frac{(12)^2}{3}$$

$$= \quad 80 \quad - 48 \quad = 32.$$

Similarly, the variation within cell ab_{21} is given by

$$\text{SS}_{21} = (14^2 + 10^2 + 6^2) - \frac{(30)^2}{3}$$

$$= \quad 332 \quad - 300 \quad = 32.$$

The other within-cell sums of squares are

$$\text{SS}_{12} = 8, \qquad \text{SS}_{22} = 8, \qquad \text{SS}_{13} = 8, \qquad \text{SS}_{23} = 18.$$

As a check on the computational work,

$$\Sigma SS_{ij} = SS_{w.cell} = 106.$$

Since the number of observations in each cell is constant, the F_{max} statistic is given by

$$F_{max} = \frac{SS(\text{largest})}{SS(\text{smallest})} = \frac{32}{8} = 4.00.$$

The critical value for a .05-level test is $F_{.95}(pq, n - 1)$, which in this case is $F_{.95}(6,2) = 266$. Since the observed F_{max} statistic does not exceed the critical value, the hypothesis of homogeneity of variance may be considered tenable. In cases in which the assumptions of homogeneity of variance cannot be considered tenable, a transformation on the scale of measurement may provide data which are amenable to the assumptions underlying the analysis model.

Approximate *F* Tests When Cell Variances Are Heterogeneous. A procedure suggested by Box (1954, p. 300) may be adapted for use in testing simple effects for factor A, even though variances may be heterogeneous. The F ratio in this case has the form

$$F = \frac{MS_{a \text{ for } b_j}}{MS_{\text{error}(b_j)}},$$

where $MS_{\text{error}(b_j)}$ is the pooled within-cell variance for all cells at level b_j. The approximate degrees of freedom for this F ratio are

$$\begin{array}{ll} 1 & \text{for numerator,} \\ n - 1 & \text{for denominator.} \end{array}$$

In testing simple effects for factor B at level a_i, the F ratio has the form

$$F = \frac{MS_{b \text{ for } a_i}}{MS_{\text{error}(a_i)}},$$

where $MS_{\text{error}(a_i)}$ is the pooled within-cell variance for all cells at level a_i. The approximate degrees of freedom for this F ratio are $(1, n - 1)$. In the usual test (assuming homogeneity of variance) the degrees of freedom for the latter F ratio are $[(q - 1), pq(n - 1)]$ if all cell variances are pooled and $[(q - 1), q(n - 1)]$ if only variances from cells at level a_i are pooled.

Alternative Notation Systems. The notation system that has been adopted for use in this and following sections is not the most widely used system but rather a slight variation on what is essentially a common theme running through several notation systems. Bennett and Franklin (1954) use a closely related notation. The equivalence between the two systems is expressed in terms of the following relationships:

$$C = \frac{G^2}{npq},$$

$$C_i = \frac{\Sigma A_i^2}{nq},$$

$$C_j = \frac{\Sigma B_j^2}{np},$$

$$C_{ij} = \frac{\Sigma (AB_{ij})^2}{n},$$

$$C_{ijk} = \Sigma X_{ijk}^2 .$$

In terms of the Bennett and Franklin notation system, the sums of squares have the following symmetric form:

$$SS_a = SS_i = C_i - C,$$

$$SS_b = SS_j = C_j - C,$$

$$SS_{ab} = SS_{ij} = C_{ij} - C_i - C_j + C,$$

$$SS_{w.cell} = C_{ijk} - C_{ij}.$$

The notation used by Kempthorne (1952) is perhaps the most widely used. In this system a single observation is designated Y_{ijk}.

$$CF = \frac{G^2}{npq},$$

$$\Sigma Y_{i..}^2 = \Sigma A_i^2,$$

$$\Sigma Y_{.j.}^2 = \Sigma B_j^2,$$

$$\sum_{i,j} Y_{ij.}^2 = \Sigma (AB_{ij})^2.$$

6.3 $p \times q$ Factorial Experiment—Unequal Cell Frequencies

Computational procedures for an unweighted-means analysis will be described in this section. The conditions under which this kind of analysis is appropriate are given in Sec. 5.22. For illustrative purposes the computational procedures are cast in terms of a 2×4 factorial experiment; these procedures may, however, be generalized to any $p \times q$ factorial experiment. Computational procedures for the least-squares solution are given in Sec. 6.14.

Suppose that the levels of factor A represent two methods for calibrating dials and levels of factor B represent four levels of background illumination. The criterion measure is an accuracy score for a series of trials. The original experiment called for five observations per cell. However, because of conditions not related to the experimental variables, the completed experiment had three to five observations per cell. The observed criterion scores are given in part i of Table 6.3-1. Summary

Table 6.3-1 Numerical Example

(i) Observed data:

	b_1	b_2	b_3	b_4
a_1	3, 4, 6, 7	5, 6, 6, 7, 7	4, 6, 8, 8	8, 10, 10, 7, 11
a_2	2, 3, 4	3, 5, 6, 3	9, 12, 12, 8	9, 7, 12, 11

(ii) Cell data:

		b_1	b_2	b_3	b_4
a_1	n_{ij}	4	5	4	5
	ΣX	20	31	26	46
	ΣX^2	110	195	180	434
	SS_{ij}	10.00	2.80	11.00	10.80
a_2	n_{ij}	3	4	4	4
	ΣX	9	17	41	39
	ΣX^2	29	79	433	395
	SS_{ij}	2.00	6.75	12.75	14.75

(iii)

$$\bar{n}_h = \frac{8}{.25 + .20 + .25 + .20 + .33 + .25 + .25 + .25} = 4.04$$

$$SS_{\text{w. cell}} = \Sigma\Sigma SS_{ij} = 10.00 + 2.80 + \cdots + 14.75 = 70.85$$

of within-cell information required in the analysis is given in part ii. The variation within cell ab_{11} is

$$SS_{11} = 110 - \frac{(20)^2}{4} = 10.00.$$

The harmonic mean of the cell frequencies is computed in part iii. The computational formula used is

$$\bar{n}_h = \frac{pq}{\Sigma\Sigma(1/n_{ij})}.$$

The pooled within-cell variation is also computed in part iii.

The data in the cells of part i of Table 6.3-2 are means of the respective n_{ij} observations in the cells. All the computational symbols in part ii are based upon these means and row and column totals of these means. In defining the computational symbols in (ii), each of the cell means is considered as if it were a single observation. Computational formulas for the main effects and interaction are given in part iii.

Table 6.3-2 Numerical Example (Continued)

Cell means:

		b_1	b_2	b_3	b_4	Total
(i)	a_1	5.00	6.20	6.50	9.20	26.90
	a_2	3.00	4.25	10.25	9.75	27.25
	Total	8.00	10.45	16.75	18.95	54.15

(ii)

$(1) = G^2/pq = (54.15)^2/8 = 366.53$

$(2) = \Sigma X^2$ (see part ii, Table 6.3-1)

$(3) = (\Sigma A_i^2)/q = (26.90^2 + 27.25^2)/4 = 366.54$

$(4) = (\Sigma B_j^2)/p = (8.00^2 + 10.45^2 + 16.75^2 + 18.95^2)/2 = 406.43$

$(5) = \Sigma(\overline{AB}_{ij})^2 = 5.00^2 + 6.20^2 + \cdots + 9.75^2 = 417.52$

(iii)

$SS_a = \tilde{n}_h[(3) - (1)] = 4.04[366.54 - 366.53] = .04$

$SS_b = \tilde{n}_h[(4) - (1)] = 4.04[406.43 - 366.53] = 161.20$

$SS_{ab} = \tilde{n}_h[(5) - (3) - (4) + (1)] = 44.76$

The analysis of variance is summarized in Table 6.3-3. The degrees

Table 6.3-3 Summary of Analysis of Variance

Source of variation	SS	df	MS	F
A (method of calibration)	.04	1	.04	
B (background illumination)	161.20	3	53.73	18.99
AB	44.76	3	14.92	5.27
Within cell	70.85	25	2.83	

of freedom for the within-cell variation are $\Sigma\Sigma n_{ij} - pq = 33 - 8 = 25$. If factors A and B are fixed, then $MS_{w.cell}$ is the proper denominator for all tests. By using the .05 level of significance, the critical value for the test on the interaction is $F_{.95}(3,25) = 2.99$. Since the observed F ratio, $F = 5.27$, is larger than the critical value, the data tend to contradict the hypothesis of zero interaction. The test on the main effects for factor B has the critical value 2.99. The observed F ratio, $F = 18.99$, is larger than the critical value for a .05-level test. Hence the data contradict the hypothesis that the main effects of factor B are zero. Inspection of profiles (Fig. 6.3-1) of the simple effects of B for

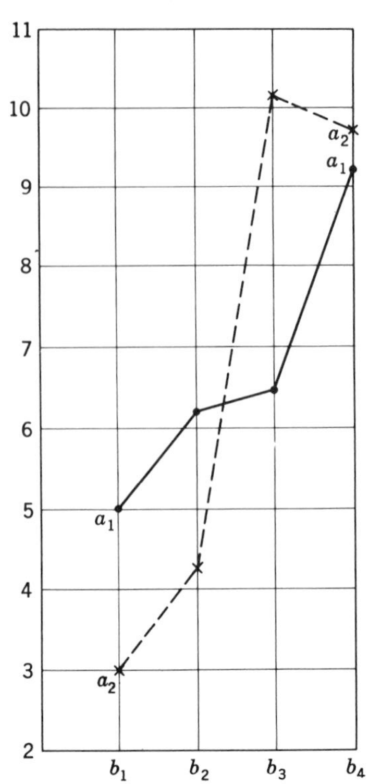

Figure 6.3-1 Profiles of simple effects for factor *B*.

levels a_1 and a_2 indicates why the effects of factor A are masked by the interaction. For the first two levels of factor B the means for level a_1 are higher than the corresponding means for level a_2, and for the other two levels the means for level a_1 are lower than the corresponding means for level a_2. Opposite algebraic signs of such differences tend to make their sum close to zero in the main effects for factor A.

To illustrate the computation of the simple effects, the variation due to the simple effects of B for level a_2 is obtained as follows (data for the computations are obtained from part ii of Table 6.3-2):

$$(5a_2) = 3.00^2 + 4.25^2 + 10.25^2 + 9.75^2 = 227.19,$$
$$(3a_2) = (27.25)^2/4 = 185.64,$$
$$\text{SS}_{b \text{ for } a_2} = \bar{n}_h[(5a_2) - (3a_2)] = 167.86,$$

$$\text{MS}_{b \text{ for } a_2} = \frac{\text{SS}_{b \text{ for } a_2}}{q-1} = 55.95.$$

A test of the hypothesis that the variance of the simple effects of factor B

at level a_2 is zero is given by the F ratio

$$F = \frac{\text{MS}_{b \text{ for } a_2}}{\text{MS}_{\text{w.cell}}} = \frac{55.95}{2.83} = 19.77.$$

The degrees of freedom for this F ratio are (3,25).

In comparing two means, the actual number of observations upon which the mean is based may be used. For example,

$$t = \frac{\overline{AB}_{14} - \overline{AB}_{11}}{\sqrt{\text{MS}_{\text{w.cell}}[(1/n_{14}) + (1/n_{11})]}} = \frac{9.20 - 5.00}{\sqrt{2.83(\frac{1}{5} + \frac{1}{4})}} = \frac{4.20}{\sqrt{1.27}} = 3.73.$$

The degrees of freedom for this t statistic are those for $\text{MS}_{\text{w.cell}}$. In making all possible tests between ordered means within a logical grouping, the procedures given in Sec. 6.2 may be followed, assuming $\tilde{n}_h$ observations per cell.

6.4 Effect of Scale of Measurement on Interaction

In Sec. 5.4 it was indicated that interactions were in part a function of the choice of the scale of measurement. When this is the case, interaction effects may be "removed" by proper choice of a scale. A numerical example will be used to illustrate this point.

The data at the left in part i of Table 6.4-1 represent observed criterion scores obtained from a 3 × 3 factorial experiment having two observations per cell. The symbols in part ii are defined in Table 6.2-1. The analysis of variance for these data is summarized in part iii. The critical value

Table 6.4-1 Analysis of Variance in Terms of Original Scale of Measurement

(i)

Observed data

	b_1	b_2	b_3
a_1	1, 0	12, 14	20, 27
a_2	9, 9	32, 30	40, 55
a_3	30, 34	64, 70	100, 96

AB summary table

	b_1	b_2	b_3	Total
a_1	1	26	47	74
a_2	18	62	95	175
a_3	64	134	196	394
Total	83	222	338	643

(ii)

(1) = 22,969.39 (3) = 31,889.50 (5) = 38,273.50
(2) = 38,449 (4) = 28,402.83

(iii)

Source of variation	SS	df	MS	F
A	8,920.11	2	4460.06	
B	5,433.44	2	2716.72	
AB	950.56	4	237.64	12.19
Within cell	175.50	9	19.50	
Total	15,479.61	17		

for a .01-level test on the interaction is $F_{.99}(4,9) = 6.42$. Since the observed F ratio exceeds the critical value, the experimental data tend to contradict the hypothesis that the interaction effects are zero. The profiles of factor B for each level of factor A are shown in Fig. 6.4-1.

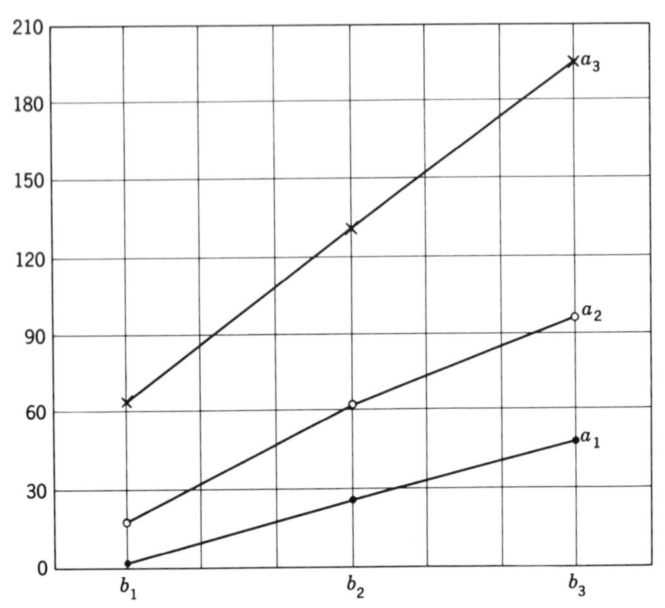

Figure 6.4-1 Profiles of factor B for levels of factor A. (Original scale of measurement.)

Within the range of the data, the profiles do not cross. Further, each profile is approximately linear in form. The major difference in these profiles is in the slopes of the respective best-fitting lines.

A square-root transformation on the original scale of measurement will make the slopes in this kind of configuration approximately equal. From the AB summary table, the ranges of the respective rows, in terms of a square-root transformation, are as follows.

Row a_1: $\sqrt{47} - \sqrt{1} \doteq 6$

Row a_2: $\sqrt{95} - \sqrt{18} \doteq 6$

Row a_3: $\sqrt{196} - \sqrt{64} \doteq 6$

The fact that these ranges are approximately equal provides partial evidence that the square-root transformation, when applied to the original observations, will yield profiles having approximately equal slopes.

In terms of the transformed scale of measurement, the observed data are given in part i of Table 6.4-2. The transformation has the following form:

$$X'_{ijk} = \sqrt{X_{ijk}}.$$

Each entry in the table at the left of part i is the square root of the corresponding entry in part i of Table 6.4-1. The analysis of variance for the transformed data is summarized in part iii of Table 6.4-2. It will be

Table 6.4-2 Analysis of Variance in Terms of Transformed Scale of Measurement

Observed data (transformed scale):

(i)

	b_1	b_2	b_3
a_1	1.0, 0.0	3.5, 3.7	4.5, 5.2
a_2	3.0, 3.0	5.7, 5.5	6.3, 7.4
a_3	5.5, 5.8	8.0, 8.4	10.0, 9.8

	b_1	b_2	b_3	Total
a_1	1.0	7.2	9.7	17.9
a_2	6.0	11.2	13.7	30.9
a_3	11.3	16.4	19.8	47.5
Total	18.3	34.8	43.2	96.3

(ii)

(1) = 515.20 (3) = 588.58 (5) = 642.38
(2) = 643.91 (4) = 568.70

(iii)

Source of variation	SS	df	MS	F
A	73.38	2	36.69	
B	53.50	2	26.75	
AB	.30	4	.075	$F < 1$
Within cell	1.53	9	.170	
Total	128.71	17		

noted that the F ratio in the test on the AB interaction is less than unity in this case. In contrast, the F ratio for the analysis in terms of the original scale was 12.19. The profiles in terms of the transformed scale of measurement are shown in Fig. 6.4-2. The magnitude of the mean squares for the main effects, relative to the within-cell mean square, is approximately constant for both scales of measurement.

Not all interaction effects can be regarded as functions of the scale of measurement. In cases where profiles cross, or in cases where the profiles have quite different shapes, transformations on the scale of measurement will not remove interaction effects. If, however, interaction effects can be removed by transformations, there are many advantages in working with a model which contains no interaction terms. This is particularly true in the mixed model, which has both fixed and random variables, since interactions between fixed and random factors form denominators for F ratios.

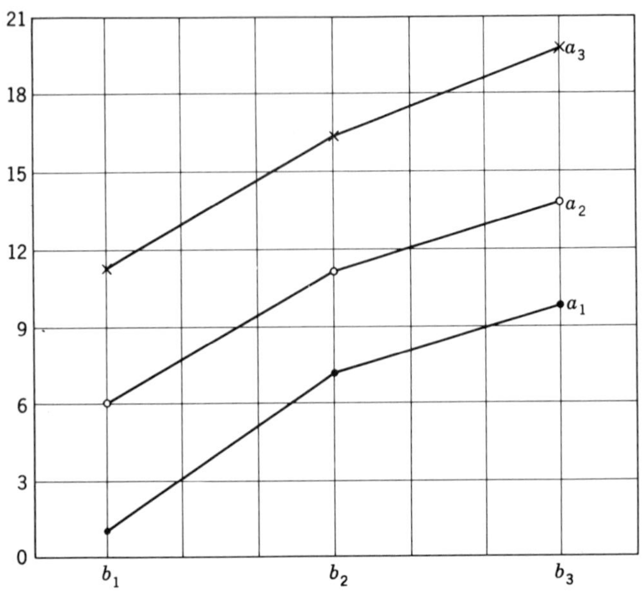

Figure 6.4-2 Profiles of factor B for levels of factor A. (Transformed scale of measurement.)

6.5 $p \times q \times r$ Factorial Experiment Having n Observations per Cell

The notation and computational procedures developed in Sec. 6.2 may be extended to three-factor experiments as well as higher-order factorial experiments. In this section the extension will be made to a $p \times q \times r$ factorial experiment. It will be assumed that there are n observations in each cell. Notation will be indicated for the special case of a $2 \times 3 \times 2$ factorial experiment. There are $pqr = 2(3)(2) = 12$ treatment combinations in this experiment. The notation for the treatment combinations is illustrated in the following table:

	c_1			c_2		
	b_1	b_2	b_3	b_1	b_2	b_3
a_1	abc_{111}	abc_{121}	abc_{131}	abc_{112}	abc_{122}	abc_{132}
a_2	abc_{211}	abc_{221}	abc_{231}	abc_{212}	abc_{222}	abc_{232}

A typical treatment combination in this experiment is designated by the notation abc_{ijk}, where i indicates the level of factor A, j the level of factor B, and k the level of factor C.

The n observations under treatment combination abc_{ijk} are represented as follows:

X_{ijk1}	X_{ijk2}	$\cdots$	X_{ijkm}	$\cdots$	X_{ijkn}

Thus the notation X_{ijkm} denotes an observation on element m under treatment combination abc_{ijk}. For the general case,

$$i = 1, 2, \ldots, p; \qquad j = 1, 2, \ldots, q; \qquad k = 1, 2, \ldots, r;$$
$$m = 1, 2, \ldots, n.$$

The sum of the n observations under treatment combination abc_{ijk} will be designated by the symbol ABC_{ijk}. Thus,

$$ABC_{ijk} = \sum_m X_{ijkm}.$$

A table of such sums will be called an ABC summary table. For the case of a $2 \times 3 \times 2$ factorial experiment, the ABC summary table has the following form:

	c_1			c_2			
	b_1	b_2	b_3	b_1	b_2	b_3	Total
a_1	ABC_{111}	ABC_{121}	ABC_{131}	ABC_{112}	ABC_{122}	ABC_{132}	A_1
a_2	ABC_{211}	ABC_{221}	ABC_{231}	ABC_{212}	ABC_{222}	ABC_{232}	A_2
Total	BC_{11}	BC_{21}	BC_{31}	BC_{12}	BC_{22}	BC_{32}	G

The column totals in this ABC summary table have the general form

$$BC_{jk} = \sum_i ABC_{ijk} = \sum_i \sum_m X_{ijkm}.$$

That is, a column total represents the sum of all observations under treatment combination bc_{jk}, the levels of factor A being disregarded. The BC summary table has the following form:

	b_1	b_2	b_3	Total
c_1	BC_{11}	BC_{21}	BC_{31}	C_1
c_2	BC_{12}	BC_{22}	BC_{32}	C_2
Total	B_1	B_2	B_3	G

Treatment combination ab_{ij} is defined to be the collection of treatment combinations $abc_{ij1}, abc_{ij2}, \ldots, abc_{ijr}$. The sum of all observations at level ab_{ij} is the sum of all observations in this collection. The sum of all observations at level ab_{ij} is thus

$$AB_{ij} = \sum_k ABC_{ijk} = \sum_k \sum_m X_{ijkm}.$$

For the case being considered,

$$AB_{11} = ABC_{111} + ABC_{112},$$
$$AB_{23} = ABC_{231} + ABC_{232}.$$

The AB summary table has the following form:

	b_1	b_2	b_3	Total
a_1	AB_{11}	AB_{12}	AB_{13}	A_1
a_2	AB_{21}	AB_{22}	AB_{23}	A_2
Total	B_1	B_2	B_3	G

The symbol AC_{ik} will be used to designate the sum of all observations at level ac_{ik}. Thus,

$$AC_{ik} = \sum_j ABC_{ijk} = \sum_j \sum_m X_{ijkm}.$$

For example, the treatment combinations at level ac_{12} for the case being considered are abc_{112}, abc_{122}, and abc_{132}. Thus,

$$AC_{12} = ABC_{112} + ABC_{122} + ABC_{132}.$$

The AC summary table has the following form:

	c_1	c_2	Total
a_1	AC_{11}	AC_{12}	A_1
a_2	AC_{21}	AC_{22}	A_2
Total	C_1	C_2	G_3

The sum of all observations at level a_i may be obtained as follows:

$$A_i = \sum_j AB_{ij} = \sum_k AC_{ik} = \sum_j \sum_k ABC_{ijk} = \sum_j \sum_k \sum_m X_{ijkm}.$$

The sum of all observations at level b_j is given by

$$B_j = \sum_i AB_{ij} = \sum_k BC_{jk} = \sum_i \sum_k ABC_{ijk} = \sum_i \sum_k \sum_m X_{ijkm}.$$

Similarly $C_k = \sum_i AC_{ik} = \sum_j BC_{jk} = \sum_i \sum_j ABC_{ijk} = \sum_i \sum_j \sum_m X_{ijkm}.$

To summarize, two-way summary tables are most readily obtained from three-way summary tables by combining levels of one of the factors. Thus the BC summary table is obtained from the ABC summary table by adding totals in the latter table which are at the same levels of factors B and C but at different levels of factor A. The AB summary table is obtained from the ABC summary table by adding totals in the latter table which are at the same levels of factors A and B but at different levels of factor C. The AC summary table is obtained in an analogous manner.

Symbols in terms of which computational formulas may be conveniently written are summarized in part i of Table 6.5-1. With the exception of symbol (2), all are obtained from either the two-way or the three-way summary tables. In each case the divisor for a computational symbol is the number of basic observations summed to obtain a term which is squared in the numerator. For example, in computational symbol (3), A_i is squared. There are nqr observations summed to obtain A_i.

There is a relatively simple method of determining the number of basic observations summed to obtain a total of the form A_i. In a

Table 6.5-1 Definition of Computational Symbols

(i)

$$
\begin{aligned}
&(1) = G^2/npqr && (6) = [\Sigma(AB_{ij})^2]/nr \\
&(2) = \Sigma X^2_{ijkm} && (7) = [\Sigma(AC_{ik})^2]/nq \\
&(3) = (\Sigma A_i^2)/nqr && (8) = [\Sigma(BC_{jk})^2]/np \\
&(4) = (\Sigma B_j^2)/npr && (9) = [\Sigma(ABC_{ijk})^2]/n \\
&(5) = (\Sigma C_k^2)/npq &&
\end{aligned}
$$

(ii)

$$
\begin{aligned}
\text{SS}_a &= nqr\Sigma(\bar{A}_i - \bar{G})^2 = (3) - (1) \\
\text{SS}_b &= npr\Sigma(\bar{B}_j - \bar{G})^2 = (4) - (1) \\
\text{SS}_c &= npq\Sigma(\bar{C}_k - \bar{G})^2 = (5) - (1) \\
\text{SS}_{ab} &= nr\Sigma(\overline{AB}_{ij} - \bar{A}_i - \bar{B}_j + \bar{G})^2 \\
&= nr\Sigma(\overline{AB}_{ij} - \bar{G})^2 - \text{SS}_a - \text{SS}_b = (6) - (3) - (4) + (1) \\
\text{SS}_{ac} &= nq\Sigma(\overline{AC}_{ik} - \bar{A}_i - \bar{C}_k + \bar{G})^2 \\
&= nq\Sigma(\overline{AC}_{ik} - \bar{G})^2 - \text{SS}_a - \text{SS}_c = (7) - (3) - (5) + (1) \\
\text{SS}_{bc} &= np\Sigma(\overline{BC}_{jk} - \bar{B}_j - \bar{C}_k + G)^2 \\
&= np\Sigma(\overline{BC}_{jk} - \bar{G})^2 - \text{SS}_b - \text{SS}_c = (8) - (4) - (5) + (1) \\
\text{SS}_{abc} &= n\Sigma(\overline{ABC}_{ijk} - \overline{AB}_{ij} - \overline{AC}_{ik} - \overline{BC}_{jk} + \bar{A}_i + \bar{B}_j + \bar{C}_k - \bar{G})^2 \\
&= n\Sigma(\overline{ABC}_{ijk} - \bar{G})^2 - \text{SS}_{ab} - \text{SS}_{ac} - \text{SS}_{bc} - \text{SS}_a - \text{SS}_b - \text{SS}_c \\
&= (9) - (6) - (7) - (8) + (3) + (4) + (5) - (1) \\
\text{SS}_{\text{w. cell}} &= \Sigma(X_{ijkm} - \overline{ABC}_{ijk})^2 = (2) - (9) \\
\text{SS}_{\text{total}} &= \Sigma(X_{ijkm} - \bar{G})^2 = (2) - (1)
\end{aligned}
$$

$p \times q \times r$ factorial experiment a basic observation is represented by the symbol X_{ijkm}. In A_i the subscripts j, k, and m are missing. The numbers of levels corresponding to these missing subscripts are, respectively, q, r, and n. The number of observations summed to obtain A_i is the product of these missing subscripts, qrn. In the total B_j, the subscripts i, k, and m are missing. The corresponding numbers of levels are p, r, and n; hence the number of observations summed is prn. In the total BC_{jk} the subscripts i and m are missing. Hence pn observations are summed in this total. In the total AC_{ik} the subscripts j and m are missing. Hence the number of observations summed in this total is qn.

Where the index of summation does not appear under the summation symbol, it is understood that the summation is over all possible terms of the form specified. For example, the notation $\Sigma(AB_{ij})^2$ indicates that the sum is over all the pq cell totals in a two-way summary table of the form AB_{ij}. Similarly, the notation $\Sigma(BC_{jk})^2$ indicates that the sum is over all the qr cell totals having the form BC_{jk}. The basic definitions of the estimates of variation due to main effects and interactions are summarized in part ii of Table 6.5-1. Corresponding computational formulas are also given.

In a three-factor factorial experiment there are various orders of simple effects. Computational formulas for these sources of variation may be obtained by specializing the symbols given in part i of Table 6.5-1. The symbol $(6a_i)$ is defined to be the equivalent of symbol (6), in which the summation is restricted to level a_i. For example,

$$(6a_1) = \frac{(AB_{11})^2 + (AB_{12})^2 + \cdots + (AB_{1q})^2}{nr},$$

$$(6a_2) = \frac{(AB_{21})^2 + (AB_{22})^2 + \cdots + (AB_{2q})^2}{nr}.$$

Similarly the symbol $(7c_k)$ is defined to be computational symbol (7), in which the summation is limited to level c_k—that is, the summation is restricted to row c_k of the BC summary table.

The computational symbol $(9a_i)$ restricts the summation in (9) to row a_i of the ABC summary table. Computational symbol $(9ab_{ij})$ restricts the summation in (9) to just those totals in which the factor A is at level a_i and factor B is at level b_j. For example,

$$(9ab_{12}) = \frac{(ABC_{121})^2 + (ABC_{122})^2 + \cdots + (ABC_{12r})^2}{n},$$

$$(9ab_{23}) = \frac{(ABC_{231})^2 + (ABC_{232})^2 + \cdots + (ABC_{23r})^2}{n}.$$

In terms of computational symbols defined in this manner, computational formulas for various orders of simple effects are given in Table 6.5-2.

The following relationships hold for the computational symbols:

$$\begin{aligned}(6) &= (6a_1) + (6a_2) + \cdots + (6a_p),\\ (7) &= (7a_1) + (7a_2) + \cdots + (7a_p),\\ (9) &= (9a_1) + (9a_2) + \cdots + (9a_p),\\ (9) &= \sum_i \sum_j (9ab_{ij}).\end{aligned}$$

Analogous relations hold for other computational symbols.

Numerical Example. The computational procedures will be illustrated by means of a $2 \times 3 \times 2$ factorial experiment. The purpose of this experiment is to evaluate the relative effectiveness of three methods of

Table 6.5-2 Computational Formulas for Simple Effects

Effects	Sum of squares
Simple interactions:	
AB for level c_k	$(9c_k) - (7c_k) - (8c_k) + (5c_k)$
AC for level b_j	$(9b_j) - (6b_j) - (8b_j) + (4b_j)$
BC for level a_i	$(9a_i) - (6a_i) - (7a_i) + (3a_i)$
Simple main effects:	
A for level b_j	$(6b_j) - (4b_j)$
A for level c_k	$(7c_k) - (5c_k)$
B for level a_i	$(6a_i) - (3a_i)$
B for level c_k	$(8c_k) - (5c_k)$
C for level a_i	$(7a_i) - (3a_i)$
C for level b_j	$(8b_j) - (4b_j)$
Simple, simple main effects:	
A for level bc_{jk}	$(9bc_{jk}) - (8bc_{jk})$
B for level ac_{ik}	$(9ac_{ik}) - (7ac_{ik})$
C for level ab_{ij}	$(9ab_{ij}) - (6ab_{ij})$

Computational checks:

$$\sum_k SS_{a \text{ for } c_k} = SS_a + SS_{ac}$$

$$\sum_k SS_{ab \text{ for } c_k} = SS_{ab} + SS_{abc}$$

$$\sum_j \sum_k SS_{a \text{ for } bc_{jk}} = SS_a + SS_{ab} + SS_{ac} + SS_{abc}$$

training (factor B). Two instructors (factor C) are used in the experiment; subjects in the experiment are classified on the basis of educational background (factor A). The plan for this experiment may be represented as follows:

	Instructor:	c_1			c_2		
	Training method:	b_1	b_2	b_3	b_1	b_2	b_3
Educational level	a_1	G_{111}	G_{121}	G_{131}	G_{112}	G_{122}	G_{132}
	a_2	G_{211}	G_{221}	G_{231}	G_{212}	G_{222}	G_{232}

In this plan G_{111} represents a group of subjects at educational level a_1 assigned to instructor c_1 to be trained under method b_1. The symbol G_{132} denotes the group of subjects at educational level a_1 assigned to instructor c_2 to be trained under method b_3. Thus each instructor teaches groups from both educational levels under each of the training methods. It will be assumed that there are 10 subjects in each of the groups, a total of 120 subjects in all.

In this experiment the methods of training (factor B) and the levels of education (factor A) will be considered fixed factors. Factor A is a classification variable included in the experiment to control potential variability in the experimental units, which is a function of level of education. (All relevant levels of the education factor must be covered if this factor is fixed.) Factor B is the treatment variable of primary interest; this factor is directly under the control of the experimenter. There is some question about whether or not factor C should be considered a fixed variable. If it is the purpose of the experiment to draw inferences about the methods of training which potentially hold for a population of instructors, of which the two instructors in the experiment can be considered a random sample, then the instructor factor is random. If inferences about the methods are to be limited to the two instructors in the experiment, then the instructor factor is fixed. Often in this type of experiment, inferences are desired about the methods over a population of specified instructors. Hence the instructor factor should be considered as a random factor, and suitable randomization procedures are required in the selection of the instructors.

The expected values for the mean squares for the case under consideration are given in Table 6.5-3. The model from which these expected values were obtained includes interaction terms with the instructor factor. According to these expected values, the test on the main effect of factor B has the form

Table 6.5-3 Expected Values of Mean Squares for Numerical Example
(A, B fixed; C random)

Effect	i	j	k	m	Expected value of mean square
α_i	0	3	2	10	$\sigma_\varepsilon^2 + 30\sigma_{\alpha\gamma}^2 + 60\sigma_\alpha^2$
β_j	2	0	2	10	$\sigma_\varepsilon^2 + 20\sigma_{\beta\gamma}^2 + 40\sigma_\beta^2$
γ_k	2	3	1	10	$\sigma_\varepsilon^2 + 60\sigma_\gamma^2$
$\alpha\beta_{ij}$	0	0	2	10	$\sigma_\varepsilon^2 + 10\sigma_{\alpha\beta\gamma}^2 + 20\sigma_{\alpha\beta}^2$
$\alpha\gamma_{ik}$	0	3	1	10	$\sigma_\varepsilon^2 + 30\sigma_{\alpha\gamma}^2$
$\beta\gamma_{jk}$	2	0	1	10	$\sigma_\varepsilon^2 + 20\sigma_{\beta\gamma}^2$
$\alpha\beta\gamma_{ijk}$	0	0	1	10	$\sigma_\varepsilon^2 + 10\sigma_{\alpha\beta\gamma}^2$
$\varepsilon_{m(ijk)}$	1	1	1	1	σ_ε^2

$$F = \frac{\text{MS}_b}{\text{MS}_{bc}}.$$

This F ratio has degrees of freedom $[(q-1), (q-1)(r-1)]$, which in this case is (2,2). When the denominator of an F ratio has only two degrees of freedom, the power of the resulting test is extremely low. This F ratio does not provide a sufficiently powerful test of the main effects of factor B to be of much practical use.

If, however, it can be assumed that interactions with the random factor (C) are negligible relative to the other uncontrolled sources of variation which are included in the experimental error, then interactions with factor C may be dropped from the original model. In terms of a model which does not include such interactions, relatively powerful tests on factor B are available. Inspection of Table 6.5-3 indicates that MS_{ac}, MS_{bc}, MS_{abc}, and $MS_{w.cell}$ are all estimates of variance due to experimental error if interactions with factor C are not included in the model. Preliminary tests on the model may be made to check on whether or not such interactions may be dropped.

Suppose that the ABC summary table for the data obtained in the experiment is that given in part i of Table 6.5-4. Each of the entries in this

Table 6.5-4 Data for Numerical Example

(i)

ABC summary table

	c_1			c_2			
	b_1	b_2	b_3	b_1	b_2	b_3	Total
a_1	20	30	12	16	33	8	119
a_2	36	38	40	40	44	42	240
Total	56	68	52	56	77	50	359

AB summary table

	b_1	b_2	b_3	Total
a_1	36	63	20	119
a_2	76	82	82	240
Total	112	145	102	359

BC summary table

	b_1	b_2	b_3	Total
c_1	56	68	52	176
c_2	56	77	50	183
Total	112	145	102	359

AC summary table

	c_1	c_2	Total
a_1	62	57	119
a_2	114	126	240
Total	176	183	359

(ii)

$$
\begin{aligned}
(1) &= (359^2)/120 &&= 1074.01\\
(2) &= \text{(not available from above data)} &&= 1360\\
(3) &= (119^2 + 240^2)/60 &&= 1196.02\\
(4) &= (112^2 + 145^2 + 102^2)/40 &&= 1099.32\\
(5) &= (176^2 + 183^2)/60 &&= 1074.42\\
(6) &= (36^2 + 63^2 + 20^2 + 76^2 + 82^2 + 82^2)/20 &&= 1244.45\\
(7) &= (62^2 + 57^2 + 114^2 + 126^2)/30 &&= 1198.83\\
(8) &= (56^2 + 68^2 + 52^2 + 56^2 + 77^2 + 50^2)/20 &&= 1101.45\\
(9) &= (20^2 + 30^2 + \cdots + 44^2 + 42^2)/10 &&= 1249.30
\end{aligned}
$$

(iii)

$$
\begin{aligned}
SS_a &= (3) - (1) = 122.01 & SS_{ab} &= (6) - (3) - (4) + (1) = 23.12\\
SS_b &= (4) - (1) = 25.31 & SS_{ac} &= (7) - (3) - (5) + (1) = 2.40\\
SS_c &= (5) - (1) = 0.41 & SS_{bc} &= (8) - (4) - (5) + (1) = 1.72
\end{aligned}
$$

$$
\begin{aligned}
SS_{abc} &= (9) - (6) - (7) - (8) + (3) + (4) + (5) - (1) = 0.32\\
SS_{w.\,cell} &= (2) - (9) = 110.70\\
SS_{total} &= (2) - (1) = 285.99
\end{aligned}
$$

table is the sum of the 10 criterion scores for the corresponding group of subjects. For example, the entry in cell abc_{132} is the sum of the 10 criterion scores for the subjects in group G_{132}, that is, the subjects at education level a_1, trained under method b_3 by instructor c_2. The two-way summary tables given in part i are obtained from the three-way summary table. For example, the entry ab_{11} in the AB summary table is given by

$$\begin{aligned} AB_{11} &= ABC_{111} + ABC_{112} \\ &= \quad 20 \quad + \quad 16 \quad = 36. \end{aligned}$$

The computational symbols defined in part i of Table 6.5-1 are obtained in part ii in Table 6.5-4. Data for all these computations, with the exception of symbol (2), are contained in part i. Symbol (2) is obtained from the individual criterion scores; the latter are not given in this table. The computation of the sums of squares is completed in part iii.

The analysis of variance is summarized in Table 6.5-5. Preliminary tests on the model will be made on the interactions with factor C (instructors) before proceeding with other tests. According to the expected values of the mean squares given in Table 6.5-3, tests on interactions with factor C all have $\text{MS}_{\text{w.cell}}$ as a denominator. These tests have the following form:

$$F = \frac{\text{MS}_{abc}}{\text{MS}_{\text{w.cell}}} = \frac{0.16}{1.02} = 0.15,$$

$$F = \frac{\text{MS}_{ac}}{\text{MS}_{\text{w.cell}}} = \frac{2.40}{1.02} = 2.33,$$

$$F = \frac{\text{MS}_{bc}}{\text{MS}_{\text{w.cell}}} = \frac{0.86}{1.02} = 0.85.$$

Only the F ratio for the test on the AC interaction is greater than unity. By use of the .10 level of significance, the critical value for the latter test is $F_{.90}(1,108) = 2.76$.

Table 6.5-5 Summary of Analysis of Variance

Source of variation	SS	df	MS
A (level of education)	122.01	$p - 1 = 1$	122.01
B (methods of training)	25.31	$q - 1 = 2$	12.66
C (instructors)	0.41	$r - 1 = 1$	0.41
AB	23.12	$(p - 1)(q - 1) = 2$	11.56
AC	2.40	$(p - 1)(r - 1) = 1$	2.40
BC	1.72	$(q - 1)(r - 1) = 2$	0.86
ABC	0.32	$(p - 1)(q - 1)(r - 1) = 2$	0.16
Within cell (experimental error)	110.70	$pqr(n - 1) = 108$	1.02
Total	285.99	119	

The outcome of these preliminary tests on the model does not contradict the hypothesis that interactions with factor C may be considered negligible. (The AC interaction is a borderline case.) On a priori grounds, if the instructors are carefully trained, interaction effects with instructors may often be kept relatively small. On these bases, the decision is made to drop interactions with factor C from the model. The expected values corresponding to this revised model are obtained from Table 6.5-3 by dropping the terms $\sigma^2_{\alpha\gamma}$, $\sigma^2_{\beta\gamma}$, and $\sigma^2_{\alpha\beta\gamma}$ from the expected values in this table. Since the degrees of freedom for the within-cell variation (108) are large relative to the degrees of freedom for the interactions with factor C, which total 5, pooling these interactions with the within-cell variation will not appreciably effect the magnitude of $\text{MS}_{\text{w.cell}}$ or the degrees of freedom for the resulting pooled error term. Hence $\text{MS}_{\text{w.cell}}$ is used under the revised model for final tests on factors A and B. These tests have the following form:

$$F = \frac{\text{MS}_{ab}}{\text{MS}_{\text{w.cell}}} = 11.33, \qquad F_{.99}(2,108) = 4.82;$$

$$F = \frac{\text{MS}_{b}}{\text{MS}_{\text{w.cell}}} = 12.41, \qquad F_{.99}(2,108) = 4.82;$$

$$F = \frac{\text{MS}_{a}}{\text{MS}_{\text{w.cell}}} = 119.62, \qquad F_{.99}(1,108) = 6.90.$$

Because of the significant AB interaction, care must be taken in interpreting the main effects due to factor B. The manner in which educational level is related to method of training is most readily shown by the profiles of the simple effects for the methods at each of the levels of education. These profiles are drawn in Fig. 6.5-1. Data for these profiles are obtained from the AB summary table. Inspection of these profiles indicates that differences between the methods of training for groups at level a_2 are not so marked as the corresponding differences for groups at level a_1.

Variation due to differences between training methods for groups at level a_1 is given by (data are obtained from row a_1 of the AB summary table)

$$\text{SS}_{b \text{ for } a_1} = \frac{36^2 + 63^2 + 20^2}{20} - \frac{119^2}{60} = 47.23.$$

This source of variation has $q - 1 = 2$ degrees of freedom. Hence the mean square of the simple effect of factor B for level a_1 is

$$\text{MS}_{b \text{ for } a_1} = \frac{47.23}{2} = 23.62.$$

To test the hypothesis of no difference between the methods of training for groups at educational level a_1,

$$F = \frac{\text{MS}_{b \text{ for } a_1}}{\text{MS}_{\text{w. cell}}} = \frac{23.62}{1.02} = 23.15.$$

The degrees of freedom for this F ratio are (2,108). The data clearly indicate a significant difference between the methods for level a_1. The test on the simple effects of methods for level a_2 indicates no significant difference between the methods; the F for this test will be found to be $F = 0.60/1.02$.

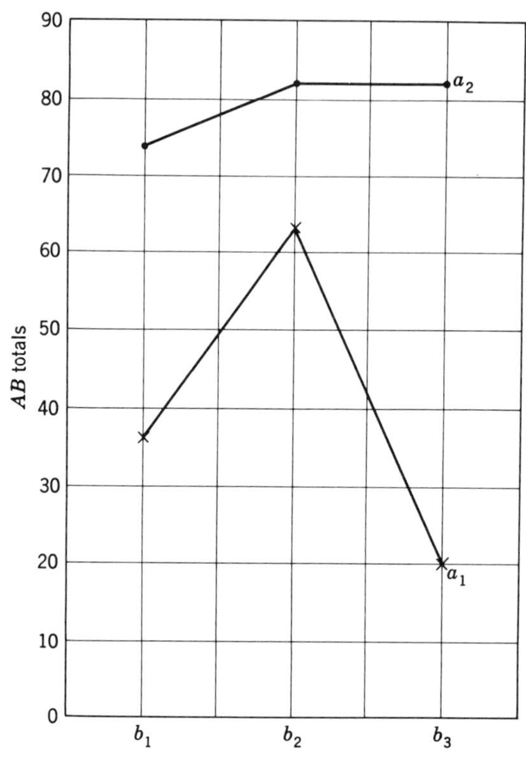

Figure 6.5-1 Profiles of simple main effects for training methods.

The levels of factor C (instructors) in this design may be considered to be replications. As indicated by the expected values for the mean squares, the proper denominator for tests on fixed effects is the corresponding interaction with replications, provided that such interactions cannot be pooled. When tests are made by using a pooled error term (this is essentially what has been done in the example that has just been considered), it is implicitly assumed that interactions with replications (instructors) do not exist.

There are potentially many reasons for the presence of interaction in this kind of experiment. Inspection of the $A \times B$ summary table indicates that groups at level a_2 are uniformly good under all the training methods. In part, this may be a function of a ceiling effect on the criterion. If the latter is the case, then the interaction is an artifact of the way in which performance is measured. Care is required in constructing the criterion to avoid this kind of artifact.

Alternative Computational Procedures for Three-factor Interactions. In Sec. 5.10 the relationship between two-factor and three-factor interactions was indicated. For a $2 \times 3 \times 2$ factorial experiment,

$$SS_{abc} = SS_{ab \text{ for } c_1} + SS_{ab \text{ for } c_2} - SS_{ab}.$$

In general, a three-factor interaction will be zero whenever the sum of the simple two-factor interactions is equal to corresponding overall two-factor interaction. Thus the three-factor interaction is a measure of the additivity of simple two-factor interactions.

The relationship given above provides a method for computing the three-factor interaction. For the data in Table 6.5-4, one may compute the following computational symbols:

$$(9c_1) = \frac{20^2 + 30^2 + 12^2 + 36^2 + 38^2 + 40^2}{10} = 578.40\,,$$

$$(9c_2) = \frac{16^2 + 33^2 + 8^2 + 40^2 + 44^2 + 42^2}{10} = 670.90\,,$$

$$(8c_1) = \frac{56^2 + 68^2 + 52^2}{20} = 523.20\,,$$

$$(8c_2) = \frac{56^2 + 77^2 + 50^2}{20} = 578.25\,,$$

$$(7c_1) = \frac{62^2 + 114^2}{30} = 561.33\,,$$

$$(7c_2) = \frac{57^2 + 126^2}{30} = 637.50\,,$$

$$(5c_1) = \frac{176^2}{60} = 516.27, \qquad (5c_2) = \frac{183^2}{60} = 558.15\,.$$

The variation due to the simple AB interaction for level c_1 is

$$(9c_1) - (7c_1) - (8c_1) + (5c_1) = 10.14.$$

The corresponding variation for level c_2 is

$$(9c_2) - (7c_2) - (8c_2) + (5c_2) = 13.30.$$

The variation due to the overall AB interaction was found to be 23.12. Hence the variation due to ABC is

$$\text{SS}_{abc} = 10.14 + 13.30 - 23.12 = 0.32.$$

6.6 Computational Procedures for Nested Factors

Factorial designs in which one or more factors are nested were discussed in Sec. 5.12. Computational procedures in which one factor is nested under a second will be illustrated from the case of a $p \times q \times r$ factorial experiment having n observations in each cell. Assume that factor B is nested under factor A. The analysis of variance for this case generally takes the following form:

Source of variation	df
A	$p - 1$
B within A	$p(q - 1)$
C	$r - 1$
AC	$(p - 1)(r - 1)$
(B within A) $\times$ C	$p(q - 1)(r - 1)$
Within cell	$pqr(n - 1)$

The variation due to B within A is defined to be

$$\text{SS}_{b(a)} = nr\sum_i\sum_j(\overline{AB}_{ij} - \bar{A}_i)^2.$$

This source of variation is actually a sum of simple main effects of factor B at each level of factor A. In terms of the computational symbols defined in Table 6.5-1,

$$\text{SS}_{b(a)} = (6) - (3).$$

For a design in which factor B is not nested under factor A,

		df
SS_b	$= (4) - (1)$	$q - 1$
SS_{ab}	$= (6) - (3) - (4) + (1)$	$(p - 1)(q - 1)$
Sum	$= (6) - (3)$	$p(q - 1)$

Thus $\text{SS}_{b(a)}$ in a design in which factor B is nested under factor A is numerically equal to $\text{SS}_b + \text{SS}_{ab}$ in the corresponding factorial design in which factor B is not nested under factor A.

The variation due to the interaction (B within A) $\times$ C is defined to be

$$SS_{b(a)c} = n\sum\sum\sum(\overline{ABC}_{ijk} - \overline{AB}_{ij} - \overline{AC}_{ik} + \bar{A}_i)^2.$$

This source of variation is actually a sum of simple interactions. In terms of the computational symbols defined in Table 6.5-1,

$$SS_{b(a)c} = (9) - (6) - (7) + (3).$$

For a $p \times q \times r$ factorial experiment in which there is no nested factor,

		df
SS_{bc}	$= (8) - (4) - (5) + (1)$	$(q-1)(r-1)$
SS_{abc}	$= (9) - (6) - (7) - (8) + (3) + (4) + (5) - (1)$	$(p-1)(q-1)(r-1)$
Sum	$= (9) - (6) - (7) + (3)$	$p(q-1)(r-1)$

Thus $SS_{b(a)c}$ is numerically equal to $SS_{bc} + SS_{abc}$. In general, if factor V is nested under factor U,

$$SS_{v(u)w} = SS_{vw} + SS_{uvw}.$$

In a four-factor experiment in which factor C is nested under both factors A and B,

$$SS_{c(ab)} = SS_c + SS_{ac} + SS_{bc} + SS_{abc},$$

$$SS_{c(ab)d} = SS_{cd} + SS_{acd} + SS_{bcd} + SS_{abcd}.$$

Returning to a three-factor factorial experiment, consider the case of a $2 \times 2 \times 3$ factorial experiment having five observations per cell. Assume that factor B is nested under factor A and that factor B is a random factor. Assume also that factors A and C are fixed factors. Under these assumptions, the expected values of the mean squares for this design are given in Table 6.6-1. From these expected values it will be noted that the test

Table 6.6-1 Expected Values of Mean Squares
(Factor B nested under factor A; A and C fixed, B random)

Effect	i	j	k	m	E(MS)
α_i	0	2	3	5	$\sigma_\varepsilon^2 + 15\sigma_{\beta(\alpha)}^2 + 30\sigma_\alpha^2$
$\beta_{j(i)}$	1	1	3	5	$\sigma_\varepsilon^2 + 15\sigma_{\beta(\alpha)}^2$
γ_k	2	2	0	5	$\sigma_\varepsilon^2 + 5\sigma_{\beta(\alpha)\gamma}^2 + 20\sigma_\gamma^2$
$\alpha\gamma_{ik}$	0	2	0	5	$\sigma_\varepsilon^2 + 5\sigma_{\beta(\alpha)\gamma}^2 + 10\sigma_{\alpha\gamma}^2$
$\beta\gamma_{j(i)k}$	1	1	0	5	$\sigma_\varepsilon^2 + 5\sigma_{\beta(\alpha)\gamma}^2$
$\varepsilon_{m(ijk)}$	1	1	1	1	σ_ε^2

on the main effects due to factor A has the form

$$F = \frac{\text{MS}_a}{\text{MS}_{b(a)}}.$$

If this denominator has relatively few degrees of freedom, the power of the test will be low. Preliminary tests on the model are often called for in this situation.

Numerical Example. A $2 \times 2 \times 3$ factorial experiment having five observations per cell will be used to illustrate the computational procedures. To make this example concrete, suppose that the experiment has the following form:

Drugs:	a_1						a_2					
Hospitals:	$b_1(a_1)$			$b_2(a_1)$			$b_1(a_2)$			$b_2(a_2)$		
Category of patients:	c_1	c_2	c_3	c_1	c_2	c_3	c_1	c_2	c_3	c_1	c_2	c_3
n:	5	5	5	5	5	5	5	5	5	5	5	5

Schematically, one has the following crossing and nesting relationships among the factors:

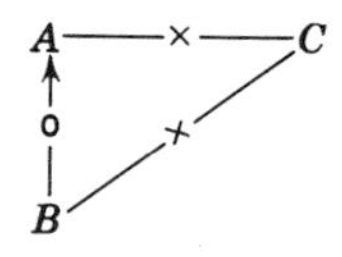

Suppose that the purpose of this experiment is to test the relative effectiveness of two drugs on patients in different diagnostic categories. Patients from hospitals $b_{1(a_1)}$ and $b_{2(a_1)}$ are given drug a_1; patients from hospitals $b_{1(a_2)}$ and $b_{2(a_2)}$ are given drug a_2. Since only one of the drugs under study is administered within a hospital, the hospital factor is nested under the drug factor. The diagnostic categories are considered to be comparable across all hospitals; hence factor C is not nested. Only a random sample of the population of hospitals about which inferences are to be drawn is included in the experiment; hence factor B is random.

Suppose that the AB summary table for the data obtained in this experiment is that in part i of Table 6.6-2. With the exception of symbol (2), which is computed from the individual observations, data for the computations are given in part i. (The definitions of these symbols appear in Table 6.5-1.) Symbols (4) and (8) are not required for this case. The analysis of variance is summarized in Table 6.6-3.

The expected values of the mean squares are given in Table 6.6-1. Before testing the main effects and interaction of the fixed factors, pre-

Table 6.6-2 Numerical Example

(i)

ABC summary table:

	a_1		a_2		Total
	$b_{1(a_1)}$	$b_{2(a_1)}$	$b_{1(a_2)}$	$b_{2(a_2)}$	
c_1	15	18	30	35	98
c_2	22	25	24	21	92
c_3	38	41	10	14	103
Total	75	84	64	70	293

AC summary table:

	a_1	a_2	Total
c_1	33	65	98
c_2	47	45	92
c_3	79	24	103
Total	159	134	293

(ii)

$(1) = (293)^2/60$	$= 1430.82$
(2) (not obtained from above summary tables)	$= 1690$
$(3) = (159^2 + 134^2)/30$	$= 1441.23$
$(5) = (98^2 + 92^2 + 103)^2/20$	$= 1433.85$
$(6) = (75^2 + 84^2 + 64^2 + 70^2)/15$	$= 1445.13$
$(7) = (33^2 + 47^2 + \cdots + 45^2 + 24^2)/10$	$= 1636.50$
$(9) = (15^2 + 22^2 + \cdots + 21^2 + 14^2)/5$	$= 1644.20$

Table 6.6-3 Summary of Analysis of Variance

(i)

$$\begin{aligned}
SS_a &= (3) - (1) &&= 10.41\\
SS_{b(a)} &= (6) - (3) &&= 3.90\\
SS_c &= (5) - (1) &&= 3.03\\
SS_{ac} &= (7) - (3) - (5) + (1) &&= 192.24\\
SS_{b(a)c} &= (9) - (6) - (7) + (3) &&= 3.80\\
SS_{w.\,cell} &= (2) - (9) &&= 45.80
\end{aligned}$$

(ii)

Source of variation	SS	df	MS
A (drugs)	10.41	1	10.41
B (hospitals within *A*)	3.90	2	1.95
C (categories)	3.03	2	1.52
AC	192.24	2	96.12
$B(A) \times C$	3.80	4	0.95
Within cell	45.80	48	0.95
Pooled error	53.50	54	0.99

liminary tests on the model are made with respect to factor B and its interaction with factor C. These preliminary tests will be made at the .10 level of significance. Inspection of the expected values of the mean squares indicates that the preliminary tests have the following form:

$$F = \frac{\text{MS}_{b(a)}}{\text{MS}_{\text{w.cell}}} = 2.05, \qquad F_{.90}(2,48) = 2.42;$$

$$F = \frac{\text{MS}_{b(a)c}}{\text{MS}_{\text{w.cell}}} = 1.00, \qquad F_{.90}(4,48) = 2.07.$$

Neither of the F ratios exceeds specified critical values. Hence variation due to $B(A)$ and $B(A) \times C$ is pooled with the within-cell variation. Thus,

$$\text{SS}_{\text{pooled error}} = \text{SS}_{b(a)} + \text{SS}_{b(a)c} + \text{SS}_{\text{w.cell}}.$$

The degrees of freedom for this term are the sum of the respective degrees of freedom for the parts.

The denominator for all final tests is $\text{MS}_{\text{pooled error}}$. For this case the final tests are

$$F = \frac{\text{MS}_{ac}}{\text{MS}_{\text{pooled error}}} = 102.26, \qquad F_{.99}(2,54) = 5.00;$$

$$F = \frac{\text{MS}_{a}}{\text{MS}_{\text{pooled error}}} = 11.07, \qquad F_{.99}(1,54) = 7.10;$$

$$F = \frac{\text{MS}_{c}}{\text{MS}_{\text{pooled error}}} = 1.62, \qquad F_{.99}(2,54) = 5.00.$$

In spite of the significant AC interaction, the main effect for factor A is significant. An analysis of the simple effects is required for an adequate interpretation of the effects of the drugs. Inspection of the AC summary table indicates that drug a_2 has the higher criterion total for category c_1; there appears to be little difference between the criterion scores for category c_2; drug a_1 has the higher criterion total for category c_3. Formal tests on these last statements have the following form:

$$F = \frac{(AC_{11} - AC_{21})^2}{2nq\text{MS}_{\text{pooled error}}} = \frac{(33 - 65)^2}{2(5)(2)(0.99)} = 54.47,$$

$$F = \frac{(AC_{12} - AC_{22})^2}{2nq\text{MS}_{\text{pooled error}}} = \frac{(47 - 45)^2}{2(5)(2)(0.99)} = 0.21,$$

$$F = \frac{(AC_{13} - AC_{23})^2}{2nq\text{MS}_{\text{pooled error}}} = \frac{(79 - 24)^2}{2(5)(2)(0.99)} = 160.90.$$

6.7 Factorial Experiment with a Single Control Group

A design closely related to one reported by Levison and Zeigler (1959) will be used to illustrate the material that will be considered in this section. The purpose of this experiment is to test the effect of amount and time of irradiation upon subsequent learning ability. Different dosages (factor B) of irradiation are administered at different age levels (factor A). When all subjects reach a specified age, they are given a series of learning tasks. Separate analyses are made of the criterion scores for each task. (A multivariate analysis of variance might also have been made.) The plan for the experiment may be represented as follows:

		Dosage of irradiation		
		b_1	b_2	
Age at which irradiation is administered	a_1	G_{11}	G_{12}	
	a_2	G_{21}	G_{22}	
	a_3	G_{31}	G_{32}	$n = 10$
	a_4	G_{41}	G_{42}	

G_{ij} represents a group of subjects under treatment ab_{ij}. Suppose that each group contains $n = 10$ subjects. In addition to these eight groups there is a group G_0, having $n_0 = 20$ subjects, which receives no irradiation treatment. G_0 represents the control group. In this case the control condition may be considered to represent level b_0 of the dosage dimension; however, the control condition cannot be classified along the age dimension.

Thus there are nine groups in all, eight groups in the cells of a 4×2 factorial experiment plus a control group. For the general case there will be $pq + 1$ groups. The analysis of variance for this experimental plan may take the following form:

Source of variation		df
Between cell		$(pq + 1) - 1$
Control vs. all others	1	
A (age)	$p - 1$	
B (dosage)	$q - 1$	
AB	$(p - 1)(q - 1)$	
Within cell		$pq(n - 1) + (n_0 - 1)$

The factorial part of this plan follows the usual computation procedures for any factorial experiment of this type. The contrast between the control group and all experimental groups is given by

$$C = pq\bar{C}_0 - \Sigma\overline{AB}_{ij} = \frac{pqC_0}{n_0} - \frac{\Sigma AB_{ij}}{n},$$

where C_0 is the sum of the observations in the control group. The corresponding mean square is

$$\text{MS}_{\text{control vs. all others}} = \frac{C^2}{[(pq)^2/n_0] + (pq/n)}.$$

The within-cell variation is obtained by pooling the within-cell variation from the factorial part of the experiment with the within-cell variation from the control group. The between-cell variation is given by

$$\text{SS}_{\text{b.cell}} = \frac{C_0^2}{n_0} + \frac{\Sigma(AB_{ij})^2}{n} - \frac{(G + C_0)^2}{npq + n_0},$$

where G is the sum of the observations in the factorial part of the experiment and C_0 is the sum of the observations in the control group.

In this case the control group may be considered as a zero-level condition of the dosage factor. Hence, as an alternative to the variation due to the main effects of factor B in the factorial part of the experiment, the following may be computed:

$$\text{SS}_b = \frac{C_0^2}{n_0} + \frac{\Sigma B_j^2}{np} - \frac{(G + C_0)^2}{npq + n_0},$$

where G is the sum of all observations in the factorial part of the experiment. In this type of experiment it is also of interest to contrast the control group with each of the experimental groups or with selected sets. The procedure described in Sec. 3.10 may be adapted for this purpose. In this case the t statistic has the form

$$t = \frac{\bar{C}_0 - \overline{AB}_{ij}}{\sqrt{\text{MS}_{\text{w.cell}}[(1/n_0) + (1/n)]}}.$$

Critical values (Dunnett) are given in Table C.6. The degrees of freedom for this statistic are those of $\text{MS}_{\text{w.cell}}$; k corresponds to the total number of groups, which in this case is $pq + 1$. If comparisons are restricted to selected sets, then k is the number of groups in a set.

Numerical Example. Suppose data obtained from the eight experimental groups and the control in the experiment which has just been described are those given in Table 6.7-1. The computational symbols at the right of part i are defined in Table 6.5-1. [Data for the computation of symbol (2) are not given.] Summary data for the control group are given in part ii. The between-cell variation (including the control) is computed in part iii.

As a partial check for homogeneity of the within-cell variance for the factorial and the control parts of the experiment, one has for the factorial part

$$\text{MS}_{\text{w.cell}} = \frac{1092.40}{8(9)} = 15.17.$$

Table 6.7-1 Numerical Example

(i) *AB* summary table ($n_0 = 20$; $n = 10$):

b_0		b_1	b_2	Total		
	a_1	380	310	690	A_1	(1) = 143,143.20
	a_2	405	340	745	A_2	(2) = 148,129
	a_3	485	470	955	A_3	(3) = 146,559.30
	a_4	504	490	994	A_4	(4) = 143,479.40
						(5) = 147,036.60
1000	Total	1774	1610	3384		
C_0		B_1	B_2	G		

Data for control group ($n_0 = 20$):

(ii) $$C_0 = \Sigma X = 1000, \quad \Sigma X^2 = 50{,}300, \quad SS_0 = 50{,}300 - \frac{(1000)^2}{20} = 300.00$$

(iii) $$SS_{b.cell} = \frac{(1000)^2}{20} + 147036.60 - \frac{(3384 + 1000)^2}{100} = 4842.04$$

(iv) $$SS_{control\ vs.\ all\ others} = \frac{[8(50) - 338.40]^2}{\frac{64}{20} + \frac{8}{10}} = 948.64$$

(v)

Source of variation	SS	df	MS	F
Between cell	4842.04	8		
Control vs. all others	948.64	1	948.64	62.00
A (age)	3416.10	3	1138.70	379.57
B (dosage)	336.20	1	336.20	21.97
AB	141.10	3	47.03	3.07
Within cell 1092.40 + 300.00 =	1392.40	91	15.30	

For the control group,

$$MS_{w.cell} = \frac{300.00}{19} = 15.79.$$

For these data, there is little question about the appropriateness of pooling the within-cell variations.

Under the definition of the variation due to main effects of factor *B* (dosage), which includes the control group as the zero level,

$$SS_b' = \frac{(1000)^2}{20} + \frac{(1774)^2}{40} + \frac{(1610)^2}{40} - \frac{(4384)^2}{100} = 1284.84.$$

Thus, $$MS_b' = \frac{1284.84}{2} = 642.42 \qquad \text{and} \qquad F = \frac{MS_b'}{MS_{w.cell}} = 41.99.$$

The corresponding *F* ratio in the analysis-of-variance table is 21.97.

The test on the interaction indicates that this source of variation is significantly greater than zero, $F_{.95}(3,91) = 2.71$. The profiles for factor A at levels b_1 and b_2 as well as the control are plotted in Fig. 6.7-1.

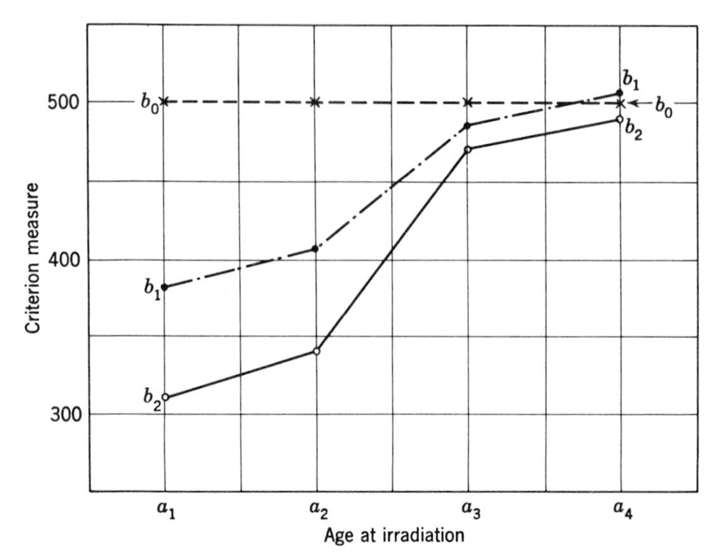

Figure 6.7-1 Profiles for different dosages.

(Data for the control group are in units which are comparable with those in the experimental groups.) Inspection of these profiles indicates relatively large differences between the control group and the groups that were irradiated at ages a_1 and a_2. There is relatively little difference between the control group and groups irradiated at ages a_3 and a_4. Further, for groups given irradiation at ages a_1 and a_2, the groups given the larger dosage (b_2) showed the greater decrement in performance. A formal test on the latter statement uses the statistic

$$F = \frac{[AB_{11} + AB_{21} - AB_{12} - AB_{22}]^2}{4n\text{MS}_{\text{w.cell}}}$$

$$= \frac{[380 + 405 - 310 - 340]^2}{40(15.30)} = 29.78.$$

The critical value for a .05-level (two-tailed) test on this comparison is $F_{.95}(1,91) = 3.95$.

To test the hypothesis that groups given irradiation at ages a_3 and a_4 do not differ from the control group, one may use the comparison

$$C = 2\bar{C}_0 - \bar{A}_3 - \bar{A}_4$$

$$= 2(50.00) - 47.75 - 49.70 = 2.55.$$

The mean square corresponding to this comparison is

$$\text{MS}_C = \frac{C^2}{(4/n_0) + (2/nq)} = \frac{(2.55)^2}{(4/20) + (2/20)} = 21.67.$$

The F ratio is given by

$$F = \frac{\text{MS}_C}{\text{MS}_{\text{w.cell}}} = \frac{21.67}{15.30} = 1.42.$$

For a (two-tailed) test at the .05-level of significance, the critical value for this statistic is $F_{.95}(1,91) = 3.95$. Thus the data indicate that there is no statistically significant difference in performance decrement between the groups irradiated at ages a_3 and a_4 and the control group with respect to the criterion of performance measured.

Additional suggestions for analyzing data of the kind discussed in this section are given in Kempthorne (1952, pp. 364–369).

6.8 Test for Nonadditivity

The assumptions underlying this test were discussed in Sec. 5.20. It should be remembered that only one of several possible sources of nonadditivity is checked in this test, namely, that source which is defined by the linear by linear cross product $\alpha_i\beta_j$. The numerical details of this test will be considered in this section.

In some cases which arise in practice, a factorial experiment may have only one observation in each cell. If the strictly additive model (no interaction effects) is appropriate, then what are computationally equivalent to interaction effects may be used as estimates of the experimental error. Computational procedures will be illustrated for the case of a 3×4 factorial experiment having one observation per cell. Suppose that the basic data for this illustration are those given at the left of part i in Table 6.8-1.

In the column headed $\bar{A}_i$, the mean of the observations in the corresponding row is entered. In the column headed c_i, an entry has the form $\bar{A}_i - \bar{G}$. For example,

$$c_1 = 14 - 11 = 3,$$
$$c_2 = 9 - 11 = -2.$$

The entries in the row headed c_j have the form $\bar{B}_j - \bar{G}$.

An entry in the column headed d_i is a weighted sum of the entries in the corresponding row, the weights being respective entries in row c_j. For example,

$$d_1 = (-6)(8) + (-5)(12) + (2)(16) + (9)(20) = 104,$$
$$d_2 = (-6)(2) + (-5)(2) + (2)(14) + (9)(18) = 168,$$
$$d_3 = (-6)(5) + (-5)(4) + (2)(9) + (9)(22) = 166.$$

Table 6.8-1 Numerical Example

		b_1	b_2	b_3	b_4	Sum	$\bar{A}_i$	$c_i = \bar{A}_i - \bar{G}$	$\sum_j c_j X_{ij} = d_i$
(i)	a_1	8	12	16	20	56	14	3	104
	a_2	2	2	14	18	36	9	−2	168
	a_3	5	4	9	22	40	10	−1	166
	Sum	15	18	39	60	132	$11 = \bar{G}$		$\sum_i c_i d_i = -190$
	$\bar{B}_j$	5	6	13	20			$\Sigma c_i^2 = 14$	
	$c_j = \bar{B}_j - \bar{G}$	−6	−5	2	9		$\Sigma c_j^2 = 146$		

(ii) (1) = 1452 (2) = 1998 (3) = 1508 (4) = 1890

$$\text{SS}_{\text{nonadd}} = \frac{(\Sigma c_i d_i)^2}{(\Sigma c_i^2)(\Sigma c_j^2)} = \frac{(-190)^2}{(14)(146)} = 17.66$$

	Source of variation	SS	df	MS	F
(iii)	A	56.00	2		
	B	438.00	3		
	AB	52.00	6		
	Nonadd 17.66		1	17.66	2.57
	Balance 34.34		5	6.87	

The numerical value of the comparison for nonadditivity is

$$\sum_i c_i(\sum_j c_j X_{ij}) = \sum_i c_i d_i$$
$$= (3)(104) + (-2)(168) + (-1)(166) = -190.$$

The sum of squares corresponding to this comparison is computed in part ii. The numerical values for the computational symbols defined in Table 6.2-1 are also given in part ii.

As a partial check on the computation of $\text{SS}_{\text{nonadd}}$, the following relationships must hold:

$$pq(\Sigma c_i^2)(\Sigma c_j^2) = (\text{SS}_a)(\text{SS}_b),$$
$$8(14)(146) = (56)(438),$$
$$24{,}528 = 24{,}528.$$

The balance, or residual variation, is given by

$$\text{SS}_{\text{bal}} = \text{SS}_{ab} - \text{SS}_{\text{nonadd}}.$$

Since $\text{SS}_{\text{nonadd}}$ is a component having a single degree of freedom, SS_{bal} has $(p - 1)(q - 1) - 1$ degrees of freedom. The F ratio in the test for nonadditivity has the form

$$F = \frac{\text{MS}_{\text{nonadd}}}{\text{MS}_{\text{bal}}} = \frac{17.66}{6.87} = 2.57.$$

If the decision rule leads to accepting the hypothesis of additivity, then the strictly additive model is used in subsequent analyses. On the other hand, if this hypothesis is rejected, the more complete model is used in subsequent analyses. The latter model is generally the more conservative in the sense that higher F ratios are required for significance at a specified level. If the level of significance is set at a numerically high value (say, $\alpha = .25$ rather than $\alpha = .05$ or $.01$), the type 2 error becomes relatively low. In this case low type 2 error implies low probability of using the additive model when in fact it is inappropriate.

If this test is made at the .25-level of significance, the critical value is $F_{.75}(1,5) = 1.69$. Since the observed value of the F statistic exceeds this critical value, the hypothesis of a strictly additive model is rejected at the .25-level of significance. If, however, MS_{ab} is used as the denominator in testing main effects, the resulting test will be biased in the direction of giving too few "significant" results.

There is an alternative method for computing SS_{nonadd} which lends itself more readily to direct generalization to higher-order interaction effects. As a first step in the computation of the comparison desired, one sets up a table in which the entry in cell ij is $c_i c_j$. For the data in Table 6.8-1 the resulting table is as follows:

	b_1	b_2	b_3	b_4	Total
a_1	−18	−15	6	27	0
a_2	12	10	−4	−18	0
a_3	6	5	−2	−9	0
Total	0	0	0	0	

The entry in cell $ab_{11} = (3)(-6) = -18$; the entry $ab_{12} = (3)(-5) = -15$. As a check on the numerical work, each row total must be zero; each column total must also be zero. Since the row sums and the column sums are zero, the entries in this table define a comparison which belongs to the interaction rather than to either of the main effects.

The comparison in the test for nonadditivity is a weighted sum of the data in the upper left-hand portion of part i of Table 6.8-1, the weight being the corresponding cell entry in the table given above. Thus,

$$\Sigma\Sigma c_i c_j X_{ij} = (-18)(8) + (-15)(12) + \cdots + (-2)(9) + (-9)(22)$$
$$= -190.$$

As a check on the numerical work,

$$\Sigma\Sigma(c_i c_j)^2 = (\Sigma c_i^2)(\Sigma c_j^2).$$

The sum on the left is given by

$$(-18)^2 + (-15)^2 + \cdots + (-2)^2 + (-9)^2 = 2044.$$

The term on the right is given by

$$(14)(146) = 2044.$$

This latter computational scheme is readily extended to the case of a $p \times q \times r$ factorial experiment. The data in Table 6.8-2 will be used to indicate the numerical details. These data represent a $3 \times 3 \times 3$ factorial experiment in which there is one observation per cell. The usual summary tables for a three-factor factorial experiment are given in part ii. Data for the latter are given in part i. Numerical values of the computational symbols defined in Table 6.5-1 are given in part ii. Since there is only one observation per cell, symbols (2) and (9) are identical, i.e.,

$$\Sigma X_{ijk}^2 = \Sigma(ABC_{ijk})^2.$$

(The letter c is used for two different concepts, but the context should make clear what is meant. In one context c_1, c_2, and c_3 represent the levels of factor C. In a second context c_i, c_j, and c_k represent deviations from the grand mean.)

Means and the deviations of the means from the grand mean are computed in part iv. In this context $c_k = \bar{C}_k - \bar{G}$. Similarly, $c_i = \bar{A}_i - \bar{G}$. The entry in cell abc_{ijk} in part v has the general form $c_i c_j c_k$. Thus the entry in cell abc_{111} is $(-3.4)(-1.8)(-5.1) = -31.2$. The entry in cell abc_{123} is $(-3.4)(1.2)(3.2) = -13.1$. As a check on the computational work,

$$\sum_i c_i c_j c_k = \sum_j c_i c_j c_k = \sum_k c_i c_j c_k = 0.$$

That is, the sum of any column in part v must be zero; also the sum over b_j within any fixed level of ac_{ik} must be zero. For example, for level ac_{11}

$$-31.2 + 20.8 + 10.4 = 0.0.$$

For level ac_{12},

$$11.6 + (-7.7) + (-3.9) = 0.0.$$

The comparison associated with nonadditivity is a weighted sum of the entries in the cell of part i, the weights being the corresponding entries in part v. Thus the comparison used in the test for nonadditivity is

$$(-31.2)(3) + (20.8)(6) + (10.4)(9) + \cdots + (1.7)(12) = 60.60.$$

The sum of squares for this comparison has the form

$$SS_{\text{nonadd}} = \frac{(60.60)^2}{\Sigma(c_i c_j c_k)^2}.$$

The divisor is given by

$$\Sigma(c_i c_j c_k)^2 = (-31.2)^2 + (20.8)^2 + \cdots + (1.7)^2 = 3742.$$

Table 6.8-2 Numerical Example

(i)

	c_1			c_2			c_3			Total
	b_1	b_2	b_3	b_1	b_2	b_3	b_1	b_2	b_3	
a_1	3	6	9	6	9	12	9	12	15	81
a_2	6	9	12	12	18	21	15	21	21	135
a_3	9	9	3	18	21	12	18	18	12	120
Total	18	24	24	36	48	45	42	51	48	336

(ii)

AB summary table

	b_1	b_2	b_3	Total
a_1	18	27	36	81
a_2	33	48	54	135
a_3	45	48	27	120
Total	96	123	117	336

AC summary table

	c_1	c_2	c_3	Total
a_1	18	27	36	81
a_2	27	51	57	135
a_3	21	51	48	120
Total	66	129	141	336

BC summary table

	c_1	c_2	c_3	Total
b_1	18	36	42	96
b_2	24	48	51	123
b_3	24	45	48	117
Total	66	129	141	336

(iii)

$(1) = 4181.33$ $(4) = 4226.00$ $(7) = 4758.00$

$(2) = 4986.00$ $(5) = 4542.00$ $(8) = 4590.00$

$(3) = 4354.00$ $(6) = 4572.00$ $(9) = 4986.00$

(iv)

$\bar{G} = 12.4$	$\bar{A}_i$	$\bar{B}_j$	$\bar{C}_k$	c_i	c_j	c_k	
1	9	10.7	7.3	−3.4	−1.8	−5.1	$\Sigma c_i^2 = 18.62$
2	15	13.7	14.3	2.5	1.2	1.9	$\Sigma c_j^2 = 5.04$
3	13.3	13	15.7	0.9	0.6	3.2	$\Sigma c_k^2 = 39.86$

(v)

	c_1			c_2			c_3			Total
	b_1	b_2	b_3	b_1	b_2	b_3	b_1	b_2	b_3	
a_1	−31.2	20.8	10.4	11.6	−7.7	−3.9	19.6	−13.1	−6.5	0.0
a_2	23.0	−15.3	−7.7	−8.6	5.7	2.9	−14.4	9.6	4.8	0.0
a_3	8.2	− 5.5	−2.7	−3.0	2.0	1.0	− 5.2	3.5	1.7	0.0
Total	0.0	0.0	0.0	0.0	0.0	0.0	0.0	0.0	0.0	

Within rounding error, the following relation must hold:

$$\Sigma(c_i c_j c_k)^2 = (\Sigma c_i^2)(\Sigma c_j^2)(\Sigma c_k^2).$$

The right-hand side of this last expression is

$$(18.62)(5.04)(39.86) = 3740.7.$$

The analysis of variance is summarized in Table 6.8-3.

Table 6.8-3 Summary of Analysis of Variance

Source of variation	SS	df	MS	F
A	172.67	2		
B	44.67	2		
C	360.67	2		
AB	173.33	4		
AC	43.33	4		
BC	3.33	4		
ABC	6.67	8		
Nonadd 0.98		1	0.98	1.21
Balance 5.69		7	0.81	

The test for nonadditivity is given by

$$F = \frac{0.98}{0.81} = 1.21.$$

The critical value of this statistic for a .25-level test is $F_{.75}(1,7) = 1.57$. Since the observed value of the F statistic does not exceed this critical value, there is no reason to reject the hypothesis of additivity. The evidence from this test supports the hypothesis that the components of the three-factor interaction are homogeneous. If the assumptions underlying the test are met, the component for nonadditivity would tend to be large relative to the other components, provided that the three-factor interaction estimated a source of variation different from experimental error. In this case, the three-factor interaction may be considered as an estimate of experimental error (granting the validity of the assumptions). Hence the three-factor interaction term may be dropped from the model. In the latter case, MS_{abc} provides an estimate of σ_ε^2.

6.9 Computation of Trend Components

Computational procedures for trends for the case of a single-factor experiment were discussed in Sec. 3.7. These procedures generalize to factorial experiments. Principles underlying this generalization were discussed in Sec. 5.18; the actual computation of trend components in a factorial experiment are considered in this section. Computational procedures will be illustrated for the case of a 3×4 factorial experiment having five observations in each cell.

It will be assumed that (1) both factors are fixed, (2) the levels of both factors represent steps along an underlying quantitative scale, and (3) the respective levels represent equally spaced steps along the respective scales. The latter assumption, which is not essential to the development, permits a simplification of the numerical work, since coefficients of orthogonal polynomials may be used to obtain desired sums of squares. For this case, the coefficients for the levels of factor A are as follows (see Table C.10):

Linear: c'_i	-1	0	1
Quadratic: c''_i	1	-2	1

The coefficients for the levels of factor B are as follows:

Linear: c'_j	-3	-1	1	3
Quadratic: c''_j	1	-1	-1	1
Cubic: c'''_j	-1	3	-3	1

The data given in part i of Table 6.9-1 will be used as a numerical example. [Assume that each entry in the AB summary table is the sum of five observations; data for the computation of symbol (2) are not given.] The analysis of variance is summarized in part iii. This particular analysis does not necessarily give the experimenter all the information he seeks. There are many other ways in which the overall variation may be analyzed.

In spite of the significant interaction, it may be of interest to study the trend components of the main effects. For illustrative purposes, the trend components of the B main effect will be obtained. The comparison associated with the linear component is a weighted sum of B_j totals, the weights being the linear coefficients. For data in part i of Table 6.9-1, the linear comparison is

$$C_{\text{lin}} = (-3)(19) + (-1)(21) + (1)(28) + (3)(37) = 61.$$

The linear component of the variation due to the main effects of factor B is

$$\text{SS}_{b(\text{lin})} = \frac{C^2_{\text{lin}}}{np\Sigma(c'_j)^2} = \frac{(61)^2}{5(3)(20)} = 12.40.$$

Computational formulas for the quadratic and cubic components of the B main effect are summarized in Table 6.9-2. It is noted that the linear component accounts for 12.40/13.25, or 94 percent, of the variation due to the main effect. This means that on the average over levels of factor A the criterion measure predominantly is a linear function of levels of factor B. A test on whether a trend component differs significantly

Table 6.9-1 Numerical Example

AB summary table ($n = 5$):

(i)

	b_1	b_2	b_3	b_4	Total
a_1	3	5	9	14	31
a_2	7	11	15	20	53
a_3	9	5	4	3	21
Total	19	21	28	37	105

(ii) (1) = 183.75 (3) = 210.55 (5) = 247.40
(2) = 280.00 (4) = 197.00

(iii)

Source of variation	SS	df	MS	F
A	26.80	2	13.40	19.71
B	13.25	3	4.42	6.50
AB	23.60	6	3.93	5.78
Within cell	32.40	48	0.68	

from zero uses the statistic

$$F = \frac{\text{MS}_{\text{trend}}}{\text{MS}_{\text{w.cell}}}.$$

For example, a test on the quadratic trend is given by

$$F = \frac{\text{MS}_{b(\text{quad})}}{\text{MS}_{\text{w.cell}}} = \frac{0.82}{0.68} = 1.21.$$

Table 6.9-2 Trends of B Main Effects

(i)

c'_j	−3	−1	1	3	$\Sigma(c'_j)^2 = 20$
c''_j	1	−1	−1	1	$\Sigma(c''_j)^2 = 4$
c'''_j	−1	3	−3	1	$\Sigma(c'''_j)^2 = 20$

B_j: 19 21 28 37

(ii)

$$\text{SS}_{b(\text{lin})} = \frac{(\Sigma c'_j B_j)^2}{np\Sigma(c'_j)^2} = \frac{(61)^2}{5(3)(20)} = 12.40$$

$$\text{SS}_{b(\text{quad})} = \frac{(\Sigma c''_j B_j)^2}{np\Sigma(c''_j)^2} = \frac{(7)^2}{5(3)(4)} = 0.82$$

$$\text{SS}_{b(\text{cubic})} = \frac{(\Sigma c'''_j B_j)^2}{np\Sigma(c'''_j)^2} = \frac{(-3)^2}{5(3)(20)} = 0.03$$

$$\text{SS}_b = 13.25$$

The critical value for a .05-level test is $F_{.95}(1,48) = 4.04$. Hence the experimental data indicate that the hypothesis of no quadratic trend in the main effect of factor B is tenable.

A significant interaction implies that the response surface for different levels of factor B (or A) is not homogeneous, i.e., that profiles are not parallel. The linear $\times$ linear, linear $\times$ quadratic, etc., components of interaction indicate the fit of variously shaped surfaces, i.e., different patterns of profiles. Computational procedures are summarized in Table 6.9-3. The weights for the linear $\times$ linear comparison are given

Table 6.9-3 Trends within *AB* Interaction

(i) Linear $\times$ Linear

	b_1	b_2	b_3	b_4
a_1	3	1	−1	−3
a_2	0	0	0	0
a_3	−3	−1	1	3

$$SS_{\text{lin}\times\text{lin}} = \frac{[\Sigma d_{ij}(AB_{ij})]^2}{n(\Sigma d_{ij}^2)} = \frac{(-56)^2}{5(40)} = 15.68$$

(ii) Quadratic $\times$ Linear

	b_1	b_2	b_3	b_4
a_1	−3	−1	1	3
a_2	6	2	−2	−6
a_3	−3	−1	1	3

$$SS_{\text{quad}\times\text{lin}} = \frac{[\Sigma d_{ij}(AB_{ij})]^2}{n(\Sigma d_{ij}^2)} = \frac{(-68)^2}{5(120)} = 7.71$$

(iii) Linear $\times$ Quadratic

	b_1	b_2	b_3	b_4
a_1	−1	1	1	−1
a_2	0	0	0	0
a_3	1	−1	−1	1

$$SS_{\text{lin}\times\text{quad}} = \frac{[\Sigma d_{ij}(AB_{ij})]^2}{n(\Sigma d_{ij}^2)} = \frac{(0)^2}{5(8)} = 0$$

in part i. An entry in cell ab_{ij} of this table has the form

$$d_{ij} = c_i'c_j'.$$

For example,

$$d_{11} = (-1)(-3) = 3, \qquad d_{12} = (-1)(-1) = 1,$$
$$d_{21} = (0)(-3) = 0, \qquad d_{22} = (0)(-1) = 0.$$

The weights for the quadratic $\times$ linear comparison are given in part ii. An entry in this latter table is given by

$$d_{ij} = c_i''c_j'.$$

(Although the symbol d_{ij} is used for the typical entry in different tables, the context will make it clear which table is meant. A more explicit, but more cumbersome, notation for the latter d_{ij} would be $d_{i''j'}$.) The entries in part ii are obtained as follows:

$$\begin{aligned} d_{11} &= (1)(3) = 3, & d_{12} &= (1)(-1) = -1, \\ d_{21} &= (-2)(3) = -6, & d_{22} &= (-2)(-1) = 2, \\ d_{31} &= (1)(3) = 3, & d_{32} &= (1)(-1) = 1. \end{aligned}$$

Computational formulas for some of the trend components of the AB interaction are summarized in Table 6.9-3. With suitable definition of d_{ij}, the other components have the same general form. In each case d_{ij} refers to an entry in a different table of weights. Of the total variation due to AB, the linear $\times$ linear component accounts for 15.68/23.60, or 66 percent. The sum of the linear $\times$ linear component and the quadratic $\times$ linear components accounts for

$$\frac{15.68 + 7.71}{23.60} = .99,$$

or 99 percent. Tests on trend components of the interaction have the following general form:

$$F = \frac{MS_{trend}}{MS_{w.cell}}.$$

It is sometimes of interest to study differences in trends for the simple effects of one factor at different levels of a second factor. For illustrative purposes, differences between trends for the simple effects of factor B at

Table 6.9-4 Difference in Trends for Simple Effects of Factor B

		b_1	b_2	b_3	b_4	Total	d'_i	d''_i	d'''_i
(i)	a_1	3	5	9	14	31	37	3	−1
	a_2	7	11	15	20	53	43	1	1
	a_3	9	5	4	3	21	−19	3	−3
	Total	19	21	28	37	105	61	7	−3
	c'_j	−3	−1	1	3	$\Sigma(c'_j)^2 = 20$			
	c''_j	1	−1	−1	1	$\Sigma(c''_j)^2 = 4$			
	c'''_j	−1	3	−3	1	$\Sigma(c'''_j)^2 = 20$			

(ii)

$$SS_{\text{diff in lin trend}} = \frac{\Sigma(d'_i)^2}{n\Sigma(c'_j)^2} - \frac{(\Sigma d'_i)^2}{np\Sigma(c'_j)^2} = 35.79 - 12.40 = 23.39$$

$$SS_{\text{diff in quad trend}} = \frac{\Sigma(d''_i)^2}{n\Sigma(c''_j)^2} - \frac{(\Sigma d''_i)^2}{np\Sigma(c''_j)^2} = .95 - .82 = .13$$

$$SS_{\text{diff in cubic trend}} = \frac{\Sigma(d'''_i)^2}{n\Sigma(c'''_j)^2} - \frac{(\Sigma d'''_i)^2}{np\Sigma(c'''_j)^2} = .11 - .03 = .08$$

$$SS_{ab} = 23.60$$

different levels of factor A will be obtained. (The degrees of freedom for such differences in trend are $p - 1 = 2$ for each trend.) Computational formulas for these sources of variation are given in Table 6.9-4.

In this table the symbol d'_i is defined as follows:

$$d'_i = \sum_j c'_j(AB_{ij}).$$

For example,

$$d'_i = (-3)(3) + (-1)(5) + (1)(9) + (3)(14) = 37.$$

The symbols d''_i and d'''_i are defined as follows:

$$d''_i = \sum_j c''_j(AB_{ij}),$$

$$d'''_i = \sum_j c'''_j(AB_{ij}).$$

The variation due to differences in linear trends in simple effects of factor B explains 23.39/23.60, or 99 percent, of the total variation of the AB interaction. This means that 99 percent of the AB interaction arises from differences between the linear trends in the profiles of factor B at the different levels of factor A. These profiles are shown in Fig. 6.9-1.

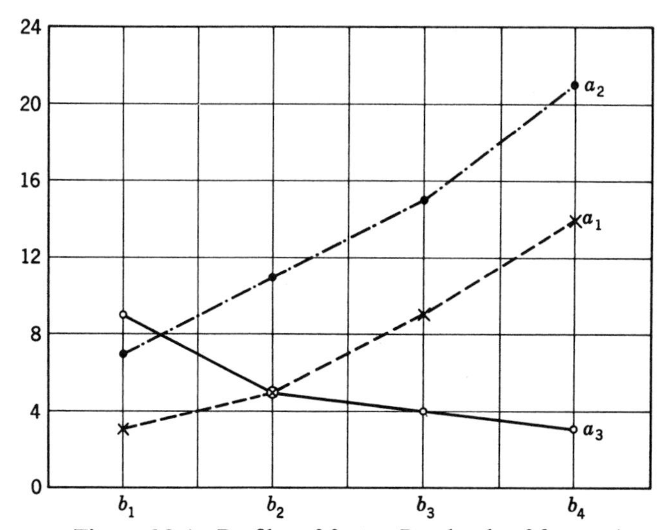

Figure 6.9-1 Profiles of factor B at levels of factor A.

If these profiles were plotted in a three-dimensional space, the response surface represented by the AB summary table would be obtained. This response surface is shown in Fig. 6.9-2. From this surface it is seen that profiles for factor A at fixed levels of factor B are predominantly quadratic in form. However, the profiles for factor B at fixed levels of factor A tend to be linear in form. Hence the shape of this surface is predominantly quadratic $\times$ linear.

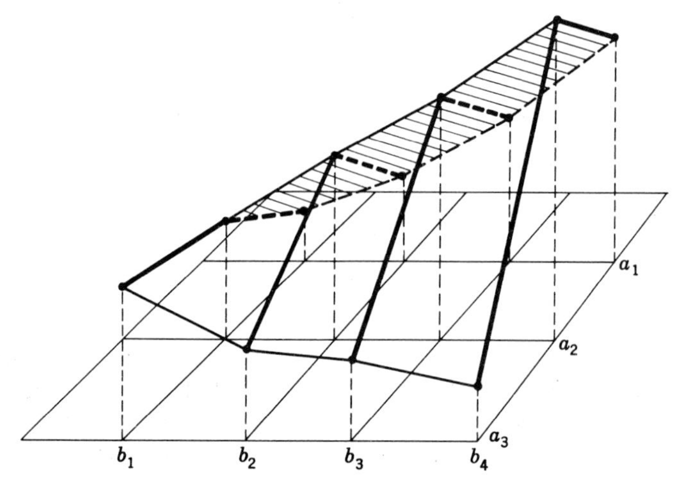

Figure 6.9-2 Response surface.

6.10 General Computational Formulas for Main Effects and Interactions

The following notation will be used in this section:

A_i = sum of all observations at level a_i.
n_a = number of observations summed to obtain A_i; n_a is assumed constant for all levels of factor A.
AB_{ij} = sum of all observations at level ab_{ij}.
n_{ab} = number of observations summed to obtain AB_{ij}; n_{ab} is assumed constant for all ab_{ij}'s.
ABC_{ijk} = sum of all observations at level abc_{ijk}.
n_{abc} = number of observations summed to obtain ABC_{ijk}; n_{abc} is assumed constant for all abc_{ijk}'s.

In terms of this notation, a general formula for the main effect due to factor A is

$$SS_a = \frac{\Sigma A_i^2}{n_a} - \frac{G^2}{n_g},$$

where G is the grand total of all observations and n_g is the number of observations summed to obtain G. (In this context an observation is a measurement on the experimental unit.) A general formula for the main effect due to factor B has the form

$$SS_b = \frac{\Sigma B_j^2}{n_b} - \frac{G^2}{n_g}.$$

The general computational formula for the variation due to the AB interaction is

$$SS_{ab} = \frac{\Sigma(AB_{ij})^2}{n_{ab}} - \frac{G^2}{n_g} - (SS_a + SS_b).$$

The general computational formula for the variation due to the AC interaction has the form

$$SS_{ac} = \frac{\Sigma(AC_{ik})^2}{n_{ac}} - \frac{G^2}{n_g} - (SS_a + SS_c).$$

(Unless otherwise indicated, the range of summation is over all possible values of the terms being summed.)

The general computational formula for variation due to the ABC interaction has the form

$$SS_{abc} = \frac{\Sigma(ABC_{ijk})^2}{n_{abc}} - \frac{G^2}{n_g} - (SS_a + SS_b + SS_c + SS_{ab} + SS_{ac} + SS_{bc}).$$

The computational formula for the variation due to the UVW interaction is

$$SS_{uvw} = \frac{\Sigma(UVW)^2}{n_{uvw}} - \frac{G^2}{n_g} - (SS_u + SS_v + SS_w + SS_{uv} + SS_{uw} + SS_{vw}).$$

Variation due to a four-factor interaction has the general formula

$$\begin{aligned} SS_{uvwx} = \frac{\Sigma(UVWX)^2}{n_{uvwx}} - \frac{G^2}{n_g} &- (SS_u + SS_v + SS_w + SS_x) \\ &- (SS_{uv} + SS_{uw} + SS_{ux} + SS_{vw} + SS_{vx} + SS_{wx}) \\ &- (SS_{uvw} + SS_{uvx} + SS_{uwx} + SS_{vwx}). \end{aligned}$$

In a factorial experiment having t factors there are

$\binom{t}{2}$ two-factor interactions,

$\binom{t}{3}$ three-factor interactions,

. . .

$\binom{t}{m}$ m-factor interactions.

For example, in a four-factor experiment, the number of possible two-factor interactions is

$$\binom{4}{2} = \frac{4(3)}{1(2)} = 6.$$

In a four-factor experiment the number of possible three-factor interactions is

$$\binom{4}{3} = \frac{4(3)(2)}{1(2)(3)} = 4.$$

Alternatively, given a t-factor interaction, the formulas that have been given above indicate the number of different two-factor, three-factor, etc., interactions that may be formed from the t letters in the t-factor interaction.

The general formula for variation due to a five-factor interaction is

$$SS_{uvwxy} = \frac{\Sigma(UVWXY)^2}{n_{uvwxy}} - \frac{G^2}{n_g} - [(i) + (ii) + (iii) + (iv)],$$

where (i) $= \Sigma SS_{\text{main effect}}$—there are five terms in this sum;

(ii) $= \Sigma SS_{\text{two-factor int}}$—there are $[5(4)]/[1(2)] = 10$ terms;

(iii) $= \Sigma SS_{\text{three-factor int}}$—there are $[5(4)(3)]/[1(2)(3)] = 10$ terms;

(iv) $= \Sigma SS_{\text{four-factor int}}$—there are $[5(4)(3)(2)]/[1(2)(3)(4)] = 5$ terms.

The general formulas given above assume that no factor is nested under any other. In case factor C is nested under factor B, the general formula for the main effect due to C within B is

$$\begin{aligned} SS_{c\text{ w. }b} &= \frac{\Sigma(BC)^2}{n_{bc}} - \frac{\Sigma B^2}{n_b} \\ &= SS_c + SS_{bc}, \end{aligned}$$

where SS_c is the variation due to the main effect of factor C if the nesting is disregarded and SS_{bc} has an analogous definition. The general formula for the $A \times (C$ within $B)$ variation is

$$\begin{aligned} SS_{a(c\text{ w. }b)} &= \frac{\Sigma(ABC)^2}{n_{abc}} - \frac{\Sigma(AB)^2}{n_{ab}} - \frac{\Sigma(BC)^2}{n_{bc}} + \frac{\Sigma B^2}{n_b} \\ &= SS_{ac} + SS_{abc}, \end{aligned}$$

where SS_{abc} is the three-factor variation computed with the nesting disregarded. In general,

$$SS_{uv(x\text{ w. }w)} = SS_{uvx} + SS_{uvxw},$$

where the terms on the right are computed as if there were no nesting.

For the case in which factor C is nested under factor B and factor D

is nested under factor C,

$$\text{SS}_{d \text{ w. } b \text{ and } c} = \frac{\Sigma(BCD)^2}{n_{bcd}} - \frac{\Sigma(BC)^2}{n_{bc}}$$
$$= \text{SS}_d + \text{SS}_{bd} + \text{SS}_{cd} + \text{SS}_{bcd}.$$

The terms in this last line are computed as if there were no nesting. Also,

$$\text{SS}_{a(d \text{ w. } b \text{ and } c)} = \frac{\Sigma(ABCD)^2}{n_{abcd}} - \frac{\Sigma(ABC)^2}{n_{abc}} - \frac{\Sigma(BCD)^2}{n_{bcd}} + \frac{\Sigma(BC)^2}{n_{bc}}$$
$$= \text{SS}_{ad} + \text{SS}_{abd} + \text{SS}_{acd} + \text{SS}_{abcd}.$$

The degrees of freedom for the variation on the left are the sum of the degrees of freedom of the variations on the right. The degrees of freedom may be checked as follows:

$$\begin{aligned} \text{df} &= pqrs - pqr - qrs + qr \\ &= pqr(s - 1) - qr(s - 1) \\ &= qr(s - 1)(p - 1). \end{aligned}$$

The terms in the first line of the above expression are obtained as follows:

$$\begin{aligned} pqrs &= \text{number of } ABCD \text{ terms}, \\ pqr &= \text{number of } ABC \text{ terms}, \\ qrs &= \text{number of } BCD \text{ terms}, \\ qr &= \text{number of } BC \text{ terms}. \end{aligned}$$

6.11 Missing Data

For a factorial experiment in which the cell frequencies are not equal but all n_{ij}'s are approximately equal, the analysis by the method of unweighted means presents no particular problems, provided that all cells contain at least one observation. In the unweighted-means analysis, the mean of each cell is estimated by the observations actually made within that cell. In cases where the number of observations within a particular cell is small relative to the number of observations in other cells in the same row and column or adjacent rows and adjacent columns, then information provided by these other cells may be used in estimating the mean of a specified cell. In an unweighted-means analysis, such information is not utilized in the estimation process.

In an experiment having no observed data in given cells, estimates of such cell means may have to be obtained from other cells in the experiment. If the form of the response surface were known, estimates of the missing entries could be obtained by using a multiple regression equation. This solution is generally not practical. One method, having somewhat limited utility, is to estimate the missing mean in cell ij by the following formula,

$$\overline{AB}'_{ij} = \bar{A}'_i + \bar{B}'_j - \bar{G}',$$

where $\bar{A}'_i$ is the mean of the observations actually made under level a_i; similar definitions hold for $\bar{B}'_j$ and $\bar{G}'$. This method of estimating missing data assumes no interaction present. Because this method of estimation does not take into account trends in the row and column of the missing entry, its utility is limited. For the case of a three-factor experiment, the method that has just been described takes the form

$$\overline{ABC}'_{ijk} = \bar{A}'_i + \bar{B}'_j + \bar{C}'_k - 2\bar{G}'.$$

The presence of interaction effects is a stumbling block in all methods for estimating missing cell entries. A review of methods that have been proposed and an extensive bibliography on the topic of missing data are given in Federer (1955, pp. 124–127, 133–134). Caution and judgment are called for in the use of any method for estimating missing data. Each of the proposed methods has a set of assumptions that must be considered with care.

Consider the numerical data given in Table 6.11-1. These data represent the available observations in a 5×4 factorial experiment. Assume the levels of the factors represent steps along a quantitative scale. There is a single measurement in 18 of the 20 cells. Entries in

Table 6.11-1 Numerical Example

	b_1	b_2	b_3	b_4	Total
a_1	28	20	11	10	69
a_2	u_{21}	16	15	8	$39 + u_{21}$
a_3	29	13	16	u_{34}	$58 + u_{34}$
a_4	27	10	18	11	66
a_5	28	11	15	10	64
Total	$112 + u_{21}$	70	75	$39 + u_{34}$	$296 + u_{21} + u_{34}$

cells ab_{21} and ab_{34} are missing. The unknown values of these cell entries are designated u_{21} and u_{34}. [The data in this table appear in Bennett and Franklin (1954, p. 383).] A method of estimating missing observations, which minimizes the interaction effect in the analysis resulting when such estimates are used in place of the missing data, is described by Bennett and Franklin (1954, pp. 382–383). Using this method, Bennett and Franklin arrive at the following estimates:

$$u_{21} = 28.6, \qquad u_{34} = 10.2.$$

If these two estimates of the missing entries are inserted in their proper places and the usual analysis of variance computed as if all data were observed, the variation due to AB will be a minimum. In essence, this principle for estimating the missing entries keeps the profiles of the simple effects as parallel as possible.

A simplified version of this latter principle utilizes information from only those cells adjacent to the cell in which data are missing. For example,

$$\frac{u_{21}}{16} = \frac{(\frac{28}{20}) + (\frac{29}{13})}{2}.$$

Solving for u_{21} gives

$$u_{21} = 29.0.$$

This method of estimation assumes that the slope of the profiles for levels b_1 and b_2 at level a_2 is the mean of corresponding slopes at levels a_1 and a_3. The latter two slopes can be obtained from the observed data. Analogously,

$$\frac{u_{34}}{16} = \frac{(\frac{8}{15}) + (\frac{11}{18})}{2},$$

from which $$u_{34} = 9.2.$$

To use the first method discussed in this section for estimating missing entries, one proceeds as follows:

$$\bar{A}_2' = \tfrac{39}{3} = 13.0, \qquad \bar{A}_3' = 19.3;$$
$$\bar{B}_1' = \tfrac{112}{4} = 28.0, \qquad \bar{B}_4' = 9.8;$$
$$\bar{G}' = \tfrac{296}{18} = 16.4.$$

In each case the denominator is the number of observations in the corresponding total. The estimates of the unknown entries are

$$u_{21} = 13.0 + 28.0 - 16.4 = 24.6,$$
$$u_{34} = 19.3 + 9.8 - 16.4 = 12.6.$$

If the analysis of variance were to be carried out with estimated values substituted for the missing values, the degrees of freedom for the resulting two-factor interaction would be

$$\text{df}_{ab} = (p - 1)(q - 1) - (\text{number of missing values}).$$

For the numerical example in Table 6.11-1,

$$\text{df}_{ab} = 4(3) - 2 = 10.$$

By way of summary, there are mathematically elegant methods for estimating missing cell entries. None of these methods is satisfactory unless the experimenter has information about the nature of the response surface being studied. There is, however, no real substitute for experimental data. In the example given in Table 6.11-1, the experimentally determined values of the missing entries were

$$u_{21} = 23, \qquad u_{34} = 24.$$

None of the methods considered yielded values relatively close to the observed value for u_{34}.

6.12 Special Computational Procedures When All Factors Have Two Levels

When each factor in a factorial experiment has two levels, the computational formulas for main effects and interactions may be simplified. For the case of a 2 × 2 factorial experiment having n observations per cell,

$$SS_a = \frac{(A_1 - A_2)^2}{4n},$$

$$SS_b = \frac{(B_1 - B_2)^2}{4n},$$

$$SS_{ab} = \frac{[(AB_{11} + AB_{22}) - (AB_{12} + AB_{21})]^2}{4n}.$$

In the expression for the interaction, note that the sum of the subscripts of each term in the first set of parentheses is an even number, whereas the sum of the subscripts for each term in the second set of parentheses is an odd number.

For the case of a 2 × 2 × 2 factorial experiment having n observations per cell,

$$SS_a = \frac{(A_1 - A_2)^2}{8n},$$

$$SS_b = \frac{(B_1 - B_2)^2}{8n},$$

$$SS_c = \frac{(C_1 - C_2)^2}{8n},$$

$$SS_{ab} = \frac{[(AB_{11} + AB_{22}) - (AB_{12} + AB_{21})]^2}{8n},$$

$$SS_{ac} = \frac{[(AC_{11} + AC_{22}) - (AC_{12} + AC_{21})]^2}{8n},$$

$$SS_{bc} = \frac{[(BC_{11} + BC_{22}) - (BC_{12} + BC_{21})]^2}{8n},$$

$$SS_{abc} = \frac{[(ABC_{111} + ABC_{122} + ABC_{212} + ABC_{221}) - (ABC_{112} + ABC_{121} + ABC_{211} + ABC_{222})]^2}{8n}.$$

Again note that in expressions for interactions the sum of the subscripts for each term within a pair of parentheses is an odd number in one case and an even number in the other case.

The computations for a three-factor experiment may be made without obtaining two-factor summary tables by following the scheme given in Table 6.12-1.

Table 6.12-1 Computational Scheme for 2 × 2 × 2 Factorial Experiment

		c_1		c_2	
		b_1	b_2	b_1	b_2
(i)	a_1	ABC_{111}	ABC_{121}	ABC_{112}	ABC_{122}
	a_2	ABC_{211}	ABC_{221}	ABC_{212}	ABC_{222}
		(1)	(2)	(3)	(4)
(i′)		s_1'	s_3'	d_1'	d_3'
		s_2'	s_4'	d_2'	d_4'
(i″)		s_1''	s_3''	d_1''	d_3''
		s_2''	s_4''	d_2''	d_4''
(i‴)		s_1'''	s_3'''	d_1'''	d_3'''
		s_2'''	s_4'''	d_2'''	d_4'''
(ii)		$SS_a = (s_2''')^2/8n$		$SS_c = (d_1''')^2/8n$	
		$SS_b = (s_3''')^2/8n$		$SS_{ac} = (d_2''')^2/8n$	
		$SS_{ab} = (s_4''')^2/8n$		$SS_{bc} = (d_3''')^2/8n$	
				$SS_{abc} = (d_4''')^2/8n$	

The ABC summary table is given in part i. The symbol s_1' in part i′ represents the sum of the entries in the first column of part i; s_2', s_3', and s_4' represent sums of entries in the respective columns of part i. These sums should be arranged as indicated in part i′. The symbol d_1' represents the difference between the two entries in the first column in part i, that is, $d_1' = ABC_{111} - ABC_{211}$. The symbols d_2', d_3', and d_4' represent corresponding differences between entries in corresponding columns in part i. For example, $d_4' = ABC_{122} - ABC_{222}$.

The entries in part i″ are obtained from the entries in part i′ in the same general manner as corresponding entries were obtained from part i. That is,

$$s_1'' = s_1' + s_2', \qquad s_3'' = d_1' + d_2',$$
$$s_2'' = s_3' + s_4', \qquad s_4'' = d_3' + d_4'.$$

Also,
$$d_1'' = s_1' - s_2', \qquad d_3'' = d_1' - d_2',$$
$$d_2'' = s_3' - s_4', \qquad d_4'' = d_3' - d_4'.$$

The entries in part i‴ are obtained from the entries in part i″ by means of the same general pattern.

$$s_1''' = s_1'' + s_2'', \qquad s_3''' = d_1'' + d_2'',$$
$$s_2''' = s_3'' + s_4'', \qquad s_4''' = d_3'' + d_4''.$$

Also
$$d_1''' = s_1'' - s_2'', \qquad d_3''' = d_1'' - d_2'',$$
$$d_2''' = s_3'' - s_4'', \qquad d_4''' = d_3'' - d_4''.$$

Computational formulas for the sums of squares are given in part ii. These computational formulas are identical to those given earlier in this section. A numerical example of this computational scheme is given in Table 6.12-2. As a check on the computational work,

$$SS_{\text{b.cells}} = \frac{\Sigma(ABC)_2}{n} - \frac{G^2}{8n}$$
$$= \Sigma(\text{main effects}) + \Sigma(\text{two-factor int}) + \Sigma(\text{three-factor int}).$$

Table 6.12-2 Numerical Example

		c_1		c_2		
(i)		b_1	b_2	b_1	b_2	$(n = 5)$
	a_1	10	15	20	10	
	a_2	20	30	30	10	
		(1)	(2)	(3)	(4)	
(i′)		30	50	−10	−10	
		45	20	−15	0	
(i″)		75	−25	−15	5	
		70	−10	30	−10	
(i‴)		145	15	5	−45	
		−35	−5	−15	15	

(ii)

$SS_a = (-35)^2/40 = 30.62$ $\qquad SS_c = (5)^2/40 = .62$

$SS_b = (15)^2/40 = 5.62$ $\qquad SS_{ac} = (-15)^2/40 = 5.62$

$SS_{ab} = (-5)^2/40 = .62$ $\qquad SS_{bc} = (-45)^2/40 = 50.62$

$SS_{abc} = (15)^2/40 = 5.62$

For the data in Table 6.12-2,

$$SS_{\text{b.cells}} = 625.00 - 525.62 = 99.38.$$

Within rounding error, this is the numerical value of the sum of the variations computed in part ii.

The computational scheme that has just been illustrated may be generalized to any 2^k factorial experiment, where k is the number of factors. The generalization to a $2^4 = 2 \times 2 \times 2 \times 2$ factorial experiment is illustrated in Table 6.12-3. In part i the $ABCD$ summary table is represented schematically. In part i′ each s' represents the sum of the entries in the corresponding columns of part i; each d' represents the difference between the elements in the corresponding columns of part i. Entries in part i″ are obtained from the entries in the columns of part i′ in an

Table 6.12-3 Computational Scheme for 2^4 Factorial Experiment

		d_1				d_2			
		c_1		c_2		c_1		c_2	
(i)		b_1	b_2	b_1	b_2	b_1	b_2	b_1	b_2
	a_1								
	a_2								
		(1)	(2)	(3)	(4)	(5)	(6)	(7)	(8)
(i′)		s'_1	s'_3	s'_5	s'_7	d'_1	d'_3	d'_5	d'_7
		s'_2	s'_4	s'_6	s'_8	d'_2	d'_4	d'_6	d'_8
(i″)		s''_1	s''_3	s''_5	s''_7	d''_1	d''_3	d''_5	d''_7
		s''_2	s''_4	s''_6	s''_8	d''_2	d''_4	d''_6	d''_8
(i‴)		s'''_1	s'''_3	s'''_5	s'''_7	d'''_1	d'''_3	d'''_5	d'''_7
		s'''_2	s'''_4	s'''_6	s'''_8	d'''_2	d'''_4	d'''_6	d'''_8
(i^{iv})		s^{iv}_1	s^{iv}_3	s^{iv}_5	s^{iv}_7	d^{iv}_1	d^{iv}_3	d^{iv}_5	d^{iv}_7
		s^{iv}_2	s^{iv}_4	s^{iv}_6	s^{iv}_8	d^{iv}_2	d^{iv}_4	d^{iv}_6	d^{iv}_8
(ii)		—	B	C	BC	D	BD	CD	BCD
		A	AB	AC	ABC	AD	ABD	ACD	$ABCD$

analogous manner. That is,

$$
\begin{aligned}
s''_1 &= s'_1 + s'_2, & d''_1 &= s'_1 - s'_2; \\
s''_2 &= s'_3 + s'_4, & d''_2 &= s'_3 - s'_4; \\
&\cdots\quad, & &\cdots\quad; \\
s''_8 &= d'_7 + d'_8, & d''_8 &= d'_7 - d'_8.
\end{aligned}
$$

This procedure continues until part i^k is completed, in this case until part i^{iv} is completed. Computational formulas for the variation due to the sources indicated in part ii may be obtained from the corresponding entry in part i^{iv}. For example,

$$SS_c = \frac{(s_5^{iv})^2}{16n}, \qquad SS_{abd} = \frac{(d_4^{iv})^2}{16n}.$$

The general formula is

$$\frac{(s^k)^2}{2^k n} \quad \text{or} \quad \frac{(d^k)^2}{2^k n}.$$

This computational procedure is particularly useful when k is 4 or larger.

6.13 Illustrative Applications

In a study reported by Wulff and Stolurow (1957) the experimental plan was a 2 × 2 factorial having 10 subjects per cell. Factor A was essentially a classification factor; its levels indicated the aptitude of the subjects as measured by a test of mechanical aptitude. The levels of factor B were the methods of instruction used in a paired-associates learning task. The criterion was the number of correct responses in a block of trials. The analysis of variance had the following form:

Source	df	MS	F
A (aptitude)	1	2175.62	17.96
B (method)	1	931.22	7.68
AB	1	9.02	
Within cell	36	121.10	

Factors A and B were both considered to be fixed; hence $MS_{w.cell}$ was the denominator for all F ratios. The factor of primary interest was B; the critical value for a .05-level test on this factor is $F_{.95}(1,36) = 4.11$. The experimental data indicated that there is a statistically significant difference between the two methods of instruction with respect to the mean number of correct responses in the specified blocks of trials.

Gordon (1959) reports a 4 × 4 factorial experiment in which both factors A and B were considered to be random. The levels of factor A represented the kind of task on which subjects were given pretraining; the levels of factor B represented the kind of task to which the subjects were transferred. The design may be represented schematically as follows:

	b_1	b_2	b_3	b_4
a_1	G_{11}	G_{12}	G_{13}	G_{14}
a_2	G_{21}	G_{22}	G_{23}	G_{24}
a_3	G_{31}	G_{32}	G_{33}	G_{34}
a_4	G_{41}	G_{42}	G_{43}	G_{44}

The symbol G_{ij} represents the group of subjects given pretraining under task i and transferred to task j. The criterion measure was the performance on the second task. The primary purpose of the study was to measure possible effects of transfer of training; the latter would be indicated by the magnitude of the interaction effects.

The analysis of variance in the Gordon study had the following form:

Source	df	MS	F
A (first task)	3	.21	1.41
B (second task)	3	4.70	32.38
AB	9	.14	3.82
Within cell	144	.038	

The analysis given above represents only part of the analysis; separate analyses were made for different stages in the learning of the second task. Since both factors A and B were considered random, the F ratios had the following form:

$$F = \frac{\text{MS}_a}{\text{MS}_{ab}}, \qquad F = \frac{\text{MS}_b}{\text{MS}_{ab}}, \qquad F = \frac{\text{MS}_{ab}}{\text{MS}_{\text{w.cell}}}.$$

These are the F ratios which would be obtained if one were to derive the expected values of the mean squares. The critical value for a .05-level test on the interaction is $F_{.95}(9,144) = 1.94$. (In the opinion of the author the factors in this kind of experiment should be considered fixed rather than random; the study did not indicate the randomization procedure whereby the tasks used in the experiment were obtained. Rather, the indications were that the tasks were deliberately chosen to meet specified objectives.)

A $6 \times 4 \times 3$ factorial having only one observation per cell is reported by Aborn et al. (1959). These writers were interested in studying the effects of contextual constraints upon recognition of words in sentences. The observation in each of the cells was a mean. A square-root transformation was used before the analysis was made. Entries in three of the cells were missing and were estimated from the data in the remaining cells. All the interactions were pooled into a single error term. The analysis of variance had the following form:

Source	df	MS	F
A (class of word)	5	4.44	26.28
B (position)	3	.47	2.76
C (sentence length)	2	.54	3.22
Pooled interactions	$61 - 3 = 58$	.17	

The pooled interaction term would have 61 degrees of freedom if there were no missing data; since 3 of the 72 cell entries were missing, degrees of freedom for the resulting pooled interaction are $61 - 3 = 58$. The separate interaction terms have the following degrees of freedom:

$$
\begin{aligned}
AB&: \quad 5 \times 3 = 15\\
AC&: \quad 5 \times 2 = 10\\
BC&: \quad 3 \times 2 = 6\\
ABC&: \quad 5 \times 3 \times 2 - 3 = 27
\end{aligned}
$$

Bamford and Ritchie (1958) report an experiment which may be represented as follows:

Subject	a_1				a_2			
	b_1	b_2	b_3	b_4	b_1	b_2	b_3	b_4
1								
2								
.								
.								
.								
9								

The levels of factor A represent control and experimental conditions for turn indicators on an airplane instrument panel. The levels of factor B represent successive trials under each condition. Each of the nine subjects was observed under each of the levels of factor A as well as each of the levels of factor B. The order in which trials were made under the level of factor A was counterbalanced.

This design, as well as the others which follow in this section, actually falls in the repeated-measure category. The usual tests on main effects and interactions are valid only under a set of highly restrictive assumptions on covariance matrices. Checks on these assumptions have, for the most part, been ignored by experimenters in the behavioral sciences. The usual tests tend to err on the side of yielding too many significant results (positive bias) when homogeneity assumptions on covariance matrices are not met. A test procedure which avoids this kind of bias is discussed in Sec. 7.2. In experiments involving a learning or practice effect, homogeneity conditions required for the usual F tests to be valid are generally not present.

In form, this plan may be considered as a special case of a $2 \times 4 \times 9$ factorial experiment. The subject factor is considered random; factors A and B are considered fixed. Assuming that all interactions with the subject factor may be pooled, i.e., that all interactions with the subject factor are independent estimates of experimental error, the analysis of variance has the following form:

Source	df	MS	F
Subjects	8	11.83	
A	1	20.77	5.82
B	3	11.42	3.19
AB	3	9.14	2.56
Pooled error	56	3.57	

Gerathewohl et al. (1957) report a study having the form of a $2 \times 9 \times 9$ factorial experiment. The purpose of this experiment was to study the effects of speed and direction of rotation upon the pattern of circular eye movements. Factor A represented the direction of rotation; factor B represented the speed of the rotation. Factor C represented a subject factor. The order in which subjects were observed under the levels of factors A and B was the same for all subjects. The subject factor was considered random, but factors A and B were considered fixed. The expected values of the mean squares for this design are analogous to those obtained in Table 6.5-3 when homogeneity assumptions discussed in Sec. 7.2 are met. The analysis of variance reported by these workers is summarized (with slight modification) in Table 6.13-1.

Table 6.13-1 Summary of Analysis of Variance

Source	df	MS	F
C (subjects)	8	30.80	
A (direction)	1	47.22	3.41
AC	8	13.84	
B (speed)	8	85.66	17.8
BC	64	4.80	
AB	8	1.75	.63
ABC	64	2.77	
Total	161		

Inspection of the interactions with the subject factor indicates that no pooling is possible. Hence the F ratios have the following form:

$$F = \frac{MS_a}{MS_{ac}}, \qquad F = \frac{MS_b}{MS_{bc}}, \qquad F = \frac{MS_{ab}}{MS_{abc}}.$$

Jerison (1959) reports a study which is a special case of a $2 \times 3 \times 4 \times 9$ factorial experiment. The purpose of this experiment was to study the effects of noise on human performance. The levels of factor A represented control and experimental noise conditions. The levels of factor B represented four successive periods of time under each of the conditions of factor A. The levels of factor C represented three clocks monitored by the subjects during the course of the experiment. The last factor represented a subject factor. The criterion was an accuracy score on each of the clocks. All factors except the subject factor were considered to be fixed. The analysis of variance is summarized in Table 6.13-2. In each case the denominator for an F ratio is the corresponding interaction with the subject factor.

Table 6.13-2 Summary of Analysis of Variance

Source	df	MS	F
Subjects	8	6544.90	
A (noise conditions)	1	8490.08	2.93
$A \times$ subjects	8	2900.09	
B (periods)	3	479.67	6.32
$B \times$ subjects	24	75.89	
C (clocks)	2	489.31	1.24
$C \times$ subjects	16	396.08	
AB	3	600.47	3.48
$AB \times$ subjects	24	172.60	
AC	2	280.52	1.18
$AC \times$ subjects	16	238.36	
BC	6	138.63	1.32
$BC \times$ subjects	48	105.09	
ABC	6	253.10	2.07
$ABC \times$ subjects	48	122.13	
Total	215		

6.14 Unequal Cell Frequencies—Least-squares Solution

The data in Table 6.14-1 will be used to illustrate the computational procedures associated with the least-squares estimates of the sums of squares. The rationale underlying these procedures was discussed in Sec. 5.23. Definitions of the basic symbols are also given in the latter section.

These data have been taken from Anderson and Bancroft (1952, p. 243). The levels of factor A represent sex of animal; levels of factor B represent four successive generations of animals. In this type of experiment, the cell frequencies are, in a real sense, an integral part of the design. Hence a least-squares analysis is more appropriate than an unweighted-means analysis. In part iii, with the exception of symbol (2), data from which the numerical values of the symbols are computed are given in parts i and ii. The raw data for symbol (2) are not given.

When one of the factors has only two levels, the simplest approach (see Rao, 1952, pp. 95–100) is to obtain $SS_{ab(\text{adj})}$ directly from part i of

Table 6.14-1 Numerical Example with Unequal Cell Frequencies ($p = 2$)

Cell frequencies:

		b_1	b_2	b_3	b_4	Total
(i)	a_1	$n_{11} = 21$	$n_{12} = 15$	$n_{13} = 12$	$n_{14} = 7$	$n_{1.} = 55$
	a_2	$n_{21} = 27$	$n_{22} = 25$	$n_{23} = 23$	$n_{24} = 19$	$n_{2.} = 94$
		$n_{.1} = 48$	$n_{.2} = 40$	$n_{.3} = 35$	$n_{.4} = 26$	$n_{..} = 149$

Cell totals:

		b_1	b_2	b_3	b_4	Total
(ii)	a_1	$AB_{11} = 3716$	$AB_{12} = 2422$	$AB_{13} = 1868$	$AB_{14} = 1197$	$A_1 = 9203$
	a_2	$AB_{21} = 2957$	$AB_{22} = 2852$	$AB_{23} = 2496$	$AB_{24} = 2029$	$A_2 = 10334$
	Total	$B_1 = 6673$	$B_2 = 5274$	$B_3 = 4364$	$B_4 = 3226$	$G = 19537$

(iii)

$$(1) = G^2/n_{..} = 2{,}561{,}707$$
$$(2) = \Sigma X^2 = 2{,}738{,}543$$
$$(3) = \Sigma(A_i^2/n_{i.}) = 1{,}539{,}913 + 1{,}136{,}080 = 2{,}675{,}993$$
$$(4) = \Sigma(B_j^2/n_{.j}) = 2{,}567{,}463$$
$$(5) = \Sigma\Sigma(AB_{ij}^2/n_{ij}) = 2{,}680{,}848$$

Table 6.14-1 and from Table 6.14-2. In the latter table,

$$d_j = \overline{AB}_{1j} - \overline{AB}_{2j} \qquad \text{and} \qquad w_j = \frac{n_{1j}n_{2j}}{n_{1j} + n_{2j}}.$$

For example,

$$w_1 = \frac{(21)(27)}{48} = 11.8125.$$

The adjusted sum of squares due to interaction is given by

$$SS_{ab(\text{adj})} = \Sigma w_j d_j^2 - \frac{(\Sigma w_j d_j)^2}{\Sigma w_j}.$$

Table 6.14-2 Direct Computation of $SS_{ab(\text{adj})}$

Cell means:

	b_1	b_2	b_3	b_4	
a_1	176.9524	161.4667	155.6667	171.0000	
a_2	109.5185	114.0900	108.5217	106.7895	
d_j	67.4339	47.3867	47.1450	64.2105	
w_j	11.8125	9.3750	7.8857	5.1154	$34.1886 = \Sigma w_j$
$w_j d_j$	796.5629	444.2344	371.7713	328.4624	$1941.0310 = \Sigma w_j d_j$
					$113{,}384.04 = \Sigma w_j d_j^2$

The first term on the right is most readily obtained from the terms $w_j d_j$, that is,

$$\Sigma w_j d_j^2 = d_1(w_1 d_1) + d_2(w_2 d_2) + \cdots + d_4(w_4 d_4)$$
$$= 113{,}384.$$

Thus $$SS_{ab(\text{adj})} = 113{,}384 - \frac{(1941.0310)^2}{34.1886} = 3183.$$

From part iii of Table 6.14-1, the unadjusted sums of squares are

$$SS_{\text{cells}} = (5) - (1) = 119{,}141,$$
$$SS_a = (3) - (1) = 114{,}286,$$
$$SS_b = (4) - (1) = 5756.$$

The adjusted sum of squares due to factor A is

$$SS_{a(\text{adj})} = SS_{\text{cells}} - SS_{ab(\text{adj})} - SS_b = 110{,}202.$$

Similarly, $$SS_{b(\text{adj})} = SS_{\text{cells}} - SS_{ab(\text{adj})} - SS_a = 1672.$$

As a partial check on the numerical work,

$$\Sigma w_j d_j^2 = R(\underline{\alpha} \mid \mu,\underline{\beta}) + R(\underline{\alpha\beta} \mid \mu,\underline{\alpha},\underline{\beta}) = SS_{a(\text{adj})} + SS_{ab(\text{adj})}$$
$$= 110{,}202 + 3183$$
$$= 113{,}385.$$

Also $$\frac{(\Sigma w_j d_j)^2}{\Sigma w_j} = SS_{a(\text{adj})} = 110{,}202.$$

From part iii of Table 6.14-1 one obtains

$$SS_{\text{error}} = (2) - (5) = 57{,}695.$$

A summary of the analysis of variance is given in Table 6.14-3. Tests follow the same pattern as that of the usual factorial experiment in which model I is appropriate.

Table 6.14-3 Summary of Analysis of Variance (Least Squares)

Source of variation	SS	df	MS
$R(\underline{\alpha} \mid \mu,\beta) = A(\text{adj})$	110,202	1	110,202
$R(\underline{\beta} \mid \mu,\alpha) = B(\text{adj})$	1,672	3	557
$R(\underline{\alpha\beta} \mid \mu,\alpha,\beta) = AB(\text{adj})$	3,183	3	1,060
Error	57,695	141	409.2

The computation of

$$SS_{b(\text{adj})} = R(\underline{\beta} \mid \mu,\underline{\alpha})$$

by means of the procedures discussed in Sec. 5.23-3 is illustrated in

Table 6.14-4. In Sec. 5.23-3 it was indicated that the reduced normal equation for $\underline{\hat{\beta}}$ was

$$\underline{\hat{\beta}} M_b = \underline{v}_b,$$

and that

$$\underline{\hat{\beta}} = E_b \underline{v}_b,$$

Table 6.14-4 Computation of $SS_{b(\text{adj})}$

(i) Cell frequencies:

	b_1	b_2	b_3	b_4	Total
a_1	21	15	12	7	55
a_2	27	25	23	19	94
Total	48	40	35	26	149

(ii)

$$D_a = \begin{bmatrix} 55 & 0 \\ 0 & 94 \end{bmatrix} \qquad D_b = \begin{bmatrix} 48 & 0 & 0 & 0 \\ 0 & 40 & 0 & 0 \\ 0 & 0 & 35 & 0 \\ 0 & 0 & 0 & 26 \end{bmatrix}$$

$$C = \begin{bmatrix} 21 & 15 & 12 & 7 \\ 27 & 25 & 23 & 19 \end{bmatrix} \qquad \underline{a} = \begin{bmatrix} 9203 \\ 10334 \end{bmatrix} \qquad \underline{b} = \begin{bmatrix} 6673 \\ 5274 \\ 4364 \\ 3226 \end{bmatrix}$$

(iii)

$$M_b = D_b - C'D_a^{-1}C = \begin{bmatrix} 32.2265 & -12.9081 & -11.1882 & -8.1302 \\ & 29.2602 & -9.3897 & -6.9623 \\ & \text{symmetric} & 26.7542 & -6.1762 \\ & & & 21.2687 \end{bmatrix}$$

$$\underline{v}_b = \underline{b} - C'D_a^{-1}\underline{a} = \begin{bmatrix} 190.85 \\ 15.69 \\ -172.46 \\ -34.08 \end{bmatrix}$$

(iv)

	b_1	b_2	b_3	b_4	$\underline{v}_b$
b_1	32.2265	−12.9081	−11.1882	−8.1302	190.85
b_2		29.2602	−9.3897	−6.9623	15.69
b_3			26.7542	−6.1762	−172.46
b_4				21.2687	−34.08
b_1'	5.6769	−2.2739	−1.9708	−1.4322	33.6187
b_2'		4.9081	−2.8262	−2.0821	18.7721
b_3'			3.8578	−3.8579	−13.7774

$$R(\underline{\beta} \mid \mu, \underline{\alpha}) = (33.6187)^2 + (18.7721)^2 + (-13.7774)^2 = 1672.43$$

where E_b was a generalized inverse of M_b. Hence

$$R(\underline{\beta} \mid \mu,\underline{\alpha}) = \underline{\beta}'\underline{v}_b.$$

One may, however, obtain $R(\underline{\beta} \mid \mu,\underline{\alpha})$ without first obtaining an explicit solution for $\hat{\underline{\beta}}$. The algorithm for doing this is outlined in Table 6.14-4.

In part i of this table are the cell and marginal frequencies corresponding to those given in Table 6.14-1. In part ii, the diagonal elements of D_a are the marginal frequencies for the levels of factor A; similarly, D_b contains the marginal frequencies for the levels of factor B. The elements of the matrix C are the cell frequencies. The elements of the vector $\underline{a}$ are the totals A_1 and A_2 as given in part ii of Table 6.14-1. The elements of the vector $\underline{b}$ are the treatment totals B_1, B_2, B_3, B_4 as given in Table 6.14-1.

The matrix M_b is computed in part iii. Note that the row totals are zero. The entries in the matrix M_b are the coefficients of $\hat{\beta}_1, \ldots, \hat{\beta}_4$ in the reduced normal equations for the levels of factor B. The reduced normal equations are those obtained from the complete set (which includes the levels of both factors A and B) when the effects of the levels of factor A are "swept out" of the complete set. What remains in the reduced set is that part of the effects of the levels of factor B which is orthogonal to the effects associated with the levels of factor A. The components of the vector $\underline{v}_b$ represent the right-hand side of the reduced normal equations.

In part iv of Table 6.14-4 the Dwyer algorithm (as described in Sec. 2.5) is applied to the first three rows of the matrix

$$[M_b \quad \underline{v}_b],$$

that is, the matrix M_b augmented by the column vector $\underline{v}_b$. Since the matrix M_b is singular (actually of rank $q - 1$), the Dwyer algorithm cannot proceed beyond the third row. The entries in column $\underline{v}_b$ in the lower half of part iv represent the following:

$$(33.6187)^2 = R(\beta_1 \mid \mu,\underline{\alpha},),$$
$$(18.7721)^2 = R(\beta_2 \mid \mu,\underline{\alpha},\beta_1),$$
$$(-13.7774)^2 = R(\beta_3 \mid \mu,\underline{\alpha},\beta_1,\beta_2).$$

Hence $R(\underline{\beta} \mid \mu,\underline{\alpha})$ may be computed as indicated in the last row of part iv. One has

$$SS_{b(\text{adj})} = R(\underline{\beta} \mid \mu,\underline{\alpha}) = 1672.$$

From the latter value, one obtains

$$SS_{a(\text{adj})} = SS_a - SS_b + SS_{b(\text{adj})}$$
$$= 114{,}286 - 5756 + 1672 = 110{,}202.$$

Also

$$SS_{ab(\text{adj})} = SS_{\text{cells}} - SS_a - SS_{b(\text{adj})}$$
$$= 119{,}141 - 114{,}286 - 1672 = 3183.$$

This last value is equal to that obtained from the direct computation in Table 6.14-2.

By way of contrast to the least-squares approach, the unweighted-means analysis for the data in Table 6.14-1 is summarized in Table 6.14-5. The cell entries in part i are the cell means. The rows and marginal totals for the cell means are also given in part i. Thus

$$A_1 = \Sigma \overline{AB}_{1j} = 665.0858,$$
$$B_1 = \Sigma \overline{AB}_{i1} = 286.4709.$$

Table 6.14-5 Unweighted-means Analysis

Cell means and marginal totals of cell means:

		b_1	b_2	b_3	b_4	Total
(i)	a_1	176.9524	161.4667	155.6667	171.0000	665.0858
	a_2	109.5185	114.0900	108.5217	106.7895	438.9197
	Total	286.4709	275.5567	264.1884	277.7895	1104.0055
		B_1	B_2	B_3	B_4	G

(ii)

$$(1) = G^2/8 = 152353.52 \qquad (4) = \Sigma B^2/2 = 152479.79$$
$$(3) = \Sigma A^2/4 = 158747.41 \qquad (5) = \Sigma(\overline{AB})^2 = 159049.19$$
$$\bar{n}_h = 15.575687$$

(iii)

$$SS_a = \bar{n}_h[(3) - (1)] = 99589$$
$$SS_b = \bar{n}_h[(4) - (1)] = 1967$$
$$SS_{ab} = \bar{n}_h[(5) - (3) - (4) + (1)] = 2744$$

The computational symbols are defined and computed in part ii of the table. The harmonic mean of the cell frequencies is also given in part ii. Thus

$$\bar{n}_h = \frac{8}{\Sigma(1/n_{ij})}.$$

In part iii the variations due to main effects and interaction are computed. A summary of the analysis of variance appears in Table 6.14-6.

Table 6.14-6 Summary of Unweighted-means Analysis of Variance

Source of variation	SS	df	MS
A	99,589	1	99,589
B	1,967	3	656
AB	2,744	3	915
Error		141	409.2

In this case the unweighted-means analysis and the least-squares analysis give roughly comparable results. The variance due to error is the same in both analyses.

The weighted-means analysis for main effects (as described in Sec. 5.23-6) is illustrated in Table 6.14-7. In part i, the entries in row h_i are the harmonic means of the cell frequencies in row a_i. The entries in row $\bar{A}_i$ are the unweighted means of the cell means at level a_i. Thus, from Table 6.14-5, one has

$$\bar{A}_1 = \frac{\sum_j \overline{AB}_{1j}}{q} = \frac{665.0858}{4} = 166.2714.$$

Table 6.14-7 Weighted-means Analysis for Main Effects

		a_1	a_2	
(i)	$h_i = \dfrac{q}{\Sigma(1/n_{ij})}$	11.7483	23.1011	
	$w_i = h_i/q$	2.9371	5.7753	$\Sigma w_i = 8.7124$
	$\bar{A}_i$	166.2714	109.7299	
	$w_i\bar{A}_i$	488.3557	633.7231	$\Sigma w_i\bar{A}_i = 1122.0788$

(ii)
$$SS_a = q^2\left[\Sigma w_i\bar{A}_i^2 - \frac{(\Sigma w_i\bar{A}_i)^2}{\Sigma w_i}\right]$$
$$= 16(150{,}737.9583 - 144{,}513.6625) = 99{,}588.73$$

		b_1	b_2	b_3	b_4	
(iii)	$h_j = \dfrac{p}{\sum_i(1/n_{ij})}$	23.6250	18.7500	15.7714	10.2308	
	$w_j = h_j/p$	11.8125	9.3700	7.8857	5.1154	$\Sigma w_j = 34.1836$
	$\bar{B}_j$	143.2354	137.7783	132.0942	138.8947	
	$w_j\bar{B}_j$	1691.9682	1290.9827	1041.6552	710.5019	$\Sigma w_j\bar{B}_j = 4735.1084$

(iv)
$$SS_b = p^2\left[\Sigma w_j\bar{B}_j^2 - \frac{(\Sigma w_j\bar{B}_j)^2}{\Sigma w_j}\right]$$
$$= 4(656{,}500.4380 - 655{,}906.5684) = 2375.48$$

Following the procedures outlined in Sec. 5.23-6, the variation due to the main effects of factor A is computed in part ii.

Since factor A has only two levels in this case, one may use the following comparison to obtain SS_a:

$$C_a = \sum_j \overline{AB}_{1j} - \sum_j \overline{AB}_{2j}.$$

Thus
$$SS_a = \frac{C_a^2}{\Sigma(1/n_{ij})}.$$

From the data in Table 6.14-5,

$$C_a = 665.0858 - 438.9197 = 226.1661,$$

$$SS_a = \frac{(226.1661)^2}{.513621} = 99589.$$

6.15 Analysis of Variance in Terms of Polynomial Regression

The sources of variation in both orthogonal and nonorthogonal experiments may be obtained by using the usual techniques of polynomial regression, provided appropriate terms are included in the polynomial. However, it is not always obvious how one should combine selected sources of variation from polynomial regression to obtain the variation due to main effects and interactions in an analysis of variance. A relatively simple numerical example will be used to illustrate some of the basic principles.

Consider a 4×4 factorial experiment in which the levels of both factors are quantitative steps along an underlying quantitative continuum. If the levels were qualitative rather than quantitative, polynomial regression might be carried out in terms of indicator variables representing the presence or absence of a treatment combination corresponding to a cell. Various coding devices have been used for such indicator variables.

Suppose factor A has four quantitative levels, namely, 0, 2, 4, 6. Suppose factor B has the four quantitative levels 1, 3, 5, 7. Further, suppose the cell totals for each of the treatment combinations are those in Table 6.15-1. The between-cell variation has been partitioned into sources

Table 6.15-1 Cell Totals

		Levels of factor B				
		1	3	5	7	
Levels of factor A	0	10	26	58	106	
	2	24	52	96	156	
	4	38	78	134	206	$n = 5$
	6	52	104	172	256	

due to main effects and interactions in Table 6.15-2. The between-cell variation for the data in this table is given by

$$SS_{b.cell} = n\Sigma(\overline{AB}_{ij} - \bar{G})^2$$
$$= \frac{\Sigma(AB_{ij})^2}{n} - \frac{G^2}{npq}.$$

For this example all the variation due to the main effects of factor A is concentrated in the linear trend. Further, all the variation due to the main effects of factor B is concentrated in the linear and quadratic trends; all the variation due to interaction is concentrated in the linear $\times$ linear component. This concentration of the total between-cell variation into four trend components simplifies the work that follows.

Table 6.15-2 Between-cell Analysis of Variance

(i)

Source of variation		SS			df
Between cell:			15,020.80		15
A			4,096.00		3
	A_{lin}	4,096.00		1	
	All other	0.00		2	
B			10,204.80		3
	B_{lin}	10,000.00		1	
	B_{quad}	204.80		1	
	B_{cubic}	0.00		1	
$A \times B$			720.00		9
	$A_{\text{lin}} \times B_{\text{lin}}$	720.00		1	
	All other	0.00		8	

(ii)

Source	r^2	r
A_{lin}	$SS_{A_{\text{lin}}}/SS_{\text{b.cell}} = .27268$	.52218
B_{lin}	$SS_{B_{\text{lin}}}/SS_{\text{b.cell}} = .66574$	.81593
B_{quad}	$SS_{B_{\text{quad}}}/SS_{\text{b.cell}} = .01363$	.11672
$A_{\text{lin}} \times B_{\text{lin}}$	$SS_{A_{\text{lin}} \times B_{\text{lin}}}/SS_{\text{b.cell}} = .04793$	.21896

In part ii of Table 6.15-2 are some of the correlations that may be computed to summarize the information about the between-cell variation. For example,

$$r^2_{B_{\text{lin}}} = \frac{SS_{B_{\text{lin}}}}{SS_{\text{b.cell}}} = .66574$$

is a measure of the proportion of the between-cell variation that is due to the linear component of the main effects of factor B. The correlations in part ii will be related to multiple and semipartial correlations obtained from polynomial regression.

Let the variables included in the regression equation be the following:

$$X_1 = \text{levels of factor } A,$$
$$X_2 = \text{levels of factor } B,$$
$$X_3 = X_2^2,$$
$$X_4 = X_1X_2.$$

The definitions of X_3 and X_4 are given by hindsight in this case merely to illustrate principles. Ordinarily one might also include such variables as X_1^2, $X_1^2X_2$, $X_1X_2^2$, and $X_1^2X_2^2$. In terms of the selected variables, the data in Table 6.15-1 may be represented in the form given in Table 6.15-3.

Table 6.15-3 Between-cell Data in Terms of Regression Variables

Cell	X_1	X_2	$X_3 = X_2^2$	$X_4 = X_1X_2$	Y
01	0	1	1	0	10
03	0	3	9	0	26
05	0	5	25	0	58
07	0	7	49	0	106
21	2	1	1	2	24
23	2	3	9	6	52
25	2	5	25	10	96
27	2	7	49	14	156
41	4	1	1	4	38
43	4	3	9	12	78
45	4	5	25	20	134
47	4	7	49	28	206
61	6	1	1	6	52
63	6	3	9	18	104
65	6	5	25	30	172
67	6	7	49	42	256

From the latter table one may obtain the matrix of intercorrelations given in Table 6.15-4.

Table 6.15-4 Intercorrelation of Regression Variables

	X_1	X_2	$X_3 = X_2^2$	$X_4 = X_1X_2$	Y
X_1	1.00000	.00000	.00000	.73209	.52218
X_2		1.00000	.97590	.54772	.81593
X_3			1.00000	.53451	.82174
X_4				1.00000	.91764
Y					1.00000

The linear regression equations given in Table 6.15-5 may be obtained from the matrix of intercorrelations. These regression equations are in terms of the standardized variables. That is,

$$X_1^* = \frac{X_1 - \bar{X}_1}{s_{X_1}}, \qquad X_2^* = \frac{X_2 - \bar{X}_2}{s_{X_2}}, \qquad \text{etc.}$$

Table 6.15-5 Regression Equations and Multiple Correlations

(1) $Y^*_{12} = .52218X^*_1 + .81593X^*_2$		
	$r^2_{Y(12)} = .93841$	
	$r^2_{Y1} = .27267$	A_{lin}
	$r^2_{Y2} = .66574$	B_{lin}
(2) $Y^*_{123} = .52218X^*_1 + .29387X^*_2 + .53495X^*_3$		
	$r^2_{Y(123)} = .95204$	
	$r^2_{Y1} = .27267$	A_{lin}
	$r^2_{Y2} = .66574$	B_{lin}
	$r^2_{Y(3.12)} = .013623$	B_{quad}
(3) $Y^*_{124} = .13055X^*_1 + .52219X^*_2 + .53629X^*_4$		
	$r^2_{Y(124)} = .98635$	
	$r^2_{Y1} = .27267$	A_{lin}
	$r^2_{Y2} = .66574$	B_{quad}
	$r^2_{Y(4.12)} = .047939$	$A_{\text{lin}} \times B_{\text{lin}}$
(4) $Y^*_{134} = .13043X^*_1 + .53501X^*_3 + .53643X^*_4$		
	$r^2_{Y(134)} = 1.00000$	
	$r^2_{Y1} = .27267$	A_{lin}
	$r^2_{Y3} = .67525$	$B_{\text{lin}} + B_{\text{quad}} + A_{\text{lin}} \times B_{\text{lin}}$
	$r^2_{Y(4.13)} = .052071$	

Equation (1) in this table, which uses only the linear information in X_1 and X_2, accounts for 93.841 percent of the between-cell variation ($r^2_{Y(12)} = .93841$). Since X_1 and X_2 are uncorrelated,

$$r^2_{Y(12)} = r^2_{Y1} + r^2_{Y2}.$$

One notes that r^2_{Y1} in the regression analysis is numerically equal to $SS_{A_{\text{lin}}}/SS_{\text{b.cell}}$ in the analysis of variance. Similarly,

$$r^2_{Y2} = \frac{SS_{B_{\text{lin}}}}{SS_{\text{b.cell}}}.$$

In equation (2) in Table 6.15-5, the regression includes the variables X_1, X_2, and X_3. One notes that X_3 is uncorrelated with X_1 but is highly correlated with X_3. Thus

$$r^2_{Y(123)} = r^2_{Y1} + r^2_{Y2} + r^2_{Y(3.12)}.$$

The contribution of X_3 above and beyond X_1 and X_2 is measured by the semipartial correlation

$$r^2_{Y(3.12)} = r^2_{Y(3.2)} = r^2_{Y(123)} - r^2_{Y1} - r^2_{Y2}$$
$$= .013623 = \frac{SS_{B_{quad}}}{SS_{b.cell}}.$$

In equation (3), $r^2_{Y(4.12)}$ represents the proportion of the between-cell variation which is a function of the $A_{lin} \times B_{lin}$ interaction. Associated with equation (4) one has

$$r^2_{Y(34.1)} = r^2_{Y3} + r^2_{Y(4.13)} = \frac{SS_{b.\ cell} - SS_{A_{lin}}}{SS_{b.cell}}.$$

An alternative way of presenting the information in Table 6.15-5 is given in Table 6.15-6. In part v of this table is an orthogonal partition of the between-cell variation in terms of semipartial correlations.

Table 6.15-6 Analysis of Multiple Regression

	Regression source	SS	df	Analysis of variance source
	X_1, X_2	$r^2_{Y(12)}SS_{b.cell} = 14{,}096$	2	$A_{lin} + B_{lin}$
(i)	X_1	$r^2_{Y1}SS_{b.cell} = 4{,}096$	1	A_{lin}
	X_2	$r^2_{Y2}SS_{b.cell} = 10{,}000$	1	B_{lin}
	X_1, X_2, X_3	$r^2_{Y(123)}SS_{b.cell} = 14{,}300$	3	$A_{lin} + B_{lin} + B_{quad}$
	X_1	$r^2_{Y1}SS_{b.cell} = 4{,}096$	1	A_{lin}
(ii)	X_2	$r^2_{Y2}SS_{b.cell} = 10{,}000$	1	B_{lin}
	$X_{3.12}$	$r^2_{Y(3.12)}SS_{b.cell} = 720$	1	B_{quad}
	X_1, X_2, X_4	$r^2_{Y(124)}SS_{b.cell} = 14{,}816$	3	$A_{lin} + B_{lin} + A_{lin} \times B_{lin}$
	X_1	$r^2_{Y1}SS_{b.cell} = 4{,}096$	1	A_{lin}
(iii)	X_2	$r^2_{Y2}SS_{b.cell} = 10{,}000$	1	B_{lin}
	X_4	$r^2_{Y(4.12)}SS_{b.cell} = 205$	1	$A_{lin} \times B_{lin}$
	X_1, X_3, X_4	$r^2_{Y(134)}SS_{b.cell} = 15{,}021$	3	$A_{lin} + B_{lin} + B_{quad} + A_{lin} \times B_{lin}$
	X_1	$r^2_{Y1}SS_{b.cell} = 4{,}096$	1	A_{lin}
(iv)	X_3	$r^2_{Y3}SS_{b.cell} = 10{,}143$	1	$B_{lin} + B_{quad} + A_{lin} \times B_{lin}$
	$X_{4.13}$	$r^2_{Y(4.13)}SS_{b.cell} = 782$	1	
	X_1, X_2, X_3, X_4	$r^2_{Y(1234)}SS_{b.cell} = 15{,}021$	4	$A_{lin} + B_{lin} + B_{quad} + A_{lin} \times B_{lin}$
	X_1	$r^2_{Y1}SS_{b.cell} = 4{,}096$	1	A_{lin}
(v)	X_2	$r^2_{Y2}SS_{b.cell} = 10{,}000$	1	B_{lin}
	$X_{3.12}$	$r^2_{Y(3.12)}SS_{b.cell} = 205$	1	B_{quad}
	$X_{4.123}$	$r^2_{Y(4.123)}SS_{b.cell} = 720$	1	$A_{lin} \times B_{lin}$

For this numerical example, it is relatively easy to set up a one-to-one correspondence between the sources of variation in the analysis of variance and the sources of variation in polynomial regression. Semipartial and multiple correlations are useful here. This simplicity is due to the orthogonality of the design. When the design is not orthogonal, it is no longer quite so simple to set up such a correspondence, although it is always possible to do so by an appropriate selection of cross-product terms which are included in the polynomial.

It is also possible to express a given polynomial in terms of the sum of a set of orthogonal polynomials. This procedure can be shown to be equivalent to transforming the variables used in the regression analysis to a suitably orthogonal set.

One of the potential advantages of the polynomial-regression approach is that it may be used to fit data in experiments (or surveys) in which there are no observations in one or more cells of what might form a factorial experiment. However, the various sources of variation in an analysis in terms of the usual polynomial regression will not be orthogonal. There is considerable potential risk in making interpretations through use of semipartial correlations based upon incomplete data.

Qualitative Levels. If the levels of a factor are qualitative (i.e., red, white, and blue) rather than quantitative, various coding devices may be used to partition the between-cell variation by means of polynomial regression. Since the qualitative levels of a factor usually cannot be ordered in any quantitative sense, any quantitative coding that is used is generally quite arbitrary insofar as interpretations are concerned. Hence the coding system that is adopted is a matter of convenience. If the cell frequencies are all equal, it is quite convenient to use the coefficients of orthogonal polynomials to build a coding system that will lead to an orthogonal partition of the between-cell variation. Interpretation of the resulting partition in terms of trend components is, of course, no longer possible. The use of the coefficients of orthogonal polynomials is merely a device for constructing a coding system in terms of which the resulting variables are uncorrelated.

To illustrate the discussion in the preceding paragraph, consider a 3×3 factorial experiment in which levels of both factors are qualitative. Suppose the variables X_1 through X_8 are defined as indicated in Table 6.15-7. (The coding system in terms of the levels of the separate factors is given at the bottom of this table.) The entries under column X_1 correspond to the linear trend due to the main effects of factor A; the entries under column X_2 correspond to the quadratic trend due to factor A. In terms of the treatment combinations, the entries in columns X_1 and X_2 define two orthogonal comparisons that belong to the main effects of factor A. Similarly, the entries in columns X_3 and X_4 define two orthogonal comparisons belonging to the main effects of factor B.

Table 6.15-7 Coding for Cells in 3×3 Factorial

Cell	X_1	X_2	X_3	X_4	$X_5 = X_1X_3$	$X_6 = X_1X_4$	$X_7 = X_2X_3$	$X_8 = X_2X_4$
ab_{11}	−1	1	−1	1	1	−1	−1	1
ab_{12}	−1	1	0	−2	0	2	0	−2
ab_{13}	−1	1	1	1	−1	−1	1	1
ab_{21}	0	−2	−1	1	0	0	2	−2
ab_{22}	0	−2	0	−2	0	0	0	4
ab_{23}	0	−2	1	1	0	0	−2	−2
ab_{31}	1	1	−1	1	−1	1	−1	1
ab_{32}	1	1	0	−2	0	−2	0	−2
ab_{33}	1	1	1	1	1	1	1	1
	A_{lin}	A_{quad}	B_{lin}	B_{quad}	$A_{\text{lin}} \times B_{\text{lin}}$	$A_{\text{lin}} \times B_{\text{quad}}$	$A_{\text{quad}} \times B_{\text{lin}}$	$A_{\text{quad}} \times B_{\text{quad}}$

	X_1	X_2
a_1	−1	1
a_2	0	−2
a_3	1	1

	X_3	X_4
b_1	−1	1
b_2	0	−2
b_3	1	1

Variables X_5 through X_8 are defined as indicated. These columns define comparisons belonging to the interaction.

The columns of Table 6.15-7 are pairwise orthogonal. Hence, if there are equal cell frequencies, the variables $X_1, \ldots, X_8$ will be uncorrelated. In terms of standardized variables, the regression equation will have the form

$$Y^* = r_{Y1}X_1^* + r_{Y2}X_2^* + \cdots + r_{Y8}X_8^*,$$

where

$$r_{Yj} = \text{product-moment correlation between } Y \text{ and } X_j.$$

The variation due to variable X_j is

$$r_{Yj}^2 \text{SS}_{\text{b.cell}}.$$

Since nine points can be fitted exactly by a regression equation having eight variables (provided the matrix of intercorrelations is nonsingular), it follows that

$$\sum_j r_{Yj}^2 = 1.$$

From the way in which the variables have been coded,

$$\text{SS}_a = (r_{Y1}^2 + r_{Y2}^2)\text{SS}_{\text{b.cell}},$$
$$\text{SS}_b = (r_{Y3}^2 + r_{Y4}^2)\text{SS}_{\text{b.cell}},$$
$$\text{SS}_{ab} = (r_{Y5}^2 + r_{Y6}^2 + r_{Y7}^2 + r_{Y8}^2)\text{SS}_{\text{b.cell}}.$$

If the design is not orthogonal, it is not always possible to make the correspondence between the analysis of variance and regression analysis in this simple manner.

If the cell frequencies are not equal, but the equivalent of an unweighted-means analysis is wanted, then the between-cell variation is defined as

$$SS_{b.cell} = \tilde{n}_h \Sigma(\overline{AB}_{ij} - \bar{G})^2.$$

If the cell means are used as the basic data (that is, if $\overline{AB}_{ij} = Y_{ij}$), then $X_1, \ldots, X_8$ will still represent an uncorrelated set of variables in the regression analysis.

If the cell frequencies are unequal and a weighted-means solution is wanted, then

$$SS_{b.cell} = \Sigma n_{ij}(\overline{AB}_{ij} - \bar{G})^2.$$

In this case, each row of Table 6.15-7 is multiplied by the corresponding n_{ij}. For example, the first row will be

$$-n_{11} \quad n_{11} \quad -n_{11} \quad n_{11} \quad n_{11} \quad -n_{11} \quad -n_{11}.$$

Again the cell means are taken as the basic data. In this case the variables $X_1, \ldots, X_8$ will not, in general, be uncorrelated. Several different orthogonal partitions of $SS_{b.cell}$ are possible in the latter case.

Instead of the coding system given in Table 6.15-7, the alternative coding system given in Table 6.15-8 might be used. The disadvantage of this coding system is that the variables $X_1, \ldots, X_8$ are not orthogonal even if the cell frequencies are equal.

Table 6.15–8 Alternative Coding System for Cells in a 3 × 3 Factorial

Cell	X_1	X_2	X_3	X_4	$X_5 = X_1X_3$	$X_6 = X_1X_4$	$X_7 = X_2X_3$	$X_8 = X_2X_4$
ab_{11}	−1	−1	−1	−1	1	1	1	1
ab_{12}	−1	−1	0	1	0	−1	0	−1
ab_{13}	−1	−1	1	0	−1	0	−1	0
ab_{21}	0	1	−1	−1	0	0	−1	−1
ab_{22}	0	1	0	1	0	0	0	1
ab_{23}	0	1	1	0	0	0	1	0
ab_{31}	1	0	−1	−1	−1	−1	0	0
ab_{32}	1	0	0	1	0	1	0	0
ab_{33}	1	0	1	0	1	0	0	0

	X_1	X_2
a_1	−1	−1
a_2	0	1
a_3	1	0

	X_3	X_4
b_1	−1	−1
b_2	0	1
b_3	1	0

Summary. Regression techniques of the kind discussed in this section have been used in the analysis of rather loosely designed experiments to identify those sources of variation that may be large. In this role, all potential variables are used in setting up the initial regression analysis;

then various selection techniques (such as the teardown procedure) are used to identify the set of variables having the largest contribution to the between-treatment variation. Conversely, one may use the information from the analysis of variance of a well-designed experiment to define the variables that are selected for inclusion in the regression equation. The latter is used for prediction purposes. What are called response-surface analysis techniques include relatively refined methods for regression analysis.

7

MULTIFACTOR EXPERIMENTS HAVING REPEATED MEASURES ON THE SAME ELEMENTS

7.1 General Purpose

Factorial experiments in which the same experimental unit (generally a subject) is observed under more than one treatment condition require special attention. Experiments of this kind will be referred to as those in which there are repeated measures. Single-factor experiments of this kind were discussed in considerable detail in Chap. 4. Just as in the case of single-factor designs having repeated measures, the designs that are discussed in this chapter are sometimes handled more efficiently through the techniques of multivariate rather than univariate analysis of variance. However, under a set of restrictions on the pattern of the parameters in the variance-covariance matrices associated with these designs, the multivariate procedures become equivalent to those used in the univariate analysis of variance.

Different approaches have been used in handling the sampling distributions of mean squares which arise in the analysis of experiments. These approaches differ to some extent in the definition of the experimental unit. Under one approach, which assumes a multivariate normal underlying distribution in the population, the experimental errors are considered to be correlated. Under an alternative approach, a nested random factor (a kind of dummy variable) is included in the model to absorb the correlation between the experimental errors. This approach has the advantage of permitting a relatively simple derivation of the expected values of the mean squares. The disadvantage of the latter approach is that it tends to mask assumptions about the pattern of the covariance matrix required to justify the form of the sampling distribution of the final F ratio used in making tests. Both approaches lead to identical F ratios when the assumptions are met.

A two-factor experiment in which there are repeated measures on factor B (that is, each experimental unit is observed under all levels of factor B) may be represented schematically as follows:

	b_1	b_2	b_3
a_1	G_1	G_1	G_1
a_2	G_2	G_2	G_2

The symbol G_1 represents a group of n subjects. The symbol G_2 represents a second group of n subjects. The subjects in G_1 are observed under treatment combinations ab_{11}, ab_{12}, and ab_{13}. Thus the subjects in G_1 are observed under all levels of factor B in the experiment, but only under one level of factor A. The subjects in G_2 are observed under treatment combinations ab_{21}, ab_{22}, and ab_{23}. Thus each subject in G_2 is observed under all levels of factor B in the experiment, but only under one level of factor A, namely, a_2.

In this design, the subjects may be considered to define a third factor having n levels. As such, the "subject" factor is crossed with factor B but nested under factor A. Schematically,

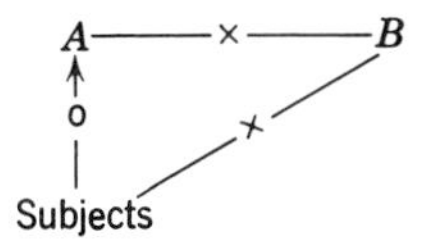

where × indicates a crossing relationship,
○ indicates a nesting relationship.

In this kind of experiment, comparisons between treatment combinations at different levels of factor A involve differences between groups as well as differences associated with factor A. On the other hand, comparisons between different levels of factor B at the same level of A do not involve differences between groups. Since measurements included in the latter comparisons are based upon the same elements, main effects associated with such elements tend to cancel. For the latter comparisons, each element serves as its own control with respect to such main effects.

In an experiment of the type represented above, the main effects of factor A are said to be completely confounded with differences between groups. On the other hand, the main effects of factor B as well as the AB interaction will be shown to be free of such confounding. Tests on B and AB will generally be more sensitive than tests on the main effects of factor A. Where no confounding with the group factor is present, there are fewer uncontrolled sources of error variance. The smaller the error variance, the more sensitive (powerful) the test.

By using the approach which assumes that the errors are correlated, the expected value of the variance due to the main effects of factor A has the general form (assuming factor B is fixed)

$$\mathrm{E}(\mathrm{MS}_a) = \sigma_{\varepsilon'}^2[1 + (q - 1)\rho] + nq\sigma_\alpha^2,$$

where ρ is the (constant) correlation between pairs of observations on the same element and q is the number of levels of factor B. The expected value of the mean square appropriate for the denominator of an F ratio used in testing the variance due to the main effect of factor A has the form

$$\sigma_{\varepsilon'}^2[1 + (q - 1)\rho].$$

(See Sec. 4.4.)

For sources of variation which are not confounded with the main effects due to differences between groups, the denominator of an F ratio has the expected value

$$\sigma_{\varepsilon'}^2(1 - \rho).$$

Thus, if correlation between pairs of measurements is positive and constant, the latter experimental error will be smaller than the former. Hence the greater sensitivity (i.e., the greater power) of tests using the latter error term.

In terms of the alternative approach, which postulates the existence of a nested random factor, the expected value for the denominator of an F ratio for a test on the main effects of factor A has the form (assuming factor B is fixed)

$$\sigma_\varepsilon^2 + q\sigma_\pi^2,$$

where σ_π^2 is the variation due to the main effects of subjects within the groups. The expected value for the denominator for tests on effects which are not confounded with main effects due to subjects within the groups has the form

$$\sigma_\varepsilon^2 + \sigma_{\beta\pi}^2,$$

where $\sigma_{\beta\pi}^2$ is the interaction between the subject and treatment factors. The magnitude of $\sigma_{\beta\pi}^2$ is generally considerably smaller than σ_π^2. (It should be noted that $\sigma_{\varepsilon'}^2$ includes more sources of variance than does σ_ε^2.)

Repeated measures on the same elements may arise in different ways. In experiments designed to study rates of learning as a function of treatment effects, repeated measures on the same subject are a necessary part of the design. Further, the order in which the observations are made is dictated by the experimental variables. On the other hand, in experiments designed to evaluate the joint effect of two or more treatments the experimenter may have his option as to whether or not the same elements are observed under more than one treatment combination. Further, the order in which the elements appear under the treatment combinations may also be under the control of the experimenter. The utility of designs calling for repeated measures is limited where carry-over effects are likely to confound results. In some cases such effects may be controlled by counterbalancing the order in which treatment combinations are given to the elements. (Designs of this kind are discussed in Sec. 10.7.)

Aside from designs having the form of learning experiments, the primary purpose of repeated measures on the same elements is the control that this kind of design provides over individual differences between experimental units. In the area of the behavioral sciences, differences between such units often are quite large relative to differences in treatment effects which the experimenter is trying to evaluate. If there are no carry-over effects, a repeated-measure design in the area of the behavioral sciences is to some degree analogous to a split-plot design in the area of agricultural experimentation. Where additive carry-over effects are present, repeated-measure designs are analogous to crossover designs.

Another (somewhat doubtful) advantage of a repeated-measure design is in terms of economy of subjects. Using different subjects under each of the treatment combinations in a factorial experiment has the *marked advantage* of providing statistically independent estimates of treatment effects from all cells in the experiment. Increasing the number of statistically independent observations is very likely to be the best way of increasing the precision of estimators. By having each subject serve as his own control, the experimenter attempts to work with a smaller sample size. However, the simple additive model underlying the usual analysis for the case of repeated measures may not be an adequate representation of the experimental phenomena. A more inclusive multivariate regression model is often required to represent fully the underlying experimental variables.

In sections that follow, special cases of factorial designs having repeated measures will be illustrated. The expected values of the mean square for these special cases will be given. The latter are obtained by means of the methods developed by Cornfield and Tukey (1956). The validity of the tests set up on the basis of such expected values rests upon assumptions about the form of the variance-covariance matrix associated with the joint multivariate normal distribution of the random variables in the model. Unless the nature of the experimental variables dictates the order in which treatments are administered to subjects, it will be assumed that the order of administration is randomized independently for each of the subjects. Further, it will be assumed that the n elements in a group are a random sample from a specified population of elements.

A strong word of warning is required in connection with order (or sequence) effects. Practice, fatigue, transfer of training, the effects of an immediately preceding success or failure illustrate what fall in the latter category. If such effects exist, randomizing or counterbalancing does not remove them; rather, such procedures completely entangle the latter with treatment effects. There is some chance that sequence effects will balance out—they generally will if a simple additive model is realistic. However, in experiments (other than those which primarily are concerned with learning or carry-over effects) where the sequence effects are likely to be marked and where primary interest lies in evaluating the effect of individual treatments in the absence of possible sequence effects, a repeated-measure design is to be avoided.

In cases where sequence effects are likely to be small relative to the treatment effects, repeated-measure designs can be used. Counterbalancing or randomizing order of administration in this case tends to prevent sequence effects from being completely confounded with one or just a selected few of the treatments. Instead such sequence effects are spread over all the treatment effects. Admittedly such sequence effects may serve to mask treatment effects; however, the potential advantages can outweigh the potential disadvantages.

7.2 Two-factor Experiment with Repeated Measures on One Factor

This kind of experiment was illustrated in the last section. The general case may be represented as follows:

	b_1	$\cdots$	b_j	$\cdots$	b_q
a_1	G_1	$\cdots$	G_1	$\cdots$	G_1
$\vdots$	$\vdots$		$\vdots$		$\vdots$
a_i	G_i	$\cdots$	G_i	$\cdots$	G_i
$\vdots$	$\vdots$		$\vdots$		$\vdots$
a_p	G_p	$\cdots$	G_p	$\cdots$	G_p

Each G represents a random sample of size n from a common population of subjects. Each of the subjects in G_i is observed under q different treatment combinations, all these treatment combinations involving factor A at level a_i. The actual observations on the subjects within group i may be represented as follows:

	Subject	b_1	$\cdots$	b_j	$\cdots$	b_q
	1	X_{i11}	$\cdots$	X_{ij1}	$\cdots$	X_{iq1}
	$\vdots$	$\vdots$		$\vdots$		$\vdots$
a_i	k	X_{i1k}	$\cdots$	X_{ijk}	$\cdots$	X_{iqk}
	$\vdots$	$\vdots$		$\vdots$		$\vdots$
	n	X_{i1n}	$\cdots$	X_{ijn}	$\cdots$	X_{iqn}

The symbol X_{ijk} denotes a measurement on subject k in G_i under treatment combination ab_{ij}. A more complete notation for subject k would be $k(i)$; the latter notation distinguishes this subject from subject k in some other group. Similarly a more complete notation for X_{ijk} would be $X_{ijk(i)}$. The latter notation is rather cumbersome; in cases in which there is no ambiguity, the symbol X_{ijk} will be used to indicate an observation on subject k in G_i made under treatment combination ab_{ij}.

For special cases the notation for a subject may be made more specific. For example, consider the case in which $p = 2$, $q = 3$, and $n = 2$. The experimental data may be represented as follows:

	Subject	b_1	b_2	b_3
a_1	1	X_{111}	X_{121}	X_{131}
	2	X_{112}	X_{122}	X_{132}
a_2	3	X_{213}	X_{223}	X_{233}
	4	X_{214}	X_{224}	X_{234}

In this notation scheme subject 3 is the first subject in G_2, and subject 4 is the second subject in G_2.

The linear model upon which the analysis will be based has the following form:

$$X_{ijk} = \mu + \alpha_i + \pi_{k(i)} + \beta_j + \alpha\beta_{ij} + \beta\pi_{jk(i)} + \varepsilon_{m(ijk)}.$$

The notation $\pi_{k(i)}$ indicates that the effect of subject k is nested under level a_i. Note that the linear model does not include any carry-over effects. From the Cornfield-Tukey algorithm as described in Sec. 5.14 the E(MS) associated with the model given above have the following form. (The dummy subscript m in the term $\varepsilon_{m(ijk)}$ is introduced in order that one may indicate that the experimental error is nested within the individual observation.)

Effect	i	j	k	m	Expected value of mean square
α_i	D_p	q	n	1	$\sigma_\varepsilon^2 + D_nD_q\sigma_{\beta\pi}^2 + nD_q\sigma_{\alpha\beta}^2 + qD_n\sigma_\pi^2 + nq\sigma_\alpha^2$
$\pi_{k(i)}$	1	q	D_n	1	$\sigma_\varepsilon^2 + D_q\sigma_{\beta\pi}^2 + q\sigma_\pi^2$
β_j	p	D_q	n	1	$\sigma_\varepsilon^2 + D_n\sigma_{\beta\pi}^2 + nD_p\sigma_{\alpha\beta}^2 + np\sigma_\beta^2$
$\alpha\beta_{ij}$	D_p	D_q	n	1	$\sigma_\varepsilon^2 + D_n\sigma_{\beta\pi}^2 + n\sigma_{\alpha\beta}^2$
$\beta\pi_{jk(i)}$	1	D_q	D_n	1	$\sigma_\varepsilon^2 + \sigma_{\beta\pi}^2$
$\varepsilon_{m(ijk)}$	1	1	1	1	σ_ε^2

Here $$D_n = 1 - \frac{n}{N}, \qquad D_p = 1 - \frac{p}{P}, \qquad D_q = 1 - \frac{q}{Q}.$$

Each D is either 0 or 1 depending upon whether the corresponding factor is fixed or random, respectively. If factors A and B are fixed, and the subject factor is random, then

$$D_n = 1, \qquad D_p = 0, \qquad D_q = 0.$$

If one makes this substitution for the D's in the E(MS) given above, the result is the E(MS) in Table 7.2-1.

Table 7.2-1 Summary of Analysis of Variance

Source of variation	df	E(MS)
Between subjects	$np - 1$	
A	$p - 1$	$\sigma_\varepsilon^2 + q\sigma_\pi^2 + nq\sigma_\alpha^2$
Subjects within groups	$p(n - 1)$	$\sigma_\varepsilon^2 + q\sigma_\pi^2$
Within subjects	$np(q - 1)$	
B	$q - 1$	$\sigma_\varepsilon^2 + \sigma_{\beta\pi}^2 + np\sigma_\beta^2$
AB	$(p - 1)(q - 1)$	$\sigma_\varepsilon^2 + \sigma_{\beta\pi}^2 + n\sigma_{\alpha\beta}^2$
$B \times$ subjects within groups	$p(n - 1)(q - 1)$	$\sigma_\varepsilon^2 + \sigma_{\beta\pi}^2$

The manner in which the total variation is partitioned in this table is quite similar to that used in a $p \times q$ factorial experiment in which there are no repeated measures. A comparison of the two partitions is shown in Table 7.2-2. It will be noted that partition of the between-cell variation is identical. However, in an experiment having repeated measures, the within-cell variation is divided into two orthogonal (nonoverlapping) parts. One part is a function of experimental error plus the main effects of subjects within groups, i.e., individual differences. The other part is a function of experimental error and $B \times$ subject-within-group interaction. If the latter interaction is negligible, then the second part of the within-cell variation is a function solely of experimental error.

Table 7.2-2 Comparison of Partitions

$p \times q$ factorial (no repeated measures)			$p \times q$ factorial (repeated measures on factor B)		
Total		$npq - 1$	Total		$npq - 1$
Between cells		$pq - 1$	Between cells		$pq - 1$
A	$p - 1$		A	$p - 1$	
B	$q - 1$		B	$q - 1$	
AB	$(p - 1)(q - 1)$		AB	$(p - 1)(q - 1)$	
Within cells		$pq(n - 1)$	Within cells		$pq(n - 1)$
			Subjects within groups		$p(n - 1)$
			$B \times$ subjects within groups		$p(n - 1)(q - 1)$

Appropriate denominators for F ratios to be used in making statistical tests are indicated by the expected values of the mean squares. Thus, to test the hypothesis that $\sigma_\alpha^2 = 0$,

$$F = \frac{\text{MS}_a}{\text{MS}_{\text{subj w.groups}}}.$$

The mean square in the denominator of the above F ratio is sometimes designated $\text{MS}_{\text{error (between)}}$. To test the hypothesis that $\sigma_\beta^2 = 0$,

$$F = \frac{\text{MS}_b}{\text{MS}_{B \times \text{subj w.groups}}}.$$

To test the hypothesis that $\sigma_{\alpha\beta}^2 = 0$, the appropriate F ratio is

$$F = \frac{\text{MS}_{ab}}{\text{MS}_{B \times \text{subj w.groups}}}.$$

The mean square in the denominator of the last two F ratios is sometimes called $\text{MS}_{\text{error (within)}}$ since it forms the denominator of F ratios used in testing effects which can be classified as part of the within-subject variation.

The mean squares used in the denominators of the above F ratios represent a pooling of different sources of variation. The variation due to subjects within groups is the sum of the following sources of variation:

Source	df
Subjects within groups	$p(n - 1)$
Subjects within G_1	$n - 1$
Subjects within G_2	$n - 1$
. . .	. . .
Subjects within G_p	$n - 1$

One of the assumptions required in order that the F ratio actually follow an F distribution is that these sources of variation be homogeneous. A partial check on this assumption may be made through use of the statistic

$$F_{\max} = \frac{\text{maximum } (\text{SS}_{\text{subj w.}G_i})}{\text{minimum } (\text{SS}_{\text{subj w.}G_i})},$$

i.e., the ratio of the largest of these sources of variation to the smallest. The critical value for this statistic in a test having level of significance equal to α is

$$F_{\max(1-\alpha)}(p, n - 1).$$

These critical values are given in Table C.7.

The variation due to $B \times$ subjects within groups represents a pooling of the following sources:

Source	df
$B \times$ subjects within groups	$p(n-1)(q-1)$
$B \times$ subjects within G_1	$(n-1)(q-1)$
$B \times$ subjects within G_2	$(n-1)(q-1)$
...	...
$B \times$ subjects within G_p	$(n-1)(q-1)$

A test on the homogeneity of these sources is given by

$$F_{\max} = \frac{\text{maximum } (\text{SS}_{B \times \text{subj w.} G_i})}{\text{minimum } (\text{SS}_{B \times \text{subj w.} G_i})}.$$

The critical value for this test is

$$F_{\max(1-\alpha)}[p, (n-1)(q-1)].$$

If the scale of measurement for the original criterion data does not satisfy these homogeneity assumptions, a transformation may often be found which will satisfy these assumptions. Indications are, however, that the F tests given above are robust with respect to minor violations of these assumptions [see Box (1954)].

For the sampling distribution of the F ratio for within-subject effects to be the F distribution with the usual degrees of freedom requires additional assumptions about the pattern of elements in $q \times q$ covariance matrices. The data at level a_i may be represented as follows:

	Subject	b_1	b_2	...	b_q
	1	X_{i11}	X_{i21}	...	X_{iq1}
	2	X_{i12}	X_{i22}	...	X_{iq2}
a_i	.	.	.		.
	.	.	.		.
	.	.	.		.
	n	X_{i1n}	X_{i2n}	...	X_{iqn}

From these data one may obtain the following $q \times q$ covariance matrix.

$$\hat{\Sigma}_{a_i} = \begin{bmatrix} \hat{\sigma}_{11} & \hat{\sigma}_{12} & \cdots & \hat{\sigma}_{1q} \\ \hat{\sigma}_{21} & \hat{\sigma}_{22} & \cdots & \hat{\sigma}_{2q} \\ \cdot & \cdot & & \cdot \\ \cdot & \cdot & & \cdot \\ \cdot & \cdot & & \cdot \\ \hat{\sigma}_{q1} & \hat{\sigma}_{q2} & \cdots & \hat{\sigma}_{qq} \end{bmatrix}.$$

Similar covariance matrices may be obtained for $i = 1, \ldots, p$. It can be shown that

$$E(\hat{\Sigma}_{a_i}) = \Sigma_{a_i}, \qquad i = 1, \ldots, p,$$

under random assignment of subjects to the levels of factor A. That is, $\hat{\Sigma}_{a_i}$ is unbiased as an estimate of Σ_{a_i}.

One assumption on the population covariance matrices is that they be homogeneous over the levels of the A factor. That is,

$$\Sigma_{a_1} = \Sigma_{a_2} = \cdots = \Sigma_{a_p} = \Sigma.$$

A check on this assumption is illustrated in Sec. 7.7. A second assumption is that the pattern of the matrix Σ be such that

$$\sigma^2_{\bar{B}_j - \bar{B}_k} = \text{constant} \qquad \text{for all } j \text{ and } k.$$

If the matrix Σ has compound symmetry, this requirement will be met. There are, however, other patterns that satisfy this requirement.

A measure of the extent to which a covariance matrix departs from this requirement is given by what Greenhouse and Geisser (1959) call ε, where

$$\varepsilon = \frac{q^2(\bar{\sigma}_{jj} - \bar{\sigma})^2}{(q-1)(\Sigma\Sigma\sigma^2_{jk} - 2q\Sigma\bar{\sigma}^2_j + q^2\bar{\sigma}^2)}$$

$\bar{\sigma}$ = mean of all entries in Σ,
$\bar{\sigma}_{jj}$ = mean of all entries of main diagonal of Σ,
$\bar{\sigma}_j$ = mean of all entries in row j of Σ,
σ_{jk} = entry in row j, column k of Σ.

ε in this case can range from 1 to $1/(q-1)$. ε will be equal to 1 if the covariance matrix meets the requirement

$$\sigma^2_{\bar{B}_j - \bar{B}_k} = \text{constant} \qquad \text{for all } j \text{ and } k.$$

The Greenhouse-Geisser procedure calls for using the critical values

$$F_{1-\alpha}[(q-1)\varepsilon, p(n-1)(q-1)\varepsilon]$$

and

$$F_{1-\alpha}[(p-1)(q-1)\varepsilon, p(n-1)(q-1)\varepsilon]$$

in making tests on the within-subject effects, when the matrix Σ does not have the required pattern of covariances.

If one lets ε equal its lower bound [in this case $1/(q-1)$], the test procedure outlined by Greenhouse and Geisser will err on the side of making the critical value in the test too large. Thus the Greenhouse-Geisser test procedure, with $\varepsilon = 1/(q-1)$, will yield a negatively biased or "conservative" test in the sense of not rejecting the hypothesis being tested as often as it should be rejected. Under this conservative procedure, the hypothesis that $\sigma_\beta^2 = 0$ has the critical value

$$F_{1-\alpha}[1, p(n-1)] \qquad \text{instead of} \qquad F_{1-\alpha}[(q-1), p(n-1)(q-1)].$$

Also the critical value for the hypothesis that $\sigma_{\alpha\beta}^2 = 0$ has the critical value

$$F_{1-\alpha}[(p-1), p(n-1)]$$

instead of $\qquad F_{1-\alpha}[(p-1)(q-1), p(n-1)(q-1)]$.

Setting ε equal to its lower bound tends to make the test procedure extremely conservative relative to using the F ratio suggested by the analysis-of-variance table. Monte Carlo studies indicate that the usual tests suggested by the analysis of variance tend to give results closer to the nominal significance levels than do results under the Greenhouse-Geisser conservative approach, provided the degree of heterogeneity of the covariances is relatively moderate. (See Collier et al., 1967.)

One notes that ε may be approximated from the sample estimate of Σ, in order to provide the basis for a critical value which does not assume that ε is at its lower bound. For example, if $\hat{\Sigma}$ $(q = 3)$ is

$$\hat{\Sigma} = \begin{bmatrix} 4.00 & 3.00 & 2.00 \\ 3.00 & 5.00 & 2.00 \\ 2.00 & 2.00 & 6.00 \end{bmatrix},$$

then
$$\hat{\varepsilon} = \frac{9(5.00 - 3.22)^2}{2[111 - 6(3.00^2 + 3.33^2 + 3.33^2) + 9(3.22)^2]} = .83.$$

The lower bound for ε in this case is .50.

As another example, if $\hat{\Sigma}$ $(q = 3)$ is

$$\hat{\Sigma} = \begin{bmatrix} 2.00 & 1.00 & 1.50 \\ 1.00 & 3.00 & 2.00 \\ 1.50 & 2.00 & 4.00 \end{bmatrix},$$

then
$$\hat{\varepsilon} = \frac{9(3.00 - 2.00)^2}{2[43.50 - 6(1.50^2 + 2.00^2 + 2.50^2) + 9(4.00)]} = 1.00.$$

The entries in the matrix $\hat{\Sigma}$ will be found to be related as follows:

$$\hat{\sigma}_{jj} + \hat{\sigma}_{kk} - 2\hat{\sigma}_{jk} = 3.00 \qquad \text{for all } j \text{ and } k, \qquad j \neq k.$$

This type of pattern for the elements of a sample covariance matrix implies

$$\hat{\sigma}^2_{\bar{B}_j - \bar{B}_k} = \text{constant} \qquad \text{for all } j \text{ and } k.$$

Computational Procedures. To illustrate the computational procedures for this plan, consider a factorial experiment in which the levels of factor A are two methods for calibrating dials and the levels of B are four shapes for the dials. Suppose that the data obtained are those given in part i of Table 7.2-3. Entries are accuracy scores on a series of trials on each

Table 7.2-3 Numerical Example

(i) Observed data:

	Subject	b_1	b_2	b_3	b_4	Total
a_1	1	0	0	5	3	$8 = P_1$
	2	3	1	5	4	$13 = P_2$
	3	4	3	6	2	$15 = P_3$
a_2	4	4	2	7	8	$21 = P_4$
	5	5	4	6	6	$21 = P_5$
	6	7	5	8	9	$29 = P_6$
	Total	23	15	37	32	$107 = G$

(ii) AB summary table:

	b_1	b_2	b_3	b_4	Total
a_1	7	4	16	9	36
a_2	16	11	21	23	71
Total	23	15	37	32	107

(iii) Computational symbols:

$$(1) = G^2/npq = (107)^2/3(2)(4) = 477.04$$
$$(2) = \Sigma X^2 = 0^2 + 0^2 + 5^2 + \cdots + 8^2 + 9^2 = 615$$
$$(3) = (\Sigma A_i^2)/nq = (36^2 + 71^2)/3(4) = 528.08$$
$$(4) = (\Sigma B_j^2)/np = (23^2 + 15^2 + 37^2 + 32^2)/3(2) = 524.50$$
$$(5) = [\Sigma(AB_{ij})^2]/n = (7^2 + 4^2 + \cdots + 21^2 + 23^2)/3 = 583.00$$
$$(6) = (\Sigma P_k^2)/q = (8^2 + 13^2 + \cdots + 21^2 + 29^2)/4 = 545.25$$

of the dials. Thus, for this experiment $p = 2$, $q = 4$, and $n = 3$. The order in which subjects are observed under the dials is randomized independently.

From the data in part i, the AB summary table given in part ii is readily obtained. Computational symbols are defined and computed in part iii. The only symbol that does not occur in a $p \times q$ factorial experiment without repeated measures is (6). This latter symbol involves P_k, which is the sum of the q observations made on subject k. These sums are given at the right of part i. In each case the divisor in a computational symbol is the number of observations summed to obtain an entry which is squared in the numerator.

Table 7.2-4 Analysis of Variance for Numerical Example

Source of variation	Computational formula	SS	df	MS	F
Between subjects	(6) − (1) =	68.21	5		
A (calibration)	(3) − (1) =	51.04	1	51.04	11.90
Subjects within groups	(6) − (3) =	17.17	4	4.29	
Within subjects	(2) − (6) =	69.75	18		
B (shape)	(4) − (1) =	47.46	3	15.82	12.76
AB	(5) − (3) − (4) + (1) =	7.46	3	2.49	2.01
$B \times$ subjects within groups	(2) − (5) − (6) + (3) =	14.83	12	1.24	

The analysis of variance is summarized in Table 7.2-4. In terms of means,

$$SS_{\text{subj w.groups}} = q\sum_k\sum_i(\bar{P}_{k(i)} - \bar{A}_i)^2,$$

$$SS_{B \times \text{subj w.groups}} = \sum_k\sum_j\sum_i(X_{ijk} - \bar{P}_{k(i)} - \overline{AB}_{ij} + \bar{A}_i)^2.$$

The computational formulas given in Table 7.2-4 are equivalent to these operations on the means.

The differential sensitivity for tests on between- and within-subject effects should be noted. The denominator of the F ratio for the between-subject effects is 4.29, whereas the denominator of the F ratios for the within-subject effects is 1.24. It is not unusual in this kind of plan to find the ratio of these two denominators to be as large as 10:1. If the level of significance for tests is set at .05, one rejects the hypothesis that $\sigma_\alpha^2 = 0$, since $F_{.95}(1,4) = 7.71$; the hypothesis that $\sigma_\beta^2 = 0$ is also rejected, since $F_{.95}(3,12) = 3.49$. However, the experimental data do not contradict the hypothesis that $\sigma_{\alpha\beta}^2 = 0$.

If the experimenter has reason (often there is) to question the pattern assumptions on the covariance matrix in the underlying population, the critical values for the within-subject tests are as follows:

Hypothesis	Conservative test	Ordinary test
$\sigma_\beta^2 = 0$	$F_{.95}[1,4] = 7.71$	$F_{.95}[3,12] = 3.49$
$\sigma_{\alpha\beta}^2 = 0$	$F_{.95}[1,4] = 7.71$	$F_{.95}[3,12] = 3.49$

Even under the more conservative test (i.e., negatively biased test) the main effects due to the shapes of the dials remain statistically significant.

If the experimenter has reason to question the homogeneity of the parts that are pooled to form the denominators of the F ratios, a check on homogeneity would logically precede the tests. Computational procedures for partitioning the relevant sums of squares are given in Table 7.2-5. A symbol of form $(6a_1)$ has the same general definition as (6), but summations are restricted to level a_1. The computational procedures

Table 7.2-5 Partition of Error Terms

(i)

$$
\begin{aligned}
(6a_1) &= (\textstyle\sum_{a_1} P_k^2)/q = (8^2 + 13^2 + 15^2)/4 & &= 114.50 \\
(6a_2) &= (\textstyle\sum_{a_2} P_k^2)/q = (21^2 + 21^2 + 29^2)/4 & &= 430.75 \\
& & (6) &= 545.25
\end{aligned}
$$

$$
\begin{aligned}
(3a_1) &= (A_1^2)/nq = 36^2/3(4) & &= 108.00 \\
(3a_2) &= (A_2^2)/nq = 71^2/3(4) & &= 420.08 \\
& & (3) &= 528.08
\end{aligned}
$$

$$
\begin{aligned}
(5a_1) &= [\textstyle\sum_{a_1} (AB_{ij})^2]/n = [7^2 + 4^2 + 16^2 + 9^2]/3 & &= 134.00 \\
(5a_2) &= [\textstyle\sum_{a_2} (AB_{ij})^2]/n = [16^2 + 11^2 + 21^2 + 23^2]/3 & &= 449.00 \\
& & (5) &= 583.00
\end{aligned}
$$

$$
\begin{aligned}
(2a_1) &= \textstyle\sum_{a_1} X^2 = 0^2 + 0^2 + \cdots + 6^2 + 2^2 & &= 150 \\
(2a_2) &= \textstyle\sum_{a_2} X^2 = 4^2 + 2^2 + \cdots + 8^2 + 9^2 & &= 465 \\
& & (2) &= 615
\end{aligned}
$$

(ii)

$$
\begin{aligned}
SS_{\text{subj w. } G_1} &= (6a_1) - (3a_1) &&= 6.50 \\
SS_{\text{subj w. } G_2} &= (6a_2) - (3a_2) &&= 10.67 \\
& && \quad 17.17 \\
SS_{B \times \text{subj w. } G_1} &= (2a_1) - (5a_1) - (6a_1) + (3a_1) &&= 9.50 \\
SS_{B \times \text{subj w. } G_2} &= (2a_2) - (5a_2) - (6a_2) + (3a_2) &&= 5.33 \\
& && \quad 14.83
\end{aligned}
$$

for parts which are pooled are given in part ii. As a check on the homogeneity of $SS_{\text{subj w.groups}}$,

$$F_{\max} = \frac{10.67}{6.50} = 1.64.$$

The critical value for a .05-level test here is

$$F_{\max .95}(2,2) = 39.00.$$

Since the computed $F_{\max}$ statistic does not exceed the critical value, the test does not contradict the hypothesis that the parts are homogeneous. Since each of the parts in this case has only two degrees of freedom, the power of a test of this kind is extremely low.

As a check on the homogeneity of the parts of $SS_{B \times \text{subj w.groups}}$,

$$F_{\max} = \frac{9.50}{5.33} = 1.78.$$

The critical value here is

$$F_{\max_{.95}}(2.6) = 5.82.$$

Again the computed value of the statistic does not exceed the critical value. Hence the hypothesis of homogeneity is not contradicted by the experimental data.

Tests on the difference between all possible pairs of means can be made in a manner similar to that given in Sec. 3.8. The procedure is illustrated in Table 7.2-6. The Newman-Keuls method is chosen for this purpose,

Table 7.2-6 Tests on Means Using Newman-Keuls Procedure

Shapes	b_2	b_1	b_4	b_3
Ordered means	2.50	3.83	5.33	6.17

(i)		b_2	b_1	b_4	b_3	r	$s_{\bar{B}}\, q_{.95}(r,12)$
	b_2		1.33	2.83	3.67	4	1.93
	b_1			1.50	2.34	3	1.73
	b_4				0.84	2	1.42

(ii) $s_{\bar{B}} = 0.46$

	$r =$	2	3	4
$q_{.95}(r,12)$:		3.08	3.77	4.20

(iii)		b_2	b_1	b_4	b_3
	b_2		—	*	*
	b_1			*	*
	b_4				—

In part i the $\bar{B}_j$'s are arranged in rank order from low to high. Differences between all possible pairs of ordered means are computed. For example,

$$6.17 - 2.50 = 3.67, \qquad 5.33 - 2.50 = 2.83, \qquad \text{etc.}$$

In part ii critical values for the ordered differences between pairs are computed. Since the main effect of factor B is a within-subject effect, the standard error of the mean for all observations at a given level of factor B is

$$s_{\bar{B}} = \sqrt{\frac{MS_{B \times \text{subj w.groups}}}{np}} = \sqrt{\frac{1.24}{6}} = \sqrt{.207} = .46.$$

The degrees of freedom associated with this standard error are those of $MS_{B \times \text{subj w.groups}}$, which in this case are 12. To obtain the critical value for the difference between two ordered means which are r steps apart in an ordered sequence, one first finds the tabled values for

$$q_{1-\alpha}(r,\mathrm{df}_{\mathrm{error}}),$$

where $\mathrm{df}_{\mathrm{error}}$ represents the degrees of freedom associated with $s_{\bar{B}}$. These values are obtained from Table C.4. For level of significance .05, the relevant values of q are given in part ii. The critical value for an ordered difference between two means r steps apart is

$$s_{\bar{B}}q_{1-\alpha}(r,\mathrm{df}_{\mathrm{error}}).$$

These critical values appear at the extreme right of part ii. For example,

$$s_{\bar{B}}q_{.95}(4,12) = 0.46(4.20) = 1.93,$$
$$s_{\bar{B}}q_{.95}(3,12) = 0.46(3.77) = 1.73,$$
$$s_{\bar{B}}q_{.95}(2,12) = 0.46(3.08) = 1.42.$$

(For the sequence in which tests on ordered pairs must be made, see Sec. 3.8.)

The pairs of means which can be considered different are indicated in part iii. The mean performance on shape b_3 is statistically different from the mean performance on shapes b_2 and b_1. The mean performance on shape b_4 is also statistically different from the mean performance on shapes b_2 and b_1. No other differences are statistically significant at the .05 level for the Newman-Keuls tests.

Tests on all possible ordered differences of the form $\bar{A}_i - \bar{A}_{i'}$ follow the same general pattern. For such tests,

$$s_{\bar{A}} = \sqrt{\frac{\mathrm{MS}_{\mathrm{subj\ w.groups}}}{nq}}.$$

The degrees of freedom associated with $s_{\bar{A}}$ are $p(n - 1)$.

If the AB interaction were significant, tests on simple main effects would be called for, rather than direct tests on main effects. The computation of the variation due to the simple main effect of factors A and B is identical to that of a two-factor factorial experiment which does not have repeated measures. To test the simple main effect of factor B, the F ratio has the form

$$F = \frac{\mathrm{MS}_{b\ \mathrm{at}\ a_i}}{\mathrm{MS}_{B \times \mathrm{subj\ w.groups}}}.$$

The denominator of this F ratio is the same as that used in testing the main effects of factor B. The F ratio for the test on the simple main effects of factor A has the form

$$F = \frac{\mathrm{MS}_{a\ \mathrm{at}\ b_j}}{\mathrm{MS}_{\mathrm{w.cell}}}.$$

The denominator of this last F ratio requires special note—it is not the denominator used in testing the main effects of factor A. For each level of factor B considered individually, this plan reduces to a single-factor experiment in which there are no repeated measures. In this latter type of experiment $MS_{w.cell}$ is the appropriate denominator for the variation due to the treatment effects.

The within-cell variation is given by

$$SS_{w.cell} = SS_{subj\ w.groups} + SS_{B \times subj\ w.groups}.$$

Within the context of a repeated-measure design, $SS_{w.cell}$ represents a pooling of what will often be heterogeneous sources of variance. Hence the F test on the simple main effects for factor A, which uses $MS_{w.cell}$ as a denominator, will tend to be biased. However, when the degrees of freedom for the within-cell variation are large (say greater than 30), the bias will be quite small. The magnitude of the bias depends in part upon the ratio of $MS_{subj\ w.groups}$ to $MS_{B \times subj\ w.groups}$.

Variation due to the simple main effects is most readily computed from the AB summary table given in part ii of Table 7.2-3. For example,

$$SS_{a\ at\ b_1} = \frac{7^2 + 16^2}{3} - \frac{23^2}{6} = 13.50;$$

$$MS_{a\ at\ b_1} = \frac{SS_{a\ at\ b_1}}{p - 1} = 13.50.$$

The denominator for the appropriate F ratio is

$$MS_{w.cell} = \frac{SS_{w.cell}}{pq(n-1)} = \frac{17.17 + 14.83}{16} = 2.00.$$

$$F = \frac{MS_{a\ at\ b_1}}{MS_{w.cell}} = 6.75.$$

In this context $MS_{w.cell}$ represents an average of heterogeneous sources of variance. Hence, for purposes of making tests, $MS_{w.cell}$ cannot be considered as having $pq(n - 1) = 16$ degrees of freedom. Further, the F ratio does not have an F distribution. However, the distribution of the F ratio (under the null hypothesis) may be approximated by an F distribution having degrees of freedom equal to $p - 1$ and f, where f is given by (Satterthwaite, 1946)

$$f = \frac{(u + v)^2}{(u^2/f_1) + (v^2/f_2)},$$

where $u = p(n - 1)MS_{subj\ w.groups}$,

$v = p(n-1)(q - 1)MS_{B \times subj\ w.groups}$,

$f_1 = p(n-1)$,

$f_2 = p(n - 1)(q - 1)$.

For the data in this example,

$$u = 17.17, \qquad v = 14.83, \qquad f_1 = 4, \qquad \text{and} \qquad f_2 = 12.$$

Hence
$$f = \frac{(17.17 + 14.83)^2}{[(17.17)^2/4] + [(14.83)^2/12]} = 11.12.$$

To the nearest integer, $f = 11$. Thus, under the appropriate null hypothesis, the F ratio

$$F = \frac{\text{MS}_{a \text{ at } b_j}}{\text{MS}_{\text{w.cell}}} \qquad \text{is distributed approximately as} \qquad F[(p-1), f].$$

It will be found that

$$p(n-1) \le f \le p(n-1) + p(n-1)(q-1).$$

For this example,

$$4 \le f \le 16.$$

f will achieve its upper bound only when

$$\text{MS}_{\text{subj w.groups}} = \text{MS}_{B \times \text{subj w.groups}}.$$

Note that

$$\text{MS}_{\text{w.cell}} = \frac{p(n-1)\text{MS}_{\text{subj w.groups}} + p(n-1)(q-1)\text{MS}_{B \times \text{subj w.groups}}}{p(n-1) + p(n-1)(q-1)}.$$

Thus $\text{MS}_{\text{w.cell}}$ is a weighted average of two mean squares. Dividing numerator and denominator of this fraction by $p(n-1)$, one obtains

$$\text{MS}_{\text{w.cell}} = \frac{\text{MS}_{\text{subj w.groups}} + (q-1)\text{MS}_{B \times \text{subj w.groups}}}{q}.$$

If one sets

$$u = \text{MS}_{\text{subj w.groups}} \qquad \text{and} \qquad v = (q-1)\text{MS}_{B \times \text{subj w.groups}}$$

in the expression for f, the numerical value will be identical with that obtained in the preceding paragraph.

The variation due to the simple main effects for factor B at level a_1 is

$$\text{SS}_{b \text{ at } a_1} = \frac{7^2 + 4^2 + 16^2 + 9^2}{3} - \frac{36^2}{12} = 26.00;$$

$$\text{MS}_{b \text{ at } a_1} = \frac{\text{SS}_{b \text{ at } a_1}}{q-1} = \frac{26.00}{3} = 8.67.$$

A test on the simple main effects of factor B at level a_1 uses the statistic

$$F = \frac{\text{MS}_{b \text{ at } a_1}}{\text{MS}_{B \times \text{subj w.groups}}} = \frac{8.67}{1.24} = 6.99.$$

The critical value for a test having level of significance .05 is

$$F_{.95}[(q-1), p(n-1)(q-1)] = F_{.95}(3,12) = 3.49.$$

Hence the experimental data tend to reject the hypothesis that there are no differences in the effects of factor B when all observations are made at level a_1.

Covariance Matrices Associated with This Design. Consider the following data from part i of Table 7.2-3:

	Subject	b_1	b_2	b_3	b_4
	1	0	0	5	3
a_1	2	3	1	5	4
	3	4	3	6	2
		7	4	16	9

The variance of the observations made under b_1 is

$$\text{var}_{b_1} = \frac{(0^2 + 3^2 + 4^2) - (7^2/3)}{2} = 4.33.$$

Similarly the variance of the observations made under b_2 is

$$\text{var}_{b_2} = \frac{(0^2 + 1^2 + 3^2) - (4^2/3)}{2} = 2.33.$$

The covariance of the observations made under b_1 and b_2 is

$$\text{cov}_{b_1b_2} = \frac{(0)(0) + (3)(1) + (4)(3) - (7)(4)/3}{2} = 2.83.$$

Similarly the covariance of the observations under b_1 and b_3 is

$$\text{cov}_{b_1b_3} = \frac{(0)(5) + (3)(5) + (4)(6) - (7)(16)/3}{2} = .83.$$

The variance-covariance matrix for level a_1 of factor A is given below (only the top half is completed since the bottom half is a duplicate of the top half):

$\hat{\Sigma}_{a_1} = B_{a_1}$		b_1	b_2	b_3	b_4
	b_1	4.33	2.83	.83	$-$ .50
	b_2		2.33	.83	-1.00
	b_3			.33	$-$ .50
	b_4				1.00

A similar variance-covariance matrix for the data at level a_2 is given below:

		b_1	b_2	b_3	b_4
$\hat{\Sigma}_{a_2} = B_{a_2}$	b_1	2.33	2.17	1.00	1.17
	b_2		2.33	.50	.33
	b_3			1.00	1.50
	b_4				2.33

If the variance-covariance matrix has compound symmetry, i.e., if Σ has the form

	b_1	b_2	b_3	b_4
b_1	σ^2	$\rho\sigma^2$	$\rho\sigma^2$	$\rho\sigma^2$
b_2		σ^2	$\rho\sigma^2$	$\rho\sigma^2$
b_3			σ^2	$\rho\sigma^2$
b_4				σ^2

then the F ratio for within-subject effects will follow an F distribution under the relevant null hypotheses. There are other patterns that the variance-covariance matrix can have and still yield an F ratio. The other patterns are often more readily handled through the use of multivariate analysis of variance.

If the underlying population variance-covariance matrices for B_{a_1} and B_{a_2} are equal but not necessarily of the form indicated above, the best estimate of the common underlying population variance-covariance matrix is the pooled sample matrices. The pooled matrix is obtained by averaging corresponding entries in the individual matrices. For the data under consideration, the pooled variance-covariance matrix is given below:

		b_1	b_2	b_3	b_4
$\hat{\Sigma} = B_{\text{pooled}}$	b_1	3.33	2.50	.92	.33
	b_2		2.33	.67	−.33
	b_3			.67	.50
	b_4				1.67

The matrix of intercorrelations as computed from the pooled covariance matrix is given below:

$$\begin{bmatrix} 1.00 & .90 & .62 & .14 \\ & 1.00 & .54 & -.17 \\ & & 1.00 & .47 \\ \text{Symmetric} & & & 1.00 \end{bmatrix}.$$

There are two stages in testing for homogeneity with respect to the variance-covariance matrices. These tests are illustrated in Sec. 7.7.

First, one is interested in finding out whether or not the B_{a_i}'s can be pooled. Second, if these matrices can be pooled, one is interested in finding out whether or not the population matrix estimated by the pooled matrix has the required symmetry, i.e., all diagonal elements equal to σ^2, and all off-diagonal elements equal to $\rho\sigma^2$. In terms of the pooled variance-covariance matrix,

$$MS_{B \times \text{subj w.groups}} = \overline{\text{var}} - \overline{\text{cov}}.$$

For the data under consideration,

$$\overline{\text{var}} - \overline{\text{cov}} = 2.00 - .76 = 1.24.$$

The term $\overline{\text{var}}$ is the mean of entries along the main diagonal of the pooled variance-covariance matrix; $\overline{\text{cov}}$ is the mean of the entries off the main diagonal in this matrix. Also,

$$\overline{\text{var}} + (q - 1)\overline{\text{cov}} = 2.00 + 3(.76) = 4.28.$$

From Table 7.2-4 it is noted that

$$MS_{\text{subj. w.groups}} = 4.29,$$

$$MS_{B \times \text{subj w.groups}} = 1.24.$$

The computational formulas for these terms, which are given in Table 7.2-4, are thus short-cut methods. The long method, however, indicates what has to be pooled along the way; the long method also produces summary data that are useful in interpreting the experimental results.

Illustrative Applications. Many applications of this form of the repeated-measure plan will be found in the literature in experimental psychology. Shore (1958) reports a study which attempts to evaluate the effects of anxiety level and muscular tension upon perceptual efficiency. Factor A in this study was the level of anxiety as measured by the Taylor Manifest Anxiety Scale. On the basis of scores made on this scale, very low (G_1), middle (G_2), and very high (G_3) groups of subjects were formed. There were six subjects in each group. Factor B in this study was the level of muscular tension exerted on a dynamometer at the time of the perceptual task. The criterion score was the number of correct symbols recognized in a series of very short exposures. The plan of this experiment may be represented as follows:

Anxiety level	Dynamometer tension					
	b_1	b_2	b_3	b_4	b_5	b_6
a_1	G_1	G_1	G_1	G_1	G_1	G_1
a_2	G_2	G_2	G_2	G_2	G_2	G_2
a_3	G_3	G_3	G_3	G_3	G_3	G_3

Each of the subjects was tested under all levels of dynamometer tension. Level b_6, the maximum tension condition, was used first for all subjects. The other tension conditions were administered under a restricted randomization procedure. The analysis of the experimental data was reported in a form equivalent to the following (this plan is a special case of a 3×6 factorial experiment with repeated measures on one of the factors, $n = 6$):

	df	MS	F
Between subjects	17		
A (anxiety level)	2	742.7	1.58
Subjects within groups	15	469.0	
Within subjects	90		
B (tension)	5	138.7	3.39**
AB	10	127.6	3.12**
$B \times$ subjects within groups	75	40.9	

The significant interaction indicated that the profiles for the groups had different shapes. Shore interprets this interaction in terms of the group profiles. Note the magnitude of the ratio of the between-subject error term (469.0) to the within-subject error term (40.9). A ratio of 10:1 is not unusual in this area of experimentation.

Another application of this same basic experimental plan is reported by Noble and McNeely (1957). This study was concerned with the relative rate of paired-associates learning as a function of degree of meaningfulness of the pairs. The experimental plan may be represented as follows:

List	Degree of meaningfulness				
	b_1	b_2	b_3	b_4	b_5
a_1	G_1	G_1	G_1	G_1	G_1
a_2	G_2	G_2	G_2	G_2	G_2
.	.	.	.	.	.
.	.	.	.	.	.
.	.	.	.	.	.
a_{18}	G_{18}	G_{18}	G_{18}	G_{18}	G_{18}

Each of the 18 lists had pairs for each of the meaningfulness categories. There were five subjects in each of the groups. Differences between groups of subjects are confounded with differences between the lists; however, primary interest is in the meaningfulness dimension of this experiment. The lists essentially represent samples of material to be learned. The analysis of variance had the following form:

	df	MS	F
Between subjects	89		
A (lists)	17	311.88	1.20
Subjects within groups	72	260.34	
Within subjects	360		
B (meaningfulness)	4	3046.72	
AB	68	82.91	1.43
$B \times$ subjects within groups	288	57.79	

The F ratio for the test on main effects for factor B, as computed by Noble and McNeely, was

$$F = \frac{\text{MS}_b}{\text{MS}_{B \times \text{subj}}} = \frac{3046.72}{57.79} = 52.72 \qquad [F_{.99}(4,288) = 3.39].$$

If, however, factor A (lists) is considered to be a random factor, then the F ratio is

$$F = \frac{\text{MS}_b}{\text{MS}_{ab}} = \frac{3046.72}{82.91} = 36.75 \qquad [F_{.99}(4,68) = 3.60].$$

If the lists used in the experiment represent a random sample from a population of lists, it is the latter F ratio that is the more appropriate for interpretation. In both cases the main effect due to factor B is statistically significant, indicating that the degree of meaningfulness of paired associates is related to the rate of learning.

A study reported by Schrier (1958) provides another illustration of the use of this basic plan. In this study the experimenter was interested in evaluating the effect of amount of reward on the performance of monkeys in a series of discrimination problems. The criterion used was percentage of correct choices in two periods—a period was a block of 20 consecutive trials. The plan of the experiment may be represented as follows (only one phase of a more extensive experiment is summarized here):

Level of reward	Period	
	b_1	b_2
a_1	G_1	G_1
a_2	G_2	G_2
a_3	G_3	G_3
a_4	G_4	G_4

There were five subjects in each group.

In designs in which there are only two measurements on each subject (and the order in which the measurements are made is not at the option of the experimenter) there is only one covariance. Hence the problem of homogeneity of covariance does not exist. The analysis of the within-subject effects in this case is equivalent to an analysis of difference scores between the two periods.

A scale of measurement in terms of percentage generally does not provide homogeneity of variance. Schrier used an arcsine transformation on the percentages, and the analysis of variance was made on the transformed criterion scale. A modified version of the analysis is given below:

	df	MS	F
Between subjects	19		
A (rewards)	3	148.6	2.52
Subjects within groups	16	58.9	
Within subjects	20		
B (periods)	1	685.8	59.12**
AB	3	1.9	
$B \times$ subjects within groups	16	1.16	

In addition to the analysis indicated above, this worker tested differences in various trends. Tests of the latter kind are discussed in a later section in this chapter.

Another example of the same basic experimental plan is reported by Denenberg and Myers (1958). This study was designed to evaluate the effects of thyroid level upon acquisition and extinction of a bar-pressing response. Schematically the acquisition phase of this study had the following form:

Thyroid level	Days			
	b_1	b_2	$\cdots$	b_{18}
a_1	G_1	G_1	$\cdots$	G_1
a_2	G_2	G_2	$\cdots$	G_2
a_3	G_3	G_3	$\cdots$	G_3

There were four subjects in each of the groups. The criterion measure was the number of responses in a Skinner box during a specified time period.

The analysis of variance had the following form:

	df	MS	F
Between subjects	11		
A (thyroid level)	2	56,665	13.32**
Subjects within groups	9	4,253	
Within subjects	204		
B (days)	17	594	2.49**
AB	34	1,190	4.98**
$B \times$ subjects within groups	153	239	

Again note the ratio of the between-subject error term (4,253) to the within-subject error term (239).

Birch, Burnstein, and Clark (1958) report an experiment which had the following form:

Deprivation interval	Blocks of trials			
	b_1	b_2	b_3	
a_1	G_1	G_1	G_1	$n_1 = 10$
a_2	G_2	G_2	G_2	$n_2 = 10$
a_3	G_3	G_3	G_3	$n_3 = 10$
a_4	G_4	G_4	G_4	$n_4 = 9$

There were 10 subjects in each of the groups except G_4; there were 9 subjects in the latter groups since 1 subject had to be discarded in the course of the experiment.

Speed of running was used as measure of response strength. The study was designed to investigate response strength as a function of deprivation interval. The analysis of variance had the following form:

	df	MS	F
Between subjects	38		
A (deprivation interval)	3	788.06	3.34*
Subjects within groups	35	236.29	
Within subjects	78		
B (trials)	2	79.92	1.17
AB	6	56.70	
$B \times$ subjects within groups	70	68.24	

In case of missing data of this kind, where the missing data are the result of conditions unrelated to the treatments, estimates of variation due to treatment effects should be obtained through use of an unweighted-means analysis.

7.3 Three-factor Experiment with Repeated Measures (Case I)

Two special cases will be considered. The first case will be that of a $p \times q \times r$ factorial experiment in which there are repeated observations on the last two factors. In the second case, repeated measures will be restricted to the last factor. A schematic representation of the first case is given below:

	b_1			$\cdots$	b_q		
	c_1	$\cdots$	c_r	$\cdots$	c_1	$\cdots$	c_r
a_1	G_1	$\cdots$	G_1	$\cdots$	G_1	$\cdots$	G_1
a_2	G_2	$\cdots$	G_2	$\cdots$	G_2	$\cdots$	G_2
$\vdots$	$\vdots$		$\vdots$		$\vdots$		$\vdots$
a_p	G_p	$\cdots$	G_p	$\cdots$	G_p	$\cdots$	G_p

There are n subjects in each group. Each subject is observed under all qr combinations of factor B and C but only under a single level of factor A. Thus there are p groups of n subjects each (np subjects in all); there are qr observations on each subject.

The observations on subjects in group i may be represented as follows:

	Subject	b_1			$\cdots$	b_q			Total
		c_1	$\cdots$	c_r	$\cdots$	c_1	$\cdots$	c_r	
	$1(i)$	X_{i111}	$\cdots$	X_{i1r1}	$\cdots$	X_{iq11}	$\cdots$	X_{iqr1}	$P_{1(i)}$
	$\vdots$	$\vdots$		$\vdots$		$\vdots$		$\vdots$	$\vdots$
a_i	$m(i)$	X_{i11m}	$\cdots$	X_{i1rm}	$\cdots$	X_{iq1m}	$\cdots$	X_{iqrm}	$P_{m(i)}$
	$\vdots$	$\vdots$		$\vdots$		$\vdots$		$\vdots$	$\vdots$
	$n(i)$	X_{i11n}	$\cdots$	X_{i1rn}	$\cdots$	X_{iq1n}	$\cdots$	X_{iqrn}	$P_{n(i)}$

The notation X_{ijkm} indicates an observation on subject $m(i)$ under treatment combination abc_{ijk}. The notation $P_{m(i)}$ denotes the sum of the qr observations on subject m in group i. Unless there is some ambiguity about which group is under discussion, the notation P_m will be used to denote this total.

In this design the subjects may be viewed as defining a fourth factor having n levels. The "subject" factor is crossed with factors B and C but is nested under factor A. Schematically,

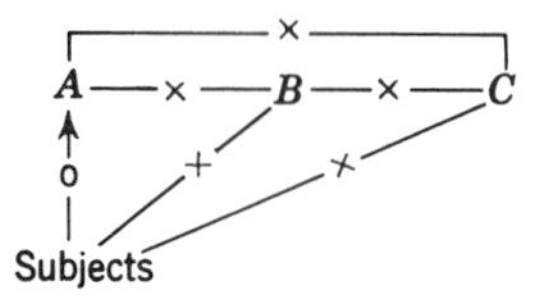

Assuming that A, B, and C are fixed factors, the analysis of variance generally takes the form shown in Table 7.3-1. The expected values of the mean squares indicate appropriate F ratios, provided that the sampling

Table 7.3-1 Summary of Analysis of Variance

Source of variation	df	E(MS)†
Between subjects	$np - 1$	
A	$p - 1$	$\sigma_\varepsilon^2 + qr\sigma_\pi^2 + nqr\sigma_\alpha^2$
Subj w. groups	$p(n - 1)$	$\sigma_\varepsilon^2 + qr\sigma_\pi^2$
Within subjects	$np(qr - 1)$	
B	$q - 1$	$\sigma_\varepsilon^2 + r\sigma_{\beta\pi}^2 + npr\sigma_\beta^2$
AB	$(p - 1)(q - 1)$	$\sigma_\varepsilon^2 + r\sigma_{\beta\pi}^2 + nr\sigma_{\alpha\beta}^2$
$B \times$ subj w. groups	$p(n - 1)(q - 1)$	$\sigma_\varepsilon^2 + r\sigma_{\beta\pi}^2$
C	$r - 1$	$\sigma_\varepsilon^2 + q\sigma_{\gamma\pi}^2 + npq\sigma_\gamma^2$
AC	$(p - 1)(r - 1)$	$\sigma_\varepsilon^2 + q\sigma_{\gamma\pi}^2 + nq\sigma_{\alpha\gamma}^2$
$C \times$ subj w. groups	$p(n - 1)(r - 1)$	$\sigma_\varepsilon^2 + q\sigma_{\gamma\pi}^2$
BC	$(q - 1)(r - 1)$	$\sigma_\varepsilon^2 + \sigma_{\beta\gamma\pi}^2 + np\sigma_{\beta\gamma}^2$
ABC	$(p - 1)(q - 1)(r - 1)$	$\sigma_\varepsilon^2 + \sigma_{\beta\gamma\pi}^2 + n\sigma_{\alpha\beta\gamma}^2$
$BC \times$ subj w. groups	$p(n - 1)(q - 1)(r - 1)$	$\sigma_\varepsilon^2 + \sigma_{\beta\gamma\pi}^2$

† Assumes A, B, and C fixed factors.

distributions of the statistics involved are actually what they are assumed to be. What is assumed will be made explicit later in this section. [The general E(MS) of which those in Table 7.3-1 are a special case are given in Table 7.3-12. If $D_n = 1$, $D_p = 0$, $D_q = 0$, $D_r = 0$ in the latter table, the results will be the E(MS) in Table 7.3-1.]

An alternative partition of the total degrees of freedom is as follows:

Source of variation	Degrees of freedom	
	General case	Special case $p = 2, q = 3, r = 4, n = 5$
Between cells	$pqr - 1$	23
Within cells	$pqr(n - 1)$	96
Subj w. groups	$p(n - 1)$	8
$B \times$ subj w. groups	$p(n - 1)(q - 1)$	16
$C \times$ subj w. groups	$p(n - 1)(r - 1)$	24
$BC \times$ subj w. groups	$p(n - 1)(q - 1)(r - 1)$	48

A cell in this context contains the n observations under treatment combination abc_{ijk}. There are pqr cells. Treatment and interaction variation are obtained by partitioning the between-cell variation. The latter partition is identical for all $p \times q \times r$ factorial experiments whether or not there are repeated measures. The manner in which the within-cell variation is partitioned depends upon the pattern of the repeated measures. For the case being considered, each of the parts of the within-cell variation forms a denominator for some F ratio. Hence the following alternative notation system:

$$\text{MS}_{\text{subj w.groups}} = \text{MS}_{\text{error}(a)},$$

$$\text{MS}_{B\times\text{subj w.groups}} = \text{MS}_{\text{error}(b)},$$

$$\text{MS}_{C\times\text{subj w.groups}} = \text{MS}_{\text{error}(c)},$$

$$\text{MS}_{BC\times\text{subj w.groups}} = \text{MS}_{\text{error}(bc)}.$$

The alternative notation system has the advantage of being more compact as well as more indicative of the role of the error term in the F tests, for the special case being considered. However, the F tests will change as a function of what model is appropriate for the data being analyzed. Thus the alternative notation has the disadvantage of not indicating how each of the terms is computed.

Each of the parts of the within-cell variation may be checked for homogeneity, the latter being one of the assumptions made when the F ratio is considered to be distributed in the form of the F distribution. For example,

Subj w. groups	$p(n - 1)$
Subj w. G_1	$n - 1$
. . .	. . .
Subj w. G_p	$n - 1$

An $F_{\max}$ test for the homogeneity of the parts would have the critical value $F_{\max(1-\alpha)}(p, n - 1)$. As another example,

$BC \times$ subj w. groups	$p(n-1)(q-1)(r-1)$
$BC \times$ subj w. G_1	$(n-1)(q-1)(r-1)$
. . .	. . .
$BC \times$ subj w. G_p	$(n-1)(q-1)(r-1)$

An $F_{\max}$ test in this last case is, in part, a check on whether or not mean covariances of a set of variance-covariance matrices are equal. The critical value for this test is $F_{\max(1-\alpha)}[(p, (n-1)(q-1)(r-1)]$. This test is only a partial check on whether or not sets of corresponding variance-covariance matrices may be pooled. More appropriate tests are generalizations of those given in Sec. 7.7.

Each of the interactions with subjects may be shown to have the general form

$$\overline{\text{var}} - \overline{\text{cov}}_i$$

with respect to a specified variance-covariance matrix. For example, corresponding to the $BC \times$ subjects within groups interaction there is a pooled $qr \times qr$ variance-covariance matrix. This matrix is assumed to estimate a variance-covariance matrix that has a pattern corresponding to that which makes the corresponding F ratio be distributed as F under the appropriate null hypothesis. If the relevant variance-covariance matrix does not have the appropriate pattern, a negatively biased (conservative) test may be constructed as indicated in the table given below.

F ratio	Critical values	
	Usual test	Conservative test
$\dfrac{MS_b}{MS_{B \times \text{subj w. group}}}$	$F_{1-\alpha}[(q-1), p(n-1)(q-1)]$	$F_{1-\alpha}[1, p(n-1)]$
$\dfrac{MS_{ab}}{MS_{B \times \text{subj w. group}}}$	$F_{1-\alpha}[(p-1)(q-1), p(n-1)(q-1)]$	$F_{1-\alpha}[(p-1), p(n-1)]$
$\dfrac{MS_c}{MS_{C \times \text{subj w. group}}}$	$F_{1-\alpha}[(r-1), p(n-1)(r-1)]$	$F_{1-\alpha}[1, p(n-1)]$
$\dfrac{MS_{ac}}{MS_{C \times \text{subj w. group}}}$	$F_{1-\alpha}[(p-1)(r-1), p(n-1)(r-1)]$	$F_{1-\alpha}[(p-1), p(n-1)]$
$\dfrac{MS_{bc}}{MS_{BC \times \text{subj w. group}}}$	$F_{1-\alpha}[(q-1)(r-1), p(n-1)(q-1)(r-1)]$	$F_{1-\alpha}[1, p(n-1)]$
$\dfrac{MS_{abc}}{MS_{BC \times \text{subj w. group}}}$	$F_{1-\alpha}[(p-1)(q-1)(r-1), p(n-1)(q-1)(r-1)]$	$F_{1-\alpha}[(p-1), p(n-1)]$

This "conservative" test is one proposed by Greenhouse and Geisser. The indications are that the conservative test (for most covariance matrices encountered in practice) is extremely conservative in the sense that a nominal level of significance of, say, .05 tends to be closer to .01—particularly when n, p, q, and r are all small.

The expected values for mean squares given in Table 7.3-1 are obtained from a model which includes interactions with the subject factor. If in fact such interactions do not exist (or are negligible relative to the magnitude of σ_ε^2), then

$$E(MS_{error(b)}) = \sigma_\varepsilon^2 \qquad \text{if } \sigma_{\beta\pi}^2 = 0,$$

$$E(MS_{error(c)}) = \sigma_\varepsilon^2 \qquad \text{if } \sigma_{\gamma\pi}^2 = 0,$$

$$E(MS_{error(bc)}) = \sigma_\varepsilon^2 \qquad \text{if } \sigma_{\beta\gamma\pi}^2 = 0.$$

In words, if all interactions with the subject factor are zero, each of the above mean squares is an estimate of the same variance, namely, that due to experimental error. Further, these estimates are independent and may be pooled to provide a single estimate of σ_ε^2. Thus,

$$MS_{error(within)} = \frac{SS_{error(b)} + SS_{error(c)} + SS_{error(bc)}}{p(n-1)(qr-1)}$$

provides an estimate of σ_ε^2 having $p(n-1)(qr-1)$ degrees of freedom.

If the experiment provides a relatively large number of degrees of freedom (say over 30) for estimating the variance due to each of the interactions with subjects, there is generally no need to consider pooling procedures. When there are relatively few degrees of freedom for such estimates, the decision about pooling should depend largely on previous experimental work. In the absence of such background information, preliminary tests on the model are useful. The purpose of such tests is to provide the experimenter with a posteriori information about whether or not certain of the interactions with random factors should be included in the model for the experiment. Such tests should be made at numerically high levels of significance (that is, $\alpha = .20$ or $\alpha = .30$). This procedure does not drop a term from the model unless the data clearly indicate that such terms can be dropped. Since terms are dropped when tests on the model do not reject the hypothesis being tested, high power is required.

Bartlett's test for homogeneity of variance may be used to indicate whether or not interactions with subjects can be pooled. (Pooling is equivalent to dropping terms from the model.) A rough but conservative test for pooling of this kind is to take the ratio of the largest to the smallest of the mean squares due to interactions with subjects. Use as a critical value for this ratio $F_{.90}(df_1, df_2)$, where the df_1 and df_2 are the respective degrees of freedom for the largest and smallest of the respective mean squares. For this procedure, the level of significance is larger than $\alpha = .20$. This test is biased in the direction of rejecting the hypothesis of homogeneity (i.e., does not lead to pooling) more often than an α-level test. The critical value errs on the side of being too small. If the hypothesis of homogeneity is rejected by the rough procedure, Bartlett's test may be used.

Tests on simple main effects have denominators of the form shown in Table 7.3-2. These denominators can be derived by considering the overall sources of variation included in these simple effects. Since

$$\sum_j \text{SS}_{a \text{ at } b_j} = \text{SS}_a + \text{SS}_{ab},$$

the F ratio for tests on the simple main effects of A at level b_j would seem to involve a mixture of the appropriate denominators for tests on the main effect of A and the AB interaction. Indeed, such is the case. Thus, MS_{error} for A at b_j is given by

$$\frac{\text{SS}_{\text{error}(a)} + \text{SS}_{\text{error}(b)}}{p(n-1) + p(n-1)(q-1)} = \frac{\text{MS}_{\text{error}(a)} + (q-1)\text{MS}_{\text{error}(b)}}{q}.$$

Similarly, since

$$\sum_k \text{SS}_{b \text{ at } c_k} = \text{SS}_b + \text{SS}_{bc},$$

the denominator of the F test on the variance due to the main effects of B at c_k is given by

$$\frac{\text{SS}_{\text{error}(b)} + \text{SS}_{\text{error}(bc)}}{p(n-1)(q-1) + p(n-1)(q-1)(r-1)} = \frac{\text{MS}_{\text{error}(b)} + (r-1)\text{MS}_{\text{error}(bc)}}{r}.$$

For tests on variance due to C at b_j,

$$\sum_j \text{SS}_{C \text{ at } b_j} = \text{SS}_c + \text{SS}_{bc},$$

both MS_c and MS_{bc} have as their denominator $\text{MS}_{\text{error}(c)}$. Hence

$$\text{MS}_{\text{error for } C \text{ at } b_j} = \text{MS}_{\text{error}(c)}.$$

A more rigorous rationale for the denominators of F ratios for simple effects is developed later in this chapter.

Tests on the difference between simple main effects take the following form:

$$t = \frac{\overline{AB}_{2j} - \overline{AB}_{1j}}{\sqrt{2[\text{MS}_{\text{error}(a)} + (q-1)\text{MS}_{\text{error}(b)}]/nrq}},$$

$$t = \frac{\overline{AC}_{2k} - \overline{AC}_{1k}}{\sqrt{2[\text{MS}_{\text{error}(a)} + (r-1)\text{MS}_{\text{error}(c)}]/nrq}}.$$

An approximate critical value for a t statistic of this kind is obtained as follows: Let t_a and t_b be the critical values for a test of level of significance equal to α for the degrees of freedom corresponding to $\text{MS}_{\text{error}(a)}$ and $\text{MS}_{\text{error}(b)}$, respectively. Then an approximate critical value for the t statistic is

$$t_{\text{critical}} = \frac{t_a\text{MS}_{\text{error}(a)} + t_b(q-1)\text{MS}_{\text{error}(b)}}{\text{MS}_{\text{error}(a)} + (q-1)\text{MS}_{\text{error}(b)}}.$$

This critical value is suggested by Cochran and Cox (1957, p. 299). In cases in which the degrees of freedom for the mean squares are both large (say over 30), the critical value may be obtained directly from tables of the normal distribution.

Table 7.3-2 Denominator of F Ratio for Simple Effects

	Simple effect	Denominator of F ratio
A at b_j	$\overline{AB}_{1j} - \overline{AB}_{2j}$	$[\text{MS}_{\text{error}(a)} + (q-1)\text{MS}_{\text{error}(b)}]/q$
A at c_k	$\overline{AC}_{1k} - \overline{AC}_{2k}$	$[\text{MS}_{\text{error}(a)} + (r-1)\text{MS}_{\text{error}(c)}]/r$
B at a_i	$\overline{AB}_{i1} - \overline{AB}_{i2}$	$\text{MS}_{\text{error}(b)}$
C at a_i	$\overline{AC}_{i1} - \overline{AC}_{i2}$	$\text{MS}_{\text{error}(c)}$
B at c_k	$\overline{BC}_{1k} - \overline{BC}_{2k}$	$[\text{MS}_{\text{error}(b)} + (r-1)\text{MS}_{\text{error}(bc)}]/r$
C at b_j	$\overline{BC}_{j1} - \overline{BC}_{j2}$	$[\text{MS}_{\text{error}(c)} + (q-1)\text{MS}_{\text{error}(bc)}]/q$
A at bc_{jk}	$\overline{ABC}_{1jk} - \overline{ABC}_{2jk}$	$\text{MS}_{\text{w. cell}}$

Alternatively, one may enter the t table with degrees of freedom given by the Satterthwaite approximation, which in this case has the form

$$f = \frac{[\text{MS}_{\text{error}(a)} + (q-1)\text{MS}_{\text{error}(b)}]^2}{\text{MS}^2_{\text{error}(a)}/p(n-1) + [(q-1)\text{MS}_{\text{error}(b)}]^2/p(n-1)(q-1)}.$$

The Satterthwaite approximation is discussed in some detail in Sec. 5.15. A summary of the degrees of freedom, as obtained from the Satterthwaite approach, is given in Table 7.3-2a.

Table 7.3-2a Approximate Degrees of Freedom for Denominators Indicated in Table 7.3-2

A at b_j	$\dfrac{[\text{MS}_{\text{error}(a)} + (q-1)\text{MS}_{\text{error}(b)}]^2}{\text{MS}^2_{\text{error}(a)}/p(n-1) + [(q-1)\text{MS}_{\text{error}(b)}]^2/p(n-1)(q-1)}$
A at c_k	$\dfrac{[\text{MS}_{\text{error}(a)} + (r-1)\text{MS}_{\text{error}(c)}]^2}{\text{MS}^2_{\text{error}(a)}/p(n-1) + [(r-1)\text{MS}_{\text{error}(c)}]^2/p(n-1)(r-1)}$
B at c_k	$\dfrac{[\text{MS}_{\text{error}(b)} + (r-1)\text{MS}_{\text{error}(bc)}]^2}{\text{MS}^2_{\text{error}(b)}/p(n-1)(q-1) + [(r-1)\text{MS}_{\text{error}(bc)}]^2 p(n-1)(r-1)(q-1)}$
C at b_j	$\dfrac{[\text{MS}_{\text{error}(c)} + (q-1)\text{MS}_{\text{error}(bc)}]^2}{\text{MS}^2_{\text{error}(c)}/p(n-1)(r-1) + [(q-1)\text{MS}_{\text{error}(bc)}]^2/p(n-1)(r-1)(q-1)}$

	Lower limit for df	Upper limit for df
A at b_j	$p(n-1)$	$pq(n-1)$
A at c_k	$p(n-1)$	$pr(n-1)$
B at c_k	$p(n-1)(q-1)$	$pr(n-1)(q-1)$
C at b_j	$p(n-1)(r-1)$	$pq(n-1)(r-1)$

Computational Procedures. With the exception of the breakdown of the within-cell variation, computational procedures are identical with those of a $p \times q \times r$ factorial experiment having n observations per cell. These procedures will be illustrated by the data in Table 7.3-3.

Suppose that the levels of factor A represent the noise background under which subjects monitor three dials. The latter define factor C. Subjects are required to make adjustments on the respective dials whenever needles swing outside a specified range. Accuracy scores are obtained for each dial during three consecutive 10-min time periods (factor B).

Table 7.3-3 Basic Data for Numerical Example

	Subjects	Periods:	b_1			b_2			b_3			Total
		Dials:	c_1	c_2	c_3	c_1	c_2	c_3	c_1	c_2	c_3	
	1		45	53	60	40	52	57	28	37	46	418
a_1	2		35	41	50	30	37	47	25	32	41	338
	3		60	65	75	58	54	70	40	47	50	519
	4		50	48	61	25	34	51	16	23	35	343
a_2	5		42	45	55	30	37	43	22	27	37	338
	6		56	60	77	40	39	57	31	29	46	435

The basic data for this experiment are given in Table 7.3-3. Subjects 1, 2, and 3 make up G_1; subjects 4, 5, and 6 make up G_2. To illustrate the meaning of the data, during the first 10-min interval (b_1) subject 1 had scores of 45, 53, and 60 on dials c_1, c_2, and c_3, respectively.

Summary tables prepared from these basic data are given in Table 7.3-4. In part i are summary tables that would be obtained for any $2 \times 3 \times 3$ factorial experiment having n observations per cell. Part ii is unique to a factorial experiment having repeated measures on factors B and C. In the $B \times$ subjects within G_1 summary table a cell entry will be denoted by the symbol BP_{jm}. For example, $BP_{11} = 158$, and $BP_{13} = 111$. Similarly an entry in a cell of the $C \times$ subjects within group summary table will be denoted by the symbol CP_{km}.

Convenient computational symbols are defined and computed in Table 7.3-5. Symbols (1) through (9) are identical to those used in any $p \times q \times r$ factorial experiment in which there are n observations in each cell. Symbols (10) through (12) are unique to a factorial experiment in which there are repeated measures on factors B and C. By using these symbols, computational formulas take the form given in Table 7.3-6. In terms of means,

$$SS_{B\times \text{subj w.groups}} = r\sum_i\sum_j\sum_j(\overline{BP}_{km(i)} - \overline{AB}_{ij} - \bar{P}_{m(i)} + \bar{A}_i)^2.$$

Table 7.3-4 Summary Tables for Numerical Example

(i)

ABC summary table

	b_1			b_2			b_3		
	c_1	c_2	c_3	c_1	c_2	c_3	c_1	c_2	c_3
a_1	140	159	185	128	143	174	93	116	137
a_2	148	153	193	95	110	151	69	79	118
Total	288	312	378	223	253	325	162	195	255

AB summary table

	b_1	b_2	b_3	Total
a_1	484	445	346	1275
a_2	494	356	266	1116
Total	978	801	612	2391

AC summary table

	c_1	c_2	c_3	Total
a_1	361	418	496	1275
a_2	312	342	462	1116
Total	673	760	958	2391

BC summary table

	c_1	c_2	c_3	Total
b_1	288	312	378	978
b_2	223	253	325	801
b_3	162	195	255	612
Total	673	760	958	2391

(ii)

$B \times$ subj w. G_1 summary table

Subject	b_1	b_2	b_3	Total
1	158	149	111	418
2	126	114	98	338
3	200	182	137	519
Total	484	445	346	1275

$B \times$ subj w. G_2 summary table

Subject	b_1	b_2	b_3	Total
4	159	110	74	343
5	142	110	86	338
6	193	136	106	435
Total	494	356	266	1116

$C \times$ subj w. G_1 summary table

Subject	c_1	c_2	c_3	Total
1	113	142	163	418
2	90	110	138	338
3	158	166	195	519
Total	361	418	496	1275

$C \times$ subj w. G_2 summary table

Subject	c_1	c_2	c_3	Total
4	91	105	147	343
5	94	109	135	338
6	127	128	180	435
Total	312	342	462	1116

Table 7.3-5 Definitions and Numerical Values of Computational Symbols

Symbol	Computation	Value
$(1) = G^2/npqr$	$= (2391)^2/3(2)(3)(3)$	$= 105{,}868.17$
$(2) = \Sigma X^2$	$= 45^2 + 53^2 + 60^2 + \cdots + 31^2 + 29^2 + 46^2$	$= 115{,}793$
$(3) = (\Sigma A_i^2)/nqr$	$= (1275^2 + 1116^2)/3(3)(3)$	$= 106{,}336.33$
$(4) = (\Sigma B_j^2)/npr$	$= (978^2 + 801^2 + 612^2)/3(2)(3)$	$= 109{,}590.50$
$(5) = (\Sigma C_k^2)/npq$	$= (673^2 + 760^2 + 958^2)/3(2)(3)$	$= 108{,}238.50$
$(6) = [\Sigma(AB_{ij})^2]/nr$	$= (484^2 + 445^2 + \cdots + 266^2)/3(3)$	$= 110{,}391.67$
$(7) = [\Sigma(AC_{ik})^2]/nq$	$= (361^2 + 418^2 + \cdots + 462^2)/3(3)$	$= 108{,}757.00$
$(8) = [\Sigma(BC_{jk})^2]/np$	$= (288^2 + 312^2 + \cdots + 255^2)/3(6)$	$= 111{,}971.50$
$(9) = [\Sigma(ABC_{ijk})^2]/n$	$= (140^2 + 159^2 + \cdots + 118^2)/3$	$= 112{,}834.33$
$(10) = (\Sigma P_m^2)/qr$	$= (418^2 + 338^2 + \cdots + 435^2)/3(3)$	$= 108{,}827.44$
$(11) = [\Sigma(BP_{jm})^2]/r$	$= (158^2 + 149^2 + \cdots + 106^2)/3$	$= 113{,}117.67$
$(12) = [\Sigma(CP_{km})^2]/q$	$= (113^2 + 142^2 + \cdots + 180^2)/3$	$= 111{,}353.67$
$(10a_1) = \left(\sum_{a_1} P_m^2\right)/qr$	$= (418^2 + 338^2 + 519^2)/3(3)$	$= 62{,}036.56$
$(10a_2) = \left(\sum_{a_2} P_m^2\right)/qr$	$= (343^2 + 338^2 + 435^2)/3(3)$	$= 46{,}790.89$
		$108{,}827.45$
$(11a_1) = \left[\sum_{a_1}\left(BP_{jm}\right)^2\right]/r$	$= (158^2 + 149^2 + \cdots + 137^2)/3$	$= 63{,}285.00$
$(11a_2) = \left[\sum_{a_2}\left(BP_{jm}\right)^2\right]/r$	$= (159^2 + 110^2 + \cdots + 106^2)/3$	$= 49{,}832.67$
		$113{,}117.67$

The formula for this source of variation given in Table 7.3-6 leads to simpler computations. This source of variation is also designated $SS_{error(b)}$.

There is an overall computational check that can be made on the sum of the error terms. The following relationship exists:

$$
\begin{aligned}
SS_{error(a)} &= (10) - (3) \\
SS_{error(b)} &= (11) - (6) - (10) + (3) \\
SS_{error(c)} &= (12) - (7) - (10) + (3) \\
SS_{error(bc)} &= (2) - (9) - (11) - (12) + (6) + (7) \\
&\qquad + (10) - (3) \\
\hline
SS_{w.\ cell} &= (2) - (9)
\end{aligned}
$$

The computational symbols on the right may be treated as algebraic symbols; that is, $(3) + (3) = 2(3)$, $(3) - (3) = 0$. The algebraic sum of the symbols on the right is $(2) - (9)$. The latter is the computational formula for $SS_{w.cell}$. The computational symbols at the bottom of Table 7.3-5 are used to partition the error variation for each of the error

Table 7.3-6 Summary of Analysis of Variance

Source of variation	Computational formula	SS	df	MS	F
Between subjects	(10) − (1)	2959.27	5		
A	(3) − (1)	468.16	1	468.16	
Subj w. groups [error (a)]	(10) − (3)	2491.11	4	622.78	
Within subjects	(2) − (10)	6965.56	48		
B	(4) − (1)	3722.33	2	1861.16	63.39*
AB	(6) − (3) − (4) + (1)	333.00	2	166.50	5.67*
$B \times$ subj w. groups [error (b)]	(11) − (6) − (10) + (3)	234.89	8	29.36	
C	(5) − (1)	2370.33	2	1185.16	89.78*
AC	(7) − (3) − (5) + (1)	50.34	2	25.17	1.91
$C \times$ subj w. groups [error (c)]	(12) − (7) − (10) + (3)	105.56	8	13.20	
BC	(8) − (4) − (5) + (1)	10.67	4	2.67	
ABC	(9) − (6) − (7) − (8) + (3) + (4) + (5) − (1)	11.32	4	2.83	
$BC \times$ subj w. groups [error (bc)]	(2) − (9) − (11) − (12) + (6) + (7) + (10) − (3)	127.11	16	7.94	

terms into parts which may be checked for homogeneity by means of an F_{max} test.

The analysis of variance is summarized in Table 7.3-6. If a model which includes interactions with subjects is appropriate for this experiment, the expected values for the mean squares are those shown in Table 7.3-1 and the structure of F ratios is determined by these expected values. However, should the existence of interactions with subjects be open to question, preliminary tests on the model are appropriate. In this case such interactions have relatively few degrees of freedom. A check on the homogeneity of such interactions is carried out in Table 7.3-7, by use of

Table 7.3-7 Test for Homogeneity of Interactions with Subjects

SS	MS	df	log MS	1/df
234.89	29.36	8	1.468	.125
105.56	13.20	8	1.121	.125
127.11	7.94	16	.900	.062
Σ SS = 467.56		Σ df = 32		Σ(1/df) = .312

$$MS_{pooled} = (\Sigma\ SS)/\Sigma\ df = 467.56/32 = 14.61$$

$$A = \Sigma[(df)_i \log MS_i] = 8(1.468) + 8(1.121) + 16(.900) = 35.112$$

$$B = (\Sigma\ df) \log MS_{pooled} = 32(1.165) = 37.280$$

$$C = 1 + \frac{1}{3(k-1)}[\Sigma(1/df) - (1/\Sigma\ df)] = 1 + \tfrac{1}{6}[.312 - .031] = 1.047$$

$$\chi^2 = \frac{2.303(B - A)}{C} = \frac{2.303(37.280 - 35.112)}{1.047} = 4.77$$

$$k = \text{number of } MS_i \qquad \chi^2_{.80}(k-1) = \chi^2_{.80}(2) = 3.22$$

Bartlett's test. Since the observed chi square (4.77) exceeds the critical value (3.22) for a test with $\alpha = .20$, the test indicates that the interactions should not be pooled. Equivalently, the test indicates that interactions with subjects should not be dropped from the model.

Thus tests on B and AB use $MS_{error(b)}$ as a denominator for F ratios; tests on C and AC use $MS_{error(c)}$ as a denominator; tests on BC and ABC use $MS_{error(bc)}$ as a denominator. The main effect of A is tested with $MS_{error(a)}$. By using $\alpha = .05$ for all tests, the main effects for factors B (periods) and C (dials) are found to be statistically significant. Inspection of the totals for levels b_1, b_2, and b_3 indicates decreasing accuracy scores for the consecutive time periods. Inspection of the totals for the dials indicates that dial c_3 is monitored with the greatest accuracy and dial c_1 monitored with the least accuracy.

The AB interaction is also noted to be statistically significant. The profiles of means corresponding to the cell totals in the AB summary table are plotted in Fig. 7.3-1. The profiles indicate a difference in the rate of decline in the accuracy scores in the three periods, the group working

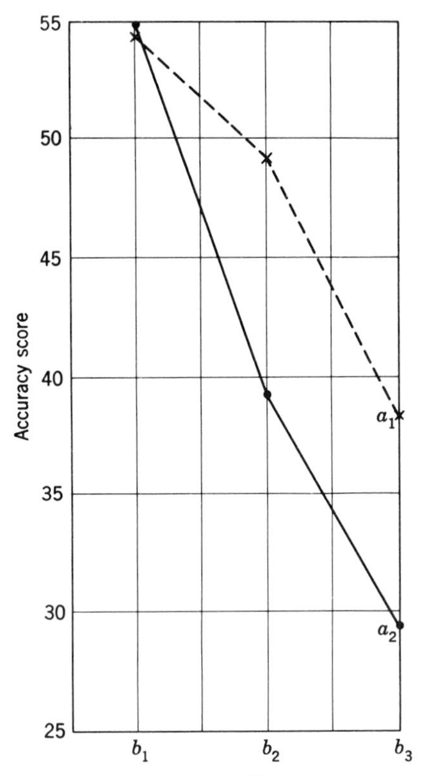

Figure 7.3-1 Profiles of means.

under noise level a_1 showing a slower decline rate than the group working under a_2. Differences between corresponding points on these profiles have the form

$$\overline{AB}_{1j} - \overline{AB}_{2j}.$$

The standard error of the difference between these two means is estimated by

$$\sqrt{\frac{2(\text{SS}_{\text{subj w.groups}} + \text{SS}_{B \times \text{subj w.groups}})}{nr[p(n-1) + p(n-1)(q-1)]}} = \sqrt{\frac{2(2491.11 + 234.89)}{9(12)}}$$
$$= \sqrt{50.48} = 7.10.$$

By way of contrast, the standard error of the difference between two means of the following form,

$$\overline{AB}_{i1} - \overline{AB}_{i2},$$

is estimated by

$$\sqrt{2\text{MS}_{\text{error}(b)}/nr} = \sqrt{2(29.36)/9} = 2.55.$$

The latter standard error is considerably smaller than that computed in the last paragraph. A difference of the form $\overline{AB}_{1j} - \overline{AB}_{2j}$ is in part confounded with between-group effects, whereas a difference of the form $\overline{AB}_{i1} - \overline{AB}_{i2}$ is entirely a within-subject effect.

For the case in which there are repeated measures on two factors, there are three sets of covariance matrices which may be constructed. The following set is obtained from the $B \times$ subj w. G_i summary tables.

$$B_{a_1} = \begin{bmatrix} 459.11 & 418.77 & 244.11 \\ & 385.44 & 220.28 \\ & & 131.44 \end{bmatrix}; \quad B_{a_2} = \begin{bmatrix} 224.77 & 122.78 & 105.78 \\ & 75.11 & 75.11 \\ & & 87.11 \end{bmatrix};$$

$$B_{\text{pooled}} = \begin{bmatrix} 341.94 & 270.78 & 174.94 \\ & 230.28 & 147.70 \\ & & 109.28 \end{bmatrix}.$$

Since n is so small, the sampling variability of the variances and covariances will be relatively large. From B_{pooled} one has

$$\overline{\text{var}} = 227.17, \qquad \overline{\text{cov}} = 197.81.$$

One notes that

$$\text{MS}_{B \times \text{subj w.groups}} = \overline{\text{var}} - \overline{\text{cov}} = 227.17 - 197.81 = 29.36.$$

The following set of covariance matrices is obtained from the $C \times$ subj w. G_i summary tables:

$$C_{a_1} = \begin{bmatrix} 398.77 & 312.45 & 327.27 \\ & 263.11 & 264.45 \\ & & 272.11 \end{bmatrix}; \quad C_{a_2} = \begin{bmatrix} 133.00 & 81.50 & 146.50 \\ & 50.33 & 87.00 \\ & & 181.00 \end{bmatrix};$$

$$C_{\text{pooled}} = \begin{bmatrix} 265.88 & 196.98 & 236.88 \\ & 156.72 & 175.72 \\ & & 226.56 \end{bmatrix}.$$

From C_{pooled},

$$\overline{\text{var}} = 216.39, \qquad \overline{\text{cov}} = 203.19.$$

One notes that

$$\text{MS}_{C \times \text{subj w.groups}} = \overline{\text{var}} - \overline{\text{cov}} = 13.20.$$

A third set of matrices may be obtained from the basic data as given in

Table 7.3-3. From the 9 × 9 matrices, it will be found that

$$32(\overline{\text{var}} - \overline{\text{cov}}) = SS_{B \times \text{subj w.groups}} + SS_{C \times \text{subj w.groups}} + SS_{BC \times \text{subj w.groups}}$$

Illustrative Applications. Many examples of this basic plan will be found in the recent experimental literature. A study by French (1959) illustrates one application. The purpose of this study was to investigate the effect of lesions in two different areas of the cortex (factor A) upon a bar-pressing task which set up two conditions of illumination (factor B). Each of the subjects was given three blocks of trials (factor C) under each illumination condition. There were four subjects per group. The criterion measure was the average length of time the bar was depressed during each block of trials. To obtain homogeneity of error variance, the scale of measurement was transformed into log $(\bar{X} + 1)$. The addition of unity to each of the means avoids the occurrence of the logarithm of zero. The plan for the experiment may be represented as follows:

Lesions	Illumination:	b_1			b_2			
	Trials:	c_1	c_2	c_3	c_1	c_2	c_3	
a_1		G_1	G_1	G_1	G_1	G_1	G_1	$n = 4$
a_2		G_2	G_2	G_2	G_2	G_2	G_2	

The analysis of variance reported by French had the form given in Table 7.3-8.

In this analysis all interactions with subjects were pooled. The resulting pooled interaction term was the denominator for all within-subject

Table 7.3-8 Analysis of Variance for French Data (1959)

Source	df	MS	F
Between subjects	7		
A (lesions)	1	.846	6.89*
Subj w. groups	6	.122	
Within subjects	40		
B (illumination)	1	.161	1.46
C (trials)	2	.945	8.61*
AB	1	.008	
AC	2	.389	3.54*
BC	2	.113	1.03
ABC	2	.078	
Pooled interactions with subjects	30	.110	

effects. The significant main effect due to factor A indicated that the groups differed in their mean overall criterion scores (on the transformed scale of measurement). The significant main effect due to trials indicated that the means changed during the trials. The significant lesion $\times$ trial (AC) interaction indicated that the rate of change during the trials differed for the two groups.

Briggs, Fitts, and Bahrick (1958) reported a series of experiments in which the experimental plan under discussion was used. In one of these studies, the effects of methods of training (factor A) upon transfer of training in a complex tracking task were studied. Performance under the last two blocks of trials in the training session and the first two blocks of the transfer session was analyzed. The plan of the experiment may be represented schematically as follows:

Method of training	Training		Transfer	
	Trial c_1	Trial c_2	Trial c_1	Trial c_2
a_1	G_1	G_1	G_1	G_1
a_2	G_2	G_2	G_2	G_2
a_3	G_3	G_3	G_3	G_3
a_4	G_4	G_4	G_4	G_4

The last two blocks of trials under training and the first two trials under transfer define the two levels of factor B. There were 17 subjects in each group. This plan may be considered as a special case of a $4 \times 2 \times 2$ factorial experiment with repeated measures on the last two factors. The criterion measure was average time on target. The analysis of variance reported by these workers had the form given in Table 7.3-9.

In the actual experiment, G_1 worked under identical conditions during the training and transfer trial. Each of the other groups had different conditions under the learning trials but a common condition under the transfer trials. The following means were obtained from the experimental data.

	b_1	b_2
a_1	2.36	2.54
a_2	1.80	2.36
a_3	1.45	2.21
a_4	1.54	2.31

Because of the significant interaction, key simple main effects were tested. The means within brackets do not differ significantly, while those means not bracketed do differ at the .05 level. It is noted that all groups were equally good, on the average, on the transfer block of trials. The significant AB interaction was due largely to the fact that the control group (a_1) showed no statistically different increase from training trials (b_1) to transfer trials (b_2), whereas all the other groups did show a statistically significant increase.

Table 7.3-9 Analysis of Variance for Briggs, Fitts, and Bahrick Data (1958)

Source	df	MS	F
Between subjects	67		
A (methods of training)	3	2,319.02	3.72*
Subj w. groups	64	623.72	
Within subjects	204		
B (training-transfer)	1	11,492.52	113.11*
AB	3	707.44	6.96*
$B \times$ subj w. groups	64	101.61	
C (blocks of trials)	1	5.95	
AC	3	9.20	
$C \times$ subj w. groups	64	90.45	
BC	1	111.90	2.06
ABC	3	24.84	
$BC \times$ subj w. groups	64	54.28	

Another study reported by Briggs, Fitts, and Bahrick (1958) may be considered as a special case of a $4 \times 3 \times 3$ factorial experiment with repeated measures on the last two factors. In this study, factor A defines the time at which transfer was made from a common training condition to a common transfer condition. Group 4 made the transfer at time zero; i.e., group 4 worked under the transfer condition during all sessions. A schematic representation of the plan is given below.

Transfer time	Sessions:	b_1				b_2				b_3			
	Trials:	c_1	c_2	c_3	c_4	c_1	c_2	c_3	c_4	c_1	c_2	c_3	c_4
a_1		G_1	G_1	G_1	G_1	G_1	G_1	G_1	G_1	G_1	G_1	G_1	G_1
a_2		G_2	G_2	G_2	G_2	G_2	G_2	G_2	G_2	G_2	G_2	G_2	G_2
a_3		G_3	G_3	G_3	G_3	G_3	G_3	G_3	G_3	G_3	G_3	G_3	G_3
a_4		G_4	G_4	G_4	G_4	G_4	G_4	G_4	G_4	G_4	G_4	G_4	G_4

$n_1 = 14, \qquad n_2 = 12, \qquad n_3 = 14, \qquad n_4 = 13$

Because of unequal group size, the equivalent of an unweighted-means analysis was made on the data. In this analysis, group means replace each of the X's and the data are considered as if n were unity in the calculation of the parts of the between-cell variation. Sums of squares computed in this manner are then multiplied by the harmonic mean of the group sizes.

The subject within groups variation is given by

$$SS_{\text{subj w.groups}} = \frac{\Sigma P_m^2}{qr} - \Sigma \frac{A_i^2}{n_i}.$$

The $B \times$ subject within groups interaction is most easily obtained by computing the separate terms $B \times$ subject within G_i and then pooling the parts. Other interactions with subjects are computed in a similar way. The analysis of variance reported by these workers had the form given in Table 7.3-10. Significant interactions led to a series of tests on simple effects as well as individual comparisons between pairs of means. Graphs of profiles corresponding to interactions were used effectively in the interpretations.

A study by Casteneda and Lipsett (1959) provides another illustrative example of the same basic experimental plan. This study was designed to evaluate the effect of level of stress (factor A) upon learning patterns of varying complexity (factor B). Four blocks of trials (factor C) were given under each level of complexity. The criterion measure was the number of correct responses in each block of trials. The plan may be represented as follows:

Level of stress	Pattern complexity:	b_1				b_2			
	Trials:	c_1	c_2	c_3	c_4	c_1	c_2	c_3	c_4
a_1		G_1	G_1	G_1	G_1	G_1	G_1	G_1	G_1
a_2		G_2	G_2	G_2	G_2	G_2	G_2	G_2	G_2

There were 54 subjects in each group. The order of presentation of the patterns was randomized, but each pattern was presented an equal number of times.

The analysis of variance reported by these workers had the form given in Table 7.3-11. It will be noted that relative to the within-subject error terms the between-subject error term is exceptionally large ($MS_{\text{subj w.groups}} = 804.66$). This observation and remarks made by the workers about the distributions suggest that the analysis of variance might more appropriately have been made in terms of a different scale of measurement, that is, log X rather than X. In the analysis as it stands, the stress $\times$ pattern interaction (AB) was found to be statistically significant. In terms of the experimental variables, this finding indicated that the stress condition interfered more in the learning of the complex pattern than it did in the learning of the simple pattern. This essentially was the hypothesis that the workers were interested in testing. Because of the extremely

Table 7.3-10 Analysis of Variance for Briggs, Fitts, and Bahrick Data (1958)

Source	df	MS	F
Between subjects	52		
A (transfer time)	3	45.79	
Subj w. groups	49	69.76	
Within subjects	583		
B (sessions)	2	426.51	17.28*
AB	6	266.12	10.78*
B × subj w. groups	98	24.68	
C (trials)	3	144.11	46.04*
AC	9	2.34	
C × subj w. groups	147	3.13	
BC	6	16.45	5.24*
ABC	18	10.61	3.38*
BC × subj w. groups	294	3.14	

high between-subject error term simple main effects of the stress conditions could not be demonstrated to be statistically significant.

Table 7.3-11 Analysis of Variance for Casteneda and Lipsett Data (1959)

Source	df	MS	F
Between subjects	107		
A (stress)	1	0.78	
Subj w. groups	106	804.66	
Within subjects	756		
B (patterns)	1	1820.04	255.26*
AB	1	148.34	20.80*
B × subj w. groups	106	7.13	
C (trials)	3	167.13	80.40*
AC	3	4.01	2.02
C × subj w. groups	318	1.98	
BC	3	7.38	2.64
ABC	3	6.48	2.32
BC × subj w. groups	318	2.79	

Table 7.3-12 General Expected Values for Mean Squares

Effect	i	j	k	m	o	E(MS)
α_i	D_p	q	r	n	1	$\sigma_\varepsilon^2 + D_qD_r\sigma_{\beta\gamma\pi}^2 + nD_qD_r\sigma_{\alpha\beta\gamma}^2 + qD_r\sigma_{\gamma\pi} + nqD_r\sigma_{\alpha\gamma}^2 + rD_q\sigma_{\beta\pi}^2 + nrD_q\sigma_{\alpha\beta}^2 + qr\sigma_\pi^2 + nqr\sigma_\alpha^2$
$\pi_{m(i)}$	1	q	r	1	1	$\sigma_\varepsilon^2 + D_qD_r\sigma_{\beta\gamma\pi}^2 + qD_r\sigma_{\gamma\pi}^2 + rD_q\sigma_{\beta\pi}^2 + qr\sigma_\pi^2$
β_j	p	D_q	r	n	1	$\sigma_\varepsilon^2 + D_r\sigma_{\beta\gamma\pi}^2 + nD_pD_r\sigma_{\alpha\beta\gamma}^2 + npD_r\sigma_{\beta\gamma}^2 + r\sigma_{\beta\pi}^2 + nrD_p\sigma_{\alpha\beta}^2 + npr\sigma_\beta^2$
$\alpha\beta_{ij}$	D_p	D_q	r	n	1	$\sigma_\varepsilon^2 + D_r\sigma_{\beta\gamma\pi}^2 + nD_r\sigma_{\alpha\beta\gamma}^2 + r\sigma_{\beta\pi}^2 + nr\sigma_{\alpha\beta}^2$
$\beta\pi_{jm(i)}$	1	D_q	r	1	1	$\sigma_\varepsilon^2 + D_r\sigma_{\beta\gamma\pi}^2 + r\sigma_{\beta\pi}^2$
γ_k	p	q	D_r	n	1	$\sigma_\varepsilon^2 + D_q\sigma_{\beta\gamma\pi}^2 + nD_pD_q\sigma_{\alpha\beta\gamma}^2 + npD_q\sigma_{\beta\gamma}^2 + q\sigma_{\gamma\pi}^2 + nqD_p\sigma_{\alpha\gamma}^2 + npq\sigma_\gamma^2$
$\alpha\gamma_{ik}$	D_p	q	D_r	n	1	$\sigma_\varepsilon^2 + D_q\sigma_{\beta\gamma\pi}^2 + nD_q\sigma_{\alpha\beta\gamma}^2 + q\sigma_{\gamma\pi}^2 + nq\sigma_{\alpha\gamma}^2$
$\gamma\pi_{km(i)}$	1	q	D_r	1	1	$\sigma_\varepsilon^2 + D_q\sigma_{\beta\gamma\pi}^2 + q\sigma_{\gamma\pi}^2$
$\beta\gamma_{jk}$	p	D_q	D_r	n	1	$\sigma_\varepsilon^2 + \sigma_{\beta\gamma\pi}^2 + nD_p\sigma_{\alpha\beta\gamma}^2 + np\sigma_{\beta\gamma}^2$
$\alpha\beta\gamma_{ijk}$	D_p	D_q	D_r	n	1	$\sigma_\varepsilon^2 + \sigma_{\beta\gamma\pi}^2 + n\sigma_{\alpha\beta\gamma}^2$
$\beta\gamma\pi_{jkm(i)}$	1	D_q	D_r	1	1	$\sigma_\varepsilon^2 + \sigma_{\beta\gamma\pi}^2$
$\varepsilon_{o(ijkm)}$	1	1	1	1	1	σ_ε^2

General Expected Values for the Mean Squares. The expected values for the mean squares given in Table 7.3-1 are for the special case in which A, B, and C are fixed factors. The general case is given in Table 7.3-12. In this table,

$$D_p = 1 - \frac{p}{P}, \qquad D_q = 1 - \frac{q}{Q}, \qquad D_r = 1 - \frac{r}{R}.$$

Thus D_p is either 0 or 1 depending upon whether factor A is fixed or random, respectively.

If $D_p = 1$, $D_q = 0$, and $D_r = 0$, then the special case given in Table 7.3-13 is obtained. There are several assumptions about the form of variance-covariance matrices that must be met in order that these expected values provide reasonable guides for setting up F ratios. A series of

Table 7.3-13 Expected Values of Mean Squares for Case in Which Factor A Is Random and Factors B and C Are Fixed

Source	E(MS)
Between subjects	
A	$\sigma_\varepsilon^2 + qr\sigma_\pi^2 + nqr\sigma_\alpha^2$
Subj w. groups	$\sigma_\varepsilon^2 + qr\sigma_\pi^2$
Within subjects	
B	$\sigma_\varepsilon^2 + r\sigma_{\beta\pi}^2 + nr\sigma_{\alpha\beta}^2 + npr\sigma_\beta^2$
AB	$\sigma_\varepsilon^2 + r\sigma_{\beta\pi}^2 + nr\sigma_{\alpha\beta}^2$
$B \times$ subj w. groups	$\sigma_\varepsilon^2 + r\sigma_{\beta\pi}^2$
C	$\sigma_\varepsilon^2 + q\sigma_{\gamma\pi}^2 + nq\sigma_{\alpha\gamma}^2 + npq\sigma_\gamma^2$
AC	$\sigma_\varepsilon^2 + q\sigma_{\gamma\pi}^2 + nq\sigma_{\alpha\gamma}^2$
$C \times$ subj w. groups	$\sigma_\varepsilon^2 + q\sigma_{\gamma\pi}^2$
BC	$\sigma_\varepsilon^2 + \sigma_{\beta\gamma\pi}^2 + n\sigma_{\alpha\beta\gamma}^2 + np\sigma_{\beta\gamma}^2$
ABC	$\sigma_\varepsilon^2 + \sigma_{\beta\gamma\pi}^2 + n\sigma_{\alpha\beta\gamma}^2$
$BC \times$ subj w. groups	$\sigma_\varepsilon^2 + \sigma_{\beta\gamma\pi}^2$

preliminary tests on the model may be made, in the absence of background information, in order to check the desirability of pooling various interactions with random factors.

7.4 Three-factor Experiment with Repeated Measures (Case II)

In the last section the case in which there were repeated measures on two of the three factors was considered. In this section the case in which there are repeated measures on only one of the three factors will be considered. This case may be represented schematically as follows:

		c_1	c_2	$\cdots$	c_r
a_1	b_1	G_{11}	G_{11}	$\cdots$	G_{11}
	$\vdots$	$\vdots$	$\vdots$		$\vdots$
	b_q	G_{1q}	G_{1q}	$\cdots$	G_{1q}
	$\vdots$				
a_p	b_1	G_{p1}	G_{p1}	$\cdots$	G_{p1}
	$\vdots$	$\vdots$	$\vdots$		$\vdots$
	b_q	G_{pq}	G_{pq}	$\cdots$	G_{pq}

Each of the groups is observed under all levels of factor C, but each group is assigned to only one combination of factors A and B. The notation G_{ij} denotes the group of subjects assigned to treatment combination ab_{ij}. A subject within group G_{ij} is identified by the subscript $m(ij)$. This notation indicates that the subject effect is nested under both factors A and B.

The structural model on which the analysis which follows is based has the following form:

$$X_{ijkm} = \mu + \alpha_i + \beta_j + \alpha\beta_{ij} + \pi_{m(ij)} + \gamma_k + \alpha\gamma_{ik} + \beta\gamma_{jk} + \alpha\beta\gamma_{ijk} + \gamma\pi_{km(ij)} + \varepsilon_{o(ijkm)}.$$

Since the subject factor is nested under both factors A and B, there can be no interaction between these latter factors and the subject factor. This model has implicit in it homogeneity assumptions on variance-covariance matrices associated with the repeated measures. The analysis of variance for this plan takes the form given in Table 7.4-1. The expected values in this table are for the special case in which A, B, and C are considered fixed factors. The general case as obtained from the Cornfield-Tukey algorithm is given in Table 7.4-8.

An alternative partition of the total variation permits a comparison

Source	df	
Total	$npqr - 1$	
Between cells	$pqr - 1$	
Within cells	$pqr(n - 1)$	
Subj w. groups		$pq(n - 1)$
$C \times$ subj w. groups		$pq(n - 1)(r - 1)$

between this plan and a $p \times q \times r$ factorial experiment in which there are no repeated measures, but n observations per cell. The main effects and

Table 7.4-1 Summary of Analysis of Variance

Source of variation	df	E(MS)†
Between subjects	$npq - 1$	
A	$p - 1$	$\sigma_\varepsilon^2 + r\sigma_\pi^2 + nqr\sigma_\alpha^2$
B	$q - 1$	$\sigma_\varepsilon^2 + r\sigma_\pi^2 + npr\sigma_\beta^2$
AB	$(p - 1)(q - 1)$	$\sigma_\varepsilon^2 + r\sigma_\pi^2 + nr\sigma_{\alpha\beta}^2$
Subj w. groups [error (between)]	$pq(n - 1)$	$\sigma_\varepsilon^2 + r\sigma_\pi^2$
Within subjects	$npq(r - 1)$	
C	$r - 1$	$\sigma_\varepsilon^2 + \sigma_{\gamma\pi}^2 + npq\sigma_\gamma^2$
AC	$(p - 1)(r - 1)$	$\sigma_\varepsilon^2 + \sigma_{\gamma\pi}^2 + nq\sigma_{\alpha\gamma}^2$
BC	$(q - 1)(r - 1)$	$\sigma_\varepsilon^2 + \sigma_{\gamma\pi}^2 + np\sigma_{\beta\gamma}^2$
ABC	$(p - 1)(q - 1)(r - 1)$	$\sigma_\varepsilon^2 + \sigma_{\gamma\pi}^2 + n\sigma_{\alpha\beta\gamma}^2$
$C \times$ subj w. groups [error (within)]	$pq(n - 1)(r - 1)$	$\sigma_\varepsilon^2 + \sigma_{\gamma\pi}^2$

† Assumes A, B, and C fixed factors.

all interactions of factors A, B, and C are part of the between-cell variation whether or not there are repeated measures. The partition of the between-cell variation is identical in the two cases. When there are repeated measures on factor C, the within-cell variation is subdivided into two parts. One of these parts is

$$\begin{aligned} SS_{\text{subj w.groups}} &= r\Sigma\Sigma\Sigma(\bar{P}_{m(ij)} - \overline{AB}_{ij})^2 \\ &= \frac{\Sigma\Sigma\Sigma P_{m(ij)}^2}{r} - \frac{\Sigma\Sigma(AB_{ij})^2}{nr}. \end{aligned}$$

(The symbol $\Sigma\Sigma\Sigma P_{m(ij)}^2$ represents the sum of the squared totals from each subject. Each total is based upon r observations.) This source of variation is a measure of the extent to which the mean of a subject differs from the mean of the group in which the subject is located. The other part of the within-cell variation is

$$SS_{C \times \text{subj w.groups}} = SS_{\text{w.cell}} - SS_{\text{subj w.groups}},$$

where

$$\begin{aligned} SS_{\text{w.cell}} &= \Sigma\Sigma\Sigma\Sigma(X_{ijkm} - \overline{ABC}_{ijk})^2 \\ &= \Sigma\Sigma\Sigma\Sigma X_{ijkm}^2 - \frac{\Sigma(ABC_{ijk})^2}{n}. \end{aligned}$$

Because of the structure of the F ratio for this plan (when A, B, and C

are fixed factors), the following notation is sometimes used:

$$SS_{\text{subj w.groups}} = SS_{\text{error(between)}},$$
$$SS_{C\times\text{subj w.groups}} = SS_{\text{error(within)}}.$$

Each of these error terms may be subdivided and tested for homogeneity by means of $F_{\max}$ tests. The first term may be subdivided into the following parts:

Source	df
Subj w. G_{11}	$n - 1$
Subj w. G_{12}	$n - 1$
...	...
Subj w. G_{pq}	$n - 1$

There will be pq terms, each having the general form

$$SS_{\text{subj w.}G_{ij}} = r\sum_m(\bar{P}_{m(ij)} - \overline{AB}_{ij})^2.$$

The critical value for an $F_{\max}$ test would be $F_{\max(1-\alpha)}(pq, n - 1)$. The second error term may be divided into the following parts:

Source	df
$C \times$ subj w. G_{11}	$(n - 1)(r - 1)$
$C \times$ subj w. G_{12}	$(n - 1)(r - 1)$
....	...
$C \times$ subj w. G_{pq}	$(n - 1)(r - 1)$

There will be pq terms; each has the general form

$$SS_{C\times\text{subj w.}G_{ij}} = \sum_m\sum_k(X_{ijkm} - \bar{P}_{m(ij)} - \overline{ABC}_{ijk} + \overline{AB}_{ij})^2.$$

The critical value for an $F_{\max}$ test in this case would be

$$F_{\max(1-\alpha)}[pq, (n - 1)(r - 1)].$$

Should either of these two error terms prove to be heterogeneous in terms of the criterion scale of measurement being used, the experimenter should consider a transformation on the scale of measurement in terms of which the analysis of variance may be carried out.

In making tests on simple main effects, denominators appropriate for F ratios are indicated in Table 7.4-2. It should be noted that a difference between two simple main effects is a mixture of main effects and interaction effects. In cases where the main effects and interaction effects have different error terms, a compromise error term is constructed. The latter is a weighted average of the different error terms, the weights being the respective degrees of freedom. Because of this pooling of heterogeneous sources of variation, the resulting F tests are potentially subject to bias.

In this design, when the pattern assumptions on the variance-covariance matrices are questionable, critical values of the conservative tests involving factor C have the form

$$F_{1-\alpha}[1, pq(n-1)] \quad \text{instead of} \quad F_{1-\alpha}[(r-1), pq(n-1)(r-1)],$$
$$F_{1-\alpha}[(p-1), pq(n-1)] \quad \text{instead of} \quad F_{1-\alpha}[(p-1)(r-1), pq(n-1)(r-1)].$$

That is, the degrees of freedom for numerator and denominator are each divided by $r - 1$.

Table 7.4-2 Denominator of F Ratio for Simple Effects

Simple effect		Denominator of F ratio
A at b_j	$\overline{AB}_{1j} - \overline{AB}_{2j}$	$\text{MS}_{\text{error(between)}}$
B at a_i	$\overline{AB}_{i1} - \overline{AB}_{i2}$	
C at a_i	$\overline{AC}_{i1} - \overline{AC}_{i2}$	$\text{MS}_{\text{error(within)}}$
C at b_j	$\overline{BC}_{j1} - \overline{BC}_{j2}$	
C at ab_{ij}	$\overline{ABC}_{ij1} - \overline{ABC}_{ij2}$	
A at c_k	$\overline{AC}_{1k} - \overline{AC}_{2k}$	$[\text{MS}_{\text{error(between)}} + (r-1)\text{MS}_{\text{error(within)}}]/r$
B at c_k	$\overline{BC}_{1k} - \overline{BC}_{jk}$	
AB at c_k	$\overline{ABC}_{12k} - \overline{ABC}_{34k}$	

Computational Procedures. A numerical example will be used to illustrate the computational procedures. [This example is a modified version of an experiment actually conducted by Meyer and Noble (1958).] Suppose that an experimenter is interested in evaluating the effect of anxiety (factor A) and muscular tension (factor B) on a learning task. Subjects who score extremely low on a scale measuring manifest anxiety are assigned to level a_1; subjects who score extremely high are assigned to level a_2. The tension factor is defined by pressure exerted on a dynamometer. One half of the subjects at level a_1 are assigned at random to tension condition b_1; the other half are assigned to level b_2. The subjects at level a_2 are divided in a similar manner. Subjects are given four blocks of trials (factor C). The criterion is the number of errors in each block of trials. Suppose that the observed data are those given in Table 7.4-3.

In this table subjects 1, 2, and 3 form group G_{11}; subjects 4, 5, and 6 form group G_{12}; etc. Subject 6 is represented symbolically as $P_{3(12)}$, that is, the third subject in G_{12}. This plan may be classified as a $2 \times 2 \times 4$ factorial experiment with repeated measures on the last factor, $n = 3$. Summary data obtained from the basic observations are given in Table

7.4-4. All these summary tables are identical to those that would be obtained for a $2 \times 2 \times 4$ factorial experiment having no repeated measures.

Table 7.4-3 Basic Data for Numerical Example

		Subjects	Blocks of trials				Total
			c_1	c_2	c_3	c_4	
a_1	b_1	1	18	14	12	6	50
		2	19	12	8	4	43
		3	14	10	6	2	32
	b_2	4	16	12	10	4	42
		5	12	8	6	2	28
		6	18	10	5	1	34
a_2	b_1	7	16	10	8	4	38
		8	18	8	4	1	31
		9	16	12	6	2	36
	b_2	10	19	16	10	8	53
		11	16	14	10	9	49
		12	16	12	8	8	44

Computational symbols are defined and evaluated in Table 7.4-5. Symbol (10) is the only one unique to a repeated-measure design. The data for the latter symbol are obtained from the total column in Table 7.4-3.

The analysis of variance is summarized in Table 7.4-6. Suppose that the .05 level of significance is used in all tests. The main effect for factor C (trials) is found to be statistically significant. This indicates that the average number of errors differed in the four blocks of trials. Inspection of the totals for the blocks indicates a decreasing number of errors from c_1 to c_4. The anxiety $\times$ tension interaction is also statistically significant. This indicates that the pattern of the number of errors in the two anxiety groups depends upon the level of muscular tension. The profiles corresponding to this interaction effect are shown in Fig. 7.4-1. These profiles indicate that the effect of muscular tension upon number of errors differs—high- and low-anxiety-level groups perform in different ways. A test on the difference between mean number of errors between the two anxiety levels under the no-tension condition (b_1) is given by

$$F = \frac{(AB_{11} - AB_{21})^2}{2nr\text{MS}_{\text{error(between)}}} = \frac{(125 - 105)^2}{2(12)(10.31)} = 1.62.$$

Table 7.4-4 Summary Table for Numerical Example

ABC summary table

		c_1	c_2	c_3	c_4	Total
a_1	b_1	51	36	26	12	125
	b_2	46	30	21	7	104
a_2	b_1	50	30	18	7	105
	b_2	51	42	28	25	146
Total		198	138	93	51	480

AB summary table

	b_1	b_2	Total
a_1	125	104	229
a_2	105	146	251
Total	230	250	480

AC summary table

	c_1	c_2	c_3	c_4	Total
a_1	97	66	47	19	229
a_2	101	72	46	32	251
Total	198	138	93	51	480

BC summary table

	c_1	c_2	c_3	c_4	Total
b_1	101	66	44	19	230
b_2	97	72	49	32	250
Total	198	138	93	51	480

Table 7.4-5 Definitions and Numerical Values of Computational Symbols

$(1) = G^2/npqr$	$= (480)^2/48$	$= 4800.00$
$(2) = \Sigma X^2$	$= 18^2 + 14^2 + \cdots + 8^2 + 8^2$	$= 6058$
$(3) = (\Sigma A_i^2)/nqr$	$= (229^2 + 251^2)/24$	$= 4810.08$
$(4) = (\Sigma B_j^2)/npr$	$= (230^2 + 250^2)/24$	$= 4808.33$
$(5) = (\Sigma C_k^2)/npq$	$= (198^2 + 138^2 + 93^2 + 51^2)/12$	$= 5791.50$
$(6) = [\Sigma(AB_{ij}^2)]/nr$	$= (125^2 + 104^2 + 105^2 + 146^2)/12$	$= 4898.50$
$(7) = [\Sigma(AC_{ik}^2)]/nq$	$= (97^2 + 66^2 + \cdots + 46^2 + 32^2)/6$	$= 5810.00$
$(8) = [\Sigma(BC_{jk}^2)]/np$	$= (101^2 + 66^2 + \cdots + 49^2 + 32^2)/6$	$= 5812.00$
$(9) = [\Sigma(ABC_{ijk}^2)]/n$	$= (51^2 + 36^2 + \cdots + 28^2 + 25^2)/3$	$= 5923.33$
$(10) = (\Sigma P_m^2)/r$	$= (50^2 + 43^2 + \cdots + 49^2 + 44^2)/4$	$= 4981.00$

Table 7.4-6 Summary of Analysis of Variance

Source of variation	Computational formula	SS	df	MS	F
Between subjects	(10) − (1)	181.00	11		
A (anxiety)	(3) − (1)	10.08	1	10.08	
B (tension)	(4) − (1)	8.33	1	8.33	
AB	(6) − (3) − (4) + (1)	80.09	1	80.09	7.77*
Subj w. groups [error (between)]	(10) − (6)	82.50	8	10.31	
Within subjects	(2) − (10)	1077.00	36		
C (trials)	(5) − (1)	991.50	3	330.50	152.30*
AC	(7) − (3) − (5) + (1)	8.42	3	2.81	1.29
BC	(8) − (4) − (5) + (1)	12.17	3	4.06	1.87
ABC	(9) − (6) − (7) − (8) + (3) + (4) + (5) − (1)	12.74	3	4.25	1.96
$C \times$ subj w. groups [error (within)]	(2) − (9) − (10) + (6)	52.17	24	2.17	

The critical value for this test is

$$F_{.95}(1,8) = 5.32.$$

Thus the data indicate no statistically significant difference between the high- and low-anxiety groups under the condition of no tension. A test on the difference between the high- and low-anxiety groups under the high-tension condition is given by

$$F = \frac{(AB_{12} - AB_{22})^2}{2nr\text{MS}_{\text{error(between)}}} = \frac{(104 - 146)^2}{2(12)(10.31)} = 7.13.$$

The critical value for this test is also 5.32. Hence the data indicate a statistically significant difference in the performance of the high- and low-anxiety groups under the high-tension conditions; the high-anxiety group tends to make significantly more errors than does the low-anxiety group.

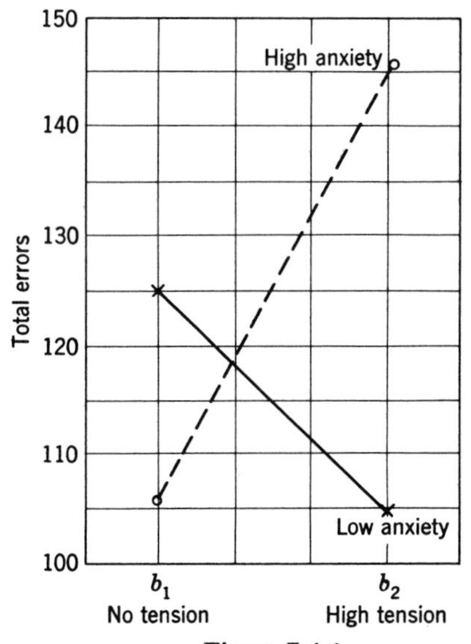

Figure 7.4-1

None of the interactions with factor C is statistically significant. Hence the data indicate that the shapes of the learning curves are essentially identical under each of the treatment combinations.

Variance-Covariance Matrices Associated with Numerical Example. Apart from the homogeneity assumptions required for the validity of the usual F tests, the variance-covariance matrices provide information which is of use in its own right in describing processes operating in the experiment.

The variance-covariance matrices associated with the data in Table 7.4-3 are given below:

Level ab_{11}

	c_1	c_2	c_3	c_4
c_1	7.00	4.00	5.00	4.00
c_2		4.00	6.00	4.00
c_3			9.33	6.00
c_4				4.00

Level ab_{12}

	c_1	c_2	c_3	c_4
c_1	9.33	4.00	0.00	−.66
c_2		4.00	4.00	2.00
c_3			7.00	4.00
c_4				2.34

Level ab_{21}

	c_1	c_2	c_3	c_4
c_1	1.33	−2.00	−2.00	−1.34
c_2		4.00	2.00	1.00
c_3			4.00	3.00
c_4				2.34

Level ab_{22}

	c_1	c_2	c_3	c_4
c_1	3.00	3.00	1.00	−.50
c_2		4.00	2.00	.00
c_3			1.33	.33
c_4				.33

The variance-covariance matrix obtained by averaging corresponding entries in each of the above matrices is given below:

Pooled variances and covariances

	c_1	c_2	c_3	c_4
c_1	5.16	2.25	1.00	.38
c_2		4.00	3.50	1.75
c_3			5.42	3.33
c_4				2.25

In this experiment, the levels of factor C represent successive blocks of trials in a learning experiment. Typically in this kind of experiment the variances tend to decrease as the learning increases. In the pooled variance-covariance matrix this trend is not clearly shown (5.16, 4.00, 5.42, 2.25). Further, the covariances between neighboring blocks of trials tend to be relatively higher than covariances between blocks which are farther apart. This trend is clearly shown in the pooled variance-covariance matrix. The pattern assumptions on the pooled variance-covariance matrix required for the strict validity of the usual F test do not, in general, hold for learning experiments. (Use of Hotelling's T^2 statistic would, however, provide an exact test even if the pattern assumptions do not hold.)

From the pooled covariance matrix, the average of the entries along the main diagonal defines $\overline{\text{var}}$, and the average of the entries off the main diagonal defines $\overline{\text{cov}}$. In this case, $\overline{\text{var}} = 4.21$, and $\overline{\text{cov}} = 2.04$. In terms of the latter quantities,

$$\text{MS}_{\text{error(between)}} = \overline{\text{var}} + (r - 1)\,\overline{\text{cov}} = 10.33,$$

$$\text{MS}_{\text{error(within)}} = \overline{\text{var}} - \overline{\text{cov}} = 2.17.$$

Within rounding error, these values are equal to those obtained for the corresponding mean squares in Table 7.4-6.

Illustrative Applications. The numerical example just considered is actually a modified version of an experiment reported by Meyer and Noble (1958). The plan for their experiment was a $2 \times 2 \times 6$ factorial experiment with repeated measures on the last factor. Factor A indicated anxiety level as measured by the Taylor Manifest Anxiety Scale. Factor B indicated the level of muscular tension exerted on a hand dynamometer during the experiment. Factor C represents blocks of trials. The criterion was the number of errors per block in the course of learning a verbal maze. The purpose of the experiment was to investigate the interaction between manifest anxiety and muscular tension during a learning task. There were 20 subjects in each of the four experimental groups.

The analysis of variance reported by Meyer and Noble is given in Table 7.4-7. The significant main effect for factor C indicated that the

Table 7.4-7 Analysis of Variance for Meyer and Noble Data

Source of variation	df	MS	F
Between subjects	79		
A (anxiety level)	1	2.33	
B (muscular tension)	1	.96	
AB	1	41.88	6.12*
Subj w. groups	76	6.84	
Within subjects	400		
C (blocks of trials)	5	216.86	309.80*
AC	5	.22	
BC	5	.29	
ABC	5	.58	
$C \times$ subj w. groups	380	.70	

mean number of errors changed during the blocks of trials. The significant AB interaction indicated that the effect of muscular tension is different in the two anxiety groups. Inspection of the profiles corresponding to this interaction indicated that the presence of muscular tension in the high-anxiety group tended to increase the number of errors relative to the no-tension condition. The effect in the low-anxiety group was just the opposite; under the tension condition there was a smaller number of errors relative to the no-test condition.

A design which, on the surface, appears identical with that being considered in this section is reported by Taylor (1958). The purpose of this experiment was to study the effect of methods of prior presentation of syllables upon the threshold value for visual recognition. The plan may be represented schematically as indicated below. This method of representation, however, will be shown to be misleading.

Method of prior presentation	Association value	Lists	
		c_1	c_2
a_1	b_1	G_{11}	G_{11}
	b_2	G_{12}	G_{12}
	b_3	G_{13}	G_{13}
a_2	b_1	G_{21}	G_{21}
	b_2	G_{22}	G_{22}
	b_3	G_{23}	G_{23}

The levels of factor A define methods of prior presentation of the syllables. The levels of factor B define the association value of the syllables. However, only the syllables in list c_1 were included in the prior representation. Syllables in list c_2 were matched with those in list c_1 for association value but were included in the experiment to serve as controls. There were 20 subjects in each group. This experimental plan does not conform to the pattern of a $2 \times 3 \times 2$ factorial experiment, since treatment combination ac_{12} is identical with treatment combination ac_{22}.

This kind of plan should not in general be analyzed as a $2 \times 3 \times 2$ factorial experiment. No meaningful interaction between factors A and C exists. One possible method of analysis is to use the difference between thresholds on syllables in list c_1 and c_2 as the criterion measure. In this case the analysis of variance takes the following form:

Source	df
A (method)	1
B (assoc. value)	2
AB	2
Within cell	114

This type of analysis in part removes differences between subjects from experimental error. As a second phase of the analysis of variance, one may combine the two levels of factor A and treat the resulting data as a

3 × 2 factorial experiment having 40 subjects in each group. This analysis takes the following form:

Source	df
Between subjects	119
B (assoc. value)	2
Subj w. groups	117
Within subjects	120
C (lists)	1
BC	2
$C \times$ subj w. groups	117

General Expected Values for Mean Squares. The expected values for the mean squares given in Table 7.4-1 are a special case of those given in Table 7.4-8. Depending upon the experimental design, D_p, D_q, and D_r are either zero or unity. The expected values in Table 7.4-1 are obtained from those in Table 7.4-8 by assuming that each of these D's is zero, i.e., that factors A, B, and C are fixed. (For Table 7.4-8, see p. 572).

7.5 Other Multifactor Repeated-measure Plans

The plans that have been considered in this chapter have the following general form:

	v_1	v_2	$\cdots$	v_h	
u_1	G_1	G_1	$\cdots$	G_1	
u_2	G_2	G_2	$\cdots$	G_2	n subjects per group
.	.	.		.	
.	.	.		.	
.	.	.		.	
u_g	G_g	G_g	$\cdots$	G_g	

The analysis of this general form may be outlined as follows:

Source	df
Between subjects	$ng - 1$
U	$g - 1$
Subjects w. groups	$g(n - 1)$
Within subjects	$ng(h - 1)$
V	$h - 1$
UV	$(g - 1)(h - 1)$
$V \times$ subj w. groups	$g(n - 1)(h - 1)$

Table 7.4-8 General Expected Values for Mean Squares

Effect	i	j	k	m	o	E(MS)
α_i	D_p	q	r	n	1	$\sigma_\varepsilon^2 + D_r\sigma_{\gamma\pi}^2 + nD_qD_r\sigma_{\alpha\beta\gamma}^2 + nqD_r\sigma_{\alpha\gamma}^2 + r\sigma_\pi^2 + nrD_q\sigma_{\alpha\beta}^2 + nqr\sigma_\alpha^2$
β_j	p	D_q	r	n	1	$\sigma_\varepsilon^2 + D_r\sigma_{\gamma\pi}^2 + nD_pD_r\sigma_{\alpha\beta\gamma}^2 + npD_r\sigma_{\beta\gamma}^2 + r\sigma_\pi^2 + nrD_p\sigma_{\alpha\beta}^2 + npr\sigma_\beta^2$
$\alpha\beta_{ij}$	D_p	D_q	r	n	1	$\sigma_\varepsilon^2 + D_r\sigma_{\gamma\pi}^2 + nD_r\sigma_{\alpha\beta\gamma}^2 + r\sigma_\pi^2 + nr\sigma_{\alpha\beta}^2$
$\pi_{m(ij)}$	1	1	r	1	1	$\sigma_\varepsilon^2 + D_r\sigma_{\gamma\pi}^2 + r\sigma_\pi^2$
γ_k	p	q	D_r	n	1	$\sigma_\varepsilon^2 + \sigma_{\gamma\pi}^2 + nD_pD_q\sigma_{\alpha\beta\gamma}^2 + npD_q\sigma_{\beta\gamma}^2 + nqD_p\sigma_{\alpha\gamma}^2 + npq\sigma_\gamma^2$
$\alpha\gamma_{ik}$	D_p	q	D_r	n	1	$\sigma_\varepsilon^2 + \sigma_{\gamma\pi}^2 + nD_q\sigma_{\alpha\beta\gamma}^2 + nq\sigma_{\alpha\gamma}^2$
$\beta\gamma_{jk}$	p	D_q	D_r	n	1	$\sigma_\varepsilon^2 + \sigma_{\gamma\pi}^2 + nD_p\sigma_{\alpha\beta\gamma}^2 + np\sigma_{\beta\gamma}^2$
$\alpha\beta\gamma_{ijk}$	D_p	D_q	D_r	n	1	$\sigma_\varepsilon^2 + \sigma_{\gamma\pi}^2 + n\sigma_{\alpha\beta\gamma}^2$
$\gamma\pi_{km(ij)}$	1	1	D_r	1	1	$\sigma_\varepsilon^2 + \sigma_{\gamma\pi}^2$
$\varepsilon_{o(ijkm)}$	1	1	1	1	1	σ_ε^2

The levels of factor U may, for example, constitute the pq treatment combinations in a $p \times q$ factorial set. In this case the following subdivisions are possible:

Source	df	Source	df
U	$g - 1$	UV	$(g - 1)(h - 1)$
A	$p - 1$	AV	$(p - 1)(h - 1)$
B	$q - 1$	BV	$(q - 1)(h - 1)$
AB	$(p - 1)(q - 1)$	ABV	$(p - 1)(q - 1)(h - 1)$

Alternatively, U may define the levels of factor A, and V may define the treatment combinations in a $q \times r$ factorial set. In this case the following subdivisions are possible:

Source	df	Source	df
V	$h - 1$	UV	$(g - 1)(h - 1)$
B	$q - 1$	AB	$(p - 1)(q - 1)$
C	$r - 1$	AC	$(p - 1)(r - 1)$
BC	$(q - 1)(r - 1)$	ABC	$(p - 1)(q - 1)(r - 1)$

Source	df
$V \times$ subj w. groups	$g(n - 1)(h - 1)$
$B \times$ subj w. groups	$p(n - 1)(q - 1)$
$C \times$ subj w. groups	$p(n - 1)(r - 1)$
$BC \times$ subj w. groups	$p(n - 1)(q - 1)(r - 1)$

As a third case, the levels of factor U may constitute pq treatment combinations in a $p \times q$ factorial set, and the levels of factor V may constitute the rs combinations in an $r \times s$ factorial set. In this case the general form specializes to the analysis summarized in Table 7.5-1. In this table, terms have been rearranged to indicate denominators for F ratios. The expected values given in this table are derived under the assumption that factors A, B, C, and D are fixed.

To illustrate the subdivision of the UV interaction for this case, one may first subdivide the U factor as follows:

Source	df
UV	$(g - 1)(h - 1)$
AV	$(p - 1)(h - 1)$
BV	$(q - 1)(h - 1)$
ABV	$(p - 1)(q - 1)(h - 1)$

Table 7.5-1 Summary of Analysis of Variance

Source of variation	df	E(MS)
Between subjects	$npq - 1$	
A	$p - 1$	$\sigma_\varepsilon^2 + rs\sigma_\pi^2 + nqrs\sigma_\alpha^2$
B	$q - 1$	$\sigma_\varepsilon^2 + rs\sigma_\pi^2 + nprs\sigma_\beta^2$
AB	$(p - 1)(q - 1)$	$\sigma_\varepsilon^2 + rs\sigma_\pi^2 + nrs\sigma_{\alpha\beta}^2$
Subjects w. groups	$pq(n - 1)$	$\sigma_\varepsilon^2 + rs\sigma_\pi^2$
Within subjects	$npq(rs - 1)$	
C	$r - 1$	$\sigma_\varepsilon^2 + s\sigma_{\gamma\pi}^2 + npqs\sigma_\gamma^2$
AC	$(p - 1)(r - 1)$	$\sigma_\varepsilon^2 + s\sigma_{\gamma\pi}^2 + nqs\sigma_{\alpha\gamma}^2$
BC	$(q - 1)(r - 1)$	$\sigma_\varepsilon^2 + s\sigma_{\gamma\pi}^2 + nps\sigma_{\beta\gamma}^2$
ABC	$(p - 1)(q - 1)(r - 1)$	$\sigma_\varepsilon^2 + s\sigma_{\gamma\pi}^2 + ns\sigma_{\alpha\beta\gamma}^2$
$C \times$ subj w. groups	$pq(n - 1)(r - 1)$	$\sigma_\varepsilon^2 + s\sigma_{\gamma\pi}^2$
D	$s - 1$	$\sigma_\varepsilon^2 + r\sigma_{\delta\pi}^2 + npqr\sigma_\delta^2$
AD	$(p - 1)(s - 1)$	$\sigma_\varepsilon^2 + r\sigma_{\delta\pi}^2 + nqr\sigma_{\alpha\delta}^2$
BD	$(q - 1)(s - 1)$	$\sigma_\varepsilon^2 + r\sigma_{\delta\pi}^2 + npr\sigma_{\beta\delta}^2$
ABD	$(p - 1)(q - 1)(s - 1)$	$\sigma_\varepsilon^2 + r\sigma_{\delta\pi}^2 + nr\sigma_{\alpha\beta\delta}^2$
$D \times$ subj w. groups	$pq(n - 1)(s - 1)$	$\sigma_\varepsilon^2 + r\sigma_{\delta\pi}^2$
CD	$(r - 1)(s - 1)$	$\sigma_\varepsilon^2 + \sigma_{\gamma\delta\pi}^2 + npq\sigma_{\gamma\delta}^2$
ACD	$(p - 1)(r - 1)(s - 1)$	$\sigma_\varepsilon^2 + \sigma_{\gamma\delta\pi}^2 + nq\sigma_{\alpha\gamma\delta}^2$
BCD	$(q - 1)(r - 1)(s - 1)$	$\sigma_\varepsilon^2 + \sigma_{\gamma\delta\pi}^2 + np\sigma_{\beta\gamma\delta}^2$
$ABCD$	$(p - 1)(q - 1)(r - 1)(s - 1)$	$\sigma_\varepsilon^2 + \sigma_{\gamma\delta\pi}^2 + n\sigma_{\alpha\beta\gamma\delta}^2$
$CD \times$ subj w. groups	$pq(n - 1)(r - 1)(s - 1)$	$\sigma_\varepsilon^2 + \sigma_{\gamma\delta\pi}^2$

Then each of the interactions with factor V may be subdivided. For example, the AV interaction may be subdivided into the following parts:

Source	df
AV	$(p - 1)(h - 1)$
AC	$(p - 1)(r - 1)$
AD	$(p - 1)(s - 1)$
ACD	$(p - 1)(r - 1)(s - 1)$

Analogous partitions may be made for the BV and ABV interactions.

Computational procedures for all treatment effects in the analysis summarized in Table 7.5-1 follow the usual procedures for a four-factor factorial experiment. Variation due to subjects within groups is part of the within-cell variation. The latter is given by

$$SS_{\text{w.cell}} = \Sigma X^2 - \frac{\Sigma(ABCD_{ijko})^2}{n}.$$

Variation due to main effects of subjects within groups is given by

$$SS_{\text{subj w.groups}} = \frac{\Sigma P_m^2}{rs} - \frac{\Sigma(AB_{ij})^2}{nrs}.$$

The pooled interaction of treatment effects with subject effects is given by

$$SS_{\text{pooled int w.groups}} = SS_{\text{w.cell}} - SS_{\text{subj w.groups}}.$$

This pooled interaction corresponds to the $V \times$ subjects within groups interaction in the general form of this plan. The $C \times$ subjects within groups part of this pooled interaction is given by

$$SS_{C \times \text{subj w.groups}} = \frac{\Sigma(CP_{km})^2}{s} - \frac{\Sigma(ABC_{ijk})^2}{ns} - SS_{\text{subj w.groups}}.$$

Still another special case of the design given at the beginning of this section is one that may be sketched as follows:

	v_1	v_2	$\cdots$	v_h	
u_1	G_1	G_1	$\cdots$	G_1	n subjects per group

One convenient analysis of variance for this design may be outlined as follows:

Source	df
Between subjects	$n - 1$
Subjects w. groups	$n - 1$
Within subjects	$n(h - 1)$
V	$h - 1$
$V \times$ subj w. groups	$(h - 1)(n - 1)$

If the levels of factor define the $h = pq$ factorial combinations of factors A and B, the analysis of variance given above takes the following form:

Source		df	
Between subjects			$n - 1$
Subjects w. groups			$n - 1$
Within subjects			$n(h - 1)$
V	A	$h - 1$	$p - 1$
	B		$q - 1$
	AB		$(p - 1)(q - 1)$
$V \times$ subj w. groups	$A \times$ subj w. groups	$(h - 1)(n - 1)$	$(p - 1)(n - 1)$
	$B \times$ subj w. groups		$(q - 1)(n - 1)$
	$AB \times$ subj w. groups		$(p - 1)(q - 1)(n - 1)$

If both factors A and B are fixed, then the within-subject variation may be subdivided as follows:

Source	df
Within subjects	$n(h-1)$
A	$p-1$
Error(a) $= A \times$ subj w. groups	$(p-1)(n-1)$
B	$q-1$
Error(b) $= B \times$ subj w. groups	$(q-1)(n-1)$
AB	$(p-1)(q-1)$
Error(ab) $= AB \times$ subj w. groups	$(p-1)(q-1)(n-1)$

F ratios for this design have the form

$$F = \frac{\text{MS}_a}{\text{MS}_{\text{error}(a)}}, \qquad F = \frac{\text{MS}_b}{\text{MS}_{\text{error}(b)}}, \qquad F = \frac{\text{MS}_{ab}}{\text{MS}_{\text{error}(ab)}}.$$

The sampling distributions of these F ratios may be approximated by the appropriate F distribution provided the pattern assumptions on the relevant covariance matrices are met.

Controlling Sequence Effects. For the plans that have been discussed in this chapter, in cases where the sequence of administration of the treatments was not dictated by the nature of the experimental variables, it was suggested that order be randomized independently for each subject. A partial control of sequence effects is provided by the use of the Latin-square principle; this principle is discussed in Chap. 9. A variety of repeated-measure designs using this principle is also discussed in Chap. 9. A more complete control of sequence effects (but one which is more costly in terms of experimental effort) is available. This more complete control is achieved by building what may be called a sequence factor into the design.

Consider a $p \times r$ factorial experiment in which there are to be repeated measures on the factor having r levels. The number of different sequences or arrangements of r levels is $r! = r(r-1)(r-2)(r-3)\cdots(1)$. For example, if r is 3, the number of possible sequences is $3! = 3 \cdot 2 \cdot 1 = 6$; if r is 5, the number of possible sequences is $5! = 5 \cdot 4 \cdot 3 \cdot 2 \cdot 1 = 120$. Each of the possible sequences may define a level of factor B in which $q = r!$. Thus, instead of the original $p \times r$ factorial experiment one has a $p \times q \times r$ factorial experiment. The analysis of the latter experiment would have the following form:

Source	df
Between subjects	$npq - 1$
A	$p - 1$
B (sequence of C)	$q - 1 = r! - 1$
AB	$(p - 1)(q - 1)$
Subjects within groups	$pq(n - 1)$
Within subjects	$npq(r - 1)$
C	$r - 1$
AC	$(p - 1)(r - 1)$
BC	$(q - 1)(r - 1)$
ABC	$(p - 1)(q - 1)(r - 1)$
$C \times$ subj w. groups	$pq(n - 1)(r - 1)$

This kind of sequence factor may be constructed in connection with any repeated-measure design in which the sequence can logically be varied. However, for designs in which the number of levels of the factor on which there are repeated measures is five or more the required number of levels of the sequence factor becomes prohibitively large.

One possible method of reducing this number is to select deliberately representative sequences from among the total possible sequences. A different approach might be to take a stratified random sample of all possible sequences, where the strata are constructed so as to assure a partial balance with respect to the order in which each of the levels appears within sequences which are used in the experiment. This kind of stratification may be achieved by using a Latin square.

7.6 Tests on Trends

Consider a $p \times q$ factorial experiment in which the levels of factor B define steps along an underlying continuum, i.e., intensity of light, dosage, blocks of trials in a learning experiment. The magnitude of the AB interaction in this kind of experiment may be regarded as a measure of global differences in the patterns or shapes of the profiles for the simple main effect of factor B. It is often of interest to study more specific aspects of such differences in patterns. Toward this end it is necessary to define dimensions in terms of which relatively irregular, experimentally determined profiles may be described. There are different methods whereby descriptive categories for this purpose may be established. In this section such categories will be defined in terms of polynomials of varying degree. Other functions, such as logarithmic or exponential, rather than polynomials, may be more appropriate for some profiles. The latter functional forms are not so readily handled as polynomials. However, polynomials may be used as first approximations to the latter forms.

Given the set of means $\overline{AB}_{i1}, \overline{AB}_{i2}, \ldots, \overline{AB}_{iq}$ in a $p \times q$ factorial experiment. The line joining these means (the profile of a simple main effect of factor B at level a_i) may have an irregular shape. As a first approximation to a quantitative description of the shape of a profile, one may obtain the best-fitting linear function (straight line). The slope of this best-fitting straight line defines the linear trend of the profile. As a second approximation to the pattern of the experimentally determined set of points, one may fit a second-degree (quadratic) function. The increase in goodness of fit over the linear fit defines what is known as the quadratic trend of the profile. As a third approximation to the pattern, one may obtain the best-fitting third-degree (cubic) function. The increase in goodness of fit of the latter function over both the linear and quadratic functions defines the cubic trend of the profile.

This process of fitting polynomial functions of increasingly higher degree can be continued up to a function of degree $q - 1$, where q is the number of points in the profile. A polynomial of degree $q - 1$ will always provide an exact fit to q points, since statistically there are only $q - 1$ degrees of freedom in this set of points. In most practical applications of the procedures to be described here, the degree of the polynomial is seldom carried beyond 3.

Global differences between shapes of profiles for simple main effects of factor B give rise to the AB interaction. Differences between the linear trends of such profiles define that part of the AB interaction which is called AB(linear). Thus the AB(linear) interaction represents a specific part of the overall AB interaction. Differences between quadratic trends in the profiles of the simple main effects define the AB(quadratic) variation. In general, the overall variation due to AB interaction may be divided into nonoverlapping, additive parts. These parts arise from specific kinds of differences in the shapes of profiles—differences in linear trends, differences in quadratic trends, etc. Symbolically, the AB interaction may be partitioned into the following parts:

Source of variation		df
AB		$(p - 1)(q - 1)$
AB (linear)	$p - 1$	
AB (quadratic)	$p - 1$	
. . .		
AB (degree $q - 1$)	$p - 1$	

These parts will sum to the overall AB variation.

The expected value of the mean square due to differences in linear trends has the following form:

$$E(MS_{ab(\text{lin})}) = \sigma_\varepsilon^2 + n\sigma_{\alpha\beta(\text{lin})}^2.$$

A test on differences in linear trends involves the hypothesis that $\sigma_{\alpha\beta(\text{lin})}^2 = 0$. This is equivalent to a test on the hypothesis that the profiles of the simple main effects have equal slopes, i.e., that the best-fitting linear functions are parallel.

The expected value of the mean square due to differences in quadratic trends has the following general form:

$$E(MS_{ab(\text{quad})}) = \sigma_\varepsilon^2 + n\sigma_{\alpha\beta(\text{quad})}^2.$$

A test on differences in these trends indicates whether or not the experimental data support the hypothesis that the profiles have equal quadratic trends.

Computational Procedures. Computational procedures will be described for the case of a $p \times q \times r$ factorial experiment in which there are repeated measures on factor C and n subjects in each group. This experimental plan has the form:

		c_1	c_2	$\cdots$	c_k	$\cdots$	c_r
a_1	b_1	G_{11}	G_{11}	$\cdots$	G_{11}	$\cdots$	G_{11}
	$\vdots$	$\vdots$	$\vdots$		$\vdots$		$\vdots$
	b_q	G_{1q}	G_{1q}	$\cdots$	G_{1q}	$\cdots$	G_{1q}
	$\vdots$	$\vdots$	$\vdots$		$\vdots$		$\vdots$
a_p	b_1	G_{p1}	G_{p1}	$\cdots$	G_{p1}	$\cdots$	G_{p1}
	$\vdots$	$\vdots$	$\vdots$		$\vdots$		$\vdots$
	b_q	G_{pq}	G_{pq}	$\cdots$	G_{pq}	$\cdots$	G_{pq}

Suppose that the levels of factor C represent equal steps along an underlying continuum. For example, suppose that the levels of factor C are r consecutive blocks of trials in a learning experiment. Under these conditions best-fitting linear, quadratic, cubic, etc., functions are most readily obtained by using the coefficients of orthogonal polynomials associated with r levels of an independent variable. Such coefficients are given in Table C.10.

The coefficients associated with the linear function will be designated

$$u_1, \quad u_2, \quad \ldots, \quad u_k, \quad \ldots, \quad u_r.$$

For example, for the case in which $r = 4$, the respective coefficients are

$$-3, \quad -1, \quad 1, \quad 3.$$

The coefficients associated with the quadratic function having r experimentally determined points will be designated

$$v_1, \quad v_2, \quad \ldots, \quad v_k, \quad \ldots, \quad v_r.$$

For the case $r = 4$, the respective coefficients are

$$1, \quad -1, \quad -1, \quad 1.$$

Note that the sum of the coefficients in each case is zero. Hence these coefficients define a comparison or contrast among the r points. The coefficients associated with the cubic function will be designated

$$w_1, \quad w_2, \quad \ldots, \quad w_k, \quad \ldots, \quad w_r.$$

In an experiment of this kind there are three sets of interactions that may be divided into parts associated with differences between trends—the AC, the BC, and the ABC interactions. Procedures for obtaining the variation due to differences in linear trends within each of these interactions will be outlined. Higher-order trends will be found to follow the same general pattern. It will also be convenient to indicate procedures for obtaining the variation due to linear and higher-order trends for the main effect of factor C. The latter indicate the goodness of fit polynomials of varying degree to the profile corresponding to the main effect of factor C.

The notation to be used in obtaining the linear part of the variation in interactions with factor C as well as the variation in the linear part of the variation in the main effect of factor C is summarized in part i of Table 7.6-1. The symbol $X'_{m(ij)}$ is a weighted sum of the r observations on subject $m(ij)$, the weights being the respective linear coefficients of the appropriate polynomial. The analysis of linear trends for this case reduces to what is essentially an analysis of variance of a $p \times q$ factorial experiment with n observations per cell, an observation being an $X'_{m(ij)}$.

Other symbols defined in part i are also weighted sums. To illustrate, ABC'_{ij} is a weighted sum of terms appearing in a row of an ABC summary table. For example, the row which corresponds to level a_1 and level b_1 has the form

$$ABC_{111} \quad ABC_{112} \quad \cdots \quad ABC_{11k} \quad \cdots \quad ABC_{11r}.$$

Each of these totals is the sum of n observations. From these totals,

$$ABC'_{11} = u_1(ABC_{111}) + u_1(ABC_{112}) + \cdots + u_r(ABC_{11r}).$$

Table 7.6-1 Notation for Analysis of Linear Trend

Coefficients for linear comparison:

$$u_1, \quad u_2, \quad \ldots, \quad u_k, \quad \ldots, \quad u_r$$

(i)

$$\begin{aligned}
X'_{m(ij)} &= \sum_k u_k(X_{ijkm}) \\
ABC'_{ij} &= \sum_k u_k(ABC_{ijk}) = \sum_m X'_{m(ij)} \\
AC'_i &= \sum_k u_k(AC_{ik}) = \sum_j ABC'_{ij} \\
BC'_j &= \sum_k u_k(BC_{jk}) = \sum_i ABC'_{ij} \\
C' &= \sum_k u_k C_k = \sum_i \sum_j ABC'_{ij} = \sum_i \sum_j \sum_m X'_{m(ij)}
\end{aligned}$$

(ii)

$$(1') = (C')^2/npq(\Sigma u_k^2) \qquad (3') = \Sigma(AC'_i)^2/nq(\Sigma u_k^2)$$

$$(2') = \Sigma(X'_{m(ij)})^2/(\Sigma u_k^2) \qquad (4') = \Sigma(BC'_j)^2/np(\Sigma u_k^2)$$

$$(5') = \Sigma(ABC'_{ij})^2/n(\Sigma u_k^2)$$

(iii)

Source	Computational formula	df
Within subjects (linear)	(2′)	npq
C (linear)	(1′)	1
AC (linear)	(3′) − (1′)	$p - 1$
BC (linear)	(4′) − (1′)	$q - 1$
ABC (linear)	(5′) − (3′) − (4′) + (1′)	$(p - 1)(q - 1)$
$C \times$ subj w. groups (linear)	(2′) − (5′)	$pq(n - 1)$

An equivalent expression for ABC'_{ij} is

$$ABC'_{ij} = X'_{1(ij)} + X'_{2(ij)} + \cdots + X'_{n(ij)} = \sum_m X'_{m(ij)}.$$

One expression serves as a computational check on the other.

Computational symbols convenient for use in the computation of the linear sources of variation are given in part ii. In the denominator of all symbols is the term Σu_k^2. The other term in the denominators is the number of observations that go into an element which is weighted in the weighted sum. For example, there are n observations in ABC_{ijk}; there are nq observations in AC_{ik}; there are np observations in BC_{jk}. Actual computational formulas are summarized in part iii. The mean square due to AC(linear) estimates an expression of the form

$$\sum_i (\beta_i - \beta)^2,$$

where the β_i's represent regression coefficients for linear profiles corresponding to simple main effects of factor C at each of the levels of factor A, and where β represents a pooled regression coefficient for all linear profiles in the set.

Computational procedures for the quadratic trend are summarized in

Table 7.6-2. Each of the entries in this table has a corresponding entry in Table 7.6-1, with v_k replacing corresponding u_k throughout. The symbol $X''_{m(ij)}$ is used to distinguish a weighted sum in terms of quadratic weights from the corresponding sum in terms of linear weights, designated by

Table 7.6-2 Notation for Analysis of Quadratic Trend

Coefficients for quadratic comparison:

$$v_1, \quad v_2, \quad \ldots, \quad v_k, \quad \ldots, \quad v_r$$

(i)

$$\begin{aligned}
X''_{m(ij)} &= \sum_k v_k(X_{ijkm}) \\
ABC''_{ij} &= \sum_k v_k(ABC_{ijk}) = \sum_m X''_{m(ij)} \\
AC''_i &= \sum_k v_k(AC_{ik}) = \sum_j ABC''_{ij} \\
BC''_j &= \sum_k v_k(BC_{jk}) = \sum_i ABC''_{ij} \\
C'' &= \sum_k v_k C_k = \sum_i \sum_j ABC''_{ij} = \sum_i \sum_j \sum_m X''_{m(ij)}
\end{aligned}$$

(ii)

$$(1'') = (C'')^2/npq(\Sigma v_k^2) \qquad (3'') = \Sigma(AC''_i)^2/nq(\Sigma v_k^2)$$

$$(2'') = \Sigma(X''_{m(ij)})^2/(\Sigma v_k^2) \qquad (4'') = \Sigma(BC''_j)^2/np(\Sigma v_k^2)$$

$$(5'') = \Sigma(ABC''_{ij})^2/n(\Sigma v_k^2)$$

(iii)

Source	Computational formula
Within subjects (quadratic)	(2″)
C (quadratic)	(1″)
AC (quadratic)	(3″) − (1″)
BC (quadratic)	(4″) − (1″)
ABC (quadratic)	(5″) − (3″) − (4″) + (1″)
$C \times$ subj w. groups (quadratic)	(2″) − (5″)

the symbol $X'_{m(ij)}$. Higher-order trends follow the same pattern, with the appropriate coefficients serving as the weights. If, for example, $r = 4$, it will be found that

$$SS_c = SS_{c(\text{lin})} + SS_{c(\text{quad})} + SS_{c(\text{cubic})}.$$

Similarly, $$SS_{ac} = SS_{ac(\text{lin})} + SS_{ac(\text{quad})} + SS_{ac(\text{cubic})}.$$

Numerical Example. The numerical data in part i of Table 7.6-3 will be used to illustrate the computational procedures. These data represent a $2 \times 2 \times 4$ factorial experiment with repeated measures on factor C, three observations in each group. Suppose that factor C represents equally spaced blocks of trials in a learning experiment. For example, subject 1 is assigned to treatment combination ab_{11} and has scores of 1, 6, 5, and 7, respectively, on a series of four blocks of trials.

Since factor C has four levels, coefficients for the case in which there are four points to be fitted are appropriate. From Table C.10 the linear coefficients are

$$-3, \quad -1, \quad 1, \quad 3.$$

These coefficients appear at the top of part i. From the data on subject 1 one obtains

$$X'_{1(11)} = (-3)(1) + (-1)(6) + (1)(5) + (3)(7) = 17,$$

the weighted sum of the observations on subject 1, the weights being the linear coefficients. Other entries in the column headed $X'_{m(ij)}$ are obtained in an analogous manner. For example,

$$X'_{4(12)} = (-3)(2) + (-1)(7) + (1)(12) + (3)(15) = 44.$$

The entries in the column headed ABC'_{ij} are obtained as follows:

$$ABC'_{11} = X'_{1(11)} + X'_{2(11)} + X'_{3(11)} = 17 + 28 + 18 = 63.$$
$$ABC'_{12} = X'_{4(12)} + X'_{5(12)} + X'_{6(12)} = 44 + 26 + 27 = 97.$$

The left-hand side of part ii represents an ABC summary table obtained in the usual manner from data on the left-hand side of part i. From the first row of the ABC summary table one obtains

$$ABC'_{11} = (-3)(4) + (-1)(20) + (1)(20) + (3)(25) = 63.$$

This entry provides a check on the entry ABC'_{11} computed in part i. The second row of the ABC summary table is used to obtain ABC'_{12}. A corresponding entry is available from part i.

The left-hand side of part iii represents an AC summary table. The entries in the column headed AC'_i are weighted sums of the entries in the corresponding rows. Checks on these entries may be obtained from part ii. Checks on entries in part iv may also be obtained from the ABC'_{ij} column of part ii. For example,

$$BC'_1 = ABC'_{11} + ABC'_{21} = 63 + 92 = 155;$$
$$BC'_2 = ABC'_{12} + ABC'_{22} = 97 + 118 = 215.$$

All the totals required in the analysis of linear trend may be obtained from the $X'_{m(ij)}$ column of part i of Table 7.6-3. Hence parts ii to iv are not actually required. In practice, however, part ii should be computed to serve as a check on the $X'_{m(ij)}$ column. Additional checks are provided by parts iii and iv.

Computational symbols defined in part ii of Table 7.6-1 are obtained in part i of Table 7.6-4. For the case in which there are four points in each profile, $\Sigma u_k^2 = 20$. Each total C_k is the sum of $npq = 12$ observations. Hence the denominator for symbol (1′) is 12(20). Since each total AC_{ik} is the sum of $nq = 6$ observations, the denominator for the symbol (3′) is 6(20).

The analysis of variance for the linear trend of the within-subject effects is summarized in part ii of Table 7.6-4. For the basic data in part i of Table 7.6-3, there are $npq = 12$ subjects and $r = 4$ observations on each

Table 7.6-3 Analysis of Linear Trend—Numerical Example

Linear coefficients: -3 -1 1 3 $\Sigma u_k^2 = 20$

(i)

		Subject	c_1	c_2	c_3	c_4	$X'_{m(ij)}$	ABC'_{ij}
a_1	b_1	1	1	6	5	7	17	
		2	0	6	7	9	28	
		3	3	8	8	9	18	63
	b_2	4	2	7	12	15	44	
		5	1	6	8	9	26	
		6	3	7	10	11	27	97
a_2	b_1	7	1	2	7	12	38	
		8	1	1	4	10	30	
		9	1	1	4	8	24	92
	b_2	10	2	2	8	12	36	
		11	3	2	10	15	44	
		12	2	2	7	13	38	118
			20	50	90	130	370	370

(ii)

		c_1	c_2	c_3	c_4	ABC'_{ij}	AC'_i
a_1	b_1	4	20	20	25	63	
	b_2	6	20	30	35	97	160
a_2	b_1	3	4	15	30	92	
	b_2	7	6	25	40	118	210
		20	50	90	130	370	370

(iii)

	c_1	c_2	c_3	c_4	AC'_i	C'
a_1	10	40	50	60	160	
a_2	10	10	40	70	210	370
	20	50	90	130	370	

(iv)

	c_1	c_2	c_3	c_4	BC'_j	C'
b_1	7	24	35	55	155	
b_2	13	26	55	75	215	370
	20	50	90	130	370	

subject. Hence the total degrees of freedom for within-subject effects are $npq(r-1) = 36$. In the analysis of variance of trend, these 36 degrees of freedom are partitioned as follows:

Within subjects	36	$npq(r-1)$
Within subjects (linear)	12	npq
Within subjects (quadratic)	12	npq
Within subjects (cubic)	12	npq

The 12 degrees of freedom for the linear trend of the within-subject effects are analyzed in part ii of Table 7.6-4.

Table 7.6-4 Analysis of Linear Trend—Numerical Example

(i)

$$
\begin{aligned}
(1') &= (370)^2/12(20) &&= 570.42 \\
(2') &= (17^2 + 28^2 + \cdots + 44^2 + 38^2)/20 &&= 616.70 \\
(3') &= (160^2 + 210^2)/6(20) &&= 580.83 \\
(4') &= (155^2 + 215^2)/6(20) &&= 585.42 \\
(5') &= (63^2 + 97^2 + 92^2 + 118^2)/3(20) &&= 596.10
\end{aligned}
$$

(ii)

Source of variation	SS	df	MS	F
Within subjects (linear)	$(2') = 616.70$	12		
C (linear)	$(1') = 570.42$	1	570.42	221.09*
AC (linear)	$(3') - (1') = 10.41$	1	10.41	4.03
BC (linear)	$(4') - (1') = 15.00$	1	15.00	5.81*
ABC (linear)	$(5') - (3') - (4') + (1') = 0.27$	1	0.27	
$C \times$ subj w. groups (linear)	$(2') - (5') = 20.60$	8	2.58	

A test on linear trend in the main effect of factor C has the form

$$F = \frac{\text{MS}_{c(\text{lin})}}{\text{MS}_{C \times \text{subj w.groups(lin)}}} = 221.09.$$

At the .05 level of significance, this test indicates that the best-fitting straight line to the profile of the C main effects has a slope which is significantly different from zero. In terms of the expected values of the mean squares, this F ratio has the structure

$$\frac{\text{E(numerator)}}{\text{E(denominator)}} = \frac{\sigma_\varepsilon^2 + \sigma_{\pi\gamma(\text{lin})}^2 + npq\sigma_{\gamma(\text{lin})}^2}{\sigma_\varepsilon^2 + \sigma_{\pi\gamma(\text{lin})}^2}.$$

The profile corresponding to the C main effects (in terms of treatment means) is shown in Fig. 7.6-1.

The profiles corresponding to the simple effects of factor C at levels b_1 and b_2 are also shown in this figure. Inspection of these profiles suggests that the best-fitting line to the profile for b_2 would have a different slope from the best-fitting line to the profile for b_1, that is, that these lines would not be parallel. A test of the hypothesis that there is no difference

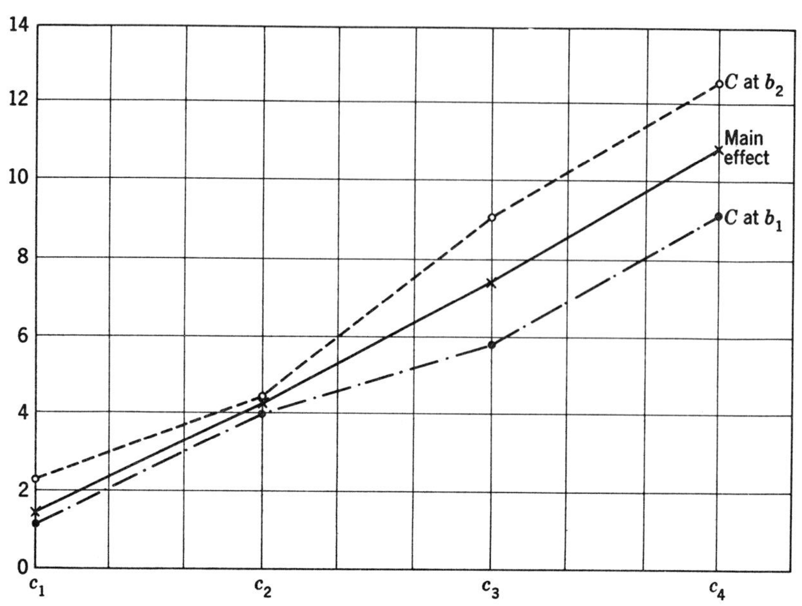

Figure 7.6-1 Profiles of *BC* interaction and *C* main effect.

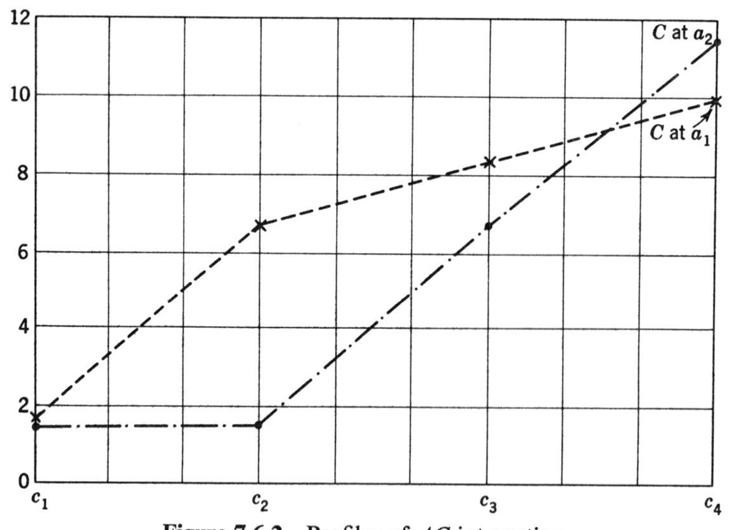

Figure 7.6-2 Profiles of *AC* interaction.

in these slopes (no difference in linear trend) has the form

$$F = \frac{MS_{bc(\text{lin})}}{MS_{C \times \text{subj w.groups(lin)}}} = 5.81.$$

At the .05 level of significance, this test indicates that the linear trends of the *BC* profiles cannot be considered to be equal.

Profiles corresponding to the *AC* interaction are shown in Fig. 7.6-2. Inspection indicates that these profiles differ in shape but that the best-fitting straight lines might be parallel. The test of the latter hypothesis is given by

$$F = \frac{MS_{ac(\text{lin})}}{MS_{C \times \text{subj w.groups(lin)}}} = 4.03.$$

The value does not exceed the critical value for a .05-level test. Hence the experimental evidence does not reject the hypothesis that the linear trends are equal. Thus differences in shapes of these profiles, if there are statistically significant differences, must be due to quadratic or high-order trends.

A summary of the analysis of variance for the quadratic trend is given in Tables 7.6-5 and 7.6-6. This analysis follows the same general procedures as those used in the analysis of the linear trend; in this case the quadratic coefficients replace the linear coefficients as weights. The test of differences in the quadratic trends of the *AC* profiles is given by

$$F = \frac{MS_{ac(\text{quad})}}{MS_{C \times \text{subj w. groups(quad)}}} = 208.36.$$

Hence the data tend to reject the hypothesis that there is no difference in the quadratic trends of the *AC* profiles. Inspection of Fig. 7.6-2 indicates that the profile at a_1 clearly has a different curvature from the profile at a_2. The significant F ratio for the main effect in Table 7.6-6 indicates that the quadratic trend in the profile of the C main effect is different from zero.

In general, the analysis of differences in trends for interaction terms is not made unless there is evidence to show that some difference in shapes exists. This evidence is provided by the usual overall tests on interactions, which indicate the presence or absence of global differences in the shapes of profiles. The overall analysis of variance for the within-subject effects is summarized in Table 7.6-8. This analysis is made by means of the usual computational procedures for a $p \times q \times r$ factorial experiment having repeated measures on factor C. The basic observational data in part i of Table 7.6-3 and the summary tables at the left of parts ii to iv are used in the overall analysis of variance. The significant *AC* and *BC*

Table 7.6-5 Analysis of Quadratic Trend—Numerical Example

Quadratic coefficients 1 −1 −1 1 $\Sigma v_k^2 = 4$

			Subject	c_1	c_2	c_3	c_4	$X''_{m(ij)}$	ABC''_{ij}
(i)	a_1	b_1	1	1	6	5	7	−3	
			2	0	6	7	9	−4	
			3	3	8	8	9	−4	−11
		b_2	4	2	7	12	15	−2	
			5	1	6	8	9	−4	
			6	3	7	10	11	−3	−9
	a_2	b_1	7	1	2	7	12	4	
			8	1	1	4	10	6	
			9	1	1	4	8	4	14
		b_2	10	2	2	8	12	4	
			11	3	2	10	15	6	
			12	2	2	7	13	6	16
				20	50	90	130	10	10

			c_1	c_2	c_3	c_4	ABC''_{ij}	AC'_i
(ii)	a_1	b_1	4	20	20	25	−11	
		b_2	6	20	30	35	−9	−20
	a_2	b_1	3	4	15	30	14	
		b_2	7	6	25	40	16	30
			20	50	90	130	10	10

Table 7.6-6 Analysis of Quadratic Trend—Numerical Example

(i)

$$
\begin{aligned}
(1'') &= (10)^2/12(4) &&= 2.08\\
(2'') &= (-3)^2 + (-4)^2 + \cdots + 6^2 + 6^2 &&= 56.50\\
(3'') &= [(-20)^2 + 30^2]/6(4) &&= 54.17\\
(4'') &= [3^2 + 7^2]/6(4) &&= 2.42\\
(5'') &= [(-11)^2 + (-9)^2 + 14^2 + 16^2]/3(4) &&= 54.50
\end{aligned}
$$

Source of variation	SS	df	MS	F
Within subjects (quadratic)	(2″) = 56.51	12		
C (quadratic)	(1″) = 2.08	1	2.08	8.32*
AC (quadratic)	(3″) − (1″) = 52.09	1	52.09	208.36*
BC (quadratic)	(4″) − (1″) = 0.34	1	0.34	
(ii) *ABC* (quadratic)	(5″) − (3″) −(4″) + (1″) = 0.00	1	0.00	
$C \times$ subj w. groups (quadratic)	(2″) − (5″) = 2.00	8	0.25	

interactions indicate that there are global differences in the trends of the corresponding profiles. The nature of such differences is explored in the analysis of the linear, quadratic, and cubic trends.

Table 7.6-7 Analysis of Cubic Trend—Numerical Example

Source of variation	SS	df	MS	F
Within subject (cubic)	20.30	12		
C (cubic)	0.42	1	0.42	
AC (cubic)	10.42	1	10.42	30.64*
BC (cubic)	6.66	1	6.66	19.59*
ABC (cubic)	0.07	1	0.07	
$C \times$ subj w. groups (cubic)	2.73	8	0.34	

Table 7.6-8 Overall Analysis of Variance for Within Subjects Effects—Numerical Example

Source of variation	SS	df	MS	F
Within subjects	693.50	36		
C	572.92	3	190.97	180.16*
AC	72.92	3	24.30	22.92*
BC	22.00	3	7.33	6.92*
ABC	0.33	3	0.11	
$C \times$ subj w. groups	25.34	24	1.06	

It is of interest to compare the overall analysis with the analysis of the individual trends. It will be noted that:

	SS	df
Within subjects (overall)	693.50	36
Within subjects (linear)	616.70	12
Within subjects (quadratic)	56.51	12
Within subjects (cubic)	20.30	12

It will also be noted that:

	SS
AC (overall)	72.92
AC (linear)	10.41
AC (quadratic)	52.09
AC (cubic)	10.42

In each case the sum of the parts will be numerically equal to the corresponding overall variation.

It is of particular interest to look at the parts of terms that go into the denominator of the F ratios. In this case:

	SS	MS
$C \times$ subj w. groups (overall)	25.34	1.06
$C \times$ subj w. groups (linear)	20.60	2.58
$C \times$ subj w. groups (quadratic)	2.00	0.25
$C \times$ subj w. groups (cubic)	2.73	0.34

The parts of the overall variation of this interaction need not be homogeneous; each of these parts does not estimate the same source of variation. The parts measure the deviation of a weighted sum about the mean of the weighted sums; the weights are different for each of the parts. The expected values of the mean squares of the parts are to a degree dependent upon the weights; the latter in turn determine the shape of the curve that is being fitted to the experimental data. Different structural models underlie the analysis of variance for linear, quadratic, and cubic trends.

Illustrative Applications. Grant (1956) gives a relatively complete account of tests for trends as well as a detailed numerical example. The plan for the experiment discussed by Grant has the following form:

		Shock	Stages of task				
			c_1	c_2	c_3	c_4	c_5
Anxiety level	a_1	b_1	G_{11}	G_{11}	G_{11}	G_{11}	G_{11}
		b_2	G_{12}	G_{12}	G_{12}	G_{12}	G_{12}
		b_3	G_{13}	G_{13}	G_{13}	G_{13}	G_{13}
	a_2	b_1	G_{21}	G_{21}	G_{21}	G_{21}	G_{21}
		b_2	G_{22}	G_{22}	G_{22}	G_{22}	G_{22}
		b_3	G_{23}	G_{23}	G_{23}	G_{23}	G_{23}

Factor A represents the level of anxiety of the subjects as measured by the Taylor Manifest Anxiety Scale. Groups G_{11}, G_{12}, and G_{13} represent subjects from one end of the scale; groups G_{21}, G_{22}, and G_{23} represent subjects from the other end of the scale. The three levels of factor B indicate the number of electric shocks received by a subject in the course of the experiment. Subjects in groups G_{11} and G_{21} were at level b_1, subjects in groups G_{12} and G_{22} at level b_2, and subjects in groups G_{13} and G_{23} at level b_3. The levels of factor C represent successive stages of a card-sorting task. There were four subjects in each group. The criterion was the square root of the number of perseverative errors. This plan may be considered as a special case of a $p \times q \times r$ factorial experiment with repeated measures on factor C.

The overall analysis, as given by Grant, had the following form:

	SS	df	MS	F
Between subjects		23		
Groups	21.91	5	4.38	2.10
Subj w. groups	37.56	18	2.09	
Within subjects		96		
C stages	16.37	4	4.09	5.37*
Stages × groups	35.26	20	1.76	2.31
Stages × subj w. groups	54.89	72	0.76	

In this analysis the six combinations of anxiety and shock are considered to be six levels of a single factor—a "group" factor. The group factor is then subdivided as follows:

	SS	df	MS	F
Groups	21.91	5		
A anxiety level	1.75	1	1.75	0.84
B shock	8.86	2	4.43	2.12
AB	11.29	2	5.65	2.71

The denominator for each of the above F ratios is $MS_{\text{subj w.groups}}$.

The 20 degrees of freedom for the stage × group interaction are associated with what Grant calls the between-group trends. This interaction is subdivided into the following parts:

	SS	df	MS	F
Stages × groups	35.26	20		
AC	17.58	4	4.40	5.79
BC	10.62	8	1.33	1.75
ABC	7.07	8	0.88	1.16

The denominator for each of the above F ratios is $MS_{\text{stages} \times \text{subj w.groups}}$. These tests indicate whether or not there are any global differences in the shapes of corresponding profiles.

The more specific analysis of differences in trends associated with factor C is summarized in Table 7.6-9. In each case the denominator for tests on differences in linear trend is the linear part of the $C \times$ subjects within groups interaction. Similarly, the quadratic part of this latter interaction serves as the denominator for differences in quadratic trend.

It will be noted that, although the overall test on the BC interaction indicates no significant difference, the test on differences in linear trends does indicate a statistically significant difference. In the absence of a significant overall interaction the experimenter would not ordinarily

Table 7.6-9 Analysis of Variance of Grant Data (1956)

Source of variation		SS		df	MS	F
Within subjects				96		
. . .				. . .		
AC		17.57		4	4.39	5.78*
Linear	16.38		1		16.38	37.99*
Quadratic	0.04		1		0.04	0.05
Cubic	0.29		1		0.29	0.26
Quartic	0.86		1		0.86	1.52
BC		10.62		8	1.33	1.75
Linear	6.84		2		3.42	7.93*
Quadratic	1.00		2		0.50	0.55
Cubic	2.29		2		1.14	1.00
Quartic	0.49		2		0.25	0.44
ABC		7.07		8	0.88	1.16
Linear	2.05		2		1.02	2.37
Quadratic	1.48		2		0.74	0.81
Cubic	0.24		2		0.12	0.10
Quartic	3.30		2		1.65	2.93
$C \times$ subj w. groups		54.89		72	0.76	
Linear	7.76		18		0.43	
Quadratic	16.43		18		0.91	
Cubic	20.55		18		1.14	
Quartic	10.15		18		0.56	

make tests on the parts—unless a priori information about the underlying sources of variation in the experimental variables indicates that certain of the trends should be more dominant than others. However, one should not hesitate to present a complete description of the experimental findings, even though some of the "tests" on parts of nonsignificant overall variation may be unduly subject to type 1 error.

Another illustrative example of the analysis of trends is given in an experiment reported by Schrier (1958). The plan for this experiment

Subjects	Trial:	a_1				a_2			
	Reward level:	b_1	b_2	b_3	b_4	b_1	b_2	b_3	b_4
1									
2									
.									
.									
.									
5									

may be represented as shown above. In this design there are repeated measures on both factors A and B. The two levels of factor A represent two blocks of trials. The four levels of factor B represent four amounts of reward. The criterion measure was the proportion of correct discriminations. To obtain homogeneity of variance, an arcsine transformation on the proportions was used in the analysis.

Since there are repeated measures on two factors, computational procedures for this design differ somewhat from those given in Table 7.6-1. General procedures for the analysis of linear trends with respect to factor B are outlined in Table 7.6-10. If the logic of the design permits, a trend analysis may also be carried out with respect to factor A. (In the case of the Schrier plan factor A has only two levels.) Higher-order trends may be analyzed by following the general pattern in Table 7.6-10, replacing the linear coefficients with appropriate higher-order coefficients throughout.

The analysis of variance reported by Schrier is summarized in Table

Table 7.6-10 Analysis of Variance of Linear Trend

Linear coefficients: $u_1, u_2, \ldots, u_j, \ldots, u_q$

(i)

$$X'_{im} = \sum_j u_j X_{ijm} \qquad AB'_i = \sum_m X'_{im}$$

$$P'_m = \sum_i X'_{im} \qquad B' = \sum_m P'_m = \sum_i AB'_i$$

(ii)

$$(1') = (B')^2/np(\Sigma u_j^2) \qquad (3') = \Sigma(AB'_i)^2/n(\Sigma u_j^2)$$

$$(2') = \Sigma(X'_{im})^2/(\Sigma u_j^2) \qquad (4') = \Sigma(P'_m)^2/p(\Sigma u_j^2)$$

(iii)

Source of variation	SS	df
B (linear)	$(1')$	1
$B \times$ subjects (linear)	$(4') - (1')$	$n - 1$
AB (linear)	$(3') - (1')$	$p - 1$
$AB \times$ subjects (linear)	$(2') - (3') - (4') + (1')$	$(n - 1)(p - 1)$

Table 7.6-11 Analysis of Variance for Schrier Data (1958)

	Source of variation		df	MS	F	Denominator
	Between subjects		4	117.8		
	Within subjects		35			
	Blocks		1	379.8	23.2*	U
U	Blocks × subjects		4	16.3		
	Rewards		3	446.1	16.2*	V
	Linear	1		1290.5	18.5*	V_1
	Quadratic	1		1.5		V_2
	Cubic	1		46.1	5.7*	V_3
V	Rewards × subjects		12	27.6		
V_1	Linear	4		69.6		
V_2	Quadratic	4		4.6		
V_3	Cubic	4		8.2		
	Blocks × rewards		3	41.2	6.2*	W
	Linear	1		75.0	6.6	W_1
	Quadratic	1		.2		W_2
	Cubic	1		48.6	16.6*	W_3
W	Blocks × rewards × subjects		12	6.6		
W_1	Linear	4		11.5		
W_2	Quadratic	4		5.6		
W_3	Cubic	4		3.0		

7.6-11. Denominators for F ratios are indicated in the right-hand column. For example, the denominator used in the test for differences in cubic trend of the blocks × rewards interaction is the entry in row W_3. The latter is $MS_{\text{blocks} \times \text{rewards} \times \text{subj(cubic)}}$. The experimental data indicate that the differences in cubic trend in the block × reward interaction cannot be considered to be zero; i.e., the shapes of corresponding profiles do not have equal cubic curvature.

7.7 Testing Equality and Symmetry of Covariance Matrices

Consider the following $p \times q$ factorial experiment in which there are repeated measures on factor B.

	b_1	b_2	$\cdots$	b_q
a_1	G_1	G_1	$\cdots$	G_1
a_2	G_2	G_2	$\cdots$	G_2
.	.	.		.
.	.	.		.
.	.	.		.
a_p	G_p	G_p	$\cdots$	G_p

Assume that there are n_1 subjects in the group (G_1) assigned to level a_1, n_2 subjects in the group (G_2) of subjects assigned to level a_2, etc. Let $N' = n'_1 + n'_2 + \cdots + n'_p$, where $n'_i = n_i - 1$.

The following additional notation will be used in this section:

$S_1 = q \times q$ matrix of covariances for level a_1.
$S_2 = q \times q$ matrix of covariances for level a_2.
. . .
$S_p = q \times q$ matrix of covariances for level a_p.
$S_{\text{pooled}} = q \times q$ matrix of pooled covariances; i.e., each entry is a weighted average of corresponding entries in S_1 through S_p, the weights being the corresponding degrees of freedom.

The variables included in the covariance matrix at level a_i are assumed to have an underlying q-variate normal distribution.

The test procedures to be described in this section are those suggested by Box (1950). To test the hypothesis that the covariance matrices $S_1, S_2, \ldots, S_p$ are random samples for populations in which the covariance matrices are $\Sigma_1 = \Sigma_2 = \cdots = \Sigma_p = \Sigma$ (that is, that the population covariance matrices are equal), one computes the following statistics:

(1) $$M_1 = N \ln |S_{\text{pooled}}| - \Sigma n_i \ln |S_i|,$$

(2) $$C_1 = \frac{2q^2 + 3q - 1}{6(q + 1)(p - 1)} \left[\Sigma\left(\frac{1}{n_i}\right) - \frac{1}{N} \right],$$

(3) $$f_1 = \frac{q(q + 1)(p - 1)}{2}.$$

Under the hypothesis that the multivariate normal populations have equal covariance matrices, the statistic

(4) $$\chi_1^2 = (1 - C_1)M_1$$

has a sampling distribution which is approximated by a chi-square distribution having f_1 degrees of freedom. Rejection of this hypothesis rules against pooling covariance matrices. If the populations have a common covariance matrix Σ, then S_{pooled} is an unbiased estimate of Σ. This test procedure is a multivariate analog of Bartlett's test for homogeneity of variance. Its power is adequate only if each n_i is large relative to q.

The model under which the usual F tests in a repeated-measure factorial experiment are valid assumes (1) that the matrix of covariances within each of the populations is Σ (the same for each of the populations) and (2) that

$$\sigma^2_{\bar{B}_j - \bar{B}_{j'}} = \text{constant} \qquad \text{for all } j \text{ and } j', \qquad \text{where } j \neq j'.$$

One pattern for Σ that satisfies this second condition is compound symmetry. A matrix having compound symmetry has the following form:

$$\Sigma = \begin{bmatrix} \sigma^2 & \rho\sigma^2 & \cdots & \rho\sigma^2 \\ \rho\sigma^2 & \sigma^2 & \cdots & \rho\sigma^2 \\ \cdot & \cdot & & \cdot \\ \cdot & \cdot & & \cdot \\ \cdot & \cdot & & \cdot \\ \rho\sigma^2 & \rho\sigma^2 & \cdots & \sigma^2 \end{bmatrix}.$$

That is, each entry on the main diagonal is equal to σ^2, and each entry off the main diagonal is equal to $\rho\sigma^2$. If, in fact, Σ has this form, then the matrix

$$S_0 = \begin{bmatrix} \overline{\text{var}} & \overline{\text{cov}} & \cdots & \overline{\text{cov}} \\ \overline{\text{cov}} & \overline{\text{var}} & \cdots & \overline{\text{cov}} \\ \cdot & \cdot & & \cdot \\ \cdot & \cdot & & \cdot \\ \cdot & \cdot & & \cdot \\ \overline{\text{cov}} & \overline{\text{cov}} & \cdots & \overline{\text{var}} \end{bmatrix},$$

where $\overline{\text{var}}$ = mean of entries on main diagonal of S_{pooled},
$\overline{\text{cov}}$ = mean of entries off main diagonal of S_{pooled},

provides an unbiased estimate of Σ.

To test the hypothesis that Σ has the form given above, one computes the following statistics:

(5) $$M_2 = -(N - p) \ln \frac{|S_{\text{pooled}}|}{|S_0|},$$

(6) $$C_2 = \frac{q(q + 1)^2(2q - 3)}{6(N - p)(q - 1)(q^2 + q - 4)},$$

(7) $$f_2 = \frac{q^2 + q - 4}{2}.$$

Under the hypothesis that Σ has the specified form, the statistic

(8) $$\chi_2^2 = (1 - C_2)M_2$$

has a sampling distribution which can be approximated by a chi-square distribution having f_2 degrees of freedom.

The computation of the statistics defined in (4) and (8) will be illustrated through use of the numerical data given in Table 7.7-1. In this table, $p = 2$, $q = 3$, and $n_1 = n_2 = 5$. The covariance matrices obtained from the data in part i of this table are given at the left in Table 7.7-2. For example, the entry 1.75 in the covariance matrix at level a_1 is the covariance between the observations in columns b_1 and b_2 at level a_1. (The symbols in part iii of Table 7.7-1 are defined in Table 7.2-3.)

Table 7.7-1 Numerical Example

(i)

	Subject	b_1	b_2	b_3	Total	
a_1	1	4	7	2	13	
	2	3	5	1	9	
	3	7	9	6	22	
	4	6	6	2	14	
	5	5	5	1	11	
	Total	25	32	12	69	
a_2	6	8	2	5	15	
	7	4	1	1	6	
	8	6	3	4	13	
	9	9	5	2	16	
	10	7	1	1	9	
	Total	34	12	13	59	$128 = G$

(ii)

	b_1	b_2	b_3	Total
a_1	25	32	12	69
a_2	34	12	13	59
Total	59	44	25	128

(iii)

$$(2a_1) = 397 \qquad (2a_2) = 333$$
$$(3a_1) = 317.40 \qquad (3a_2) = 232.07$$
$$(5a_1) = 358.60 \qquad (5a_2) = 293.80$$
$$(6a_1) = 350.33 \qquad (6a_2) = 255.67$$

(iv)

$$SS_{B \times \text{subj w. } a_1} = (2a_1) - (5a_1) - (6a_1) + (3a_1) = 5.47$$
$$SS_{B \times \text{subj w. } a_2} = (2a_2) - (5a_2) - (6a_2) + (3a_2) = 15.60$$
$$SS_{B \times \text{subj}} \text{ (pooled)} = 21.07$$
$$MS_{B \times \text{subj w. } a_1} = 5.47/8 = 0.684$$
$$MS_{B \times \text{subj w. } a_2} = 15.60/8 = 1.950$$
$$MS_{B \times \text{subj}} \text{ (pooled)} = 21.04/16 = 1.315$$

Table 7.7-2 Covariance Matrices Associated with Data in Table 7.7-1

a_1	b_1	b_2	b_3
b_1	2.50	1.75	2.50
b_2	1.75	2.80	3.30
b_3	2.50	3.30	4.30

$\overline{\text{var}} - \overline{\text{cov}} = 0.68$

a_2	b_1	b_2	b_3
b_1	3.70	2.10	1.15
b_2	2.10	2.80	0.70
b_3	1.15	0.70	3.30

$\overline{\text{var}} - \overline{\text{cov}} = 1.95$

Pooled	b_1	b_2	b_3
b_1	3.10	1.92	1.82
b_2	1.92	2.80	2.00
b_3	1.82	2.00	3.80

$\overline{\text{var}} - \overline{\text{cov}} = 1.32$

The matrix corresponding to S_{pooled} is given at the right of Table 7.7-2. The entry 3.10 is the mean of corresponding entries at levels a_1 and a_2, that is, $(2.50 + 3.70)/2 = 3.10$. Similarly, the entry 1.92 is the mean of corresponding entries at a_1 and a_2. The numerical values of the determinants corresponding to S_1, S_2, and S_{pooled} are

$$|S_1| = 1.08, \qquad |S_2| = 17.51, \qquad |S_{\text{pooled}}| = 11.28.$$

The statistic defined in (4) is obtained from

$$\begin{aligned} M_1 &= 8 \ln (11.28) - 4 \ln (1.08) - 4 \ln (17.51) \\ &= 8(2.421) \quad - 4(0.077) \quad - 4(2.863) \\ &= 7.608 \end{aligned}$$

$$C_1 = \frac{18 + 9 - 1}{6(4)(1)} \left(\frac{1}{4} + \frac{1}{4} + \frac{1}{8} \right) = .406$$

$$f_1 = \frac{3(4)(1)}{2} = 6.$$

Hence $$\chi_1^2 = (1 - .406)(7.608) = 4.52.$$

If $\alpha = .05$, the critical value for the test of homogeneity of population covariance matrices is $\chi^2_{.95}(6) = 12.6$. Since the observed chi-square statistic does not exceed the latter value, the hypothesis of homogeneity of covariances may be considered tenable. An unbiased estimate of Σ is given by S_{pooled}.

The matrix S_0 is given by

$$S_0 = \begin{bmatrix} 3.23 & 1.91 & 1.91 \\ 1.91 & 3.23 & 1.91 \\ 1.91 & 1.91 & 3.23 \end{bmatrix}.$$

Note that

$$\overline{\text{var}} - \overline{\text{cov}} = 3.23 - 1.91 = 1.32.$$

Within rounding error, this is the value of $MS_{B \times \text{subj}}$ computed in part iv of Table 7.7-1. The numerical value of the determinant corresponding to S_0 is

$$|S_0| = 12.30.$$

To obtain the statistic defined in (8),

$$M_2 = -(10 - 3) \ln \frac{11.28}{12.30} = -7(-.0866) = .692,$$

$$C_2 = \frac{3(16)(3)}{6(8)(2)(8)} = 0.19,$$

$$f_2 = \frac{9 + 3 - 4}{2} = 4.$$

Hence $\chi_2^2 = (1 - .19)(.692) = 0.56.$

The critical value for a .05-level test on the hypothesis that all the diagonal values of Σ are σ^2 and all the off-diagonal entries are $\rho\sigma^2$ is $\chi^2_{.95}(4) = 9.5$. These data do not contradict this hypothesis.

It is of interest to note the following relationships:

$$\text{MS}_{\text{subj w. } a_1} = 8.23,$$
$$\overline{\text{var}}_{a_1} + (q-1)\,\overline{\text{cov}}_{a_1} = 3.20 + (2)(2.52) = 8.24;$$
$$\text{MS}_{\text{subj w. } a_2} = 5.90,$$
$$\overline{\text{var}}_{a_2} + (q-1)\,\overline{\text{cov}}_{a_2} = 3.27 + (2)(1.32) = 5.91.$$

In general, within rounding error,

$$\text{MS}_{\text{subj w. } a_i} = \overline{\text{var}}_{a_i} + (q-1)\,\overline{\text{cov}}_{a_i}.$$

7.8 Unequal Group Size

Consider the following design:

	b_1	$\cdots$	b_q	Group size
a_1	G_1	$\cdots$	G_1	n_1
a_2	G_2	$\cdots$	G_2	n_2
.	.		.	.
.	.		.	.
.	.		.	.
a_p	G_p	$\cdots$	G_p	n_p

Total number of subjects $= N$

If the levels of factor A represent different strata within a specified population, then the n_i may be proportional to the number of individuals actually in each of these strata in the population. In this case, a least-squares solution for the effects and the sums of squares is appropriate. However, if the original plan for an experiment calls for equal group size, but the completed experiment does not have equal group size because of conditions unrelated to the treatments per se, then an unweighted-means solution is the more appropriate. Both types of solution are considered in this section.

Because the cell frequencies will be proportional by columns, the least-squares solution can be obtained quite simply. The case of proportional cell frequencies is discussed in Sec. 5.23-8. What simplifies the computational procedures is the orthogonality of SS_a and SS_b. If one defines

$$\text{SS}_{ab} = \text{SS}_{\text{b.cell}} - \text{SS}_a - \text{SS}_b,$$

one will have the following additivity:

$$SS_{b.cell} = SS_a + SS_b + SS_{ab}.$$

This type of additivity holds for equal cell frequencies as well as proportional cell frequencies. In general, for unequal cell frequencies

$$SS_{a+b} \neq SS_a + SS_b;$$

however, for proportional cell frequencies

$$SS_{a+b} = SS_a + SS_b.$$

Computational symbols for this case are defined in Table 7.8-1. The sums of squares are obtained by the relations given in Table 7.3-6, using the symbols as defined in Table 7.8-1. The degrees of freedom for the

Table 7.8-1 Unequal Group Size—Least-squares Solution

$(1) = \dfrac{G^2}{Nqr}$	$(6) = \dfrac{\Sigma[(AB_{ij})^2/n_i]}{r}$
$(2) = \Sigma X^2$	$(7) = \dfrac{\Sigma[(AC_{ik})^2/n_i]}{q}$
$(3) = \dfrac{\Sigma(A_i^2/n_i)}{qr}$	$(8) = \dfrac{\Sigma(BC_{jk})^2}{N}$
$(4) = \dfrac{\Sigma B_j^2}{Nr}$	$(9) = \Sigma\left[\dfrac{(ABC_{ijk})^2}{n_i}\right]$
$(5) = \dfrac{\Sigma C_k^2}{Nq}$	$(10) = \dfrac{\Sigma P_m^2}{qr}$

sums of squares are obtained from those given in Table 7.3-1 by replacing np with N throughout, where $N = \Sigma n_i$. For example,

$$p(n-1) \quad \text{becomes} \quad N - p,$$
$$p(n-1)(q-1) \quad \text{becomes} \quad (N-p)(q-1).$$

The starting point for an unweighted-means solution is an ABC' summary table in which a cell entry is a mean; that is,

$$ABC'_{ijk} = \frac{ABC_{ijk}}{n_i}.$$

From this summary table, AB' and AC' summaries are computed in the usual manner; that is,

$$AB'_{ij} = \sum_k ABC'_{ijk},$$

$$A'_1 = \sum_j AB'_{ij}.$$

Computational symbols appropriate for this case are given in Table 7.8-2. Only those sums of squares which do not involve the subject

Table 7.8-2 Unequal Group Size—Unweighted-means Solution

(i)	$(1') = G'^2/pqr$ $(3') = (\Sigma A_i'^2)/qr$ $(4') = (\Sigma B_j'^2)/pr$ $(5') = (\Sigma C_k'^2)/pq$	$(6') = [\Sigma(AB_{ij}')^2]/r$ $(7') = [\Sigma(AC_{ik}')^2]/q$ $(8') = [\Sigma(BC_{jk}')^2]/p$ $(9') = \Sigma(ABC_{ijk}')^2$
(ii)	$SS_a = \bar{n}_h[(3') - (1')]$ $SS_b = \bar{n}_h[(4') - (1')]$ $SS_c = \bar{n}_h[(5') - (1')]$	$SS_{ab} = \bar{n}_h[(6') - (3') - (4') + (1')]$ $SS_{ac} = \bar{n}_h[(7') - (3') - (5') + (1')]$ $SS_{bc} = \bar{n}_h[(8') - (4') - (5') + (1')]$
	$SS_{abc} = \bar{n}_h[(9') - (6') - (7') - (8') + (3') + (4') + (5') - (1')]$	

factor are given here. Sums of squares which do involve the subject factor are identical with those in a least-squares analysis.

In a least-squares solution,

$$SS_{\text{total}} = SS_{\text{b.cell}} + SS_{\text{w.cell}}.$$

This relationship, however, does not hold for an unweighted-means solution. If the n_i do not differ markedly, both types of solution lead to numerically similar final products.

The computational procedures that have been discussed above can be specialized to the case of a $p \times q$ factorial experiment with repeated measures on factor B. This is done by setting $r = 1$ and dropping all terms involving factor C. The starting point for an unweighted-means analysis in this case is a summary table in which

$$AB'_{ij} = \frac{AB_{ij}}{n_i}.$$

A numerical example of this case is given in Table 7.8-3.

Table 7.8-3 Numerical Example

	Subject	b_1	b_2	b_3	Total	
a_1	1	3	6	9	18	
	2	6	10	14	30	$n_1 = 3$
	3	10	15	18	43	
a_2	4	8	12	16	36	
	5	3	5	8	16	
	6	1	3	8	12	$n_2 = 5$
	7	12	18	26	56	
	8	9	10	18	37	
a_3	9	10	22	16	48	
	10	3	15	8	26	
	11	7	16	10	33	$n_3 = 4$
	12	5	20	12	37	
	Total	77	152	163	392	$N = 12$

A summary of the computational steps in obtaining both the least-squares and the unweighted-means solutions appears in Table 7.8-4. Symbols associated with the least-squares solution appear at the left. The unweighted-means analysis is summarized in Table 7.8-5. The two solutions are compared in Table 7.8-6.

Table 7.8-4 Computational Procedures

(i) *AB* summary table

	b_1	b_2	b_3	Total
a_1	19	31	41	91
a_2	33	48	76	157
a_3	25	73	46	144
Total	77	152	163	392

(i) *AB′* summary table

	b_1	b_2	b_3	Total
a_1	6.33	10.33	13.67	30.33
a_2	6.60	9.60	15.20	31.40
a_3	6.25	18.25	11.50	36.00
Total	19.18	38.18	40.37	97.73

(ii)

$(1) = G^2/Nq = 4268.44$

$(2) = \Sigma X^2 = 5504$

$(3) = \Sigma(A_i^2/n_i q) = 4291.38$

$(4) = (\Sigma B_j^2)/N = 4633.50$

$(5) = \Sigma[(AB_{ij})^2/n_i] = 4852.30$

$(6) = (\Sigma P_m^2)/q = 4904.00$

$(1') = G'^2/pq = 1061.24$

$(3') = (\Sigma A_i'^2)/q = 1067.29$

$(4') = (\Sigma B_j'^2)/p = 1151.77$

$(5') = \Sigma(AB'_{ij})^2 = 1204.78$

(iii)

$$\tilde{n}_h = \frac{p}{\Sigma(1/n_i)} = \frac{3}{(1/3) + (1/5) + (1/4)} = 3.830$$

Table 7.8-5 Analysis of Variance for Numerical Example—Unweighted-means Solution

Source of variation	Computational formula	SS	df	MS	F
Between subjects			11		
A	$\tilde{n}_h[(3') - (1')]$	23.17	2	11.29	
Subjects w. groups	$(6) - (3)$	612.62	9	68.06	
Within subjects			24		
B	$\tilde{n}_h[(4') - (1')]$	346.73	2	173.36	79.89
AB	$\tilde{n}_h[(5') - (3') - (4') + (1')]$	179.86	4	44.97	20.72
$B \times$ subjects w. groups	$(2) - (5) - (6) + (3)$	39.08	18	2.17	

Individual comparisons for the case of a $p \times q$ factorial experiment having repeated measures on factor B will be outlined in what follows. For the case of the least-squares solution,

$$\bar{A}_i = \frac{A_i}{n_i q} \qquad \text{and} \qquad \bar{B}_j = \frac{B_j}{N}.$$

Table 7.8-6 Comparison of Solutions

Source	Unweighted means	Least squares
A	23.17	22.94
B	346.73	365.06
AB	179.86	195.86

F ratios in tests on individual comparisons take the form

$$F = \frac{(\bar{A}_i - \bar{A}_{i'})^2}{\text{MS}_{\text{subj w.groups}}[(1/n_i q) + (1/n_{i'} q)]},$$

$$F = \frac{(\bar{B}_j - \bar{B}_{j'})^2}{\text{MS}_{B \times \text{subj w.groups}}(2/N)}.$$

For the case of the unweighted-means solution,

$$\bar{A}_i = \frac{A'_i}{q} \qquad \text{and} \qquad \bar{B}_j = \frac{B'_j}{p}.$$

F ratios in tests on individual comparisons take the form

$$F = \frac{(\bar{A}_i - \bar{A}_{i'})^2}{\text{MS}_{\text{subj w.groups}}[(1/n_i q) + (1/n_{i'} q)]},$$

$$F = \frac{(\bar{B}_j - \bar{B}_{j'})^2}{\text{MS}_{B \times \text{subj w.groups}}(2/\bar{n}_h p)}.$$

8

FACTORIAL EXPERIMENTS IN WHICH SOME OF THE INTERACTIONS ARE CONFOUNDED

8.1 General Purpose

Precision of estimation requires that treatment effects be free of between-block variation. In this context, a block is a person, a group of people, a period of time, a source of experimental material, etc. Since the number of treatment combinations in a factorial experiment increases rapidly with either the number of factors or the number of levels of the factors, observing all treatment combinations within the same block often demands unworkably large block capacity. In this chapter, techniques will be considered for assigning a relatively small number of the possible treatment combinations to blocks in a way that permits within-block estimates of the most important sources of variation.

Small block size in this context is equivalent to homogeneity of the conditions under which the treatment effects are measured. There can be considerable variation among blocks; this source of variation does not affect the precision of the within-block information. When practical working conditions rule against having a complete replication within a single block and still assure homogeneity of the uncontrolled sources of error, then balanced incomplete-block designs provide the next best alternative.

Most of the plans to be developed use all the treatment combinations required for the complete factorial experiment. Within any one block only a fraction of all possible treatment combinations appear; the number of treatment combinations per block will be called the block size. The primary object of these plans is to control experimental error (1) by keeping the block size small and (2) by eliminating block differences from experimental error. The cost of this added control on experimental error will be the loss of some information on higher-order interactions. Differences between blocks will form a part of some interaction; hence components of such interactions will be confounded with the block effects. Whether or not the sacrifice of information on components of higher-order interactions is worth the added control depends upon the magnitude of the block effects.

Admittedly there is some danger in using designs in which any effect is confounded. This is particularly true in exploratory studies. However, the potential advantage of these designs—increased precision with respect to effects of primary interest—provides the experimenter with a potent source of motivation for their use.

In this chapter the general principles underlying designs involving confounding of interactions will be considered; the analysis and applications of these designs will also be illustrated. In balanced designs some information will be available on all components of the interactions; in other designs some of the components of interactions will be completely confounded with between-block effects. The designs in Chap. 9, which use the Latin-square principle, are actually special cases of the designs in this chapter.

Construction of the designs to be presented in this chapter depends upon techniques for analyzing interaction terms into parts. The first step is to divide the treatment combinations into sets which are balanced with respect to the main effects of the factors involved. For example, in a 2×2 factorial experiment the following sets are balanced with respect to the main effects:

Set I	Set II
ab_{11}	ab_{12}
ab_{22}	ab_{21}

Note that a_1 appears once and only once in each set. Similarly, a_2, b_1, and b_2 each appear once and only once in each set. Thus there is balance with respect to the main effects of both factors A and B. Any difference between the sums for the sets is *not* a function of the main effects, but rather a function of the AB interaction. These results follow only if factors A and B are fixed; blocks are assumed to be random. Further, the underlying model is strictly additive with respect to block effects; i.e., no interactions with block effects appear in the model.

$$X_{ijkm} = \mu + \alpha_i + \beta_j + \alpha\beta_{ij} + (\text{block})_k + \varepsilon_{m(ijk)}.$$

In obtaining the components of a three-factor interaction one divides all treatment combinations into balanced sets. In this case, however, the balance is with respect to all two-factor interactions as well as all main effects. For example, in a $2 \times 2 \times 2$ factorial experiment balanced sets with respect to all effects except the ABC interaction are:

Set I	Set II
abc_{111}	abc_{112}
abc_{122}	abc_{211}
abc_{212}	abc_{121}
abc_{221}	abc_{222}

Note that each level of each factor occurs an equal number of times within a set. Note also that all possible pairs of treatments occur once and only once with each set. Hence each set is balanced with respect to the A, B, and C main effects as well as AB, AC, and BC interactions. Therefore, the difference between the sum of all observations in set I and the sum of all observations in set II is a function of the ABC interaction.

Procedures for constructing balanced sets of treatment combinations are given in later sections. Comparisons between the sums over balanced sets define components of the interactions.

The sets of treatment combinations which are balanced with respect to main effects and lower-order interactions will be assigned to separate blocks. Hence the between-block differences, though free of main effects and lower-order interactions, will be confounded with components of the higher-order interactions. By replicating the experiment an appropriate number of times, different components of higher-order interactions can be used to form the sets of treatment combinations that make up the blocks. Hence some information on all components of the higher-order interactions will often be obtainable from the experiment as a whole.

8.2 Modular Arithmetic

The material to be presented in later sections is simplified through use of modular arithmetic. By definition an integer I modulus an integer m is the remainder obtained by dividing I by m. For example, the integer 18 to the modulus 5 is 3, since the remainder when 18 is divided by 5 is 3. This result is usually written

$$18 \pmod 5 = 3$$

and is read "18 modulo 5 is 3." Other examples are:

$$20 \pmod 5 = 0,$$
$$7 \pmod 5 = 2,$$
$$3 \pmod 5 = 3.$$

Alternatively, 18 is said to be congruent to 3, modulus 5; 20 is said to be congruent to 0, modulus 5. Thus all integers are congruent to one of the integers 0, 1, 2, 3, or 4, modulus 5.

If the modulus 3 is used, all integers are congruent to 0, 1, or 2. For the modulus 3,

$$18 \pmod 3 = 0,$$
$$7 \pmod 3 = 1,$$
$$20 \pmod 3 = 2.$$

For purposes of the work in the following sections, the moduli will be limited to prime numbers, i.e., numbers divisible by no number smaller than it except unity. For example, 1, 2, 3, 5, 7, 11, etc., are prime numbers.

The operation of modular addition is shown in the following examples:

$$2 + 1 = 0 \pmod 3,$$
$$0 + 2 = 2 \pmod 3,$$
$$2 + 2 = 1 \pmod 3;$$

$$2 + 2 = 4 \pmod 5,$$
$$4 + 4 = 3 \pmod 5,$$
$$1 + 4 = 0 \pmod 5.$$

To add two integers, one obtains the ordinary sum, and then one expresses this sum in terms of the modulus. For example, $4 + 4 = 8$; $8 \pmod 5 = 3$. Hence, $4 + 4 = 3 \pmod 5$. Unless the modulus is understood from the context, it is written after the operation as (mod m). The operation of multiplication is illustrated by the following examples:

$$2 \cdot 2 = 1 \pmod 3,$$
$$2 \cdot 0 = 0 \pmod 3;$$

$$4 \cdot 2 = 3 \pmod 5,$$
$$3 \cdot 3 = 4 \pmod 5.$$

The product of two numbers is formed as in ordinary multiplication; then the product is expressed in terms of the modulus.

Algebraic equations may be solved in terms of a modular system. For example, by using the modulus 3, the equation

$$2x = 1 \pmod 3$$

has the solution $x = 2$. To obtain the solution, both sides of this equation are multiplied by a number which will make the coefficient of x equal to unity, modulus 3. Thus

$$2 \cdot 2x = 2 \pmod 3.$$

Since $2 \cdot 2 = 1 \pmod 3$, the last equation becomes $x = 2$. As another example, the equation

$$4x = 3 \pmod 5$$

has the solution $x = 2$. To obtain this solution, both sides of the equation are multiplied by an integer that will make the coefficient of x equal to unity. If both sides of this equation are multiplied by 4,

$$4 \cdot 4x = 4 \cdot 3 \pmod{5}.$$

Expressing the respective products to the modulus 5, one has

$$16x = 12 \qquad \text{or} \qquad x = 2 \pmod{5}.$$

Equations of the form $ax_1 + bx_2 = c \pmod{m}$ may always be reduced to the form $x_1 + dx_2 = k \pmod{m}$. For example, the equation

$$2x_1 + x_2 = 1 \pmod{3}$$

becomes, after multiplying both sides by 2,

$$x_1 + 2x_2 = 2 \pmod{3}.$$

As another example,

$$2x_1 + 4x_2 = 2 \pmod{5}$$

becomes, after multiplying both sides by 3,

$$x_1 + 2x_2 = 1 \pmod{5}.$$

The equation $2x_1 + 4x_2 = 2 \pmod{5}$ and the equation $x_1 + 2x_2 = 1 \pmod{5}$ have the same roots. It may be verified that, when $x_1 = 0$, $x_2 = 3$; hence one pair of roots for these equations is (0,3). Other roots are (1,0) and (2,2). To show that (2,2) is a root of the equation $x_1 + 2x_2 = 1 \pmod{5}$, substituting $x_1 = 2$ and $x_2 = 2$ in this equation yields

$$2 + 2(2) = 1 \pmod{5}.$$

The numerical value of the left-hand side of the equation is 6, which to the modulus 5 is 1.

8.3 Revised Notation for Factorial Experiments

The introduction of a modified notation for the treatment combinations in a factorial experiment will permit more convenient application of modular arithmetic. This revised notation is illustrated for the case of a $3 \times 3 \times 2$ factorial experiment in Table 8.3-1. The three levels of factor

Table 8.3-1 Notation for a $3 \times 3 \times 2$ Factorial Experiment

	c_0			c_1		
	b_0	b_1	b_2	b_0	b_1	b_2
a_0	(000)	(010)	(020)	(001)	(011)	(021)
a_1	(100)	(110)	(120)	(101)	(111)	(121)
a_2	(200)	(210)	(220)	(201)	(211)	(221)

A are designated by the subscripts 0, 1, and 2. The treatment combination consisting of level a_1, level b_0, and level c_1 is designated by the symbol (101). The digit in the first position indicates the level of factor A, the digit in the second position indicates the level of factor B, and the digit in the third position indicates the level of factor C. Thus, the symbol (ijk) represents the treatment combination abc_{ijk}.

To illustrate how modular arithmetic may be used to define sets of treatment combinations, consider a 3×3 factorial experiment. All the treatment combinations in this experiment may be expressed in the form (ij) by suitable choice of i and j. Let x_1 stand for the digit in the first position of this symbol, and let x_2 stand for the digit in the second position of this symbol. The relation $x_1 + x_2 = 0 \pmod 3$ is satisfied by the symbols (00), (12), and (21). To show, for example, that the symbol (12) satisfies this relation, substituting $x_1 = 1$ and $x_2 = 2$ yields $1 + 2 = 0 \pmod 3$. Equivalently, the relation $x_1 + x_2 = 0$ is said to define or generate the set of treatment combinations (00), (12), (21). By similar reasoning, the relation

$$x_1 + x_2 = 1 \pmod 3 \text{ defines the set (01), (10), (22);}$$

and the relation

$$x_1 + x_2 = 2 \pmod 3 \text{ defines the set (02), (11), (20).}$$

Each of these sets is balanced with respect to main effects.

8.4 Method for Obtaining the Components of Interactions

The method of subdividing the degrees of freedom for interactions to be described in this section applies only to factorial experiments of the form $p \times p \times \cdots \times p$, where p is a prime number. Thus, this method applies to experiments of the form 2×2, $2 \times 2 \times 2, \ldots ; 3 \times 3$, $3 \times 3 \times 3, \ldots ; 5 \times 5$, $5 \times 5 \times 5, \ldots$; etc. A 3×3 factorial experiment will be used to illustrate the method. The $A \times B$ interaction, which has four degrees of freedom, consists of an I component having two degrees of freedom and a J component having two degrees of freedom. The sums of squares for $A \times B$ may be partitioned as follows:

SS	df
$A \times B$	4
$AB(I)$	2
$AB(J)$	2

The symbol $AB(I)$ denotes the I component of the $A \times B$ interaction. Such components may have no meaning in terms of the levels of the factors; their purpose is merely to provide a convenient method for subdividing the degrees of freedom.

An alternative, but more convenient, notation for the components of the interaction uses the symbol AB for the J component and the symbol AB^2 for the I component. The rationale underlying this notation scheme will become clear when the computational procedure for these components is described. The nine treatment combinations in a 3×3 factorial experiment may be divided into three nonoverlapping sets by means of the following relations:

$x_1 + x_2 = 0 \pmod 3$	$x_1 + x_2 = 1 \pmod 3$	$x_1 + x_2 = 2 \pmod 3$
(00)	(01)	(02)
(12)	(10)	(11)
(21)	(22)	(20)

The digits in the symbols for the treatments satisfy the relations under which the symbols appear. No treatment combination appears in more than one set; all treatment combinations of the 3×3 factorial experiment are included. Each of these sets is balanced with respect to the main effects of both factors (assuming that factors A and B are both fixed factors).

To illustrate what is meant by balance in this context, assume that the following linear model holds:

$$X_{ij} = \mu + \alpha_i + \beta_j + (\alpha\beta)_{ij} + \varepsilon_{ij},$$

where $\Sigma_i \alpha_i = 0$, $\Sigma_j \beta_j = 0$, and $\Sigma_i(\alpha\beta)_{ij} = \Sigma_j(\alpha\beta)_{ij} = 0$. When A and B are fixed factors, these restrictions on the parameters in the model follow from the definition of the effects. An observation made on a treatment combination on the left-hand side of the following equations estimates the sum of the parameters indicated on the right-hand side,

$$\begin{aligned}(00) &= \mu + \alpha_0 + \beta_0 + (\alpha\beta)_{00} + \varepsilon_{00},\\ (12) &= \mu + \alpha_1 + \beta_2 + (\alpha\beta)_{12} + \varepsilon_{12},\\ (21) &= \mu + \alpha_2 + \beta_1 + (\alpha\beta)_{21} + \varepsilon_{21}.\end{aligned}$$

The sum of the observations on this set of treatments will not contain effects associated with either of the main effects, since $\alpha_0 + \alpha_1 + \alpha_2 = 0$ and $\beta_0 + \beta_1 + \beta_2 = 0$. The sum of the observations in this set of treatments will, however, involve an interaction effect, since $(\alpha\beta)_{00} + (\alpha\beta)_{12} + (\alpha\beta)_{21} \neq 0$. Thus the set (00), (12), (21), which is defined by the relation $x_1 + x_2 = 0 \pmod 3$, is said to be balanced with respect to both main effects but not balanced with respect to the interaction effect. The sets defined by the relations $x_1 + x_2 = 1$ and $x_1 + x_2 = 2$ are also balanced with respect to both main effects but unbalanced with respect to the interaction effect.

Differences between the sums of the observations for each of these three sets define two of the four degrees of freedom of the $A \times B$ interaction. These two degrees of freedom are associated with what is called the $AB(J)$

component (or the AB component) of the $A \times B$ interaction. The numerical example in Table 8.4-1 illustrates the computation of this com-

Table 8.4-1 Computation of Components of Interaction in 3 × 3 Factorial Experiment Having Four Observations per Cell

(i)

	b_0	b_1	b_2	Total	
a_0	45	20	20	$85 = A_0$	
a_1	25	50	20	$95 = A_1$	$(n = 4)$
a_2	20	50	80	$150 = A_2$	
Total	90	120	120	$330 = G$	
	B_0	B_1	B_2		

(ii)

$$SS_{A\times B} = \frac{45^2 + 20^2 + \cdots + 50^2 + 80^2}{4} - \frac{85^2 + 95^2 + 150^2}{12} - \frac{90^2 + 120^2 + 120^2}{12} + \frac{330^2}{36}$$

$$= 633.33$$

(iii)

$x_1 + x_2 = 0$	$x_1 + x_2 = 1$	$x_1 + x_2 = 2$
$(00) = 45$	$(01) = 20$	$(02) = 20$
$(12) = 20$	$(10) = 25$	$(11) = 50$
$(21) = 50$	$(22) = 80$	$(20) = 20$
$(AB)_0 = 115$	$(AB)_1 = 125$	$(AB)_2 = 90$

$$SS_{AB} = \frac{115^2 + 125^2 + 90^2}{12} - \frac{330^2}{36} = 54.17$$

(iv)

$x_1 + 2x_2 = 0$	$x_1 + 2x_2 = 1$	$x_1 + 2x_2 = 2$
$(00) = 45$	$(02) = 20$	$(01) = 20$
$(11) = 50$	$(10) = 25$	$(12) = 20$
$(22) = 80$	$(21) = 50$	$(20) = 20$
$(AB^2)_0 = 175$	$(AB^2)_1 = 95$	$(AB^2)_2 = 60$

$$SS_{AB^2} = \frac{175^2 + 95^2 + 60^2}{12} - \frac{330^2}{36} = 579.17$$

ponent of the interaction. The entries in the cells of part i of this table are assumed to be the sums over four observations; i.e., the data in part i represent a summary of a 3 × 3 factorial experiment in which there are four observations per cell.

The sum of squares for the overall $A \times B$ interaction, which has four degrees of freedom, is computed in part ii. The sum of squares for the AB component, which has two degrees of freedom, is computed in part iii. The sum of all observations in the set of treatment combinations defined by the relation $x_1 + x_2 = 0$ is denoted by the symbol $(AB)_0$. Similarly the sum of all observations in the set defined by $x_1 + x_2 = 1$ is denoted by the symbol $(AB)_1$. The sum of squares corresponding to the AB component is given by

$$\text{SS}_{AB} = \frac{(AB)_0^2 + (AB)_1^2 + (AB)_2^2}{3n} - \frac{G^2}{9n},$$

where n is the number of observations in each of the cell entries in part i. For the data in part i, $\text{SS}_{AB} = 54.17$.

The AB^2 component of the interaction is obtained from sets of treatment combinations defined by the following relations (modulus 3 is understood):

$x_1 + 2x_2 = 0$	$x_1 + 2x_2 = 1$	$x_1 + 2x_2 = 2$
(00)	(02)	(01)
(11)	(10)	(12)
(22)	(21)	(20)

These sets are also balanced with respect to the main effects. Hence differences among the sums of the observations in these three sets will define a component of the overall interaction—in this case the AB^2 component. The numerical value of this component is computed in part iv of Table 8.4-1. The symbol $(AB^2)_0$ in this part denotes the sum of all observations in the set of treatments defined by the relation $x_1 + 2x_2 = 0$. The sum of squares for this component is given by

$$\text{SS}_{AB^2} = \frac{(AB^2)_0^2 + (AB^2)_1^2 + (AB^2)_2^2}{3n} - \frac{G^2}{9n}.$$

The numerical value for this sum of squares is 579.17.

The partition of the overall interaction may be summarized as follows:

Source	SS	df
$A \times B$	633.33	4
AB	54.17	2
AB^2	579.17	2

Within rounding error $\text{SS}_{A\times B} = \text{SS}_{AB} + \text{SS}_{AB^2}$. In this case the AB^2 component of the interaction is considerably larger than the AB component. Inspection of part i of Table 8.4-1 indicates that the large totals are on the diagonal running from upper left to lower right. The treatment totals on this diagonal are in the same set for the AB^2 component but in different sets for the AB component. Hence the AB^2 component is large relative to the AB component.

The computation of the components of the $A \times B$ interaction may be simplified by use of the procedure illustrated in Table 8.4-2. The cell entries at the top of this table are the same as those in part i of Table 8.4-1, with rows a_0 and a_1 appearing twice. Totals of the form $(AB)_i$ are obtained by summing the entries on diagonals running from upper right to lower left; totals of the form $(AB^2)_i$ are obtained by summing entries on diagonals running from upper left to lower right. From these totals the components of the interaction are computed by the same formulas as those used in Table 8.4-1. The treatment combinations falling along these diagonals actually form the balanced sets defined by modular relations.

Table 8.4-2 Simplified Computation of the Components of the $A \times B$ Interaction

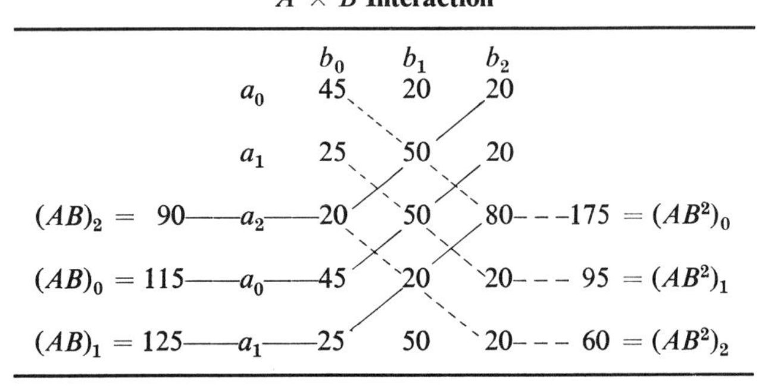

The use of modular arithmetic in defining balanced sets of treatment combinations is readily extended to 5×5 factorial experiments. In this latter type of experiment the 16 degrees of freedom for the $A \times B$ interaction may be partitioned as follows:

Source	df
$A \times B$	16
AB	4
AB^2	4
AB^3	4
AB^4	4

The balanced sets of treatment combinations from which the AB^3 component of the interaction is obtained are defined by the following relations, all with respect to the modulus 5:

$x_1 + 3x_2 = 0$	$x_1 + 3x_2 = 1$	$x_1 + 3x_2 = 2$	$x_1 + 3x_2 = 3$	$x_1 + 3x_2 = 4$
(00)	(02)	(04)	(01)	(03)
(13)	(10)	(12)	(14)	(11)
(21)	(23)	(20)	(22)	(24)
(34)	(31)	(33)	(30)	(32)
(42)	(44)	(41)	(43)	(40)

Each of these sets is balanced with respect to both main effects. The balanced sets defining the AB^4 component of the interaction are obtained from the relation $x_1 + 4x_2 = i$ (modulus 5), where $i = 0, \ldots, 4$.

The computation of the components of the $A \times B$ interaction in a 5×5 factorial experiment may be carried out by an extension of the method used in Table 8.4-2. This extension is illustrated by the data in Table 8.4-3. In part i, the totals required for the computation of the AB component appear at the left, and the totals required for the AB^4 component appear at the right. In part ii, totals required for the AB^2 and the AB^3 components are obtained. The arrangement of the b's in part ii is given by multiplying subscripts to b's in part i by 2 and then reducing the resulting numbers to the modulus 5. By performing this operation, the sequence b_0, b_1, b_2, b_3, b_4 becomes the sequence b_0, b_2, b_4, b_1, b_3; the latter sequence appears in part ii. The AB^2 component of the interaction is obtained from the totals in the lower right of part ii. Assuming one observation per cell,

$$SS_{AB^2} = \frac{65^2 + 70^2 + 65^2 + 80^2 + 100^2}{5} - \frac{380^2}{25} = 174.00.$$

Other components are obtained in an analogous manner. The numerical values of the components of the interaction for the data in Table 8.4-3 are as follows:

Source	SS	df
$A \times B$	1136.00	16
AB	414.00	4
AB^2	174.00	4
AB^3	254.00	4
AB^4	294.00	4

In a $3 \times 3 \times 3$ factorial experiment, the three-factor interaction may be partitioned into the following components:

Source	df
$A \times B \times C$	8
ABC	2
ABC^2	2
AB^2C	2
AB^2C^2	2

Table 8.4-3 Computation of Components of Interaction in a 5 × 5 Factorial Experiment

				b_0	b_1	b_2	b_3	b_4		
			a_0	5	10	15	20	25		
			a_1	10	15	20	25	30		
			a_2	15	15	15	15	20		
			a_3	20	15	10	10	5		
(i)	$(AB)_4$ =	110	a_4	30	20	5	5	5	50	$= (AB^4)_0$
	$(AB)_0$ =	80	a_0	5	10	15	20	25	65	$= (AB^4)_1$
	$(AB)_1$ =	55	a_1	10	15	20	25	30	85	$= (AB^4)_2$
	$(AB)_2$ =	55	a_2	15	15	15	15	20	100	$= (AB^4)_3$
	$(AB)_3$ =	80	a_3	20	15	10	10	5	80	$= (AB^4)_4$
		380							380	

				b_0	b_2	b_4	b_1	b_3		
			a_0	5	15	25	10	20		
			a_1	10	20	30	15	25		
			a_2	15	15	20	15	15		
			a_3	20	10	5	15	10		
(ii)	$(AB^3)_4$ =	95	a_4	30	5	5	20	5	65	$= (AB^2)_0$
	$(AB^3)_0$ =	55	a_0	5	15	25	10	20	70	$= (AB^2)_1$
	$(AB^3)_1$ =	60	a_1	10	20	30	15	25	65	$= (AB^2)_2$
	$(AB^3)_2$ =	90	a_2	15	15	20	15	15	80	$= (AB^2)_3$
	$(AB^3)_3$ =	80	a_3	20	10	5	15	10	100	$= (AB^2)_4$
		380							380	

The sets of treatment combinations from which the AB^2C^2 component is computed are defined by the following relations:

$x_1 + 2x_2 + 2x_3 = 0 \pmod 3$	$x_1 + 2x_2 + 2x_3 = 1 \pmod 3$	$x_1 + 2x_2 + 2x_3 = 2 \pmod 3$
(000)	(002)	(001)
(012)	(011)	(010)
(021)	(020)	(022)
(101)	(100)	(102)
(110)	(112)	(111)
(122)	(121)	(120)
(202)	(201)	(200)
(211)	(210)	(212)
(220)	(222)	(221)

The sum of the observations within each of these sets is balanced with respect to all the main effects and all the two-factor interactions. There is no balance for the three-factor interaction. Hence differences between sums over the sets form one of the components of the three-factor interaction, in this case the AB^2C^2 component. Treatment combinations from which the AB^2C component is computed are defined by relations of the form $x_1 + 2x_2 + x_3 = i \pmod 3$.

The computation of the components of the three-factor interaction is illustrated in Table 8.4-4. In part i summary data for observations made at level c_0 are given. Summary data for levels c_1 and c_2 appear in parts ii and iii, respectively. Totals corresponding to the AB component for level c_0 appear at the left in part i. These totals also appear in part iv under the column headed c_0. The totals at the left of parts ii and iii make up the columns c_1 and c_2 in part iv. The ABC component of the three-factor interaction is computed from totals at the left of part iv; the ABC^2 component is obtained from the totals at the right of part iv. For

Table 8.4-4 Computation of the Components of a Three-factor Interaction in a 3 × 3 × 3 Factorial Experiment

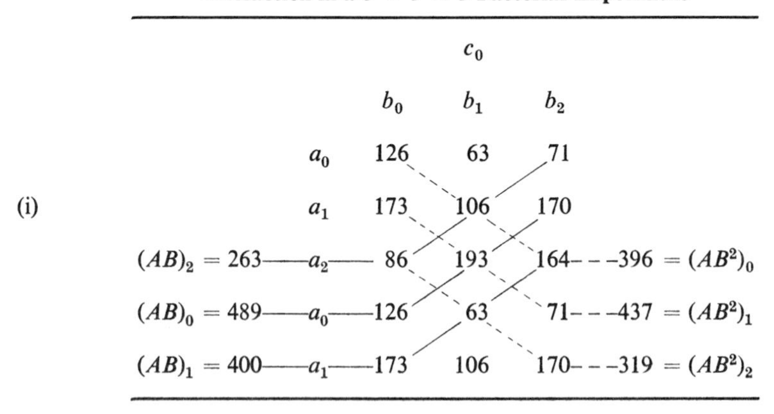

(ii) c_1

			b_0	b_1	b_2	
		a_0	127	158	106	
(ii)		a_1	80	109	123	
	$(AB)_2 = 307$	a_2	92	139	104	$340 = (AB^2)_0$
	$(AB)_0 = 389$	a_0	127	158	106	$325 = (AB^2)_1$
	$(AB)_1 = 342$	a_1	80	109	123	$373 = (AB^2)_2$

(iii) c_2

			b_0	b_1	b_2	
		a_0	103	80	54	
(iii)		a_1	123	52	109	
	$(AB)_2 = 198$	a_2	92	101	60	$215 = (AB^2)_0$
	$(AB)_0 = 313$	a_0	103	80	54	$278 = (AB^2)_1$
	$(AB)_1 = 263$	a_1	123	52	109	$281 = (AB^2)_2$

(iv)

			c_0	c_1	c_2	
		$(AB)_0$	489	389	313	
		$(AB)_1$	400	342	263	
(iv)	$(ABC)_2 = 918$	$(AB)_2$	263	307	198	$1029 = (ABC^2)_0$
	$(ABC)_0 = 1059$	$(AB)_0$	489	389	313	$1020 = (ABC^2)_1$
	$(ABC)_1 = 987$	$(AB)_1$	400	342	263	$915 = (ABC^2)_2$
	2964					2964

(v)

			c_0	c_1	c_2	
		$(AB^2)_0$	396	340	215	
		$(AB^2)_1$	437	325	278	
(v)	$(AB^2C)_2 = 859$	$(AB^2)_2$	319	373	281	$1002 = (AB^2C^2)_0$
	$(AB^2C)_0 = 1047$	$(AB^2)_0$	396	340	215	$1025 = (AB^2C^2)_1$
	$(AB^2C)_1 = 1058$	$(AB^2)_1$	437	325	278	$937 = (AB^2C^2)_2$

example, if there are r observations in each cell of a $3 \times 3 \times 3$ factorial experiment, the ABC component is given by

$$SS_{ABC} = \frac{918^2 + 1059^2 + 987^2}{9r} - \frac{2964^2}{27r}.$$

In part v of the table, the entries under column c_0 are obtained from the totals at the right of part i. The entries in columns c_1 and c_2 are obtained, respectively, from the totals to the right of parts ii and iii. Totals at the left of part v are used to obtain the AB^2C component of the three-factor interaction; totals at the right are used to obtain the AB^2C^2 component.

The treatment combinations that are summed to obtain the total $(AB^2C^2)_0$ in part v may be shown to be those belonging to set $x_1 + 2x_2 + 2x_3 = 0$. Similarly the total $(AB^2C^2)_1$ is the sum of all observations on the treatment set $x_1 + 2x_2 + 2x_3 = 1$. The scheme presented in Table 8.4-4 provides a convenient method for obtaining the sum over balanced sets. In general the symbol $(AB^iC^j)_k$ represents the sum of all treatment combinations satisfying the relation

$$x_1 + ix_2 + jx_3 = k \pmod 3).$$

Rather than using the scheme presented in Table 8.4-4, it is sometimes more convenient to form the sets of treatment combinations by means of their defining relations and obtain the sums directly from the sets.

In the notation system used by Yates as well as Cochran and Cox, the following components of the three-factor interaction in a $3 \times 3 \times 3$ factorial experiment are equivalent:

Modular notation	Yates notation
ABC	$ABC(Z)$
ABC^2	$ABC(Y)$
AB^2C	$ABC(X)$
AB^2C^2	$ABC(W)$

The modular-notation system is the more convenient and lends itself to generalization beyond the $3 \times 3 \times 3$ factorial experiment. For example, in a $5 \times 5 \times 5$ factorial experiment, the three-factor interaction may be partitioned into the following parts:

Source	df
$A \times B \times C$	64
ABC	4
ABC^2	4
ABC^3	4
ABC^4	4
AB^2C	4
...	...
AB^4C^4	4

There are 16 components in all, each having four degrees of freedom. The balanced sets used to obtain the AB^4C^4 component are defined by the relations $x_1 + 4x_2 + 4x_3 = 0, 1, 2, 3, 4 \pmod 5$. For example, the sum $(AB^4C^4)_2$ is obtained from the treatment combinations belonging to the set defined by the relation $x_1 + 4x_2 + 4x_3 = 2$; there will be 25 treatment combinations in this set.

The notation system for the components of the interaction is not unique unless the convention is adopted that the exponent of the first letter is always unity. For example, in a 3×3 factorial experiment A^2B and AB^2 define the same component, since

$$2x_1 + x_2 = i \pmod 3$$

defines the same set of treatment combinations as

$$2(2)x_1 + 2x_2 = 2i \pmod 3,$$

$$x_1 + 2x_2 = 2i \pmod 3.$$

Multiplying both sides of a defining relation by a constant is algebraically equivalent to raising the symbol for the corresponding component to a power equal to that constant. In this case the constant is 2. Hence

$$(A^2B)^2 = A^4B^2 = AB^2,$$

upon expressing the exponents modulo 3.

As another example, in a 5×5 factorial experiment A^3B^2 is equivalent to AB^4. To show this,

$$3x_1 + 2x_2 = i \pmod 5.$$

Multiplying both sides of this equation by 2,

$$6x_1 + 4x_2 = 2i \pmod 5,$$

$$x_1 + 4x_2 = 2i \pmod 5.$$

Alternatively, $\quad A^3B^2 = (A^3B^2)^2 = A^6B^4 = AB^4.$

8.5 Designs for $2 \times 2 \times 2$ Factorial Experiments in Blocks of Size 4

The eight treatment combinations in a $2 \times 2 \times 2$ factorial experiment may be divided into two balanced sets in such a way that all main effects and two-factor interactions are balanced within each set. The plan in Table 8.5-1 provides such sets. The treatment combinations in block 1 satisfy the relation $x_1 + x_2 + x_3 = 0 \pmod 2$, and those in block 2 satisfy the relation $x_1 + x_2 + x_3 = 1 \pmod 2$. For r replications of this experiment the analysis takes the form given in Table 8.5-2. It is assumed that the blocks are random but that factors A, B, and C are fixed. It is also assumed that the treatment by block interaction is negligible relative to other effects in the experiment.

The tests on main effects and two-factor interactions use the within-block residual in the denominator of F ratios. The test on the three-factor interaction uses the between-block residual as the denominator. This latter test will generally be considerably less sensitive than tests which are based upon within-block information. The within-block residual in this design is the pooled interaction of the replications with the main effects and with the two-factor interactions.

The plan in Table 8.5-1 does not provide any within-block information on the three-factor interaction. By way of contrast the plan in Table 8.5-3

Table 8.5-1 2 × 2 × 2 Factorial Experiment in Blocks of Size 4

Block 1	Block 2
(000)	(001)
(011)	(010)
(101)	(100)
(110)	(111)

provides within-block information on the three-factor interaction as well as the two-factor interactions. A minimum of eight blocks is required for this latter design. In the first replication, AB is confounded with differences between blocks 1 and 2. However, within-block information on AB is available from all other replications. The treatment combinations in block 1 satisfy the relation $x_1 + x_2 = 0 \pmod 2$; the treatment combinations appearing in block 2 satisfy the relation $x_1 + x_2 = 1 \pmod 2$.

Table 8.5-2 Analysis of 2 × 2 × 2 Factorial Experiment in Blocks of Size 4

Source of variation	df	E(MS)
Between blocks	$2r - 1$	
Replications	$r - 1$	
Blocks within reps	r	
ABC	1	$\sigma_\varepsilon^2 + 4\sigma_{\text{blocks}}^2 + r\sigma_{\alpha\beta\gamma}^2$
Residual (between blocks)	$r - 1$	$\sigma_\varepsilon^2 + 4\sigma_{\text{blocks}}^2$
Within blocks	$6r$	
A	1	$\sigma_\varepsilon^2 + 4r\sigma_\alpha^2$
B	1	$\sigma_\varepsilon^2 + 4r\sigma_\beta^2$
C	1	$\sigma_\varepsilon^2 + 4r\sigma_\gamma^2$
AB	1	$\sigma_\varepsilon^2 + 2r\sigma_{\alpha\beta}^2$
AC	1	$\sigma_\varepsilon^2 + 2r\sigma_{\alpha\gamma}^2$
BC	1	$\sigma_\varepsilon^2 + 2r\sigma_{\beta\gamma}^2$
Residual (within block)	$6r - 6$	σ_ε^2

Since within-block information on AB is available from six of the eight blocks, the relative within-block information on the AB interaction is $\frac{6}{8} = \frac{3}{4}$.

To show that block 3, for example, provides within-block information on AB, two of the four treatment combinations in this block satisfy the relation $x_1 + x_2 = 0$ and hence belong to what will be called the set J_0,

Table 8.5-3 Balanced Design with Partial Within-block Information on All Interactions

Rep 1		Rep 2		Rep 3		Rep 4	
Block 1	Block 2	Block 3	Block 4	Block 5	Block 6	Block 7	Block 8
(000)	(010)	(000)	(001)	(000)	(001)	(000)	(001)
(001)	(011)	(010)	(011)	(011)	(010)	(011)	(010)
(110)	(100)	(101)	(100)	(100)	(101)	(101)	(100)
(111)	(101)	(111)	(110)	(111)	(110)	(110)	(111)
AB		AC		BC		ABC	

that is, (000) and (111). The other two treatment combinations in block 3 belong to the set J_1, that is, satisfy the relation $x_1 + x_2 = 1$. The difference between the totals for the J_0 set and the J_1 set provides information on the AB component, which is free of block effects. In addition to block 3, each of blocks 4 through 8 contains two treatment combinations belonging to set J_0 and two belonging to set J_1.

The AC interaction is completely confounded with the difference between blocks 3 and 4, but within-block information on AC is available from all other blocks. Hence the relative within-block information on AC is $\frac{6}{8}$ or $\frac{3}{4}$. In block 3, the treatment combinations satisfy the relation $x_1 + x_3 = 0 \pmod 2$; in block 4 the treatment combinations satisfy the relation $x_1 + x_3 = 1 \pmod 2$. Blocks 7 and 8 in Table 8.5-3 are identical to blocks 1 and 2 of the design in Table 8.5-1. In the latter design, no within-block information is available on the ABC interaction; the design in Table 8.5-3 provides $\frac{3}{4}$ relative within-block information on ABC, such information being available from blocks 1 through 6.

The analysis of the design in Table 8.5-3 takes the form given in Table 8.5-4. The symbol $(AB)'$ indicates that only partial within-block information is available for AB. In the computation of $(AB)'$, only information from blocks 3 through 8 is used. The AB comparison given by the difference between block 1 and block 2 is confounded by differences between these blocks; this comparison represents one of the four degrees of freedom for blocks within replications. In computing $(AC)'$, only information from blocks 1, 2, and 5 through 8 is used. Similarly, $(ABC)'$ is based upon information from blocks 1 through 6. The difference between the total for block 7 and the total for block 8 gives rise to the between-block component of ABC.

Table 8.5-4 Analysis of Balanced Design

Source of variation	df	E(MS)
Between blocks	7	
Replications	3	
Blocks within reps	4	
AB	1	
AC	1	
BC	1	
ABC	1	
Within blocks	24	
A	1	$\sigma_\varepsilon^2 + 16\sigma_\alpha^2$
B	1	$\sigma_\varepsilon^2 + 16\sigma_\beta^2$
C	1	$\sigma_\varepsilon^2 + 16\sigma_\gamma^2$
$(AB)'$	1	$\sigma_\varepsilon^2 + (\frac{3}{4})8\sigma_{\alpha\beta}^2$
$(AC)'$	1	$\sigma_\varepsilon^2 + (\frac{3}{4})8\sigma_{\alpha\gamma}^2$
$(BC)'$	1	$\sigma_\varepsilon^2 + (\frac{3}{4})8\sigma_{\beta\gamma}^2$
$(ABC)'$	1	$\sigma_\varepsilon^2 + (\frac{3}{4})4\sigma_{\alpha\beta\gamma}^2$
Residual	17	σ_ε^2

Each of the main effects is determined from data obtained from all four replications; the within-block information on the interactions is based upon data obtained in each case from three of the four replications. Hence there are three *effective* replications for the interactions, but four *effective* replications for the main effects. The degrees of freedom for the within-block residual are made up of the following parts:

Residual	17
$A \times$ reps	3
$B \times$ reps	3
$C \times$ reps	3
$(AB)' \times$ reps	2
$(AC)' \times$ reps	2
$(BC)' \times$ reps	2
$(ABC)' \times$ reps	2

Since each of the interactions has only three effective replications, their interaction with replications has only two degrees of freedom.

The expected values for the mean squares given in Table 8.5-4 assume that there is one observation per cell. Suppose that each block represents a group of n subjects and that the groups are assigned at random to one of the eight blocks. Also suppose each subject within the groups is observed under each of the four treatment combinations in the block to which the group is assigned. Further assume that terms of the form $\sigma^2_{\alpha\pi}$ are equal to zero. Under these assumptions the analysis takes the form given in Table 8.5-5. In this analysis the individual subject is considered

Table 8.5-5 Analysis of Balanced Design with Repeated Measures

Source of variation	df	E(MS)
Between subjects	$8n - 1$	
Groups	7	
Reps	3	
AB	1	$\sigma^2_\varepsilon + 4\sigma^2_{\text{people}} + 2n\sigma^2_{\alpha\beta}$
AC	1	$\sigma^2_\varepsilon + 4\sigma^2_{\text{people}} + 2n\sigma^2_{\alpha\gamma}$
BC	1	$\sigma^2_\varepsilon + 4\sigma^2_{\text{people}} + 2n\sigma^2_{\beta\gamma}$
ABC	1	$\sigma^2_\varepsilon + 4\sigma^2_{\text{people}} + n\sigma^2_{\alpha\beta\gamma}$
Subjects within groups	$8(n - 1)$	$\sigma^2_\varepsilon + 4\sigma^2_{\text{people}}$
Within subjects	$24n$	
A	1	$\sigma^2_\varepsilon + 16n\sigma^2_\alpha$
B	1	$\sigma^2_\varepsilon + 16n\sigma^2_\beta$
C	1	$\sigma^2_\varepsilon + 16n\sigma^2_\gamma$
$(AB)'$	1	$\sigma^2_\varepsilon + (\frac{3}{4})8n\sigma^2_{\alpha\beta}$
$(AC)'$	1	$\sigma^2_\varepsilon + (\frac{3}{4})8n\sigma^2_{\alpha\gamma}$
$(BC)'$	1	$\sigma^2_\varepsilon + (\frac{3}{4})8n\sigma^2_{\beta\gamma}$
$(ABC)'$	1	$\sigma^2_\varepsilon + (\frac{3}{4})4n\sigma^2_{\alpha\beta\gamma}$
Residual	$24n - 7$	σ^2_ε

to be the experimental unit. Tests on main effects and interactions use the residual mean square in the denominator of the F ratio. There is usually no interest in a test on differences between blocks or the components of such differences. In some cases between-block estimates of the interactions may be combined with the corresponding within-block estimates. Such pooled estimates are given by a weighted sum of the between-subject and within-subject components, the weights being the respective reciprocals of the between- and within-person residuals. F tests on such combined estimates require a weighted pooling of the respective residuals.

The cell totals for the design in Table 8.5-3 may be designated by the following notation:

	b_0		b_1	
	c_0	c_1	c_0	c_1
a_0	$X_{000.}$	$X_{001.}$	$X_{010.}$	$X_{011.}$
a_1	$X_{100.}$	$X_{101.}$	$X_{110.}$	$X_{111.}$

Each of these totals will be based upon four observations. Since all the treatment combinations do not appear in all the blocks, means based upon these totals will not be free of block effects. If, for example, the cell total for treatment combination (101) were free of block effects, it would estimate

$$X_{101.} = 4\mu + 4\alpha_1 + 4\beta_0 + 4\gamma_1 + 4(\alpha\beta)_{10} + 4(\alpha\gamma)_{11} + 4(\beta\gamma)_{01} + 4(\alpha\beta\gamma)_{101} + (\text{sum of all block effects}).$$

In this last expression, the sum of all block effects would be a constant (under the assumptions made); hence differences between any two X's would be free of block effects. It is readily verified that the following quantity will estimate the sum of the parameters on the right-hand side of the last expression:

$$X'_{101.} = \frac{A_1 + B_0 + C_1}{4} + \frac{(AB)'_1 + (AC)'_0 + (BC)'_1 + (ABC)'_0}{3} - \frac{3G}{4},$$

where $(AB)'_1$ = total of all treatment combinations which satisfy the relation $x_1 + x_2 = 1 \pmod 2$ and appear in blocks 3 through 8; similarly $(AC)'_0$ = total of all treatment combinations which satisfy the relation $x_1 + x_3 = 0 \pmod 2$ and appear in blocks 1, 2, 4 through 8. The quantity $X'_{101.}$ is called an adjusted cell total. The adjusted cell mean for treatment combination (101) is

$$\bar{X}'_{101} = \frac{X'_{101.}}{4}.$$

For the general case, an adjusted cell total for the design in Table 8.5-3 has the form

$$X'_{ijk.} = \frac{A_i + B_j + C_k}{4} + \frac{(AB)'_{i+j} + (AC)'_{i+k} + (BC)'_{j+k}}{3} + \frac{(ABC)'_{i+j+k}}{3} - \frac{3G}{4},$$

where $(AB)'_{i+j}$ = sum of all treatment combinations which satisfy the relation $x_1 + x_2 = (i + j) \pmod 2$ and which appear in blocks providing within-block information on AB. Similarly $(ABC)'_{i+j+k}$ = sum of all treatment combinations which satisfy the relation $x_1 + x_2 + x_3 = (i + j + k) \pmod 2$ and which appear in blocks providing within-block information on ABC.

Adjusted totals having the form $X'_{ij..}$ are obtained from the relation

$$X'_{ij..} = X'_{ij0.} + X'_{ij1.}.$$

Similarly

$$X'_{i.k.} = X'_{i0k.} + X'_{i1k.}.$$

Individual comparisons among the treatment combinations use the adjusted cell totals. The effective number of observations in the adjusted cell total $X'_{ijk.}$ is actually somewhat less than four since there is only $\frac{3}{4}$ relative within-block information on the interaction effects. For most practical purposes a comparison between two adjusted cell totals is given by

$$F = \frac{(X'_{ijk.} - X'_{pqr.})^2}{8\text{MS}_{\text{res}}}.$$

$$F = \frac{(X'_{ij..} - X'_{pq..})^2}{16\text{MS}_{\text{res}}}.$$

The squared standard error of a difference between two adjusted cell means which takes into account the effective number of replications is readily obtained. For the design given in Table 8.5-3, consider the difference

$$\overline{ABC}'_{011} - \overline{ABC}'_{111} = [\bar{A}_0 - \bar{A}_1] + [(\overline{AB})'_1 - (\overline{AB})'_0] + [(\overline{AC})'_1 - (\overline{AC})'_0] + [(\overline{ABC})'_0 - (\overline{ABC})'_1].$$

Since there is complete information on each $\bar{A}_i$, there are 16 basic observations contributing to this mean. Since there is only $\frac{3}{4}$ information on each of the interactions, there are only 12 basic observations contributing to each of the other means on the right. Hence the squared standard errors of the terms in brackets are, respectively,

$$\frac{2\text{MS}_{\text{res}}}{16}, \quad \frac{2\text{MS}_{\text{res}}}{12}, \quad \frac{2\text{MS}_{\text{res}}}{12}, \quad \frac{2\text{MS}_{\text{res}}}{12}.$$

Since the terms in brackets on the right are distributed independently, the squared standard error of the difference between the adjusted cell means is given by

$$\hat{\sigma}^2_{(\overline{ABC}'_{011} - \overline{ABC}'_{111})} = \frac{2\text{MS}_{\text{res}}}{16} + \frac{3(2\text{MS}_{\text{res}})}{12} = \frac{5\text{MS}_{\text{res}}}{8}.$$

If there were no confounding,

$$\hat{\sigma}^2_{(\overline{ABC}'_{011} - \overline{ABC}'_{111})} = \frac{8\text{MS}_{\text{res}}}{16} = \frac{\text{MS}_{\text{res}}}{2}.$$

The *relative* effective number of replications on the comparison in question is given by

$$\frac{\frac{1}{2}\text{MS}_{\text{res}}}{\frac{5}{8}\text{MS}_{\text{res}}} = \frac{4}{5}.$$

A numerical example of the design given in Table 8.5-1 appears in Sec. 8.7. Computational procedures are simplified by means of special procedures for 2^k factorial experiments given in the next section.

8.6 Simplified Computational Procedures for 2^k Factorial Experiments

Computational procedures for 2×2, $2 \times 2 \times 2$, $2 \times 2 \times 2 \times 2$, etc., factorial experiments may be simplified by use of the device employed in forming balanced sets. This simplified procedure was presented in a different form in Sec. 6.12. The simplified computational procedures will be illustrated by the numerical example given in Table 8.6-1. Cell totals in part i are the sum of five observations. The sums of squares obtained by the procedures to be outlined are the *unadjusted* sums of squares if the experiment involves any confounding.

The patterns of the algebraic signs in part ii are determined as follows (the x's refer to digits in the treatment combinations at the top of each

Table 8.6-1 $2 \times 2 \times 2$ **Factorial Experiment with Five Observations per Cell**

(i)

	b_0		b_1		
	c_0	c_1	c_0	c_1	Total
a_0	5	10	15	15	45
a_1	10	20	20	5	55
	15	30	35	20	100

(ii)

	Treat comb:	(000)	(001)	(010)	(011)	(100)	(101)	(110)	(111)	Comparison
	Cell total:	5	10	15	15	10	20	20	5	100
G		+	+	+	+	+	+	+	+	100
A		−	−	−	−	+	+	+	+	10
B		−	−	+	+	−	−	+	+	10
C		−	+	−	+	−	+	−	+	0
AB		+	+	−	−	−	−	+	+	−20
AC		+	−	+	−	−	+	−	+	−10
BC		+	−	−	+	+	−	−	+	−30
ABC		−	+	+	−	+	−	−	+	−20

(iii)

$$SS_A = \frac{(10)^2}{5(8)} = 2.50 \qquad SS_{AC} = \frac{(-10)^2}{5(8)} = 2.50$$

$$SS_B = \frac{(10)^2}{5(8)} = 2.50 \qquad SS_{BC} = \frac{(-30)^2}{5(8)} = 22.50$$

$$SS_C = 0 \qquad SS_{ABC} = \frac{(-20)^2}{5(8)} = 10.00$$

$$SS_{AB} = \frac{(-20)^2}{5(8)} = 10.00$$

column; all addition is modulo 2):

G:	All positive			
A:	$-$ if x_1	$= 0$,	$+$ if x_1	$= 1$
B:	$-$ if x_2	$= 0$,	$+$ if x_2	$= 1$
C:	$-$ if x_3	$= 0$,	$+$ if x_3	$= 1$
AB:	$+$ if $x_1 + x_2$	$= 0$,	$-$ if $x_1 + x_2$	$= 1$
AC:	$+$ if $x_1 + x_3$	$= 0$,	$-$ if $x_1 + x_3$	$= 1$
BC:	$+$ if $x_2 + x_3$	$= 0$,	$-$ if $x_2 + x_3$	$= 1$
ABC:	$-$ if $x_1 + x_2 + x_3$	$= 0$,	$+$ if $x_1 + x_2 + x_3$	$= 1$

In general, if the number of factors which are interacting is even, then the zero modular sum receives the positive sign; if the number of factors interacting is odd, then the zero modular sum receives a negative sign. In this context main effects are classified with the interaction of an odd number of factors.

The entry at the right of each row in part ii is a weighted sum of the cell totals at the top of each column, the weights being ± 1 as determined by the pattern of signs in each row. For example, the entry at the right of row A is given by

$$-5 - 10 - 15 - 15 + 10 + 20 + 20 + 5 = 10.$$

As another example, the entry at the right of row ABC is given by

$$-5 + 10 + 15 - 15 + 10 - 20 - 20 + 5 = -20.$$

The pattern of the signs in the last expression is that in row ABC. The weighted sum of the cell totals for row ABC actually corresponds to the difference $(ABC)_1 - (ABC)_0$, where $(ABC)_1$ represents the sum of all observations on treatment combinations which satisfy the relation

$$x_1 + x_2 + x_3 = 1 \pmod{2}.$$

Alternatively, the totals at the right of the rows represent the comparisons corresponding to the effects at the left of the rows. The general form of a comparison (assuming the same number of observations in each cell total) is

$$\text{Comparison} = c_1T_1 + c_2T_2 + \cdots + c_kT_k, \qquad \Sigma c_j = 0,$$

where the T's represent cell totals. The sum of squares corresponding to a comparison is given by

$$\text{SS}_{\text{comparison}} = \frac{(\text{comparison})^2}{n\Sigma c^2},$$

where n is the number of observations summed to obtain the cell totals. The c's in part ii are either $+1$ or -1, depending upon the sign; for each of the rows $\Sigma c^2 = 8$. The sums of squares for comparisons corresponding to main effects and interactions are computed in part iii. In this example, n is assumed to be equal to 5.

The extension of these computational procedures to any factorial experiment of the form 2^k is direct. Once the pattern of signs corresponding to a row in part iii is determined, the weighted sum of the cell totals corresponding to a row gives the number value of the comparison. For any comparison determined in this manner, $\Sigma c^2 = 2^k$.

In 2^k factorial experiments a specialized notation is frequently used to designate the treatment combinations. This specialized notation for a $2 \times 2 \times 2$ factorial experiment is as follows:

	b_0		b_1	
	c_0	c_1	c_0	c_1
a_0	(1)	c	b	bc
a_1	a	ac	ab	abc

The relationship between the notation system that has been used so far in this chapter and the specialized notation system is as follows:

Treatment combination	(000)	(010)	(001)	(011)	(100)	(110)	(101)	(111)
Specialized notation	(1)	b	c	bc	a	ab	ac	abc

In the specialized notation system, the symbol for a treatment combination contains only those letters for which the factor is at level 1. Conversely, if a factor is at level 0 in a treatment combination, then the letter corresponding to that factor does not appear in the symbol.

In terms of the latter notation system sign patterns for the comparisons may be determined by a relatively simple rule. In the sign pattern for an interaction of the form XY, a symbol receives a positive sign if it contains both of the letters X and Y or neither of these letters; a symbol receives a negative sign otherwise. In the sign pattern for an interaction of the form XYZ, a symbol receives a positive sign if it contains all three of the letters X, Y, and Z or just one of the three letters; a symbol receives a negative sign otherwise. For the general case, in the interaction of an *even* number of factors, a symbol receives a positive sign if it contains an even number of the interacting factors (zero being considered an even number); otherwise it receives a negative sign. In the interaction of an *odd* number of factors, a symbol receives a positive sign if it contains an odd number of the interacting factors; otherwise the symbol receives a minus sign. A main effect may be considered included under the latter case. The following example illustrates the general case:

	Symbols:	(1)	*b*	*c*	*bc*	*a*	*ab*	*ac*	*abc*
A		−	−	−	−	+	+	+	+
C		−	−	+	+	−	−	+	+
AC		+	+	−	−	−	−	+	+

For the case of 2^4 factorial experiment, the sign patterns for some of the comparisons are illustrated below:

	(1)	*a*	*b*	*ab*	*c*	*ac*	*bc*	*abc*	*d*	*ad*	*bd*	*abd*	*cd*	*acd*	*bcd*	*abcd*
A	−	+	−	+	−	+	−	+	−	+	−	+	−	+	−	+
AB	+	−	−	+	+	−	−	+	+	−	−	+	+	−	−	+
ABC	−	+	+	−	+	−	−	+	−	+	+	−	+	−	−	+
ABCD	+	−	−	+	−	+	+	−	−	+	+	−	+	−	−	+

Given the pattern of signs corresponding to a comparison, the sum of squares for that comparison is readily obtained from the treatment totals.

A systematic computing scheme for obtaining the comparisons corresponding to main effects and interactions is illustrated in Table 8.6-2.

Table 8.6-2 Simplified Computational Procedures

	Effect:					
	Treatment combination	Total	(1)	(2)	(3)	Comparison
(i)	(1)	5	15	50	100	*G*
	a	10	35	50	10	*A*
	b	15	30	10	10	*B*
	ab	20	20	0	−20	*AB*
	c	10	5	20	0	*C*
	ac	20	5	−10	−10	*AC*
	bc	15	10	0	−30	*BC*
	abc	5	−10	−20	−20	*ABC*
(ii)	Upper half	...	100	110	100	
	Lower half	...	10	−10	−60	
	Odds	45	60	80		
	Evens	55	50	20		

This scheme is quite useful when the number of factors is large. The basic data in Table 8.6-2 are obtained from part i of Table 8.6-1. The treatment combinations must be arranged in the order given in this table. (If a fourth factor were present, the order for the treatment combinations having this factor at level 1 would be *d*, *ad*, *bd*, *abd*, *cd*, *acd*, *bcd*, *abcd*.)

The entries in the total column are the cell totals obtained from part i of Table 8.6-1. The entries in the upper half of column 1 are the sums of successive pairs of the entries in the total column, i.e., $5 + 10 = 15$, $15 + 20 = 35$, $10 + 20 = 30$, and $15 + 5 = 20$. The entries in the lower half of column 1 are differences between successive pairs in the total column, i.e., $10 - 5 = 5$, $20 - 15 = 5$, $20 - 10 = 10$, and $5 - 15 = -10$.

The entries in the upper half of column 2 are the sums of successive pairs of entries in column 1, that is, $15 + 35 = 50$, $30 + 20 = 50$. The entries in the lower half of column 2 are the differences between successive pairs of entries in column 1, that is, $35 - 15 = 20$, $20 - 30 = -10$.

The entries in column 3 are obtained from the entries in column 2 by a procedure analogous to that by which the entries in column 2 were obtained from column 1. This procedure is continued until column k is reached, where k is the number of factors. In this example $k = 3$; hence the procedure is terminated at column 3. The entries in column 3 give the numerical values of the comparisons corresponding to the treatment effects to the left of each row. These values are identical to those obtained in part ii of Table 8.6-1.

Checks on the numerical work are given in part ii of Table 8.6-2. The respective sums for the upper and lower halves for each of the columns are obtained. One then obtains the sum of every other entry in each column, starting with the first entry, i.e., the sum of odd-numbered entries. One also obtains the sum of the even-numbered entries in each of the columns. These sums are shown in part ii. The sum of the upper half of column 1 is checked by the sum of the odds and evens under the total column; the sum of the lower half of column 1 is checked by the difference between the evens and odds in the total column. The sum of the upper half of column 2 is checked by the sum of the odds and the evens under column 1; the sum of the lower half of column 2 is checked by the difference between evens and odds for column 1. Analogous checks are made for column 3.

8.7 Numerical Example of $2 \times 2 \times 2$ Factorial Experiment in Blocks of Size 4

The purpose of this experiment was to evaluate the effects of various treatment combinations upon the progress of mental patients in specified diagnostic categories. The treatments and their levels are given in Table 8.7-1. It was desired to have 10 patients in each cell of a factorial experiment, necessitating a total of 80 patients for the experiment. However, no more than 20 patients meeting specifications were available from a single hospital, but four hospitals were available for the study. It was anticipated that there would be large differences among hospitals. In order to prevent such differences from confounding main effects and two-factor interactions, the design in Table 8.7-2 was used.

In the construction of this design, the relation $x_1 + x_2 + x_3 = i$ (mod 2) was used to divide the eight treatment combinations into blocks of

Table 8.7-1 Definition of Factors

Factor	Level	Definition
Drug A	a_0 a_1	No drug A Drug A administered
Drug B	b_0 b_1	No drug B Drug B administered
Psychotherapy	c_0 c_1	No psychotherapy Psychotherapy administered

size 4. By using the specialized notation, this procedure is equivalent to assigning symbols containing an even number of letters to one set of blocks and symbols containing an odd number of letters to a second set of blocks. (In terms of the design in Table 8.7-2, symbols having an even number of letters appear in blocks 2 and 4.) Hospitals were assigned at random to blocks. Hospitals 1 and 2 make up one replication; hospitals 3 and 4 make up the second replication. Within each replication, the ABC interaction is completely confounded with between-hospital differences. From each hospital, 20 patients meeting the specifications were selected and assigned at random to subgroups of 5 patients each. The subgroups were then assigned at random to one of the treatment conditions allocated to the hospital to which the subgroup belonged.

Each patient was rated by a panel of judges before and after the treatment combinations were administered. The difference between these two ratings was taken as the criterion of progress. Since some of the treatments were administered to the subgroups as a unit, the subgroup of five

Table 8.7-2 Design and Data for Numerical Example

Hospital 1		Hospital 2		Hospital 3		Hospital 4	
a	6	(1)	2	a	14	(1)	3
b	10	ab	4	b	15	ab	6
c	6	ac	15	c	9	ac	25
abc	8	bc	18	abc	12	bc	22
	30		39		50		56

$\Sigma\Sigma X^2 = 2605$

patients was considered to be the experimental unit, rather than the individual patient. The mean criterion score for each subgroup is given in Table 8.7-2 to the right of the corresponding symbol for the treatment combination. Each of these means is considered to be a single observation for purposes of the analysis of variance.

Summary tables and details of the computations are given in Table 8.7-3. Cell totals for the treatment combinations appear in part i; these totals are obtained by combining the data from the two replications. Each of these cell totals is considered to be the sum of two observations. The numerical values for individual comparisons corresponding to the main effects and the interactions are computed in part ii. Here the entries in the upper half of column 1 are obtained from the total column by summing successive pairs of values in the total column; the entries in the lower half of column 1 are obtained by taking the difference between successive pairs of values in the total column. Column 2 is obtained from column 1 by

Table 8.7-3 Summary of Numerical Analysis

(i)

	b_0		b_1		Total
	c_0	c_1	c_0	c_1	
a_0	5	15	25	40	85
a_1	20	40	10	20	90
	25	55	35	60	175

(ii)

Treatment combinations	Total	(1)	(2)	(3)	
(1)	5	25	60	175	G
a	20	35	115	5	$A = A_1 - A_0$
b	25	55	0	15	$B = B_1 - B_0$
ab	10	60	5	−75	AB
c	15	15	10	55	$C = C_1 - C_0$
ac	40	−15	5	5	AC
bc	40	25	−30	−5	BC
abc	20	−20	−45	−15	ABC

(iii)

$SS_A = (5)^2/16 = 1.57$ $\qquad SS_{AC} = (5)^2/16 = 1.57$

$SS_B = (15)^2/16 = 14.06$ $\qquad SS_{BC} = (-5)^2/16 = 1.57$

$SS_C = (55)^2/16 = 189.06$ $\qquad SS_{ABC} = (-15)^2/16 = 14.06$

$SS_{AB} = (-75)^2/16 = 351.57$

$SS_{\text{hospitals}} = (30^2 + 39^2 + 50^2 + 56^2)/4 - (175)^2/16 = 100.19$

$SS_{\text{total}} = 2605 - (175)^2/16 = 690.94$

an analogous procedure, and similarly column 3 is obtained from column 2. The entries in column 3 are the values of the comparisons. The corresponding sums of squares are given in part iii. The sum of squares for the main effect of factor A has the general definition

$$SS_A = \frac{C_A^2}{2^k r},$$

where C_A = comparison corresponding to main effect of factor A,
k = number of factors,
r = number of replications.

In this case C_A is the entry in column 3 in row a, which is 5. For this case $r = 2$, and $k = 3$; hence

$$SS_A = \frac{(5)^2}{2^k(2)} = \frac{25}{16} = 1.57.$$

The analysis of variance for these data is given in Table 8.7-4. The

Table 8.7-4 Analysis of Variance

Source	SS	df	MS	F
Between hospitals	100.19	3		
Replications	85.56	1		
Residual (b)	14.63	2		
Within hospitals	590.75	12		
A Drug A	1.57	1	1.57	—
B Drug B	14.06	1	14.06	2.69
C Psychotherapy	189.06	1	189.06	36.22**
AB	351.57	1	351.57	67.35**
AC	1.57	1	1.57	—
BC	1.57	1	1.57	—
Residual (w)	31.35	6	5.22	

$^{**}F_{.99}(1,6) = 13.74$

within-hospital sum of squares is obtained by subtracting the between-hospital sum of squares from the total sum of squares. The within-hospital residual is obtained from the relation

$$SS_{res} = SS_{w.\ hospital} - \text{(sum of main effects and two-factor interactions)}$$
$$= 590.75 - 559.40 = 31.35.$$

Although the ABC interaction is completely confounded with between-hospital differences within a single replication, by including information from the two replications the ABC component may be estimated.

The within-hospital residual mean square is used as the denominator for all tests on within-hospital effects. The F tests show no significant interactions involving factor C (psychotherapy), but a significant main effect. It may be concluded from this information that the effect of psychotherapy is independent of the effect of the drugs; further, the groups given psychotherapy showed significantly greater improvement than did groups which were not given psychotherapy.

The interaction between the drugs is seen to be highly significant. The following summary data obtained from part i of Table 8.7-3 are useful in the interpretation of this interaction:

	b_0	b_1
a_0	20	65
a_1	60	30

These data indicate that the use of both drugs simultaneously is not better than the use of either drug alone. The test on the comparison between the levels of drug B in the absence of drug A is given by

$$F = \frac{(AB_{00} - AB_{01})^2}{2nr\,\text{MS}_{\text{res}(w)}} = \frac{(20 - 65)^2}{2(4)(5.22)} = 48.49.$$

If this comparison is considered to belong in the a priori category, the critical value for a .05-level test is $F_{.95}(1,6) = 5.99$. If this comparison is considered to belong in the a posteriori category, the critical value for a .05-level test is, as given by Scheffé, $3F_{.95}(3,6) = 3(4.76) = 14.28$. (In the present context this test would be considered as being in the a priori category.) In either instance, it may be concluded that drug B has a significant effect upon progress in the absence of drug A.

A test on the effect of drug A in the absence of drug B is given by

$$F = \frac{(20 - 60)^2}{2(4)(5.22)} = 38.31.$$

Clearly drug A has a significant effect upon progress in the absence of drug B. To compare the relative effectiveness of the two drugs when each is used in the absence of the other,

$$F = \frac{(60 - 65)^2}{2(4)(5.22)} = 0.60.$$

Thus the data indicate that drug A and drug B are equally effective.

The denominator of the F tests made in the last paragraph has the general form $nr(\Sigma c^2)(\text{MS}_{\text{error}})$; nr is the number of experimental units summed to obtain the total in the comparison, Σc^2 is the sum of the squares of the coefficients in the comparison, and MS_{error} is the appropriate error term for the comparison.

In Table 8.7-4 the within-hospital residual sum of squares was obtained by subtraction. This term is actually the pooled interaction of treatment effects (with the exception of ABC) with the hospitals receiving the same set of treatment combinations. Direct computation of the residual sum of squares is illustrated in Table 8.7-5. Since hospitals 1 and 3 had the same set of treatment conditions, the data at the left in part i provide three of the six degrees of freedom of the residual term. If the four treatment combinations common to hospitals 1 and 3 are considered as four levels of a single factor, then the interaction of the levels of this factor with the hospital factor (defined by hospitals 1 and 3) is part of the sum of squares for residuals. This interaction is computed at the left in part ii. An analogous interaction term is obtained from the data at the right in part i.

Table 8.7-5 Computation of the Components of the Residual (w)

	Treat	Hosp. 1	Hosp. 3	Total	Treat	Hosp. 2	Hosp. 4	Total
(i)	a	6	14	20	(1)	2	3	5
	b	10	15	25	ab	4	6	10
	c	6	9	15	ac	15	25	40
	abc	8	12	20	bc	18	22	40
		30	50	80		39	56	95
(ii)	Treat × hosp. interaction $= (6^2 + 10^2 + \cdots + 9^2 + 12^2) - (20^2 + \cdots + 20^2)/2 - (30^2 + 50^2)/4 + 80^2/8 = 7.00$				Treat × hosp. interaction $= (2^2 + 4^2 + \cdots + 25^2 + 22^2) - (5^2 + \cdots + 40^2)/2 - (39^2 + 56^2)/4 + 95^2/8 = 24.37$			
(iii)	$SS_{res} = 7.00 + 24.37 = 31.37$							

8.8 Numerical Example of $2 \times 2 \times 2$ Factorial Experiment in Blocks of Size 4 (Repeated Measures)

The purpose of this experiment was to evaluate the preferences for advertisements made up by varying the size (factor A), the style of type (factor B), and the color (factor C). The definitions of the levels of these three factors are given in Table 8.8-1.

Table 8.8-1 Definition of Factors

Factor	Level	Definition
Size (A)	a_0	Small
	a_1	Large
Style (B)	b_0	Gothic
	b_1	Roman
Color (C)	c_0	Green
	c_1	Blue

The experimenter desired to have within-subject estimates on all main effects and interactions; however, the task of having each subject judge all eight combinations was not considered experimentally feasible. Loss of interest on the part of the subjects and excessive time demands (as indicated by a pilot study) ruled against the procedure of having each subject judge all combinations. The experimenter was willing to sacrifice precision with respect to the three-factor interaction in order to keep the number of judgments an individual had to make down to four. The plan outlined in Table 8.8-2 was chosen for use.

In order to keep this illustrative example simple, it will be assumed that a sample of six subjects was used in the experiment. (In practice, this sample size would be too small.) The subjects were divided at random into two groups. Individuals within each group judged only four of the eight different make-ups. The combinations of factors judged by group I satisfied the relation $x_1 + x_2 + x_3 = 1 \pmod 2$, and the combinations judged by group II satisfied the relation $x_1 + x_2 + x_3 = 0 \pmod 2$. In terms of the specialized notation system for 2^k factorial experiments, group I judged treatment combinations represented by symbols having an odd number of letters; group II judged the remaining treatment combinations. In this design the ABC comparison is completely confounded with differences between groups; within-subject estimates are available on all other factorial effects.

Table 8.8-2 Outline of Plan and Basic Data

(i)

Group I						Group II					
Person	*a*	*b*	*c*	*abc*	Total	Person	(1)	*ab*	*ac*	*bc*	Total
1	16	8	2	8	34	4	10	12	8	3	33
2	10	4	3	7	24	5	11	16	10	5	42
3	9	3	0	5	17	6	4	7	7	2	20
Total	35	15	5	20	75	Total	25	35	25	10	95

(ii)

	Treat comb.:	*a*	*b*	*c*	*abc*	(1)	*ab*	*ac*	*bc*	Comparison
	Cell totals:	35	15	5	20	25	35	25	10	
G		+	+	+	+	+	+	+	+	170
A		+	−	−	+	−	+	+	−	60
B		−	+	−	+	−	+	−	+	−10
C		−	−	+	+	−	−	+	+	−50
AB		−	−	+	+	+	+	−	−	0
AC		−	+	−	+	+	−	+	−	0
BC		+	−	−	+	+	−	−	+	10
ABC		+	+	+	+	−	−	−	−	−20

(iii)

$SS_A = (60)^2/24 = 150.00$ $\quad SS_{AB} = 0 \quad SS_{BC} = (10)^2/24 = 4.17$

$SS_B = (-10)^2/24 = 4.17$ $\quad SS_{AC} = 0$

$SS_C = (-50)^2/24 = 104.17$ $\quad SS_{ABC} = (-20)^2/24 = 16.67$

$SS_{subj} = (34^2 + 24^2 + 17^2 + 33^2 + 42^2 + 20^2)/4 - (170)^2/24 = 114.33$

$SS_{total} = (16^2 + 8^2 + \cdots + 7^2 + 2^2) - (170)^2/24 = 409.83$

The order in which a subject judged a particular combination was randomized independently for each subject. Order could, however, have been controlled by means of a Latin square. (Plan 9.7-7 utilizes the Latin-square principle with what is essentially the design given in Table 8.8-2. The number of subjects per group would have to be a multiple of 4 in order to use the Latin square.)

The comparisons corresponding to the main effects and interactions of the factors are obtained at the right in part ii of Table 8.8-2. The method by which the *ABC* comparison is obtained shows clearly that this comparison is completely confounded with differences between the two groups. The sums of squares corresponding to the factorial effects, between-subject variation, and total variation are computed in part iii. The overall analysis of variance is summarized in Table 8.8-3.

Table 8.8-3 Summary of Analysis

Source of variation	SS	df	MS	F
Between subjects	114.33	5		
ABC (groups)	16.67	1	16.67	
Subjects within groups	97.67	4	24.42	
Within subjects	295.50	18		
A Size	150.00	1	150.00	54.54**
B Type style	4.17	1	4.17	1.51
C Color	104.17	1	104.17	37.88**
AB	0	1	0	
AC	0	1	0	
BC	4.17	1	4.17	1.51
Residual	32.99	12	2.75	

** $F_{.99}(1,12) = 9.33$

The sum of squares due to the within-subject variation is obtained from the relation

$$SS_{w.\ subj} = SS_{total} - SS_{b.\ subj}$$
$$= 409.83 - 114.33 = 295.50.$$

The residual sum of squares is given by

$$SS_{res} = SS_{w.\ subj} - (\text{sum of main effects and 2-factor interactions})$$
$$= 295.50 - 262.51 = 32.99.$$

Assuming that factors A, B, and C are fixed, the residual term is the proper denominator for all within-subject effects. The proper denominator for between-subject effects is the mean square for subjects within groups. In this design the ABC interaction is a between-subject effect—but this interaction is completely confounded with differences among groups.

Generally there would be little interest in testing the ABC interaction. (In this case the F ratio would be less than 1.)

For the within-subject data in Table 8.8-3, none of the interactions is statistically significant. This result implies that the main effects (if any) operate independently; i.e., the main effects are additive. The tests made in Table 8.8-3 indicate that the size and color main effects are statistically significant. Inspection of the summary data in Table 8.8-2 indicates that large size (a_1) is preferred no matter which of the styles or colors is used. Similarly, green (c_0) is preferred to blue no matter which size or style of type is used. There is no statistically significant difference between the two styles of type used in the experiment; there is, however, a slight preference for the Gothic (b_0).

If one were to make an overall recommendation with respect to the most preferred make-up of the advertising copy, the large size in the green color would be the best combination. The two type styles are, statistically, equally good, but there is a slight observed difference in favor of the Gothic.

Returning to structural considerations underlying the analysis of this design, the residual term in the analysis consists of the following pooled interactions:

Subject × treatments (within group I)	6
Subject × treatments (within group II)	6
Residual	12

The residual term in Table 8.8-3 may be computed directly from the interaction terms given above. The four treatment combinations assigned to the subjects within group I are considered to be four levels of a single

Table 8.8-4 Analysis of General Case of Design in Table 8.8-2

Source of variation		df
Between subjects		$2nr - 1$
Replications		$r - 1$
Groups within replications		r
ABC	1	
Residual (groups)	$r - 1$	
Subjects within groups		$2r(n - 1)$
Within subjects		$6nr$
A		1
B		1
C		1
AB		1
AC		1
BC		1
Residual (within subject)		$6(nr - 1)$

factor; the subject by treatment interaction is computed by means of the usual computational formulas for a two-factor interaction.

The analysis of the general case of a $2 \times 2 \times 2$ factorial experiment in blocks of size 4 with repeated measures is outlined in Table 8.8-4. In this analysis there are n subjects in each group, and the experiment is replicated r times. The analysis of the design in Table 8.8-3 is a special case of the more general design in Table 8.8-4. In the special case $n = 3$, and $r = 1$. In a design having more than one replication, the ABC interaction may be tested by using the subjects within groups in the denominator of an F ratio. This test will generally be considerably less sensitive than tests on the main effects and two-factor interactions.

8.9 Designs for 3 × 3 Factorial Experiments

The nine treatment combinations in this experiment may be partitioned into three sets in such a way that differences between sets form one of the components of the $A \times B$ interaction. The design outlined in Table 8.9-1 gives two such partitions. In replication 1, the AB (or J) component is used to define the sets (blocks); in replication 2, the AB^2 (or I) component is used to define the sets (blocks).

Table 8.9-1 3 × 3 Factorial Experiment in Blocks of Size 3

	Replication 1			Replication 2		
	Block 1	Block 2	Block 3	Block 4	Block 5	Block 6
	(00) (12) (21)	(01) (10) (22)	(02) (20) (11)	(00) (11) (22)	(02) (10) (21)	(01) (12) (20)
Component confounded		AB (or J)			AB^2 (or I)	

The treatment combinations in block 1 satisfy the relation $x_1 + x_2 = 0$ (mod 3); the treatment combinations in blocks 2 and 3 satisfy the respective relations $x_1 + x_2 = 1$ (mod 3) and $x_1 + x_2 = 2$ (mod 3). Hence differences between blocks within replication 1 are completely confounded with the AB (or J) component of $A \times B$. That is, a difference between two block totals in replication 1 is simultaneously an estimate of block differences as well as differences which form part of the AB component. (A formal proof of this is given later in this section.)

In replication 2, each of the blocks contains treatment combinations which satisfy the relation $x_1 + 2x_2 = i$ (mod 3). For example, the treatment combinations in block 6 all satisfy the relation $x_1 + 2x_2 = 2$ (mod 3). Hence differences between blocks within replication 2 are completely confounded with the AB^2 (or I) component of $A \times B$.

Assuming one observation per cell, the analysis takes the form given in Table 8.9-2. Within-block information on the main effects is available

Table 8.9-2 Analysis of Design in Table 8.9-1

Source		df	E(MS)
Between blocks		5	
Replications		1	
AB (from rep 1)		2	
AB^2 (from rep 2)		2	
Within blocks		12	
A		2	$\sigma_\varepsilon^2 + 6\sigma_\alpha^2$
B		2	$\sigma_\varepsilon^2 + 6\sigma_\beta^2$
$A \times B$		4	
AB (from rep 2)	2		$\sigma_\varepsilon^2 + (\frac{1}{2})(2)\sigma_{\alpha\beta_J}^2$
AB^2 (from rep 1)	2		$\sigma_\varepsilon^2 + (\frac{1}{2})(2)\sigma_{\alpha\beta_I}^2$
Residual		4	σ_ε^2
$A \times$ rep	2		
$B \times$ rep	2		

from both replications. Within-block information on the AB (or J) component is available only from replication 2; within-block information on the AB^2 (or I) component is available only from replication 1. Since within-block information on the components of $A \times B$ is available from only one of the two replications, the relative within-block information on $A \times B$ is said to be $\frac{1}{2}$. For main effects, however, the relative within-block information is $\frac{2}{2} = 1$, since both replications provide within-block information on the main effects. The design in Table 8.9-1 is balanced in the sense that the same amount of relative within-block information is provided on each of the components of $A \times B$.

The denominator of F ratios for within-block effects is the pooled interaction of the main effects with the replications. The two components of the $A \times B$ interaction are generally combined, and the combined interaction (with four degrees of freedom) is tested as a unit. If, however, the individual components are meaningful in terms of the experimental variables, separate tests on the components may be made. In most cases the components of $A \times B$ that are confounded with block effects are not tested. There are, however, techniques for combining the between-block information on the interaction with the within-block information to obtain an overall estimate of the interaction. Such techniques are said to recover the between-block information on the interaction effects. These techniques will not be considered in this section.

The block in this design may be an individual subject—in this case each subject is observed under all treatment combinations assigned to a given block. The block may be a group of three subjects—in this case each subject is observed under the three treatment combinations assigned to the block. The block may be a group of n subjects—in this case the n subjects may be observed under all treatment combinations within a given block. The analysis of this latter design takes the form given in Table 8.9-3.

Table 8.9-3 Analysis of 3 × 3 Factorial Experiment in Blocks of Size 3 (Repeated Measurements)

Source		df
Between subjects		$6n - 1$
Groups		5
Replications	1	
AB (from rep 1)	2	
AB^2 (from rep 2)	2	
Subjects within groups		$6(n - 1)$
Within subjects		$12n$
A		2
B		2
AB (from rep 2)		2
AB^2 (from rep 1)		2
Residual		$12n - 8$

The expected values of the mean squares have the same general form as those given in Table 8.9-2. The residual again provides an estimate of the within-block experimental error. The $12n - 8$ degrees of freedom for the residual is the pooled interaction of the main effects with the replications and the interaction of the main effects with the subjects within groups. The breakdown of these degrees of freedom is as follows:

Residual	$12n - 8$
$A \times$ rep	2
$B \times$ rep	2
Treat × subj w. group	$12n - 12$

A formal algebraic proof will now be outlined to demonstrate that the AB component in the design in Table 8.9-1 is confounded with block effects within replication 1 but is free of such confounding within replication 2. This proof assumes that an observation under treatment combination ab_{ij} in block k provides an estimate (disregarding the experimental error) of the sum of the parameters on the right-hand side of the following expression,

$$X_{ijk} \doteq \mu + \alpha_i + \beta_j + (\alpha\beta)_{ij} + \pi_k,$$

where the symbol π_k designates the effect of block k. Assuming that factors A and B are fixed, $\Sigma\alpha_i = 0$, $\Sigma\beta_j = 0$, $\sum_i(\alpha\beta)_{ij} = 0$, and $\sum_j(\alpha\beta)_{ij} = 0$. These restrictions on the above model follow directly from the basic definition of treatment effects. Interactions with block effects do not appear on the right-hand side of the model; hence the implicit assumption that such effects either do not exist or that such effects are negligible relative to the magnitude of the other effects.

The set of treatment combinations which satisfy the relation $x_1 + x_2 = i$ define the set J_i. It will be convenient to define the sum of the interaction effects in the set J_0 by the symbol $3(\alpha\beta)_{J_0}$. Thus

$$3(\alpha\beta)_{J_0} = (\alpha\beta)_{00} + (\alpha\beta)_{12} + (\alpha\beta)_{21}.$$

Similarly, by definition,

$$3(\alpha\beta)_{J_1} = (\alpha\beta)_{01} + (\alpha\beta)_{10} + (\alpha\beta)_{22},$$
$$3(\alpha\beta)_{J_2} = (\alpha\beta)_{02} + (\alpha\beta)_{11} + (\alpha\beta)_{20}.$$

In terms of the basic model and the specialized definitions, for observations on treatment combinations in the set J_0 in replication 1 of the design in Table 8.9-1,

$$\begin{aligned} X_{00} &\doteq \mu + \alpha_0 + \beta_0 + (\alpha\beta)_{00} + \pi_1 \\ X_{12} &\doteq \mu + \alpha_1 + \beta_2 + (\alpha\beta)_{12} + \pi_1 \\ X_{21} &\doteq \mu + \alpha_2 + \beta_1 + (\alpha\beta)_{21} + \pi_1 \\ \hline J_{0.} &\doteq 3\mu + 3(\alpha\beta)_{J_0} + 3\pi_1 \quad . \end{aligned}$$

The symbol $J_{0.}$ is used to designate the sum of all observations on treatment combinations in the set J_0. Similarly

$$J_{1.} \doteq 3\mu + 3(\alpha\beta)_{J_1} + 3\pi_2,$$
$$J_{2.} \doteq 3\mu + 3(\alpha\beta)_{J_2} + 3\pi_3.$$

The sum of squares for the AB component of $A \times B$ is obtained from differences between these J totals. For example,

$$J_{0.} - J_{1.} \doteq 3(\alpha\beta)_{J_0} - 3(\alpha\beta)_{J_1} + 3\pi_1 - 3\pi_2.$$

Each of the other possible comparisons among the J's also includes block effects. Hence, from the data in replication 1, SS_{AB} is completely confounded with differences among blocks.

In contrast, for replication 2,

$$\begin{aligned} X_{00} &\doteq \mu + \alpha_0 + \beta_0 + (\alpha\beta)_{00} + \pi_4 \\ X_{12} &\doteq \mu + \alpha_1 + \beta_2 + (\alpha\beta)_{12} + \pi_6 \\ X_{21} &\doteq \mu + \alpha_2 + \beta_1 + (\alpha\beta)_{21} + \pi_5 \\ \hline J_{0.} &\doteq 3\mu + 3(\alpha\beta)_{J_0} + \pi_4 + \pi_5 + \pi_6. \end{aligned}$$

Similarly, for replication 2,

$$J_{1.} \doteq 3\mu + 3(\alpha\beta)_{J_1} + \pi_4 + \pi_5 + \pi_6,$$
$$J_{2.} \doteq 3\mu + 3(\alpha\beta)_{J_2} + \pi_4 + \pi_5 + \pi_6.$$

Differences between J totals in replication 2 are free of block effects. For example,

$$J_{0.} - J_{1.} \doteq 3(\alpha\beta)_{J_0} - 3(\alpha\beta)_{J_1}.$$

Hence within-block information on the AB components of $A \times B$ may be obtained from replication 2.

To show that replication 1 provides within-block information on the AB^2 (or I) component of $A \times B$,

$$\begin{aligned} X_{00} &\doteq \mu + \alpha_0 + \beta_0 + (\alpha\beta)_{00} + \pi_1 \\ X_{11} &\doteq \mu + \alpha_1 + \beta_1 + (\alpha\beta)_{11} + \pi_3 \\ X_{22} &\doteq \mu + \alpha_2 + \beta_2 + (\alpha\beta)_{22} + \pi_2 \\ \hline I_{0.} &\doteq 3\mu + 3(\alpha\beta)_{I_0} + \pi_1 + \pi_2 + \pi_3. \end{aligned}$$

Similarly, for replication 1,

$$I_{1.} \doteq 3\mu + 3(\alpha\beta)_{I_1} + \pi_1 + \pi_2 + \pi_3,$$
$$I_{2.} \doteq 3\mu + 3(\alpha\beta)_{I_2} + \pi_1 + \pi_2 + \pi_3.$$

Since each I total contains the same block effects, differences between these totals will be free of block effects. Hence information on the I component of $A \times B$ obtained from replication 1 will not be confounded with block effects.

For the general case of a design having r replications (one observation per cell) of the design in Table 8.9-1, with $r/2$ replications of the form of replication 1 and $r/2$ of the form of replication 2, the sum of squares for the J component of $A \times B$ is

$$SS'_{AB(J)} = \frac{\Sigma J_{i.}^2}{3(r/2)} - \frac{(\Sigma J_{i.})^2}{9(r/2)},$$

where J totals are restricted to the replications in which the J component of $A \times B$ is free of block effects. Similarly, the sum of squares for the I components of $A \times B$ is given by

$$SS'_{AB(I)} = \frac{\Sigma I_{i.}^2}{3(r/2)} - \frac{(\Sigma I_{i.})^2}{9(r/2)},$$

where the I totals are restricted to replications in which the I component of $A \times B$ is free of block effects.

If the symbol AB_{ij} represents the sum of all observations on treatment combination ab_{ij}, then this sum will not be free of block effects, since all treatment combinations do not appear in each of the blocks. An adjusted sum which is free of block effects is given by

$$AB'_{ij} = \frac{A_i + B_j}{3} + \frac{J_{(i+j).} + I_{(i+2j).}}{\frac{3}{2}} - \frac{G}{3},$$

where A_i = sum of all observations at level a_i,
B_j = sum of all observations at level b_j,
$J_{(i+j).}$ = sum of all observations on treatment combinations which satisfy the relation $x_1 + x_2 = i + j \pmod 3$ in replication in which differences between J's are free of block effects,
$I_{(i+2j).}$ = sum of all observations on treatment combinations which satisfy the relation $x_1 + 2x_2 = i + 2j \pmod 3$ in replications in which differences between I's are free of block effects.

For example, the adjusted total for all observations made under treatment combination ab_{02} is

$$AB'_{02} = \frac{A_0 + B_2}{3} + \frac{J_{2.} + I_{1.}}{\frac{3}{2}} - \frac{G}{3}.$$

In terms of the general linear model, this last expression estimates the parameters on the right-hand side of the following expression:

$$AB'_{02} \doteq r\mu + r\alpha_0 + r\beta_2 + r(\alpha\beta)_{J_2} + r(\alpha\beta)_{I_1} + \frac{\text{sum of all block effects}}{3}.$$

Since all adjusted totals will contain this last term, differences between the adjusted total will be free of block effects. To demonstrate that AB'_{02} actually does estimate the parameters on the right-hand side,

$$\frac{A_0}{3} \doteq r\mu + r\alpha_0 + \tfrac{1}{3}(\text{sum of all block effects})$$

$$\frac{B_2}{3} \doteq r\mu + r\beta_2 + \tfrac{1}{3}(\text{sum of all block effects})$$

$$\frac{J_{2.}}{\frac{3}{2}} \doteq r\mu + r(\alpha\beta)_{J_2} + \tfrac{3}{2}(\text{sum of blocks in which } J \text{ is free of block effects})$$

$$\frac{I_{1.}}{\frac{3}{2}} \doteq r\mu + r(\alpha\beta)_{I_2} + \tfrac{3}{2}(\text{sum of blocks in which } I \text{ is free of block effects})$$

$$\frac{-G}{3} \doteq -3r\mu - (\text{sum of all block effects})$$

$$AB'_{02} \doteq r\mu + r\alpha_0 + r\beta_2 + r(\alpha\beta)_{J_2} + r(\alpha\beta)_{I_1} + \frac{\text{sum of all block effects}}{3}.$$

In general it can be shown that

$$(\bar{J}_{(i+j).} - \bar{G}) + (\bar{I}_{(i+2j).} - \bar{G}) = \widehat{\alpha\beta}_{ij}$$

or

$$(\widehat{\alpha\beta})_{J_{i+j}} + (\widehat{\alpha\beta})_{I_{i+2j}} = \widehat{\alpha\beta}_{ij}.$$

For example, if $i = 0$ and $j = 2$,

$$\begin{aligned}
3\bar{J}_{2.} &= \overline{AB}_{02} + \overline{AB}_{11} + \overline{AB}_{20} \\
3\bar{I}_{1.} &= \overline{AB}_{02} + \overline{AB}_{10} + \overline{AB}_{21} \\
0 &= \overline{AB}_{02} + \overline{AB}_{12} + \overline{AB}_{22} - (\overline{AB}_{02} + \overline{AB}_{12} + \overline{AB}_{22}) \\
\hline
3(\bar{J}_{2.} + \bar{I}_{1.}) &= 3\overline{AB}_{02} + 3\bar{A}_1 + 3\bar{A}_2 - 3\bar{B}_2 \\
&= 3\overline{AB}_{02} + 3\bar{G} - 3\bar{A}_0 - 3\bar{B}_2.
\end{aligned}$$

Hence

$$\begin{aligned}
(\bar{J}_{2.} - \bar{G}) + (\bar{I}_{1.} - \bar{G}) &= \overline{AB}_{02} - \bar{A}_0 - \bar{B}_2 + \bar{G} \\
&= \widehat{\alpha\beta}_{02}.
\end{aligned}$$

Similarly, if $i = 1$ and $j = 1$, then

$$\begin{aligned}
(\bar{J}_{2.} - \bar{G}) + (\bar{I}_{0.} - \bar{G}) &= \overline{AB}_{11} - \bar{A}_1 - \bar{B}_1 + \bar{G} \\
&= \widehat{\alpha\beta}_{11}.
\end{aligned}$$

In making comparisons among the means of treatment combinations, one uses the adjusted totals to obtain the means. The effective number of observations in each of these adjusted means is less than r, since information on the I and J components of the $A \times B$ interaction is obtained from only one-half of the replications. The number of *effective* replications in this adjusted total is given by

$$\frac{1 + 1 + 1 + 1 + 1}{1 + 1 + (1/\frac{1}{2}) + (1/\frac{1}{2}) + 1}\, r = \frac{5}{7}\, r.$$

The terms in the numerator of this expression represent the relative information in each of the totals entering into AB'_{ij}, assuming there was no confounding. The terms in the denominator are the reciprocals of the relative information in each of the totals actually given by the design.

For most practical purposes, however, the adjusted cell means may be handled as if they were based upon r replications. The adjusted mean for cell ab_{ij} is

$$\overline{AB}'_{ij} = \frac{AB'_{ij}}{r}.$$

If this design had r replications with n observations per cell, AB'_{ij}

would have the definition given above but

$$\overline{AB}'_{ij} = \frac{AB'_{ij}}{nr}.$$

If there were no confounding in a design having r replications,

$$\sigma^2_{\overline{AB}} = \frac{\sigma^2_\varepsilon}{r}.$$

For the design given in Table 8.9-1, in which one has only $\frac{1}{2}$ information on the interaction,

$$\sigma^2_{\overline{AB'}} = \frac{\sigma^2_\varepsilon}{(\frac{5}{7})r} = \frac{7\sigma^2_\varepsilon}{5r} = \frac{\sigma^2_\varepsilon}{r_{\text{effective}}}.$$

When the effective number of replications is close to the actual number of replications, the latter is often used in place of $r_{\text{effective}}$ in the denominator.

8.10 Numerical Example of 3 × 3 Factorial Experiment in Blocks of Size 3

In the experiment to be described it was desired to have within-subject information on all factorial effects in a 3 × 3 factorial experiment. However, it was not experimentally feasible to have each subject observed under each of the nine treatment combinations. Each person could be observed under three treatment combinations. The design outlined in Table 8.10-1 was selected for use.

The experimenter wished to have two observations under each treatment combination. A random sample of 12 subjects was obtained from a specified population; the sample was divided at random into six groups of 2 subjects each. The groups were then assigned at random to the blocks in Table 8.10-1. The 2 subjects within each group were observed under each of the three treatment combinations in the assigned block. The order in which a subject was observed under a treatment condition was randomized for each subject. The data obtained from this experiment are summarized in Table 8.10-1.

In the row having person 1 at the left, the entries 14, 7, and 15 are the respective observations (with order randomized) on person 1 under treatment combinations (00), (12), and (21). The sum of these three observations on person 1 is denoted by the symbol P_1. The symbols G_i denote the totals for the six observations made within a group (block). The symbols R_j denote the totals of the 18 observations made within a single replication. Additional data required for the computation of the sums of squares in the analysis are given in Table 8.10-2.

The sums of squares for all effects except the $A \times$ B interaction are computed in the same manner as that used in any factorial experiment having

Table 8.10-1 Summary of Observed Data

Block	Person				Total	
		(00)	(12)	(21)		
1	1	14	7	15	$36 = P_1$	
	2	6	3	5	$14 = P_2$	$50 = G_1$
		(01)	(10)	(22)		
2	3	3	4	10	$17 = P_3$	
	4	7	6	30	$43 = P_4$	$60 = G_2$
		(02)	(11)	(20)		
3	5	5	15	7	$27 = P_5$	
	6	5	5	3	$13 = P_6$	$40 = G_3$
						$150 = R_1$
		(00)	(11)	(22)		
4	7	10	10	15	$35 = P_7$	
	8	15	20	25	$60 = P_8$	$95 = G_4$
		(02)	(10)	(21)		
5	9	3	5	12	$20 = P_9$	
	10	7	10	18	$35 = P_{10}$	$55 = G_5$
		(01)	(12)	(20)		
6	11	6	7	5	$18 = P_{11}$	
	12	4	3	5	$12 = P_{12}$	$30 = G_6$
						$180 = R_2$

repeated measures. Data required for computation of all effects except the components of $A \times B$ are summarized in parts ii and iii of Table 8.10-2. From the data in part iii, the between-subject sum of squares is given by

$$\frac{\Sigma P^2}{3} - \frac{G^2}{36} = 3815.33 - 3025.00 = 790.33.$$

The within-subject sum of squares is given by

$$\Sigma\Sigma X^2 - \frac{\Sigma P^2}{3} = 4494 - 3815.33 = 678.67.$$

Data required for the sums of squares for groups, replications, and the main effects of factors A and B are also given in part iii.

From the data in part i, both the between- and within-subject components of the $A \times B$ interaction may be computed. From the totals at the right of replication 1, the within-subject information on the I component of $A \times B$ is given by

$$SS_{AB(I)} = SS_{AB^2} = \frac{\Sigma I_{i.}^2}{3n} - \frac{(\Sigma I_{i.})^2}{9n}$$

$$= \frac{80^2 + 40^2 + 30^2}{3(2)} - \frac{(150)^2}{18} = 233.33.$$

Table 8.10-2 Summary Data Required for Analysis

(i)

Replication 1

		b_0	b_1	b_2	
	a_0	20	10	10	
	a_1	10	20	10	
40	a_2	10	20	40	80 = $I_{0.}$
50	a_0	20	10	10	40 = $I_{1.}$
60	a_1	10	20	10	30 = $I_{2.}$
150					150

Replication 2

		b_0	b_1	b_2	
	a_0	25	10	10	
	a_1	15	30	10	
$J_{2.}$ = 50	a_2	10	30	40	95
$J_{0.}$ = 65	a_0	25	10	10	55
$J_{1.}$ = 65	a_1	15	30	10	30
180					180

(ii)

Combined replications

	b_0	b_1	b_2	Total
a_0	45	20	20	85 = A_0
a_1	25	50	20	95 = A_1
a_2	20	50	80	150 = A_2
Total	90	120	120	330 = G
	B_0	B_1	B_2	

(iii)

$G^2/9nr = 3025.00$	$(\Sigma P^2)/3 = 3815.33$
$\Sigma\Sigma X^2 = 4494$	$(\Sigma G_i^2)/3n = 3441.67$
$(\Sigma A^2)/3nr = 3229.17$	$(\Sigma R^2)/9n = 3050.00$
$(\Sigma B^2)/3nr = 3075.00$	

From the totals at the left of replication 2, the within-subject information on the J component of $A \times B$ is given by

$$SS_{AB(J)} = SS_{AB} = \frac{\Sigma J_{i.}^2}{3n} - \frac{(\Sigma J_{i.})^2}{9n}$$

$$= \frac{50^2 + 65^2 + 65^2}{6} - \frac{(180)^2}{18} = 25.00.$$

Thus within-subject information on the components of the $A \times B$ interaction is obtained separately from each of the replications—one component from replication 1, the other from replication 2.

The between-subject information on the J component of $A \times B$ is obtained from the totals at the left of replication 1.

$$SS_{AB(J)} = SS_{AB} = \frac{40^2 + 50^2 + 60^2}{6} - \frac{(150)^2}{18} = 33.33.$$

Table 8.10-3 Summary of Analysis

Source		SS		df	MS	F
Between subjects		790.33		11		
Groups (blocks)		416.67		5		
Replications	25.00		1			
AB (from rep 1)	33.33		2			
AB^2 (from rep 2)	358.33		2			
Subjects within groups		373.66		6	62.28	
Within subjects		678.67		24		
A		204.17		2	102.08	9.82**
B		50.00		2	25.00	2.41
$A \times B$ (adjusted)		258.33		4	64.58	6.22**
AB (from rep 2)	25.00		2			
AB^2 (from rep 1)	233.33		2			
Residual		166.17		16	10.39	

** $F_{.99}(2,16) = 6.23$; $F_{.99}(4,16) = 4.77$

The between-subject information on the I component of $A \times B$ is obtained from the totals to the right of replication 2.

$$SS_{AB(I)} = SS_{AB^2} = \frac{95^2 + 55^2 + 30^2}{6} - \frac{(180)^2}{18} = 358.33.$$

The residual sum of squares in Table 8.10-3 is obtained from the relation

$$SS_{res} = SS_{w.\ subj} - SS_A - SS_B - SS_{AB}(\text{within}) - SS_{AB^2}(\text{within}).$$

The residual sum of squares may also be computed directly from the interactions of factor A and B with the replications and subjects within groups.

Unless there is an a priori reason for handling the components of the $A \times B$ interaction separately, a single test is made on the combined within-subject components of this interaction. (In cases in which the large cell values in a two-way summary table fall along the diagonal running from upper left to lower right, the AB^2 component will be large relative to the AB component.) In this numerical example the $A \times B$ interaction is statistically significant. The simple effects for both factor A and factor B are between-subject effects; however, approximations to tests on simple effects may be obtained by working with the adjusted cell totals. Differences among the latter totals may be considered to be within-subject effects.

To illustrate the computation of the adjusted cell totals, for treatment combination ab_{02},

$$AB'_{02} = \frac{A_0 + B_2}{3} + \frac{I_{1.} + J_{2.}}{\frac{3}{2}} - \frac{G}{3}$$

$$= \frac{85 + 120}{3} + \frac{40 + 50}{\frac{3}{2}} - \frac{330}{3} = 18.33.$$

(Note that in part i of Table 8.10-2 cell ab_{02} contributes to the $I_{1.}$ total in replication 1 and to the $J_{2.}$ total in replication 2.) The unadjusted cell total is 20. As another example, the adjusted total for cell ab_{11} has the form

$$AB'_{11} = \frac{A_1 + B_1}{3} + \frac{I_{0.} + J_{2.}}{\frac{3}{2}} - \frac{G}{3}.$$

Comparisons among the adjusted cell totals use the within-subject residual mean square as an error term. For approximate tests, the number of effective observations in each cell is considered to be $nr = 4$.

To illustrate the procedures for tests on adjusted cell totals,

$$F = \frac{(AB'_{02} - AB'_{22})^2}{2nr\,\text{MS}_{\text{res}}}.$$

The adjusted total AB'_{02} was found to be 18.33. The adjusted total AB'_{22} is given by

$$AB'_{22} = \frac{A_2 + B_2}{3} + \frac{I_{0.} + J_{1.}}{\frac{3}{2}} - \frac{G}{3}.$$

(Note that ab_{22} contributes to the $I_{0.}$ total in replication 1 and to the $J_{1.}$ total in replication 2.)

$$AB'_{22} = \frac{150 + 120}{3} + \frac{80 + 65}{\frac{3}{2}} - \frac{330}{3}$$

$$= 76.57.$$

Thus $$F = \frac{(18.33 - 76.67)^2}{2(2)(2)(10.39)} = 40.95.$$

The critical value for a .05-level (a priori) test is $F_{.95}(1,16) = 4.49$.

8.11 Designs for 3 × 3 × 3 Factorial Experiments

The 27 treatment combinations in a 3 × 3 × 3 factorial experiment may be divided into sets in several different ways and balance still be maintained with respect to main effects and two-factor interactions. Each of the following components of the $A \times B \times C$ interactions can serve to define such balanced sets:

	df
$A \times B \times C$	8
ABC	2
ABC^2	2
AB^2C	2
AB^2C^2	2

For example, the sets formed by using the ABC component are as follows:

$x_1 + x_2 + x_3 = 0$ (mod 3)	$x_1 + x_2 + x_3 = 1$ (mod 3)	$x_1 + x_2 + x_3 = 2$ (mod 3)
(000)	(001)	(002)
(012)	(010)	(011)
(021)	(022)	(020)
(102)	(100)	(101)
(111)	(112)	(110)
(120)	(121)	(122)
(201)	(202)	(200)
(210)	(211)	(212)
(222)	(220)	(221)

A design consisting of these three blocks would provide within-block information on all effects except the ABC component of the three-factor interaction. With only a single replication (assuming only one observation per cell) no estimate of experimental error is available.

If the experiment were to be replicated by using the same block structure as that given above, the design would lack balance with respect to the components of the three-factor interaction. That is, within-block information would be available on three of the four components of this interaction, but no within-block information would be available on the ABC component. A design which has balance with respect to the components of the three-factor interaction may be constructed by means of the scheme outlined in Table 8.11-1. A minimum of four replications is required for balance. The blocks within each of the replications are constructed by means of relations associated with a component of the three-factor interaction; each of the replications involves a different component. Hence within-block information on all the three-factor components is available from some three of the four replications.

Table 8.11-1 Construction of Blocks of Size 9

Replication	Block	Component confounded	Defining relation (mod 3)
1	1	$(ABC)_0$	$x_1 + x_2 + x_3 = 0$
	2	$(ABC)_1$	$x_1 + x_2 + x_3 = 1$
	3	$(ABC)_2$	$x_1 + x_2 + x_3 = 2$
2	4	$(ABC^2)_0$	$x_1 + x_2 + 2x_3 = 0$
	5	$(ABC^2)_1$	$x_1 + x_2 + 2x_3 = 1$
	6	$(ABC^2)_2$	$x_1 + x_2 + 2x_3 = 2$
3	7	$(AB^2C)_0$	$x_1 + 2x_2 + x_3 = 0$
	8	$(AB^2C)_1$	$x_1 + 2x_2 + x_3 = 1$
	9	$(AB^2C)_2$	$x_1 + 2x_2 + x_3 = 2$
4	10	$(AB^2C^2)_0$	$x_1 + 2x_2 + 2x_3 = 0$
	11	$(AB^2C^2)_1$	$x_1 + 2x_2 + 2x_3 = 1$
	12	$(AB^2C^2)_2$	$x_1 + 2x_2 + 2x_3 = 2$

The structure of the blocks defined by the design in Table 8.11-1 is given in Table 8.11-2. Blocks in replication 1 are balanced with respect to all

Table 8.11-2 Design Corresponding to Table 8.11-1

Replication 1			Replication 2		
Block 1	Block 2	Block 3	Block 4	Block 5	Block 6
(000)	(001)	(002)	(000)	(002)	(001)
(012)	(010)	(011)	(011)	(010)	(012)
(021)	(022)	(020)	(022)	(021)	(020)
(102)	(100)	(101)	(101)	(100)	(102)
(111)	(112)	(110)	(112)	(111)	(110)
(120)	(121)	(122)	(120)	(122)	(121)
(201)	(202)	(200)	(202)	(201)	(200)
(210)	(211)	(212)	(210)	(212)	(211)
(222)	(220)	(221)	(221)	(220)	(222)
ABC			ABC^2		
Replication 3			Replication 4		
Block 7	Block 8	Block 9	Block 10	Block 11	Block 12
(000)	(001)	(002)	(000)	(002)	(001)
(011)	(012)	(010)	(012)	(011)	(010)
(022)	(020)	(021)	(021)	(020)	(022)
(102)	(100)	(101)	(101)	(100)	(102)
(110)	(111)	(112)	(110)	(112)	(111)
(121)	(122)	(120)	(122)	(121)	(120)
(201)	(202)	(200)	(202)	(201)	(200)
(212)	(210)	(211)	(211)	(210)	(212)
(220)	(221)	(222)	(220)	(222)	(221)
AB^2C			AB^2C^2		

main effects and all two-factor interactions. These blocks are also balanced with respect to all components of the three-factor interaction except the ABC component. To illustrate the balance with respect to the ABC^2 component within block 1, the nine treatment combinations in this block may be divided into the following sets:

$(ABC^2)_0$	$(ABC^2)_1$	$(ABC^2)_2$
(000)	(021)	(012)
(120)	(111)	(102)
(210)	(201)	(222)

Differences between the totals for these sets form part of the ABC^2 component.

Assuming one observation per cell, the analysis of this design has the form given in Table 8.11-3. The computation of the sums of squares for

Table 8.11-3 Analysis of Design Having Block Size 9 (One Observation per Cell)

Source of variation		df	E(MS)
Between blocks		11	
Replications		3	
Blocks within reps		8	
ABC (from rep 1)	2		
ABC^2 (from rep 2)	2		
AB^2C (from rep 3)	2		
AB^2C^2 (from rep 4)	2		
Within blocks		96	
A		2	$\sigma_\varepsilon^2 + 36\sigma_\alpha^2$
B		2	$\sigma_\varepsilon^2 + 36\sigma_\beta^2$
C		2	$\sigma_\varepsilon^2 + 36\sigma_\gamma^2$
$A \times B$		4	$\sigma_\varepsilon^2 + 12\sigma_{\alpha\beta}^2$
$A \times C$		4	$\sigma_\varepsilon^2 + 12\sigma_{\alpha\gamma}^2$
$B \times C$		4	$\sigma_\varepsilon^2 + 12\sigma_{\beta\gamma}^2$
$(A \times B \times C)'$ $(\frac{3}{4})$		8	$\sigma_\varepsilon^2 + (\frac{3}{4})(4)\sigma_{\alpha\beta\gamma}^2$
ABC (omit rep 1)	2		
ABC^2 (omit rep 2)	2		
AB^2C (omit rep 3)	2		
AB^2C^2 (omit rep 4)	2		
Residual		70	σ_ε^2
Main effects × reps	18		
2-factor int × reps	36		
3-factor int × reps	16		

main effects and two-factor interactions follows the same rules as those of the replicated factorial experiment. In computing the within-block components of $A \times B \times C$, data from replications in which a component is confounded with block effects are not used. For example, to compute the within-block component of ABC^2, data from all replications except replication 2 are combined to form a three-way summary table. From this summary table one obtains the totals $(ABC^2)'_0$, $(ABC^2)'_1$, and $(ABC^2)'_2$, where $(ABC^2)'_i$ represents the sum of observations in all cells which satisfy the relation $x_1 + x_2 + 2x_3 = i \pmod 3$. Then

$$SS'_{ABC^2} = \frac{\Sigma(ABC^2)'^2_i}{27} - \frac{[\Sigma(ABC^2)'_i]^2}{81}.$$

As a partial check on the numerical work, $\Sigma(ABC^2)'_i$ is equal to the sum of all observations in replications 1, 3, and 4.

The within-cell residual is most conveniently obtained by subtracting the within-cell treatment effects from the total within-cell variation. The sources of variation which are pooled to form the residual term are shown in Table 8.11-3. These sources may be computed separately. In this design there are only three effective replications on each of the components of the three-factor interaction; thus the $ABC \times$ replications interaction, for example, has four degrees of freedom. Hence the three-factor $\times$ replications interaction has a total of 16 degrees of freedom—4 degrees of freedom for each of the four components of the three-factor interaction.

In making tests on differences between adjusted cell means, adjusted cell totals are required. Such totals are given by

$$\begin{aligned} 9ABC'_{ijk} = {} & A_i + B_j + C_k + (AB)_{i+j} + (AC)_{i+k} + (BC)_{j+k} + (AB^2)_{i+2j} \\ & + (AC^2)_{i+2k} + (BC^2)_{j+2k} + \tfrac{4}{3}[(ABC)'_{i+j+k} + (ABC^2)'_{i+j+2k} \\ & + (AB^2C)'_{i+2j+k} + (AB^2C^2)'_{i+2j+2k}] - 4G, \end{aligned}$$

where the symbol $(ABC)'_{i+j+k}$ represents a total obtained from those replications in which this part of the three-factor interaction is not confounded with block totals. For example,

$$\begin{aligned} 9ABC'_{012} = {} & A_0 + B_1 + C_2 + (AB)_1 + (AC)_2 + (BC)_0 + (AB^2)_2 + (AC^2)_1 \\ & + (BC^2)_2 + \tfrac{4}{3}[(ABC)'_0 + (ABC^2)'_2 + (AB^2C)'_1 + (AB^2C^2)'_0] \\ & - 4G, \end{aligned}$$

$$\begin{aligned} 9ABC'_{001} = {} & A_0 + B_0 + C_1 + (AB)_0 + (AC)_1 + (BC)_1 + (AB^2)_0 \\ & + (AC^2)_2 + (BC^2)_2 + \tfrac{4}{3}[(ABC)'_1 + (ABC^2)'_2 + (AB^2C)'_1 \\ & + (AB^2C^2)'_2] - 4G. \end{aligned}$$

The difference between the means corresponding to these adjusted totals is given by

$$\begin{aligned} \overline{ABC}'_{012} - \overline{ABC}'_{001} = {} & (\bar{B}_1 - \bar{B}_0) + (\bar{C}_2 - \bar{C}_1) + [(\overline{AB})_1 - (\overline{AB})_0] \\ & + [(\overline{AC})_2 - (\overline{AC})_2] + [(\overline{BC})_0 - (\overline{BC})_1] \\ & + [(\overline{AB^2})_2 - (\overline{AB^2})_0] + [(\overline{AC^2})_1 - (\overline{AC^2})_2] \\ & + [(\overline{ABC})'_0 - (\overline{ABC})'_1] + [(\overline{AB^2C^2})'_0 - (\overline{AB^2C^2})'_2]. \end{aligned}$$

The effective information in this difference relative to a design in which there is no confounding is given by

$$\frac{9}{7 + 2(4/3)} = \frac{27}{29}.$$

Since the two-factor interactions in the design given in Table 8.11-2 are not confounded with block effects, no adjustments are required for cell totals in two-way summary tables. Differences between means computed from such totals will be free of block effects.

Block Size = 3. In dividing the 27 treatment combinations into blocks of size 9, a relation of the general form

$$u = x_1 + u_2x_2 + u_3x_3 = i \pmod 3$$

was used for each of the replications. The relation u corresponded to some component of the three-factor interaction. In the analysis of the resulting design, the u component was confounded with block effects within the replication having blocks defined by the u relation. The 27 treatment combinations in a $3 \times 3 \times 3$ factorial experiment may be divided into blocks of size 3 by requiring that the treatment combinations within a given block simultaneously satisfy the two relations

$$u = i \pmod 3,$$

$$v = j \pmod 3,$$

where v is some relation other than u.

In the analysis of a replication having blocks formed in this way, the components corresponding to u and v will both be confounded with differences between blocks. In addition the components corresponding to the relations $u + v$ and $u + 2v$ will also be confounded with block effects. The latter components are known as the *generalized interactions*, or *aliases*, of the components corresponding to u and v. To illustrate, suppose that the blocks within a replication are defined by the relations

$$u = x_1 + x_2 + 2x_3 = i \pmod 3,$$

$$v = x_1 + 2x_2 \qquad = j \pmod 3.$$

These relations correspond, respectively, to the ABC^2 and the AB^2 components. One of the generalized interactions, or aliases, is given by

$$\begin{aligned} u + v &= 2x_1 + 3x_2 + 2x_3 = (i + j) \pmod 3 \\ &= 2x_1 + 2x_3 = (i + j) \pmod 3 \\ &= x_1 + x_3 = 2(i + j) \pmod 3. \end{aligned}$$

(The last line is obtained from the one above by multiplying both sides of the equation by 2 and then reducing coefficients to the modulus 3.) This last relation corresponds to the AC component. The second generalized interaction of u and v is given by

$$\begin{aligned} u + 2v &= 3x_1 + 5x_2 + 2x_3 = (i + 2j) \pmod 3 \\ &= 2x_2 + 2x_3 = (i + 2j) \pmod 3 \\ &= x_2 + x_3 = 2(i + 2j) \pmod 3. \end{aligned}$$

This last relation corresponds to the BC component.

To summarize this illustrative example, blocks are defined by relations corresponding to the ABC^2 and AB^2 components. The generalized interactions, or aliases, of these components are the AC and BC components. Hence, in a replication defined by these relations, the analysis would have the following form:

Between blocks	8
AB^2	2
AC	2
BC	2
ABC^2	2
Within blocks	18
A, B, C	2 each
AB, AC^2, BC^2	2 each
ABC, AB^2C, AB^2C^2	2 each

The defining relations in this example may be symbolized by $(ABC^2)_i$ and $(AB^2)_j$; the aliases may be symbolized by $(AC)_{2(i+j)}$ and $(BC)_{2(i+2j)}$. The block defined by $(ABC^2)_0$ and $(AB^2)_1$ is

$$(022), \qquad (101), \qquad (210).$$

It may be verified that each of the treatment combinations in the blocks also satisfies the relations $(AC)_{2(0+1)} = (AC)_2$ and $(BC)_{2(0+2)} = (BC)_1$.

In general, if two relations are used simultaneously to define blocks, one needs to know the generalized interactions of the components corresponding to these relations in order to carry out the analysis. There is a relatively simple rule for determining the aliases of any two components. Given two components having the general form W and X, their aliases will have the general forms WX and WX^2. To illustrate the use of this rule, given

$$W = ABC^2 \qquad \text{and} \qquad X = AB^2,$$

one alias has the form

$$WX = (ABC^2)(AB^2) = A^2B^3C^2 = A^2C^2 = AC,$$

upon reduction of the exponents to the modulus 3. The second alias has the form

$$WX^2 = (ABC^2)(AB^2)^2 = (ABC^2)(A^2B^4) = A^3B^5C^2 = B^2C^2 = BC.$$

It is immaterial which of the components is designated W and which is designated X, since the aliases are also given by the general forms W^2X and WX. As another example, the aliases of the components AB and AC are

$$WX = (AB)(AC) = A^2BC = AB^2C^2,$$
$$WX^2 = (AB)(AC)^2 = A^3BC^2 = BC^2.$$

To construct a design having within-block information on all components of the interactions, and yet provide complete information on all main effects, the scheme presented in Table 8.11-4 may be used. The aliases associated with the pairs of defining relations are indicated. The design represented by this scheme will provide some within-block information with respect to all interaction components. For example, the AB^2 component is confounded with block effects in replications 1 and 2

Table 8.11-4 Construction of Blocks of Size 3

Replication	Defining relations	Aliases
1	ABC, AB^2	AC^2, BC^2
2	ABC^2, AB^2	AC, BC
3	AB^2C, AB	AC^2, BC
4	AB^2C^2, AB	AC, BC^2

but is free of block effects in replications 3 and 4. Hence the relative within-block information on this component is $\frac{1}{2}$. The AC^2 component is confounded in replications 1 and 3 but is not confounded in blocks 2 and 4. It will be found that all the components of two-factor interactions are confounded with blocks in two of the four replications but are free of such confounding in two of the four replications. Each of the components of the three-factor interaction is confounded with block effects in one of the replications but is free of block effects in three of the four replications. Hence the relative information on the components of the three-factor interaction is $\frac{3}{4}$.

The blocks corresponding to the design outlined in Table 8.11-4 are obtained by subdividing the blocks in Table 8.11-2. Block 1 in the latter table is subdivided as follows:

$(AB^2)_0$	$(AB^2)_1$	$(AB^2)_2$
(000)	(021)	(012)
(111)	(102)	(120)
(222)	(210)	(201)

Each of the treatment combinations within a block satisfies the relation $(ABC)_0$ as well as the relation heading the block. Block 2 in Table 8.11-2 is subdivided as follows:

$(AB^2)_0$	$(AB^2)_1$	$(AB^2)_2$
(001)	(022)	(010)
(112)	(100)	(121)
(220)	(211)	(202)

Block 3 in Table 8.11-2 is subdivided as follows:

$(AB^2)_0$	$(AB^2)_1$	$(AB^2)_2$
(002)	(020)	(011)
(110)	(101)	(122)
(221)	(212)	(200)

The nine blocks of size 3 that have just been constructed form replication 1 of the design in Table 8.11-4.

The overall analysis of the latter design takes the form given in Table 8.11-5. Within-block information on the main effects is complete; hence

Table 8.11-5 Analysis of Design in Table 8.11-4

Source of variation	df	
Between blocks	35	
Replications	3	
Blocks within reps	32	
Within blocks	72	
Main effects: A, B, C	6	(2 each)
Two-factor interaction:		
AB, AB^2 ($\frac{1}{2}$ information)	4	(2 each)
AC, AC^2 ($\frac{1}{2}$ information)	4	(2 each)
BC, BC^2 ($\frac{1}{2}$ information)	4	(2 each)
Three-factor interaction:		
ABC, ABC^2, AB^2C, AB^2C^2 ($\frac{3}{4}$ information)	8	(2 each)
Residual	46	

no special computational procedures are required. In computing the AB component, only information from blocks in which this component is not confounded is used. Blocks of this kind are in replications 1 and 2; the blocks in replications 3 and 4 do not provide within-block information on AB. All within-block information on AB^2 is obtained from replications 3 and 4. Hence the sum of squares for AB^2 is computed from a summary table prepared from replications 3 and 4. Given this summary table,

$$SS'_{AB^2} = \frac{(AB^2)_0'^2 + (AB^2)_1'^2 + (AB^2)_2'^2}{18} - \frac{[\Sigma(AB^2)_i']^2}{54},$$

where $(AB^2)_i'$ is the sum of observations on all treatment combinations in replications 3 and 4 which satisfy the relation $x_1 + 2x_2 = i \pmod 3$. As a partial check on the numerical work $\Sigma(AB^2)_i'$ is equal to the sum of all observations in replications 3 and 4.

Other components of the two-factor interactions are computed in an analogous manner. The within-block information on AC is obtained from replications 1 and 3; similar information on AC^2 is obtained from replications 2 and 4.

Within-block information on the three-factor interaction components is obtained from some three of the four replications. For example, information from replications 2, 3, and 4 is combined to obtain the ABC

component. Given a summary table of observations from these three replications,

$$\text{SS}'_{ABC} = \frac{(ABC)_0'^2 + (ABC)_1'^2 + (ABC)_2'^2 + (ABC)_3'^2}{27} - \frac{[\Sigma(ABC)_i']^2}{81},$$

where $(ABC)_i'$ is the sum of observations in replications 2, 3, and 4 satisfying the relation $x_1 + x_2 + x_3 = i \pmod 3$. The other components of the three-factor interaction are computed in an analogous manner, the ABC^2 component, for example, being obtained from replications 1, 3, and 4. Since components of both two-factor and three-factor interactions are partially confounded with block effects, estimates of cell means require adjustment for block effects. The latter adjustment is most readily made by adjusting cell totals. Adjusted cell totals are given by

$$\begin{aligned} ABC'_{ijk} &= \tfrac{1}{9}(A_i + B_j + C_k) \\ &\quad + \tfrac{2}{9}[(AB)'_{i+j} + (AB^2)'_{i+2j} + (AC)'_{i+k} + (AC^2)'_{i+2k} + (BC)'_{j+k} + (BC^2)'_{j+2k}] \\ &\quad + \tfrac{4}{27}[(ABC)'_{i+j+k} + (ABC^2)'_{i+j+2k} + (AB^2C)'_{i+2j+k} + (AB^2C^2)'_{i+2j+2k}] \\ &\quad - \tfrac{4}{9}G. \end{aligned}$$

The primes indicate that summations are restricted to replications in which a component is not confounded with block effects. The sums required for obtaining an adjusted cell total are obtained in the course of computing the within-block components of the interactions. In terms of the basic linear model, it may be shown that

$$\begin{aligned} ABC'_{ijk} \doteq 4[\mu + \alpha_i + \beta_j + \gamma_k + (\alpha\beta)_{ij} + (\alpha\gamma)_{ik} + (\beta\gamma)_{jk} \\ + (\alpha\beta\gamma)_{ijk}] + 4(\text{sum of all block effects}). \end{aligned}$$

The right-hand side of this last expression represents the parameters that would be estimated by a cell total if each block contained all treatment combinations. Differences between two adjusted totals will be free of block effects, since the last term on the right-hand side is a constant for all cells. With the exception of the block effects, each of the parameters on the right-hand side is estimated independently of all others. Estimates of the parameters associated with main effects are each based upon four effective replications. However, estimates of parameters for two-factor interactions are based upon two effective replications, and estimates of parameters for three-factor interactions are based upon three effective replications. In making comparisons among adjusted cell totals in the three-way table, for most practical purposes one may consider each of these adjusted cell totals as being based upon four replications. Exact methods for taking into account the difference in the effective number of replications for the separate parts of the adjusted cell totals are, however, available [cf. Federer (1955, chap. 9)].

An adjusted total for a cell in a two-way summary table, the cell ab_{ij}, for example, has the form

$$AB'_{ij} = \tfrac{1}{3}(A_i + B_j) + \tfrac{2}{3}[(AB)'_{i+j} + (AB^2)'_{i+2j}] - \tfrac{1}{3}G.$$

Similarly an adjusted total for the cell ac_{ik} has the form

$$AC'_{ik} = \tfrac{1}{3}(A_i + C_k) + \tfrac{2}{3}[(AC)'_{i+k} + (AC^2)'_{i+2k}] - \tfrac{1}{3}G.$$

For comparisons among adjusted totals in a two-way summary table, the effective number of observations in a total varies as a function of the nature of the comparison. The difference between adjusted totals which are in the same row or column of a two-way summary table is based upon $(\frac{3}{5})(12)$ effective observations, whereas the difference between adjusted totals which are not in the same row or column is based upon $(\frac{2}{3})(12)$ effective observations. Comparisons involving mixtures of these two kinds of differences have effective numbers of observations which may be determined by the general expression for the variance of a linear combination of independent terms.

So far the design outlined in Table 8.11-4 has been considered for the case in which there is one observation per cell. Several variations on this basic design are possible. One variation is to have the block represent a group of n subjects. Each of the groups would be assigned at random to the blocks, and each of the n subjects within a group would be observed, in a random order, under each of the treatment combinations within a given block. The analysis of the resulting design would have the form shown in Table 8.11-6. Thus, with only three observations per subject, within-subject estimates of all factorial effects are available.

To summarize the principles underlying the construction of designs for

Table 8.11-6 Analysis of Design in Table 8.11-4 with Repeated Measures

Source of variation	df	
Between subjects	$36n - 1$	
Groups	35	
Subjects within groups	$36(n - 1)$	
Within subjects	$72n$	
Main effects: A, B, C	6	(2 each)
Two-factor interaction:		
AB, AB^2 ($\frac{1}{2}$ information)	4	(2 each)
AC, AC^2 ($\frac{1}{2}$ information)	4	(2 each)
BC, BC^2 ($\frac{1}{2}$ information)	4	(2 each)
Three-factor interaction:		
ABC, ABC^2, AB^2C, AB^2C^2 ($\frac{3}{4}$ information)	8	(2 each)
Residual	$72n - 26$	

$3 \times 3 \times 3$ factorial experiments, blocks of size 9 may be constructed by utilizing relations corresponding to three-factor interactions. If this is done, components of three-factor interactions are either completely or partially confounded with block effects. To construct blocks of size 3, the blocks of size 9 are subdivided by means of relations corresponding to two-factor interactions. In this case, in addition to the interaction corresponding to the defining relations, the aliases also become confounded with block effects.

The principles developed in this section may be extended to include factorial experiments of the form $p \times p \times p$, where p is a prime number. Use of components of the three-factor interaction reduces the block size to p^2. Then blocks of size p^2 are reduced to size p through use of relations corresponding to the components of two-factor interactions. For example, in a $5 \times 5 \times 5$ factorial experiment, the relations u and v may be used to obtain blocks of size 5. In addition to components corresponding to the relations u and v, components corresponding to the following relations will also be confounded with block effects:

$$u + v, \qquad u + 2v, \qquad u + 3v, \qquad u + 4v.$$

For a single replication with block size 5, the number of blocks required would be 25. Hence the degrees of freedom for blocks within a single replication would be 24. Within a single replication, six components of interactions are confounded with block effects; each component has 4 degrees of freedom. Hence 24 degrees of freedom corresponding to the six components of interactions are confounded with differences between blocks, the latter differences also having 24 degrees of freedom.

In terms of the multiplicative scheme, the aliases of the components W and X in a $5 \times 5 \times 5$ factorial experiment have the general form

$$WX, \qquad WX^2, \qquad WX^3, \qquad WX^4.$$

To make the notation system unique, the exponent of the first letter in an interaction term is made unity by raising the whole term to an appropriate power and then reducing exponents modulo 5. It is not difficult to show that WX^k and $(WX^k)^n$, where n is any number modulo 5, correspond to equivalent relations. For example, the following relations are equivalent in the sense that they define the same set of treatment combinations:

$$AC^2: \qquad x_1 + 2x_3 = i \pmod 5,$$
$$A^2C^4: \qquad 2x_1 + 4x_3 = 2i \pmod 5.$$

8.12 Balanced $3 \times 2 \times 2$ Factorial Experiment in Blocks of Size 6

A balanced design for $3 \times 2 \times 2$ factorial experiments in which the block size is 6 can be constructed at the cost of partial confounding of the BC and ABC interactions. A minimum of three replications is required for balance; in this balanced design the relative information on BC is $\frac{8}{9}$, and the relative information on ABC is $\frac{5}{9}$. The balanced design is given in schematic form in Table 8.12-1. In this table the symbol K_0 designates the set of treatment combinations satisfying the relation $x_2 + x_3 = 0$ (mod 2), and the symbol K_1 designates the set of treatment combinations satisfying the relation $x_2 + x_3 = 1$ (mod 2). In terms of the symbols for individual treatment combinations, the design is given in Table 8.12-2.

Table 8.12-1

Level of A	Rep 1		Rep 2		Rep 3	
	Block 1	Block 2	Block 3	Block 4	Block 5	Block 6
a_0	K_0	K_1	K_1	K_0	K_1	K_0
a_1	K_1	K_0	K_0	K_1	K_1	K_0
a_2	K_1	K_0	K_1	K_0	K_0	K_1

Table 8.12-2

Rep 1		Rep 2		Rep 3	
Block 1	Block 2	Block 3	Block 4	Block 5	Block 6
(000)	(001)	(001)	(000)	(001)	(000)
(011)	(010)	(010)	(011)	(010)	(011)
(101)	(100)	(100)	(101)	(101)	(100)
(110)	(111)	(111)	(110)	(110)	(111)
(201)	(200)	(201)	(200)	(200)	(201)
(210)	(211)	(210)	(211)	(211)	(210)

The spacings within the blocks designate treatment combinations belonging to different K sets.

The computational procedures for all factorial effects, except the BC and ABC interactions, are identical to those for a factorial experiment in which there is no confounding. The BC and ABC interactions require adjustment for block effects. A simplified procedure for computing the BC interaction will be outlined for the case in which there is no confounding. The adjustments required for the partial confounding with block effects will then be obtained.

Assuming that the BC interaction were not confounded with blocks, the total of all observations in the set K_0, which will be designated by the symbol $K_{0.}$, may be shown to be an estimate of the following parameters of the linear model (assuming one observation per cell):

$$K_{0.} \doteq 18\mu + 18(\beta\gamma)_{K_0}.$$

The symbol $(\beta\gamma)_{K_0}$ denotes the K_0 component of the BC interaction. Similarly the total of all observations in set K_1, designated by the symbol $K_{1.}$, may be shown to estimate

$$K_{1.} \doteq 18\mu + 18(\beta\gamma)_{K_1}.$$

Hence the difference between these two totals provides an estimate of

$$K_{0.} - K_{1.} \doteq 18(\beta\gamma)_{K_0} - 18(\beta\gamma)_{K_1}.$$

Since the BC interaction has only these two components,

$$(\beta\gamma)_{K_0} + (\beta\gamma)_{K_1} = 0, \qquad \text{and hence} \qquad (\beta\gamma)_{K_0} = -(\beta\gamma)_{K_1}.$$

(The sum of all components of an interaction term in the general linear model is assumed to be zero, since these terms are measured in deviation units about the grand mean.)

Suppose that the symbol $(\beta\gamma)$ is used to denote $(\beta\gamma)_{K_0}$. Then

$$-(\beta\gamma) = (\beta\gamma)_{K_1}.$$

Hence $$K_{0.} - K_{1.} \doteq 18(\beta\gamma) - [-18(\beta\gamma)] = 36(\beta\gamma).$$

Thus, if there were no confounding with blocks, the difference between the sum of all observations in set K_0 and all observations in set K_1 estimates a parameter which is equal in absolute value to the difference between the K components of the BC interaction. Since the BC interaction has only one degree of freedom, there is only one independent parameter associated with it. In terms of the parameters of the model, disregarding the experimental error, the sum of squares for the BC interaction would be

$$\text{SS}_{BC} \doteq 18(\beta\gamma)^2_{K_0} + 18(\beta\gamma)^2_{K_1} = 36(\beta\gamma)^2.$$

Thus a computational formula for the BC interaction, if there were no confounding with blocks, would be

$$\text{SS}_{BC} = \frac{(K_{0.} - K_{1.})^2}{36} \doteq 36(\beta\gamma)^2.$$

In the design in Table 8.12-2, however, the BC interaction is partially confounded with block effects. The difference between the total of all observations in the set K_0 and all observations in the set K_1 actually estimates a mixture of $(\beta\gamma)$ and block effects, i.e.,

$$K_{0.} - K_{1.} \doteq 36(\beta\gamma) + 2(\pi_2 + \pi_4 + \pi_6 - \pi_1 - \pi_3 - \pi_5),$$

where the π's represent block effects. If P_i represents the total of all observations in block i,

$$(\tfrac{1}{3})(P_2 + P_4 + P_6 - P_1 - P_3 - P_5)$$
$$\doteq 4(\beta\gamma) + 2(\pi_2 + \pi_4 + \pi_6 - \pi_1 - \pi_3 - \pi_5).$$

The quantity on the left-hand side of this last equation may be used to adjust the difference $K_{0.} - K_{1.}$ for block effects. The adjusted difference takes the form

$$Q = K_{0.} - K_{1.} - (\tfrac{1}{3})(P_2 + P_4 + P_6 - P_1 - P_3 - P_5)$$
$$\doteq 32(\beta\gamma).$$

Thus the adjusted difference between the K totals provides 32 effective observations in the estimation of $(\beta\gamma)$; but if there were no confounding with blocks, the unadjusted difference between the K totals would provide 36 effective observations. Hence the relative effective information on BC is $\frac{32}{36}$, or $\frac{8}{9}$. The adjusted sum of squares for BC takes the form

$$SS'_{BC} = \frac{Q^2}{32} \doteq 32(\beta\gamma)^2.$$

The problem of finding the appropriate adjustment for the ABC interaction follows the same general pattern as that which has just been indicated for the BC interaction. A simplified procedure for finding the ABC interaction will first be outlined for the case in which there is no confounding. The symbol K_{0i} will be used to designate the subset of all treatment combinations in the set K_0 which are at level a_i; similarly the symbol K_{1i} will be used to designate the subset of treatment combinations in the set K_1 which are at level a_i. The totals for all observations in the respective subsets will be denoted by the symbols $K_{0i.}$ and $K_{1i.}$. If the symbol $(\alpha\beta\gamma)_{K_{0i}}$ denotes an effect associated with the ABC interaction,

$$K_{0i.} - K_{1i.} \doteq 6(\alpha\beta\gamma)_{K_{1i}} - 6(\alpha\beta\gamma)_{K_{1i}} + 12(\beta\gamma).$$

Since the sum of the components of an interaction at a fixed level of one of the factors is zero,

$$(\alpha\beta\gamma)_{K_{0i}} + (\alpha\beta\gamma)_{K_{1i}} = 0; \qquad \text{hence} \qquad (\alpha\beta\gamma)_{K_{0i}} = -(\alpha\beta\gamma)_{K_{1i}}.$$

If the symbol $(\alpha\beta\gamma)_i$ designates either $(\alpha\beta\gamma)_{K_{0i}}$ or $-(\alpha\beta\gamma)_{K_{1i}}$, assuming there were no confounding with blocks,

$$K_{0i.} - K_{1i.} \doteq 12(\alpha\beta\gamma)_i + 12(\beta\gamma).$$

Specifically, for each level of factor A,

$$K_{00.} - K_{10.} \doteq 12(\alpha\beta\gamma)_0 + 12(\beta\gamma),$$
$$K_{01.} - K_{11.} \doteq 12(\alpha\beta\gamma)_1 + 12(\beta\gamma),$$
$$K_{02.} - K_{12.} \doteq 12(\alpha\beta\gamma)_2 + 12(\beta\gamma).$$

Thus, if no confounding were present,

$$\frac{\Sigma(K_{0i.} - K_{1i.})^2}{12} \doteq 12(\alpha\beta\gamma)_0^2 + 12(\alpha\beta\gamma)_1^2 + 12(\alpha\beta\gamma)_2^2 + 36(\beta\gamma)^2.$$

The sum of the first three terms on the right-hand side of the last equation defines SS_{ABC} in terms of the parameters of the model; the last term defines SS_{BC}. (In both cases the error component has not been included.) Hence

$$\frac{\Sigma(K_{0i.} - K_{1i.})^2}{12} = SS_{ABC} + SS_{BC}.$$

Thus, if there were no confounding, a computational formula for SS_{ABC} would be given by

$$SS_{ABC} = \frac{\Sigma(K_{0i.} - K_{1i.})^2}{12} - \frac{(K_{0.} - K_{1.})^2}{36}.$$

In words, the ABC sum of squares is obtained by summing the BC interaction at each level of factor A and then subtracting the overall BC interaction.

$$SS_{ABC} = \Sigma SS_{BC \text{ at } a_i} - SS_{BC}.$$

For factorial experiments in which some of the factors are at two levels, this general computational procedure is simpler than direct caclulation of the three-factor interaction.

In the design under consideration the difference $K_{00} - K_{10}$ is not free of block effects. For this design

$$K_{00} - K_{10} \doteq 12(\alpha\beta\gamma)_0 + 12(\beta\gamma) + 2(\pi_1 + \pi_4 + \pi_6 - \pi_2 - \pi_3 - \pi_5).$$

The blocks in which K_0 appears at level a_0 have positive signs; the blocks in which K_1 appears at level a_0 have negative signs. In obtaining the adjustment for block effects it is more convenient to work with the expression

$$3(K_{00} - K_{10}) \doteq 36(\alpha\beta\gamma)_0 + 36(\beta\gamma) + 6(\pi_1 + \pi_4 + \pi_6 - \pi_2 - \pi_3 - \pi_5).$$

The adjustment for block effects requires the term

$$\begin{aligned}(\text{adj } a_0) &= P_1 + P_4 + P_6 - P_2 - P_3 - P_5\\ &\doteq 16(\alpha\beta\gamma)_0 + 4(\beta\gamma) + 6(\pi_1 + \pi_4 + \pi_6 - \pi_2 - \pi_3 - \pi_5).\end{aligned}$$

The adjusted difference used to obtain the sum of squares for the three-factor interaction is

$$3R_0 = 3(K_{00.} - K_{10.}) - (\text{adj } a_0) - Q \doteq 20(\alpha\beta\gamma)_0,$$

where Q is the quantity used in the computation of SS'_{BC}. Other adjustments for the levels of factor A are

$$\begin{aligned}(\text{adj } a_1) &= P_2 + P_3 + P_6 - P_1 - P_4 - P_5,\\ (\text{adj } a_2) &= P_2 + P_4 + P_5 - P_1 - P_3 - P_6.\end{aligned}$$

The adjusted differences used in the computation of SS'_{ABC} are

$$3R_1 = 3(K_{01.} - K_{11.}) - (\text{adj } a_1) - Q \doteq 20(\alpha\beta\gamma)_1,$$

$$3R_2 = 3(K_{02.} - K_{12.}) - (\text{adj } a_2) - Q \doteq 20(\alpha\beta\gamma)_2.$$

An adjusted difference of the form $3R_i$ provides 20 effective observations in the estimation of the parameter $(\alpha\beta\gamma)_i$, whereas the corresponding unadjusted difference provides 36 effective observations. Hence the relative information for the ABC interaction in this design is $\frac{20}{36} = \frac{5}{9}$. A computational formula for the adjusted sum of squares for the ABC interaction is given by

$$SS'_{ABC} = \frac{\Sigma(3R_i)^2}{9(\frac{20}{3})} \doteq (\tfrac{20}{3})[(\alpha\beta\gamma)_0^2 + (\alpha\beta\gamma)_1^2 + (\alpha\beta\gamma)_2^2].$$

If there were no confounding with blocks,

$$SS_{ABC} \doteq 12[(\alpha\beta\gamma)_0^2 + (\alpha\beta\gamma)_1^2 + (\alpha\beta\gamma)_2^2],$$

where the error component has been disregarded. The ratio $(\frac{20}{3})/12 = \frac{5}{9}$, the relative information for ABC. The overall analysis of the balanced design in Table 8.12-2 is given in Table 8.12-3. In the underlying model, all factors are assumed to be fixed, blocks are assumed to be random, and the block $\times$ treatment interactions are assumed to be zero or negligible. Only under these stringent assumptions are the adjustments for block effects valid. In making F tests on all factorial effects, MS_{res} forms the denominator of all F ratios. In making comparisons between adjusted cell means, the effective number of observations on the component parts should be taken into account.

Table 8.12-3

Source of variation	SS	df	MS	E(MS)
Between blocks		5		
Replications		2		
Blocks within reps		3		
Within blocks		30		
A		2		$\sigma_\varepsilon^2 + nqr\sigma_\alpha^2$
B		1		$\sigma_\varepsilon^2 + npr\sigma_\beta^2$
C		1		$\sigma_\varepsilon^2 + npq\sigma_\gamma^2$
AB		2		$\sigma_\varepsilon^2 + nr\sigma_{\alpha\beta}^2$
AC		2		$\sigma_\varepsilon^2 + nq\sigma_{\alpha\gamma}^2$
$(BC)'$		1		$\sigma_\varepsilon^2 + (\frac{8}{9})np\sigma_{\beta\gamma}^2$
$(ABC)'$		2		$\sigma_\varepsilon^2 + (\frac{5}{9})n\sigma_{\alpha\beta\gamma}^2$
Residual		19		σ_ε^2

The adjustment for $\overline{BC}_{00}$ is obtained by considering the parameters actually estimated by the total of all observations under treatment combination bc_{00}, the levels of factor A being disregarded. There are nine such observations.

$$BC_{00} \doteq 9\mu + 9\beta_0 + 9\gamma_0 + 9(\beta\gamma)_{00} + (\pi_1 + \pi_3 + \pi_5) + 2(\pi_2 + \pi_4 + \pi_6).$$

To obtain an adjustment for BC_{00}, one must construct an expression which estimates the block effects on the right-hand side of this last equation. This expression is given by

$$(\text{adj } BC_{00}) = (\tfrac{1}{6})(P_1 + P_3 + P_5) + (\tfrac{1}{3})(P_2 + P_4 + P_6) - (\tfrac{1}{32})Q - (\tfrac{1}{4})G$$
$$\doteq (\pi_1 + \pi_3 + \pi_5) + 2(\pi_2 + \pi_4 + \pi_6).$$

The adjusted total for the cell bc_{00} is thus

$$BC'_{00} = BC_{00} - (\text{adj } BC_{00}) \doteq 9\mu + 9\beta_0 + 9\gamma_0 + 9(\beta\gamma)_{00}.$$

The adjusted mean for cell bc_{00} is given by

$$\overline{BC}'_{00} = \frac{BC'_{00}}{9} \doteq \mu + \beta_0 + \gamma_0 + (\beta\gamma)_{00}.$$

Since the treatment combination bc_{00} and bc_{11} both belong to the set K_0, and since the design is symmetrical with respect to the parts of K_0, the adjustment for $\overline{BC}_{11}$ is the same as that used for $\overline{BC}_{00}$.

By similar arguments and use of the relation that $(\beta\gamma)_{K_0} = -(\beta\gamma)_{K_1}$, the adjustment for both $\overline{BC}_{01}$ and $\overline{BC}_{10}$ may be shown to be

$$(\text{adj } BC_{01}) = (\text{adj } BC_{10}) = (\tfrac{1}{3})(P_1 + P_3 + P_5)$$
$$+ (\tfrac{1}{6})(P_2 + P_4 + P_6) + (\tfrac{1}{32})Q - (\tfrac{1}{4})G.$$

In the arguments that have been used in the course of arriving at various adjustments, certain assumptions about restrictions on the parameters in the general linear model have been invoked. These restrictions are made explicit in Table 8.12-4.

Table 8.12-4

	c_0		c_1	
	b_0	b_1	b_0	b_1
a_0	$(\alpha\beta\gamma)_0$	$-(\alpha\beta\gamma)_0$	$-(\alpha\beta\gamma)_0$	$(\alpha\beta\gamma)_0$
a_1	$(\alpha\beta\gamma)_1$	$-(\alpha\beta\gamma)_1$	$-(\alpha\beta\gamma)_1$	$(\alpha\beta\gamma)_1$
a_2	$(\alpha\beta\gamma)_2$	$-(\alpha\beta\gamma)_2$	$-(\alpha\beta\gamma)_2$	$(\alpha\beta\gamma)_2$
Sum	0	0	0	0

	b_0	b_1
c_0	$(\beta\gamma)$	$-(\beta\gamma)$
c_1	$-(\beta\gamma)$	$(\beta\gamma)$
Sum	0	0

8.13 Numerical Example of 3 × 2 × 2 Factorial Experiment in Blocks of Size 6

Data for this type of design will generally take the form given in Table 8.13-1. The observation is given opposite the symbol for the treatment

Table 8.13-1

Rep 1		Rep 2		Rep 3		
Block 1	Block 2	Block 3	Block 4	Block 5	Block 6	
(000) 4	(001) 10	(001) 5	(000) 3	(001) 5	(000) 3	
(011) 5	(010) 20	(010) 10	(011) 4	(010) 10	(011) 1	
(101) 15	(100) 5	(100) 15	(101) 5	(101) 10	(100) 0	
(110) 20	(111) 10	(111) 15	(110) 5	(110) 5	(111) 5	
(201) 20	(200) 10	(201) 10	(200) 10	(200) 10	(201) 20	
(210) 10	(211) 1	(210) 5	(211) 5	(211) 4	(210) 5	
$P_1 = 74$	$P_2 = 56$	$P_3 = 60$	$P_4 = 32$	$P_5 = 44$	$P_6 = 34$	$G = 300$

combination. The make-up of the blocks is that given in Table 8.12-1. In the behavioral sciences the blocks may correspond to an individual or to a group of individuals.

Summary data for the factorial effects are given in Table 8.13-2. From

Table 8.13-2

(i)

	b_0		b_1		
	c_0	c_1	c_0	c_1	Total
a_0	10	20	40	10	80
a_1	20	30	30	30	110
a_2	30	50	20	10	110
Total	60	100	90	50	300

(ii)

	b_0	b_1
a_0	30	50
a_1	50	60
a_2	80	30
Total	160	140

(iii)

	c_0	c_1
a_0	50	30
a_1	50	60
a_2	50	60
Total	150	150

(iv)

	c_0	c_1
b_0	60	100
b_1	90	50
Total	150	150

part iv of this table one obtains $K_{0.}$ and $K_{1.}$:

$$K_{0.} = 60 + 50 = 110,$$
$$K_{1.} = 90 + 100 = 190.$$

From the block totals in Table 8.13-1, one finds

$$P_2 + P_4 + P_6 - P_1 - P_3 - P_5 = -56.$$

Hence $$Q = K_{0.} - K_{1.} - \frac{P_2 + P_4 + P_6 - P_1 - P_3 - P_5}{3}$$

$$= 110 - 190 - \frac{(-56)}{3}$$

$$= -61.33.$$

Having the numerical value of Q, one obtains the adjusted sum of squares for BC from the relation

$$SS'_{BC} = \frac{Q^2}{32} = \frac{(-61.33)^2}{32} = 117.54.$$

Expressions of the form $3(K_{0i.} - K_{1i.})$ are obtained from part i of Table 8.13-2.

$$\begin{aligned} 3(K_{00.} - K_{10.}) &= 3[(10 + 10) - (20 + 40)] = -120, \\ 3(K_{01.} - K_{11.}) &= 3[(20 + 30) - (30 + 30)] = -30, \\ 3(K_{02.} - K_{12.}) &= 3[(30 + 10) - (50 + 20)] = -90. \end{aligned}$$

Adjustments for the above expressions are, respectively,

$$\begin{aligned} (\text{adj } a_0) &= P_1 + P_4 + P_6 - P_2 - P_3 - P_5 = -20 \\ (\text{adj } a_1) &= P_2 + P_3 + P_6 - P_1 - P_4 - P_5 = 0 \\ (\text{adj } a_2) &= P_2 + P_4 + P_5 - P_1 - P_3 - P_6 = -36 \\ \hline \text{Sum} &= P_2 + P_4 + P_6 - P_1 - P_3 - P_5 = -56 \end{aligned}$$

The R's required for computing SS'_{ABC} are given by

$$\begin{aligned} 3R_0 &= 3(K_{00.} - K_{10.}) - (\text{adj } a_0) - Q = -38.67 \\ 3R_1 &= 3(K_{01.} - K_{11.}) - (\text{adj } a_1) - Q = 31.33 \\ 3R_2 &= 3(K_{02.} - K_{12.}) - (\text{adj } a_2) - Q = 7.33 \\ \hline \Sigma(3R) &= 3(-80) - (-56) - 3(-61.33) = -\ .01 \end{aligned}$$

Within rounding error, $\Sigma(3R) = 0$.

The adjusted sum of squares for ABC is

$$SS'_{ABC} = \frac{\Sigma(3R_i)^2}{60} = 42.18.$$

A summary of the computation of all sums of squares required in the analysis of variance is given in Table 8.13-3. The residual term in this analysis may be obtained either by subtracting the total of the adjusted treatment sums of squares from $SS_{\text{w. block}}$ or by the method indicated in the table.

The relative efficiency of a design involving small block size as compared with a completely randomized design depends in large part upon the relative magnitudes of $MS_{\text{block w. rep}}$ and MS_{res} (adj). For the data under consideration these mean squares are, respectively, 33.55 and 20.37. The smaller the latter mean square relative to the former, the more efficient the design involving the smaller block size.

Table 8.13-3

(1) = $G^2/36$	= 2500	(7) = $[\Sigma(AC)^2]/6$	= 2600.00
(2) = ΣX^2	= 3618	(8) = $[\Sigma(BC)^2]/9$	= 2688.89
(3) = $(\Sigma A^2)/12$	= 2550.00	(9) = $[\Sigma(ABC)^2]/3$	= 3066.67
(4) = $(\Sigma B^2)/18$	= 2511.11	(10) = $(\Sigma P^2)/6$	= 2721.33
(5) = $(\Sigma C^2)/18$	= 2500.00	(11) = $(\Sigma\ \text{Rep}^2)/12$	= 2620.67
(6) = $[\Sigma(AB)^2]/6$	= 2800.00		

SS_{blocks}	= (10) − (1)	= 221.33
SS_{reps}	= (11) − (1)	= 120.67
$SS_{\text{blocks w. rep}}$	= (10) − (11)	= 100.66
$SS_{\text{w. block}}$	= (2) − (10)	= 896.67
SS_A	= (3) − (1)	= 50.00
SS_B	= (4) − (1)	= 11.11
SS_C	= (5) − (1)	= 0
SS_{AB}	= (6) − (3) − (4) + (1)	= 238.89
SS_{AC}	= (7) − (3) − (5) + (1)	= 50.00
SS_{BC}(unadj)	= (8) − (4) − (5) + (1)	= 177.78
SS_{BC}(adj)	= SS'_{BC}	= (117.54)
SS_{ABC}(unadj)	= (9) − (6) − (7) − (8) + (3) + (4) + (5) − (1)	= 38.89
SS_{ABC}(adj)	= SS'_{ABC}	= (42.18)
SS_{res}(unadj)	= (2) − (9) − (10) + (1)	= 330.00
SS_{res}(adj)	= SS_{res}(unadj) + SS_{BC}(unadj) + SS_{ABC}(unadj) − SS_{BC}(adj) − SS_{ABC}(adj)	= 386.95

To illustrate the computation of the adjustments for the BC means,

$$\begin{aligned}(\text{adj } BC_{00}) &= (\tfrac{1}{6})(P_1 + P_3 + P_5) + (\tfrac{1}{3})(P_2 + P_4 + P_6) - (\tfrac{1}{32})Q - (\tfrac{1}{4})G \\ &= (\tfrac{1}{6})(178) + (\tfrac{1}{3})(122) - (-1.92) - 75 \\ &= 29.67 + 40.67 + 1.92 - 75 = -2.74.\end{aligned}$$

The adjusted mean is

$$\overline{BC}'_{00} = \overline{BC}_{00} - \frac{(-2.74)}{9} = 6.67 - \frac{(-2.74)}{9} = 6.97.$$

The adjustment for $\overline{BC}_{01}$ will be numerically equal but opposite in sign to the adjustment for $\overline{BC}_{00}$. Thus

$$\overline{BC}'_{01} = \overline{BC}_{01} - \frac{2.74}{9} = 11.11 - \frac{2.74}{9} = 10.81.$$

The restrictions on the parameters underlying the linear model for a $3 \times 2 \times 2$ factorial experiment as shown in Table 8.12-3 may be illustrated numerically through use of the data in Table 8.13-2. These restrictions also hold for estimates of these parameters. For example, assuming no confounding,

$$(\alpha\beta\gamma)_{000} \doteq \frac{ABC_{000}}{3} - \frac{AB_{00}}{6} - \frac{AC_{00}}{6} - \frac{BC_{00}}{9} + \frac{A_0}{12} + \frac{B_0}{18} + \frac{C_0}{18} - \frac{G}{36}$$

$$\doteq -1.11.$$

According to Table 8.12-3, $(\alpha\beta\gamma)_{000} = -(\alpha\beta\gamma)_{000}$. The latter parameter is estimated by

$$(\alpha\beta\gamma)_{001} \doteq \frac{ABC_{001}}{3} - \frac{AB_{00}}{6} - \frac{AC_{01}}{6} - \frac{BC_{01}}{9} + \frac{A_0}{12} + \frac{B_0}{18} + \frac{C_1}{18} - \frac{G}{36}$$

$$\doteq 1.11.$$

Thus $$(\alpha\beta\gamma)_{000} = -(\alpha\beta\gamma)_{001} = (\alpha\beta\gamma)_0 \doteq -1.11.$$

8.14 3 × 3 × 3 × 2 Factorial Experiment in Blocks of Size 6

The principles used in the construction of the 3 × 2 × 2 factorial design for block size 6 may be extended to cover a 3 × 3 × 3 × 2 factorial experiment with block size 6. To illustrate the construction of this design, the following notation will be used:

L_{ij} = set of all treatment combinations satisfying both the relations

$$x_1 + x_2 = i \pmod 3 \qquad \text{and} \qquad x_1 + x_3 = j \pmod 3).$$

For example, the fourth factor being disregarded, the treatment combinations in the set L_{00} have the form (000−), (122−), and (211−). As another example, the treatment combinations in the set L_{12} have the form (012−), (101−), and (220−). For each of the treatment combinations in the set L_{12} the relations $x_1 + x_2 = 1 \pmod 3$ and $x_1 + x_3 = 2 \pmod 3$ are satisfied. Each of the sets L_{ij} consists of three treatment combinations at level d_0 and three at level d_1.

The design in Table 8.14-1 represents a partially balanced design. The actual treatment combinations in this design are given in Table 8.14-2. A

Table 8.14-1

Level of D	Block:	1	2	3	4	5	6	7	8	9
d_0		L_{00}	L_{01}	L_{02}	L_{10}	L_{11}	L_{12}	L_{20}	L_{21}	L_{22}
d_1		L_{12}	L_{22}	L_{11}	L_{01}	L_{20}	L_{21}	L_{02}	L_{00}	L_{10}

minimum of four replications is required for complete balance. To construct the latter, the defining relations for i and j may be changed; i.e., the relations

$$x_1 + x_2 = i \pmod 3), \qquad x_1 + 2x_3 = j \pmod 3)$$

Table 8.14-2

Block:	1	2	3	4	5	6	7	8	9
	(0000)	(0010)	(0020)	(0100)	(0110)	(0120)	(0200)	(0210)	(0220)
	(1220)	(1200)	(1210)	(1020)	(1000)	(1010)	(1120)	(1100)	(1110)
	(2110)	(2120)	(2100)	(2210)	(2220)	(2200)	(2010)	(2020)	(2000)
	(0121)	(0221)	(0111)	(0011)	(0201)	(0211)	(0021)	(0001)	(0101)
	(1011)	(1111)	(1001)	(1201)	(1121)	(1101)	(1211)	(1221)	(1021)
	(2201)	(2001)	(2221)	(2121)	(2011)	(2021)	(2101)	(2111)	(2211)

will define sets of L's which could make up a second replication. The assignment of the L's defined in this manner would be identical to that shown in Table 8.14-1.

The design given in this table has the following restrictions imposed upon the assignment of the L's to the blocks: (1) All the possible L's occur once and only once at each level of factor D. (2) Within each block a subscript does not occur twice in the same position; that is, L_{01} and L_{02} cannot occur within the same block, since the subscript zero would be repeated within the same position within the same block. (3) The sum of the subscripts (modulo 3) for L's within the same block must be equal. For example, in block 9 one finds L_{22} and L_{10} in the same block; the sum of the subscripts for L_{22} is $2 + 2 = 1$ (mod 3), and the sum of the subscripts for L_{10} is $1 + 0 = 1$ (mod 3).

In the original definition of L_{ij}, the relations used correspond to the AB and AC components of the $A \times B$ and $A \times C$ interactions, respectively. The relative information on these components is $\frac{3}{4}$. The generalized interactions of the AB and AC components are BC^2 and AB^2C^2. The relative information on BC^2 is $\frac{3}{4}$, but there is no within-block information on the AB^2C^2 component. The ABC, ACD, and B^2C^2D components are also partially confounded with block effects; the relative within-block information on each of these components is $\frac{1}{4}$. However, the AB^2C^2D component is not confounded with block effects.

To show that the design in Table 8.14-1 provides no within-block information on the AB^2C^2 component, it is necessary to show that no comparisons which belong to this component can be obtained from information provided by a single block. The treatment combinations defining the AB^2C^2 sets are located in the following blocks:

	Blocks
$x_1 + 2x_2 + 2x_3 = 0 \pmod 3$	1, 6, 8
$x_1 + 2x_2 + 2x_3 = 1 \pmod 3$	3, 5, 7
$x_1 + 2x_2 + 2x_3 = 2 \pmod 3$	2, 4, 9

Thus no single block contains treatment combinations which belong to more than one of the AB^2C^2 sets. Hence any comparison between such sets will involve differences between blocks.

The picture with respect to the AB^2C^2 component may be contrasted with that presented by the AB component, for which there is $\frac{3}{4}$ relative within-block information. The blocks which contain treatment combinations belonging to the AB sets are the following:

		Blocks
J_0:	$x_1 + x_2 = 0 \pmod 3$	1, 2, 3, 4, 7, 8
J_1:	$x_1 + x_2 = 1 \pmod 3$	1, 3, 4, 5, 6, 9
J_2:	$x_1 + x_2 = 2 \pmod 3$	2, 5, 6, 7, 8, 9

Thus within-block information on the AB component is available on the following comparisons in the blocks indicated:

	Blocks
J_0 versus J_1	1, 3, 4
J_0 versus J_2	2, 7, 8
J_1 versus J_2	5, 6, 9

In the design given in Table 8.14-1 each J set occurs in the same block with each of the other J sets three times.

Procedures for obtaining the adjusted sums of squares for the two-factor interactions which are partially confounded with block effects will be illustrated by working with the AB (or J) component. The symbol J_0 will be used to represent the set of treatment combinations which satisfy the relation $x_1 + x_2 = 0 \pmod 3$. For the design in Table 8.14-1, these treatment combinations occur in cells of the form L_{0j}, that is, cells in which the first subscript of an L is zero. The symbol $J_{0.}$ will be used to designate the sum of all treatment combinations in the set J_0. In terms of the parameters of the general linear model,

$$2J_{0.} = 2\sum_j L_{0j} \doteq 36\mu + 36(\alpha\beta)_{J_0} + 6(\pi_1 + \pi_2 + \pi_3 + \pi_4 + \pi_7 + \pi_8).$$

In words, the right-hand side of the last expression indicates that the total $J_{0.}$ is partially confounded with block effects. The symbol $\Sigma P_{L_{0j}}$ will be used to designate the sum of totals for blocks in which treatment combinations belonging to J_0 are located. Thus

$$\Sigma P_{L_{0j}} = P_1 + P_2 + P_3 + P_4 + P_7 + P_8.$$

In terms of the parameters of the general linear model,

$$\Sigma P_{L_{0j}} \doteq 36\mu + 9(\alpha\beta)_{J_0} + 6(\pi_1 + \pi_2 + \pi_3 + \pi_4 + \pi_7 + \pi_8).$$

A quantity convenient for use in obtaining the adjusted sum of squares for the J component of $A \times B$ is

$$2Q_{J_0} = 2J_{0.} - \Sigma P_{L_{0j}} \doteq 27(\alpha\beta)_{J_0}.$$

The coefficient of $(\alpha\beta)_{J_0}$ is 36 in the $2J_{0.}$ total and 27 in the $2Q_{J_0}$ total. The ratio $\frac{27}{36} = \frac{3}{4}$ gives the relative within-block information for the AB component.

The other quantities required for the computation of the adjusted sum of squares for the AB component are

$$2Q_{J_1} = 2J_{1.} - \Sigma P_{L_{1j}} \doteq 27(\alpha\beta)_{J_1},$$

where $P_{L_{1j}}$ represents a total for a block containing treatment combinations belonging to the set J_1, and

$$2Q_{J_2} = 2J_{2.} - \Sigma P_{L_{2j}} \doteq 27(\alpha\beta)_{J_2}.$$

The adjusted sum of squares for the AB component is given by

$$SS'_{J(AB)} = \frac{\Sigma(2Q_{J_i})^2}{108} = (\tfrac{27}{2})\Sigma(\alpha\beta)^2_{J_i}.$$

(The error component has been disregarded in this last expression.) If the AB component were not partially confounded with block effects, the sum of squares for this component would be given by

$$SS_{J(AB)} = \frac{\Sigma(2J_{i.})^2}{144} - \frac{[\Sigma(2J_i)]^2}{432} \doteq 18\Sigma(\alpha\beta)^2_{J_i}.$$

[The ratio of the coefficients of $\Sigma(\alpha\beta)^2$ provides the formal definition of relative within-block information. In this case the ratio is $(\frac{27}{2})/18 = \frac{3}{4}$.]

Adjustments for the AC component of $A \times C$ are obtained in an analogous manner. If K_j designates the set of treatment combinations which satisfy the relation $x_1 + x_3 = j \pmod 3$, and if $K_{j.}$ designates the sum of all observations on treatment combinations belonging to the set K_j, then

$$2Q_{K_j} = 2K_{j.} - \sum_i P_{L_{ij}} \doteq 27(\alpha\gamma)_{K_j}.$$

For example, $$2Q_{K_0} = 2K_{0.} - \sum_i P_{L_{i0}} \doteq 27(\alpha\gamma)_{K_0}.$$

The adjusted sum of squares for the AC component is

$$SS'_{K(AC)} = \frac{\Sigma(2Q_{K_j})^2}{108} \doteq (\tfrac{27}{2})\Sigma(\alpha\gamma)^2_{K_j}.$$

Adjustments for the BC^2 component may be cast in the following form:

$$2Q_{(BC^2)_m} = 2(BC^2)_{m.} + \Sigma P_{(BC^2)_m} \doteq 27(\beta\gamma^2)_m.$$

In this expression $(BC^2)_{m.}$ designates the sum of all observations on treatment combinations which satisfy the relation $x_2 + 2x_3 = m \pmod 3$, and $P_{(BC^2)_m}$ designates a total for a block containing treatment combinations which belong to the set $(BC^2)_m$. Blocks which have treatment combinations in the latter set are those containing L_{ij}'s which satisfy the relation $i + 2j = m \pmod 3$. For example, treatment combinations in the set $(BC^2)_0$ are included in L_{00}, L_{22}, and L_{11}; these L's are located in blocks 1, 2, 3, 5, 8, and 9. Hence

$$\Sigma P_{(BC^2)_0} = P_1 + P_2 + P_3 + P_5 + P_8 + P_9.$$

The process of obtaining the adjustments for the three-factor interactions which are partially confounded with block effects is simplified by utilizing the following restrictions on the underlying parameters,

$$(\alpha\beta\delta)_{J_0 \text{ at } d_0} + (\alpha\beta\delta)_{J_0 \text{ at } d_1} = 0,$$

and hence

$$(\alpha\beta\delta)_{J_0 \text{ at } d_0} = -(\alpha\beta\delta)_{J_0 \text{ at } d_1} = (\alpha\beta\delta)_{J_0}.$$

Analogous restrictions hold for J_1 and J_2.

$$(\alpha\beta\delta)_{J_1 \text{ at } d_0} = -(\alpha\beta\delta)_{J_1 \text{ at } d_1} = (\alpha\beta\delta)_{J_1},$$
$$(\alpha\beta\delta)_{J_2 \text{ at } d_0} = -(\alpha\beta\delta)_{J_2 \text{ at } d_1} = (\alpha\beta\delta)_{J_2}.$$

If there were no confounding with blocks, the difference between the totals $J_{0.\text{ at } d_0}$ and $J_{0.\text{ at } d_1}$ would be a function only of the term $(\alpha\beta\delta)_{J_0}$. This result follows from the basic definition of a three-factor interaction; in this case the ABD interaction is a measure of the difference between AB profiles at the two levels of factor D. Since the AB component is partially confounded with block effects, the $AB \times D$ component will also be partially confounded with block effects.

$$2J_{0.\text{ at } d_0} - 2J_{0.\text{ at } d_1} \doteq 36(\alpha\beta\delta)_{J_0} + 6(\pi_1 + \pi_2 + \pi_3 - \pi_4 - \pi_7 - \pi_8).$$

The last term in the above expression represents the block effects with which the three-factor interaction is confounded. Blocks containing L_{0j} at level d_0 appear with positive signs; blocks containing L_{0j} at level d_1 appear with negative signs. To adjust for block effects,

$$\Sigma P_{L_{0j} \text{ at } d_0} - \Sigma P_{L_{0j} \text{ at level } d_1} \doteq 27(\alpha\beta\delta)_{J_0} + 6(\pi_1 + \pi_2 + \pi_3 - \pi_4 - \pi_7 - \pi_8).$$

The expression $\Sigma P_{L_{0j} \text{ at } d_0}$ denotes the sum of totals for all blocks which contain L_{0j} at level d_0. The right-hand side of the above expression makes use of the fact that $(\alpha\beta\delta)_{J_0} + (\alpha\beta\delta)_{J_1} + (\alpha\beta\delta)_{J_2} = 0$. An estimate of $(\alpha\beta\delta)_{J_0}$, which is free of block effects, is given by

$$2R_{J_0} = 2J_{0.\text{ at } d_0} - 2J_{0.\text{ at } d_1} - (\Sigma P_{L_{0j} \text{ at } d_0} - \Sigma P_{L_{0j} \text{ at } d_1})$$
$$\doteq 9(\alpha\beta\delta)_{J_0}.$$

If there were no confounding with blocks, the coefficient of $(\alpha\beta\delta)_{J_0}$ would be 36. The ratio $\frac{9}{36} = \frac{3}{4}$ gives the relative within-block information on the $AB \times D$ component of the $A \times B \times D$ interaction. Other quantities required in order to compute $SS'_{AB \times D}$ are

$$
\begin{aligned}
2R_{J_1} &= 2J_{1.\,\text{at}\,d_0} - 2J_{1.\,\text{at}\,d_1} - (\Sigma P_{L_{1j}\,\text{at}\,d_0} + \Sigma P_{L_{1j}\,\text{at}\,d_1}) \\
&\doteq 9(\alpha\beta\delta)_{J_1}, \\
2R_{J_2} &= 2J_{2.\,\text{at}\,d_0} - 2J_{2.\,\text{at}\,d_1} - (\Sigma P_{L_{2j}\,\text{at}\,d_0} + \Sigma P_{L_{2j}\,\text{at}\,d_1}) \\
&\doteq 9(\alpha\beta\delta)_{J_2}.
\end{aligned}
$$

The adjusted sum of squares for $AB \times D$ is

$$
SS'_{AB \times D} = \frac{\Sigma 2R^2_{J_i}}{36} \doteq 9\Sigma(\alpha\beta\delta)^2_{J_i}.
$$

Quantities required for the adjusted sum of squares for the $AC \times D$ component have the form

$$
\begin{aligned}
2R_{K_j} &= 2K_{j.\,\text{at}\,d_0} - 2K_{j.\,\text{at}\,d_1} - \left(\sum_i P_{L_{ij}\,\text{at}\,d_0} + \sum_i P_{L_{ij}\,\text{at}\,d_1}\right) \\
&\doteq 9(\alpha\gamma\delta)_{K_j}.
\end{aligned}
$$

The corresponding quantities for the $BC^2 \times D$ component have the form

$$
\begin{aligned}
2R_{(BC^2)_m} &= (BC^2)_{m\,\text{at}\,d_0} - (BC^2)_{m\,\text{at}\,d_1} - (\Sigma P_{(BC^2)_m\,\text{at}\,d_0} + \Sigma P_{(BC^2)_m\,\text{at}\,d_1}) \\
&\doteq 9(\beta\gamma^2\delta)_m.
\end{aligned}
$$

8.15 Fractional Replication

In the designs considered up to this point, all the possible treatment combinations in the factorial set were included in the actual experiment. The number of treatment combinations in a complete factorial set becomes quite large as the number of factors increases. For example, a 2^8 factorial experiment requires a total of 256 treatment combinations; a 2^{16} factorial experiment requires 65,536 treatment combinations. If higher-order interactions can be considered negligible relative to main effects and lower-order interactions, only a selected fraction of the complete factorial set needs to be included in an experiment. The cost of the experimenter is the confounding of higher-order interactions with main effects and lower-order interactions. The gain is usually a substantial reduction in experimental effort, accompanied by a somewhat broader scope for the inferences. In cases where ambiguity is present because of confounding, the initial experiment may be supplemented by follow-up experiments specifically designed to clarify such ambiguities. There are situations in which it is highly desirable to run a sequence of fractional replications, the choice of the successive fractions being determined by the results of the preceding fractions. In the designs to be discussed in this section, all factors are considered to be fixed.

The principles for selecting the set of treatments which will provide maximum information on main effects and lower-order interactions are essentially those followed in assigning treatments to blocks in a complete factorial experiment. To illustrate the kind of confounding which arises in a fractional replication, consider a one-half replication of a 2^3 factorial experiment. Suppose that the treatments in this one-half replication correspond to the treatments in the set $(ABC)_0$. The latter are (000), (011), (101), (110). Comparisons corresponding to the main effects and interactions may be indicated schematically as follows (the columns indicate the weights for a comparison):

	A	B	C	AB	AC	BC	ABC
(000) = (1)	−	−	−	−	−	−	−
(011) = bc	−	+	+	+	+	−	−
(101) = ac	+	−	+	+	−	+	−
(110) = ab	+	+	−	−	+	+	−

Note that the pattern of the comparison corresponding to the main effect of A (the column headed A) and the pattern of the comparison corresponding to the BC interaction (the column headed BC) are identical. (These two patterns would not continue to be identical if the remaining treatments in the complete factorial were added.) Hence, if only information on these four treatment combinations is available, variation due to the main effect of A is completely confounded with variation due to the BC interaction. Similarly, by using information from the one-half replication given above, B and AC are completely confounded; C and AB are also completely confounded. The sign pattern of ABC is not that of a comparison, it will be found that variation due to ABC cannot be estimated.

The effects which cannot be distinguished in a fractional replication are called aliases—the same source of variation is, in essence, called by two different names. The aliases in a fractional replication may be determined by means of a relatively simple rule. Consider a one-half replication of a 2^4 factorial experiment defined by the relation $(ABCD)_0$. The alias of the main effect of A is given by

$$A \times ABCD = A^2BCD = BCD.$$

BCD is the generalized interaction of A and $ABCD$; it is also the alias of A in a one-half replication defined by $(ABCD)_0$ or $(ABCD)_1$. The alias of AB is

$$AB \times ABCD = A^2B^2CD = CD.$$

The alias of ABC is the generalized interaction of ABC and the defining relation, i.e.,

$$ABC \times ABCD = D.$$

In general, the alias of an effect in a fractional replication is the generalized interaction of that effect with the defining relation or relations, should there be more than one.

The analysis of variance for a one-half replication of a 2^4 factorial experiment, by using $(ABCD)_0$ or $(ABCD)_1$ as the defining relation, is given in Table 8.15-1. The aliases are indicated. (It will be noted that, if BCD is the generalized interaction of A and $ABCD$, then A will be the

Table 8.15-1 One-half Replication of a 2^4 Factorial Experiment

Source	df
A (BCD)	1
B (ACD)	1
C (ABD)	1
D (ABC)	1
AB (CD)	1
AC (BD)	1
AD (BC)	1
Within cell	$8(n - 1)$
Total	$8n - 1$

generalized interaction of BCD and $ABCD$.) Main effects are aliased with three-factor interactions, and two-factor interactions are aliased with other two-factor interactions. The four-factor interaction cannot be estimated. If three-factor and higher-order interactions may be considered negligible, then estimates of the variance due to the main effects are given by this fractional replication. There is considerable ambiguity about what interpretation should be made if two-factor interactions should prove to be significant, since pairs of two-factor interactions are completely confounded.

By way of contrast, consider a one-half replication of a 2^5 factorial experiment. Suppose that the treatments are selected through use of the relation $(ABCDE)_0$ or $(ABCDE)_1$. The analysis of variance may be outlined as indicated below (aliases are indicated in parentheses):

Source	df
Main effects (4-factor interactions)	5
Two-factor interactions (3-factor interactions)	10
Within cell	$16(n - 1)$
Total	$16n - 1$

If three-factor and higher-order interactions are negligible, this fractional replication provides information for tests on main effects as well as two-factor interactions.

A single defining relation for a 2^4 factorial experiment will select a one-half replication. If k is large, it is desirable to have a one-quarter or even a one-eighth replication. In order to select a one-fourth replication, the treatments are required to satisfy, simultaneously, two defining relations.

Suppose that these defining relations are designated U and V. The aliases of an effect E will have the form

$$E \times U, \qquad E \times V, \qquad E \times (U \times V).$$

In words, the aliases of an effect in a one-fourth replication are the generalized interactions of that effect with each of the defining relations as well as the generalized interaction of the effect with the interaction of the defining relations.

For example, consider a one-fourth replication of a 2^6 factorial experiment. Suppose that the defining relations are $(ABCE)_0$ and $(ABDF)_0$. The generalized interaction of these two relations is $(CDEF)_0$. A partial list of the aliases is as follows:

Effect	Aliases
A	*BCE, BDF, ACDEF*
B	*ACE, ADF, BCDEF*
AB	*CE, DF, ABCDEF*
CD	*ABED, ABCF, EF*

In this case, main effects are aliased with three-factor and higher-order interactions; two factor interactions are aliased with other two-factor as well as higher-order interactions.

As another example, consider a one-fourth replication of a 2^7 factorial experiment. Suppose that the defining relations for the selected treatments are $(ABCDE)_0$ and $(ABCFG)_0$. The generalized interaction of these relations is $(DEFG)_0$. A partial listing of the aliases is as follows:

Effect	Aliases
A	*BCDE, BCFG, ADEFG*
B	*ACDE, ACFG, BDEFG*
AB	*CDE, CFG, ABDEFG*
DE	*ABC, ABCDEFG, FG*

In this design main effects are aliased with four-factor and higher-order interactions. With the exception of $DE = FG$, $DG = EF$, $DF = EG$, all other two-factor interactions are aliased with three-factor or higher-order interactions. The two-factor interactions which are equated are aliased. This design yields unbiased tests on main effects and several of the two-factor interactions, provided that three-factor and higher-order interactions are negligible.

For a one-eighth replication of a 2^6 factorial experiment, three defining relations are required to select the set of treatments. If these relations are designated U, V, and W, the aliases of an effect E are given by the following interactions:

$$E \times U, \quad E \times V, \quad E \times W;$$

$$E \times (U \times V), \quad E \times (U \times W), \quad E \times (V \times W), \quad E \times (U \times V \times W).$$

In general, if m is the number of defining relations, then the number of aliases of an effect is

$$m + \binom{m}{2} + \binom{m}{3} + \cdots + \binom{m}{m}.$$

For example, when $m = 3$,

$$3 + \binom{3}{2} + \binom{3}{3} = 3 + 3 + 1 = 7.$$

Fractional Replication in Blocks. To illustrate the general method for arranging the treatments in a fractional replication into blocks,

Table 8.15-2 One-half Replication of 2^6 Factorial [Defining Relation: $(ABCDEF)_0$]

	Block 1 $(ABC)_0$				Block 2 $(ABC)_1$			
(i)	(1)	*ab*	*ac*	*bc*	*ae*	*af*	*ad*	*bd*
	abef	*ef*	*de*	*df*	*bf*	*be*	*ce*	*cf*
	acde	*acdf*	*abdf*	*acef*	*cd*	*abcd*	*abcf*	*abce*
	bcdf	*bcde*	*bcef*	*abde*	*abcdef*	*cdef*	*bdef*	*adef*

	Block 1′ $(ABD)_0$		Block 1″ $(ABD)_1$		Block 2′ $(ABD)_1$		Block 2″ $(ABD)_0$	
(ii)	(1)	*ab*	*ac*	*bc*	*ae*	*af*	*ad*	*bd*
	abef	*ef*	*de*	*df*	*bf*	*be*	*ce*	*cf*
	acde	*acdf*	*abdf*	*acef*	*cf*	*abcd*	*abcf*	*abce*
	bcdf	*bcde*	*bcef*	*abde*	*abcdef*	*cdef*	*bdef*	*adef*

consider a 2^6 factorial experiment. Suppose that a one-half replication of 64 treatments is selected by means of the relation $(ABCDEF)_0$. Now suppose that the resulting 32 treatments are subdivided into two sets of 16 treatments by means of the relation ABC, one set being $(ABC)_0$ and the other $(ABC)_1$. The resulting sets of 16 treatments are given in part i of Table 8.15-2. Suppose that the latter sets define the blocks. In the analysis of the resulting experiment, main effects will be aliased with five-factor interactions. Two-factor interactions will be aliased with four-factor interactions. Three-factor interactions will be aliased in pairs as follows:

Effect	Alias
ABC	DEF
ABD	CEF
ABE	CDF

The pair $ABC = DEF$ is confounded with differences between blocks.

Suppose that the blocks of 16 are further subdivided into blocks of size 8 by means of the relations $(ABD)_0$ and $(ABD)_1$. The resulting blocks are given in part ii of the table. The aliases remain the same as in the previous analysis, but now there is additional confounding with blocks. The three degrees of freedom confounded with between-block differences are

$$ABC = DEF, \qquad ABD = CEF,$$

as well as

$$ABC \times ABD = CD.$$

Note that the generalized interaction of DEF and CEF is also CD.

Thus in a one-half replication of a 2^6 factorial experiment in blocks of size 8, if three-factor and higher-order interactions are negligible, main effects and all two-factor interactions except CD may be tested. Some of the three-factor interactions as well as CD are confounded with block effects, but other pairs of three-factor interactions are clear of block effects, provided that the appropriate model is strictly additive with respect to block effects.

Computational Procedures. Computational procedures for a one-half replication of a 2^k factorial experiment are identical to those of a complete replication of a 2^{k-1} factorial experiment. For example, the eight treatments in a one-half replication of a 2^4 factorial experiment actually form a complete replication of a 2^3 factorial experiment if the levels of one of the factors are disregarded. If the one-half replication is defined by the relation $(ABCD)_0$, the treatments are

$$(1), ab, ac, ad, bc, bd, cd, abcd.$$

If the levels of factor D are disregarded, the treatments are

$$(1), ab, ac, a, bc, b, c, abc;$$

these are the eight treatment combinations in a 2^3 factorial experiment. Corresponding effects in the three-factor and four-factor experiments are given below:

Three-factor experiment	Four-factor experiment (one-half rep)
A	$A = BCD$
B	$B = ACD$
C	$C = ABD$
AB	$AB = CD$
AC	$AC = BD$
BC	$BC = AD$
ABC	$ABC = D$

Suppose that a one-quarter replication of a 2^5 factorial experiment is defined by the relations $(ABE)_0$ and $(CDE)_0$. The treatments in this fractional replication are

$$(1),\ ab,\ cd,\ ace,\ bce,\ ade,\ bde,\ abcde.$$

If factors D and E are disregarded, the treatments are

$$(1),\ ab,\ c,\ ac,\ bc,\ a,\ b,\ abc;$$

these are the treatments in a complete replication of a 2^3 factorial experiment. The aliases of these effects are their generalized interactions with ABE, CDE, and $ABE \times CDE = ABCD$. Thus there is the following correspondence between this 2^3 factorial experiment and the one-quarter replication of a 2^5 factorial experiment:

Five-factor experiment (one-quarter rep)
$A = BE, ACDE, BCD$
$B = AE, BCDE, ACD$
$C = ABCE, DE, ABD$
$AB = E, ABCDE, CD$
$AC = BCE, ADE, BD$
$BC = ACE, BDE, AD$
$ABC = CE, ABDE, D$

That is, if a one-quarter replication of a 2^5 factorial is analyzed as if it were a complete replication of a 2^3 factorial, the corresponding effects are indicated—thus the analysis reduces to the analysis of a complete factorial having two fewer factors.

Extensive tables of fractional replications of experiments in the 2^k series are given in Cochran and Cox (1957, pp. 276–289). The plans in these tables permit the arrangement of the treatments into blocks of various sizes. To avoid having main effects and two-factor interactions aliased with lower-order interactions, care must be taken in selecting the defining relations for fractional replications and blocks. The plans tabulated in Cochran and Cox are those which tend to minimize undesirable aliases.

Fractional Replication for Designs in the 3^k Series. Principles used in the 2^k series may be generalized to factorial experiments in the p^k series, where p is a prime number. In a 3^3 factorial experiment, a one-third replication may be constructed by any of the components of the three-factor interaction: ABC, ABC^2, AB^2C, or AB^2C^2. Each of these components will subdivide the 27 treatments in a 3^3 factorial set into three sets of 9 treatments each. If one of the sets is defined by $(ABC)_i$, where $i = 0$, 1, or 2 is used, the aliases of the main effect of factor A are

$$A \times ABC = A^2BC = AB^2C^2,$$

$$A^2 \times ABC = BC.$$

The aliases of AB^2 are

$$AB^2 \times ABC = A^2B^3C = AC^2,$$

$$(AB^2)^2 \times ABC = A^3B^5C = B^2C = BC^2.$$

In general, if the defining relation for a one-third replication in a 3^k experiment is R, then the aliases of an effect E are

$$E \times R \qquad \text{and} \qquad E^2 \times R.$$

The following correspondence may be established between a one-third replication of a 3^3 factorial and a complete replication of a 3^2 factorial; assume that the defining relation is $(ABC)_i$:

3^2 factorial experiment	3^3 factorial experiment (one-third rep)
A	$A = AB^2C^2 = BC$
B	$B = AB^2C = AC$
AB	$AB = ABC^2 = C$
AB^2	$AB^2 = AC^2 = BC^2$

From the point of view of computational procedures, a one-third replication of a 3^3 factorial is equivalent to a complete replication of a 3^2 factorial experiment. Since two-factor interactions are aliased with main effects, this plan is of little practical use unless it can be assumed that all interactions are negligible.

A one-third replication of a 3^4 factorial experiment may be constructed from any one of the components of the four-factor interaction. If one of

Table 8.15-3 One-third Replication of a 3^4 Factorial Experiment (Defining Relation: $ABCD$)

Source	df
$A = AB^2CD^2 = BCD$	2
$B = AB^2CD = ACD$	2
$C = ABC^2D = ABD$	2
$AB = ABC^2D^2 = CD$	2
$AB^2 = AC^2D^2 = BC^2D^2$	2
$AC = AB^2CD^2 = BD$	2
$AC^2 = AB^2D^2 = BC^2D$	2
$BC = AB^2C^2D = AD$	2
$BC^2 = AB^2D = AC^2D$	2
$ABC = ABCD^2 = D$	2
$ABC^2 = ABD^2 = CD^2$	2
$AB^2C = ACD^2 = BD^2$	2
$AB^2C^2 = AD^2 = BCD^2$	2
Within cell	$27(n-1)$
Total	$27n - 1$

the sets of 27 treatments defined by $(ABCD)_i$ is used, then the correspondence given in Table 8.15-3 may be established between the fractional replication of a 3^4 factorial and a complete replication of a 3^4 factorial. Main effects are seen to be aliased with three-factor and higher-order interactions. Some of the two-factor interactions are also aliased with other two-factor interactions. If, however, the two-factor interactions with factor D are negligible, then the other two-factor interactions are clear of two-factor aliases.

Assignment of treatments to blocks follows the same general principles as those given in earlier sections of this chapter. Plans for fractional replications in the 3^k series are given in Cochran and Cox (1957, pp. 290–291). The following notation systems are equivalent:

$$AB(I) = AB^2, \qquad ABC(W) = AB^2C^2, \qquad ABC(Y) = ABC^2,$$

$$AB(J) = AB, \qquad ABC(X) = AB^2C, \qquad ABC(Z) = ABC.$$

Other Fractional-replication Designs. Latin squares and Greco-Latin squares may be considered as fractional replications of factorial experiments. Experimental designs in these categories will be considered in Chap. 9.

9

LATIN SQUARES AND RELATED DESIGNS

9.1 Definition of Latin Square

A Latin square, as used in experimental design, is a balanced two-way classification scheme. Consider the following 3×3 arrangement:

$$\begin{matrix} a & b & c \\ b & c & a \\ c & a & b \end{matrix}$$

In this arrangement, each letter occurs just once in each row and just once in each column. The following arrangement also exhibits this kind of balance:

$$\begin{matrix} b & c & a \\ a & b & c \\ c & a & b \end{matrix}$$

The latter arrangement was obtained from the first by interchanging the first and second rows.

Use of this type of balance may be incorporated into a variety of designs. For example, consider an experiment involving the administration of three treatments to each of three subjects. The order in which subjects receive the treatments may be either completely randomized or randomized under the restriction of balance required for the Latin square. If the treatments are designated a_1, a_2, and a_3, a balanced arrangement with respect to order of administration is obtained by employing the following plan:

	Order 1	Order 2	Order 3
Subject 1	a_1	a_2	a_3
Subject 2	a_3	a_1	a_2
Subject 3	a_2	a_3	a_1

If this plan is followed, each treatment is administered first once, second once, and third once. If there is a systematic additive effect associated with the order of administration, this effect can be evaluated. Had order of administration been randomized independently for each subject, this balance would not in general be achieved, and evaluation of the order effect could not be made readily. (In the design outlined above, instead of individuals, groups of subjects may be assigned to the rows of the Latin square.)

Two Latin squares are orthogonal if, when they are combined, the same pair of symbols occurs no more than once in the composite square. For example, consider the following 3×3 Latin squares:

(1)			(2)			(3)		
a_1	a_2	a_3	b_2	b_3	b_1	c_1	c_2	c_3
a_2	a_3	a_1	b_3	b_1	b_2	c_3	c_1	c_2
a_3	a_1	a_2	b_1	b_2	b_3	c_2	c_3	c_1

Combining squares (1) and (2) yields the composite square

$$\begin{array}{ccc} a_1b_2 & a_2b_3 & a_3b_1 \\ a_2b_3 & a_3b_1 & a_1b_2 \\ a_3b_1 & a_1b_2 & a_2b_3 \end{array}$$

In this composite the treatment combination a_1b_2 occurs more than once; hence squares (1) and (2) are not orthogonal.

Combining squares (1) and (3) yields the following composite:

$$\begin{array}{ccc} a_1c_1 & a_2c_2 & a_3c_3 \\ a_2c_3 & a_3c_1 & a_1c_2 \\ a_3c_2 & a_1c_3 & a_2c_1 \end{array}$$

In this composite no treatment combination is repeated. There are nine possible treatment combinations that may be formed from three levels of factor A and three levels of factor C. Each of these possibilities appears in the composite. Hence squares (1) and (3) are orthogonal.

Extensive tables of sets of orthogonal Latin squares are given in Fisher and Yates (1953) and Cochran and Cox (1957). The composite square obtained by combining two orthogonal Latin squares is called a Greco-Latin square. The 3×3 Greco-Latin square obtained by combining squares (1) and (3) above may be represented schematically as follows:

11	22	33
23	31	12
32	13	21

This representation uses only the subscripts that appear with the a's and c's. Interchanging any two rows of a Greco-Latin square will still yield a Greco-Latin square. For example, interchanging the first and third rows of the above square yields the following Greco-Latin square:

32	13	21
23	31	12
11	22	33

Any two columns of a Greco-Latin square may also be interchanged without affecting the required balance.

A collection of $p - 1$ Latin squares of size $p \times p$ is said to form a *complete set* if each square in the collection is orthogonal to every other square. There can be at most $p - 1$ squares in a complete set. For the case of 4×4 squares, the following collection defines a complete set:

1	2	3	4	1	2	3	4	1	2	3	4
2	1	4	3	3	4	1	2	4	3	2	1
3	4	1	2	4	3	2	1	2	1	4	3
4	3	2	1	2	1	4	3	3	4	1	2

There are many such sets for 4×4 squares. However each set cannot contain more than three squares.

It is not always possible to find a Latin square orthogonal to a given Latin square. For example, no orthogonal squares exist for 6×6 squares. If the dimension of a square is capable of being expressed in the form (prime number)^n, where n is any integer, then a complete set of squares exists. For example, complete sets exist for squares of the following dimensions:

$$3 = 3^1, \qquad 4 = 2^2, \qquad 5 = 5^1, \qquad 8 = 2^3, \qquad 9 = 3^2.$$

In addition, there are cases in which complete sets exist even though the dimension of the square is not of the form (prime number)^n, particularly when the dimension is divisible by 4. For example, a complete set may be constructed for squares of dimension 12×12.

9.2 Enumeration of Latin Squares

The standard form of a Latin square is, by definition, that square obtained by rearranging the rows and columns until the letters in the first row and the letters in the first column are in alphabetical order. For example, the square

$$\begin{matrix} b & a & c \\ c & b & a \\ a & c & b \end{matrix} \tag{1}$$

has the standard form

$$\begin{matrix} a & b & c \\ b & c & a \\ c & a & b \end{matrix} \tag{2}$$

The standard form (2) is obtained from (1) by interchanging columns 1 and 2. All 3×3 Latin squares may be reduced to this standard form. From this standard form $(3!) \cdot (2!) - 1 = (3 \cdot 2 \cdot 1)(2 \cdot 1) - 1 = 11$ different nonstandard 3×3 Latin squares may be constructed. Hence, including the standard form, there are 12 different 3×3 Latin squares.

For 4×4 Latin squares there are four different standard forms. One of these four is given in (3).

$$\begin{matrix} a & b & c & d \\ b & a & d & c \\ c & d & b & a \\ d & c & a & b \end{matrix} \tag{3}$$

A second standard form is given by (4).

$$\begin{matrix} a & b & c & d \\ b & c & d & a \\ c & d & a & b \\ d & a & b & c \end{matrix} \tag{4}$$

From each of these standard forms $(4!)(3!) - 1 = 143$ different nonstandard 4×4 Latin squares may be constructed by the process of interchanging rows and columns. Nonstandard squares constructed from standard form (3) will be different from nonstandard squares constructed from standard form (4). Thus (3) potentially represents 144 Latin squares (143 nonstandard squares plus 1 standard square). Standard form (4) also represents potentially 144 different 4×4 Latin squares. Since there are 4 different standard forms, the potential number of different 4×4 Latin squares is $4(144) = 576$.

The number of possible standard forms of a Latin square increases quite rapidly as the dimension of the square increases. For example, a

6×6 Latin square has 9408 standard forms. Each of these forms represents potentially $(6!)(5!) = 86{,}400$ different squares. The total number of different 6×6 Latin squares that may be constructed is

$$(9408)(86{,}400) = 812{,}851{,}200.$$

In most tables, only standard forms of Latin squares are given. To obtain a Latin square for use in an experimental design, one of the standard squares of suitable dimension should be selected at random. The rows and columns of the selected square are then randomized independently. The levels of the factorial effects are then assigned at random to the rows, columns, and Latin letters of the square, respectively. To illustrate the procedure for randomizing the rows and columns of a 4×4 Latin square, suppose that square (3) is obtained from a table of Latin squares. Two random sequences of digits 1 through 4 are then obtained from tables of random sequences. Suppose that the sequences obtained are (2,4,1,3) and (3,4,1,2). The columns of square (3) are now rearranged in accordance with the first sequence. That is, column 2 is moved into the first position, column 4 into the second position, column 1 into the third position, and column 3 into the fourth position. The resulting square is

$$\begin{matrix} b & d & a & c \\ a & c & b & d \\ d & a & c & b \\ c & b & d & a \end{matrix}$$

The rows of the resulting squares are now rearranged in accordance with the second random sequence. That is, row 3 is moved to the first position, row 4 moved to the second position, row 1 moved to the third position, and row 2 moved to the fourth position. After these moves have been made, the resulting square is

$$\begin{matrix} d & a & c & b \\ c & b & d & a \\ b & d & a & c \\ a & c & b & d \end{matrix}$$

This last square represents a random rearrangement of the rows and columns of the original 4×4 standard form. This last square may be considered as a random choice from among the 144 squares represented by the standard form (3). Since (3) was chosen at random from the standard forms, the square that has just been constructed may be considered a random choice from the 576 possible different 4×4 Latin squares.

Some additional definitions of types of Latin squares will be helpful in using tables. The *conjugate* of a Latin square is obtained by interchanging the rows and columns. For example, the following squares are conjugates:

$$\begin{array}{ccc} a & b & c \\ c & a & b \\ b & c & a \end{array} \qquad \begin{array}{ccc} a & c & b \\ b & a & c \\ c & b & a \end{array}$$

The square on the right is obtained from the square on the left by writing the columns as rows. That is, the first column of the square on the left is the first row of the square on the right, the second column of the square on the left is the second row of the square on the right, etc. Conversely, the square on the left may be obtained from the square on the right by writing the rows of the latter as columns.

A one-step *cyclic permutation* of a sequence of letters is one which moves the first letter in the sequence to the extreme right, simultaneously moving all other letters in one position to the left. For example, given the sequence *abcd*, a one-step cyclic permutation yields *bcda*; a second cyclic permutation yields *cdab*; a third cyclic permutation yields *dabc*; and a fourth cyclic permutation yields *abcd*—the latter is the starting sequence. Given a sequence of p letters, a $p \times p$ Latin square may be constructed by $p - 1$ one-step cyclic permutations of these letters. In the case of the sequence *abcd*, the Latin square formed by cyclic permutations is

$$\begin{array}{cccc} a & b & c & d \\ b & c & d & a \\ c & d & a & b \\ d & a & b & c \end{array}$$

A *balanced set* of Latin squares is a collection in which each letter appears in each possible position once and only once. For a $p \times p$ square, there will be p squares in a balanced set. Given any $p \times p$ square, a balanced set may be constructed by cyclic permutations of the columns. For example, the squares given below form a balanced set:

$$\begin{array}{ccc} & \text{I} & \\ a & b & c \\ c & a & b \\ b & c & a \end{array} \qquad \begin{array}{ccc} & \text{II} & \\ b & c & a \\ a & b & c \\ c & a & b \end{array} \qquad \begin{array}{ccc} & \text{III} & \\ c & a & b \\ b & c & a \\ a & b & c \end{array}$$

Square II is obtained from square I by moving the first column of square I to the extreme right. Square III is obtained from square II by moving the first column of square II to the extreme right. The resulting set is balanced in the sense that each of the nine positions within the squares contains any given letter once and only once.

One square may be obtained from another by substitution of one letter for another. For example, given the following square:

a	*b*	*c*	*d*
b	*a*	*d*	*c*
c	*d*	*b*	*a*
d	*c*	*a*	*b*

Suppose that each of the *a*'s is replaced by a *b*, each of the *b*'s replaced by a *c*, each of the *c*'s by an *a*, and each of the *d*'s is not changed. This kind of replacement procedure is called a *one-to-one transformation.* The square resulting from this transformation is given below:

b	*c*	*a*	*d*
c	*b*	*d*	*a*
a	*d*	*c*	*b*
d	*a*	*b*	*c*

This last square has the standard form given below:

a	*b*	*c*	*d*
b	*d*	*a*	*c*
c	*a*	*d*	*b*
d	*c*	*b*	*a*

Since this latter square has a different standard form from that of the original square, the one-to-one transformation yields a square which is different from any square which can be constructed by permutation of the rows or columns of the original square.

9.3 Structural Relation between Latin Squares and Three-factor Factorial Experiments

There is an interesting structural relation between a $p \times p \times p$ factorial experiment and a $p \times p$ Latin square. This relation will be illustrated by means of a $3 \times 3 \times 3$ factorial experiment. It will be shown that the latter may be partitioned into a balanced set of 3×3 Latin squares. In the following representation of a $3 \times 3 \times 3$ factorial experiment, those treatment combinations marked *X* form one Latin square, those marked *Y* form a second, and those marked *Z* form a third. The set of three squares formed in this manner constitutes a balanced set:

	c_1			c_2			c_3		
	b_1	b_2	b_3	b_1	b_2	b_3	b_1	b_2	b_3
a_1	*X*	*Y*	*Z*	*Z*	*X*	*Y*	*Y*	*Z*	*X*
a_2	*Z*	*X*	*Y*	*Y*	*Z*	*X*	*X*	*Y*	*Z*
a_3	*Y*	*Z*	*X*	*X*	*Y*	*Z*	*Z*	*X*	*Y*

The treatment combinations marked X may be grouped as follows:

$$\begin{array}{ccc} abc_{111} & abc_{122} & abc_{133} \\ abc_{221} & abc_{232} & abc_{213} \\ abc_{331} & abc_{312} & abc_{323} \end{array}$$

This set of treatment combinations may also be represented in the following schematic form:

(X)

	b_1	b_2	b_3
a_1	c_1	c_2	c_3
a_2	c_3	c_1	c_2
a_3	c_2	c_3	c_1

In this latter arrangement the c's form a Latin square.

The treatment combinations marked Y and Z in the factorial experiment may be represented by the following Latin squares:

(Y)

	b_1	b_2	b_3
a_1	c_3	c_1	c_2
a_2	c_2	c_3	c_1
a_3	c_1	c_2	c_3

(Z)

	b_1	b_2	b_3
a_1	c_2	c_3	c_1
a_2	c_1	c_2	c_3
a_3	c_3	c_1	c_2

In terms of the modular notation used in Chap. 8, there is the following correspondence between the Latin-square fractions of the $3 \times 3 \times 3$ factorial given above and the components of the $A \times B \times C$ interaction.

Latin square	Component of $A \times B \times C$
(X)	$(AB^2C)_0$
(Y)	$(AB^2C)_2$
(Z)	$(AB^2C)_1$

That is, if one were to express the levels of the factors as a_0, a_1, a_2, etc., then the Latin square given by (X) consists of those treatment combinations which satisfy the relation

$$x_1 + 2x_2 + x_3 = 0 \pmod 3).$$

In a complete factorial experiment there is balance in the sense that treatment a_i occurs in combination with each b_j and each c_k as well as in combination with all possible pairs bc_{jk}. In a Latin square there is only partial balance; each a_i occurs in combination with each b_j and each c_k, but each a_i does not occur in combination with all possible pairs bc_{jk}. For example, in square (Y) only the following combinations of bc occur with level a_1:

$$bc_{13} \quad bc_{21} \quad bc_{32}$$

The following combinations of bc do *not* occur with level a_1:

$$bc_{11} \quad bc_{12} \quad bc_{22} \quad bc_{23} \quad bc_{31} \quad bc_{33}$$

A Latin square is balanced with respect to main effects but only partially balanced with respect to two-factor interactions.

It is of interest to note the parameters estimated by the sum of all observations at level a_1 in Latin square (Y). Assume for the moment that the cell means have the following form:

$$\begin{aligned}
\overline{ABC}_{113} &= \mu + \alpha_1 + \beta_1 + \gamma_3 + \varepsilon_{113} + \alpha\beta_{11} + \alpha\gamma_{13} + \beta\gamma_{13} \\
\overline{ABC}_{121} &= \mu + \alpha_1 + \beta_2 + \gamma_1 + \varepsilon_{121} + \alpha\beta_{12} + \alpha\gamma_{11} + \beta\gamma_{21} \\
\overline{ABC}_{132} &= \mu + \alpha_1 + \beta_3 + \gamma_2 + \varepsilon_{132} + \alpha\beta_{13} + \alpha\gamma_{12} + \beta\gamma_{32} \\
\hline
\bar{A}_1 &= \mu + \alpha_1 + 0 \ + 0 \ + \bar{\varepsilon}_1 \ \ + 0 \quad + 0 \quad + \overline{\beta\gamma}_{\alpha_1}
\end{aligned}$$

Assuming further that factors A, B, and C are fixed, it follows that

$$\sum_j \beta_j = 0, \qquad \sum_k \gamma_k = 0, \qquad \sum_j \alpha\beta_{1j} = 0, \qquad \sum_k \alpha\gamma_{1k} = 0.$$

However, $$\beta\gamma_{13} + \beta\gamma_{21} + \beta\gamma_{32} \neq 0.$$

Thus, in addition to sources of variation due to α and $\bar{\varepsilon}$, $\bar{A}$ includes a source of variation due to $\beta\gamma$. Unless $\sigma^2_{\beta\gamma} = 0$, the main effect due to factor A will be confounded with the BC interaction.

In general, when a Latin square is considered as a fractional replication of a three-factor experiment, unless $\sigma^2_{\beta\gamma} = 0$, the main effect due to factor A will be confounded with the BC interaction. Similarly, it may be shown that unless $\sigma^2_{\alpha\gamma} = 0$, the main effect due to factor B will be confounded with the AC interaction; unless $\sigma^2_{\alpha\beta} = 0$, the main effect due to factor C will be confounded with the AB interaction. The variance due to the three-factor interaction, $\sigma^2_{\alpha\beta\gamma}$, is assumed to be zero throughout.

9.4 Uses of Latin Squares

In agricultural experimentation, the cells of a $p \times p$ Latin square define p^2 experimental units. In this case, the selection of a Latin square defines the manner in which the treatments are assigned to the experimental units. That is, the procedure followed in selecting the Latin square specifies the randomization procedure necessary for the validity of statistical tests made in subsequent analyses. In its classic agricultural setting, the Latin-square design represents a single-factor experiment with restricted randomization with respect to row and column effects associated with the experimental units. In this design it is assumed that treatment effects do not interact with the row and column effects. Fisher (1951) and Wilk and Kempthorne (1957) carefully distinguish between the use of a Latin square to provide the design for the randomization procedure and the use of a Latin square for other purposes.

In biological experimentation, the Latin square provides a method for controlling individual differences among experimental units. For example, suppose that it is desired to control differences among litters as well as differences among sizes within litter for animals assigned to given treatment conditions. If there are four treatments, and if the size of the litter is four, then a 4 × 4 Latin square may be used to obtain the required balance. Use of this plan provides the experimenter with a dual balance. That is, the experimental units under any treatment are balanced

Litter	Size within litter			
	1	2	3	4
1	t_2	t_1	t_4	t_3
2	t_3	t_4	t_1	t_2
3	t_1	t_2	t_3	t_4
4	t_4	t_3	t_2	t_1

with respect to both litter and size within litter. This plan may be replicated as many times as is required. An alternative plan, not having this dual balance, would assign the experimental units at random to the treatment conditions. This plan is to be preferred to the balanced plan when the variables under control by restricted randomization do not in fact reduce the experimental error sufficiently to offset the loss in the degrees of freedom for estimating the experimental error.

In the behavioral sciences, the more usual application of the Latin square is in selecting a balanced fractional replication from a complete factorial experiment. To illustrate this procedure, consider an experiment conducted in accordance with the following plan (in this plan c_1, c_2, and c_3 represent three categories of patients):

	Drug 1	Drug 2	Drug 3
Hospital 1	c_2	c_1	c_3
Hospital 2	c_1	c_3	c_2
Hospital 3	c_3	c_2	c_1

Suppose that 15 patients are sampled from each of the hospitals, 5 patients from each of the three categories of patients. According to this design:

Patients from hospital 1 in category 2 are given drug 1.

Patients from hospital 1 in category 1 are given drug 2.

Patients from hospital 1 in category 3 are given drug 3.

Upon completion of the experiment, drug 1 is administered to some patients from each of the categories as well as some patients from each of the hospitals. This same kind of balance also holds for the other drugs.

In order to have five observations in each cell, the complete factorial would require 45 patients from each hospital, 15 patients in each category from each of the hospitals, that is, 135 patients in all. The Latin-square design requires a total of 45 patients. In this context, a Latin square represents one-third of the complete factorial experiment. Under some conditions, however, the Latin-square design may provide the experimenter with all the important information. If interactions with the hospital factor may be considered negligible relative to the hospital main effects and the drug by category interaction, then this design provides the experimenter with complete information on the main effects of drugs and categories as well as partial information on the main effects of hospitals and drug by category interaction. Prior experimentation in an area and familiarity with the subject matter are the best guides as to whether assumptions about negligible interactions are warranted. Pilot studies are also a valuable source of information about underlying assumptions.

Another important use of the Latin square in the area of the behavioral sciences is to counterbalance order effects in plans calling for repeated measures. For example, consider the following experimental plan:

	Order 1	Order 2	Order 3	Order 4
Group 1	a_3	a_1	a_2	a_4
Group 2	a_2	a_3	a_4	a_1
Group 3	a_4	a_2	a_1	a_3
Group 4	a_1	a_4	a_3	a_2

Suppose that each of the groups represents a random subsample from a larger random sample. Each of the subjects within group 1 is given treatments a_1 through a_4 in the order indicated by the columns of the Latin square. Upon completion of the experiment, each of the levels of factor A will be administered once in each of the orders. Hence there is a kind of balance with respect to the order effect. If an order effect exists, it is "controlled" in the weak sense that each treatment appears equally often in each of the orders. Implicit in this type of "control" are the assumptions of a strictly additive model with respect to the order factor. That is, the order factor is assumed to be additive with respect to the treatment and group factors; also, order is assumed not to interact with the latter factors. (Further, the homogeneity-of-covariance assumptions underlying a repeated-measure design are still required for the validity of the final tests.)

In another sense, however, order effects are not under control. Consider the administration of a_2 under order 3. For the plan given above, a_2 is preceded by a_3 and a_1 in that order. Should the sequence in which a_1 and a_3 precede a_2 have any appreciable effect, this plan does not provide an adequate control on such sequence (carry-over) effects. (The term sequence effect will be used to indicate differential effects associated with the order in which treatments precede other treatments.) Differences among the groups in this plan essentially represent differences among the four sequences called for by this plan.

A modification of the plan that has just been considered leads to another important use of the Latin square. Suppose that each of the groups is assigned to a different level of factor *B*. If the sequence effects are negligible relative to the effects of factor *B*, differences among the groups measure the effects due to factor *B*. If interactions with order effects are negligible, then partial information with respect to the *AB* interaction may be obtained. Assuming *n* subjects in each of the groups, the analysis of this plan takes the following form:

Source of variation	df
Between subjects	$4n - 1$
Groups (*B*)	3
Subjects within groups	$4(n - 1)$
Within subjects	$12n$
Order	3
A	3
$(AB)'$ (partial information)	6
Residual	$12n - 12$

In many applications, the Latin square forms a building block within a more comprehensive design. Use of a Latin square provides economy of experimental effort only in cases where certain of the interactions (to be specified more completely in the following section) are zero. Complete factorial experiments result in unnecessary experimental effort in areas where the assumptions underlying the analysis of Latin squares are appropriate.

9.5 Analysis of Latin-square Designs—No Repeated Measures

Plan 1. Consider the following 3×3 Latin square as a fractional replication of a $3 \times 3 \times 3$ factorial experiment:

	b_1	b_2	b_3
a_1	c_2	c_1	c_3
a_2	c_3	c_2	c_1
a_3	c_1	c_3	c_2

Assume there are *n* observations in each cell. Assume further that two-factor and three-factor interactions are negligible relative to main effects. The model for an observation made in cell *ijk* may be expressed as

$$X_{ijkm} = \mu + \alpha_{i(s)} + \beta_{j(s)} + \gamma_{k(s)} + \text{res}_{(s)} + \varepsilon_{m(ijk)}.$$

The subscript (*s*) is used to indicate that the effect in question is estimated from data obtained from a Latin square. The term $\text{res}_{(s)}$ includes all sources of variation due to treatment effects which are not predictable from the sum of the main effects. Under this model, the analysis and the expected values of the mean square are outlined below:

Source of variation	df	df for general case	E(MS)
A	2	$p - 1$	$\sigma_\varepsilon^2 + np\sigma_\alpha^2$
B	2	$p - 1$	$\sigma_\varepsilon^2 + np\sigma_\beta^2$
C	2	$p - 1$	$\sigma_\varepsilon^2 + np\sigma_\gamma^2$
Residual	2	$(p - 1)(p - 2)$	$\sigma_\varepsilon^2 + \sigma_{\text{res}}^2$
Within cell	$9(n - 1)$	$p^2(n - 1)$	σ_ε^2

If, indeed, interactions are negligible, the variance due to the residual sources should not differ appreciably from the variance due to experimental error. According to the expected values of the mean squares, when $\sigma_{\text{res}}^2 = 0$, the expected values for the residual and the within-cell variances are both estimates of variance due to experimental error. A partial test on the appropriateness of the model is therefore given by the F ratio

$$F = \frac{\text{MS}_{\text{res}}}{\text{MS}_{\text{w. cell}}}.$$

The magnitude of this F ratio indicates the extent to which the observed data conform to the model (which postulates no interactions). However, a test of this kind is somewhat unsatisfactory. Decisions as to the appropriateness of the model should, in general, be based upon evidence independent of that obtained in the actual experiment. Pilot studies, guided by subject-matter knowledge, should, wherever possible, serve as the basis for formulating the model under which the analysis is to be made.

Granting the appropriateness of the model, tests on main effects can readily be carried out. If interaction is present, some of the tests on main effects will no longer be possible. If, for example, all interactions with factor A may be considered negligible but the BC interaction not negligible, then the main effect of factor A will no longer be clear of interaction effects. However, main effects of B and C will be clear of interaction effects; i.e., tests on these latter main effects can be carried out.

Numerical Example. Computational procedures for the plan to be described are similar to those used in a two-factor factorial experiment.

The data in Table 9.5-1 will be used for illustrative purposes. Suppose that an experimenter is interested in evaluating the relative effectiveness of three drugs (factor B) on three categories (factor C) of patients. Patients for the experiment are obtained from three different hospitals (factor A).

Table 9.5-1 Numerical Example

(i)

Design:

	b_1	b_3	b_2
a_2	c_3	c_2	c_1
a_1	c_2	c_1	c_3
a_3	c_1	c_3	c_2

Observed data:

	b_1	b_3	b_2
a_2	6, 8, 12, 7	0, 0, 1, 4	0, 2, 2, 5
a_1	2, 5, 3, 1	2, 2, 4, 6	9, 10, 12, 12
a_3	0, 1, 1, 4	2, 1, 1, 5	0, 1, 1, 4

(ii)

Cell totals

	b_1	b_3	b_2	Total
a_2	33	5	9	$47 = A_2$
a_1	11	14	43	$68 = A_1$
a_3	6	9	6	$21 = A_3$
Total	50	28	58	$136 = G$
	B_1	B_3	B_2	

$n = 4$

$p = 3$

29	22	85
C_1	C_2	C_3

$$(1) = G^2/np^2 = 513.78 \qquad (4) = \Sigma B^2/np = 554.00$$
$$(2) = \Sigma X^2 = 978.00 \qquad (5) = \Sigma C^2/np = 712.50$$
$$(3) = \Sigma A^2/np = 606.17 \qquad (6) = \Sigma(ABC)^2/n = 878.50$$

(iii)

$$SS_a = (3) - (1) = 92.39 \qquad SS_{b.\ cells} = (6) - (1) = 364.72$$
$$SS_b = (4) - (1) = 40.22 \qquad SS_{w.\ cells} = (2) - (6) = 99.50$$
$$SS_c = (5) - (1) = 198.72$$
$$SS_{res} = (6) - (3) - (4) - (5) + 2(1) = 33.39$$

The experimenter obtains 12 patients from each of the hospitals. Of these 12 patients, 4 belong to category c_1, 4 to category c_2, and 4 to category c_3. The total number of patients in the experiment is 36; the total number of patients in any one category is 12.

The design for the experiment is the Latin square given at the left of part i. The first row of this Latin square indicates the manner in which the 12 patients from hospital a_2 are assigned to the drugs. The four patients in category c_3 are given drug b_1; the four patients in category c_2 are given drug b_3; the four patients in category c_1 are given drug b_2. The criterion scores for these three sets of four patients are given in the first row of the table of observed data.

In part ii are the cell totals for the four observations within each cell of the design. The entry in row a_2, column b_1, which is 33, is the sum of the criterion scores for the four patients in category c_3 from hospital a_2—these four patients were given drug b_1. This total will be designated by the symbol ABC_{213}. Similarly, the entry in row a_1, column b_2, which is 43, will be designated by the symbol ABC_{123}—this total is the sum of the four observations under treatment conditions abc_{123}. Row totals and column totals in part ii define, respectively, the A_i's and the B_j's. The C_k's are also obtained from the cell totals. From the design, one locates cells at specified levels of factor C; thus

$$C_1 = 9 + 14 + 6 = 29,$$
$$C_2 = 5 + 11 + 6 = 22,$$
$$C_3 = 33 + 43 + 9 = 85.$$

Computational symbols are defined in part iii. In terms of the symbols, the variation due to the main effect of factor A is estimated by

$$SS_a = np\Sigma(\bar{A}_i - \bar{G})^2 = (3) - (1) = 92.39.$$

The variation due to the main effect of factor B is estimated by

$$SS_b = np\Sigma(\bar{B}_j - \bar{G})^2 = (4) - (1) = 40.22.$$

Similarly the variation due to the main effect of factor C is estimated by

$$SS_c = np\Sigma(\bar{C}_k - G)^2 = (5) - (1) = 198.72.$$

The residual sum of squares is computed in a somewhat indirect manner. The variation between the nine cells in the Latin square is

$$SS_{\text{b. cells}} = n\Sigma(\overline{ABC}_{ijk} - \bar{G})^2 = (6) - (1) = 364.72.$$

This source of variation represents a composite of all factorial effects, main effects as well as interactions. There are $8 = p^2 - 1$ degrees of freedom in this source of variation. The residual variation is given by

$$\begin{aligned} SS_{\text{res}} &= SS_{\text{b. cells}} - SS_a - SS_b - SS_c \\ &= (6) - (3) - (4) - (5) + 2(1) = 33.39. \end{aligned}$$

From this point of view, the residual is that part of the between-cell variation which cannot be accounted for by additivity of the three main effects. If no interactions exist, this source of variation provides an estimate of experimental error. If interactions do exist, the residual variation is, in part, an estimate of these interactions.

The within-cell variation is estimated by

$$SS_{\text{w. cell}} = \sum_m (X_{ijkm} - \overline{ABC}_{ijk})^2 = (2) - (6) = 99.50.$$

A summary of the analysis of variance is given in Table 9.5-2.

Table 9.5-2 Analysis of Variance for Numerical Example

Source of variation	SS	df	MS	F
Hospitals (A)	92.39	2	46.20	12.52*
Drugs (B)	40.22	2	20.11	5.45*
Categories (C)	198.72	2	99.36	26.93*
Residual	33.39	2	16.70	4.53*
Within cell	99.50	27	3.69	
Total	464.22	35		

* $F_{.95}(2,27) = 3.35$

A partial test of the hypothesis that all interactions are negligible is given by

$$F = \frac{MS_{res}}{MS_{w.\ cell}} = \frac{16.70}{3.69} = 4.53.$$

For a .05-level test, the critical value is $F_{.95}(2,27) = 3.35$. Hence the experimental data tend to contradict the hypothesis that the interactions are negligible. Under these conditions, the adequacy of the Latin-square design is questionable—estimates of the main effects will be confounded by interaction terms.

If all interactions with the hospital factor are negligible, then the variation due to the residual represents partial information on the drug by category interaction. Under this latter assumption, the main effects of the drug factor and category factor will not be confounded with interaction effects. If the assumption is made that all interactions with the hospital factor are negligible, the experimental data indicate that the drug by category interaction is statistically significant. Tests on main effects for drugs and categories also indicate statistically significant variation. The test on the main effects due to hospitals cannot adequately be made since this source of variation is partially confounded with the drug by category interaction. If the hospital factor is a random factor, there will ordinarily be little intrinsic interest in this test.

The experimental data do indicate statistically significant differences between the relative effectiveness of the drugs; however, in view of the presence of drugs by category interaction, the interpretation of these differences requires an analysis of the simple effects. Information for this latter type of analysis is not readily obtained from a Latin-square design. Had the existence of an interaction effect been anticipated, a complete factorial experiment with respect to the interacting factors would have been the more adequate design.

To show the relationship between the analysis of variance for a Latin square and the analysis for a two-factor factorial experiment, suppose factor A in Table 9.5-1 is disregarded. Then the analysis of variance is as follows:

Source	SS	df
B	$(4) - (1) = 40.22$	2
C	$(5) - (1) = 198.72$	2
BC	$\frac{\Sigma(BC)^2}{n} - (4) - (5) + (1) = 125.78$	4
Within cell	$(2) - (6) = 99.50$	27

The two degrees of freedom for the main effect of factor A are part of the BC interaction of the two-factor factorial experiment. Thus in a Latin square the interaction term is partitioned as follows:

BC	SS	df
A	92.39	2
Residual	$SS_{bc} - SS_a = 33.39$	2
	125.78	4

Plan 2. Use of a Latin square as part of a larger design is illustrated in the following plan:

		b_1	b_2	b_3
d_1	a_1	c_3	c_2	c_1
	a_2	c_1	c_3	c_2
	a_3	c_2	c_1	c_3

		b_1	b_2	b_3
d_2	a_1	c_1	c_3	c_2
	a_2	c_2	c_1	c_3
	a_3	c_3	c_2	c_1

Assume that there are n independent observations within each cell. All observations in a single square are made at the same level of factor D. The separate squares are at different levels of factor D. In some plans it is desirable to choose a balanced set of squares, in others use of the same square throughout is to be preferred, while in still other plans independent randomization of the squares is to be preferred. Guides in choosing the squares will be discussed after the analysis has been outlined.

A model for which this plan provides adequate data is the following:

$$E(X_{ijkmo}) = \mu + \alpha_{i(s)} + \beta_{j(s)} + \gamma_{k(s)} + \delta_m + \alpha\delta_{i(s)m} + \beta\delta_{j(s)m} + \gamma\delta_{k(s)m} + \text{res}_{(s)}.$$

The subscript (s) indicates that an effect forms one of the dimensions of the square. It will be noted that no interactions are assumed to exist between factors that form part of the same Latin square (i.e., factors A, B, and C). However, interactions between the factor assigned to the whole square (factor D) and the factors that form the parts are included in the model.

Assuming that all factors are fixed and that there are n observations in each cell, the appropriate analysis and expected values for the mean squares (as obtained from the above model) are outlined in Table 9.5-3. A partial check on the appropriateness of the model is provided by the F ratio

$$F = \frac{\text{MS}_{\text{res}}}{\text{MS}_{\text{w. cell}}}.$$

When the model is appropriate, this ratio should not differ appreciably from 1.00.

Use of this experimental plan requires a highly restrictive set of assumptions with respect to some of the interactions. Should these assumptions be violated, main effects will be partially confounded with interaction effects. If, for example, it may be assumed that interactions with factor A are negligible, but that the BC interaction cannot be considered negligible, then the main effect of factor A will be confounded with the BC interaction. However, main effects of factors B and C will not be confounded. When interactions with factor A are negligible, partial information on the BC interaction is available from the within-square residuals. When such information is to be used to make inferences about the BC interaction, it is generally advisable (if the design permits) to work with a balanced set of Latin squares rather than to work with independently randomized squares. Use of a balanced set of squares will provide partial

Table 9.5-3 Analysis of Plan 2

Source of variation	df	df for general case	E(MS)
A	2	$p - 1$	$\sigma_\varepsilon^2 + npq\sigma_\alpha^2$
B	2	$p - 1$	$\sigma_\varepsilon^2 + npq\sigma_\beta^2$
C	2	$p - 1$	$\sigma_\varepsilon^2 + npq\sigma_\gamma^2$
D	1	$q - 1$	$\sigma_\varepsilon^2 + np^2\sigma_\delta^2$
AD	2	$(p - 1)(q - 1)$	$\sigma_\varepsilon^2 + np\sigma_{\alpha\delta}^2$
BD	2	$(p - 1)(q - 1)$	$\sigma_\varepsilon^2 + np\sigma_{\beta\delta}^2$
CD	2	$(p - 1)(q - 1)$	$\sigma_\varepsilon^2 + np\sigma_{\gamma\delta}^2$
Residual	4	$q(p - 1)(p - 2)$	$\sigma_\varepsilon^2 + n\sigma_{\text{res}}^2$
Within cell	$18(n - 1)$	$p^2q(n - 1)$	σ_ε^2

information on all the components of the interaction term. Use of the same square throughout is to be avoided in this context, since information on only a limited number of the interaction components will be available.

Computational Procedures for Plan 2. The computational procedures assume that there are q levels of factor D and n observations in each cell of the $p \times p$ Latin squares included in the plan. The observed data will consist of q squares, one for each level of factor D; each square will contain np^2 observations. Hence the total number of observations in the experiment is np^2q. From the observed data, a summary table of the following form may be prepared (for illustrative purposes, assume that $p = 3$ and $q = 2$):

	d_1	d_2	Total
a_1	AD_{11}	AD_{12}	A_1
a_2	AD_{21}	AD_{22}	A_2
a_3	AD_{31}	AD_{32}	A_3
Total	D_1	D_2	G

Each of the AD_{im}'s in this summary table is the sum of the np observations in the experiment which were made under treatment combination ad_{im}. From this summary table, one may compute the following sums of squares:

$$\text{SS}_a = \frac{\Sigma A_i^2}{npq} - \frac{G^2}{np^2q},$$

$$\text{SS}_d = \frac{\Sigma D_m^2}{np^2} - \frac{G^2}{np^2q},$$

$$\text{SS}_{ad} = \frac{\Sigma(AD_{im})^2}{np} - \frac{\Sigma A_i^2}{npq} - \frac{\Sigma D_m^2}{np^2} + \frac{G^2}{np^2q}.$$

In a manner analogous to that by which the AD summary table is constructed, one may construct a BD summary table. From this latter table one computes SS_b and SS_{bd}. From a CD summary table one can compute SS_d and SS_{cd}.

To obtain the variation due to the residual sources, one proceeds as follows: The Latin square assigned to level d_m will be called square m. The residual variation within square m is given by

$$\text{SS}_{\text{res}(m)} = \frac{\sum_{(m)} (\text{cell totals})^2}{n} - \frac{\sum_{(m)} A_i^2}{np} - \frac{\sum_{(m)} B_j^2}{np} - \frac{\sum_{(m)} C^2}{np} + \frac{2G_m^2}{np^2}.$$

The notation $\sum_{(m)}$ indicates that the summation is restricted to square m. The degrees of freedom for residual variation within square m are

$$(p - 1)(p - 2).$$

The residual variation for the whole experiment is

$$\text{SS}_{\text{res}} = \Sigma\text{SS}_{\text{res}(m)}.$$

The degrees of freedom for SS_{res} are

$$df_{res} = \Sigma(p - 1)(p - 2) = q(p - 1)(p - 2).$$

As a partial check on the assumptions underlying the analysis, the within-square residuals for each of the squares should be homogeneous. One may use an F_{max} test for this purpose.

The within-cell variation is given by

$$SS_{w.\ cell} = \Sigma X^2 - \frac{\Sigma(\text{cell total})^2}{n},$$

where the summation is over the entire experiment. To check the assumption of homogeneity of error variation, the summation in the last expression may be restricted to the individual cells. The qp^2 sums of squares obtained in this manner may then be checked for homogeneity by means of an F_{max} test.

As an additional check on the assumption of no interactions between the factors which form the dimensions of the squares, one has the F ratio

$$F = \frac{MS_{res}}{MS_{w.\ cell}}.$$

When $\sigma^2_{res} = 0$, this F ratio should be approximately unity. The residual variation is sometimes pooled with the within-cell variation to provide a pooled estimate of the experimental error, i.e.,

$$SS_{pooled\ error} = SS_{res} + SS_{w.\ cell}.$$

The degrees of freedom for $SS_{pooled\ error}$ is the sum of the degrees of freedom for the parts.

Plan 3. This plan resembles Plan 2 in form, but the assumptions underlying it are quite different. Here the treatment combinations in a $p \times p \times p$ factorial experiment are divided into a balanced set of $p \times p$ Latin squares. The levels of factors A, B, and C are then assigned at random to the symbols defining the Latin square. Then the levels of factor D are assigned at random to the whole squares. Hence the number of levels for each factor must be p. An illustration in terms of a balanced set of 3×3 Latin squares is given below:

		b_1	b_2	b_3
d_1	a_1	c_1	c_3	c_2
	a_2	c_3	c_2	c_1
	a_3	c_2	c_1	c_3

		b_1	b_2	b_3
d_2	a_1	c_2	c_1	c_3
	a_2	c_1	c_3	c_2
	a_3	c_3	c_2	c_1

		b_1	b_2	b_3
d_3	a_1	c_3	c_2	c_1
	a_2	c_2	c_1	c_3
	a_3	c_1	c_3	c_2

For the special case in which p is a prime number, a balanced set of squares may be constructed by means of the modular arithmetic discussed in Chap. 8. The squares given above correspond to the relationship

$$x_1 + x_2 + x_3 = 0, 1, 2 \qquad (\text{mod } 3)$$

when one expresses the levels of the factors in terms of the subscripts 0, 1, 2, that is, when a_1, a_2, a_3 are replaced by a_0, a_1, a_2, etc. If p is not a prime number, one may obtain a balanced set of squares by means of a series of one-step cyclic permutations.

In this plan all the 27 possible combinations of the factors A, B, and C are present. However, only one-third of the 81 possible combinations of the factors A, B, C, and D that would be present in the complete four-factor factorial experiment are present in the above plan. If interactions with factor D are negligible, then this plan will provide complete information with respect to the main effects of factors A, B, and C, as well as all two-factor interactions between these factors. The main effect of factor D will be partially confounded with the ABC interaction; however, partial information with respect to the ABC interaction may be obtained.

This design is particularly useful when the experimenter is interested primarily in a three-factor factorial experiment in which control with respect to the main effects of a fourth factor is deemed desirable. The model appropriate for this plan is the following:

$$E(X_{ijkmo}) = \mu + \alpha_i + \beta_j + \gamma_k + \alpha\beta_{ij} + \alpha\gamma_{ik} + \beta\gamma_{jk} + \delta_m + \alpha\beta\gamma'_{ijk}.$$

The prime symbol on the three-factor interaction indicates only partial information. (This plan can perhaps be more accurately classified as a balanced incomplete-block design.) Note that no interactions with factor D are included in the model; should such interactions exist, estimates of the sources of variation due to factors A, B, C, and their two-factor interactions will be confounded with interactions with factor D.

Assuming (1) that the model is appropriate for the experimental data, (2) that there are n observations in each of the p^3 cells in the experiment, and (3) that A, B, and C are fixed factors, an outline of the analysis and the expected values of the mean squares is given in Table 9.5-4.

Table 9.5-4 Analysis of Plan 3

Source of variation	df	df for general case	E(MS)
A	2	$p-1$	$\sigma_\varepsilon^2 + np^2\sigma_\alpha^2$
B	2	$p-1$	$\sigma_\varepsilon^2 + np^2\sigma_\beta^2$
C	2	$p-1$	$\sigma_\varepsilon^2 + np^2\sigma_\gamma^2$
AB	4	$(p-1)^2$	$\sigma_\varepsilon^2 + np\sigma_{\alpha\beta}^2$
AC	4	$(p-1)^2$	$\sigma_\varepsilon^2 + np\sigma_{\alpha\gamma}^2$
BC	4	$(p-1)^2$	$\sigma_\varepsilon^2 + np\sigma_{\beta\gamma}^2$
D	2	$(p-1)$	$\sigma_\varepsilon^2 + np^2\sigma_\delta^2$
$(ABC)'$	6	$(p-1)^3 - (p-1)$	$\sigma_\varepsilon^2 + n\sigma_{\alpha\beta\gamma}^2$
Within cell	$27(n-1)$	$p^3(n-1)$	σ_ε^2

Computational Procedures for Plan 3. Because this plan includes a balanced set of Latin squares, one may construct an ABC summary table of the following form:

	b_1			b_2			b_3		
	c_1	c_2	c_3	c_1	c_2	c_3	c_1	c_2	c_3
a_1	ABC_{111}			ABC_{121}			ABC_{131}		
a_2									
a_3									

Each total of the form ABC_{ijk} is based upon n observations. By means of the usual computational formulas for a three-factor factorial experiment having n observations per cell, one may obtain the sums of squares for all the factorial effects, including SS_{abc}. For example,

$$\text{SS}_a = \frac{\Sigma A_i^2}{np^2} - \frac{G^2}{np^3}.$$

The sum of squares due to the three-factor interaction, SS_{abc}, includes the variation due to the main effect of factor D. The adjusted sum of squares for this three-factor interaction is given by

$$\text{SS}'_{abc} = \text{SS}_{abc} - \left(\frac{\Sigma D_m^2}{np^2} - \frac{G^2}{np^3}\right).$$

SS'_{abc} includes that part of the ABC interaction which is not confounded with the main effect due to factor D. The degrees of freedom for SS'_{abc} are

$$\begin{aligned} \text{df}_{\text{SS}'_{abc}} &= \text{df}_{\text{SS}_{abc}} - \text{df}_{\text{SS}_d} \\ &= (p-1)^3 - (p-1). \end{aligned}$$

The within-cell variation is given by

$$\text{SS}_{\text{w. cell}} = \Sigma X^2 - \frac{\Sigma(\text{cell total})^2}{n},$$

where the notation (cell total) represents the sum of the n observations in the experiment made under a unique combination of factors A, B, C, and D. The summation is over all cells in the experiment.

Plan 4. This plan uses the treatment combinations making up a factorial set as one or more dimensions of a Latin square. From many points of view, Plan 4 may be regarded as a special case of Plan 1. A square of dimension $p^2 \times p^2$ is required for this plan. It will be illustrated by the following 4×4 square:

	cd_{11}	cd_{12}	cd_{21}	cd_{22}
ab_{11}	t_2	t_1	t_3	t_4
ab_{12}	t_3	t_2	t_4	t_1
ab_{21}	t_1	t_4	t_2	t_3
ab_{22}	t_4	t_3	t_1	t_2

The treatment combinations along the rows are those forming a 2×2 factorial experiment; the treatment combinations across the columns also form a 2×2 factorial experiment. The letters in the square may or may not form a factorial set. For illustrative purposes suppose that they do not. Assume that there are n observations in each cell. If the treatment sets which define the dimensions of the Latin square do not interact, then an appropriate analysis is given in Table 9.5-5. The expected values of the mean squares assume that A, B, C, and D are fixed factors. (Note that the factors within a dimension of the Latin square may interact.)

Table 9.5-5 Analysis of Plan 4

Source of variation	df		df for general case		E(MS)
Row effects		3		$p^2 - 1$	
A	1		$p - 1$		$\sigma_\varepsilon^2 + np^2\sigma_\alpha^2$
B	1		$p - 1$		$\sigma_\varepsilon^2 + np^2\sigma_\beta^2$
AB	1		$(p - 1)^2$		$\sigma_\varepsilon^2 + np\sigma_{\alpha\beta}^2$
Column effects		3		$p^2 - 1$	
C	1		$p - 1$		$\sigma_\varepsilon^2 + np^2\sigma_\gamma^2$
D	1		$p - 1$		$\sigma_\varepsilon^2 + np^2\sigma_\delta^2$
CD	1		$(p - 1)^2$		$\sigma_\varepsilon^2 + np\sigma_{\gamma\delta}^2$
Letters in cells		3		$p^2 - 1$	
T	3		$p^2 - 1$		$\sigma_\varepsilon^2 + np^2\sigma_\tau^2$
Residual		6		$(p^2 - 1)(p^2 - 2)$	$\sigma_\varepsilon^2 + n\sigma_{\text{res}}^2$
Within cell		$16(n - 1)$		$p^4(n - 1)$	σ_ε^2

Computational Procedures for Plan 4. To obtain SS_a, SS_b, and SS_{ab}, the p^2 row totals are arranged in the form of an AB summary table.

	b_1	b_2	Total
a_1	AB_{11}	AB_{12}	A_1
a_2	AB_{21}	AB_{22}	A_2
Total	B_1	B_2	G

Each AB_{ij} is the sum of np^2 observations. Hence each A_i total is based upon np^3 observations. The variation due to the main effects of factor A is

$$\text{SS}_a = \frac{\Sigma A_i^2}{np^3} - \frac{G^2}{np^4}.$$

The variation due to the AB interaction is

$$\text{SS}_{ab} = \frac{\Sigma(AB_{ij})^2}{np^2} - \frac{\Sigma A_i^2}{np^3} - \frac{\Sigma B_j^2}{np^3} + \frac{G^2}{np^4}.$$

The sums of squares SS_c, SS_d, and SS_{cd} are computed in an analogous manner from the column totals.

If the sum of the np^2 observations at level t_o is designated by the symbol T_o, then the variation due to the treatments assigned to the letters of the square is

$$\text{SS}_t = \frac{\Sigma T_o^2}{np^2} - \frac{G^2}{np^4}.$$

The variation due to residual sources is

$$\text{SS}_{\text{res}} = \frac{\Sigma(\text{cell total})^2}{n} - \frac{\Sigma(AB_{ij})^2}{np^2} - \frac{\Sigma(CD_{km})^2}{np^2} - \frac{\Sigma T_o^2}{np^2} + \frac{2G^2}{np^4}.$$

The within-cell variation is

$$\text{SS}_{\text{w. cell}} = \Sigma X^2 - \frac{\Sigma(\text{cell total})^2}{n},$$

where the summation is over all cells in the experiment.

Summary of Plans in Sec. 9.5

Plan 1

	b_1	b_2	b_3
a_1	c_2	c_1	c_3
a_2	c_3	c_2	c_1
a_3	c_1	c_3	c_2

Plan 2

		b_1	b_2	b_3
d_1	a_1	c_3	c_2	c_1
	a_2	c_1	c_3	c_2
	a_3	c_2	c_1	c_3

		b_1	b_2	b_3
d_2	a_1	c_1	c_3	c_2
	a_2	c_2	c_1	c_3
	a_3	c_3	c_2	c_1

Plan 3

		b_1	b_2	b_3
d_1	a_1	c_1	c_2	c_3
	a_2	c_2	c_3	c_1
	a_3	c_3	c_1	c_2

		b_1	b_2	b_3
d_2	a_1	c_2	c_3	c_1
	a_2	c_3	c_1	c_2
	a_3	c_1	c_2	c_3

		b_1	b_2	b_3
d_3	a_1	c_3	c_1	c_2
	a_2	c_1	c_2	c_3
	a_3	c_2	c_3	c_1

Plan 4

	cd_{11}	cd_{12}	cd_{21}	cd_{22}
ab_{11}	t_2	t_1	t_3	t_4
ab_{12}	t_3	t_2	t_4	t_1
ab_{21}	t_1	t_4	t_2	t_3
ab_{22}	t_4	t_3	t_1	t_2

9.6 Analysis of Greco-Latin Squares

In its classic context, a Latin-square arrangement permits a two-way control in variation of the experimental units, i.e., control of row and column effects. In a similar context, a Greco-Latin square permits a three-way control in the variation of the experimental units, i.e., row effects, column effects, and "layer" effects. Thus, a Greco-Latin square defines a restricted randomization procedure whereby p treatments are assigned to p^2 experimental units so as to obtain balance along three dimensions.

From the point of view of construction, a Greco-Latin square is a composite of two orthogonal Latin squares. (Independent randomization of rows and columns is required before the resulting composite square is used in practice.) In order to maintain the identity of the squares which are combined to form the composite, the cells of one square are often designated by Latin letters, and the cells of the second square by Greek letters. This procedure is demonstrated for two 3×3 squares.

I			II			Composite		
a	b	c	α	β	γ	$a\alpha$	$b\beta$	$c\gamma$
b	c	a	γ	α	β	$b\gamma$	$c\alpha$	$a\beta$
c	a	b	β	γ	α	$c\beta$	$a\gamma$	$b\alpha$

An equivalent representation in terms of numbers rather than letters is given below. In the resulting composite square the first digit in a pair represents the level of one effect, and the second digit the level of a second effect.

I			II			Composite		
1	2	3	1	2	3	11	22	33
2	3	1	3	1	2	23	31	12
3	1	2	2	3	1	32	13	21

In a Greco-Latin square, there are in reality four variables—namely, row, column, Latin-letter, and Greek-letter variables. From this point of view, a Greco-Latin square may be regarded as a kind of four-factor experiment. There are p^2 cells in this square; hence there are p^2 treatment combinations. In a four-factor factorial experiment in which each factor has p levels, there are p^4 treatment combinations. A Greco-Latin square may also be regarded as a p_2/p^4 or $1/p^2$ fractional replication of a $p \times p \times p \times p$ factorial experiment.

In terms of the modular notation introduced in Chap. 8, a Greco-Latin square is a balanced fraction of a p^4 factorial experiment, if p is a prime number. For example, when $p = 3$, the nine treatment combinations which simultaneously satisfy the relations

$$x_2 + x_3 + x_4 = 0(\text{mod } 3),$$
$$x_1 + x_2 + 2x_3 = 0(\text{mod } 3),$$

are the following:

$$\begin{array}{ccc} (0\ 0\ 0\ 0) & (0\ 1\ 1\ 1) & (0\ 2\ 2\ 2) \\ (1\ 0\ 1\ 2) & (1\ 1\ 2\ 0) & (1\ 2\ 0\ 1) \\ (2\ 0\ 2\ 1) & (2\ 1\ 0\ 2) & (2\ 2\ 1\ 0) \end{array}$$

These nine treatments may be arranged as follows:

	b_0	b_1	b_2
a_0	00	11	22
a_1	12	20	01
a_2	21	02	10

The entries in the table represent the levels of factor C and D. The modular condition given above may be expressed in the form

$$BCD = 0, \qquad ABC^2 = 0.$$

The generalized interaction is

$$AB^2D = 0, \qquad ACD^2 = 0.$$

For the Greco-Latin square given above one has

$$ABC^2 = AB^2D = ACD^2 = BCD = \mathrm{I},$$

where I represents what is called an identity element. That is, in this Greco-Latin square, considered as a fraction of a 3^4 factorial experiment, the components of the interactions corresponding to the symbols given above are completely confounded.

As a fractional replication of a factorial experiment, main effects of each of the factors will be confounded with two-factor and higher-order interaction effects. For example, the main effects of factor A will be confounded with the BC, CD, and BCD interactions. In general, the utility of a single Greco-Latin square is limited to experimental situations in which the four dimensions of the square have negligible interactions. However, Greco-Latin squares may be used to good advantage as part of more inclusive designs. The latter are illustrated in Sec. 9.7. If all interactions between factors defining the dimensions of a Greco-Latin square are negligible, then the analysis of variance takes the form given in Table 9.6-1. This analysis assumes that there are n independent observations in each of the cells. The specific degrees of freedom are for the case of a 3×3 square. For this special case, the degrees of freedom of the residual variation are zero; only for the case $p > 3$ will the residual variation be estimable. A partial check on the assumptions made about negligible interactions is given by

$$F = \frac{\mathrm{MS_{res}}}{\mathrm{MS_{w.\ cell}}}.$$

Depending upon the outcome of this test, the residual sum of squares is sometimes pooled with the within-cell variation to provide an overall estimate of the variation due to experimental error.

Table 9.6-1 Analysis of Variance for Greco-Latin Square

Source of variation	df	df for general case	E(MS)
A (rows)	2	$p - 1$	$\sigma_\varepsilon^2 + np\sigma_\alpha^2$
B (columns)	2	$p - 1$	$\sigma_\varepsilon^2 + np\sigma_\beta^2$
C (Latin letters)	2	$p - 1$	$\sigma_\varepsilon^2 + np\sigma_\gamma^2$
D (Greek letters)	2	$p - 1$	$\sigma_\varepsilon^2 + np\sigma_\delta^2$
Residual	–	$(p - 1)(p - 3)$	$\sigma_\varepsilon^2 + n\sigma_{\text{res}}^2$
Within cell	$9(n - 1)$	$p^2(n - 1)$	σ_ε^2
Total	$9n - 1$	$np^2 - 1$	

A summary of the computational procedures is given in Table 9.6-2. These procedures differ only slightly from those appropriate for the Latin square. The residual variation is obtained from the relation

$$SS_{\text{res}} = SS_{\text{b. cells}} - SS_a - SS_b - SS_c - SS_d.$$

Residual variation includes interaction terms if these are not negligible. Otherwise the residual variation provides an estimate of experimental error.

Table 9.6-2 Computational Procedures for Greco-Latin Square

(i)	$(1) = G^2/np^2$ $(2) = \Sigma X^2$ $(3) = (\Sigma A_i^2)/np$ $(4) = (\Sigma B_j^2)/np$	$(5) = (\Sigma C_k^2)/np$ $(6) = (\Sigma D_m^2)/np$ $(7) = [\Sigma(\text{cell total})^2]/n$
(ii)	A	$(3) - (1)$
	B	$(4) - (1)$
	C	$(5) - (1)$
	D	$(6) - (1)$
	Residual	$(7) - (3) - (4) - (5) - (6) + 3(1)$
	Within cell	$(2) - (7)$
	Total	$(2) - (1)$

9.7 Analysis of Latin Squares—Repeated Measures

In the plans that are discussed in this section, all the restrictions on the model underlying a repeated-measure design for a factorial experiment are necessary in order that the final F tests be valid. These restrictions were discussed in Chap. 7. Special attention should be given to the possible presence of nonadditive sequence effects in experiments which do not involve learning. In particular, a repeated-measure design assumes that all pairs of observations on the same subjects have a constant correlation. If this assumption is violated, resulting tests on within-subject effects tend to be biased in the direction of yielding too many significant results.

The equivalent of the conservative test proposed by Box (1954) and by Greenhouse and Geisser (1959) can be adapted for use in connection with the designs that follow. For example, if the test for a within-subject main effect (as indicated by the model requiring the homogeneity-of-correlation condition) requires the critical value

$$F_{1-\alpha}[(p-1), p(n-1)(p-1)],$$

then the corresponding critical value under the conservative test procedure is

$$F_{1-\alpha}[1, p(n-1)].$$

In principle, the degrees of freedom for the numerator and denominator of the F distribution required in the usual test (as given by the expected values of the mean squares obtained from the restricted model) are divided by the degrees of freedom for the factor on which there are repeated measures.

Plan 5. Consider the following 3 × 3 Latin square in which groups of n subjects are assigned at random to the rows:

	a_1	a_2	a_3
G_1	b_1	b_1	b_1
G_2	b_2	b_2	b_2
G_3	b_3	b_3	b_3

In this plan each of the n subjects in G_1 is observed under all treatment combinations in row 1, that is, ab_{11}, ab_{21}, ab_{31}, For example, the levels of factor A may represent three kinds of targets, and the levels of factor B may represent three distances. There are nine possible treatment combinations in the complete 3 × 3 factorial experiment; each of these nine appears in the Latin square. However, the individuals within any one group are observed only under three of the nine possibilities. Suppose the order in which a subject is observed under each treatment combination is randomized independently for each subject.

If the interactions with the group factor are negligible, the following model will be appropriate for the analysis (this assumption is reasonable if the groups represent random subsamples from a common population):

$$E(X_{ijkm}) = \mu + \delta_k + \pi_{m(k)} + \alpha_i + \beta_j + \alpha\beta'_{ij}.$$

In this model δ_k represents effects associated with the groups and $\pi_{m(k)}$ effects associated with subjects within the groups. The symbol $\alpha\beta'_{ij}$

indicates that only partial information is available on this source of variation. Assuming that factors A and B are fixed factors, the analysis and the expected values of the mean squares are given in Table 9.7-1.

In this analysis, only $(p - 1)(p - 2)$ degrees of freedom for the AB interaction appear as within-subject effects. The missing $p - 1$ degrees of freedom define the variation among the groups. Since differences among the groups, in part, reflect differences due to the effects of various combinations of A and B (which are balanced with respect to main effects), such differences define part of the AB interaction. It is readily shown that

$$SS_{ab} = SS_{groups} + SS'_{ab},$$

where SS_{ab} is the variation due to the AB interaction as computed in a two-factor factorial experiment. From some points of view SS_{groups} may be regarded as the between-subject component of the AB interaction. For most practical purposes, only SS'_{ab} (the within-subject component) is tested; tests on the latter component will generally be the more powerful.

Table 9.7-1 Analysis of Plan 5

Source of variation	df	df for general case	E(MS)
Between subjects	$3n - 1$	$np - 1$	
Groups	2	$p - 1$	$\sigma_\varepsilon^2 + p\sigma_\pi^2 + np\sigma_\delta^2$
Subjects within groups	$3(n - 1)$	$p(n - 1)$	$\sigma_\varepsilon^2 + p\sigma_\pi^2$
Within subjects	$6n$	$np(p - 1)$	
A	2	$p - 1$	$\sigma_\varepsilon^2 + np\sigma_\alpha^2$
B	2	$p - 1$	$\sigma_\varepsilon^2 + np\sigma_\beta^2$
$(AB)'$	2	$(p - 1)(p - 2)$	$\sigma_\varepsilon^2 + n\sigma_{\alpha\beta}^2$
Error (within)	$6n - 6$	$p(n - 1)(p - 1)$	σ_ε^2

The appropriateness of the model should be given serious consideration before this plan is used for an experiment. A possible alternative plan, which requires the same amount of experimental effort but does not utilize the Latin-square principle, is the following:

	a_1	a_2	a_3
G_1	b_3	b_1	b_2
G_2	b_1	b_2	b_3
G_3	b_2	b_3	b_1

Again the groups are assumed to be random samples of size n from a common population. In this plan each of the n subjects in group 1 is observed under the treatment combinations ab_{13}, ab_{21}, and ab_{32}. Assume that the order in which subjects are observed under each treatment combination is randomized independently for each subject. In this plan there are repeated measures on factor A but no repeated measures on factor B. A model appropriate for estimation and tests under this plan is

$$E(X_{ijkm}) = \mu + \beta_j + \pi_{k(j)} + \alpha_i + \alpha\beta_{ij} + \alpha\pi_{ik(j)}.$$

For this model, assuming that factors A and B are fixed, the analysis is outlined in Table 9.7-2. In contrast to Plan 5, under this plan the AB interaction is not confounded with between-group differences; however, all components of the main effect of factor B are between-subject components. That is, differences between levels of factor B are simultaneously differences between groups of people. Hence use of the Latin square permits a more sensitive test on the main effects of factor B, provided that the model under which the analysis is made is appropriate. In general the $A \times$ subjects-within-groups interaction under the alternative plan will be of the same order of magnitude as the error (within) term in Plan 5.

If the experimenter's primary interest is in factor A and its interaction with factor B, and if there is little interest in the main effects of factor B,

Table 9.7-2 Analysis of Alternative to Plan 5

Source of variation	df	df for general case	E(MS)
Between subjects	$3n - 1$	$np - 1$	
B	2	$p - 1$	$\sigma_\varepsilon^2 + p\sigma_\pi^2 + np\sigma_\beta^2$
Subjects within groups	$3(n - 1)$	$p(n - 1)$	$\sigma_\varepsilon^2 + p\sigma_\pi^2$
Within subjects	$6n$	$np(p - 1)$	
A	2	$p - 1$	$\sigma_\varepsilon^2 + \sigma_{\alpha\pi}^2 + np\sigma_\alpha^2$
AB	4	$(p - 1)(p - 1)$	$\sigma_\varepsilon^2 + \sigma_{\alpha\pi}^2 + n\sigma_{\alpha\beta}^2$
$A \times$ subjects within groups	$6n - 6$	$p(n - 1)(p - 1)$	$\sigma_\varepsilon^2 + \sigma_{\alpha\pi}^2$

use of the Latin-square plan is not recommended. However, if the main effects of factor B are also of primary interest to the experimenter, use of the Latin square has much to recommend it. By suitable replication, some within-subject information may be obtained on all components of the AB interaction. Plan 5 forms a building block out of which other plans may be constructed.

Computational Procedures for Plan 5. Computational procedures for this plan are similar to those given for Plan 1. The sources of variation SS_a, SS_b, and SS_{groups} are computed in a manner analogous to those used in Plan 1 for corresponding effects. The variation due to the within-subject components of the AB interaction is given by

$$SS'_{ab} = SS_{ab} - SS_{groups},$$

where SS_{ab} is computed from an AB summary table by means of the usual computational formulas for this interaction.

The variation due to subjects within groups is given by

$$SS_{\text{subj w. groups}} = \frac{\Sigma P^2_{m(k)}}{p} - \frac{\Sigma G_k^2}{np},$$

where $P_{m(k)}$ represents the sum of the p observations on person m in group k, and where G_k represents the sum of the np observations in group k. The summation is over the whole experiment. If the summation is limited to a single group, one obtains $SS_{\text{subj w. group } k}$. There will be p such sources of variation. It is readily shown that

$$SS_{\text{subj w. groups}} = \Sigma SS_{\text{subj w. group } k}.$$

That is, the sum of squares on the left represents a pooling of the sources of variation on the right; the latter may be checked for homogeneity by means of an $F_{\max}$ test.

It is convenient to compute $SS_{\text{error(within)}}$ from $SS_{\text{w. cell}}$. The latter is given by

$$SS_{\text{w. cell}} = \Sigma X^2 - \frac{\Sigma(AB_{ij})^2}{n},$$

where AB_{ij} represents a cell total. The summation is over the whole experiment. From this source of variation,

$$SS_{\text{error(within)}} = SS_{\text{w. cell}} - SS_{\text{subj w. groups}}.$$

A somewhat modified computational procedure provides the parts of the error term that may be checked for homogeneity. One first computes

$$SS_{\text{w. cell for } G_k} = \Sigma X^2 - \frac{\Sigma(AB_{ij})^2}{n},$$

where the summation is limited to those cells in which group k participates. The degrees of freedom for this source of variation are $p(n - 1)$. Then

$$SS_{\text{error(within) for } G_k} = SS_{\text{w. cell for } G_k} - SS_{\text{subj w. } G_k}.$$

Degrees of freedom for this source of variation are $p(n - 1) - (n - 1)$, which is equal to $(n - 1)(p - 1)$. It is readily shown that

$$SS_{\text{error(within)}} = \Sigma SS_{\text{error(within) for } G_k};$$

that is, the variation on the left is a pooling of the sources of variation on the right. The degrees of freedom for the pooled error (within) and the degrees of freedom of the parts are related as follows:

$$df_{\text{error(within)}} = \sum_k (n - 1)(p - 1) = p(n - 1)(p - 1).$$

Thus error (within) is partitioned into p parts; each of the parts has $(n - 1)(p - 1)$ degrees of freedom.

Illustrative Applications. Staats et al. (1957) report an experiment which may be represented schematically as follows:

	a_1	a_2
G_1	b_1	b_2
G_2	b_2	b_1

In this experiment the levels of factor A represent two different words. The levels of factor B represent two different kinds of meaning associated with the words. Subjects were assigned at random to each of the groups. The criterion measure was a rating of the "pleasantness" of the word after a series of associations defined by the level of factor B. The analysis of variance had the following form:

	df	MS	F
Between subjects	159		
Groups	1	0.00	
Subj w. groups	158	2.30	
Within subjects	160		
A Words	1	45.75	18.67
B Meaning	1	22.58	9.22
Error (within)	158	2.45	

In this design, differences between groups may be considered as part of the AB interaction.

Martindale and Lowe (1959) report an experiment which is also a special case of Plan 5. In this case, a 5×5 Latin square is central to the plan. The levels of factor A represent different angles for a television camera; the levels of factor B represent trials. There were three subjects in each of the groups. Subjects were required to track a target which was visible to them only through a television screen. The criterion was time on target. The analysis of variance had the following form:

		df		MS	F
Between subjects		14			
Groups		4		5,000	5.03
Subj.w. groups		10		995	
Within subjects		60			
A Angles		4		15,102	38.82
B Trials		4		167	
Pooled error		52		389	
$(AB)'$	12		597		
Error (w)	40		326		

The authors refer to $(AB)'$ as the residual from the Latin square; the error (within) is called the residual within the Latin square. These latter sources of variation are pooled to form the within-subject error.

Plan 6. In Plan 5, the groups assigned to the rows of the Latin square form a quasi factor. Essentially, Plan 5 may be regarded as one complete replication of a two-factor factorial experiment arranged in incomplete blocks. From this point of view, the plan to be discussed in this section may be regarded as a fractional replication of a three-factor factorial experiment arranged in incomplete blocks.

In this plan each subject within G_1 is assigned to the treatment combinations abc_{111}, abc_{231}, and abc_{321}. Thus each subject in G_1 is observed under all levels of factors A and B but under only one level of factor C. For each subject there is balance with respect to the main effects of factors A and B, but there is no balance with respect to any of the interactions. Similarly, each subject in G_2 is assigned to the treatment combinations abc_{122}, abc_{212}, and abc_{332}. Again there is balance with respect to the main effects of factors A and B, but there is no balance with respect to interactions.

		a_1	a_2	a_3
G_1	c_1	b_1	b_3	b_2
G_2	c_2	b_2	b_1	b_3
G_3	c_3	b_3	b_2	b_1

If all interactions are negligible relative to main effects (a highly restrictive assumption), the following model is appropriate for making estimates and tests in the analysis of variance:

$$E(X_{ijkm}) = \mu + \gamma_{k(s)} + \pi_{m(k)} + \alpha_{i(s)} + \beta_{j(s)} + \text{res}_{(s)}.$$

The analysis of variance and the expected values of the mean squares are summarized in Table 9.7-3.

Table 9.7-3 Analysis of Plan 6

Source of variation	df	df for general case	E(MS)
Between subjects	$3n - 1$	$np - 1$	
C	2	$p - 1$	$\sigma_\varepsilon^2 + p\sigma_\pi^2 + np\sigma_\gamma^2$
Subjects within groups	$3(n - 1)$	$p(n - 1)$	$\sigma_\varepsilon^2 + p\sigma_\pi^2$
Within subjects	$6n$	$np(p - 1)$	
A	2	$p - 1$	$\sigma_\varepsilon^2 + np\sigma_\alpha^2$
B	2	$p - 1$	$\sigma_\varepsilon^2 + np\sigma_\beta^2$
Residual	2	$(p - 1)(p - 2)$	$\sigma_\varepsilon^2 + n\sigma_{\text{res}}^2$
Error (within)	$6n - 6$	$p(n - 1)(p - 1)$	σ_ε^2

The analysis of Plan 6 is quite similar to the analysis of Plan 5. Differences among groups in the latter plan correspond to differences due to the main effects of factor C in the present plan; what was part of the AB interaction in Plan 5 is a residual term in the present plan. It is of interest to note the relationship between the analysis of a two-factor factorial experiment which does not have repeated measures and the analysis of Plan 6. The total number of observations in Plan 6 is the same as that of a two-factor factorial experiment in which there are n observations per cell. The Latin-square arrangement in Plan 6 permits the partition of what is formally a two-factor interaction into a main effect and a residual effect. The latter is a mixture of interaction effects if these exist. The fact that there are repeated measures in Plan 6 permits the within-cell variation to be partitioned into one part involving differences between subjects and one part which does not involve differences between subjects, provided of course that homogeneity of covariances exists.

df	Two-factor experiment	Plan 6	df
$p - 1$	A	A	$p - 1$
$p - 1$	B	B	$p - 1$
$(p - 1)(p - 1)$	AB	C	$p - 1$
		Residual	$(p - 1)(p - 2)$
$p^2(n - 1)$	Within cell	Subj w. gp	$p(n - 1)$
		Error (within)	$p(n - 1)(p - 1)$

Because of the highly restrictive assumptions with respect to the interactions, use of Plan 6 is appropriate only when experience has shown that such interactions are negligible. A partial check on this assumption is given by the ratio

$$F = \frac{\text{MS}_{\text{res}}}{\text{MS}_{\text{error(within)}}}.$$

According to the model under which the analysis of variance is made, σ^2_{res} should be zero when the assumptions with respect to the interactions are satisfied. In cases in which one of the factors, say A, is the experimental variable of primary interest and factors B and C are of the nature of control factors (i.e., replications or order of presentation) this plan is potentially quite useful. With this type of experimental design interactions can frequently be assumed to be negligible relative to main effects.

In exploratory studies in which interaction effects may be of primary interest to the experimenter, there is generally no substitute for the complete factorial experiment. The complete factorial analog of Plan 6 is represented schematically in Table 9.7-4. In order to have complete within-subject information on the main effects of factors A and B as well as complete within-subject information on all interactions, including interactions with factor C, p^2 observations on each subject are required. In

Table 9.7-4 Complete Factorial Analog of Plan 6

		a_1	a_2	a_3	a_1	a_2	a_3	a_1	a_2	a_3
G_1	c_1	b_1	b_1	b_1	b_2	b_2	b_2	b_3	b_3	b_3
G_2	c_2	b_1	b_1	b_1	b_2	b_2	b_2	b_3	b_3	b_3
G_3	c_3	b_1	b_1	b_1	b_2	b_2	b_2	b_3	b_3	b_3

Plan 6, only p observations are made on each of the subjects. Use of Plan 6 reduces the overall cost of the experiment in terms of experimental effort—at the possible cost, however, of inconclusive results. Should the experimenter find evidence of interaction effects which invalidate the analysis of Plan 6, he may, if his work is planned properly, enlarge the experiment into Plan 8a. It will be noted that the latter may be constructed from a series of plans having the form of Plan 6. The assumptions in Plan 8a are less restrictive with respect to interactions than are those underlying Plan 6.

Computational Procedures for Plan 6. Computational procedures for this plan are identical to those outlined for Plan 5. Here $SS_c = SS_{groups}$ and $SS'_{ab} = SS_{res}$. All other sums of squares are identical. Tests for homogeneity of error variance are also identical.

Plan 7. This plan is related to Plan 5 as well as to Plan 6. Plan 7 may be regarded as being formed from Plan 5 by superimposing an orthogonal Latin square. This plan may also be viewed as a modification of Plan 6 in which the C factor is converted into a within-subject effect. The combinations of factors B and C which appear in the cells of the following plan are defined by a Greco-Latin square:

	a_1	a_2	a_3
G_1	bc_{11}	bc_{23}	bc_{32}
G_2	bc_{22}	bc_{31}	bc_{13}
G_3	bc_{33}	bc_{12}	bc_{21}

The groups of subjects are assigned at random to the rows of the square; the levels of factor A are also assigned at random to the columns of the square. The subjects within G_1 are observed under treatment combinations abc_{111}, abc_{223}, and abc_{332}. (Assume that the order in which a subject is observed under a particular treatment combination is randomized independently for each subject.) Provided that all interactions are negligible, unbiased estimates of the differential main effects of factors A, B, and C can be obtained. Further, the expected values of the mean squares for these main effects will not involve a between-subject component. The model under which the analysis of Plan 7 can be carried out is

$$E(X_{ijkmo}) = \mu + \delta_{m(s)} + \pi_{o(m)} + \alpha_{i(s)} + \beta_{j(s)} + \gamma_{k(s)}.$$

The symbol $\delta_{m(s)}$ designates the effect of group m, and the symbol $\pi_{o(m)}$ designates the additive effect of person o within group m.

An outline of the analysis, assuming that the model is appropriate, is given in Table 9.7-5. For a 3×3 square, the variation due to the residual cannot be estimated, since there are zero degrees of freedom for this source of variation. However, for squares of higher dimension a residual term will be estimable. The latter represents the variation due to interactions, if these exist. The variation between groups may also be regarded as part of the interaction variation.

If the levels of factor C define the order in which subjects are observed under combinations of factors A and B, then Plan 7 becomes a special

Table 9.7-5 Analysis of Plan 7

Source of variation	df	df for general case	E(MS)
Between subjects	$3n - 1$	$np - 1$	
Groups	2	$p - 1$	$\sigma_\varepsilon^2 + p\sigma_\pi^2 + np\sigma_\delta^2$
Subjects within groups	$3(n - 1)$	$p(n - 1)$	$\sigma_\varepsilon^2 + p\sigma_\pi^2$
Within subjects	$6n$	$np(p - 1)$	
A	2	$p - 1$	$\sigma_\varepsilon^2 + np\sigma_\alpha^2$
B	2	$p - 1$	$\sigma_\varepsilon^2 + np\sigma_\beta^2$
C	2	$p - 1$	$\sigma_\varepsilon^2 + np\sigma_\gamma^2$
Residual	0	$(p - 1)(p - 3)$	$\sigma_\varepsilon^2 + n\sigma_{\text{res}}^2$
Error (within)	$6n - 6$	$p(n - 1)(p - 1)$	σ_ε^2

case of Plan 5. In terms of factor C as an order factor, the schematic representation of Plan 7 given earlier in this section may be reorganized as follows:

	Order 1	Order 2	Order 3
G_1	ab_{11}	ab_{33}	ab_{22}
G_2	ab_{23}	ab_{12}	ab_{31}
G_3	ab_{32}	ab_{21}	ab_{13}

Thus the subjects within G_1 are observed under treatment combinations ab_{11}, ab_{33}, and ab_{22} in that order. With one of the factors representing an order factor, the variation between groups may be regarded as representing a sequence effect; this latter source of variation may also be regarded as part of the confounded interaction effects.

Computational Procedures for Plan 7. The computational procedures for this plan do not differ appreciably from those of the Greco-Latin square. Following the computational procedures for the latter, one obtains SS_a, SS_b, SS_c, and SS_{groups}. The between-cell variation is given by

$$SS_{b.\ cells} = \frac{\Sigma(\text{cell total})^2}{n} - \frac{G^2}{np^2}.$$

From this sum of squares one obtains

$$SS_{res} = SS_{b.\ cells} - (SS_a + SS_b + SS_c + SS_{groups}).$$

From this last relationship, the degrees of freedom for the residual source of variation are

$$\begin{aligned} df_{res} = (p^2 - 1) - 4(p - 1) &= (p - 1)(p + 1 - 4) \\ &= (p - 1)(p - 3). \end{aligned}$$

The variation due to subjects within groups is

$$SS_{subj\ w.\ groups} = \frac{\Sigma P^2_{m(k)}}{p} - \frac{\Sigma G_k^2}{np},$$

where $P_{m(k)}$ is the sum of the p observations on person m in group k, and where G_k is the sum of the np observations in group k. The error (within) variation is obtained indirectly from the within-cell variation. The latter is given by

$$SS_{w.\ cell} = \Sigma X^2 - \frac{\Sigma(\text{cell total})^2}{n},$$

where the summation is over the whole experiment. From this one obtains

$$SS_{error(within)} = SS_{w.\ cell} - SS_{subj\ w.\ groups}.$$

This method of computing this source of variation shows that the degrees of freedom for error (within) are

$$\begin{aligned} df_{error(within)} &= p^2(n - 1) - p(n - 1) \\ &= p(n - 1)(p - 1). \end{aligned}$$

The sum of squares for error (within) may be partitioned by the procedures given under Plan 5. That is, by limiting the summations for $SS_{w.\ cells}$ and $SS_{subj\ w.\ groups}$ to single groups, one may obtain p terms each of the general form $SS_{error\ (within)\ for\ G_k}$. The degrees of freedom for each of these terms are $(n - 1)(p - 1)$.

Illustrative Application. Leary (1958) reports a study using a plan closely related to Plan 7. This experiment may be represented schematically as follows:

Square I

Subject	Day			
	1	2	3	4
1	(11)	(22)	**(33)**	(44)
2	(23)	(14)	**(41)**	(32)
3	(34)	(43)	**(12)**	(21)
4	(42)	(31)	**(24)**	(13)

Square II

Subject	Day			
	1	2	3	4
5	(33)	(41)	(14)	(22)
6	(12)	(24)	(31)	(43)
7	(21)	(13)	(42)	(34)
8	(44)	(32)	(23)	(11)

The basic plan consists of two different Greco-Latin squares. The symbol (23), for example, represents an experimental condition having reward level 2 for list 3—a list being a set of pairs of objects. The analysis of variance for this experiment had the following form:

Source		df
Between subjects		7
Squares		1
Subjects within squares		6
Within subjects		24
Days		3
Lists		3
Rewards		3
Pooled residual		15
Days × squares	3	
Lists × squares	3	
Rewards × squares	3	
Res from square I	3	
Res from square II	3	

In this design there is only one subject per cell, but there are two replications. Since the squares are considered to define a random factor, the interactions with the squares are pooled to form part of the experimental error. The residual within square I is formally equivalent to the interaction between any two of the dimensions of square I minus the sum of the main effects of the other factors in the square. Thus

$$SS_{\text{res from I}} = SS_{\text{days} \times \text{lists}} - (SS_{\text{rewards}} + SS_{\text{subj w. I}}).$$

In each case the summations are limited to square I. This residual term is an estimate of experimental error if the assumptions underlying the model are valid. Similarly the residual from square II is an estimate of the experimental error. An $F_{\max}$ test may be used as a partial check on the pooling of the various estimates of experimental error.

In terms of the analysis outlined in Table 9.7-5, the degrees of freedom for the pooled residual in this design are

$$(p - 1)(p - 3) + p(n - 1)(p - 1) = (3)(1) + 4(1)(3) = 15,$$

where n in this case is the number of replications.

Plan 8. This plan uses Plan 5 as a building block. Disregarding the repeated-measure aspect, this plan also resembles Plan 2. A schematic representation of Plan 8 is given below:

Square I

		a_1	a_2	a_3
	G_1	b_1	b_2	b_3
c_1	G_2	b_2	b_3	b_1
	G_3	b_3	b_1	b_2

Square II

		a_1	a_2	a_3
	G_4	b_2	b_3	b_1
c_2	G_5	b_1	b_2	b_3
	G_6	b_3	b_1	b_2

In general there will be q squares, one for each level of factor C. Different squares are used for each level of factor C. In square I, all observations are at level c_1; in square II, all observations are at level c_2. There is no restriction on the number of levels of factor C.

Depending upon what can be assumed about the interactions, different analyses are possible. If the interactions between factors A, B, and groups (the dimensions of the squares) are negligible, then unbiased tests can be made on all main effects and the two-factor interactions with factor C. The model under which the latter analysis is made is the following:

$$E(X_{ijkmo}) = \mu + \gamma_k + \delta_{m(k)} + \pi_{o(km)} + \alpha_{i(s)} + \beta_{j(s)} + \alpha\gamma_{ik} + \beta\gamma_{jk}.$$

The analysis outlined in Table 9.7-6 follows from this model; the expected values for the mean squares assume that factor C is fixed. It is also

Table 9.7-6 Analysis of Plan 8

Source of variation	df	df for general case	E(MS)
Between subjects	$6n - 1$	$npq - 1$	
C	1	$q - 1$	$\sigma_\varepsilon^2 + p\sigma_\pi^2 + np\sigma_\delta^2 + np^2\sigma_\gamma^2$
Groups within C	4	$q(p - 1)$	$\sigma_\varepsilon^2 + p\sigma_\pi^2 + np\sigma_\delta^2$
Subjects within groups	$6(n - 1)$	$pq(n - 1)$	$\sigma_\varepsilon^2 + p\sigma_\pi^2$
Within subjects	$12n$	$npq(p - 1)$	
A	2	$p - 1$	$\sigma_\varepsilon^2 + npq\sigma_\alpha^2$
B	2	$p - 1$	$\sigma_\varepsilon^2 + npq\sigma_\beta^2$
AC	2	$(p - 1)(q - 1)$	$\sigma_\varepsilon^2 + np\sigma_{\alpha\gamma}^2$
BC	2	$(p - 1)(q - 1)$	$\sigma_\varepsilon^2 + np\sigma_{\beta\gamma}^2$
Residual	4	$q(p - 1)(p - 2)$	$\sigma_\varepsilon^2 + n\sigma_{\text{res}}^2$
Error (within)	$12n - 12$	$pq(n - 1)(p - 1)$	σ_ε^2

assumed that the order of administration of the treatment combinations is randomized independently for each subject.

The assumptions underlying Plan 8, which lead to the analysis summarized in Table 9.7-6, are different from the assumptions used in the analysis of Plan 5. A set of assumptions, consistent with those in Plan 5, leads to the analysis for Plan 8*a*. The latter plan has the same schematic representation as Plan 8. However, in Plan 8*a* only the interactions involving the group factor are assumed to be negligible. Under these latter assumptions, the appropriate model is the following:

$$E(X_{ijkmo}) = \mu + \gamma_k + \delta_{m(k)} + \pi_{o(km)} + \alpha_i + \beta_j + \alpha\beta'_{ij} + \alpha\gamma_{ik} + \beta\gamma_{jk}.$$

The analysis resulting from this model, assuming factors A and B fixed, is outlined in Table 9.7-7.

Table 9.7-7 Analysis of Plan 8*a*

Source of variation	df	df for general case	E(MS)
Between subjects	$6n - 1$	$npq - 1$	
C	1	$q - 1$	$\sigma_\varepsilon^2 + p\sigma_\pi^2 + np\sigma_\delta^2 + np^2\sigma_\gamma^2$
Groups within C	4	$q(p - 1)$	$\sigma_\varepsilon^2 + p\sigma_\pi^2 + np\sigma_\delta^2$
Subjects within groups	$6(n - 1)$	$pq(n - 1)$	$\sigma_\varepsilon^2 + p\sigma_\pi^2$
Within subjects	$12n$	$npq(p - 1)$	
A	2	$p - 1$	$\sigma_\varepsilon^2 + npq\sigma_\alpha^2$
B	2	$p - 1$	$\sigma_\varepsilon^2 + npq\sigma_\beta^2$
AC	2	$(p - 1)(q - 1)$	$\sigma_\varepsilon^2 + np\sigma_{\alpha\gamma}^2$
BC	2	$(p - 1)(q - 1)$	$\sigma_\varepsilon^2 + np\sigma_{\beta\gamma}^2$
AB' from square I	2	$(p - 1)(p - 2)$	$\sigma_\varepsilon^2 + n\sigma_{\alpha\beta}^2$
AB' from square II	2	$(p - 1)(p - 2)$	$\sigma_\varepsilon^2 + n\sigma_{\alpha\beta}^2$
Error (within)	$12n - 12$	$pq(n - 1)(p - 1)$	σ_ε^2

If the number of levels of factor C is equal to p, and if a balanced set of squares is used, then the components of the AB interaction which are confounded will be balanced. However, if the same Latin square is used for each of the levels of factor C, the same set of $p - 1$ degrees of freedom of the AB interaction is confounded in each square. It is of interest to compare the partition of the total degrees of freedom in a $3 \times 3 \times 3$ factorial experiment with the partition made under Plan 8*a*, assuming that a balanced set of squares is used in the construction of the design. These partitions are presented in Table 9.7-8.

The 12 degrees of freedom corresponding to the AB and ABC interactions in the factorial experiment appear in Plan 8*a* as the sum of 6 degrees of

Table 9.7-8 Partitions in Plan 8a and 3 × 3 × 3 Factorial Experiment

3 × 3 × 3 factorial		Plan 8*a*	
df	Source of variation	Source of variation	df
2	*A*	*A*	2
2	*B*	*B*	2
2	*C*	*C*	2
4	*AC*	*AC*	4
4	*BC*	*BC*	4
4	*AB* }	{Groups within *C*	6
8	*ABC* }	{$\Sigma(AB'$ for each square)	6
$27(n-1)$	Within cell	{Subjects within groups	$9(n-1)$
		{Error (within)	$18(n-1)$

freedom for groups within *C* and 6 degrees of freedom for what is partial information on the simple interaction of *AB* at each level of c_k. In the general case, the relationship is as follows:

df			df
$(p-1)^2$	*AB* }	{Groups within *C*	$p(p-1)$
$(p-1)^3$	*ABC* }	{$\Sigma(AB'$ for each square)	$p(p-1)(p-2)$
$p(p-1)^2$			$p(p-1)^2$

In order to have complete within-subject information on all effects, except the main effects of factor *C*, p^2 observations on each subject are required. In Plan 8*a* (constructed with a balanced set of squares) only *p* observations are required on each subject. What is lost in reducing the number of observations per subject from p^2 to p is partial information on the *AB* and *ABC* interactions, specifically $p(p-1)$ of these degrees of freedom. This loss in information is, in many cases, a small price to pay for the potential saving in experimental feasibility and effort.

In Plan 8*a*, the source of variation SS'_{ab}, which is actually part of the simple *AB* interaction for each of the levels of factor *C*, may be tested separately or as a sum. As a sum, this source of variation is a mixture of the *AB* and *ABC* interactions.

Computational Procedures for Plan 8. Computational procedures for this plan are summarized in Table 9.7-9. These procedures may also be followed for Plan 8*a*. In the latter plan, SS_{res} becomes $\Sigma SS'_{ab}$ for each square.

In part i of this table symbols convenient for use in the computational formulas are defined. Computational formulas are given in part ii. The order in which the sums of squares are given in part ii corresponds to the order in Table 9.7-6. For purposes of checking homogeneity of the parts of $SS_{error(within)}$, this source of variation may be partitioned into *pq* nonoverlapping parts—one part for each of the pq groups in the experiment. The part corresponding to G_m is

$$SS_{\text{error(within) for } G_m} = (2g_m) - (9g_m) - (8g_m) + (10g_m),$$

Table 9.7-9 Summary of Computational Procedures for Plan 8

(i)	$(1) = G^2/np^2q$	$(6) = [\Sigma(AC_{ik})^2]/np$
	$(2) = \Sigma X^2$	$(7) = [\Sigma(BC_{jk})^2]/np$
	$(3) = (\Sigma A_i^2)/npq$	$(8) = \Sigma(P_{o(m)}^2)/p$
	$(4) = (\Sigma B_j^2)/npq$	$(9) = [\Sigma(\text{cell total})^2]/n$
	$(5) = (\Sigma C_k^2)/np^2$	$(10) = (\Sigma G_m^2)/np$
	(In each case the summation is over the whole experiment.)	
(ii)	$SS_{\text{b. subjects}}$	$= (8) - (1)$
	SS_c	$= (5) - (1)$
	$SS_{\text{groups w. } C}$	$= (10) - (5)$
	$SS_{\text{subj w. groups}}$	$= (8) - (10)$
	$SS_{\text{w. subj}}$	$= (2) - (8)$
	SS_a	$= (3) - (1)$
	SS_b	$= (4) - (1)$
	SS_{ac}	$= (6) - (3) - (5) + (1)$
	SS_{bc}	$= (7) - (4) - (5) + (1)$
	SS_{res}	$= (9) - (10) - (6) - (7) + 2(5)$
	$SS_{\text{error (within)}}$	$= (2) - (9) - (8) + (10)$

where $(2g_m)$ designates the numerical value of (2) if the summation in the latter is restricted to G_m. Analogous definitions hold for the other computational symbols containing g_m. It is readily shown that

$$SS_{\text{error(within)}} = \Sigma SS_{\text{error(within) for } G_m}.$$

The $F_{\max}$ statistic may be used to check on the homogeneity of the parts that are pooled to provide the overall error term for the within-subject effects. It is readily verified that

$$\Sigma(2g_m) = (2), \qquad \Sigma(9g_m) = (9), \qquad \text{etc.}$$

The variation due to subjects within groups may also be checked for homogeneity. The sum of squares $SS_{\text{subj w. groups}}$ may be partitioned into pq parts, one part for each group. Each part has the general form

$$SS_{\text{subj w. } G_m} = (8g_m) - (10g_m),$$

where $(8g_m)$ is the numerical value of (8) if the summation is restricted to group m.

If a balanced set of squares is used in the construction of Plan 8a, the computational procedures may be simplified. One may use the regular computational procedures for a three-factor factorial experiment having

n observations per cell as the starting point. In addition one computes $SS_{\text{groups w. } C}$ and $SS_{\text{subj w. groups}}$ by means of the procedures given for Plan 8. Then

$$\Sigma(SS'_{ab \text{ for each square}}) = SS_{ab} + SS_{abc} - SS_{\text{groups w. } C},$$

$$SS_{\text{error(within)}} = SS_{\text{w. cell}} - SS_{\text{subj w. groups}}.$$

Illustrative Application. Conrad (1958) reports an experiment that may be considered a special case of Plan 8. The purpose of this experiment was to investigate the accuracy with which subjects performed under four methods for dialing telephone numbers. There were 24 subjects in the experiment. Rather than assigning groups of 6 subjects to the rows of a single 4 × 4 square, individual subjects were assigned at random to six different 4 × 4 squares. The squares thus represented different sets of sequences under which the subjects used the methods. The analysis of variance for the Conrad data was as follows:

	df	MS	F
Between subjects	23		
C Squares	5	2,446	
Subj w. squares	18	28,651	
Within subjects	72		
A Methods	3	3,018	9.61
B Order	3	1,106	3.52
AC	15	197	
BC	15	215	
Residual	36	314	

Plan 9. This plan may be viewed as a special case of Plan 8a in which the same square is used for all levels of factor C. Hence the same components of the AB interaction are confounded within each of the squares, and the same components of the ABC interaction are confounded in differences between squares. Full information will be available on some components of both the AB and ABC interactions. The number of levels of factors A and B must be equal; there is no restriction on the number of levels of factor C. Factors A, B, and C are considered to be fixed; interactions with the group factor are assumed to be negligible. A schematic representation of this plan for the case in which $p = 3$ and $q = 3$ is given below:

I

		a_1	a_2	a_3
	G_1	b_2	b_3	b_1
c_1	G_2	b_1	b_2	b_3
	G_3	b_3	b_1	b_2

II

		a_1	a_2	a_3
	G_4	b_2	b_3	b_1
c_2	G_5	b_1	b_2	b_3
	G_6	b_3	b_1	b_2

III

		a_1	a_2	a_3
	G_7	b_2	b_3	b_1
c_3	G_8	b_1	b_2	b_3
	G_9	b_3	b_1	b_2

It is of considerable interest to indicate the manner in which the AB and ABC interactions are confounded. A given row in each of the above squares represents the same combination of treatments A and B. For example, the first row in each square involves ab_{12}, ab_{23}, and ab_{31}. A summary table of the following form will be shown to have special meaning in the interpretation of the AB and ABC interactions:

Row	c_1	c_2	c_3	Total
1	G_1	G_4	G_7	R_1
2	G_2	G_5	G_8	R_2
3	G_3	G_6	G_9	R_3
Total	C_1	C_2	C_3	Grand total

In this summary table each G designates the sum of the np observations made within a group.

The total R_1 represents the sum of the npq observations in row 1 of all squares; this sum is also $G_1 + G_4 + G_7$. This total is balanced with respect to the main effects of factors A, B, and C; it is also balanced with respect to the ABC interaction. The latter balance follows from the fact that sums of the form

$$\alpha\beta\gamma_{121} + \alpha\beta\gamma_{122} + \alpha\beta\gamma_{123} = 0,$$

$$\alpha\beta\gamma_{231} + \alpha\beta\gamma_{232} + \alpha\beta\gamma_{233} = 0.$$

However, the total R_1 is not balanced with respect to the AB interaction, since

$$\alpha\beta_{12} + \alpha\beta_{23} + \alpha\beta_{31} \neq 0.$$

By similar reasoning each of the other R's may be shown to be balanced with respect to all main effects as well as the ABC interaction; however, each of the other R's is not balanced with respect to the AB interaction. Thus variation due to differences between the R's represents two of the four degrees of freedom of the AB interaction. In general such differences represent $p - 1$ of the $(p - 1)^2$ degrees of freedom of the AB interaction.

Since differences between the rows define $p - 1$ degrees of freedom of the AB interaction, the row $\times$ C interaction will define $(p - 1)(q - 1)$ degrees of freedom of the $AB \times C$ interaction. The latter interaction is equivalent to the ABC interaction. Hence $(p - 1)(q - 1)$ of the $(p - 1)(p - 1)(q - 1)$ degrees of freedom for the ABC interaction are confounded with row effects. The remaining degrees of freedom are not confounded with row effects; that is, $(p - 1)(p - 2)(q - 1)$ degrees of freedom of the ABC interaction are within-subject effects.

In some texts the following notation has been used:

$$\text{SS}_{\text{rows}} = \text{SS}_{ab(\text{between})},$$

$$\text{SS}_{\text{row} \times C} = \text{SS}_{abc(\text{between})}.$$

In terms of this notation,

$$\text{SS}_{ab} = \text{SS}_{ab(\text{between})} + \text{SS}_{ab(\text{within})}.$$

The corresponding degrees of freedom are

$$(p-1)(p-1) = (p-1) + (p-1)(p-2).$$

The three-factor interaction takes the following form:

$$\text{SS}_{abc} = \text{SS}_{abc(\text{between})} + \text{SS}_{abc(\text{within})},$$

$$(p-1)(p-1)(q-1) = (p-1)(q-1) + (p-1)(p-2)(q-1).$$

In each case the "between" component is part of the between-subject variation, and the "within" component is part of the within-subject effects.

The model under which the analysis of variance for this plan may be carried out is

$$\text{E}(X_{ijkmo}) = \mu + \gamma_k + (\text{row})_m + (\gamma \times \text{row})_{km} + \pi_{o(m)} + \alpha_i + \beta_j + \alpha\beta'_{ij} + \alpha\gamma_{ik} + \beta\gamma_{jk} + \alpha\beta\gamma'_{ijk}.$$

The analysis corresponding to this model is summarized in Table 9.7-10. The expected values of the mean squares in this table are derived under the assumptions that factors A, B, and C are fixed; the group factor and

Table 9.7-10 Analysis of Plan 9

Source of variation	df	df for general case	E(MS)
Between subjects	$9n - 1$	$npq - 1$	
C	2	$q - 1$	$\sigma_\varepsilon^2 + p\sigma_\pi^2 + np^2\sigma_\gamma^2$
Rows [AB (between)]	2	$p - 1$	$\sigma_\varepsilon^2 + p\sigma_\pi^2 + nq\sigma_{\alpha\beta}^2$
$C \times$ row [ABC (between)]	4	$(p-1)(q-1)$	$\sigma_\varepsilon^2 + p\sigma_\pi^2 + n\sigma_{\alpha\beta}^2$
Subjects w. groups	$9(n-1)$	$pq(n-1)$	$\sigma_\varepsilon^2 + p\sigma_\pi^2$
Within subjects	$18n$	$npq(p-1)$	
A	2	$p - 1$	$\sigma_\varepsilon^2 + npq\sigma_\alpha^2$
B	2	$p - 1$	$\sigma_\varepsilon^2 + npq\sigma_\beta^2$
AC	4	$(p-1)(q-1)$	$\sigma_\varepsilon^2 + np\sigma_{\alpha\gamma}^2$
BC	4	$(p-1)(q-1)$	$\sigma_\varepsilon^2 + np\sigma_{\beta\gamma}^2$
$(AB)'$	2	$(p-1)(p-2)$	$\sigma_\varepsilon^2 + nq\sigma_{\alpha\beta}^2$
$(ABC)'$	4	$(p-1)(p-2)(q-1)$	$\sigma_\varepsilon^2 + n\sigma_{\alpha\beta\gamma}^2$
Error (within)	$18n - 18$	$pq(p-1)(n-1)$	σ_ε^2

subjects within groups are considered random. All interactions with the group and subject effects are considered negligible.

The analysis given in this table indicates that Plan 9 may be considered as a special case of a $p \times p \times q$ factorial experiment arranged in blocks of size p, the groups having the role of blocks. The differences between the groups may be partitioned as follows:

Between groups		$pq - 1$
C		$q - 1$
Groups within C		$q(p - 1)$
Rows	$p - 1$	
$C \times$ rows	$(p - 1)(q - 1)$	

What corresponds to the within-cell variation in a $p \times p \times q$ factorial experiment having n observations per cell is partitioned as follows in Plan 9.

Within cell	$p^2q(n - 1)$
Subjects w. groups	$pq(n - 1)$
Error (within)	$pq(n - 1)(p - 1)$

Numerical Example of Plan 9. Suppose that a research worker is interested in evaluating the relative effectiveness of three variables in the design of a package. The variables to be studied are kind of material, style of printing to be used, and color of the material. Each variable constitutes a factor; there are three levels under each factor. That is, there are three kinds of material, three styles of printing, and three colors. A total of $3 \times 3 \times 3 = 27$ different packages can be constructed. Suppose that all 27 packages are constructed.

If the research worker selects Plan 9 for use, each subject is required to judge only 3 of the 27 packages. Under the conditions of the study, this number is considered to be the largest number feasible for any one subject. One of the main effects in Plan 9 (that of factor C) is wholly a between-subject effect, but within-subject information is available with respect to the interactions with that factor. Anticipating interactions with the color factor, and desiring within-subject information on interactions with the color factor, the research worker decides to let color correspond to factor C in Plan 9. This plan is symmetrical with respect to factors A and B; hence there is no difference whether the material factor corresponds to factor A or B. Suppose that the material factor is made to correspond to factor A.

The number of subjects to be used in each of the groups depends upon the precision that is desired. In related studies of this kind a minimum of 10 to 20 subjects per group is generally required to obtain a satisfactory power. Depending upon the degree of variability of the judgments, a larger or smaller number of subjects will be required. For purposes of illustrating the computational procedures, suppose that there are only two subjects in each group. The basic observational data for this illustration are given in Table 9.7-11. The design is that given earlier in this section; the data have been rearranged for computational convenience.

The subjects are asked to rate the packages assigned to them on a 15-point criterion scale. Subjects are assigned at random to the groups; the order in which a subject judges a set of packages is randomized independently for each subject. Subjects in groups 1, 4, and 7 each judge combinations ab_{12}, ab_{23}, and ab_{31}. However, these combinations

Table 9.7-11 Numerical Example of Plan 9

	Group	Person	a_1	a_2	a_3	Person total	Group total	Row total
			b_2	b_3	b_1			
c_1	G_1	1	2	2	3	$7 = P_1$		
		2	1	1	7	9	$16 = G_1$	
c_2	G_4	3	5	8	1	14		
		4	9	12	7	28	42	
c_3	G_7	5	5	4	6	15		
		6	7	5	9	21	36	$94 = R_1$
			b_1	b_2	b_3			
c_1	G_2	7	5	4	7	16		
		8	8	5	8	21	37	
c_2	G_5	9	8	10	4	22		
		10	10	14	6	30	52	
c_3	G_8	11	10	10	8	28		
		12	7	3	9	19	47	136
			b_3	b_1	b_2			
c_1	G_3	13	3	2	5	10		
		14	6	4	9	19	29	
c_2	G_6	15	8	9	6	23		
		16	10	10	5	25	48	
c_3	G_9	17	12	6	10	28		
		18	8	8	2	18	46	123
	Total		124	117	112	$353 = G$	353	353

appear in a different color (factor C) for each of the three groups. Person 1 made the ratings 2, 2, and 3 for the respective combinations, all of which are at color c_1. Person 6 made the ratings 7, 5, and 9 for the respective combinations—these combinations are in color c_3. The assignment of the treatment combinations to the groups follows the schematic representation of Plan 9 given earlier in this section. For ease in computation the data in Table 9.7-11 have been arranged. In addition to the basic observed data, totals needed in the analysis of variance are given at the right.

Summary data obtained from Table 9.7-11 appear in parts i and ii of Table 9.7-12. An ABC summary table appears in part i. Entries in part i

Table 9.7-12 Summary Data for Numerical Example

(i)

	c_1			c_2			c_3			Total
	a_1	a_2	a_3	a_1	a_2	a_3	a_1	a_2	a_3	
b_1	13	6	10	18	19	8	17	14	15	120
b_2	3	9	14	14	24	11	12	13	12	112
b_3	9	3	15	18	20	10	20	9	17	121
										353

(ii)

	a_1	a_2	a_3	Total
b_1	48	39	33	120
b_2	29	46	37	112
b_3	47	32	42	121
Total	124	117	112	353

	a_1	a_2	a_3	Total
c_1	25	18	39	82
c_2	50	63	29	142
c_3	49	36	44	129
Total	124	117	112	353

	b_1	b_2	b_3	Total
c_1	29	26	27	82
c_2	45	49	48	142
c_3	46	37	46	129
Total	120	112	121	353

	Row 1	Row 2	Row 3	Group total
c_1	16	37	29	82
c_2	42	52	48	142
c_3	36	47	46	129
Total	94	136	123	353

(iii)

$(1) = G^2/np^2q$	$= 2307.57$	$(7) = [\Sigma(AC_{ik})^2]/np$	$= 2568.83$
$(2) = \Sigma X^2$	$= 2811$	$(8) = [\Sigma(BC_{jk})^2]/np$	$= 2429.50$
$(3) = (\Sigma A_i^2)/npq$	$= 2311.61$	$(9) = [\Sigma(ABC_{ijk})^2]/n$	$= 2654.50$
$(4) = (\Sigma B_j^2)/npq$	$= 2310.28$	$(10) = (\Sigma P_o^2)/p$	$= 2575.00$
$(5) = (\Sigma C_k^2)/np^2$	$= 2418.28$	$(11) = (\Sigma G_m^2)/np$	$= 2476.50$
$(6) = [\Sigma(AB_{ij})^2]/nq$	$= 2372.83$	$(12) = (\Sigma R_s^2)/npq$	$= 2358.94$

have the general symbol ABC_{ijk} and are each based upon two observations. For example, the entry ABC_{111} is the sum of all observations made under treatment combination abc_{111}. Only subjects 7 and 8 are observed under this treatment combination. Hence as

$$ABC_{111} = 5 + 8 = 13.$$

As another example, only subjects 11 and 12 are observed under treatment combination abc_{333}; hence as

$$ABC_{333} = 8 + 9 = 17.$$

From the data in part i, *AB*, *AC*, and *BC* summary tables may be prepared; these appear in part ii. Data for the $C \times$ row summary table in part ii are obtained from the column headed Group total in Table 9.7-11. Computational symbols convenient for use in obtaining the required sums of squares are defined in part iii. The numerical values of these symbols for the data are also given.

The analysis of variance is summarized in Table 9.7-13. Factors *A*, *B*, and *C* are considered to be fixed. Tests on the between-subject effects use $MS_{\text{subj w. groups}}$ in the denominator of *F* ratios. Tests on within-subject effects use $MS_{\text{error(within)}}$. It will be noted that the denominator for the within-subject effects (3.22) is considerably smaller than the denominator for the between-subject effects (10.94).

The *F* tests indicate a highly significant *AC* interaction. To help interpret this interaction effect, the profiles for simple main effects of factor *A* (materials) at the various levels of factor *C* (colors) are given in Fig. 9.7-1. These profiles were prepared from the *AC* summary table in part ii of Table 9.7-12. Although color c_2 has the highest overall average rating, c_2 in combination with material a_3 has one of the lower average

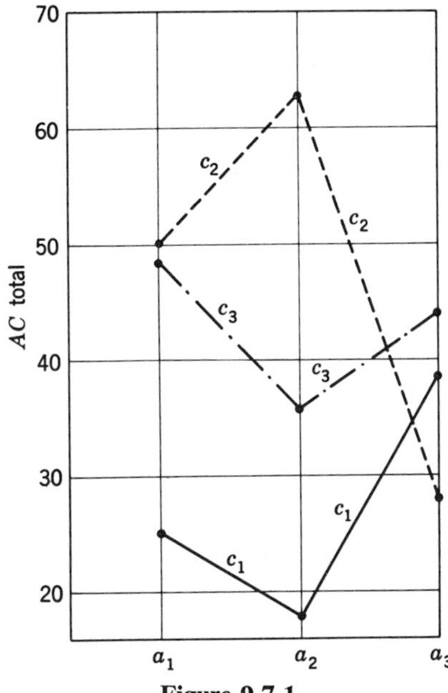

Figure 9.7-1

Table 9.7-13 Analysis of Variance for Numerical Example

Source of variation	Computational formula	SS	df	MS	F
Between subjects	(10) − (1)	267.43	17		
C (color)	(5) − (1)	110.71	2	55.36	5.06*
Rows	(12) − (1)	51.37	2	25.68	2.35
$C \times$ row	(11) − (5) − (12) + (1)	6.85	4	1.71	
Subjects within group	(10) − (11)	98.50	9	10.94	
Within subjects	(2) − (10)	236.00	36		
A (material)	(3) − (1)	4.04	2	2.02	
B (printing)	(4) − (1)	2.71	2	1.36	
AC	(7) − (3) − (5) + (1)	146.51	4	36.63	11.38**
BC	(8) − (4) − (5) + (1)	8.51	4	2.13	
AB'	[(6) − (3) − (4) + (1)] − [(12) − (1)]	7.14	2	3.57	
ABC'	[(9) − (6) − (7) − (8) + (3) + (4) + (5) − (1)] − [(11) − (5) − (12) + (1)]	9.09	4	2.27	
Error (within)	(2) − (10) − (9) + (11)	58.00	18	3.22	

ratings. Material a_2 in combination with color c_2 has the highest average rating. An analysis of the simple main effects of factor A for the levels of factor C will permit statistical tests of the differences among the set of points in the same profile. These tests are summarized in Table 9.7-14.

Table 9.7-14 Analysis of Simple Effects for Factor A

Source of variation	Computational formula	SS	df	MS	F
A at level c_1	$(3c_1) - (1c_1)$	38.11	2	19.06	5.92*
A at level c_2	$(3c_2) - (1c_2)$	98.11	2	49.06	15.24**
A at level c_3	$(3c_3) - (1c_3)$	14.33	2	7.16	2.22
Error (within)	(From Table 9.7-13)	58.00	18	3.22	

Differences between the points on the profile for c_2 are statistically significant ($F = 15.24$). A test on the difference between a_1 and a_2 at level c_2 is given by

$$F = \frac{(AC_{12} - AC_{22})^2}{2np\,\text{MS}_{\text{error(within)}}}$$
$$= \frac{(50 - 63)^2}{2(6)(3.22)} = 4.37.$$

If the latter test is considered as being in the post hoc category for differences between the levels of factor A at c_2, the .05-level Scheffé critical value is

$$(p - 1)F_{.95}(p - 1, \text{df}_{\text{error(within)}}) = 2F_{.95}(2,18)$$
$$= 2(3.55) = 7.10.$$

However, if this test is considered as a single comparison in the post hoc partition of A and AC, then the .05-level Scheffé critical value is

$$[(p + 1) + (p - 1)(q - 1)]F_{.95}[(p - 1) + (p - 1)(q - 1), \text{df}_{\text{error(within)}}]$$
$$= 6F_{.95}(6,18) = 6(2.66) = 15.96.$$

Tests on simple main effects are generally made along the dimension of greatest interest to the experimenter. This dimension, by design, is most frequently the within-subject dimension. However, should the experimenter desire to make tests on the simple main effects of what corresponds to factor C, the denominator of such tests is not $\text{MS}_{\text{subj w. groups}}$. Rather the denominator is $\text{MS}_{\text{w. cell}}$ for the appropriate level of the simple main effect. For example, in making tests on the simple main effects of factor C at level a_2, the appropriate denominator is $\text{MS}_{\text{w. cell(for } a_2)}$. Computationally,

$$\text{MS}_{\text{w. cell (for } a_2)} = \frac{(2a_2) - (9a_2)}{pq(n - 1)}.$$

Illustrative Application. A study reported by Michels, Bevan, and Strassel (1958) represents a special case of Plan 9. Individual animals were assigned at random to the rows of a 3×3 Latin square. Problems (factor A) were assigned to the columns of the square; the order in which the animals were assigned to the problems was determined by the levels of factor B. Nine replications (factor C) of the experiment were run; the same Latin square was used throughout. The criterion was the number of trials required to learn the problems.

Since the same square was used in each of the replications, corresponding rows of the squares represent a sequence effect. The experiment was actually conducted in distinct replications. Factor C, the replication factor, was considered to be a random factor. The analysis of variance reported by the authors is summarized in Table 9.7-15. Since the inter-

Table 9.7-15 Analysis of Variance for Michels et al. Experiment

Source		df	MS	F
Between subjects		26		
Sequences (rows)		2	3.05	
C (replications)		8	5.71	
Sequence $\times$ reps		16	7.59	
Within subjects		54		
A (order)		2	56.64	5.12
B (problem)		2	417.64	37.73
Pooled error		50	11.07	
Order $\times$ reps	16		12.89	
Problem $\times$ reps	16		9.43	
Residuals within squares	18		10.90	

actions with the replication factor were homogeneous and did not differ from the residuals within the squares, a single pooled error term was used in making all within-subject tests.

Plan 10. This plan is essentially an extension of Plan 6; more explicitly, Plan 6 is the building block from which Plan 10 is constructed. The latter is a series of such building blocks, each at a different level of factor D. Alternatively, Plan 10 may be viewed as a fractional replication of a $p \times p \times p \times q$ factorial experiment. As an illustration, a $3 \times 3 \times 3 \times 2$ factorial experiment will be used. A schematic representation of this plan is given below:

			a_1	a_2	a_3
	G_1	c_1	b_3	b_1	b_2
d_1	G_2	c_2	b_2	b_3	b_1
	G_3	c_3	b_1	b_2	b_3

			a_1	a_2	a_3
	G_4	c_1	b_2	b_1	b_3
d_2	G_5	c_2	b_1	b_3	b_2
	G_6	c_3	b_3	b_2	b_1

As shown above, a different square is used for the two levels of factor D. In cases where partial information on confounded interactions is to be recovered, it is desirable to use the same square for each level of factor D.

The complete factorial experiment for the illustrative example includes $3 \times 3 \times 3 \times 2 = 54$ treatment combinations. Only 18 treatment combinations appear in the above plan (i.e., one-third of the total). If the factors forming the dimensions of the square (factors A, B, and C) do not interact with each other (these dimensions may, however, interact with factor D), then an outline of the analysis of variance is given in Table 9.7-16. The expected values given in this table are derived under

Table 9.7-16 Analysis of Plan 10

Source of variation	df	df for general case	E(MS)
Between subjects	$6n - 1$	$npq - 1$	
C (AB')	2	$p - 1$	$\sigma_\varepsilon^2 + p\sigma_\pi^2 + npq\sigma_\gamma^2$
D	1	$q - 1$	$\sigma_\varepsilon^2 + p\sigma_\pi^2 + np^2\sigma_\delta^2$
CD ($AB' \times D$)	2	$(p - 1)(q - 1)$	$\sigma_\varepsilon^2 + p\sigma_\pi^2 + np\sigma_{\gamma\delta}^2$
Subjects within groups	$6(n - 1)$	$pq(n - 1)$	$\sigma_\varepsilon^2 + p\sigma_\pi^2$
Within subjects	$12n$	$npq(p - 1)$	
A	2	$p - 1$	$\sigma_\varepsilon^2 + npq\sigma_\alpha^2$
B	2	$p - 1$	$\sigma_\varepsilon^2 + npq\sigma_\beta^2$
AD	2	$(p - 1)(q - 1)$	$\sigma_\varepsilon^2 + np\sigma_{\alpha\delta}^2$
BD	2	$(p - 1)(q - 1)$	$\sigma_\varepsilon^2 + np\sigma_{\beta\delta}^2$
$(AB)''$ } residual	2	$(p - 1)(p - 2)$	$\sigma_\varepsilon^2 + nq\sigma_{\alpha\beta}^2$
$(AB)'' \times D$ } residual	2	$(p - 1)(p - 2)(q - 1)$	$\sigma_\varepsilon^2 + n\sigma_{\alpha\beta\delta}^2$
Error (within)	$12n - 12$	$pq(n - 1)(p - 1)$	σ_ε^2

the assumption that factors A, B, C, and D are fixed. Groups and subjects within groups are considered random.

If, for example, the AB interaction were not negligible, then the main effects of factor C and the CD interaction would be confounded with this interaction. A partial check on the assumption that dimensions of the Latin square do not interact with each other is provided by the F ratio

$$F = \frac{MS_{res}}{MS_{error\,(within)}}.$$

A somewhat better check on these assumptions is obtained by computing the residual and error (within) separately for each level of factor D as well as the corresponding pooled terms. Individual tests are made for the separate levels as well as for the pooled terms.

Relative to what would be the case in a complete factorial experiment, the within-cell variation is partitioned as follows:

Within cell	$p^2q(n-1)$
Subjects w. groups	$pq(n-1)$
Error (within)	$pq(n-1)(p-1)$

All other sources of variation are part of the between-cell variation; the degrees of freedom for the latter are $p^2q - 1$.

Plan 10 is useful in situations illustrated by the following example: Suppose that a research worker is interested in evaluating the effectiveness of q different methods of training (factor D) on marksmanship. There are p different kinds of targets (factor B) to be used in the evaluation process. Each target is to be placed at p different distances (factor C). Subjects are to fire at each of the p targets, but subjects are assigned to only one distance. The order (factor A) in which subjects fire at the targets is balanced by means of a Latin square. In this kind of experiment, primary interest is generally in the main effects of factor D and the interactions with factor D, particularly the BD and CD interactions. The BD interaction is a within-subject effect; the CD interaction is a between-subject effect. If the experimenter has the choice of what variables are assigned as factors B and C, factor B should be the one on which the more precise information is desired. The experimenter may not always have his choice in such matters—the dictates of experimental feasibility frequently force a decision of this kind in a given direction.

Computational Procedures. Computational procedures for this plan are summarized in Table 9.7-17. Since all treatment combinations in the

Table 9.7-17 Computational Procedures for Plan 10

(i)	$(1) = G^2/np^2q$	$(7) = [\Sigma(AD_{im})^2]/np$
	$(2) = \Sigma X^2$	$(8) = [\Sigma(BD_{jm})^2]/np$
	$(3) = (\Sigma A_i^2)/npq$	$(9) = [\Sigma(CD_{km})^2]/np$
	$(4) = (\Sigma B_j^2)/npq$	$(10) = (\Sigma P_o^2)/p$
	$(5) = (\Sigma C_k^2)/npq$	$(11) = [\Sigma(\text{cell total})^2]/n$
	$(6) = (\Sigma D_m^2)/np^2$	
(ii)	Between subjects	$(10) - (1)$
	C	$(5) - (1)$
	D	$(6) - (1)$
	CD	$(9) - (5) - (6) + (1)$
	Subjects within groups	$(10) - (9)$
	Within subjects	$(2) - (10)$
	A	$(3) - (1)$
	B	$(4) - (1)$
	AD	$(7) - (3) - (6) + (1)$
	BD	$(8) - (4) - (6) + (1)$
	Residual	$(11) - (7) - (8) - (9) + 2(6)$
	Error (within)	$(2) - (10) - (11) + (9)$

factorial experiment involving factors C and D appear in this plan, a CD summary table may be prepared. From the latter summary table one may compute sums of squares of the main effects of C and D as well as the sum of squares for their interaction. One may also prepare AD and BD summary tables, since all the treatment combinations in corresponding two-factor factorial experiments occur for these two sets of variables.

For the basic observations one may obtain each of the P_o, where the latter symbol denotes the sum of the p observations on an individual subject. The expression (cell total) used in computational symbol (11) denotes the sum of the n observations made under a specified treatment combination. Another symbol for this total is $ABCD_{ijkm}$.

The residual variation is that part of the variation between cells remaining after (presumably) all the unconfounded sources of variation due to treatment effects have been taken into account. In symbols,

$$SS_{res} = SS_{b.\ cells} - (SS_a + SS_b + SS_c + SS_d + SS_{cd} + SS_{ad} + SS_{bd}).$$

The homogeneity of the residual variation may be checked by computing separate residual terms for each Latin square. Since each of the Latin squares represents a different level of factor D, the latter procedure is equivalent to computing a separate residual term for each level of factor D.

$$SS_{\text{res for level } d_m} = (11d_m) - (7d_m) - (8d_m) - (9d_m) + 2(6d_m),$$

where the symbol $(11d_m)$ is (11) with the summation restricted to level d_m.

The variation due to error (within) is that part of the within-cell variation which remains after variation due to differences between subjects within cells is removed. In symbols,

$$SS_{\text{error(within)}} = SS_{\text{w. cell}} - SS_{\text{subj w. groups}}.$$

The latter term may be checked for homogeneity by computing a separate $SS_{\text{error (within)}}$ for each group. Computationally,

$$SS_{\text{error(within) for } G_r} = (2g_r) - (10g_r) - (11g_r) + (9g_r).$$

Plan 11. In Plan 7 only a fraction of the treatment combinations in a $p \times p \times p$ factorial experiment actually appears in the experiment. However, in Plan 11 all the treatment combinations in this factorial experiment are used. The primary purpose of this plan is to obtain complete within-subject information on all main effects, as well as partial within-subject information on all interaction effects in a $p \times p \times p$ factorial experiment. Yet only p observations on each subject are required.

The construction of this plan will be illustrated by means of a $3 \times 3 \times 3$ factorial experiment. The starting point for this plan is a balanced set of 3×3 Latin squares.

I			II			III		
1	3	2	3	2	1	2	1	3
3	2	1	2	1	3	1	3	2
2	1	3	1	3	2	3	2	1

It is noted that square II is obtained from square I by means of a one-step cyclic permutation of the rows. Similarly, square III is obtained from square II by a one-step cyclic permutation. One now constructs a square orthogonal to square I. The following square has this property:

I′

2	1	3
3	2	1
1	3	2

This square is also orthogonal to squares II and III.

In the design given in part i of Table 9.7-18, the subscripts for factor B

Table 9.7-18 Schematic Representation of Plan 11

		a_1	a_2	a_3		a_1	a_2	a_3		a_1	a_2	a_3
(i)	G_1	bc_{12}	bc_{31}	bc_{23}	G_4	bc_{32}	bc_{21}	bc_{13}	G_7	bc_{22}	bc_{11}	bc_{33}
	G_2	bc_{33}	bc_{22}	bc_{11}	G_5	bc_{23}	bc_{12}	bc_{31}	G_8	bc_{13}	bc_{32}	bc_{21}
	G_3	bc_{21}	bc_{13}	bc_{32}	G_6	bc_{11}	bc_{33}	bc_{22}	G_9	bc_{31}	bc_{23}	bc_{12}
(ii)	G_1	(112)	(231)	(323)	G_4	(132)	(221)	(313)	G_7	(122)	(211)	(333)
	G_2	(133)	(222)	(311)	G_5	(123)	(212)	(331)	G_8	(113)	(232)	(321)
	G_3	(121)	(213)	(332)	G_6	(111)	(233)	(322)	G_9	(131)	(223)	(312)

are determined by the corresponding numbers in squares I, II, and III. The subscripts for factor C are determined by the numbers in square I′. A different notation system is used in part ii of this table—the numbers in parentheses represent the respective subscripts for the treatment combinations.

The n subjects in each group are observed under the treatment combinations in each of the rows. These sets are balanced with respect to main effects but only partially balanced with respect to interaction effects. For purposes of illustrating the manner in which the AB interaction is partially confounded with differences between groups, consider the sets of treatment combinations assigned to the following groups:

G_1	(112)	(231)	(323)
G_6	(111)	(233)	(322)
G_8	(113)	(232)	(321)

Within each of these groups there are repeated measures on the same set of combinations of factor A and B, namely, ab_{11}, ab_{23}, and ab_{32}. There is balance with respect to main effects as well as the AC, BC, and ABC interactions when the sum of all observations in the three groups is obtained. If all factors are fixed, it may be shown by direct substitution in the basic linear model that

$$E(G_{1+6+8}) = 9n\mu + 3n(\alpha\beta_{11} + \alpha\beta_{23} + \alpha\beta_{32}) + 3n(\delta_1 + \delta_6 + \delta_8).$$

The symbol G_{1+6+8} denotes the sum of the $9n$ observations in these three groups. The effects δ_1, δ_6, and δ_8 designate group effects. Since G_2, G_4, and G_9 are each observed under the set of treatment combinations ab_{13}, ab_{22}, and ab_{31}, and since there is balance with respect to the main effects as well as the AC and BC interactions,

$$E(G_{2+4+9}) = 9n\mu + 3n(\alpha\beta_{13} + \alpha\beta_{22} + \alpha\beta_{31}) + 3n(\delta_2 + \delta_4 + \delta_9).$$

It may also be shown that

$$E(G_{3+5+7}) = 9n\mu + 3n(\alpha\beta_{12} + \alpha\beta_{21} + \alpha\beta_{33}) + 3n(\delta_3 + \delta_5 + \delta_7).$$

Thus differences between the three totals G_{1+6+8}, G_{2+4+9}, and G_{3+5+7} are in part due to the AB interaction and in part due to group effects. Hence two of the four degrees of freedom of the AB interaction are partially confounded with differences between groups.

In Plan 11, as constructed in Table 9.7-18, two-factor interactions are partially confounded with the sets of group totals given below:

Interaction	Sets of group totals
AB	G_{1+6+8}, G_{2+4+9}, G_{3+5+7}
AC	G_{1+4+7}, G_{2+5+8}, G_{3+6+9}
BC	G_{1+5+9}, G_{2+6+7}, G_{3+4+8}

In each case two of the four degrees of freedom for the respective two-factor interactions are confounded with differences between groups. The remaining degrees of freedom for the variation due to differences between groups are confounded with the three-factor interaction. The totals involved in the latter are G_{1+2+3}, G_{4+5+6}, and G_{7+8+9}. An outline of the analysis of variance appears in Table 9.7-19.

This plan may be improved by constructing a replication in which different components of the interactions are confounded with group effects. This replication may be obtained from a second set of balanced squares. The following balanced set is different from the original set:

I			II			III		
1	2	3	2	3	1	3	1	2
2	3	1	3	1	2	1	2	3
3	1	2	1	2	3	2	3	1

The following square is orthogonal to square I:

I′		
1	2	3
3	1	2
2	3	1

Table 9.7-19 Analysis of Plan 11

Source of variation	df	df for general case	E(MS)
Between subjects	$9n - 1$	$np^2 - 1$	
Groups	8	$p^2 - 1$	
$(AB)'$	2	$p - 1$	
$(AC)'$	2	$p - 1$	
$(BC)'$	2	$p - 1$	
$(ABC)'$	2	$(p - 1)(p - 2)$	
Subjects within groups	$9n - 9$	$p^2(n - 1)$	
Within subjects	$18n$	$np^2(p - 1)$	
A	2	$p - 1$	$\sigma_\varepsilon^2 + np^2\sigma_\alpha^2$
B	2	$p - 1$	$\sigma_\varepsilon^2 + np^2\sigma_\beta^2$
C	2	$p - 1$	$\sigma_\varepsilon^2 + np^2\sigma_\gamma^2$
$(AB)''$	2	$(p - 1)(p - 2)$	$\sigma_\varepsilon^2 + np\sigma_{\alpha\beta}^2$
$(AC)''$	2	$(p - 1)(p - 2)$	$\sigma_\varepsilon^2 + np\sigma_{\alpha\gamma}^2$
$(BC)''$	2	$(p - 1)(p - 2)$	$\sigma_\varepsilon^2 + np\sigma_{\beta\gamma}^2$
$(ABC)''$	6	$(p - 1)^3 - (p - 1)(p - 2)$	$\sigma_\varepsilon^2 + n\sigma_{\alpha\beta\gamma}^2$
Error (within)	$18n - 18$	$p^2(p - 1)(n - 1)$	σ_ε^2

A plan constructed from this set of squares is given in Table 9.7-20. For this replication, it may be shown that

$$E(G_{1+6+8}) = 9n + 3n(\alpha\beta_{11} + \alpha\beta_{22} + \alpha\beta_{33}) + 3n(\delta_1 + \delta_6 + \delta_8).$$

The set of $\alpha\beta$'s included in this expected value is different from that included in the original plan. For this replication it may be shown that variation among the totals

$$G_{1+6+8}, \qquad G_{2+4+9}, \qquad \text{and} \qquad G_{3+5+7}$$

represents two of the four degrees of freedom of the AB interaction; however, the two degrees of freedom that are confounded here are not identical with those confounded in the original plan.

In terms of the notation of Chap. 8 on components of interaction, in the original plan the AB^2 components of the $A \times B$ interaction are confounded with group effects; in the replication the AB components are confounded with group effects. Hence the original plan provided within-subject information on the AB^2 components, and the replication provides

Table 9.7-20 Replication of Plan 11

G_1	(111)	(222)	(333)	G_4	(121)	(232)	(313)	G_7	(131)	(212)	(323)
G_2	(123)	(231)	(312)	G_5	(133)	(211)	(322)	G_8	(113)	(221)	(332)
G_3	(132)	(213)	(321)	G_6	(112)	(223)	(331)	G_9	(122)	(233)	(311)

within-subject information on the AB components. Similarly, in the original plan the AC components of $A \times C$ are confounded with group effects; in the replication the AC^2 components are confounded. A comparable condition holds for the $B \times C$ interaction. Use of the replication will provide some within-subject information on all components of the two-factor interactions. However, additional replications are required to obtain within-subject information on all components of the three-factor interaction. The original plan provides within-subject information on the AB^2C^2 component; the replication provides within-subject information on the ABC component. No within-subject information is available on the ABC^2 or the AB^2C^2 components.

Computational Procedures for Plan 11. Computational procedures here are similar to those used in a $p \times p \times p$ factorial experiment in which there are n observations per cell. The first steps actually duplicate the latter procedures. The nonreplicated version of Plan 11 will be considered first; the replicated plan will be considered later.

Since all the treatment combinations in a $p \times p \times p$ factorial experiment appear in Plan 11, an ABC summary table may be prepared from the basic observations. From the latter summary table one obtains AB, AC, and BC summary tables. Most of the computational symbols in part i of Table 9.7-21 are obtained from these summary tables. Symbol (10)

Table 9.7-21 Definition of Computational Symbols

(i)	$(1) = G^2/np^3$	$(6) = [\Sigma(AB_{ij})^2]/np$
	$(2) = \Sigma X^2$	$(7) = [\Sigma(AC_{ik})^2]/np$
	$(3) = (\Sigma A_i^2)/np^2$	$(8) = [\Sigma(BC_{jk})^2]/np$
	$(4) = (\Sigma B_j^2)/np^2$	$(9) = [\Sigma(ABC_{ijk})^2]/n$
	$(5) = (\Sigma C_k^2)/np^2$	$(10) = (\Sigma P_o^2)/p$
(ii)	$(11) = (\Sigma G_m^2)/np$	
	$(12) = (\Sigma G_{ab}^2)/np^2 = (G_{1+6+8}^2 + G_{2+4+9}^2 + G_{3+5+7}^2)/np^2$	
	$(13) = (\Sigma G_{ac}^2)/np^2 = (G_{1+4+7}^2 + G_{2+5+8}^2 + G_{3+6+9}^2)/np^2$	
	$(14) = (\Sigma G_{bc}^2)/np^2 = (G_{1+5+9}^2 + G_{2+6+7}^2 + G_{3+4+8}^2)/np^2$	

involves the totals P_o; the latter are the sums of the set of observations on an individual subject. The computational symbols in part ii require special comment.

The symbol G_m denotes the sum of the np observations in group m. The symbol G_{ab} is the sum of those G_m's which are assigned to the same set of ab_{ij}. For assignments made in accordance with the principles given in the last section, in a $3 \times 3 \times 3$ experiment the G_{ab}'s are

$$G_{1+6+8}, \qquad G_{2+4+9}, \qquad \text{and} \qquad G_{3+5+7},$$

where $G_{1+6+8} = G_1 + G_6 + G_8$. In general there will be np^2 observations in each of such totals. The G_{ac}'s are made up of the following totals:

$$G_{1+4+7}, \qquad G_{2+5+8}, \qquad \text{and} \qquad G_{3+6+9}.$$

Each of the latter G_m's which are combined into a single total is assigned to the same set of ac_{ik}. For example, inspection of Table 9.7-18 indicates that groups 1, 4, and 7 are assigned to the sets ac_{12}, ac_{21}, and ac_{33}. The level of factor B changes for the different groups, but the set of ac_{ik} remains the same.

Computational formulas for the sum of squares are summarized in Table 9.7-22. The parts of $SS_{error\ (within)}$ may be checked for homogeneity

Table 9.7-22 Computational Formulas

Between subjects	(10) − (1)
Between groups	(11) − (1)
AB'	(12) − (1)
AC'	(13) − (1)
BC'	(14) − (1)
ABC'	(11) − (12) − (13) − (14) + 2(1)
Subjects within groups	(10) − (11)
Within subjects	(2) − (10)
A	(3) − (1)
B	(4) − (1)
C	(5) − (1)
AB''	[(6) − (3) − (4) + (1)] − [(12) − (1)]
AC''	[(7) − (3) − (5) + (1)] − [(13) − (1)]
BC''	[(8) − (4) − (5) + (1)] − [(14) − (1)]
ABC''	[(9) − (6) − (7) − (8) + (3) + (4) + (5) − (1)] − [(11) − (12) − (13) − (14) + 2(1)]
Error (within)	[(2) − (9)] − [(10) − (11)]

by computing a separate sum of squares for each of the groups. Thus

$$SS_{error\ (within)\ for\ G_m} = [(2g_m) - (9g_m)] - [(10g_m) - (11g_m)],$$

where the symbol $(2g_m)$ represents the symbol (2), in which the summation is restricted to G_m.

If Plan 11 is replicated, it is generally wise to carry out separate computations for each replication; those terms which are homogeneous may then be combined. Since different components of the two-factor interactions are estimated in each of the two replications, the two-factor interactions are in a sense nested within the replications. Thus,

$$SS''_{ab} = SS''_{ab\ \text{from rep 1}} + SS''_{ab\ \text{for rep 2}}.$$

Similar relationships hold for other two-factor interactions as well as the three-factor interaction. The error (within) term for the combined replications is given by

$$\text{SS}_{\text{error (within)}} = \text{SS}_{\text{error (within) from rep 1}} + \text{SS}_{\text{error (within) from rep 2}}.$$

In a sense, the error (within) effects are also nested within each replication.

Main effects, however, are not nested within replications. The latter are computed by obtaining $A \times$ replication, $B \times$ replication, and $C \times$ replication summary tables from the basic data. From such tables main effects and corresponding interactions may be computed. The latter interactions are pooled with error (within) if the components prove to be homogeneous with error (within).

Plan 12. This plan resembles Plan 5 as well as Plan 8 but yet has features that neither of the latter designs has. A schematic representation of Plan 12 is given below:

	c_1			c_2		
	b_1	b_2	b_3	b_1	b_2	b_3
G_1	a_2	a_1	a_3	a_2	a_1	a_3
G_2	a_3	a_2	a_1	a_3	a_2	a_1
G_3	a_1	a_3	a_2	a_1	a_3	a_2

In general, there will be p levels of factor A and p levels of factor B. The same $p \times p$ Latin square is used at each of the q levels of factor C. Plan 12 may be regarded as a fractional replication of a $p \times p \times q$ factorial experiment. If the interaction with the group factor is negligible, complete within-subject information is available for the main effects of factors A, B, and C. Complete within-subject information is also available on the AC and BC interactions; partial within-subject information is available on the AB and ABC interactions.

The analysis of variance is outlined in Table 9.7-23. In obtaining the expected values of the mean squares it is assumed that factors A, B, and C are fixed; subjects within the groups define a random variable. The terms σ_u^2, σ_v^2, and σ_w^2, appearing in the expected values, represent interactions with subject effects. Because factors A and B are dimensions of a Latin square, the $A \times$ subject within group interaction cannot be distinguished from the $B \times$ subject within group interaction. However, the $C \times$ subject within group interaction can be distinguished from the others, since factor C is not one of the dimensions of the Latin square. In general, the three residual terms should be pooled into a single error term if there is neither a priori nor experimental evidence for heterogeneity of these components.

The computational procedures for this plan are simplified if the analysis of variance is carried out in two stages. In the first stage one of the dimensions of the Latin square, say factor A, is disregarded. Then the plan reduces to a $p \times q$ factorial experiment with repeated measures on

Table 9.7-23 Analysis of Plan 12

Source	df	E(MS)
Between subjects	$np - 1$	
Groups	$p - 1$	
Subj w. groups	$p(n - 1)$	
Within subjects	$np(pq - 1)$	
A	$p - 1$	$\sigma_\varepsilon^2 + \sigma_u^2 + npq\sigma_\alpha^2$
B	$p - 1$	$\sigma_\varepsilon^2 + \sigma_u^2 + npq\sigma_\beta^2$
$(AB)'$	$(p - 1)(p - 2)$	$\sigma_\varepsilon^2 + \sigma_u^2 + nq\sigma_{\alpha\beta'}^2$
Residual (1)	$p(n - 1)(p - 1)$	$\sigma_\varepsilon^2 + \sigma_u^2$
C	$q - 1$	$\sigma_\varepsilon^2 + \sigma_v^2 + np^2\sigma_\gamma^2$
$C \times$ groups	$(p - 1)(q - 1)$	$\sigma_\varepsilon^2 + \sigma_v^2 + np\sigma_{\gamma\delta}^2$
Residual (2)	$p(n - 1)(q - 1)$	$\sigma_\varepsilon^2 + \sigma_v^2$
AC	$(p - 1)(q - 1)$	$\sigma_\varepsilon^2 + \sigma_w^2 + np\sigma_{\alpha\gamma}^2$
BC	$(p - 1)(q - 1)$	$\sigma_\varepsilon^2 + \sigma_w^2 + np\sigma_{\beta\gamma}^2$
$(AB)'C$	$(p - 1)(p - 2)(q - 1)$	$\sigma_\varepsilon^2 + \sigma_w^2 + n\sigma_{\alpha\beta'\gamma}^2$
Residual (3)	$p(n - 1)(p - 1)(q - 1)$	$\sigma_\varepsilon^2 + \sigma_w^2$

Pooled error = residual (1) + residual (2) + residual (3)
$\text{df}_{\text{pooled error}} = p(n - 1)(pq - 1)$

both factors. The analysis of variance for the first stage is outlined in Table 9.7-24. The detailed computational procedures given in Sec. 7.3 may be adapted for use to obtain the first stage of the analysis.

In the second stage of the analysis, the presence of factor A as a dimension of the Latin square is taken into account. The $B \times$ group interaction is partitioned as follows:

$B \times$ group	$(p - 1)^2$
A	$p - 1$
AB'	$(p - 1)(p - 2)$

The latter interaction term may be obtained by subtraction or by the relation

$$SS_{ab'} = SS_{ab} - SS_{\text{groups}},$$

where SS_{ab} is obtained from an AB summary table for the combined levels of factor C. The $BC \times$ group interaction is partitioned as follows:

$BC \times$ group	$(p - 1)^2(q - 1)$
AC	$(p - 1)(q - 1)$
$AB'C$	$(p - 1)(p - 2)(q - 1)$

The latter interaction term may be obtained by subtraction from the following relation,

$$SS_{ab'c} = SS_{abc} - SS_{c \times \text{group}},$$

where SS_{abc} is obtained from an ABC summary table.

Table 9.7-24 First Stage in the Analysis of Plan 12

Source	df
Between subjects	$np - 1$
Groups	$p - 1$
Subjects w. group	$p(n - 1)$
Within subjects	$np(pq - 1)$
B	$p - 1$
$B \times$ group	$(p - 1)^2$
$B \times$ subj w. group	$p(n - 1)(p - 1)$
C	$(q - 1)$
$C \times$ group	$(p - 1)(q - 1)$
$C \times$ subj w. group	$p(n - 1)(q - 1)$
BC	$(p - 1)(q - 1)$
$BC \times$ group	$(p - 1)^2(q - 1)$
$BC \times$ subj w. group	$p(n - 1)(p - 1)(q - 1)$

The residual terms in Table 9.7-23 are equivalent to the following interactions:

$$SS_{\text{res}(1)} = SS_{b \times \text{subj w. group}},$$

$$SS_{\text{res}(2)} = SS_{c \times \text{subj w. group}},$$

$$SS_{\text{res}(3)} = SS_{bc \times \text{subj w. group}}.$$

If only the pooled error is obtained, the latter is given by

$$SS_{\text{pooled error}} = SS_{\text{w. cell}} - SS_{\text{subj w. group}}.$$

Illustrative Application. An experiment reported by Briggs, Fitts, and Bahrick (1957) represents a special case of Plan 12. In this study the dimensions of a 4 × 4 Latin square were noise level (factor A) and blocks of trials (factor B). Factor C represented three experimental sessions. There were three subjects in each of the groups. The groups were assigned at random to the rows of the Latin square. The groups were required to track a visual target under various levels of visual noise. There were three experimental sessions; within each session there were four blocks of trials—each block was under a noise level determined by the letters of a randomly selected Latin square. The same square was used in all sessions. The rows of the Latin square define the sequence in which the noise levels were given. Hence the group factor may be considered as a sequence factor.

The analysis of variance reported by the authors differs slightly from that given in Table 9.7-23. Their analysis is outlined in Table 9.7-25.

Table 9.7-25 Analysis of Variance for Briggs et al. Experiment

Source	df	MS	F
Between subjects	11		
G Groups (sequence)	3	160	
Subj w. group	8	1369	
Within subjects	132		
C Sessions	2	2546	13.35
CG	6	16	
$C \times$ subj w. group	16	191	
A Noise level	3	5289	31.31
B Blocks of trials	3	330	1.95
AC	6	293	1.73
BC	6	109	
$(AB)'$ Square uniqueness	6	148	
$(AB)' \times C$($C \times$ square uniq.)	12	95	
Residual	72	169	

The residual term in this analysis corresponds to residual (1) + residual (3). The $C \times$ subjects within group term corresponds to residual (2). No attempt was made to interpret the partial information on the AB and ABC interactions. Instead, these sources of variation were considered to be functions of the particular Latin square selected for use. The latter interpretation is the preferred one if the sequence (group) factor can be considered as contributing toward these interactions. If there is no reason for considering residual (2) as potentially different from residuals (1) and (3), the experimental data do not rule against pooling the equivalent of residuals (1), (2), and (3) into a single pooled error term to be used in testing all within-subject effects.

Plan 13. This plan resembles Plan 9; in this case, however, a Greco-Latin square replaces the Latin square. A schematic representation of Plan 13 is given below:

		c_1	c_2	c_3
d_1	G_1	ab_{12}	ab_{31}	ab_{23}
	G_2	ab_{33}	ab_{22}	ab_{11}
	G_3	ab_{21}	ab_{13}	ab_{32}

		c_1	c_2	c_3
d_2	G_4	ab_{12}	ab_{31}	ab_{23}
	G_5	ab_{33}	ab_{22}	ab_{11}
	G_6	ab_{21}	ab_{13}	ab_{32}

The same Greco-Latin square is used for each level of factor D.

The analysis of variance for this plan is most easily understood if it is made in two stages. In the first stage, two of the four dimensions of the Greco-Latin square are disregarded. Suppose that factors A and B are disregarded. The resulting plan may be considered as a $q \times p \times p$ factorial experiment with repeated measures on one of the factors. The analysis of variance for the first stage appears in the upper part of Table 9.7-26.

Table 9.7-26 Analysis of Variance for Plan 13

	Source	df
(i)	Between subjects	$npq - 1$
	Rows	$p - 1$
	D	$q - 1$
	$D \times$ row	$(p - 1)(q - 1)$
	Subjects within groups	$pq(n - 1)$
	Within subjects	$npq(p - 1)$
	C	$p - 1$
	$C \times$ row	$(p - 1)^2$
	CD	$(p - 1)(q - 1)$
	$CD \times$ row	$(p - 1)^2(q - 1)$
	$C \times$ subj w. group (Error)	$pq(n - 1)(p - 1)$
(ii)	$C \times$ row	$(p - 1)^2$
	A	$p - 1$
	B	$p - 1$
	$(AB)'$	$(p - 1)(p - 3)$
	$CD \times$ row	$(p - 1)^2(q - 1)$
	AD	$(p - 1)(q - 1)$
	BD	$(p - 1)(q - 1)$
	$(AB)'D$	$(p - 1)(p - 3)(q - 1)$

In the second stage, the interactions which involve two dimensions of the square (C and rows) are partitioned into main effects and interactions associated with factors A and B. This stage of the analysis is shown in the lower part of Table 9.7-26. It should be noted that $(AB)'$ cannot be distinguished from $(AC)'$ or $(BC)'$. This latter source of variation is sometimes called the uniqueness of the Greco-Latin square. In the partition of the $CD \times$ row interaction, $(AB)'D$ cannot be distinguished from $(AC)'D$ or $(BC)'D$.

If all factors are fixed, all within-subject effects are tested by means of F ratios having $C \times$ subjects within group as a denominator. If interactions between the dimensions of the square are negligible, $(AB)'$ and $(AB)'D$ may be pooled with the experimental error. The F ratios for between-subject effects have subjects within groups as a denominator.

Illustrative Example. Briggs (1957) reports an experiment which is a special case of Plan 13. A 4 × 4 Greco-Latin square formed the basis for the plan. Groups of four subjects each were assigned to the rows of a Greco-Latin square. The letters of the square represented the lists (factor A) of syllables to be learned and the degree of overlearning (factor B). The columns of the square (factor C) represented the experimental session. There were five squares (factor D) in all—subjects in different squares received different amounts of initial learning. The same Greco-Latin square was used throughout; hence corresponding rows of the squares represent the same sequence of combinations of factors A and B.

Table 9.7-27 Analysis of Variance for Briggs Data

Source	df	MS	F
Between subjects	79		
R Rows (sequences)	3	1,088	2.29
D Squares (initial learning)	4	8,975	18.85
DR	12	427	
Subjects w. groups	60	476	
Within subjects	240		
A Lists	3	1,668	10.80
B Overlearning	3	22,389	144.94
C Session	3	755	4.88
AD	12	122	
BD	12	346	2.24
CD	12	127	
$(AB)'$	3	152	
$(AB)'D$	12	114	
Error	180	154	

A slightly modified analysis of variance for the Briggs data is given in Table 9.7-27.

Summary of Plans in Sec. 9.7

Plan 5

	a_1	a_2	a_3
G_1	b_3	b_1	b_2
G_2	b_1	b_2	b_3
G_3	b_2	b_3	b_1

Plan 6

		a_1	a_2	a_3
G_1	c_1	b_1	b_3	b_2
G_2	c_2	b_2	b_1	b_3
G_3	c_3	b_3	b_2	b_1

Plan 7

	a_1	a_2	a_3
G_1	bc_{11}	bc_{23}	bc_{32}
G_2	bc_{22}	bc_{31}	bc_{13}
G_3	bc_{33}	bc_{12}	bc_{21}

Plan 8

		a_1	a_2	a_3
c_1	G_1	b_1	b_2	b_3
	G_2	b_2	b_3	b_1
	G_3	b_3	b_1	b_2

		a_1	a_2	a_3
c_2	G_4	b_2	b_3	b_1
	G_5	b_1	b_2	b_3
	G_6	b_3	b_1	b_2

Plan 9

		a_1	a_2	a_3
c_1	G_1	b_2	b_3	b_1
	G_2	b_1	b_2	b_3
	G_3	b_3	b_1	b_2

		a_1	a_2	a_3
c_2	G_4	b_2	b_3	b_1
	G_5	b_1	b_2	b_3
	G_6	b_3	b_1	b_2

		a_1	a_2	a_3
c_3	G_7	b_2	b_3	b_1
	G_8	b_1	b_2	b_3
	G_9	b_3	b_1	b_2

Plan 10

			a_1	a_2	a_3
d_1	G_1	c_1	b_3	b_1	b_2
	G_2	c_2	b_2	b_3	b_1
	G_3	c_3	b_1	b_2	b_3

			a_1	a_2	a_3
d_2	G_4	c_1	b_2	b_1	b_3
	G_5	c_2	b_1	b_3	b_2
	G_6	c_3	b_3	b_2	b_1

Plan 11

	a_1	a_2	a_3
G_1	bc_{12}	bc_{31}	bc_{23}
G_2	bc_{33}	bc_{22}	bc_{11}
G_3	bc_{21}	bc_{13}	bc_{32}

	a_1	a_2	a_3
G_4	bc_{32}	bc_{21}	bc_{13}
G_5	bc_{23}	bc_{12}	bc_{31}
G_6	bc_{11}	bc_{33}	bc_{22}

	a_1	a_2	a_3
G_7	bc_{22}	bc_{11}	bc_{33}
G_8	bc_{13}	bc_{32}	bc_{21}
G_9	bc_{31}	bc_{23}	bc_{12}

Plan 12

	c_1			c_2		
	b_1	b_2	b_3	b_1	b_2	b_3
G_1	a_2	a_1	a_3	a_2	a_1	a_3
G_2	a_3	a_2	a_1	a_3	a_2	a_1
G_3	a_1	a_3	a_2	a_1	a_3	a_2

Plan 13

		c_1	c_2	c_3
d_1	G_1	ab_{12}	ab_{31}	ab_{23}
	G_2	ab_{33}	ab_{22}	ab_{11}
	G_3	ab_{21}	ab_{13}	ab_{32}

		c_1	c_2	c_3
d_2	G_4	ab_{12}	ab_{31}	ab_{23}
	G_5	ab_{33}	ab_{22}	ab_{11}
	G_6	ab_{21}	ab_{13}	ab_{32}

10
ANALYSIS OF COVARIANCE

10.1 General Purpose

There are two general methods for controlling variability due to experimental error—direct and statistical. Direct control includes such methods as grouping the experimental units into homogeneous strata or blocks, increasing the uniformity of the conditions under which the experiment is run, and increasing the accuracy of the measurements. Replicated experiments, randomized block designs, repeated-measure designs, split-plot designs, incomplete-block designs—these designs use the direct-control principle to increase the precision of the experiment.

In this chapter, designs which use an indirect, or statistical, control (1) to increase the precision of the experiment and (2) to remove potential sources of bias in the experiment will be discussed. The latter objective is one that is particularly important in situations where the experimenter cannot assign individual units at random to the experimental conditions. Statistical control is achieved by measuring one or more concomitant variates in addition to the variate of primary interest. The latter variate will be termed the criterion, or simply the variate; the concomitant variates will be called covariates. Measurements on the covariates are made for the purpose of adjusting the measurements on the variate.

For example, suppose that the purpose of an experiment is to determine the effect of various methods of extinction upon some kind of learned response. The variate in this experiment may be a measure of extinction; the covariate may be a measure associated with the degree of learning at the start of the extinction trials. As another example, suppose that the purpose of an experiment is to measure the effect of various stress situations upon blood pressure. In this case a measure of blood pressure under a condition of no stress may be the covariate. As still another example, suppose that the purpose of an experiment is to evaluate the effect of electrical stimulation on the weight of denervated muscle. The weight of the corresponding normal muscle may serve as a covariate in this type of experiment.

Considerable care is needed in using the analysis of covariance in "adjusting" for initial or final differences between groups of experimental units. Lord (1969) gives some good examples of how this type of adjustment may result in highly misleading interpretations. Interpretations and possible misinterpretations of adjusted means are discussed later in this chapter. The intrinsic nature of the concomitance between the variate and the covariate requires examination. If this concomitance is predominantly due to a dimension along which an adjustment is desired, such an adjustment may be made. However, should this concomitance be due largely to factors which are not those for which control is wanted, then making an adjustment for the covariate may lead to relatively meaningless results.

Direct control and statistical control may, of course, be used simultaneously within the same experiment. One or more variates may be under direct control; one or more variates may be under statistical control.

Suppose that the means on the variate in a single-factor experiment are denoted

$$\bar{Y}_1, \bar{Y}_2, \ldots, \bar{Y}_k,$$

and the means on the covariate are denoted

$$\bar{X}_1, \bar{X}_2, \ldots, \bar{X}_k.$$

Primary interest lies in differences among the $\bar{Y}_j$. Suppose that differences in the $\bar{X}_j$ are due to sources of variation related to the $\bar{Y}_j$ but not directly related to the treatment effects. If this is the case, then more precise information on the treatment effects may be obtained by adjusting the $\bar{Y}_j$ for the association with the $\bar{X}_j$. Suppose the adjusted variate means are denoted

$$\bar{Y}'_1, \bar{Y}'_2, \ldots, \bar{Y}'_k.$$

There may be several different ways of making the adjustment. In some cases, the adjustment may take the form of a simple difference between variate and covariate; that is,

$$\bar{Y}'_j = \bar{Y}_j - \bar{X}_j.$$

In other cases, the adjusted mean may take the form

$$\bar{Y}'_j = \bar{Y}_j / \bar{X}_j.$$

The appropriate form of the adjustment is usually determined from prior knowledge about the interrelationship between variate and covariate. Often the average effect of an increase of 1 unit in the covariate upon the variate is given by some form of regression analysis.

Although the form of the regression need not be linear, only the linear case will be considered in this section. In terms of a linear adjustment,

$$\bar{Y}'_j = \bar{Y}_j - b(\bar{X}_j - \bar{X}),$$

where b is the regression coefficient for the regression of Y on X.

The change in the experimental error under linear adjustment of the variate depends upon the magnitude of the linear correlation between the variate and the covariate. If this correlation is denoted by ρ, and if the experimental error per unit, disregarding the covariate, is σ_ε^2, then the experimental error after the adjustment is

$$\sigma_\varepsilon^2(1 - \rho^2)\frac{f_e}{f_e - 1},$$

where f_e represents the degrees of freedom for the estimation of σ_ε^2.

If, instead of using a covariance adjustment, the experimental units are grouped into blocks (equivalently, into strata or classes) which are homogeneous with respect to the covariate, and if the relationship between X and Y is linear, then the effect of this kind of blocking is to reduce the experimental error from

$$\sigma_\varepsilon^2 \quad \text{to} \quad \sigma_\varepsilon^2(1 - \rho^2).$$

Thus, when the regression is linear, covariance adjustment is approximately as effective as stratification with respect to the covariate. However, if the regression is not linear, and a linear adjustment is used, then stratification will generally provide greater reduction in the experimental error. In a real sense, stratification is a function-free regression scheme. (Stratification with respect to the covariate converts the covariate into a factor—a single-factor experiment is converted into a two-factor experiment, the factors being the treatment and the covariate.)

When the covariate is actually affected by the treatment, the adjustment process removes more than what can be considered an error component from the variate. It also may remove part of the treatment effect on Y.

If the measurements on both the variate and the covariate are made after the administration of the treatment, it is possible that the covariate may have been affected by the treatment. An analysis of variance on the covariate may throw some light on this issue. In some experimental work where both the variate and covariate are affected by the treatments, the purpose of such experiments has been to investigate the underlying process by which the treatment actually affects the variate. There are some difficulties that are encountered in the interpretation of the results of such experiments. These difficulties are not that the model used in the analysis necessarily breaks down but rather that the adjusted data may not correspond to any realistic experimental situation.

If the measurements on the covariate are made before the treatments are administered, the covariate cannot be affected by the treatments. When an experimenter is forced by the nature of the real world to work with intact groups, the covariate means under treatments j and m, denoted $\bar{X}_j$ and $\bar{X}_m$, may differ. The adjusted difference between the variate means, using a linear adjustment, takes the form

$$\bar{Y}'_j - \bar{Y}'_m = \bar{Y}_j - \bar{Y}_m - b(\bar{X}_j - \bar{X}_m).$$

If one defines the difference

$$\tau_j - \tau_m = \mu_{y_j} - \mu_{y_m} - \beta(\bar{X}_j - \bar{X}_m),$$

then

$$\begin{aligned} E(\bar{Y}'_j - \bar{Y}'_m) &= E(\bar{Y}_j - \bar{Y}_m) - E[b(\bar{X}_j - \bar{X}_m)] \\ &= \mu_{y_j} - \mu_{y_m} - \beta(\bar{X}_j - \bar{X}_m) \\ &= \tau_j - \tau_m. \end{aligned}$$

That is, the expected value of the difference between the adjusted means is an unbiased estimator of the corresponding difference between the treatment effects. The expected value of the difference between the unadjusted treatment means is

$$E(\bar{Y}_j - \bar{Y}_m) = \mu_{y_j} - \mu_{y_m} = \tau_j - \tau_m + \beta(\bar{X}_j - \bar{X}_m);$$

that is, this difference will be biased as an estimate of the corresponding treatment effects by a factor that depends upon β and the magnitude of $\bar{X}_j - \bar{X}_m$. When this latter difference is relatively large, the standard error of the adjusted difference will tend to be quite large. At best, covariance adjustments for large initial biases on the covariate are poor substitutes for direct control. A more complete discussion of this point is given by Cochran (1957).

An excellent as well as quite readable summary of the wide variety of uses of the analysis of covariance is contained in a special issue of *Biometrics* (1957, **13,** no. 3), which is devoted entirely to this topic. In this issue is a rather lengthy discussion by H. F. Smith on the interpretation of the adjusted treatment means, particularly for the case in which the treatment affects the covariate.

10.2 Single-factor Experiment

The covariate will be denoted by the symbol X, the variate by the symbol Y. The notation that will be used is summarized in Table 10.2-1. Thus

T_{x_j} = sum of measurements on covariate under treatment j.
T_{y_j} = sum of measurements on variate under treatment j.
$\bar{X}_j$ = mean of measurements on covariate under treatment j.

Table 10.2-1 Notation for the Analysis of Covariance

	Treatment 1		...	Treatment j		...	Treatment k		
	Y_{11}	X_{11}		Y_{1j}	X_{1j}		Y_{1k}	X_{1k}	
	Y_{21}	X_{21}		Y_{2j}	X_{2j}		Y_{2k}	X_{2k}	
	.	.		.	.		.	.	
	.	.		.	.		.	.	
	.	.		.	.		.	.	
	Y_{n1}	X_{n1}		Y_{nj}	X_{nj}		Y_{nk}	X_{nk}	
Sum	T_{y_1}	T_{x_1}	...	T_{y_j}	T_{x_j}	...	T_{y_k}	T_{x_k}	$\Sigma T_{y_j} = G_y$ $\Sigma T_{x_j} = G_x$
Mean	$\bar{Y}_1$	$\bar{X}_1$	...	$\bar{Y}_j$	$\bar{X}_j$	...	$\bar{Y}_k$	$\bar{X}_k$	$\bar{Y}$ $\bar{X}$

$T_{xx} = n\Sigma(\bar{X}_j - \bar{X})^2$	$T_{xy} = n\Sigma(\bar{X}_j - \bar{X})(\bar{Y}_j - \bar{Y})$	$T_{yy} = n\Sigma(\bar{Y}_j - \bar{Y})^2$
$E_{xx_j} = \sum_i(X_{ij} - \bar{X}_j)^2$	$E_{xy_j} = \sum_i(X_{ij} - \bar{X}_j)(Y_{ij} - \bar{Y}_j)$	$E_{yy_j} = \sum_i(Y_{ij} - \bar{Y}_j)^2$
$E_{xx} = \sum_j E_{xx_j}$	$E_{xy} = \sum_j E_{xy_j}$	$E_{yy} = \sum_j E_{yy_j}$
$S_{xx} = T_{xx} + E_{xx} = \Sigma\Sigma(X_{ij} - \bar{X})^2$	$S_{xy} = T_{xy} + E_{xy} = \Sigma\Sigma(X_{ij} - \bar{X})(Y_{ij} - \bar{Y})$	$S_{yy} = T_{yy} + E_{yy} = \Sigma\Sigma(Y_{ij} - \bar{Y})^2$

$\bar{Y}_j$ = mean of measurements on variate under treatment j.
E_{xx_j} = variation on covariate under treatment j.
E_{yy_j} = variation on variate under treatment j.
E_{xy_j} = covariation between variate and covariate under treatment j.
$E_{xx} = \Sigma E_{xx_j}$, $E_{yy} = \Sigma E_{yy_j}$, $E_{xy} = \Sigma E_{xy_j}$.
T_{xx} = between-treatment variation on covariate.
T_{yy} = between-treatment variation on variate.
T_{xy} = between-treatment covariation.
S_{xx} = overall variation on covariate $= T_{xx} + E_{xx}$.
S_{yy} = overall variation on variate $= T_{yy} + E_{yy}$.
S_{xy} = overall covariation $= T_{xy} + E_{xy}$.

Least-squares Estimators of the Parameters in Fixed Model. The model for an observation on element i under treatment j has the form

$$Y_{ij} = \mu + \beta(X_{ij} - \bar{X}) + \tau_j + \varepsilon_{ij}. \tag{1}$$

For the fixed model, the only random variable is ε_{ij}.

Assume that the ε_{ij} are uncorrelated with mean zero and variance σ_ε^2, where σ_ε^2 does not depend upon j. The least-squares estimators of the parameters μ, β, and the τ_j are obtained by making

$$\Sigma\hat{\varepsilon}_{ij}^2 = \Sigma[Y_{ij} - \hat{\mu} - \hat{\beta}(X_{ij} - \bar{X}) - \hat{\tau}_j]^2 = \text{minimum}, \tag{2}$$

under the constraint that $\Sigma\hat{\tau}_j = 0$.

Taking the partial derivative of (2) with respect to $\hat{\mu}$ and setting the result equal to zero yield the normal equation

$$-2\sum_{i,j}[Y_{ij} - \hat{\mu} - \hat{\beta}(X_{ij} - \bar{X}) - \hat{\tau}_j] = 0. \tag{3}$$

Solving (3) for $\hat{\mu}$ gives

$$\sum_{i,j}\hat{\mu} = \sum_{i,j}Y_{ij} - \hat{\beta}\sum_{i,j}(X_{ij} - \bar{X}) - \sum_{i,j}\hat{\tau}_j,$$

$$\hat{\mu} = \bar{Y}. \tag{4}$$

The normal equation for $\hat{\tau}_j$ has the form

$$-2\sum_{i}[Y_{ij} - \hat{\mu} - \hat{\beta}(X_{ij} - \bar{X}) - \hat{\tau}_j] = 0. \tag{5}$$

Solving (5) for $\hat{\tau}_j$ and using the result in (4) give

$$\hat{\tau}_j = \bar{Y}_j - \bar{Y} - \hat{\beta}(\bar{X}_j - \bar{X}). \tag{6}$$

The least-squares estimator of τ_j in the analysis of variance is that special case of (6) in which $\hat{\beta} = 0$.

The normal equation for estimating β is

$$-2\sum_{i,j}[Y_{ij} - \hat{\mu} - \hat{\beta}(X_{ij} - \bar{X}) - \hat{\tau}_j](X_{ij} - \bar{X}) = 0. \tag{7}$$

Using the results in (4) and (6), one may write (7) in the form

$$\sum_{i,j}[Y_{ij} - \bar{Y} - \hat{\beta}(X_{ij} - \bar{X}) - (\bar{Y}_j - \bar{Y}) + \hat{\beta}(\bar{X}_j - \bar{X})](X_{ij} - \bar{X}) = 0,$$

or

$$(8)\quad \hat{\beta}[\sum_{i,j}(X_{ij} - \bar{X})^2 - \sum_{i,j}(\bar{X}_j - \bar{X})(X_{ij} - \bar{X})] = \sum_{i,j} Y_{ij}(X_{ij} - \bar{X}) - \sum_{i,j}\bar{Y}_j(X_{ij} - \bar{X}).$$

The following relationships are readily established.

$$T_{xx} = n\sum_j(\bar{X}_j - \bar{X})^2 = \sum_{i,j}(\bar{X}_j - \bar{X})(X_{ij} - \bar{X}),$$

$$S_{xy} = \sum_{i,j}(X_{ij} - \bar{X})(Y_{ij} - \bar{Y}) = \sum_{i,j} Y_{ij}(X_{ij} - \bar{X}),$$

$$T_{xy} = n\sum_j(\bar{X}_j - \bar{X})(\bar{Y}_j - \bar{Y}) = \sum_{i,j}\bar{Y}_j(X_{ij} - \bar{X}).$$

Using these relationships in (8) yields

$$\hat{\beta}(S_{xx} - T_{xx}) = S_{xy} - T_{xy}.$$

Hence the least-squares estimator of β is given by

$$(9)\qquad \hat{\beta} = \frac{S_{xy} - T_{xy}}{S_{xx} - T_{xx}} = \frac{E_{xy}}{E_{xx}} = b_E.$$

The regression coefficient b_E is called the pooled within-class regression coefficient. One notes that

$$E_{xy} = \Sigma E_{xy_j} \qquad \text{and} \qquad E_{xx} = \Sigma E_{xx_j}.$$

Hence

$$(10)\qquad b_E = \frac{\Sigma E_{xy_j}}{\Sigma E_{xx_j}}.$$

If one considers only the information from within treatment class j, the least-squares estimator of β may be shown to be

$$\hat{\beta}_j = b_{E_j} = \frac{E_{xy_j}}{E_{xx_j}}.$$

Since it was assumed as part of model (1) that σ_ε^2 did not depend upon j, implicit in this derivation is the assumption that

$$\beta_1 = \beta_2 = \cdots = \beta_k = \beta,$$

that is, homogeneity of the within-class regressions.

The minimum value of $\sum_{i,j}\hat{\varepsilon}_{ij}^2$ may be obtained from (2) by replacing the estimators by the corresponding expressions in (4), (6), and (9). After considerable algebraic manipulations one obtains the result

$$\text{(11)} \qquad \text{Minimum} \sum_{i,j} \hat{\varepsilon}_{ij}^2 = E'_{yy} = E_{yy} - \frac{E_{xy}^2}{E_{xx}}.$$

The normal equations developed above may be represented schematically in the relatively simple form given below.

$$\begin{aligned}
\hat{\mu}&: \quad kn\hat{\mu} + \qquad n\Sigma\hat{\tau}_j + \sum_{i,j}(X_{ij} - \bar{X})\hat{\beta} = G_y, \\
\hat{\tau}_j&: \quad n\hat{\mu} + \qquad n\hat{\tau}_j + \sum_{i}(X_{ij} - \bar{X})\hat{\beta} = T_{y_j}, j = 1, \ldots, k \\
\hat{\beta}&: \quad \sum_{i,j}(X_{ij} - \bar{X})\hat{\mu} + \sum_{i,j}(X_{ij} - \bar{X})\hat{\tau}_j + \qquad S_{xx}\hat{\beta} = S_{xy}.
\end{aligned}$$

From the general linear model, the predictable variation is given by

$$\begin{aligned}
R(\tau_j,\beta \mid \mu) &= \sum_j \hat{\tau}_j T_{y_j} + \hat{\beta} S_{xy} \\
&= \sum_j [(\bar{Y}_j - \bar{Y}) - \hat{\beta}(\bar{X}_j - \bar{X})]T_{y_j} + \hat{\beta}(T_{xy} + E_{xy}) \\
&= T_{yy} - \hat{\beta}T_{xy} + \hat{\beta}T_{xy} + \hat{\beta}E_{xy} \\
&= T_{yy} + b_E E_{xy} \\
&= T_{yy} + \frac{E_{xy}^2}{E_{xx}}.
\end{aligned}$$

In the analysis of variance, the predictable variation is T_{yy}. The increase in predictability due to the covariate is thus $b_E E_{xy}$ or E_{xy}^2/E_{xx}.

The nonpredictable or error variation in the analysis of covariance is thus

$$\begin{aligned}
E'_{yy} = S_{yy} - \left(T_{yy} + \frac{E_{xy}^2}{E_{xx}}\right) &= S_{yy} - T_{yy} - \frac{E_{xy}^2}{E_{xx}} \\
&= E_{yy} - \frac{E_{xy}^2}{E_{xx}} = E_{yy} - b_E E_{xy}.
\end{aligned}$$

This last result is the expression given in (11).

Alternative Approach. Consider the three models in Table 10.2-2. Model (*a*) represents the regression of Y on X, ignoring the treatment effects. Under model (*a*), the regression coefficient is estimated by

$$\hat{\beta} = b_S = \frac{S_{xy}}{S_{xx}}.$$

Hence the predictable variation in Y is

$$R(\beta \mid \mu) = b_S S_{xy} = \frac{S_{xy}^2}{S_{xx}}.$$

Table 10.2-2 Model Underlying the Analysis of Covariance

Model

(*a*) $Y_{ij} = \mu + \beta(X_{ij} - \bar{X}) + \varepsilon_{ij}$

(*b*) $Y_{ij} = \mu + \tau_j + \varepsilon_{ij}$

(*c*) $Y_{ij} = \mu + \beta(X_{ij} - \bar{X}) + \tau_j + \varepsilon_{ij}$
$= \mu + \beta(X_{ij} - \bar{X}_j) + \beta(\bar{X}_j - \bar{X}) + \tau_j + \varepsilon_{ij}$

Model	Predictable variation	Error variation
(*a*)	$R(\beta \mid \mu) = S_{xy}^2/S_{xx}$	$S'_{yy} = S_{yy} - (S_{xy}^2/S_{xx})$
(*b*)	$R(\tau_j \mid \mu) = T_{yy}$	E_{yy}
(*c*)	$R(\beta,\tau_j \mid \mu) = S_{yy} - [E_{yy} - (E_{xy}^2/E_{xx})]$ $= T_{yy} + (E_{xy}^2/E_{xx})$	$E'_{yy} = E_{yy} - (E_{xy}^2/E_{xx})$

Model (*b*) represents the usual analysis-of-variance model when there is no covariate or, equivalently, when the covariate is ignored. Under this model the predictable variation in Y is

$$R(\tau_j \mid \mu) = T_{yy},$$

which is the between-class variation in the usual analysis of variance.

Model (*c*) combines the regression model with the analysis-of-variance model to produce the analysis-of-covariance model. The predictable variation here is given by

$$R(\beta, \tau_j \mid \mu) = T_{yy} + \frac{E_{xy}^2}{E_{xx}} = T_{yy} + b_E E_{xy}.$$

Thus the increase in predictability of model (*c*) over model (*b*) is a function only of the within-class regression. Since

$$T_{yy} = S_{yy} - E_{yy},$$

the predictable variation under model (*c*) takes the form

$$R(\beta, \tau_j \mid \mu) = S_{yy} - \left[E_{yy} - \frac{E_{xy}^2}{E_{xx}}\right].$$

Hence the term in brackets represents variation due to error under model (*c*).

Since model (*c*) does not contain a term of the form $X_{ij}\tau_j$, implicit in this model is the assumption that the expected value (over a population of experiments) of the covariation between X and τ will be zero. Within a single experiment this covariation will not necessarily be zero.

That part of the predictable variation associated with model (*c*) which cannot be predicted from model (*a*) is given by

$$(12) \qquad R(\tau_j \mid \mu,\beta) = R(\beta,\tau_j \mid \mu) - R(\beta \mid \mu)$$

$$= S_{yy} - \left[E_{yy} - \frac{E_{xy}^2}{E_{xx}}\right] - \frac{S_{xy}^2}{S_{xx}}$$

$$= \left[S_{yy} - \frac{S_{xy}^2}{S_{xx}}\right] - \left[E_{yy} - \frac{E_{xy}^2}{E_{xx}}\right]$$

$$= S'_{yy} - E'_{yy},$$

where S'_{yy} and E'_{yy} are the terms in brackets. The expression (12) represents what is called the *reduced* variation due to treatments and is denoted by the symbol T_{yyR}. Thus,

$$T_{yyR} = R(\tau_j \mid \mu, \beta) = S'_{yy} - E'_{yy}.$$

Since $T_{yy} = S_{yy} - E_{yy}$, (12) may be written in the form

$$(13) \qquad T_{yyR} = S_{yy} - E_{yy} + \frac{E_{xy}^2}{E_{xx}} - \frac{S_{xy}^2}{S_{xx}}$$

$$= T_{yy} + \frac{E_{xy}^2}{E_{xx}} - \frac{S_{xy}^2}{S_{xx}}$$

$$= T_{yy} + b_E E_{xy} - b_S S_{xy}.$$

Consider now the pairs of means

$$(\bar{X}_1, \bar{Y}_1), (\bar{X}_2, \bar{Y}_2), \ldots, (\bar{X}_k, \bar{Y}_k).$$

From these pairs one may compute the between-class regression coefficient, given by

$$b_T = \frac{T_{xy}}{T_{xx}}.$$

The variation of the $\bar{Y}_j$ is T_{yy}. That part of T_{yy} that is predictable from the between-class regression is

$$b_T T_{xy} = \frac{T_{xy}^2}{T_{xx}}.$$

The corresponding error variation (that is, the deviations of the $\bar{Y}_j$ about the between-class regression line) is thus

$$T'_{yy} = T_{yy} - \frac{T_{xy}^2}{T_{xx}}.$$

It may be shown that

$$T_{yyR} = T'_{yy} + (b_T - b_E)^2 \frac{T_{xx} E_{xx}}{T_{xx} + E_{xx}}.$$

Thus, if

$$b_T = b_E, \qquad \text{then} \qquad T_{yyR} = T'_{yy};$$

if

$$b_T \neq b_E, \qquad \text{then} \qquad T_{yyR} > T'_{yy}.$$

It is noted that, if $b_T = b_E$, then it must be that $b_T = b_E = b_S$. This follows from the fact that

$$b_S = \frac{T_{xy} + E_{xy}}{T_{xx} + E_{xx}} = \frac{S_{xy}}{S_{xx}}.$$

An unbiased estimate of β in model (*c*), which will not be influenced by the magnitude of the between-class covariation between the covariate and the variate, is given by the within-class regression coefficient b_E. The adjusted means for the variate are defined by

$$\text{(14)} \qquad \bar{Y}'_j = \bar{Y}_j - b_E(\bar{X}_j - \bar{X}).$$

This adjustment "removes" or "partials out" from $\bar{Y}_j$ that part which may be considered a linear function of $\bar{X}_j$. Thus, if the model for a treatment mean is

$$\bar{Y}_j = \mu + \beta(\bar{X}_j - \bar{X}) + \tau_j + \bar{\varepsilon}_j,$$

then

$$\bar{Y}'_j = \bar{Y}_j - \beta(\bar{X}_j - \bar{X}) = \mu + \tau_j + \bar{\varepsilon}_j.$$

Replacing β by its estimator b_E,

$$\bar{Y}'_j = \bar{Y}_j - b_E(\bar{X}_j - \bar{X}) = \mu + \tau_j + \bar{\varepsilon}'_j,$$

where $\bar{\varepsilon}'_j$ includes a part which is a function of the sampling error of the estimator b_E. Since

$$\hat{\tau}_j = \bar{Y}_j - \bar{Y} - b_E(\bar{X}_j - \bar{X}),$$

one has

$$\bar{Y}'_j = \hat{\tau}_j + \bar{Y}.$$

The variation due to the adjusted treatment means for the variate is defined by

$$T_{yyA} = n\Sigma(\bar{Y}'_j - \bar{Y})^2.$$

The reduced treatment variation is related to the variation of the adjusted treatment means as follows:

$$T_{yyR} = T_{yyA} - (b_T - b_E)^2 \frac{T_{xx}^2}{T_{xx} + E_{xx}}.$$

Thus, if

$$b_T = b_E, \qquad \text{then} \qquad T_{yyR} = T_{yyA};$$

if

$$b_T \neq b_E, \qquad \text{then} \qquad T_{yyR} < T_{yyA}.$$

In terms of model (*c*) it may be shown that

$$b_T - b_E = \frac{\Sigma(\bar{X}_j - \bar{X})\tau_j}{\Sigma(\bar{X}_j - \bar{X})^2} + f(\varepsilon),$$

where $f(\varepsilon)$ is a term that is a function of ε. The expected value of this term is zero. If the covariate is uncorrelated with the τ_j, then

$$\mathrm{E}[\Sigma(\bar{X}_j - \bar{X})\tau_j] = 0.$$

Hence

$$\mathrm{E}(b_T - b_E) = 0.$$

A summary of some of the possible regression coefficients that may be obtained from the data in the analysis of covariance is given in Table 10.2-3. The residual or error variation associated with the corresponding regression lines are represented by terms having a prime symbol. From the middle part of this table, one notes that the variation of the residuals about the overall regression line [as obtained from model (*a*)] is partitioned into three parts:

$$E'_{yy}, \qquad T'_{yy}, \qquad \text{and} \qquad (b_T - b_E)^2 \frac{T_{xx}E_{xx}}{T_{xx} + E_{xx}}.$$

The sum of the last two parts is equal to the reduced variation due to treatments. The third part, which involves the term $(b_T - b_E)^2$, is a

Table 10.2-3 Summary of Various Regressions

	Pooled within-class	Between class	Overall
(i)	$b_E = E_{xy}/E_{xx}$	$b_T = T_{xy}/T_{xx}$	$b_S = S_{xy}/S_{xx}$
	$E'_{yy} = E_{yy} - b_E E_{xy}$	$T'_{yy} = T_{yy} - b_T T_{xy}$	$S'_{yy} = S_{yy} - b_S S_{xy}$
	$= E_{yy} - (E^2_{xy}/E_{xx})$	$= T_{yy} - (T^2_{xy}/T_{xx})$	$= S_{yy} - (S^2_{xy}/S_{xx})$

(ii) $$S'_{yy} = E'_{yy} + T'_{yy} + (b_T - b_E)^2 \frac{T_{xx}E_{xx}}{T_{xx} + E_{xx}} = E'_{yy} + T_{yyR}$$

Alternative expressions for the reduced treatment variation:

(iii)
$$T_{yyR} = T_{yy} + b_E E_{xy} - b_S S_{xy} = T_{yy} + \frac{E^2_{xy}}{E_{xx}} - \frac{S^2_{xy}}{S_{xx}}$$

$$= T'_{yy} + (b_T - b_E)^2 \frac{T_{xx}E_{xx}}{T_{xx} + E_{xx}}$$

$$= T_{yyA} - (b_T - b_E)^2 \frac{T^2_{xx}}{T_{xx} + E_{xx}}$$

function of the covariation between the covariate and the treatment effects in the experimental data; the expected value of this covariation over a population of experiments is assumed to be zero.

The within-class correlation coefficient is defined by

$$r_{\text{w. class}} = \frac{E_{xy}}{\sqrt{E_{xx}}\sqrt{E_{yy}}}.$$

Similarly the between-class correlation coefficient is defined by

$$r_{\text{b. class}} = \frac{T_{xy}}{\sqrt{T_{xx}}\sqrt{T_{yy}}}.$$

Thus $$b_E = \frac{E_{xy}}{E_{xx}} = r_{\text{w. class}} \frac{\sqrt{E_{yy}}}{\sqrt{E_{xx}}}.$$

Also $$b_T = \frac{T_{xy}}{T_{xx}} = r_{\text{b. class}} \frac{\sqrt{T_{yy}}}{\sqrt{T_{xx}}}.$$

Hence $$E'_{yy} = E_{yy} - b_E E_{xy} = E_{xy} - r_{\text{w. class}} \frac{E_{xy}\sqrt{E_{yy}}}{\sqrt{E_{xx}}}$$

$$= E_{yy} - r^2_{\text{w. class}} E_{yy}$$

$$= (1 - r^2_{\text{w. class}}) E_{yy}.$$

Also, by the same reasoning,

$$T'_{yy} = (1 - r^2_{\text{b. class}}) T_{yy}.$$

Summary of Assumptions Underlying the Analysis of Covariance. From the various parts of the model given in Table 10.2-2, it follows that there is only a single regression coefficient. This implies that the within-class regression coefficients are homogeneous, that is,

$$\beta_{E_1} = \beta_{E_2} = \cdots = \beta_{E_k} = \beta_E.$$

Further, the implication is that

$$\beta_E = \beta_T = \beta_S.$$

Implicit also in the model is that the correct form of the relationship between the variate and the covariate is linear.

In addition to these assumptions about the regression part of the model are the usual assumptions about the analysis-of-variance part of the model, namely, that the ε_{ij} are distributed independently as $N(0,\sigma^2_\varepsilon)$ within each of the treatment classes.

Upon combining the regression model with the analysis-of-variance model to obtain the complete analysis-of-covariance model, an assumption with respect to additivity of treatment and regression effects is implied. Further, the error term in the complete model is assumed to be $N(0,\sigma^2_{\varepsilon'})$ within each of the treatment classes, where

$$\hat{\varepsilon}'_{ij} = Y_{ij} - \hat{\mu} - \hat{\beta}(X_{ij} - \bar{X}) - \hat{\tau}_j$$

$$= Y_{ij} - \bar{Y} - b_E(X_{ij} - \bar{X}) - [\bar{Y}_j - \bar{Y} - b_E(\bar{X}_j - \bar{X})]$$

$$= (Y_{ij} - \bar{Y}_j) - b_E(X_{ij} - \bar{X}_j).$$

Hence

$$\begin{aligned}\Sigma(\hat{\varepsilon}')^2 &= E_{yy} + b_E^2 E_{xx} - 2b_E E_{xy} \\ &= E_{yy} + \frac{E_{xy}^2}{E_{xx}} - 2\frac{E_{xy}^2}{E_{xx}} \\ &= E_{yy} - \frac{E_{xy}^2}{E_{xx}}.\end{aligned}$$

Bivariate Normal Model. The analysis of covariance need not be restricted to the case in which X is a fixed variable. Assume X and Y are bivariate normal with parameters

$$\mu_x, \quad \mu_y, \quad \sigma_x^2, \quad \sigma_y^2, \quad \text{and} \quad \rho = \frac{\sigma_{xy}}{\sigma_x \sigma_y}.$$

The regression of Y on X has the form

$$Y = \mu_y + \beta(X - \mu_x), \qquad \text{where} \qquad \beta = \frac{\sigma_{xy}}{\sigma_x^2}.$$

For a given X, the distribution of Y is normal in form with mean

$$\mu_{y \mid x} = \mu_y + \beta(X - \mu_x)$$

and variance

$$\sigma_{y \mid x}^2 = \sigma_y^2 - \beta\sigma_{xy} = \sigma_y^2(1 - \rho^2).$$

Within treatment class j, assume that X and Y are also bivariate normal with parameters

$$\mu_{x_j}, \quad \mu_{y_j}, \quad \sigma_x^2, \quad \sigma_y^2, \quad \text{and} \quad \rho = \frac{\sigma_{xy}}{\sigma_x \sigma_y}.$$

Note that the variances and covariance do not depend upon j. Within treatment class j, the regression of Y on X has the form

$$Y_j = \mu_{y_j} + \beta(X - \mu_{x_j}).$$

For a given X, the distribution of Y is normal in form with mean

$$\mu_{y_j \mid x} = \mu_{y_j} + \beta(X - \mu_{x_j})$$

and variance

$$\sigma_{y_j \mid x}^2 = \sigma_y^2(1 - \rho^2).$$

Within treatment class m (where m is any treatment class other than class j), it is also assumed the X and Y are bivariate normal with parameters

$$\mu_{x_m}, \quad \mu_{y_m}, \quad \sigma_x^2, \quad \sigma_y^2, \quad \text{and} \quad \rho = \frac{\sigma_{xy}}{\sigma_x \sigma_y}.$$

The regression of Y on X within this treatment class is

$$Y_m = \mu_{y_m} + \beta(X - \mu_{x_m}).$$

For a given X, the distribution of Y in this treatment class will be normal in form with mean

$$\mu_{y_m \mid x} = \mu_{y_m} + \beta(X - \mu_{x_m})$$

and variance

$$\sigma^2_{y_m \mid x} = \sigma^2_y(1 - \rho^2).$$

Since β is a constant for all treatment classes, the within-class regression lines will be parallel. For a given X, the difference between Y_j and Y_m is

$$Y_j - Y_m = \mu_{y_j} - \mu_{y_m} - \beta(\mu_{x_j} - \mu_{x_m}).$$

If the two regression lines are not only parallel but also identical, then

$$Y_j - Y_m = \mu_{y_j} - \mu_{y_m} - \beta(\mu_{x_j} - \mu_{x_m}) = 0$$

or

$$\mu_{y_j} - \mu_{y_m} = \beta(\mu_{x_j} - \mu_{x_m}). \tag{15}$$

Consider now the following model for an observation on element i in treatment class j.

$$Y_{ij} = \mu_y + \beta(X_{ij} - \mu_x) + \tau_j + \varepsilon_{ij},$$

where $\quad \mathrm{E}(\varepsilon_{ij}) = 0.$

Taking the expected value of both sides with respect to i gives

$$\mu_{y_j} = \mu_y + \beta(\mu_{x_j} - \mu_x) + \tau_j.$$

Hence τ_j may be expressed in the form

$$\tau_j = \mu_{y_j} - \mu_y - \beta(\mu_{x_j} - \mu_x).$$

By analogy, the treatment effect for treatment class m is given by

$$\tau_m = \mu_{y_m} - \mu_y - \beta(\mu_{x_m} - \mu_x).$$

Hence the difference between the treatment effects may be expressed in the form

$$\tau_j - \tau_m = \mu_{y_j} - \mu_{y_m} - \beta(\mu_{x_j} - \mu_{x_m}).$$

Under the hypothesis that $\tau_j = 0$ for $j = 1, \ldots, k$,

$$\tau_j - \tau_m = 0 \qquad \text{for all } j \text{ and } m.$$

Hence $\quad 0 = \mu_{y_j} - \mu_{y_m} = \beta(\mu_{x_j} - \mu_{x_m})$

or

$$\mu_{y_j} - \mu_{y_m} = \beta(\mu_{x_j} - \mu_{x_m}). \tag{16}$$

But this last relationship is the same as (15) which was obtained under the hypothesis that the within-class regression lines were identical. Hence the hypothesis of no treatment effects in the analysis of covariance is equivalent to the hypothesis that the within-class regression lines are identical.

Nothing has been said in this development about whether or not the covariate is influenced by the treatments. As long as the model holds, statistically rigorous tests can be made. However, if the treatments do influence the magnitude of the covariate, the hypothesis that is tested may have little or no meaning in terms of an empirical experiment. That is, it may not be possible, experimentally, to vary Y when X is held constant.

Geometric Representation. In Fig. 10.2-1, the data points for an experiment having $k = 3$ treatments and $n = 7$ observations under

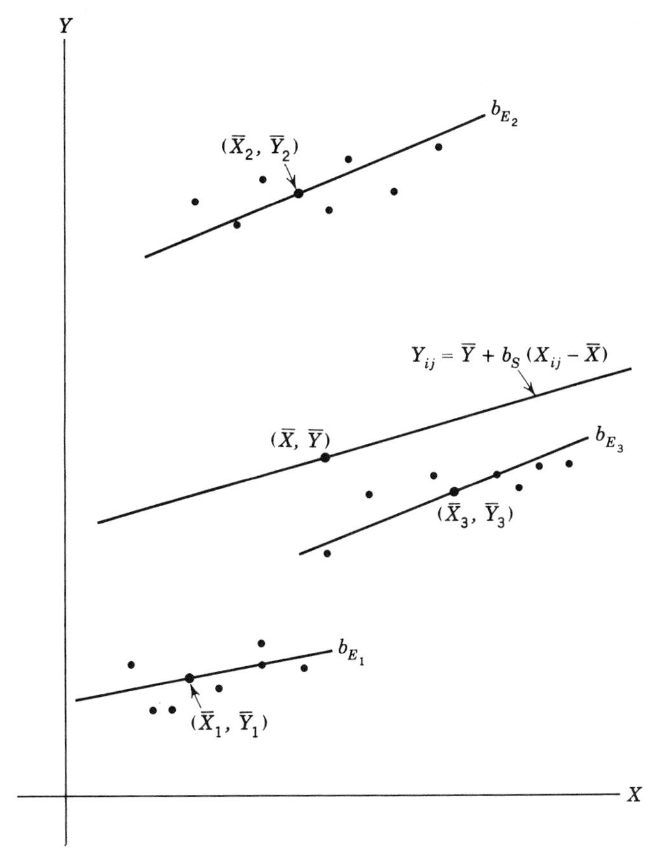

Figure 10.2-1 Within-class regression lines with slopes b_{E_j} and overall regression line.

each treatment are shown. The regression lines computed from the separate treatment classes are indicated. That is, the regression line for the $n = 7$ observations in treatment class 1 has slope

$$b_{E_1} = \frac{E_{xy_1}}{E_{xx_1}}.$$

The point $(\bar{X}_1, \bar{Y}_1)$ is on this regression line. The overall regression line, obtained by disregarding the treatment classes, is also shown. It is noted that the overall regression line does not provide a good fit to the data points since the $\bar{Y}_j$ tend to fall relatively far from this line.

In Fig. 10.2-2, the lines with common slope b_E are drawn through the points $(\bar{X}_j, \bar{Y}_j)$. These lines may be considered to define the within-class

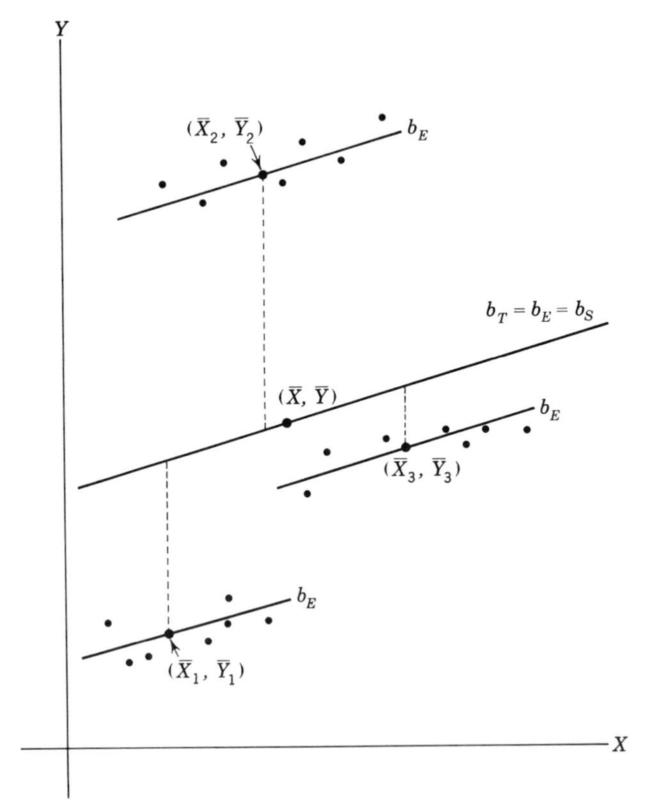

Figure 10.2-2 Within-class regression lines each with slope b_E and between-class regression line with slope $b_T = b_E$.

regression under the assumption of homogeneous regression. The between-class regression line is also shown in this figure. Since this line is drawn parallel to the within-class regression lines, implicit is the assumption that $b_T = b_E$. (If this equality holds then $b_E = b_S$.)

The adjusted treatment means are shown in Fig. 10.2-3. The adjusted means are obtained from the intersection of the within-class regression lines with common slope b_E and the line $X = \bar{X}$.

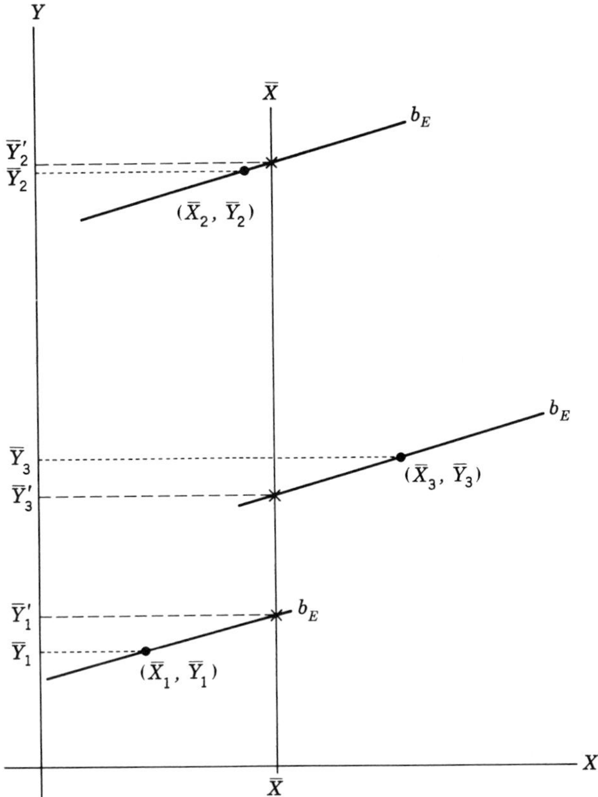

Figure 10.2-3 Adjusted treatment means, $\bar{Y}_j' = \bar{Y}_j - b_E(\bar{X}_j - \bar{X})$.

Test Procedure. In the analysis of regression, to test the hypothesis that $\sigma^2_{u \mid v} = 0$, an F ratio has the general structure

$$F = \frac{(D' - D)/f_1}{D/f_2}, \tag{17}$$

where $D' = S_{yy} - R(u)$,

$D = S_{yy} - R(u,v)$,

f_1 = degrees of freedom for $D' - D$,

f_2 = degrees of freedom for D.

Hence the term in the numerator of the F ratio is

$$\begin{aligned} D' - D &= [S_{yy} - R(u)] - [S_{yy} - R(u,v)] \\ &= R(u,v) - R(u) = R(u \mid v). \end{aligned}$$

In the context of the analysis of covariance,

$$R(u,v) = R(\beta,\tau_j \mid \mu),$$
$$R(u) = R(\beta \mid \mu),$$
$$R(u \mid v) = R(\tau_j \mid \mu,\beta).$$

Hence
$$D' - D = T_{yyR} = S'_{yy} - E'_{yy},$$
$$D = E'_{yy} = E_{yy} - b_E E_{xy}.$$

The degrees of freedom for E'_{yy} are equal to the degrees of freedom for E_{yy}, which are $k(n-1)$, less the one degree of freedom associated with the term $b_E E_{xy}$. Thus

$$f_2 = k(n-1) - 1.$$

The degrees of freedom for $S'_{yy} - E'_{yy}$ are the difference between the degrees of freedom for the parts. Thus

$$f_1 = [kn-2] - [k(n-1)-1] = k-1.$$

Thus, to test the hypothesis that $\sigma^2_{\tau|\beta} = 0$ in the context of the analysis of covariance, the F ratio in (17) takes the form

$$F = \frac{T_{yyR}/(k-1)}{E'_{yy}/[k(n-1)-1]}. \tag{18}$$

Under the hypothesis that $\sigma^2_{\tau|\beta} = 0$ (and under the assumptions about the form of sampling distributions which will be indicated later in this section), the F ratio in (18) will be distributed as $F[k-1, k(n-1)-1]$.

A comparison of the analysis of variance and the analysis of covariance is presented in Table 10.2-4. If one ignores the data on the covariate,

Table 10.2-4 Comparison of Analysis of Variance and Analysis of Covariance

Analysis of variance

	Source of variation	SS	df	MS	E(MS)
(i)	Treatments	T_{yy}	$k-1$	MS_{treat}	$\sigma^2_\varepsilon + n\sigma^2_\tau$
	Error	E_{yy}	$k(n-1)$	MS_{error}	σ^2_ε
	Total	S_{yy}	$kn-1$		

Analysis of covariance

	Source of variation	SS	df	MS	E(MS)
(ii)	Treatments	T_{yyR}	$k-1$	$MS_{treat(R)}$	$\sigma^2_{\varepsilon\mid\beta} + n\sigma^2_{\tau\mid\beta}$
	Error	E'_{yy}	$k(n-1)-1$	MS'_{error}	$\sigma^2_{\varepsilon\mid\beta}$
	Total	S'_{yy}	$kn-2$		

one obtains the analysis of variance given in part i. Here the error variance, σ_ε^2, is estimated from the within-cell variation. In the analysis of covariance, the error variance, $\sigma_{\varepsilon'}^2 = \sigma_{\varepsilon \mid \beta}^2$, is estimated from the within-cell variation less that part of the within-cell variation which is a function of the covariate.

In the analysis of variance, the variation due to treatments is considered to be all the between-class variation. However, in the analysis of covariance, the reduced treatment variation is related to the between-class variation as shown in part iii of Table 10.2-3. To differentiate between the case in which the covariate is ignored and the case in which adjustment is made for the covariate, the notation σ_τ^2 and $\sigma_{\tau \mid \beta}^2$ is used. The structure of the F ratio is indicated by the expected values of the mean squares.

In terms of the regression lines in Fig. 10.2-2, the test procedure which uses the F ratio in (14) is equivalent to testing the hypothesis that the overall regression line with slope b_S fits the data as well as the within-class regression lines with slope b_E. The numerator of (18) is

$$\frac{\begin{bmatrix}\text{Variation due to residuals}\\ \text{about overall regression}\\ \text{lines}\end{bmatrix} - \begin{bmatrix}\text{Variation due to residuals about}\\ \text{regression lines with common slope}\\ b_E \text{ fitted through each of the}\\ \text{class means}\end{bmatrix}}{k-1}$$

Equivalently, this numerator has the form

$$\frac{\text{Error }(\omega) - \text{Error }(\Omega)}{(kn-1) - k(n-1)},$$

where Error (ω) = variation due to error assuming H_1 is true (i.e., no difference due to treatments)

$= S_{yy} - (S_{xy}^2/S_{xx})$,

Error (Ω) = variation due to error with no restriction other than homogeneity of within-class regression

$= E_{yy} - (E_{xy}^2/E_{xx})$.

Adjusted Treatment Means. The estimate of τ_j as given by (6) is

$$\hat{\tau}_j = (\bar{Y}_j - \bar{Y}) - b_E(\bar{X}_j - \bar{X}).$$

The adjusted treatment mean $\bar{Y}_j'$ has been defined as

$$\bar{Y}_j' = \hat{\tau}_j + \bar{Y} = \bar{Y}_j - b_E(\bar{X}_j - \bar{X}).$$

An estimate of the square of the standard error of an adjusted mean is given by

$$s_{\bar{Y}_j'}^2 = \text{MS}_{\text{error}}'\left[\frac{1}{n} + \frac{(\bar{X}_j - \bar{X})^2}{E_{xx}}\right].$$

This estimate is that obtained from the regression model which assumes

X to be fixed. An estimate of the square of the standard error of the difference $\bar{Y}'_j - \bar{Y}'_m$ has the form

$$s^2_{\bar{Y}'_j - \bar{Y}'_m} = \text{MS}'_{\text{error}}\left[\frac{2}{n} + \frac{(\bar{X}_j - \bar{X}_m)^2}{E_{xx}}\right].$$

A statistic that may be used to test the difference between two adjusted treatment means has the form

$$F = \frac{(\bar{Y}'_j - \bar{Y}'_m)^2}{\text{MS}'_{\text{error}}\left[\frac{2}{n} + \frac{(\bar{X}_j - \bar{X}_m)^2}{E_{xx}}\right]}.$$

Under the hypothesis of no difference between τ_j and τ_m, this test statistic is distributed as F with degrees of freedom 1 and $k(n - 1) - 1$, the latter being the degrees of freedom associated with $\text{MS}'_{\text{error}}$.

Note that the variance of the difference between two adjusted treatment means depends upon the magnitude of $(\bar{X}_j - \bar{X}_m)^2$. The average variance between the difference between two adjusted means, averaged over all values of $(\bar{X}_j - \bar{X}_m)^2$, is

$$\frac{2\text{MS}'_{\text{error}}}{n}\left[1 + \frac{T_{xx}/(k - 1)}{E_{xx}}\right].$$

This average may be used in making tests between several pairs of adjusted means or all possible pairs [as in the Tukey (*a*) procedure].

Tests on Model. Although there is evidence to indicate that the analysis of covariance is robust with respect to homogeneity assumptions on within-class variances and regression coefficients, the test procedures described here can serve as useful checks on these assumptions.

The residual variation about the overall regression line with slope b_S is partitioned in Table 10.2-5. The major partition is

$$S'_{yy} = E'_{yy} + T_{yyR}.$$

In the upper half of the table, E'_{yy} is partitioned as follows:

$$E'_{yy} = S_1 + S_2,$$

where S_1 = variation due to residuals about w. class regression lines with slopes b_{E_j},

$$S_2 = \left[\begin{array}{l}\text{variation due to residuals} \\ \text{about w. class regression} \\ \text{lines with common slope } b_E\end{array}\right] - S_1.$$

Table 10.2-5 Partition of Residual Variation about Overall Regression

Source of variation	SS	df
Within-class residual	$E'_{yy} = E_{yy} - (E^2_{xy}/E_{xx}) = E_{yy} - b_E E_{xy}$	$k(n-1)-1$
Residual w. class variation using individual w. class regression	$S_1 = E_{yy} - \Sigma b_{E_j} E_{xy_j}$	$k(n-1)-k$
Difference between the b_{E_j} and b_E	$S_2 = [E_{yy} - b_E E_{xy}] - [E_{yy} - \Sigma b_{E_j} E_{xy_j}] = \Sigma b_{E_j} E_{xy_j} - b_E E_{xy}$	$k-1$
Reduced treatment	$T_{yyR} = T_{yy} + b_E E_{xy} - b_S S_{xy}$	$k-1$
Residual b. class variation using b. class regression	$S_3 = T_{yy} - b_T T_{xy} = T'_{yy}$	$k-2$
Difference between b_T and b_E	$S_4 = [S_{yy} - b_S S_{xy}] - [S_{yy} - (b_T T_{xy} + b_E E_{xy})] = b_T T_{xy} + b_E E_{xy} - b_S S_{xy}$	1
Total	$S'_{yy} = S_{yy} - b_S S_{xy}$	$kn-2$

To test the hypothesis of homogeneity of the within-class regression coefficients, one uses the F ratio:

$$F = \frac{(E'_{yy} - S_1)/(k-1)}{S_1/[k(n-1)-k]}$$

$$= \frac{S_2/(k-1)}{S_1/[k(n-1)-k]}.$$

This F ratio has the general form

$$\frac{(D' - D)/f_1}{D/f_2}.$$

Under the hypothesis of homogeneity of the within-class regressions, the sampling distribution of this statistic is given by the F distribution having $k - 1$ and $k(n - 1) - k$ degrees of freedom.

In the lower half of Table 10.2-5, one has the following partition of the reduced treatment variation:

$$T_{yyR} = S_3 + S_4,$$

where $S_3 = T'_{yy}$ = variation of residuals about b. class regression line,

S_4 = variation related to difference between b_T and b_E

$$= (b_T - b_E)^2 \frac{T_{xx} + E_{xx}}{T_{xx}E_{xx}}.$$

One notes that

$$(b_T - b_E)^2 \frac{T_{xx}E_{xx}}{T_{xx} + E_{xx}} = \frac{1}{S_{xx}}\left[\frac{E_{xx}T_{xy}^2}{T_{xx}} + \frac{T_{xx}E_{xy}^2}{E_{xx}} - 2T_{xy}E_{xy}\right].$$

The term on the right may be written

$$\frac{1}{S_{xx}}\left[\frac{E_{xx}T_{xy}^2 + T_{xx}T_{xy}^2}{T_{xx}} + \frac{T_{xx}E_{xy}^2 + E_{xx}E_{xy}^2}{E_{xx}} - T_{xy}^2 - E_{xy}^2 - 2T_{xy}E_{xy}\right]$$

$$= \frac{T_{xy}^2}{T_{xx}} + \frac{E_{xy}^2}{E_{xx}} - \frac{(T_{xy} + E_{xy})^2}{S_{xx}} = S_4.$$

Under the hypothesis that the between-class regression is linear, E'_{yy} and S_3 would both be estimates of variation due to experimental error. Hence the statistic

$$F_3 = \frac{S_3/(k-2)}{E'_{yy}/[k(n-1)-1]}$$

is distributed as the F distribution with degrees of freedom $k - 2$ and $k(n - 1) - 1$, when the hypothesis of linearity of regression is true (assuming homogeneity of the within-class regression).

Unless k, the number of treatments, is relatively large, the estimate of b_T, which is based upon the k pairs $(\bar{X}_j, \bar{Y}_j)$, will be relatively unstable. A test for the equality of β_T and β_E is somewhat more complex than the tests which have just been described. The term $[S_{yy} - (b_T T_{xy} + b_E E_{xy})]$ in S_4 is not the variation due to residuals about any single regression line. Hence the appropriate F ratio does not have a simple form.

An estimate of $(\beta_T - \beta_E)^2$ is given by

$$(b_T - b_E)^2 = S_4 \frac{T_{xx} + E_{xx}}{T_{xx}E_{xx}}.$$

An estimate of the square of the standard error of $b_T - b_E$ is given by

$$s_{b_T - b_E}^2 = s_{b_T}^2 + s_{b_E}^2$$

$$= \frac{v_T}{T_{xx}} + \frac{v_E}{E_{xx}} = \frac{v_T E_{xx} + v_E T_{xx}}{T_{xx}E_{xx}},$$

where $$v_T = \frac{T'_{yy}}{k-2}, \qquad v_E = \frac{E'_{yy}}{k(n-1)-1}.$$

Hence, to test the hypothesis that $\beta_T = \beta_E$, one may use the F' statistic

$$F' = \frac{S_4(T_{xx} + E_{xx})/T_{xx}E_{xx}}{s_{b_T - b_E}^2} = \frac{S_4(T_{xx} + E_{xx})}{v_T E_{xx} + v_E T_{xx}}.$$

The sampling distribution of this F' statistic may be approximated by an F distribution having one degree of freedom for the numerator, and the degrees of freedom for the denominator, as given by the Satterthwaite approximation, are

$$\mathrm{df}_{\text{denom}} = \frac{(v_T E_{xx} + v_E T_{xx})^2}{[(v_T E_{xx})^2/(k-2)] + \{(v_E T_{xx})^2/[k(n-1)-1]\}}.$$

10.3 Numerical Example of Single-factor Experiment

Computational symbols that will be used in the numerical example are summarized in Table 10.3-1. The symbols in parts i and ii are required

Table 10.3-1 Computational Formulas for Analysis of Covariance

(i)	$(2x_j) = \sum_i X_{ij}^2$ $(3x_j) = T_{x_j}^2/n$	$(2xy_j) = \sum_i X_{ij} Y_{ij}$ $(3xy_j) = T_{x_j} T_{y_j}/n$	$(2y_j) = \sum_i Y_{ij}^2$ $(3y_j) = T_{y_j}^2/n$
(ii)	$E_{xx_1} = (2x_1) - (3x_1)$ . . . $E_{xx_k} = (2x_k) - (3x_k)$	$E_{xy_1} = (2xy_1) - (3xy_1)$ . . . $E_{xy_k} = (2xy_k) - (3xy_k)$	$E_{yy_1} = (2y_1) - (3y_1)$ . . . $E_{yy_k} = (2y_k) - (3y_k)$
(iii)	$(1x) = G_x^2/kn$ $(2x) = \Sigma(2x_j)$ $= \Sigma X^2$ $(3x) = \Sigma(3x_j)$ $= (\Sigma T_{x_j}^2)/n$	$(1xy) = G_x G_y/kn$ $(2xy) = \Sigma(2xy_j)$ $= \Sigma XY$ $(3xy) = \Sigma(3xy_j)$ $= (\Sigma T_{x_j} T_{y_j})/n$	$(1y) = G_y^2/kn$ $(2y) = \Sigma(2y_j)$ $= \Sigma Y^2$ $(3y) = \Sigma(3y_j)$ $= (\Sigma T_{y_j}^2)/n$
(iv)	$S_{xx} = (2x) - (1x)$ $E_{xx} = (2x) - (3x)$ $T_{xx} = (3x) - (1x)$	$S_{xy} = (2xy) - (1xy)$ $E_{xy} = (2xy) - (3xy)$ $T_{xy} = (3xy) - (1xy)$	$S_{yy} = (2y) - (1y)$ $E_{yy} = (2y) - (3y)$ $T_{yy} = (3y) - (1y)$
(v)		$S'_{yy} = S_{yy} - (S_{xy}^2/S_{xx})$ $E'_{yy} = E_{yy} - (E_{xy}^2/E_{xx})$ $T_{yyR} = S'_{yy} - E'_{yy}$	

only if tests on the assumptions underlying the analysis of covariance are made. Symbols required for the overall analysis of covariance are summarized in parts iii and iv.

The numerical data in Table 10.3-2 will be used to illustrate the computational procedures. Suppose that the $k = 3$ treatments represent different methods of training. The experimenter is not at liberty to assign the subjects at random to the different methods of training; he is required to use groups that are already formed. However, from all available information, the groups were not originally selected on the basis of variables that are considered directly relevant to the study; for most practical purposes the groups can be considered as random samples from a common population.

There are $n = 7$ subjects in each of the groups. The groups are assigned at random to training methods. *Before* the subjects are trained under the method to which they are assigned, they are given a common aptitude test. The scores on this test define the covariate measure X. After the training is completed, the subjects are given a common achievement

Table 10.3-2 Numerical Example

		Treat 1		Treat 2		Treat 3		Totals	
		X	Y	X	Y	X	Y		
(i)		3	6	4	8	3	6		
		1	4	5	9	2	7		
		3	5	5	7	2	7		
		1	3	4	9	3	7		
		2	4	3	8	4	8		
		1	3	1	5	1	5		
		4	6	2	7	4	7		
	$\Sigma(\ )$	15	31	24	53	19	47	58	131
	$\Sigma(\ )^2$	41	147	96	413	59	321	196	881
	ΣXY	75		191		132		398	

		Treat 1	Treat 2	Treat 3	Total
(ii)	E_{xx_j}	8.86	13.71	7.43	$30.00 = E_{xx}$
	E_{xy_j}	8.57	9.29	4.43	$22.29 = E_{xy}$
	E_{yy_j}	9.72	11.71	5.43	$26.86 = E_{yy}$

(iii)

$$(1x) = (58)^2/21 = 160.19 \qquad (1y) = (131)^2/21 = 817.19$$
$$(2x) = 196 \qquad (2y) = 881$$
$$(3x) = (15^2 + 24^2 + 19^2)/7 = 166.00 \qquad (3y) = (31^2 + 53^2 + 47^2)/7 = 854.14$$
$$(1xy) = (58)(131)/21 = 361.81$$
$$(2xy) = 398$$
$$(3xy) = [(15)(31) + (24)(53) + (19)(47)]/7 = 375.71$$

(iv)

$$S_{xx} = (2x) - (1x) = 35.81 \qquad S_{xy} = (2xy) - (1xy) = 36.19$$
$$E_{xx} = (2x) - (3x) = 30.00 \qquad E_{xy} = (2xy) - (3xy) = 22.29$$
$$T_{xx} = (3x) - (1x) = 5.81 \qquad T_{xy} = (3xy) - (1xy) = 13.90$$
$$S_{yy} = (2y) - (1y) = 63.81$$
$$E_{yy} = (2y) - (3y) = 26.86$$
$$T_{yy} = (3y) - (1y) = 36.95$$

test over the material covered in the training. The score on the latter test is the criterion measure Y. For example, the first subject in treatment class 1 made a score of $X = 3$ on the aptitude test and a score of $Y = 6$ on the achievement test.

In part i of this table, the entries under treatment 1 are as follows:

$$\sum_i X_{i1} = 15 \qquad \sum_i Y_{i1} = 31$$
$$\sum_i X_{i1}^2 = 41 \qquad \sum_i Y_{i1}^2 = 147$$
$$\sum_i X_{i1} Y_{i1} = 75.$$

In words, these entries are the sum, the sum of the squares, and the sum of the products of the pairs of observations under treatment 1. The totals to the right of part i are as follows:

$$\sum_j \left(\sum_i X_{ij} \right) = 58 \qquad \sum_j \left(\sum_i Y_{ij} \right) = 131$$
$$\sum_j \left(\sum_i X_{ij}^2 \right) = 196 \qquad \sum_j \left(\sum_i Y_{ij}^2 \right) = 881$$
$$\sum_j \left(\sum_i X_{ij} Y_{ij} \right) = 398.$$

These entries are the sums of the corresponding entries under the treatment headings.

The entries in part ii are within-class error terms for each of the classes. For example,

$$E_{xx_1} = 41 - (15^2/7) = 8.86,$$
$$E_{xy_1} = 75 - [(15)(31)/7] = 8.57,$$
$$E_{yy_1} = 147 - (31^2/7) = 9.72.$$

At the right of part ii are the totals for the entries in the corresponding rows.

Computational symbols used in the overall analysis are given in part iii. These symbols are defined in part iii of Table 10.3-1.

To test the hypothesis of homogeneity of within-class regression, from part ii of Table 10.3-2 one computes

$$\Sigma \frac{E_{xy_j}^2}{E_{xx_j}} = \frac{8.57^2}{8.86} + \frac{9.29^2}{13.71} + \frac{4.43^2}{7.43}$$
$$= 8.29 + 6.29 + 2.64 = 17.22.$$

The variation of the individual observations about the unpooled within-class regression lines is

$$S_1 = E_{yy} - \Sigma \frac{E_{xy_j}^2}{E_{xx_j}} = 26.86 - 17.22 = 9.64.$$

The variation of the individual within-class regression coefficients about the pooled within-class regression coefficient is

$$S_2 = \Sigma \frac{E^2_{xy_j}}{E_{xx_j}} - \frac{E^2_{xy}}{E_{xx}} = 17.22 - 16.56 = .66.$$

The statistic used in the test for homogeneity of within-class regression is

$$F = \frac{S_2/(k-1)}{S_1/k(n-2)} = \frac{(.66)/2}{(9.64)/15} = .52.$$

The critical value for a .10-level test is $F_{.90}(2,15) = 2.70$. The experimental data do not contradict the hypothesis of homogeneity of within-class regression. The outcome of this test, in part, does not rule against pooling the within-class regressions. The appropriateness of the covariance model for the data justifies, in the long run, such pooling procedures.

An analysis of variance on the criterion variable alone is summarized in Table 10.3-3. Although the F test here indicates statistically significant

Table 10.3-3 Analysis of Variance

Source	SS	df	MS	F
Total	$S_{yy} = 63.81$	20		
Error	$E_{yy} = 26.86$	18	1.49	
Treatments	$T_{yy} = 36.95$	2	18.48	12.40

differences in the treatment means, $F_{.99}(2,18) = 6.01$, direct interpretation of this test is difficult since the treatment classes appear to have different covariate means. A question might be raised about differences between treatments after adjustment is made for the linear trend in the relationship between the criterion and the covariate. In other words, do differences between the criterion means remain after a statistical adjustment has been made for the effects of the covariate? In a sense the analysis of covariance attempts to approximate the situation in which each of the treatment groups is equated on the covariate; this approximation procedure assumes that a linear model is appropriate.

The analysis of covariance is summarized in Table 10.3-4. The F ratio in this table provides a test of the hypothesis that $\sigma^2_\tau = 0$, after the criterion data have been adjusted for the linear trend on the covariate.

Table 10.3-4 Analysis of Covariance

Source	SS	df	MS	F
Total	$S'_{yy} = 27.24$	19		
Error	$E'_{yy} = 10.30$	17	.61	
Treatments	$T_{yyR} = 16.94$	2	8.47	13.89

The critical value for a .01-level test in this case is $F_{.99}(2,17) = 6.11$. Thus the experimental data indicate statistically significant differences between the criterion scores for the groups even after adjustment is made for the linear effect of the covariate.

The adjusted means are given in Table 10.3-5. In making multiple

Table 10.3-5 Adjusted Treatment Means

$b = E_{xy}/E_{xx} = .743$

$\bar{X}_j \ldots$	2.14	3.43	2.71	$\bar{X} = 2.76$
$\bar{X}_j - \bar{X} \ldots$	$-.62$	.67	$-.05$	
$\bar{Y}_j \ldots$	4.43	7.57	6.71	$\bar{Y} = 6.24$
$\bar{Y}_j' = \bar{Y}_j - b(\bar{X}_j - \bar{X}) \ldots$	4.89	7.07	6.75	$\bar{Y}' = 6.24$

$$\text{MS}'_{\text{error(effective)}} = \text{MS}'_{\text{error}}\left[1 + \frac{T_{xx}/(k-1)}{E_{xx}}\right]$$

$$= .61\left[1 + \frac{5.81/2}{30.00}\right] = .67$$

tests between these means, the average effective error per unit is that given in this table. In making tests which fall in the a priori category, a somewhat different error term is used. For example, to test the hypothesis that $\tau_2 = \tau_3$,

$$F = \frac{(7.07 - 6.75)^2}{.61\left[\frac{2}{7} + \frac{(3.43 - 2.71)^2}{30.00}\right]} = \frac{.1024}{.1848} = .55.$$

The critical value for a .01-level test in this case is $F_{.99}(1,17) = 8.40$. Had the average effective error been used, this ratio would be

$$F = \frac{(7.07 - 6.75)^2}{2(.67)/7} = \frac{.1024}{.1914} = .54.$$

The critical value used in the previous test would also be appropriate in this latter test.

It is of interest to examine the magnitude of the individual within-class correlations. The squares of these correlations are as follows:

$$r_1^2 = \frac{E_{xy_1}^2/E_{xx_1}}{E_{yy_1}} = .85, \qquad r_2^2 = \frac{E_{xy_2}^2/E_{xx_2}}{E_{yy_2}} = .54, \qquad r_3^2 = \frac{E_{xy_3}^2/E_{xx_3}}{E_{yy_3}} = .49.$$

The square of the pooled within-class correlation is

$$r_w^2 = \frac{E_{xy}^2/E_{xx}}{E_{yy}} = .617.$$

In terms of this latter correlation, the error term in the analysis of covariance is given by

$$\text{MS}'_{\text{error}} = \text{MS}_{\text{error}}(1 - r_w^2)\left[\frac{k(n-1)}{k(n-1)-1}\right] = 1.49(.383)(\tfrac{18}{17}) = .61,$$

where MS_{error} is obtained from Table 10.3-3. Thus the reduction in the error term (increase in precision) is primarily a function of the magnitude of r_w^2. The larger the latter correlation, the greater the reduction in the error term.

The reduced sum of squares for treatments, T_{yyR}, depends in part upon the magnitude of r_w^2 and in part upon the magnitude of the squared between-class correlation. The latter is

$$r_b^2 = \frac{T_{xy}^2/T_{xx}}{T_{yy}} = .90.$$

When r_b is large relative to r_w, the reduction in the treatment variation can be relatively larger than the reduction in the error variation. When this occurs, the F ratio in the analysis of covariance will actually be smaller than the corresponding F ratio in the analysis of variance. The latter finding would generally indicate that bias due to the covariate had inflated the F ratio in the original analysis of variance.

When r_b is negative and r_w is positive, T_{yyA} will always be larger than T_{yy}. The reduced sum of squares, T_{yyR}, may in this case also be larger than T_{yy}. When this happens, the F ratio in the original analysis of variance is deflated because of the bias due to the covariate.

If the treatment classes are disregarded, the square of the overall correlation between the covariate and the criterion is

$$r_{\text{total}}^2 = \frac{S_{xy}^2/S_{xx}}{S_{yy}} = .57.$$

Rather than using a covariance analysis for the data in Table 10.3-2, the experimenter might have attempted to use the covariate as a classification or stratification factor. If this were done, the experiment would be analyzed as a two-factor experiment. The data would have the form given in Table 10.3-6. In this particular case the cell frequencies would be quite small; further, there would be no entries in some of the cells. If each of the resulting cell frequencies is relatively large, say 5 or more, this type of stratification on the covariate is generally to be preferred to the analysis of covariance. Often neighboring ordered categories having small frequencies may be grouped in order to build up the cell frequencies.

Unequal Cell Frequencies. If the number of subjects in each treatment class is not constant, the only changes in the computational formulas are

Table 10.3-6 Use of Covariate as a Classification Factor

Covariate classification	Treatment 1	Treatment 2	Treatment 3
5		9,7	
4	6	8,9	8,7
3	6,5	8	6,7
2	4	7	7,7
1	4,3,3	5	5

as follows:

$$(3x_j) = \frac{T_{x_j}^2}{n_j} \qquad (3xy_j) = \frac{T_{x_j}T_{y_j}}{n_j} \qquad (3y_j) = \frac{T_{y_j}^2}{n_j}$$

$$(3x) = \Sigma(3x_j) \qquad (3xy) = \Sigma(3xy_j) \qquad (3y) = \Sigma(3y_j)$$

The degrees of freedom for this case are as follows:

Source	df
S'_{yy}	$(\Sigma n_j) - 2$
E'_{yy}	$[\Sigma(n_j - 1)] - 1$
T'_{yy}	$k - 1$

With these changes, the computational procedures outlined in Table 10.3-1 may be used.

10.4 Factorial Experiment

The analysis of covariance for a factorial experiment is a direct generalization of the corresponding analysis for a single-factor experiment. Assuming a $p \times q$ factorial experiment having n observations in each cell, the model is as follows:

$$Y'_{ijk} = Y_{ijk} - \beta(X_{ijk} - \bar{X}) = \mu + \alpha_j + \beta_k + \alpha\beta_{jk} + \varepsilon_{ijk}.$$

That is, an observation, adjusted for the effect of the covariate, estimates the parameters in the usual analysis of variance. If the covariate is ignored, its effect augments variation due to experimental error; the latter includes all uncontrolled sources of variation. It will be assumed throughout this section that A and B represent fixed factors.

The observations on the covariate and the criterion within cell ab_{jk} of the experiment are represented as follows:

X	Y
X_{1jk}	Y_{1jk}
X_{2jk}	Y_{2jk}
. . .	. . .
X_{njk}	Y_{njk}

In each of the pq cells in the experiment there are n pairs of observations. The following notation denotes the various sums needed in the analysis:

$$
\begin{aligned}
AB_{x_{jk}} &= \text{sum of covariate measures in cell } ab_{jk}.\\
AB_{y_{jk}} &= \text{sum of criterion measures in cell } ab_{jk}.\\
A_{x_j} &= \text{sum of all covariate measures at level } a_j.\\
A_{y_j} &= \text{sum of all criterion measures at level } a_j.\\
B_{x_k} &= \text{sum of all covariate measures at level } b_k.\\
B_{y_k} &= \text{sum of all criterion measures at level } b_k.\\
G_x &= \text{sum of all covariate measures.}\\
G_y &= \text{sum of all criterion measures.}
\end{aligned}
$$

The means corresponding to the sets of sums defined above are obtained by dividing the respective sums by the number of experimental units over which the sum is taken. For example,

$$\overline{AB}_{x_{jk}} = \frac{AB_{x_{jk}}}{n}, \qquad \bar{A}_{y_j} = \frac{A_{y_j}}{nq}, \qquad \bar{G}_y = \frac{G_y}{npq}.$$

The variation due to main effects and interactions are defined as follows. For the variate:

$$
\begin{aligned}
A_{yy} &= nq\Sigma(\bar{A}_{y_j} - \bar{G}_y)^2,\\
B_{yy} &= np\Sigma(\bar{B}_{y_k} - \bar{G}_y)^2,\\
AB_{yy} &= n\Sigma(\overline{AB}_{y_{jk}} - \bar{A}_{y_j} - \bar{B}_{y_k} + \bar{G}_{y_j})^2.
\end{aligned}
$$

For the covariate:

$$
\begin{aligned}
A_{xx} &= nq\Sigma(\bar{A}_{x_j} - \bar{G}_{x_j})^2,\\
B_{xx} &= np\Sigma(\bar{B}_{x_k} - \bar{G}_x)^2,\\
AB_{xx} &= n\Sigma(\overline{AB}_{x_{jk}} - \bar{A}_{x_j} - \bar{B}_{x_k} + \bar{G}_x)^2.
\end{aligned}
$$

For the cross products:

$$
\begin{aligned}
A_{xy} &= nq\Sigma(\bar{A}_{x_j} - \bar{G}_x)(\bar{A}_{y_j} - \bar{G}_y),\\
B_{xy} &= np\Sigma(\bar{B}_{x_j} - \bar{G}_x)(\bar{B}_{y_j} - \bar{G}_y),\\
AB_{xy} &= n\Sigma(\overline{AB}_{x_{jk}} - \bar{A}_{x_j} - \bar{B}_{x_k} + \bar{G}_x)(\overline{AB}_{y_{jk}} - \bar{A}_{y_j} - \bar{B}_{y_k} + \bar{G}_y).
\end{aligned}
$$

The variation of the covariate within cell ab_{jk} is

$$E_{xx_{jk}} = \sum_i (X_{ijk} - \overline{AB}_{x_{jk}})^2.$$

The variation of the variate within this cell is

$$E_{yy_{jk}} = \sum_i (Y_{ijk} - \overline{AB}_{y_{jk}})^2.$$

The covariation for this cell is

$$E_{xy_{jk}} = \sum_i (X_{ijk} - \overline{AB}_{x_{jk}})(Y_{ijk} - \overline{AB}_{y_{jk}}).$$

The pooled within-cell variations are given by

$$E_{xx} = \Sigma E_{xx_{jk}}, \qquad E_{yy} = \Sigma E_{yy_{jk}}, \qquad E_{xy} = \Sigma E_{xy_{jk}}.$$

From the data within cell ab_{jk}, the regression coefficient for the regression of Y on X is

$$b_{E_{jk}} = \frac{E_{xy_{jk}}}{E_{xx_{jk}}}.$$

The residual about this regression line within cell ab_{jk} is

$$E'_{yy_{jk}} = E_{yy_{jk}} - \frac{E^2_{xy_{jk}}}{E_{xx_{jk}}}.$$

Under the hypothesis that

$$\beta_{E_{11}} = \beta_{E_{12}} = \cdots = \beta_{E_{pq}} = \beta_E,$$

that is, homogeneity of the within-cell regressions, an estimate of β_E is

$$b_E = \frac{E_{xy}}{E_{xx}}.$$

The within-class regression lines, based upon the pooled within-class estimate of the regression coefficient, have the form

$$\hat{Y}_{ijk} = b_E(X_{ijk} - \overline{AB}_x) + \overline{AB}_{y_{jk}}.$$

The variation of the residuals about these lines is

$$D = E'_{yy} = E_{yy} - b_E E_{xy} = E_{yy} - \frac{E^2_{xy}}{E_{xx}}.$$

If factors A and B are fixed, this source of variation is an estimate of experimental error in the analysis of covariance; its form is identical to the corresponding term in a single-factor experiment.

Equivalently, E'_{yy} is the error variation associated with the model

$$Y_{ijk} = (\alpha + \beta X_{ijk}) + \tau_{jk} + \varepsilon_{ijk},$$

where $$\tau_{jk} = \alpha_j + \beta_k + \alpha\beta_{jk}.$$

If one restricts this model to the case in which

$$\alpha_j = 0, \qquad j = 1, 2, \ldots, p,$$

then the variation due to error in the restricted model may be shown to be

$$D'_a = (A_{yy} + E_{yy}) - \frac{(A_{xy} + E_{xy})^2}{A_{xx} + E_{xx}}.$$

The restricted model is the one which corresponds to the hypothesis that all $\alpha_j = 0$. A measure of deviation from hypothesis is given by

$$A'_{yy} = D'_a - D = A_{yy} - \frac{(A_{xy} + E_{xy})^2}{A_{xx} + E_{xx}} + \frac{E^2_{xy}}{E_{xx}}.$$

In terms of the notation system used for a single-factor experiment,

$$A'_{yy} = A_{yyR}.$$

The "prime" symbol will be used throughout in this section and those that follow on factorial experiments to imply the reduced variation.

If one restricts the model in the preceding paragraph so that $\beta_k = 0$ (for all k), then the variation due to error is

$$D'_b = (B_{yy} + E_{yy}) - \frac{(B_{xy} + E_{xy})^2}{B_{xx} + E_{xx}}.$$

A measure of deviation from the hypothesis that $\beta_k = 0$ is thus

$$B'_{yy} = D'_b - D = B_{yy} - \frac{(B_{xy} + E_{xy})^2}{B_{xx} + E_{xx}} + \frac{E^2_{xy}}{E_{xx}}.$$

Similarly, if the model is restricted to the case in which all $\alpha\beta_{jk}$ are set equal to zero, the variation due to error is

$$D'_{ab} = (AB_{yy} + E_{yy}) - \frac{(AB_{xy} + E_{xy})^2}{AB_{xx} + E_{xx}}.$$

A measure of deviation from hypothesis with respect to the interaction effects is

$$AB'_{yy} = D'_{ab} - D = AB_{yy} - \frac{(AB_{xy} + E_{xy})^2}{AB_{xx} + E_{xx}} + \frac{E^2_{xy}}{E_{xx}}.$$

Each of the variations due to deviation from hypothesis has the general form

$$S_\omega - S_\Omega,$$

where S_ω = variation due to error under model restricted to correspond to case in which a specified hypothesis is true,
S_Ω = variation due to error under no restrictions on model.

The analysis of variance is contrasted with the analysis of covariance in Table 10.4-1. Assuming all factors fixed, MS'_{error} is the proper denominator for F tests on main effects and interaction. An approximation to the reduced sum of squares for main effects and interactions is

$$A''_{yy} = A_{yy} - 2b_E A_{xy} + b^2_E A_{xx},$$
$$B''_{yy} = B_{yy} - 2b_E B_{xy} + b^2_E B_{xx},$$
$$AB''_{yy} = AB_{yy} - 2b_E AB_{xy} + b^2_E AB_{xx}.$$

Table 10.4-1 Summary of Analysis of Variance and Analysis of Covariance

Source	Analysis of variance			Analysis of covariance		
	SS	df	MS	SS	df	MS
A	A_{yy}	$p - 1$	MS_a	A'_{yy}	$p - 1$	MS'_a
B	B_{yy}	$q - 1$	MS_b	B'_{yy}	$q - 1$	MS'_b
AB	AB_{yy}	$(p - 1)(q - 1)$	MS_{ab}	AB'_{yy}	$(p - 1)(q - 1)$	MS'_{ab}
Error	E_{yy}	$pq(n - 1)$	MS_{error}	E'_{yy}	$pq(n - 1) - 1$	MS'_{error}

In cases where the treatments do not affect the covariate and the bias in the treatment means associated with the covariate is not large, these approximations will be quite close.

Adjusted cell means have the following form:

$$\overline{AB}'_{y_j} = \overline{AB}_{y_j} - b_E(\overline{AB}_{x_j} - \overline{AB}_x).$$

The average effective error for the difference between two adjusted cell means is

$$\frac{2MS'_{error}}{n}\left[1 + \frac{\text{Cell}_{xx}/(pq - 1)}{E_{xx}}\right],$$

where $\quad \text{Cell}_{xx} = n\Sigma(\overline{AB}_{x_{jk}} - \bar{G}_x)^2 = A_{xx} + B_{xx} + AB_{xx}.$

In making multiple comparisons among cell means, the effective error per experimental unit is

$$MS'_{error}\left[1 + \frac{\text{Cell}_{xx}/(pq - 1)}{E_{xx}}\right].$$

The estimate of the variance of a comparison among the cell means of the form $\Sigma c_{jk}\overline{AB}_{y_{jk}}$, where $\Sigma c_{jk} = 0$, is given by

$$MS'_{error}\left[\frac{\Sigma c_{jk}^2}{n} + \frac{(\Sigma c_{ij}\overline{AB}_{x_{jk}})^2}{E_{xx}}\right].$$

For the special case of a simple difference between two cell means, $\overline{AB}_{y_{jk}} - \overline{AB}_{y_{mp}}$, the estimate of the variance of a comparison specializes to

$$MS'_{error}\left[\frac{2}{n} + \frac{(\overline{AB}_{x_{jk}} - \overline{AB}_{x_{mp}})^2}{E_{xx}}\right].$$

Since $\quad \bar{A}_{y_j} = \dfrac{\overline{AB}_{y_{j1}} + \overline{AB}_{y_{j2}} + \cdots + \overline{AB}_{y_{jq}}}{q},$

$$\bar{A}_{y_m} = \frac{\overline{AB}_{y_{m1}} + \overline{AB}_{y_{m2}} + \cdots + \overline{AB}_{y_{mq}}}{q},$$

the difference
$$\bar{A}_{y_j} - \bar{A}_{y_m}$$
represents a comparison for which $\Sigma c_{ij}^2 = \Sigma(1/q)^2 = 2q/q^2 = 2/q$. Hence an estimate of the variance of the difference between two adjusted means of the form $\bar{A}'_{y_j} - \bar{A}'_{y_m}$ is given by

$$\text{MS}_{\text{error}}\left[\frac{2}{nq} + \frac{(\bar{A}_{x_j} - \bar{A}_{x_m})^2}{E_{xx}}\right].$$

In this case,

$$\bar{A}'_{y_j} = \bar{A}_{y_j} - b_E(\bar{A}_{x_j} - \bar{G}_x).$$

It will be found that

$$\bar{A}'_{y_j} = \frac{\sum_k \overline{AB}'_{jk}}{q}.$$

In making multiple tests among the $\bar{A}'_j$'s, the average effective error per experimental unit may be taken as

$$\text{MS}'_{\text{error}}\left[1 + \frac{A_{xx}/(p-1)}{E_{xx}}\right].$$

An adjusted mean for level b_k is

$$\bar{B}'_{y_k} = \bar{B}_{y_k} - b_E(\bar{B}_{x_k} - \bar{G}_x).$$

The corresponding average experimental error per unit is

$$\text{MS}'_{\text{error}}\left[1 + \frac{B_{xx}/(q-1)}{E_{xx}}\right].$$

In making a test for homogeneity of within-cell regression, the error term may be partitioned as follows:

Source	df
$E_{yy} = E_{yy} - (E_{xy}^2/E_{xx})$	$pq(n-1) - 1$
$S_1 = E_{yy} - \Sigma\Sigma(E_{xy_{jk}}^2/E_{xx_{jk}})$	$pq(n-2)$
$S_2 = \Sigma\Sigma(E_{xy_{jk}}^2/E_{xx_{jk}}) - (E_{xy}^2/E_{xx})$	$pq - 1$

S_1 represents the variation of the observations about the individual within-cell regression lines. S_2 represents the variation of the individual cell regression coefficients about the pooled within-class regression coefficient. The test for homogeneity of within-class regression uses the statistic

$$F = \frac{S_2/(pq-1)}{S_1/pq(n-2)}.$$

It should be noted that the pooled within-class regression coefficient is used in obtaining $\bar{A}'_y$, $\bar{B}'_y$, and $\overline{AB}'_y$. Implicit in this adjustment process is the assumption that b_E is the appropriate regression coefficient for all these adjustments. In most experimental situations in the behavioral sciences this assumption very probably oversimplifies what is a more complex model. Caution in the use of covariates in factorial experiments is advised. The model for the analysis outlined in this section is a highly restrictive one. The experimenter should be aware that such designs exist, but he also should be aware of both their strengths and their weaknesses.

10.5 Computational Procedures for Factorial Experiment

Computational procedures for the analysis of covariance for a $p \times q$ factorial experiment having n observations per cell will be considered first. The case of unequal cell frequencies is considered later in this section, where the equivalent of an unweighted-means analysis is outlined. Procedures in this latter case require relatively small changes from those to be given for the case of equal cell frequencies.

Assuming n pairs of observations in each of the pq cells, computational formulas for the sums of squares needed in the analysis of covariance are given in Table 10.5-1. In each case the range of summation is over all possible values of the total that is squared and summed. Symbols $(1x)$ through $(5x)$ are those used in an analysis of variance on the covariate. Symbols $(1y)$ through $(5y)$ are those used in the usual analysis of variance

Table 10.5-1 Computational Formulas for the Analysis of Covariance in a Factorial Experiment

(i)	$(1x) = G_x^2/npq$	$(1xy) = G_xG_y/npq$	$(1y) = G_y^2/npq$
	$(2x) = \Sigma X^2$	$(2xy) = \Sigma XY$	$(2y) = \Sigma Y^2$
	$(3x) = (\Sigma A_x^2)/nq$	$(3xy) = (\Sigma A_xA_y)/nq$	$(3y) = (\Sigma A_y^2)/nq$
	$(4x) = (\Sigma B_x^2)/np$	$(4xy) = (\Sigma B_xB_y)/np$	$(4y) = (\Sigma B_y^2)/np$
	$(5x) = (\Sigma AB_x^2)/n$	$(5xy) = (\Sigma AB_xAB_y)/n$	$(5y) = (\Sigma AB_y^2)/n$
(ii)	$A_{xx} = (3x) - (1x)$	$A_{xy} = (3xy) - (1xy)$	$A_{yy} = (3y) - (1y)$
	$B_{xx} = (4x) - (1x)$	$B_{xy} = (4xy) - (1xy)$	$B_{yy} = (4y) - (1y)$
	$AB_{xx} = (5x) - (3x) - (4x) + (1x)$	$AB_{xy} = (5xy) - (3xy) - (4xy) + (1xy)$	$AB_{yy} = (5y) - (3y) - (4y) + (1y)$
	$E_{xx} = (2x) - (5x)$	$E_{xy} = (2xy) - (5xy)$	$E_{yy} = (2y) - (5y)$

(iii)

$$E'_{yy} = E_{yy} - (E_{xy}^2/E_{xx})$$

$$(A + E)'_{yy} = (A_{yy} + E_{yy}) - \frac{(A_{xy} + E_{xy})^2}{A_{xx} + E_{xx}}, \qquad A'_{yy} = (A + E)'_{yy} - E'_{yy}$$

$$(B + E)'_{yy} = (B_{yy} + E_{yy}) - \frac{(B_{xy} + E_{xy})^2}{B_{xx} + E_{xx}}, \qquad B'_{yy} = (B + E)'_{yy} - E'_{yy}$$

$$(AB + E)'_{yy} = (AB_{yy} + E_{yy}) - \frac{(AB_{xy} + E_{xy})^2}{(AB_{xx} + E_{xx})}, \qquad AB'_{yy} = (AB + E)'_{yy} - E'_{yy}$$

on the criterion. Symbols $(1xy)$ through $(5xy)$ are used to estimate the covariances needed in the adjustment process.

A 2×3 factorial experiment having $n = 5$ observations per cell will be used to illustrate the computational procedures. Suppose that the $p = 2$ levels of factor A represent methods of instructing in teaching map reading, and suppose that the $q = 3$ levels of factor B represent three instructors. For purposes of the present analysis, both factors are considered to be fixed. The covariate measure in this experiment is the score on an achievement test on map reading prior to the training; the criterion measure is the score on a comparable form of the achievement test after training is completed. Assume that intact groups of $n = 5$ subjects each are assigned at random to the cells of the experiment. Suppose that data obtained from this experiment are those given in the upper part of Table 10.5-2.

In this table, for example, the first subject under method a_1 and instructor b_2 has scores of 30 and 85, respectively, on the covariate and the criterion.

Table 10.5-2 Numerical Example

Method	b_1 Instr. 1		b_2 Instr. 2		b_3 Instr. 3	
	X	Y	X	Y	X	Y
a_1	40	95	30	85	50	90
	35	80	40	100	40	85
	40	95	45	85	40	90
	50	105	40	90	30	80
	45	100	40	90	40	85
a_2	50	100	50	100	45	95
	30	95	30	90	30	85
	35	95	40	95	25	75
	45	110	45	90	50	105
	30	88	40	95	35	85

AB summary:

	b_1		b_2		b_3		Total	
	X	Y	X	Y	X	Y	X	Y
a_1	210	475	195	450	200	430	605	1355
a_2	190	488	205	470	185	445	580	1403
Total	400	963	400	920	385	875	1185	2758

$(1x) = 46{,}808$	$(1xy) = 108{,}941$	$(1y) = 253{,}552$
$(2x) = 48{,}325$	$(2xy) = 110{,}065$	$(2y) = 255{,}444$
$(3x) = 46{,}828$	$(3xy) = 108{,}901$	$(3y) = 253{,}629$
$(4x) = 46{,}822$	$(4xy) = 109{,}008$	$(4y) = 253{,}939$
$(5x) = 46{,}895$	$(5xy) = 108{,}979$	$(5y) = 254{,}019$

In symbols,

$$X_{121} = 30, \qquad Y_{121} = 85.$$

An AB summary table appears under the observed data. There are two entries in each cell of this table—one represents the sum of the observations on the covariate, the other the sum for the criterion. For example, the sum of the $n = 5$ observations on the covariate under treatment combination ab_{12} is

$$AB_{x_{12}} = 30 + 40 + 45 + 40 + 40 = 195.$$

The corresponding sum for the criterion data is

$$AB_{y_{12}} = 85 + 100 + 85 + 90 + 90 = 450.$$

The entries in the total columns at the right of the AB summary table are the sums of corresponding entries in the rows. For example,

$$A_{x_1} = \sum_k AB_{x_{1k}} = 210 + 195 + 200 = 605.$$

The corresponding sum for the criterion data is

$$A_{y_1} = \sum_k AB_{y_{1k}} = 475 + 450 + 430 = 1355.$$

The total of the first column in the summary table is

$$B_{x_1} = \sum_j AB_{x_{j1}} = 210 + 190 = 400.$$

The corresponding sum for the criterion is

$$B_{y_1} = \sum_j AB_{y_{j1}} = 475 + 488 = 963.$$

The grand totals for the covariate and the criterion are

$$G_x = \Sigma A_{x_j} + \Sigma B_{x_k} = 1185,$$
$$G_y = \Sigma A_{y_j} + \Sigma B_{y_k} = 2758.$$

The computational symbols in the lower part of Table 10.5-2 are defined in part i of Table 10.5-1. The only symbols requiring special comment are those in the center column. These entries are obtained as follows:

$$(1xy) = \frac{(1185)(2758)}{30},$$

$$(2xy) = (40)(95) + (35)(80) + \cdots + (50)(105) + (35)(85),$$

$$(3xy) = \frac{(605)(1355) + (580)(1403)}{15},$$

$$(4xy) = \frac{(400)(963) + (400)(920) + (385)(875)}{10},$$

$$(5xy) = \frac{(210)(475) + (190)(488) + \cdots + (185)(445)}{5}.$$

The basic data for all these symbols except $(2xy)$ are obtained from the AB summary table.

Sums of squares and sums of products are given in Table 10.5-3. Computational formulas for these terms are given in parts ii and iii of Table 10.5-1. Note that it is possible for the entries that are used to obtain co-

Table 10.5-3 Summary Data for Numerical Example

$A_{xx} = 21$	$A_{xy} = -40$	$A_{yy} = 77$
$B_{xx} = 15$	$B_{xy} = 67$	$B_{yy} = 387$
$AB_{xx} = 52$	$AB_{xy} = 11$	$AB_{yy} = 3$
$E_{xx} = 1430$	$E_{xy} = 1086$	$E_{yy} = 1425$
1518	1124	1892

$$E'_{yy} = 1425 - (1086^2/1430) = 600$$

$(A + E)'_{yy} = 748$	$A'_{yy} = 748 - 600 = 148$
$(B + E)'_{yy} = 892$	$B'_{yy} = 892 - 600 = 292$
$(AB + E)'_{yy} = 616$	$AB'_{yy} = 616 - 600 = 16$

variances to be either positive or negative. In this case the between-class covariation of the totals corresponding to the main effects of factor A is negative (-40). Inspection of the total columns at the right of the AB summary in Table 10.5-2 indicates that the higher criterion total is paired with the lower covariate total; hence the negative covariation.

The analysis of variance for the criterion data is summarized in Table 10.5-4. This analysis disregards the presence of the covariate. Differences between the methods of training are tested by means of the statistic

$$F = \frac{77}{59.4} = 1.30.$$

(The instructor factor is considered to be fixed.) This test indicates no statistically significant difference between the methods insofar as the mean of the groups is concerned.

Table 10.5-4 Analysis of Variance

Source	SS	df	MS	F
A Methods	$A_{yy} = 77$	1	77	1.30
B Instructors	$B_{yy} = 387$	2	193.5	3.26
AB	$AB_{yy} = 3$	2	1.5	
Error	$E_{yy} = 1425$	24	59.4	
Total	1892	29		

$F_{.95}(2,24) = 3.40$

The analysis of covariance is summarized in Table 10.5-5. Note that the error mean square in this case is 26.1, compared with 59.4 in the case of the analysis of variance. Further note that the adjusted method mean square is 148, compared with 77 in the analysis of variance. This increase in the adjusted method variance is a function of the negative covariance for the between-method totals. A .05-level test on the methods in the analysis of covariance indicates statistically significant differences between the criterion means. Thus, when a linear adjustment is made for the effect of variation due to differences in prior experience in map reading, as measured by the covariate, there are statistically significant differences between the training methods.

Table 10.5-5 Analysis of Covariance

Source	SS	df	MS	F
A Methods	$A'_{yy} = 148$	1	148.0	5.67
B Instructors	$B'_{yy} = 292$	2	146.0	5.59
AB	$AB'_{yy} = 16$	2	8.0	
Error	$E'_{yy} = 600$	23	26.1	
		28		

$F_{.95}(1,23) = 4.28;\ F_{.95}(2,23) = 3.44$

An estimate of the square of the within-cell correlation is

$$r^2_{\text{within}} = \frac{E^2_{xy}/E_{xx}}{E_{yy}} = \frac{(1086)^2/1430}{1425} = .5788.$$

The mean square due to experimental error in the analysis of covariance is

$$\begin{aligned} \text{MS}'_{\text{error}} &= \text{MS}_{\text{error}}(1 - r^2_{\text{within}}) \frac{pq(n-1)}{pq(n-1)-1} \\ &= 59.375(.4212)(24/23) = 26.10. \end{aligned}$$

The adjusted criterion means for factor A are given in Table 10.5-6. Note that the difference between the adjusted means is larger than the

Table 10.5-6 Adjusted Means

$b_E = E_{xy}/E_{xx} = .76$

	Method 1	Method 2	Mean
$\bar{A}_{x_j}$	40.3	38.7	$39.5 = \bar{G}_x$
$\bar{A}_{x_j} - \bar{G}_x$	.8	−.8	
$\bar{A}_y$	90.3	93.5	$91.9 = \bar{G}_y$
$\bar{A}'_y = \bar{A}_y - .76(\bar{A}_{x_j} - \bar{G}_x)$	89.7	94.1	91.9

corresponding difference between the unadjusted means. Had the covariance in this case been positive rather than negative, the difference between the adjusted means would have been smaller rather than larger than the difference between the unadjusted means.

It is of interest to compare the adjusted mean squares given in Table 10.5-5 with those that would be obtained by the approximation method described in the last section. The latter mean squares are as follows:

Source of variation	SS	MS
$A''_y = A_{yy} - 2b_E^2 A_{xy} + b_E^2 A_{xx}$ $= 77 - 2(.76)(-40) + (.76^2)(21)$	150	150
$B''_y = B_{yy} - 2b_E B_{xy} - b_E^2 B_{xx}$	309	154
$AB''_y = AB_{yy} - 2b_E AB_{xy} - b_E^2 AB_{xx}$	16	8

These mean squares are slightly larger than the corresponding mean squares in Table 10.5-5.

Unequal Cell Frequencies—Unweighted-means Analysis. Under the conditions for which the unweighted-means analysis is appropriate in a factorial design which does not include a covariate, there is an equivalent unweighted-means analysis for the analysis of covariance. The computational procedures for a two-factor factorial experiment having one covariate are outlined in Table 10.5-7. These procedures are readily generalized to higher-order factorial experiments having one covariate.

If one considers the variate separately from the covariate, the procedures in Table 10.5-7 will produce an unweighted-means analysis of variance. Similarly, if the covariate were considered separately, one would have an unweighted-means analysis on the covariate.

The cell means for the variate and the covariate are given at the top of Table 10.5-7. The row and column totals for these cell means are also given. For example,

$$A_{x_1} = \sum_j \overline{AB}_{x_{1j}}, \qquad A_{y_1} = \sum_j \overline{AB}_{y_{1j}}.$$

The computational symbols are defined in part i. Here

$$\tilde{n} = \frac{pq}{\Sigma(1/n_{ij})} = \text{harmonic mean of cell frequencies.}$$

The sums of squares and sums of products needed in the analysis of covariance are defined in terms of the computational symbols in part ii. Part iii in Table 10.5-7 is the same as part iii in Table 10.5-1.

A numerical example is worked in Table 10.5-8. In part i of this

Table 10.5-7 Computational Formulas (Unequal Cell Frequencies)

Notation:		b_1		b_2		b_3		Total	
$\overline{AB}_{x_{jk}} = AB_{x_{jk}}/n_{jk}$	a_1	$\overline{AB}_{x_{11}}$	$\overline{AB}_{y_{11}}$	$\overline{AB}_{x_{12}}$	$\overline{AB}_{y_{12}}$	$\overline{AB}_{x_{13}}$	$\overline{AB}_{y_{13}}$	A_{x_1}	A_{y_1}
$\overline{AB}_{y_{jk}} = AB_{y_{jk}}/n_{jk}$	a_2	$\overline{AB}_{x_{21}}$	$\overline{AB}_{y_{21}}$	$\overline{AB}_{x_{22}}$	$\overline{AB}_{y_{22}}$	$\overline{AB}_{x_{23}}$	$\overline{AB}_{y_{23}}$	A_{x_2}	A_{y_2}
	Total	B_{x_1}	B_{y_1}	B_{x_2}	B_{y_2}	B_{x_3}	B_{y_3}	G_x	G_y

(i)	$(1x) = \tilde{n}G_x^2/pq$	$(1xy) = \tilde{n}G_xG_y/pq$	$(1y) = \tilde{n}G_y^2/pq$
	$(2x) = \Sigma X^2$	$(2xy) = \Sigma XY$	$(2y) = \Sigma Y^2$
	$(3x) = \tilde{n}\Sigma(A_x^2)/q$	$(3xy) = \tilde{n}\Sigma(A_xA_y)/q$	$(3y) = \tilde{n}\Sigma(A_y^2)/q$
	$(4x) = \tilde{n}\Sigma(B_x^2)/p$	$(4xy) = \tilde{n}\Sigma(B_xB_y)/p$	$(4y) = \tilde{n}\Sigma(B_y^2)/p$
	$(5x) = \tilde{n}\Sigma(\overline{AB}_x^2)$	$(5xy) = \tilde{n}\Sigma(\overline{AB}_x\overline{AB}_y)$	$(5y) = \tilde{n}\Sigma(\overline{AB}_y^2)$
	$(5'x) = \Sigma(AB_{x_{jk}}^2/n_{jk})$	$(5'xy) = \Sigma(AB_{x_{jk}}AB_{y_{jk}}/n_{jk})$	$(5'y) = \Sigma(AB_{y_{jk}}^2/n_{jk})$
(ii)	$A_{xx} = (3x) - (1x)$	$A_{xy} = (3xy) - (1xy)$	$A_{yy} = (3y) - (1y)$
	$B_{xx} = (4x) - (1x)$	$B_{xy} = (4xy) - (1xy)$	$B_{yy} = (4y) - (1y)$
	$AB_{xx} = (5x) - (3x) - (4x) + (1x)$	$AB_{xy} = (5xy) - (3xy) - (4xy) + (1xy)$	$AB_{yy} = (5y) - (3y) - (4y) + (1y)$
	$E_{xx} = (2x) - (5'x)$	$E_{xy} = (2xy) - (5'xy)$	$E_{yy} = (2y) - (5'y)$
(iii)	Same as part iii of Table 10.5-1		

table are the basic data. The cell frequencies and the harmonic mean of the cell frequencies are given in part ii. The cell totals are also given in part ii. In the lower half of part ii are the cell means and the marginal totals for the cell means. It is from this part of the table that one computes most of the computational symbols defined in part i of Table 10.5-7. The numerical values of these symbols for the data in part i of Table 10.5-8 appear in part iii. For example,

$$(1x) = \frac{\tilde{n}G_x^2}{pq} = \frac{4.961(25.73)^2}{2(3)} = 547.39,$$

$$(1xy) = \frac{\tilde{n}G_xG_y}{pq} = \frac{4.961(25.73)(93.90)}{2(3)} = 1997.67.$$

The analysis of covariance is summarized in Table 10.5-9. Part i of this table follows from part ii of Table 10.5-7. Part ii follows from part iii of Table 10.5-7. The resulting analysis of variance and analysis of covariance are summarized in part iii. The large increase in the F ratio for the main effects of factor B is due to the relatively large negative between-class covariance between the variate and covariate means for the levels of factor B.

Table 10.5-8 Unequal Cell Frequencies—Numerical Example

(i)

	b_1		b_2		b_3	
	X	Y	X	Y	X	Y
a_1	3	8	2	14	3	16
	5	16	1	11	2	10
	1	10	8	20	1	14
	9	24	7	15	2	14
			4	12	6	22
					2	16
a_2	7	18	0	8	0	10
	0	7	4	16	1	15
	4	10	8	20	9	26
	6	15	5	18	4	18
	9	23			4	18
					7	26
					8	24

n_{jk}:	b_1	b_2	b_3
a_1	4	5	6
a_2	5	4	7

$\tilde{n} = 4.961$

AB_{jk}:	b_1		b_2		b_3		Total	
	X	Y	X	Y	X	Y	X	Y
a_1	18	58	22	72	16	92	56	222
a_2	26	73	17	62	33	137	76	272
Total	44	131	39	134	49	229	132	494

(ii)

$\overline{AB}_{jk}$:	b_1		b_2		b_3		Total	
	X	Y	X	Y	X	Y	X	Y
a_1	4.50	14.50	4.40	14.40	2.67	15.33	11.57	44.23
a_2	5.20	14.60	4.25	15.50	4.71	19.57	14.16	49.67
Total	9.70	29.10	8.65	29.90	7.38	34.90	25.73	93.90

(iii)

$(1x) = 547.39$	$(1xy) = 1997.67$	$(1y) = 7290.36$
$(2x) = 822$	$(2xy) = 2500$	$(2y) = 8742$
$(3x) = 552.94$	$(3xy) = 2009.32$	$(3y) = 7314.83$
$(4x) = 554.09$	$(4xy) = 1980.60$	$(4y) = 7339.38$
$(5x) = 565.68$	$(5xy) = 2001.82$	$(5y) = 7387.00$
$(5'x) = 583.49$	$(5'xy) = 2112.09$	$(5'y) = 7996.55$

Illustrative Applications. Learning and conditioning experiments have made relatively extensive use of the analysis of covariance. Prokasy, Grant, and Myers (1958) report a typical application of the analysis of

Table 10.5-9 Unequal Cell Frequencies—Numerical Example
(Continued from Table 10.5-8)

(i)	$A_{xx} = 5.55$	$A_{xy} = 11.65$	$A_{yy} = 24.47$
	$B_{xx} = 6.70$	$B_{xy} = -17.07$	$B_{yy} = 49.02$
	$AB_{xx} = 6.04$	$AB_{xy} = 9.57$	$AB_{yy} = 23.15$
	$E_{xx} = 238.51$	$E_{xy} = 387.91$	$E_{yy} = 745.45$

(ii)

$$E'_{yy} = 745.45 - \frac{(387.91)^2}{238.51} = 114.56$$

$$(A + E)'_{yy} = 769.92 - \frac{(399.56)^2}{244.06} = 115.78 \qquad A'_{yy} = 1.22$$

$$(B + E)'_{yy} = 794.47 - \frac{(370.84)^2}{245.21} = 233.64 \qquad B'_{yy} = 119.08$$

$$(AB + E)'_{yy} = 768.60 - \frac{(397.48)^2}{244.55} = 122.55 \qquad AB'_{yy} = 7.99$$

	Source of variation	Analysis of variance				Analysis of covariance			
		SS	df	MS	F	SS	df	MS	F
(iii)	A	24.47	1	24.47	.82	1.22	1	1.22	.26
	B	49.02	2	24.51	.82	119.08	2	59.54	12.48**
	AB	23.15	2	11.58	.39	7.99	2	4.00	.84
	Within cell	745.45	25	29.82		114.56	24	4.77	

covariance in this context. The purpose of their experiment was to study the effect of stimulus intensity (four levels) and intertrial interval (three levels) upon acquisition and extinction of eyelid conditioning. The criterion was the proportion of conditioned responses in units of the arcsine transformation. In the acquisition phase of the study, the score on the first day's trials served as the covariate for analyzing the results of the second day's trials. In the extinction phase, the combined scores during the first and second days' trials during acquisition served as the covariate. The analysis of covariance during the acquisition phase was used to clarify interpretation of the trend of the effects. It was found that the intertrial interval had an effect during day 1, and the effect continued during day 2; however, the latter effect was almost entirely predictable from that observed on day 1. Zimbardo and Miller (1958) reported an experiment in which covariance analysis is employed for essentially the same purpose.

Payne (1958) reports an experiment that is a special case of a $2 \times 2 \times 2$

factorial experiment. The plan of this experiment may be represented as follows:

Method of initial learning	Method of relearning	Drugs	
		c_1	c_2
a_1	b_1 b_2	G_{111} G_{121}	G_{112} G_{122}
a_2	b_1 b_2	G_{211} G_{221}	G_{212} G_{222}

The covariate in this study was the logarithm of the number of trials required for initial learning; the criterion was the logarithm of the number of trials required for relearning. For example, the subjects in group G_{121} learned initially under method a_1. Relearning was done under method b_2 while the subjects were under the influence of drug c_1. The use of covariance analysis in this case is to adjust the relearning data for the linear effect of the initial learning.

Cotton, Lewis, and Metzger (1958) report an experiment which is a $3 \times 2 \times 2$ factorial. The type of apparatus in which a behavior pattern was acquired defined one factor. A second factor was defined by the type of apparatus in which the behavior pattern was extinguished. The time of restriction in a goal box defined the third factor. The covariate for the extinction data was the score on the last five trials of the acquisition phase.

10.6 Factorial Experiment—Repeated Measures

The design to be considered in this section is the analog of the split-plot design in agricultural research. The design may be represented schematically as shown at the left of Table 10.6-1. Assume that there are n sub-

Table 10.6-1 Factorial Experiment, Repeated Measures

					Source	SS	df
	b_1	b_2	$\cdots$	b_q	Between subjects		$np - 1$
					A	A_{yy}	$p - 1$
a_1	G_1	G_1	$\cdots$	G_1	Subj w. gp	P_{yy}	$p(n - 1)$
a_2	G_2	G_2	$\cdots$	G_2	Within subjects		$np(q - 1)$
$\cdot$	$\cdot$	$\cdot$		$\cdot$	B	B_{yy}	$q - 1$
$\cdot$	$\cdot$	$\cdot$		$\cdot$	AB	AB_{yy}	$(p - 1)(q - 1)$
a_p	G_p	G_p	$\cdots$	G_p	Residual	E_{yy}	$p(q - 1)(n - 1$
							$npq - 1$

jects in each of the groups. At the right is the usual analysis of variance for this design when there is no covariate.

Adjustment procedures depend upon whether or not the between- and within-subject regression coefficients can be considered homogeneous. In the first part of the discussion that follows these regression coefficients are not assumed homogeneous.

Aside from the matter of homogeneity of the between- and within-subject regressions, two cases of this design need to be distinguished. In case (1) there is a single covariate measure associated with all the criterion scores for an individual. The data for subject i may be represented as follows:

		b_1	b_2	$\cdots$	b_q
Subject i	X_i	Y_{i1}	Y_{i2}	$\cdots$	Y_{iq}

Here the covariate score X_i is a measure taken before the administration of any of the treatments. Hence the same X_i is paired with all criterion scores on subject i. In contrast, for case (2) the covariate measure is taken just before, just after, or simultaneously with the criterion measure. The data for subject i in this case may be represented as follows:

	b_1		b_2		$\cdots$	b_q	
Subject i	X_{i1}	Y_{i1}	X_{i2}	Y_{i2}	$\cdots$	X_{iq}	Y_{iq}

Thus for case (2) each criterion measure on subject i is paired with a unique covariate measure. Case (1) may be considered as a special case of case (2) in which all the X_{ij}'s for subject i are equal. Hence computational procedures for case (2) may be used for case (1).

Under case (2), both the between-subject (whole-plot) comparisons and the within-subject (split-plot) comparisons are adjusted for the effect of the covariate. Under case (1), only the between-subject (whole-plot) comparisons are adjusted for the effect of the covariate—the within-subject (split-plot) comparisons will all have adjustments which are numerically equal to zero.

The notation that will be used is essentially that defined in Sec. 10.4. Two additional symbols are required.

$P_{x_{i(j)}}$ = sum of the q observations on the covariate for subject i in group G_j.
$P_{y_{i(j)}}$ = sum of the q observations on the criterion for subject i in group G_j.

The following variations and covariations are associated with differences between subjects within the groups:

$$P_{xx} = q\Sigma\Sigma(\bar{P}_{x_{i(j)}} - \bar{A}_{x_j})^2,$$
$$P_{xy} = q\Sigma\Sigma(\bar{P}_{x_{i(j)}} - \bar{A}_{x_j})(\bar{P}_{y_{i(j)}} - \bar{A}_{y_j}),$$
$$P_{yy} = q\Sigma\Sigma(\bar{P}_{y_{i(j)}} - \bar{A}_{y_j})^2.$$

The linear regression equation for the subjects-within-group data has the following form:

$$\bar{P}'_{y_{i(j)}} = b_p(\bar{P}_{x_{i(j)}} - \bar{A}_{x_j}) + \bar{A}_{y_j},$$

where b_p is given by

$$b_p = \frac{P_{xy}}{P_{xx}}.$$

The variation of the residuals about this regression line is

$$P'_{yy} = P_{yy} - \frac{P_{xy}^2}{P_{xx}}.$$

The mean square corresponding to this latter variation is the adjusted error for between-subject effects.

The variations and covariations associated with the treatment effects are estimated by the same procedures as those used in Sec. 10.4. The error terms for within-subject effects in this design have the following form:

$$\text{Residual } (x) = \text{within cell } (x) - P_{xx}.$$

Thus

$$E_{xx} = \Sigma(X_{ijk} - \overline{AB}_{x_{jk}})^2 - P_{xx},$$
$$E_{xy} = \Sigma[X_{ijk}Y_{ijk} - (\overline{AB}_{x_{jk}})(\overline{AB}_{y_{jk}})] - P_{xy},$$
$$E_{yy} = \Sigma(Y_{ijk} - \overline{AB}_{y_{jk}})^2 - P_{yy}.$$

The regression coefficient for the within-subject effects is given by

$$b_w = \frac{E_{xy}}{E_{xx}}.$$

The residuals about the regression line for the within-subject effects have variation equal to

$$E'_{yy} = E_{yy} - \frac{E_{xy}^2}{E_{xx}}.$$

An outline of the analysis of covariance is given in Table 10.6-2.

Computational procedures for this design differ only slightly from those given in Table 10.5-1. One additional set of computational symbols is required for the case of repeated measures.

$$(6x) = \frac{(\Sigma P_x^2)}{q}, \qquad (6xy) = \frac{\Sigma(P_xP_y)}{q}, \qquad (6y) = \frac{(\Sigma P_y^2)}{q}.$$

In each case the summation is over all possible values of the totals. A more complete notation for these totals is $P_{x_{i(j)}}$, each total being over q observations. In terms of the symbols defined in Table 10.5-1 and the

Table 10.6-2 Covariance Analysis for Design in Table 10.6-1

Source	X^2	XY	Y^2	Adjusted variation	df
A	A_{xx}	A_{xy}	A_{yy}	$A'_{yy} = (A_{yy} + P_{yy}) - \frac{(A_{xy} + P_{xy})^2}{A_{xx} + P_{xx}} - P'_{yy}$	$p - 1$
Subj w. gp	P_{xx}	P_{xy}	P_{yy}	$P'_{yy} = P_{yy} - (P^2_{xy}/P_{xx})$	$p(n - 1) - 1$
B	B_{xx}	B_{xy}	B_{yy}	$B'_{yy} = (B_{yy} + E_{yy}) - \frac{(B_{xy} + E_{xy})^2}{B_{xx} + E_{xx}} - E'_{yy}$	$q - 1$
AB	AB_{xx}	AB_{xy}	AB_{yy}	$AB'_{yy} = (AB_{yy} + E_{yy}) - \frac{(AB_{xy} + E_{xy})^2}{AB_{xx} + E_{xx}} - E'_{yy}$	$(p - 1)(q - 1)$
Residual	E_{xx}	E_{xy}	E_{yy}	$E'_{yy} = E_{yy} - (E^2_{xy}/E_{xx})$	$p(q - 1)(n - 1) - 1$

Note: If a single covariate measure is associated with all scores on a given subject, the following sums of squares will be equal to zero:

$$B_{xx}, \quad AB_{xx}, \quad E_{xx}, \quad B_{xy}, \quad AB_{xy}, \quad E_{xy}.$$

set of symbols defined here,

$$\begin{aligned}
P_{xx} &= (6x) - (3x), \\
P_{xy} &= (6xy) - (3xy), \\
E_{xx} &= (2x) - (5x) - (6x) + (3x), \\
E_{xy} &= (2xy) - (5xy) - (6xy) + (3xy), \\
P_{yy} &= (6y) - (3y), \\
E_{yy} &= (2y) - (5y) - (6y) + (3y).
\end{aligned}$$

The variation associated with the treatment effects is defined in Table 10.5-1.

The adjusted mean for level a_j has the form

$$\bar{A}'_{y_j} = \bar{A}_{y_j} - b_p(\bar{A}_{x_j} - \bar{G}_x).$$

The effective error variance for a difference between two adjusted means in this case is

$$s_p^2\left[\frac{2}{nq} + \frac{(\bar{A}_{x_j} - \bar{A}_{x_m})^2}{P_{xx}}\right],$$

where

$$s_p^2 = \frac{P'_{yy}}{p(n-1) - 1}.$$

An adjusted mean for level b_j has the form

$$\bar{B}'_{y_k} = \bar{B}_{y_k} - b_w(\bar{B}_{x_k} - \bar{G}_x).$$

The effective error variance for the difference between two adjusted means in this case is

$$s_w^2\left[\frac{2}{np} + \frac{(\bar{B}_{x_k} - \bar{B}_{x_m})^2}{E_{xx}}\right],$$

where

$$s_w^2 = \frac{E'_{yy}}{p(q-1)(n-1) - 1}.$$

An adjusted cell mean is given by

$$\overline{AB}'_{y_{jk}} = \overline{AB}_{y_{jk}} - b_p(\bar{A}_{x_j} - \bar{G}_x) - b_w(\overline{AB}_{x_{jk}} - \bar{A}_{x_j}).$$

The effective error for the difference between two adjusted cell means which are at the same level of factor A is estimated by

$$s_w^2\left[\frac{2}{n} + \frac{(\overline{AB}_{x_{jk}} - \overline{AB}_{x_{jm}})^2}{E_{xx}}\right].$$

The difference between two adjusted cell means which are not at the same level of factor A has the following form:

$$\begin{aligned}
\overline{AB}'_{y_{jk}} - \overline{AB}'_{y_{ms}} = \overline{AB}_{y_{jk}} - \overline{AB}_{y_{ms}} &- b_p(\bar{A}_{x_j} - \bar{A}_{x_m}) \\
&- b_w(\overline{AB}_{x_{jk}} - \bar{A}_{x_j} - \overline{AB}_{x_{ms}} + \bar{A}_{x_m}).
\end{aligned}$$

Since this difference involves both between- and within-subject effects, the effective error variance is somewhat more complex. The latter variance is estimated by

$$\frac{2[s_p^2 + (q-1)s_w^2]}{nq} + \frac{(\bar{A}_{x_j} - \bar{A}_{x_m})^2 s_p^2}{P_{xx}} + \frac{(\overline{AB}_{x_{jk}} - \bar{A}_{x_j} - \overline{AB}_{x_{ms}} + \bar{A}_{x_m})^2 s_w^2}{E_{xx}}.$$

If the regression coefficients for between-subject (β_p) and within-subject (β_w) effects are equal, the within-subject regression may be used throughout in making the adjustments. When the treatments do not affect the covariate, it is reasonable to expect that the between- and within-subject regressions will be equal. A test on the hypothesis that $\beta_p = \beta_w$ is given by

$$t' = \frac{b_p - b_w}{\sqrt{s_1^2 + s_2^2}},$$

where $s_1^2 = s_p^2/P_{xx}$ and $s_2^2 = s_w^2/E_{xx}$. s_1^2 and s_2^2 are the respective error variances for b_p and b_w. Since the variances in the denominator of this t' statistic will not in general be homogeneous, the sampling distribution of t' is not that of the usual t statistic. If the degrees of freedom for s_p^2 and s_w^2 are both larger than 20, the normal distribution $N(0,1)$ may be used to approximate the sampling distribution of t'. In other cases the sampling distribution of t' may be approximated by the usual t distribution with degrees of freedom f,

$$f = \frac{(s_1^2 + s_2^2)^2}{(s_1^4/f_p) + (s_2^4/f_w)},$$

where f_p and f_w are the respective degrees of freedom for s_p^2 and s_w^2.

When it can be assumed that $\beta_p = \beta_w$, the analysis of covariance has the form given in Table 10.6-3. All the adjustments for within-subject effects are identical to those given in Table 10.6-2. The adjustment procedures for the between-subject effects are, however, different. For purposes of making overall tests, there is some indication that the adjustments for between-subject effects given at the bottom of this table are to be preferred to those indicated at the top. The two adjustment procedures are not algebraically equivalent.

Adjusted means for within-subject effects are identical to those given earlier in this section. The adjusted mean for level a_j now has the form

$$\bar{A}'_{y_j} = \bar{A}_{y_j} - b_w(\bar{A}_{x_j} - \bar{G}_x).$$

The error variance for the difference between $\bar{A}'_{y_j}$ and $\bar{A}'_{y_m}$ in this case is approximately

$$s_p'^2\left[\frac{2}{nq} + \frac{(\bar{A}_{x_j} - \bar{A}_{x_m})^2}{E_{xx}}\right].$$

Table 10.6-3 Covariance Analysis When $\beta_p = \beta_w$

Source	Adjusted variation	df	MS
A	$A''_{yy} = (A_{yy} + E_{yy}) - \dfrac{(A_{xy} + E_{xy})^2}{A_{xx} + E_{xx}} - E'_{yy}$	$p - 1$	
Subj w. gp	$P''_{yy} = (P_{yy} + E_{yy}) - \dfrac{(P_{xy} + E_{xy})^2}{P_{xx} + E_{xx}} - E'_{yy}$	$p(n - 1)$	s'^2_p
B	$B'_{yy} = (B_{yy} + E_{yy}) - \dfrac{(B_{xy} + E_{xy})^2}{(B_{xx} + E_{xx})} - E'_{yy}$	$q - 1$	
AB	$AB'_{yy} = (AB_{yy} + E_{yy}) - \dfrac{(AB_{xy} + E_{xy})^2}{(AB_{xx} + E_{xx})} - E'_{yy}$	$(p - 1)(q - 1)$	
Residual	$E'_{yy} = E_{yy} - (E^2_{xy}/E_{xx})$	$p(q - 1)(n - 1) - 1$	s^2_w
A	$A''_{yy} = A_{yy} - 2b_w A_{xy} + b^2_w A_{xx}$	$p - 1$	
Subj w. A	$P''_{yy} = P_{yy} - 2b_w P_{xy} + b^2_w P_{xx}$	$p(n - 1)$	s'^2_p

An adjusted cell mean in this case is

$$\overline{AB}'_{y_{jk}} = \overline{AB}_{y_{jk}} - b_w(\overline{AB}_{x_{yk}} - \bar{G}_x).$$

The error variance for the difference between two adjusted cell means which are at the same level of factor A is identical to that given earlier in this section—this difference is a within-subject effect. The error variance for the difference between two adjusted cell means which are not at the same level of A is approximately

$$\frac{2[s_p'^2 + (q-1)s_w^2]}{nq} + \frac{(\overline{AB}_{x_{jk}} - \overline{AB}_{x_{ms}})^2 s_w^2}{E_{xx}}.$$

Numerical Example. The data in Table 10.6-4 will be used to illustrate the computational procedures for this design. Disregarding the covariate, part i represents data that would be obtained in a 2 × 2 factorial experiment having repeated measures on factor B. There are four subjects under each level of factor A. Suppose that the covariate measure on a given sub-

Table 10.6-4 Numerical Example

		Person	b_1		b_2		Total	
			X	Y	X	Y	X	Y
(i)	a_1	1	3	10	3	8	6	18
		2	5	15	5	12	10	27
		3	8	20	8	14	16	34
		4	2	12	2	6	4	18
	a_2	5	1	15	1	10	2	25
		6	8	25	8	20	16	45
		7	10	20	10	15	20	35
		8	2	15	2	10	4	25
		Total	39	132	39	95	78	227

		b_1		b_2		Total	
		X	Y	X	Y	X	Y
(ii)	a_1	18	57	18	40	36	97
	a_2	21	75	21	55	42	130
	Total	39	132	39	95	78	227

(iii)	(1x) = 380.25	(1xy) = 1106.62	(1y) = 3220.56
	(2x) = 542	(2xy) = 1282	(2y) = 3609
	(3x) = 382.50	(3xy) = 1119.00	(3y) = 3288.62
	(4x) = 380.25	(4xy) = 1106.62	(4y) = 3306.12
	(5x) = 382.50	(5xy) = 1119.00	(5y) = 3374.75
	(6x) = 542.00	(6xy) = 1282.00	(6y) = 3516.50

ject is obtained before the administration of any of the treatments; then the covariate measure for a given subject will be a constant for both levels of factor B. Thus the data in Table 10.6-4 represent case (1) of this design.

An AB summary table for both the variate and covariate appears in part ii. With the exception of symbols $(6x)$, $(6xy)$, and $(6y)$, the computational symbols given in part iii are defined in part i of Table 10.5-1. Symbols containing the number (6) were defined earlier in this section. Not all the symbols in part iii are required in the analysis of data for case (1), since within-subject (split-plot) adjustments will be zero; to show, however, that such adjustments will be zero, the complete analysis will be illustrated.

The computation of the variation for the between- and within-subject effects is illustrated in Table 10.6-5. By inserting x, xy, or y in the symbols given at the left of part i, one obtains the variations given under the headings X^2, XY, and Y^2, respectively. For example,

$$\begin{aligned} A_{xx} &= (3x) - (1x) = 2.25, \\ A_{xy} &= (3xy) - (1xy) = 12.38, \\ A_{yy} &= (3y) - (1y) = 68.06. \end{aligned}$$

The adjusted value for the between-subject error term is given by

$$P'_{yy} = P_{yy} - \frac{P^2_{xy}}{P_{xx}} = 227.88 - \frac{163.00^2}{159.50} = 61.30.$$

The adjusted value for the variation due to the main effect of factor A is given by

$$\begin{aligned} A'_{yy} &= (A_{yy} + P_{yy}) - \frac{(A_{xy} + P_{xy})^2}{A_{xx} + P_{xx}} - P'_{yy} \\ &= (68.06 + 227.88) - \frac{(12.38 + 163.00)^2}{2.25 + 159.50} - 61.30 \\ &= 44.48. \end{aligned}$$

The analysis of variance as well as the analysis of covariance are summarized in part ii of Table 10.6-5. Only the between-subject (whole-plot) comparisons will have nonzero adjustments. If one were to compute the adjusted value for the main effect due to factor B, the adjusted value would be

$$\begin{aligned} B'_{yy} &= (B_{yy} + E_{yy}) - \frac{(B_{xy} + E_{xy})^2}{B_{xx} + E_{xx}} - E'_{yy} \\ &= (85.56 + 6.37) - \frac{(0 + 0)^2}{0 + 0} - 6.37 = 85.56 = B_{yy}. \end{aligned}$$

That these adjustments should be zero follows from the fact that the covariate measure is constant for all criterion measures on the same subject.

It should be noted that the error mean square for the between-subject (whole-plot) comparisons is 37.98 in the analysis of variance. In the analysis of covariance the corresponding error mean square is 12.26. Hence the covariate adjustment reduces the error mean square for between-subject comparisons from 37.98 to 12.26. This relatively large reduction in the error mean square is a function of the magnitude of the within-class correlation. The square of this latter correlation is estimated by

$$r_p^2 = \frac{P_{xy}^2/P_{xx}}{P_{yy}} = .73.$$

The mean square due to the main effect of factor A is reduced from 68.06 to 44.48 in the adjustment process.

Under the analysis of variance, the F ratio for the test on the main effect of factor A is 1.79, with degrees of freedom 1 and 6; the corresponding F ratio under the analysis of covariance is 3.63, with degrees of freedom 1 and 5.

Table 10.6-5 Analysis of Variance and Covariance

			X^2	XY	Y^2	$(Y')^2$
(i)	A	(3) − (1)	2.25	12.38	68.06	44.48
	Subj w. A	(6) − (3)	159.50	163.00	227.88	61.30
	B	(4) − (1)	0	0	85.56	
	AB	(5) − (3) − (4) + (1)	0	0	.57	
	Residual	(2) − (5) − (6) + (3)	0	0	6.37	

	Source of variation	SS	df	MS	F
(ii)	A	68.06	1	68.06	1.79
	Subj w. A	227.88	6	37.98	
	B	85.56	1	85.56	80.72
	AB	.57	1	.57	
	Residual	6.37	6	1.06	
	A (adj)	44.48	1	44.48	3.63
	Subj w. A (adj)	61.30	5	12.26	

Another Numerical Example. The data in Table 10.6-6 will be used to illustrate a design in which there are distinct covariate measures for each observation. The design in this table may be considered as a special case of a 3×2 factorial experiment with repeated measures on factor B. There are $n = 3$ subjects in each group.

Table 10.6-6 Numerical Example

	Subject	b_1		b_2		Total	
		X	Y	X	Y	X	Y
	1	3	8	4	14	7	22
a_1	2	5	11	9	18	14	29
	3	11	16	14	22	25	38
		19	35	27	54	46	89
	4	2	6	1	8	3	14
a_2	5	8	12	9	14	17	26
	6	10	9	9	10	19	19
		20	27	19	32	39	59
	7	7	10	4	10	11	20
a_3	8	8	14	10	18	18	32
	9	9	15	12	22	21	37
		24	39	26	50	50	89
Total		63	101	72	136	135	237

$(1x) = 1012.50$	$(1xy) = 1777.50$	$(1y) = 3120.50$
$(2x) = 1233$	$(2xy) = 2004$	$(2y) = 3495$
$(3x) = 1022.83$	$(3xy) = 1807.50$	$(3y) = 3220.50$
$(4x) = 1017.00$	$(4xy) = 1795.00$	$(4y) = 3188.56$
$(5x) = 1034.33$	$(5xy) = 1835.67$	$(5y) = 3305.00$
$(6x) = 1207.50$	$(6xy) = 1964.00$	$(6y) = 3397.50$

Computational symbols obtained in part ii are defined in part i of Table 10.5-1. Symbol (6) has the following definition:

$$(6x) = \frac{(\Sigma P_x^2)}{q} = \frac{7^2 + 14^2 + \cdots + 18^2 + 21^2}{2} = 1207.50,$$

$$(6xy) = \frac{(\Sigma P_x P_y)}{q} = \frac{7(22) + 14(19) + \cdots + 21(37)}{2} = 1964.00,$$

$$(6y) = \frac{(\Sigma P_y^2)}{q} = \frac{22^2 + 29^2 + \cdots + 32^2 + 37^2}{2} = 3397.50.$$

The sums of squares and sums of products for the variations and covariations are summarized in part i of Table 10.6-7. The analysis of variance is summarized in part ii. The analysis of covariance is summarized in part iii. The adjusted variation is defined in Table 10.6-2. For example,

$$E'_{yy} = E_{yy} - \frac{E^2_{xy}}{E_{xx}} = 13.00 - \frac{(11.83)^2}{14.00} = 3.00,$$

$$B'_{yy} = (B_{yy} + E_{yy}) - \frac{(B_{xy} + E_{xy})^2}{B_{xx} + E_{xx}} - E'_{yy}$$

$$= (68.06 + 13.00) - \frac{(17.50 + 11.83)^2}{4.50 + 14.00} - 3.00 = 31.56.$$

The between-subject effects in part iii are also adjusted in accordance with the procedures given in Table 10.6-2. The estimates of β_p and β_w are

Table 10.6-7 Analysis of Variance and Covariance

			X^2	XY	Y^2
(i)	A	(3) − (1)	10.33	30.00	100.00
	Subj w. A	(6) − (3)	184.67	156.50	177.00
	B	(4) − (1)	4.50	17.50	68.06
	AB	(5) − (3) − (4) + (1)	7.00	10.67	16.44
	B × (subj w. A)	(2) − (5) − (6) + (3)	14.00	11.83	13.00
	Total	(2) − (1)	220.50	226.50	374.50

	Source	SS	df	MS	F
(ii)	Between subjects				
	A	100.00	2	50.00	1.69
	Subj w. A	177.00	6	29.50	
	Within subjects				
	B	68.06	1	68.06	31.36
	AB	16.44	2	8.22	3.78
	B × (subj w. A)	13.00	6	2.17	

	Source (adjusted)	SS	df	MS	F
(iii)	Between subjects				
	A'	54.26	2	27.13	3.06
	Subj w. A (P'_{yy})	44.37	5	8.87	
	Within subjects				
	B'	31.56	1	31.56	52.60
	AB'	2.33	2	1.16	1.93
	B × (subj w. A) (E'_{yy})	3.00	5	0.600	

$$b_p = \frac{P_{xy}}{P_{xx}} = \frac{156.50}{184.67} = .847,$$

$$b_w = \frac{E_{xy}}{E_{xx}} = \frac{11.83}{14.00} = .845.$$

In this case, inspection indicates that b_w may be used throughout in the adjustment process. If the adjustment of the between-subject effects is made in accordance with procedures given at the bottom of Table 10.6-3, the adjusted sums of squares are as follows:

$$\begin{aligned} A'_{yy} &= A_{yy} - 2b_w A_{xy} + b_w^2 A_{xx} \\ &= 100.00 - 2(.845)(30.00) + (.845)^2(10.33) \\ &= 56.68, \\ P'_{yy} &= P_{yy} - 2b_w P_{xy} + b_w^2 P_{xx} \\ &= 177.00 - 2(.845)(156.50) + (.845)^2(184.67) \\ &= 44.36. \end{aligned}$$

The overall F test for the hypothesis $\sigma_\alpha^2 = 0$ has the form

$$F = \frac{A'_{yy}/(p-1)}{P'_{yy}/p(n-1)} = \frac{56.68/2}{44.36/6} = 3.83.$$

It is noted that the error term for the within-subject effects is 2.17 in the analysis of variance and 0.600 in the analysis of covariance. For the between-subject effects the corresponding error terms are 29.50 and 44.36/6 = 7.39.

Illustrative Application. Suppose that an experimenter is interested in evaluating the effect of two drugs upon blood pressure. Suppose further that the experimenter desires to evaluate effect upon blood pressure when a linear adjustment is made for the effect of the drugs upon pulse rate. The covariate, pulse rate, is in part affected by the drugs. In the design, a random sample of $2n$ subjects is divided at random into two groups, each of size n. Subjects in group 1 receive drug 1 first. After the drug has had a chance to act, blood pressure and pulse rate are taken on each subject. After the effect of drug 1 has dissipated, subjects in group 1 are given drug 2. Blood pressure and pulse rate are then obtained under the latter drug. Subjects in group 2 are given the drugs in the reverse order. The design may be represented schematically as follows:

	Drug	
	b_1	b_2
a_1	G_1	G_1
a_2	G_2	G_2

The analysis of covariance for this design will permit the experimenter to evaluate the effect of the drugs upon blood pressure after a linear adjustment has been made for the effect of the drugs upon pulse rate. In this type of experiment, one or more levels of the drug factor may represent control conditions.

10.7 Multiple Covariates

The principles developed for the case of a single covariate generalize to more than one covariate. In this section this generalization will be considered in the setting of a single-factor experiment. However, multiple covariates may be used with any of the experimental designs for which there is a corresponding analysis of variance.

Suppose that there are two covariates X and Z, and a single criterion Y. The observations under treatment j is a single-factor experiment may be represented as follows:

Treatment j		
X_{1j}	Z_{1j}	Y_{1j}
X_{2j}	Z_{2j}	Y_{2j}
$\cdot$	$\cdot$	$\cdot$
$\cdot$	$\cdot$	$\cdot$
$\cdot$	$\cdot$	$\cdot$
X_{ij}	Z_{ij}	Y_{ij}
$\cdot$	$\cdot$	$\cdot$
$\cdot$	$\cdot$	$\cdot$
$\cdot$	$\cdot$	$\cdot$
X_{nj}	Z_{nj}	Y_{nj}

In the analysis of covariance, the criterion data are adjusted for the linear effects of both X and Z through use of a multiple regression equation. The multiple regression equation obtained from the data in class j has the form

$$Y'_{ij} = b_{x_j}(X_{ij} - \bar{X}_j) + b_{z_j}(Z_{ij} - \bar{Z}_j) + \bar{Y}_j.$$

If the within-class regressions are homogeneous, pooled within-class regression coefficients may be computed. The multiple regression equation in terms of the pooled within-class regression coefficients has the form

$$Y'_{ij} = b_x(X_{ij} - \bar{X}_j) + b_z(Z_{ij} - \bar{Z}_j) + \bar{Y}_j.$$

The pooled within-class variances and covariances are defined as follows:

$$\begin{aligned} E_{xx} &= \Sigma\Sigma(X_{ij} - \bar{X}_j)^2, & E_{zz} &= \Sigma\Sigma(Z_{ij} - \bar{Z}_j)^2, \\ E_{xy} &= \Sigma\Sigma(X_{ij} - \bar{X}_j)(Y_{ij} - \bar{Y}_j), & E_{zy} &= \Sigma\Sigma(Z_{ij} - \bar{Z}_j)(Y_{ij} - \bar{Y}_j), \\ E_{yy} &= \Sigma\Sigma(Y_{ij} - \bar{Y}_j)^2, & E_{xz} &= \Sigma\Sigma(X_{ij} - \bar{X}_j)(Z_{ij} - \bar{Z}_j). \end{aligned}$$

In terms of these variances and covariances, the pooled within-class regression coefficients are given by

$$b_x = \frac{E_{zz}E_{xy} - E_{xz}E_{zy}}{d},$$

$$b_z = \frac{E_{xx}E_{zy} - E_{xz}E_{xy}}{d},$$

where $d = E_{xx}E_{zz} - E^2_{xz}$. The sum of the squares of the residuals about this regression line is

$$E'_{yy} = E_{yy} - b_xE_{xy} - b_zE_{zy}.$$

The treatment variations and covariations are defined as follows:

$$T_{xx} = n\Sigma(\bar{T}_{x_j} - \bar{G}_x)^2, \qquad T_{zz} = n\Sigma(\bar{T}_{z_j} - \bar{G}_z)^2,$$

$$T_{xy} = n\Sigma(\bar{T}_{x_j} - \bar{G}_x)(\bar{T}_{y_j} - \bar{G}_y), \qquad T_{zy} = n\Sigma(\bar{T}_{z_j} - \bar{G}_z)(\bar{T}_{y_j} - \bar{G}_y),$$

$$T_{yy} = n\Sigma(\bar{T}_{y_j} - \bar{G}_y)^2, \qquad T_{xz} = n\Sigma(\bar{T}_{x_j} - \bar{G}_x)(\bar{T}_{z_j} - \bar{G}_z).$$

To obtain the reduced treatment sum of squares, one combines corresponding treatment and error variations and covariations, computes a multiple regression equation for the combined data, and computes the variation of the residuals about this equation. The reduced treatment variation is obtained by subtracting the adjusted error variation from the variation of the residuals in the latter regression equation. The combined variations and covariations are as follows:

$$S_{xx} = T_{xx} + E_{xx}, \qquad S_{xz} = T_{xz} + E_{xz}, \qquad S_{xy} = T_{xy} + E_{xy},$$

$$S_{yy} = T_{yy} + E_{yy}, \qquad S_{zz} = T_{zz} + E_{zz}, \qquad S_{zy} = T_{zy} + E_{zy}.$$

The multiple regression equation associated with these combined data has the following form:

$$Y''_{ij} = b''_x(X_{ij} - \bar{X}) + b''_z(Z_{ij} - \bar{Z}) + \bar{Y}.$$

For the case of a single-factor experiment, this regression line is the one that would be obtained if the treatment classes were disregarded and an overall least-squares fit obtained for the entire set of experimental data. The regression coefficients are given by

$$b''_x = \frac{S_{zz}S_{xy} - S_{xz}S_{zy}}{d''},$$

$$b''_z = \frac{S_{xx}S_{zy} - S_{xz}S_{xy}}{d''},$$

where $d'' = S_{xx}S_{zz} - S^2_{xz}$.

The sum of the squares of the residuals about this regression line is

$$S'_{yy} = (T + E)'_{yy} = S_{yy} - b''_xS_{xy} - b''_zS_{zy}.$$

The reduced sum of squares for treatments is

$$T_{yyR} = (T + E)'_{yy} - E'_{yy}.$$

The analysis of covariance for this type of experiment is summarized in Table 10.7-1. Since two regression coefficients are estimated from the

Table 10.7-1 Analysis of Covariance, Two Covariates

Source	SS	df	MS	F
Treatments	T_{yyR}	$k - 1$	$\text{MS}'_{\text{treat}}$	$F = \dfrac{\text{MS}'_{\text{treat}}}{\text{MS}'_{\text{error}}}$
Error	E'_{yy}	$k(n - 1) - 2$	$\text{MS}'_{\text{error}}$	
Total	S'_{yy}	$kn - 3$		

experimental data to obtain the adjusted error, the degrees of freedom for the latter source of variation are $k(n - 1) - 2$.

To compute the variations and covariations, the procedures in parts i through iv of Table 10.3-1 may be extended to cover any number of covariates. For example,

$$E_{zz} = (2z) - (3z), \qquad E_{xz} = (2xz) - (3xz),$$
$$T_{zz} = (3z) - (1z), \qquad T_{xz} = (3xz) - (1xz).$$

In order to obtain the adjusted criterion variations, the pooled within-class as well as the overall regression coefficients are required. Given the variations and covariations, the regression coefficients are obtained from the relations given in this section.

The adjusted mean for treatment j is

$$\bar{Y}'_j = \bar{Y}_j - b_x(\bar{X}_j - \bar{X}) - b_z(\bar{Z}_j - \bar{Z}).$$

The average effective error variance per experimental unit in making comparisons among adjusted treatment means is

$$s'^2_y = s'^2_{\text{error}}\left[1 + \frac{T_{xx}E_{zz} - 2T_{xz}E_{xz} + T_{zz}E_{xx}}{(k - 1)(E_{xx}E_{zz} - E^2_{xz})}\right],$$

where s'^2_{error} is the mean square for error in the analysis of covariance. The average effective error variance for the difference between two adjusted treatment means is

$$s'^2_{\bar{Y}'_j - \bar{Y}'_k} = \frac{2s'^2_y}{n}.$$

The effective error variance for the difference between two adjusted treatment means for a test in the a priori category is

$$s'^2_{\text{error}}\left[\frac{2}{n} + \frac{(\bar{X}_j - \bar{X}_k)^2E_{zz} - 2(\bar{X}_j - \bar{X}_k)(\bar{Z}_j - \bar{Z}_k)E_{xz} + (\bar{Z}_j - \bar{Z}_k)^2E_{xx}}{E_{xx}E_{zz} - E^2_{xz}}\right].$$

When the treatments affect the covariate, the last expression for the error variance is the more appropriate for use in testing differences between adjusted treatment means.

For the case of p covariates, expressions for the regression coefficients are most easily written in matrix notation. Suppose that the criterion is designated X_0 and the covariates $X_1, X_2, \ldots, X_p$. Let

$$\underline{e}_0' = [E_{01} \quad E_{02} \quad \cdots \quad E_{0p}], \qquad \underline{s}_0' = [S_{01} \quad S_{02} \quad \cdots \quad S_{0p}],$$

$$E = \begin{bmatrix} E_{11} & E_{12} & \cdots & E_{1p} \\ E_{21} & E_{22} & \cdots & E_{2p} \\ \cdot & \cdot & & \cdot \\ \cdot & \cdot & & \cdot \\ \cdot & \cdot & & \cdot \\ E_{p1} & E_{p2} & \cdots & E_{pp} \end{bmatrix}, \qquad S = \begin{bmatrix} S_{11} & S_{12} & \cdots & S_{1p} \\ S_{21} & S_{22} & \cdots & S_{2p} \\ \cdot & \cdot & & \cdot \\ \cdot & \cdot & & \cdot \\ \cdot & \cdot & & \cdot \\ S_{p1} & S_{p2} & \cdots & S_{pp} \end{bmatrix},$$

$$\underline{b}_w' = [b_1 \quad b_2 \quad \cdots \quad b_p], \qquad \underline{b}_s' = [b_1'' \quad b_2'' \quad \cdots \quad b_p''].$$

In terms of these matrices, the regression coefficients for the within-class data are

$$\underline{b}_w = E^{-1}\underline{e}_0.$$

The regression coefficients for the overall data are

$$\underline{b}_s = S^{-1}\underline{s}_0.$$

The error and treatment variations are as follows:

$$\begin{aligned} E_{yy}' &= E_{yy} - \underline{b}_w'\underline{e}_0 \\ &= E_{yy} - \underline{e}_0'E^{-1}\underline{e}_0, \\ T_{yyR} &= S_{yy} - \underline{b}_s'\underline{s}_0 - E_{yy}'. \end{aligned}$$

Assuming n observations under each of k treatments, the analysis of covariance has the following form:

Source	SS	df
Total	$S_{yy}' = S_{yy} - \underline{b}_s' \underline{s}_0$	$kn - p - 1$
Error	$E_{yy}' = E_{yy} - \underline{b}_w' \underline{e}_0$	$k(n-1) - p$
Treatments	$T_{yyR} = S_{yy}' - E_{yy}'$	$k - 1$

APPENDIX A

RANDOM VARIABLES

The concept of a *random* variable is basic in modern statistics. In mathematics, a function of a variable is usually a rule whereby one or more values (usually numbers) are associated with different values of the variable. For example, consider the function

$$f(x) = 2x + 5.$$

When

$$x = 0, \qquad f(x) = 5;$$
$$x = 1, \qquad f(x) = 7;$$
$$x = 2, \qquad f(x) = 9.$$

If x is a random variable (or *variate*), the probability density function on x, say $f(x)$, gives the probability that the variate assumes the value x. That is,

$$f(x) = \Pr(x).$$

If one distinguishes between the name of the variate (x) and the values (say x') that the variate can assume, then

$$f(x = x') = \Pr(x = x').$$

In some contexts, $f(x)$ represents a model for the relative frequency (in a series of experiments) with which the variate x will assume specified values. For example, if $x = 0$, 1, or 2, but no other values, a probability density function may be formulated as follows:

$$f(x = 0) = .50 = p_0,$$
$$f(x = 1) = .20 = p_1,$$
$$f(x = 2) = .30 = p_2.$$

If this density function is an appropriate model for a series of n independent experiments, the expected frequencies of the outcomes are as indicated below.

Outcome	Expected frequency
$x = 0$	$np_0 = .50n$
$x = 1$	$np_1 = .20n$
$x = 2$	$np_2 = .30n$

One of the basic problems in statistics is to evaluate the goodness of fit of various models to experimental data.

A.1 Random Variables and Probability Distributions

Associated with a *random* variable is a probability distribution. If the symbol x denotes a random variable, then the symbol $f(x)$ will denote the *probability density* for x. For example, the probability that the random variable x assumes the specific value x_j is given by

$$\Pr(x = x_j) = f(x_j).$$

A probability *measure* on the random variable x is defined by a function having the following properties:

(i) $$0 \le \Pr(x \le x_j) \le 1.$$

(ii) $$\Pr(-\infty \le x \le \infty) = 1.$$

(iii) For $x_k > x_j$, $$\Pr(x_j \le x \le x_k) = \Pr(x \le x_k) - \Pr(x \le x_j).$$

The *distribution function* for a random variable x will be denoted by the symbol $F(x)$. The distribution function, as a probability law, is interpreted as follows:

$$\Pr(x \le x_j) = F(x_j) = \int_{-\infty}^{x_j} f(x)\,dx.$$

That is, the distribution function of a random variable x represents a cumulative probability. In some contexts the "product" $f(x)\,dx$, which can be considered to define the area of a rectangle with height $f(x)$ and width dx, is called the *probability element*.

Any nonnegative function, whose integral over the entire range of the variable in the function is unity, will define a probability density. Conversely, a random variable x is said to have a density function $f(x)$ if

$$\int_{-\infty}^{x_j} f(x)\,dx = \Pr(x \le x_j).$$

The term *distribution* is used rather loosely to refer either to the probability density or the distribution function. The context will generally make it clear which one is implied. The distribution which defines a random variable is called its probability *law*.

It is convenient to define the range of a random variable to be from $-\infty$ to $+\infty$. If, for example, the density function is zero for all values of $x \leq 0$, the probability density may be defined as

$$\begin{aligned} \Pr(x) &= f(x), \qquad & x > 0, \\ &= 0, & x \leq 0. \end{aligned}$$

As another example, if the random variable x has zero probability outside the interval 0 to 1,

$$\begin{aligned} \Pr(x) &= f(x), \qquad & 0 < x < 1, \\ &= 0, & \text{otherwise.} \end{aligned}$$

A random variable may be either continuous or discrete. Only continuous variables are considered in this section. Associated with the distribution of a random variable are one or more *parameters*. The latter are constants which determine certain characteristics of the distribution. For example, suppose

$$\begin{aligned} f(x) &= \gamma e^{-\gamma x}, \qquad & x > 0, \gamma > 0, \\ &= 0, & \text{otherwise.} \end{aligned}$$

For this probability density, the random variable is x, and the parameter is γ. To make explicit the distinction between the variable and the parameter one may write

$$f(x \mid \gamma) = \gamma e^{-\gamma x}, \qquad x > 0, \gamma > 0.$$

For $\gamma = 1$, the probability density takes the form

$$\begin{aligned} f(x \mid \gamma = 1) &= e^{-x}, \qquad & x > 0, \\ &= 0, & \text{otherwise.} \end{aligned}$$

This probability law is shown geometrically in Fig. A.1-1. This example defines what is called the exponential probability law or the exponential distribution for a random variable x. The random variable defined by this density is called an "exponential variable." The exponential distribution is actually a family of distributions—a member of the family is specified by assigning a numerical value to the parameter γ. The distribution function for an exponential variable has the form

$$\Pr(x \leq x_j) = \int_0^{x_j} f(x)\, dx = \int_0^{x_j} \gamma e^{-\gamma x}\, dx.$$

It is sometimes convenient to introduce a dummy variable, say t, and represent the distribution function in the form

$$F(x) = \Pr(t \le x) = \int_0^x f(t)\,dt = \int_0^x \gamma e^{-\gamma t}\,dt.$$

The dummy variable t is called the variable of integration. This notation makes explicit the fact that the value of this integral depends upon the upper limit.

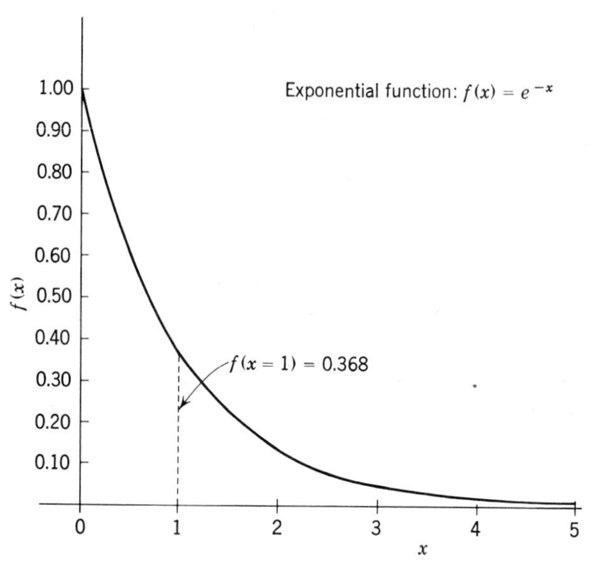

Figure A.1-1 Exponential probability law when $\gamma = 1$.

If a probability density for a random variable x contains the parameters $\theta_1, \ldots, \theta_p$, the following alternative notation systems are used:

$$f(x \mid \theta_1, \ldots, \theta_p), \qquad f_x(\theta_1, \ldots, \theta_p), \qquad f(\theta_1, \ldots, \theta_p),$$

or simply

$$f(x).$$

The random variable itself is called the argument of the function. Which one of the notation systems is used depends upon what the context is trying to emphasize. The corresponding distribution is written in the alternative forms

$$F(x \mid \theta_1, \ldots, \theta_p), \qquad F_x(\theta_1, \ldots, \theta_p), \qquad F(x).$$

Expected Value. Let $F(x)$ be the distribution function for a random variable x. Let $g(x)$ be an arbitrary function on x. If the Riemann-Stieltjes integral

$$\int_{-\infty}^{\infty} g(x)\,dF(x)$$

exists, its value is said to be the *expected value* of $g(x)$. That is,

$$\mathrm{E}[g(x)] = \int_{-\infty}^{\infty} g(x)\, dF(x).$$

The existence of this expected value requires that $g(x)$ be integrable with respect to $F(x)$. Under additional regularity assumptions with respect to $F(x)$, $\mathrm{E}[g(x)]$ may be expressed in terms of the density function. Thus,

$$\mathrm{E}[g(x)] = \int_{-\infty}^{\infty} g(x) f(x)\, dx.$$

If the random variable x is discrete, then the Riemann-Stieltjes integral becomes the sum of products. A requirement for the existence of the expected value is that

$$\int_{-\infty}^{\infty} |g(x)|\, dF(x) < \infty.$$

The *expected value* of a random variable (or, equivalently, the first *moment* of a probability distribution) is, by definition,

$$\alpha_1 = \mu_1 = \mathrm{E}(x) = \int_{-\infty}^{\infty} x f(x)\, dx.$$

The mth *noncentral* (or raw) moment of a probability distribution is

$$\alpha_m = \mathrm{E}(x^m) = \int_{-\infty}^{\infty} x^m f(x)\, dx, \qquad m \geq 1.$$

(m is usually an integer.) The mth *central* moment is

$$\mu_m = \mathrm{E}[x - \mathrm{E}(x)]^m = \int_{-\infty}^{\infty} [x - \mathrm{E}(x)]^m f(x)\, dx, \qquad m > 1.$$

The second central moment is the variance of the distribution.

For example, if

$$f(x) = \gamma e^{-\gamma x},$$

it can be shown that

$$\mu_1 = \mathrm{E}(x) = \int_{-\infty}^{\infty} x f(x)\, dx = \frac{1}{\gamma}.$$

The second central moment is

$$\mu_2 = \mathrm{E}\left[x - \frac{1}{\gamma}\right]^2 = \int_{-\infty}^{\infty} \left[x - \frac{1}{\gamma}\right]^2 f(x)\, dx = \frac{1}{\gamma^2}.$$

For many probability densities, the first and second moments are either equal to corresponding parameters of the distribution or are some relatively simple function of the parameters of the distribution. In general, the parameters of a probability density determine its moments.

Vice versa, a distribution is uniquely determined by its moments. In practice the parameters of a distribution are estimated from the moments of an empirical distribution. If a distribution is symmetric about $\mathrm{E}(x)$, then all odd moments will be zero.

Associated with some probability distributions is what is called a *moment-generating* function. The latter, if it exists, is given by

$$M_x(t) = \mathrm{E}(e^{tx}) = \int_{-\infty}^{\infty} e^{tx} f(x)\, dx,$$

where $f(x)$ is the density function for the random variable x. If $M_x(t)$ is finite for $|t| \leq T$ for some $T > 0$, then $M_x(t)$ may be expressed in the following form (valid for $|t| < T$):

$$M_x(t) = 1 + \mathrm{E}(x)t + \mathrm{E}(x^2)\frac{t^2}{2!} + \cdots + \mathrm{E}(x^m)\frac{t^m}{m!} + \cdots.$$

Thus the mth raw moment is the coefficient of t^m in the series expansion of $M_x(t)$. Equivalently,

$$\mathrm{E}(x^m) = M_x^{(m)}(t = 0), \qquad \text{where } M_x^{(m)}(t = 0)$$

is the mth derivative of $M_x(t)$ with respect to t, evaluated for $t = 0$. It will be found that

$$\begin{aligned} M_x^{(1)}(t) &= \mathrm{E}(xe^{tx}) \qquad \text{which, for } t = 0, \text{ is } \mathrm{E}(x), \\ M_x^{(2)}(t) &= \mathrm{E}(x^2e^{tx}) \qquad \text{which, for } t = 0, \text{ is } \mathrm{E}(x^2), \\ M_x^{(3)}(t) &= \mathrm{E}(x^3e^{tx}) \qquad \text{which, for } t = 0, \text{ is } \mathrm{E}(x^3). \end{aligned}$$

For example, the moment-generating function for the exponential distribution has the relatively simple form

$$g(t) = \left(1 - \frac{t}{\gamma}\right)^{-1}.$$

The first derivative of this function is

$$\frac{dg}{dt} = \frac{1}{\gamma}\left(1 - \frac{t}{\gamma}\right)^{-2}.$$

For $t = 0$,

$$\frac{dg}{dt} = \frac{1}{\gamma} = \mathrm{E}(x).$$

The second derivative of $g(t)$ is

$$\frac{d^2g}{dt} = \frac{2}{\gamma^2}\left(1 - \frac{t}{\gamma}\right)^{-3}.$$

For $t = 0$,

$$\frac{d^2g}{dt^2} = \frac{2}{\gamma^2} = \mathrm{E}(x^2) = \alpha_2.$$

To obtain the second central moment,

$$\mu_2 = \mathrm{E}[x - \mathrm{E}(x)]^2 = \mathrm{E}[x - \mu_1]^2 = \mathrm{E}(x^2) - \mu_1^2$$
$$= \frac{2}{\gamma^2} - \frac{1}{\gamma^2} = \frac{1}{\gamma^2}.$$

The mathematical principle underlying the moment-generating function is a relatively simple one. For suitable choice of t and x, one has the following converging power series:

$$e^{tx} = 1 + \frac{(tx)}{1!} + \frac{(tx)^2}{2!} + \cdots + \frac{(tx)^m}{m!} + \cdots.$$

Hence

$$M_x(t) = \int_{-\infty}^{\infty} e^{tx} f(x)\, dx$$
$$= 1\int_{-\infty}^{\infty} f(x)\, dx + \frac{t}{1!}\int_{-\infty}^{\infty} x f(x)\, dx + \frac{t^2}{2!}\int_{-\infty}^{\infty} x^2 f(x)\, dx$$
$$+ \cdots + \frac{t^m}{m!}\int_{-\infty}^{\infty} x^m f(x)\, dx + \cdots$$
$$= 1 + \frac{t}{1!}\mathrm{E}(x) + \frac{t^2}{2!}\mathrm{E}(x^2) + \cdots + \frac{t^m}{m}\mathrm{E}(x^m) + \cdots$$
$$= \sum_{j=0}^{\infty} \frac{t^j}{j!}\alpha_j, \qquad \text{where } \alpha_j = \mathrm{E}(x^j).$$

For discrete (in contrast to continuous) variables it is convenient to work with the *probability-generating function* rather than with the moment-generating function. For a discrete distribution in which $x = 0, 1, 2, \ldots,$ the probability-generating function is defined to be

$$P(t) = \sum_{j=0}^{\infty} t^j f(x_j)$$
$$= t^0 f(0) + t^1 f(1) + t^2 f(2) + t^3 f(3) + \cdots.$$

If one takes the first and second derivatives of $P(t)$ with respect to t one has

$$\frac{dP}{dt} = f(1) + 2t f(2) + 3t^2 f(3) + 4t^3 f(4) + \cdots,$$
$$\frac{d^2P}{dt^2} = 2f(2) + 6t f(3) + 12t^2 f(4) + \cdots.$$

If one sets $t = 1$ in the expressions given above and defines ϕ_1 and ϕ_2 as indicated below, one has

$$\phi_1 = \left.\frac{dP}{dt}\right|_{t=1} = f(1) + 2f(2) + 3f(3) + 4f(4) + \cdots,$$
$$\phi_2 = \left.\frac{d^2P}{dt^2}\right|_{t=1} = 2f(2) + 6f(3) + 12f(4) + 20f(5) + \cdots.$$

One notes that

$$\phi_1 = \mathrm{E}(x) \qquad \text{and} \qquad \phi_2 = \mathrm{E}[x(x-1)].$$

In general, the mth *factorial* moment of a distribution is

$$\phi_m = \mathrm{E}[x(x-1)(x-2)\cdots(x-m+1)] = \left.\frac{d^mP}{dt^m}\right|_{t=1}.$$

The factorial moments are related to the raw moments as follows:

$$\phi_1 = \alpha_1, \qquad \phi_2 = \alpha_2 - \alpha_1, \qquad \phi_3 = \alpha_3 - 3\alpha_2 + 2\alpha_1,$$
$$\phi_4 = \alpha_4 - 6\alpha_3 + 11\alpha_2 - 6\alpha_1.$$

Rather than working with the moment-generating function, it is often convenient to work with the *characteristic function*, which is defined by

$$g(t) = \mathrm{E}(e^{itx}) = \int_{-\infty}^{\infty} e^{itx} f(x)\, dx.$$

The integral on the right exists for all distribution functions. If $\mathrm{E}(x^m)$ exists, then $g(t)$ may be expanded in terms of a Taylor series as follows:

$$g(t) = \sum_{j=0}^{m} \alpha_j \frac{(it)^j}{j!} + 0(t^m) \qquad \text{as } t \to 0.$$

The mth derivative (with respect to t) of $g(t)$ may be shown to be

$$\frac{d^mg}{dt^m} = i^m \int_{-\infty}^{\infty} e^{itx} x^m f(x)\, dx.$$

If one sets $t = 0$ in this derivative one has

$$\frac{d^mg}{dt^m} = i^m \int_{-\infty}^{\infty} x^m f(x)\, dx = i^m \mathrm{E}(x^m).$$

Hence it follows that

$$\mathrm{E}(x^m) = \alpha_m,$$

where α_m is the coefficient of $(it)^m/m!$ in the Taylor series expansion of $g(t)$. Thus the function $g(t)$ may be considered a special case of a moment-generating function.

In addition to the moments, there is another set of parameters which are useful in characterizing a distribution function; the latter set are called *cumulants*. The cumulants are obtained from the log of the characteristic function. Let

$$h(t) = \ln g(t).$$

The function $h(t)$ is called the second characteristic of the random variable x. If $\mathrm{E}(x^m)$ exists, then $h(x)$ may be expanded in a Taylor series as follows:

$$h(t) = \sum_{j=0}^{m} \kappa_j \frac{(it)^j}{j!} + 0(t^m) \qquad \text{as } t \to 0.$$

κ_j in this series is the jth cumulant of the distribution. Alternatively. the mth cumulant may be obtained from

$$\left. \frac{d^m h}{dt^m} \right|_{t=0} = \kappa_m.$$

Lower-order cumulants, raw moments, and central moments are related as follows:

$$\begin{aligned}
\kappa_1 &= \alpha_1 = \mu_1, \\
\kappa_2 &= \alpha_2 - \alpha_2^2 = \mu_2, \\
\kappa_3 &= \alpha_3 - 3\alpha_1\alpha_2 + 2\alpha_1^3 = \mu_3, \\
\kappa_4 &= \alpha_4 - 3\alpha_2^2 - 4\alpha_1\alpha_3 + 12\alpha_1^2\alpha_2 - 6\alpha_1^4 = \mu_4 - 3\mu_2^2.
\end{aligned}$$

The *skewness* of a distribution is measured by

$$\sqrt{\beta_1} = \frac{\mu_3}{\mu_2^{3/2}} = \frac{\kappa_3}{\kappa_2^{3/2}}.$$

The *kurtosis* of a distribution is measured by

$$\beta_2 = \frac{\mu_4}{\mu_2^2} = \frac{\kappa_4}{\kappa_2^2} + 3.$$

Independent Random Variables. Consider two random variables x and y. The *joint* probability density of x and y gives

$$\Pr(x,y) = f(x,y).$$

The joint distribution function gives

$$\Pr(x \le x_i, y \le y_j) = \int_{-\infty}^{x_i} \int_{-\infty}^{y_j} f(x,y)\, dx\, dy.$$

The probability density function of the random variable x considered by itself (or the marginal density for x) is

$$\Pr(x) = \int_{-\infty}^{\infty} f(x,y)\, dy = g(x).$$

Similarly the probability density function for the random variable y considered by itself is

$$\Pr(y) = \int_{-\infty}^{\infty} f(x,y)\, dx = h(y).$$

If the joint probability density of x and y is the product of the marginal densities, i.e., if

$$f(x,y) = g(x)\, h(y),$$

then the random variables x and y are said to be *independent*. Equivalently, the random variables x and y are said to be independently distributed. For example, if the joint density of x and y is

$$f(x,y) = kme^{x+y},$$

and if the marginal densities are

$$g(x) = ke^{x} \qquad \text{and} \qquad h(x) = me^{y},$$

then
$$f(x,y) = g(x)\,h(y) = (ke^{x})(me^{y}) = kme^{x+y}.$$

For this example, x and y are distributed independently.

A.2 Normal Distribution

A random variable x has a normal distribution with parameters μ and σ^2 [symbolized $N(\mu,\sigma^2)$] if its density function is given by

$$f(x) = \frac{1}{\sqrt{2\pi}\,\sigma} \exp\left[-\frac{(x-\mu)^2}{2\sigma^2}\right].$$

The distribution function is given by

$$\Pr(x \le x_j) = F(x_j) = \int_{-\infty}^{x_j} f(x)\,dx.$$

The density function and its relationship to the distribution function are illustrated in Fig. A.2-1.

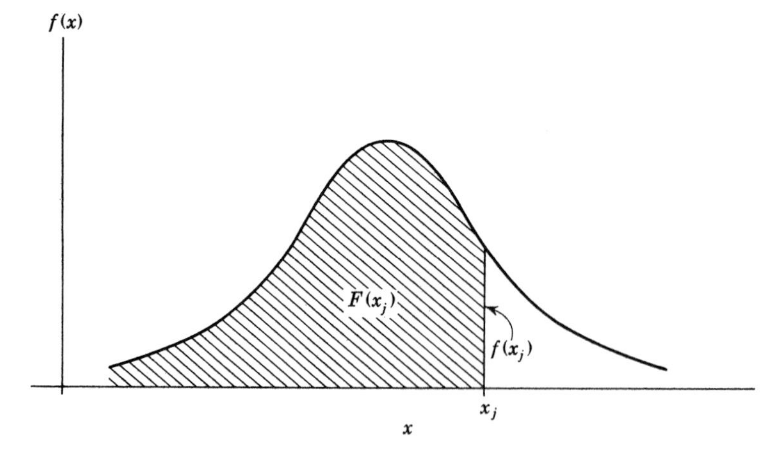

Figure A.2-1 Normal probability distribution.

The density function $f(x_j)$ defines the height of the curve at the point $x = x_j$. The distribution function $F(x_j)$ gives the area under the curve from $-\infty$ to x_j. Area in this case corresponds to probability. Thus

$$\Pr(x \le x_j) = F(x_j).$$

In Fig. A.2-2, the density at the point $x = x_i$ is given by the height $f(x_i)$.

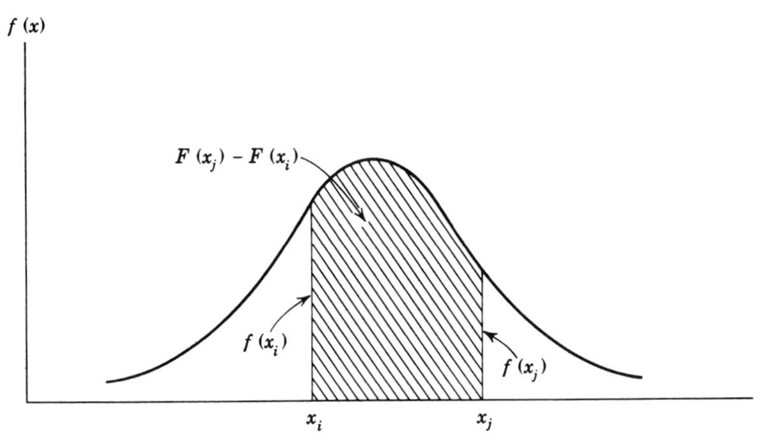

Figure A.2-2 Normal probability distribution.

The density at the point $x = x_j$ is given by $f(x_j)$. The area under the curve between the points x_i and x_j corresponds to the probability

$$\Pr(x_i \leq x \leq x_j) = F(x_j) - F(x_i),$$

that is, the difference between the areas under the curve from $-\infty$ to x_j and from $-\infty$ to x_i.

The expected value of the random variable x is

$$E(x) = \mu.$$

Thus the parameter μ of the normal distribution is the expected value of the distribution. The variance of the random variable x is

$$E(x - \mu)^2 = \sigma^2.$$

Thus the parameter σ^2 in the normal distribution is actually the variance of the distribution.

One also has

$$E(x - \mu)^3 = 0.$$

All odd central moments will be zero. Further

$$E(x - \mu)^4 = 3\sigma^4.$$

All even central moments may be expressed as functions of σ^2.

The normal distribution has the following reproductive property. If $x_1, \ldots, x_n$ are independently distributed as $N(\mu_i, \sigma_i^2)$, $i = 1, \ldots, n$, then the random variable

$$y = x_1 + \cdots + x_n$$

is distributed as $N(\mu = \Sigma\mu_i, \sigma^2 = \Sigma\sigma_i^2)$. That is, the random variable y is normally distributed with parameters $\Sigma\mu_i$ and $\Sigma\sigma_i^2$. This reproductive property of the normal distribution has important consequences. In particular, if $x_1, \ldots, x_n$ are independently, identically, and normally distributed as $N(\mu,\sigma^2)$, then the random variable

$$y = \frac{x_1 + \cdots + x_n}{n}$$

is distributed as

$$N\left(\mu, \frac{\Sigma\sigma^2}{n^2}\right) = N\left(\mu, \frac{n\sigma^2}{n^2}\right) = N\left(\mu, \frac{\sigma^2}{n}\right).$$

Variables which are Normal and Independent, with Identical Distributions, are said to be NIID variables. In this case, the mean of the distribution of the random variable y is

$$\mathrm{E}(y) = \mu,$$

and the variance of the random variable y is

$$\mathrm{E}(y - \mu)^2 = \sigma_y^2 = \frac{\sigma^2}{n}.$$

A.3 Gamma and Chi-square Distributions

A random variable x has a *gamma* distribution if its density function has the form

$$g(\alpha,p) = \frac{\alpha^p}{\Gamma(p)} e^{-\alpha x} x^{p-1}, \qquad x, \alpha, p > 0, \tag{1}$$

where α and p are parameters of the distribution. The corresponding distribution function is

$$G(\alpha,p) = \int_0^x g(\alpha,p)\, dx. \tag{2}$$

A random variable having this distribution function is called a gamma variable. The expected value of a gamma variable is

$$\mathrm{E}(x) = \mu = \frac{p}{\alpha};$$

its variance is

$$\sigma^2 = \frac{p}{\alpha^2}.$$

A geometric representation of the gamma probability law is shown in Fig. A.3-1. For each of the three densities shown in this figure, $\alpha = \frac{1}{2}$. In this case, the larger p is, the flatter the appearance of the density function. Also, the larger the value of the parameter p, the larger is the mean. In all cases the total area under the curve is unity.

The gamma distribution has the following *reproductive* property. If the random variables $x_1, \ldots, x_k$ are each independently distributed as

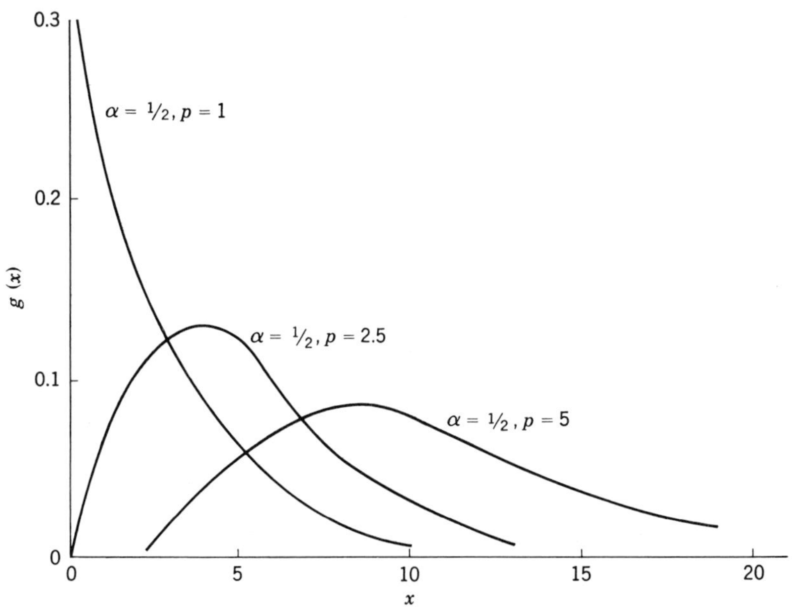

Figure A.3-1 Gamma probability distribution.

gamma variables with parameters α, p_i, then the random variable y, defined by

$$y = \Sigma x_i,$$

is a gamma variable with parameters α, Σp_i. That is, the probability density of the random variable y has the form

$$f(y) = g(\alpha, \Sigma p_i).$$

The chi-square distribution is that special case of the gamma distribution for which the parameters are

$$\alpha = \frac{1}{2} \quad \text{and} \quad p = \frac{k}{2}.$$

Thus a random variable x is distributed as chi square if the density function has the form

$$g\left(\frac{1}{2}, \frac{k}{2}\right) = \frac{(\frac{1}{2})^{k/2}}{\Gamma(k/2)} e^{-x/2} x^{(k/2)-1}.$$

The density function corresponding to this special case of a gamma variable is denoted

$$\chi^2(k) = g\left(\frac{1}{2}, \frac{k}{2}\right).$$

The parameter k is called the degrees of freedom of the chi-square distribution. The expected value of a chi-square variable is

$$\mu = \mathrm{E}(x) = \frac{p}{\alpha} = \frac{k/2}{\frac{1}{2}} = k.$$

The variance of a chi-square variable is

$$\sigma^2 = \frac{p}{\alpha^2} = \frac{k/2}{(\frac{1}{2})^2} = 2k.$$

A chi-square variable has the same sort of reproductive property as a gamma variable. If $x_1, \ldots, x_p$ are independent chi-square variables distributed as $\chi^2(k_1), \ldots, \chi^2(k_p)$, then the random variable

$$y = x_1 + \cdots + x_p$$

has the density function given by $\chi^2(k = \Sigma k_i)$.

A chi-square variable is related to a normally distributed variable. If x is a random variable with the distribution $N(0,1)$, then the random variable $y = x^2$ is distributed as chi square with one degree of freedom; symbolically, y is $\chi^2(1)$. Further, if the random variables $x_1, \ldots, x_k$ are independently distributed as $N(0,1)$, then the random variable y defined by

$$y = x_1^2 + \cdots + x_k^2$$

will be distributed as chi square with k degrees of freedom. Symbolically,

$$y\colon \chi^2(k).$$

This last statement is read: y is distributed as chi square with k degrees of freedom.

If $x_1, \ldots, x_k$ are independently distributed as $N(\mu_i,\sigma^2)$, where $i = 1, \ldots, k$, then the random variable y defined by

$$y = \frac{x_1^2 + \cdots + x_k^2}{\sigma^2}$$

will be distributed as a *noncentral* chi-square variable with degrees of freedom equal to k and noncentrality parameter λ given by

$$\lambda = \frac{\mu_1^2 + \cdots + \mu_k^2}{\sigma^2} = \frac{\mu^2}{\sigma^2}, \qquad \text{where } \mu^2 = \Sigma\mu_i^2.$$

[Scheffé (1959) defines the noncentrality parameter as

$$\delta^2 = \frac{\Sigma\mu_i^2}{\sigma^2} \qquad \text{or} \qquad \delta = \sqrt{\frac{\Sigma\mu_i^2}{\sigma^2}}.$$

Rao (1965) defines the noncentrality parameter as

$$\lambda_R = \frac{\Sigma\mu_i^2}{\sigma^2}.$$

Graybill (1961) defines the noncentrality parameter as

$$\lambda_G = \frac{\Sigma\mu_i^2}{2\sigma^2}.$$

The notation system adopted by Rao is the one followed here.]

If $x_1, \ldots, x_k$ are independently distributed as $N(\mu_i,\sigma^2/n)$, then the random variable

$$y = \frac{x_1^2 + \cdots + x_k^2}{\sigma^2/n} = \frac{n\Sigma x_i^2}{\sigma^2}$$

will be distributed as a noncentral chi square with noncentrality parameter

$$\lambda = \frac{n\Sigma\mu_i^2}{\sigma^2}.$$

This is the form of the noncentrality parameter that one encounters in the analysis of variance. Each x_i in this case corresponds to a mean based upon n observations. It should be noted that the noncentrality parameter as defined here is obtained by replacing each x_i in y by $\mathrm{E}(x_i)$.

The noncentral chi-square distribution is denoted by the symbol $\chi^2(k;\lambda)$. The density function for this distribution has the form

$$\chi^2(k;\lambda) = e^{-\lambda/2}\sum_{r=0}^{\infty}\frac{1}{r!}\left(\frac{\lambda}{2}\right)^r g\left(\frac{1}{2}, r + \frac{k}{2}\right).$$

When $\lambda = 0$, the only nonzero term in this summation is that corresponding to $r = 0$. Hence, when $\lambda = 0$,

$$\chi^2(k;0) = g\left(\frac{1}{2}, \frac{k}{2}\right) = \chi^2(k).$$

The chi-square distribution is that special case of the noncentral chi-square distribution corresponding to $\lambda = 0$. To distinguish between the former and the latter distributions, the former is sometimes called the *central* chi-square distribution. When the term "chi-square distribution" is used, it is the central distribution that is implied unless otherwise indicated.

The expected value of a variate having a noncentral chi-square distribution is

$$\mathrm{E}[\chi^2(k;\lambda)] = k + \lambda.$$

The variance of a noncentral chi-square variate is

$$2k + 4\lambda.$$

The noncentral chi-square distribution has the same kind of additive properties as does the central chi-square distribution. If u_1 is distributed as $\chi^2(k_1;\lambda_1)$ and u_2 is independently distributed as $\chi^2(k_2;\lambda_2)$, then the random variable $u_1 + u_2$ is distributed as

$$\chi^2(k_1 + k_2; \lambda_1 + \lambda_2).$$

The noncentral chi square may be approximated by a central chi-square distribution as follows:

$$\chi^2_{1-\alpha}(k;\lambda) = c\chi^2_{1-\alpha}(k'),$$

where c and k' are determined from the following relationships:

$$ck' = k + \lambda,$$
$$c^2k' = k + 2\lambda.$$

These relationships equate the means and variances of the central and noncentral chi-square distributions. For example, if $k = 10$ and $\lambda = 2$,

$$c = 1.17 \qquad \text{and} \qquad k' = 10.29.$$

A.4 Beta and F Distributions

Let x_1 and x_2 be independently distributed random variables with

$$x_1\colon g(\alpha,p_1) \qquad \text{and} \qquad x_2\colon g(\alpha,p_2).$$

Thus x_1 and x_2 are independent gamma variables. Let the random variable y be defined as

$$y = \frac{x_1}{x_1 + x_2} = \frac{x_1/x_2}{(x_1/x_2) + 1}.$$

The density function for y is given by what is called a beta distribution having parameters p_1 and p_2 and denoted $b(p_1,p_2)$. Thus

$$y\colon b(p_1,p_2) = \frac{\Gamma(p_1 + p_2)}{\Gamma(p_1)\Gamma(p_2)} y^{p_1-1}(1 - y)^{p_2-1}$$
$$= \frac{1}{\beta(p_1,p_2)} y^{p_1-1}(1 - y)^{p_2-1}, \qquad 0 < y < 1.$$

The expected value of a beta variable is

$$\mu = \mathrm{E}(y) = \frac{p_1}{p_1 + p_2}.$$

The variance of a beta variable is

$$\sigma^2 = \frac{p_1 p_2}{(p_1 + p_2)^2(p_1 + p_2 + 1)}.$$

The form of the density function for various values of the parameters is shown in Fig. A.4-1.

There is a quasi-symmetry to the beta distribution in the sense that

$$\int_0^{y_i} b(p_1,p_2)\,dy = 1 - \int_0^{1-y_i} b(p_2,p_1)\,dy,$$

or, equivalently,

$$\int_{y_i}^{1} b(p_1,p_2)\,dy = \int_0^{1-y_i} b(p_2,p_1)\,dy.$$

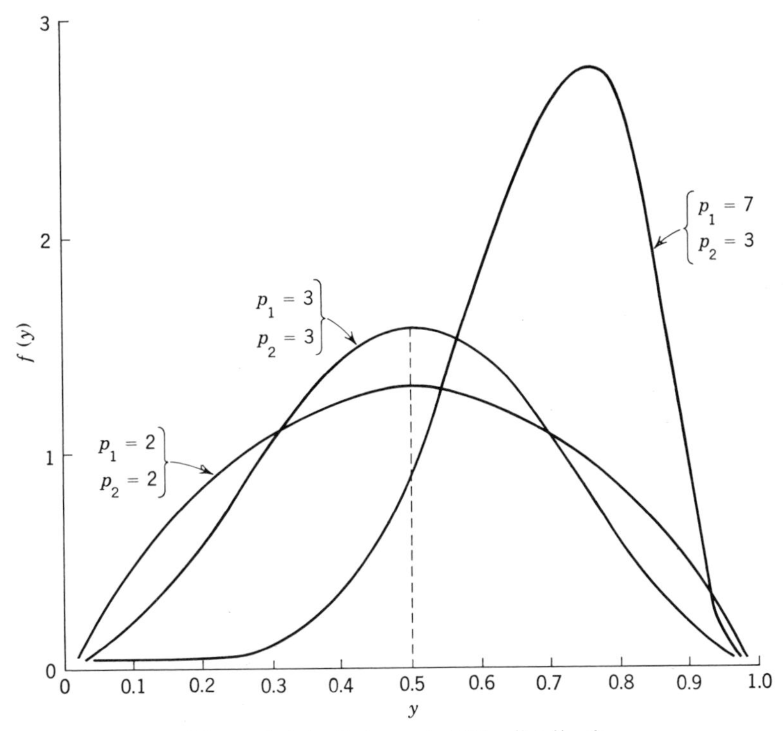

Figure A.4-1 Beta probability distribution.

This quasi-symmetry is illustrated in Fig. A.4-2.

Consider now the independent random variables

$$x_1: \chi^2(k_1) \qquad \text{and} \qquad x_2: \chi^2(k_2).$$

Define the random variable F as

$$F = \frac{x_1/k_1}{x_2/k_2}.$$

Thus F is the ratio of two independent chi-square variables. The density function for F is that of the F distribution with parameters k_1 and k_2.

Symbolically,

$$F\colon F(k_1,k_2) = \frac{(k_1/k_2)^{k_1/2}}{\beta(k_1/2,k_2/2)} \frac{F^{(k_1/2)-1}}{[1 + (k_1/k_2)F]^{(k_1+k_2)/2}}.$$

The variable y, which was defined above, may be cast in the form

$$y = \frac{x_1}{x_1 + x_2} = \frac{x_1/x_2}{(x_1/x_2) + 1} = \frac{(k_1/k_2)F}{(k_1/k_2)F + 1} = \frac{k_1 F}{k_1 F + k_2},$$

where

$$F = \frac{x_1/k_1}{x_2/k_2}.$$

Solving this last equation for F gives

$$F = \frac{k_2}{k_1}\frac{y}{1 - y}.$$

Thus the random variable F is a transformed beta variable. The relationship between the F and beta distributions may be expressed as follows:

$$\int_{y_i}^{1} b\left(\frac{k_1}{2}, \frac{k_2}{2}\right) dy = 1 - \int_0^{y_i} b\left(\frac{k_1}{2}, \frac{k_2}{2}\right) dy = \int_{F_i}^{\infty} F(k_1,k_2)\, dF,$$

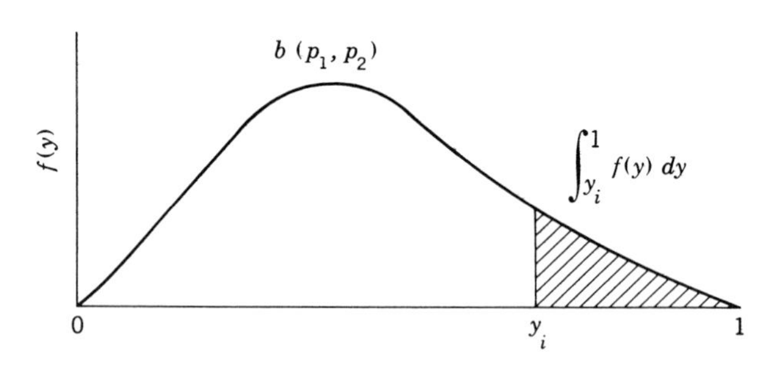

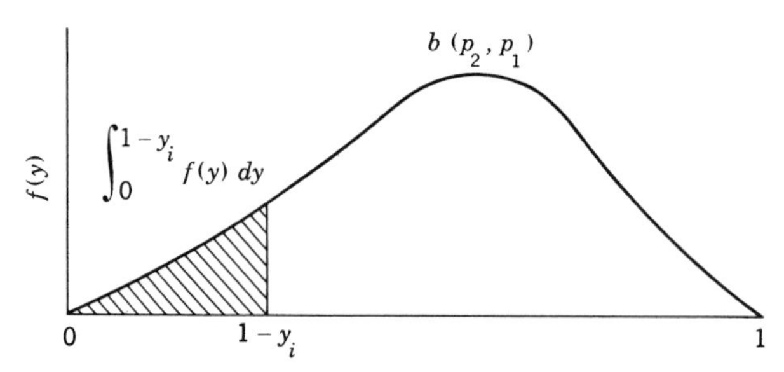

Figure A.4-2 Quasi-symmetry of beta distribution.

where $$y_i = \frac{k_1 F_i}{k_1 F_i + k_2} \qquad \text{or} \qquad F_i = \frac{k_2}{k_1} \frac{y_i}{1 - y_i}.$$

Thus $$\Pr(y \geq y_i) = \Pr(F \geq F_i).$$

Note that

$$1 - y_i = \frac{k_2}{k_2 + k_1 F_i} = \frac{1}{1 + (k_1/k_2) F_i}.$$

Tables of the F distribution are actually computed from tables of the beta distribution. The relationship between probabilities (areas) of the beta and F distributions is illustrated in Fig. A.4-3.

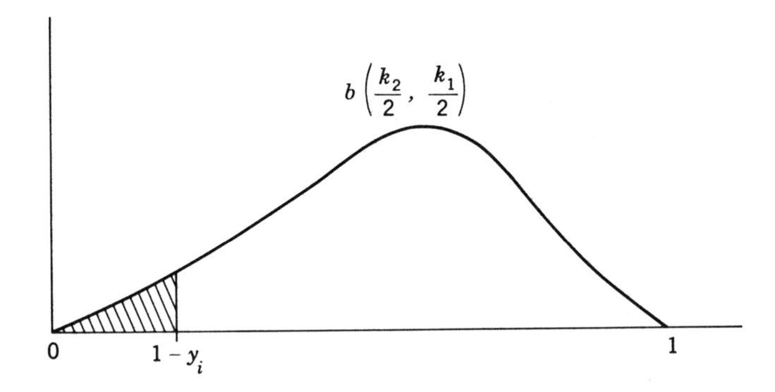

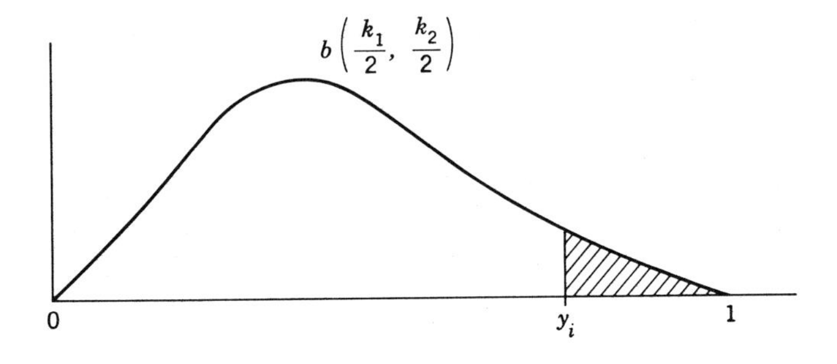

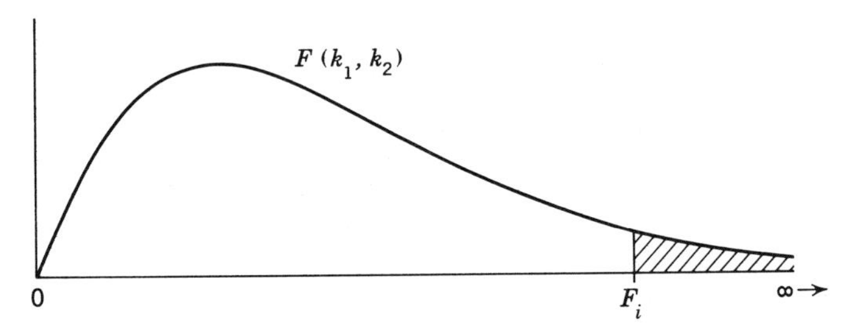

Figure A.4-3 Relationship between probabilities obtained from beta and F distributions.

The first moment of the F distribution is

$$E(F) = \frac{k_2}{k_2 - 2}, \qquad k_2 > 2.$$

The mode of the F distribution, for $k_1 > 2$, is at the point

$$\frac{k_1 - 2}{k_1} \frac{k_2}{k_2 + 2}.$$

The second central moment of the F distribution is

$$E[F - E(F)]^2 = \frac{2k_2^2(k_1 + k_2 - 2)}{k_1(k_2 - 2)^2(k_2 - 4)}, \qquad k_2 > 4.$$

Consider now the independent random variables

$$x_1\colon \chi^2(k_1;\lambda) \qquad \text{and} \qquad x_2\colon \chi^2(k_2).$$

That is, x_1 is distributed as noncentral chi square, and x_2 is distributed as central chi square. Let the random variable F be defined as

$$F = \frac{x_1/k_1}{x_2/k_2}.$$

This random variable has a *noncentral* F distribution with noncentrality parameter λ. The symbol $F(k_1,k_2;\lambda)$ is used to denote the noncentral F distribution. The noncentral F is closely related to the noncentral beta distribution; the latter has the probability density

$$b\left(\frac{k_1}{2}, \frac{k_2}{2}; \lambda\right) = e^{-\lambda/2} \sum_{r=0}^{\infty} \left(\frac{\lambda}{2}\right)^r \left(\frac{1}{r!}\right) b\left(\frac{k_1}{2} + r, \frac{k_2}{2}\right).$$

When $\lambda = 0$, the only nonzero term in this summation is that corresponding to $r = 0$. In this case, the term corresponding to $r = 0$ is

$$b\left(\frac{k_1}{2}, \frac{k_2}{2}; 0\right) = b\left(\frac{k_1}{2}, \frac{k_2}{2}\right).$$

Thus the ordinary (or central) beta distribution is that special case of the noncentral beta in which the noncentrality parameter is zero. (See Fig. A.4-4.)

The expected value of the noncentral F distribution is

$$\frac{k_2(k_1 + \lambda)}{(k_2 - 2)k_1} = \frac{k_2}{k_2 - 2}\left(1 + \frac{\lambda}{k_1}\right), \qquad k_2 > 2.$$

The second raw moment is

$$\frac{k_2^2}{(k_2 - 2)(k_2 - 4)k_1^2} [(k_1 + \lambda)^2 + 2(k_2 + 2\lambda)], \qquad k_2 > 4.$$

The relationship between the noncentral F and the noncentral beta

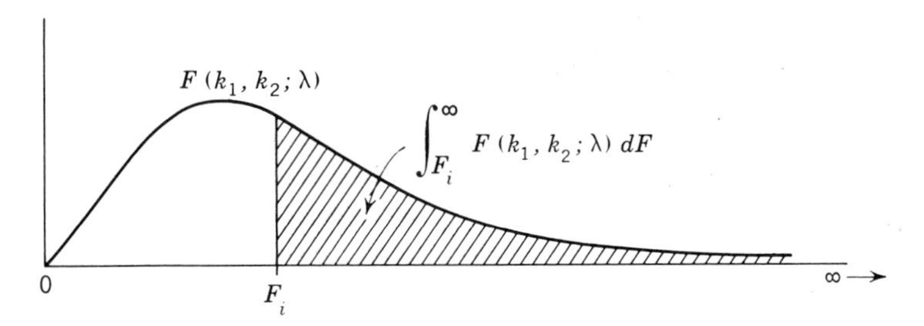

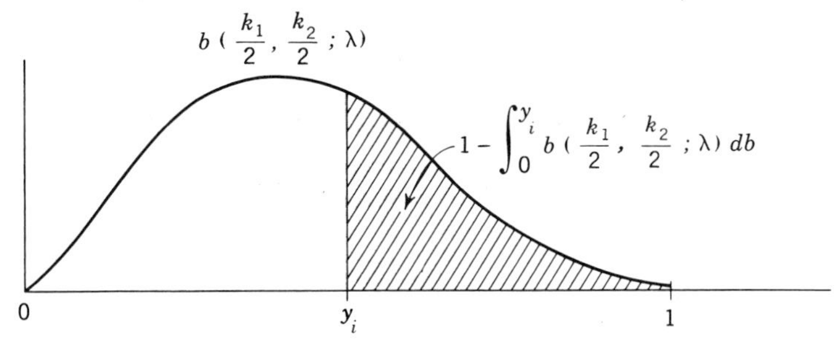

Figure A.4-4 Relationship between noncentral F and noncentral beta.

distributions is the same as the corresponding relationship between the central distributions. Tables of the noncentral beta distribution as prepared by Tang appear in Graybill (1961, pp. 444–459). In terms of the noncentrality parameter as defined here, the noncentrality parameter as used in these tables is

$$\phi = \sqrt{\frac{\lambda}{k_1 + 1}}.$$

From the tables as constructed by Tang,

$$1 - \int_0^{y_i} b\left(\frac{k_1}{2}, \frac{k_2}{2}; \phi\right) dy = \int_{F_i}^{\infty} F(k_1,k_2;\lambda)\, dF,$$

where
$$y_i = \frac{k_1 F_i}{k_1 F_i + k_2}.$$

A portion of the tables of the noncentral F constructed by Tiku (1967) appears in Table C.14.

Explicitly the probability density for the noncentral F is

$$F(k_1,k_2;\lambda) = e^{(-\lambda/2)} \sum_{r=0}^{\infty} \frac{(\lambda/2)^r (k_1/k_2)^{f_r}}{\beta(f_r,k_2/2)} \frac{F^{f_r - 1}}{[1 + (k_1/k_2)F]^{f_r + (k_2/2)}},$$

where
$$f_r = \frac{k_1}{2} + r.$$

The noncentral F distribution may be approximated quite well by a central F distribution as follows:

$$\int_{F_{1-\alpha}}^{\infty} F(k_1,k_2;\lambda)\,dF \doteq \int_{(1/c)F_{1-\alpha}}^{\infty} F(k_1',k_2)\,dF,$$

where

$$F_{1-\alpha} = F_{1-\alpha}(k_1,k_2), \qquad c = \frac{k_1+\lambda}{k_1}, \qquad k_1' = \frac{(k_1+\lambda)^2}{k_1+2\lambda}.$$

For example, if

$$\alpha = .01, \qquad k_1 = 4, \qquad k_2 = 6, \qquad \text{and} \qquad \lambda = 16.2,$$

then

$$F_{.99}(4,6) = 9.15.$$

From tables of the noncentral F distribution one will find that

$$\int_{9.15}^{\infty} F(4,6;\,16.2) = .24.$$

(Note: $\phi = \sqrt{\dfrac{16.2}{5}} = 1.80.$)

To use the approximation by means of the central F distribution,

$$c = \frac{4+16.2}{4} = 5.05, \qquad k_1' = \frac{(4+16.2)^2}{4+2(16.2)} = 8.99,$$

$$\frac{1}{c}F_{.99}(4,6) = \frac{1}{5.05}(9.15) = 1.81.$$

By interpolation in tables of the central F distribution,

$$\int_{1.81}^{\infty} F(8.99,\,6)\,dF = .24.$$

Relatively extensive tables of the noncentral F have been prepared by Tiku (1967).

The F distribution has a quasi-symmetry which is similar to that of the beta distribution. This symmetry is illustrated for the case of the central F distribution in Fig. A.4-5. From this relationship one may compute left-hand areas of $F(k_2,k_1)$ from right-hand areas of $F(k_1,k_2)$. The latter areas are found in tables of the central F distribution.

A.5 Student's *t* Distribution

Let the random variables x and y be independently distributed as

$$x\colon N(0,1), \qquad y\colon \chi^2(k).$$

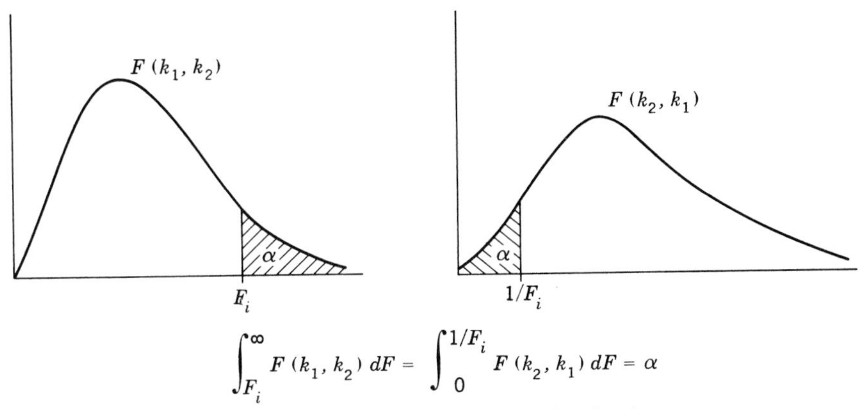

$$\int_{F_i}^{\infty} F(k_1, k_2)\, dF = \int_0^{1/F_i} F(k_2, k_1)\, dF = \alpha$$

Figure A.4-5 Quasi-symmetry of F distribution.

That is, x is a unit normal variate, and y is a chi-square variate. Define the random variable t as

$$t = \frac{x}{\sqrt{y/k}}.$$

The random variable t will be distributed as Student's t with parameter k. Symbolically,

$$t\colon t(k).$$

The probability density function for Student's t distribution has the form

$$\begin{aligned} f(t) &= \frac{\Gamma[\frac{1}{2} + (k/2)]}{\sqrt{k}\,\Gamma(\frac{1}{2})\Gamma(k/2)}\left(1 + \frac{t^2}{k}\right)^{-(k+1)/2} \\ &= \frac{1}{\sqrt{k}\,\beta(\frac{1}{2},k/2)}\left(1 + \frac{t^2}{k}\right)^{-(k+1)/2}. \end{aligned}$$

Student's t distribution is symmetric about $E(t)$, which is

$$E(t) = 0.$$

The second moment of the t distribution is

$$E(t^2) = \sigma_t^2 = \frac{k}{k-2}.$$

In general, since the t distribution is symmetric about zero, all odd moments will be zero. For even moments, if $m < k$, the mth moment is given by

$$\mu_m = \frac{1(3)(5)\cdots(m-1)k^{m/2}}{(k-2)(k-4)\cdots(k-m)}.$$

In particular, if $m = 2$,

$$\mu_2 = \mathrm{E}(t^2) = \frac{k}{k-2}.$$

Student's t distribution is closely related to the F distribution. Specifically, if one defines the random variable $v = t^2$, where t: $t(k)$, then the distribution of the random variable v is $F(1,k)$. This relationship follows from the fact that

$$t^2 = \frac{x^2}{y/k} = \frac{\chi^2(1)}{\chi^2(k)/k} = F(1,k).$$

One notes that

$$\mathrm{E}(t^2) = \sigma_t^2 = \mathrm{E}[F(1,k)] = \frac{k}{k-2}.$$

That is, the variance of the t distribution is the mean of the F distribution that corresponds to the t distribution.

Consider now the independently distributed random variables

$$x\colon N(\mu,\sigma^2) \quad \text{and} \quad \frac{y}{\sigma^2} = \chi^2(k).$$

Let $$\sqrt{\lambda} = \sqrt{\frac{\mu^2}{\sigma^2}} \quad \text{and} \quad t = \frac{x/\sigma}{\sqrt{y/k\sigma^2}} = \frac{x}{\sqrt{y/k}}.$$

The random variable defined above has a *noncentral* t distribution with noncentrality parameter $\sqrt{\lambda}$. The density function for the noncentral t distribution has the form

$$t(k;\sqrt{\lambda}) = \frac{k^{k/2}}{\Gamma(k/2)} \frac{e^{-\lambda/2}}{(k+t^2)^{(k+1)/2}} h(t),$$

where $$h(t) = \sum_{s=0}^{\infty} \Gamma\left(\frac{k+s+1}{2}\right)\left(\frac{\lambda^{s/2}}{s!}\right)\left(\frac{2t^2}{k+t^2}\right)^{s/2}.$$

When $\sqrt{\lambda} = 0$, the only nonzero term in the summation is that corresponding to $s = 0$. It will be found that $t(k;0)$ reduces to the ordinary (or central) Student's t distribution.

The noncentral t distribution may be approximated by

$$\Pr(t \le u) = \Pr\left[z \le (u - \sqrt{\lambda})\left(1 + \frac{u^2}{2k}\right)^{-1/2}\right],$$

where z is $N(0,1)$. When $\sqrt{\lambda} = 0$, one has an approximation to the central t distribution. For example, if $k = 20$ and $\lambda = 0$, from tables of the central t distribution one has

$$\Pr(t \le 1.32) \doteq .90.$$

The approximation gives

$$\Pr(t \le 1.32) \doteq \Pr\left\{z \le 1.32\left[1 + \frac{(1.32)^2}{20}\right]^{-1/2}\right\} = \Pr(z \le 1.29)$$
$$= .90.$$

Just as the square of a central t variate is a central F variate, so the square of a noncentral t variate will be a noncentral F variate. Thus, $t^2(k;\sqrt{\lambda}) = F(1,k;\lambda)$. Hence one may also use the method outlined under the noncentral F to approximate the noncentral t.

A.6 Bivariate Normal Distribution

If one is considering, simultaneously, two random variables, say x and y, the joint distribution function has the general form

$$F(x',y') = \Pr(x \le x', y \le y') = \int_{-\infty}^{x'}\int_{-\infty}^{y'} f(x,y)\,dx\,dy,$$

where $f(x,y)$ is the joint density function. Geometrically, the probability element $f(x,y)\,dx\,dy$ represents the volume of a solid rectangle having a base area equal to $dx\,dy$ and height equal to $f(x,y)$.

The bivariate normal distribution has the density function

$$f(x,y) = ke^{-Q^2/2},$$

where $$Q^2 = \frac{1}{1-\rho^2}\left[\frac{(x-\mu_x)^2}{\sigma_x^2} + \frac{(y-\mu_y)^2}{\sigma_y^2} - \frac{2\rho(x-\mu_x)(y-\mu_y)}{\sigma_x\sigma_y}\right],$$

$$k = \frac{1}{2\pi\sigma_x\sigma_y\sqrt{1-\rho^2}}.$$

To make the parameters of this distribution explicit, the density function may be written in the form

$$f(x,y;\mu_x,\mu_y,\sigma_x^2,\sigma_y^2,\rho).$$

The bivariate normal distribution may also be represented in an alternative form. Let

$$\underline{u} = \begin{bmatrix} x \\ y \end{bmatrix}, \quad \underline{\mu} = \begin{bmatrix} \mu_x \\ \mu_y \end{bmatrix}, \quad \Sigma = \begin{bmatrix} \sigma_x^2 & \rho\sigma_x\sigma_y \\ \rho\sigma_x\sigma_y & \sigma_y^2 \end{bmatrix}, \quad |\Sigma| = \sigma_x^2\sigma_y^2(1-\rho^2).$$

The density function for the bivariate normal is

$$f(\underline{u}) = k\exp\left[-\tfrac{1}{2}(\underline{u}-\underline{\mu})'\Sigma^{-1}(\underline{u}-\underline{\mu})\right],$$

where $$k = \frac{1}{2\pi\,|\Sigma|^{1/2}}.$$

Associated with a bivariate distribution is a set of marginal distributions. The marginal distribution of the y variable is

$$f_1(y) = \int_{-\infty}^{\infty} f(x,y)\,dx.$$

For the bivariate normal distribution, this marginal distribution is the univariate normal distribution with parameters μ_y and σ_y^2. The marginal distribution for the x variable is

$$f_2(x) = \int_{-\infty}^{\infty} f(x,y)\,dy.$$

For the bivariate normal distribution, this marginal distribution is the univariate normal distribution with parameters μ_x and σ_x^2. Thus the parameters μ_x, μ_y, σ_x^2, σ_y^2 of the bivariate distribution are the parameters of the corresponding univariate marginal distributions.

Also associated with a bivariate distribution is a set of conditional distributions, that is, the distribution of one variate for a fixed value of the second variate. The conditional probability density of y for fixed x is denoted by the symbol $f(y \mid x)$. This conditional density is related to the joint and marginal densities as follows:

$$f(y \mid x) = \frac{f(x,y)}{f_2(x)}.$$

In terms of probabilities,

$$\Pr(y \mid x) = \frac{\Pr(x,y)}{\Pr(x)} \qquad \text{or} \qquad \Pr(x,y) = \Pr(y \mid x)\Pr(x).$$

That is, the joint probability of x and y is the product of the conditional probability of y given x and the probability of x. The distribution function for a conditional distribution has the form

$$F(y = y' \mid x = x') = \int_{-\infty}^{y'} f(y \mid x = x')\,dy = \int_{-\infty}^{y'} \frac{f(x',y)}{f_2(x')}\,dy.$$

For the bivariate normal distribution, the probability density $f(y \mid x)$ is the univariate normal distribution with parameters

$$\mu_{y|x} = \alpha_{yx} + \beta_{yx}x \qquad \text{and} \qquad \sigma_{y|x}^2 = \sigma_y^2(1 - \rho^2),$$

where

$$\beta_{yx} = \rho\left(\frac{\sigma_y}{\sigma_x}\right) \qquad \text{and} \qquad \alpha_{yx} = \mu_y - \beta_{yx}\mu_x.$$

The parameter ρ is defined by

$$\rho = \frac{\sigma_{xy}}{\sigma_x\sigma_y}, \qquad \text{where } \sigma_{xy} = \text{covariance between } x \text{ and } y.$$

One notes that the conditional variance $\sigma_{y|x}^2$ does not depend upon x.

In general, the conditional distribution of x for fixed y is given by

$$f(x \mid y) = \frac{f(x,y)}{f_1(y)}.$$

For the special case of the bivariate normal distribution, the probability density $f(x \mid y)$ is the univariate normal with parameters

$$\mu_{x|y} = \alpha_{xy} + \beta_{xy} y \qquad \text{and} \qquad \sigma^2_{x|y} = \sigma^2_x(1 - \rho^2),$$

where

$$\beta_{xy} = \rho\left(\frac{\sigma_x}{\sigma_y}\right) \qquad \text{and} \qquad \alpha_{xy} = \mu_x - \beta_{xy}\mu_y.$$

One notes that the conditional variance $\sigma_{x|y}$ does not depend upon y.

In general, the r, s joint central moment of a bivariate distribution is defined by

$$\mu_{r,s} = \mathrm{E}[(x - \mu_x)^r(y - \mu_y)^s] = \int_{-\infty}^{\infty}\int_{-\infty}^{\infty} (x - \mu_x)^r(y - \mu_y)^s f(x,y)\,dx\,dy.$$

For the special case of the bivariate normal distribution,

$$\mu_{10} = \mu_x \qquad \text{and} \qquad \mu_{01} = \mu_y.$$

Thus μ_x and μ_y, which are moments of the marginal distributions, are also the indicated moments of the joint distribution. One also has

$$\mu_{20} = \sigma^2_x \qquad \text{and} \qquad \mu_{02} = \sigma^2_y.$$

It may also be shown that (for the special case of the bivariate normal)

$$\mu_{11} = \sigma_{xy} = \rho\sigma_x\sigma_y.$$

Thus the covariance between x and y, σ_{xy}, is a moment of the joint distribution. In terms of the joint-moment notation, the parameter ρ has the form

$$\rho = \frac{\mu_{11}}{\sqrt{\mu_{10}\mu_{01}}}.$$

Other joint moments of the bivariate normal distribution are as follows:

$$\begin{aligned}
&\mu_{30} = \mu_{12} = \mu_{21} = 0,\\
&\mu_{40} = 3\sigma^4_x, \qquad \mu_{31} = 3\rho\sigma^3_x\sigma_y, \qquad \mu_{22} = (1 + 2\rho^2)\sigma^2_x\sigma^2_y,\\
&\mu_{04} = 3\sigma^4_y, \qquad \mu_{13} = 3\rho\sigma_x\sigma^3_y.
\end{aligned}$$

A.7 Multivariate Normal Distribution

Let

$$\underline{x}' = [x_1 \quad x_2 \quad \cdots \quad x_p]$$

be a vector of random variables whose joint probability density function is

$$f(\underline{x}) = \frac{1}{(2\pi)^{p/2}\,|\Sigma|^{1/2}} \exp\left[-\tfrac{1}{2}(\underline{x} - \underline{\mu})'\Sigma^{-1}(\underline{x} - \underline{\mu})\right],$$

where

$$\underset{p,1}{\underline{\mu}} = \begin{bmatrix} \mu_1 \\ \mu_2 \\ \cdot \\ \cdot \\ \cdot \\ \mu_p \end{bmatrix}, \qquad \underset{p,p}{\Sigma} = \begin{bmatrix} \sigma_1^2 & \sigma_{12} & \cdots & \sigma_{1p} \\ \sigma_{21} & \sigma_2^2 & \cdots & \sigma_{2p} \\ \cdot & \cdot & & \cdot \\ \cdot & \cdot & & \cdot \\ \cdot & \cdot & & \cdot \\ \sigma_{p1} & \sigma_{p2} & \cdots & \sigma_p^2 \end{bmatrix}, \qquad \text{rank}(\Sigma) = p.$$

A vector random variable $\underline{x}$ having this density function is said to be distributed as a p-variate normal distribution with parameters $\underline{\mu}$ and Σ. In symbols,

$$\underline{x}: N_p(\underline{\mu},\Sigma).$$

The parameter Σ is called the *dispersion* (or covariance) matrix for the vector variable $\underline{x}$. The distribution function $F(\underline{x}_j)$ represents

$$F(\underline{x}_j) = \Pr(\underline{x} \le \underline{x}_j), \qquad \text{where } \underline{x}_j' = [x_{1j} \quad x_{2j} \quad \cdots \quad x_{pj}].$$

The marginal distribution for each of the component variables x_i can be shown to be

$$x_i: N(\mu_i,\sigma_i^2), \qquad i = 1, 2, \ldots, p.$$

That is, each x_i is univariate normal with the parameters indicated. For the special case in which

$$\Sigma = \begin{bmatrix} \sigma_1^2 & 0 & \cdots & 0 \\ 0 & \sigma_2^2 & \cdots & 0 \\ \cdot & \cdot & & \cdot \\ \cdot & \cdot & & \cdot \\ \cdot & \cdot & & \cdot \\ 0 & 0 & \cdots & \sigma_p^2 \end{bmatrix},$$

the joint density function becomes

$$f(\underline{x}) = f_1(x_1)f_2(x_2)\cdots f_p(x_p),$$

where $f_i(x_i)$ is the univariate density $N(\mu_i,\sigma_i^2)$. Thus, for a multivariate normal distribution, if the dispersion matrix is diagonal, the component variables are *statistically independent;* that is, the joint distribution is the product of the marginal distributions.

Let the random variable u be defined by

$$u = \underset{1,p}{\underline{k}'} \; \underset{p,1}{\underline{x}},$$

where $\underline{k}$ is a vector of known constants. The random variable u will be distributed as a univariate normal with parameters

$$\mu_u = \underline{k}'\underline{\mu} \qquad \text{and} \qquad \sigma_u^2 = \underline{k}'\Sigma\underline{k}.$$

In general, any linear function of the components of $\underline{x}$ will be normally distributed.

Let the vector variable $\underline{x}$ be partitioned as follows:

$$\underset{p,1}{\underline{x}} = \begin{bmatrix} x_1 \\ \cdot \\ \cdot \\ \cdot \\ x_{p-1} \\ \hline x_p \end{bmatrix} = \begin{matrix} p-1 \\ 1 \end{matrix}\begin{bmatrix} \underline{x}_{p-1} \\ x_p \end{bmatrix}.$$

Also let $\underset{p,1}{\underline{\mu}} = \begin{matrix} p-1 \\ 1 \end{matrix}\begin{bmatrix} \underline{\mu}_{p-1} \\ \mu_p \end{bmatrix}, \qquad \underset{p,p}{\Sigma} = \begin{matrix} \\ p-1 \\ 1 \end{matrix}\begin{matrix} p-1 \qquad 1 \\ \begin{bmatrix} \Sigma_{p-1} & \underline{\sigma}_{p-1} \\ \underline{\sigma}'_{p-1} & \sigma_p^2 \end{bmatrix} \end{matrix},$

where

$$\underline{\sigma}_{p-1} = \begin{bmatrix} \sigma_{1p} \\ \sigma_{2p} \\ \cdot \\ \cdot \\ \cdot \\ \sigma_{p-1,p} \end{bmatrix}.$$

Thus the random variable $\underline{x}_{p-1}$ consists of the first $p - 1$ components of the vector $\underline{x}$. Hence,

$$\underline{x}_{p-1}\colon N_{p-1}(\underline{\mu}_{p-1},\Sigma_{p-1}), \qquad x_p\colon N_1(\mu_p,\sigma_p^2).$$

The conditional distribution of x_p for a fixed set of values for $\underline{x}_{p-1}$, say $\underline{x}_{p-1,\,j}$, will be univariate normal in form with mean

$$\mu_p \mid \underline{x}_{p-1,\,j} = \alpha + \underline{\beta}'\underline{x}_{p-1,\,j},$$

where $\qquad \underline{\beta} = \Sigma_{p-1}^{-1}\,\underline{\sigma}_{p-1} \qquad \text{and} \qquad \alpha = \mu_p - \underline{\beta}'\underline{\mu}_{p-1}.$

The variance of this conditional distribution is

$$\begin{aligned} \sigma_p^2 \mid \underline{x}_{p-1,\,j} &= \sigma_p^2 - \underline{\beta}'\Sigma_{p-1}\,\underline{\beta} \\ &= \sigma_p^2 - \underline{\sigma}'_{p-1}\,\Sigma_{p-1}^{-1}\,\underline{\sigma}_{p-1}. \end{aligned}$$

The last line follows from the one above by replacing $\underline{\beta}$ with $\Sigma_{p-1}^{-1}\,\underline{\sigma}_{p-1}$. One notes that the conditional variance does not depend upon $\underline{x}_{p-1,j}$. One also notes that the conditional means are linear functions of the components of $\underline{x}_{p-1}$. When

$$\Sigma_{p-1} = \text{diagonal matrix},$$

that is, when the components of $\underline{x}_{p-1}$ are statistically independent, then

$$\beta_j = \frac{\sigma_{jp}}{\sigma_j^2}, \qquad j = 1, \ldots, p-1.$$

The square of the correlation between x_p and a linear function of the components of $\underline{x}_{p-1}$, say $m = \underline{k}'\underline{x}_{p-1}$, is given by

$$\rho^2_{x_p m} = \frac{\text{cov}^2\,(x_p,m)}{\text{var}\,(x_p)\,\text{var}\,(m)} = \frac{(\underline{k}'\underline{\sigma}_{p-1})^2}{\sigma_p^2(\underline{k}'\Sigma_{p-1}\,\underline{k})}.$$

The numerator follows from the relationship

$$\text{cov}\,(x_p, \underline{k}'\underline{x}_{p-1}) = \sum_{j=1}^{p-1} k_j\sigma_{jp} = \underline{k}'\underline{\sigma}_{p-1}.$$

The maximum possible squared correlation is attained when

$$\underline{k} = c\underline{\beta} \qquad \text{or} \qquad m = c\underline{\beta}'\underline{x}_{p-1},$$

where

$$c = \text{arbitrary constant},$$

$$\underline{\beta} = \Sigma_{p-1}^{-1}\,\underline{\sigma}_{p-1}.$$

For this choice of the vector $\underline{k}$ one has

$$\rho^2_{x_p m} = \frac{(c\underline{\beta}'\underline{\sigma}_{p-1})^2}{\sigma_p^2(c^2\underline{\beta}'\Sigma_{p-1}\,\underline{\beta})}$$

$$= \frac{(\underline{\beta}'\underline{\sigma}_{p-1})^2}{\sigma_p^2(\underline{\beta}'\Sigma_{p-1}\Sigma_{p-1}^{-1}\,\underline{\sigma}_{p-1})}$$

$$= \frac{(\underline{\beta}'\underline{\sigma}_{p-1})^2}{\sigma_p^2(\underline{\beta}'\underline{\sigma}_{p-1})} = \frac{\underline{\beta}'\underline{\sigma}_{p-1}}{\sigma_p^2}.$$

This maximum correlation is called the *multiple correlation.* $\rho_{x_p m}$ is taken to be $+\sqrt{\rho^2_{x_p m}}$.

If one drops the arbitrary constant c and lets

$$m = \underline{\beta}'\underline{x}_{p-1},$$

then the random variable m is univariate normal with expected value

$$\text{E}(m) = \underline{\beta}'\underline{\mu}_{p-1}$$

and variance

$$\sigma_m^2 = \underline{\beta}'\Sigma_{p-1}\underline{\beta} = \underline{\sigma}_{p-1}'\Sigma_{p-1}^{-1}\,\underline{\sigma}_{p-1} = \underline{\beta}'\underline{\sigma}_{p-1}.$$

Let the vector $\underline{x}$ now be partitioned as follows:

$$\underline{x} = \begin{bmatrix} x_1 \\ \cdot \\ \cdot \\ \cdot \\ x_k \\ \hline x_{k+1} \\ \cdot \\ \cdot \\ \cdot \\ x_p \end{bmatrix} = \begin{bmatrix} \underline{x}_1 \\ \underline{x}_2 \end{bmatrix},$$

where $\underline{x}_1$ contains the first k components of $\underline{x}$, and $\underline{x}_2$ contains the last $p - k$ components of $\underline{x}$. Let $\underline{\mu}$ and Σ be partitioned in a similar manner.

$$\underline{\mu} = \begin{matrix} k \\ p-k \end{matrix}\begin{bmatrix} \underline{\mu}_1 \\ \underline{\mu}_2 \end{bmatrix}, \qquad \Sigma = \begin{matrix} \\ k \\ p-k \end{matrix}\begin{matrix} k & p-k \\ \begin{bmatrix} \Sigma_{11} & \Sigma_{12} \\ \Sigma_{21} & \Sigma_{22} \end{bmatrix} \end{matrix},$$

where Σ_{11} = dispersion matrix of $\underline{x}_1$,
Σ_{22} = dispersion matrix of $\underline{x}_2$,
Σ_{12} = covariance matrix between $\underline{x}_1$ and $\underline{x}_2$.

The vector variables $\underline{x}_1$ and $\underline{x}_2$ are distributed as follows:

$$\underline{x}_1\colon N_k(\underline{\mu}_1, \Sigma_{11}), \qquad \underline{x}_2\colon N_{p-k}(\underline{\mu}_2, \Sigma_{22}).$$

The joint conditional distribution of the vector variable $\underline{x}_1$ given that $\underline{x}_2$ is equal to some fixed value, say $\underline{x}_{2j}$, will be

$$f(\underline{x}_1 \mid \underline{x}_2 = \underline{x}_{2j}) = N_k(\underline{\mu}_{1|2j}, \Sigma_{1|2}),$$

where

$$\Sigma_{1|2} = \Sigma_{11} - B'\Sigma_{22}B,$$
$$B = \Sigma_{22}^{-1}\Sigma_{21}.$$

An alternative expression for $\Sigma_{1|2}$, the matrix of conditional variances and covariances, is obtained if B is replaced by its definition:

$$\Sigma_{1|2} = \Sigma_{11} - \Sigma_{12}\Sigma_{22}^{-1}\Sigma_{21}.$$

A typical diagonal element in this matrix is

$$\sigma^2(x_j \mid \underline{x}_2) = \sigma^2_{x_j} - \underline{\beta}_j'\Sigma_{22}\,\underline{\beta}_j, \qquad j = 1, 2, \ldots, k,$$

where

$$\underline{\beta}_j = j\text{th column of } B.$$

$\sigma^2(x_j \mid \underline{x}_2)$ is called the conditional variance of x_j given that $\underline{x}_2$ is some fixed value. A typical off-diagonal element in the matrix $\Sigma_{1|2}$ has the form

$$\sigma(x_i,x_j \mid \underline{x}_2) = \sigma_{x_ix_j} - \underline{\beta}_i'\Sigma_{22}\,\underline{\beta}_j,$$

where

$$\underline{\beta}_i = i\text{th column of } B,$$

$$\underline{\beta}_j = j\text{th column of } B.$$

The expected value of this joint conditional distribution is

$$\mathrm{E}(\underline{x}_1 \mid \underline{x}_{2j}) = \underline{\alpha} + B'\underline{x}_{2j},$$

where

$$\underset{p-k,k}{B} = \Sigma_{22}^{-1}\Sigma_{21}$$

$$\underset{k,1}{\underline{\alpha}} = \underline{\mu}_1 - B'\underline{\mu}_2.$$

Consider now two populations of elements. In both populations the random variable is $\underline{x}$. Assume that in population 1 the distribution of $\underline{x}$ is $N_p(\underline{\mu}_1,\Sigma)$; in population 2, assume that the distribution of $\underline{x}$ is $N_p(\underline{\mu}_2,\Sigma)$. Note that Σ is the same for both distributions. The ratio of the two probability densities for the vector $\underline{x}$ is

$$\lambda(\underline{x}) = \frac{\exp\left[-\frac{1}{2}(\underline{x} - \underline{\mu}_1)'\Sigma^{-1}(\underline{x} - \underline{\mu}_1)\right]}{\exp\left[-\frac{1}{2}(\underline{x} - \underline{\mu}_2)'\Sigma^{-1}(\underline{x} - \underline{\mu}_2)\right]}.$$

If one lets

$$y = L(\underline{x}) = \ln \lambda(\underline{x}),$$

then

$$\begin{aligned} y = L(\underline{x}) &= \left[-\tfrac{1}{2}(\underline{x} - \underline{\mu}_1)'\Sigma^{-1}(\underline{x} - \underline{\mu}_1)\right] - \left[-\tfrac{1}{2}(\underline{x} - \underline{\mu}_2)'\Sigma^{-1}(\underline{x} - \underline{\mu}_2)\right] \\ &= (\underline{\mu}_1' - \underline{\mu}_2')\Sigma^{-1}\underline{x} - \tfrac{1}{2}(\underline{\mu}_1'\Sigma^{-1}\underline{\mu}_1 - \underline{\mu}_2'\Sigma^{-1}\underline{\mu}_2) \\ &= \underline{\beta}'\underline{x} - c, \end{aligned}$$

where $\quad \underline{\beta} = \Sigma^{-1}(\underline{\mu}_1 - \underline{\mu}_2), \qquad c = \frac{1}{2}(\underline{\mu}_1'\Sigma^{-1}\underline{\mu}_1 - \underline{\mu}_2'\Sigma^{-1}\underline{\mu}_2).$

$\lambda(\underline{x})$ is called the likelihood ratio. $L(x)$ is called the linear discriminant function. Thus the linear discriminant function is the natural logarithm of the likelihood ratio.

The random variable y defined in the preceding paragraph is a linear function of the components of $\underline{x}$. Hence y is normally distributed. One has

$$L(\underline{\mu}_1) = \underline{\beta}'\underline{\mu}_1 - c,$$

$$L(\underline{\mu}_2) = \underline{\beta}'\underline{\mu}_2 - c.$$

Hence $\quad L(\underline{\mu}_1) - L(\underline{\mu}_2) = \underline{\beta}'(\underline{\mu}_1 - \underline{\mu}_2) = (\underline{\mu}_1' - \underline{\mu}_2')\Sigma^{-1}(\underline{\mu}_1 - \underline{\mu}_2)$

$$= \underline{\beta}'\Sigma\underline{\beta}.$$

From the definition of $L(\underline{x})$, it follows that

$$\sigma_y^2 = \underline{\beta}'\Sigma\underline{\beta} = L(\underline{\mu}_1) - L(\underline{\mu}_2).$$

hold is that any *one* of the following be true:

(i) P_j be idempotent, $j = 1, \ldots, m,$

(ii) $P_j P_{j'} = 0,$ $j \neq j',$

(iii) $\text{rank}\,(\Sigma P_j) = \Sigma(\text{rank}\,P_j).$

An example of the application of this last theorem is as follows: Suppose

$$Q_j\colon \chi^2(k_j;\lambda_j) \quad \text{where } \lambda_j = \underline{\mu}' P_j\,\underline{\mu},$$
$$Q_{j'}\colon \chi^2(k_{j'};0) \quad \text{where } \underline{\mu}' P_{j'}\,\underline{\mu} = 0,\ P_j P_{j'} = 0.$$

Then $F = \dfrac{Q_j/k_j}{Q_{j'}/k_{j'}}$ is distributed as $F(k_j, k_{j'};\,\lambda_j)$.

APPENDIX B

TOPICS CLOSELY RELATED TO THE ANALYSIS OF VARIANCE

B.1 Kruskal-Wallis *H* Test

Analogous to a single classification analysis of variance in which there are no repeated measures is the analysis of variance for ranked data. Suppose that there are k treatment classes having n_j observations in each class. Suppose further that the observations are in the form of ranks. That is, the criterion scores are ranks assigned irrespective of the treatment class to which an observation belongs. The data given below illustrate what is meant:

Treatment 1	Treatment 2	Treatment 3
1	3	6
2	5	9
4	8	12
7	10	13
	11	14

To test the hypothesis that the ranks within the treatment classes are a random sample from a common population of ranks, the following statistic may be used:

$$H = \frac{SS_{treat}}{MS_{total}}.$$

Numerator and denominator of the H statistic have the usual analysis-of-variance definitions.

When the hypothesis being tested is true, and when each n_j is larger than 5, the sampling distribution of this statistic may be approximated by a chi-square distribution having $k - 1$ degrees of freedom. For small values of n_j and k, special tables for the H statistic are available. Computational procedures for this test duplicate the procedures for a single classification analysis of variance. The latter procedures correct for tied ranks, whereas the specialized formulas for the H statistic require corrections for tied ranks, if these occur.

Wallace (1959) compared various methods of approximating the exact sampling distribution of the H statistic for the special case of $k = 3$ and each $n_j < 6$. When all $n_j = 5$, the chi-square approximation gave results which differed only in the third decimal place from the exact probability. A typical example is the following: When the exact probability was .0094, the chi-square approximation was .0089.

The Mann-Whitney U statistic is closely related to the H statistic when $k = 2$. Extensive tables for the U statistic for small n_j are available [see Siegel (1956, pp. 271–277)]. Individual comparisons between two treatments following an overall H test may be made by means of the U statistic. An application of this procedure will be found in Lewis and Cotton (1958). If one of the treatments represents a control group, the nonparametric analog of the Dunnett procedure is described in Sec. B.3 of this appendix.

A different approach for handling data which are in terms of ranks is to transform the ranks into normalized scores. Tables for doing this are given in Walker and Lev (1953, p. 480). More extensive tables of expected values of normal order statistics are in Harter (1960). In the latter form the data may be handled by means of the usual analysis of variance. The latter approach may lead to somewhat different conclusions. If the population to which inferences are to be made is considered to be one in which the criterion scores are normally distributed, then the analysis of variance in terms of the transformed scores is the more appropriate. On the other hand, if inferences are limited to ordinal measurement on the criterion scale, then the Kruskal-Wallis H statistic provides the more appropriate type of analysis.

B.2 Contingency Table with Repeated Measures

Consider an experiment in which n judges are asked to assign ranks to r products. Data obtained from this experiment may be summarized as follows:

Product	Rank						Total
	1	2	...	j	...	r	
1	n_{11}	n_{12}	...	n_{1j}	...	n_{1r}	n
2	n_{21}	n_{22}	...	n_{2j}	...	n_{2r}	n
.	.	.		.		.	.
.	.	.		.		.	.
.	.	.		.		.	.
i	n_{i1}	n_{i2}	...	n_{ij}	...	n_{ir}	n
.	.	.		.		.	.
.	.	.		.		.	.
.	.	.		.		.	.
r	n_{r1}	n_{r2}	...	n_{rj}	...	n_{rr}	n
Total	n	n	...	n	...	n	nr

In this summary, n_{ij} represents the number of times product i receives a rank of j. (The sampling distributions to be discussed in this section are obtained by limiting procedures which assume n to be large. The approximations have been shown to be reasonably close for $nr = 30$ and larger. In similar limiting procedures, $n = 5$ and larger provide adequate approximations.)

To test the hypothesis of no differences between the products with respect to the frequency with which the products receive the rank of j, the statistic

$$Q_j^2 = \frac{r\sum_i [n_{ij} - (n/r)]^2}{n} = \frac{(r\sum_i n_{ij}^2) - n^2}{n}$$

may be used. When there are no differences between the frequencies in column j, except those due to sampling error, Q_j has a sampling distribution which is approximated by a chi-square distribution having $r - 1$ degrees of freedom.

To test the overall hypothesis of no differences between the ranks for the products, the statistic

$$Q^2 = \Sigma Q_j^2$$

may be used. Under the hypothesis of no difference between the ranks assigned to the products, Anderson (1959) has shown that the statistic

$$\frac{(r-1)Q^2}{r} \doteq \chi^2 \qquad \text{with df} = (r-1)^2.$$

If the statistic $(r - 1)Q^2/r$ exceeds the critical value for a test having level of significance α, as determined from the appropriate sampling distribution, the hypothesis of no difference between the frequencies within the columns is rejected.

Anderson (1959) has shown that the Friedman statistic discussed in Sec. 4.7 provides a test on $r - 1$ components of the overall chi square. The latter may be partitioned into individual comparisons, or contrasts, each having a single degree of freedom. In making tests on such comparisons, one uses the following estimates of the variances and covariances for the cell frequencies:

$$\text{var}\,(n_{ij}) = \frac{n(r-1)}{r^2},$$

$$\text{cov}\,(n_{ij},n_{ik}) = \frac{-n}{r^2},$$

where n_{ij} and n_{ik} are two frequencies in the same row,

$$\text{cov}\,(n_{ij},n_{kj}) = \frac{-n}{r^2},$$

where n_{ij} and n_{kj} are two frequencies in the same column, and

$$\text{cov}\,(n_{ij},n_{km}) = \frac{n}{r^2(r-1)},$$

where n_{ij} and n_{km} are two frequencies in different rows and columns.

A 3×3 contingency table will be used for illustrative purposes. The cell frequencies are the column headings.

	n_{11}	n_{12}	n_{13}	n_{21}	n_{22}	n_{23}	n_{31}	n_{32}	n_{33}
C_1	0	0	0	−1	0	1	0	0	0
C_2	1	0	0	0	0	0	−1	0	0
C_3	1	0	−1	0	0	0	−1	0	1

The coefficients in row C_1 represent a linear comparison among the ranks assigned to product 2. The numerical value of the chi-square statistic corresponding to this comparison is

$$\chi^2_{C_1} = \frac{(n_{23} - n_{21})^2}{\text{var}\,(n_{23} - n_{21})}.$$

The denominator of this statistic is

$$\begin{aligned}\text{var}\,(n_{23} - n_{21}) &= \text{var}\,(n_{23}) + \text{var}\,(n_{21}) - 2\,\text{cov}\,(n_{23},n_{21}) \\ &= \frac{n(r-1)}{r^2} + \frac{n(r-1)}{r^2} - \frac{2(-n)}{r^2} \\ &= \frac{2n}{r}.\end{aligned}$$

The above chi-square statistic has one degree of freedom. Should this statistic exceed the critical value for an α-level test, the data would indicate a statistically significant difference between the rankings assigned to product 2.

The coefficients in row C_2 represent a linear comparison among the products for rank 1. (This comparison is not orthogonal to C_1.) The chi-square statistic corresponding to this comparison is

$$\chi^2_{C_2} = \frac{(n_{11} - n_{31})^2}{\text{var}\,(n_{11} - n_{31})} = \frac{(n_{11} - n_{31})^2}{2n/r}$$

$$= \frac{r(n_{11} - n_{31})^2}{2n}.$$

This chi-square statistic has one degree of freedom.

Table B.2-1 Numerical Example

(i)

Judge	Product a	Product b	Product c	Total
1	2	1	3	$6 = P_1$
2	1	2	3	6
3	1	3	2	6
4	1	2	3	6
5	2	1	3	6
6	1	2	3	6
7	1	3	2	6
8	1	2	3	6
	$T_a = 10$	16	22	48

$G^2/nr = (48^2)/8(3) = 96.00$ $\qquad (\Sigma T_j^2)/n = 105.00$

$\Sigma X^2 = 112$ $\qquad (\Sigma P_i^2)/r = 96.00$

$SS_{\text{products}} = 105 - 96.00 = 9.00$

$SS_{\text{w. judge}} = 112 - 96.00 = 16.00$ $\qquad MS_{\text{w. judge}} = 16.00/16 = 1.00$

$$\chi^2_{\text{ranks}} = \frac{SS_{\text{products}}}{MS_{\text{w. judge}}} = 9.00$$

(ii)

Product	Rank 1	Rank 2	Rank 3	Total
a	6	2	0	8
b	2	4	2	8
c	0	2	6	8
	8	8	8	

$\chi^2_{C_1} = 3(2 - 2)^2/2(8) = 0$

$\chi^2_{C_3} = 2(6 - 0 + 6 - 0)/4(8) = 9$

$\chi^2_{C_1} + \chi^2_{C_3} = 9$

The coefficients in row C_3 represent a comparison between the differences in linear rankings for products 1 and 3. (Comparison C_3 is orthogonal to comparison C_1.) The chi-square statistic corresponding to this comparison is

$$\chi^2_{C_3} = \frac{(n_{11} - n_{13} + n_{33} - n_{31})^2}{\text{var}\,(n_{11} - n_{13} + n_{33} - n_{31})}.$$

The individual variances and covariances required to obtain the term in the denominator are given by

$$\begin{aligned}\text{var}\,(n_{11} - n_{13} + n_{33} - n_{31}) &= 4\,\text{var}\,(n_{ij}) - 4\,\text{cov}\,(n_{ij},n_{ik}) \\ &\quad - 4\,\text{cov}\,(n_{ij},n_{kj}) + 4\,\text{cov}\,(n_{ij},n_{km}) \\ &= \frac{4n}{r-1}.\end{aligned}$$

Thus $$\chi^2_{C_3} = \frac{(r-1)(n_{11} - n_{13} + n_{33} - n_{31})^2}{4n}.$$

The chi-square statistic used in the Friedman test is equivalent to the sum of $r - 1$ orthogonal comparisons among the products. For the case of a 3×3 contingency table, C_1 and C_3 are orthogonal comparisons of this kind. Hence

$$\chi^2_{C_1} + \chi^2_{C_3} = \chi^2_{\text{ranks}} = \frac{\text{SS}_{\text{products}}}{\text{MS}_{\text{w. judge}}},$$

where the last term on the right is the statistic used in the Friedman test.

The numerical example given in Table B.2-1 illustrates this last relationship. Basic data are given in part i. There are $n = 8$ judges and $r = 3$ products. The rankings assigned by each judge are shown. The variations obtained in part i are defined as follows:

$$\text{SS}_{\text{products}} = \frac{\Sigma T_j^2}{n} - \frac{G^2}{nr},$$

$$\text{SS}_{\text{w. judge}} = \Sigma X^2 - \frac{\Sigma P_i^2}{r}.$$

Computation of this latter source of variation may be simplified when no tied ranks are permitted. The critical value for a .05-level test is $\chi^2_{.95}(2) = 6.00$. Hence the test in part i indicates that the differences in ranks assigned to the products are statistically significant.

Data from part i are rearranged to form a contingency table in part ii. The comparison C_1, which was defined earlier in this section, indicates no difference in the linear ranking for product b. The critical value associated with a .05-level test on C_3 is $\chi^2_{.95}(1) = 3.8$. Hence the data indicate that there is a statistically significant difference between the linear rankings for products a and c. Note that

$$\chi^2_{C_1} + \chi^2_{C_3} = \chi^2_{\text{ranks}}.$$

B.3 Comparing Treatment Effects with a Control

Procedures for comparing all treatments with a control were discussed in Sec. 3.10. A nonparametric analog of these procedures has been developed by Steel (1959). A numerical example will be used to illustrate the procedures for comparing all treatments with a control when data are in terms of ranks. In a sense, these comparisons are part of the overall hypothesis tested by the Kruskal-Wallis H statistic.

The basic data for this numerical example are given in part i of Table B.3-1. Suppose that only the rank order of these measurements is

Table B.3-1 Numerical Example

	Control		Treatment a	Treatment b	Treatment c
(i)	45		35	58	75
	50		40	62	78
	60	$n = 5$	45	70	80
	62		48	78	80
	75		50	80	84

	Control			Treatment a	Treatment b	Treatment c
	a	b	c			
(ii)	3.5	1	1	1	3	5.5
	6.5	2	2	2	6	7
	8	4	3	3.5	8	8.5
	9	5	4	5	9	8.5
	10	7	5.5	6.5	10	10
	37.0	19.0	15.5	18.0	36.0	39.5
	T'_a	T'_b	T'_c	T_a	T_b	T_c

considered meaningful. The data in part ii are in terms of ranks. To obtain these ranks, the control scores and the treatment a scores are combined; then ranks 1 to $2n$ are assigned to the combined set of scores. In case of ties, the mean of the tied ranks is used. The combined sets of scores for the control and treatment a groups are as follows:

Scores . . .	35	40	45	<u>45</u>	48	50	<u>50</u>	<u>60</u>	<u>62</u>	<u>75</u>
Ranks . . .	1	2	3.5	<u>3.5</u>	5	6.5	<u>6.5</u>	<u>8</u>	<u>9</u>	<u>10</u>

Data from the control condition are underscored. The combined sets of scores from the control and treatment c conditions are as follows:

Scores ...	45	50	60	62	75	75	78	80	80	84
Ranks ...	1	2	3	4	5.5	5.5	7	8.5	8.5	10

The sum of the ranks for the control group and each of the treatment groups is then computed. T'_a represents the sum of ranks for the control condition when the scores are ranked with reference to treatment a. The test statistic used in the decision rule about the difference between treatment a and the control condition is $\min(T'_a,T_a)$, that is, the smaller of T'_a and T_a. In this case,

$$\min(T'_a,T_a) = \min(37,18) = 18.$$

As a partial check on the numerical work,

$$T'_i + T_i = n(2n + 1).$$

Steel (1959) constructed tables of the sampling distribution of the statistic $\min(T'_i,T_i)$. Probabilities in these tables are in terms of an experimentwise error rate. By definition, the latter is the ratio of the number of experiments with one or more false significance statements to the total number of experiments. For the case $n = 5$ and $k = 3$, where k is the number of treatments (excluding the control), the critical value for the rank sum statistic for a two-tailed test with error rate .05 is 16. The decision is made to reject the hypothesis of no difference between treatment i and the control if

$$\min(T'_i,T_i) \leq 16.$$

For the data in Table B.3-1, treatment c is statistically different from the control, but none of the other differences between the treatments and the control is statistically significant, with a .05-level experimentwise error rate. Had the direction of the differences between the control and the experimental groups been predicted prior to the experiment, one-tailed rather than two-tailed tests would be appropriate. The critical value for a .05-level one-tailed test in which $n = 5$ and $k = 3$ is 18.

B.4 General Partition of Degrees of Freedom in a Contingency Table

To illustrate the procedures to be discussed in this section, consider the three-dimensional contingency table having the following form (all observations are assumed to be independent):

	c_1		c_2	
	b_1	b_2	b_1	b_2
a_1	n_{111}	n_{121}	n_{112}	n_{122}
a_2	n_{211}	n_{221}	n_{212}	n_{222}
a_3	n_{311}	n_{321}	n_{312}	n_{322}

In general there will be p classes for category A, q classes for category B, and r classes for category C. The frequency in cell abc_{ijk} will be designated by the symbol n_{ijk}.

If the B category in the above contingency table is disregarded, the resulting AC summary table will have the following form:

	c_1	c_2	Total
a_1	$n_{1.1}$	$n_{1.2}$	$n_{1..}$
a_2	$n_{2.1}$	$n_{2.2}$	$n_{2..}$
a_3	$n_{3.1}$	$n_{3.2}$	$n_{3..}$
Total	$n_{..1}$	$n_{..2}$	$n_{...}$

In general the following notation will be used:

$$\sum_i n_{ijk} = n_{.jk}, \qquad \sum_j n_{ijk} = n_{i.k}, \qquad \sum_k n_{ijk} = n_{ij.};$$

$$\sum_i \sum_j n_{ijk} = \sum_i n_{i.k} = n_{..k}, \qquad \sum_i \sum_k n_{ijk} = \sum_i n_{ij.} = n_{.j.};$$

$$\sum_i \sum_j \sum_k n_{ijk} = \sum_i \sum_j n_{ij.} = \sum_i n_{i..} = n_{...}.$$

If sampling is random with respect to all categories, $pqr - 1$ parameters are necessary to specify the population from which the sample of size $n_{...}$ was drawn. These parameters may be specified in terms of the following proportions:

$$P_{ijk} = \text{proportion of population frequency in cell } abc_{ijk}.$$

The expected frequency in cell abc_{ijk}, which will be designated by the symbol n'_{ijk}, is

$$n'_{ijk} = P_{ijk}n_{...}.$$

The expected frequencies for the marginal totals of category A would be

$$n'_{i..} = \sum_j \sum_k n'_{ijk};$$

alternatively,

$$n'_{i..} = \sum_j \sum_k P_{ijk}n_{...} = P_{i..}n_{...}.$$

The symbol $P_{i..}$ designates the population proportion for the category a_i. The other expected marginal frequencies are

$$n'_{.j.} = P_{.j.}n_{...},$$

$$n'_{..k} = P_{..k}n_{...}.$$

The expected frequency for a cell in the AB summary table is given by

$$n'_{ij.} = \sum_k n'_{ijk}$$

$$= \sum_k P_{ijk}n_{...} = P_{ij.}n_{...}.$$

Other expected frequencies for two-way summary tables are

$$n'_{i.k} = P_{i.k}n_{...},$$
$$n'_{.jk} = P_{.jk}n_{...}.$$

If all the $pqr - 1$ parameters in the population are specified by an a priori model, and if the sampling is random with respect to all categories, then the total chi square indicated in Table B.4-1 may be partitioned in the manner shown in this table. This partition bears a marked resemblance to an analysis-of-variance table.

Tests with respect to conformity with the specified model may be made, provided that the sampling distributions for the statistics indicated may be approximated by chi-square distributions. If each of the expected cell frequencies is greater than 5, the chi-square distributions will provide good approximations. If a relatively small number of expected frequencies are less than 5, the chi-square approximations will still be good.

A review of some of the work done on the partition of chi square in contingency tables appears in Sutcliffe (1957). A comprehensive review of methods for testing interactions is summarized in Goodman (1970). If the model for the population can be completely specified on a priori grounds, and if the sampling is random with respect to all categories, then the method of partition indicated in Table B.4-1 may be carried out quite

Table B.4-1 Partition of Chi Square

Source	Chi square	df
Total	$\chi^2_{\text{total}} = \Sigma\Sigma\Sigma[(n_{ijk} - n'_{ijk})^2/n'_{ijk}]$	$pqr - 1$
A	$\chi^2_a = \Sigma[(n_{i..} - n'_{i..})^2/n'_{i..}]$	$p - 1$
B	$\chi^2_b = \Sigma[(n_{.j.} - n'_{.j.})^2/n'_{.j.}]$	$q - 1$
C	$\chi^2_c = \Sigma[(n_{..k} - n'_{..j})^2/n'_{..k}]$	$r - 1$
AB	$\chi^2_{ab} = \Sigma\Sigma[(n_{ij.} - n'_{ij.})^2/n'_{ij.}] - \chi^2_a - \chi^2_b$	$(p - 1)(q - 1)$
AC	$\chi^2_{ac} = \Sigma\Sigma[(n_{i.k} - n'_{i.k})^2/n'_{i.k}] - \chi^2_a - \chi^2_c$	$(p - 1)(r - 1)$
BC	$\chi^2_{bc} = \Sigma\Sigma[(n_{.jk} - n'_{.jk})^2/n'_{.jk}] - \chi^2_b - \chi^2_c$	$(q - 1)(r - 1)$
ABC	$\chi^2_{abc} = \chi^2_{\text{total}} - \chi^2_a - \chi^2_b - \chi^2_c - \chi^2_{ab} - \chi^2_{ac} - \chi^2_{bc}$	$(p - 1)(q - 1)(r - 1)$

readily. In practice, however, certain of the parameters in the model are often estimated from the observed data. For example, the parameters $P_{i..}$, $P_{.j.}$, and $P_{..k}$ may be estimated from the marginal frequencies of the sample data. Under the hypothesis of no interactions of any order (i.e., no two-category or no three-category interactions), the expected proportion for cell abc_{ijk} is

$$P_{ijk} = P_{i..}P_{.j.}P_{..k},$$

and the expected frequency in cell abc_{ijk} is

$$n'_{ijk} = P_{ijk}n_{...}.$$

Under this method for specifying the model for the population, the total chi square may be partitioned as shown in Table B.4-2. In this case,

Table B.4-2 Partition of Chi Square When Probabilities Are Estimated from Marginal Totals

Source	Chi square	df
Total	$\chi^2_{\text{total}} = \Sigma\Sigma\Sigma[(n_{ijk} - n'_{ijk})^2/n'_{ijk}]$	$(pqr - 1) - (p - 1) - (q - 1) - (r - 1)$
AB	$\chi^2_{ab} = \Sigma\Sigma[(n_{ij.} - n'_{ij.})^2/n'_{ij.}]$	$(p - 1)(q - 1)$
AC	$\chi^2_{ac} = \Sigma\Sigma[(n_{i.k} - n'_{i.k})^2/n'_{i.k}]$	$(p - 1)(r - 1)$
BC	$\chi^2_{bc} = \Sigma\Sigma[(n_{.jk} - n'_{.jk})^2/n'_{.jk}]$	$(q - 1)(r - 1)$
ABC	$\chi^2_{abc} = \chi^2_{\text{total}} - \chi^2_{ab} - \chi^2_{ac} - \chi^2_{bc}$	$(p - 1)(q - 1)(r - 1)$

note that the degrees of freedom for the total chi square are

$$(pqr - 1) - (p - 1) - (q - 1) - (r - 1) = pqr - p - q - r + 2.$$

Should the three-factor interaction be statistically significant in this type of analysis, the two-way summary tables should be studied separately within a fixed level of the third category. In these latter tables, the marginal totals may be used in some cases to estimate the cell frequencies. For example, if the AB data for level c_1 are being studied under the hypothesis of no interaction between categories A and B for level c_1,

$$n'_{ij1} \doteq \frac{n_{i.1}n_{.j1}}{n_{..1}}.$$

This expected value for cell $ij1$ will not in general be the same as that obtained under the hypothesis of no interactions of any order.

Another case which arises in practice is one in which the sampling is restricted with respect to the number of observations in each of the cells of the form ab_{ij} but random with respect to the category C. If the marginal totals are used in the estimation of $P_{..k}$, then

$$P_{..k} = \frac{n_{..k}}{n_{...}}, \qquad P_{ij.} = \frac{n_{ij.}}{n_{...}}.$$

Under the hypothesis of no interactions,

$$n'_{ijk} = P_{ij.}P_{..k}n_{...}.$$

In this case, chi square may be partitioned in the following manner:

Source	df
Total	$(pq - 1)(r - 1)$
AC	$(p - 1)(r - 1)$
BC	$(q - 1)(r - 1)$
ABC	$(p - 1)(q - 1)(r - 1)$

Should the three-factor interaction prove to be statistically significant in this case, it is advisable to study the equivalent of simple effects for category C at each of the separate levels of factors A and B.

In analyzing contingency tables into main effects and interactions of various orders, Goodman (1970) defines such effects in terms of $\ln P_{ijk}$, $\ln P_{ij.}$, $\ln P_{i..}$, etc. Distribution problems associated with tests of hypotheses are more readily handled in terms of this transformation.

APPENDIX C

TABLES

Table C.1 Unit Normal Distribution†

$[P(z \le z_{1-\alpha}) = 1 - \alpha]$

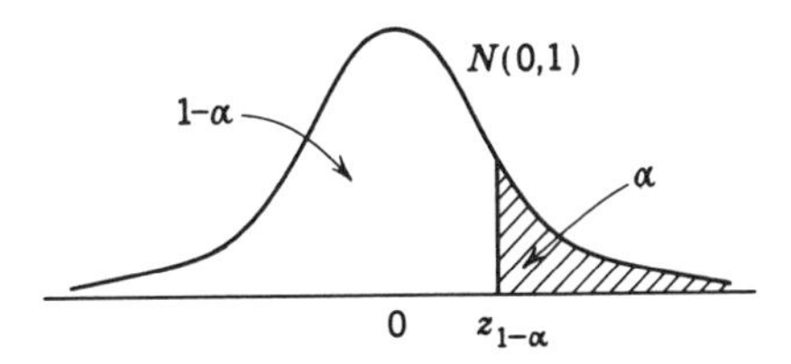

$z_{1-\alpha}$	$1-\alpha$	$z_{1-\alpha}$	$1-\alpha$	$z_{1-\alpha}$	$1-\alpha$	$z_{1-\alpha}$	$1-\alpha$
.00	.500	.35	.637	.70	.758	1.05	.853
.01	.504	.36	.641	.71	.761	1.06	.855
.02	.508	.37	.644	.72	.764	1.07	.858
.03	.512	.38	.648	.73	.767	1.08	.860
.04	.516	.39	.652	.74	.770	1.09	.862
.05	.520	.40	.655	.75	.773	1.10	.864
.06	.524	.41	.659	.76	.776	1.11	.867
.07	.528	.42	.663	.77	.779	1.12	.869
.08	.532	.43	.666	.78	.782	1.13	.871
.09	.536	.44	.670	.79	.785	1.14	.873
.10	.540	.45	.674	.80	.788	1.15	.875
.11	.544	.46	.677	.81	.791	1.16	.877
.12	.548	.47	.681	.82	.794	1.17	.879
.13	.552	.48	.684	.83	.797	1.18	.881
.14	.556	.49	.688	.84	.800	1.19	.883
.15	.560	.50	.691	.85	.802	1.20	.885
.16	.564	.51	.695	.86	.805	1.21	.887
.17	.567	.52	.698	.87	.808	1.22	.889
.18	.571	.53	.702	.88	.811	1.23	.891
.19	.575	.54	.705	.89	.813	1.24	.893
.20	.579	.55	.709	.90	.816	1.25	.894
.21	.583	.56	.712	.91	.819	1.26	.896
.22	.587	.57	.716	.92	.821	1.27	.898
.23	.591	.58	.719	.93	.824	1.28	.900
.24	.595	.59	.722	.94	.826	1.29	.901
.25	.599	.60	.726	.95	.829	1.30	.903
.26	.603	.61	.729	.96	.831	1.31	.905
.27	.606	.62	.732	.97	.834	1.32	.907
.28	.610	.63	.736	.98	.836	1.33	.908
.29	.614	.64	.739	.99	.839	1.34	.910
.30	.618	.65	.742	1.00	.841	1.35	.911
.31	.622	.66	.745	1.01	.844	1.36	.913
.32	.626	.67	.749	1.02	.846	1.37	.915
.33	.629	.68	.752	1.03	.848	1.38	.916
.34	.633	.69	.755	1.04	.851	1.39	.918

† This table is abridged from Table 9 in *Biometrika Tables for Statisticians*, vol. 1. (2d ed.) New York: Cambridge, 1958. Edited by E. S. Pearson and H. O. Hartley. Reproduced with kind permission of E. S. Pearson and the trustees of *Biometrika*.

Table C.1 (*Continued*)

$[P(z \leq z_{1-\alpha}) = 1 - \alpha]$

$z_{1-\alpha}$	$1 - \alpha$	$z_{1-\alpha}$	$1 - \alpha$	$z_{1-\alpha}$	$1 - \alpha$	$z_{1-\alpha}$	$1 - \alpha$
1.40	.919	1.75	.960	2.10	.982	2.45	.993
1.41	.921	1.76	.961	2.11	.983	2.46	.993
1.42	.922	1.77	.962	2.12	.983	2.47	.993
1.43	.924	1.78	.962	2.13	.983	2.48	.993
1.44	.925	1.79	.963	2.14	.984	2.49	.994
1.45	.926	1.80	.964	2.15	.984	2.50	.994
1.46	.928	1.81	.965	2.16	.985	2.51	.994
1.47	.929	1.82	.966	2.17	.985	2.52	.994
1.48	.931	1.83	.966	2.18	.985	2.53	.994
1.49	.932	1.84	.967	2.19	.986	2.54	.994
1.50	.933	1.85	.968	2.20	.986	2.55	.995
1.51	.934	1.86	.969	2.21	.986	2.56	.995
1.52	.936	1.87	.969	2.22	.987	2.57	.995
1.53	.937	1.88	.970	2.23	.987	2.58	.995
1.54	.938	1.89	.971	2.24	.987	2.59	.995
1.55	.939	1.90	.971	2.25	.988		
1.56	.941	1.91	.972	2.26	.988	2.60	.9953
1.57	.942	1.92	.973	2.27	.988		
1.58	.943	1.93	.973	2.28	.989	2.70	.9965
1.59	.944	1.94	.974	2.29	.989		
						2.80	.9974
1.60	.945	1.95	.974	2.30	.989		
1.61	.946	1.96	.975	2.31	.990	2.90	.9981
1.62	.947	1.97	.976	2.32	.990		
1.63	.948	1.98	.976	2.33	.990	3.00	.9987
1.64	.949	1.99	.977	2.34	.990		
						3.20	.9993
1.65	.951	2.00	.977	2.35	.991		
1.66	.952	2.01	.978	2.36	.991	3.40	.9997
1.67	.953	2.02	.978	2.37	.991		
1.68	.954	2.03	.979	2.38	.991	3.60	.9998
1.69	.954	2.04	.979	2.39	.992		
						3.80	.99993
1.70	.955	2.05	.980	2.40	.992		
1.71	.956	2.06	.980	2.41	.992	4.00	.999968
1.72	.957	2.07	.981	2.42	.992	4.50	.999997
1.73	.958	2.08	.981	2.43	.992	5.00	.9999997
1.74	.959	2.09	.982	2.44	.993	5.50	.9999999
1.645	.950	2.326	.990	3.090	.999	3.891	.99995
1.960	.975	2.576	.995	3.291	.9995	4.417	.999995

Table C.2 Student's t Distribution

df	Percentile point								
	70	75	80	87.5	90	95	97.5	99	99.5
1	.73	1.00	1.38	2.41	3.08	6.31	12.71	31.82	63.66
2	.62	.81	1.06	1.60	1.89	2.92	4.30	6.96	9.92
3	.58	.79	.98	1.42	1.64	2.35	3.18	4.54	5.84
4	.57	.77	.94	1.34	1.53	2.13	2.78	3.75	4.60
5	.56	.75	.92	1.30	1.48	2.02	2.57	3.36	4.03
6	.55	.74	.91	1.27	1.44	1.94	2.45	3.14	3.71
7	.55	.73	.90	1.25	1.42	1.89	2.36	3.00	3.50
8	.55	.72	.89	1.24	1.40	1.86	2.31	2.90	3.36
9	.54	.71	.88	1.23	1.38	1.83	2.26	2.82	3.25
10	.54	.70	.88	1.22	1.37	1.81	2.23	2.76	3.17
11	.54	.70	.88	1.21	1.36	1.80	2.20	2.72	3.11
12	.54	.69	.87	1.21	1.36	1.78	2.18	2.68	3.05
13	.54	.69	.87	1.21	1.35	1.77	2.16	2.65	3.01
14	.54	.69	.87	1.20	1.34	1.76	2.14	2.62	2.98
15	.54	.69	.87	1.20	1.34	1.75	2.13	2.60	2.95
16	.54	.69	.86	1.20	1.34	1.75	2.12	2.58	2.92
17	.53	.69	.86	1.19	1.33	1.74	2.11	2.57	2.90
18	.53	.69	.86	1.19	1.33	1.73	2.10	2.55	2.88
19	.53	.69	.86	1.19	1.33	1.73	2.09	2.54	2.86
20	.53	.69	.86	1.18	1.32	1.72	2.09	2.53	2.85
21	.53	.69	.86	1.18	1.32	1.72	2.08	2.52	2.83
22	.53	.69	.86	1.18	1.32	1.72	2.07	2.51	2.82
23	.53	.68	.86	1.18	1.32	1.71	2.07	2.50	2.81
24	.53	.68	.86	1.18	1.32	1.71	2.06	2.49	2.80
25	.53	.68	.86	1.18	1.32	1.71	2.06	2.49	2.79
26	.53	.68	.86	1.18	1.32	1.71	2.06	2.48	2.78
27	.53	.68	.86	1.18	1.31	1.70	2.05	2.47	2.77
28	.53	.68	.86	1.17	1.31	1.70	2.05	2.47	2.76
29	.53	.68	.85	1.17	1.31	1.70	2.05	2.46	2.76
30	.53	68	.85	1.17	1.31	1.70	2.04	2.46	2.75
40	.53	.68	.85	1.17	1.30	1.68	2.02	2.42	2.70
50	.53	.67	.85	1.16	1.30	1.68	2.01	2.40	2.68
60	.53	.67	.85	1.16	1.30	1.67	2.00	2.39	2.66
80	.53	.67	.85	1.16	1.29	1.66	1.99	2.37	2.64
100	.53	.67	.84	1.16	1.29	1.66	1.98	2.36	2.63
200	.52	.67	.84	1.15	1.29	1.65	1.97	2.35	2.60
500	.52	.67	.84	1.15	1.28	1.65	1.96	2.33	2.59
∞	.52	.67	.84	1.15	1.28	1.64	1.96	2.33	2.58

Table C.3

df for denom.	$1-\alpha$	1	2	3	4	5	6	7	8	9	10	11	12
		df for numerator											
1	.75	5.83	7.50	8.20	8.58	8.82	8.98	9.10	9.19	9.26	9.32	9.36	9.41
	.90	39.9	49.5	53.6	55.8	57.2	58.2	58.9	59.4	59.9	60.2	60.5	60.7
	.95	161	200	216	225	230	234	237	239	241	242	243	244
2	.75	2.57	3.00	3.15	3.23	3.28	3.31	3.34	3.35	3.37	3.38	3.39	3.39
	.90	8.53	9.00	9.16	9.24	9.29	9.33	9.35	9.37	9.38	9.39	9.40	9.41
	.95	18.5	19.0	19.2	19.2	19.3	19.3	19.4	19.4	19.4	19.4	19.4	19.4
	.99	98.5	99.0	99.2	99.2	99.3	99.3	99.4	99.4	99.4	99.4	99.4	99.4
3	.75	2.02	2.28	2.36	2.39	2.41	2.42	2.43	2.44	2.44	2.44	2.45	2.45
	.90	5.54	5.46	5.39	5.34	5.31	5.28	5.27	5.25	5.24	5.23	5.22	5.22
	.95	10.1	9.55	9.28	9.12	9.10	8.94	8.89	8.85	8.81	8.79	8.76	8.74
	.99	34.1	30.8	29.5	28.7	28.2	27.9	27.7	27.5	27.3	27.2	27.1	27.1
4	.75	1.81	2.00	2.05	2.06	2.07	2.08	2.08	2.08	2.08	2.08	2.08	2.08
	.90	4.54	4.32	4.19	4.11	4.05	4.01	3.98	3.95	3.94	3.92	3.91	3.90
	.95	7.71	6.94	6.59	6.39	6.26	6.16	6.09	6.04	6.00	5.96	5.94	5.91
	.99	21.2	18.0	16.7	16.0	15.5	15.2	15.0	14.8	14.7	14.5	14.4	14.4
5	.75	1.69	1.85	1.88	1.89	1.89	1.89	1.89	1.89	1.89	1.89	1.89	1.89
	.90	4.06	3.78	3.62	3.52	3.45	3.40	3.37	3.34	3.32	3.30	3.28	3.27
	.95	6.61	5.79	5.41	5.19	5.05	4.95	4.88	4.82	4.77	4.74	4.71	4.68
	.99	16.3	13.3	12.1	11.4	11.0	10.7	10.5	10.3	10.2	10.1	9.96	9.89
6	.75	1.62	1.76	1.78	1.79	1.79	1.78	1.78	1.77	1.77	1.77	1.77	1.77
	.90	3.78	3.46	3.29	3.18	3.11	3.05	3.01	2.98	2.96	2.94	2.92	2.90
	.95	5.99	5.14	4.76	4.53	4.39	4.28	4.21	4.15	4.10	4.06	4.03	4.00
	.99	13.7	10.9	9.78	9.15	8.75	8.47	8.26	8.10	7.98	7.87	7.79	7.72
7	.75	1.57	1.70	1.72	1.72	1.71	1.71	1.70	1.70	1.69	1.69	1.69	1.68
	.90	3.59	3.26	3.07	2.96	2.88	2.83	2.78	2.75	2.72	2.70	2.68	2.67
	.95	5.59	4.74	4.35	4.12	3.97	3.87	3.79	3.73	3.68	3.64	3.60	3.57
	.99	12.2	9.55	8.45	7.85	7.46	7.19	6.99	6.84	6.72	6.62	6.54	6.47
8	.75	1.54	1.66	1.67	1.66	1.66	1.65	1.64	1.64	1.64	1.63	1.63	1.62
	.90	3.46	3.11	2.92	2.81	2.73	2.67	2.62	2.59	2.56	2.54	2.52	2.50
	95	5.32	4.46	4.07	3.84	3.69	3.58	3.50	3.44	3.39	3.35	3.31	3.28
	.99	11.3	8.65	7.59	7.01	6.63	6.37	6.18	6.03	5.91	5.81	5.73	5.67
9	.75	1.51	1.62	1.63	1.63	1.62	1.61	1.60	1.60	1.59	1.59	1.58	1.58
	.90	3.36	3.01	2.81	2.69	2.61	2.55	2.51	2.47	2.44	2.42	2.40	2.38
	.95	5.12	4.26	3.86	3.63	3.48	3.37	3.29	3.23	3.18	3.14	3.10	3.07
	.99	10.6	8.02	6.99	6.42	6.06	5.80	5.61	5.47	5.35	5.26	5.18	5.11
10	.75	1.49	1.60	1.60	1.59	1.59	1.58	1.57	1.56	1.56	1.55	1.55	1.54
	.90	3.28	2.92	2.73	2.61	2.52	2.46	2.41	2.38	2.35	2.32	2.30	2.28
	.95	4.96	4.10	3.71	3.48	3.33	3.22	3.14	3.07	3.02	2.98	2.94	2.91
	.99	10.0	7.56	6.55	5.99	5.64	5.39	5.20	5.06	4.94	4.85	4.77	4.71
11	.75	1.47	1.58	1.58	1.57	1.56	1.55	1.54	1.53	1.53	1.52	1.52	1.51
	.90	3.23	2.86	2.66	2.54	2.45	2.39	2.34	2.30	2.27	2.25	2.23	2.21
	.95	4.84	3.98	3.59	3.36	3.20	3.09	3.01	2.95	2.90	2.85	2.82	2.79
	.99	9.65	7.21	6.22	5.67	5.32	5.07	4.89	4.74	4.63	4.54	4.46	4.40
12	.75	1.46	1.56	1.56	1.55	1.54	1.53	1.52	1.51	1.51	1.50	1.50	1.49
	.90	3.18	2.81	2.61	2.48	2.39	2.33	2.28	2.24	2.21	2.19	2.17	2.15
	.95	4.75	3.89	3.49	3.26	3.11	3.00	2.91	2.85	2.80	2.75	2.72	2.69
	.99	9.33	6.93	5.95	5.41	5.06	4.82	4.64	4.50	4.39	4.30	4.22	4.16

F Distribution†

df for numerator												1 − α	df for denom.
15	20	24	30	40	50	60	100	120	200	500	∞		
9.49	9.58	9.63	9.67	9.71	9.74	9.76	9.78	9.80	9.82	9.84	9.85	.75	
61.2	61.7	62.0	62.3	62.5	62.7	62.8	63.0	63.1	63.2	63.3	63.3	.90	1
246	248	249	250	251	252	252	253	253	254	254	254	.95	
3.41	3.43	3.43	3.44	3.45	3.45	3.46	3.47	3.47	3.48	3.48	3.48	.75	
9.42	9.44	9.45	9.46	9.47	9.47	9.47	9.48	9.48	9.49	9.49	9.49	.90	2
19.4	19.4	19.5	19.5	19.5	19.5	19.5	19.5	19.5	19.5	19.5	19.5	.95	
99.4	99.4	99.5	99.5	99.5	99.5	99.5	99.5	99.5	99.5	99.5	99.5	.99	
2.46	2.46	2.46	2.47	2.47	2.47	2.47	2.47	2.47	2.47	2.47	2.47	.75	
5.20	5.18	5.18	5.17	5.16	5.15	5.15	5.14	5.14	5.14	5.14	5.13	.90	3
8.70	8.66	8.64	8.62	8.59	8.58	8.57	8.55	8.55	8.54	8.53	8.53	.95	
26.9	26.7	26.6	26.5	26.4	26.4	26.3	26.2	26.2	26.2	26.1	26.1	.99	
2.08	2.08	2.08	2.08	2.08	2.08	2.08	2.08	2.08	2.08	2.08	2.08	.75	
3.87	3.84	3.83	3.82	3.80	3.80	3.79	3.78	3.78	3.77	3.76	3.76	.90	
5.86	5.80	5.77	5.75	5.72	5.70	5.69	5.66	5.66	5.65	5.64	5.63	.95	4
14.2	14.0	13.9	13.8	13.7	13.7	13.7	13.6	13.6	13.5	13.5	13.5	.99	
1.89	1.88	1.88	1.88	1.88	1.88	1.87	1.87	1.87	1.87	1.87	1.87	.75	
3.24	3.21	3.19	3.17	3.16	3.15	3.14	3.13	3.12	3.12	3.11	3.10	.90	5
4.62	4.56	4.53	4.50	4.46	4.44	4.43	4.41	4.40	4.39	4.37	4.36	.95	
9.72	9.55	9.47	9.38	9.29	9.24	9.20	9.13	9.11	9.08	9.04	9.02	.99	
1.76	1.76	1.75	1.75	1.75	1.75	1.74	1.74	1.74	1.74	1.74	1.74	.75	
2.87	2.84	2.82	2.80	2.78	2.77	2.76	2.75	2.74	2.73	2.73	2.72	.90	6
3.94	3.87	3.84	3.81	3.77	3.75	3.74	3.71	3.70	3.69	3.68	3.67	.95	
7.56	7.40	7.31	7.23	7.14	7.09	7.06	6.99	6.97	6.93	6.90	6.88	.99	
1.68	1.67	1.67	1.66	1.66	1.66	1.65	1.65	1.65	1.65	1.65	1.65	.75	
2.63	2.59	2.58	2.56	2.54	2.52	2.51	2.50	2.49	2.48	2.48	2.47	.90	7
3.51	3.44	3.41	3.38	3.34	3.32	3.30	3.27	3.27	3.25	3.24	3.23	.95	
6.31	6.16	6.07	5.99	5.91	5.86	5.82	5.75	5.74	5.70	5.67	5.65	.99	
1.62	1.61	1.60	1.60	1.59	1.59	1.59	1.58	1.58	1.58	1.58	1.58	.75	
2.46	2.42	2.40	2.38	2.36	2.35	2.34	2.32	2.32	2.31	2.30	2.29	.90	8
3.22	3.15	3.12	3.08	3.04	3.02	3.01	2.97	2.97	2.95	2.94	2.93	.95	
5.52	5.36	5.28	5.20	5.12	5.07	5.03	4.96	4.95	4.91	4.88	4.86	.99	
1.57	1.56	1.56	1.55	1.55	1.54	1.54	1.53	1.53	1.53	1.53	1.53	.75	
2.34	2.30	2.28	2.25	2.23	2.22	2.21	2.19	2.18	2.17	2.17	2.16	.90	9
3.01	2.94	2.90	2.86	2.83	2.80	2.79	2.76	2.75	2.73	2.72	2.71	.95	
4.96	4.81	4.73	4.65	4.57	4.52	4.48	4.42	4.40	4.36	4.33	4.31	.99	
1.53	1.52	1.52	1.51	1.51	1.50	1.50	1.49	1.49	1.49	1.48	1.48	.75	
2.24	2.20	2.18	2.16	2.13	2.12	2.11	2.09	2.08	2.07	2.06	2.06	.90	10
2.85	2.77	2.74	2.70	2.66	2.64	2.62	2.59	2.58	2.56	2.55	2.54	.95	
4.56	4.41	4.33	4.25	4.17	4.12	4.08	4.01	4.00	3.96	3.93	3.91	.99	
1.50	1.49	1.49	1.48	1.47	1.47	1.47	1.46	1.46	1.46	1.45	1.45	.75	
2.17	2.12	2.10	2.08	2.05	2.04	2.03	2.00	2.00	1.99	1.98	1.97	.90	11
2.72	2.65	2.61	2.57	2.53	2.51	2.49	2.46	2.45	2.43	2.42	2.40	.95	
4.25	4.10	4.02	3.94	3.86	3.81	3.78	3.71	3.69	3.66	3.62	3.60	.99	
1.48	1.47	1.46	1.45	1.45	1.44	1.44	1.43	1.43	1.43	1.42	1.42	.75	
2.10	2.06	2.04	2.01	1.99	1.97	1.96	1.94	1.93	1.92	1.91	1.90	.90	12
2.62	2.54	2.51	2.47	2.43	2.40	2.38	2.35	2.34	2.32	2.31	2.30	.95	
4.01	3.86	3.78	3.70	3.62	3.57	3.54	3.47	3.45	3.41	3.38	3.36	.99	

Table C.3

df for denom.	1 − α	df for numerator											
		1	2	3	4	5	6	7	8	9	10	11	12
13	.75	1.45	1.54	1.54	1.53	1.52	1.51	1.50	1.49	1.49	1.48	1.47	1.47
	.90	3.14	2.76	2.56	2.43	2.35	2.28	2.23	2.20	2.16	2.14	2.12	2.10
	.95	4.67	3.81	3.41	3.18	3.03	2.92	2.83	2.7[illegible]	2.71	2.67	2.63	2.60
	.99	9.07	6.70	5.74	5.21	4.86	4.62	4.44	4.30	4.19	4.10	4.02	3.96
14	.75	1.44	1.53	1.53	1.52	1.51	1.50	1.48	1.48	1.47	1.46	1.46	1.45
	.90	3.10	2.73	2.52	2.39	2.31	2.24	2.19	2.15	2.12	2.10	2.08	2.05
	.95	4.60	3.74	3.34	3.11	2.96	2.85	2.76	2.70	2.65	2.60	2.57	2.53
	.99	8.86	6.51	5.56	5.04	4.69	4.46	4.28	4.14	4.03	3.94	3.86	3.80
15	.75	1.43	1.52	1.52	1.51	1.49	1.48	1.47	1.46	1.46	1.45	1.44	1.44
	.90	3.07	2.70	2.49	2.36	2.27	2.21	2.16	2.12	2.09	2.06	2.04	2.02
	.95	4.54	3.68	3.29	3.06	2.90	2.79	2.71	2.64	2.59	2.54	2.51	2.48
	.99	8.68	6.36	5.42	4.89	4.56	4.32	4.14	4.00	3.89	3.80	3.73	3.67
16	.75	1.42	1.51	1.51	1.50	1.48	1.48	1.47	1.46	1.45	1.45	1.44	1.44
	.90	3.05	2.67	2.46	2.33	2.24	2.18	2.13	2.09	2.06	2.03	2.01	1.99
	.95	4.49	3.63	3.24	3.01	2.85	2.74	2.66	2.59	2.54	2.49	2.46	2.42
	.99	8.53	6.23	5.29	4.77	4.44	4.20	4.03	3.89	3.78	3.69	3.62	3.55
17	.75	1.42	1.51	1.50	1.49	1.47	1.46	1.45	1.44	1.43	1.43	1.42	1.41
	.90	3.03	2.64	2.44	2.31	2.22	2.15	2.10	2.06	2.03	2.00	1.98	1.96
	.95	4.45	3.59	3.20	2.96	2.81	2.70	2.61	2.55	2.49	2.45	2.41	2.38
	.99	8.40	6.11	5.18	4.67	4.34	4.10	3.93	3.79	3.68	3.59	3.52	3.46
18	.75	1.41	1.50	1.49	1.48	1.46	1.45	1.44	1.43	1.42	1.42	1.41	1.40
	.90	3.01	2.62	2.42	2.29	2.20	2.13	2.08	2.04	2.00	1.98	1.96	1.93
	.95	4.41	3.55	3.16	2.93	2.77	2.66	2.58	2.51	2.46	2.41	2.37	2.34
	.99	8.29	6.01	5.09	4.58	4.25	4.01	3.84	3.71	3.60	3.51	3.43	3.37
19	.75	1.41	1.49	1.49	1.47	1.46	1.44	1.43	1.42	1.41	1.41	1.40	1.40
	.90	2.99	2.61	2.40	2.27	2.18	2.11	2.06	2.02	1.98	1.96	1.94	1.91
	.95	4.38	3.52	3.13	2.90	2.74	2.63	2.54	2.48	2.42	2.38	2.34	2.31
	.99	8.18	5.93	5.01	4.50	4.17	3.94	3.77	3.63	3.52	3.43	3.36	3.30
20	.75	1.40	1.49	1.48	1.46	1.45	1.44	1.42	1.42	1.41	1.40	1.39	1.39
	.90	2.97	2.59	2.38	2.25	2.16	2.09	2.04	2.00	1.96	1.94	1.92	1.89
	.95	4.35	3.49	3.10	2.87	2.71	2.60	2.51	2.45	2.39	2.35	2.31	2.28
	.99	8.10	5.85	4.94	4.43	4.10	3.87	3.70	3.56	3.46	3.37	3.29	3.23
22	.75	1.40	1.48	1.47	1.45	1.44	1.42	1.41	1.40	1.39	1.39	1.38	1.37
	.90	2.95	2.56	2.35	2.22	2.13	2.06	2.01	1.97	1.93	1.90	1.88	1.86
	.95	4.30	3.44	3.05	2.82	2.66	2.55	2.46	2.40	2.34	2.30	2.26	2.23
	.99	7.95	5.72	4.82	4.31	3.99	3.76	3.59	3.45	3.35	3.26	3.18	3.12
24	.75	1.39	1.47	1.46	1.44	1.43	1.41	1.40	1.39	1.38	1.38	1.37	1.36
	.90	2.93	2.54	2.33	2.19	2.10	2.04	1.98	1.94	1.91	1.88	1.85	1.83
	.95	4.26	3.40	3.01	2.78	2.62	2.51	2.42	2.36	2.30	2.25	2.21	2.18
	.99	7.82	5.61	4.72	4.22	3.90	3.67	3.50	3.36	3.26	3.17	3.09	3.03
26	.75	1.38	1.46	1.45	1.44	1.42	1.41	1.40	1.39	1.37	1.37	1.36	1.35
	.90	2.91	2.52	2.31	2.17	2.08	2.01	1.96	1.92	1.88	1.86	1.84	1.81
	.95	4.23	3.37	2.98	2.74	2.59	2.47	2.39	2.32	2.27	2.22	2.18	2.15
	.99	7.72	5.53	4.64	4.14	3.82	3.59	3.42	3.29	3.18	3.09	3.02	2.96
28	.75	1.38	1.46	1.45	1.43	1.41	1.40	1.39	1.38	1.37	1.36	1.35	1.34
	.90	2.89	2.50	2.29	2.16	2.06	2.00	1.94	1.90	1.87	1.84	1.81	1.79
	.95	4.20	3.34	2.95	2.71	2.56	2.45	2.36	2.29	2.24	2.19	2.15	2.12
	.99	7.64	5.45	4.57	4.07	3.75	3.53	3.36	3.23	3.12	3.03	2.96	2.90

F **Distribution** (*Continued*)†

df for numerator												1 − α	df for denom.
15	20	24	30	40	50	60	100	120	200	500	∞		
1.46	1.45	1.44	1.43	1.42	1.42	1.42	1.41	1.41	1.40	1.40	1.40	.75	
2.05	2.01	1.98	1.96	1.93	1.92	1.90	1.88	1.88	1.86	1.85	1.85	.90	13
2.53	2.46	2.42	2.38	2.34	2.31	2.30	2.26	2.25	2.23	2.22	2.21	.95	
3.82	3.66	3.59	3.51	3.43	3.38	3.34	3.27	3.25	3.22	3.19	3.17	.99	
1.44	1.43	1.42	1.41	1.41	1.40	1.40	1.39	1.39	1.39	1.38	1.38	.75	
2.01	1.96	1.94	1.91	1.89	1.87	1.86	1.83	1.83	1.82	1.80	1.80	.90	
2.46	2.39	2.35	2.31	2.27	2.24	2.22	2.19	2.18	2.16	2.14	2.13	.95	14
3.66	3.51	3.43	3.35	3.27	3.22	3.18	3.11	3.09	3.06	3.03	3.00	.99	
1.43	1.41	1.41	1.40	1.39	1.39	1.38	1.38	1.37	1.37	1.36	1.36	.75	
1.97	1.92	1.90	1.87	1.85	1.83	1.82	1.79	1.79	1.77	1.76	1.76	.90	15
2.40	2.33	2.29	2.25	2.20	2.18	2.16	2.12	2.11	2.10	2.08	2.07	.95	
3.52	3.37	3.29	3.21	3.13	3.08	3.05	2.98	2.96	2.92	2.89	2.87	.99	
1.41	1.40	1.39	1.38	1.37	1.37	1.36	1.36	1.35	1.35	1.34	1.34	.75	
1.94	1.89	1.87	1.84	1.81	1.79	1.78	1.76	1.75	1.74	1.73	1.72	.90	16
2.35	2.28	2.24	2.19	2.15	2.12	2.11	2.07	2.06	2.04	2.02	2.01	.95	
3.41	3.26	3.18	3.10	3.02	2.97	2.93	2.86	2.84	2.81	2.78	2.75	.99	
1.40	1.39	1.38	1.37	1.36	1.35	1.35	1.34	1.34	1.34	1.33	1.33	.75	
1.91	1.86	1.84	1.81	1.78	1.76	1.75	1.73	1.72	1.71	1.69	1.69	.90	17
2.31	2.23	2.19	2.15	2.10	2.08	2.06	2.02	2.01	1.99	1.97	1.96	.95	
3.31	3.16	3.08	3.00	2.92	2.87	2.83	2.76	2.75	2.71	2.68	2.65	.99	
1.39	1.38	1.37	1.36	1.35	1.34	1.34	1.33	1.33	1.32	1.32	1.32	.75	
1.89	1.84	1.81	1.78	1.75	1.74	1.72	1.70	1.69	1.68	1.67	1.66	.90	18
2.27	2.19	2.15	2.11	2.06	2.04	2.02	1.98	1.97	1.95	1.93	1.92	.95	
3.23	3.08	3.00	2.92	2.84	2.78	2.75	2.68	2.66	2.62	2.59	2.57	.99	
1.38	1.37	1.36	1.35	1.34	1.33	1.33	1.32	1.32	1.31	1.31	1.30	.75	
1.86	1.81	1.79	1.76	1.73	1.71	1.70	1.67	1.67	1.65	1.64	1.63	.90	19
2.23	2.16	2.11	2.07	2.03	2.00	1.98	1.94	1.93	1.91	1.89	1.88	.95	
3.15	3.00	2.92	2.84	2.76	2.71	2.67	2.60	2.58	2.55	2.51	2.49	.99	
1.37	1.36	1.35	1.34	1.33	1.33	1.32	1.31	1.31	1.30	1.30	1.29	.75	
1.84	1.79	1.77	1.74	1.71	1.69	1.68	1.65	1.64	1.63	1.62	1.61	.90	20
2.20	2.12	2.08	2.04	1.99	1.97	1.95	1.91	1.90	1.88	1.86	1.84	.95	
3.09	2.94	2.86	2.78	2.69	2.64	2.61	2.54	2.52	2.48	2.44	2.42	.99	
1.36	1.34	1.33	1.32	1.31	1.31	1.30	1.30	1.30	1.29	1.29	1.28	.75	
1.81	1.76	1.73	1.70	1.67	1.65	1.64	1.61	1.60	1.59	1.58	1.57	.90	22
2.15	2.07	2.03	1.98	1.94	1.91	1.89	1.85	1.84	1.82	1.80	1.78	.95	
2.98	2.83	2.75	2.67	2.58	2.53	2.50	2.42	2.40	2.36	2.33	2.31	.99	
1.35	1.33	1.32	1.31	1.30	1.29	1.29	1.28	1.28	1.27	1.27	1.26	.75	
1.78	1.73	1.70	1.67	1.64	1.62	1.61	1.58	1.57	1.56	1.54	1.53	.90	24
2.11	2.03	1.98	1.94	1.89	1.86	1.84	1.80	1.79	1.77	1.75	1.73	.95	
2.89	2.74	2.66	2.58	2.49	2.44	2.40	2.33	2.31	2.27	2.24	2.21	.99	
1.34	1.32	1.31	1.30	1.29	1.28	1.28	1.26	1.26	1.26	1.25	1.25	.75	
1.76	1.71	1.68	1.65	1.61	1.59	1.58	1.55	1.54	1.53	1.51	1.50	.90	26
2.07	1.99	1.95	1.90	1.85	1.82	1.80	1.76	1.75	1.73	1.71	1.69	.95	
2.81	2.66	2.58	2.50	2.42	2.36	2.33	2.25	2.23	2.19	2.16	2.13	.99	
1.33	1.31	1.30	1.29	1.28	1.27	1.27	1.26	1.25	1.25	1.24	1.24	.75	
1.74	1.69	1.66	1.63	1.59	1.57	1.56	1.53	1.52	1.50	1.49	1.48	.90	28
2.04	1.96	1.91	1.87	1.82	1.79	1.77	1.73	1.71	1.69	1.67	1.65	.95	
2.75	2.60	2.52	2.44	2.35	2.30	2.26	2.19	2.17	2.13	2.09	2.06	.99	

Table C.3

df for denom.	1 − α	df for numerator 1	2	3	4	5	6	7	8	9	10	11	12
30	.75	1.38	1.45	1.44	1.42	1.41	1.39	1.38	1.37	1.36	1.35	1.35	1.34
	.90	2.88	2.49	2.28	2.14	2.05	1.98	1.93	1.88	1.85	1.82	1.79	1.77
	.95	4.17	3.32	2.92	2.69	2.53	2.42	2.33	2.27	2.21	2.16	2.13	2.09
	.99	7.56	5.39	4.51	4.02	3.70	3.47	3.30	3.17	3.07	2.98	2.91	2.84
40	.75	1.36	1.44	1.42	1.40	1.39	1.37	1.36	1.35	1.34	1.33	1.32	1.31
	.90	2.84	2.44	2.23	2.09	2.00	1.93	1.87	1.83	1.79	1.76	1.73	1.71
	.95	4.08	3.23	2.84	2.61	2.45	2.34	2.25	2.18	2.12	2.08	2.04	2.00
	.99	7.31	5.18	4.31	3.83	3.51	3.29	3.12	2.99	2.89	2.80	2.73	2.66
60	.75	1.35	1.42	1.41	1.38	1.37	1.35	1.33	1.32	1.31	1.30	1.29	1.29
	.90	2.79	2.39	2.18	2.04	1.95	1.87	1.82	1.77	1.74	1.71	1.68	1.66
	.95	4.00	3.15	2.76	2.53	2.37	2.25	2.17	2.10	2.04	1.99	1.95	1.92
	.99	7.08	4.98	4.13	3.65	3.34	3.12	2.95	2.82	2.72	2.63	2.56	2.50
120	.75	1.34	1.40	1.39	1.37	1.35	1.33	1.31	1.30	1.29	1.28	1.27	1.26
	.90	2.75	2.35	2.13	1.99	1.90	1.82	1.77	1.72	1.68	1.65	1.62	1.60
	.95	3.92	3.07	2.68	2.45	2.29	2.17	2.09	2.02	1.96	1.91	1.87	1.83
	.99	6.85	4.79	3.95	3.48	3.17	2.96	2.79	2.66	2.56	2.47	2.40	2.34
200	.75	1.33	1.39	1.38	1.36	1.34	1.32	1.31	1.29	1.28	1.27	1.26	1.25
	.90	2.73	2.33	2.11	1.97	1.88	1.80	1.75	1.70	1.66	1.63	1.60	1.57
	.95	3.89	3.04	2.65	2.42	2.26	2.14	2.06	1.98	1.93	1.88	1.84	1.80
	.99	6.76	4.71	3.88	3.41	3.11	2.89	2.73	2.60	2.50	2.41	2.34	2.27
∞	.75	1.32	1.39	1.37	1.35	1.33	1.31	1.29	1.28	1.27	1.25	1.24	1.24
	.90	2.71	2.30	2.08	1.94	1.85	1.77	1.72	1.67	1.63	1.60	1.57	1.55
	.95	3.84	3.00	2.60	2.37	2.21	2.10	2.01	1.94	1.88	1.83	1.79	1.75
	.99	6.63	4.61	3.78	3.32	3.02	2.80	2.64	2.51	2.41	2.32	2.25	2.18

† This table is abridged from Table 18 in *Biometrika Tables for Statisticians*, vol. 1. Reproduced with the kind permission of E. S. Pearson and the trustees of *Biometrika*.

Table C.3*a*

df for denom.	1 − α	df for numerator 1	2	3	4	5	6	7	8	9	10
2	.995	198.5	199.0	199.2	199.2	199.3	199.3	199.4	199.4	199.4	199.4
	.999	998.5	999.0	999.2	999.2	999.3	999.3	999.4	999.4	999.4	999.4
3	.995	55.55	49.80	47.47	46.20	45.39	44.84	44.43	44.13	43.88	43.69
	.999	167.0	148.5	141.1	137.1	134.6	132.8	131.6	130.6	129.9	129.2
4	.995	31.33	26.28	24.26	23.16	22.46	21.98	21.62	21.35	21.14	20.97
	.999	74.14	61.25	56.18	53.44	51.71	50.52	49.66	49.00	48.48	48.05
5	.995	22.79	18.31	16.53	15.56	14.94	14.51	14.20	13.96	13.77	13.62
	.999	47.18	37.12	33.20	31.08	29.75	28.83	28.16	27.65	27.24	26.92
6	.995	18.64	14.54	12.92	12.03	11.46	11.07	10.79	10.57	10.39	10.15
	.999	35.51	27.00	23.70	21.92	20.80	20.03	19.46	19.03	18.69	18.41
7	.995	16.24	12.40	10.88	10.05	9.52	9.16	8.89	8.68	8.51	8.38
	.999	29.25	21.69	18.77	17.20	16.21	15.52	15.02	14.63	14.33	14.28
8	.995	14.69	11.04	9.60	8.81	8.30	7.95	7.69	7.50	7.34	7.21
	.999	25.42	18.49	15.83	14.39	13.48	12.86	12.40	12.05	11.77	11.54

F **Distribution** (*Continued*)†

df for numerator													
15	20	24	30	40	50	60	100	120	200	500	∞	1 − α	df for denom.
1.32	1.30	1.29	1.28	1.27	1.26	1.26	1.25	1.24	1.24	1.23	1.23	.75	
1.72	1.67	1.64	1.61	1.57	1.55	1.54	1.51	1.50	1.48	1.47	1.46	.90	30
2.01	1.93	1.89	1.84	1.79	1.76	1.74	1.70	1.68	1.66	1.64	1.62	.95	
2.70	2.55	2.47	2.39	2.30	2.25	2.21	2.13	2.11	2.07	2.03	2.01	.99	
1.30	1.28	1.26	1.25	1.24	1.23	1.22	1.21	1.21	1.20	1.19	1.19	.75	
1.66	1.61	1.57	1.54	1.51	1.48	1.47	1.43	1.42	1.41	1.39	1.38	.90	40
1.92	1.84	1.79	1.74	1.69	1.66	1.64	1.59	1.58	1.55	1.53	1.51	.95	
2.52	2.37	2.29	2.20	2.11	2.06	2.02	1.94	1.92	1.87	1.83	1.80	.99	
1.27	1.25	1.24	1.22	1.21	1.20	1.19	1.17	1.17	1.16	1.15	1.15	.75	
1.60	1.54	1.51	1.48	1.44	1.41	1.40	1.36	1.35	1.33	1.31	1.29	.90	60
1.84	1.75	1.70	1.65	1.59	1.56	1.53	1.48	1.47	1.44	1.41	1.39	.95	
2.35	2.20	2.12	2.03	1.94	1.88	1.84	1.75	1.73	1.68	1.63	1.60	.99	
1.24	1.22	1.21	1.19	1.18	1.17	1.16	1.14	1.13	1.12	1.11	1.10	.75	
1.55	1.48	1.45	1.41	1.37	1.34	1.32	1.27	1.26	1.24	1.21	1.19	.90	120
1.75	1.66	1.61	1.55	1.50	1.46	1.43	1.37	1.35	1.32	1.28	1.25	.95	
2.19	2.03	1.95	1.86	1.76	1.70	1.66	1.56	1.53	1.48	1.42	1.38	.99	
1.23	1.21	1.20	1.18	1.16	1.14	1.12	1.11	1.10	1.09	1.08	1.06	.75	
1.52	1.46	1.42	1.38	1.34	1.31	1.28	1.24	1.22	1.20	1.17	1.14	.90	200
1.72	1.62	1.57	1.52	1.46	1.41	1.39	1.32	1.29	1.26	1.22	1.19	.95	
2.13	1.97	1.89	1.79	1.69	1.63	1.58	1.48	1.44	1.39	1.33	1.28	.99	
1.22	1.19	1.18	1.16	1.14	1.13	1.12	1.09	1.08	1.07	1.04	1.00	.75	
1.49	1.42	1.38	1.34	1.30	1.26	1.24	1.18	1.17	1.13	1.08	1.00	.90	∞
1.67	1.57	1.52	1.46	1.39	1.35	1.32	1.24	1.22	1.17	1.11	1.00	.95	
2.04	1.88	1.79	1.70	1.59	1.52	1.47	1.36	1.32	1.25	1.15	1.00	.99	

(2d ed.) New York: Cambridge, 1958. Edited by E. S. Pearson and H. O. Hartley.

F **Distribution (Supplement)**

df for denom.	1 − α	df for numerator									
		1	2	3	4	5	6	7	8	9	10
9	.995	13.61	10.11	8.72	7.96	7.47	7.13	6.88	6.69	6.54	6.42
	.999	22.86	16.39	13.90	12.56	11.71	11.13	10.70	10.37	10.11	9.89
10	.995	12.83	9.43	8.08	7.34	6.87	6.54	6.30	6.12	5.97	5.85
	.999	21.04	14.91	12.56	11.28	10.48	9.93	9.52	9.20	8.96	8.75
12	.995	11.75	8.51	7.23	6.52	6.07	5.76	5.52	5.35	5.20	5.09
	.999	18.64	12.97	10.80	9.63	8.89	8.38	8.00	7.71	7.48	7.29
20	.995	9.94	6.99	5.82	5.17	4.76	4.47	4.26	4.09	3.96	3.85
	.999	14.82	9.53	8.10	7.10	6.46	6.02	5.69	5.44	5.24	5.08
40	.995	8.83	6.07	4.98	4.37	3.99	3.71	3.51	3.35	3.22	3.12
	.999	12.61	8.25	6.59	5.70	5.13	4.73	4.44	4.21	4.02	3.87
60	.995	8.49	5.80	4.73	4.14	3.76	3.49	3.29	3.13	3.01	2.90
	.999	11.97	7.77	6.17	5.31	4.76	4.37	4.09	3.86	3.69	3.54
120	.995	8.18	5.54	4.50	3.92	3.55	3.28	3.09	2.93	2.81	2.71
	.999	11.38	7.32	5.78	4.95	4.42	4.04	3.77	3.55	3.38	3.24

Table C.4 Distribution of the Studentized Range Statistic†

df for $s_{\bar{X}}$	$1-\alpha$	r = number of steps between ordered means													
		2	3	4	5	6	7	8	9	10	11	12	13	14	15
1	.95	18.0*	27.0	32.8	37.1	40.4	43.1	45.4	47.4	49.1	50.6	52.0	53.2	54.3	55.4
	.99	90.0	135	164	186	202	216	227	237	246	253	260	266	272	277
2	.95	6.09	8.3	9.8	10.9	11.7	12.4	13.0	13.5	14.0	14.4	14.7	15.1	15.4	15.7
	.99	14.0	19.0	22.3	24.7	26.6	28.2	29.5	30.7	31.7	32.6	33.4	34.1	34.8	35.4
3	.95	4.50	5.91	6.82	7.50	8.04	8.48	8.85	9.18	9.46	9.72	9.95	10.2	10.4	10.5
	.99	8.26	10.6	12.2	13.3	14.2	15.0	15.6	16.2	16.7	17.1	17.5	17.9	18.2	18.5
4	.95	3.93	5.04	5.76	6.29	6.71	7.05	7.35	7.60	7.83	8.03	8.21	8.37	8.52	8.66
	.99	6.51	8.12	9.17	9.96	10.6	11.1	11.5	11.9	12.3	12.6	12.8	13.1	13.3	13.5
5	.95	3.64	4.60	5.22	5.67	6.03	6.33	6.58	6.80	6.99	7.17	7.32	7.47	7.60	7.72
	.99	5.70	6.97	7.80	8.42	8.91	9.32	9.67	9.97	10.2	10.5	10.7	10.9	11.1	11.2
6	.95	3.46	4.34	4.90	5.31	5.63	5.89	6.12	6.32	6.49	6.65	6.79	6.92	7.03	7.14
	.99	5.24	6.33	7.03	7.56	7.97	8.32	8.61	8.87	9.10	9.30	9.49	9.65	9.81	9.95
7	.95	3.34	4.16	4.69	5.06	5.36	5.61	5.82	6.00	6.16	6.30	6.43	6.55	6.66	6.76
	.99	4.95	5.92	6.54	7.01	7.37	7.68	7.94	8.17	8.37	8.55	8.71	8.86	9.00	9.12
8	.95	3.26	4.04	4.53	4.89	5.17	5.40	5.60	5.77	5.92	6.05	6.18	6.29	6.39	6.48
	.99	4.74	5.63	6.20	6.63	6.96	7.24	7.47	7.68	7.87	8.03	8.18	8.31	8.44	8.55
9	.95	3.20	3.95	4.42	4.76	5.02	5.24	5.43	5.60	5.74	5.87	5.98	6.09	6.19	6.28
	.99	4.60	5.43	5.96	6.35	6.66	6.91	7.13	7.32	7.49	7.65	7.78	7.91	8.03	8.13
10	.95	3.15	3.88	4.33	4.65	4.91	5.12	5.30	5.46	5.60	5.72	5.83	5.93	6.03	6.11
	.99	4.48	5.27	5.77	6.14	6.43	6.67	6.87	7.05	7.21	7.36	7.48	7.60	7.71	7.81
11	.95	3.11	3.82	4.26	4.57	4.82	5.03	5.20	5.35	5.49	5.61	5.71	5.81	5.90	5.99
	.99	4.39	5.14	5.62	5.97	6.25	6.48	6.67	6.84	6.99	7.13	7.26	7.36	7.46	7.56
12	.95	3.08	3.77	4.20	4.51	4.75	4.95	5.12	5.27	5.40	5.51	5.62	5.71	5.80	5.88
	.99	4.32	5.04	5.50	5.84	6.10	6.32	6.51	6.67	6.81	6.94	7.06	7.17	7.26	7.36

13	.95	3.06	3.73	4.15	4.45	4.69	4.88	5.05	5.19	5.32	5.43	5.53	5.63	5.71	5.79
	.99	4.26	4.96	5.40	5.73	5.98	6.19	6.37	6.53	6.67	6.79	6.90	7.01	7.10	7.19
14	.95	3.03	3.70	4.11	4.41	4.64	4.83	4.99	5.13	5.25	5.36	5.46	5.55	6.64	5.72
	.99	4.21	4.89	5.32	5.63	5.88	6.08	6.26	6.41	6.54	6.66	6.77	6.87	6.96	7.05
16	.95	3.00	3.65	4.05	4.33	4.56	4.74	4.90	5.03	5.15	5.26	5.35	5.44	5.52	5.59
	.99	4.13	4.78	5.19	5.49	5.72	5.92	6.08	6.22	6.35	6.46	6.56	6.66	6.74	6.82
18	.95	2.97	3.61	4.00	4.28	4.49	4.67	4.82	4.96	5.07	5.17	5.27	5.35	5.43	5.50
	.99	4.07	4.70	5.09	5.38	5.60	5.79	5.94	6.08	6.20	6.31	6.41	6.50	6.58	6.65
20	.95	2.95	3.58	3.96	4.23	4.45	4.62	4.77	4.90	5.01	5.11	5.20	5.28	5.36	5.43
	.99	4.02	4.64	5.02	5.29	5.51	5.69	5.84	5.97	6.09	6.19	6.29	6.37	6.45	6.52
24	.95	2.92	3.53	3.90	4.17	4.37	4.54	4.68	4.81	4.92	5.01	5.10	5.18	5.25	5.32
	.99	3.96	4.54	4.91	5.17	5.37	5.54	5.69	5.81	5.92	6.02	6.11	6.19	6.26	6.33
30	.95	2.89	3.49	3.84	4.10	4.30	4.46	4.60	4.72	4.83	4.92	5.00	5.08	5.15	5.21
	.99	3.89	4.45	4.80	5.05	5.24	5.40	5.54	5.56	5.76	5.85	5.93	6.01	6.08	6.14
40	.95	2.86	3.44	3.79	4.04	4.23	4.39	4.52	4.63	4.74	4.82	4.91	4.98	5.05	5.11
	.99	3.82	4.37	4.70	4.93	5.11	5.27	5.39	5.50	5.60	5.69	5.77	5.84	5.90	5.96
60	.95	2.83	3.40	3.74	3.98	4.16	4.31	4.44	4.55	4.65	4.73	4.81	4.88	4.94	5.00
	.99	3.76	4.28	4.60	4.82	4.99	5.13	5.25	5.36	5.45	5.53	5.60	5.67	5.73	5.79
120	.95	2.80	3.36	3.69	3.92	4.10	4.24	4.36	4.48	4.56	4.64	4.72	4.78	4.84	4.90
	.99	3.70	4.20	4.50	4.71	4.87	5.01	5.12	5.21	5.30	5.38	5.44	5.51	5.56	5.61
∞	.95	2.77	3.31	3.63	3.86	4.03	4.17	4.29	4.39	4.47	4.55	4.62	4.68	4.74	4.80
	.99	3.64	4.12	4.40	4.60	4.76	4.88	4.99	5.08	5.16	5.23	5.29	5.35	5.40	5.45

† This table is abridged from Table II.2 in *The Probability Integrals of the Range and of the Studentized Range*, prepared by H. Leon Harter, Donald S. Clemm, and Eugene H. Guthrie. These tables are published in WADC tech. Rep. 58–484, vol. 2, 1959, Wright Air Development Center, and are reproduced with the kind permission of the authors.

Table C.5 Arcsin Transformation ($\phi = 2 \arcsin \sqrt{X}$)

X	ϕ	X	ϕ	X	ϕ	X	ϕ	X	ϕ
.001	.0633	.041	.4078	.36	1.2870	.76	2.1177	.971	2.7993
.002	.0895	.042	.4128	.37	1.3078	.77	2.1412	.972	2.8053
.003	.1096	.043	.4178	.38	1.3284	.78	2.1652	.973	2.8115
.004	.1266	.044	.4227	.39	1.3490	.79	2.1895	.974	2.8177
.005	.1415	.045	.4275	.40	1.3694	.80	2.2143	.975	2.8240
.006	.1551	.046	.4323	.41	1.3898	.81	2.2395	.976	2.8305
.007	.1675	.047	.4371	.42	1.4101	.82	2.2653	.977	2.8371
.008	.1791	.048	.4418	.43	1.4303	.83	2.2916	.978	2.8438
.009	.1900	.049	.4464	.44	1.4505	.84	2.3186	.979	2.8507
.010	.2003	.050	.4510	.45	1.4706	.85	2.3462	.980	2.8578
.011	.2101	.06	.4949	.46	1.4907	.86	2.3746	.981	2.8650
.012	.2195	.07	.5355	.47	1.5108	.87	2.4039	.982	2.8725
.013	.2285	.08	.5735	.48	1.5308	.88	2.4341	.983	2.8801
.014	.2372	.09	.6094	.49	1.5508	.89	2.4655	.984	2.8879
.015	.2456	.10	.6435	.50	1.5708	.90	2.4981	.985	2.8960
.016	.2537	.11	.6761	.51	1.5908	.91	2.5322	.986	2.9044
.017	.2615	.12	.7075	.52	1.6108	.92	2.5681	.987	2.9131
.018	.2691	.13	.7377	.53	1.6308	.93	2.6062	.988	2.9221
.019	.2766	.14	.7670	.54	1.6509	.94	2.6467	.989	2.9315
.020	.2838	.15	.7954	.55	1.6710	.95	2.6906	.990	2.9413
.021	.2909	.16	.8230	.56	1.6911	.951	2.6952	.991	2.9516
.022	.2978	.17	.8500	.57	1.7113	.952	2.6998	.992	2.9625
.023	.3045	.18	.8763	.58	1.7315	.953	2.7045	.993	2.9741
.024	.3111	.19	.9021	.59	1.7518	.954	2.7093	.994	2.9865
.025	.3176	.20	.9273	.60	1.7722	.955	2.7141	.995	3.0001
.026	.3239	.21	.9521	.61	1.7926	.956	2.7189	.996	3.0150
.027	.3301	.22	.9764	.62	1.8132	.957	2.7238	.997	3.0320
.028	.3363	.23	1.0004	.63	1.8338	.958	2.7288	.998	3.0521
.029	.3423	.24	1.0239	.64	1.8546	.959	2.7338	.999	3.0783
.030	.3482	.25	1.0472	.65	1.8755	.960	2.7389		
.031	.3540	.26	1.0701	.66	1.8965	.961	2.7440		
.032	.3597	.27	1.0928	.67	1.9177	.962	2.7492		
.033	.3654	.28	1.1152	.68	1.9391	.963	2.7545		
.034	.3709	.29	1.1374	.69	1.9606	.964	2.7598		
.035	.3764	.30	1.1593	.70	1.9823	.965	2.7652		
.036	.3818	.31	1.1810	.71	2.0042	.966	2.7707		
.037	.3871	.32	1.2025	.72	2.0264	.967	2.7762		
.038	.3924	.33	1.2239	.73	2.0488	.968	2.7819		
.039	.3976	.34	1.2451	.74	2.0715	.969	2.7876		
.040	.4027	.35	1.2661	.75	2.0944	.970	2.7934		

Table C.6 Distribution of t Statistic in Comparing Treatment Means with a Control†‡

df for MS_{error}	$1 - \alpha$	k = number of means (including control)								
		2	3	4	5	6	7	8	9	10
5	.95	2.02	2.44	2.68	2.85	2.98	3.08	3.16	3.24	3.30
	.975	2.57	3.03	3.29	3.48	3.62	3.73	3.82	3.90	3.97
	.99	3.36	3.90	4.21	4.43	4.60	4.73	4.85	4.94	5.03
	.995	4.03	4.63	4.98	5.22	5.41	5.56	5.69	5.80	5.89
6	.95	1.94	2.34	2.56	2.71	2.83	2.92	3.00	3.07	3.12
	.975	2.45	2.86	3.10	3.26	3.39	3.49	3.57	3.64	3.71
	.99	3.14	3.61	3.88	4.07	4.21	4.33	4.43	4.51	4.59
	.995	3.71	4.21	4.51	4.71	4.87	5.00	5.10	5.20	5.28
7	.95	1.89	2.27	2.48	2.62	2.73	2.82	2.89	2.95	3.01
	.975	2.36	2.75	2.97	3.12	3.24	3.33	3.41	3.47	3.53
	.99	3.00	3.42	3.66	3.83	3.96	4.07	4.15	4.23	4.30
	.995	3.50	3.95	4.21	4.39	4.53	4.64	4.74	4.82	4.89
8	.95	1.86	2.22	2.42	2.55	2.66	2.74	2.81	2.87	2.92
	.975	2.31	2.67	2.88	3.02	3.13	3.22	3.29	3.35	3.41
	.99	2.90	3.29	3.51	3.67	3.79	3.88	3.96	4.03	4.09
	.995	3.36	3.77	4.00	4.17	4.29	4.40	4.48	4.56	4.62
9	.95	1.83	2.18	2.37	2.50	2.60	2.68	2.75	2.81	2.86
	.975	2.26	2.61	2.81	2.95	3.05	3.14	3.20	3.26	3.32
	.99	2.82	3.19	3.40	3.55	3.66	3.75	3.82	3.89	3.94
	.995	3.25	3.63	3.85	4.01	4.12	4.22	4.30	4.37	4.43
10	.95	1.81	2.15	2.34	2.47	2.56	2.64	2.70	2.76	2.81
	.975	2.23	2.57	2.76	2.89	2.99	3.07	3.14	3.19	3.24
	.99	2.76	3.11	3.31	3.45	3.56	3.64	3.71	3.78	3.83
	.995	3.17	3.53	3.74	3.88	3.99	4.08	4.16	4.22	4.28
11	.95	1.80	2.13	2.31	2.44	2.53	2.60	2.67	2.72	2.77
	.975	2.20	2.53	2.72	2.84	2.94	3.02	3.08	3.14	3.19
	.99	2.72	3.06	3.25	3.38	3.48	3.56	3.63	3.69	3.74
	.995	3.11	3.45	3.65	3.79	3.89	3.98	4.05	4.11	4.16
12	.95	1.78	2.11	2.29	2.41	2.50	2.58	2.64	2.69	2.74
	.975	2.18	2.50	2.68	2.81	2.90	2.98	3.04	3.09	3.14
	.99	2.68	3.01	3.19	3.32	3.42	3.50	3.56	3.62	3.67
	.995	3.05	3.39	3.58	3.71	3.81	3.89	3.96	4.02	4.07
13	.95	1.77	2.09	2.27	2.39	2.48	2.55	2.61	2.66	2.71
	.975	2.16	2.48	2.65	2.78	2.87	2.94	3.00	3.06	3.10
	.99	2.65	2.97	3.15	3.27	3.37	3.44	3.51	3.56	3.61
	.995	3.01	3.33	3.52	3.65	3.74	3.82	3.89	3.94	3.99
14	.95	1.76	2.08	2.25	2.37	2.46	2.53	2.59	2.64	2.69
	.975	2.14	2.46	2.63	2.75	2.84	2.91	2.97	3.02	3.07
	.99	2.62	2.94	3.11	3.23	3.32	3.40	3.46	3.51	3.56
	.995	2.98	3.29	3.47	3.59	3.69	3.76	3.83	3.88	3.93

Table C.6 (*Continued*)† ‡

df for MS_{error}	$1 - \alpha$	k = number of means (including control)								
		2	3	4	5	6	7	8	9	10
16	.95	1.75	2.06	2.23	2.34	2.43	2.50	2.56	2.61	2.65
	.975	2.12	2.42	2.59	2.71	2.80	2.87	2.92	2.97	3.02
	.99	2.58	2.88	3.05	3.17	3.26	3.33	3.39	3.44	3.48
	.995	2.92	3.22	3.39	3.51	3.60	3.67	3.73	3.78	3.83
18	.95	1.73	2.04	2.21	2.32	2.41	2.48	2.53	2.58	2.62
	.975	2.10	2.40	2.56	2.68	2.76	2.83	2.89	2.94	2.98
	.99	2.55	2.84	3.01	3.12	3.21	3.27	3.33	3.38	3.42
	.995	2.88	3.17	3.33	3.44	3.53	3.60	3.66	3.71	3.75
20	.95	1.72	2.03	2.19	2.30	2.39	2.46	2.51	2.56	2.60
	.975	2.09	2.38	2.54	2.65	2.73	2.80	2.86	2.90	2.95
	.99	2.53	2.81	2.97	3.08	3.17	3.23	3.29	3.34	3.38
	.995	2.85	3.13	3.29	3.40	3.48	3.55	3.60	3.65	3.69
24	.95	1.71	2.01	2.17	2.28	2.36	2.43	2.48	2.53	2.57
	.975	2.06	2.35	2.51	2.61	2.70	2.76	2.81	2.86	2.90
	.99	2.49	2.77	2.92	3.03	3.11	3.17	3.22	3.27	3.31
	.995	2.80	3.07	3.22	3.32	3.40	3.47	3.52	3.57	3.61
30	.95	1.70	1.99	2.15	2.25	2.33	2.40	2.45	2.50	2.54
	.975	2.04	2.32	2.47	2.58	2.66	2.72	2.77	2.82	2.86
	.99	2.46	2.72	2.87	2.97	3.05	3.11	3.16	3.21	3.24
	.995	2.75	3.01	3.15	3.25	3.33	3.39	3.44	3.49	3.52
40	.95	1.68	1.97	2.13	2.23	2.31	2.37	2.42	2.47	2.51
	.975	2.02	2.29	2.44	2.54	2.62	2.68	2.73	2.77	2.81
	.99	2.42	2.68	2.82	2.92	2.99	3.05	3.10	3.14	3.18
	.995	2.70	2.95	3.09	3.19	3.26	3.32	3.37	3.41	3.44
60	.95	1.67	1.95	2.10	2.21	2.28	2.35	2.39	2.44	2.48
	.975	2.00	2.27	2.41	2.51	2.58	2.64	2.69	2.73	2.77
	.99	2.39	2.64	2.78	2.87	2.94	3.00	3.04	3.08	3.12
	.995	2.66	2.90	3.03	3.12	3.19	3.25	3.29	3.33	3.37
120	.95	1.66	1.93	2.08	2.18	2.26	2.32	2.37	2.41	2.45
	.975	1.98	2.24	2.38	2.47	2.55	2.60	2.65	2.69	2.73
	.99	2.36	2.60	2.73	2.82	2.89	2.94	2.99	3.03	3.06
	.995	2.62	2.85	2.97	3.06	3.12	3.18	3.22	3.26	3.29
∞	.95	1.64	1.92	2.06	2.16	2.23	2.29	2.34	2.38	2.42
	.975	1.96	2.21	2.35	2.44	2.51	2.57	2.61	2.65	2.69
	.99	2.33	2.56	2.68	2.77	2.84	2.89	2.93	2.97	3.00
	.995	2.58	2.79	2.92	3.00	3.06	3.11	3.15	3.19	3.22

† Entries in rows .975 and .995 are for two-sided simultaneous confidence intervals with $\alpha = .05$ and .01, respectively. Entries in rows .95 and .99 are for one-sided confidence simultaneous intervals with $\alpha = .05$ and 0.1, respectively.

‡ This table is reproduced from: A multiple comparison procedure for comparing several treatments with a control. *Journal of the American Statistical Association*, 1955, **50**, 1096–1121, and New tables for multiple comparisons with a control. *Biometrics*, 1964, **20**, 482–491, with the permission of the author, C. W. Dunnett, and the editors.

Table C.7 Distribution of F_{max} Statistic†

df for s_X^2	$1-\alpha$	k = number of variances								
		2	3	4	5	6	7	8	9	10
4	.95	9.60	15.5	20.6	25.2	29.5	33.6	37.5	41.4	44.6
	.99	23.2	37.	49.	59.	69.	79.	89.	97.	106.
5	.95	7.15	10.8	13.7	16.3	18.7	20.8	22.9	24.7	26.5
	.99	14.9	22.	28.	33.	38.	42.	46.	50.	54.
6	.95	5.82	8.38	10.4	12.1	13.7	15.0	16.3	17.5	18.6
	.99	11.1	15.5	19.1	22.	25.	27.	30.	32.	34.
7	.95	4.99	6.94	8.44	9.70	10.8	11.8	12.7	13.5	14.3
	.99	8.89	12.1	14.5	16.5	18.4	20.	22.	23.	24.
8	.95	4.43	6.00	7.18	8.12	9.03	9.78	10.5	11.1	11.7
	.99	7.50	9.9	11.7	13.2	14.5	15.8	16.9	17.9	18.9
9	.95	4.03	5.34	6.31	7.11	7.80	8.41	8.95	9.45	9.91
	.99	6.54	8.5	9.9	11.1	12.1	13.1	13.9	14.7	15.3
10	.95	3.72	4.85	5.67	6.34	6.92	7.42	7.87	8.28	8.66
	.99	5.85	7.4	8.6	9.6	10.4	11.1	11.8	12.4	12.9
12	.95	3.28	4.16	4.79	5.30	5.72	6.09	6.42	6.72	7.00
	.99	4.91	6.1	6.9	7.6	8.2	8.7	9.1	9.5	9.9
15	.95	2.86	3.54	4.01	4.37	4.68	4.95	5.19	5.40	5.59
	.99	4.07	4.9	5.5	6.0	6.4	6.7	7.1	7.3	7.5
20	.95	2.46	2.95	3.29	3.54	3.76	3.94	4.10	4.24	4.37
	.99	3.32	3.8	4.3	4.6	4.9	5.1	5.3	5.5	5.6
30	.95	2.07	2.40	2.61	2.78	2.91	3.02	3.12	3.21	3.29
	.99	2.63	3.0	3.3	3.4	3.6	3.7	3.8	3.9	4.0
60	.95	1.67	1.85	1.96	2.04	2.11	2.17	2.22	2.26	2.30
	.99	1.96	2.2	2.3	2.4	2.4	2.5	2.5	2.6	2.6
∞	.95	1.00	1.00	1.00	1.00	1.00	1.00	1.00	1.00	1.00
	.99	1.00	1.00	1.00	1.00	1.00	1.00	1.00	1.00	1.00

† This table is abridged from Table 31 in *Biometrika Tables for Statisticians*, vol. 1. (2d ed.) New York: Cambridge, 1958. Edited by E. S. Pearson and H. O. Hartley. Reproduced with the kind permission of E. S. Pearson and the trustees of *Biometrika*.

Table C.8 Critical Values for Cochran's Test for Homogeneity of Variance†

$C = (\text{largest } s^2)/(\Sigma s_j^2)$

df for s_j^2	$1-\alpha$	k = number of variances										
		2	3	4	5	6	7	8	9	10	15	20
1	.95	.9985	.9669	.9065	.8412	.7808	.7271	.6798	.6385	.6020	.4709	.3894
	.99	.9999	.9933	.9676	.9279	.8828	.8376	.7945	.7544	.7175	.5747	.4799
2	.95	.9750	.8709	.7679	.6838	.6161	.5612	.5157	.4775	.4450	.3346	.2705
	.99	.9950	.9423	.8643	.7885	.7218	.6644	.6152	.5727	.5358	.4069	.3297
3	.95	.9392	.7977	.6841	.5981	.5321	.4800	.4377	.4027	.3733	.2758	.2205
	.99	.9794	.8831	.7814	.6957	.6258	.5685	.5209	.4810	.4469	.3317	.2654
4	.95	.9057	.7457	.6287	.5441	.4803	.4307	.3910	.3584	.3311	.2419	.1921
	.99	.9586	.8335	.7212	.6329	.5635	.5080	.4627	.4251	.3934	.2882	.2288
5	.95	.8772	.7071	.5895	.5065	.4447	.3974	.3595	.3286	.3029	.2195	.1735
	.99	.9373	.7933	.6761	.5875	.5195	.4659	.4226	.3870	.3572	.2593	.2048
6	.95	.8534	.6771	.5598	.4783	.4184	.3726	.3362	.3067	.2823	.2034	.1602
	.99	.9172	.7606	.6410	.5531	.4866	.4347	.3932	.3592	.3308	.2386	.1877
7	.95	.8332	.6530	.5365	.4564	.3980	.3535	.3185	.2901	.2666	.1911	.1501
	.99	.8988	.7335	.6129	.5259	.4608	.4105	.3704	.3378	.3106	.2228	.1748
8	.95	.8159	.6333	.5175	.4387	.3817	.3384	.3043	.2768	.2541	.1815	.1422
	.99	.8823	.7107	.5897	.5037	.4401	.3911	.3522	.3207	.2945	.2104	.1646
9	.95	.8010	.6167	.5017	.4241	.3682	.3259	.2926	.2659	.2439	.1736	.1357
	.99	.8674	.6912	.5702	.4854	.4229	.3751	.3373	.3067	.2813	.2002	.1567
16	.95	.7341	.5466	.4366	.3645	.3135	.2756	.2462	.2226	.2032	.1429	.1108
	.99	.7949	.6059	.4884	.4094	.3529	.3105	.2779	.2514	.2297	.1612	.1248
36	.95	.6602	.4748	.3720	.3066	.2612	.2278	.2022	.1820	.1655	.1144	.0879
	.99	.7067	.5153	.4057	.3351	.2858	.2494	.2214	.1992	.1811	.1251	.0960
144	.95	.5813	.4031	.3093	.2513	.2119	.1833	.1616	.1446	.1308	.0889	.0675
	.99	.6062	.4230	.3251	.2644	.2229	.1929	.1700	.1521	.1376	.0934	.0709

† Reproduced with permission from C. Eisenhart, M. W. Hastay, and W. A. Wallis, *Techniques of Statistical Analysis*, chap. 15. New York: McGraw-Hill, 1947.

Table C.9 Chi-square Distribution†

df	Percentile point						
	50	75	90	95	97.5	99	99.5
1	.46	1.3	2.7	3.8	5.0	6.6	7.9
2	1.4	2.8	4.6	6.0	7.4	9.2	10.6
3	2.4	4.1	6.3	7.8	9.4	11.3	12.8
4	3.4	5.4	7.8	9.5	11.1	13.3	14.9
5	4.4	6.6	9.2	11.1	12.8	15.1	16.7
6	5.4	7.8	10.6	12.6	14.4	16.8	18.5
7	6.4	9.0	12.0	14.1	16.0	18.5	20.3
8	7.3	10.2	13.4	15.5	17.5	20.1	22.0
9	8.3	11.4	14.7	16.9	19.0	21.7	23.6
10	9.3	12.5	16.0	18.3	20.5	23.2	25.2
11	10.3	13.7	17.3	19.7	21.9	24.7	26.8
12	11.3	14.8	18.5	21.0	23.3	26.2	28.3
13	12.3	16.0	19.8	22.4	24.7	27.7	29.8
14	13.3	17.1	21.1	23.7	26.1	29.1	31.3
15	14.3	18.2	22.3	25.0	27.5	30.6	32.8
16	15.3	19.4	23.5	26.3	28.8	32.0	34.3
17	16.3	20.5	24.8	27.6	30.2	33.4	35.7
18	17.3	21.6	26.0	28.9	31.5	34.8	37.2
19	18.3	22.7	27.2	30.1	32.9	36.2	38.6
20	19.3	23.8	28.4	31.4	34.2	37.6	40.0
21	20.3	24.9	29.6	32.7	35.5	38.9	41.4
22	21.3	26.0	30.8	33.9	36.8	40.3	42.8
23	22.3	27.1	32.0	35.2	38.1	41.6	44.2
24	23.3	28.2	33.2	36.4	39.4	43.0	45.6
25	24.3	29.3	34.4	37.7	40.6	44.3	46.9
26	25.3	30.4	35.6	38.9	41.9	45.6	48.3
27	26.3	31.5	36.7	40.1	43.2	47.0	49.6
28	27.3	32.6	37.9	41.3	44.5	48.3	51.0
29	28.3	33.7	39.1	42.6	45.7	49.6	52.3
30	29.3	34.8	40.3	43.8	47.0	50.9	53.7
40	39.3	45.6	51.8	55.8	59.3	63.7	66.8
60	59.3	67.0	74.4	79.1	83.3	88.4	92.0
100	99.3	109.1	118.5	124.3	129.6	135.8	140.2

For df > 30, $\chi^2_{1-\alpha} \doteq [\sqrt{2(\text{df}) - 1} + z_{1-\alpha}]^2/2$.
For example, when df = 60,

$$\chi^2_{.95} = [\sqrt{2(60) - 1} + 1.645]^2/2 = 79.$$

† This table is abridged from Table 8 in *Biometrika Tables for Statisticians*, vol. 1. (2d ed.) New York: Cambridge, 1958. Edited by E. S. Pearson and H. O. Hartley. Reproduced with the kind permission of E. S. Pearson and the trustees of *Biometrika*.

Table C.10 Coefficients of Orthogonal Polynomials

k	Polynomial	$X = 1$	2	3	4	5	6	7	8	9	10	$\Sigma\xi'^2$	λ
3	Linear	−1	0	1								2	1
	Quadratic	1	−2	1								6	3
4	Linear	−3	−1	1	3							20	2
	Quadratic	1	−1	−1	1							4	1
	Cubic	−1	3	−3	1							20	10⁄3
5	Linear	−2	−1	0	1	2						10	1
	Quadratic	2	−1	−2	−1	2						14	1
	Cubic	−1	2	0	−2	1						10	5⁄6
	Quartic	1	−4	6	−4	1						70	35⁄12
6	Linear	−5	−3	−1	1	3	5					70	2
	Quadratic	5	−1	−4	−4	−1	5					84	3⁄2
	Cubic	−5	7	4	−4	−7	5					180	5⁄3
	Quartic	1	−3	2	2	−3	1					28	7⁄12
7	Linear	−3	−2	−1	0	1	2	3				28	1
	Quadratic	5	0	−3	−4	−3	0	5				84	1
	Cubic	−1	1	1	0	−1	−1	1				6	1⁄6
	Quartic	3	−7	1	6	1	−7	3				154	7⁄12
8	Linear	−7	−5	−3	−1	1	3	5	7			168	2
	Quadratic	7	1	−3	−5	−5	−3	1	7			168	1
	Cubic	−7	5	7	3	−3	−7	−5	7			264	2⁄3
	Quartic	7	−13	−3	9	9	−3	−13	7			616	7⁄12
	Quintic	−7	23	−17	−15	15	17	−23	7			2184	7⁄10
9	Linear	−4	−3	−2	−1	0	1	2	3	4		60	1
	Quadratic	28	7	−8	−17	−20	−17	−8	7	28		2772	3
	Cubic	−14	7	13	9	0	−9	−13	−7	14		990	5⁄6
	Quartic	14	−21	−11	9	18	9	−11	−21	14		2002	7⁄12
	Quintic	−4	11	−4	−9	0	9	4	−11	4		468	3⁄20
10	Linear	−9	−7	−5	−3	−1	1	3	5	7	9	330	2
	Quadratic	6	2	−1	−3	−4	−4	−3	−1	2	6	132	1⁄2
	Cubic	−42	14	35	31	12	−12	−31	−35	−14	42	8580	5⁄3
	Quartic	18	−22	−17	3	18	18	3	−17	−22	18	2860	5⁄12
	Quintic	−6	14	−1	−11	−6	6	11	1	−14	6	780	1⁄10

Table C.11 Curves of Constant Power for Tests on Main Effects†

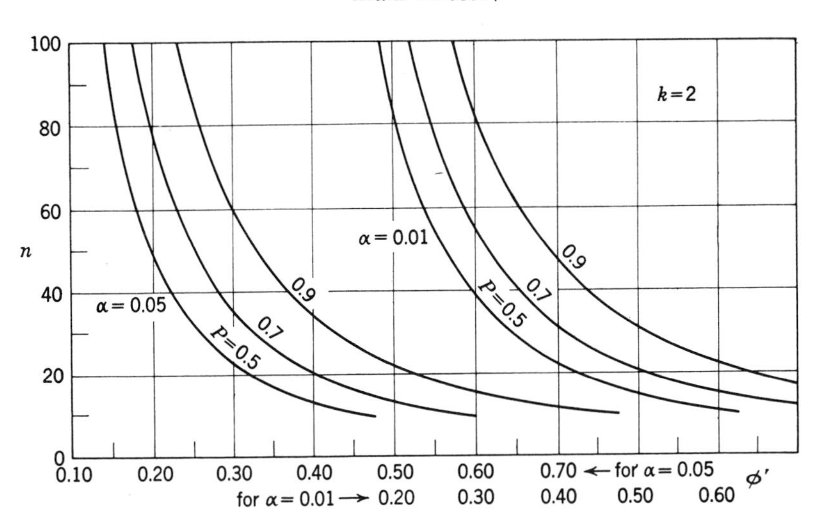

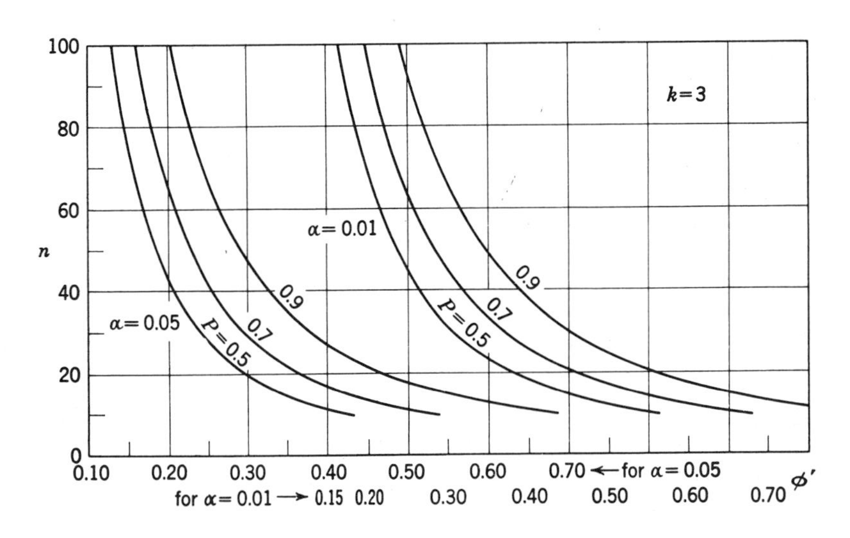

Table C.11 Curves of Constant Power for Tests on Main Effects (*Continued*)†

100
80
60
n
40
20
0
$k=4$
$\alpha=0.01$
$\alpha=0.05$
$P=0.5$ 0.7 0.9
0.10 0.20 0.30 0.40 0.50 0.60 0.70 ← for $\alpha=0.05$
for $\alpha=0.01$ → 0.15 0.20 0.30 0.40 0.50 0.60 0.70 ϕ'

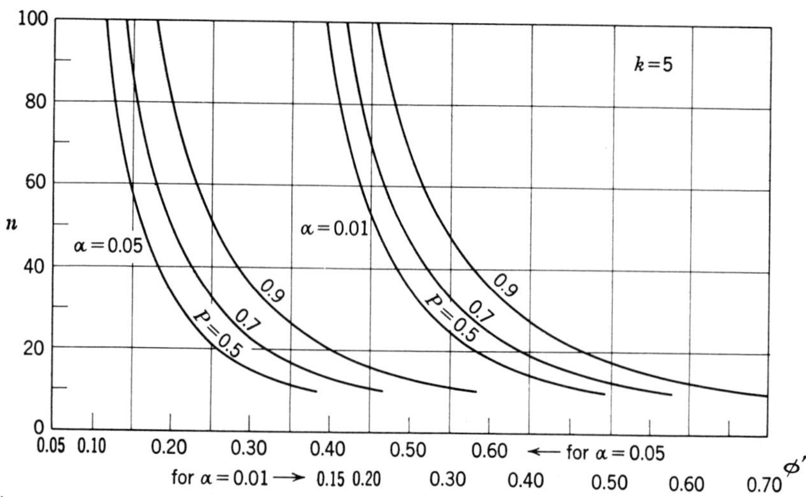

† Reproduced from L. S. Feldt and M. W. Mahmoud, Power function charts for specification of sample size in analysis of variance. *Psychometrika*, 1958, 23, 201–210, with permission of the editor.

Table C.12 Random Permutations of 16 Numbers

12	11	16	11	7	4	13	7	16	16	14	15	7	13	13	10	6	1	8	10	15	16	1	2	1	3	12	1	4	3	2	9	5	13	8	15	10	15	13	3
5	9	9	3	1	2	5	6	3	5	11	7	13	16	7	7	5	13	2	14	4	13	3	4	5	6	10	16	1	8	8	14	4	8	14	6	6	10	5	16
13	15	2	5	6	8	2	11	6	10	13	2	9	15	6	2	8	4	5	8	7	2	9	15	15	14	11	8	6	5	7	2	12	6	2	13	2	13	15	10
16	4	5	16	9	9	16	4	8	15	8	5	12	3	9	13	4	3	10	4	1	15	4	12	9	9	8	12	15	2	12	10	13	10	1	12	14	8	7	5
2	2	13	1	11	11	12	13	5	3	5	8	14	7	3	15	13	11	4	7	9	11	7	1	13	4	16	9	13	12	4	11	3	11	10	8	8	9	12	4
10	12	10	7	14	15	14	10	12	4	3	12	5	14	12	16	2	2	9	15	16	3	14	7	6	10	1	13	7	16	9	15	14	7	11	14	9	4	8	15
8	14	15	14	12	7	1	1	7	13	7	10	6	9	11	9	7	6	16	11	3	12	13	16	16	1	14	15	2	11	15	4	11	16	6	4	11	7	14	6
3	16	3	6	16	6	8	3	9	11	4	16	2	6	5	1	12	10	6	9	5	4	10	13	7	7	9	7	3	10	6	1	16	15	9	16	5	1	4	11
1	1	8	9	2	16	10	14	14	2	10	1	15	5	14	12	14	12	3	2	13	9	5	14	4	5	15	10	12	15	10	16	2	3	13	10	7	2	1	12
15	3	12	8	3	10	11	8	2	14	15	4	11	8	16	8	9	14	14	1	8	5	15	3	8	2	7	5	8	7	13	7	15	14	4	1	4	14	10	7
6	6	1	10	13	13	9	5	15	1	6	13	16	12	15	4	11	8	12	6	12	6	2	8	11	13	2	2	10	14	1	6	7	4	7	9	1	12	16	1
4	10	4	4	15	3	15	2	1	12	1	9	8	1	8	14	16	5	13	5	6	1	8	6	2	16	5	11	16	6	3	13	9	12	5	5	3	16	6	8
7	8	7	15	10	5	3	9	11	7	9	3	10	4	4	3	10	9	1	3	10	7	16	5	10	15	6	14	9	13	16	8	6	1	15	11	15	11	11	14
14	13	6	12	5	12	4	16	4	9	12	14	3	10	1	6	15	16	15	12	11	8	6	11	14	12	13	3	5	9	14	3	1	9	3	2	12	3	3	2
11	5	14	2	4	14	6	15	10	8	16	11	1	2	10	5	1	15	7	13	14	14	11	10	12	11	4	4	14	1	11	12	8	2	16	7	13	5	9	13
9	7	11	13	8	1	7	12	13	6	2	6	4	11	2	11	3	7	11	16	2	10	12	9	3	8	3	6	11	4	5	5	10	5	12	3	16	6	2	9

1	10	1	15	16	10	11	4	6	4	14	12	9	8	7	5	13	3	13	3	12	12	10	2	14	15	3	10	9	2	8	9	12	13	10	3	16	8	9	8
3	3	5	3	6	7	7	8	5	12	10	14	15	1	2	11	4	5	15	9	7	15	8	15	10	4	16	12	13	14	14	8	11	14	2	10	6	11	6	6
10	8	11	11	7	16	12	15	9	9	12	6	16	15	16	9	10	12	16	15	9	9	11	6	2	9	8	1	15	4	6	3	16	12	14	1	12	16	11	2
9	2	4	7	1	3	15	10	14	11	3	3	12	14	15	13	6	4	1	16	4	2	16	4	8	8	15	13	10	8	11	13	4	1	6	7	15	12	14	11
6	16	9	4	8	12	8	2	8	6	2	9	8	16	4	6	5	15	7	8	3	4	3	14	12	2	5	6	14	10	9	12	10	7	13	8	2	1	8	9
2	12	10	9	12	14	9	5	13	5	4	13	7	7	9	12	14	8	8	11	16	10	15	8	3	10	14	2	7	3	15	6	6	3	9	2	3	9	1	13
12	1	15	5	15	6	4	12	4	14	9	5	14	11	10	1	3	13	3	5	8	5	2	7	7	16	11	14	16	13	16	10	3	8	7	9	1	13	7	12
13	6	8	6	13	11	3	1	10	7	11	15	6	12	5	7	11	1	14	4	2	6	4	16	9	5	2	11	2	12	7	5	2	10	16	15	14	5	5	3
14	5	3	16	4	13	10	13	15	3	7	8	3	13	11	8	7	7	12	7	15	14	13	11	6	3	12	3	4	7	2	11	15	16	8	4	10	2	10	4
8	15	2	14	9	1	1	7	3	13	8	1	13	6	3	3	15	9	9	12	6	13	9	9	1	11	6	15	11	5	3	14	5	11	15	6	9	6	3	15
11	13	16	12	10	15	2	9	2	16	5	11	10	10	12	15	16	14	5	2	11	8	1	13	5	1	1	5	12	9	5	16	7	4	4	5	13	15	12	16
16	9	6	10	14	5	6	6	7	15	16	4	5	5	13	4	9	16	2	6	10	3	12	12	15	7	13	8	3	16	12	15	14	9	12	11	8	7	15	5
7	14	12	8	3	4	5	16	12	8	15	10	1	4	6	16	2	6	11	1	1	7	14	10	4	13	4	16	1	15	10	4	8	15	5	13	4	10	13	7
4	4	13	2	5	8	14	3	1	1	6	7	11	9	14	10	8	11	4	13	14	1	6	5	13	14	10	4	5	1	4	7	13	5	11	12	5	4	2	1
5	7	14	1	2	2	16	11	16	2	13	2	4	2	1	2	12	2	10	14	5	16	5	1	16	6	9	7	6	6	1	2	1	2	3	14	7	14	16	14
15	11	7	13	11	9	13	14	11	10	1	16	2	3	8	14	1	10	6	10	13	11	7	3	11	12	7	9	8	11	13	1	9	6	1	16	11	3	4	10

Table C.12 Random Permutations of 16 Numbers (*Continued*)

12	13	16	6	12	10	5	12	1	10	4	6	6	1	4	5	7	13	2	10	16	11	15	9	2	12	4	10	10	1	9	6	15	15	1	13	1	1	16	2
6	1	13	2	7	12	16	10	11	13	8	2	15	7	9	8	16	1	14	3	14	12	14	1	7	16	7	12	8	6	4	10	6	3	10	16	7	10	12	12
10	14	11	12	4	15	13	16	9	3	7	16	11	8	3	3	12	2	3	4	8	10	13	6	5	5	9	5	1	7	3	12	11	2	15	3	16	16	6	7
7	7	4	10	10	6	1	9	5	9	15	5	12	5	7	16	5	11	8	1	5	6	2	8	3	7	13	15	4	14	1	16	10	13	9	1	5	8	15	14
2	16	10	8	11	14	4	3	8	15	14	9	2	16	1	12	6	14	4	13	13	8	8	12	10	4	2	6	14	11	6	8	4	6	4	10	6	7	1	16
11	3	6	4	15	1	12	4	6	1	16	14	3	15	11	11	3	9	12	5	2	5	1	3	9	2	1	16	2	5	16	7	9	10	5	12	2	4	14	5
9	11	1	3	6	2	15	14	7	4	9	10	16	2	10	2	10	7	10	16	6	4	10	5	14	8	8	8	6	13	15	1	13	7	12	9	9	13	9	8
8	15	3	16	14	16	14	2	2	7	10	12	5	12	5	14	13	16	5	6	3	7	11	14	6	1	16	1	3	4	10	2	2	14	13	8	4	6	13	13
16	5	14	14	8	5	7	15	10	12	3	13	14	4	2	1	14	8	6	12	7	3	16	16	16	6	6	11	9	12	14	5	7	12	2	6	10	2	11	1
13	10	7	7	9	3	3	13	12	6	2	4	13	3	16	9	9	3	7	14	4	14	3	7	1	11	10	3	5	2	11	4	3	8	11	4	11	9	10	15
4	12	12	5	16	9	11	6	3	16	11	7	8	14	6	4	4	4	15	11	11	13	12	10	4	3	11	2	7	15	7	14	14	11	16	2	14	15	4	11
5	9	8	9	3	13	8	8	14	11	13	1	9	10	12	10	11	10	9	8	15	15	4	11	12	15	5	14	12	10	8	11	1	9	3	14	8	3	7	3
15	6	9	13	5	11	2	11	13	5	6	8	4	9	8	15	8	6	11	9	1	9	7	15	11	9	14	4	15	16	12	9	5	1	6	15	3	5	3	10
3	4	15	15	13	7	6	5	4	14	1	11	7	11	13	6	1	15	13	15	9	1	5	13	15	10	12	13	11	8	13	13	12	5	14	11	15	14	8	9
1	2	5	11	1	8	10	1	15	8	12	3	10	6	14	7	2	12	16	7	12	16	6	2	13	13	15	9	16	3	5	15	16	4	8	7	13	11	5	6
14	8	2	1	2	4	9	7	16	2	5	15	1	13	15	13	15	5	1	2	10	2	9	4	8	14	3	7	13	9	2	3	8	16	7	5	12	12	2	4

5	7	15	11	10	14	10	16	4	16	6	14	1	16	8	7	2	3	3	12	12	11	5	11	16	7	14	5	4	7	14	12	6	15	10	7	7	9	8	10
8	6	1	2	5	1	11	13	13	15	16	4	15	4	3	12	12	1	4	7	10	5	11	12	7	4	11	3	5	8	2	10	2	2	15	15	3	11	5	12
6	13	6	5	13	9	15	4	9	10	10	15	14	9	10	1	14	8	8	16	7	9	12	7	1	12	5	6	9	15	8	15	9	3	9	5	14	10	13	8
16	4	3	16	2	8	9	8	7	7	5	13	16	1	7	5	11	2	9	3	3	6	1	8	14	9	7	8	8	9	1	1	5	13	12	8	2	7	10	6
14	9	5	10	15	5	13	9	3	4	12	1	12	12	14	16	3	11	11	8	8	2	10	14	8	16	8	12	12	13	12	5	12	10	14	12	1	1	2	5
10	10	10	3	4	12	1	3	2	11	7	11	6	3	11	4	4	6	6	9	13	15	15	15	6	14	16	14	10	3	16	16	8	4	7	1	11	15	1	3
7	11	4	7	3	2	7	14	11	1	9	3	3	7	13	11	10	4	5	1	14	13	2	13	9	3	1	2	14	12	3	3	1	7	1	10	10	8	7	14
1	2	2	15	7	4	14	15	8	8	4	9	13	13	6	8	15	9	1	14	6	1	4	1	13	1	13	16	2	5	4	13	10	9	3	14	6	16	4	7
15	16	8	9	9	11	16	12	12	3	1	8	4	11	5	2	16	10	12	4	9	8	8	9	5	10	6	15	15	11	13	11	7	16	6	3	16	2	11	1
2	3	13	1	8	10	5	10	5	12	2	16	5	8	1	9	5	12	16	6	15	12	6	3	3	5	4	13	11	14	7	6	4	8	8	11	12	6	9	16
3	14	14	6	12	3	8	1	15	5	15	10	7	10	9	10	6	14	10	11	16	14	9	5	2	8	3	7	7	1	11	14	3	12	2	16	13	5	16	2
9	15	11	14	6	7	6	2	6	6	14	2	8	2	4	13	8	5	15	5	5	16	14	4	12	2	9	9	3	6	15	2	14	1	11	4	5	3	3	9
13	1	7	13	1	16	3	11	16	13	13	12	9	6	2	15	7	15	7	13	4	3	7	6	10	11	15	4	1	16	6	8	16	6	16	2	4	12	6	11
12	8	16	8	11	15	2	6	14	2	11	5	10	15	12	3	1	13	13	10	2	4	16	16	15	13	2	10	13	10	9	7	11	5	5	6	9	14	12	15
4	5	9	4	14	13	12	5	1	14	8	6	11	14	15	6	9	16	2	2	1	10	3	2	11	15	10	1	16	2	10	4	15	11	13	9	15	13	15	13
11	12	12	12	16	6	4	7	10	9	3	7	2	5	16	14	13	7	14	15	11	7	13	10	4	6	12	11	6	4	5	9	13	14	4	13	8	4	14	4

Table C.13 Noncentral t Distribution†

Values of the noncentrality parameter for a one-sided test with $\alpha = .050$

f	β = type 2 error										
	.01	.05	.10	.20	.30	.40	.50	.60	.70	.80	.90
1	16.47	12.53	10.51	8.19	6.63	5.38	4.31	3.35	2.46	1.60	.64
2	6.88	5.52	4.81	3.98	3.40	2.92	2.49	2.07	1.63	1.15	.50
3	5.47	4.46	3.93	3.30	2.85	2.48	2.13	1.79	1.43	1.02	.46
4	4.95	4.07	3.60	3.04	2.64	2.30	1.99	1.67	1.34	.96	.43
5	4.70	3.87	3.43	2.90	2.53	2.21	1.91	1.61	1.29	.92	.42
6	4.55	3.75	3.33	2.82	2.46	2.15	1.86	1.57	1.26	.90	.41
7	4.45	3.67	3.26	2.77	2.41	2.11	1.82	1.54	1.24	.89	.40
8	4.38	3.62	3.21	2.73	2.38	2.08	1.80	1.52	1.22	.88	.40
9	4.32	3.58	3.18	2.70	2.35	2.06	1.78	1.51	1.21	.87	.39
10	4.28	3.54	3.15	2.67	2.33	2.04	1.77	1.49	1.20	.86	.39
11	4.25	3.52	3.13	2.65	2.31	2.02	1.75	1.48	1.19	.86	.39
12	4.22	3.50	3.11	2.64	2.30	2.01	1.74	1.47	1.19	.85	.38
13	4.20	3.48	3.09	2.63	2.29	2.00	1.74	1.47	1.18	.85	.38
14	4.18	3.46	3.08	2.62	2.28	2.00	1.73	1.46	1.18	.84	.38
15	4.17	3.45	3.07	2.61	2.27	1.99	1.72	1.46	1.17	.84	.38
16	4.16	3.44	3.06	2.60	2.27	1.98	1.72	1.45	1.17	.84	.38
17	4.14	3.43	3.05	2.59	2.26	1.98	1.71	1.45	1.17	.84	.38
18	4.13	3.42	3.04	2.59	2.26	1.97	1.71	1.45	1.16	.83	.38
19	4.12	3.41	3.04	2.58	2.25	1.97	1.71	1.44	1.16	.83	.38
20	4.12	3.41	3.03	2.58	2.25	1.97	1.70	1.44	1.16	.83	.38
21	4.11	3.40	3.03	2.57	2.24	1.96	1.70	1.44	1.16	.83	.38
22	4.10	3.40	3.02	2.57	2.24	1.96	1.70	1.44	1.16	.83	.37
23	4.10	3.39	3.02	2.56	2.24	1.96	1.69	1.43	1.15	.83	.37
24	4.09	3.39	3.01	2.56	2.23	1.95	1.69	1.43	1.15	.83	.37
25	4.09	3.38	3.01	2.56	2.23	1.95	1.69	1.43	1.15	.83	.37
26	4.08	3.38	3.01	2.55	2.23	1.95	1.69	1.43	1.15	.82	.37
27	4.08	3.38	3.00	2.55	2.23	1.95	1.69	1.43	1.15	.82	.37
28	4.07	3.37	3.00	2.55	2.22	1.95	1.69	1.43	1.15	.82	.37
29	4.07	3.37	3.00	2.55	2.22	1.94	1.68	1.42	1.15	.82	.37
30	4.07	3.37	3.00	2.54	2.22	1.94	1.68	1.42	1.15	.82	.37
40	4.04	3.35	2.98	2.53	2.21	1.93	1.67	1.42	1.14	.82	.37
60	4.02	3.33	2.96	2.52	2.19	1.92	1.66	1.41	1.13	.81	.37
100	4.00	3.31	2.95	2.50	2.18	1.91	1.66	1.40	1.13	.81	.37
∞	3.97	3.29	2.93	2.49	2.17	1.90	1.64	1.39	1.12	.80	.36

$$\Pr[\text{noncentral } t > t_{1-\alpha} \mid \delta = (\mu_1 - \mu_0)(\sqrt{n}/\sigma)] = 1 - \beta.$$

† This table is reproduced from: The power of Student's t test. *Journal of the American Statistical Association*, 1965, **60**, 320–333, with the permission of the author, D. B. Owen, and the editors.

Table C.13 (*Continued*)
Values of the noncentrality parameter for a one-sided test with $\alpha = .025$

f	β = type 2 error										
	.01	.05	.10	.20	.30	.40	.50	.60	.70	.80	.90
1	32.83	24.98	20.96	16.33	13.21	10.73	8.60	6.68	4.91	3.22	1.58
2	9.67	7.77	6.80	5.65	4.86	4.21	3.63	3.07	2.50	1.88	1.09
3	6.88	5.65	5.01	4.26	3.72	3.28	2.87	2.47	2.05	1.57	.94
4	5.94	4.93	4.40	3.76	3.31	2.93	2.58	2.23	1.86	1.44	.87
5	5.49	4.57	4.09	3.51	3.10	2.75	2.43	2.11	1.76	1.37	.82
6	5.22	4.37	3.91	3.37	2.98	2.64	2.34	2.03	1.70	1.32	.80
7	5.06	4.23	3.80	3.27	2.89	2.57	2.27	1.98	1.66	1.29	.78
8	4.94	4.14	3.71	3.20	2.83	2.52	2.23	1.94	1.63	1.27	.77
9	4.85	4.07	3.65	3.15	2.79	2.48	2.20	1.91	1.60	1.25	.76
10	4.78	4.01	3.60	3.11	2.75	2.45	2.17	1.89	1.59	1.23	.75
11	4.73	3.97	3.57	3.08	2.73	2.43	2.15	1.87	1.57	1.22	.74
12	4.69	3.93	3.54	3.05	2.70	2.41	2.13	1.85	1.56	1.21	.74
13	4.65	3.91	3.51	3.03	2.69	2.39	2.12	1.84	1.55	1.21	.73
14	4.62	3.88	3.49	3.01	2.67	2.38	2.11	1.83	1.54	1.20	.73
15	4.60	3.86	3.47	3.00	2.66	2.37	2.09	1.82	1.53	1.19	.72
16	4.58	3.84	3.46	2.98	2.65	2.36	2.09	1.81	1.53	1.19	.72
17	4.56	3.83	3.44	2.97	2.64	2.35	2.08	1.81	1.52	1.18	.72
18	4.54	3.82	3.43	2.96	2.63	2.34	2.07	1.80	1.52	1.18	.72
19	4.52	3.80	3.42	2.95	2.61	2.33	2.06	1.80	1.51	1.17	.71
20	4.51	3.79	3.41	2.95	2.61	2.33	2.06	1.79	1.51	1.17	.71
21	4.50	3.78	3.40	2.93	2.60	2.32	2.05	1.79	1.50	1.17	.71
22	4.49	3.77	3.39	2.93	2.60	2.32	2.05	1.78	1.50	1.17	.71
23	4.48	3.77	3.39	2.93	2.59	2.31	2.05	1.78	1.50	1.17	.71
24	4.47	3.76	3.38	2.92	2.59	2.31	2.04	1.78	1.50	1.16	.71
25	4.46	3.75	3.37	2.92	2.58	2.30	2.04	1.77	1.49	1.16	.71
26	4.46	3.75	3.37	2.92	2.58	2.30	2.04	1.77	1.49	1.16	.70
27	4.45	3.74	3.36	2.91	2.58	2.30	2.03	1.77	1.49	1.16	.70
28	4.44	3.73	3.36	2.90	2.57	2.29	2.03	1.77	1.49	1.16	.70
29	4.44	3.73	3.35	2.90	2.57	2.29	2.03	1.77	1.48	1.16	.70
30	4.43	3.73	3.35	2.90	2.57	2.29	2.02	1.76	1.48	1.16	.70
40	4.39	3.69	3.32	2.87	2.55	2.27	2.01	1.75	1.47	1.15	.69
60	4.36	3.66	3.29	2.85	2.53	2.25	1.99	1.73	1.46	1.14	.69
100	4.33	3.64	3.27	2.83	2.51	2.23	1.98	1.73	1.45	1.12	.68
∞	4.29	3.60	3.24	2.80	2.48	2.21	1.96	1.71	1.44	1.12	.68

Pr [noncentral $t > t_{1-\alpha} \mid \delta = (\mu_1 - \mu_0)(\sqrt{n}/\sigma)] = 1 - \beta$.

Table C.13 (*Continued*)
Values of the noncentrality parameter for a one-sided test with $\alpha = .010$

f	β = type 2 error										
	.01	.05	.10	.20	.30	.40	.50	.60	.70	.80	.90
1	82.00	62.40	52.37	40.80	33.00	26.79	21.47	16.69	12.27	8.07	4.00
2	15.22	12.26	10.74	8.96	7.73	6.73	5.83	4.98	4.12	3.20	2.08
3	9.34	7.71	6.86	5.87	5.17	4.59	4.07	3.56	3.03	2.44	1.66
4	7.52	6.28	5.64	4.88	4.34	3.88	3.47	3.06	2.63	2.14	1.48
5	6.68	5.62	5.07	4.40	3.93	3.54	3.17	2.81	2.42	1.98	1.38
6	6.21	5.25	4.74	4.13	3.70	3.33	2.99	2.66	2.30	1.88	1.32
7	5.91	5.01	4.53	3.96	3.55	3.20	2.88	2.56	2.22	1.82	1.27
8	5.71	4.85	4.39	3.84	3.44	3.11	2.80	2.49	2.16	1.77	1.24
9	5.56	4.72	4.28	3.75	3.37	3.04	2.74	2.43	2.11	1.74	1.22
10	5.45	4.63	4.20	3.68	3.31	2.99	2.69	2.39	2.08	1.71	1.20
11	5.36	4.56	4.14	3.63	3.26	2.94	2.65	2.36	2.05	1.69	1.18
12	5.29	4.50	4.09	3.58	3.22	2.91	2.62	2.33	2.03	1.67	1.17
13	5.23	4.46	4.04	3.55	3.19	2.88	2.60	2.31	2.01	1.65	1.16
14	5.18	4.42	4.01	3.51	3.16	2.86	2.57	2.29	1.99	1.64	1.15
15	5.14	4.38	3.98	3.49	3.14	2.84	2.56	2.28	1.98	1.63	1.14
16	5.11	4.35	3.95	3.47	3.12	2.82	2.54	2.26	1.97	1.62	1.14
17	5.08	4.33	3.93	3.45	3.10	2.80	2.53	2.25	1.96	1.61	1.13
18	5.05	4.31	3.91	3.43	3.09	2.79	2.52	2.24	1.95	1.60	1.13
19	5.03	4.29	3.89	3.42	3.07	2.78	2.50	2.23	1.94	1.60	1.12
20	5.01	4.27	3.88	3.40	3.06	2.77	2.50	2.22	1.93	1.59	1.12
21	4.99	4.25	3.86	3.39	3.05	2.76	2.49	2.22	1.92	1.59	1.11
22	4.97	4.24	3.85	3.38	3.04	2.75	2.48	2.21	1.92	1.58	1.11
23	4.96	4.23	3.84	3.37	3.03	2.74	2.47	2.20	1.91	1.58	1.11
24	4.94	4.22	3.83	3.36	3.02	2.73	2.47	2.20	1.91	1.57	1.11
25	4.93	4.20	3.82	3.35	3.02	2.73	2.46	2.19	1.90	1.57	1.10
26	4.92	4.19	3.81	3.34	3.01	2.72	2.45	2.19	1.90	1.57	1.10
27	4.91	4.19	3.80	3.34	3.00	2.72	2.45	2.18	1.90	1.56	1.10
28	4.90	4.18	3.79	3.33	3.00	2.71	2.44	2.18	1.89	1.56	1.10
29	4.89	4.17	3.79	3.32	2.99	2.71	2.44	2.17	1.89	1.56	1.10
30	4.88	4.16	3.78	3.32	2.99	2.70	2.44	2.17	1.89	1.55	1.09
40	4.82	4.11	3.74	3.28	2.95	2.67	2.41	2.15	1.86	1.54	1.08
60	4.76	4.06	3.69	3.24	2.92	2.64	2.38	2.12	1.84	1.52	1.07
100	4.72	4.03	3.66	3.21	2.89	2.62	2.36	2.10	1.83	1.51	1.06
∞	4.65	3.97	3.61	3.17	2.85	2.58	2.33	2.07	1.80	1.48	1.04

Pr [noncentral $t > t_{1-\alpha} \mid \delta = (\mu_1 - \mu_0)(\sqrt{n}/\sigma)] = 1 - \beta$.

Table C.14 Noncentral F Distribution†

Tabled entries are $\Pr\,[F(f_1,f_2;\phi) < F_{1-\alpha}(f_1,f_2)]$

Power = 1 − (tabled entry)

f_2	$1-\alpha$	ϕ .50	1.0	1.2	1.4	1.6	1.8	2.0	2.2	2.6	3.0	4.0
							$f_1 = 1$					
2	.95	.93	.86	.83	.78	.74	.69	.64	.59	.49	.40	.20
	.99	.99	.97	.96	.95	.94	.93	.91	.90	.87	.83	.72
4	.95	.91	.80	.74	.67	.59	.51	.43	.35	.22	.12	.02
	.99	.98	.95	.94	.92	.89	.86	.82	.78	.67	.56	.23
6	.95	.91	.78	.70	.62	.52	.43	.34	.26	.14	.06	.00
	.99	.98	.93	.90	.86	.81	.75	.69	.61	.46	.31	.08
8	.95	.90	.76	.68	.59	.49	.39	.30	.22	.11	.04	.00
	.99	.98	.92	.89	.84	.78	.70	.62	.54	.37	.22	.03
10	.95	.90	.75	.66	.57	.47	.37	.28	.20	.09	.03	.00
	.99	.98	.92	.87	.82	.75	.67	.58	.49	.31	.17	.02
12	.95	.90	.74	.65	.56	.45	.35	.26	.19	.08	.03	
	.99	.97	.91	.87	.81	.73	.65	.55	.46	.28	.14	
16	.95	.90	.74	.64	.54	.43	.33	.24	.17	.07	.02	
	.99	.97	.90	.85	.79	.71	.61	.52	.42	.24	.11	
20	.95	.90	.73	.63	.53	.42	.32	.23	.16	.06	.02	
	.99	.97	.90	.85	.78	.69	.59	.49	.39	.21	.10	
30	.95	.89	.72	.62	.52	.40	.31	.22	.15	.06	.02	
	.99	.97	.89	.83	.76	.67	.57	.46	.36	.19	.08	
∞	.95	.89	.71	.70	.49	.38	.28	.19	.12	.04	.01	
	.99	.97	.88	.81	.72	.62	.51	.40	.30	.14	.05	
f_2	$1-\alpha$						$f_1 = 2$					
2	.95	.93	.88	.85	.82	.78	.75	.70	.66	.57	.48	.29
	.99	.99	.98	.97	.96	.95	.94	.93	.92	.89	.86	.78
4	.95	.92	.82	.77	.70	.62	.54	.46	.38	.24	.14	.02
	.99	.98	.96	.94	.92	.89	.85	.81	.76	.66	.54	.27
6	.95	.91	.79	.71	.63	.53	.43	.34	.26	.13	.05	.00
	.99	.98	.94	.91	.87	.82	.76	.70	.62	.46	.31	.07
8	.95	.91	.77	.68	.58	.48	.37	.28	.20	.08	.03	.00
	.99	.98	.93	.89	.84	.78	.70	.61	.52	.34	.19	.02
10	.95	.91	.75	.66	.55	.44	.34	.24	.16	.06	.02	.00
	.99	.98	.92	.88	.82	.74	.65	.55	.45	.26	.13	.01
12	.95	.90	.74	.64	.53	.42	.31	.22	.14	.05	.01	
	.99	.98	.91	.86	.80	.71	.61	.51	.40	.22	.09	
16	.95	.90	.73	.62	.51	.39	.28	.19	.12	.04	.01	
	.99	.97	.90	.84	.77	.67	.57	.45	.34	.16	.06	
20	.95	.90	.72	.61	.49	.37	.26	.17	.11	.03	.01	
	.99	.97	.90	.83	.75	.65	.53	.42	.31	.14	.04	
30	.95	.90	.71	.59	.47	.35	.24	.15	.09	.02	.00	
	.99	.97	.88	.82	.72	.61	.49	.37	.26	.10	.03	
∞	.95	.89	.68	.56	.43	.30	.20	.12	.06	.01	.00	
	.99	.97	.86	.77	.66	.53	.40	.28	.18	.05	.01	

† This table is abridged from: Tables of the power of the F test. *Journal of the American Statistical Association*, 1967, **62**, 525–539, with the permission of the author, M. L. Tiku, and the editors.

Table C.14 *(Continued)*

f_2	$1-\alpha$	ϕ .50	1.0	1.2	1.4	1.6	1.8	2.0	2.2	2.6	3.0	4.0
		$f_1 = 3$										
2	.95	.93	.89	.86	.83	.80	.76	.73	.69	.60	.52	.32
	.99	.99	.98	.97	.96	.96	.95	.94	.93	.90	.88	.80
4	.95	.92	.83	.77	.71	.63	.55	.47	.39	.25	.14	.02
	.99	.98	.96	.94	.92	.89	.86	.82	.77	.67	.55	.28
6	.95	.91	.79	.71	.62	.52	.42	.33	.24	.11	.04	.00
	.99	.98	.94	.91	.87	.82	.76	.69	.61	.44	.29	.06
8	.95	.91	.76	.67	.57	.46	.35	.25	.17	.06	.02	.00
	.99	.98	.93	.89	.84	.77	.68	.59	.49	.30	.16	.01
10	.95	.91	.75	.65	.53	.41	.30	.21	.13	.04	.01	
	.99	.98	.92	.87	.80	.72	.62	.52	.41	.22	.09	
12	.95	.90	.73	.62	.51	.38	.27	.18	.11	.03	.01	
	.99	.98	.91	.85	.78	.69	.58	.46	.35	.17	.06	
16	.95	.90	.71	.60	.47	.34	.23	.14	.08	.02	.00	
	.99	.97	.90	.83	.74	.64	.51	.39	.28	.11	.03	
20	.95	.90	.70	.58	.45	.32	.21	.13	.07	.01	.00	
	.99	.97	.89	.82	.72	.60	.47	.35	.24	.08	.02	
30	.95	.89	.68	.55	.42	.29	.18	.10	.05	.01	.00	
	.99	.97	.87	.79	.68	.55	.42	.29	.18	.05	.01	
∞	.95	.88	.64	.50	.36	.23	.13	.07	.03	.00	.00	
	.99	.97	.84	.73	.59	.44	.30	.18	.10	.02	.00	
f_2	$1-\alpha$	$f_1 = 4$										
2	.95	.94	.89	.87	.84	.81	.77	.74	.70	.62	.54	.34
	.99	.99	.98	.97	.97	.96	.95	.94	.93	.91	.88	.81
4	.95	.92	.83	.78	.71	.64	.55	.47	.39	.25	.14	.02
	.99	.98	.96	.94	.92	.89	.86	.82	.78	.67	.56	.28
6	.95	.92	.79	.71	.62	.52	.41	.31	.23	.10	.04	.00
	.99	.98	.94	.91	.87	.82	.76	.68	.60	.43	.28	.05
8	.95	.91	.76	.66	.55	.44	.33	.23	.15	.05	.01	.00
	.99	.98	.93	.89	.83	.76	.67	.57	.47	.28	.14	.01
10	.95	.91	.74	.63	.51	.39	.27	.18	.11	.03	.01	
	.99	.98	.92	.87	.79	.70	.60	.49	.37	.19	.07	
12	.95	.90	.72	.61	.48	.35	.24	.15	.08	.02	.00	
	.99	.98	.91	.85	.76	.66	.55	.42	.31	.13	.04	
16	.95	.90	.70	.57	.44	.31	.19	.11	.06	.01	.00	
	.99	.97	.89	.82	.72	.60	.47	.34	.23	.08	.02	
20	.95	.89	.68	.55	.41	.28	.17	.09	.04	.01	.00	
	.99	.97	.88	.80	.69	.56	.42	.29	.18	.05	.01	
30	.95	.89	.66	.52	.37	.24	.14	.07	.03	.00	.00	
	.99	.97	.86	.77	.64	.50	.35	.22	.13	.03	.00	
∞	.95	.88	.60	.45	.29	.17	.08	.04	.01	.00	.00	
	.99	.96	.81	.68	.53	.36	.22	.11	.05	.01	.00	

CONTENT REFERENCES

Addelman, S. (1967). *The selection of sequences of two-level fractional factorial plans.* ARL 67-0013, Office of Aerospace Research, USAF, Wright-Patterson AFB, Ohio.

Alexander, H. W. (1946). A general test for trend. *Psychological Bulletin,* **43,** 533–557.

Anderson, R. L. (1959). Use of contingency tables in the analysis of consumer preference studies. *Biometrics,* **15,** 582–590.

Anderson, R. L., and E. E. Houseman (1942). *Tables of orthogonal polynomial values extended to* $N = 104$. *Research Bulletin* 297, Ames, Iowa.

Anderson, R. L., and T. A. Bancroft (1952). *Statistical theory in research.* New York: McGraw-Hill.

Anderson, T. W. (1958). *Introduction to multivariate statistical analysis.* New York: Wiley.

Aspin, A. A. (1949). Tables for use in comparisons whose accuracy involves two variances separately estimated. *Biometrika,* **36,** 290–293.

Balaam, L. N., and W. T. Federer (1965). Error base rates. *Technometrics,* **7,** 260–262.

Bennett, C. A., and N. L. Franklin (1954). *Statistical analysis in chemistry and the chemical industry.* New York: Wiley.

Bock, R. D. (1963). Programming univariate and multivariate analysis of variance. *Technometrics,* **5,** 95–118.

Box, G. E. P. (1950). Problems in the analysis of growth and wear curves. *Biometrics,* **6,** 362–389.

Box, G. E. P. (1953). Non-normality and tests on variance. *Biometrika,* **40,** 318–335.

Box, G. E. P. (1954). Some theorems on quadratic forms applied in the study of analysis of variance problems. *The Annals of Mathematical Statistics*, **25**, 290–302, 484–498.

Box, G. E. P., and J. S. Hunter (1957). Multi-factor experimental designs for exploring response surfaces. *The Annals of Mathematical Statistics*, **28**, 195–241.

Bozivich, H., T. A. Bancroft, and H. O. Hartley (1956). Power of analysis of variance procedures for certain incompletely specified models. *The Annals of Mathematical Statistics*, **27**, 1017–1043.

Cochran, W. G. (1947). Some consequences when assumptions for the analysis of variance are not satisfied. *Biometrics*, **3**, 22–38.

Cochran, W. G. (1950). The comparison of percentages in matched samples. *Biometrika*, **37**, 256–266.

Cochran, W. G. (1951). Testing a linear relation among variances. *Biometrics*, **7**, 17–32.

Cochran, W. G. (1957). Analysis of covariance: Its nature and use. *Biometrics*, **13**, 261–281.

Cochran, W. G., and G. M. Cox (1957). *Experimental designs*. (2d ed.) New York: Wiley.

Collier, R. O., Jr., and F. B. Baker (1963). The randomization distribution of *F*-ratios for the split-plot design—an empirical investigation. *Biometrika*, **50**, 431–438.

Collier, R. O., Jr., F. B. Baker, G. K. Mandeville, and T. F. Hayes (1967). Estimates of test size for several test procedures based on variance ratios in the repeated measures design. *Psychometrika*, **32**, 339–353.

Cornfield, J., and J. W. Tukey (1956). Average values of mean squares in factorials. *The Annals of Mathematical Statistics*, **27**, 907–949.

Cornish, E. A. (1957). An application of the Kronecker product of matrices in multiple regression. *Biometrics*, **13**, 19–27.

Cox, D. R. (1958). *Planning of experiments*. New York: Wiley.

Crump, S. L. (1951). The present status of variance component analysis. *Biometrics*, **7**, 1–16.

Dixon, W. J., and J. W. Tukey (1968). Approximate behavior of the distribution of winsorized *t* (trimming/winsorization 2). *Technometrics*, **10**, 83–98.

Duncan, D. B. (1955). Multiple range and multiple *F* tests. *Biometrics*, **11**, 1–42.

Duncan, D. B. (1957). Multiple range tests for correlated and heteroscedastic means. *Biometrics*, **13**, 164–176.

Dunnett, C. W. (1955). A multiple comparison procedure for comparing several treatments with a control. *Journal of the American Statistical Association*, **50**, 1096–1121.

Dunnett, C. W. (1964). New tables for multiple comparisons with a control. *Biometrics*, **20**, 482–491.

Eisenhart, C. (1947). The assumptions underlying the analysis of variance. *Biometrics*, **3**, 1–21.

Federer, W. T. (1955). *Experimental design*. New York: Macmillan.

Federer, W. T. (1963). Relationships between a three-way classification disproportionate numbers analysis of variance and several two-way classification and nested analyses. *Biometrics*, **19**, 629–637.

Federer, W. T., and M. Zelen (1966). Analysis of multifactor classifications with unequal numbers of observations. *Biometrics*, **22**, 525–552.

Feldt, L. S., and M. W. Mahmoud (1958). Power function charts for specification of sample size in analysis of variance. *Psychometrika*, **23**, 201–210.

Fisher, R. A. (1951). *The design of experiments.* (6th ed.) Edinburgh: Oliver & Boyd.

Fisher, R. A., and F. Yates (1953). *Statistical tables for biological, agricultural, and medical research.* (4th ed.) Edinburgh: Oliver & Boyd.

Fleiss, J. L. (1969). Estimating the magnitude of experimental effects. *Psychological Bulletin*, **72**, 273–276.

Gayen, A. K. (1949). The distribution of Student's t in random samples of any size drawn from non-normal universes. *Biometrika*, **36**, 353–369.

Gaylor, D. W., and F. N. Hopper (1969). Estimating the degrees of freedom for linear combinations of mean squares by Satterthwaite's formula. *Technometrics*, **11**, 691–706.

Geisser, S. (1959). A method for testing treatment effects in the presence of learning. *Biometrics*, **15**, 389–395.

Geisser, S., and S. W. Greenhouse (1958). An extension of Box's results on the use of the F distribution in multivariate analysis. *The Annals of Mathematical Statistics*, **29**, 885–891.

Goodman, L. A. (1970). The multivariate analysis of qualitative data: Interactions among multiple classifications. *Journal of the American Statistical Association*, **65**, 226–256.

Gosslee, G. D., and H. L. Lucas (1965). Analysis of variance of disproportionate data when interaction is present. *Biometrics*, **21**, 115–133.

Grant, D. A. (1956). Analysis of variance tests in the analysis and comparison of curves. *Psychological Bulletin*, **53**, 141–154.

Graybill, F. A. (1961). *An introduction to linear statistical models.* (Vol. I.) New York: McGraw-Hill.

Graybill, F. A., and R. B. Deal (1959). Combining unbiased estimators. *Biometrics*, **15**, 543–550.

Green, B. F., and J. Tukey (1960). Complex analysis of variance: General problems. *Psychometrika*, **25**, 127–152.

Greenhouse, S. W., and S. Geisser (1959). On methods in the analysis of profile data. *Psychometrika*, **24**, 95–112.

Harter, H. L. (1957). Error rates and sample sizes for range tests in multiple comparisons. *Biometrics*, **13**, 511–536.

Harter, H. L. (1960). *Expected values of normal order statistics.* ARL Tech. Rep. 60-292, Wright-Patterson AFB, Ohio.

Henderson, C. R. (1953). Estimation of variance and covariance components. *Biometrics*, **9**, 226–252.

Hughes, H. M., and M. B. Danford (1958). *Repeated measurement designs, assuming equal variances and covariances.* Rep. 59-40, Air University, School of Aviation Medicine, USAF, Randolph AFB, Texas.

Johnson, N. L., and B. L. Welch (1940). Applications of the noncentral t distribution. *Biometrika*, **31**, 362–389.

Kempthorne, O. (1952). *The design and analysis of experiments.* New York: Wiley.

Lord, F. M. (1969). Statistical adjustments when comparing pre-existing groups. *Psychological Bulletin*, **72**, 336–337.

Mielke, P. W., Jr., and R. B. McHugh (1965). Non-orthogonality in the two-way classification for the mixed effects finite population model. *Biometrics*, **21**, 308–323.

Morrison, D. F. (1967). *Multivariate statistical methods.* New York: McGraw-Hill.

Odeh, R. E., and E. G. Olds (1959). *Notes on the analysis of variance of logarithms of variances.* WADC Tech. Note 59-82, Wright-Patterson AFB, Ohio.

Olds, E. G., T. B. Mattson, and R. E. Odeh (1956). *Notes on the use of transformations in the analysis of variance.* WADC Tech. Rep. 56-308, Wright Air Development Center, Ohio.

Olkin, I., and J. W. Pratt (1958). Unbiased estimation of certain correlation coefficients. *The Annals of Mathematical Statistics*, **29**, 201–211.

Overall, J. E., and D. K. Spiegel (1969). Concerning least squares analysis of experimental data. *Psychological Bulletin*, **72**, 311–322.

Owen, D. B. (1965). The power of Student's t test. *Journal of the American Statistical Association*, **60**, 320–333.

Patnaik, P. B. (1949). The noncentral χ^2- and F-distributions and their approximations. *Biometrika*, **36**, 202–232.

Paull, A. E. (1950). On preliminary tests for pooling mean squares in the analysis of variance. *The Annals of Mathematical Statistics*, **21**, 539–556.

Plackett, R. L., and J. P. Burman (1946). The design of optimum multifactorial experiments. *Biometrika*, **33**, 305–325.

Rao, C. R. (1952). *Advanced statistical methods in biometric research.* New York: Wiley.

Rao, C. R. (1958). Some statistical methods for the comparison of growth curves. *Biometrics*, **14**, 1–17.

Rao, C. R. (1965). *Linear statistical inference and its applications.* New York: Wiley.

Rider, P. R., H. L. Harter, and M. D. Lum (1956). *An elementary approach to the analysis of variance.* WADC Tech. Rep. 56-20, Wright Air Development Center, Ohio.

Robson, D. S. (1959). A simple method for construction of orthogonal polynomials when the independent variable is unequally spaced. *Biometrics*, **15**, 187–191.

Satterthwaite, F. E. (1946). An approximate distribution of estimates of variance components. *Biometrics Bulletin*, **2**, 110–114.

Scheffé, H. A. (1953). A method for judging all possible contrasts in the analysis of variance. *Biometrika*, **40**, 87–104.

Scheffé, H. A. (1959). *The analysis of variance.* New York: Wiley.

Seal, K. C. (1951). On errors of estimates in double sampling procedure. *Sankhya*, **11**, (pt. 2), 125–144.

Searle, S. R. (1968). Another look at Henderson's methods of estimating variance components. *Biometrics*, **24**, 749–788.

Siegel, S. (1956). *Nonparametric statistics.* New York: McGraw-Hill.

Smith, R. A. (1968). An empirical analysis of the effect of unequal sample size on the Tukey Studentized Range Technique. Unpublished doctoral dissertation, University of Colorado, Boulder.

Steel, R. G. D. (1959). A multiple comparison rank sum test: Treatments versus control. *Biometrics*, **15,** 560–572.

Sutcliffe, J. P. (1957). A general method of analysis of frequency data for multiple classification designs. *Psychological Bulletin*, **54,** 134–137.

Tiku, M. L. (1967). Tables of the power of the *F* test. *Journal of the American Statistical Association*, **62,** 525–539.

Tukey, J. W. (1949). One degree of freedom for nonadditivity. *Biometrics*, **5,** 232–242.

Tukey, J. W. (1955). Answer to query. *Biometrics*, **11,** 111–113.

Tukey, J. W. (1956). Variance of variance components. I. Balanced designs. *The Annals of Mathematical Statistics*, **27,** 722–736.

Tukey, J. W. (1957). The comparative anatomy of transformations. *The Annals of Mathematical Statistics*, **28,** 602–632.

Tukey, J. W. (1962). The future of data analysis. *The Annals of Mathematical Statistics*, **33,** 1–67.

Walker, H., and J. Lev (1953). *Statistical inference.* New York: Holt.

Wallace, D. L. (1959). Simplified beta-approximations to the Kruskal-Wallis *H* test. *Journal of the American Statistical Association*, **54,** 225–230.

Welch, B. L. (1947). The generalization of Student's problem when several different population variances are involved. *Biometrika*, **34,** 28–35.

Wilk, M. B., and O. Kempthorne (1955). Fixed, mixed, and random models. *Journal of the American Statistical Association*, **50,** 1144–1167.

Wilk, M. B., and O. Kempthorne (1957). Nonadditivities in a Latin square design. *Journal of the American Statistical Association*, **52,** 218–236.

REFERENCES TO EXPERIMENTS

Aborn, M., H. Rubenstein, and T. D. Sterling (1959). Sources of contextual constraints upon words in sentences. *Journal of Experimental Psychology*, **57,** 171–180.

Bamford, H. E., and M. L. Ritchie (1958). Complex feedback displays in a man-machine system. *Journal of Applied Psychology*, **42,** 141–146.

Birch, D., E. Burnstein, and R. A. Clark (1958). Response strength as a function of hours of deprivation under a controlled maintenance schedule. *Journal of Comparative and Physiological Psychology*, **51,** 350–354.

Briggs, G. E. (1957). Retroactive inhibition as a function of the degree of original and interpolated learning. *Journal of Experimental Psychology*, **54,** 60–67.

Briggs, G. E., P. M. Fitts, and H. P. Bahrick (1957). Learning and performance in a complex tracking task as a function of visual noise. *Journal of Experimental Psychology*, **53,** 379–387.

Briggs, G. E., P. M. Fitts, and H. P. Bahrick (1957). Effects of force and amplitude in a complex tracking task. *Journal of Experimental Psychology*, **54,** 262–268.

Briggs, G. E., P. M. Fitts, and H. P. Bahrick (1958). Transfer effects from a double integral tracking system. *Journal of Experimental Psychology*, **55,** 135–142.

Castenada, A., and L. P. Lipsett (1959). Relation of stress and differential position habits to performance in motor learning. *Journal of Experimental Psychology*, **57,** 25–30.

Conrad, R. (1958). Accuracy of recall using keyset and telephone dial, and the effect of a prefix digit. *Journal of Applied Psychology*, **42,** 285–288.

Cotton, J. W., D. J. Lewis, and R. Metzger (1958). Running behavior as a function of apparatus and of restriction of goal box activity. *Journal of Comparative and Physiological Psychology*, **51,** 336–341.

Denenberg, V. H., and R. D. Myers (1958). Learning and hormone activity. I: The effects of thyroid levels upon the acquisition and extinction of an operant response. *Journal of Comparative and Physiological Psychology*, **51,** 213–219.

French, G. M. (1959). A deficit associated with hypermobility in monkeys with lesions of the dorsolateral frontal granular cortex. *Journal of Comparative and Physiological Psychology*, **52,** 25–28.

Gerathewohl, S. J., H. Strughold, and W. F. Taylor (1957). The oculomotoric pattern of circular eye movements during increasing speed of rotation. *Journal of Experimental Psychology*, **53,** 249–256.

Gordon, N. B. (1959). Learning a motor task under varied display conditions. *Journal of Experimental Psychology*, **57,** 65–73.

Jerison, H. J. (1959). Effects of noise on human performance. *Journal of Applied Psychology*, **43,** 96–101.

Leary, R. W. (1958). Homogeneous and heterogeneous reward of monkeys. *Journal of Comparative and Physiological Psychology*, **51,** 706–710.

Levison, B., and H. P. Zeigler (1959). The effects of neonatal X irradiation upon learning in the rat. *Journal of Comparative and Physiological Psychology*, **52,** 53–55.

Lewis, D. J., and J. W. Cotton (1958). Partial reinforcement and nonresponse acquisition. *Journal of Comparative and Physiological Psychology*, **51,** 251–257.

Martindale, R. L., and W. F. Lowe (1959). Use of television for remote control: a preliminary study. *Journal of Applied Psychology*, **43,** 122–124.

Meyer, D. R., and M. E. Noble (1958). Summation of manifest anxiety and muscular tension. *Journal of Experimental Psychology*, **55,** 599–602.

Michels, K. M., W. Bevan, and H. C. Strassel (1958). Discrimination learning and interdimensional transfer under conditions of systematically controlled visual experience. *Journal of Comparative and Physiological Psychology*, **51,** 778–781.

Noble, C. E., and D. A. McNeely (1957). The role of meaningfulness in paired-associate verbal learning. *Journal of Experimental Psychology*, **53,** 16–22.

Payne, R. B. (1958). An extension of Hullian theory to response decrements resulting from drugs. *Journal of Experimental Psychology*, **55,** 342–346.

Prokasy, W. F., D. A. Grant, and N. Myers (1958). Eyelid conditioning as a function of unconditioned stimulus intensity and intertrial interval. *Journal of Experimental Psychology*, **55,** 242–246.

Schrier, A. M. (1958). Comparison of two methods of investigating the effect of amount of reward on performance. *Journal of Comparative and Physiological Psychology*, **51,** 725–731.

Shore, M. F. (1958). Perceptual efficiency as related to induced muscular effort and manifest anxiety. *Journal of Experimental Psychology*, **55,** 179–183.

Staats, A. W., et al. (1957). Language conditioning of meaning to meaning using a semantic generalization paradigm. *Journal of Experimental Psychology*, **57,** 187–192.

Taylor, J. A. (1958). Meaning, frequency, and visual duration threshold. *Journal of Experimental Psychology*, **55**, 329–334.

Wulff, J. J., and L. M. Stolurow (1957). The role of class-descriptive cues in paired-associates learning. *Journal of Experimental Psychology*, **53**, 199–206.

Zimbardo, P. G., and N. E. Miller (1958). Facilitation of exploration by hunger in rats. *Journal of Comparative and Physiological Psychology*, **51**, 43–46.

INDEX